# Food Quality and Shelf Life

# Food Quality and Shelf Life

*Edited by*

*Charis M. Galanakis*
Research & Innovation Department, Galanakis Laboratories, Chania, Greece
Food Waste Recovery Group, ISEKI Food Association, Vienna, Austria

ELSEVIER

ACADEMIC PRESS
An imprint of Elsevier

Academic Press is an imprint of Elsevier
125 London Wall, London EC2Y 5AS, United Kingdom
525 B Street, Suite 1650, San Diego, CA 92101, United States
50 Hampshire Street, 5th Floor, Cambridge, MA 02139, United States
The Boulevard, Langford Lane, Kidlington, Oxford OX5 1GB, United Kingdom



## Notices

Knowledge and best practice in this field are constantly changing. As new research and experience broaden our understanding, changes in research methods, professional practices, or medical treatment may become necessary.

Practitioners and researchers must always rely on their own experience and knowledge in evaluating and using any information, methods, compounds, or experiments described herein. In using such information or methods they should be mindful of their own safety and the safety of others, including parties for whom they have a professional responsibility.

To the fullest extent of the law, neither the Publisher nor the authors, contributors, or editors, assume any liability for any injury and/or damage to persons or property as a matter of products liability, negligence or otherwise, or from any use or operation of any methods, products, instructions, or ideas contained in the material herein.

**Library of Congress Cataloging-in-Publication Data**
A catalog record for this book is available from the Library of Congress

**British Library Cataloguing-in-Publication Data**
A catalogue record for this book is available from the British Library

ISBN: 978-0-12-817190-5

For information on all Academic Press publications
visit our website at https://www.elsevier.com/books-and-journals

*Publisher:* Charlotte Cockle
*Acquisition Editor:* Nina Bandeira
*Editorial Project Manager:* Katerina Zaliva
*Production Project Manager:* James Selvam
*Cover Designer:* Christian J. Bilbow

Typeset by SPi Global, India

# Contents

# Contributors

**A. Sibel Akalin** Department of Dairy Technology Faculty of Agriculture, Ege University, Bornova-İzmir, Turkey

**Vildan Akdeniz** Department of Dairy Technology Faculty of Agriculture, Ege University, Bornova-İzmir, Turkey

**Monica Anese** Department of Agriculture, Food, Environmental and Animal Sciences, University of Udine, Udine, Italy

**Gastón Ares** Sensometrics & Consumer Science, Instituto Polo Tecnológico de Pando, Universidad de la República, Canelones, Uruguay

**Zinash A. Belay** Postharvest Technology Research Laboratory, South African Research Chair in Postharvest Technology, Department of Horticultural Science, Faculty of AgriSciences, Stellenbosch University, Stellenbosch, South Africa

**Hulya Cakmak** Department of Food Engineering, Hitit University, Corum, Turkey

**Oluwafemi J. Caleb** Post-harvest and Agro-Processing Technologies (PHATs), Agricultural Research Council (ARC), Infruitec-Nietvoorbij, Stellenbosch, South Africa

**Sonia Calligaris** Department of Agriculture, Food, Environmental and Animal Sciences, University of Udine, Udine, Italy

**Irma Caro** Department of Food Science and Nutrition, Faculty of Medicine, University of Valladolid, Valladolid, Spain

**Rosen Milanov Chochkov** Department Technology of Cereals, Fodder, Bread and Confectionary Products, Technological Faculty, University of Food Technologies, Plovdiv, Bulgaria

**Rui M.S. Cruz** Department of Food Engineering, Institute of Engineering; Centre for Mediterranean Bioresources and Food (MeditBio), Faculty of Sciences and Technology, University of Algarve, Faro, Portugal

**Petko Nedyalkov Denev** Laboratory of Biologically Active Substances, Institute of Organic Chemistry With Centre of Phytochemistry, Bulgarian Academy of Sciences, Plovdiv, Bulgaria

**Zapryana Rangelova Denkova** Department of Microbiology, Technological Faculty, University of Food Technologies, Plovdiv, Bulgaria

**Rositsa Stefanova Denkova-Kostova** Department of Biochemistry and Molecular Biology, Technological Faculty, University of Food Technologies, Plovdiv, Bulgaria

**Ivelina Nikolaeva Deseva** Department of Analytical Chemistry and Physicochemistry, Technological Faculty, University of Food Technologies, Plovdiv, Bulgaria

**Nayil Dinkçi** Department of Dairy Technology Faculty of Agriculture, Ege University,Bornova-İzmir, Turkey

**E.H. Drosinos** Laboratory of Food Quality Control and Hygiene; Department of Food Science and Human Nutrition, Agricultural University of Athens, Athens, Greece

**Ana Giménez** Sensometrics & Consumer Science, Instituto Polo Tecnológico de Pando, Universidad de la República, Canelones, Uruguay

**Hamed Hosseini** Food Additives Department, Food Science and Technology Research Institute, Research Center for Iranian Academic Center for Education, Culture and Research (ACECR), Khorasan Razavi Branch, Mashhad, Iran

**Seid Mahdi Jafari** Faculty of Food Science and Technology, Gorgan University of Agricultural Sciences and Natural Resources, Gorgan, Iran

**William Kerr** Department of Food Science and Technology, University of Georgia, Athens, GA, United States

**Georgi Atanasov Kostov** Department of Wine and Beer Technology, Technological Faculty, University of Food Technologies, Plovdiv, Bulgaria

**Vera Lavelli** DeFENS, Department of Food, Environmental and Nutritional Sciences, University of Milan, Milano, Italy

**Nadia Lotti** Department of Civil, Chemical, Environmental and Materials Engineering, University of Bologna, Bologna, Italy

**Lara Manzocco** Department of Agriculture, Food, Environmental and Animal Sciences, University of Udine, Udine, Italy

**Javier Mateo** Department of Food Hygiene and Food Technology, University of León, León, Spain

**G. Moatsou** Laboratory of Dairy Research; Department of Food Science and Human Nutrition, Agricultural University of Athens, Athens, Greece

**E. Moschopoulou** Laboratory of Dairy Research; Department of Food Science and Human Nutrition, Agricultural University of Athens, Athens, Greece

**G.V. Nevárez-Moorillon** Faculty of Chemical Sciences, Autonomous University of Chihuahua, Chihuahua, Mexico

**Maria Cristina Nicoli** Department of Agriculture, Food, Environmental and Animal Sciences, University of Udine, Udine, Italy

**Umezuruike Linus Opara** Postharvest Technology Research Laboratory, South African Research Chair in Postharvest Technology, Department of Horticultural Science, Faculty of AgriSciences, Stellenbosch University, Stellenbosch, South Africa

**S. Paramithiotis** Laboratory of Food Quality Control and Hygiene; Department of Food Science and Human Nutrition, Agricultural University of Athens, Athens, Greece

**Emiliano J. Quinto** Department of Food Science and Nutrition, Faculty of Medicine, University of Valladolid, Valladolid, Spain

**Bruna P.M. Rico** Department of Food Engineering, Institute of Engineering, University of Algarve, Faro, Portugal

**Valentina Siracusa** Department of Chemical Science (DSC), University of Catania, Catania, Italy

**Anton Minchev Slavov** Department of Organic and Inorganic Chemistry, Technological Faculty, University of Food Technologies, Plovdiv, Bulgaria

**M.K. Syrokou** Laboratory of Food Quality Control and Hygiene; Department of Food Science and Human Nutrition, Agricultural University of Athens, Athens, Greece

**Zhila Tajiani** Agriculture Faculty, Ferdowsi University of Mashhad, Mashhad, Iran

**Desislava Georgieva Teneva** Laboratory of Biologically Active Substances, Institute of Organic Chemistry With Centre of Phytochemistry, Bulgarian Academy of Sciences, Plovdiv, Bulgaria

**Margarida C. Vieira** Department of Food Engineering, Institute of Engineering; Centre for Mediterranean Bioresources and Food (MeditBio), Faculty of Sciences and Technology, University of Algarve, Faro, Portugal

**Luz H. Villalobos-Delgado** Institute of Agroindustry, Technological University of the Mixteca, Huajuapan de León, Mexico

# Preface

The quality of food products is directly connected not only to production stages and manufacture, but also to their shelf life. This connection is nowadays more important than ever, as the preference for minimally processed and tailor-made foods with fresh-like properties is continuously growing. However, this preference is accompanied by concerns surrounding efficacy of the available sanitizing methods to appropriately deal with food-borne diseases. Furthermore, ensuring reliable shelf life of foods is essential in order to assure food quality and protect consumers from detrimental effects during preservation and storage. Current consumer demands, tailor-made applications, and newly strict legislative criteria are driving the field of food science and technology to update and change perspectives on food shelf life. Nowadays, modern food technologists, manufacturers, and retailers often deal with the development and maintenance of new products, and more integral information is needed in this direction. Thus, there is a need for a new reference connecting aspects of shelf life with the quality of the final food products.

Over recent years, the Food Waste Recovery Group (www.foodwasterecovery.group) of ISEKI Food Association has organized a series of activities (webinars, workshops, e-course, etc.) and published handbooks dealing with sustainable food systems, innovations in the food industry and traditional foods, food waste recovery and nonthermal processing, targeting alternative food ingredients, and functional compounds such as polyphenols and proteins. The current book aspires to fill in the gap existing in the current literature by providing information with regard to food quality in all stages between production and consumers' plates. The ultimate goal is to support the scientific community, professionals, and enterprises by highlighting insights regarding the new trends in the field.

The book consists of 12 chapters. Chapter 1 deals with the various quality changes that occur during shelf life of foods and which tend to limit it. It explains how intrinsic factors (e.g., originated from food, like water activity, pH, chemical, and microbiological composition) and extrinsic factors (e.g., storage conditions and packaging), may affect the changes of food quality during shelf life. In Chapter 2, parameters that are key predictors of the stability of dried and intermediate moisture food ingredients are discussed. In addition, the moisture-dependent degradation of bioactive compounds is illustrated.

Chapter 3 addresses and critically evaluates important emerging cold pasteurization technologies (ionizing and ultraviolet irradiation, pulsed electric field application, cold plasma, pulsed light, high pressure dense phase $CO_2$ treatments, etc.), and their combinations to ensure high food quality, increased shelf life, and convenience while keeping the food chemical and biochemical changes as low as possible. Chapter 4 describes the range of natural antimicrobials potentially useful for food preservation and food safety, as well as their action mechanisms. The different methods used for the integration of natural antimicrobials in food are also explained. In Chapter 5, the main types of food spoilage, namely chemical and microbial spoilage, are introduced, followed by an explanation of the significance of different encapsulated ingredients on limiting or preventing the major spoilage routs, as well as comparing the performance of encapsulated preservatives with free ones. Chapter 6 deals with the survival of probiotics in functional foods during their shelf life, highlighting the role of alternative processing and storage conditions as well as the use of modified storage environments and packaging materials for this purpose.

Chapter 7 examines the recent advances in new and emerging packaging systems such as high-oxygen, controlled, and intelligent modified atmosphere packaging. Respective applications for handling and marketing fresh and fresh-cut fruit and vegetables, mushrooms, meat, and aquatic products are presented. Future prospects for modified atmosphere packaging include the integration of novel information and communication technologies, biosensors, and intelligent packaging materials are discussed. Chapter 8 provides an overview of recent technological progress in active and intelligent packaging food systems. Chapter 9 presents different plastic packaging materials, their properties, stimulants, and testing conditions, as well as the migration and release kinetics of compounds into food from those packages and from new packaging systems. In Chapter 10, fruit and vegetable preharvest and postharvest quality assessment with nondestructive techniques are summarized. Chapter 11 discusses current methodological approaches for sensory shelf life estimation. Implementation, applications, advantages, and disadvantages of quality-based methods, acceptability limit, cut-off point methodology, and survival analysis are also described. Finally, the basic principles for the application of accelerated shelf life testing are reviewed in Chapter 12, with special emphasis on the application of temperature and light as accelerating factors. Possible uncertainties, pitfalls, and future research needed on this topic are also discussed.

Conclusively, the book addresses food professionals, food scientists, and technologists working with food processing, preservation, and quality, as well as those who are interested in the development of innovative food products. It could be used by university libraries and institutes all around the world as a textbook and/or ancillary reading in under-graduate and post-graduate level multi-discipline courses dealing with food preservation, food quality, food science and technology, and food processing, especially in post-graduate programs.

I would like to thank all the authors for their fruitful cooperation, bringing together different topics of food quality that affect shelf life. Accepting my invitation, following editorial guidelines, and meeting the deadlines are highly appreciated. Indeed, I consider myself fortunate to have had the opportunity to collaborate with so many experts from Bulgaria, Greece, Italy, Iran, Mexico, Portugal, South Africa, Spain, Turkey, Uruguay, and the United States. I would also like to thank the acquisition editor, Nina Bandeira, the book manager, Katerina Zaliva, and Elsevier's production team for their assistance during the editing and publishing process.

Last but not least, a message for you, the reader. This book comprises a collaborative effort involving many colleagues and more than one hundred thousand words. It is almost impossible not to contain errors and gaps. Consequently, any comments, suggestions, or criticism are welcome. Please do not hesitate to contact me in order to discuss relevant issues of the book.

*Charis M. Galanakis*
Research & Innovation Department, Galanakis Laboratories, Chania, Greece
Food Waste Recovery Group, ISEKI Food Association, Vienna, Austria

# Food quality changes during shelf life

1

*E. Moschopoulou**, *G. Moatsou**, *M.K. Syrokou*†, *S. Paramithiotis*†, *E.H. Drosinos*†

*Laboratory of Dairy Research; Department of Food Science and Human Nutrition, Agricultural University of Athens, Athens, Greece, †Laboratory of Food Quality Control and Hygiene; Department of Food Science and Human Nutrition, Agricultural University of Athens, Athens, Greece

## Chapter outline



**Food Quality and Shelf Life. https://doi.org/10.1016/B978-0-12-817190-5.00001-X**

# 1 Introduction

According to European Union legislation (EC Regulation No. 1169/, 2011) shelf life is referred to as the "date of minimum durability" and requires that the shelf life of a foodstuff be indicated by either a "best before" date or a "use by" date. The "best before" date, is the date until which a foodstuff retains its specific properties, for example, taste, aroma, appearance, and any specific qualities that are related to the product, providing the product has been stored appropriately and the package is unopened. The "use by" date is related to the safety and is the date that afterwards a foodstuff may not be safe.

In practice, the term "shelf life" describes the duration of consumer's acceptance of a foodstuff. From this point of view, the end of shelf life differs among products and classifies them into those having intermediate or long shelf life and those having short shelf life. Therefore, all foods either with the "use by" or with the "best before" labeling have a certain shelf life, which depends firstly on their specific quality characteristics and secondly on their environmental conditions. In other words, labeling reflects the fact that during shelf life, quality characteristics are not stable and, consequently, foods deteriorate. The quality changes that may occur in food are presented in Table 1.

Factors in food quality changes may be physicochemical, like pH and $a_w$, microbiological, like spore-forming bacteria, or organoleptic, referred to as intrinsic factors; or they may be the storage conditions, the packaging etc., referred to as extrinsic factors.

# 2 Intrinsic factors affecting food shelf life

## 2.1 *Ingredients*

Proteins, fats, carbohydrates, and vitamins are the food components that exhibit the most important changes during shelf life under the influence of microflora, enzymes, and processing conditions. Shelf life of fatty foods is much affected by their concentration in unsaturated fatty acids, since these acids are prone to oxidation during production and storage, especially in the absence of antioxidant substances such as phenolic compounds, alpha-tocopherol, or ascorbic acid. For example, the oxidation degree of olive oil is greatly affected by the ratio of saturated/unsaturated fatty acids and the total phenolic compounds. To extend or limit shelf life of fatty foods, antioxidants are used. Other substances that may be added in other foods to inhibit quality deterioration are sweeteners that can act as humectants or cryoprotectants, that is, for frozen products.

## 2.2 *Water activity ($a_w$)*

Water is the substrate for chemical and biochemical reactions and is an important constituent not only of those products in liquid form but also of those in powdered form. The latter deteriorate quickly if water content exceeds 5%, becoming subject to textural and sensory changes.

The water activity value ($a_w$) is the ratio of the vapor pressure in a solution or a food material and that of pure water at the same temperature. In practice, $a_w$ expresses the

**Table 1** Changes of food quality occurred during shelf life

| Type of change | | Examples of foods |
|---|---|---|
| Physical changes | Moisture gain or loss | Fresh fruits and vegetables, cheeses |
| | Water vapor transfer, ice sublimation | Frozen products, e.g., ice cream |
| | Loss of characteristic flavor, development of undesired odor or flavor ("taint") | Juices, chocolate, tea leaves |
| | Loss of carbon dioxide | Carbonated drinks |
| | Crystallization and textural changes | Milk powders, toffee |
| | Emulsion destabilizing and breakdown | Butter, margarine |
| | Syneresis | Yoghurt products, jams, sauces |
| Chemical and biochemical changes | Chilling injury | Fruits and vegetables |
| | Hydrolysis of fatty acids | Butter, whole milk powder, cheeses, cereal grains |
| | Oxidation of fats, proteins, pigments, vitamins | Cereals (wheat, barley, maize), oilseeds (soya), fish, meat, milk |
| | Nonenzymic browning | Dehydrated fruits and vegetables, evaporated and dry milk products |
| | Enzymic browning | Precut fruits and vegetables |
| | Off-flavors, oxidative discoloration, interactions between food and packaging | Foods in tinplate cans |
| Light-induced changes | Photo-oxidation of vitamins, light induced oxidation, color fading | Milk, snacks, bakery products |
| Microbiological changes | Growth of pathogenic microorganisms, microbial toxins | Fish, meat products, poultry, milk, cheeses |
| | Microbial spoilage by Gram-negative, spore-forming bacteria, Enterobacteriaceae, NSLAB, yeasts, and molds | Fish, meat products, poultry, milk cheeses |

Based on Man, D.C.M., 2015. Shelf Life, second ed. Wiley Blackwell, UK.

availability of water in the food. High-moisture foods that usually contain >50% w/w water have an $a_w$ from 0.9 to 0.999, intermediate moisture foods containing water from 10% to 50% have an $a_w$ from 0.60 to 0.90. Foods with $a_w$ below 0.60 are low-moisture foods and are seldom spoiled by microorganisms. In Table 2, water activity and pH values of different foods are presented. Water activity, pH, and temperature are the most important factors that control the rates of food deterioration caused by microbial and enzymic activity (Roos, 2001).

During storage, many foods tend to gain, lose, or exchange moisture depending on their composition and environmental conditions. In addition, water transfer may

**Table 2** pH value and water activity ($a_w$) of some foods

| pH range | Food | pH | Water activity |
|---|---|---|---|
| Low acid (pH 7.0–5.5) | Milk (whole or skim) | 6.5–6.7 | 0.988–0.996 |
| | Cheddar cheese | 5.9 | 0.950 |
| | Red meat | 5.4–6.2 | 0.970–0.990 |
| | Ham | 5.9–6.1 | 0.967 |
| | Poultry | 5.6–6.4 | 0.979–0.982 |
| | Fish | 6.6–6.8 | 0.938–0.990 |
| | Butter (salted or unsalted) | 6.1–6.4 | 0.894–0.980 |
| | Potatoes | 5.6–6.2 | 0.997–0.988 |
| | Bread | 5.3–5.8 | 0.939–0.960 |
| Medium acid (pH 5.5–4.5) | Cottage cheese | 4.5 | 0.95–0.99 |
| | Feta cheese | 4.6–4.8 | 0.95 |
| | Bananas | 4.5–5.2 | 0.964–0.987 |
| | Green beans | 4.6–5.5 | 0.987–0.996 |
| Acid (pH 4.5–3.7) | Mayonnaise | 3.0–4.1 | 0.930–0.96 |
| | Tomatoes | 4.0 | 0.998 |
| High acid (pH < 3.7) | Fruit juice | 3.5–3.9 | 0.970–0.988 |
| | Citrus juice | 3.0–3.5 | 0.983 |
| | Apples | 2.9–3.3 | 0.975–0.988 |

Based on Man, D.C.M., 2015. Shelf Life, second ed. Wiley Blackwell, UK.

occur at freezer temperature, that is, water vapor transfer. Loss of moisture results in diffusion and evaporation of desirable flavors of liquid products and changes in textural characteristics of solid foods. Transfer of moisture into food may cause sensory changes, for example, unpleasant odor or flavor; textural changes, for example, plasticization, graining, softness, stickiness, or crystallization due to absorption of moisture on the food surface; or biochemical changes, for example, spore germination in powdered foods (Roos, 2001; Man, 2015).

## 2.3 pH

pH value is a critical factor that may induce or stop microbial growth. This is why foods are classified according to their pH value as low-acid, medium-acid, acid, and high-acid foods. Pathogens and spoilage bacteria cannot grow in an acidic environment, but most foods have low- and medium-acid pH, as shown in Table 2.

## 2.4 Microflora

Microflora is a key factor for maintaining food quality during shelf life. Undesirable microorganisms, that is, pathogens and spore-forming bacteria, are eliminated by heat treatment, at least by pasteurization during food manufacture. In contrast, selective desirable microorganisms that play important roles in some foods like yoghurt and fermented meat products are added. Among the potential spoilage microorganisms present in a perishable food, mesophilic microorganisms at counts higher than $10^6$ cfu/g or ml have been associated with the end of shelf life (Notermans, 1993).

### 2.5 Enzymes

Enzymes found naturally in food, that is, indigenous enzymes, may cause deterioration when food is exposed to favorable conditions for their activity. For example, in the presence of oxygen, polyphenol oxidase causes enzymic oxidation of phenolic compounds and limits the shelf life of precut fruits and vegetables resulting in enzymic browning. In addition, enzymes produced from microbial growth can deteriorate the food inducing lipolysis and/or proteolysis and resulting in off-flavors like rancidity and bitter peptides, respectively.

### 2.6 Redox potential ($E_h$)

The redox (reduction and oxidation) reaction is a coupled reaction, in which oxidation (loss of electrons) and the corresponding reduction (gain of electrons) take place simultaneously. Redox potential influences the microflora as well as many chemical and biochemical reactions of a food (Man, 2015).

## 3 Extrinsic factors affecting food shelf life

### 3.1 Processing

Several different processing procedures are applied to different foods. Some of these aim to alter physical properties of foods, for example, fat homogenization; some aim to separate components, for example, milk fat centrifugation; and others, to kill the microorganisms that cause food deterioration. Heat treatment can be applied to almost any kind of food and results in shelf life extension. However, since heat treatment may also affect sensitive food compounds like vitamins, novel processing methods are used as alternatives to, or in combination with, heat treatment to protect food from intrinsic factors extending, in parallel, the shelf life. High pressure, ultra-high pressure homogenization, cold plasma, ohmic heating, and electric pulse fields are some of the novel methods that have been studied for food production. Nevertheless, even in the case of a novel process, deterioration of quality occurs. For example, Gavahian et al. (2018) recently reported that cold plasma processing may induce lipid oxidation in cereals, edible oils, dairy, and meat products, due to the oxygen present as a component of the gas affecting their acceptability and shelf life.

### 3.2 Storage conditions

Relative humidity, temperature, oxygen, and light are critical parameters of storage conditions affecting the shelf life of perishable foods. For example, quality of fats and oils is sensitive to these factors because rancidity may be caused. High storage temperature accelerates the food deterioration process by inducing microbial growth or spore germination. Therefore, cooling is the main method to slow the spoilage of foods during storage. Increase of relative humidity in storage rooms favors enzyme activity and growth of fungi that can spoil nonpackaged food. Regarding other parameters of storage conditions, insects and mites are not discussed here, as they should be completely absent.

### 3.3 Packaging

The main role of packaging is to protect food from environmental spoilage and mechanical destruction, keeping its quality and hence its shelf life. However, biological, chemical, and physical interactions between food, package, and environment may occur. For example, deterioration of food, that is, fat oxidation or vitamin loss, may occur via light transmission through transparent bottles. In recent decades, food-packaging systems have been designed to determine food shelf life in real time, forecasting accurately food demand and avoiding food waste. The best-known smart packaging systems are time temperature indicators (TTIs), which have the potential to monitor or predict safety and shelf life of refrigerating foods (Taoukis and Labuza, 1989; Zhang et al., 2016), and active packaging. Active packaging systems have incorporated components that can release antioxidant, antimicrobial substances, or carbon dioxide into the packaged food, or absorb oxygen, moisture, or ethylene from the packaged food or the environment surrounding the food (Yildirim et al., 2018). Therefore, the active packaging intends to maintain or to improve the conditions of the packaged food, extending the shelf life.

### 3.4 Logistics and retail issues

For perishable foods like fish, meat products, dairy products, fruit, and vegetable products, storage, transportation, distribution, and retail should follow the "cold chain." During cold storage, continuous monitoring of temperature, humidity, and their fluctuations supports the real-time assessment of product quality and determines the rest of its shelf life. To control the cold chain several systems based on sensor devices such as wireless sensor networks and radio frequency identification are applied (Spagnol et al., 2018). In addition, consumers' handling after pack opening plays a critical role in food quality maintenance, at least up to the indicated date. Consumers should follow the guides written on food package regarding food handling, either before or after opening of package, for example, store in cool and dry place, after opening keep refrigerating and use within 3 days, etc. After pack opening, the deterioration of food quality accelerates, and for this reason, Nicoli and Calligaris (2018) reported the concept of "secondary shelf life": the period after pack opening, during which a food product maintains an acceptable quality level.

## 4 Quality changes during shelf life of dairy products

It is worth discussing the shelf life of dairy products in detail, since they have different properties and so, different behavior during storage.

Milk is a nutritious product rich in fat, protein, lactose, minerals, and vitamins. It contains, however, microorganisms and indigenous enzymes that may spoil it quickly depending on the storage conditions before processing. Indigenous and contaminant microflora of raw milk are classified as spoilage and pathogenic. The most important spoilage microorganisms that may affect milk products are: (i) psychrotrophs, which can grow during milk storage at 6°C or less, (ii) thermoduric bacteria, which can survive pasteurization conditions, (iii) spore-forming bacteria, which produce heat resistant spores, and (iv) yeasts and molds (Moatsou and Moschopoulou, 2014). Mesophilic and thermophilic aerobic spore-forming bacteria, such as *Paenibacillus* spp. and *Geobacillus stearothermophilus* may limit the shelf life of milk powder, whereas psychrotrophic aerobic spore-forming bacteria like *Bacillus cereus* may limit the shelf life of fluid milk. Anaerobic spore-forming bacteria, such as *Clostridium tyrobutyricum* may cause defects in hard cheeses. The heat stable protease plasmin, the lipoprotein lipase, and catalase correlated with an increased number of somatic cells counts of raw milk, influence the raw milk quality causing protein and fat degradation, producing off-flavors. The practice of cold storage of raw milk immediately after milking until its processing causes growth of psychrotrophic bacteria that produce extracellular or intracellular thermo-resistant proteolytic and lipolytic enzymes spoiling the final product (Barbano et al., 2006). A deep cooling to 2°C for 48h controls the growth of psychrotrophic microbiota and increases raw milk shelf life (Moatsou and Moschopoulou, 2014).

## 4.1 Processed fluid milk

Fluid milk for direct consumption is manufactured by heat treatment of milk at least at 72°C for 15s, that is, HTST pasteurization, in combination with clean packaging; at temperatures higher than 100°C for a few seconds, that is, ultra-pasteurized milk; or at a temperature of 135–145°C for a few seconds, that is, ultra-high heated (UHT) milk in combination with aseptic packaging. Shelf life of pasteurized milk is about 7–10days, providing that post pasteurization contamination is absent. Depending on processing design, pasteurization may be combined with bactofugation or microfiltration for shelf life extension. Quality of stored pasteurized milk is equally affected by the surviving microflora, and by storage conditions like temperature, oxygen, and light.

Microbial counts, somatic cells count, and spores may affect shelf life detrimentally (Barbano et al., 2006; Boor, 2001) and post pasteurization contamination may introduce to the processed milk psychrotolerant Enterobacteriaceae and *Pseudomonas* spp., which can grow at 6°C, causing sensory defects within 7–10days after processing (Martin et al., 2017). *B. cereus* spores can survive pasteurization and cause also sensory defects such as sweet curdling of milk and formation of flakes in cream known as "bitty cream" (Moatsou and Moschopoulou, 2014). For this reason, a limit for *B. cereus* $10^4$cfu/mL has been proposed (Schoder et al., 2007). In addition, *Paenibacillus* bacteria from spores that survive pasteurization can grow in cold storage and finally spoil pasteurized milk (Ziarno and Zaręba, 2010). According to a recent systematic review using metaanalysis of data from previous studies, spore-forming bacteria may increase within the range of 0.58–2.41 logs cfu/mL from raw to pasteurized milk indicating the significance of lowering spore forming bacteria levels and preventing post pasteurization contamination (Ortuzar et al., 2018). In pasteurized whole milk, high bacteria counts have been correlated with increased volatile compounds and increased protease and lipase activities throughout shelf life (Ziyaina et al., 2018). However, the correlation between microbial counts and volatile compounds is not linear and the concentration of volatiles changes considerably when microbial counts are $10^6$–$10^8$ cfu/mL (Silcock et al., 2014). Moreover, Alothman et al. (2018) showed that there is a weak relationship between total microbial counts, volatile compounds, and sensory characteristics, indicating that the kind of microflora plays a critical role.

Another critical factor affecting the shelf life of pasteurized milk is the storage temperature (Brown et al., 1984; Griffiths and Phillips, 1988). Storage temperature below 4°C delays microbial growth, while above 6°C favors it. At 4°C, spoilage is mainly due to *Pseudomonas* spp., whereas Enterobacteriaceae and Gram-positive bacteria are involved in spoilage at temperatures over 10°C (Griffiths and Phillips, 1988). Stability of vitamin A and vitamin E in infant milk is also affected by the storage temperature, decreasing up to 66% and 35%, respectively, when milk is stored at 37°C for 9 months (Haro-Vicente et al., 2013). Moreover, storage temperature affects the rate of deterioration of sensory characteristics of pasteurized skimmed milk kept at 10°C for 8 days and flavors such as vegetable, sour, cardboard, and chalky have been attributed to it (Watson and McEwan, 1995).

Although pasteurization is the minimum heat treatment for fluid milk consumption, it affects heat sensitive vitamins. It causes loss of ascorbic acid and folate between 0%–17% and 10%–12%, respectively, and of vitamin B1 and vitamin B12 between 5% and 10%. However, during storage, B12 and folate remain intact, while 25%–45% of the residual ascorbic acid may be lost (Andersson and Osten, 1994). Vitamin D2 is not affected by heat treatment, but it is affected by the material packaging. It significantly decreases when milk is kept in plastic bottles compared to milk kept in glass bottles at refrigeration temperatures. In addition, it significantly decreases when milk is kept in polyethylene pouches under certain light intensities (Kaushik et al., 2014). In general, packaging material influence the stability of sensitive compounds of pasteurized milk when combined with certain storage conditions in terms of light and oxygen. In a study assessing different packaging materials it was clearly shown that microbial counts and chemical composition were not affected over a 7-day period, whereas Vitamin A was affected. More specifically, loss of vitamin A was 11%, 16%, and 31% in case of multilayer colored HDPE and PET bottles, paperboard cartons, and clear PET bottles, respectively (Moyssiadi et al., 2004). In pasteurized whole milk, vitamin A and riboflavin, the main vitamin of milk, are the most sensitive ingredients during storage and losses from 50% to 73% and from 34% to 45%, respectively, have been reported for milk kept either in clear or in LDPE bottles (Vassila et al., 2002). Therefore, packaging with a specific combination of materials must be designed to prevent excessive losses of vitamins.

Shelf life of sterilized milk, known as long life milk, may be up to 12 months, although in practice it is consumed much earlier. Deterioration of UHT milk is mainly due to the indigenous protease plasmin and proteases from *Pseudomonas* spp. (Stoeckel et al., 2016) and to heat-resistant spores and heat-stable enzymes from the spore-forming bacteria belonging to the genera *Bacillus* spp. and *Clostridium* spp. Recently it was shown that commercial Brazilian UHT milk was contaminated with aerobic mesophilic and spore forming microorganisms at considerable counts suggesting that this product was at potential risk (Pinto et al., 2018).

In addition, oxidation of fatty acids may occur during storage of UHT milk. This phenomenon is influenced by the fatty acid profile and the presence of metal ions, as well as the presence of antioxidant substances, either naturally occurring or added. On the other hand, presence of light is a more critical factor for stability of UHT milk during storage, compared to the fatty acid profile. Light degrades riboflavin, hydrophilic antioxidants found in the milk serum, and vitamin E (Smet et al., 2009). For this reason, modern materials for UHT packaging contain light barriers. Finally, somatic cell counts as high as 800,000 cells/mL do not affect lipolysis and apparent viscosity of UHT milk up to 120 days of storage, but increase proteolysis (Fernandes et al., 2008).

## 4.2 Yoghurt

Yoghurt exhibits a self-protection against quality changes during storage due to its low pH value, that is, from 4.4 to 4.6, which is achieved by the accumulation of lactic acid produced from lactose fermentation during manufacture. Consequently, lactic acid is the main component of flavor and taste of yoghurt products, while acetaldehyde, diacetyl, and ethanol are the most abundant volatile compounds, the concentration of which are affected by the type of starter culture used as well as by the storage period (Pinto et al., 2009).

Shelf life of yoghurt products depends on the type of product, production conditions, and storage temperature. For example, Greek strain yoghurt and Greek traditional yoghurt may be kept for about 40 days and 15 days, respectively, at 4°C. The nominal shelf life of labneh, a concentrated yoghurt produced in Middle Eastern countries, is about 18, 14, 9, and 3 days when kept at 5°C, 7°C, 15°C, and 25°C, respectively (Al-Kadamany et al., 2003; Nsabimana et al., 2005).

Yoghurt deterioration from bacteria is limited, but yeasts, fungi, and some lactic acid bacteria can spoil it, producing off-flavors. In addition, when postacidification occurs during storage, the excess acidity results in syneresis, which yields serum on the top of the product, affecting its sensory quality. Moreover, when the fermentation continues after the fermentation process, organic acid derivatives and proteolysis off-flavor compounds are produced. The latter problem is usually solved by using starter cultures that stop growing at pH 4.4.

The shelf life of yoghurt and fermented milk products is usually extended by pasteurization and addition of preservatives, while programs to estimate shelf life, integrated planning, and schedule of yoghurt production have been proposed (Lutke-Entrup et al., 2005).

## 4.3 Cheese

Cheese is a very diverse food group in terms of composition, functionality, and sensory properties. Variability results from different cheese making conditions and ripening treatments. As a result, the main shelf life limiting factors and mechanisms can differ within cheese varieties.

Intrinsic factors that affect the shelf life of cheese are: (i) the microbiological profile and indigenous enzymes of cheese milk; (ii) the gross composition of curd and cheese, that is, residual lactose, salt, protein, and fat content; and (iii) physicochemical conditions in cheese, that is, pH, water activity ($a_w$), and redox potential ($E_h$).

In addition, crucial extrinsic factors for the shelf life of cheese are related to processes and treatments applied during cheese making and throughout ripening and storage, that is, temperature/time combinations, hygiene, storage, and distribution conditions are very important.

Dehydration of cheese curd by means of whey removal is an essential step of every cheese making procedure. However, various degrees of dehydration are applied to various cheese varieties resulting in variable physicochemical composition. Most cheese varieties undergo ripening. In general, the changes of the main constituents and microflora that take place during ripening enhance the shelf life of cheese. However, under particular circumstances, they can shorten its shelf life, for example, by inducing spoilage.

Salt, pH, water activity ($a_w$), and redox potential ($E_h$) prevent the growth of pathogens in cheese and minimize spoilage (Guinee and Fox, 2004). Examples of the physicochemical conditions in several cheese varieties are shown in Table 3.

The $a_w$ expresses the available water in cheese, that is, water not bound to its components, for example, in Cheddar coagulum, the bound water is 0.2–0.3 g/g of solids, whereas in harder Emmental cheese coagulum it is 0.45 g/g of solids (Hardy, 1986). It depends on cheese moisture and on the concentration of low molecular mass components in the first days/h after manufacture, and is determined almost entirely by the concentration of salt in the aqueous phase (Marcos, 1999; Guinee and Fox, 2004). The $a_w$ is not constant during cheese ripening because, (i) dehydration of cheese takes place, and (ii) low molecular mass compounds are produced, such as numerous peptides that bind water in their amino- and carboxyl-groups (Hardy, 1986; Marcos, 1999).

**Table 3** Examples of physicochemical conditions that influence cheese shelf life

| Cheese | % moisture | $a_w$ | pH | % NaCl |
|---|---|---|---|---|
| Unripened | 70–85 | 0.980–0.995 | 4.3–4.7 | |
| Soft | 50–60 | 0.970–0.990 | 4.5–5.0 | 2 |
| Mold-ripened | 45–55 | 0.940–0.970 | 4.7–5.0 | 2 |
| Uncooked hard | 45–55 | 0.970 | 4.8–5.3 | 1.8–2.0 |
| Cooked hard | 25–35 | 0.885–0.905 | 5.0–5.3 | 1.8 |
| Cheddar | 37 | 0.950 | 5.5 | 1.5 |
| Gouda | 41 | 0.950 | 5.8 | 2.0 |
| Gruyère | 33 | 0.950 | 5.7 | 1.1 |
| Parmesan | 31 | 0.920 | 5.4 | 2.6 |
| Camembert | 52 | 0.970 | 6.9 | 2.5 |
| Roquefort | 40 | 0.910 | 6.4 | 3.5 |

Based on Veillet-Poncet, L., 1986. Manufacturing hygiene. In: Eck, A. (Eds.), Cheesemaking-Science and Technology, second ed. translated by C.D. Thomson. Lavoisier Publishing Inc., New York, pp. 148–149. and Guinee, T.P., Fox, P.F., 2004. Salt in cheese: physical, chemical and biological aspects. In: Fox, P.F., McSweeney, P.L.H., Cogan, T.M., Guinee, T.P., (Eds.), Cheese: Chemistry, Physics and Microbiology, vol. 1. Elsevier Ltd, London, pp. 207–259.

Bacteria are the most sensitive in the reduction of cheese $a_w$ and, except for halophilic bacteria, they cannot tolerate $a_w < 0.85$. Most yeasts and molds, with the exception of some *Saccharomyces* species, are inhibited at $a_w < 0.80$, and only osmophilic species can tolerate $a_w$ 0.55; on the other hand, the minimum $a_w$ for the growth of pathogens is close to 0.85, for example, for *Staphylococcus aureus* (Hardy, 1986; Veillet-Poncet, 1986; Guinee and Fox, 2004).

Pathogens and other microbiological factors related to food borne diseases are beyond the scope of this chapter. In brief, (Veillet-Poncet, 1986; Marcos, 1999; Donnelly, 2007) the greatest safety risk for cheeses are the pathogens *Salmonella* spp., *Listeria monocytogenes*, *Escherichia coli*, and *St. aureus*. The later may produce toxin in cheese manufactured without starters. Bloomy rind cheeses, washed rind cheeses, Hispanic-style cheeses, and Tomme-style cheeses from raw milk can be liable to pathogens. For raw milk cheeses, 90 days of ripening are necessary to ensure safety. Hard-type Swiss and Italian cheeses are of low risk due to high scalding of curd and long ripening time, along with low moisture. Metabolites and bactericidal substances produced during cheese ripening, along with a decrease of $a_w$, may eliminate pathogenic bacteria, especially in cooked hard cheeses. Since pH values close to neutral are consistent with the maximum growth of pathogenic bacteria, the acidification of the cheese mass does not favor their growth. This effect is species specific for both the lactic acid bacteria and the pathogens. The extrinsic factors that affect pathogen growth are heat treatments, ripening temperatures, environmental conditions, and packaging. Cheese making is a hurdles technology, among which the most important is $\alpha_w$ decreasing with time. In this respect, semihard and hard cheeses have very rarely been associated with food poisoning, whereas brucellosis, listeriosis, and shigellosis have been associated with surface-ripened Brie, and Camembert, fresh or ripened for less than 2 months.

Various microbiological aspects are the most common reason for the limitation of cheese shelf life. A very important factor for the appearance of microbial-type defects is the content of residual lactose at the end of cheese making. Development of early or late gas by nondesirable fermentations occurring in the cheese mass during ripening results in eyes, cracks, and holes in the cheese mass, or gas within packaging, and reduces dramatically the shelf life of cheese.

### 4.3.1 Gas production

Early blowing of cheese due to early gas production will be presented in brief, according to Choisy et al. (1986b) and Alichanidis (2007). It takes place shortly after manufacture, that is, within a few hours or days after cheese making and it is most probable in raw milk cheeses and in fresh cheeses with high $a_w$, when residual lactose content is high and salt absorbance low or nonuniform. It results in numerous small holes in the cheese mass, which in turn becomes rather "spongy." If cheese milk has been pasteurized, mainly coliforms, but also yeasts or heterofermentative lactic acid bacteria coming from contamination, provoke early gas production. Coliforms are mesophilic, but they can grow also at 10°C and 40°C, they are inhibited at low pH 4.1–3.85, and include psychrotrophic species. They are capable of fermenting residual lactose in fresh cheese producing, among others, lactic acid, acetic acid, ethanol, $CO_2$, and $H_2$. $H_2$ is very weakly soluble in water and favors blowing. Gas produced by yeasts is $CO_2$ from the metabolism of lactate or lactose. Holes from coliforms are very small, with a diameter of 1–2 mm, and when they co-exist with larger ones, yeasts are also present. Other microorganisms that cause gas defects are citrate positive lactococci or Leuconostocs producing $CO_2$, and salt tolerant lactobacilli producing $CO_2$ and unclean flavors being present in "weak" brines. Early blowing is the most common defect of white brined cheeses. Sources of contamination are:

**(i)** poor hygiene of plant and equipment;
**(ii)** post pasteurization contamination of cheese milk; and
**(iii)** contaminated starter.

For the appearance of early gas, either a high count of coliforms and yeasts ($>10^5$–$10^6$ per mL milk), insufficient (slow) rate of acidification by the starters, or recontamination of cheese milk with antibiotics is necessary.

Although the coliform counts decline dramatically during ripening, high moisture and high residual lactose, along with rather high temperature and pH of fresh curd, enhance their multiplication in fresh cheese. Therefore, rapid fermentation of lactose by starters, proper draining, heat treatment of cheese milk, hygiene in manufacture, and low temperatures after curd formation inhibit early gas production.

Late gas production results in late blowing and its characteristics are summarized according to Choisy et al. (1986b), Alichanidis (2007), McSweeney (2007), and Sheehan (2007a). It is due to the anaerobic fermentation of lactate by *Clostridium* spp., particularly *C. tyrobutyricum*, producing butyric acid and gasses. The accumulation of hydrogen and carbon dioxide in the cheese mass occurs weeks or months after manufacture, during ripening or storage, and results in the appearance of cracks, slits, or abnormal eyes in semihard and hard scalded cheeses, such as Swiss and Dutch-type cheeses.

*C. tyrobutyricum* is acid tolerant, growing anaerobically within pH 4.5–7.5, and metabolizes lactate producing butyric acid, acetic acid, carbon dioxide, and hydrogen. At optimum pH 5.8, *C. tyrobutyricum tolerates* 5.5%–6% salt. Therefore, rapid pH decreases at values lower than pH 5.0 and rapid salt intake (due to dry salting) at >5.5% salt-in-moisture occurring in white brined cheeses do not favor late blowing in these cheese varieties. Late gas in dry-salted cheese, such as Cheddar cheese, can be due to fermentation of residual lactose and galactose to $CO_2$ by heterofermentative lactobacilli (NSLAB) if the starter activity is low. Other sources of late gas could be the decarboxylation of amino acids during ripening.

Measures to avoid late gas in cheese are related to both cheese milk quality and cheese ripening conditions. Good quality silage and good milking practices limit the initial milk spore counts. They can be removed from cheese milk by physical methods, such as bactofugation or microfiltration. Their germination and the growth of vegetative cells may be prevented by using $KNO_3$, $NaNO_3$ (10–20 g/100 L milk), or lysozyme, if allowed. Coliform, lactobacilli surface microflora, and the enzyme xanthine oxidase of cheese reduce nitrate to nitrite. The later delay the spore germination until a uniform salt distribution within the cheese mass is achieved. Opposite to the white dry-salted cheeses-in-brine (brined), the appropriate salt intake throughout the mass of brine-salted cheeses is slow. However, nitrates cannot be used in propionic fermentation cheeses due to the inhibitory effect of nitrite on propionibacteria. Moreover, under low pH conditions, the nitrite reacts with aromatic amino acids producing nitrosamines, which are potentially carcinogenic substances. In addition, the use of bacteriosinogenic starters does not seem effective due to their susceptibility to bacteriophages and their inhibitory activity on useful starter and nonstarter bacteria (Brändle et al., 2016). Interestingly, D'Incecco et al. (2018) proved that not only spores in the cheese milk but also vegetative cells of *C. tyrobutyricum* can be risk factors for late blowing due to the ability of the later to sporulate during the vat processing of Grana-Padano cheese.

Ripening and storage conditions may inhibit or induce late blowing. For example, the conditions during the hot room ripening of Swiss-type propionic fermentation cheeses are favorable for spore germination and clostridia growth. According to Silvetti et al. (2018), ripening temperatures below 10°C can prevent late blowing. In particular, at temperatures lower than 15°C, the gas formation by vegetative cells of Clostridia is prevented for over 70 days, but a combination of temperature ≤15°C, pH ≤5.0, and NaCl at 2%, or a temperature ≤10°C is necessary to avoid spores outgrowth and gas production. However, at 20°C, gas production resulting from *C. tyrobutyricum* spore germination and growth is not affected by acidity and salt concentration. Emerging technologies such as HHP treatment of young cheese at >300 MPa could inhibit spore germination, as reported by Ávila et al. (2016).

### 4.3.2 Cheese surface deterioration

Cheese rind is the outer part of cheese and it is a feature specific for each formatted cheese variety. For example, the rind of surface-ripened cheeses like Brie and Camembert is bloomy, the wash-rind of brine-salted Swiss-type cheeses is firm and dry, whereas white cheeses in brine do not have any particular rind. The external part of cheese rind is the cheese surface, which is vulnerable to spoilage due to its direct contact with the environment and, in particular, with air. Although protection of the cheese surface by means of packaging is a widespread practice, previously started spoilage can limit the shelf life of cheese. Moreover, packaging during ripening may influence the biochemical pathways that are involved in cheese ripening.

The composition of cheese rind is different from that of the core. Cheese rind is characterized by low $a_w$ and high salt content, in particular at the early stages of ripening, which, along with the availability of the oxygen, configures a particular microflora. Cheese surface microbiota is noninoculated microflora consisting of both prokaryotes (Actinobacteria, Firmicutes, Proteobacteria) and eukaryotes (mainly yeasts and molds) that contribute to flavor development of cheeses due to their various enzymatic activities. Interestingly, commercial ripening cultures, used for smear-rind cheese manufactured from pasteurized milk, are not always the dominant surface microorganisms. The microflora of cheese surface is a barrier against pathogen and spoilage microorganisms and a carrier of various enzymatic activities that affect cheese flavor (Irlinger et al., 2015).

In general, the conditions on the cheese surface favor the growth of yeasts and molds, and of aerobic and salt-tolerant micrococci and corynebacteria. Corynebacteria include

*Corynebacterium* spp. and *Arthrobacter* spp. with two subgroups, *Brevibacterium* and the heat-resistant *Microbacterium.* The highly proteolytic *Bacterium linens* is very important for the ripening of smear ripened cheeses (Choisy et al., 1986b).

Wolfe et al. (2014) considered the rind developed on the cheese surface during ripening as a biofilm and, according to their findings, >60% of bacteria and >25% of the fungi present in the rind of various cheeses originate from environmental sources. They also report that the microbial composition of the cheese rind is configured by rind moisture and type and by interactions among microbes that result on succession of species on the cheese rind. Indeed, yeasts and molds metabolize lactic acid and produce ammonia by proteolysis. In addition, Wolfe et al. (2014) reported that marine associated γ-proteobacteria are abundant in both washed and bloomy rind cheeses and they suggest sea salt as a possible source, in accordance to the findings of Ishikawa et al. (2007). Similarly, Garnier et al. (2017) report also the isolation from the rind of hard cheese of *Penicillium antarcticum,* originally described in marine and subglacial environments.

Studies on smear-ripened cheeses have highlighted the importance and the complexity of cheese surface microflora that can result in changes that limit cheese shelf life. Yeasts are very important for the growth of bacteria on cheese surface. Initially, yeasts such as *Debaryomyces hansenii, Kluyveromyces marxians, Yarrowia lipolytica,* and the yeast-like *Geotrichum candidum,* followed by molds, grow on the surface and utilize lactic acid causing deacidification. Thus, surface pH rises from 4.8–5.2 to up to 6–8.2, which favors the growth of salt-tolerant and acid-sensitive bacteria such as *Arthrobacter arilaitensis, Brevibacterium aurantiacum, Brevibacterium linens,* and *Corynebacterium casei* (Mounier et al., 2008; Irlinger and Mounier, 2009; Montel et al., 2014; Sutthiwong et al., 2018).

Moisture loss from the cheese surface during ripening regulates the growth of various microbial groups on the cheese surface. Milk can be a source of molds when raw cheese milk is used. However, they are not expected to persist in cheese environment. The same holds true for molds coming from the cheese-making plant. Many mold species isolated from dry-ripened foods are mycotoxigenic under particular biotic and abiotic conditions, reviewed by Hymery et al. (2014). The main sources of fungal contamination of cheeses are the processing equipment, shelves, air, and brining baths, while the raw materials do not contribute. Brining baths may contain up to $10^6$ salt-tolerant yeasts per mL, resulting in a contamination of $10^6$ yeasts/dm$^2$ of the cheese surface. In addition, molds developing on the side of the brining bath may contaminate the brine up to $10^3$ cfu/mL. Molds are always present inside wooden shelves, despite the application of cleaning and disinfection. Moreover, Mycelium fragments and mold spores may be circulated by the air-conditioning systems of the ripening rooms (Stark, 2007).

Fungal development on the cheese surface is a very common shelf life-limiting factor for cheese, especially when vacuum packaging is not applied, or fails during ripening. This brief description of this defect is based on Sheehan (2007b) and Choisy et al. (1986b). Blue-mold defect is due to *Penicillium* spp., and black spot is due to the genera *Mucor* spp. and *Rhizopus* spp. The means of dispersal of mold spores depends on species. The spores of *Penicillium* spp. are dry and, therefore, they are spread through air, unlike the moist spores of Mucor spp., which are spread through wet surfaces in humid environments. Blue mold is associated with insufficient acid formation. The growth of *Penicillium roqueforti* under the low oxygen concentration within the mechanical openings of the cheese mass of soft or white-mold cheeses is a serious defect. Insufficient acidity, poor hygiene practices, and excess humidity during production and ripening are responsible for the appearance of black spots. The growth of acid tolerant *Cladosporium herbarum* and *Scopulariopsis fiscal* on mature cheese surface and on paper and packaging materials also limits shelf life. The former forms small dark-green to black spots on cheeses ripened or stored at low temperatures without appropriate acidification, or when pH increase occurs during ripening. It can be expanded to any part of ripening or storage rooms. *S. fiscal* forms dry, floury, brownish spots on the cheese surface and, due to its acid-intolerance, succeeds *Penicillium*. The "thread-mold" defect in the folds and creases of vacuum-packed Cheddar cheese blocks is usually due to free whey from the cheese blocks and is associated with the growth of *Penicillium* spp., *Cladosporium* spp., and *Phoma* spp. species. Finally, although the presence of Aspergillus in cheese is not common, nonsterilized pepper can contaminate cheese with *Aspergillus flavus* (Choisy et al., 1986b).

Lund et al. (1995) found that 91% of fungi isolates from cheeses belonging to various groups were *Penicillium* spp., 45% of them were *Penicillium commune*, and 24% *Penicillium nalgiovense*. The abundance of *P. commune* among the isolates from 48 spoiled hard cheeses has been also reported by Garnier et al. (2017), although at 22% of the total isolates, followed by *Penicillium bialowiezense* at 18% of total isolates. A few spoilage yeasts were isolated (3 isolates) from 48 spoiled hard cheeses (*Y. lipolytica*, *Sporodiobolus salmonicolor*, *D. hansenii*) as opposed to 14 isolates from 6 fresh cheese samples. In the latter, the most abundant were *Trichosporon asahii*, at 35% of the isolates, followed by *Pichia fermentans*, *Rhodotorula mucilagilosa*, and *Kluyveromyces marxianus*. They also observed that many species were identified in both spoiled products and their corresponding dairy plant environment, suggesting that the latter is an important source of contamination.

Garnier et al. (2017) present in brief the methods that can be applied in combination to avoid or delay fungal spoilage of cheeses. Prevention methods (HACCP, good manufacturing and hygiene practice, air filtration) along with hurdle technologies (heat treatments, cold storage, appropriate packaging, sodium chloride, fermentation) are used. Moreover, natamycin and organic acids and their salts (acetic, lactic, propionic, sorbic, benzoic) can be used for the treatment of the cheese surface. However, they reported that a few of their isolates presented very high resistance to various preservatives; for example, *Y. lipolytica* significantly resisted natamycin, calcium propionate, and sodium benzoate. A biocontrol of molds on cheese surface has been proposed by Delgado et al. (2016).

### 4.3.3 Flavor defects

Flavor defects of microbial origin also limit the shelf life of cheese (Choisy et al., 1986a; McSweeney, 2007). Bitter taste can develop during the ripening of cheese due to the accumulation of <6 kDa hydrophobic peptides from the hydrolysis of caseins by rennet or by cell-bound proteases of bacteria. Low fat and low salt content in cheeses enhance the production of bitter peptides, which accumulate when peptidase activity in the cheese mass is weak. For this reason, bitterness can be considered as a defect of microbial origin. The development of bitterness has been extensively reviewed by Lemieux and Simard (1991, 1992). The low content of polyunsaturated fatty acids and the low redox potential in cheese does not favor lipid oxidation of fat in milk during ripening. However, the accumulation of FAA due to hydrolysis of triaglycerols may result in undesirable rancid flavor in particular cheese varieties. Use of raw milk or milk with high psychrotrophic counts, prolonged ripening, molds or other surface microorganisms, or specific enzyme preparations can induce rancid flavor. More information about lipolysis is given by Collins et al. (2003). Some examples of cheese aroma defects given by Hardy (1986) are: the production of styrene by *Penicillium caseicolum*, potato taste due to 2-methoxy-3-isopropylpyrazine produced by certain *Pseudomonas* species, rancidity due to psychrotrophs in soft cheese, fat oxidation due to exposure to light or due to the use of frozen curd, and rancidity due to excessive free fatty acid content.

### 4.3.4 Formation of biogenic amines

O'Brien and O'Connor (2007) briefly reviewed the formation of biogenic amines due to decarboxylation of amino acids during cheese ripening. The principal biogenic amines detected in cheese are histamine, tyramine, tryptamine, putrescine, cadaverine, and phenylethylamine. Their presence is consistent with heavy contamination by spoilage microorganisms and with unpasteurized cheese milk. Bacteria of *Enterobacteriaceae* family and nonstarter lactobacilli are the main cause of biogenic amine formation following proteolysis during cheese ripening.

## 4.4 Butter

Butter contains over 80% milk fat; because of this, oxidation is the most common deterioration during shelf life, and can be prevented by exclusion of oxygen from packaging. Microbial spoilage resulting from lipolysis at low temperature may be caused by psychrotrophic bacteria and yeasts, while growth of pathogens on this product is limited.

Butter has limited shelf life at room temperature. It is best stored at 4°C for 3 months and at freezing conditions for longer time. Traditional Tunisian butter can only be kept for 2 days and 12 days at 10°C and 4°C,respectively, due to growth of yeasts and fungi, whereas the produced butter oil can be kept up to 14 days (Samet-Bali et al., 2009). Finally, the use of irradiated fermented cream from sheep milk does not extend the shelf life of produced butter (Carvalho Santos et al., 2017).

## 4.5 Milk powders

Milk powders, because of their low $a_w$, can be kept for several months, but it is important for their moisture content to range between 2% and 4% to avoid deterioration during shelf life. Shelf life of whole milk powder depends on preheated treatment of milk, the $a_w$, and the storage conditions (Stapelfeldt et al., 1997). It is also influenced by changes in fatty acid profile caused by microbial thermostable lipases. Sensory quality, lipid oxidation, and free radicals are more affected in low-heated powder than in medium- or high-heated powder during storage at 45°C. Water activities between 0.11 and 0.23 assure the quality of the whole milk powder independently of the heat treatment.

Finally, skimmed milk powder stored in barrier bags at normal ambient temperatures has a shelf life for over one year. Quality defects may be caused by spores and heat-stable enzymes from *Bacillus* spp. and *Clostridium* spp. (McHugh et al., 2017; Sadiq et al., 2018).

## 4.6 Quality changes during shelf life of meat products

The most frequent quality changes that may occur in meat and meat products are the development of off-odors, pigments, and slime. All these have been attributed to microbial growth. Depending on the type of product, in terms of composition, processing, and storage conditions, different microorganisms may dominate the respective microecosystem, leading to specific defects.

Meat is an excellent substrate for microbial growth, as it can provide adequate amounts of glucose, proteins, amino acids, nucleotides, and lipids. As a result, nutrient supply may not be considered as a limiting factor. On the other hand, the factors that affect microbial development and, consequently, the extent of quality changes in the final meat product are the intrinsic characteristics of the product (e.g., pH value, $a_w$, salt and nitrite contents, etc.), the storage conditions (temperature, packaging atmosphere), and the initial microbial load.

The initial microbial load is determined by the microbiological quality of the meat, casings, additives, premises, and utensils. The microecosystem of fresh meat under chill aerobic storage is dominated by Gram-negative rods, especially pseudomonads. Members of the Enterobacteriaceae family, yeasts-molds as well as Gram-positive bacteria, including lactic acid bacteria, *Brochothrix thermosphacta*, and spore formers constitute the secondary fresh meat microbiota (von Holy et al., 1992; Gill, 2005; Nychas et al., 2008) and dominate under modified atmosphere packaging conditions. The importance of casings and additives has been adequately highlighted by Houben (2005) and by Paramithiotis and Drosinos (2010), respectively. In the first study, 138 casing samples were analyzed and 53 were found positive in sulfite-reducing clostridia. In the latter study, a wide variety of spices, spice mixtures, and additives were analyzed. Although absence of foodborne pathogens was reported, opposing previous studies (Schwab et al., 1982; De Boer et al., 1985; Garcia et al., 2001; Banerjee and Sarkar, 2003; Sagoo et al., 2009), total aerobic mesophilic counts exceeding 7 log CFU/g were reported in many cases. Finally, sufficient cleaning and disinfection of premises and utensils is necessary to avoid cross-contamination of the raw materials or even the final products (Chevallier et al., 2006; Lebert et al., 2007).

Microbial metabolic activity is detectable as off-odors, when the respective population reaches $10^7$ CFU/g, or $cm^2$ in the case of surfaces. Slime is observable at populations near $10^8$ CFU/g or $cm^2$ (Garcia-Lopez et al., 1998). In general, pseudomonads dominate the microecosystem upon storage under aerobic conditions and at low temperatures, usually below 20°C; above that, mesophilic Enterobacteriaceae or lactic acid bacteria may dominate. However, specific characteristics of the product may alter the spoilage potential of some species by affecting the respective growth rate and, consequently, the microecosystem composition and development.

A wide variety of meat products currently exist. They may be classified into fresh, cured, dried, and fermented meat products, which, from a microbiological perspective, present many common properties. In order to facilitate understanding of the parameters that affect the microecosystem development and, consequently, the quality changes that take place during their shelf life, the aforementioned categories are separately discussed.

## 4.7 Fresh meat

The factors that affect the development of fresh meat microecosystem and, consequently, the observed quality changes during shelf life, are the characteristics of the product (nutrient composition, pH value), storage temperature, and packaging. In the scientific literature, a classic paper by Ingram and Dainty (1971), almost half a century ago, provided at that time the foundation to our understanding of this topic.

Meat stored aerobically will soon demonstrate off-odor development and discoloration due to the rapid growth of pseudomonads. *Pseudomonas fragi*, *P. fluorescens*, and *P. lundensis* have been reported as the pseudomonads that most frequently drive spoilage (Drosinos and Board, 1994, 1995a; Labadie, 1999; Ercolini et al., 2007) under aerobic conditions and at temperatures below 20°C. At temperatures near or above 30°C, Enterobacteriaceae will dominate the microecosystem (Fernandez, 2009), resulting in the same defects. At chilled temperatures, and provided that surface drying is avoided, pseudomonads will also dominate. *Shewanella putrefaciens* may also grow and contribute to discoloration and off-odor development, but only when the pH value of the meat is above 6.0 (dark, firm, dry meat (DFD)). On the other hand, if the meat surface is dried, the development of yeasts and molds will take place (Fernandez, 2009).

Meat stored under vacuum may also develop off-odors and gas production during shelf life. The latter may be due to the development of psychrophilic and psychrotrophic clostridia species, such as *Clostridium algidicarnis*, *C. algidixylanolyticum*, *C. estertheticum*, *C. frigidicarnis*, *C. gasigenes*, *C. laramiense*, and *C. putrefaciens* (Dainty et al., 1989; Lawson et al., 1994; Broda et al., 1996, 2000, 2002; Spring et al., 2003). Off-odors may be produced by lactic acid bacteria that dominate the microecosystem under these conditions, as well as by members of the Enterobacteriaceae family, such as *Hafnia alvei*, *Serratia* spp., and *Enterobacter* spp. (Hernandez-Macedo et al., 2011).

Storage under increased $CO_2$ concentration inhibits growth of pseudomonads, but favors growth of members of the Enterobacteriaceae family such as *Serratia* spp., *Proteus* spp., *H. alvei*, *Rahnella* spp., and lactic acid bacteria (Drosinos and Board, 1995b; Doulgeraki et al., 2012). This results in defects such as sour smell due to production of D-lactic acid, acetic acid, and drop in pH value. In an effort to delay detrimental changes due to microbial activity, efforts were also made with supplementation of meat ecosystems with energy sources, mainly glucose (Lambropoulou et al., 1996).

## 4.8 Cured meat products

Cured meat products, that is, products that have been treated with sodium chloride and nitrate/nitrite along with several other ingredients, may be subdivided into uncooked (e.g., bacon, dry ham) and cooked cured products (e.g., luncheon meats, frankfurters). In general, microbial stability during shelf life is ensured by the addition of sodium chloride, nitrite, packaging, and refrigerated storage. However, a wide diversity of products belongs to this category made by different recipes and procedures, altering the relative importance of the aforementioned hurdles. For that purpose and for reasons of clarity, the quality changes during shelf life of representative ones will be discussed in more detail.

Stability of bacon is ensured by the presence of sodium chloride, nitrite, packaging under vacuum or modified atmosphere, and refrigerated storage. In addition, in some products, cooking and/or smoking may be also applied. When the pH value of the product is below 5.4, lactic acid bacteria may develop during storage resulting in souring of the product (Vignolo et al., 1988). When the pH value is above 5.5–5.6 or the storage temperature exceeds 30°C, Enterobacteriaceae may develop and the product becomes putrid. Organoleptic deviations may also occur upon growth of yeast and molds due to their proteolytic and lipolytic activities when packaging is compromised.

Stability of dry-cured ham relies on the low water activity (<0.90) and the use of sodium chloride and nitrite. Growth of molds on the surface of the ham as well as "bone taint" or "deep spoilage" are the major problems that may be encountered both during production and shelf life. The first may result in organoleptic deviations through proteolytic and lipolytic activities; it may also pose a safety concern due to the production of mycotoxins (Chiavaro et al., 2002; Pietri et al., 2006; Toscani et al., 2007; Pizzolato Montanha et al., 2018). Bone-taint, on the other hand, results from growth of proteolytic Gram-positive catalase, positive cocci, and/or members of the Enterobacteriaceae family. Proteolysis results in the production of off-flavors, due to the production of several metabolic end products including ammonia and sulfides, as well as a pasty texture (Paarup et al., 1999; Garcia et al., 2000; Martin et al., 2008).

The stability of cooked cured meat products is based on the addition of curing agents, thermal treatment, packaging, and storage temperature. In the case of shelf-stable products (e.g., luncheon meats), canning and subsequent sterilization take place and, provided that integrity of the package is maintained, no quality changes are expected during shelf life. As far as the perishable products are concerned (e.g., frankfurters), with water activity values above 0.94 and pH value around 6.0, they may be characterized as ideal substrates for microbial growth. An important change that is noted in these products is growth of lactic acid bacteria that lead to discoloration and pH reduction, which in turn results in water exudation in the packaged slides of these products (Anifantaki et al., 2002; Pexara et al., 2002). This is also depicted by the large number of outbreaks that have been reported (Table 4). Thus, stability of these products is ensured by the addition of antimicrobial agents, vacuum or modified

**Table 4** Foodborne outbreaks in the United States in the years 1998–2016 having as food vehicle perishable cooked cured meat products

| Year | State | Etiology | Food vehicle | I/D |
|---|---|---|---|---|
| 1998 | South Carolina | *St. aureus* | Multiple deli meat | 35/0 |
| 1998 | New York | *Y. enterocolitica* | Multiple deli meat | 9/0 |
| 1999 | Florida | *L. monocytogenes* | Multiple deli meat | 2/1 |
| 2000 | Multistate | *L. monocytogenes* | Sliced turkey | 29/7 |
| 2001 | Florida | Norovirus Genogroup II | Multiple deli meat | 9/0 |
| 2002 | Florida | *S. enterica*, Norovirus Genogroup II | Multiple deli meat | 18/0 |
| 2002 | Multistate | *L. monocytogenes* | Sliced turkey | 54/8 |
| 2003 | Nevada | *S. Reading* | Sliced turkey | 6/0 |
| 2003 | California | *S. Enteritidis* | Multiple deli meat | 76/0 |
| 2003 | North Dakota | Norovirus Genogroup I | Sliced turkey | 34/0 |
| 2005 | Connecticut | Norovirus Genogroup I | Sliced turkey | 6/0 |
| 2005 | Multistate | *L. monocytogenes* | Sliced turkey | 13/1 |
| 2006 | New York | *S. Saintpaul* | Sliced turkey | 3/0 |
| 2008 | Minnesota | *S. Enteritidis* | Sliced turkey ham | 46/0 |
| 2008 | Iowa | *S. Hadar* | Sliced turkey | 3/0 |
| 2009 | Minnesota | Norovirus | Sliced turkey | 6/0 |
| 2011 | Multistate | *E. coli* O157:H7 | Bologna | 14/0 |

*I/D*, Illnesses/Deaths; *E.*: *Escherichia*; *L.*: *Listeria*; *S. Salmonella*; *St.*: *Staphylococcus*; *Y.*: *Yersinia*.
Data from www.cdc.gov.

atmosphere packaging, and refrigerated storage. However, gas production, as well as the development of ropy slime, discolorations, as mentioned previously, and off-flavors is very often observed. Depending on the compromised antimicrobial agent, Gram-positive or Gram-negative bacteria, or even yeasts and molds, may develop, according to the principles mentioned in the previous paragraph.

## 4.9 Dried meat products

Dried meat products, like jerky and biltong, can be made from a variety of species (such as beef, poultry, game, etc.), marinating recipes, and techniques, as well as drying procedures (Prior, 1984; Calicioglu et al., 2002, 2003; Ingham et al., 2004, 2006; Boles et al., 2007; Burnham et al., 2008; Porto-Fett et al., 2008, 2009; Naidoo and Lindsay, 2010). Their common characteristic is the low water activity that is usually reported between 0.60 and 0.80 (Holley, 1985; Harrison et al., 1998; Ingham et al., 2004, 2006). However, depending on the drying process, lower or higher $a_w$ values can also be retrieved from the literature (Albright et al., 2003; Yoon et al., 2005, 2006; Buege et al., 2006; Allen et al., 2007; Boles et al., 2007; Borowski et al., 2009a,b). Water activity is a critical property for both quality and safety of the product. For that purpose, USDA has recommended a maximum of 0.85, accompanied by a maximum moisture to protein ratio of 0.75:1 (USDA, 2014). However, the low $a_w$ may only inhibit growth of spoilage or pathogenic microorganisms. Thus, USDA has also recommended a cooking step before dehydration and have proposed time, temperature, and humidity combinations that may achieve a 5-log population reduction in *Salmonella* and *E. coli* O157:H7 (USDA, 2014). These recommendations, as well as their regular update, are the result of several outbreaks linked to the consumption of jerky (Table 5) as well as to the reported prevalence of foodborne pathogens (Eidson et al., 2000).

The cooking step may eliminate the majority of the microorganisms that have contaminated the raw materials. Then, the low $a_w$ obtained by the dehydration step may inhibit growth of the remaining ones. If the $a_w$ is equal to, or less than, 0.85, the product may be stored in an oxygen-containing environment, provided that surface mold growth is prevented. If the $a_w$ is equal to, or less than, 0.91, the product should be stored in an anaerobic environment. There recommendations were provided by the USDA and were based on the growth limits of *St. aureus* under the respective conditions (USDA, 2014).

During shelf life, if the integrity of the packaging is maintained, no microbial growth and, consequently, no quality changes are expected to occur. On the other hand, rancidity may develop when fatty tissues instead of lean meat are used.

## 4.10 Fermented meat products

A wide variety of fermented meat products has been studied worldwide (Kumar et al., 2017). These products may differ by the type and amount of the raw materials and seasonings used as well as fermentation and ripening conditions. Their quality, safety, and stability are ensured by a series of hurdles that are applied sequentially and their relative importance may differ, depending on the production procedure. More accurately, nitrite, sodium chloride, and fermentation temperature are the first hurdles that are applied, and these inhibit growth of a significant proportion of the initial batter microbiota. The members of the microecosystem that are not inhibited utilize the available oxygen constraining growth of aerobic microorganisms. In these conditions, lactic acid bacteria prevail, resulting in the reduction of the pH value and accumulation of acidity. In addition, drying reduces water activity, which constitutes the last hurdle

**Table 5** Foodborne outbreaks in the United States in the years 1998–2016 having as food vehicle various types of jerky

| Year | Product | State | Etiology | I/D |
|---|---|---|---|---|
| 2001 | Beef jerky | California | *S. Typhimurium* var. Cope | 6/0 |
| 2001 | Rabbit jerky | Ohio | *St. aureus* | 6/0 |
| 2002 | Jerky | South Dakota | Norovirus Genogroup I | 19/0 |
| 2003 | Beef jerky | New Mexico | *S. Kiambu* | 23/0 |
| 2005 | Beef jerky | Arkansas | *E. coli* O157:H7 | 3/0 |
| 2007 | Jerky | Pennsylvania | *E. coli* O157:H7 | 8/0 |
| 2012 | Turkey jerky | Minnesota | *S. enterica* | 5/5 |

*I/D*, Illnesses/Deaths; *E.*: *Escherichia*; *S. Salmonella*; *St.*: *Staphylococcus*.
Data from www.cdc.gov.

**Table 6** Foodborne outbreaks in the years 1998–2016 having as food vehicle various types of fermented meat products

| Year | Country/State | Etiology | I/D |
|---|---|---|---|
| 2001 | Alaska | *C. botulinum* toxin | 14 |
| 2001 | Arkansas | *St. aureus* | 32/0 |
| 2002 | Sweden | *E. coli* O157:H7 | 38/0 |
| 2004 | Italy | *S. Typhimurium* DT104A | 63/0 |
| 2005 | Sweden | *S. Typhimurium* NST | 15/0 |
| 2006 | Norway | *S. Kedougou* | 54/1 |
| 2007 | Italy | *E. coli* 0157 | 2/0 |
| 2009–10 | United States-multistate | *S. Montevideo* | 272/0 |
| 2010 | Denmark | *S. Typhimurium* | 20/ |
| 2010 | Oregon | Norovirus Genogroup I | 12/0 |

*C., Clostridium; E., Escherichia; S., Salmonella; St., Staphylococcus.*
Data from www.cdc.gov and www.eurosurveillance.org.

that is applied. All these will occur when fermentation, ripening, and drying are effectively controlled, which is not always the case. The latter is particularly exhibited by the number of the outbreaks linked to fermented meat consumption (Table 6).

During production and shelf life, uncontrolled growth of yeasts-molds, pseudomonads and members of the Enterobacteriaceae family may result in a series of quality changes including physicochemical and sensorial defects, discoloration, and slime production.

An important technological role has been attributed to yeasts and molds. Molds require oxygen for their growth and thus they grow on the surface of the product, forming a mycelium that allows homogeneous and not excessive dehydration to occur during production. Yeasts, on the other hand, apart from oxygen, may also utilize organic molecules for electron acceptors, which increases the depth to which they may develop. The latter action depletes oxygen, enhances curing reactions, and delays lipid oxidation. Both yeasts and molds may also contribute significantly to flavor development through their proteolytic and lipolytic activities. Uncontrolled growth may result in significant defects, both during production and shelf life. In the first case, early mycelium development may prevent dehydration. Surface growth during storage, which is the result of compromised packaging, may result in discoloration, slime formation, development of off-flavors, and even mycotoxin production (Iacumin et al., 2009; Markov et al., 2013; Pleadin et al., 2015).

The population of Gram-negative bacteria in the initial batter depends upon the quality of the raw materials; they have been reported from below detection limit to 5 log cfu/g (Samelis et al., 1998; Aymerich et al., 2003; Drosinos et al., 2005; Comi et al., 2005; Fontana et al., 2005; Rantsiou et al., 2005; Lebert et al., 2007). Their fate during fermentation depends upon the speed with which lactic acid bacteria dominate. If LAB fail to grow fast and develop an acidic environment, and/or the quality of the raw materials is poor, Gram-negative bacteria may remain in high population and their metabolic activity becomes organoleptically detectable. Pseudomonads, being oxidative microorganisms, may form a biofilm on the surface of the fermenting batter, slowing dehydration rate. During shelf life, pseudomonads may grow on the surface if the packaging integrity is compromised leading to discoloration and off-odor development. Enterobacteriaceae, on the other hand, may grow and result in the same defects only if the acidification of the product is not adequate.

Lactic acid bacteria and staphylococci are considered as the major technological biota. During shelf life, no defects resulting from their uncontrolled growth are expected to occur. However, during production several problems may be encountered, especially in the case of spontaneous fermentations. Discoloration is one such problem that may result from ineffective staphylococcal nitrate reductase activity; in that case, metmyoglobin instead of nitrosylmyoglobin is formed, leading to products with brown instead of bright red color. In addition, greening, gas production, and sensorial deviations may be observed in spontaneous fermentations due to the production of peroxides, carbon dioxide, and a mixture of organic acids, respectively, when heterofermentative instead of the technologically desired homofermentative LAB prevail (Anifantaki et al., 2002).

## 5 Conclusion

Concluding, it is obvious that shelf life of foods, and especially of dairy and meat foods, is fully depended on both the intrinsic and the extrinsic factors, which, in many cases, interact with each other. Consequently, maintenance of food quality is a complex procedure, especially when the food is perishable. For this reason, good manufacturing practice and good hygienic practice should be applied during food manufacture. In addition, further knowledge of the interactions between the factors that affect food quality is needed in order to design new food processing methods and materials to maintain food quality for a longer time.

## References

Albright, S.N., Kendall, P.A., Avens, J.S., Sofos, J.N., 2003. Pretreatment effect on inactivation of *Escherichia coli* O157:H7 inoculated beef jerky. Lebensm. Wiss. Technol. 36 (4), 381–389.

Alichanidis, E., 2007. What causes early and late gas blowing in white-brined cheeses? In: McSweeney, P.L.H. (Ed.), Cheese Problems Solved. Woodhead Publishing, Cambridge, pp. 161–162.

Al-Kadamany, E., Khattar, M., Haddad, T., Toufeili, I., 2003. Estimation of shelf-life of concentrated yogurt by monitoring selected microbiological and physicochemical changes during storage. Lebensm. Wiss. Technol. 36, 407–414.

Allen, K., Cornforth, D., Whittier, D., Vasavada, M., Nummer, B., 2007. Evaluation of high humidity and wet marinade methods for pasteurization of jerky. J. Food Sci. 72 (7), C351–C355.

Alothman, M., Lusk, K., Silcock, P., Bremer, P., 2018. Relationship between total microbial numbers, volatile organic compound composition, and the sensory characteristics of whole fresh chilled pasteurized milk. Food Packag. Shelf Life 15, 69–75.

Andersson, I., Osten, R., 1994. Nutritional quality of pasteurized milk. Vitamin $B_{12}$, folate and ascorbic acid content during storage. Int. Dairy J. 4, 161–172.

Anifantaki, K., Metaxopoulos, J., Kammenou, M., Drosinos, E.H., Vlasi, M., 2002. The effect of smoking, packaging and storage temperature on the bacterial greening of frankfurters caused by *Leuconostoc mesenteroides* subsp. *mesenteroides*. Ital. J. Food Sci. 14, 135–144.

Ávila, M., Gómez-Torres, N., Delgado, D., Gaya, P., Garde, S., 2016. Application of high pressure processing for controlling *Clostridium tyrobutyricum* and late blowing defect on semi-hard cheese. Food Microbiol. 60, 165–173.

Aymerich, T., Martin, B., Garriga, M., Hugas, M., 2003. Microbial quality and direct PCR identification of lactic acid bacteria and nonpathogenic Staphylococci from artisanal low-acid sausages. Appl. Environ. Microbiol. 69, 4583–4594.

Banerjee, M., Sarkar, P.K., 2003. Microbiological quality of some retail spices in India. Food Res. Int. 36, 469–474.

Barbano, D.M., Ma, Y., Santos, M.V., 2006. Influence of raw milk quality on fluid milk shelf life. J. Dairy Sci. 89 (Suppl 1), E15–E19.

Boles, J.A., Neary, K., Clawson, K., 2007. Survival of *Listeria monocytogenes* on jerky contaminated postprocessing. J. Muscle Foods 18 (2), 186–193.

Boor, K.J., 2001. Fluid dairy product quality and safety: looking to the future. J. Dairy Sci. 84 (1), 1–11.

Borowski, A.G., Ingham, S.C., Ingham, B.H., 2009a. Lethality of home-style dehydrator processes against *Escherichia coli* O157:H7 and *Salmonella* serovars in the manufacture of ground-and-formed beef jerky and the potential for using a pathogen surrogate in process validation. J. Food Prot. 72 (10), 2056–2064.

Borowski, A.G., Ingham, S.C., Ingham, B.H., 2009b. Validation of ground-and-formed beef jerky processes using commercial lactic acid bacteria starter cultures as pathogen surrogates. J. Food Prot. 72 (6), 1234–1247.

Brändle, J., Domig, K.J., Kneifel, W., 2016. Relevance and analysis of butyric acid producing clostridia in milk and cheese. Food Control 67, 96–113.

Broda, D.M., Bell, R.G., Boerema, J.A., Musgrave, D.R., 2002. The abattoir source of culturable psychrophilic *Clostridium* spp causing "blown pack" spoilage of vacuum-packed chilled venison. J. Appl. Microbiol. 93 (5), 817–824.

Broda, D.M., De Lacy, K.M., Bell, R.G., Bragiins, T.J., Cook, R.L., 1996. Psychrotrophic *Clostridium* spp. associated with "blown pack" spoilage of chilled vaccum-packed red meats and dog rolls in gas impermeable plastic casings. Int. J. Food Microbiol. 29 (1–2), 335–352.

Broda, D.M., Saul, D.J., Lawson, P.A., Bell, R.G., Musgrave, D.R., 2000. *Clostridium gasigenes* sp. Nov. a psychrophile causing spoilage of vacuum-packed meat. Int. J. Syst. Evol. Microbiol. 50 (1), 107–118.

Brown, J.V., Ranjith, H.M.P., Prentice, G.A., 1984. Comparative shelf-lives of skimmed, semi-skimmed and whole milks. J. Soc. Dairy Technol. 37 (4), 132–135.

Buege, D.R., Searls, G., Ingham, S.C., 2006. Lethality of commercial whole-muscle beef jerky manufacturing processes against *Salmonella* serovars and *Escherichia coli* O157:H7. J. Food Prot. 69 (9), 2091–2099.

Burnham, G.M., Hanson, D.J., Koschick, C.M., Ingham, S.C., 2008. Death of *Salmonella* serovars, *Escherichia coli* O157:H7, *Staphylococcus aureus* and *Listeria monocytogenes* during the drying of meat: a case study using biltong and droewors. J. Food Saf. 28 (2), 198–209.

Calicioglu, M., Sofos, J.N., Kendall, P.A., 2003. Influence of marinades on survival during storage of acid-adapted and nonadapted *Listeria monocytogenes* inoculated post-drying on beef jerky. Int. J. Food Microbiol. 86 (3), 283–292.

Calicioglu, M., Sofos, J.N., Samelis, J., Kendall, P.A., Smith, G.C., 2002. Destruction of acid- and non-adapted *Listeria monocytogenes* during drying and storage of beef jerky. Food Microbiol. 19 (6), 545–559.

Carvalho Santos, I., Pinto, J., Pimenta, A.I., Madureira, J., Matos, P., Viegas, C., Raposo, A., Margaça, F.M.A., Cabo Verde, S., 2017. Use of gamma radiation in sheep butter manufacturing process for shelf-life extension. Int. Dairy J. 71, 43–49.

Chevallier, I., Ammor, S., Laguet, A., Labayle, S., Castanet, V., Dufour, E., Talon, R., 2006. Microbial ecology of a small-scale facility producing traditional dry sausage. Food Control 17, 446–453.

Chiavaro, E., Lepiani, A., Colla, F., Bettoni, P., Pari, E., Spotti, E., 2002. Ochratoxin A determination in ham by immunoaffinity clean-up and a quick fluorometric method. Food Addit. Contam. 19, 575–581.

Choisy, C., Desmazeaud, M., Gripon, J.C., Lamberet, G., Lenoir, J., Tourneur, C., 1986a. Ripening, a complex biochemical ripening. In: Eck, A. (Ed.), translated by C.D. Thomson. Cheesemaking-Science and Technology, second ed. Lavoisier Publishing Inc., New York, pp. 62–96.

Choisy, C., Gugeuen, M., Lenoir, J., Schmidt, J.L., Tourneur, C., 1986b. Microbiological aspects. In: Eck, A. (Ed.), translated by C.D. Thomson. Cheesemaking-Science and Technology, second ed. Lavoisier Publishing Inc., New York, pp. 259–292.

Collins, Y.F., McSweeney, P.L.H., Wilkinson, M.G., 2003. Lipolysis and free fatty acid catabolism in cheese: a review of current knowledge. Int. Dairy J. 13, 841–866.

Comi, G., Urso, R., Iacumin, L., Rantsiou, K., Cattaneo, P., Cantoni, C., Cocolin, L., 2005. Characterisation of naturally fermented sausages produced in the north east of Italy. Meat Sci. 69, 381–392.

Dainty, R.H., Edwards, R.A., Hibbarb, C.M., 1989. Spoilage of vacuum-packed beef by a *Clostridium* sp. J. Sci. Food Agric. 49 (4), 473–486.

De Boer, E., Spiegelenberg, W.M., Janssen, F.W., 1985. Microbiology of spices and herbs. Antonie Van Leeuwenhoek 51, 435–438.

Delgado, J., Peromingo, B., Núñez, F., Asensio, M.A., 2016. Use of molds and their antifungal proteins for biocontrol of toxigenic molds on dry-ripened cheese and meats. Curr. Opin. Food Sci. 11, 40–45.

D'Incecco, P., Pellegrino, L., Hogenboom, J.A., Cocconcelli, P.S., Bassi, D., 2018. The late blowing defect of hard cheeses: behaviour of cells and spores of *Clostridium tyrobutyricum* throughout the cheese manufacturing and ripening. LWT- Food Sci. Technol. 87, 134–141.

Donnelly, C.W., 2007. What cheeses are the most liable to pathogens? Do pathogens grow during cheese ripening? In: McSweeney, P.L.H. (Ed.), Cheese Problems Solved. Woodhead Publishing, Cambridge, pp. 134–139.

Doulgeraki, A.I., Ercolini, D., Villani, F., Nychas, G.-J.E., 2012. Spoilage microbiota associated to the storage of raw meat in different conditions. Int. J. Food Microbiol. 157, 130–141.

Drosinos, E.H., Board, R.G., 1995b. A survey of minced lamb packaged in modified atmospheres. Fleischwirtschaft 75, 281–284.

Drosinos, E.H., Board, R.G., 1994. Metabolic activities of pseudomonads in batch cultures in extract of minced lamb. J. Appl. Bacteriol. 77, 613–620.

Drosinos, E.H., Board, R.G., 1995a. Microbial and physicochemical attributes of minced lamb: sources of contamination with pseudomonads. Food Microbiol. 12, 189–197.

Drosinos, E.H., Mataragas, M., Xiraphi, N., Moschonas, G., Gaitis, F., Metaxopoulos, J., 2005. Characterization of the microbial flora from a traditional Greek fermented sausage. Meat Sci. 69, 307–317.

Eidson, M., Sewell, C.M., Graves, G., Olson, R., 2000. Beef jerky gastroenteritis outbreaks. J. Environ. Health 62, 9–13.

Ercolini, D., Russo, F., Blaiotta, G., Pepe, I., Mauriello, G., Villani, F., 2007. Simultaneous detection of *Pseudomonas fragi*, *P. lundensis*, and *P. putida* from meat by use of a multiplex PCR assay targeting the *carA* gene. Appl. Environ. Microbiol. 73, 2354–2359.

European Commission, Regulation (EU) No. 1169/2011 on the provision of food information to consumers, amending Regulations (EC) No. 1924/2006 and (EC) No. 1925/2006 of the European Parliament and of the Council, and repealing Commission Directive 87/250/EEC, Council Directive 90/496/EEC, Commission Directive 1999/10/EC, Directive 2000/13/EC of the European Parliament and of the Council, Commission Directives 2002/67/EC and 2008/5/EC and Commission Regulation (EC) No 608/2004, Off. J. Eur. Union. L304, 18–63.

Fernandes, A.M., Moretti, T.S., Bovo, F., Lima, C.G., Oliveira, C.A.F., 2008. Effect of somatic cell counts on lipolysis, proteolysis and apparent viscosity of UHT milk during storage. Int. J. Dairy Technol. 61 (4), 327–332.

Fernandez, R., 2009. Cured meats and poultry, including cooked cured meats. In: Fernandez, P. (Ed.), Microbiology Handbook-Meat Products. Leatherhead Food Research, Surrey, pp. 1–52.

Fontana, C., Cocconcelli, P.S., Vignolo, G., 2005. Monitoring the bacterial population dynamics during fermentation of artisanal Argentinean sausages. Int. J. Food Microbiol. 103, 131–142.

Garcia, C., Martin, A., Timon, M.L., Cordoba, J.J., 2000. Microbial populations and volatile compounds in the "bone taint" spoilage of dry-cured ham. Lett. Appl. Microbiol. 30, 61–66.

Garcia, S., Iracheta, F., Galvan, F., Heredia, N., 2001. Microbiological survey of retail herbs and spices from Mexican markets. J. Food Prot. 64, 99–103.

Garcia-Lopez, M.L., Prieto, M., Otero, A., 1998. The physiological attributes of gram-negative bacteria associated with spoilage of meat and meat products. In: Davies, A., Board, R. (Eds.), The Microbiology of Meat and Poultry. Blackie Academic & Professional, London, pp. 1–34.

Garnier, L., Valence, F., Pawtowski, A., Auhustsinava-Galerne, L., Frotté, N., Baroncelli, R., Deniel, F., Coton, E., Mounier, J., 2017. Diversity of Spoilage Fungi Associated with Various French Dairy Products. Int. J. Food Microbiol. 241, 191–197.

Gavahian, M., Chu, Y.-H., Mousavi Khaneghah, A., Barba, F.J., Misra, N.N., 2018. A critical analysis of the cold plasma induced lipid oxidation in foods. Trends Food Sci. Technol. 77, 32–41.

Gill, C.O., 2005. Sources of bacterial contamination at slaughtering plants. In: Sofos, J.N., (Ed.). Improving the Safety of Fresh Meat, CRC/Woodhead Publishing Limited, Cambridge, pp. 231–243.

Griffiths, M.W., Phillips, J.D., 1988. Modelling the relation between bacterial growth and storage temperature in pasteurized milks of varying hygienic quality. J. Soc. Dairy Technol. 41 (4), 96–102.

Guinee, T.P., Fox, P.F., 2004. Salt in cheese: physical, chemical and biological aspects. In: Fox, P.F., McSweeney, P.L.H., Cogan, T.M., Guinee, T.P. (Eds.), Cheese: Chemistry, Physics and Microbiology. vol. 1. Elsevier Ltd, London, pp. 207–259.

Hardy, J., 1986. Water activity and the salting of cheese. In: Eck, A. (Ed.), translated by C.D. Thomson Cheesemaking-Science and Technology, second ed. Lavoisier Publishing Inc., New York, pp. 37–61.

Haro-Vicente, J.F., Frontelaseta, K., Romeo-Braquehais, F., Ros Berruezo, G., 2013. Changes in content of vitamins A and E in growing-up milk throughout its shelf life. Int. J. Dairy Technol. 66 (1), 31–36.

Harrison, J.A., Harrison, M.A., Rose, R.A., 1998. Survival of *Escherichia coli* O157:H7 in ground beef jerky assessed on two plating media. J. Food Prot. 61 (1), 11–13.

Hernandez-Macedo, M.L., Barancelli, G.V., Contreras-Castillo, C.J., 2011. Microbial deterioration of vacuum-packaged chilled beef cuts and techniques for microbiota detection and characterization: a review. Braz. J. Microbiol. 42, 1–11.

Holley, R.A., 1985. Beef jerky: fate of *Staphylococcus aureus* in marinated and corned beef during jerky manufacture and 2.5 degrees C storage. J. Food Prot. 48 (2), 107–111.

Houben, J.H., 2005. A survey of dry-salted natural casings for the presence of *Salmonella* spp., *Listeria monocytogenes* and sulphite-reducing *Clostridium* spores. Food Microbiol. 22, 221–225.

Hymery, N., Vasseur, V., Coton, M., Mounier, J., Jany, J.-L., Barbier, G., Coton, E., 2014. Filamentous fungi and mycotoxins in cheese: a review. Compr. Rev. Food Sci. Food Saf. 13, 437–456.

Iacumin, L., Chiesa, L., Boscolo, D., Manzano, M., Cantoni, C., Orlic, S., Comi, G., 2009. Moulds and ochratoxin A on surfaces of artisanal and industrial dry sausages. Food Microbiol. 26, 65–70.

Ingham, S.C., Buege, D.R., Dropp, B.K., Losinski, J.A., 2004. Survival of *Listeria monocytogenes* during storage of ready-to-eat meat products processed by drying, fermentation, and/or smoking. J. Food Prot. 67 (12), 2698–2702.

Ingham, S.C., Searls, G., Mohanan, S., Buege, D.R., 2006. Survival of *Staphylococcus aureus* and *Listeria monocytogenes* on vacuum-packaged beef jerky and related products stored at 21 degrees C. J. Food Prot. 69 (9), 2263–2267.

Ingram, M., Dainty, R.H., 1971. Changes caused by microbes in spoilage of meats. J. Appl. Bacteriol. 34, 21–39.

Irlinger, F., Mounier, J., 2009. Microbial interactions in cheese: implications for cheese quality and safety. Curr. Opin. Biotechnol. 20, 142–148.

Irlinger, F., Layec, S., Hélinck, S., Dugat-Bony, E., 2015. Cheese rind microbial communities: diversity, composition and origin. FEMS Microbiol. Lett. 362 (2). Article number fnu015.

Ishikawa, M., Kodama, K., Yasuda, H., Okamoto-Kainuma, A., Koizumi, K., Yamasato, K., 2007. Presence of halophilic and alkaliphilic lactic acid bacteria in various cheeses. Lett. Appl. Microbiol. 44, 308–313.

Kaushik, R., Sachdeva, B., Arora, S., 2014. Vitamin D2 stability in milk during processing, packaging and storage. LWT- Food Sci. Technol. 56 (2), 421–426.

Kumar, P., Chatli, M.K., Verma, A.K., Mehta, N., Malav, O.P., Kumar, D., Sharma, N., 2017. Quality, functionality, and shelf life of fermented meat and meat products: a review. Crit. Rev. Food Sci. Nutr. 57 (13), 2844–2856.

Labadie, J., 1999. Consequences of packaging on bacterial growth. Meat is an ecological niche. Meat Sci. 52, 299–305.

Lambropoulou, K.A., Drosinos, E.H., Nychas, G.J.E., 1996. The effect of glucose supplementation on the spoilage microflora and chemical composition of minced beef stored aerobically or under a modified atmosphere at 4°C. Int. J. Food Microbiol. 30, 281–291.

Lawson, P., Dainty, R.H., Kristiansen, N., Berg, J., Collins, M.D., 1994. Characterisation of a psychrotrophic *Clostridium* causing spoilage in vacuum-packed cooked pork: description of *Clostridium algidicarnis* sp. nov. Lett. Appl. Microbiol. 19 (3), 153–157.

Lebert, I., Leroy, S., Giammarinaro, P., Lebert, A., Chacornac, J.P., Bover-Cid, S., Vidal-Carou, M.C., Talon, R., 2007. Diversity of microorganisms in environments and dry fermented sausages of French traditional small units. Meat Sci. 76, 112–122.

Lemieux, L., Simard, R.E., 1991. Bitter flavour in dairy products. I. A review of the factors likely to influence its development, mainly in cheese manufacture. Lait 71, 599–636.

Lemieux, L., Simard, R.E., 1992. Bitter flavour in dairy products (II): a review of bitter peptides from caseins: their formation, isolation and identification, structure, masking and inhibition. Lait 72, 335–382.

Lund, F., Filtenborg, O., Frisvad, J.C., 1995. Associated microflora in cheese. Food Microbiol. 12, 173–180.

Lutke-Entrup, M.L., Günther, H.-O., Van Beek, P., Grunow, M., Seiler, T., 2005. Mixed-integer linear programming approaches to shelf-life-integrated planning and scheduling in yoghurt production. Int. J. Prod. Res. 43 (23), 5071–5100.

Man, D.C.M., 2015. Shelf Life, second ed. Wiley Blackwell, UK.

Marcos, A., 1999. Water activity in cheese in relation to composition, stability and safety. In: Fox, P.F. (Ed.), Cheese: Chemistry, Physics and Microbiology, second ed. vol. 1. Aspen Publishers Inc., Gaittensburg, Maryland, pp. 439–470.

Markov, K., Pleadin, J., Bevardi, M., Vahcic, N., Sokolic-Mihalak, D., Frece, J., 2013. Natural occurrence of aflatoxin B1, ochratoxin A and citrinin in Croatian fermented meat products. Food Control 34, 312–317.

Martin, A., Benito, M.J., Hernandez, A., Perez-Nevado, F., Cordoba, J.J., Cordoba, M.G., 2008. Characterisation of microbial deep spoilage in Iberian dry cured ham. Meat Sci. 78, 475–484.

Martin, N.H., Boor, K.J., Martin Wiedmann, M., 2017. Effect of post pasteurization contamination on fluid milk quality. J. Diary Sci. 101 (1), 861–870.

McHugh, A.J., Feehily, C., Hill, C., Cotter, P.D., 2017. Detection and enumeration of spore-forming bacteria in powdered dairy products. Front. Microbiol. 8, 109.

McSweeney, P.L.H., 2007. Flavour, texture and flavour defects in hard and semi-hard cheeses. In: McSweeney, P.L.H. (Ed.), Cheese Problems Solved. Woodhead Publishing, Cambridge, pp. 189–207.

Moatsou, G., Moschopoulou, E., 2014. Microbiology of raw milk. In: Ozer, B., Akdemir-Evrendilek, G. (Eds.), Dairy Microbiology and Biochemistry: Recent Developments. CRC Press, Boca Raton, Florida, pp. 1–38.

Montel, M.-C., Buchin, S., Mallet, A., Delbes-Paus, C., Vuitton, D.A., Desmasures, N., Berthier, F., 2014. Traditional cheeses: rich and diverse microbiota with associated benefits. Int. J. Food Microbiol. 177, 136–154.

Mounier, J., Monnet, C., Vallaeys, T., Arditi, R., Sarthou, A.S., Hélias, A., Irlinger, F., 2008. Microbial interactions within a cheese microbial community. Appl. Environ. Microbiol. 74, 72–181.

Moyssiadi, M., Badeka, A., Kondyli, E., Vakirtzi, T., Savvaidis, I., Kontominas, M.G., 2004. Effect of light transmittance and oxygen permeability of various packaging materials on keeping quality of low fat pasteurized milk: chemical and sensorial aspects. Int. Dairy J. 14, 429–436.

Naidoo, K., Lindsay, D., 2010. Survival of *Listeria monocytogenes* and enterotoxin producing *Staphylococcus aureus* and *Staphylococcus pasteuri* during two types of biltong manufacturing processes. Food Control 21 (7), 1042–1050.

Nicoli, M.C., Calligaris, S., 2018. Secondary shelf life: an underestimated issue. Food Eng. Rev. 10 (2), 57–65.

Notermans, S., 1993. A user guide to microbial challenge testing for ensuring the safety and stability of food products. Int. J. Food Microbiol. 10, 145–147.

Nsabimana, C., Jiang, B., Kossah, R., 2005. Manufacturing, properties and shelf life of labneh: a review. Int. J. Dairy Technol. 58 (3), 129–137.

Nychas, G.J.E., Skandamis, P.N., Tassou, C.C., Koutsoumanis, K.P., 2008. Meat spoilage during distribution. Meat Sci. 78, 77–89.

O'Brien, N.M., O'Connor, T.P., 2007. What are biogenic amines and how are they produced? In: McSweeney, P.L.H. (Ed.), Cheese Problems Solved. Woodhead Publishing, Cambridge, pp. 148–149.

Ortuzar, J., Martinez, B., Bianchini, A., Stratton, J., Rupnow, J., Wang, B., 2018. Quantifying changes in spore-forming bacteria contamination along the milk production chain from farm to packaged pasteurized milk using systematic review and meta-analysis. Food Control 86, 319–331.

Paarup, T., Nieto, J.C., Pelaez, C., Reguera, J.I., 1999. Microbial and physicochemical characterisation of deep spoilage in Spanish dry cured hams and characterisation of isolated Enterobacteriaceae with regard to salt and temperature tolerance. Eur. Food Res. Technol. 209, 366–371.

Paramithiotis, S., Drosinos, E.H., 2010. Microbiological quality and aflatoxin B1 content of some spices and additives used in meat. Qual. Assur. Saf. Crops Food 2, 41–45.

Pexara, E.S., Metaxopoulos, J., Drosinos, E.H., 2002. Evaluation of shelf life of cured, cooked, sliced turkey fillets and cooked pork sausages—"piroski"—stored under vacuum and modified atmospheres at + 4° and + 10°C. Meat Sci. 62, 33–43.

Pietri, A., Bertuzzi, T., Gualla, A., Piva, G., 2006. Occurrence of ochratoxin A in raw ham muscles and in pork products from northern Italy. Ital. J. Food Sci. 18, 99–106.

Pinto, C., Souza, L.V., Meloni, V.S., Batista, C.S., Silva, R., Martins, E.M.F., Gomes-Cruz, A., Martins, M., 2018. Microbiological quality of Brazilian UHT milk: identification and spoilage potential of spore-forming bacteria. Int. J. Dairy Technol. 71 (1), 20–26.

Pinto, S.M., Clemente, M.D.G., Ronaldo de Abreu, L., 2009. Behaviour of volatile compounds during the shelf life of yoghurt. Int. J. Dairy Technol. 62 (2), 215–223.

Pizzolato Montanha, F., Anater, A., Burchard, J.F., Luciano, F.B., Meca, G., Manyes, L., Pimpão, C.T., 2018. Mycotoxins in dry-cured meats: a review. Food Chem. Toxicol. 111, 494–502.

Pleadin, J., Staver, M.M., Vahcic, N., Kovacevic, D., Milone, S., Saftic, L., Scortichini, G., 2015. Survey of aflatoxin B1 and ochratoxin A occurrence in traditional meat products coming from Croatian households and markets. Food Control 52, 71–77.

Porto-Fett, A.C.S., Call, J.E., Hwang, C.-A., Juneja, V., Ingham, S., Ingham, B., Luchansky, J.B., 2009. Validation of commercial processes for inactivation of *Escherichia coli* 0157:H7, *Salmonella* typhimurium and *Listeria monocytogenes* on the surface of whole-muscle Turkey jerky. Poult. Sci. 88 (6), 1275–1281.

Porto-Fett, A.C.S., Call, J.E., Luchansky, J.B., 2008. Validation of a commercial process for inactivation of *Escherichia coli* O157:H7, *Salmonella* typhimurium, and *Listeria monocytogenes* on the surface of whole muscle beef jerky. J. Food Prot. 71 (5), 918–926.

Prior, B.A., 1984. Role of micro-organisms in biltong flavour development. J. Appl. Bacteriol. 56 (1), 41–45.

Rantsiou, K., Urso, R., Iacumin, L., Cantoni, C., Cattaneo, P., Comi, G., Cocolin, L., 2005. Culture-dependent and -independent methods to investigate the microbial ecology of Italian fermented sausages. Appl. Environ. Microbiol. 71, 1977–1986.

Roos, Y.H., 2001. Water activity and plasticization. In: Eskin, N.A.M., Robinson, D.S. (Eds.), Food Shelf Life Stability: Chemical, Biochemical and Microbiological Changes. CRC Press, Boca Raton, Florida, pp. 3–36.

Sadiq, F.A., Flint, S., He, G., 2018. Microbiota of milk powders and the heat resistance and spoilage potential of aerobic spore-forming bacteria. Int. Dairy J. 85, 159–168.

Sagoo, S.K., Little, C.L., Greenwood, M., Mithani, V., Grant, K.A., McLauchlin, J., de Pinna, E., Threlfall, E.J., 2009. Assessment of the microbiological safety of dried spices and herbs from production and retail premises in the United Kingdom. Food Microbiol. 26, 39–43.

Samelis, J., Metaxopoulos, J., Vlassi, M., Pappa, A., 1998. Stability and safety of traditional Greek salami—a microbiological ecology study. Int. J. Food Microbiol. 44, 69–82.

Samet-Bali, O., Ayadi, M.A., Attia, H., 2009. Traditional Tunisian butter: physicochemical and microbial characteristics and storage stability of the oil fraction. LWT- Food Sci. Technol. 42 (4), 899–905.

Schoder, D., Zangerl, P., Manafi, M., Wagner, M., Lindner, G., Foissy, H., 2007. *Bacillus cereus*—a foodborne pathogen of different risk potential for the dairy industry [*Bacillus cereus*—Ein problemkeim der milchwirtschaft mit unterschiedlich eingeschätztem risikopotenzial]. Wien. Tierarztl. Monatsschr. 94 (1–2), 25–33.

Schwab, A.H., Harpestad, A.D., Swartzentruber, A., Lanier, J.M., Wentz, B.A., Duran, A.P., Barnard, R.J., Read Jr., R.B., 1982. Microbiological quality of some spices and herbs in retail markets. Appl. Environ. Microbiol. 44, 627–630.

Sheehan, J.J., 2007a. What causes the development of gas during ripening? In: McSweeney, P.L.H. (Ed.), Cheese Problems Solved. Woodhead Publishing, Cambridge, pp. 131–133.

Sheehan, J.J., 2007b. Why does mould develop under the packaging? In: McSweeney, P.L.H. (Ed.), Cheese Problems Solved. Woodhead Publishing, Cambridge, pp. 161–162.

Silcock, P., Alothman, M., Zardin, E., 2014. Microbially induced changes in the volatile constituents of fresh chilled pasteurized milk during storage. Food Packag. Shelf Life 2, 81–90.

Silvetti, T., Morandi, S., Brasca, M., 2018. Growth factors affecting gas production and reduction potential of vegetative cell and spore inocula of dairy-related Clostridium species. LWT- Food Sci. Technol. 92, 32–39.

Smet, K., De Block, J., De Campeneere, S., De Brabander, D., Herman, L., Raes, K., Dewettinck, K., Coudijzer, K., 2009. Oxidative stability of UHT milk as influenced by fatty acid composition and packaging. Int. Dairy J. 19, 372–379.

Spagnol, W.A., Silveira Jr., V., Pereira, E., Filho, N.G., 2018. Monitoring the cold chain: new technologies and recent advances [Monitoramento da cadeia do frio: Novas tecnologias e recentes avanços]. Braz. J. Food Technol. 21, 2–8.

Spring, S., Merkhoffer, B., Weiss, N., Kroppenstedt, R.M., Hippe, H., Stackebrandt, E., 2003. Characterization of novel psychrophilic clostridia from an Antarctic microbial mat: description of *Clostridium frigoris* sp. nov., *Clostridium lacusfryxellense* sp. nov., *Clostridium bowmanii* sp. nov. and *Clostridium psychrophilum* sp. nov. and reclassification of *Clostridium laramiense* as *Clostridium estertheticum* subsp. *laramiense* subsp. nov Int. J. Syst. Evol. Microbiol. 53 (4), 1019–1029.

Stapelfeldt, H., Nielsen, B.R., Skibsted, L.H., 1997. Effect of heat treatment, water activity and storage temperature on the oxidative stability of whole milk powder. Int. Dairy J. 7, 331–339.

Stark, J., 2007. Cheese and fermented sausages. In: Dijksterhius, J., Sanson, R.A. (Eds.), Food Mycology: A Multifaced Approach to Fungi. CRC Press Taylor and Francis Group, Boca Raton, Florida, pp. 319–331.

Stoeckel, M., Lidolt, M., Stressler, T., Fischer, L., Wenning, M., Hinrichs, J., 2016. Heat stability of indigenous milk plasmin and proteases from *Pseudomonas*: a challenge in the production of ultra-high temperature milk products. Int. Dairy J. 61, 250–261.

Sutthiwong, N., Fouillaud, M., Dufossé, L., 2018. The influence of pH, NaCl, and the deacidifying yeasts *Debaryomyces hansenii* and *Kluyveromyces marxianus* on the production of pigments by the cheese-ripening bacteria *Arthrobacter arilaitensis*. Foods 7 (11), 190.

Taoukis, P.S., Labuza, T.P., 1989. Applicability of time-temperature indicators as shelf-life monitors of food-products. J. Food Sci. 54, 783–788.

Toscani, T., Moseriti, A., Dossena, A., Dall'Asta, C., Simoncini, N., Virgili, R., 2007. Determination of ochratoxin a in dry-cured meat products by a HPLCFLD quantitative method. J. Chromatogr. 855, 242–248.
USDA, E., 2014. FSIS Compliance Guideline for Meat and Poultry Jerky Produced by Small and Very Small Establishments. 2014 Compliance Guideline. FSIS, Omaha, NE. https://www.fsis.usda.gov/wps/wcm/connect/5fd4a01d-a381-4134-8b91-99617e56a90a/Compliance-Guideline-Jerky-2014.pdf?MOD=AJPERES.
Vassila, E., Badeka, A., Kondyli, E., Savvaidis, I., Kontominas, M.G., 2002. Chemical and microbiological changes in fluid milk as affected by packaging conditions. Int. Dairy J. 12, 715–722.
Veillet-Poncet, L., 1986. Manufacturing hygiene. In: Eck, A. (Ed.), translated by C.D. Thomson. Cheesemaking-Science and Technology, Second ed. Lavoisier Publishing Inc., New York, pp. 148–149.
Vignolo, G.M., Pesce de Ruiz Holgado, A., Oliver, G., 1988. Some physiological, biochemical and tecnological characteristics of gram-positive cocci isolated from cured meat products. Microbiol. Aliments Nutrit. 6, 323–327.
von Holy, A., Holzapfel, W.H., Dykes, G.A., 1992. Bacterial populations associated with Vienna sausage packaging. Food Microbiol. 9, 45–53.
Watson, M., McEwan, J.A., 1995. Sensory changes in liquid milk during storage and the effect on consumer acceptance. J. Soc. Dairy Technol. 48 (1), 1–8.
Wolfe, B.E., Button, J.E., Santarelli, M., Dutton, R.J., 2014. Cheese rind communities provide tractable systems for in situ and in vitro studies of microbial diversity. Cell 158, 422–433.
Yildirim, S., Röcker, B., Pettersen, M.K., Nilsen-Nygaard, J., Ayhan, Z., Rutkaite, R., Radusin, T., Suminska, P., Marcos, B., Coma, V., 2018. Active packaging applications for food. Compr. Rev. Food Sci. Food Saf. 17 (1), 165–199.
Yoon, Y., Calicioglu, M., Kendall, P.A., Smith, G.C., Sofos, J.N., 2005. Influence of inoculum level and acidic marination on inactivation of *Escherichia coli* O157:H7 during drying and storage of beef jerky. Food Microbiol. 22 (5), 423–431.
Yoon, Y., Skandamis, P.N., Kendall, P.A., Smith, G.C., Sofos, J.N., 2006. A predictive model for the effect of temperature and predrying treatments in reducing *Listeria monocytogenes* populations during drying of beef jerky. J. Food Prot. 69 (1), 62–70.
Zhang, X., Sun, G., Xiao, X., Liu, Y., Zheng, X., 2016. Application of microbial TTIs as smart label for food quality: response mechanism, application and research trends. Trends Food Sci. Technol. 51, 12–23.
Ziarno, M., Zaręba, D., 2010. Characteristics and occurrence of *Paenibacillus* bacteria in milk and milk products [Charakterystyka i występowanie bakterii rodzaju Paenibacillus w mleku i przetworach mlecznych]. Med. Weter. 66 (9), 600–603.
Ziyaina, M., Rasco, B., Coffey, T., Mattinson, D.S., Salbani, S., 2018. Correlation of volatile compound concentrations with bacterial counts in whole pasteurized milk under various storage conditions. Int. J. Dairy Technol. 70 (1), 1–11.

## Further reading

McSweeney, P.L.H., 2004. Biochemistry of cheese ripening. Int. J. Dairy Technol. 57, 127–144.

# Moisture properties and stability of novel bioactive ingredients

**2**

*Vera Lavelli*, William Kerr†*
*DeFENS, Department of Food, Environmental and Nutritional Sciences, University of Milan, Milano, Italy, †Department of Food Science and Technology, University of Georgia, Athens, GA, United States

**Chapter outline**



## 1 Predictive parameters for moisture-dependent food degradation

Drying technologies are often applied to inhibit microbial growth and prolong stability of foods stored at ambient temperature. In contrast to cold-chain supply, "dry chain" products require an initial energy input for drying, but no further energy input is required to maintain product quality as long as the integrity of the storage container is preserved (Bradford et al., 2018). It is of great importance to determine the moisture content associated with maximum product stability. To this end, the thermodynamic property of water activity ($a_w$) has proven to be a useful marker for assessing quality changes. Water activity can be defined as:

$$a_w = p / p_o = \text{relative humidity} / 100 \qquad (1)$$

where $p$ is the partial pressure of water in the food and $p_0$ the vapor pressure of pure water at the same temperature. For a food at equilibrium with its environment, $a_w$ is related to the relative humidity of the surrounding air (Rahman, 1995).

One of the most common methods of measuring $a_w$ is to equilibrate the sample with air in a closed vessel and then determine the humidity of the air with a hygrometer. Furthermore, it is often desirable to determine the unique relationship between moisture content and water activity of a given food. At a specific temperature, this is known as the moisture sorption isotherm. While five classifications (I–V) of sorption

Food Quality and Shelf Life. https://doi.org/10.1016/B978-0-12-817190-5.00002-1

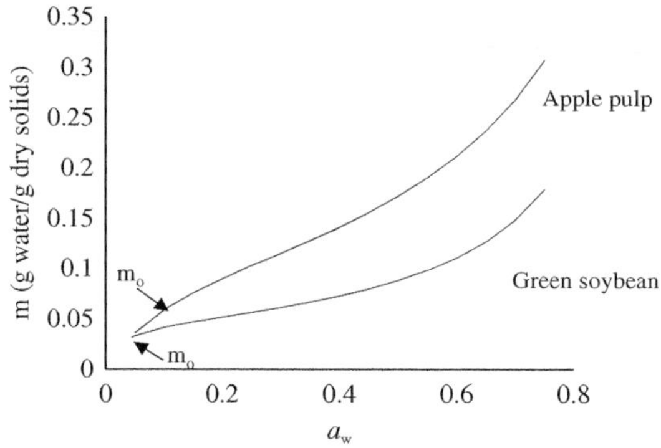


**Fig. 1** Moisture sorption isotherms corresponding to the models proposed by Brunauer et al. (1940): type II (apple pulp, moisture absorption at 30°C, data from Corey et al., 2011) and type III (green soybean, absorption at 25°C, data from Yang et al., 2012). *Arrows* indicate the GAB monolayer moisture content.

isotherms have been proposed (Brunauer et al., 1940), most foods are of Type II, although some crystalline materials are of Type III, and some anticaking agents are of Type I (Fig. 1). Both empirical and theoretical equations have been applied to model moisture sorption isotherms (Rahman, 1995). The original thermodynamic model proposed by Brunauer et al. (1940) was modified by Guggenheim-Anderson-de Boer. The semiempirical GAB model is considered one of the best for fitting food isotherms over a wide-range of water activity. The GAB model can be written as:

$$m = m_0 Cka_w / \left[ (1 - ka_w)(1 - ka_w + Cka_w) \right] \tag{2}$$

where $m$ is the equilibrium moisture content (kg water/kg dry solids); $m_o$ is the GAB monolayer moisture content; and $C$ and $k$ are constants related to the heat of sorption of the first and subsequent sorbed layers of water. While $k$ is obtained empirically, it is believed to have a physico-chemical meaning.

Many studies have related $a_W$ to the microbial and chemical stability of foods. There is usually a critical $a_W$ level below which microorganisms do not grow or produce toxins; however, the specific value can vary with pH and type of solutes present. Hence, $a_W$ has been a useful marker for constructing general rules regarding microbial stability (Rahman, 1995). It should be noted, however, that while most microorganisms do not grow at low $a_W$, some can survive low $a_W$ conditions and result in food-borne illnesses. For example, *Salmonella* spp. has been reported in a variety of dried milk, cereal, and spice products (Gurtler et al., 2014)

The GAB monolayer moisture content, which signifies moisture that is strongly bound to specific sites on the solids, has been related to the maximum stability of food toward lipid oxidation. Lipid oxidation generally displays a U-shaped curve with respect to $a_W$ level, with the minimum oxidation rate occurring when sample moisture is near the monolayer value. Minimum lipid oxidation typically occurs when $a_W$ is between 0.2 and 0.4 (Rahman, 1995). In comparison, reactions such as nonenzymatic browning generally have minimal rates at relatively low and relatively high $a_w$ levels, and their rate is greatest at intermediate moisture levels corresponding to $a_w$ between 0.4 and 0.7 (Rahman, 1995). However, specifying an $a_w$ at which deteriorative chemical reactions occur at the greatest rate is difficult as that depends on the particular reaction, available substrates, the food matrix and pH.

The glass transition temperature ($T_g$) is another parameter that has been used to predict food stability. $T_g$ is the temperature above which a hard amorphous solid will transform into a soft, rubbery material. Materials in the glassy state have very high viscosity and heating to $T > T_g$ results in a marked reduction of viscosity and increase in molecular mobility (Harnkarnsujarit et al., 2012). This phenomenon is generally studied by differential scanning calorimetry (DSC), differential thermal analysis (DTA), or thermomechanical analysis (TMA). Thermogravimetric analysis (TGA), dynamic mechanical analysis (DMA), and electron spin resonance (ESR) have also been applied (Rahman, 1995).

The glass transition temperature depends on molecular weight and structural characteristics of the solutes, as well as the moisture content. As water content increases, $T_g$ decreases, thus $T_g$ also decreases with increasing $a_w$ (Fig. 2). The Gordon and Taylor equation has been proposed to predict the glass transition temperature of a polymer blend (Bell, 2007). It can be written as:

$$T_g = w_s T_{gs} + k w_m T_{gm} / w_s + k w_m \tag{3}$$

where $T_g$ is the observed glass transition temperature, $T_{gs}$ the glass transition temperature of amorphous dry solid, $T_{gm}$ the glass transition temperature of amorphous water, $k$ a constant, $w_s$ is the weight fraction of solids, and $w_m$ is the weight fraction of water. Although difficult to measure, $T_{gm}$ has been found to be −135°C (Welti-Chanes et al., 1999).

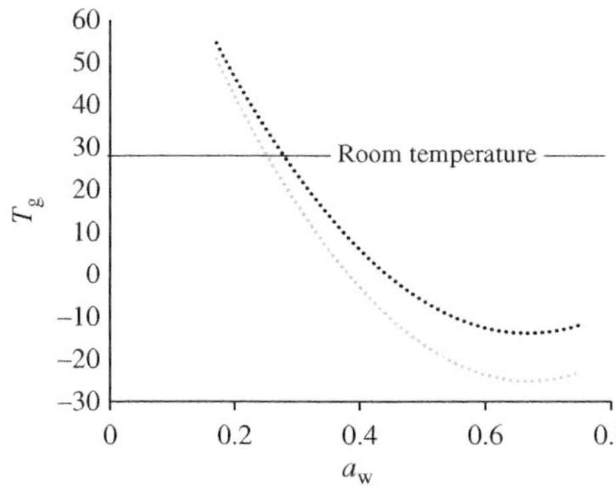


**Fig. 2** $T_g$ values for tomato peel (*gray dotted line*) and tomato pulp (*black dotted line*) as a function of $a_w$.
Data from Lavelli et al. (2013).

It has been suggested that in glassy states various chemical reactions may become diffusion-limited. In general, it has been observed that a given food is most stable at and below its glass transition temperature ($T_g$ or $T_g'$ in frozen foods), and that as $T-T_g$ (or $T/T_g$) increase above the glass transition, the greater the rate of deteriorative reactions (Rahman, 2010). Foods are often very stable in the glassy state, as compounds involved in deteriorative reactions take many months or even years to diffuse over molecular distances and approach close enough to each other to react. However, some reactions were found to occur even in the glassy state (Hung et al., 2007). Additionally, the glassy systems are more porous than rubbery systems and thus allow easier diffusion of oxygen. Conversely, at $T>T_g$, the material could undergo structural collapse. Collapse is usually defined as the readily visible deformation (or shrinkage) of the sample under the force of gravity during storage of a dried amorphous matrix. Collapse of freeze-dried materials is associated with the disappearance of micropores and cavities, which limits oxygen access to oxidizable targets (Harnkarnsujarit et al., 2012).

The molecular mobility of water has also been proposed as a food stability indicator, as water itself may be a reactant, or diminished mobility may hamper the diffusion of other chemical reactants. However, there is limited data to address the complexity of this subject and indeed it can be difficult to ascertain just what "water mobility" is and how best to measure it. A variety of techniques have been used to help researchers understand molecular mobility and relaxation phenomena below and above $T_g$. Nuclear magnetic resonance (NMR) is a common means to characterize molecular mobility in foods. In this approach, paramagnetic nuclei in a static magnetic field are excited by an oscillating magnetic field pulse, and the rate at which magnetic spins return to thermal equilibrium is monitored. In particular, proton NMR has been used to monitor the environments in which water resides. Water with restricted mobility is better able to exchange nuclear spins with other molecules, resulting in a more rapid decay of the NMR signal.

Thermomechanical methods measure mobility as a sample is placed under a mechanical stress, then store that energy in molecular alignments, or release it as increased thermal motion. Calorimetric methods heat the sample while precisely measuring the amount of energy required to raise the temperature. For systems undergoing phase changes, more energy is required, as the system gains greater degrees of motional freedom. In all cases, these motions occur over characteristic times and distances. Thermal and mechanical methods typically measure mobility over 20–300 nm, while NMR measures molecular relaxation in a 1–2 nm range. In the time domain, NMR can detect molecular motions in the ps-ms range, while DSC and DMA detect relaxations in the ms-s range (McBrierty and Packer, 1993). Therefore, data on bulk properties such as viscosity, elastic modulus, or DSC-glass transitions do not characterize all of the possible molecular changes in a sample (Vittadini et al., 2002). As $a_w$, $T_g$, and proton relaxation are distinct methods of assessing water binding and molecular mobility, it may be useful to combine the approaches to study chemical reactions that occur in foods (Wang et al., 2004; Venturi et al., 2007). In many instances, these parameters alone or in combination explain clearly the complexity of food stability. Hence, they can be used as a tool to design stability maps for dehydrated and intermediate moisture foods.

As noted above, substantial research has examined the influence of moisture on lipid oxidation, browning reactions, and changes in nutrients in common foods. However, the food supply chains are constantly changing in order to address either the requirements for healthier foods, the need for better resource management, or enhanced environmental protection (Galanakis, 2015; Lavelli et al., 2018). Promising opportunities come from the recovery of phytochemicals and proteins from food byproducts and their conversion into novel food ingredients. Even if consumers are concerned about possible safety risks associated with food byproducts, suitable management and traceability systems have become key tools to increase consumers' trust (Luning et al., 2011; Lavelli, 2013). Little is known, however, about the storage stability of phytochemicals and proteins as a function of moisture content. There is a need to set predictive models for quality degradation in order to find out formulation strategies aimed at the extension of the stability of these novel ingredients at ambient temperature. In this context, in the following sections, recent research on moisture-dependent stability of anthocyanins, flavonols, carotenoids and proteins, or partially hydrolyzed proteins will be discussed.

## 2 Kinetics of moisture-dependent anthocyanin degradation

Anthocyanins are plant secondary metabolites from the flavonoid family. Many red to blue fruits, and their derived products, are major dietary sources of anthocyanins. In addition to the benefits of direct consumption, studies have shown that the residual skins and peels left over from the processing of grape and other juices are a rich source of anthocyanins (Lavelli et al., 2017a,b; Galanakis, 2015). Anthocyanins are glycosylated forms of anthocyanidins, and 31 anthocyanidin aglycones have been identified. However, 90% of all naturally occurring anthocyanins are based on six aglycones: cyanidin (~50%); delphinidin, pelargodin, and peonidin (~12% each); and petunidin and malvidin (~7% each). Anthocyanins differ in the type and number of bonded sugar moieties. These sugars may also be modified with aliphatic or aromatic carboxylates, with the most common being 3-monosides (mainly glucosides); 3-biosides; and 3,5- and 3,7-diglucosides from the 3 nonmethylated aglycones (cyanidin, delphinidin, and pelargodin) (Olivas-Aguirre et al., 2016).

In recent years, there have been numerous studies that indicate that anthocyanins have gastroprotective, antiinflammatory, antithrombotic, and chemo-preventive properties. Anthocyanins can also influence gene expression, protect against *Helicobacter pylori* infection, and play a role in ameliorating Type 2 diabetes, cardiovascular disease, metabolic syndrome, and oral cancer. Thus, anthocyanidins are of interest as functional ingredients that provide natural coloring and health-promoting properties (Sigurdson et al., 2017; Olivas-Aguirre et al., 2016). When compared to other phenolics, anthocyanins display the lowest stability under intermediate moisture conditions.

The degradation rate of several anthocyanins and phenolics was studied in apple pomace at $a_w$ 0.75 and 30°C; conditions that would allow these compounds to take part in chemical reactions. The order of stability was: phloridzin > chlorogenic acid > quercetin 3-*O*-galactoside > epicatechin > procyanidin B2 > cyanidin galactoside (Lavelli and Corti, 2011). A detailed insight into the effect of $a_w$ on anthocyanin stability was provided by Gradinaru et al. (2003) in a study on the anthocyanins from roselle (*Hibiscus sabdariffa* L.), which are mainly comprised of delphinidin-3-sambubioside and cyanidin-3-sambubioside. Aqueous extracts containing anthocyanins were either lyophilized directly or co-lyophilized with pullulan, and their stability investigated during storage at different relative humidity levels corresponding to the $a_w$ range 0.33–0.80, for 16 days at 40°C. Increasing the $a_w$ decreased the $T_g$ of anthocyanin powders from about 60 to −10°C, and that of pullulan from about 110°C to −5°C. Hence, the powders stored at the lowest $a_w$ were in the glassy state, while those stored at the highest $a_w$ would have substantial molecular mobility.

Pullulan was expected to increase the stability of anthocyanins due to its contribution to higher $T_g$. During storage, anthocyanin degradation followed first-order kinetics, with rate constants increasing at greater $a_w$, particularly above 0.54 (Table 1). For instance, the half-lives at 40°C were 317 and 3.9 days for the powder stored at $a_w$ levels of 0.33 and 0.84, respectively. In the presence of pullulan, the half-lives were 321 and 6 days for the powders stored at the same conditions (Table 1). A plot of reciprocal rate constants against $T-T_g$ showed a linear response for both glassy and nonglassy states. Hence, anthocyanin degradation rates diminished as $T-T_g$ increased, but degradation occurred also below $T_g$; that is, in the glassy state. Macroscopic heterogeneities in the glassy matrix, nonhomogeneous distribution of water, and phase separation phenomena (demixing of reactants and inert matrix) are likely to influence the apparent reaction rates and further explain why reactions do not cease below the measured $T_g$. Pullulan, which shows good film barrier properties (impermeability to oxygen) improved anthocyanin stability by a factor of 1.5. However, pullulan was found to be more efficient in improving stability of other water-soluble pigments (e.g., crocins) against oxidation, decreasing the degradation rate constant up to 20 times (Selim et al., 2000).

Detailed information on the effects of various carriers on anthocyanin stability was provided by Tonon et al. (2010) using the juice of açai (*Euterpe oleracea* M.), which is mainly comprised of cyanidin-3-rutinoside. Acai juice was spray dried using four types of carrier agents: maltodextrin 10DE, maltodextrin 20DE, gum arabic, and tapioca starch. The resulting powders were stored at $a_w$ levels of 0.328 and 0.529 for 120 days at 25°C and 35°C, respectively. The sorption isotherms of these matrices were modeled by the GAB equation. The powders produced with tapioca starch showed the lowest water adsorption, followed by that produced with maltodextrin 10DE, while samples with maltodextrin 20DE and gum arabic were the most hygroscopic. Such differences in water adsorption were explained by the chemical structure of each agent. Maltodextrin 20DE and gum arabic have a greater number of hydrophilic groups and, therefore, can more easily adsorb moisture from the ambient air. Maltodextrin 10DE is less hydrolyzed, with fewer hydrophilic groups and thus adsorbs less water. Tapioca starch is a native starch that is not hydrolyzed, which explains its lower hygroscopicity. The powder produced with maltodextrin 10DE had a higher $T_g$ compared to that produced with maltodextrin 20DE, which was related to

**Table 1** First-order rate constants ($day^{-1}$) and half-lives (day) for the degradation of anthocyanins during storage in the dark as a function of the individual molecules involved, matrix and storage temperature (°C), time (day), and water activity conditions

| Main anthocyanin studied | Matrix | $T$ | $t$ | $a_w$ | $k \times 10^3$ | $t_{1/2}$ | References |
|---|---|---|---|---|---|---|---|
| Delphinidin-3-sambubioside<br>Cyanidin-3-sambubioside | Freeze-dried roselle water extract | 40 | 16 | 0.33 | 2.2 ± 0.3 | 317 | Gradinaru et al. (2003) |
| | | | | 0.54 | 23.2 ± 1.2 | 30 | |
| | | | | 0.75 | 145.2 ± 8.0 | 4.7 | |
| | | | | 0.84 | 174.6 ± 8.0 | 3.9 | |
| | Freeze-dried roselle water extract encapsulated in pullulan | 40 | 16 | 0.33 | 2.2 ± 0.4 | 321 | |
| | | | | 0.54 | 12.7 ± 0.7 | 54.6 | |
| | | | | 0.75 | 83.3 ± 4.1 | 8.3 | |
| | | | | 0.84 | 115.3 ± 5.5 | 6.0 | |
| Cyanidin-3-rutinoside | Spray-dried açai juice encapsulated in maltodextrin D10 | 35 | 120 | 0.33 | | 880 | Tonon et al. (2010) |
| | | | | 0.54 | | 411 | |
| | Spray-dried açai juice encapsulated in maltodextrin D20 | 35 | 120 | 0.33 | | 516 | |
| | | | | 0.54 | | 260 | |
| | Spray-dried açai juice encapsulated in gum arabic | 35 | 120 | 0.33 | | 767 | |
| | | | | 0.54 | | 328 | |
| | Spray-dried açai juice encapsulated in tapioca starch | 35 | 120 | 0.33 | | 655 | |
| | | | | 0.54 | | 248 | |
| Delphinidin-3-glucoside<br>Cyanidin-3-glucoside<br>Petunidin-3-glucoside<br>Peonidin-3-glucoside<br>Malvidin-3-glucoside<br>Malvidin-3-*p*-coumaroyl-glucoside | Air-dried grape skins | 30 | 180 | 0.11 | 0.77 ± 0.07 | 896 | Lavelli et al. (2017b)[a] |
| | | | | 0.22 | 0.83 ± 0.10 | 837 | |
| | | | | 0.33 | 1.11 ± 0.44 | 627 | |
| | | | | 0.54 | 2.27 ± 0.42 | 305 | |
| | | | | 0.75 | 4.30 ± 0.02 | 161 | |
| | Spray-dried ethanol extract of grape skin extract encapsulated in maltodextrin D10 | 30 | 180 | 0.11 | n.d. | n.d. | |
| | | | | 0.22 | 0.68 ± 0.31 | 1026 | |
| | | | | 0.33 | 1.02 ± 0.03 | 679 | |
| | | | | 0.54 | 3.65 ± 0.00 | 190 | |
| | | | | 0.75 | 7.56 ± 0.10 | 92 | |

[a] Rate constants and half-lives refer to Mv-3-glucoside.

the smaller molecular weight of the 20DE sample. At $a_w$ 0.328, the $T_g$ of maltodextrin 20DE and gum arabic was 60°C, while that of maltodextrin 10DE and tapioca starch was 53°C. At an $a_w$ of 0.529, the $T_g$ of maltodextrin and gum arabic was 40°C, while that of maltodextrin DE20 and tapioca starch was about 32°C (Tonon et al., 2009).

When anthocyanins were prepared with these polysaccharides and stored at $a_w$ levels 0.328 and 0.529 at 25°C and 35°C, respectively, anthocyanin degradation occurred even below the $T_g$ and exhibited two first-order kinetic regimes. The first, with higher reaction rate constant, occurred up to 45–60 days of storage, while the second, with lower degradation rate, occurred after 60 days of storage. The higher degradation rate in the first stage was attributed to the nonencapsulated material, which had greater contact with oxygen, or even to the material in contact with the oxygen present in the interior of pores. As observed previously (Gradinaru et al., 2003), anthocyanin stability decreased with increasing $a_w$ and was greater for powders stored at 35°C (Table 1). Maltodextrin 10DE was the carrier agent that showed the best pigment protection for all the conditions studied, followed by tapioca starch, especially at $a_w$ 0.54. In fact, the half-lives for cyanidin-3-rutinoside degradation at $a_w$ 0.54 and 35°C were 411, 260, 328, and 248 days for the pigment encapsulated in maltodextrin 10DE, maltodextrin 20DE, gum arabic, and tapioca starch, respectively (Table 1).

Lavelli et al. (2017b) studied anthocyanin stability in dried grape skins (*Vitis vinifera* L.) as compared to ethanolic grape skin extracts encapsulated with maltodextrin DE10. The hygroscopicity of these samples was assessed after fitting moisture isotherms with the GAB equation. Due to the presence of numerous hydrophilic sites, hygroscopicity was greater for encapsulated grape skin extracts than for grape skins. At 0.56 $a_w$ the encapsulated phenolic powder formed lumps, while at 0.75 $a_w$ the product was a viscous liquid with 0.20 g $H_2O$/g d.w. In contrast, at 0.75 $a_w$ the grape skins absorbed only 0.08 g water/g d.w. According to the reported $T_g$ values for maltodextrins with DE10 (Tonon et al., 2009), at 30°C this polysaccharide is in the rubbery state for $a_w > 0.56$ and in the glassy state for $a_w < 0.56$. The $T_g$ of grape skins were above 50°C, even at $a_w$ 0.75 (Georget et al., 1997). Hence, at room temperature the dried grape skins were in the glassy state. The decrease in anthocyanins during storage at various $a_w$ levels followed first-order kinetics, as observed previously (Gradinaru et al., 2003). At $a_w$ 0.11, no degradation occurred in maltodextrin DE10 encapsulated grape skin extract, while in grape skins degradation occurred at a slower rate than samples at higher $a_w$, even if the storage temperature was below $T_g$ (Table 1). Storage at $a_w$ levels of 0.22 and 0.32 resulted in a slight increase in anthocyanin degradation rates, while storage at $a_w > 0.54$ caused a much greater increase in anthocyanin degradation. The half-life for the primary anthocyanin in grape extract (malvidin 3-glucoside) at $a_w$ 0.75 was ~92 days for the encapsulated extract and 161 days for grape skins (Table 1). The half-life varied among individual compounds. In fact, delphinidin-3-*O*-glucoside was the least stable while malvidin-*p*-coumaroyl-glucoside most stable (not shown).

As a general rule, for all sources and matrices, anthocyanin stability decreases with increasing $a_w$, especially at $a_w > 0.54$. While reactants are more concentrated at lower $a_w$, decreasing moisture increases viscosity and it is thought that this limits molecular diffusion and the ability of reactants to participate in chemical reactions. Degradation may also occur in the glassy state, but the rate is much lower. Encapsulation with polysaccharides, especially maltodextrin DE10, decreases the $T_g$ and increases anthocyanin stability with respect to the free extract. Anthocyanins are also stable in the dehydrated fruit skins, which have high $T_g$.

# 3 Kinetics of moisture-dependent flavanol degradation

Flavanols are a structurally complex subclass of phenolics ranging from simple monomers (such as catechin and its isomer epicatechin) to oligomers (from dimers to decamers), polymers (>10mers), and other derived compounds (such as theaflavins and thearubigins). The oligomers and polymers of flavan-3-ols are also referred to as condensed tannins or proanthocyanidins (PA), named for their ability to yield anthocyanidins when heated in acidic media. Flavanols are found ubiquitously in plants as secondary metabolites. These compounds are abundant in various food byproducts such as apple pomace (Virot et al., 2010), grape seeds and skins (Sri Harsha et al., 2013, 2014), cocoa byproducts (Cádiz-Gurrea et al., 2017), and green tea byproducts (Lee et al., 2006). Total and individual compounds of flavan-3-ols have been studied extensively for their antioxidant, antiinflammatory, immunomodulatory, and anticarcinogenic effects (Suzuki et al., 2009; Yang et al., 2009; Hooper et al., 2008). Hence, flavanol-rich sources are being studied as ingredients for functional foods.

With this aim, some studies have investigated the stability of flavanols in intermediate moisture and dried conditions as a function of $a_w$. Corey et al. (2011) studied flavanol degradation in dehydrated apple pulp stored in the $a_w$ range 0.11–0.75 at 30°C. As moisture content and $a_w$ increased, $T_g$ decreased from 8.72°C at 0.11 $a_w$ 0.11 to −38.76°C at $a_w$ 0.75. All $T_g$ values were below the storage temperature of 30°C, indicating that none of the samples were truly in the glassy state. However, the temperature difference (30°C − $T_g$) determines how much the material exhibits solid-like behavior. Indeed, flavanol degradation occurred following first-order kinetics, with increasing rate constants as $a_w$ increased and $T_g$ decreased (Corey et al., 2011).

In a related study, Lavelli and Kerr (2012) compared the stability of dehydrated apple pomace to that of dehydrated apple pulp and found that flavanol degradation in the pulp and pomace of apple occurred on a very different time scale. In apple pomace, degradation of epicatechin was significant for $a_w > 0.32$. The half-life decreased from 384 to 40 days when the $a_w$ was increased from 0.32 to 0.75. In comparison, in apple pulp the half-life of epicatechin decreased from 65 to 20 days when the $a_w$ increased from 0.11 to 0.75 (Table 2). Similarly, for the dimer procyanidin B2, the half-life decreased from 414 to 26 days in apple pomace and from 43 to 13 days in apple pulp when $a_w$ was increased from 0.32 to 0.75 (not shown). This result was related to the relatively lower water mobility in apple pomace than in the pulp. There are compositional differences between the two dried materials. As it encompasses more peel, pomace contains a greater fraction of cell wall polysaccharides, while the pulp has a greater fraction of low molecular weight sugars. Thus, the pulp has a greater tendency to bind water, promoting a higher molecular mobility environment. Relatively low molecular weight sugars can also act as plasticizers, and this also contributes to greater mobility. $^1$H NMR revealed that apple pulp at $a_w$ 0.11 or 0.22 had signals that decayed

**Table 2** First-order rate constants ($day^{-1}$) and half-lives (day) for the degradation of flavanols during storage in the dark as a function of the individual molecules involved, matrix and storage temperature (°C), time (day), and water activity conditions

| Main flavanol studied | Matrix | $T$ | $t$ | $a_w$ | $k \times 10^3$ | $t_{1/2}$ | References |
|---|---|---|---|---|---|---|---|
| Epicatechin, catechin Procyanidin B2[a] | Freeze-dried apple pulp | 30 | 45 | 0.32 | 10.6 ± 0.8 | 65 | Corey et al. (2011) |
| | | | | 0.56 | 21 ± 2 | 33 | |
| | | | | 0.75 | 34 ± 2 | 20 | |
| Epicatechin, catechin Procyanidin B2[a] | Freeze-dried apple pomace | 30 | 270 | 0.32 | 1.8 ± 0.7 | 384 | Lavelli and Kerr (2012) |
| | | | | 0.56 | 6.6 ± 2.4 | 104 | |
| | | | | 0.75 | 17 ± 3 | 40 | |
| Epicatechin, catechin[a] | Air-dried grape skin | 30 | 180 | 0.32 | n.s. | | Lavelli et al. (2017b) |
| | | | | 0.56 | 4.47 ± 0.55 | 155 | |
| | | | | 0.75 | 8.33 ± 0.07 | 83 | |
| Proanthocyanidins | Air-dried grape skin | 30 | 180 | 0.32 | 2.70 ± 1.8 | 253 | |
| | | | | 0.56 | 2.32 ± 1.40 | 299 | |
| | | | | 0.75 | 3.88 ± 0.16 | 179 | |
| Epicatechin, catechin[a] | Spray-dried ethanol extract of grape skin extract encapsulated in maltodextrin D10 | 30 | 180 | 0.32 | n.s. | | |
| | | | | 0.56 | 5.51 ± 0.16 | 126 | |
| | | | | 0.75 | 11.18 ± 0.13 | 63 | |
| Proanthocyanidins | Spray-dried ethanol extract of grape skin extract encapsulated in maltodextrin D10 | 30 | 180 | 0.32 | n.s. | | |
| | | | | 0.56 | 0.61 ± 0.36 | 1127 | |
| | | | | 0.75 | 1.20 ± 0.56 | 580 | |
| Green tea flavanols:[b] epicatechin, catechin, epicatechin gallate, epigallo catechin, epigallo catechin gallate | Supercritical $CO_2$ extracted fresh green tea leaves encapsulated in gum arabic | 25 | 84 | 0.26 | | 392 | Zokti et al. (2016) |
| | Supercritical $CO_2$ extracted fresh green tea leaves encapsulated in maltodextrin D10-16 | 25 | 84 | 0.28 | | 350 | |
| | Supercritical $CO_2$ extracted green tea leaves | 25 | 84 | 0.45 | | 238 | |

[a] Rate constants and half-lives refer to epicatechin.
[b] Rate constants and half-lives refer to total flavanols.

within 40 μs, indicative of solid-like behavior. At $a_w > 0.32$, the signal decayed over several hundred microseconds. While this is not the highly mobile domain associated with bulk water, it shows enhanced mobility within a viscous domain. Interestingly, even at $a_w \geq 0.32$, water mobility in apple pomace was much lower than that found in apple pulp. In fact, for apple pulp held at $a_w \geq 0.32$, the $^1H$ NMR free induction decay curves showed decay over thousands of microseconds (Lavelli and Kerr, 2012).

Flavanol stability was also investigated in grape skins in the $a_w$ range 0.11–0.75 at 30°C (Lavelli et al., 2017b). No degradation was observed at $a_w < 0.32$, whereas above this $a_w$, flavanol degradation occurred at a slightly lower rate than in apple pomace (Table 2). The differences observed between apples and grapes could be due to the occurrence of Maillard reactions in apple, as demonstrated by the formation of hydroxymethylfurfural, which can lead to antioxidant degradation (Lavelli and Vantaggi, 2009). In contrast, hydroxymethylfurfural was not detected in grape skins.

In another approach, Zokti et al. (2016) applied polysaccharides such as gum arabic and maltodextrin DE10-16 to encapsulate flavanols of green tea and proved that encapsulated flavanols were more stable than the free extract during storage at $a_w$ 0.26–0.4 and 25°C. Lavelli et al. (2017b) compared the stability of flavanols in dehydrated grape skins and encapsulated in maltodextrin DE10. Epicatechin was found to be more stable in grape skins, while high molecular weight polymers (proanthocyanidins) were found to be more stable when encapsulated in maltodextrin DE10. However, the degradation rates were, in any case, relatively low (Table 2).

As a general rule, monomeric, dimeric, and polymeric flavanols showed similar stability to anthocyanins; that is, their stability decreased with increasing $a_w$ level. Moreover, polysaccharides increased flavanol stability, both when present as a matrix component, and when used as encapsulating agents.

# 4 Kinetics of moisture-dependent carotenoid degradation

Carotenoids are 40-carbon tetraterpenoids, constituted of isoprenoid units joined in a head-to-tail pattern. They have colors ranging from yellow to orange to red. Carotenoids are widely distributed in nature, including higher plants, bacteria, fungi, yeast, and algae. They are abundant in various food by-products such as tomato peel (lycopene, β-carotene, and lutein), carrot pomace (α-carotene, β-carotene, and lutein), mango peel (β-carotene), and shrimp waste (astaxantin) (Galanakis, 2015).

They primarily exert antioxidant effects, but individual carotenoids may also act through other mechanisms; for example, β-carotene has a pro-vitamin A function, while lutein/zeaxanthin is found in macular eye pigments. There is evidence that carotenoids, in addition to beneficial effects on eye health, also produce improvements in cognitive function and cardiovascular health, and may help to prevent some types of cancer. Carotenoids are also used as natural colorants in foods (Eggersdorfer and Wyss, 2018).

The effect of moisture on the stability of crocins, the polar carotenoids of saffron, was studied at 35°C in the $a_w$ range 0.41–0.75 (Selim et al., 2000). A natural water extract including the crocins (all-*trans*-crocetin di-(β-D-gentiobiosyl) ester, all-*trans*-crocetin β-D-gentiobiosyl-β-D-glucosyl ester, all-*trans*-crocetin di-(β-D-glucosyl) ester, all-*trans*-crocetin mono-(β-D-gentiobiosyl) ester, 13-*cis*-crocetin di-(β-D-gentiobiosyl) ester, and 13-*cis*-crocetin β-D-gentiobiosyl-β-D-glucosyl ester) was considered in parallel with an encapsulated extract. The moisture dependent oxidation rate of saffron extract was similar to that of phenolics discussed previously. Saffron carotenoid stability was the highest at the lowest $a_w$ (0.43) studied, with a half-life of 60days at 35°C. Again, this was most likely due to limited molecular mobility (Table 3). When samples were held at $a_w$ 0.75, the half-life decreased to 2days.

Encapsulation of saffron carotenoids with either polyvinylpyrrolidone (PVP40), average MW 40,000, or pullulan, both having high $T_g$ values, greatly improved crocins' stability to oxidation. Oxidation was still moisture-dependent and the reaction rate increased with increasing $a_w$. By using PVP40, the half-life at 35°C was 257days at $a_w$ 0.43 and 154 days at $a_w$ 0.75, respectively (Table 3). However, pullulan, which has a similar $T_g$ to PVP40, was less efficient in improving crocins' stability (data not shown). That is, the $T_g$ of the polymer used as wall material was not a useful predictor of carotenoid stability in this case. It is plausible that when in presence of these polymers, factors other than molecular mobility, such as microstructure and porosity of the polymeric matrix, may affect reaction kinetics. Indeed, the stabilizing effect of PVP40 was attributed to higher structural collapse of the matrix with PVP40 than with pullulan.

The degradation of nonpolar carotenoids has a different dependence on moisture than that found for polar carotenoids and phenolics. Prado et al. (2006) investigated the stability of purified β-carotene encapsulated in PVP40 during storage at 25°C and $a_w$ between 0.11 and 0.75. The glass transition temperature of this matrix ranged between 86°C at $a_w$ 0.11 and −9°C at $a_w$ 0.75. At $a_w$ ≤0.53, the storage temperature ($T$) was lower than $T_g$ and hence the samples were in the glassy state, while at $a_w$ >0.53 the temperature was higher than $T_g$ and samples were in the rubbery state. β-Carotene degradation was found to be minimal in systems that were fully plasticized, that is, where $T > T_g$. For instance, the half-life for β-carotene degradation at 25°C was 2 days at $a_w$ 0.11, and 99 days at $a_w$ 0.75 (Table 3). This suggested that more than just molecular mobility is important to the reaction rate when lipophilic compounds are involved. Prado et al. (2006) concluded that the glassy systems were more porous and thus allowed easier diffusion of oxygen. Conversely, at $T > T_g$, the material underwent structural collapse, which collapsed the micropores and limited oxygen access to β-carotene.

Researchers have also studied the potential of encapsulation for improving carotenoid stability. Lim et al. (2014) designed freeze-dried single-layer (SL) and layer-by-layer (LBL) emulsions using trehalose as a wall material to protect the carotenoids. For the SL emulsions, whey protein isolate (2%, w/w, in oil) was used as an emulsifier and the samples were homogenized with sunflower oil containing β-carotene (0.05%, w/w) and lutein (0.05%, w/w). To create LBL emulsions, the SL emulsion was mixed with 0.005% w/w gum arabic solution before mixing with trehalose (20%, w/w) and then freeze-drying. Compared to the degradation rate found previously (Prado et al., 2006), the loss of β-carotene was lower, especially in the LBL system. Indeed, the

**Table 3** First-order rate constants (day$^{-1}$) and half-lives (day) for the degradation of carotenoids during storage in the dark as a function of the individual molecules involved, matrix and storage temperature (°C), time (day), and water activity conditions

| Main carotenoids studied | Matrix | $T$ | $t$ | $a_w$ | $k \times 10^3$ | $t_{1/2}$ | References |
|---|---|---|---|---|---|---|---|
| Crocins: all-*trans*-crocetin di-(β-D-gentiobiosyl) ester, all-*trans*-crocetin β-D-gentiobiosyl-β-D-glucosyl ester, all-*trans*-crocetin di-(β-D-glucosyl) ester, all-*trans*-crocetin mono-(β-D-gentiobiosyl) ester, 13-*cis*-crocetin di-(β-D-gentiobiosyl) ester, and 13-*cis*-crocetin β-D-gentiobiosyl-β-D-glucosyl ester[a] | Freeze-dried saffron water extract | 35 | 75 | 0.43 | 11.6 ± 0.5 | 60 | Selim et al. (2000) |
| | | | | 0.53 | 43.0 ± 2.8 | 9.3 | |
| | Freeze-dried saffron water extract encapsulated in poly(vinyl pyrrolidone) of average MW 40,000 | 35 | 75 | 0.64 | 417 ± 11 | 1.7 | |
| | | | | 0.75 | 347 ± 5.1 | 2.0 | |
| | | | | 0.43 | 2.7 ± 0.1 | 257 | |
| | | | | 0.53 | 4.1 ± 0.1 | 169 | |
| | | | | 0.64 | 4.6 ± 0.2 | 151 | |
| | | | | 0.75 | 4.5 ± 0.2 | 154 | |
| All-*trans*-β-carotene | Freeze-dried emulsion of β-carotene encapsulated in poly(vinyl pyrrolidone) of average MW 40,000 | 25 | 60 | 0.11 | 270 ± 50 | 2 | Prado et al. (2006) |
| | | | | 0.44 | 220 ± 60 | 3 | |
| | | | | 0.53 | 100 ± 10 | 6 | |
| | | | | 0.64 | 10 ± 0.8 | 69 | |
| | | | | 0.75 | 7 ± 0.2 | 99 | |
| All-*trans*-β-carotene and all-*trans*-Lutein[b] | Freeze-dried single layer emulsion (sunflower oil, whey protein and trehalose) | 37 | 70 | 0.33 | 14 | 48 | Lim et al. (2014) |
| | Freeze-dried layer-by-layer emulsion (gum arabic and the above-indicated primary layer) | 37 | 70 | 0.33 | 8.7 | 80 | |
| All-*trans*-Lycopene | Freeze-dried supercritical carbon dioxide extract of tomato pomace encapsulated in poly-γ-glutamic acid | 35 | 30 | n.d. | 24.7 | 28 | Chiu et al. (2007) |
| All-*trans*-Lycopene | Freeze-dried tomato peel | 30 | 139 | 0.17 | 17 ± 1.2 | 41 | Lavelli et al. (2013) |
| | | | | 0.22 | 11 ± 1.5 | 63 | |
| | Freeze-dried tomato pulp | 30 | 139 | 0.32 | 8.5 ± 1.1 | 81 | |
| | | | | 0.56 | 6.0 ± 0.8 | 115 | |
| | | | | 0.17 | 8.1 ± 0.5 | 86 | |
| | | | | 0.22 | 7.7 ± 0.5 | 90 | |
| | | | | 0.32 | 5.8 ± 0.2 | 119 | |
| | | | | 0.56 | 6.2 ± 0.3 | 111 | |

[a] Rate constants and half-lives refer total crocin.
[b] Rate constants and half-lives refer all-*trans*-β-carotene.

half-life for β-carotene in the LBL system at $a_w$ 0.33 and 37°C was 80 days (Table 3). The loss of lutein occurred at a similar rate to that of β-carotene.

The degradation of nonpolar carotenoids in food matrices has been found to be more complex. The stability of α- and β-carotene and lutein in dehydrated carrots (Lavelli et al., 2007) and β-carotene in dehydrated mango (Harnkarnsujarit et al., 2012) was modeled as a function of $a_w$. When plotted against $a_w$, the first-order rate constants for all these carotenoids showed a U-shape curve typical of most oxidative reactions. Indeed, in the $a_w$ range 0.05–0.32, carotenoid oxidation rates decreased with increasing $a_w$, as was also observed by Prado et al. (2006) for encapsulated β-carotene. This effect was probably due to the occurrence of collapse phenomenon that resulted in diminished oxygen access. Additionally, metal catalysts are present in food matrices. Hence, increasing water from the dry state may slow oxidation by hydrating or diluting heavy metal catalysts or precipitating them as hydroxides. Indeed, Boon et al. (2009) found that transition metal induced oxidation may be the predominant mechanism of carotenoid degradation. In contrast, in the $a_w$ range of 0.32–0.54 carotenoid degradation in the food matrices increased with $a_w$. These findings diverge somewhat from the previously mentioned study. Such behavior was attributed to the increase of water in the matrix, which may augment the action of oxidative enzymes such as polyphenol oxidase and peroxidase (Harnkarnsujarit et al., 2012; Lavelli and Torresani, 2011).

Some other nonpolar carotenoids have shown very low stability in the dry state. Chiu et al. (2007) used gelatin and poly($\gamma$-glutamic acid) as coating materials for the encapsulation of lycopene. However, during storage of the microencapsulated powder, the concentrations of *cis*-, *trans*-, and total lycopene decreased quickly. The $a_w$ level was not specified, however, and the half-life was found to be 28 days at 35°C (Chiu et al., 2007). Lavelli et al. (2013) studied lycopene stability in dried tomato pulp and peel stored at 30°C with $a_w$ in the range 0.17–0.56. At $a_w$ 0.17, $T_g$ values were 52°C for the dried peel and 57°C for the pulp. As expected, $T_g$ decreased with increasing $a_w$/moisture content, as water plasticizes the molecules of the powder matrix, and at $a_w$ 0.56, $T_g$ was −12°C for the peel and −24°C for the pulp. At $a_w$ <0.32, the values for $T_g$ were >30°C for both the pulp and peel, indicating that they were in the glassy state and with little molecular mobility. At $a_w$ ≥0.32, $T_g$ was <30°C, indicating the samples were in the rubbery state, with decreased viscosity, and increased molecular mobility. In the entire $a_w$ range, lycopene stability in these natural matrices was generally better than that found for model systems (Table 3). However, lycopene was very unstable for samples in the glassy state (Table 3). α-Tocopherol behaved similarly to lycopene, while the hydrophilic antioxidants hydroxycinnamic acid and naringenin were stable for samples in the glassy state, but were unstable for samples in the rubbery state (data not shown).

In the glassy state, lycopene was more unstable in the dried peel, which is a less hydrophilic matrix than the pulp. For instance, at $a_w$ 0.17 the half-life for lycopene was 41 and 86 days in the peel and pulp, respectively (Table 3). This effect was attributed to the presence of sugars in the pulp that affect the microstructure. In fact, mono- and disaccharides were shown to increase β-carotene stability at $a_w$ levels of 0.11 and 0.33 (Desorby et al., 1997). In related work, Harnkarnsujarit et al. (2012) studied the microstructural formations of maltodextrin and sugar matrices in freeze-dried systems by scanning electron microscopy and demonstrated that the presence of sugars (glucose, fructose, and sucrose) resulted in smaller pore sizes in the freeze-dried solids. As sugars cause a decrease in freezing temperature, melting of ice crystals can occur during freeze-drying; this leads to structural collapse and causes the formation of small pores in the dried matrices.

In conclusion, the moisture-dependent degradation of polar carotenoids increases with increasing $a_w$, as expected for diffusion controlled reactions. Conversely, nonpolar carotenoids are highly unstable in the glassy state and their stability increases with increasing $a_w$ when the rubbery state is reached, due to structure collapse that collapses the micropores, and also to the dilution of metal catalysts. After an optimal $a_w$, where nonpolar carotenoid stability is maximum, further increases in $a_w$ lead to increasing degradation rates, most probably due to activation of oxidative enzymes. Strategies to improve carotenoid stability include encapsulation through layer-by-layer systems and freeze-drying in the presence of di- and monosaccharides, which cause a decrease in the pore size of the matrix.

## 5 Kinetics of moisture-dependent protein and protein hydrolizate degradation

Considering the rapid increase in the world's population along with limited land, water, and food resources, it is important to find efficient protein sources to meet human nutritional needs. Whey proteins that are obtained from cheese making are characterized by their high nutritional quality, predominantly determined by amino acid composition and physiological functionality. As reviewed by Smithers (2015), whey proteins and peptides have been implicated in a range of biological and physiological effects. These include: (i) improved muscle recovery after exercise and general muscle health; (ii) increased satiety, weight management, and control of obesity; (iii) enhanced wound management and repair; (iv) improved cardiovascular health and reduction of risk factors; (v) control of microbial and viral infections; (vi) better infant growth and nutrition; and (vii) anticancer activity. Whey has been established as a rich and unique source of functional and bioactive peptides, either naturally occurring or isolated from whey proteins through the targeted use of enzymes (Smithers, 2008). Moreover, proteins in foods exert various functional properties such as gelling, foaming, water binding, and emulsifying capacity, which justify their inclusion as ingredients in various formulations. To this aim, both whey and hen egg protein are applied either as such or after hydrolysis (Rao et al., 2016).

On the other hand, protein can undergo degradation through the Maillard reaction, especially at intermediate $a_w$ levels. In the initial stage, the Maillard reaction forms an Amadori adduct from the condensation of glucose with an amino group of proteins. Loss of essential amino acids, especially lysine, is the first effect of the Maillard reaction in its initial stage. Next, Amadori products can form cross-links between adjacent proteins or with other amino groups, giving rise to fluorescent polymeric aggregates—so called advanced glycation end products (AGEs) (Lavelli and Sri Harsha, 2019).

Upon glycation, proteins lose their functionality and have increased resistance to removal by proteolytic enzymes. In addition, AGE receptors are expressed on various cells. The binding of AGEs to these receptors mediate intracellular signaling and promote oxidative stress, thus causing tissue damage (Engelen et al., 2013).

Changes in available lysine were studied in infant rice cereal stored at 25°C and 55°C for 12 months. Lysine losses were greatest at relatively high $a_w$ (0.65) and temperature (55°C), and were diminished by packaging with nitrogen. At $a_w$ 0.65, lysine losses after 7 days were similar to those produced after 12 months of storage at lower $a_w$ (Ramirez-Jimenez et al., 2003). Nonenzymatic browning of whey protein at intermediate $a_w$ was studied by McMahon et al. (2009), who also outlined the effects of hydrolysis on the reaction rate. Whey protein and partially hydrolyzed whey proteins were separately formulated with high-fructose corn syrup or sorbitol syrup, such that the initial $a_w$ was 0.60. During storage for 42 days at 32°C, the extent of Maillard reactions was much greater in the hydrolyzed whey protein than in the intact protein, both in the presence of high-fructose corn syrup and sorbitol syrup. However, the degradation rate was not kinetically modeled. In a related study, the moisture sorption isotherms of the hydrolysates of casein, hen egg white, meat, and fish protein were modeled using the GAB equation, giving monolayer moisture values in the $a_w$ range 0.06–0.3, corresponding to 4.7–6.8 g water/g solid. These values were similar to those found for the corresponding intact protein, indicating that protein hydrolysis exposes few, if any, new adsorption sites (Rao et al., 2016). However, at $a_w > 0.7$ it was observed that the greater the hydrolysis degree, the greater was the moisture holding capacity in the powdered protein hydrolysates. It was concluded that at the higher hydrolysis degree, more hydrophilic groups in the hydrolyzed protein are exposed. In general, the increase in hydrolysis resulted in smaller average molecular weight than for the intact protein, causing a decrease in $T_g$, especially at high water contents. Thus, the decrease in $T_g$ was related to the faster rate of the Maillard reaction (Rao et al., 2016).

The relevance of $T_g$ for predicting protein stability was also tested in milk powder stored at 37–60°C (Pereyra Gonzales et al., 2010). The Maillard reaction kinetics were analyzed by following the loss of available lysine at the first stage of this reaction. Lysine loss was fitted to a pseudo-first-order model at all temperatures and $a_w$ values considered. At 37°C and the lowest $a_w$ (0.32), the Maillard reaction occurred at a relatively slow rate, with a half-life of 35 days. Conversely, at $a_w$ levels of 0.47–0.70, extensive lysine losses were noticed after a few days of storage (Table 4). It should be noted that 37°C is very close to the $T_g$ (33°C) reported by Jouppila and Roos (1994) for skim milk stored at $a_w$ 0.33. Although the Maillard reaction did not stop in the glassy state, the rate constant decreased considerably at temperatures close to $T_g$. Hence, it was concluded that the rate of nonenzymatic browning was low in the vicinity of the glass transition temperature, because of limited molecular mobility. Conversely, at $a_w$ 0.43, Jouppila and Roos (1994) reported a $T_g$ of 9°C, much lower than 37°C. Hence, at this temperature, the increase in the mobility of the reactants caused a dramatic rise in reaction rate.

Schmitz-Schug et al. (2013) applied low resolution $^1$H NMR to investigate the relationship between lysine loss and molecular mobility in dairy powders, and to complement the stability concept based on the physical state. The loss of available lysine increased with increasing molecular mobility in the glassy and rubbery state and was

**Table 4** First-order rate constants ($day^{-1}$), first-order hyperbolic model rate constants ($day^{-1}$) and half-lives (day) for the Maillard reaction occurring in protein and protein hydrolysates during storage in the dark as a function matrix and storage temperature (°C), time (day), and water activity conditions

| Parameter studied | Matrix | $T$ | $t$ | $a_w$ | $k \times 10^3$ | $t_{1/2}$ | References |
|---|---|---|---|---|---|---|---|
| Lysin[a] | Rice cereal | 25 | 28 | 0.65 | 4.55 | 152 | Ramirez-Jimenez et al. (2003) |
| | | 55 | 28 | 0.65 | 23.9 | 29 | |
| Lysin[a] | Skim milk powder | 37 | 27 | 0.33 | 0.50 ± 0.1 | 1459 | Pereyra Gonzales et al. (2010) |
| | | | | 0.43 | 26.8 ± 1.8 | 26 | |
| | | | | 0.52 | 18.3 ± 2.3 | 38 | |
| | | | | 0.69 | 19.2 ± 1.1 | 36 | |
| | | | | 0.85 | 12.9 ± 1.2 | 54 | |
| | | | | 0.98 | 5.9 ± 0.1 | 118 | |
| Advanced-glycation-endproducts[b] | White hen egg protein | 45 | 50 | 0.05 | n.s. | | Rao and Labuza (2012) |
| | | | | 0.31 | n.s. | | |
| | | | | 0.43 | 11–12 | 56.8–65.1 | |
| | | | | 0.50 | 8–21 | 32.4–87.8 | |
| | | | | 0.53 | 17–28 | 24.8–41.0 | |
| | | | | 0.73 | 18–38 | 18.1–38.1 | |
| | | | | 0.79 | 16–40 | 17.4–42.2 | |

[a] Degradation was modeled according to a first-order equation.
[b] Formation was modeled according to a hyperbolic model.

decelerated by crystallization. Lysine in lactose-hydrolyzed skim milk powder is less stable that in skim milk powder. In fact, the kinetics of lysine loss was compared in model systems made of casein and lactose, casein and glucose, or casein and galactose. The lysine loss in the system with lactose was 10-fold lower than in those with the monosaccharides. $T_g$ values determined for casein-glucose, casein-galactose, and casein-lactose systems equilibrated at $a_w$ 0.33 were 33°C, 25°C, and 5°C, respectively. Hence, the molecular mobility was higher in the matrix with the monosaccharides, leading to higher reaction rates (Naranjo et al., 2013).

In a study on egg white protein, Rao and Labuza (2012) confirmed that at intermediate and high moisture levels, the effect of moisture on hydrolyzed egg protein was more severe than that of intact egg protein. For the intact protein, no significant change in lightness ($L^*$) was observed during storage for 4 months at 23°C within the $a_w$ range 0.05–0.85. For hydrolyzed egg protein, the $L^*$ value deceased slightly when the $a_w$ was below 0.54 (moisture content below 12.0%, db), while above this value, the lightness started to decrease significantly. The Maillard reaction occurred in the hydrolyzed egg protein even if the glucose content was relatively low (0.07%). Indeed, Rao and Labuza (2012) found that at $a_w$ values of 0.43, 0.54, and 0.64, the $T_g$ values for ovoalbumin were 113°C, 106°C, and 98°C, while those of egg protein hydrolyzates (hydrolysis degree 7%–14%) were 50°C, 31°C, and 11°C. In this study, kinetics of the Maillard reaction were followed by measuring the increase in fluorescence ($k_{ex}$ = 340–370 nm and $k_{em}$ = 420–440 nm) due to the formation of AGEs (Table 4).

Rao and Labuza (2012) found that after about 2 months of storage at 45°C, the fluorescence did not increase significantly for hydrolyzed hen egg powder stored at $a_w$ levels from 0.05 to 0.31. Under these latter conditions, the hydrolyzed hen egg protein powder was below the glass transition temperature. However, when the storage $a_w$ was from 0.43 to 0.79 the fluorescence increased markedly, as expected from the increase in $T_g$. A first-order hyperbolic model was applied to fit this increase. Interestingly, after 1 month of storage at 45°C, the free amino groups at $a_w$s of 0.50 and 0.79 decreased by about 5% and 6%, respectively, while there was no significant change in the sample at lower $a_w$ values. On the other hand, previous studies showed that when ovalbumin was stored with different reducing sugars (1:1, g/g) in a solid system under accelerated conditions (50°C and $a_w$ 0.65), the remaining free amino groups of ovalbumin were <30% of the original within 1 month (Kato et al., 1988). Hence, Rao and Labuza (2012) concluded that significant amino acid loss can be expected in food products where hydrolyzed hen egg protein are added as ingredients.

As a general rule, during storage at intermediate $a_w$ levels, and above the $T_g$, proteins are involved in the Maillard reaction. The extent of lysine loss and increase in AGEs depends also on the composition of the matrix, and could be very relevant in the presence of free sugars.

# 6 Conclusion

As food supply chains are constantly evolving toward healthier food production and better resource management, novel food ingredients are needed that combine promising bioactive potential with sustainable processing technologies. From a sustainability perspective, drying, which is notably an energy-consuming operation, should be followed by storage at ambient temperature; that is, with no further energy input. For all the bioactive compounds analyzed in this Chapter, $a_w$, $T_g$, and the water mobility concepts proved to be useful predictors of the kinetics of degradation phenomena, even if the complexity of food stability still remains to be fully understood.

Moisture-dependent degradation of anthocyanin flavanols and carotenoids is related to their polarity. For polar anthocyanins and flavanols, the degradation rate significantly decreases with decreasing $a_w$ level, especially when the matrix is in the glassy state, even if it is not completely ended. Conversely, the lipid-soluble carotenoids are highly unstable in the glassy state and their degradation decreased in the rubbery state. For all these compounds, the degradation kinetics also depends on the matrix composition. In particular fibers, either when present as major components (i.e., in fruit peels) or when applied as encapsulating agents, have been shown to be the most promising matrix constituents to be modulated to enhance bioactive compound stability. Regarding protein-based ingredients, the hydrolysis degree affects the polarity, and consequently, the moisture sorption properties and $T_g$, ultimately resulting in decreased stability with decreasing the molecular weight of the components. Reducing sugar content is also a crucial factor affecting degradation kinetics of protein.

## References

Bell, L.N., 2007. Moisture effects on food's chemical stability. In: Barbosa-Canovas, G.V., Anthony, J., Fontana, J., Schmidt, S.J., Labuza, T.P. (Eds.), Water Activity in Foods: Fundamentals and Applications. IFT Press, Blackwell Publishing, Ames, Iowa, pp. 173–198.

Boon, C.S., McClements, D.J., Weiss, J., Decker, E.A., 2009. Role of iron and hydroperoxides in the degradation of lycopene in oil-in-water emulsions. J. Agric. Food Chem. 57 (7), 2993–2998.

Bradford, K.J., Dahala, P., Van Asbrouck, J., Kunusothc, K., Bello, P., Thompsond, J., Wue, F., 2018. The dry chain: reducing postharvest losses and improving food safety in humid climates. Trends Food Sci. Technol. 71, 84–93.

Brunauer, S., Deming, L.S., Deming, W.E., Troller, E., 1940. On the theory of Van der Waals adsorption of gases. J. Am. Chem. Soc. 62 (7), 1723–1732.

Cádiz-Gurrea, M.L., Borrás-Linares, I., Lozano-Sánchez, J., Joven, J., Fernández-Arroyo, S., Segura-Carretero, A., 2017. Cocoa and grape seed byproducts as a source of antioxidant and anti-inflammatory proanthocyanidins. Int. J. Mol. Sci. 18 (2), 376.

Chiu, Y.T., Chiu, C.P., Chien, J.T., Ho, G.H., Yang, J., Chien, B.H., 2007. Encapsulation of lycopene extract from tomato pulp waste with gelatin and poly($\gamma$-glutamic acid) as carrier. J. Agric. Food Chem. 55 (13), 5123–5130.

Corey, M.E., Kerr, W.L., Mulligan, J.H., Lavelli, V., 2011. Phytochemical stability in dried apple and green tea functional products as related to moisture properties. LWT- Food Sci. Technol. 44 (1), 67–74.

Desorby, S.A., Netto, F.M., Labuza, T.P., 1997. Comparison of spray-drying, drum-drying and freeze-drying for β-carotene encapsulation and preservation. J. Food Sci. 62 (6), 1158–1162.

Eggersdorfer, M., Wyss, A., 2018. Carotenoids in human nutrition and health. Arch. Biochem. Biophys. 652, 18–26.

Engelen, L., Stehouwer, C.D., Schalkwijk, C.G., 2013. Current therapeutic interventions in the glycation pathway: evidence from clinical studies. Diabetes. Obes. Metab. 15 (8), 677–689.

Galanakis, C.M. (Ed.), 2015. Food Waste Recovery: Processing Technologies and Industrial Techniques. Elsevier Inc., London.

Georget, D.M.R., Guardo, M., Ng, A., Smith, A.C., Waldron, K.W., 1997. Mechanical and thermal analysis of raisin components. Thermochim. Acta 294 (1), 71–79.

Gradinaru, G., Biliaderis, C.G., Kallithraka, S., Kefalas, P., Garcia-Viguera, C., 2003. Thermal stability of *Hibiscus sabdariffa* L. anthocyanins in solution and in solid state: effects of copigmentation and glass transition. Food Chem. 83 (3), 423–436.

Gurtler, J.B., Doyle, M.P., Kornacki, J.L. (Eds.), 2014. The Microbiological Safety of Low Water Activity Foods and Spices. Springer, New York.

Harnkarnsujarit, N., Charoenrein, S., Roos, Y.H., 2012. Microstructure formation of maltodextrin and sugar matrices in freeze-dried systems. Carbohydr. Polym. 88 (2), 734–742.

Hooper, L., Kroon, P.A., Rimm, E.B., Cohn, J.S., Harvey, I., Le Cornu, K.A., Ryder, J.J., Hall, W.L., Cassidy, A., 2008. Flavonoids, flavonoid- rich foods, and cardiovascular risk: a meta-analysis of randomized controlled trials. Am. J. Clin. Nutr. 88 (1), 38–50.

Hung, L., Horagai, Y., Kimura, Y., Adachi, S., 2007. Decomposition and discoloration of L-ascorbic acid freeze-dried with saccharides. Innovative Food Sci. Emerg. Technol. 8 (4), 500–506.

Jouppila, K., Roos, Y.H., 1994. Glass transition and crystallization in milk powders. J. Dairy Sci. 77 (10), 2907–2915.

Kato, Y., Matsuda, T., Kato, N., Nakamura, R., 1988. Browning and protein polymerization induced by amino carbonyl reaction of ovalbumin with glucose and lactose. J. Agric. Food Chem. 36 (4), 806–809.

Lavelli, V., 2013. High-warranty traceability system in the poultry meat supply chain: a medium-sized enterprise case study. Food Control 33, 148–156.

Lavelli, V., Corti, S., 2011. Phloridzin and other phytochemicals in apple pomace: stability evaluation upon dehydration and storage of dried product. Food Chem. 129 (4), 1578–1583.

Lavelli, V., Kerr, W., 2012. Apple pomace is a good matrix for phytochemical retention. J. Agric. Food Chem. 60 (22), 5660–5666.

Lavelli, V., Sri Harsha, P.S.C., 2019. Microencapsulation of grape skin phenolics for pH controlled release of antiglycation agents. Food Res. Int. 119, 822–828.

Lavelli, V., Torresani, M.C., 2011. Modelling the stability of lycopene-rich by-products of tomato processing. Food Chem. 125 (2), 529–535.

Lavelli, V., Vantaggi, C., 2009. Rate of antioxidant degradation and color variations in dehydrated apples as related to water activity. J. Agric. Food Chem. 57 (11), 4733–4738.

Lavelli, V., Zanoni, B., Zaniboni, A., 2007. Effect of water activity on carotenoid degradation in dehydrated carrots. Food Chem. 104 (4), 1705–1711.

Lavelli, V., Kerr, W., Sri Harsha, P.S.C., 2013. Phytochemical stability in dried tomato pulp and peel as affected by moisture properties. J. Agric. Food Chem. 61 (3), 700–707.

Lavelli, V., Kerr, W.L., García-Lomillo, J., González-SanJosé, M.L., 2017a. Applications of recovered bioactive compounds in food products. In: Galanakis, C. (Ed.), Handbook of Grape Processing by-Products: Sustainable Solutions. Elsevier Inc., pp. 233–266.

Lavelli, V., Sri Harsha, P.S.C., Laureati, M., Pagliarini, E., 2017b. Degradation kinetics of encapsulated grape skin phenolics and micronized grape skins in various water activity environments and criteria to develop wide-ranging and tailor-made food applications. Innov. Food Sci. Emerg. Technol. 39, 156–164.

Lavelli, V., Proserpio, C., Gallotti, F., Laureati, M., Pagliarini, E., 2018. Circular reuse of bioresources: the role of Pleurotus spp. in the development of functional foods. Food Funct. 9, 1353–1372.

Lee, N.Y., Jo, C., Sohn, S.H., Kim, J.K., Byun, M.W., 2006. Effects of gamma irradiation on the biological activity of green tea byproduct extracts and a comparison with green tea leaf extracts. J. Food Sci. 71 (4), C269–C274.

Lim, A.S.L., Griffin, C., Roos, Y.H., 2014. Stability and loss kinetics of lutein and β-carotene encapsulated in freeze-dried emulsions with layered interface and trehalose as glass former. Food Res. Int. 62, 403–409.

Luning, P.A., Marcelis, W.J., Rovira, J., van Boekel, M.A.J.S., Uyttendaele, M., Jacxsens, L., 2011. A tool to diagnose context riskiness in view of food safety activities and microbiological safety output. Trends Food Sci. Technol. 22, S67–S79.

McBrierty, V.J., Packer, K.J. (Eds.), 1993. Nuclear Magnetic Resonance in Solid Polymers. Cambridge University Press, Cambridge.

McMahon, D.J., Adams, S.L., McManus, W.R., 2009. Hardening of high-protein nutrition bars and sugar/polyol-protein phase separation. J. Food Sci. 74 (6), E312–E321.

Naranjo, G.B., Pereyra Gonzales, A.S., Leiva, G.E., Malec, L.S., 2013. The kinetics of Maillard reaction in lactose-hydrolysed milk powder and related systems containing carbohydrate mixtures. Food Chem. 141 (4), 3790–3795.

Olivas-Aguirre, F.J., Rodrigo-García, J., Martínez-Ruiz, N., Cárdenas-Robles, A.I., Mendoza-Díaz, S.O., Álvarez-Parrilla, E., González-Aguilar, G.A., de la Rosa, L.A., Ramos-Jiménez, A., Wall-Medrano, A., 2016. Cyanidin-3-O-glucoside: physical-chemistry, foodomics and health effects. Molecules 21, 1264.

Pereyra Gonzales, A.S., Naranjo, G.B., Leiva, G.E., Malec, L.S., 2010. Maillard reaction kinetics in milk powder: effect of water activity at mild temperatures. Int. Dairy J. 20 (1), 40–45.

Prado, S.M., Buera, M.P., Elizalde, B.E., 2006. Structural collapse prevents β-carotene loss in a supercooled polymer matrix. J. Agric. Food Chem. 54 (1), 79–85.

Rahman, S., 1995. Water activity and sorption properties of food. In: Clydesdale, F.M. (Ed.), Food Properties Handbook. CRC Press, Boca Raton, pp. 1–83.

Rahman, M.S., 2010. Food stability determination by macro–micro region concept in the state diagram and by defining a critical temperature. J. Food Eng. 99 (4), 402–416.

Ramirez-Jimenez, A., Guerra-Hernandez, E., Garcia-Villanova, B., 2003. Evolution of non-enzymatic browning during storage of infant rice cereal. Food Chem. 83 (2), 219–225.

Rao, Q.C., Labuza, T.P., 2012. Effect of moisture content on selected physicochemical properties of two commercial hen egg white powders. Food Chem. 132 (1), 373–384.

Rao, Q., Klaassen Kamdar, A., Labuza, T.P., 2016. Storage stability of food protein hydrolysates—a review. Crit. Rev. Food Sci. Nutr. 56 (7), 1169–1192.

Schmitz-Schug, I., Gianfrancesco, A., Kulozik, U., Foerst, P., 2013. Physical state, molecular mobility and chemical stability of powdered dairy formulations. Food Res. Int. 53 (1), 268–277.

Selim, K., Tsimidou, M., Biliaderis, C.G., 2000. Kinetic studies of degradation of saffron carotenoids encapsulated in amorphous polymer matrices. Food Chem. 71 (2), 199–206.

Sigurdson, G.T., Tang, P., Giusti, M.M., 2017. Natural colorants: food colorants from natural sources. Annu. Rev. Food Sci. Technol. 8, 261–280.

Smithers, G.W., 2008. Whey and whey proteins—from "gutter-to-gold". Int. Dairy J. 18 (7), 695–704.

Smithers, G.W., 2015. Whey-ing up the options—yesterday, today and tomorrow. Int. Dairy J. 48, 2–14.

Sri Harsha, P.S.C., Gardana, C., Simonetti, P., Spigno, G., Lavelli, V., 2013. Characterization of phenolics, in vitro reducing capacity and anti-glycation activity of red grape skins recovered from winemaking by-products. Bioresour. Technol. 140, 263–268.

Sri Harsha, P.S.C., Lavelli, V., Scarafoni, A., 2014. Protective ability of phenolics from white grape vinification by-products against structural damage of bovine serum albumin induced by glycation. Food Chem. 156, 220–226.

Suzuki, J., Isobe, M., Morishita, R., Nagai, R., 2009. Tea polyphenols regulate key mediators on inflammatory cardiovascular diseases. Mediat. Inflamm. 2009, 494928.
Tonon, R.V., Baroni, A.F., Brabet, C., Gibert, O., Pallet, D., Hubinger, M.D., 2009. Water sorption and glass transition temperature of spray dried acai (*Euterpe oleracea* Mart.) juice. J. Food Eng. 94 (3–4), 215–221.
Tonon, R.V., Brabet, C., Hubinger, M.D., 2010. Anthocyanin stability and antioxidant activity of spray-dried açai (*Euterpe oleracea* Mart.) juice produced with different carrier agents. Food Res. Int. 43, 907–914.
Venturi, L., Rocculi, P., Cavani, C., Placucci, G., Dalla Rosa, M., Cremonini, M.A., 2007. Water absorption of freeze-dried meat at different water activities: a multianalytical approach using sorption isotherm, differential scanning calorimetry, and nuclear magnetic resonance. J. Agric. Food Chem. 55 (26), 10572–10578.
Virot, M., Tomao, V., Le Bourvellec, C., Renard, C.M.C.G., Chemat, F., 2010. Towards the industrial production of antioxidants from food processing by-products with ultrasound-assisted extraction. Ultrason. Sonochem. 17 (6), 1066–1074.
Vittadini, E., Dickinson, L.C., Chinachoti, P., 2002. NMR water mobility in xanthan and locust bean gum mixtures: possible explanation of microbial response. Carbohydr. Polym. 49 (3), 261–269.
Wang, X., Choi, S.G., Kerr, W.L., 2004. Water dynamics in white bread and starch gels as affected by water and gluten content. LWT- Food Sci. Technol. 37 (3), 377–384.
Welti-Chanes, J., Guerrero, J.A., Barcenas, M.E., Aguilera, J.M., Vergara, F., Barbosa-Canovas, G.V., 1999. Glass transition temperature (Tg) and water activity (aw) of dehydrated apple products. J. Food Process Eng. 22 (2), 91–101.
Yang, C.S., Wang, X., Lu, G., Picinich, S.C., 2009. Cancer prevention by tea: animal studies, molecular mechanisms and human relevance. Nat. Rev. Cancer 9 (6), 429–439.
Yang, Z., Zhu, E., Zhu, Z., 2012. Moisture sorption isotherms and net isosteric heats of sorption of green soybean. Int. J. Food Eng. 8 (3). 15.
Zokti, J.A., Baharin, B.S., Mohammed, A.S., Abas, F., 2016. Green tea leaves extract: microencapsulation, physicochemical and storage stability study. Molecules 21 (8), 940.

## Further reading

Ayranci, E., Ayranci, G., Dogantan, Z., 1990. Moisture sorption isotherms of dried apricot, fig and raisins at 20°C and 36°C. J. Food Sci. 55 (6), 1591–1593.
Lavelli, V., Corey, M., Kerr, W., Vantaggi, C., 2011. Stability and anti-glycation properties of intermediate moisture apple products fortified with green tea. Food Chem. 127 (2), 589–595.

# Emerging cold pasteurization technologies to improve shelf life and ensure food quality

3

*Anton Minchev Slavov*, Petko Nedyalkov Denev*[†], *Zapryana Rangelova Denkova*[‡], *Georgi Atanasov Kostov*[§], *Rositsa Stefanova Denkova-Kostova*[¶], *Rosen Milanov Chochkov*[||], *Ivelina Nikolaeva Deseva*[#], *Desislava Georgieva Teneva*[†]

*Department of Organic and Inorganic Chemistry, Technological Faculty, University of Food Technologies, Plovdiv, Bulgaria, [†]Laboratory of Biologically Active Substances, Institute of Organic Chemistry With Centre of Phytochemistry, Bulgarian Academy of Sciences, Plovdiv, Bulgaria, [‡]Department of Microbiology, Technological Faculty, University of Food Technologies, Plovdiv, Bulgaria, [§]Department of Wine and Beer Technology, Technological Faculty, University of Food Technologies, Plovdiv, Bulgaria, [¶]Department of Biochemistry and Molecular Biology, Technological Faculty, University of Food Technologies, Plovdiv, Bulgaria, [||]Department Technology of Cereals, Fodder, Bread and Confectionary Products, Technological Faculty, University of Food Technologies, Plovdiv, Bulgaria, [#]Department of Analytical Chemistry and Physicochemistry, Technological Faculty, University of Food Technologies, Plovdiv, Bulgaria

**Chapter Outline**



## 1 Introduction

How to collect enough food in order to ensure survival? Why does food spoil? How to preserve food longer? These are among the most important questions ancient humans had to solve along with development. The deeper the knowledge mankind

Food Quality and Shelf Life. https://doi.org/10.1016/B978-0-12-817190-5.00003-3

gained with time, the better these issues were addressed. Nowadays, most of the problems related to microbial contamination, food preservation, and foodborne illnesses are solved to a large extent, or a reasonable compromise has been made. However, all of the "traditional" methods for ensuring food safety suffer from some disadvantages. Thermal treatment (sterilization, pasteurization, etc.), although the modern technologies rely on short exposure to elevated temperatures, leads to some loss of valuable nutrients, chemical and physico-chemical changes, and could initiate Maillard reaction (Young et al., 2018). The decrease of water activity is another method of control and extending shelf life. Either performed by water content reduction or by addition of compounds binding water strongly, the process significantly changes final products, and does not ensure freshness and texture compared to untreated foodstuffs.

Freezing and cold storage are among the most popular methods for microbial development delay. The usual cold storage temperature is around 4°C. However, some nonproteolytic strains of *Clostridium botulinum* could even germinate at 2–3°C (Selby, 2017), and *Listeria monocytogenes* and *Yerstnia enterocoltttca* could grow below 1°C (Gould, 2000). This is particularly important when an improper temperature control (as the domestic one very often is) is applied. Freezing below −18°C ensures prevention of microbial growth but care should be taken to avoid large ice crystal formation, which could damage the cell walls and lead to valuable nutrients loss (Chemat et al., 2017). Chemical preservatives and antimicrobial agents usage is one of the earliest and easiest ways to control microbial growth. Nevertheless, it is a controversial issue, does not always completely fulfil the aim, and lately, consumers' opinion tends towards utilization of fewer chemicals. For example, application of chlorine-based treatment could lead to formation of chlorinated derivatives with possible carcinogenic activities (Rico et al., 2007). Poultry treatment with chlorine has been banned in the EU since 1997 due to uncertainties of successful disinfection of species of *Salmonella* and *Campylobacter* depending on how heavily the birds were initially soiled. A recent study questioned the complete eradication of *L. monocytogenes* and *Salmonella enterica* if they were induced by chlorine treatment to enter a viable-but-nonculturable state (Highmore et al., 2018). Synthetic antioxidants, such as butylated hydroxyanisole, butylated hydroxytoluene, tertiary butylhydroquinone, etc., and higher amounts of salt and sugar in the diet are often blamed for provoking various socially significant diseases.

Contemporary world trends, however, are oriented towards higher and higher consumption of ready-to-eat and minimally processed foods. Consumers expect foodstuffs to exhibit a combination of the convenience of semiprepared products and, at the same time, the quality of freshly picked fruit, vegetables, or freshly prepared meat, milk, or other type of meal. Besides, concerns of the public health risks, the possibility of bioterrorism threats has to be taken into account. All these expectations and attitudes raise an additional question: how to ensure longer shelf life of food while keeping it fresh, minimally processed, and without food quality deterioration? In order to adequately answer this question, new "emerging" technologies and their combined use have entered into practice and technology.

# 2 Emerging technologies

All of the new technologies rely heavily on physical treatments of the products while trying to keep the temperature as low as possible. For this reason, the emerging technologies are also often referred to as "cold pasteurization." Some of the methods employ electromagnetic fields, such as pulsed light (PL), pulsed electric field (PEF), high voltage arc discharge (HVAD), cold plasma (CP), or microwaving (MW). Others use a combination of high-energy electromagnetic waves and particles (ionizing radiation, IR), sound waves (ultrasound treatment, US), high pressure (high pressure technologies, which could vary depending on the effect achieved and techniques employed), pressure and carbon dioxide (dense phase carbon dioxide treatment, DPCD), and mechanical waves (shockwave treatment). In general, the effects of some of the physical processes were well known from the past but, due to the lack of proper equipment at that time, these technologies only recently became practically applicable.

The main driving forces of the introduction and commercialization of the new technologies, according to Jermann et al. (2015), could be summarized and ordered by significance as follows:

- enhanced quality and/or added value of the food products;
- safety issues concerns;
- extending shelf life of food systems;
- convenience;
- lowering the price, increase in competitiveness, running costs decrease;
- regulations and legislation;
- better waste management and eco-friendly procedures;
- globalization trends;
- financial issues;
- research support; and
- installation quality and operability.

## 2.1 *High pressure treatment*

The application of high pressure on food systems has been known since the end of the 19th century. The first attempts were made with raw milk with dubious success. The major limitation was lack of proper equipment. At the end of the 20th century, advancement of the technologies allowed construction of machinery able to achieve high enough pressures. Nowadays, several major approaches for utilization of high pressure in the food industry could be distinguished: high pressure processing (HPP, high hydrostatic pressure processing); high pressure homogenization (HPH); microfluidization (MF); and high-pressure jet (HPJ) processing. In general, microfluidization is performed in the 20–280 MPa range, HPH in the range 100–400 MPa, HPJ uses pressures up to 600 MPa, and HPP, in the range 400–1000 MPa (Tobin et al., 2015; Tran et al., 2018) (Fig. 1).

The European Commission, after several major producers submitted, according to Regulation (EC) 258/97, applications for placing on the market food products treated by high pressure, decided in May 2001 that "the use of high-pressure pasteurization

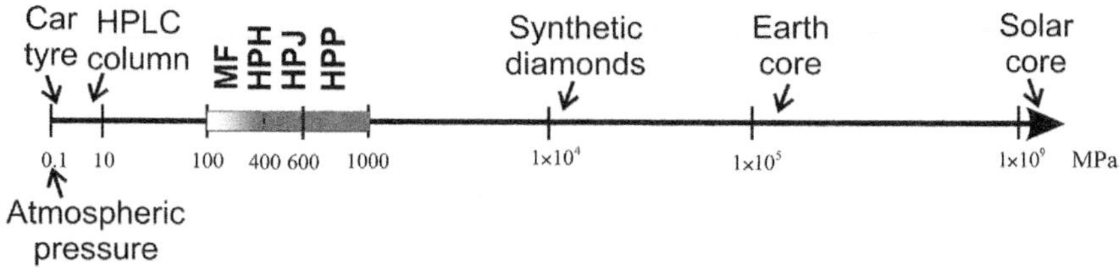


**Fig. 1** Comparative scheme of the high pressure technologies and the pressures utilized. *MF*, microfluidization; *HPH*, high pressure homogenization; *HPJ*, high pressure jet processing; *HPP*, high pressure processing.

in the production of fruit preparations is not likely to have an effect on public health so that a decision can be taken without consultation of the Scientific Committee for Food" (EU Commission decision, 2001). Similarly, the FDA ruled that HPP of fruit juices does not necessarily require specific FDA approval (US Food and Drug Administration, 2004), but the processors have to include in their plan the process time and pressure as critical limits. The Canadian authorities (Canadian Food Inspection Agency, 2016), based on a position from December 2016 of Health Canada (Health Canada, 2016), also stated that the results and proofs of safety issues with HPP treated food systems are already numerous and the novel status should be removed. However, the differences in the structure, microbial load, and texture for each product have to be carefully examined and taken into account.

The primary application of high pressure in the food industry is for nonthermal pasteurization and enzymes inactivation (Tran et al., 2018). Among the methods employing high pressure, HPP is the most investigated approach for food preservation, while at the same time retaining unchanged the nutrients, biologically active and aroma substances, and appearance of the treated products. HPJ, HPH, and MF are suitable for fluid or semifluid type food systems and mostly applied for emulsification, and droplet size reduction, especially for milk and creamy texture products (Mohan et al., 2016), although the higher pressures in HPH and HPJ also result in substantial microorganism reduction. HPP has several important advantages: it could be applied for a variety of products—liquid, solid and semisolid, the temperature could be easily controlled, and the product could be "cold" pasteurized after proper packing. Among the main disadvantages is the batch processing. HPJ, HPH, and MF are exclusively applied on liquids and the advantage is that they could be performed as continuous processes. The main disadvantages are related to packing after the treatment and the temperature control.

### 2.1.1 *Effect of high pressure on microorganisms*

Microorganism destruction at high pressure goes through various stages and depends on the type and age of microorganisms, the pressure applied, the pH, the food matrix, etc. Initially high pressure affects the transport functions of the membrane. As the pressure rises, the membrane begins to deform, lose its integrity, and ultimately, under the effect of the difference in external and internal pressure, it breaks down. Depending on the cell type and the processing conditions, the effect on the transport systems is due to protein denaturation in the cell membrane, modification of some protein structures, disruption of the membrane bonds, and lipid bilayer destruction. The pressurizing stage induces firmer packing of bilayered membrane, and gel-like structures are formed, which then, by diminishing the applied pressure at the end of the cycle/process, leading to puncturing and permeability of the cells (Barba et al., 2017).

Since there are a number of types of cellular structures in nature, the effect of high pressure in cell inactivation occurs differently and under different boundary pressure values (Abe, 2007; Hsiao-Wen et al., 2014). It should be noted that there is a correlation between the pressure and the processing temperature that can be used to enhance the treatment effect (Hsiao-Wen et al., 2014; Mohamed et al., 2012; Wang et al., 2013). Another important feature is the use of alternating pressure, i.e., the pressure is increased and released periodically, and the result is an increase in the final effect (Hsiao-Wen et al., 2014).

Simultaneously with membrane damaging, high pressure disturbs the physiological functions of the cellular structures. Firstly, the pressure disturbs the transposition mechanisms of substrates and metabolic products, which directly affects the cellular metabolic pathways naturally occurring in microbial cells. There is also evidence that high pressure treatment disturbs the synthesis of adenosine triphosphate. The proton flow in the cells changes and the intracellular pH, which is responsible for the physiological functions of the cellular structures, is affected (Hsiao-Wen et al., 2014; Tholozan et al., 2000). The enzyme functions are also impaired (Moreirinha et al., 2016).

High pressure changes the tertiary and quaternary structure of the cell proteins. Covalent bonds in proteins are not disturbed but breakdown of hydrogen bonds occurs (Hsiao-Wen et al., 2014). These variations are observed above 200 MPa and with pressure increase, the degree of change in the protein structure increases. Different authors, examining the influence of pressure on protein structure, state that at pressure values of up to 300 MPa changes in protein structures are reversible, but above this limit the changes become irreversible. Therefore, in order to substantially disturb the physiological functions of the cell, the pressure must be above 300 MPa (Hsiao-Wen et al., 2014; Yaldagard et al., 2008).

HPP affects the genetic material of the cell by disturbing DNA replication and transcription and also causes changes in the proteins of the nuclear cell structure. Stress factors such as high pressure may increase the resistance of some Gram-positive bacteria due to changes in their DNA structure. It is interesting to note that if the cells are processed at lower pressures they become resistant to shock conditions such as temperature and pressure due to genetic mutations (Hsiao-Wen et al., 2014; Wemekamp-Kamphuis et al., 2004). In addition, some bacteria could exhibit a recovery after HPP and survive in the food product (Muñoz-Cuevas et al., 2013).

Gram-positive bacteria, in general, showed better survivability compared to Gram-negative bacteria. Gram-positive bacteria have a thick cell wall composed of several layers of peptidoglycan (mucopeptide responsible for the thick solid structure) and two types of teihoic acids. Some species also have a layer on the cell surface called surface layer protein.The wall of Gram-positive bacteria contains several grid layers, while that of Gram-negative bacteria contains only one layer (Denkova et al., 2018;

Hsiao-Wen et al., 2014). Cells showed higher resistance in the presence of more unsaturated acids in the membrane because of greater membrane fluidity. Certain specific protein groups in the cellular membrane also increase cell stability (Allen et al., 1999; Denkova et al., 2018; Hsiao-Wen et al., 2014; Wemekamp-Kamphuis et al., 2004).

The age of the cell also affects its stability. In general, cells that are in an exponential growth phase are more sensitive due to the ongoing fission processes and the incomplete cellular structure. Cells in the stationary growth phase have well-formed membrane structures and are more resistant. However, there are exceptions, mostly related to the presence of RNA polymerase proteins in the cell wall. The spore-forming ability of some bacteria in the stationary phase is also important. Some spores can withstand pressure of 1000 MPa at room temperature. Spores have a multilayered structure and this leads to increased resistance. The best inactivation is application of periodically changing pressure, which induces spore germination and, further, their destruction (Hsiao-Wen et al., 2014; Minh et al., 2010).

The pH of the medium is important for cellular resistance. Data suggests that, at neutral pH, bacteria have a higher resistance than at lower (Bayindirli et al., 2006; Rastogi et al., 2007). The same applies to spore-forming bacteria, such as some *Bacillus* species. But this is not a rule, and the choice of a combination of pH-temperature-pressure processing conditions should be optimized for each particular type of microorganism. Islam et al. (2006) found that relatively low pressures and pH 7 did not affect significantly the resistance of *Bacillus coagulans*, but the pH reduction of three units increased the pasteurization effect even at lower pressures. Other authors investigated the condition of *B. coagulans* spores in model conditions and tomato juice applying a constant pressure of 800 MPa and a variable temperature of 25–75°C, and found that bacterial spores are much more sensitive to the combination of high temperature and high pressure under acidic conditions (Hsiao-Wen et al., 2014;Vercammen, et al., 2012).

For products with pH below 4.5 (most juices) it was suggested that pressures of 580–600 MPa and duration of 4–6 min, in general, are sufficient (Augusto et al., 2018). However, it had to be taken into account that even at conditions as harsh as 1200 MPa, the molds and bacterial spores could survive and germinate later in a better environment (Datta and Deeth, 2011), and spores are also resistant to heat, irradiation, and some chemical treatment (Hsiao-Wen et al., 2014). Increasing the pressure to 1000–1200 MPa at low temperatures does not significantly affect the spores of the species of *Bacillus* and *Clostridium.* As the temperature increases, the spore resistance of the two genera decreases, thus reducing the pressure needed (Aouadhi et al., 2013; Hsiao-Wen et al., 2014; Shao and Ramaswamy, 2011). Typically, the inactivation of spores occurs by combining the effects of temperature and pressure, quite often combined with other methods, including the addition of chemicals (Heinz and Buckow, 2010; Hsiao-Wen et al., 2014).

Destruction of protozoa and parasites is achieved at relatively low pressures (100–400 MPa), whereas inhibition of fungi and yeast requires pressure of more than 400 MPa. Furthermore, fungal spores are more resistant to HPP than vegetative cells (Rendueles et al., 2011). Viruses have different sensitivity to high pressure (Rendueles et al., 2011).

An important element in high pressure processing is the type of food matrix. Data suggested that solid food matrices protect cells better than liquid media (Tassou et al., 2007). The presence of inhibitory components such as bacteriocins and antimicrobial agents reduces cellular resistance, especially if these agents affect the cellular membrane structure (López-Pedemonte et al., 2003).

In conclusion, the reduction of the vegetative forms of microorganisms by 4–5 log $N$ is obtained at pressures of about 400–600 MPa at room temperature. Study data by a number of authors can be summarized as follows:

- Prokaryotic cells are more resistant to high pressure than eukaryotic ones;
- Gram-positive bacteria have higher resistance than Gram-negative bacteria;
- Cells with a spherical shape, in comparison with rod-shaped bacteria, are more resistant; and
- Spores of spore-forming microorganisms are also more resistant than vegetative forms (Hsiao-Wen et al., 2014).

Increasing the effect of the treatment could be achieved by combining with other factors: changes in temperature, changes in pH, presence of chemical inhibitors, use of pulsating pressure, etc. (Rendueles et al., 2011). A nonexhaustive list of applications for inactivation of microorganism and potential structural, chemical, physicochemical, and biochemical consequences from the high pressure treatment of various food products is presented in Table 1.

### *2.1.2 Effect of high pressure on low molecular substances, biopolymers and food structure*

The effective conditions for microbial reduction of applied preservation technology very often have negative effects on food structure. The emulsions are important biphasic systems in different food and cosmetic products. The emulsion stability determines the proper physical structure and appearance of the formulation and often could be disrupted by application of preservation methods. A recent study (Young et al., 2018) compared the stability of emulsion systems and biologically active substances after thermal treatment: short time-high temperature (90°C, 3 min), low temperature-long time (70°C, 30 min), and sterilization (121°C, 15 min), with HPP pasteurization (600 MPa, 5 min). The results implied that the disruption of the emulsion systems treated by HPP was the lowest and the recovery of retinol (used as a model of biologically active substance encapsulated in the micelles) was the highest. It should be noted, however, that the results depend on the type of matrix (e.g., the emulsifier used) and duration of the treatment, and the results could not be unambiguously interpreted. Other studies have found jellification of the food systems or food components treated by HPP (Ribeiro et al., 2018; Yang et al., 2016), which contributed to the better textural and functional properties of the food products. It has been found also that HPP leads to an improvement in the organoleptic profile of meat as some enzymes such as proteases are activated (Hsiao-Wen et al., 2014; Simonin, et al., 2012).

The volatile aroma compounds play a key role for the quality of food products. The effect of HPP (600 MPa, 6 min, 21°C) on volatile substances in Iberian dry-cured ham was studied (Martínez-Onandi et al., 2017b) and a significant change was found in the amount of the key odorants. In addition, an increase of substances with unpleasant

**Table 1** High pressure application for treatment of food systems

| Pressure/ technology | Product/medium | Microorganisms reduction | Optimum conditions | Co-processing; additives; storage | Structure related changes | Chemical changes | References |
|---|---|---|---|---|---|---|---|
| HPP | Fresh cheese | *Listeria monocytogenes* | 600 MPa, 10 min, (20°C) | Refrigeration, 4°C | Not investigated | Not investigated | Tomasula et al. (2014) |
| HPP | Whole milk<br>Skim milk | *Cronobacter sakazakii* | 400 MPa, 3 min | No | Not investigated | Not investigated | Jiao et al. (2016) |
| HPP | Korla pear (*Pyrus bretschneideri rehd*) juice | Total microorganisms count | 500 MPa, 10 min | Pretreatment by ultrafiltration | Clarity, color and degree of browning | Ascorbic acid, total polyphenols, aroma compounds | Zhao et al. (2016) |
| HPP | Mixed sweet pepper–orange juice | Total aerobic bacteria, molds and yeasts | 550 MPa, 5 min | Storage (4°C, 25 days) | Color, sensory analysis | Ascorbic acid, total polyphenols, antioxidant activity | Xu et al. (2015) |
| HPP | Strawberry (*Fragaria×ananassa Duch.*) puree and juice | Total aerobic bacteria, molds and yeasts | 500 MPa, 3 min | Storage (6°C, 49 days) | Color, pH, soluble matters, sensory analysis | Ascorbic acid, total, antocyanins | Aaby et al. (2018) |
| HPP | Carrot (*Daucus carota* L.) juice | Total aerobic bacteria, molds and yeasts | 550 MPa, 6 min | Storage (4°C, 20 days) | Color, pH, soluble matters, viscosity, clarity, sensory analysis | Total phenols, carotenoids, sugars, antioxidant activity | Zhang et al. (2016) |

| | | | | | | | |
|---|---|---|---|---|---|---|---|
| HPP | Clear cucumber juice | Total aerobic bacteria, molds and yeasts | 500 MPa, 5 min | Storage (4°C, 20 days) | Color, soluble matters, clarity, pH, acidity, sensory analysis | Key odorants retained: (*Z*)-6-nonenal, (*E*,*Z*)-2,6-nonadienal, (*E*)-2-nonenal, and (*E*,*Z*)-3,6-nonadien-1-ol | Liu et al. (2016) |
| Thermally assisted HPP | Mango (*Mangifera indica* L.) pulp | Aerobic mesophiles count, yeast and mold, total coliforms, lactic acid bacteria, psychrotrophs | 600 MPa, 10 min (52°C) | No | Color, acidity, soluble matters | Ascorbic acid, total phenols and flavonoids, antioxidant activity | Kaushik et al. (2016) |
| Ultra HPH | Orange juice | *Alicyclobacillus acidoterrestris*, *Alicyclobacillus hesperidum* spores | 300 MPa | Inlet temperature 70°C | pH | Not investigated | Roig-Sagués et al. (2015) |
| Ultra HPH | Tiger nut (*Cyperus esculentus* L.) milk | Psychrotrophs, Aerobic mesophiles, Lactobacilli, Enterobacteriaceae, faecal coliforms, total mesophilic spores, molds and yeasts | 300 MPa, 40°C | Storage (4°C, 57 days) | Color (slightly changed), pH, sensory analysis | Not investigated | Codina-Torrella et al. (2018) |

*Continued*

**Table 1** Continued

| Pressure/technology | Product/medium | Microorganisms reduction | Optimum conditions | Co-processing; additives; storage | Structure related changes | Chemical changes | References |
|---|---|---|---|---|---|---|---|
| HPP | Black tiger shrimp | Aerobic mesophiles, Psychrotrophs, *Escherichia coli*, *Staphylococcus aureus* | 300 MPa, 3–9 min; 400 MPa, 3 min | No | Color: loss in transparency; decreased redness, increased of yellow. Texture: higher hardness, springiness, cohesiveness, chewiness and gumminess | Trimethylamine nitrogen content: decreased in all conditions. Lipid oxidation: significant increase, but below the critical limits | Kaur et al. (2016) |
| HPP | Chilean jack mackerel (*Trachurus murphyi*) | Mesophilic and psychrophilic aerobic microorganisms; *Shewanella putrefaciens* ($H_2S$ producing bacteria) | 450 MPa, 3 min; 550 MPa 4 min (20°C) | No | Not investigated | Trimethylamine nitrogen content: decreased and remained for 26 days below 5 mg/100g pH increase slowly compared to control | Reyes et al. (2015) |
| HPP | Horse mackerel (*Trachurus trachurus*) | – | 150 MPa, 5 min | Frozen storage (1 and 3 months) | Texture and sensory analysis. Color change: loss in transparency | Water content decrease | Torres et al. (2014) |

| | | | | | | | |
|---|---|---|---|---|---|---|---|
| HPP | Sea bass (*Dicentrarchus labrax*) fillets | Total microorganisms count | 600 MPa, 30 min | No | Appearance: cooked-like; whither and loss of translucency; WHC[a] decreased | Changes in the protein profiles | Teixeira et al. (2014) |
| HPP | Haddock (*Merlangius euxinus*) | Total mesophilic aerobic bacteria | 400 MPa, 5 min (15°C) | Refrigeration, 4°C, 15 days | Color changes: decreased *L* value (*L*a*b* system). Sensory analysis (odor): 15 days shelf life; control–7days | Increased TBA[b]; Decreased TMA-N[c], DMA-N[d] and TVB-N[e] values compared to control | Alpas and Akhan (2012) |
| HPP | Fresh farmed salmon (*Salmo salar*) and cod (*Gadus morhua*) Fresh wild mackerel (*Scomber scombrus*) | Total aerobic count; Hydrogen sulphide producing bacteria | 500 MPa, 2 min | Refrigeration, 0.5°C, 26 days | Not investigated | Increased levels of lipid oxidation (TBA[b] method). Decreased phosphatase activity | Rode and Hovda (2016) |
| HPP | Marinated herring (*Clupea harengus*) fillet | *Morganella psychrotolerans*; $H_2S$-producing bacteria; Total psychrophilic bacteria | 300 MPa, 10 min; 500 MPa, 5 or 10 min | 4% acetic acid, up to 90 days (for 500 MPa) | Not investigated | Histamine and cadaverine not observed for the storage period | Ucak et al. (2018) |
| HPH | Hen egg white lysozyme activity | Antimicrobial activity against *Escherichia coli*, *Enterococcus faecalis*, *Geobacillus stearothermophilus*, *Bacillus cereus* | 120 MPa, 20°C | Combined action of lysozyme and HPH | Not investigated | Not investigated | Tribst et al. (2017) |

*Continued*

**Table 1** Continued

| Pressure/ technology | Product/medium | Microorganisms reduction | Optimum conditions | Co-processing; additives; storage | Structure related changes | Chemical changes | References |
|---|---|---|---|---|---|---|---|
| HPP | Serrano ham | Aerobic mesophiles<br>Psychrotrophs<br>Lactic acid bacteria<br>*Micrococcaceae*<br>Moulds and yeasts | 600 MPa, 6 min (21°C) | No | Not investigated | Not investigated | Martínez-Onandi et al. (2017a) |
| HPP | Spanish chorizo sausage | *Listeria monocytogenes* | 550 MPa, 12.3 min | Maintaining $a_w$ at 0.86 | Not investigated | Not investigated | Rubio et al. (2018) |
| HPP | Saline solution; meat slurry | *Staphylococcus aureus* | 350 MPa, 6 min | Nisin (500 ppm) | Not investigated | Not investigated | Yao et al. (2015) |
| HPP | Ready-to-eat sliced dry-cured ham | *Listeria monocytogenes* | 600 MPa, 5 min | Nisin | Not investigated | Not investigated | Hereu et al. (2012) |
| HPP | Cooked ham mortadella | *Listeria monocytogenes* | 400 MPa, 5 min | No | Not investigated | Not investigated | Hereu et al. (2014) |
| HPP | Ground chicken | *Listeria monocytogenes* | 600 MPa, 3 min | NaCl, KCl, $CaCl_2$ | Not investigated | Not investigated | Balamurugan et al. (2016) |
| HPP | Ground beef | *Escherichia coli* (Shiga toxin-producing) | 350 MPa, up to 40 min (4°C) | Observed some resistance | Not investigated | Not investigated | Sheen et al. (2015) |

[a] WHC: water holding capacity.
[b] TBA: thiobarbituric acid value (mg malondialdehyde/kg sample).
[c] TMA-N: trimethylamine-nitrogen.
[d] DMA-N: dimethylamine-nitrogen.
[e] TVB-N: total volatile basic nitrogen.

odor, such as nonanal, methanethiol, and dimethyl trisulfide, alter the sensorial profile of the product. Similar observations were made by Martínez-Onandi et al. (2016a,b) investigating Serrano dry-cured ham. The mechanism of formation of the volatiles in meat products containing higher fat levels is based on the lipolysis and further oxidative destruction of fatty acids. The increased formation of free radicals, especially at higher pressures (600–800 MPa), could also be related to the higher rates of formation and breakdown of hydroperoxides, and direct the reaction towards formation of less branched aldehydes (Fuentes et al., 2010). The upper limit of applied pressure that resulted in less radical formation, according to Bolumar et al. (2012), was 400 MPa (25°C) and 500 MPa (5°C). An additional advantage of HPP is inactivation of enzymes able to produce biogenic amines. Treatments at 350–400 MPa for 10–15 min significantly lowered the levels of cadaverine, putrescine, and tyramine in fermented meat products (Nuñez et al., 2016).

Although HPP was proven to ensure preservation with minimal heating, in certain cases a final product with the consumers' preferred textural characteristics and flavor cannot be achieved. Pingen et al. (2016) investigated ham production, trying to replace the traditional application of heating in the technological scheme by HPP (500 and 600 MPa). As a result, the authors failed to produce ham and the result might be described as a "ham-like" product with a softer texture and color diverging from the traditionally prepared product. Combined pretreatment with HPP and mild heating at 53°C gave better results, but in this case the advantage of shorter production time is lost.

HPP induces fewer changes compared to heat treatment but significant effects are observed and their influence on human health have to be carefully studied. In a recent study, Bogahawaththa et al. (2018) compared the impact of HPP and HTST on raw milk proteins. The results revealed that HTST has preserved to a large extent the natural conformations of proteins and this could suggest almost unchanged immunogenicity of milk. On the contrary, HPP at 400, 500, and 600 MPa (15 min, 30°C) led to denaturation of β-lactoglobulin and immunoglobulin G, and formation of protein clusters with participation of κ-casein. In addition, the authors observed alteration of immunogenicity of the treated milk (for pressures of 400 and 500 MPa) by measuring changes in the ratio of different types of T helper cells.

Today, it is roughly estimated that around 28% of vegetables and vegetable/fruit based products, 26% of meat products, 15% of seafood and fish, 14% of juices and beverages, and around 17% of other foods, are treated with high pressure (Liepa et al., 2016; Paterson, 2005). Fig. 2 presents an example of HPP treated meat products sold in Greece.

No doubt the application of high pressure technologies offers substantial advantages over traditional preservation methods, such as extending the shelf life, packaging before treatment, avoiding changes related to heating (even for a short time), etc. Nevertheless, limitations and shortcomings exist and often make the practical application of HPP difficult. Among the main disadvantages are the high initial investment, inability to utilize on all food systems, and structural and chemical changes, which need further research. It seems that the best way of applying HPP is in combination with other methods in order to ensure high quality longer shelf life foods and keeping processing at minimum levels.

**Fig. 2** Meat products treated by HPP and sold in the retail network of Greece.

## 2.2 Ultrasound treatment

Ultrasound (US) consists of sound waves with frequencies higher than the upper limit of human hearing—approximately 20kHz. US techniques are simple, reliable, energy-saving, and cheap, and find application in many spheres of food processing. Based on the frequency, the applications of US in food processing and technology could be divided into low and high energy. Low power US, with frequencies higher than 100kHz, at intensities below $1\,W\,cm^{-2}$, is among the most widely used nondestructive analytical methods for analysis and monitoring of foods, to ensure high quality during processing and storage. It is used for quality control of fresh vegetables and fruits, detection of honey adulteration, and assessment of the aggregation state, size, and type of protein, etc. High power ultrasound (HPUS) uses intensities higher than $1\,W\,cm^{-2}$ and frequencies between 20 and 500kHz. These are disruptive and have effects on the physical, mechanical, or chemical/biochemical properties of microorganisms and organic matter. HPUS is recognized as an emerging technology with various application for controlling microstructure of foods, modifying textural characteristics of fat products, emulsification, defoaming; inactivation or acceleration of enzymatic activity, microbial inactivation, freezing, drying, and extraction of various food and bioactive components (Awad et al., 2012; Knorr et al., 2004).

### 2.2.1 Microbicidal effect of ultrasound

The antimicrobial effect of US on vegetative forms has been known since 1929, when Harvey and Loomis (1929) proposed the use of US for food processing and preservation. Since then, many studies have been performed to clarify the mechanisms behind the microbicidal effect of ultrasound. US-mediated inactivation of microorganisms occurs through physical, mechanical, and/or chemical mechanisms. Cavitation resulting from US treatment appears to be the main promoter of US-assisted microbial death. Cavitation can be classified into acoustic, hydrodynamic, optic, and particle, based on the mode of generation (Gogate et al., 2006). The chemical effects of US are due mainly to acoustic cavitation, which includes formation, growth, and implosive collapse of gas bubbles in liquids (Kentish and Ashokkumar, 2011). During bubble explosion, a considerable amount of energy is generated and resulting pressure causes mechanical weakening and destruction of cell walls (Maleki et al., 2013). In addition, hydroxyl radicals are generated from the dissociation of the water molecules

(Gogate and Pandit, 2011). Generated free radicals attack the bacterial cell wall and cause weakening and destruction (Kentish and Ashokkumar, 2011; Raso and Barbosa-Canovas, 2003). It has been proposed also that US could deagglomerate bacterial clusters or flocs (Joyce et al., 2003).

### Factors affecting US microbicidal efficiency

Several factors affect the microbicidal efficiency of US treatment: type of microorganism and its vegetative form, degree of contamination, US frequency, heating, pressure, or other treatments. For example, US treatment is more efficient in inactivation of Gram-negative and rod-shaped bacteria, whereas spores, Gram-positive, and coccal bacteria are more resistant to US-mediated rupturing (Feng et al., 2008). Arroyo et al. (2011) revealed that US resistance of *Cronobacter sakazakii* in standard conditions was similar to that of other Gram-negative species and lower than that of Gram-positives. This effect is related to the higher amount of peptidoglucans in Gram-positive bacteria cell walls, which determines their higher strength and stability. Different strains differ in their resistance to US, as well. For example applying the same US parameters, *Escherichia coli* and *Saccharmyces cerevisiae* were reduced by more than 99% after ultrasonication, whereas *Lactobacillus acidphilus* was reduced by 72%–84% (Cameron et al., 2008). Usually a frequency of about 20kHz is applied for microbial inactivation, but mortality rate is highly dependent on US frequency and wave amplitude (Anaya-Esparza et al., 2017). As already mentioned, cavitation occurs in water media and, therefore, moisture content of foods is crucial for microbicidal effectiveness. Therefore, US is more efficient in liquid foods in comparison to solid ones. Arroyo et al. (2011) demonstrated that decreasing water activity of the treatment medium resulted in a significant increase in the resistance of *C. sakazakii*. In general, HPUS is not very effective in killing bacteria in foods alone, and is often combined with other preservation methods, such as antibiotics, hypochlorite, mild heat, or pressure (Baumann et al., 2005, Furuta et al., 2004; Lee et al., 2003; Piyasena et al., 2003a; Villamiel and de Jong, 2000).

## 2.2.2 Application of US for inactivation of food spoilage microflora and improving food quality

### Application for liquid foods

The most widely US-treated liquid foods are fruit and vegetable juices. In general US treatment increases juices' microbial stability and retains their nutritional components better than conventional thermal treatments. Experimental results suggested that thermosonication leads to a significant reduction of the microbial count (1–9 logs) of different microorganisms such as *S. cerevisiae* (Bermúdez-Aguirre and Barbosa-Canovas, 2012; Char et al., 2010; Ferrario et al., 2015; Marx et al., 2011; Wordon et al., 2012), *E. coli* and *Salmonella enteritidis* (Char et al., 2010; Kiang et al., 2013; Muñoz et al., 2011, 2012; Sagong et al., 2011), *C. sakazakii* (Adekunte et al., 2010; Arroyo et al., 2011, 2012), *Alicyclobacillus acidophilus* and *Alicyclobacillus acidoterrestris* (Evelyn and Silva, 2016; Ferrario et al., 2015; Wang et al., 2010), and *Staphylococcus aureus* (Walkling-Ribeiro et al., 2009). For example, manothermosonication (4bar,

43°C, 750 W, 20 kHz) significantly reduced microbial *E. coli* and *Pichia fermentans* counts inoculated in a fresh blend of apple and cranberry juice, compared to the untreated sample (Palgan et al., 2011). Thermosonication (24 kHz, 120 µm amplitude) at different temperatures (50°C, 54°C, and 58°C) was used to increase the shelf life of carrot juice. Samples sonicated at 58°C had the best quality, while microbial growth remained low at around 3-log for mesophiles, 4.5-log for yeasts and molds, and 2-log for enterobacteria after 20 days of storage. Furthermore, sonicated juice at 58°C retained more than 98% of its carotenoids and 100% of ascorbic acid.

Khandpur and Gogate (2015, 2016) investigated the effect of US (20 kHz, 100 W, 15 min) on the nutritional quality of different juices (orange, sweet lime, carrot, and spinach) in comparison to pasteurization (80°C, 10 min). US-processed juices retained most of the nutrients (polyphenols, vitamin C, carbohydrates, etc.) to a higher extent. Moreover, a scale up of the process with spinach juice confirmed that the technology could retain the required nutrient contents and microbial safety in refrigerated storage conditions for 10 weeks. The survival tests of *L. monocytogenes* in apple cider indicated the possibility of using a mild treatment condition in combination with ultrasound to achieve a 5-log reduction in pathogen populations (Baumann et al., 2005). Bermúdez-Aguirre and Barbosa-Cánovas (2008) used US to reduce *Listeria innocua* in milk and suggested that US-treatment efficiency depended on the fat content of the milk. Inactivation of *Listeria* cells occurred first in fat free milk, and rate of inactivation decreased with increasing fat content. No degradation of protein, a slight decrease in pH, and a minor increase of the milk's lactic acid content was observed.

## Application to solid foods

Fruit and vegetable are among the most widely US-treated foods. US was used by Aday et al. (2013), who investigated the effect of different ultrasound powers (30, 60, 90 W) and treatment times on quality of fresh strawberries. The authors revealed that power levels between 30 and 60 W improved quality of the fruits, resulting in higher oxygen concentration inside packages. The pH, total soluble solid content, and color (30 and 60 W treatments) were better than untreated control and 90 W treatment. Authors concluded that US power as high as 90 W is detrimental for the quality of strawberries. However, combining US with the chemical sanitizers (ozone and chlorine dioxide) was more beneficial for quality factors such as pH, total soluble solids, electrical conductivity, and texture, compared to the individual treatment or untreated fruit (Aday and Caner, 2014). In a similar study, freshly harvested strawberries were treated with 0, 25, 28, 40 or 59 kHz ultrasound at 20°C for 10 min. The results showed that 40 kHz US treatment significantly reduced decay incidence and numbers of microorganisms. US treatment also inhibited the decrease of firmness and maintained significant higher levels of total soluble solids, total titratable acidity, and vitamin C (Cao et al., 2010).

In a more recent study, strawberries were treated with ultrasound (33 kHz, 60 W) for different time periods (0, 10, 20, 30, 40, 60 min) and analyzed for the storage period of 15 days at 4°C. 20 and 30 min treatments were optimum for the firmness and color of the fruits. The pH, vitamin C, acidity, and total soluble solids were better retained between 30 and 40 min, while at 40 min treatment time, the bacterial count decreased from 5.91 to 3.91 log CFU $g^{-1}$, while yeast and mold count decreased from 4.80 to 3.58 log CFU $g^{-1}$. 30–40 min US treated fruits were shelf stable for the entire storage period (15 days). The US treatment was found to be effective for color retention, texture loss, preserving sensorial quality, and microbial load reduction of whole tomatoes (Pinheiro et al., 2015). Vivek et al. (2016) used response surface methodology for optimization of US treatment combined with sodium hypochlorite solution of kiwifruit (*Actinidia deliciosa*). The optimal parameters obtained were US intensity (368 W $cm^{-2}$), temperature (25°C), treatment time (8 min), and concentration of the solvent (30 ppm). US combined with NaOCl was found to be the most effective treatment in inhibiting the microbial growth (bacteria, yeast, and mold) and preserving the quality of kiwifruits, suggesting that US treatment may provide an alternative for extending the shelf life of whole kiwifruit.

In a recent study, Amaral et al. (2017) demonstrated that sonication of potato strips (40 kHz, 5 min) coupled with an acid medium gives better results in comparison to blanching or coating with edible films, in respect to their quality and microbial inactivation. Thermosonication combined with slightly acidic electrolyzed water effectively prolonged the shelf life of fresh-cut kale during refrigeration storage (Mansur and Oh, 2015). Chen and Zhu (2011) investigated the effect of the simultaneous US treatment (100 W for 10 min) and aqueous chlorine dioxide (40 mg $L^{-1}$ $ClO_2$ for 10 min) on postharvest storage quality of plum (*Prunus salicina* L.). Combination of the two processes was effective in reducing the initial microflora and retaining sensory qualities of plum fruit, extending shelf life from 35 days (for controls) to 60 days after the treatment.

The first studies on the effect of US on the quality and shelf life of meat (dating back to the early 1990s) did not find any significant effect of US on microbial inactivation (Pohlman et al., 1996: Smith et al., 1991). More recently, Kordowska-Wiater and Stasiak (2011) investigated the removal of Gram-negative bacteria (*Salmonella anatum, E. coli, Proteus* sp., and *Pseudomonas fluorescens*) from the surface of chicken skin after treatment with US (40 kHz, 2.5 W $cm^{-2}$ for 3 or 6 min) in water and 1% lactic acid. Sonication in water alone or in lactic acid for 3 min resulted in a reduction of microorganisms on the skin surface by 1.0 CFU $cm^{-2}$; 6 min treatment reduced more than 1.0 CFU $cm^{-2}$ in the water samples, and 1.5 log CFU $cm^{-2}$ in the lactic acid samples. Hanieh et al. (2014) showed that steam treatment and US applied to chicken carcasses can significantly reduce the number of *Campylobacte*r on contaminated birds. The total viable count was reduced by approximately three logs by applying steam and US immediately after slaughter (Alarcon-Rojo et al., 2015).

More often, combined techniques are applied for microbial reduction of meats. US treatment inhibited the growth of psychrotrophic and lactic bacteria, reduced the lipid oxidation, and promoted only minor pH and texture modifications during storage, improving the pasteurization of hot dog sausages (Cichoski et al., 2015). Ferrentino and Spilimbergo (2016) revealed the feasibility and potential of combining US with DPCD for the microbial stabilization of cooked ham. The authors demonstrated that US with DPCD was more effective in comparison to DPCD alone, reducing the treatment time for inactivation of lactic acid bacteria, yeasts, and molds to undetectable levels. In another case, the same authors revealed that HPUS alone was not able to induce any microbial inactivation of *L. monocytogenes*, while in combination with DPCD there was significant inactivation, depending on the parameters of the treatment (Spilimbergo et al., 2014). Morild et al. (2011) evaluated the inactivation of pathogens by the application of pressurized steam simultaneously with HPUS (30–40 kHz, 0.5–4.0 s) on the surface of pig skin and meat. Authors observed reduction of viable bacteria (*Salmonella typhimurium*, *Salmonella derby*, *Salmonella infantis*, *Yersinia enterocolitica*, and *E. coli*) with 1.1 log CFU $cm^{-2}$ after treatment for 1 s and 3.3 log CFU $cm^{-2}$ after treatment for 4 s. The reduction of microorganisms in pig skin was significantly greater than the reduction in meat.

Pedrós-Garrido et al. (2017) assessed the applicability of HPUS for surface decontamination of salmon (S. *salar*), mackerel (*S. scombrus*), cod (*G. morhua*), and hake (*M. merluccius*) fillets, and evaluated its impact on fish quality. US treatment reduced microbiological counts in oily fish species (salmon and mackerel) with 1.5 and 1.1 log CFU $g^{-1}$ for psychrophilic and mesophilic, respectively. In white fish species, the reduction was 0.5 log CFU $g^{-1}$. Lipid content did not change significantly by US treatment, whereas there was significant reduction of thiobarbituric acid reactive substance (TBARS) values.

HPUS is a cost-efficient and environmentally friendly technology to reduce microbial count and preserve nutritional components of various foods, thus extending their shelf life. However, its low efficiency in inactivation of bacterial spores impairs its application as a sterilization procedure. In combination with heat and/or pressure, HPUS can accelerate the rate of food sterilization. Among these combined techniques, HPUS under pressure at nonlethal and lethal temperatures appears to have great potential for industrial applications as they are more energy-efficient and effective in killing microorganisms. Furthermore, these techniques have shown several advantages compared to heat pasteurization, in terms of sensory and nutritional quality of treated foods.

## 2.3 Dense phase carbon dioxide

Fraser (1951) and Foster et al. (1962) studied the effect of DPCD on microorganisms and found that rapid decompression of $CO_2$ from high to atmospheric pressure resulted in the destruction of microbial cells. In 1969, the American company Swift & Co. received the first patent for food pasteurization by using $CO_2$ in "Super atmospheric" pressure. Various microorganisms had been shown to exhibit a different sensitivity to DPCD by 2002, when Spilimbergo et al. (2002) proposed the potential use of DPCD as an alternative nonthermal technique for pasteurization of foods.

### 2.3.1 Microbicidal effect of DPCD

Several mechanisms elucidate the microbicidal effect of DPCD: extracellular and intracellular pH decrease; inhibition of biological activities; cell rupture; modifications of cell walls; precipitation of proteins and ions; extraction of cellular substances; and disordering of the intracellular electrolyte balance (Erkmen, 2012). High pressure facilitates saturation of water with carbon dioxide, which in aqueous media transforms to $H^+$ and carbonate ions. The accumulation of $H^+$ decreases pH in both the media (extracellular pH) and the microbial cells. The low extracellular pH does not have a significant direct microbicidal effect on the cells. Instead, it acts indirectly by increasing the permeability of membranes, which further facilitates the penetration of $CO_2$ in cells (Lin et al., 1994; Erkmen, 2012). Due to the high affinity of $CO_2$ to membrane phospholipids, $CO_2$ is accumulated fast in membranes, which modifies them and leads to an increased membrane fluidity and permeability (Hong and Pyun 2001; Damar and Balaban, 2006; Garcia-Gonzalez et al., 2007). Due to the high pressure, $CO_2$ accumulates in cells to a critical level that leads to a removal of important cellular constituents, compromised structure of membrane, and imbalance in biological systems (Erkmen, 2012).

Other studies indicate cytoplasmic leakage, rather than membrane modification, as a result of the elevated $CO_2$ in microbial cell (Hong and Pyun, 1999, 2001). Constant cytoplasmic pH is vital for microbial cells and, therefore, is maintained via different mechanisms. Several studies indicate that reduction of intracellular pH is the main microbicidal tool of DPCD (Dillow et al., 1999). Acidification of cytoplasm leads to denaturation of numerous enzymes, which actually is among the main reasons for the microbial inactivation (Erkmen and Bozoglu, 2008). For example, treatment with DPCD of *E. coli* leads to a lack of activity of its alkaline phosphatase and galactosidase (Zhang et al., 2006). Hong and Pyun (2001) revealed that DPCD inhibits the activity of 13 enzymes from *Lactobacillus plantarum*, while the activity of others remained unaltered. This indicates that DPCD-induced inactivation of enzymes is selective. The concentration of the dissolved bicarbonate is crucial for the inhibition of enzyme activity since it regulate the general anion-sensitive site on the enzymes (Damar and Balaban, 2006). Furthermore, it affects the carboxylation and decarboxylation reactions in microbial cells (Erkmen, 2012). Increased permeability and fluidity of cell membrane, as a result of DPCD treatment, could lead to extraction of cellular components. Depressurization could also facilitate the transfer of membrane and intracellular materials out of the cell. DPCD treatment triggers series of subsequent and/or parallel morphological or metabolic changes in microbial cells which could lead to either inactivation of the cells or rupturing (Hong and Pyun, 1999; Dillow et al., 1999; White et al., 2006).

The effectiveness of DPCD for destroying microbial cells, and hence the effectiveness of pasteurization, depends primarily on the sensitivity of microorganisms to DPCD. In turn, microbial sensitivity depends on many genetic and/or environmental factors such as type of microorganisms, their vegetative form, initial microbial number, culture conditions and growth phase, presence of injured microorganisms, etc. Technological parameters such as water content, physical and chemical properties, $CO_2$ physical state and concentration, temperature, pressure, agitation, pressurization and depressurization rate, pressure cycling, and treatment time are even more important (Erkmen. 2012).

## Genetic and environmental factors affecting DPCD efficiency

The cell walls of Gram-positive bacteria are made up of a thick layer of peptidoglucans, which determines their strength and stability. In the cell walls of Gram-negative bacteria, the layer of peptidoglucans is thinner, making their cell walls more susceptible to impacts and more permeable to DPCD (Erkmen, 2012). Different studies investigated the effect of DPCD on *E. coli* in different nutrient media, applying various pressures (10–31 MPa), temperatures (34–40°C), and contact times from 10 to 120 min (Chen et al., 2017; Erkmen, 2001; Kobayashi et al., 2009; Spilimbergo et al., 2003a,b; Wu et al., 2007) and reduction in the number of viable cells by 2.5–9 logs was observed. Similar results were reported for *L. monocytogenes* (21.7 MPa, 45°C for 60 min) in milk (Lin et al., 1994), *Listeria dunnifii* (20.5 MPa, 34°C for 36 min) and *Proteus vulgaris* (20.5 MPa, 34°C for 36 min) (Dillow et al., 1999), *Pseudomonas aeruginosa* (7.4 MPa, 38°C for 2.5 min and 20.5 MPa, 34°C for 36 min) (Dillow et al., 1999; Spilimbergo et al., 2002), *Serratia mercescens* (7.4 MPa, 38°C for 10 min) (Spilimbergo et al., 2003a), and *Salmonella* spp. (Dillow et al. 1999).

The effect of DPCD on Gram-positive bacteria (*Enterococcus faecium*, *Lactobacillus casei*, *Lactobacillus brevis*, *Lactobacillus innocua*, *Lactobacillus dextranicum*, *L. plantarum*, *S. aureus*, etc.) showed either higher or at least equal effect in comparison to Gram-negative microorganisms (Dillow et al., 1999; Hong and Pyun, 2001; Spilimbergo et al., 2002, 2003a,b; Wu et al., 2007). DPCD has been successfully used to inactivate vegetative forms of molds and yeasts (Erkmen, 2012; Ferrentino et al., 2017; Gunes et al., 2005; Isenschmid et al., 1995; Parton et al., 2007). All these studies revealed that DPCD is a promising processing alternative for pasteurization of foods at moderate temperature and pressure. Moreover, the inactivation effect was strongly dependent on the technological parameters such as pressure, temperature, and contact time. Isenschmid et al. (1995) revealed that DPCD treatment led to a removal of vital components from the cell membrane, rather than to rupture of the cell. The initial number of microorganisms also affects the effectiveness of DPCD, as a higher number would require a longer exposure to DPCD (Erkmen, 2012). In general, bacteria are more susceptible to DPCD at lower temperatures, which is very important in regards to the practical application of this technology for food processing. The same is valid also for the pH of the media, since DPCD is more effective at pH levels different from the optimal acidity levels for a given microorganism. Additionally, young cells are more sensitive to DPCD in comparison to mature cells. For example, Hong and Pyun (1999) revealed that *L. plantarum* cells in stationary phase are more resistant to DPCD than the culture in logarithmic phase.

Besides microorganisms, DPCD could be used as an alternative to the common fumigations techniques since it is very effective in eradicating eggs, larvae, or beetles after a short exposure (10–20 min). Even a pressure of 2.5 MPa killed completely the most common insects in rice, including their eggs (Capilla et al., 2003).

## Technological factors affecting DPCD efficiency

Probably, pressure and temperature are the most important parameters for the effectiveness of DPCD. Moreover, pressure and temperature determine the state of $CO_2$ and, hence, influence greatly its microbicidal activity (Illera et al., 2019). In supercritical conditions, the liquid-like state of $CO_2$ determines its increased density and solubility. On the other hand, the gas-like state is related to a decreased viscosity and elevated diffusivity, which allows quicker diffusion and effective penetration in cells (Garcia-Gonzalez et al., 2007). Therefore, supercritical $CO_2$ is a more effective antimicrobial agent than pressurized liquid or gaseous $CO_2$. In general, the higher the pressure, the stronger is the effectiveness of DPCD. On one hand, pressure could act directly on cells but most probably its effect on $CO_2$ state and therefore its penetrating ability is more important for the microbicidal significance of this factor. However, the solubility of $CO_2$ in liquids and food matrices is limited and Spilimbergo et al. (2002) indicated that an increase of $CO_2$ pressure over 10 MPa does not influence significantly its solubility in water, which is the most common liquid in foods.

Temperature has a profound effect on microbial cells by destructing biological systems, which is used in many conventional technologies for sterilization, based on elevated temperatures. However, due to the milder temperatures (30–50°C) during DPCD treatment, the effect of this factor is related again mainly with its influence on $CO_2$ state. Additionally, elevation in temperature stimulates the diffusion of $CO_2$ into cells. To summarize, the microbicidal effect of DPCD could be increased by elevating pressure, temperature, and saturation of system with $CO_2$. However, technological optimum should be sought for every food matrix, since higher pressures and temperatures are related to more resistant equipment and higher production costs. Agitation of the system could also enhance the effectiveness of DPCD treatment, since it enhances the transfer of pressurized $CO_2$ and its solubilization rate, resulting in easier penetration of $CO_2$ through cell membranes (Oulé et al., 2006; Tsuji et al., 2005). The effectiveness of DPCD could be increased by the rate of pressurization and depressurization. Pressurized $CO_2$ rapidly expands through cells, leading to a mechanical rupturing of cell structures. The effectiveness of an explosive depressurization was demonstrated by Nakamura et al. (1994) with yeast cells. Conversely, Enomoto et al. (1997) revealed that even slower depressurization inactivates *S. cerevisia* cells similarly to the explosive depressurization of $CO_2$. Similar results were obtained by Kumagai et al. (1997), indicating that depressurization rate is not so essential in cell destruction.

The effect of pressure cycling (i.e., applying pressure repeatedly in cycles) on the effectiveness of DPCD processing is contradictory. Physical and chemical properties of food matrix strongly affect the effectiveness of DPCD treatment. pH is very important, since higher acidity could enhance the effect of DPCD (Damar and Balaban, 2006). Water content, viscosity, protein, and fat content are other factors affecting the treatment (Hong and Pyun, 1999). For example, under the same conditions *L. monocytogenes* in eggs is eliminated for twice the time in comparison to broth, and *E. faecalis* are eliminated for 24 h in whole milk, 16 h in skimmed milk, 3 h in orange juice, and 4 h in carrot juice (Erkmen, 2012). High moisture content is essential for an effective inactivation of microorganisms, since $CO_2$ must dissolve in an aqueous solution at a sufficient concentration, and dissociate into corresponding ions to reduce intracellular pH. Microorganisms in foods with low water content can show low or no inactivation behavior with DPCD treatment (Nakamura et al., 1994). From a technological point of view, batch systems require a longer time for microbial inactivation than continuous systems (Spilimbergo et al., 2003b). In general, inactivation increases with elongated exposure to DPCD, but the efficiency of the treatment depends on all the above-mentioned factors.

## 2.3.2 Application of DPCD for pasteurization of foods

As already mentioned DPCD is used mainly for nonthermal treatment of liquid foods due to their high moisture content, but recently it is used more often for pasteurization of solid foods. DPCD is used not only for inactivation of food microflora but also for inactivation of some enzymes that improve food sensory properties and flavor stability.

### Application for the treatment of liquid foods

There are many studies on the use of DPCD for reduction of microbes and enzyme activities in liquid foods and its effect on their nutritional components, quality parameters, and sensory properties (Balaban and Duong, 2014; Damar and Balaban, 2006; Garcia-Gonzalez et al., 2007). Fruit juices, and particularly orange and apple juices, are among the most investigated food matrices. Niu et al. (2010a) treated orange juice at 40MPa and 55°C for 10–60min and compared with thermally treated (90°C for 60s) juice. Authors found that cloud and stability of the juice were improved due to the DPCD treatment. Similar findings were made by Kincal et al. (2005) in a continuous-flow system. Gui et al. (2006) reported significant reduction of the browning degree of cloudy apple juice after DPCD treatment. Gasperi et al. (2009) did not find significant differences in the sensory properties after DPCD treatment of apple juice. Similarly, Amaral et al. (2018a,b) did not find differences among physicochemical parameters of DPCD-treated whey-grape juice and untreated control. Niu et al. (2010b) reported that polyphenol oxidase in cloudy apple juice was completely inactivated by DPCD and pectin methylesterase activity was significantly reduced when compared to mildly heated control. More recently, Manzocco et al. (2017) reported that DPCD treatment (12MPa, 35°C, 10min) led to 80% inhibition of polyphenoloxidase activity and higher microbial stability of freshly squeezed apple juice in comparison to a control, while keeping a sensory fresh-likelihood higher than thermally pasteurized. Ferrentino et al. (2009a) revealed that DPCD is equally efficient in microbial inactivation of apple juice as a conventional thermal treatment (85°C, 60min).

In another study, *A. acidoterrestris* spores in apple juice were completely inactivated by DPCD and, again, no influence on physical and sensory properties of the juice were observed (Bae et al., 2009). Lim et al. (2006) achieved maximal microbial reduction in mandarin juice at 35°C, 41.1MPa, for 9min at 7% $CO_2$. Moreover, cloud was enhanced by 38% and color parameters of juice improved. Five log reduction for yeasts and molds and total aerobic microorganisms occurred at 34.5MPa and 7min of treatment of fresh-squeezed grapefruit juice (Ferrentino et al., 2009b). Pineapple juice was completely free of microbiota after DPCD treatment (10MPa, 31.5°C) at different contact times (3.06–4.6min) and differences in quality attributes were minimal (Martínez et al., 2018). During storage, microorganisms were not able to recover and the vitamin C decrease could be limited to 8.2% after 4 weeks.

The treatment of watermelon juice at 34.4MPa, 40°C, 10% $CO_2$ for 5min led to reduction of the native aerobic microflora of the juice by 6logs, which allowed subsequent refrigeration storage of the product for 8 weeks without spoilage (Lecky and Balaban, 2004). In a comparative study, Damar et al. (2009) revealed that there was no difference in the sensory properties of continuous-flow DPCD-processed (34.5MPa,

25°C, 13% $CO_2$, 6 min) coconut water in comparison to that of fresh, untreated samples, whereas a heat-processed (74°C, 15 s) sample was liked less. DPCD can effectively inactivate yeasts (by 6 logs) in grape juice (Gunes et al., 2005). As the $CO_2$-juice ratio, temperature, and pressure increased, the inactivation rate increased, without detectable flavor degradation. In another study, Del Pozo-Insfran et al. (2006) showed that thermal pasteurization decreased anthocyanins (by 16%), polyphenols (by 26%), and antioxidant capacity (by 10%) of muscadine grape juice, whereas DPCD-treated grape juice retained fresh-like levels of these. As revealed by Zhou et al. (2009b), total soluble solids and carotenoid content of DPCD-treated carrot juice were stable. In a further study with peach juice, the same authors observed that DPCD significantly altered product particle size distribution. They observed decrease of protein and polyphenols, while pectin and calcium levels remained unchanged. The DPCD treatment of melon juice led to significant inactivation of microorganisms and enzymes (polyphenol oxidase, peroxidase and lipoxygenase), while retaining the flavor (Chen et al., 2010).

Besides fruit juices, DPCD was used for pasteurization of beer in a continuous system (27.6 MPa, 21°C, 5% $CO_2$, 5 min). DPCD-treated beer showed unaltered sensory properties in comparison to fresh beer after 1 month of storage, whereas those of heat-treated beer (74°C, 30 s) were significantly reduced. Haze in DPCD-treated beer was reduced, as well foam capacity and foam stability (Dagan and Balaban, 2006).

## Application for treatment of solid foods

There are several difficulties in the treatment of solid foods with DPCD: the continuous processing is difficult; the penetration of $CO_2$ in most of them is slow; and there are concerns about cellular damage leading to texture and/or surface changes, affecting food quality. DPCD was used for treatment of different solid foods, such as fruit and vegetables, meat, seafood, herbs and spices, etc. Continuous $CO_2$ flow was used by Valverde et al. (2010) to inactivate *S. cerevisiae* in fresh-cut pears. Using DPCD, at 55°C, the authors achieved inactivation of yeast by 5 logs. pH and °Brix of pears were not affected by DPCD treatment and peroxidase activity was partially reduced. However, some texture loss and darkening was observed. Less effective inactivation of *E. coli* and total aerobic bacterial count was achieved by Mazzoni et al. (2001) treating alfalfa seeds with DPCD (27.6 MPa, 50°C for 60 min). Jung et al. (2009) investigated the effect of DPCD with different intensity on the contamination of alfalfa seeds with *E. coli*, *Listeria* and *Salmonella*. DPCD treatment for 15 min (20 MPa, 45°C) reduced the pathogen levels by 7 logs, but impaired their germination capability. Seeds were able to germinate if treated for 10 min at 15 MPa and 35°C, but the reduction of *E. coli* was about 3.5 logs. DPCD (30 MPa, 65°C for 40 min) did not have any microbicidal effect on dry cocoa powder. However, after addition of 5%–10% water, microbes were fully inactivated (Calvo et al., 2007).

Similarly, Dehghani et al. (2008) achieved complete inactivation of bacteria and fungi on ginseng root, moisturized with water/ethanol/$H_2O_2$, after applying DPCD (10 MPa, 60°C, 6 h). However, when using dry ginseng, treatment was not effective and inactivation of microorganisms did not reach the required regulatory levels. Zhong et al. (2008) reported 5 logs reduction of *E. coli* on fresh spinach leaves treated with

DPCD (7.5–10 MPa, 40°C) for 40 min, which was accompanied by undesirable discoloration of the leaves and changes in texture. Calvo and Torres (2010) observed that water content and temperature were the influential parameters for microbial reduction of paprika powder. 35% water content and 6 MPa led to a 1.5 logs microbial reduction, while 30 MPa resulted in 5 logs reductions and retention of paprika color. Ferrentino et al. (2012) treated fresh-cut coconut pieces with DPCD (45°C, 12 MPa, 15 min) resulting in 4 logs CFU/g reduction of mesophilic microorganisms, lactic acid bacteria, total coliforms, yeasts, and molds. However, some microstructure irregularities occurred after the DPCD treatment. Beside microorganism inactivation, Zhou et al. (2009a) revealed that pectin methylesterase activity in carrots is inactivated by HPCD, whereas Manzocco et al. (2016) reported polyphenoloxydase inactivation in mushrooms.

Application of DPCD to meat products is less effective in microorganism reduction in comparison to plant foods. Erkmen (2000) revealed that DPCD at 6 MPa, 45°C for 150 min inactivates 5 and 1 log microorganisms in minced and skinned beef, respectively. Applying 10 MPa for 15 min resulted in 1 log reduction of *E. coli* in beef trimmings (Meurehg, 2006). Choi et al. (2008, 2009a,b) investigated the inactivation of *E. coli*, *Listeria* and *Salmonella* in fresh, treated, and marinated pork and observed 1.9–2.5 log reduction of *Listeria*. In a more recent study, Ferrentino et al. (2013) totally inactivated *Listeria* in dry cured ham pieces at 50°C, 12 MPa, 15 min. DPCD treatment slightly affected the sensory attributes of the product. Ji et al. (2012) found that 15 MPa, for 26 min at 55°C reduced the microbial load of shrimps by 3.5 logs. Meujo et al. (2010) succeeded in reducing total aerobic plate counts by 3 logs and *Vibrio* by 6 logs, in the digestive system of oysters by exposing them to 17.2 MPa, 60°C for 60 min. In a more recent study, myosin and shrimp surimi from *Litopenaeus vannamei* were treated with DPCD at 5–25 MPa and 40–60°C for 20 min. DPCD had a greater effect on secondary structure of myosyn than heating, increased the hardness, springiness, and weight loss, meanwhile decreasing the redness of sausages.

De Matos et al. (2013) used pressure cycles and different depressurization rates to treat oysters with DPCD to reduce *Vibrio parahaemolyticus*. The authors observed that after 30 min, 75% of the initial count was inactivated and 100% inactivation was observed after 3 h. Investigating the inactivation of several bacteria on solid food matrices through supercritical $CO_2$, Galvanin et al. (2014) developed mathematical models to represent the DPCD pasteurization process at different experimental conditions. A more recent study investigated DPCD applicability to a solid-liquid product consisting of apples in syrup. The optimal process conditions (12 MPa, 55°C, 15 min) led to inactivation to undetectable levels of mesophilic microorganisms, total coliforms, yeasts, and molds. Polyphenol oxidase was totally inactivated, while pH, color, and total acidity were slightly affected. The shelf life study of DPCD treated products revealed their microbial, enzymatic, and quality stability.

DPCD is a nonthermal pasteurization technology, which is mainly applied to thermosensitive foods. $CO_2$ under pressure (either in supercritical, liquid, or gaseous state) is employed to destroy microbial cells and/or inactivate enzyme systems. There are different technological solutions that rely on a batch, semicontinuous, or continuous process. These solutions provide contact of the processed feedstock with $CO_2$, with a pressure usually less than 50 MPa for a certain time period. The process temperature is low, generally from 30°C to 50°C. Therefore, DPCD largely retains the quality and nutrients in the food matrix. It is mainly used for liquid foods, but recently the accumulated knowledge and experience in the field is directed to DPCD application to many solid foods, as well. There is also great interest in combining DPCD with other nonthermal food processing technologies. Unfortunately, DPCD equipment is available mostly at a laboratory scale (Ferrentino and Balaban, 2012). With its potential for safe food without loss of nutrients or quality, and scientific background proving its efficiency, it is not surprising that DPCD treatment is listed among the emerging technologies for food processing with the most commercial importance in the next 5–10 years (Jermann et al., 2015).

## 2.4 CP application

CP processing is a new perspective and alternative method of sterilization that does not require high temperature. The nonthermal nature of the method allows treatment of thermosensitive materials (plastic foils) and heat-labile food systems (fresh fruits and vegetables). The substances formed during processing (atomic oxygen, ozone, ions) have a reactive character, and a short life span (Thirumdas et al., 2015; Yanakiev, 2017). CP is a type of nonthermal plasma, based on generation of ionized gas composed of positively and negatively charged particles (ions), chemical reactants such as atomic oxygen and ozone, UV rays, and accelerated electrons. Two types of CP are used in the food industry—CP generated under vacuum and CP generated at atmospheric pressure. The first type is generated in a hermetically sealed chamber filled with gas (Ar, N, $N_2O$, $N_2$, $O_2$) or a gas mixture in vacuum (0.01–0.02 MPa), via a high frequency electric field source, or source of microwave energy. Atmospheric CP is generated at atmospheric pressure through an electric field, generating a high potential difference between the two electrodes placed in air or other gas environment (Coutinho et al., 2018; Yanakiev, 2017).

Depending on the electric discharge and the generator design, CP may be classified as plasma generated via corona discharge, generated via a dielectric barrier discharge (DBD), and generated via an atmospheric pressure plasma jet (APPJ). According to the source of energy, CP is AC, Pulsed DC, and radio frequency (RF). Based on the composition of the feed gas, CP can be classified as atmospheric plasma or named after the gas used for generation: $Ar_2$, $He_2$, $N_2$, etc. (Pankaj et al., 2014b; Pervez et al., 2014; Shakila et al., 2012; Yanakiev, 2017). The most commonly used sources of CP are corona discharge, plasma jet, and DBD (Yanakiev, 2017).

The antimicrobial effect of CP is due to generation of reactive oxygen species (ROS) and reactive nitrogen species, such as: $H_2O_2$, $O_3$, $O_2^{\bullet -}$, $HO_2^{\bullet}$, $RO^{\bullet}$, $ROO^{\bullet}$, $^1O_2$, $^{\bullet}OH$, $NO^{\bullet}$, $^{\bullet}NO_2$, $ONOO^{\bullet}$, ROONO, etc. formed during discharge (Arjunan et al., 2015; Gaens and Bogaerts, 2013; Lu et al., 2016; Misra and Jo, 2017). ROS interact with various cellular structures. First of all, ROS are responsible for changing the DNA structure and bind to different proteins and enzymes in the microbial cell. ROS cause peroxidation of the lipids in the cell membrane, thus changing its biological functions. At the same time, the cells are subjected to intense tension from the electric field leading to their destruction. ROS can attack the peptidoglycan in the membrane of the Gram-positive bacteria, which alters its functionality. In Gram-negative bacteria, ROS interacts chemically with lipopolysaccharides in the cell membrane. At the same time, the tension that the cells experienced leads to leakage of the membrane and loss of cell integrity. The cell growth inhibition during CP treatment depends on a number of factors. For example, it has been found that some fungal spores are not affected by the ozone and NO formed during discharge, but predominantly by atomic oxygen (Hashizume et al., 2013, 2014; Iseki et al., 2011; Ishikawa and Hori, 2014; Misra and Jo, 2017; Misra et al., 2016; Yusupov et al., 2013).

Of great importance for growth inhibition are the UV rays generated during discharge. These rays are largely adsorbed by the cell, resulting in a change in the DNA structure and chemical reactions in the cytoplasm. UV rays generation strongly depends on the method of obtaining (Misra and Jo, 2017; Schneider et al., 2011, 2012). The influence of CP on different groups of microorganisms largely depends on the generation source and the experimental conditions (Misra and Jo, 2017; Coutinho et al., 2018).

Recently, a comparison of different CP approaches in the treatment of fresh and cooked meat against the major groups of food pathogens was performed (Misra and Jo, 2017). Depending on the CP source, the gas phase present reduces microbial contamination by 0.5–3 log*N* (relative to the total bacterial abundance (TBA) of the samples tested) (Ulbin-Figlewicz et al., 2015a,b). The processing time of samples in order to kill the main psychotropic microorganisms is about 10 min. CP also has a strong impact on the two major pathogenic microorganisms in meat—*E. coli* and *L. monocytogenes*. Depending on the plasma source and the gas phase (He, $N_2$, $O_2$, $NO_2$), the reduction in the number of viable *L. monocytogenes* cells is more than 3 log*N*. The trend with *E. coli* is similar. The structure of the food matrix is important in CP processing of meat. In general, it has been found that treatment in surface layers was more efficient (there was a reduction in the number of viable pathogen cells by more than 3 log*N*), whereas in the inner layers, the reduction is up to 1 log*N*. The processing time varies from 2 to 10 min (Misra and Jo, 2017).

Depending on the plasma generation, there are also differences in cell growth inhibition. Korachi et al. (2009) showed that corona-generated plasma is a suitable method for inhibiting the growth of *E. coli*, *S. aureus*, *P. aeruginosa*, *Streptococcus mutans*, *Candida albicans*, and *Aspergillus niger*, but the method did not show satisfactory results in the processing of *Bacillus subtilis*, most likely due to its inherent spore formation.

It has been found that plasma jet processing (20 min, 400–600 V/20–35 mA) significantly reduces the number of viable *S. aureus*. The pH medium was also important for the inactivation of all cells, for example, it can lead to 100% inhibition in combination with the CP. This is likely to be due to $HOO^{\bullet}$, which initiates oxidation of fatty acids in the membrane (Liu, et al., 2010). It was found that the treatment with CP generated by plasma jet resulted in the inactivation of *L. monocytogenes*, *E. coli,* and *Salmonella* sp. in nuts, fruits and vegetables by up to 3 log*N* or up to 90%, depending on the initial contamination quantity (Lee et al., 2011; Niemira, 2012; Wang et al., 2012).

Jayasena et al. (2015) investigated the effect of CP (generated by DBD) treatment in meat packaging. They determined its inactivation ability against several microorganisms (*L. monocytogenes, E. coli* O157: H7, and *S. typhimurium*) and the influence of plasma on the physicochemical properties of the product. A 10 min treatment reduced the number of viable cells of the three microorganisms tested by 2–3 log*N*. Miao and Yun (2011) investigated the bactericidal efficacy of CP generated by DBD in the treatment of samples infected with *E. coli* ATCC8099 at a different cell titer. Ziuzina et al. (2014) explored the possibilities for inactivation of *E. coli, Salmonella* sp., and *L. monocytogenes* on cherry tomatoes and strawberries by CP. Their experiments were conducted under different conditions, but it was found that processing up to 120 s provided a reduction in the number of viable cells from 3 to 7 log*N*. It has been found that *Salmonella* sp. and *L. monocytogenes* were more sensitive than coliforms. Patil et al. (2014a,b) applied direct treatment with DBD-CP when sterilizing the surfaces of packaged products. The achieved result was a reduction of *Bacillus atrophaeus* spores by 6 log*N* after treatment with 70 kV/50 Hz CP for 60 s.

Yanakiev (2017) examined the possibilities of applying CP treatment to surfaces coming into contact with food products. Antimicrobial activity of CP was found to be higher at low water activity on the treated surface. It has been demonstrated that the application of a longer treatment duration led to an increase in the effect and a higher level of reduction of the viable microorganisms was observed. The *Salmonella abony* and *S. aureus* had different resistance against the antimicrobial action of CP. The differences in the resistance of the nonspore-producing bacterial cells, *S. abony* and *S. aureus*, may be due to the different cell wall structure of the two microorganisms, being representatives of Gram-negative and Gram-positive bacteria, respectively. Duration of 4.8 s, a discharge voltage of 10 kV, and a discharge frequency of 550 Hz was the most energy efficient for a reduction by 1 log*N* of living cells number in both microorganisms. The highest reduction of the tested microorganisms was achieved (power required $141.75 \times 10^{-3}$ W h) for *S. aureus* (3.47 log*N* reduction) and for *S. abony*: 3.35 log*N*.

Besides decontamination, atmospheric pressure CP has shown promising potential for a number of innovative applications including surface hydrophobicity enhancement in biscuits (Misra et al., 2014), modification of dough rheology and mixing properties (Misra et al., 2015), modulation of whey protein functionality (Segat et al., 2015), technological performance of rice flour (Pal et al., 2016), enhancement of food grain (seed) germination (Jiafeng et al., 2014), and enhancement of mass transfer (Kodama et al., 2014).

The influence of CP on enzyme functionality in food systems was also assessed. Bernard et al. (2006) observed the degradation of immobilized lysozyme under the action of the afterglow of plasma in nitrogen-oxygen mixture, operating in a flowing mode. They proposed a model including destruction of functional sites under the influence of reactive species. Kylián et al. (2008) demonstrated etching of protein deposits (including lysozyme) on a range of substrates (e.g., silicon wafers, polystyrene, glass) using a low pressure plasma source. Surowsky et al. (2013) reported inactivation of peroxidase and polyphenol oxidase in a model system using a CP jet with various gas mixtures. The changes in tryptophan emission fluorescence measurements brought by the duration of plasma exposure further corroborated the structural modifications in proteins. The inactivation of pectin methyl esterase and peroxidase in fresh cut fruit systems has also been studied by Tappi et al. (2014, 2016). CP was reported to inactivate a range of enzymes such as lysozyme (Takai et al., 2012), polyphenol oxidase, and peroxidase (Pankaj et al., 2014a; Surowsky et al., 2013).

CP technology is an emerging, green process offering many potential applications for food packaging. While it was originally developed to increase the surface energy of polymers, enhancing adhesion and printability, it has recently emerged as a powerful tool for surface decontamination of both foodstuffs and food packaging materials. New trends aim to develop in-package decontamination, offering nonthermal treatment of foods postpackaging. Food packaging materials retain very low barrier properties after plasma treatment and transparency is not affected even after 5 minute treatment (Pankaj et al., 2014a). Key limitations of CP are the relatively early state of technology development, the variety and complexity of the necessary equipment, and the largely unexplored impacts of CP treatment on the sensory and nutritional qualities of treated foods. Also, the antimicrobial modes of action for various CP systems vary depending on the type of CP generated. Optimization and scale up to commercial treatment levels require a more complete understanding of these chemical processes. Nevertheless, this area of technology is promising and is the subject of active research to enhance efficacy (Niemira, 2012).

## 2.5 Radiation methods for food decontamination

Radiation energy refers to energy sources that pass from one place to another through the air or other space. It can consist of particles or electromagnetic waves that are characterized by their length and have a different impact on microorganisms. The lower the wavelength is, the more energy is accumulated in them, and the stronger their effect on the organisms is. The basis of their action is different physical or chemical changes occurring in cells, microorganisms, and the environment (Fig. 3). Changes can only be the result of the absorption of the rays, that is, their effective penetration is important for effective irradiation.

Electromagnetic waves with a wavelength of 800–1000 nm in the infrared spectrum are used as an energy source by photosynthetic bacteria. Visible light with a wavelength of 380–760 nm is also used as an energy source by photosynthetic bacteria and higher plants. Electromagnetic waves with a wavelength of 200–380 nm are ultraviolet rays. Both infrared and ultraviolet rays are invisible to the human eye (Murgov and Denkova, 2012).

If the treatment includes the use of a source of radiation, for example, UV irradiation or pulsed light, FDA approval of the means of treatment for the control of microorganisms is required. Both UV radiation and pulsed light have been approved by FDA for the control of microorganisms, and the regulations specifying the conditions under which they may be safely used are at 21 CFR 179.39 (UV radiation) and 21 CFR 179.41 (pulsed light).

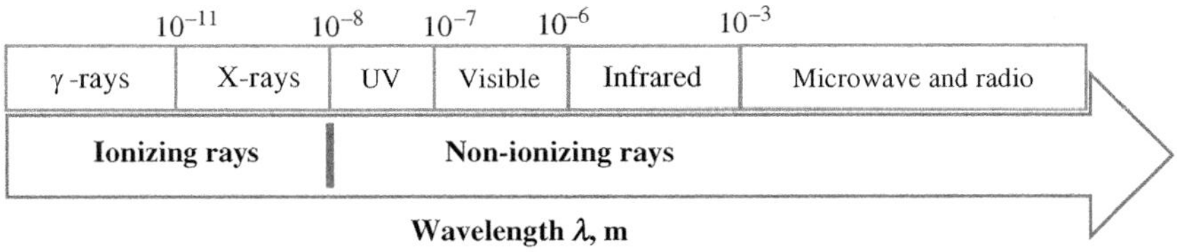


**Fig. 3** Radiation treatment methods in the electromagnetic spectrum according to the wavelength.

### 2.5.1 *UV treatment*

Ultraviolet rays are the most active rays of the solar spectrum, which also determines their bactericidal effect. They are characterized by high energy, with the greatest activity being at wavelengths of 250–260nm (253.7nm). The effect of the treatment depends on the intensity ($I$, in $\mu W\,cm^{-2}$), the treatment duration ($\tau$, in s), and the UV permeability of the medium in which they propagate (UVT, in %) (Smith, 2016):

$$\text{UV} - \text{dose} = \frac{I}{\text{UVT}}, \text{W s cm}^{-2}$$

The effective action of UV rays on microorganisms depends on the irradiation dose (amount of energy absorbed), the substrate properties, the pH, the degree of contamination by microorganisms, and the temperature. Low irradiation doses stimulate the individual functions of microorganisms, including hereditary ones. This is used, in practice, for obtaining antibiotics, enzymes, amino acids, and other biologically active substances.

Purine and pyrimidine bases of DNA and RNA absorb at their maximum UV rays at 260nm while proteins absorb at 280nm. The influence of UV rays on microorganisms is related to their effect on DNA and the most lethal effects of UV light are at a wavelength of 253.7nm (Murgov and Denkova, 2012; Smith, 2016). They induce dimerization in adjacent pyrimidine bases in the DNA molecule. The dimers formed lead to changes in the genetic structure and to genetic mutations. There is evidence suggesting that UV rays induce mutations indirectly as well, by the formation of free radicals and peroxides having mutagenic properties (Bolton, 2001). Higher doses of UV rays cause microbial cell death. It should be noted that different microorganisms have different sensitivity to similar irradiation doses (Figs. 4–7).

Experimental data, as suggested by the figures, clearly indicate that bacteria that do not emit pigments into the environment are more sensitive to UV radiation. The carotenoid pigments contained in the cells have protective properties against UV rays (Fig. 4). Bacterial spores are significantly more resistant than vegetative cells. In order to destroy spores, more energy and longer exposure is needed (Fig. 5). The susceptibility of microorganisms to UV rays is as follows: Gram-positive bacteria < Gram-negative bacteria < yeasts < bacterial spores < mold spores (Figs. 6 and 7). Similar results were reported by Akshay (2013).

Microbial cell death can be due both to the direct effects of UV rays and to adverse changes in the substrate structure. UV rays inactivate enzymes, and being absorbed by important cellular substances (proteins, nucleic acids, etc.), cause damage to their

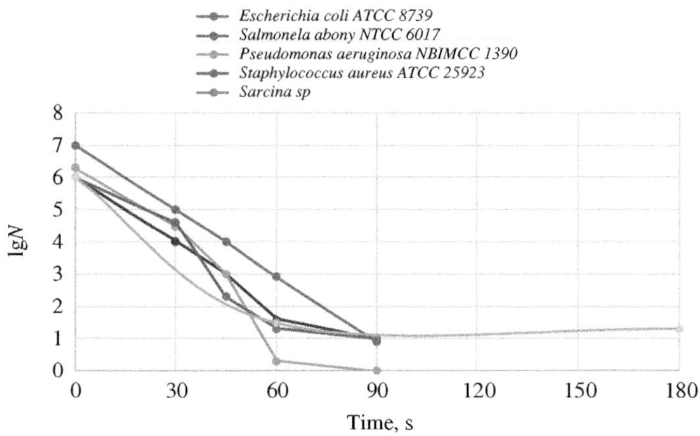


**Fig. 4** Death curves of various bacteria under the action of UV rays.

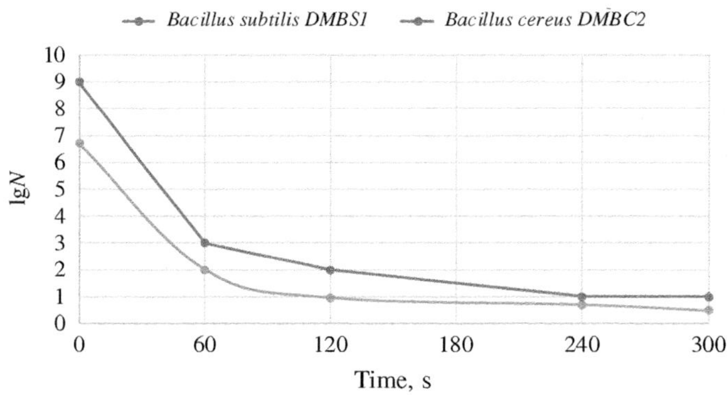


**Fig. 5** Destruction of spores of spore-forming bacteria under the action of UV rays.

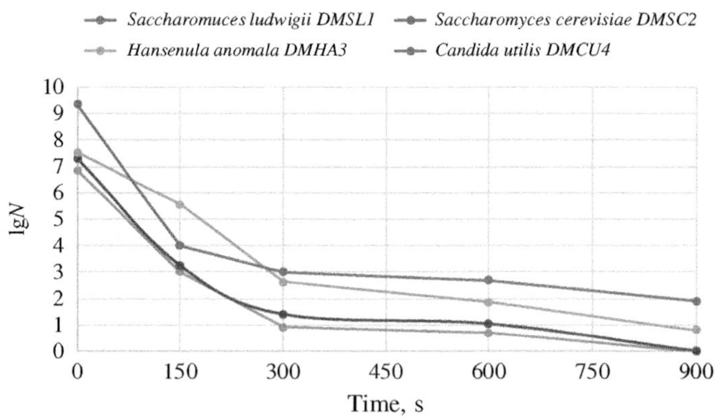


**Fig. 6** Death curves of yeast cells under the action of UV rays.

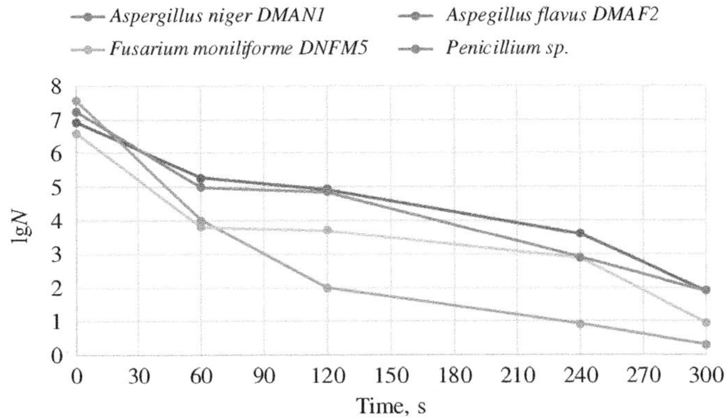


**Fig. 7** Destruction of fungal spores under the action of UV rays.

molecules. Substances ($H_2O_2$, $O_3$, etc.) that destroy microorganisms can also be formed in the irradiated environment. An often used artificial source of ultraviolet radiation is a low-pressure argon-mercury lamp (Bolton, 2001). UV rays are used to destroy microorganisms in the air and on the food surface (Fig. 8). Refrigeration chambers, healing, and production premises are disinfected by their application. Up to 100% of bacteria and fungi found in the air were destroyed after 3 hours (Table 2). UV rays are also applied in the packaging of foodstuffs, medical preparations, and for decontamination of trays, packaging materials, equipment, and utensils in public catering.

Food sterilization with UV rays is obstructed due to their low penetration ability (Sattar, 2011). Therefore, their effect is only on the surface or in a very thin layer. Positive results have been achieved in water disinfection by exposure to UV rays,

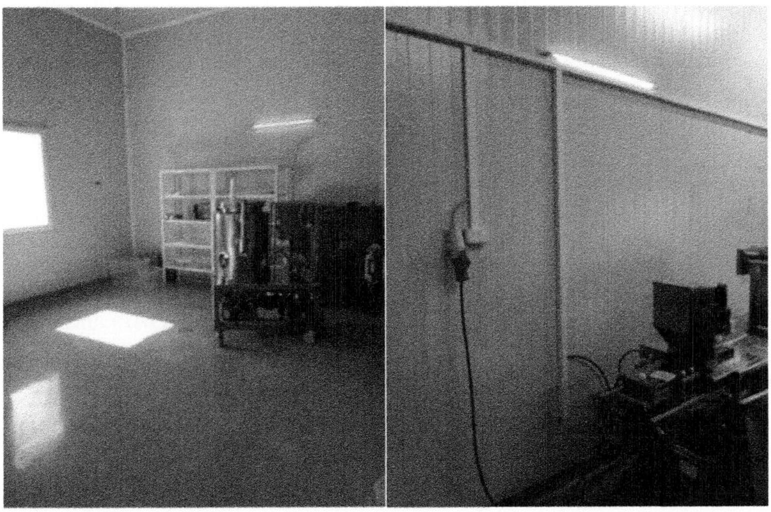

**Fig. 8** UV sterilization (90W lamps) facilities at the Scientific laboratory of fermentation and lyophilization (Plovdiv, Bulgaria).

**Table 2** Influence of UV rays on microflora of air in production premises and equipment

| Air in production premises (Protey-M Ltd, Bulgaria; Fermentation hall; 2 UV lamps 90 W); method of determination—Sedimentation method (Murgov and Denkova, 2012) | | | | | | |
|---|---|---|---|---|---|---|
| Time, min | 0 | 30 | 60 | 120 | 180 | 360 |
| TBA[a], CFU/m$^3$ | 2.2.10$^2$ | 2.0.10$^2$ | 1.76.10$^2$ | 1.76.10$^2$ | 44 | 0 |
| Freeze-drier equipment (Protey-M Ltd; 2 UV lamps 90 W); method of determination—Sedimentation method (Murgov and Denkova, 2012) | | | | | | |
| Time, min | 0 | 30 | 60 | 120 | 180 | 360 |
| TBA[a], CFU/m$^3$ | 44 | 0 | 0 | 0 | 0 | 0 |
| Air in production premises (Protey-M Ltd, Bulgaria; Packaging hall; 2 UV lamps 90 W); method of determination—Sedimentation method (Murgov and Denkova, 2012) | | | | | | |
| Time, min | 0 | 20 | 40 | 60 | 120 | 180 |
| TBA[a], CFU/m$^3$ | 1.76.10$^2$ | 880 | 44 | 44 | 44 | 0 |

[a] Total bacteria abundance.

when viruses, bacteria, yeasts, and other pathogenic microorganisms were destroyed (Masschelein, 2002).

UV rays are successfully used in the processing of liquid-type foods with high transparency, such as sugar solutions and juices (Caminiti et al., 2011a; Choi and Nielsen, 2005; Fan, 2007; Franz et al., 2009; Koutchma et al., 2007). Application of UV rays for sterilization of fruit juices and wines (in a thin layer) allows their storage for longer periods than pasteurized ones.

Results have also been obtained in the treatment of solid food products in a thin layer, such as sand sugar, resulting in complete decontamination of *Bacillus stearothermophilus*, and thermal, ultrasonic, and ultraviolet inactivation of *Salmonella* in thin films of aqueous media and chocolate (Lee et al., 1989; Stermer et al., 1987).

For some foods (oil, milk) sterilization with UV rays leads to deterioration of taste and nutritional qualities of the food. When treating such products with UV rays, a decrease in the vitamin content (vitamins K, A, $B_{12}$, $B_2$, folic acid, and tocopherols) (Koutchma et al., 2009), deterioration in quality indicators such as appearance, color, flavor, and taste (Azzara, 1992; Borle et al., 2001; Frankel, 1991; Min and Boff, 2002), as well as physico-chemical changes in the composition of the food and the raw materials occurred (Kehoe et al., 2008; Wong and Kitts, 2001).

### 2.5.2 Pulsed light treatment

Among the existing nonthermal technologies, PL irradiation is one of the promising due to its ease of use, application to most types of microorganisms, nongeneration of chemical residues, and low cost of equipment compared with other sterilization methods

(Gómez-López et al., 2007; Ramos et al., 2013; Heinrich et al., 2015). The mechanism of generating intense PL can be described as follows: electric energy is stored in a high power capacitor and then discharged to a specially designed xenon lamp unit. The lamp converts the high voltage pulses to light pulses of the same duration and frequency as that of the received DC electric pulses. The high-voltage electric pulses (up to $70\,kV\,cm^{-1}$) applied to the lamp causes it to emit an intense PL focused on the treatment area between two electrodes, which typically lasts a few hundred microseconds (1 ms–0.1 s). This produces several high-energy flashes, typically pulsed at 1–20 flashes per second, and an energy density in the range of $0.01–50\,J\,cm^{-2}$ at the irradiated surface; each flash has intensity almost 20,000 times that of sunlight at sea level.

The spectrum of PL ranges from UV to near-NIR and at least 2/3 of the emitted electromagnetic radiation falls within the wavelength range from 170 to 2600 nm, with a peak emission between 400 and 500 nm (Gómez-López et al., 2005). The wavelength range of the PL spectrum includes the sun-light spectrum wavelengths and particularly about 25% UV wavelengths in the range 180–300 nm that do not normally reach the earth's surface, since they are filtered by the ozone layer (Gómez-López et al., 2005, 2007). When PL at wavelengths under 300 nm is blocked, the overall effectiveness of sterilization decreases dramatically. This indicates that the sterilization function of PL is primarily dependent on light in the UV range. Two main effects could be distinguished: photochemical and photothermal.

*Photochemical effect*: Although details of the mechanism by which PL inactivates microbial organisms is still being developed, it is generally believed that, basically, the germicidal effects of PL on microbial organisms are attributed mainly to DNA mutations induced from the absorption of UV energy of PL by proteins, DNA, and RNA, causing damage, which kills microorganisms (Cheigh et al., 2012; Muñoz et al., 2012; Turtoi and Nicolau, 2007; Wang et al., 2005).

According to Wuytack et al. (2003), a major reason for inactivation is structural changes in DNA, with minor reasons including membrane damage, and protein and other macromolecules modifications. In general, the UV light causes dimerization of the DNA thymine, resulting in mutations, damage to the genetic information, failure of replication, and gene transcription (Turtoi and Nicolau, 2007).

Another factor influencing the efficacy of PL treatment is the damage recovery of some microbes. According to Kao et al. (2005), there are generally two mechanisms through which microorganisms can recover from damage. The first is photo reactivation mechanism that occurs if products are exposed to harmless blue or near UV light during storage. The other is called dark repair, which does not need light to catalyze; however, it does not always lead to complete repair, which makes it less efficient than photo reactivation (Hijnen et al., 2006). It was also found that enzymatic repair of DNA does not occur after damaging by PL (Slieman and Nicholson, 2000).

*Photothermal effect*: When a PL lamp emits electromagnetic radiation, a portion of it lies in the visible and NIR range. When absorbed by the irradiated surfaces, these wavelengths may generate heat. According to Bang-Yuan et al. (2015), this heat may increase surface temperature of the objects up to 50–150°C. The generated heat is mainly concentrated on the surface of the objects, with a maximal penetration depth up to 10 mm (depends on the optical properties of the irradiated surface and moisture content). The heat is evenly conducted to the center of the irradiated object so no considerable increase in the internal temperature can be expected. However, these temperature differences may generate thermal stresses that can eventually damage the cell wall. Thus, it can be suggested that the lethal effect of PL can also be due to photothermal effect. Wekhof (2000) hypothesized that the PL microbial lethal effect is achieved through a rupture/explosion of bacteria during their temporary overheating caused by absorption of NIR light with a fluence exceeding $0.5\,J\,cm^{-2}$, which was confirmed by SEM photographs of irradiated *A. niger* spores presenting severe deformation and rupture.

A possible coexistence of both photochemical and photothermal mechanisms could be expected also, along with other structural damages induced in cell walls, membranes, and internal structures (Artíguez et al., 2011; Elmnasser et al., 2007; Wang et al., 2005). Ohlsson and Bengtsson (2002) reported changes in ion flow, increased cell membrane permeability, and depolarization of cell membranes. The exact origin of these minor effects is not yet well understood, but experiments (Takeshita et al., 2003) confirmed that they were observed after treatment of yeast with pulsed UV light, but not observed after similar treatment with continuous UV light.

Parameters affecting the efficiency of PL on microbial inactivation can be grouped into two categories: process and target parameters. The process parameters include: fluence (light intensity), fluence rate, exposure (treatment) time, peak wavelength (spectral distribution), peak power, pulse frequency (pulse width and pulse duration), total number of pulses applied, and the distance between the lamp and sample. Fluence (or energy density) is the energy received by the sample per unit surface irradiated area during the PL treatment (measured in $J\,m^{-2}$). Fluence is believed to be the most important parameter when characterizing PL treatment (Gómez-López et al., 2007). There is an agreement in the published literature that effectiveness of microbial inactivation increases with the total fluence delivered to the object (Pataro et al., 2011). Cho et al. (2002) studied the inactivation effects by PL on lactic acid bacteria (*L. plantarum*, *L. brevis*, *Leuconostoc mesenteroides*, and *Pediococcus pentosaceus*) for 0–2500 s treatment at 15–25 kV and found an exponential relationship between the electric power and inactivation of microorganisms. After irradiating *A. niger* spores inoculated into a corn meal suspension, Jun et al. (2003) also reported that inactivation increases with both the electric power and treatment time.

The fluence rate can be defined as the received energy from the lamp by the sample per unit area per unit time ($W\,m^{-2}$). Exposure time is the time period in seconds during which treatment is given. Spectral distribution refers to the width in wavelength distribution of the PL. The lamp spectral distribution depends strongly on lamp characteristics and design, where the desired peak wavelength can be obtained by proper adjustment of the electrical pulse current by closed loop system equipped with photo-detectors. Alternatively, appropriate light filters can be used to block unwanted wavelengths in the PL before reaching the target or a laser with defined wavelength of emittance.

Pulse frequency (known also as pulse duration or pulse repetition rate) has similar meaning from a physical point of view—the number of pulses per unit time or the time interval during energy delivery. For a fixed input current density, fluence is determined by the pulse duration. High fluence (pulse power) can be achieved with a low current density if the pulse duration is shortened so that the generated light energy can be dissipated within a short time. The maximum frequency of a pulse light system that can be effectively achieved is usually limited by the design characteristics of the flash lamp. When higher pulse frequency must be achieved, two or more lamps arranged in sequence can be used.

The total fluence delivered on a given surface depends on the pulse power, as well as the total number of pulses delivered. From a practical point of view, the so-called Bunsen-Roscoe law states that the product of intensity of light multiplied by the time of exposure equals a constant. Therefore high energy density can be attained by using either high fluence rate with low time of exposure or low fluence rate with high time of exposure. However, in some experiments, especially where the pulse fluence is low, a threshold for number of pulses exists, below which no inactivation can be achieved.

The distance from the light source has an impact on the light intensity and, therefore, on the penetration depth. As the distance from light source and depth of the substrate increases, the absorption and scattering diminishes. The term optical penetration depth represents the distance over which light decreases in fluence rate to 37% of its initial value. The optical penetration varies with wavelength, with shorter wavelengths providing deeper penetration into the food than longer wavelengths (Abida et al., 2014). The target parameters include the optical properties of the objects, such as scattering and absorption, color, opacity, thickness, and surface topography of the irradiated object, viscosity, and flow conditions of the liquid samples, as well as moisture content (Pataro et al., 2011) The chemical composition and physical properties of the irradiated cells and the wavelength of the incident light determine to a great extend whether light will be reflected, scattered and absorbed or transmitted. In the case of biological tissues and photothermal effect, absorption and scattering are the most relevant types of light-substrate interaction (Abida et al., 2014).

Esbelin et al. (2013) found that the color of the fungal spores (*A. niger* vs. *Fusarium culmorum* spores) played a significant role in PL disinfection, i.e., darker *A. niger* spores are more resistant than *F. culmorum* spores. This was explained by the presence of protective dark pigments in the wall layers in *A. niger* spores that absorb more UV-C light than the pigment in *F. culmorum* spores. Turtoi and Nicolau (2007) also reported similar dependence and found that pigment extracts of *A. niger* absorb strongly UV in the range 240–480 nm.

Bang-Yuan et al. (2015) and Ringus and Moraru (2013) hypothesized that the physical properties of the treated surfaces, including surface topography and reflectivity, affect the efficacy of PL. For smooth surfaces, the incident light on the surface comes out at the same angle as the incident beam, with the same spectral distribution of energy. For rough surfaces, light travels through the outer layers of the material, where the incident light is partly absorbed, this phenomenon is called diffuse reflection. Surface roughness and crevices have also been suggested to shield microbial cells during treatment as they can shadow some microorganisms from the intensive light and may reduce the effectiveness of inactivation. Therefore, samples subjected to microbial inactivation with PL should have a smooth surface and must be placed in containers that are transparent and not reflective to at least the UV range of the spectrum.

Hillegas and Demirci (2003) indicated that the efficacy of PL is also closely related to the moisture content of the product. Higher moisture content typically increases UV and NIR absorbance by the material and, therefore, the lethal effect of the treatment. However, Fine and Gervais (2004) demonstrated that PL can also be successfully applied for food powders, such as wheat flour and black pepper, and suggested that PL has a potential to inactivate *S. cerevisiae* and *L. monocytogenes* cell population in powdered infant foods.

PL has been widely used for the decontamination of different food products: meat products, eggs, milk and dairy products, mushrooms, juices, baby foods, etc. Hierro et al. (2011) demonstrated inactivation of *L. monocytogenes* in ready-to-eat cooked processed meat products. Paskeviciute et al. (2011) reduced viability of *S. typhimurium* and *L. monocytogenes* inoculated on the surface of chicken, and organoleptic properties were not affected. It has been found that PL at a dose of $11.9\,\mathrm{J\,cm^{-2}}$ effectively reduces the number of viable cells of *L. monocytogenes* and *S. typhimurium* in ready-to-eat cured meat products (Ganan et al., 2013). Lasagabaster et al. (2011) irradiated eggs and reported that $2.1\,\mathrm{J\,cm^{-2}}$ caused death of *Salmonella* cells on the egg surface with slight increase in temperature. When fluence was increased to $10.5\,\mathrm{J\,cm^{-2}}$ no penetration of *Salmonella* cells to the egg contents from the shell was observed.

Continuous flow PL was also used for the inactivation of *S. aureus* in milk and of brine during Mozzarella cheese production (Li and Farid, 2016). Matak et al. (2005) reported reduction of *L. monocytogenes* in goat milk. However, according to Krishnamurthy et al. (2007), the milk temperature increased during PL treatment in a continuous flow system, causing a fouling effect and undesired changes in milk. Fresh slices of mushrooms were exposed to PL treatment varying between 4.8 and $28\,\mathrm{J\,cm^{-2}}$ and was found to increase in shelf life by 2–3 days in comparison to untreated samples (Oms-Oliu et al., 2010). However, the authors reported side effects of the treatment at higher fluence levels (12 and $28\,\mathrm{J\,cm^{-2}}$), such as affected texture and decrease of phenolic compounds and vitamin C.

Pataro et al. (2011) inoculated apple and orange juice with Gram-positive (*Listeria innocua* 11288) and Gram-negative (*E. coli* DH5-α) bacteria and then irradiated with continuous flow PL system. The authors reported microbial load reductions in apple and orange juices for *E. coli* to be 4.00 and 2.90 log-cycle, respectively, and for *Listeria innocua* were 2.98 and 0.93 log-cycles, respectively, when fluence of $4\,\mathrm{J\,cm^{-2}}$ was applied.

The sterilization effects of PL treatment on commercial infant foods (infant beverages, an infant meal and infant powdered milk) inoculated with *L. monocytogenes* were investigated at 15 kV, which level showed higher energy efficiency for the inactivation of *L. monocytogenes* (Choi et al., 2010). PL was also successfully used for packaging and surface decontamination (Rajkovic et al., 2010; Turtoi and Nicolau, 2007).

PL treatment delivers to the irradiated objects high energy density but without generating significant thermal effects, which can affect quality and nutrients (Abida et al., 2014; Ohlsson and Bengtsson, 2002). The bactericidal effect is based mainly on the photochemical and photothermal effects. In contrast to the continuous UV radiation where the effects of inactivation are reversible due to the presence of repairing enzymes, the absence of such enzymes in the damaged DNA molecules of cells irradiated with PL makes it much more effective (Palmieri and Cacace, 2005). Moreover, PL has shown potential in reducing allergies associated with some foods, that is, peanuts and soy extracts (Yang et al., 2010, 2012).

The major challenge for PL is possible resistance of some strains (Caminiti et al., 2011a; Manzocco et al., 2011; Rajkovic et al., 2010; Uysal and Toklucu, 2011). Another limitation of PL is the depth of penetration of light. It is usually limited by the power and wavelength distribution of the PL and by the optical properties and chemical composition of the object and, in general, it hardly exceeds 10mm. When very rough surfaces are irradiated, folds or fissures in the food may protect microbes from exposure to the PL (Brown, 2008). In general PL is useful mostly for liquids and solid foods surfaces.

### 2.5.3 Ionizing radiation (food irradiation)

At the end of the 19th century, two discoveries, of X-rays by Röntgen (1895) and natural radioactivity by Becquerel (1896), paved the way for the introduction of new physical effects of high energy particles on living organisms. The application of ionizing radiation for food treatments dates back to the beginning of the 20th century, when the first patents were issued. Commercially, the method was introduced in 1957 (Germany) for treatment of spices (Harder et al., 2016) and nowadays, more than 60 countries allow food irradiation, with more than 500,000 tons irradiated yearly (Eurofins, 2016). International organizations (the UN International Atomic Energy Agency, WHO, FAO) have approved foodstuffs that can undergo radiation treatment and the radiation doses. The most commonly irradiated food items in the United States are spices/herbs (around 78%), fruit and vegetables (14.6%), and meat and poultry (7.8%). In the EU, 46% of the irradiated products are frog legs, followed by dried spices and herbs (around 25%). In comparison with other methods for pasteurization, the labeling for irradiated foods is compulsory. It has to be strictly pointed out that the food product is treated by ionizing radiation and a specific symbol, called Radura, has to be on the packing. The food legislation in the EU is regulated by two directives, 1999/2/EC and 1999/3/EC, which determine the type of the food systems, the sources of radiation used, and the facilities (which also need prior registration and approval). In France, food irradiation is regulated by Decree 2001/1097 of 16 November 2001 and the Decree of 20 August 2002, and the following irradiated products may be marketed: frozen aromatic herbs, onions, garlic, shallots, pulses, dried fruit, flakes and germs of cereals, rice flour, meat and offal of poultry, frozen shrimps, shelled or headstocks, frozen frog legs, egg white, casein, and animal blood caseinate.

Ionizing rays are more powerful than nonionizing rays and include corpuscular (flow of accelerated elemental particles, α-rays, β-rays, high-speed electrons, and neutrons) and electromagnetic (γ-rays and short-wave X-rays). Characteristic for radioactive rays is their ability to induce ionization of atoms and molecules, which in turn leads to the destruction of molecular structures. They cause ionization of water and other substances. The mutation effect is indirectly due to ionization, which generates free radicals, hydroxyl radicals ($OH^{\bullet}$) being the most significant among them. Free radicals react with DNA, which is probably as sensitive to ionizing rays as other macromolecules, but since each gene is present in only one copy in it, its inactivation has severe consequences for the microbial cell. The effectiveness of ionizing radiation on microorganisms depends on the dose of radiation absorbed and other factors. At low doses of radiation, some vital processes in the microbial cell are activated, triggering enzyme systems. They induce hereditary changes in the properties of the microorganisms leading to mutations. At high doses, metabolism is significantly impaired, pathological changes occur in the cell and they lead to its destruction. It should be noted that microbial cells are significantly more resistant to radioactive radiation than the cells of the higher organisms. The lethal dose for them is hundreds to thousands times higher than for animals and plants. Microorganisms most sensitive to radioactive irradiation are those with higher carbohydrate and proteolytic activity, containing more $Mn^{2+}$ and $Ca^{2+}$ in their cells, as well as those with high levels of protective metabolic processes. *E. coli*, *Proteus* sp., and *Pseudomonas* sp., which typically cause microbial spoilage of meat and fish foods, are sensitive to irradiation. Micrococci are characterized by remarkable resistance. *Micrococcus radiodurans* is particularly resistant to ionizing rays and is easily isolated from foods that are subjected to ionizing radiation (Murgov and Denkova, 2012).

Vegetative cells are more susceptible to radiation than spores of bacteria and fungi. The persistence of microorganisms is enhanced by reduced humidity (dry products). Some substances, mainly organic, containing sulfhydryl groups protect microorganisms against the detrimental action of ionizing rays.

Many enzyme systems, membrane structures, the nuclear apparatus, especially DNA, are sensitive to ionizing radiation, which affects cell proliferation. Different microorganisms are distinguished by their resistance to ionizing radiation (e.g., the radiation dose that reduces the number of live *E. coli* cells 10 times is within the range of 5–10kGy, for spore-forming bacteria of the genera *Bacillus* and *Clostridium*; it ranges from 10 to 50kGy for vegetative cells and from 5 to 10kGy for sporulation forms; for yeast and fungal cells it is between 5 and 10kGy, and for viruses, in the 5–10kGy range. Depending on the radiation resistance, microorganisms can be divided into four groups (Murgov and Denkova, 2012):

- radio-sensitive—with lethal dose of up to 5kGy;
- relatively radio-resistant—up to 10kGy;
- meso-radio resistant—up to 15kGy; and
- radio-resistant—up to and over 20kGy.

The absorption dose is influenced by the age of the culture, the composition of the medium, the temperature, and the dose power (irradiation dose for a certain period of time) (Farkas and Mohácsi-Farkas, 2011; Smith and Pillai, 2004). The microbial cells have the ability to repair radiation damage. The nature of the repairing process is determined by the species specificity, physiological state, the absorption dose, and the $\gamma$-radiation power.

Ionizing rays, unlike UV rays, have high penetration through solids such as glass, plastic, paper, etc. (Silindir and Ozer, 2009). This penetrating ability makes them dangerous for people and they are, therefore, used in shielded facilities. This actually necessitates proper handling of the irradiation equipment, isolation of the radiation source, and well-trained personnel. γ-Rays are the most frequently applied form of radiation as they have the greatest penetrating ability and upon irradiation they do not cause the appearance of so-called induced radiation in the product, but minor chemical changes in the composition of the food, as well as in their flavor might occur (Aquino, 2012).

In food processing, the radiation effect depends on the microflora composition, food matrix, absorbance dose (the time for the reduction of the number of viable cells in the population by 1 log*N* at a given exposure), and power.

The International Atomic Energy Agency has proposed a specific terminology for the radiation treatment of foodstuffs (Arvanitoyannis et al., 2009; Farkas and Mohácsi-Farkas, 2011; Smith and Pillai, 2004) with doses sufficient to destroy human pathogenic microorganisms:

- radiation (6–10kGy) applied to reduce the number of microorganisms causing microbial spoilage; and
- radarization (10–50kGy) performed in industrial sterilization of food under conditions excluding re-contamination with microorganisms.

The main concerns of consumers still remain in the sphere of safety of the irradiated foods. Although the effect of the process on food quality and safety has been investigated over more than 100 years, there are no significant reports raising awareness of the ionizing radiation application. There were some issues reported related to possible negative effect of the irradiated foods but the studies seem not to have adequately addressed public concerns (Ehlermann, 2016). An experiment in India with malnourished children was performed with only 20 persons and few of them were given irradiated wheat (Bhaskaram and Sadasivan, 1975). The authors claimed that the patients receiving irradiated food developed polypoid and certain abnormal cells and the effect disappeared after stopping irradiated wheat feeding. However, the experiment was set up in a manner that could be difficult to repeat, the number of the patients was too small, and the initial severity of the children's health conditions were not taken into account.

Another isolated incident in Australia with irradiated cat food was again used as an occasion to raise issues about the safety of irradiated food (Ehlermann, 2016). This incident, however, was isolated and related to the sole importer of irradiated cat food. It is well know that irradiation generates free radicals, but in presence of water the radicals are neutralized very fast; some sensitive substances in food are partially destroyed, such as anthocyanins and vitamins. There were also concerns about generation of potential toxic substances in the irradiated foods (Ehlermann, 2016). It was found that in most cases the compounds generated during irradiation could also be found in thermally treated foods. The particular case with 2-alkylcyclobutanones (2-ACBs) led, in the beginning, to the opinion that these substances were generated only in the irradiated foods but later they were discovered in untreated products, altough in minimal amounts. Studies, however, till now did not find any evidence for potential harmful effects and, for this reason, The Scientific Committee on Food (SCF, today EFSA) of the EU had already concluded that the occurrence of 2-ACBs is of no concern (Statement of the Scientific Committee on Food on a report on 2-alkylcyclo-butanones) (expressed on 3 July 2002); European Commission; https://ec.europa.eu/food/sites/food/files/safety/docs/sci-com_scf_out135_en.pdf (last visited 20 September 2018). Nevertheless, ionizing radiation has some clear disadvantages, such as high investments cost, precautions with the source of radiation, packaging safety and migration characteristics, etc., which to a large extent hamper the wider application of the method (Luksiene and Paskeviciute, 2011).

## 2.6 Other methods

Other emerging and promising methods were also proposed. Pulsed electric field processing (PEF) has a dual advantage: prolonging the shelf life of food with minimal changes in nutrient content and sensory quality. PEF technology involves food processing by placing food between electrodes and processing with high voltage electrical pulses of 20–80 $\mathrm{kV\,cm^{-1}}$ (several microseconds). PEF has a lethal effect on the vegetative cells of bacteria, yeasts, and molds (Pan et al., 2017). A series of short high-voltage impulses destroy cytoplasmic membranes of cells in liquid media by expanding existing pores or forming new ones. The vegetative cell membranes become permeable to small molecules, permeability causes swelling, penetration of macromolecular components within the cell, and eventually, cell membrane rupture. The PEF affects the proteins and the phospholipid bilayer of the cell membrane, resulting in leakage of cellular contents (Jaeger et al., 2014). The other way to inactivate cells is the loss of transport functions due to the electric field. In the latter case, electroporation of the membrane and leakage of the cellular content is observed. The stability of the microorganisms in this case depends on the medium conditions: type of microorganisms, electric field, impulse size, medium permeability, pH, and temperature. In general, vegetative forms of microorganisms are more unstable, while bacterial spores are more resistant. Data shows that yeasts are more resistant than bacteria, which is quite natural, as they are eukaryotes. Due to the presence of a thicker peptide glycan layer, Gram-positive bacteria are more resistant than Gram-negative bacteria (Pal, 2017).

PEF processing can also be used to inactivate some groups of food enzymes, but in this case, the electric field should be more powerful than the one used for the inactivation of microorganisms. When treating dairy and meat products with a field of 21.5 $\mathrm{kV\,cm^{-1}}$, a decrease in lipase and peroxidase activity in the range of 25%–65% in raw milk was observed. At the same time, there was no loss of alkaline phosphatase activity. PEF processing is extremely useful in treating thermally unstable products such as juices and soups. It has been shown that PEF processing does not result in losses of vitamins B1, B2, D, and E, whereas some losses are seen in vitamin C content (Pal, 2017). PEF was also found to have a beneficial effect on bioactive compounds, such as carotenoids (García-Parra et al., 2018; Vallverdú-Queralt et al., 2013), polyphenols (Vallverdú-Queralt et al., 2012), etc.

Shockwave treatment (hydrodynamic pressure processing) was initially used as a novel approach for meat tenderization (Bowker et al., 2011; Ha et al., 2017; Warner et al., 2017). The first attempts were made at the end of the 20th century by Solomon and coworkers (Solomon et al., 1997, 1998) using controlled explosions in sealed vessels. Due to difficulties in controlling the explosions, safety issues with explosives, and packaging problems, nowadays electric discharge processes are used (Ha et al., 2017; Long and Ayers, 1999; Long and Waits, 2000). Although the main purpose of shockwave treatment is related to tenderization of meat and meat products, there is also observed clear bactericidal effect (Hazell et al., 2010; Leighs et al., 2014; Schiffer et al., 2017; Udagawa and Suzuki, 2013). Studies related to changes in the appearance of the treated meat suggested that no substantial change in the meat color in response to shockwave were observed (Ha et al., 2017). Considering the cooking loss of treated meat, there was a small, but significant increase compared with the control (untreated) samples. Data for the disruptive effect of the shockwaves on the collagen microfibril network were reported (Zuckerman et al., 2013). In generall, neverthelss, the technology offers some advantages for modification of physico-chemical and microbiological properties of meat products; however, the technology met difficulties with industrial commercialization, especially the variant using explosions (Warner et al., 2017).

Photosensitization (photodynamic treatment) is a method relying on the action of visible light on the microbial cells loaded with compound (photosensitizer—accumulating preferentially in the microbial cells), which can absorb the light and provoke photochemical reactions detrimental for the host organisms (Luksiene and Paskeviciute, 2011). Various photosensitizers were explored: porphyrin derivatives, phtalocyanins, phenothiazinium dyes, xanthylium dyes, fullerens, etc. The main requirements for the substances used are efficiency in formation of triplet excited state when light with specific wavelength is used, and preferential affinity to the microbial cells (Brovko, 2010). According to Luksiene and Zukauskas (2009), photodynamic treatment has the advantages that there is no bacterial resistance (or mutagenicity and antibiotic resistance influence on efficacy), no harmful effect on the surrounding observed, the cost is comparatively low, and it is a generally environmentally friendly procedure. Nevertheless, there are some points of concerns that should be mentioned: the necessity for presence of oxygen, addition of compounds (although natural and allowed for food purposes), which usually are dyes and could interfere or change the appearance of the surface, the penetration ability of the light, etc.

Ultrahigh-frequency electric field technology (UHF), relying on waves with short wavelengths of 10–50 m and ultra-short wavelengths (from 10 m to mm) also had a sterilization effect. This is due to the high frequency (HF) and ultrahigh frequency (UHF) current changes during the passage of waves with short and ultra-short wavelengths through the medium. In the electromagnetic field, the electrical energy is transformed into thermal energy. The nature of heating in the UHF field is different from conventional heating techniques and has a number of advantages: the object is heated rapidly and evenly throughout the whole volume/mass, and heating may be selective, as individual parts of a nonuniform object have different electrophysical properties and are heated to different degrees.

The death of microorganisms in the UHF field occurs as a result of the thermal effect, but the mechanism of action of the UHF field is still not fully understood. Some researchers believe that electromagnetic waves have a specific impact. It has been found that a low-intensity UHF field does not cause medium heating and affects some physiological and biochemical properties of microbial cells. Data on the death of certain bacteria and yeasts at temperatures of 35–40°C were reported (Youn et al., 2011;

Ziuzina et al., 2018). Due to its specific features, this treatment method is used for pasteurization and sterilization of foods, including canned fruits. Compared to ordinary steam treatment in autoclaves, the heating time with UHF energy to the same temperature is reduced, thus completely preserving the taste and nutritional properties of the product while the effect on its microflora is the same. It has been found that UHF energy in the processing of spices, whole meats and meat products has a detrimental effect on *Salmonella* sp., *Staphylococcus aureus*, and *Escherichia* sp. These microorganisms, as well as their toxins, are destroyed under different modes of treatment at power of 410–540 W (Awuah et al., 2005; Youn et al., 2011; Ziuzina et al., 2018). UHF electromagnetic food processing is widely used in the food and catering industry for cooking, drying, baking, thawing, etc. (Piyasena et al., 2003b).

## 3 Hurdle approach and combinations of methods

In most cases, a failure of single technologies to achieve proper microbial control with minimal and mild processing has led to the introduction of the idea of combining several approaches to ensure high-quality safe foods. This approach became known as "hurdle technology" (Leistner and Gorris, 1995). More than 50 hurdles were proposed; some are well-known from the past, some are recently introduced. Based on the type of influence on microorganisms and products, they can be categorized as physical (temperature, electromagnetic energy, pressure, ultrasound, packaging, etc.), physicochemical (water activity, pH, redox potential, etc.), chemical (nitrite, polyphenols, spices, oxygen, organic acids, etc.), and microbial (competitive microorganisms, bacteriocins, protective cultures, etc.) (Khan et al., 2017).

Ideally, the technological processes and steps necessary to obtain the desired product have to serve as treatments ensuring microbiological safety. Unfortunately, in most cases, additional steps (hurdles) have to be introduced, depending on the type of food system. The lower the additional hurdles the better, from a technological point of view, from a sensory point of view, for consumers, and for economical reasons, etc.

The inability of HPP alone to ensure complete eradication of bacterial spores gave ground for utilization of combinations of two or more methods. Concepts like mild pressure assisted thermal sterilization (Wimalaratne and Farid, 2008), relying on thermal expansion generated pressure, and pressure assisted thermal sterilization (Wilson et al., 2008), in which heating is applied before HPP treatment, has proven inactivation of microorganisms and spores (Li and Farid, 2016). Similarly, combined action of HPP and bioprotective microorganisms on cooked ham was investigated (Ramaroson et al., 2018) and it was found that, from 63 investigated *Lactobacillus* strains, three gave promising results to regrow after HPP and for suppression of *Bacillus* and *Clostridium* spore-forming bacteria. Combinations of HPP and US with mild heat treatments were also very often proposed for microorganism reduction (Evelyn et al., 2017). The data suggested that heat assisted HPP (600 MPa, 70–75°C) ensured better bacterial and mold spore inactivation than thermosonication and heat treatment. Similarly, García-Parra et al. (2018) concluded that combine action of PEF and HPT could ensure food safety of pumpkin puree.

Spilimbergo et al. (2003b) investigated the synergistic effect of PEF and DPCD on *E. coli*, *S. aureus*, and *Bacillus cereus* and found that it could be an alternative approach for treatment of thermally unstable substances. Various combinations were also tested and promising results were achieved: US followed by mild heating (Wordon et al., 2012); thermosonication and PEF for *S. aureus* inactivation (Walkling-Ribeiro et al., 2009); UV and PEF and PL plus PEF for *E. coli* and *P. fermentans* inactivation (Palgan et al., 2011); US and steam treatment of broilers (Musavian et al., 2014); DPCD and US of cooked ham and dry cured ham (Ferrentino and Spilimbergo, 2016; Spilimbergo et al., 2014), etc.

High intensity PL with PEF was used to inactivate *E. coli* in apple juice with no significant effect on the juice quality (Caminiti et al., 2011b). Caminiti et al. (2011a) applied UV light (5.3 J $cm^{-2}$), high intensity PL (3.3 J $cm^{-2}$), PEF (34 kV $cm^{-1}$, 18 Hz, 93 ms), and manothermosonication (5 bar, 43°C, 750 W, 20 kHz) to a blend of apple and cranberry juice, and found that the high intensity PL and PEF combination was the most effective.

Combining US with pressure at nonlethal (manosonication) and lethal (manothermosonication) temperatures appears to have great potential (Condón et al., 2005, 2010). Knorr et al. (2004) inactivated *E. coli* more easily when US was combined with heat and Álvarez et al. (2006) used heat and US under pressure for inactivation of *Salmonella senftenberg* 775 W. Other works have demonstrated the microbicidal efficiency of US with chemical antimicrobials against *Salmonella* and *E. coli* on alfalfa seed (Scouten and Beuchat, 2002), and for reduction of bacteria in contaminated water (Phull et al., 1997).

Promising results were obtained using combinations of physical and chemical treatments of food systems: US with ozone and $ClO_2$ (Aday and Caner, 2014), US with NaOCl (Vivek et al., 2016), thermosonication with acidic electrolyzed water (Mansur and Oh, 2015), US with $ClO_2$ (Chen and Zhu, 2011), and US combined with organic acids (Sagong et al., 2011), etc. Terpens present in essential oils of many herbs and spices have a pronounced antibacterial activity against different microorganisms. Utilization of both carvacrol and PEF was investigated (Ait-Ouazzou et al., 2013) and it was found that the combination was able to reduce 5 $\log_{10}$ of *E. coli* O157:H7 in mango juice. The study also revealed that 20–50 pulses (30 kV $cm^{-1}$ electric field strength and 1 Hz frequency of repetitions) were necessary for mango, apple, tomato, and orange juice preservation. The inclusion of carvacrol in food formulations is an excellent example of consumers' demand for utilization of a naturally occurring chemical preservative (Wang et al., 2016). Additionally, various plant extracts or wastes of the essential oil industry were also investigated as antimicrobial agents. However, very often the concentration of chemical natural constituents either isolated or as part of natural extracts, is not sufficient for effective microbial preservation. Besides, the higher amount of such natural extracts very often leads to negative alteration of the product taste and flavor (Vasileva et al., 2018).

## 4 Future trends and perspectives

A recent survey on the importance and application of the novel "cold" pasteurization technologies outlined the current status, development, major advancements, limitations, trends, and future perspectives for the next 10 years (Jermann et al., 2015).

Specialists from the industry, government, and academic institutions from North America and Europe (two separate surveys) were interviewed and the main purpose was to identify which of the emerging technologies (commercially exploited now or with potential for commercialization) will be the most utilized in the near future. With regard to the present situation the most utilized technologies seems to be HPP and microwave heating (as alternative to the traditional thermal treatments). Nowadays more than 300 HPP units are used globally and the HPP market is estimated to be more than $10 billion (Huang et al., 2017). The UV (in North America) and PEF (for Europe) pasteurization methods were outlined also as currently often used. The predictions for future implementations and applications of the technologies revealed HPP as the most utilized method for the next 5–10 years. For Europe PEF and microwave heating will be the second and the third mostly used technologies, while for North Americans the prognoses suggested microwave heating, thermal sterilization assisted by applying pressure and UV to be the next after HPP important techniques. It is interesting to note also that according to the opinions of the participants the HPP was the most researched and developed method for the past 10 years. More than 200 HPP installations are industrially available and intensively used in North America (Higgins, 2016).

With regard to potential for application in different food sectors, HPP was the most promising and potentially applicable technology for the meat, poultry, and seafood industries, and UV was considered to be the best alternative for treatment of liquid state food systems and the fresh produce industry. No doubt the major limitation of further spread of the new technologies is the initial investment needed, which actually reflects later on production costs and final products prices. Additionally, lack of funding is also a limitation for some of the food processors. The applications of the emerging technologies will remain mainly for shelf life extension and preservation but also trends toward obtaining new functional and value added foods with reduced salt content are observed (O'Flynn et al., 2014; Yang et al., 2015). Another survey (Lindell, 2018), conducted in October 2017 with participation of food producers and retailers, revealed growing interest on HPP application and the main reasons are shelf life increase and food waste reduction.

From another point of view it is very important to consider consumers' preferences and expectations. The conclusions of a survey organized in Bourg-en-Bresse, Rennes, and Strasbourg (Bourgade et al., 2015) on the technologies for amelioration of food quality and safety, reduction of food wastes, and consumers' awareness, showed that a novel technology is better accepted if it allows substantial reduction or elimination of some additives (especially preservatives), and if it is more understandable by the consumers and closer to the "natural." Similarly, consumers react negatively if a technology is too complex, introduces too many changes in the final product, and relies on materials which could lead to usage of too much packaging (from the point of view of waste reduction, and for minimization of possibility for chemical compounds migration into foodstuffs). Based on the above-mentioned, the consumers ranked as accepted technologies like HPP, microwaving, usage of biopreservatives, etc., while technologies like pulsed light and especially food irradiation were more "suspicious" to consumers.

Generally, the overall impression is that these new technologies for now are spread out mostly in the developed countries from where the consumers' demand for fresher, healthier, minimally processed products mainly comes. Underdeveloped, poorer countries, where the hygienic conditions are not at the highest levels often, could benefit from the emerging "cold" pasteurization technologies with development of research, improvement of the techniques, and mostly with decrease of the installation and final products costs.

# 5 Conclusions

Conventional techniques for thermal sterilization of food have certain disadvantages that may affect the quality of food, the organoleptic properties, and nutrients. The advantages and merits of the new emerging technologies, such as HPP, UV, US, PEF, PL treatment, CP application, ionizing radiation, etc., have led to a rapid spread in the production facilities. Nevertheless, with the continuous growth of utilization and commercialization of the emerging technologies and the new solutions come many more questions and challenges.

How to ensure longer shelf life of food products and in the same time keeping food fresh, minimally processed, and without food quality deterioration? Can the emerging technologies answer this important question? In general, the solutions offered were closer to yes, but yet with some restrictions and limitations. At this point no single technology could offer a perfect solution, and combinations, and applications based on case by case studies should be regarded as the best possible solution.

# References


Aaby, K., Grimsbo, I.H., Hovda, M.B., Rode, T.M., 2018. Effect of high pressure and thermal processing on shelf life and quality of strawberry purée and juice. Food Chem. 260, 115–123. https://doi.org/10.1016/j.foodchem.2018.03.100.

Abe, F., 2007. Exploration of the effects of high hydrostatic pressure on microbial growth, physiology and survival: perspectives from piezophysiology. Biosci. Biotechnol. Biochem. 71 (10), 2347–2357. https://doi.org/10.1271/bbb.70015.

Abida, J., Rayees, B., Masoodi, F.A., 2014. Pulsed light technology: a novel method for food preservation. Int. Food Res. J. 21 (3), 839–848.

Aday, M.S., Caner, C., 2014. Individual and combined effects of ultrasound, ozone and chlorine dioxide on strawberry storage life. LWT- Food Sci. Technol. 57 (1), 344–351. https://doi.org/10.1016/j.lwt.2014.01.006.

Aday, M.A., Temizkan, R., Büyükcan, M.B., Caner, C., 2013. An innovative technique for extending shelf life of strawberry: ultrasound. LWT- Food Sci. Technol. 52 (2), 93–101. https://doi.org/10.1016/j.lwt.2012.09.013.

Adekunte, A., Valdramidis, V.P., Tiwari, B.K., Slone, N., Cullen, P.J., Donnell, C.P., Scannell, A., 2010. Resistance of *Cronobacter sakazakii* in reconstituted powdered infant formula during ultrasound at controlled temperatures: a quantitative approach on microbial responses. Int. J. Food Microbiol. 142 (1–2), 53–59. https://doi.org/10.1016/j.ijfoodmicro.2010.05.028.

Ait-Ouazzou, A., Espina, L., García-Gonzalo, D., Pagán, R., 2013. Synergistic combination of physical treatments and carvacrol for *Escherichia coli* O157:H7 inactivation in apple, mango, orange, and tomato juices. Food Control 32 (1), 159–167.

Akshay, K.A., 2013. Microbial Inactivation and Allergen Mitigation of Food Matrix by Pulsed Ultraviolet Light. http://ufdc.ufl.edu/UFE0045406/00001. [(Accessed August 29, 2018)].

Alarcon-Rojo, A.D., Janacua, H., Rodriguez, J.C., Paniwnyk, L., Mason, T.J., 2015. Power ultrasound in meat processing. Meat Sci. 107, 86–93. https://doi.org/10.1016/j.meatsci.2015.04.015.

Allen, E.E., Facciotti, D., Bartlett, D.H., 1999. Monounsaturated but not polyunsaturated fatty acids are required for growth of the deep-sea bacterium *Photobacterium profundum* SS9 at high pressure and low temperature. Appl. Environ. Microbiol. 65 (4), 1710–1720.

Alpas, H., Akhan, C., 2012. Shelf-life extension and safety concerns about haddock (*Merlangius euxinus*) under high hydrostatic pressure. J. Food Saf. 32, 517–527. https://doi.org/10.1111/jfs.12015.

Álvarez, I., Mañas, P., Virto, R., Condón, S., 2006. Inactivation of *Salmonella* Senftenberg 775W by ultrasonic waves under pressure at different water activities. Int. J. Food Microbiol. 108 (2), 218–225. https://doi.org/10.1016/j.ijfoodmicro.2005.11.011.

Amaral, R.D.A., Achaerandio, I., Benedetti, B.C., Pujola, M., 2017. The influence of edible coatings, blanching and ultrasound treatments on quality attributes and shelf-life of vacuum packaged potato strips. LWT- Food Sci. Technol. 85, 449–455. https://doi.org/10.1016/j.lwt.2017.03.062. (B).

Amaral, G.V., Silva, E.K., Cavalcanti, R.N., Martins, C.P.C., Andrade, L.G.Z.S., Moraes, J., Alvarenga, V.O., Guimarães, J.T., Esmerino, E.A., Freitas, M.Q., Silva, M.C., Raices, R.S.L., Sant'Ana, A.S., Meireles, M.A.A., Cruz, A.G., 2018a. Whey-grape juice drink processed by supercritical carbon dioxide technology: physicochemical characteristics, bioactive compounds and volatile profile. Food Chem. 239, 697–703. https://doi.org/10.1016/j.foodchem.2017.07.003.

Amaral, G.V., Silva, K.E., Costa, A.L.R., Alvarenga, V.O., Cavalcanti, R.N., Esmerino, E.A., Guimarães, J.T., Freitas, M.Q., Sant'Ana, A.S., Cunha, R.L., Moraes, J., Silva, M.C., Meireles, M.A., Cruz, A.G., 2018b. Whey-grape juice drink processed by supercritical carbon dioxide technology: physical properties and sensory acceptance. LWT 92, 80–86. https://doi.org/10.1016/j.lwt.2018.02.005.

Anaya-Esparza, L.M., Velázquez-Estrada, R.M., Roig, A.X., García-Galindo, H.S., Sayago-Ayerdi, S.G., Montalvo-Gonázlez, E., 2017. Thermosonication: an alternative processing for fruit and vegetable juices. Trends Food Sci. Technol. 61, 26–37. https://doi.org/10.1016/j.tifs.2016.11.020.

Aouadhi, C., Simonin, H., Prévost, H., Lamballerie, M.D., Maaroufi, A., Mejri, S., 2013. Inactivation of *Bacillus sporothermodurans* LTIS27 spores by high hydrostatic pressure and moderate heat studied by response surface methodology. LWT- Food Sci. Technol. 50 (1), 50–56. https://doi.org/10.1016/j.lwt.2012.07.015.

Aquino, K.A.S., 2012. Sterilization by gamma irradiation. In: Adrovic, F. (Ed.), Gamma Radiation. IntechOpen, https://doi.org/10.5772/34901.

Arjunan, K., Sharma, V., Ptasinska, S., 2015. Effects of atmospheric pressure plasmas on isolated and cellular DNA—a review. Int. J. Mol. Sci. 16 (2), 2971–3016. https://doi.org/10.3390/ijms16022971.

Arroyo, C., Cebrián, G., Pagán, R., Condón, S., 2011. Inactivation of *Cronobacter sakazakii* by ultrasonic waves under pressure in buffer and foods. Int. J. Food Microbiol. 144 (3), 446–454. https://doi.org/10.1016/j.ijfoodmicro.2010.10.033.

Arroyo, C., Cebrian, G., Pagan, R., Condon, S., 2012. Synergistic combination of heat and ultrasonic waves under pressure for *Cronobacter sakazakii* inactivation in apple juice. Food Control 25, 342–348. https://doi.org/10.1016/j.foodcont.2011.10.056.

Artíguez, M.L., Lasagabaster, A., de Marañón, I.M., 2011. Factors affecting microbial inactivation by pulsed light in a continuous flow-through unit for liquid products treatment. Procedia Food Sci. 1, 786–791. https://doi.org/10.1016/j.profoo.2011.09.119.

Arvanitoyannis, I.S., Stratakos, A.C., Tsarouhas, P., 2009. Irradiation applications in vegetables and fruits: a review. Crit. Rev. Food Sci. Nutr. 49 (5), 427–462. https://doi.org/10.1080/10408390802067936.

Augusto, P.E.D., Tribst, A.A.L., Cristianini, M., 2018. Chapter 20–High hydrostatic pressure and high-pressure homogenization processing of fruit juices. In: Rajauria, G., Tiwari, B.K. (Eds.), Fruit Juices Extraction, Composition and Analysis. Academic Press, pp. 393–421. https://doi.org/10.1016/B978-0-12-802230-6.00020-5. Elsevier, Oxford.

Awad, S., Moharram, H.A., Shaltout, O.E., Asker, D., Youssef, M.M., 2012. Applications of ultrasound in analysis, processing and quality control of food: a review. Food Res. Int. 48 (2), 410–427. https://doi.org/10.1016/j.foodres.2012.05.004.

Awuah, G.B., Ramaswamy, H.S., Economides, A., Mallikarjuanan, K., 2005. Inactivation of *Escherichia coli* K-12 and *Listeria innocua* in milk using radio frequency (RF) heating. Innovative Food Sci. Emerg. Technol. 6 (4), 396–402. https://doi.org/10.1016/j.ifset.2005.06.002.

Azzara, C.D., 1992. Off-flavors of dairy products. In: Azzara, C.D., Campbell, L.B., Charalambous, G. (Eds.), Off-Flavors in Foods and Beverages. Elsevier Science Publishers, Amsterdam, The Netherlands, pp. 329–374.

Becquerel, H., 1896. Sur les radiations invisibles émises par les corps phosphorescents. Comptes Rendus 122, 501–503.

Bae, Y.Y., Lee, H.J., Kim, S.A., Rhee, M.S., 2009. Inactivation of *Alicyclobacillus acidoterrestris* spores in apple juice by supercritical carbon dioxide. Int. J. Food Microbiol. 136, 95–100. https://doi.org/10.1016/j.ijfoodmicro.2009.09.015.

Balaban, M.O., Duong, T., 2014. Dense phase carbon dioxide research: current focus and directions. Agric. Agric. Sci. Procedia 2, 2–9. https://doi.org/10.1016/j.aaspro.2014.11.002.

Balamurugan, S., Ahmeda, R., Chibeu, A., Gao, A., Koutchma, T., Strange, P., 2016. Effect of salt types and concentrations on the high-pressure inactivation of *Listeria monocytogenes* in ground chicken. Int. J. Food Microbiol. 218, 51–56. https://doi.org/10.1016/j.ijfoodmicro.2015.11.010.

Bang-Yuan, C., Hsiang-Mei, L., Binghuei, B.Y., Chung-Yi, W., 2015. Pulsed light sterilization of packaging materials. Food Packag. Shelf Life 5, 1–9. https://doi.org/10.1016/j.fpsl.2015.04.002.

Barba, F.J., Koubaa, M., Silva, L.P., Orlien, V., Sant'Ana, A.S., 2017. Mild processing applied to the inactivation of the main foodborne bacterial pathogens: a review. Trends Food Sci. Technol. 66, 20–35. https://doi.org/10.1016/j.tifs.2017.05.011.

Baumann, A.R., Martin, S.E., Feng, H., 2005. Power ultrasound treatment of *Listeria monocytogenes* in apple cider. J. Food Prot. 68 (11), 2333–2340.

Bayindirli, A., Alpas, H., Bozoğlu, F., Hizal, M., 2006. Efficiency of high pressure treatment on inactivation of pathogenic microorganisms and enzymes in apple, orange, apricot and sour cherry juices. Food Control 17 (1), 52–58. https://doi.org/10.1016/j.foodcont.2004.09.002.

Bermúdez-Aguirre, D., Barbosa-Cánovas, G.V., 2008. Study of butter fat content in milk on the inactivation of *Listeria innocua* ATCC 51742 by thermo-sonication. Innov. Food Sci. Emerg. Technol. 9 (2), 176–185. https://doi.org/10.1016/j.ifset.2007.07.008.

Bermúdez-Aguirre, D., Barbosa-Canovas, G.V., 2012. Inactivation of *Saccharomyces* in pineapple, grape and cranberry juices under pulsed and continuous thermo-sonication treatments. J. Food Eng. 108 (3), 383–392. https://doi.org/10.1016/j.jfoodeng.2011.06.038.

Bernard, C., Leduc, A., Barbeau, J., Saoudi, B., Yahia, L.H., De Crescenzo, G., 2006. Validation of cold plasma treatment for protein inactivation: a surface plasmon resonance-based biosensor study. J. Phys. D Appl. Phys. 39 (16), 3470. https://doi.org/10.1088/0022-3727/39/16/S04.

Bhaskaram, C., Sadasivan, G., 1975. Effects of feeding irradiated wheat to malnourished children. Am. J. Clin. Nutr. 28 (2), 130–135. https://doi.org/10.1093/ajcn/28.2.130.

Bogahawaththa, D., Buckow, R., Chandrapala, J., Vasiljevic, T., 2018. Comparison between thermal pasteurization and high pressure processing of bovine skim milk in relation to denaturation and immunogenicity of native milk proteins. Innov. Food Sci. Emerg. Technol. 47, 301–308. https://doi.org/10.1016/j.ifset.2018.03.016.

Bolton, J.R., 2001. Ultraviolet Applications Handbook. Bolton Photosciences Inc., Ayr, ON.

Bolumar, T., Skibsted, L.H., Orlien, V., 2012. Kinetics of the formation of radicals in meat during high pressure processing. Food Chem. 134, 2114–2120. https://doi.org/10.1016/j.foodchem.2012.04.013.

Borle, F., Sieber, R., Bosset, J.O., 2001. Photo-oxidation and photoprotection of foods, with particular reference to dairy products. Sci. Aliment. 21 (6), 571–590. https://doi.org/10.3166/sda.21.571-590.

Bourgade, H., Chartier, O., Cluzel, E., Hémon, N., Dole, P., Zuber, F., 2015. Technological Innovations, Levers for Reducing Waste in the Agri-Food Sector: Challenges for Consumers and Businesses. http://agriculture.gouv.fr/les-innovations-technologiques-leviers-de-reduction-du-gaspillage-dans-le-secteur-agroalimentaire. [(Accessed August 25, 2018)].

Bowker, B.C., Schaefer, R.B., Grapperhaus, M.J., Solomon, M.B., 2011. Tenderization of beef loins using a high efficiency sparker. Innov. Food Sci. Emerg. Technol. 12, 135–141. https://doi.org/10.1016/j.ifset.2011.01.010.

Brovko, L., 2010. Photodynamic treatment: a new efficient alternative for surface sanitation. Adv. Food Nutr. Res. 61, 119–147. https://doi.org/10.1016/B978-0-12-374468-5.00003-9.

Brown, A.C., 2008. In: Brown, A.C. (Ed.), Understanding Food: Principles and Preparation, third ed. Thompson/Wadsworth Publishing, Belmont, CA.

Calvo, L., Torres, E., 2010. Microbial inactivation of paprika using high-pressure $CO_2$. J. Supercrit. Fluids 52 (1), 134–141. https://doi.org/10.1016/j.supflu.2009.11.002.

Calvo, L., Muguerza, B., Cienfuegos-Jovellanos, E., 2007. Microbial inactivation and butter extraction in a cocoa derivative using high pressure $CO_2$. J. Supercrit. Fluids 42 (1), 80–87. https://doi.org/10.1016/j.supflu.2007.01.009.

Cameron, M., McMaster, L.D., Britz, T.J., 2008. Electron microscopic analysis of dairy microbes inactivated by ultrasound. Ultrason. Sonochem. 15 (6), 960–964. https://doi.org/10.1016/j.ultsonch.2008.02.012.

Caminiti, I.M., Noci, F., Muñoz, A., Whyte, P., Morgan, D.J., Cronin, D.A., Lyng, J.G., 2011a. Impact of selected combinations of non-thermal processing technologies on the quality of an apple and cranberry juice blend. Food Chem. 124 (4), 1387–1392. https://doi.org/10.1016/j.foodchem.2010.07.096.

Caminiti, I.M., Palgan, I., Noci, F., Muñoz, A., Whyte, P., Cronin, D.A., Morgan, D.J., Lyng, J.G., 2011b. The effect of pulsed electric fields (PEF) in combination with high intensity light pulses (HILP) on *Escherischia coli* and quality attributes in apple juice. Innov. Food Sci. Emerg. Technol. 12 (2), 118–123. https://doi.org/10.1016/j.ifset.2011.01.003.

Canadian Food Inspection Agency, 2016. Guidance on Food Products Treated With High Pressure Processing (HPP). http://www.inspection.gc.ca/food/safe-food-production-systems/technical-references/high-pressure-processing/eng/1498504011314/1498504256677. [(Accessed July 22, 2018)].

Cao, S., Hu, Z., Pang, B., Wang, H., Xie, H., Wu, F., 2010. Effect of ultrasound treatment on fruit decay and quality maintenance in strawberry after harvest. Food Control 21 (4), 529–532. https://doi.org/10.1016/j.foodcont.2009.08.002.

Capilla, V., Manez, M., Moreno, M.J., Jimenez, R., 2003. Disinfection and disinfection effect of $CO_2$ under pressure on food matrix. In: Proceedings of the 5th International Smposium on Supercritical Fluids, pp. 2095–2100. Versailles, France.

Char, C.D., Mitilinaki, E., Guerrero, S.N., Alzamora, S.M., 2010. Use of highintensity ultrasound and UV-C light to inactivate some microorganisms in fruit juices. Food Bioproc. Tech. 3, 797–803. https://doi.org/10.1007/s11947-009-0307-7.

Cheigh, C.I., Park, M.H., Chung, M.S., Shin, J.K., Park, Y.S., 2012. Comparison of intense pulsed light- and ultraviolet (UVC)-induced cell damage in *Listeria Monocytogenes* and *Escherichia coli* O157:H7. Food Control 25 (2), 654–659. https://doi.org/10.1016/j.foodcont.2011.11.032.

Chemat, F., Rombaut, N., Meullemiestre, A., Turk, M., Perino, S., Tixier, A.S.F., Vian, M.A., 2017. Review of green food processing techniques. Preservation, transformation, and extraction. Innov. Food Sci. Emerg. Technol. 41, 357–377. https://doi.org/10.1016/j.ifset.2017.04.016.

Chen, Z., Zhu, C., 2011. Combined effects of aqueous chlorine dioxide and ultrasonic treatments on postharvest storage quality of plum fruit (*Prunus salicina* L.). Postharvest Biol. Technol. 61 (2-3), 117–123. https://doi.org/10.1016/j.postharvbio.2011.03.006.

Chen, J.L., Zhang, J., Song, L., Jiang, Y., Wu, J., Hu, X.S., 2010. Changes in microorganism, enzyme, aroma of hami melon (*Cucumis melo* L.) juice treated with dense phase carbon dioxide and stored at 4°C. Innov. Food Sci. Emerg. Technol. 11 (4), 623–629. https://doi.org/10.1016/j.ifset.2010.05.008.

Chen, Y.Y., Temelli, F., Gänzle, M.G., 2017. Mechanisms of inactivation of dry *Escherichia coli* by high pressure carbon dioxide. Appl. Environ. Microbiol. 83 (10). https://doi.org/10.1128/AEM.00062-17. e00062-17.

Cho, H.Y., Shin, J.K., Song, Y.A., Yoon, S.J., Kim, J.M., Pyun, Y.R., 2002. Nonthermal pasteurization of lactic acid bacteria by high intensity light pulse. Korean J. Food Sci. Technol. 34 (4), 631–636.

Choi, L.H., Nielsen, S.S., 2005. The effects of thermal and nonthermal processing methods on apple cider quality and consumer acceptability. J. Food Qual. 28 (1), 13–29. https://doi.org/10.1111/j.1745-4557.2005.00002.x.

Choi, Y.M., Ryu, Y.C., Lee, S.H., Go, G.W., Shin, H.G., Kim, K.H., Rhee, M.S., Kim, B.C., 2008. Effects of supercritical carbon dioxide treatment for sterilization purpose on meat quality of porcine *longissimus dorsi* muscle. Lebensm. Wiss. Technol. 41 (2), 317–322. https://doi.org/10.1016/j.lwt.2007.02.020.

Choi, Y.M., Bae, Y.Y., Kim, K.H., Kim, B.C., Rhee, M.S., 2009a. Effects of supercritical carbon dioxide treatment against generic *Escherichia coli*, *Listeria monocytogenes*, *Salmonella typhimurium*, and *Escherichia coli* O157:H7 in marinades and marinated pork. Meat Sci. 82 (4), 419–424. https://doi.org/10.1016/j.meatsci.2009.02.016.

Choi, Y.M., Kim, O.Y., Kim, K.H., Kim, B.C., Rhee, M.S., 2009b. Combined effect of organic acids and supercritical carbon dioxide treatments against nonpathogenic *Escherichia coli*, *Listeria monocytogenes*, *Salmonella typhimurium* and *E. coli* O157:H7 in fresh pork. Lett. Appl. Microbiol. 49, 510–515. https://doi.org/10.1111/j.1472-765X.2009.02702.x.

Choi, M.S., Cheigh, C.I., Jeong, E.A., Shin, J.K., Chung, M.S., 2010. Nonthermal sterilization of *Listeria monocytogenes* in infant foods by intense pulsed-light treatment. J. Food Eng. 97 (4), 504–509. https://doi.org/10.1016/j.jfoodeng.2009.11.008.

Cichoski, A.J., Rampelotto, C., Silva, M.S., de Moura, H.C., Terra, N.N., Wagner, R., de Menezes, C.R., Flores, E.M.M., Barin, H.S., 2015. Ultrasound-assisted post-packaging pasteurization of sausages. Innov. Food Sci. Emerg. Technol. 30, 132–137. https://doi.org/10.1016/j.ifset.2015.04.011.

Codina-Torrella, I., Guamis, B., Zamora, A., Quevedo, J.M., Trujillo, A.J., 2018. Microbiological stabilization of tiger nuts' milk beverage using ultra-high pressure homogenization. A preliminary study on microbial shelf-life extension. Food Microbiol. 69, 143–150. https://doi.org/10.1016/j.fm.2017.08.002.

Condón, S., Raso, J., Pagán, R., 2005. Microbial inactivation by ultrasound. In: Barbosa-Cánovas, G.V., Tapia, M.S., Cano, M.P. (Eds.), Novel Food Processing Technologies. CRC Press, Boca Ratón, pp. 423–442.

Condón, S., Mañas, P., Cebrián, G., 2010. Manothermosonication for microbial inactivation. In: Feng, H., Weiss, J. (Eds.), Ultrasound Technologies for Food and Bioprocessing. Springer Science, New York. Chapter 11.

Coutinho, N.M., Silveira, M.R., Rocha, R.S., Moraes, J., Ferreira, M.V.S., Pimentel, T.C., Freitas, M.Q., Silva, M.C., Raices, R.C.L., Ranadheera, C.S., Borges, F.O., Mathias, S.P., Fernandes, F.A.N., Rodrigues, S., Cruz, A.G., 2018. Cold plasma processing of milk and dairy products. Trends Food Sci. Technol. 74, 56–68. https://doi.org/10.1016/j.tifs.2018.02.008.

Dagan, G.F., Balaban, M.O., 2006. Pasteurization of beer by a continuous dense-phase $CO_2$ system. J. Food Sci. 71 (3), E164–E169. https://doi.org/10.1111/j.1365-2621.2006.tb15630.x.

Damar, S., Balaban, M.O., 2006. Review of dense phase $CO_2$ technology: microbial and enzyme inactivation, and effects on food quality. J. Food Sci. 71 (1), R1–R11. https://doi.org/10.1111/j.1365-2621.2006.tb12397.x.

Damar, S., Balaban, M.O., Sims, C.A., 2009. Continuous dense phase $CO_2$ processing of coconut water. Int. J. Food Sci. Technol. 44 (4), 666–673. https://doi.org/10.1111/j.1365-2621.2008.01784.x.

Datta, N., Deeth, H.C., 2011. Heat treatment of milk. Non-thermal technologies: high pressure processing. In: Fuquay, J.W., Fox, P.F., McSweeney, P.L. (Eds.), Encyclopedia of Dairy Sciences. Academic Press. Elsevier, San Diego, CA, pp. 732–737. https://doi.org/10.1016/B978-0-12-374407-4.00219-3.

De Matos, K.H.O., Lerin, L.A., Soares, D., Smania, A., de Oliveira, J.V., Monteiro, A.R., 2013. Inactivation Kinetics of *Vibrio parahaemolyticus* Naturally Occurring in Oysters Using Supercritical $CO_2$ Treatment. Ibero American Conference on Supercritical fluids Cartagena de Indias, Colombia.

Dehghani, F., Annabi, N., Titus, M., Valtchev, P., Tumilar, A., 2008. Sterilization of ginseng using a high pressure $CO_2$ at moderate temperatures. Biotechnol. Bioeng. 102 (2), 569–576. https://doi.org/10.1002/bit.22059.

Del Pozo-Insfran, D., Balaban, M.O., Talcott, S.T., 2006. Microbial stability, phytochemical retention, and organoleptic attributes of dense phase $CO_2$ processed muscadine grape juice. J. Agric. Food Chem. 54 (15), 5468–5473. https://doi.org/10.1021/jf060854o.

Denkova, Z., Denkova-Kostova, R., Kostov, G., Teneva, D., Goranov, B., 2018. Food Microbiology. Academic Publisher of UFT, p. 302.

Dillow, A.K., Dehghani, F., Hrkach, J.S., Foster, N.R., Langer, R., 1999. Bacterial inactivation by using near- and supercritical carbon dioxide. Proc. Natl. Acad. Sci. U. S. A. 96 (18), 10344–10348. https://doi.org/10.1073/pnas.96.18.10344.

Ehlermann, D.A.E., 2016. Wholesomeness of irradiated food. Radiat. Phys. Chem. 129, 24–29. https://doi.org/10.1016/j.radphyschem.2016.08.014.

Elmnasser, N., Guilou, S., Leroi, F., Orange, N., Bakhrouf, A., Federighi, M., 2007. Pulsed-light system as a novel food decontamination technology: a review. Can. J. Microbiol. 53 (7), 813–821. https://doi.org/10.1139/W07-042.

Enomoto, A., Nakamura, K., Hakoda, M., Amaya, N., 1997. Lethal effect of high-pressure $CO_2$ on a bacterial spore. J. Ferment. Bioeng. 83 (3), 305–307. https://doi.org/10.1016/S0922-338X(97)80999-3.

Erkmen, O., 2000. Antimicrobial effects of pressurisec carbon dioxide on *Brochothrix thermosphacta* in broth and foods. J. Sci. Food Agric. 80 (9), 1365–1370. 10.1002/1097-0010(200007)80:9<1365::AID-JSFA652>3.0.CO;2-P.

Erkmen, O., 2001. Mathematical modeling of *Escherichia coli* inactivation under high-pressure carbon dioxide. J. Biosci. Bioeng. 92 (1), 39–43. https://doi.org/10.1016/S1389-1723(01)80196-1.

Erkmen, O., 2012. Effects of dense phase carbon dioxide on vegetative cells. In: Balaban, M.O., Ferrentino, G. (Eds.), Dense Phase Carbon Dioxide Food and Pharmaceutical Applications, pp. 67–98. https://doi.org/10.1002/9781118243350.ch4.

Erkmen, O., Bozoglu, T.F., 2008. Food Microbiology III: Food Presevation. Like Publishing Company, Ankara, pp. 187–189.

Esbelin, J., Mallea, S., Ram, A.F., Carlin, F., 2013. Role of pigmentation in protecting *Aspergillus niger* conidiospores against pulsed light radiation. Photochem. Photobiol. 89 (3), 758–761. https://doi.org/10.1111/php.12037.

EU Commission decision, 2001. Commission decision of 23 May 2001 authorising the placing on the market of pasteurised fruit based preparations produced using high pressure pasteurisation under Regulation (EC) No 258/97 of the European Parliament and of the Council. Off. J. Eur. Communities L151, 42–43. 7.6.2001, https://eur-lex.europa.eu/legal-content/EN/TXT/?uri=CELEX%3A32001D0424. [(Accessed July 22, 2018)].

Eurofins, 2016. Irradiation Testing for Correct Labeling You Can Trust. https://www.eurofins.com/food-and-feed-testing/food-testing-services/food-irradiation/. [(Accessed August 10, 2018)].

Evelyn, E., Silva, F.V., 2016. High pressure processing pretreatment enhanced the thermosonication inactivation of *Alicyclobacillus acidoterrestris* spores in orange juice. Food Control 62, 365–372. https://doi.org/10.1016/j.foodcont.2015.11.007.

Evelyn, Milani, E., Silva, F.V.M., 2017. Comparing high pressure thermal processing and thermosonication with thermal processing for the inactivation of bacteria, moulds, and yeasts spores in foods. J. Food Eng. 214, 90–96. https://doi.org/10.1016/j.jfoodeng.2017.06.027.

Fan, X., 2007. Furan formation in sugar solution and apple cider upon ultraviolet treatment. J. Agric. Food Chem. 55 (19), 7816–7821. https://doi.org/10.1021/jf071366z.

Farkas, J., Mohácsi-Farkas, C., 2011. History and future of food irradiation. Trends Food Sci. Technol. 22, 121–126. https://doi.org/10.1016/j.tifs.2010.04.002.

Feng, H., Yang, W., Hiescher, T., 2008. Power ultrasound. Food Sci. Technol. Int. 14 (5), 433–436. https://doi.org/10.1177/1082013208098814.

Ferrario, M., Alzamora, S.M., Guerrero, S., 2015. Study of the inactivation of spoilage microorganisms in apple juice by pulsed light and ultrasound. Food Microbiol. 46, 635–642. https://doi.org/10.1016/j.fm.2014.06.017.

Ferrentino, G., Balaban, M.O., 2012. Introduction to dense phase carbon dioxide technology. In: Balaban, M.O., Ferrentino, G. (Eds.), Dense Phase Carbon Dioxide Food and Pharmaceutical Applications. Wiley-Blackwell, IA, USA, pp. 1–4. https://doi.org/10.1002/9781118243350.ch1.

Ferrentino, F., Spilimbergo, S., 2016. A combined high pressure carbon dioxide and high power ultrasound treatment for the microbial stabilization of cooked ham. J. Food Eng. 174, 47–55. https://doi.org/10.1016/j.jfoodeng.2015.11.013.

Ferrentino, G., Bruno, M., Ferrari, G., Poletto, M., Balaban, M.O., 2009a. Microbial inactivation and shelf life of apple juice treated with high pressure carbon dioxide. J. Biol. Eng. 3, 3. https://doi.org/10.1186/1754-1611-3-3.

Ferrentino, G., Plaza, M.L., Ramirez-Rodrigues, M., Ferrari, G., Balaban, M.O., 2009b. Effects of dense phase carbon dioxide pasteurization on the physical and quality attributes of a red grapefruit juice. J. Food Sci. 74 (6), E333–E341. https://doi.org/10.1111/j.1750-3841.2009.01250.x.

Ferrentino, G., Balzan, S., Dorigato, A., Pegoretti, A., Spilimbergo, S., 2012. Effect of supercritical $CO_2$ pasteurization on natural microbiota, texture and microstructure of fresh-cut coconut. J. Food Sci. 77 (5), E137–E143. https://doi.org/10.1111/j.1750-3841.2012.02669.x.

Ferrentino, G., Balzan, S., Spilimbergo, S., 2013. Supercritical $CO_2$ processing of dry cured ham spiked with *Listeria monocytogenes*: inactivation kinetics, color, and sensory evaluations. Food Bioproc. Tech. 6 (5), 1164–1174. https://doi.org/10.1007/s11947-012-0819-4.

Ferrentino, G., Schuster, J., Braeuer, A., Spilimbergo, S., 2017. *In situ* Raman quantification of the dissolution kinetics of carbon dioxide in liquid solutions during a dense phase and ultrasound treatment for the inactivation of *Saccharomyces cerevisiae*. J. Supercrit. Fluids 111, 104–111. https://doi.org/10.1016/j.supflu.2016.01.019.

Fine, F., Gervais, P., 2004. Efficiency of pulsed ultravoilet light for microbial decontamination of food powders. J. Food Prot. 67 (4), 787–792.

Foster, J.W., Cowan, R.M., Maag, T.A., 1962. Rupture of bacteria by explosive decomposition. J. Bacteriol. 83 (2), 330–334.

Frankel, E.N., 1991. Recent advances in lipid oxidation. J. Sci. Food Agric. 54 (4), 495–511. https://doi.org/10.1002/jsfa.2740540402.

Franz, C., Specht, I., Cho, G.S., Gräf, V., Stahl, M., 2009. UV-C-inactivation of microorganisms in naturally cloudy apple juice using novel inactivation equipment based on Dean vortex technology. Food Control 20 (12), 1103–1107. https://doi.org/10.1016/j.foodcont.2009.02.010.

Fraser, D., 1951. Bursting bacteria by release of gas pressure. Nature 167, 33–34.

Fuentes, V., Ventanas, J., Morcuende, D., Estévez, M., Ventanas, S., 2010. Lipid and protein oxidation and sensory properties of vacuum-packaged dry-cured ham subjected to high hydrostatic pressure. Meat Sci. 85, 506–514. https://doi.org/10.1016/j.meatsci.2010.02.024.

Furuta, M., Yamaguchi, M., Tsukamoto, T., Yim, B., Stavarache, C.E., Hasiba, K., Maeda, Y., 2004. Inactivation of *Escherichia coli* by ultrasonic irradiation. Ultrason. Sonochem. 11 (2), 57–60. https://doi.org/10.1016/S1350-4177(03)00136-6.

Gaens, W.V., Bogaerts, A., 2013. Kinetic modelling for an atmospheric pressure argon plasma jet in humid air. J. Phys. D Appl. Phys. 46 (53pp). https://doi.org/10.1088/0022-3727/46/27/275201.

Galvanin, F., De Luca, R., Ferrentino, G., Barolo, M., Spilimbergo, S., Bezzo, F., 2014. Bacterial inactivation on solid food matrices through supercritical $CO_2$: a correlative study. J. Food Eng. 120, 146–157. https://doi.org/10.1016/j.jfoodeng.2013.07.027.

Ganan, M., Hierro, E., Hospital, X.F., Barroso, E., Fernández, M., 2013. Use of pulsed light to increase the safety of ready-to-eat cured meat products. Food Control 32 (2), 512–517. https://doi.org/10.1016/j.foodcont.2013.01.022.

Garcia-Gonzalez, L., Geeraerd, A.H., Spilimbergo, S., Elst, K., Van Ginneken, L.V., Debevere, J., Van Impe, J.F., Devlieghere, F., 2007. High pressure carbon dioxide inactivation of microorganisms in foods: the past, the present and the future. Int. J. Food Microbiol. 117 (1), 1–28. https://doi.org/10.1016/j.ijfoodmicro.2007.02.018.

García-Parra, J., González-Cebrino, F., Delgado-Adámez, J., Cava, R., Martín-Bellosoc, O., Elez-Martínez, P., Ramírez, R., 2018. Application of innovative technologies, moderate-intensity pulsed electric fields and high-pressure thermal treatment, to preserve and/or improve the bioactive compounds content of pumpkin. Innov. Food Sci. Emerg. Technol. 45, 53–61. https://doi.org/10.1016/j.ifset.2017.09.022.

Gasperi, F., Aprea, E., Biasioli, F., Carlin, S., Endrizzi, I., Pirretti, G., Spilimbergo, S., 2009. Effects of supercritical $CO_2$ and $N_2O$ pasteurisation on the quality of fresh apple juice. Food Chem. 115 (1), 129–136. https://doi.org/10.1016/j.foodchem.2008.11.078.

Gogate, P.R., Pandit, A.B., 2011. Sonocrystallization and its application in food and bioprocessing. In: Feng, H., Barbosa-Canovas, G., Weiss, J. (Eds.), Ultrasound Technologies for Food and Bioprocessing. Springer, New York, pp. 467–493.

Gogate, P.R., Tayal, R.K., Pandit, A.B., 2006. Cavitation: a technology on the horizon. Curr. Sci. 91 (1), 35–46.

Gómez-López, V.M., Devlieghere, F., Bonduelle, V., Debevere, J., 2005. Factors affecting the inactivation of micro-organisms by intense light pulses. J. Appl. Microbiol. 99, 460–470. https://doi.org/10.1111/j.1365-2672.2005.02641.x.

Gómez-López, V.M., Ragaert, P., Debevere, J., Devlieghere, F., 2007. Pulsed light for food decontamination: a review. Trends Food Sci. Technol. 18, 464–473. https://doi.org/10.1016/j.tifs.2007.03.010.

Gould, G.W., 2000. Preservation: past, present and future. Br. Med. Bull. 56 (1), 84–96.

Gui, F.Q., Wu, J.H., Chen, F., Liao, X.J., Hu, X.S., Zhang, Z.H., Wang, Z.F., 2006. Change of polyphenol oxidase activity, color, and browning degree during storage of cloudy apple juice treated by supercritical carbon dioxide. Eur. Food Res. Technol. 223 (3), 427–432. https://doi.org/10.1007/s00217-005-0219-3.

Gunes, G., Blum, L.K., Hotchkiss, J.H., 2005. Inactivation of yeasts in grape juice using a continuous dense phase carbon dioxide processing system. J. Sci. Food Agric. 85 (14), 2362–2368. https://doi.org/10.1002/jsfa.2260.

Ha, M., Dunshea, F.R., Warner, R.D., 2017. A meta-analysis of the effects of shockwave and high pressure processing on color and cook loss of fresh meat. Meat Sci. 132, 107–111. https://doi.org/10.1016/j.meatsci.2017.04.016.

Hanieh, S., Niels, H., Nonboe, K.U., Corry, J.E.L., Purnell, G., 2014. Combined steam and ultrasound treatment of broilers at slaughter: a promising intervention to significantly reduce numbers of naturally occurring campylobacters on carcasses. Int. J. Food Microbiol. 176, 23–28. https://doi.org/10.1016/j.ijfoodmicro.2014.02.001.

Harder, M.N.C., Arthur, V., Arthur, P.B., 2016. Irradiation of foods: processing technology and effects on nutrients: effect of ionizing radiation on food components. In: Caballero, B., Finglas, P.M., Toldrá, F. (Eds.), Encyclopedia of Food and Health. Academic Press, Oxford, pp. 476–481.

Harvey, E.N., Loomis, A.L., 1929. The destruction of luminuous bacteria by high frequency sound waves. J. Bacteriol. 17, 373–376.

Hashizume, H., Ohta, T., Fengdong, J., Takeda, K., Ishikawa, K., Hori, M., 2013. Inactivation effects of neutral reactive-oxygen species on *Penicillium digitatum* spores using non-equilibrium atmospheric-pressure oxygen radical source. Appl. Phys. Lett. 103, 153–708. https://doi.org/10.1063/1.4824892.

Hashizume, H., Ohta, T., Takeda, K., Ishikawa, K., Hori, M., Ito, M., 2014. Oxidation mechanism of *Penicillium digitatum* spores through neutral oxygen radicals. Jpn. J. Appl. Phys. 53 (1).

Hazell, P.J., Beveridge, C., Groves, K., Appleby-Thomas, G., 2010. The shock compression of microorganism-loaded broths and emulsions: experiments and simulations. Int. J. Impact Eng. 37 (4), 433–440. https://doi.org/10.1016/j.ijimpeng.2009.08.007.

Health Canada, 2016. Health Canada Position - High Pressure Processing (HPP) is no longer a novel process. < https://www.canada.ca/en/health-canada/services/food-nutrition/legislation-guidelines/guidance-documents/position-high-pressure-processing-no-longer-novel-process-treated-food-products-treated-food-products-2013.html > (accessed 23.07.18).

Heinrich, V., Zunabovic, M., Varzakas, T., Bergmair, J., Kneifel, W., 2015. Pulsed light treatment of different food types with a special focus on meat: a critical review. Crit. Rev. Food Sci. Nutr. 56 (4), 591–613. https://doi.org/10.1080/10408398.2013.826174.

Heinz, V., Buckow, R., 2010. Food preservation by high pressure. J. Consum. Prot. Food Saf. 5, 73–81. https://doi.org/10.1007/s00003-009-0311-x.

Hereu, A., Cid, S.B., Garriga, M., Aymerich, T., 2012. High hydrostatic pressure and biopreservation of dry-cured ham to meet the food safety objectives for *Listeria monocytogenes*. Int. J. Food Microbiol. 154, 107–112. https://doi.org/10.1016/j.ijfoodmicro.2011.02.027.

Hereu, A., Dalgaard, P., Garriga, M., Aymerich, T., Cid, S.B., 2014. Analysing and modelling the growth behaviour of *Listeria monocytogenes* on RTE cooked meat products after a high pressure treatment at 400 MPa. Int. J. Food Microbiol. 186, 84–94. https://doi.org/10.1016/j.ijfoodmicro.2014.06.020.

Hierro, E., Barroso, E., de la Hoz, L., Ordonez, J.A., Manzano, S., Fernandez, M., 2011. Efficacy of pulsed light for shelf-life extension and inactivation of *Listeria monocytogenes* on ready-to-eat cooked meat products. Innov. Food Sci. Emerg. Technol. 12 (3), 275–281. https://doi.org/10.1016/j.ifset.2011.04.006.

Higgins, K.T., 2016. New pasteurization processes give food companies options. Food Process. J. 77 (8), 59–63. https://www.foodprocessing.com/articles/2016/pasteurization-processes/?show=all. [(Accessed July 23, 2018)].

Highmore, C.J., Warner, J.C., Rothwell, S.D., Wilks, S.A., Keevil, C.W., 2018. Viable-but-nonculturable *Listeria monocytogenes* and *Salmonella enterica* serovar Thompson induced by chlorine stress remain infectious. MBio 9 (2), e00540-18. https://doi.org/10.1128/mBio.00540-18.

Hijnen, W.A.M., Beerendonk, E.F., Medema, G.J., 2006. Inactivation credit of UV radiation for viruses, bacteria and protozoan (oo) cysts in water: a review. Water Res. 40 (1), 3–22. https://doi.org/10.1016/j.watres.2005.10.030.

Hillegas, S.L., Demirci, A., 2003. Inactivation of *Clostridium sporogenes* in clover honey by pulsed UV-light treatment. Agric. Eng. Int. CIGR J. Sci. Res. Dev 5, 1–7. Manuscript FP 03 009, http://hdl.handle.net/1813/10327.

Hong, S.I., Pyun, Y.R., 1999. Inactivation kinetics of *Lactobacillus plantarum* by high pressure carbon dioxide. J. Food Sci. 64 (4), 728–733. https://doi.org/10.1111/j.1365-2621.1999.tb15120.x.

Hong, S.I., Pyun, Y.R., 2001. Membrane damage and enzyme inactivation of *Lactobacillus plantarum* by high pressure $CO_2$ treatment. Int. J. Food Microbiol. 63 (1-2), 19–28. https://doi.org/10.1016/S0168-1605(00)00393-7.

Hsiao-Wen, H., Hsiang-Mei, L., Binghuei, B.Y., Chung-Yi, W., 2014. Responses of microorganisms to high hydrostatic pressure processing. Food Control 40, 250–259. https://doi.org/10.1016/j.foodcont.2013.12.007.

Huang, H.W., Wu, S.J., Lu, J.K., Shyu, Y.T., Wang, C.Y., 2017. Current status and future trends of high-pressure processing in food industry. Food Control 72A, 1–8. https://doi.org/10.1016/j.foodcont.2016.07.019.

Illera, E., Sanz, M.T., Beltrán, S., Melgosa, R., 2019. High pressure $CO_2$ solubility in food model solutions and fruit juices. J. Supercrit. Fluids 143, 120–125. https://doi.org/10.1016/j.supflu.2018.07.009.

Iseki, S., Hashizume, H., Jia, F., Takeda, K., Ishikawa, K., Ohta, T., 2011. Inactivation of *Penicillium digitatum* spores by a high-density ground-state atomic oxygen-radical source employing an atmospheric-pressure plasma. Appl. Phys. Express 4 (11), 116201. https://doi.org/10.1143/APEX.4.116201.

Isenschmid, A., Marison, I.W., von Stockar, U., 1995. The influence of pressure and temperature of compressed $CO_2$ on the survival of yeast cells. J. Biotechnol. 39 (3), 229–237. https://doi.org/10.1016/0168-1656(95)00018-L.

Ishikawa, K., Hori, M., 2014. Diagnostics of plasma-biological surface interactions in low pressure and atmospheric pressure plasmas. Int. J. Mod. Phys.: Conf. Ser. 32, 1460318. https://doi.org/10.1142/S2010194514603184.

Islam, M.S., Inoue, A., Igura, N., Shimoda, M., Hayakawa, I., 2006. Inactivation of *Bacillus* spores by the combination of moderate heat and low hydrostatic pressure in ketchup and potage. Int. J. Food Microbiol. 107 (2), 124–130. https://doi.org/10.1016/j.ijfoodmicro.2005.08.021.

Jaeger, H., Meneses, N., Knorr, D., 2014. Food technologies: pulsed electric field technology. In: Motarjemi, Y. (Ed.), Encyclopedia of Food Safety, third ed. Academic Press, London, pp. 239–244. https://doi.org/10.1016/B978-0-12-378612-8.00260-2.

Jayasena, D.D., Kim, H.J., Yong, H.I., Park, S., Kim, K., Choe, W., Jo, C., 2015. Flexible thin-layer dielectric barrier discharge plasma treatment of pork butt and beef loin: effects on pathogen inactivation and meat-quality attributes. Food Microbiol. 46, 51–57. https://doi.org/10.1016/j.fm.2014.07.009.

Jermann, C., Koutchma, T., Margas, E., Leadley, C., Polski, V.R., 2015. Mapping trends in novel and emerging food processing technologies around the world. Innov. Food Sci. Emerg. Technol. 31, 14–27. https://doi.org/10.1016/j.ifset.2015.06.007.

Ji, H., Zhang, L., Liu, S., Qu, X., Zhang, C., Gao, J., 2012. Optimization of microbial inactivation of shrimp by dense phase $CO_2$. Int. J. Food Microbiol. 156 (1), 44–49. https://doi.org/10.1016/j.ijfoodmicro.2012.02.020.

Jiafeng, J., Xin, H., Ling, L., Jiangang, L., Hanliang, S., Qilai, X., Renhong, Y., Yuanhua, D., 2014. Effect of cold plasma treatment on seed germination and growth of wheat. Plasma Science and Technology 16 (1), 54–58. https://doi.org/10.1088/1009-0630/16/1/12.

Jiao, R., Gao, J., Li, Y., Zhang, X., Zhang, M., Ye, Y., Wu, Q., Fan, H., 2016. Short communication: effects of high-pressure processing on the inactivity of *Cronobacter sakazakii* in whole milk and skim milk samples. J. Dairy Sci. 99, 1–5. https://doi.org/10.3168/jds.2016-11418.

Joyce, E., Phull, S.S., Lorimer, J.P., Mason, T.J., 2003. The development and evaluation of ultrasound for the treatment of bacterial suspensions. A study of frequency, power and sonication time on cultured *Bacillus* species. Ultrason. Sonochem. 10 (6), 315–318. https://doi.org/10.1016/S1350-4177(03)00101-9.

Jun, S.J., Irudayaraj, J., Demirci, A., Geiser, D., 2003. Pulsed UV-light treatment of corn meal for inactivation of *Aspergillus niger* spores. Int. J. Food Sci. Technol. 38 (8), 883–888. https://doi.org/10.1046/j.0950-5423.2003.00752.x.

Jung, W.Y., Choi, Y.M., Rhee, M.S., 2009. Potential use of supercritical carbon dioxide to decontaminate *Escherichia coli* 0157:H7, *Listeria monocytogenes* and *Salmonella typhimurium* in alfalfa sprouted seeds. Int. J. Food Microbiol. 136 (1), 66–70. https://doi.org/10.1016/j.ijfoodmicro.2009.08.014.

Kao, Y.T., Saxena, C., Wang, L., Sancar, A., Zhong, D., 2005. Direct observation of thymine dimer repair in DNA by photolyase. Proc. Natl. Acad. Sci. U. S. A. 102 (45), 16128–16132. https://doi.org/10.1073/pnas.0506586102.

Kaur, B.P., Rao, P.S., Nema, P.K., 2016. Effect of hydrostatic pressure and holding time on physicochemical quality and microbial inactivation kinetics of black tiger shrimp (*Penaeus monodon*). Innov. Food Sci. Emerg. Technol. 33, 47–55. https://doi.org/10.1016/j.ifset.2015.12.002.

Kaushik, N., Rao, P.S., Mishra, H.N., 2016. Process optimization for thermal-assisted high pressure processing of mango (*Mangifera indica* L.) pulp using response surface methodology. LWT- Food Sci. Technol. 69, 372–381. https://doi.org/10.1016/j.lwt.2016.01.062.

Kehoe, J.J., Remondetto, G.E., Subirade, M., Morris, E.R., Brodkorb, A., 2008. Tryptophan-mediated denaturation of beta-lactoglobulin A by UV irradiation. J. Agric. Food Chem. 56 (12), 4720–4725. https://doi.org/10.1021/jf0733158.

Kentish, S., Ashokkumar, M., 2011. The physical and chemical effects of ultrasound. In: Feng, H., Barbosa-Canovas, G., Weiss, J. (Eds.), Ultrasound Technologies for Food and Bioprocessing. Food Engineering Series. Springer, New York, NY. https://doi.org/10.1007/978-1-4419-7472-3_1.

Khan, I., Tango, C.N., Miskeen, S., Lee, B.H., Oh, D.H., 2017. Hurdle technology: a novel approach for enhanced food quality and safety—A review. Food Control 73, 1426–1444. https://doi.org/10.1016/j.foodcont.2016.11.010.

Khandpur, P., Gogate, P.R., 2015. Effect of novel ultrasound based processing on the nutrition quality of different fruit and vegetable juices. Ultrason. Sonochem. 27, 125–136. https://doi.org/10.1016/j.ultsonch.2015.05.008.

Khandpur, P., Gogate, P.R., 2016. Evaluation of ultrasound based sterilization approaches in terms of shelf life and quality parameters of fruit and vegetable juices. Ultrason. Sonochem. 29, 337–353. https://doi.org/10.1016/j.ultsonch.2015.10.008.

Kiang, W.S., Bhat, R., Rosma, A., Cheng, L.H., 2013. Effects of thermosonication on the fate of *Escherichia coli* O157:H7 and *Salmonella enteritidis* in mango juice. Lett. Appl. Microbiol. 56 (4), 251–257. https://doi.org/10.1111/lam.12042.

Kincal, D., Hill, W.S., Balaban, M.O., Portier, K.M., Wei, C.I., Marshall, M.R., 2005. A continuous high pressure $CO_2$ system for microbial reduction in orange juice. J. Food Sci. 70 (5), M249–M254. https://doi.org/10.1111/j.1365-2621.2005.tb09979.x.

Knorr, D., Zenker, M., Heinz, V., Lee, D., 2004. Applications and potential of ultrasonics in food processing. Trends Food Sci. Technol. 15 (5), 261–266. https://doi.org/10.1016/j.tifs.2003.12.001.

Kobayashi, F., Yazama, F., Ikeura, H., Hayata, Y., Muto, N., Osajima, Y., 2009. Inactivation of microorganisms in untreated water by a continuous flow system with supercritical $CO_2$ bubbling. J. Water Environ. Nanotechnol. 7 (4), 241–250. https://doi.org/10.2965/jwet.2009.241.

Kodama, S., Thawatchaipracha, B., Sekiguchi, H., 2014. Enhancement of essential oil extraction for steam distillation by dbd surface treatment. Plasma Processes Polym. 11 (2), 126–132. https://doi.org/10.1002/ppap.201300047.

Korachi, M., Turan, Z., Senturk, K., Sahin, F., Aslan, N., 2009. An investigation into the biocidal effect of high voltage AC/DC atmospheric corona discharges on bacteria, yeasts, fungi and algae. J. Electrostat. 67 (4), 678–685. https://doi.org/10.1016/j.elstat.2009.03.002.

Kordowska-Wiater, M., Stasiak, D.M., 2011. Effect of ultrasound on survival of gramnegative bacteria on chicken skin surface. Bull. Vet. Inst. Pulawy 55, 207–210.

Koutchma, T., Parisi, B., Patazca, E., 2007. Validation of UV coiled tube reactor for fresh juices. J. Environ. Eng. Sci. 6 (3), 319–328. https://doi.org/10.1139/s06-058.

Koutchma, T.N., Forney, L.J., Moraru, C.I., 2009. Ultraviolet Light in Food Technology. Principles and Applications. CRC Press, USA, p. 279.

Krishnamurthy, K., Demirci, A., Irudayaraj, J.M., 2007. Inactivation of *Staphylococcus aureus* in milk using flow-through pulsed UV-light treatment system. J. Food Sci. 72 (7), 233–239. https://doi.org/10.1111/j.1750-3841.2007.00438.x.

Kumagai, H., Hata, C., Nakamura, K., 1997. $CO_2$ sorption by microbial cells and sterilization by high-pressure $CO_2$. Biosci. Biotechnol. Biochem. 61 (6), 931–935. https://doi.org/10.1271/bbb.61.931.

Kylián, O., Rauscher, H., Gilliland, D., Brétagnol, F., Rossi, F., 2008. Removal of model proteins by means of low-pressure inductively coupled plasma discharge. J. Phys. D Appl. Phys. 41 (9), 095201. https://doi.org/10.1088/0022-3727/41/9/095201.

Lasagabaster, A., Arboleya, J.C., de Marañón, I.M., 2011. Pulsed light technology for surface decontamination of eggs, impact on *Salmonella* inactivation and egg quality. Innov. Food Sci. Emerg. Technol. 12 (2), 124–128. https://doi.org/10.1016/j.ifset.2011.01.007.

Lecky, M., Balaban, M.O., 2004. Continuous high pressure carbon dioxide processing of watermelon juice. 49H-3, page 133. In: Institute of Food Technologists Annual Meeting, Las Vegas, Nevada, July 12–16, 2004.

Lee, B.H., Kermasha, B.E., Baker, B.E., 1989. Thermal, ultrasonic and ultraviolet inactivation of *Salmonella* in thin films of aqueous media and chocolate. Food Microbiol. 6 (3), 143–152. https://doi.org/10.1016/S0740-0020(89)80022-X.

Lee, D., Heinz, V., Knorr, D., 2003. Effects of combination treatments of nisin and high-intensity ultrasound with high pressure on the microbial inactivation in liquid whole egg. Innov. Food Sci. Emerg. Technol. 4 (4), 387–393. https://doi.org/10.1016/S1466-8564(03)00039-0.

Lee, H.J., Jung, H., Choe, W., Ham, J.S., Lee, J.H., Jo, C., 2011. Inactivation of *Listeria monocytogenes* on agar and processed meat surfaces by atmospheric pressure plasma jets. Food Microbiol. 28 (8), 1468–1471. https://doi.org/10.1016/j.fm.2011.08.002.

Leighs, J.A., Appleby-Thomas, G.J., Wood, D.C., Goff, M.J., Hameed, A., Hazell, P.J., 2014. The bactericidal effect of shock waves. J. Phys. Conf. Ser. 500, 182026. https://doi.org/10.1088/1742-6596/500/18/182026.

Leistner, L., Gorris, G.M., 1995. Food preservation by hurdle technology. Trends Food Sci. Technol. 6 (2), 41–46.

Li, X., Farid, M., 2016. A review on recent development in non-conventional food sterilization technologies. J. Food Eng. 182, 33–45. https://doi.org/10.1016/j.jfoodeng.2016.02.026.

Liepa, M., Zagorska, J., Galoburda, R., 2016. High-pressure processing as novel technology in dairy industry: a review. In: Treija, S., Skujeniece, S. (Eds.), Proceedings 22nd Annual International Scientific Conference "Research for Rural Development 2016" 18–20 May, 2016. vol. 1, pp. 76–83. ISSN 2255-923X.

Lim, S., Yagiz, Y., Balaban, M.O., 2006. Continuous high pressure carbon dioxide processing of mandarin juice. Food Sci. Biotechnol. 15 (1), 13–18.

Lin, H.M., Cao, N., Chen, L.F., 1994. Antimicrobial effect of pressurized carbon dioxide on *Listeria monocytogenes*. J. Food Sci. 59 (3), 657–659. https://doi.org/10.1111/j.1365-2621.1994.tb05587.x.

Lindell, C., 2018. Food Safety Strategies. Universal Pure Survey: Food Safety Helps Drive Increase of HPP. https://www.foodsafetystrategies.com/articles/270-universal-pure-survey-food-safety-drives-increase-of-hpp. [(Accessed August 29, 2018)].

Liu, F., Sun, P., Bai, N., Tian, Y., Zhou, H., Wei, S., Zhou, Y., Zhang, J., Zhu, W., Becker, K., Fang, J., 2010. Inactivation of bacteria in an aqueous environment by a direct-current, cold-atmospheric-pressure air plasma microjet. Plasma Processes Polym. 7, 231–236. https://doi.org/10.1002/ppap.200900070.

Liu, F., Zhang, X., Zhao, L., Wang, Y., Liao, X., 2016. Potential of high-pressure processing and high-temperature/short-time thermal processing on microbial, physicochemical and sensory assurance of clear cucumber juice. Innov. Food Sci. Emerg. Technol. 34, 51–58. https://doi.org/10.1016/j.ifset.2015.12.030.

Long, J.B., Ayers, R.A. 1999. Shock-Wave Food Processing With Acoustic Converging Wave Guide. US Patent US6224476B1.

Long, J.B., Waits, D. 2000. Shock-wave meat treatment. Patent CA2431547C.

López-Pedemonte, T.J., Roig-Sagués, A.X., Trujillo, A.J., Capellas, M., Guamis, B., 2003. Inactivation of spores of *Bacillus cereus* in cheese by high hydrostatic pressure with the addition of nisin or lysozyme. J. Dairy Sci. 86 (10), 3075–3081. https://doi.org/10.3168/jds.S0022-0302(03)73907-1.

Lu, X., Naidis, G.V., Laroussi, M., Reuter, S., Graves, D.B., Ostrikov, K., 2016. Reactive species in non-equilibrium atmospheric-pressure plasmas: generation, transport, and biological effects. Phys. Rep. 630, 1–84. https://doi.org/10.1016/j.physrep.2016.03.003.

Luksiene, Z., Paskeviciute, E., 2011. Microbial control of food-related surfaces: Na-chlorophyllin-based photosensitization. J. Photochem. Photobiol. B Biol. 105, 69–74. https://doi.org/10.1016/j.jphotobiol.2011.06.011.

Luksiene, Z., Zukauskas, A., 2009. Prospects of photosensitization in control of pathogenic and harmful micro-organisms. J. Appl. Microbiol. 107, 1415–1424. https://doi.org/10.1111/j.1365-2672.2009.04341.x.

Maleki, A., Shahmoradiorcid, B., Daraei, H., Kalantar, E., 2013. Assessment of ultrasound irradiation on inactivation of gram negative and positive bacteria isolated from hospital in aqueous solution. J. Adv. Environ. Health Res. 1 (1), 9–14. https://doi.org/10.22102/JAEHR.2013.40119.

Mansur, A.R., Oh, D.H., 2015. Combined effects of thermosonication and slightly acidic electrolyzed water on the microbial quality and shelf life extension of fresh-cut kale during refrigeration storage. Food Microbiol. 51, 154–162. https://doi.org/10.1016/j.fm.2015.05.008.

Manzocco, L., Da Pieve, S., Maifreni, M., 2011. Impact of ultraviolet (UV)-C light on safety and quality of fresh-cut melon. Innov. Food Sci. Emerg. Technol. 12 (1), 13–17. https://doi.org/10.1016/j.ifset.2010.11.006.

Manzocco, L., Ignat, A., Valoppi, F., Burrafato, K.R., Lippe, G., Spilimbergo, S., Nicoli, M.C., 2016. Inactivation of mushroom polyphenoloxidase in model systems exposed to high-pressure carbon dioxide. J. Supercrit. Fluids 107, 669–675. https://doi.org/10.1016/j.supflu.2015.07.029.

Manzocco, L., Plazzotta, S., Spilimbergo, S., Nicoli, M.C., 2017. Impact of high-pressure carbon dioxide on polyphenoloxidase activity and stability of fresh apple juice. LWT- Food Sci. Technol. 85 (B), 363–371. https://doi.org/10.1016/j.lwt.2016.11.052.

Martínez, I.P., Mulet, A., García-Alvarado, M.A., Benedito, J., 2018. Inactivation of the microbiota and effect on the quality attributes of pineapple juice using a continuous flow ultrasound-assisted supercritical carbon dioxide system. Food Sci. Technol. J. 24 (7), 547–554. https://doi.org/10.1177/1082013218774694.

Martínez-Onandi, N., Rivas-Cañedo, A., Nuñez, M., Picon, A., 2016a. Effect of chemical composition and high pressure processing on the volatile fraction of Serrano dry cured ham. Meat Sci. 111, 130–138. https://doi.org/10.1016/j.meatsci.2015.09.004.

Martínez-Onandi, N., Rivas-Cañedo, A., Picon, A., Nuñez, M., 2016b. Influence of physicochemical parameters and high pressure processing on the volatile compounds of Serrano dry-cured ham after prolonged refrigerated storage. Meat Sci. 122, 101–108. https://doi.org/10.1016/j.meatsci.2016.07.027.

Martínez-Onandi, N., Castioni, A., San Martín, E., Cañedo, A.R., Nuñez, M., Torriani, S., Picon, A., 2017a. Microbiota of high-pressure-processed Serrano ham investigated by culture-dependent and culture-independent methods. Int. J. Food Microbiol. 241, 298–307. https://doi.org/10.1016/j.ijfoodmicro.2016.11.001.

Martínez-Onandi, N., Rivas-Cañedo, A., Ávila, M., Garde, S., Nuñez, M., Picon, A., 2017b. Influence of physicochemical characteristics and high pressure processing on the volatile fraction of Iberian dry-cured ham. Meat Sci. 131, 40–47. https://doi.org/10.1016/j.meatsci.2017.04.233.

Marx, G., Moody, A., Bermúdez-Aguirre, D., 2011. A comparative study on the structure of *Saccharomyces cerevisiae* under nonthermal technologies: high hydrostatic pressure, pulsed electric fields and thermo-sonication. Int. J. Food Microbiol. 151 (3), 327–337. https://doi.org/10.1016/j.ijfoodmicro.2011.09.027.

Masschelein, W.J., 2002. Ultraviolet Light in Water and Wastewater Sanitation. CRC Press, Boca Raton, p. 192.

Matak, K.E., Churey, J.J., Worobo, R.W., Sumner, S.S., Hovingh, E., Hackney, C.R., Pierson, M.D., 2005. Efficacy of UV light for the reduction of *Listeria monocytogenes* in goat's milk. J. Food Prot. 68 (10), 2212–2216. https://doi.org/10.4315/0362-028X-68.10.2212.

Mazzoni, A.M., Sharma, R.R., Demirci, A., Ziegler, G.R., 2001. Supercritical carbon dioxide treatment to inactivate aerobic microorganisms on alfalfa seeds. J. Food Saf. 21 (4), 215–223. https://doi.org/10.1111/j.1745-4565.2001.tb00320.x.

Meujo, D.A.F., Kevin, D.A., Peng, J., Bowling, J.J., Liu, J., Hamann, M.T., 2010. Reducing oyster-associated bacteria levels using supercritical fluid $CO_2$ as an agent of warm pasteurization. Int. J. Food Microbiol. 138 (1-2), 63–70. https://doi.org/10.1016/j.ijfoodmicro.2009.11.012.

Meurehg, C.A.T., 2006. Control of *Escherichia coli* O157:H7, generic *Escherichia coli*, and *Salmonella spp.* on beef trimmings prior to grinding using a controlled phase carbon dioxide ($cpCO_2$) system. PhD Thesis, Manhattan, Kansas: Kansas State University.

Miao, H., Yun, G., 2011. The sterilization *of Escherichia coli* by dielectric-barrier discharge plasma at atmospheric pressure. Appl. Surf. Sci. 257 (16), 7065–7070. https://doi.org/10.1016/j.apsusc.2011.03.014.

Min, D.B., Boff, J.M., 2002. Chemistry and reaction of singlet oxygen in foods. Compr. Rev. Food Sci. Food Saf. 1 (2), 58–72. https://doi.org/10.1111/j.1541-4337.2002.tb00007.x.

Minh, H.N.T., Dantigny, P., Perrier-Cornet, J.M., Gervais, P., 2010. Germination and inactivation of *Bacillus subtilis* spores induced by moderate hydrostatic pressure. Biotechnol. Bioeng. 107, 876–883. https://doi.org/10.1002/bit.22849.

Misra, N.N., Jo, C., 2017. Applications of cold plasma technology for microbiological safety in meat industry. Trends Food Sci. Technol. 64, 74–86. https://doi.org/10.1016/j.tifs.2017.04.005.

Misra, N., Sullivan, C., Pankaj, S.K., Alvarez-Jubete, L., Cama, R., Jacoby, F., Cullen, P.J., 2014. Enhancement of oil spreadability of biscuit surface by nonthermal barrier discharge plasma. Innov. Food Sci. Emerg. Technol. 26, 456–461. https://doi.org/10.1016/j.ifset.2014.10.001.

Misra, N.N., Kaur, S., Tiwari, B.K., Kaur, A., Singh, N., Cullen, P.J., 2015. Atmospheric pressure cold plasma (ACP) treatment of wheat flour. Food Hydrocoll. 44, 115–121. https://doi.org/10.1016/j.foodhyd.2014.08.019.

Misra, N.N., Pankaj, S.K., Segat, A., Ishikawa, K., 2016. Cold plasma interactions with enzymes in foods and model systems. Trends Food Sci. Technol. 55, 39–47. https://doi.org/10.1016/j.tifs.2016.07.001.

Mohamed, H.M.H., Diono, B.H.S., Yousef, E.Y., 2012. Structural changes in *Listeria monocytogenes* treated with gamma radiation, pulsed electric field and ultra-high pressure. J. Food Saf. 32 (1), 66–73. https://doi.org/10.1111/j.1745-4565.2011.00345.x.

Mohan, M.S., Ye, R., Harte, F., 2016. Initial study on high pressure jet processing using a modified waterjet on physicochemical and rennet coagulation properties of pasteurized skim milk. Int. Dairy J. 55, 52–58. https://doi.org/10.1016/j.idairyj.2015.12.004.

Moreirinha, C., Almeida, A., Saraiva, J.A., Delgadillo, I., 2016. High-pressure processing effects on foodborne bacteria by mid-infrared spectroscopy analysis. LWT- Food Sci. Technol. 73, 212–218. https://doi.org/10.1016/j.lwt.2016.05.041.

Morild, R.K., Christiansen, P.S., Anders, H., Nonboe, U., Aabo, S., 2011. Inactivation of pathogens on pork by steam-ultrasound treatment. J. Food Prot. 74 (5), 769–775. https://doi.org/10.4315/0362-028X.JFP-10-338.

Muñoz, A., Palgan, I., Noci, F., Morgan, D.J., Cronin, D.A., Whyte, P., Lyng, J.G., 2011. Combinations of high intensity light pulses and thermosonication for the inactivation of *Escherichia coli* in orange juice. Food Microbiol. 28 (6), 1200–1204. https://doi.org/10.1016/j.fm.2011.04.005.

Muñoz, A., Caminiti, I.M., Palgan, I., Pataro, G., Noci, F., Morgan, D.J., Cronin, D.A., Whyte, P., Ferrari, G., Lyng, J.G., 2012. Effects on *Escherichia coli* inactivation and quality attributes in apple juice treated by combinations of pulsed light and thermosonication. Food Res. Int. 45 (1), 299–305. https://doi.org/10.1016/j.foodres.2011.08.020.

Muñoz-Cuevas, M., Guevara, L., Aznar, A., Martínez, A., Periago, P.M., Fernández, P.S., 2013. Characterisation of the resistance and the growth variability of *Listeria monocytogenes* after high hydrostatic pressure treatments. Food Control 29 (2), 409–415. https://doi.org/10.1016/j.foodcont.2012.05.047.

Murgov, I., Denkova, Z., 2012. Microbiology. Academic Publisher of UFT, p. 250.

Musavian, H.S., Krebs, N.H., Nonboe, U., Corry, J.E.L., Purnell, G., 2014. Combined steam and ultrasound treatment of broilers at slaughter: a promising intervention to significantly reduce numbers of naturally occurring campylobacters on carcasses. Int. J. Food Microbiol. 176, 23–28. https://doi.org/10.1016/j.ijfoodmicro.2014.02.001.

Nakamura, K., Enomoto, A., Fukushima, H., Nagai, K., Hakoda, M., 1994. Disruption of microbial cells by the flash discharge of high-pressure carbon dioxide. Biosci. Biotechnol. Biochem. 58 (7), 1297–1301. https://doi.org/10.1271/bbb.58.1297.

Niemira, B.A., 2012. Cold plasma reduction of *Salmonella* and *Escherichia coli* O157:H7 on almonds using ambient pressure gases. J. Food Sci. 77 (3), 53–62. https://doi.org/10.1111/j.1750-3841.2011.02594.x.

Niu, L., Hu, X., Wu, J., Liao, X., Chen, F., Zhao, G., Wang, Z., 2010a. Effect of dense phase carbon dioxide process on physicochemical properties and flavor compounds of orange juice. J. Food Process. Preserv. 34 (s2), 530–548. https://doi.org/10.1111/j.1745-4549.2009.00369.x.

Niu, S., Xu, Z., Fang, Y., Zhang, L., Yang, Y., Liao, X., Hu, X., 2010b. Comparative study on cloudy apple juice qualities from apple slices treated by high pressure carbon dioxide and mild heat. Innov. Food Sci. Emerg. Technol. 11 (1), 91–97. https://doi.org/10.1016/j.ifset.2009.09.002.

Nuñez, A., del Olmo, A., Calzada, J., 2016. Biogenic amines. In: Caballero, B., Finglas, P.M., Toldrá, F. (Eds.), Encyclopedia of Food and Health. Academic press, Oxford, pp. 416–423.

O'Flynn, C.C., Cruz-Romero, M.C., Troy, D., Mullen, A.M., Kerry, J.P., 2014. The application of high-pressure treatment in the reduction of salt levels in reduced-phosphate breakfast sausages. Meat Sci. 96, 1266–1274. https://doi.org/10.1016/j.meatsci.2013.11.010.

Ohlsson, T., Bengtsson, N., 2002. In: Ohlsson, T., Bengtsson, N. (Eds.), Minimal processing technologies in the food industry. Woodhead Publishing, Cambridge, England, pp. 112–121.

Oms-Oliu, G., Aguiló-Aguayo, I., Martin-Belloso, O., Soliva-Fortuny, R., 2010. Effects of pulsed light treatments on quality and antioxidant properties of fresh-cut mushrooms (*Agaricus bisporus*). Postharvest Biol. Technol. 56 (3), 216–222. https://doi.org/10.1016/j.postharvbio.2009.12.011.

Oulé, M.K., Tano, K., Bernier, A.M., Arul, J., 2006. *Escherichia coli* inactivation mechanism by pressurized $CO_2$. Can. J. Microbiol. 52 (12), 1208–1217. https://doi.org/10.1139/w06-078.

Pal, M., 2017. Pulsed electric field processing: an emerging technology for food preservation. J. Exp. Food Chem. 3 (2), 126. https://doi.org/10.4172/2472-0542.1000126.

Pal, P., Kaur, P., Singh, N., Kaur, A., Misra, N.N., Tiwari, B.K., Cullen, P.J., Virdi, A.S., 2016. Effect of nonthermal plasma on physico-chemical, amino acid composition, pasting and protein characteristics of short and long grain rice flour. Food Res. Int. 81, 50–57. https://doi.org/10.1016/j.foodres.2015.12.019.

Palgan, I., Caminiti, I.M., Muñoz, A., Noci, F., Whyte, P., Morgan, D.J., Cronin, D.A., Lyng, J.G., 2011. Combined effect of selected non-thermal technologies on *Escherichia coli* and *Pichia fermentans* inactivation in an apple and cranberry juice blend and on product shelf life. Int. J. Food Microbiol. 151 (1), 1–6. https://doi.org/10.1016/j.ijfoodmicro.2011.07.019.

Palmieri, L., Cacace, D., 2005. High intensity pulsed light technology. In: Da-Wen, S. (Ed.), Emerging Technologies for Food Processing. Academic Press, London, pp. 279–306.

Pan, Y., Sun, D., Han, Z., 2017. Applications of electromagnetic fields for nonthermal inactivation of microorganisms in foods: an overview. Trends Food Sci. Technol. 64, 13–22. https://doi.org/10.1016/j.tifs.2017.02.014.

Pankaj, S.K., Bueno-Ferrer, C., Misra, N.N., Cullen, P.J., 2014a. Zein Film: effects of dielectric barrier discharge atmospheric cold plasma. J. Appl. Polym. Sci. 131 (18). https://doi.org/10.1002/app.40803.

Pankaj, S.K., Bueno-Ferrer, C., Misra, N.N., Milosavljevic, V., O'Donnell, C.P., Bourke, P., Keener, K.M., Cullen, P.J., 2014b. Applications of cold plasma technology in food packaging. Food Sci. Technol. 35 (1), 5–17. https://doi.org/10.1016/j.tifs.2013.10.009.

Parton, T., Bertucco, A., Bertoloni, G., 2007. Pasteurisation of grape must and tomato paste by dense-phase $CO_2$. Ital. J. Food Sci. 19 (4), 425–437.

Paskeviciute, E., Buchovec, I., Luksiene, Z., 2011. High-power pulsed light for decontamination of chicken from food pathogens: a study on antimicrobial efficiency and organoleptic properties. J. Food Saf. 31 (1), 61–68. https://doi.org/10.1111/j.1745-4565.2010.00267.x.

Pataro, G., Muñoz, A., Palgan, I., Noci, F., Ferrari, G., Lyng, J.G., 2011. Bacterial inactivation in fruit juices using a continuous flow pulsed light (PL) system. Food Res. Int. 44 (6), 1642–1648. https://doi.org/10.1016/j.foodres.2011.04.048.

Paterson, M.F., 2005. A review: microbiology of pressure treated foods. J. Appl. Microbiol. 98 (6), 1400–1409. https://doi.org/10.1111/j.1365-2672.2005.02564.x.

Patil, S., Keener, K.M., Moiseev, T., Mosnier, J.P., Misra, N.N., Cullen, P.J., Bourke, P., 2014a. In package inactivation of Bacillus atrophaeus spores using high voltage atmospheric cold plasma. In: Society of Engineering Science, 51st Annual Technical Meeting, Purdue University, Indiana, USA, 1–3 October 2014.

Patil, S., Moiseev, T., Misra, N.N., Cullen, P.J., Mosnier, J.P., Keener, K.M., Bourke, P., 2014b. Influence of high voltage atmospheric cold plasma process parameters and role of relative humidity on inactivation of *Bacillus atrophaeus spores* inside a sealed package. J. Hosp. Infect. 88 (3), 162–169. https://doi.org/10.1016/j.jhin.2014.08.009.

Pedrós-Garrido, S., Condón-Abanto, S., Beltrán, J.A., Lyng, J.G., Brunton, N.P., Bolton, D., Whyte, P., 2017. Assessment of high intensity ultrasound for surface decontamination of salmon (*S. salar*), mackerel (*S. scombrus*), cod (*G. morhua*) and hake (*M. merluccius*) fillets, and its impact on fish quality. Innov. Food Sci. Emerg. Technol. 41, 64–70. https://doi.org/10.1016/j.ifset.2017.02.006.

Pervez, M.R., Begum, A., Laroussi, M., 2014. Plasma based sterilization: overview and the stepwise inactivation process of microbial by non-thermal atmospheric pressure plasma jet. Int. J. Eng. Technol. 14 (5), 7–16.

Phull, S.S., Newman, A.P., Lorimer, J.P., Pollet, B., Mason, T.J., 1997. The development and evaluation of ultrasound in the biocidal treatment of water. Ultrason. Sonochem. 4 (2), 157–164. https://doi.org/10.1016/S1350-4177(97)00029-1.

Pingen, S., Sudhaus, N., Becker, A., Krischek, C., Klein, G., 2016. High pressure as an alternative processing step for ham production. Meat Sci. 118, 22–27. https://doi.org/10.1016/j.meatsci.2016.03.014.

Pinheiro, J., Alegria, C., Abreu, M., Gonsalves, E.M., Silva, C.L.M., 2015. Influence of postharvest ultrasounds treatments on tomato (*Solanum lycopersicum*, cv. Zinac) quality and microbial load during storage. Ultrason. Sonochem. 27, 552–559. https://doi.org/10.1016/j.ultsonch.2015.04.009.

Piyasena, P., Mohareb, E., McKellar, R.C., 2003a. Inactivation of microbes using ultrasound: a review. Int. J. Food Microbiol. 87 (3), 207–216. https://doi.org/10.1016/S0168-1605(03)00075-8.

Piyasena, P., Dussault, C., Koutchma, T., Ramaswamy, H.S., Awuah, G.B., 2003b. Radio frequency heating of foods: principles, applications and related properties—a review. Crit. Rev. Food Sci. Nutr. 43 (6), 587–606. https://www.doi.org/10.1080/10408690390251129.

Pohlman, F.W., Dikeman, M.E., Zayasq, J.F., 1996. The effect of low-intensity ultrasound treatment on shear properties, color stability and shelf-life of vacuum-packaged beef *semitendinosus* and *biceps femoris* muscles. Meat Sci. 45 (3), 329–337. https://doi.org/10.1016/S0309-1740(96)00106-4.

Rajkovic, A., Tomasevic, I., Smigic, N., Uyttendaele, M., Radovanovic, R., Devlieghere, F., 2010. Pulsed ultraviolet light as an intervention strategy against *Listeria monocytogenes* and *Escherichia coli* O157:H7 on the surface of a meat slicing knife. J. Food Eng. 100 (3), 446–451. https://doi.org/10.1016/j.jfoodeng.2010.04.029.

Ramaroson, M., Guillou, S., Rossero, A., Rezé, S., Anthoine, V., Moriceau, N., Martin, J.L., Duranton, F., Zagorec, M., 2018. Selection procedure of bioprotective cultures for their combined use with high pressure processing to control spore-forming bacteria in cooked ham. Int. J. Food Microbiol. 276, 28–38. https://doi.org/10.1016/j.ijfoodmicro.2018.04.010.

Ramos, B., Miller, F.A., Brandão, T.R.S., Teixeira, P., Silva, C.L.M., 2013. Fresh fruits and vegetables—An overview on applied methodologies to improve its quality and safety. Innov. Food Sci. Emerg. Technol. 20, 1–15. https://doi.org/10.1016/j.ifset.2013.07.002.

Raso, J., Barbosa-Canovas, G.V., 2003. Nonthermal preservation of foods using combined processing techniques. Crit. Rev. Food Sci. Nutr. 43 (3), 265–285. https://doi.org/10.1080/10408690390826527.

Rastogi, N.K., Raghavarao, K.S.M.S., Balasubramaniam, V.M., Niranjan, K., Knorr, D., 2007. Opportunities and challenges in high pressure processing of foods. Crit. Rev. Food Sci. Nutr. 47 (1), 69–112. https://doi.org/10.1080/10408390600626420.

Rendueles, E., Omer, M.K., Alvseike, O., Alonso-Calleja, C., Capita, R., Prieto, M., 2011. Microbiological food safety assessment of high hydrostatic pressure processing: a review. LWT- Food Sci. Technol. 44 (5), 1251–1260. https://doi.org/10.1016/j.lwt.2010.11.001.

Reyes, J.E., Munizaga, G.T., Won, M.P., Maluenda, D., Roco, T., 2015. Effect of high hydrostatic pressure (HHP) treatments on microbiological shelf-life of chilled Chilean jack mackerel (*Trachurus murphyi*). Innov. Food Sci. Emerg. Technol. 29, 107–112. https://doi.org/10.1016/j.ifset.2015.01.010.

Ribeiro, A.T., Elias, M., Teixeira, B., Pires, C., Duarte, R., Saraiva, J.A., Mendes, R., 2018. Effects of high pressure processing on the physical properties of fish ham prepared with farmed meagre (*Argyrosomus regius*) with reduced use of microbial transglutaminase. LWT- Food Sci. Technol. 96, 296–306. https://doi.org/10.1016/j.lwt.2018.05.044.

Rico, D., Martin-Diana, A.B., Barat, J., Barry-Ryan, C., 2007. Extending and measuring the quality of fresh-cut fruit and vegetables: a review. Trends Food Sci. Technol. 18 (7), 373–386. https://doi.org/10.1016/j.tifs.2007.03.011.

Ringus, D.L., Moraru, C.I., 2013. Pulsed light inactivation of *Listeria innocua* on food packaging materials of different surface roughness and reflectivity. J. Food Eng. 114 (3), 331–337. https://doi.org/10.1016/j.jfoodeng.2012.08.022.

Rode, T.M., Hovda, M.B., 2016. High pressure processing extend the shelf life of fresh salmon, cod and mackerel. Food Control 70, 242–248. https://doi.org/10.1016/j.foodcont.2016.05.045.

Roig-Sagués, A.X., Asto, E., Engers, I., Hernández-Herrero, M.M., 2015. Improving the efficiency of ultra-high pressure homogenization treatments to inactivate spores of *Alicyclobacillus* spp. in orange juice controlling the inlet temperature. LWT- Food Sci. Technol. 63, 866–871. https://doi.org/10.1016/j.lwt.2015.04.056.

Röntgen, W., 1895. Ueber eine neue Art von Strahlen. Vorläufige Mitteilung. Aus den Sitzungsberichten der Würzburger Physik.-medic. Gesellschaft Würzburg. 137–147.

Rubio, B., Possas, A., Rincón, F., Gímeno, R.M.G., Martínez, B., 2018. Model for *Listeria monocytogenes* inactivation by high hydrostatic pressure processing in Spanish chorizo sausage. Food Microbiol. 69, 18–24. https://doi.org/10.1016/j.fm.2017.07.012.

Sagong, H.G., Lee, S.Y., Chang, P.S., Heu, S., Ryu, S., Choi, Y.J., Kang, D.H., 2011. Combined effect of ultrasound and organic acids to reduce *Escherichia coli* O157:H7, *Salmonella typhimurium*, and *Listeria monocytogenes* on organic fresh lettuce. Int. J. Food Microbiol. 145 (1), 287–292. https://doi.org/10.1016/j.ijfoodmicro.2011.01.010.

Sattar, S., 2011. Cleaning, disinfection and sterilisation. In: IFIC Basic Concepts of Infection Control. IFIC, Northern Ireland. Chapter 12).

Schiffer, A., Gardner, M.N., Lynn, R.H., Tagarielli, V.L., 2017. A new apparatus to induce lysis of planktonic microbial cells by shock compression, cavitation and spray. R. Soc. Open Sci. 4 (3), 160939. https://doi.org/10.1098/rsos.160939.

Schneider, S., Lackmann, J.W., Narberhaus, F., Bandow, J.E., Denis, B., Benedikt, J., 2011. Separation of VUV/UV photons and reactive particles in the effluent of a He/$O_2$ atmospheric pressure plasma jet. J. Phys. D Appl. Phys. 44 (29). https://doi.org/10.1088/0022-3727/44/29/295201. 295201(9pp).

Schneider, S., Lackmann, J.W., Ellerweg, D., Denis, B., Narberhaus, F., Bandow, J.E., 2012. The role of VUV radiation in the inactivation of bacteria with an atmospheric pressure plasma jet. Plasma Processes Polym. 9 (6), 561–568. https://doi.org/10.1002/ppap.201100102.

Scouten, A.J., Beuchat, L.R., 2002. Combined effects of chemical, heat and ultrasound treatments to kill *Salmonella* and *Escherichia coli* O157:H7 on alfalfa seeds. J. Appl. Microbiol. 92 (4), 668–674. https://doi.org/10.1046/j.1365-2672.2002.01571.x.

Segat, A., Misra, N.N., Cullen, P., Innocente, N., 2015. Atmospheric pressure cold plasma (ACP) treatment of whey protein isolate model solution. Innov. Food Sci. Emerg. Technol. 29, 247–254. https://doi.org/10.1016/j.ifset.2015.03.014.

Selby, K., 2017. Growth Temperature Variation and Heat Stress Response of *Clostridium botulinum*. PhD thesis, University of Helsinki, Finland. ISBN 978-951-51-3005-1 (paperback) Unigrafia, Helsinki. ISBN 978-951-51-3006-8 (PDF) http://ethesis.helsinki.fi.

Shakila, B.M., Sasikala, P., Dhanapal, A., Kavitha, V., Yazhini, G., Rajamani, L., 2012. Cold plasma as a novel food processing technology. Int. J. Emerg. Trends Eng. Dev. 2 (4), 803–818.

Shao, Y., Ramaswamy, H.S., 2011. *Clostridium sporogenes* ATCC 7955 spore destruction kinetics in milk under high pressure and elevated temperature treatment conditions. Food Bioproc. Tech. 4, 458–468. https://doi.org/10.1007/s11947-008-0165-8.

Sheen, S., Cassidy, J., Scullen, B., Sommers, C., 2015. Inactivation of a diverse set of shiga toxin-producing *Escherichia coli* in ground beef by high pressure processing. Food Microbiol. 52, 84–87. https://doi.org/10.1016/j.fm.2015.07.001.

Silindir, M., Ozer, A., 2009. Sterilization methods and the comparison of e-beam sterilization with gamma radiation sterilization. FABAD J. Pharm. Sci. 34, 43–53.

Simonin, H., Duranton, F., de Lamballerie, M., 2012. New insights into the high-pressure processing of meat and meat products. Compr. Rev. Food Sci. Food Saf. 11 (3), 285–306. https://doi.org/10.1111/j.1541-4337.2012.00184.x.

Slieman, T.A., Nicholson, W.L., 2000. Artificial and solar UV radiation induces strand breaks and cyclobutane dimers in *Bacillus subtilis* spore DNA. Appl. Environ. Microbiol. 66 (1), 199–205.

Smith, G., 2016. UV Intensity and UV Dose. http://www.fluidquip.com.au/contractors/uv-intensity-and-uv-dose/. [(Accessed August 23, 2018)].

Smith, J.S., Pillai, S., 2004. Irradiation and food safety. Food Technol. 58 (11), 48–55.

Smith, N.B., Cannon, J.E., Novakofski, J.E., McKeith, F.K., O'Brien, W.D., 1991. Tenderization of semitendinosus muscle using high intensity ultrasound. Ultrasonics Symposium Proceedings IEEE, 1371–1374.

Solomon, M.B., Long, J.B., Eastridge, J.S., 1997. The hydrodyne—A new process to improve beef tenderness. J. Anim. Sci. 75, 1534–1537.

Solomon, M.B., Carpenter, C.E., Snowder, G.D., Cockett, N.E., 1998. Tenderizing callipyge lamb with the hydrodyne process and electrical stimulation. J. Muscle Foods 9, 305–311. https://doi.org/10.1111/j.1745-4573.1998.tb00663.x.

Spilimbergo, S., Elvassore, N., Bertucco, A., 2002. Microbial inactivation by high pressure. J. Supercrit. Fluids 22 (1), 55–63. https://doi.org/10.1016/S0896-8446(01)00106-1.

Spilimbergo, S., Elvassore, N., Bertucco, A., 2003a. Inactivation of microorganisms by supercritical $CO_2$ in a semi-continuous process. Ital. J. Food Sci. 15, 115–124.

Spilimbergo, S., Dehghani, F., Bertucco, A., Foster, N.R., 2003b. Inactivation of bacteria and spores by pulse electric field and high pressure $CO_2$ at low temperature. Biotechnol. Bioeng. 82 (1), 118–125. https://doi.org/10.1002/bit.10554.

Spilimbergo, S., Cappelletti, M., Ferrentino, G., 2014. High pressure carbon dioxide combined with high power ultrasound processing of dry cured ham spiked with *Listeria monocytogenes*. Food Res. Int. 66, 264–273. https://doi.org/10.1016/j.foodres.2014.09.024.

Stermer, R.A., Lasater-Smith, M., Brasington, C., 1987. Ultraviolet radiation—An effective bactericide for fresh meat. J. Food Prot. 20, 108–111. https://doi.org/10.4315/0362-028X-50.2.108.

Surowsky, B., Fischer, A., Schlüter, O., Knorr, D., 2013. Cold plasma effects on enzyme activity in a model food system. Innov. Food Sci. Emerg. Technol. 19, 146–152. https://doi.org/10.1016/j.ifset.2013.04.002.

Takai, E., Kitano, K., Kuwabara, J., Shiraki, K., 2012. Protein inactivation by low-temperature atmospheric pressure plasma in aqueous solution. Plasma Processes Polym. 9 (1), 77–82. https://doi.org/10.1002/ppap.201100063.

Takeshita, K., Shibato, J., Sameshima, T., Fukunaga, S., Isobe, S., Arihara, K., Itoh, M., 2003. Damage of yeast cells induced by pulsed light irradiation. Int. J. Food Microbiol. 85, 151–158. https://doi.org/10.1016/S0168-1605(02)00509-3.

Tappi, S., Berardinelli, A., Ragni, L., Dalla Rosa, M., Guarnieri, A., Rocculi, P., 2014. Atmospheric gas plasma treatment of fresh-cut apples. Innov. Food Sci. Emerg. Technol. 21, 114–122. https://doi.org/10.1016/j.ifset.2013.09.012.

Tappi, S., Gozzi, G., Vannini, L., Berardinelli, A., Romani, S., Ragni, L., Rocculi, P., 2016. Cold plasma treatment for fresh-cut melon stabilization. Innov. Food Sci. Emerg. Technol. 33, 225–233. https://doi.org/10.1016/j.ifset.2015.12.022.

Tassou, C.C., Galiatsatou, P., Samaras, F.J., Mallidis, C.G., 2007. Inactivation kinetics of a piezotolerant *Staphylococcus aureus* isolated from high-pressure-treated sliced ham by high pressure in buffer and in a ham model system: evaluation in selective and non-selective medium. Innov. Food Sci. Emerg. Technol. 8 (4), 478–484. https://doi.org/10.1016/j.ifset.2007.04.002.

Teixeira, B., Fidalgo, L., Mendes, R., Costa, G., Cordeiro, C., Marques, A., Saraiva, J.A., Nunes, M.L., 2014. Effect of high pressure processing in the quality of sea bass (*Dicentrarchus labrax*) fillets: pressurization rate, pressure level and holding time. Innov. Food Sci. Emerg. Technol. 22, 31–39. https://doi.org/10.1016/j.ifset.2013.12.005.

Thirumdas, R., Sarangapani, C., Annapure, U.S., 2015. Cold plasma: a novel non-thermal technology for food processing. Food Biophys. 10 (1), 1–11. https://doi.org/10.1007/s11483-014-9382-z.

Tholozan, J.L., Ritz, M., Jugiau, F., Federighi, M., Tissier, J.P., 2000. Physiological effects of high hydrostatic pressure treatments on *Listeria monocytogenes* and *Salmonella typhimurium*. J. Appl. Microbiol. 88 (2), 202–212. https://doi.org/10.1046/j.1365-2672.2000.00917.x.

Tobin, J., Heffernan, S.P., Mulvihill, D.M., Huppertz, T., Kelly, A.L., 2015. Applications of high-pressure homogenization and microfluidization for milk and dairy products. In: Datta, N., Tomasula, P.M. (Eds.), Emerging Dairy Processing Technologies: Opportunities for the Dairy Industry. Wiley-Blackwell, Ames, IA, pp. 93–114.

Tomasula, P.M., Renye, J.A., Van Hekken, D.L., Tunick, M.H., Kwoczak, R., Toht, M., Leggett, L.N., Luchansky, J.B., Porto-Fett, A.C.S., Phillips, J.G., 2014. Effect of high-pressure processing on reduction of *Listeria monocytogenes* in packaged Queso Fresco. J. Dairy Sci. 97 (3), 1281–1295. https://doi.org/10.3168/jds.2013-7538.

Torres, J.A., Saraiva, J.A., Rodríguez, E.G., Aubourg, S.P., Vázquez, M., 2014. Effect of combining high-pressure processing and frozen storage on the functional and sensory properties of horse mackerel (*Trachurus trachurus*). Innov. Food Sci. Emerg. Technol. 21, 2–11. https://doi.org/10.1016/j.ifset.2013.12.001.

Tran, M., Roberts, R., Felix, T.L., Harte, F.M., 2018. Effect of high-pressure-jet processing on the viscosity and foaming properties of pasteurized whole milk. J. Dairy Sci. 101 (5), 3887–3899. https://doi.org/10.3168/jds.2017-14103.

Tribst, A.A.L., Ribeiro, L.R., Cristianini, M., 2017. Comparison of the effects of high pressure homogenization and high pressure processing on the enzyme activity and antimicrobial profile of lysozyme. Innov. Food Sci. Emerg. Technol. 43, 60–67. https://doi.org/10.1016/j.ifset.2017.07.026.

Tsuji, M., Sato, Y., Komiyama, Y., 2005. Inactivation of microorganisms and enzymes in juices by supercritical carbon dioxide method with continous flow system. Nippon Shokuhin Kagaku Kogaku Kaishi 52 (11), 528–531. https://doi.org/10.3136/nskkk.52.528.

Turtoi, M., Nicolau, A., 2007. Intense light pulse treatment as alternative method for mould spores destruction on paper–polyethylene packaging material. J. Food Eng. 83 (1), 47–53. https://doi.org/10.1016/j.jfoodeng.2006.11.017.

U.S. Food and Drug Administration, 2004. Guidance for Industry: Juice HACCP Hazards and Controls Guidance First Edition. https://www.fda.gov/food/guidanceregulation/ucm072557.htm. [(Accessed July 23, 2018)].

Ucak, I., Gokoglu, N., Kiessling, M., Toepfl, S., Galanakis, C.M., 2018. Inhibitory effects of high pressure treatment on microbial growth and biogenic amine formation in marinated herring (*Clupea harengus*) inoculated with *Morganella psychrotolerans*. LWT- Food Sci. Technol. https://doi.org/10.1016/j.lwt.2018.09.058. accepted.

Udagawa, Y., Suzuki, M., 2013. A study on treatment of microorganisms using underwater shock waves by gas imploding detonation. Trans. Jpn. Soc. Mech. Eng., B 79 (808), 2818–2825. https://doi.org/10.1299/kikaib.79.2818.

Ulbin-Figlewicz, N., Brychcy, E., Jarmoluk, A., 2015a. Effect of low-pressure cold plasma on surface microflora of meat and quality attributes. J. Food Sci. Technol. 52, 1228–1232. https://doi.org/10.1007/s13197-013-1108-6.

Ulbin-Figlewicz, N., Jarmoluk, A., Marycz, K., 2015b. Antimicrobial activity of low-pressure plasma treatment against selected foodborne bacteria and meat microbiota. Ann. Microbiol. 65 (3), 1537–1546. https://doi.org/10.1007/s13213-014-0992-y.

Uysal, P.C., Toklucu, A.K., 2011. Effect of ultravoilet-C light on anthocyanin content and other quality parameters of pomegranate juice. J. Food Compos. Anal. 24 (6), 790–795. https://doi.org/10.1016/j.jfca.2011.01.003.

Vallverdú-Queralt, A., Odriozola-Serrano, I., Oms-Oliu, G., Lamuela-Raventós, R.M., Elez-Martínez, P., Martín-Belloso, O., 2012. Changes in the polyphenol profile of tomato juices processed by pulsed electric fields. J. Agric. Food Chem. 60 (38), 9667–9672. https://doi.org/10.1021/jf302791k.

Vallverdú-Queralt, A., Odriozola-Serrano, I., Oms-Oliu, G., Lamuela-Raventós, R.M., Elez-Martínez, P., Martín-Belloso, O., 2013. Impact of high-intensity pulsed electric fields on carotenoids profile of tomato juice made of moderate-intensity pulsed electric field-treated tomatoes. Food Chem. 141 (3), 3131–3138. https://doi.org/10.1016/j.foodchem.2013.05.150.

Valverde, M.T., Marin-Iniesta, F., Calvo, L., 2010. Inactivation of *Saccharomyces cerevisiae* in conference pear with high pressure carbon dioxide and effects on pear quality. J. Food Eng. 98 (4), 421–428. https://doi.org/10.1016/j.jfoodeng.2010.01.022.

Vasileva, I., Denkova, R., Chochkov, R., Teneva, D., Denkova, Z., Dessev, T., Denev, P., Slavov, A., 2018. Effect of lavender (*Lavandula angustifolia*) and melissa (*Melissa Officinalis*) waste on quality and shelf life of bread. Food Chem. 253, 13–21. https://doi.org/10.1016/j.foodchem.2018.01.131.

Vercammen, A., Vivijs, B., Lurquin, I., Michiels, C.W., 2012. Germination and inactivation of *Bacillus coagulans* and *Alicyclobacillus acidoterrestris* spores by high hydrostatic pressure treatment in buffer and tomato sauce. Int. J. Food Microbiol. 152 (3), 162–167. https://doi.org/10.1016/j.ijfoodmicro.2011.02.019.

Villamiel, M., de Jong, P., 2000. Inactivation of *Pseudomonas fluorescens* and *Streptococcus thermophilus* in trypticase soy broth and total bacteria in milk by continuous-flow ultrasonic treatment and conventional heating. J. Food Eng. 45 (3), 171–179. https://doi.org/10.1016/S0260-8774(00)00059-5.

Vivek, K., Subbarao, K.V., Srivastava, B., 2016. Optimization of postharvest ultrasonic treatment of kiwifruit using RSM. Ultrason. Sonochem. 32, 328–335. https://doi.org/10.1016/j.ultsonch.2016.03.029.

Walkling-Ribeiro, M., Noci, F., Riener, J., Cronin, D.A., Lyng, J.G., Morgan, D.J., 2009. The impact of thermosonication and pulsed electric fields on *Staphylococcus aureus* inactivation and selected quality parameters in orange juice. Food Bioproc. Tech. 2, 422–430. https://doi.org/10.1007/s11947-007-0045-7.

Wang, T., MacGregor, S.J., Anderson, J.G., Woolsey, G.A., 2005. Pulsed ultra-violet inactivation spectrum of *Escherichia coli*. Water Res. 39 (13), 2921–2925. https://doi.org/10.1016/j.watres.2005.04.067.

Wang, J., Hu, X., Wang, Z., 2010. Kinetics models for the inactivation of *Alicyclobacillus acidiphilus* DSM14558 T and *Alicyclobacillus acidoterrestris* DSM 3922 T in apple juice by ultrasound. Int. J. Food Microbiol. 139 (3), 177–181. https://doi.org/10.1016/j.ijfoodmicro.2010.02.029.

Wang, R.X., Nian, W.F., Wu, H.Y., Feng, H.Q., Zhang, K., Zhang, J., Zhu, W.D., Becker, K.H., Fang, J., 2012. Atmospheric-pressure cold plasma treatment of contaminated fresh fruit and vegetable slices: inactivation and physiochemical properties evaluation. Eur. Phys. J. D Atom. Mol. Opt. Phys. 66, https://doi.org/10.1140/epjd/e2012-30053-1. 276(7 pp).

Wang, C.Y., Huang, H.W., Hsu, C.P., Shyu, Y.T., Yang, B.B., 2013. Inactivation and morphological damage of *Vibrio parahaemolyticus* treated with high hydrostatic pressure. Food Control 32, 348–353. https://doi.org/10.1016/j.foodcont.2012.12.023.

Wang, L.H., Wang, M.S., Zeng, X.A., Zhang, Z.H., Gong, D.M., Huang, Y.B., 2016. Membrane destruction and DNA binding of *Staphylococcus aureus* cells induced by carvacrol, and its combined effect with pulsed electric field. J. Agric. Food Chem. 64 (32), 6355–6363. https://doi.org/10.1021/acs.jafc.6b02507.

Warner, R.D., McDonnell, C.K., Bekhit, A.E.D., Claus, J., Vaskoska, R., Sikes, A., Dunshea, F.R., Ha, M., 2017. Systematic review of emerging and innovative technologies for meat tenderisation. Meat Sci. 132, 72–89. https://doi.org/10.1016/j.meatsci.2017.04.241.

Wekhof, A., 2000. Disinfection with flash lamps. PDA J. Pharm. Sci. Technol. 54 (3), 264–276.

Wemekamp-Kamphuis, H.H., Wouters, J.A., de Leeuw, P.P., Hain, T., Chakraborty, T., Abee, T., 2004. Identification of sigma factor $\sigma^B$-controlled genes and their impact on acid stress, high hydrostatic pressure, and freeze survival in *Listeria monocytogenes* EGD-e. Appl. Environ. Microbiol. 70 (6), 3457–3466. https://doi.org/10.1128/AEM.70.6.3457-3466.2004.

White, A., Burns, D., Christensen, T.W., 2006. Effective terminal sterilization using supercritical carbon dioxide. J. Biotechnol. 123 (4), 504–515. https://doi.org/10.1016/j.jbiotec.2005.12.033.

Wilson, D.R., Dabrowski, L., Stringer, S., Moezelaar, R., Brocklehurst, T.F., 2008. High pressure in combination with elevated temperature as a method for the sterilisation of food. Trends Food Sci. Technol. 19, 289–299. https://doi.org/10.1016/j.tifs.2008.01.005.

Wimalaratne, S.K., Farid, M.M., 2008. Pressure assisted thermal sterilization. Food Bioprod. Process. 86, 312–316. https://doi.org/10.1016/j.fbp.2007.08.001.

Wong, P.Y.Y., Kitts, D.D., 2001. Factors influencing ultraviolet and electron beam irradiation-induced free radical damage of ascorbic acid. Food Chem. 74 (1), 75–84. https://doi.org/10.1016/S0308-8146(01)00101-7.

Wordon, B.A., Mortimer, B., Mcmaster, L.D., 2012. Comparative real-time analysis of *Saccharomyces cerevisiae* cell viability, injury and death induced by ultrasound (20 kHz) and heat for the application of hurdle technology. Food Res. Int. 47 (2), 134–139. https://doi.org/10.1016/j.foodres.2011.04.038.

Wu, Y., Yao, S.J., Guan, Y.X., 2007. Inactivation of microorganisms in carbon dioxide at elevated pressure and ambient temperature. Ind. Eng. Chem. Res. 46 (19), 6345–6352. https://doi.org/10.1021/ie0702330.

Wuytack, E.Y., Phuong, L.D.T., Aertsen, A., Reyns, K.M.F., Marquenie, D., De Ketelaere, B., Masschalck, B., Van Opstal, I., Diels, A.M.J., Michiels, C.W., 2003. Comparison of sublethal injury induced in *Salmonella enterica* serovar Typhimurium by heat and by different nonthermal treatments. J. Food Prot. 66 (1), 31–37.

Xu, Z., Lina, T., Wang, Y., Liao, X., 2015. Quality assurance in pepper and orange juice blend treated by high pressure processing and high temperature short time. Innov. Food Sci. Emerg. Technol. 31, 28–36. https://doi.org/10.1016/j.ifset.2015.08.001.

Yaldagard, M., Mortazavi, S.A., Tabatabaie, F., 2008. The principles of ultra high pressure technology and its application in food processing/preservation: a review of microbiological and quality aspects. Afr. J. Biotechnol. 7 (16), 2739–2767.

Yanakiev, C., 2017. Application of Cold Plasma for the Treatment of Food Contact Surfaces, PhD Thesis, University of Food Technologies, Plovdiv, p.150.

Yang, W.W., Chung, S.Y., Ajayi, O., Krishnamurthy, K., Konan, K., Goodrich-Schneider, R., 2010. Use of pulsed ultraviolet light to reduce the allergenic potency of soybean extracts. Int. J. Food Eng. 6 (3), 11–14. https://doi.org/10.2202/1556-3758.1876.

Yang, W.W., Mwakatage, N.R., Goodrich-Schneider, R., Krishnamurthy, K., Rababah, T.M., 2012. Mitigation of major peanut allergens by pulsed ultraviolet light. Food Bioproc. Tech. 5 (7), 2728–2738. https://doi.org/10.1007/s11947-011-0615-6.

Yang, H., Han, M., Bai, Y., Han, Y., Xu, X., Zhou, G., 2015. High pressure processing alters water distribution enabling the production of reduced-fat and reduced-salt pork sausages. Meat Sci. 102, 69–78. https://doi.org/10.1016/j.meatsci.2014.10.010.

Yang, H., Khan, M.A., Yu, X., Zheng, H., Han, M., Xu, X., Zhou, G., 2016. Changes in protein structures to improve the rheology and texture of reduced-fat sausages using high pressure processing. Meat Sci. 121, 79–87. https://doi.org/10.1016/j.meatsci.2016.06.004.

Yao, J., Zhou, B., Wang, R., Wang, T., Hu, X., Liao, X., Zhang, Y., 2015. Inactivation of *Staphylococcus aureus* by high hydrostatic pressure in saline solution and meat slurry with different initial inoculum levels. Food Bioprod. Process. 94, 592–600. https://doi.org/10.1016/j.fbp.2014.06.005.

Youn, K.S., Sagong, H.G., Choi, H.S., Ryu, S., Kang, D.H., 2011. Radio-frequency heating to inactivate *Salmonella typhimurium* and *Escherichia coli* O157:H7 on black and red pepper spice. Int. J. Food Microbiol 153, 171–175. https://doi.org/10.1016/j.ijfoodmicro.2011.11.004.

Young, S., Basiana, E., Nitin, N., 2018. Effects of interfacial composition on the stability of emulsion and encapsulated bioactives after thermal and high pressure processing. J. Food Eng. 231, 22–29. https://doi.org/10.1016/j.jfoodeng.2018.02.022.

Yusupov, M., Bogaerts, A., Huygh, S., Snoeckx, R., van Duin, A.C.T., Neyts, E.C., 2013. Plasma-induced destruction of bacterial cell wall components: a reactive molecular dynamics simulation. J. Phys. Chem. C 117 (11), 5993–5998. https://doi.org/10.1021/jp3128516.

Zhang, J., Burrows, S., Gleason, C., Matthews, M.A., Drews, M.J., LaBerge, M., An, Y.H., 2006. Sterilizing *Bacillus pumilus* spores using supercritical carbon dioxide. J. Microbiol. Methods 66 (3), 479–485. https://doi.org/10.1016/j.mimet.2006.01.012.

Zhang, Y., Liu, X.C., Wang, Y., Zhao, F., Sun, Z., Liao, X., 2016. Quality comparison of carrot juices processed by high-pressure processing and high-temperature short-time processing. Innov. Food Sci. Emerg. Technol. 33, 135–144. https://doi.org/10.1016/j.ifset.2015.10.012.

Zhao, L., Wang, Y., Hu, X., Sun, Z., Liao, X., 2016. Korla pear juice treated by ultrafiltration followed by high pressure processing or high temperature short time. LWT- Food Sci. Technol. 65, 283–289. https://doi.org/10.1016/j.lwt.2015.08.011.

Zhong, Q., Black, D.G., Davidson, P.M., Golden, D.A., 2008. Non thermal inactivation of *Escherichia coli* K-12 on spinach leaves, using dense phase carbon dioxide. J. Food Prot. 71 (5), 1015–1017.

Zhou, L., Zhang, Y., Hu, X., Liao, X., He, J., 2009a. Comparison of the inactivation kinetics of pectin methylesterases from carrot and peach by high-pressure carbon dioxide. Food Chem. 115 (2), 449–455. https://doi.org/10.1016/j.foodchem.2008.12.028.

Zhou, L., Wang, Y., Hu, X., Wu, J., Liao, X., 2009b. Effect of high pressure carbon dioxide on the quality of carrot juice. Innov. Food Sci. Emerg. Technol. 10 (3), 321–327. https://doi.org/10.1016/j.ifset.2009.01.002.

Ziuzina, D., Patil, S., Cullen, P.J., Keener, K.M., Bourke, P., 2014. Atmospheric cold plasma inactivation of *Escherichia coli, Salmonella enterica serovar typhimurium* and *Listeria monocytogenes* inoculated on fresh produce. Food Microbiol. 42, 109–116. https://doi.org/10.1016/j.fm.2014.02.007.

Ziuzina, D., Los, A., Bourke, P., 2018. Inactivation of *Staphylococcus aureus* in foods by thermal and nonthermal control strategies. In: Fetsch, A. (Ed.), Staphylococcus aureus. Academic Press, London, pp. 235–255. https://doi.org/10.1016/B978-0-12-809671-0.00012-7.

Zuckerman, H., Bowker, B.C., Eastridge, J.S., Solomon, M.B., 2013. Microstructure alterations in beef intramuscular connective tissue caused by hydrodynamic pressure processing. Meat Sci. 95 (3), 603–607. https://doi.org/10.1016/j.meatsci.2013.05.041.

## Further reading

European Commission, Statement of the Scientific Committee on Food on a Report on 2-Alkylcyclo-Butanones. (expressed on 3 July2012) 2002; https://ec.europa.eu/food/sites/food/files/safety/docs/sci-com_scf_out135_en.pdf (Accessed 30 August 2018).

Solomon, M.B., Sharma, M., Patel, J.R., 2010. Hydrodynamic pressure processing of meat products. In: Zhang, H.Q., Barbosa-Cánovas, G.V., Balasubramaniam, V.M.B., Dunne, C.P., Farkas, D.F., Yuan, J.T.C. (Eds.), Nonthermal Processing Technologies for Food. Wiley-Blackwell, Chichester, West Sussex, pp. 98–108.

Suslick, K.S., 1989. The chemical effects of ultrasound. Sci. Am. 260 (2), 80–86.

# Natural antimicrobial agents to improve foods shelf life

4

*Luz H. Villalobos-Delgado*, G.V. Nevárez-Moorillon†, Irma Caro‡, Emiliano J. Quinto‡, Javier Mateo§*

*Institute of Agroindustry, Technological University of the Mixteca, Huajuapan de León, Oaxaca, Mexico, †Faculty of Chemical Sciences, Autonomous University of Chihuahua, Chihuahua, Mexico, ‡Department of Food Science and Nutrition, Faculty of Medicine, University of Valladolid, Valladolid, Spain, §Department of Food Hygiene and Food Technology, University of León, León, Spain

## Chapter outline



## 1 Introduction

Food spoilage includes physical damage and chemical changes, such as lipid oxidation and color changes, or the appearance of off-flavors and off-odors resulting from microbial growth and the product's associated metabolism (Zheng et al., 2013). Pathogens, especially bacteria, are responsible for causing a wide variety of infections and intoxications through contaminated foods that result in moderate to severe illness and deaths (Fu et al., 2016). Therefore, foodborne illnesses are a major concern for consumers, the food industry, and food safety authorities (Gram et al., 2002).

Different methods, including thermal processing, lowering of water activity, various methods of packaging, irradiation, plasma processing, high pressure, high-intensity

Food Quality and Shelf Life. https://doi.org/10.1016/B978-0-12-817190-5.00004-5

pulsed electric field processing, and the addition of chemical preservatives have been utilized to produce safer foodstuffs (Fu et al., 2016; Khorshidian et al., 2018; Tajkarimi et al., 2010). Nonetheless, control of food spoilage and pathogenic bacteria has been achieved mainly by heat treatment and the addition of chemicals such as synthetic antimicrobials (Calo et al., 2015; Zheng et al., 2013). Among the chemical antimicrobials, several organic acids and salts could be included, such as sodium benzoates and propionates, potassium sorbates, parabens, sorbic acid, sulphites, sodium nitrites and nitrates, chlorides, phosphates, sulfur dioxide, ascorbic acid, citric acid, tartaric acid, etc. (Carocho et al., 2015; Gutiérrez-del-Río et al., 2018; Pisoschi et al., 2018). However, the use of these chemicals is limited due to undesirable effects including carcinogenicity, acute toxicity, teratogenicity, and slow degradation periods, which could lead to environmental problems such as pollution (Calo et al., 2015). For example, sulphites cause degradation of vitamin B1 (thiamine) in food (Gutiérrez-del-Río et al., 2018) as well as causing various allergic reactions in the case of sulfite hypersensitive consumers (Pisoschi et al., 2018), while sodium sorbate is genotoxic on in vitro blood lymphocytes, depending on the dose used (Carocho et al., 2015).

In recent years, considerable effort has been made to find natural antimicrobials that can inhibit bacterial and fungal growth in foods in order to improve quality and shelf life. This, in turn, has led to a search for antimicrobials derived from a variety of natural sources (Gyawalli and Ibrahim, 2014). Natural antimicrobials are produced and isolated from various sources, namely plant and animal materials, bacteria, algae, and mushrooms (Gyawalli and Ibrahim, 2014; Pisoschi et al., 2018). Among them, high relevance has been recognized for lysozyme, lactoferrin, and other iron-binding proteins, bacteriocins, natamycin, spices, and their essential oils (EOs), onions and garlic antimicrobials, isothiocyanates, and phenolic compounds. Moreover, it should also be noted that by-products generated from fruits and vegetables, as well as crustacean waste such as chitosan, can also be used as sources of antibacterial compounds (Gyawalli and Ibrahim, 2014; Khorshidian et al., 2018; Nazzaro et al., 2018; Pisoschi et al., 2018; Tajkarimi et al., 2010), as the waste by-products often present equivalent or even higher bioactive compounds contents than the final product itself (Silva et al., 2014). Several reports have examined the microbiological activities of a small number of secondary metabolites usually present in plant-food by-products, such as tannins, terpenoids, alkaloids, and phenolics (Guil-Guerrero et al., 2016).

On the other hand, the incorporation of antimicrobials in food systems can be performed in different forms to control the growth of pathogenic and spoilage microorganisms. The natural antimicrobials can be added directly to the food, or incorporated in edible films and coatings or plastic packaging, so that they are released during transportation and/or storage of food and contribute to increasing food shelf life (Ayala-Zavala et al., 2013; Campos et al., 2011; Gyawalli and Ibrahim, 2014; Hamed et al., 2016; Lopes and Brandelli, 2017; Ribeiro-Santos et al., 2017; Shah et al., 2012; Shokri and Ehsani, 2017; Sokovic et al., 2010; Zou et al., 2012). Thus, antimicrobials originating from natural sources, including their by-products, provide new modalities to ensure microbial safety and increase the shelf life of foodstuffs (Gyawalli and Ibrahim, 2014; Pisoschi et al., 2018). In addition to improving shelf life and food safety, natural antimicrobial agents may allow novel food products with enhanced quality and nutritional properties to be introduced to the market (Tiwari et al., 2009).

This chapter describes the range of types of natural antimicrobials with potential to be used for food preservation and their mechanism of action. The application of natural antimicrobials in food is then explained. Finally, the limitations and perspectives of the use of natural antimicrobials as food preservatives are discussed.

# 2 Types of natural antimicrobials and their mechanisms of action

Biopreservation is focused on the utilization of natural preservatives from sources such as bacteria, fungi, algae, plants, and animals, with the ability to ensure food safety due to their antimicrobial activity exerted against a wide spectrum of foodborne pathogens (Calo et al., 2015; Pisoschi et al., 2018). Moreover, secondary metabolites with antimicrobial activity are produced by fruits, vegetables, seeds, herbs and spices, milk, eggs, animal tissues, and microorganisms such as bacteria and fungi (Fig. 1) (Pisoschi et al., 2018).

## 2.1 *Animal sources*

### 2.1.1 *Lactoferrin*

Bovine lactoferrin (Lf), which shares about 70% sequence homology with human Lf, is a multifunctional protein from milk involved in bacteriostatic, bactericidal, and fungistatic activities (Liu et al., 2018). The antimicrobial properties of Lf against foodborne microorganisms such as *Listeria monocytogenes*, enteropathogenic *Escherichia coli*, and *Klebsiella pneumoniae* have been reported.

There are several explanations that may account for Lf's antimicrobial action. One is that it limits microbial access to iron via Fe-chelation, which thereby produces an iron deficient medium. Another possibility is its capability of destabilizing the outer membrane of Gram-negative bacteria, thus increasing the membrane permeability, which results in the liberation of lipopolysaccharides (Gyawalli and Ibrahim, 2014; Hugo and Hugo, 2015). Additionally, the antimicrobial effect of Lf can be potentiated in food containing *Bifidobacterium* spp. strains. Under low iron concentrations, these probiotic bacteria appear to be better at iron scavenging than pathogenic bacteria and, therefore, contribute to limit the already low Fe availability (Gyawalli and Ibrahim, 2014).

### 2.1.2 *Lysozyme*

Lysozyme is an enzyme that is naturally present in avian eggs and mammalian milk. The lysozyme used for direct addition to foods is derived from eggs (Carocho et al., 2015; Gyawalli and Ibrahim, 2014). The antimicrobial activity of lysozyme is due to its ability to hydrolyze the β-1,4 linkage between *N*-acetylmuramic acid and *N*-acetyl glucosamine present in peptidoglycan of the microbial cell wall (Barbiroli et al., 2012;

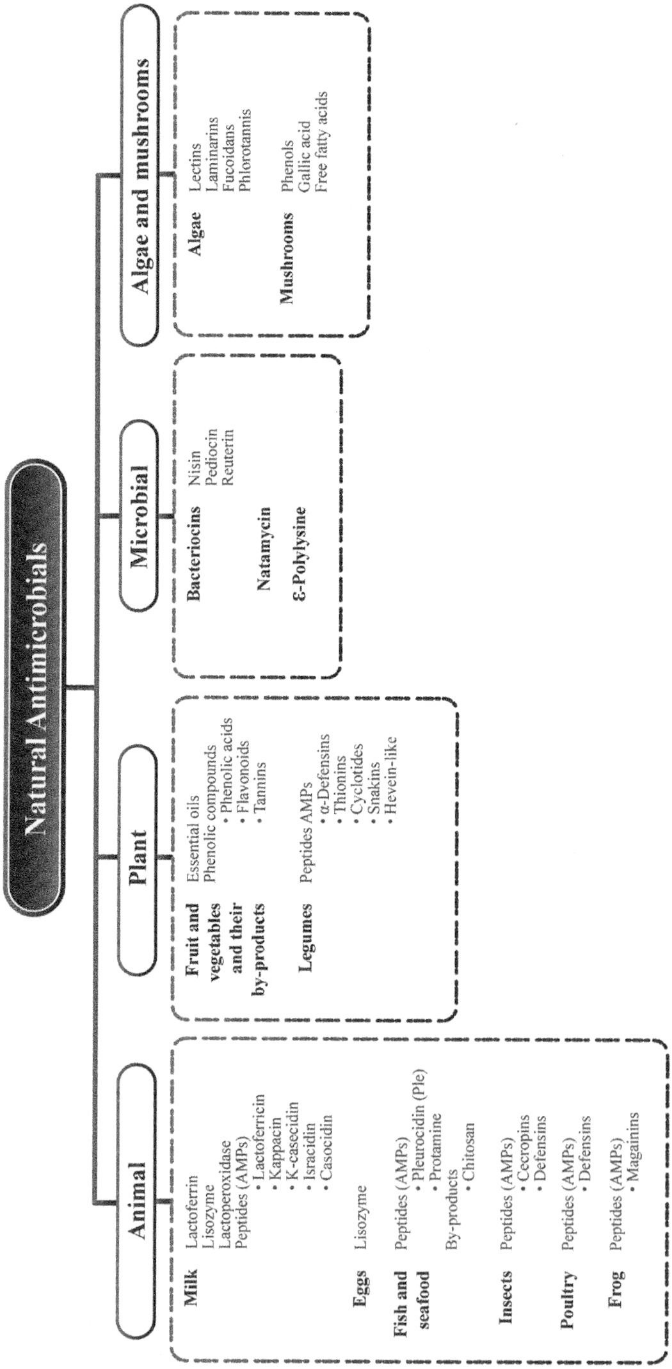


**Fig. 1** Natural antimicrobials with potential application in food systems.

Gyawali and Ibrahim, 2014). Gram-positive bacteria are very susceptible to lysozyme, as peptidoglycan can represent up to 90% of their cell wall. In Gram-negatives, peptidoglycan counts for only for 5%–10% of the cell wall and lies beneath the outer membrane of the cell envelope. Nevertheless, lysozyme is more effective against Gram-negative bacteria when used in combination with detergents and chelators (e.g., EDTA), nisin, and lactoferrin, as these compounds can initiate membrane disruption (Barbiroli et al., 2012; Hugo and Hugo, 2015; Pisoschi et al., 2018). The antimicrobial spectrum of lysozyme can be broadened, for example, by pretreating the target microbial cell in a number of ways. Conjugation of lysozyme with dextran increased its activity against Gram-positive and Gram-negative bacteria, particularly if the temperature was raised during treatment. Freeze-thaw treatments sensitized *E. coli* to lysozyme. Gram-negative bacteria, including salmonella, become lysozyme-sensitive following an osmotic downshift (e.g., achieved by sudden dilution of a salt solution) (Gould, 1996).

### 2.1.3 Lactoperoxidase

Lactoperoxidase belongs to the peroxidases family, widespread throughout nature. It is an enzyme that occurs naturally in milk and other secretions such as saliva and tears (Aprodu et al., 2014; Montiel et al., 2016). The lactoperoxidase system (LPS) consists of three components: lactoperoxidase, thiocyanate, and hydrogen peroxide ($H_2O_2$). The enzyme catalyzes the oxidation of thiocyanate ($SCN^-$) by the use of $H_2O_2$ and produces hypothiocyanite ($OSCN^-$) and hypothiocyanous acid (HOSCN). These products inhibit microorganisms by the oxidation of sulphydryl (–SH) groups in their enzyme systems and proteins (Campos et al., 2011). This system has been shown to be bactericid or bacteriostatic for a wide range of spoilage and food-poisoning microorganisms, including species of *Listeria*, *Staphylococcus*, *Campylobacter*, *Salmonella*, and *Bacillus cereus*, as well as some yeasts. Consequently, the use of LPS has been promoted as an effective method of extending the shelf life of raw milk, especially in places where refrigeration is not readily available. Moreover, the effectiveness of incorporation of LPS into whey proteins and alginate films has been demonstrated (Aprodu et al., 2014; Carocho et al., 2015; Cissé et al., 2015; Montiel et al., 2016).

### 2.1.4 Antimicrobial peptides (AMPs)

#### Pleurocidin

Pleurocidin (Ple), a novel antimicrobial peptide with 25 amino acids, is derived from the skin-secreted mucous of the winter flounder (*Pleuronectes americanus*) and exhibits broad antimicrobial activity against both Gram-positive and Gram-negative bacteria (Tiwari et al., 2009). Additionally, the benefits of insensitivity to salt, adaptation in a wide range of pH, thermal resistance, low cytotoxicity, and hemolytic activity allow Ple to be recognized as a potentially excellent preservative for various food systems (Wang et al., 2015). Ple has been found to be effective against foodborne organisms including *Vibrio parahaemolyticus*, *L. monocytogenes*, *E. coli* O157:H7, and against yeast and molds, such as *Saccharomyces cerevisiae*, and *Penicillium expansum*.

As an action mechanism, Pleurocidin exhibits strong membrane translocation and pore-formation ability, reacting with both neutral and acidic anionic phospholipid membranes. However, magnesium and calcium can inhibit their antimicrobial activity, which may limit their use as an antibacterial peptide (Burrowes et al., 2004; Pisoschi et al., 2018; Rai et al., 2016).

### Protamine

Protamine is the most cationic naturally occurring cationic antimicrobial peptide (CAP). It consists of 31 amino acids with 20 arginine residues. The peptide may be isolated from spermatic cells of fish, birds, and mammals and is commercially recovered from herring (clupeine) and salmon (salmine) milt (Potter et al., 2005). Protamine has a broad range of antimicrobial activity against Gram-positive and Gram-negative bacteria, as well as yeasts and molds (Gill et al., 2006; Tiwari et al., 2009). Moreover, protamine is heat resistant and able to withstand steam sterilization without the loss of antimicrobial activity. It supposedly kills susceptible bacteria by cell envelope lysis and condensation of the cytoplasm. A number of Japanese patents exist for the use of protamine as food preservatives (Gill et al., 2006; Potter et al., 2005).

### Cecropins

Cecropins are positively charged AMPs that were originally isolated from insects and cecropia moths, while a mammalian homologue, cecropin P1 (CP1), was isolated from pig intestines. Insect cecropins are highly potent against both Gram-negative and Gram-positive bacteria (Fitriyanti and Narsimhan, 2018). The peptides are thought to integrate into the acidic cell membranes of bacteria, either by forming voltage-dependent ion channels or causing a general disruption of the membrane by a "carpet-like" mechanism (Ekengren and Hultmark, 1999; Li et al., 2012). Cecropin P1 is able destroy microorganisms by forming an α-helical chain when penetrating the cell membrane to make it leaky. Moreover, it is not cytotoxic to mammalian cells and can, therefore, find potential applications in food preservation, for example in food packaging (Fitriyanti and Narsimhan, 2018).

### Defensins

Defensins belong to cysteine-rich cationic peptides isolated from plants, mammals, and insects (Rai et al., 2016). Insect defensins are mainly effective against Gram-positive bacteria. They probably act through membrane permeabilization and their antimicrobial activity is salt-dependent (Faruck et al., 2016; Li et al., 2012). On the other hand, defensins produced by vertebrates such as mammalian phagocytes and epithelial cells of chickens, turkeys, etc., have been confirmed to possess antimicrobial potential exerted over Gram-positive and Gram-negative bacteria, fungi, and enveloped viruses (Pisoschi et al., 2018; Tiwari et al., 2009). Mammalian defensins participate in the nonoxidative microbicidal mechanisms through membrane permeabilization of ingested microorganisms (Hoffmann and Hetru, 1992).

### Magainins

Magainin is a polycationic peptide that has been isolated from frogs. This peptide consists of 23 amino acids and produces an amphipatic helical structure that appears to be responsible for the formation of transmembrane pores, resulting in an alteration of cell membrane permeability that leads to cell death. Magainin peptides have a broad spectrum of antimicrobial activity and low toxicity to normal eukaryotic cells (Cruz-Chamorro et al., 2006). It has been found to be effective against a range of food-related pathogens implying their application as food preservative (Tiwari et al., 2009), and has been used in meat and cheese (Rai et al., 2016).

### Milk-derived peptides

Lactoferricin, kappacin and κ-casecidin are examples of antimicrobial peptides derived from milk. They possess remarkable antibacterial activity and are, therefore, used as food preservatives (Rai et al., 2016). Lactoferricin, a peptide released by bovine lactoferrin pepsin digestion, acts on microbial cells by means that do not depend on iron binding. The antimicrobial potency of this peptide is much higher than that of an equimolar amount of intact bovine lactoferrin. A potential advantage of lactoferricin in food preservation is its relative heat resistance (Gould, 1996; Martinez-Hernández et al., 2017). Kappacin corresponds to the nonglycosylated, phosphorylated form of bovine caseinomacropeptide (CMP) f(106–169), which exhibit growth-inhibitory activity in Gram-positive and Gram-negative bacteria, while κ-casecidin from the tryptic digest of bovine κ-casein f(17–21) is able to inhibit the growth of pathogenic bacteria such as *Staphylococcu aureus*, *E. coli*, and *Salmonella typhimurium* (Sibel, 2014).

On the other hand, isracidin, that was the first antimicrobial peptide identified in the sequence of bovine $\alpha_{s1}$-casein (Sibel, 2014), has been found to inhibit the in vitro growth of lactobacilli, as well as being effective against *S. aureus*, *L. monocytogenes*, and *Candida albicans* (Gyawalli and Ibrahim, 2014; López and Recio, 2006). Casocidin is the hydrolysis product of $\alpha_{S2}$-casein, by the enzyme chymosin, found in rennet (Pisoschi et al., 2018). It is a heat and acid-stable peptide, corresponding to residues f150-188, isolated from boiled and acidified milk (Sibel, 2014). It possesses antibacterial activity, acting against *Staphylococcus* spp., *Sarcina* spp., *Bacillus subtilis*, *Diplococcus pneumonia*, and *Streptococcus pyogenes* (Gyawalli and Ibrahim, 2014; Pisoschi et al., 2018). Lactoferrampin, isolated as a fragment of lactoferrin, displays inhibitory activity against *Streptococcus mutans*, *E. coli*, *B. subtilis*, and *Pseudomonas aeruginosa* while isfracidin and lactoferrin B are effective against *Candida albicans* (Mohanty et al., 2016). Moreover, peptides released during the digestion of β-lactoglobulin with trypsin have demonstrated antimicrobial activity against foodborne pathogens such as *S. aureus*, *L. monocytogenes*, *Salmonella* spp., and *E. coli* O157 (Gyawalli and Ibrahim, 2014). Antimicrobial peptides generally have two common physical features: a cationic charge and a significant proportion of hydrophobic residues. In this way, most of the antimicrobial peptides form cationic amphipathic secondary structures, typically α-helices and β-sheet, which can selectivity interact with anionic bacterial membranes with the help of electrostatic interactions (Sibel, 2014).

## 2.2 Plant sources

Plants have been found to be rich sources of antimicrobials, many of which are phenolic compounds found in leaves, seeds, flowers, and bulbs (Fu et al., 2016) as well as in twigs, herbs, wood, fruits, and roots (Tajkarimi et al., 2010). Raw fruit, vegetables as well as legumes are a good source of polyphenols (Tylewicz et al., 2018) which have been extensively studied and associated with antimicrobial potential against a wide spectrum of microorganisms (Pina-Pérez and Ferrús, 2018).

### 2.2.1 Polyphenols as antimicrobials

Polyphenols are secondary metabolites of plants and are generally involved in the defense against ultraviolet radiation or aggression by pathogens (Pandey and Rizci, 2009). These are classified into two major classes: flavonoids and nonflavonoids. The latter includes structurally simple molecules such as phenolic acids among stand up derivatives of benzoic acid (e.g., gallic, protocatehuic, syringic, vanillic, hydroxybenzoic) and derivatives of cinnamic acid (e.g., caffeic, chlorogenic, ferulic, *p*-coumaric, sinapic), and highly complex molecules such as stilbenes (e.g., resveratrol), tannins (e.g., proanthocyanidins, gallotannins, and ellagitannins), and lignins (e.g., secoisolariciresinol). Flavonoids can be divided into six subclasses: flavonols, flavones, flavanones, flavanols, anthocyanins, and isoflavones. Quercetin, myricetin, catechins, etc., are some of the most common flavonoids (Daglia, 2012; Gutiérrez-del-Río et al., 2018; Khan et al., 2018; Pandey and Rizci, 2009).

Low-molecular-weight phenolic compounds are among the most common components of plant extracts with antimicrobial activity. These are generally recognized as being more efficient against Gram-positive than against Gram-negative (Gutiérrez-Larraínzar et al., 2012) and exert antimicrobial effects by the diffusion of the undissociated acid across the membrane, leading to the acidification of the cytoplasm and, in some cases, cell death. However, factors related to lipophilicity such as pH, ring substitutions (hydroxyl and methoxy groups), and the saturation of the side chain are determinants for activity of cinnamic acids. Among these, caffeic acid is reportedly active against viruses, bacteria, and fungi (Guil-Guerrero et al., 2016). Furthermore, direct antibacterial activity of flavonoids has been attributed to three mechanisms: cytoplasmic membrane damage (caused by perforation and/or a reduction in membrane fluidity), inhibition of nucleic acid synthesis (caused by topoisomerase inhibition), and inhibition of energy metabolism (caused by NADH-cytochrome *c* reductase inhibition) (Cushnie and Lamb, 2011).

Among polyphenols, flavan-3-ols, flavanols, and tannins have received most attention due to their wide spectrum and higher antimicrobial activity in comparison with other polyphenols. On the other hand, several mechanisms could explain the effect of the A-type proanthocyanidin in bacterial growth inhibition, such as the destabilization of the cytoplasmic membrane, the permeabilization of the cell membrane, the inhibition of extracellular microbial enzymes, direct actions on microbial metabolism, or the deprivation of the substrates required for microbial growth, especially essential mineral micronutrients such as iron and zinc (via proanthocyanidin chelation with metals), whose depletion can severely limit bacterial growth (Daglia, 2012). In general, among the mechanisms most accepted as being responsible for the antimicrobial capability of polyphenols are the following: destabilization of the cytoplasmic membrane, permeabilization of the cell membrane, deprivation of essential mineral micronutrients such as iron and zinc by means of chelating and inhibition of extracellular microbial enzymes, as well as direct interference in microbial metabolism (Pina-Pérez and Ferrús, 2018).

### 2.2.2 Essential oils (EOs)

EOs consist of a mixture of various compounds characterized by an aromatic smell, generally colorless liquid to slightly yellowish color, and are insoluble in water, but soluble in organic solvents. About 90%–95% of the whole EO comprises volatile components consisting of monoterpenes and sesquiterpene hydrocarbons and their oxygenated derivatives besides aliphatic aldehydes, alcohols, and esters. The nonvolatile portion constitutes about 5%–10% of the whole EO that includes mainly hydrocarbons, fatty acids, sterols, carotenoids, waxes, cumarines, and flavonoids. The most active antimicrobial compounds present in EOs can be divided into four groups: terpenes (e.g., *p*-cymene, limonene), terpenoids (e.g., thymol, carvacrol), phenylpropenes (e.g., eugenol, vanillin), and other compounds such as allicin or isothiocyanates (Khorshidian et al., 2018). Most EOs have wide antimicrobial spectra against bacteria, yeasts, and molds (Tiwari et al., 2009). Gram-positive bacteria are more susceptible to EOs than Gram-negative bacteria due to the direct interaction of cells with hydrophobic components of the EOs (Calo et al., 2015; Tajkarimi et al., 2010). The mechanism of action may relate to the ability of active compounds to alter microbial cell permeability, damage cytoplasmic membranes, interfere with the cellular energy (ATP) generation system, and disrupt the proton-motive force. The disrupted permeability of the cytoplasmic membrane can result in cell death (Calo et al., 2015; Fu et al., 2016; Pisoschi et al., 2018; Tiwari et al., 2009).

### 2.2.3 Plant antimicrobial peptides

Plant AMPs (pAMPs) are predominantly cysteine-rich compounds that have been extracted and isolated from different plants and plant organs such as stems, roots, seeds, flowers, and leaves (Tang et al., 2018). Based on amino acid sequence homology, AMPs are mostly classified as *a*-defensins, thionins, lipid transfer proteins, cyclotides, snakins, and hevein-like (Hintz et al., 2015; Rai et al., 2016; Said et al., 2018), which have a broad spectrum of activity against bacteria, fungi, and viruses (de Souza et al., 2014). Their antibacterial action involves cell membranes of targeted organisms and is driven by net positive charge, flexibility, and hydrophobicity to enable interaction with bacterial membranes (Hintz et al., 2015). AMPs in hydrolysate form can be used as a food additive for beverages and infant formulae, as a food texture enhancer, and as a pharmaceutical ingredient (Tang et al., 2018).

## 2.3 Microbial sources

### 2.3.1 Bacteriocins

Bacteriocins consist of ribosomally synthesized antimicrobial peptides or complex proteins secreted by various Gram-positive and Gram-negative bacteria (Ahmad et al., 2017). Among the Gram-positive bacteria, lactic acid bacteria (LAB) have gained particular attention, due to the production of bacteriocins. These tolerate high thermal stress, are active over a wide pH range, remain effective at fairly low concentrations (Woraprayote et al., 2016), and demonstrate the ability to preserve a food without affecting its sensory qualities. These can be broadly divided into two classes: class I, of which the lantibiotics (posttranslationally modified peptides containing unusual amino acids) are the best known example, and class II, containing unmodified peptides (O'Connor et al., 2015). Moreover, bacteriocins can kill bacteria or inhibit bacterial growth; however, they do not generally inhibit Gram-negative bacteria as they cannot penetrate the bacterial outer membrane (Fu et al., 2016). Their action mechanism is mainly due to disruption of the integrity of the cell wall or inhibition of protein or nucleic acid synthesis. Bacteriocins bind to cell wall components, including lipid or surface molecular binding sites, via specific or nonspecific receptor binding, which facilitates pore formation or direct cell lysis, resulting in cell death via dissipation of the proton motive force of a bacterial system (Ahmad et al., 2017). Although there are various bacteriocins in nature (e.g., nisin, pediocin, reuterin, plantaricin, helveticin, etc.), nisin is the only FDA registered GRAS bacteriocin extensively used in the food processing industry (Fu et al., 2016).

### Nisin

Nisin has been considered as a food additive in the UE (E-234) and as a constituent of coatings and films (Carocho et al., 2015). It is produced by strains of *Lactococcus lactis* subsp. *Lactis* and acts against Gram-positive (*S. aureus*, *Micrococcus luteus*) and spore-forming bacteria (*B. cereus* and *Clostridium*) (Hugo and Hugo, 2015), while Gram-negative bacteria are generally resistant as the outer membrane blocks the access of nisin to the cytoplasmic membrane (Modugno et al., 2018). There are two bactericidal mechanisms of nisin. One mechanism is that nisin forms pores in membranes that disrupt the proton-motive force and pH equilibrium, thus causing leakage of ions and hydrolysis of ATP leading to cell death. The other mechanism involves nisin interfering with the cell wall biosynthesis, thus preventing the use of lipid II from synthesizing peptidoglycans, the main component of membranes, and causing cell disruption and death (Balciunas et al., 2013; Liao et al., 2018). To date, eight natural nisin variants have been discovered (A, Z, F, Q, H, U, U2, and P) (Woraprayote et al., 2016).

### Pediocin

Pediocin is produced by *Pediococcus* strains such as *P. acidilactici* (Ach, PA-1, JD) and *P. pentosaceus* (A, N5p, ST18 and PD1) and has GRAS status in certain food applications (Hugo and Hugo, 2015; Tiwari et al., 2009). They are characterized by thermostability and functionality over a broad pH range. Pedicions exhibit efficiency against spoilage and pathogenic organisms, such as *L. monocytogenes*, *E. faecalis*, *S. aureus*, and *Clostridium perfringens* (Pisoschi et al., 2018). Pediocin PA-1 performs its function by permeabilizing the cytoplasmic membrane of the receptor bacteria, resulting in a leakage of ions and small molecules (Woraprayote et al., 2016).

### Reuterin

Reuterin or β-hydroxypropionaldehyde (β-HPA) is produced as an intermediate step by some *Lactobacillus reuteri* strains in the conversion of glycerol to 1,3-propanediol. It is water-soluble, active in a wide range of pH values and resistant to proteolytic and lipolytic enzymes, with a broad spectrum of antimicrobial activity towards a wide range of pathogens and food spoilage organisms including Gram-positive and Gram-negative bacteria, yeasts, and molds (Garde et al., 2016; Montiel et al., 2016). Reuterin induces oxidative stress in cells by modifying thiol groups in protein and small molecules, ultimately resulting in cell death (Gómez-Torres et al., 2016; Gyawalli and Ibrahim, 2014).

### 2.3.2 Natamycin

Natamycin, also known as pimaricin, is a macrolide polyene antifungal agent that is produced during fermentation by the bacteria *Streptomyces natalensis*, *Streptomyces chmanovgensis*, and *Streptomyces gilvosporeus* (Duchateau and van Scheppingen, 2018). It is especially active against yeasts and molds and virtually without effect on bacteria, protozoa and viruses. It has been employed in dairy-based food products to control spoilage by fungi, especially in cheese (Carocho et al., 2015). Natamycin is a very effective antifungal agent as it binds irreversibly to fungal ergosterol, thus disrupting the fungal cell membrane leading to a loss of solutes from the cytoplasm (Carocho et al., 2015; Hondrodimou et al., 2011).

### 2.3.3 ε-Polylysine (ε-PL)

As a natural antimicrobial food additive, ε-polylysine was approved by the FDA as safe for food in 2003 and has been widely used in Japan (Li et al., 2014). It has strong antimicrobial activity against a wide range of Gram-positive and Gram-negative bacteria, yeasts, molds, and bacteriophages (Lopez-Pena and McClements, 2015). Besides *Streptomyces albulus*, a number of other ε-PL producing species belonging to *Streptomyces*, *Kitasatospora*, and an *Ergot fungi*, *Epichloë*, have been isolated (Shukla et al., 2012). Their antimicrobial efficacy has been attributed to its high cationic charge density. The proposed mechanism for ε-PL is that it is absorbed onto anionic microbial cell surfaces due to electrostatic interactions, leading to the stripping of the outer membrane, disruption of the cytoplasm, and eventually, cell death. Nevertheless, its widespread use within the food and beverage industry may be limited because of its potential to interact with various components within the compositionally complex matrix found in these products (Chang et al., 2014; Lopez-Pena and McClements, 2015).

## 2.4 Algae and mushrooms

The algae group is divided into multicellular organisms, macroalgae or seaweed (often fast-growing, reaching sizes of up 60 m in length), and unicellular organisms, known as microalgae (measuring from 1 μm to several cm). A standard classification has been established, dividing seaweed organisms into three groups on the basis of their pigmentation:

**(i)** brown seaweed (*Phaeophyceae*);
**(ii)** red seaweed (*Rhodophyceae*) and
**(iii)** green seaweed (*Chlorophyceae*) (Pina-Pérez et al., 2017).

Among the major bioactive constituents of algae with demonstrated antimicrobial potential are: proteins (lectins), polysaccharides (laminarins and fucoidans), fatty acids (myristic, palmitic, oleic, and eicosapentaenoic acids), terpenoids (diterpenes, sesquiterpenes, and related compounds), and antioxidants such as polyphenols (phlorotannis), flavonoids, and carotenoids. In addition to those mentioned above, bromophenols are present in red algae, which have been reported to function as antimicrobial compounds (Gupta and Abu-Ghannam, 2011; Pina-Pérez et al., 2017). The presence of these compounds suggests a number of alternative mechanisms for antimicrobial action (Gyawalli and Ibrahim, 2014). For example, the mechanism of action of fatty acids as antimicrobials may be due to cell leakage derived from membrane damage. However, most antimicrobial compounds in algae are polyphenols and polysaccharides that act by inhibiting microbial growth, or in a more direct manner, by destroying the living structures of microorganisms (Pina-Pérez et al., 2017; Pisoschi et al., 2018). The phenolic compounds attack the organism's cell walls and cell membranes, which results in the release of intracellular constituents. Moreover, interference with the membrane function such as disruption of electron transport, nutrient uptake, protein, nucleic acid synthesis, and enzyme activity has also been reported to be a mode to action (Gupta and Abu-Ghannam, 2011).

On the other hand, among fungi, mushrooms are commonly eaten as food and have been shown to possess antimicrobial properties (Gyawalli and Ibrahim, 2014). A total of 158 species from 88 mushroom genera have been reported to possess antimicrobial properties compared with important foodborne pathogens. The most common mushrooms genera with antimicrobial properties include *Lentinus*, *Pleurotus*, *Dictyophora*, *Cordyceps*, *Ganoderma*, and *Tremella*. Their antimicrobial activity may be attributed to the presence of various bioactive secondary metabolites, volatile compounds, some phenols, gallic acids, free fatty acids, and their derivatives with inhibitory activity against targeted microorganisms. In addition, some nutritional metabolites, including polysaccharides, proteins, and enzymes, have also been shown to have antimicrobials activities (Gyawalli and Ibrahim, 2014; Shen et al., 2017).

## 2.5 Processing by-products

Food industries are currently focusing on solving the problems of waste management and recycling by utilization of by-products that contain valuable nutrients or bioactive compounds that can be used for developing novel value-added products (Silvan and Martinez-Rodriguez, 2018). The waste streams that are originated by various branches of the food industry can be divided into two main groups: plant (cereals, roots and tubers, oil crops, and pulses, as well as fruits and vegetables) or animal origin (meat products, fish, and seafood, as well as dairy products) (Galanakis, 2012, 2016; Nazzaro et al., 2018). These wastes generally comprise dietary fibers, proteins and peptides, lipids, fatty acids, and phenolic compounds (Silvan and Martinez-Rodriguez, 2018). Therefore, fruit and vegetable by-products are potentially good sources of phenolic compounds, which have demonstrated remarkable antimicrobial activity in different foods (Ayala-Zavala et al., 2011; Gyawalli and Ibrahim, 2014; Kowalska et al., 2017; Qin et al., 2013). Furthermore, in the case of animal origin by-products, specifically fish and seafood, the application of crustacean waste is mainly focused on the production of chitin/chitosan, with widespread use in the food industry. Chitosan-based films are very effective in food preservation; chitosan polymers structured as films, fibers, and beads can be used for the packaging of various foods such as fruits and vegetables (Campos et al., 2011; Nazzaro et al., 2018; Riaz et al., 2018) (Fig. 2).

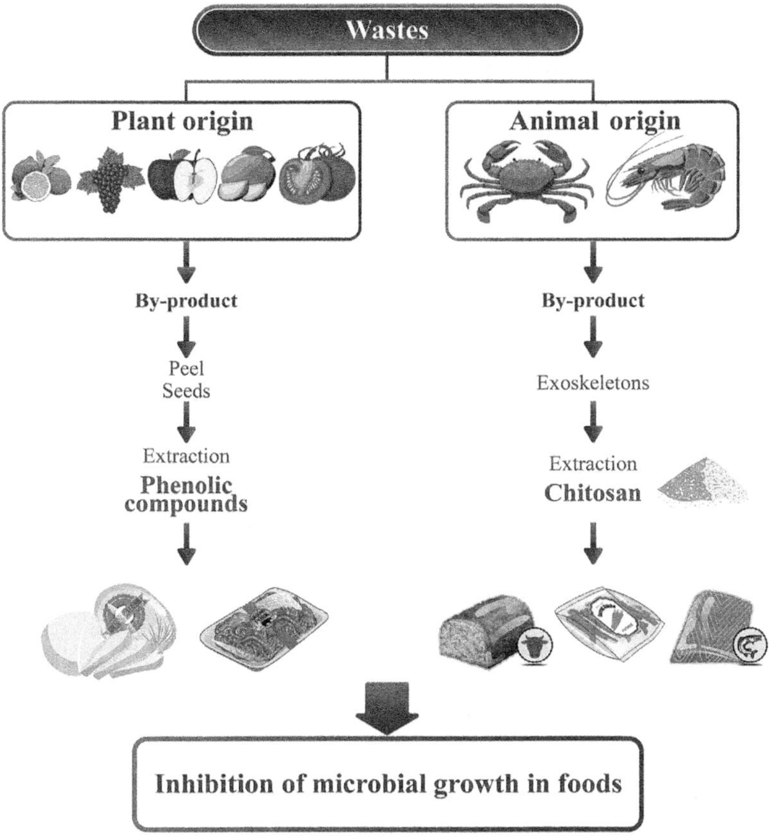


**Fig. 2** Inhibition of microbial growth of food from animal and plant by-products.

### *2.5.1 Fruit and vegetables as plant by-products*

By-products such as peel, pulp, seeds, stones, and stems from raw fruit and vegetables contain valuable bioactive compounds, such as flavonols, flavanols, anthocyanins, and phenolic acids (ferulic acid, vanillic acid, caffeic acid, etc.) which have several functionalities including antimicrobial activity (Guil-Guerrero et al., 2016; Gyawalli and Ibrahim, 2014; Kowalska et al., 2017; Mirabella et al., 2014; Tylewicz et al., 2018). Furthermore, the phenolic compounds from legume by-products (chickpeas, beans, lentils, and peas) are mainly represented by tannins, phenolic acids, and flavonoids (Tylewicz et al., 2018). These bioactive compounds from plant by-products (e.g., avocado, pomegranate, bergamota, red-grape, quince, citrus, apple and potato peels, cacao bean husk, coconut palm fronds, husks of the coconut fruits, sugarcane bagasse, mango kernel, stink bean pod, mustard seed meal, tomato seeds, etc.) have shown antibacterial activity against a wide range of microorganisms such as Gram-positive and Gram-negative bacteria, and sometimes yeast and mold (Agourram et al., 2013; Ayala-Zavala et al., 2011; Guil-Guerrero et al., 2016; Gyawalli and Ibrahim, 2014; Kowalska et al., 2017; Silvan and Martinez-Rodriguez, 2018; Taveira et al., 2010).

### 2.5.2 Fish and seafood as animal by-products

The huge amount of processed aqua-cultured and open water species has led to large amounts of marine by-product waste, such as crustacean shells, viscera, heads, skins, fins, trimmings, and shellfish waste. The application of crustacean waste is mainly focused on the production of chitin/chitosan, with wide application in the food industry (Nazzaro et al., 2018).

#### Chitosan

Chitosan is the *N*-deacetylated derivative of chitin. It is found in the shells of crustaceans, such as crab, shrimp, squid pen, and crawfish (Hamed et al., 2016; Hosseinnejad and Jafari, 2016) as well as insects and the cell wall of mushrooms (Hugo and Hugo, 2015; Perinelli et al., 2018; Poverenov et al., 2018). Chitosan inhibits the growth of a wide variety of fungi, yeasts, and bacteria (Hosseinnejad and Jafari, 2016), and is more effective against Gram-negative than Gram-positive bacteria (Pisoschi et al., 2018). The main antimicrobial effect of chitosan is related to its positively charged amine groups' ability to bind to the negatively charged surface of bacteria walls or plasma membrane. This causes a modification of the cell permeability, leading to an osmotic damage with the efflux of ions and proteins from the cytoplasm to the extracellular space (Hamed et al., 2016; Perinelli et al., 2018). Moreover, other mechanisms have been proposed, such as the interaction of diffused hydrolysis products with microbial DNA, which leads to the inhibition of the mRNA and protein synthesis. Chitosan also inhibits the microbial growth by the chelation of nutrients and essential metals. Finally, a polymer membrane can form on the surface of the cell, which prevents nutrients from entering the cell or acts as an oxygen barrier, which can inhibit the growth of aerobic bacteria (Hosseinnejad and Jafari, 2016; Verlee et al., 2017).

### 2.5.3 Bovine whey as animal by-products

Bovine whey is a by-product of cheese and casein manufacture in the dairy industry. Its main proteins include immunoglobulin, bovine serum albumin (BSA), alpha-lactalbumin, glycomacropeptide, beta-lactoglobulin, lactoferrin, and lactoperoxidase (LP). Among various valuable whey proteins, LP is of growing interest due to its distinctive biological functions, such as biocidal and biostatic activities (Pan et al., 2015). It can be used in combination with other materials for food packaging. Moreover, films and coatings based on whey protein are interesting, because they are edible and biodegradable (Shokri and Ehsani, 2017).

# 3 Application of natural antimicrobials in foods

## 3.1 Generalities

Plant-derived raw materials rich in natural antimicrobials have long been used for food preparation. An example of this is the addition of selected spices and herbs, for example, clove, oregano, thyme, cinnamon, garlic, or cumin, among others, to traditional food (Liu et al., 2017). Preservative effects of extracts from many types of plants that are used as seasoning agents in foods and beverages have been claimed for a long time (Elgayyar et al., 2001). These authors studied the effectiveness of selected herb and spice EOs for control of growth and survival of microorganisms. EOs from anise, angelica, basil, carrot, celery, cardamom, coriander, dill weed, fennel, oregano, parsley, and rosemary were studied. The grades of inhibition ranged from complete with oregano to no inhibition with carrot oil for *L. monocytogenes*, *S. aureus*, *E. coli* O:157:H7, *Yersinia enterocolitica*, *P. aeruginosa*, *Lactobacillus plantarum*, *Aspergillus niger*, *Geotrichum*, and *Rhodotorula*. Therefore, spices and herbs have been used in foods not only as flavoring compounds but also for their antimicrobial activity, although this activity is moderate at the used concentrations (Blaszyk and Holley, 1998).

On the other hand, it is also known that natural antimicrobials can be formed in lactic-acid fermented foods, which can contain antimicrobial compounds derived from microbial metabolism, such as organic acids and bacteriocins, among others (Bintsis, 2018). Recently, Turgis et al. (2012) studied bacteriocins from LAB as useful to the food industry. These authors studied the effect of nisin, pediocin, and MT104 and MT162 bacteriocins produced by *Enterococcus faecium*, and of EOs from *Cinnamomum cassia*, *Cymbopogon nardus*, *Brassica hirta*, *Origanum vulgare*, *Thymus vulgaris*, and *Satureja montagna*, alone and in combination. Each agent was tested against seven bacterial foodborne pathogens (*B. cereus*, *E. coli*, *L. monocytogenes*, *S. typhimurium*, and *S. aureus*) and food spoilage microorganisms (*Lactobacillus sakei* and *Pseudomonas putida*) in vitro. The results from these authors demonstrated that the use of EOs in combination with nisin, pediocin, or bacteriocins could cause synergistic or additive effects on several food pathogens. The results would allow the use of lower concentrations of EOs without affecting the reduction of foodborne pathogens. The increasing knowledge of the potential antimicrobial effect of natural antimicrobials has been the starting point for the development of recent approaches in food processing practices to increase the shelf life of perishable fresh or heat-treated food products (Patrignani et al., 2015).

## 3.2 Natural antimicrobial preparation before use

Nowadays, there is a growing interest in new approaches on the use of natural antimicrobials for food preservation. In this context, current literature and practices show that the bioactive components that are present in natural antimicrobial-rich raw materials often have to be extracted and eventually purified before being used as additives in processed foods. Extraction and purification allow the use of higher and standardized concentrations of bioactive compounds (Valdés et al., 2017).

Animal-derived peptides or proteins, some of them showing antimicrobial properties, are usually purified before use. The methods for this include precipitation, gel filtration, and chromatographic methods. In many cases, the physicochemical properties of proteins or peptides represent a challenge and at the same time, make it possible to develop extraction and purification approaches. In the case of bacteriocins, it is necessary to consider their affinity to organic solvents and solubility in concentrate salt solutions at a given pH value (Parada et al., 2007; Jamaluddin et al., 2017). Affinity chromatography has been developed for purification and concentration of peptides and proteins based on their physicochemical or specific binding properties. As an example, Hirsch et al. (2018) have proposed a purification process for lysozyme from egg white using sulphanilic acid-modified chitosan microbeads in a cation exchange chromatography.

On the other hand, one traditional method for obtaining EOs has been based on hydrodistillation. Plant material is immersed in water in a vessel that is connected to a condenser and then a decanter. The EOs' components have high boiling point temperatures, but in co-distillation with water, the heterogeneous mixture needs a temperature lower than 100°C for distillation. Once the mixture is condensed and retained in the decanter, the hydrophobic compounds are separated from the carrying water by simple decantation (El Asbahani et al., 2015). The Clevenger type hydro-distillers are usually used in labs for this purpose. The main disadvantage of hydrodistillation is the time required for the EOs extraction and that the high temperatures used can have detrimental effects on bioactive compounds (Gavahian et al., 2015). Steam distillation is a modification of the hydrodistillation process, which is faster and less deleterious to bioactive components.

Another traditional method for the extraction of EOs is the use of organic solvents. The plant is macerated in the solvent in small pieces or particles to increase the contact area between solvent and the extracting plant material for at least 24h. Since the extraction temperatures are lower than the ones achieved in hydrodistillation, degradation or problems associated with the use of boiling water are avoided (El Asbahani et al., 2015). The use of solvents, however, has the disadvantage that many of them are not allowed to be used in food preparation. Furthermore, when organic solvents are used for extraction of EOs, its chemical composition is different from the one obtained using the distillation methods. On the other hand, the extraction of polyphenols, polysaccharides, and other antimicrobial nonvolatile compounds different to those found in EOs have commonly been carried out using polar solvents including water or alcohol (Coss et al., 2018).

New methods for the extraction and purification of EOs and polyphenols have been proposed in the later decades, including the use of supercritical fluids, and solvent free microwave extraction. These new methodologies reduce the time, solvents, and/or temperatures used and increase the recovery (García-Salas et al., 2010; El Asbahani et al., 2015). In supercritical fluid extraction, the solvent—the most commonly used solvent is $CO_2$—is applied under a critical pressure and critical temperature. Wenquiang et al. (2007) compared four methods for the extraction of clove EO, including hydrodistillation, steam distillation, Soxhlet extraction, and supercritical fluid extraction. The authors described the differences in chemical composition of the EOs obtained, and proposed supercritical fluid extraction as the method of choice. Microwave-assisted extraction has been studied in recent years, since it reduces the time required for extraction, requires less solvent for extraction, and improves product yield and reproducibility. For EOs there are several types of microwave-assisted extraction, including vacuum microwave hydrodistillation, microwave steam distillation, and solvent free microwave extraction (El Asbahani et al., 2015; Gavahian et al., 2015). The latter has been extensively studied since it was first proposed by Lucchesi et al. (2004).

The application of the biorefinery concept, using environmentally friendly solvents and the step-by-step progressive obtention of different useful components, has been suggested for the extraction and purification of antimicrobials from plants. An example of a biorefinery is presented by Boukroufa et al. (2015), where an agroindustrial residue is subjected to further processing to obtain valuable products. Orange peels, a residue from the preparation of orange juice, are used for extraction of EOs by a microwave-assisted method, and the residue is then processed using ultrasound-assisted extraction for obtaining polyphenols. Pectin is finally obtained from the remaining solids using the residual water as a solvent for the extraction of EOs.

## 3.3 Incorporation of natural antimicrobials into food

### 3.3.1 Direct incorporation

The direct incorporation into food matrices of a variety of natural antimicrobial agents such as peptides (bacteriocins, lactoferrin), plant-derived substances (EOs, polyphenols, and isothiocyanates), and enzymes (lysozyme) as ingredients has shown a decreasing or inhibitory effect on the growth of spoilage and pathogenic microorganisms (Atarés and Chiralt, 2016; Calo et al., 2015; Lopes and Brandelli, 2017; Meira et al., 2014; Rhim et al., 2013). The incorporation of these antimicrobials into processed food can be carried out in different forms. The simplest way for this consists of mixing the natural antimicrobials with the rest of the food ingredients. In case the antimicrobial needs to be placed on the food surface (i.e., a whole-piece solid food) the addition should be performed by spraying, dipping, or brushing on the surface of food pieces (Galotto et al., 2015).

The effectiveness of the antimicrobials used in food depends on many factors such as the food product, the storage conditions, its handling, and the target microorganisms (Davidson, 1997). Gould (1989) and Davidson (1997) classified the factors that affect the antimicrobials activity into the following groups:

- *Microbial factors*. They include the inherent resistance of the microorganisms, their initial number, lag time and maximal growth rate, interspecies interactions and intraspecies interactions, and the physiological status.
- *Intrinsic factors*. All are related with the food product, such as composition, pH, water activity, etc.
- *Extrinsic factors*. Environmental factors including temperature, atmosphere, humidity, etc.
- *Processing factors*. Including the changes suffered by the food along their processing steps.

The direct incorporation of natural antimicrobials into food mixes or food surfaces presents the disadvantage of a loss of inhibitory efficacy due to their interaction with food components. Another disadvantage is related to the difficulty of achieving a homogeneous distribution of the antimicrobial in the food system or to reaching the correct amount of antimicrobial where its action is needed. Solutions to overcome these problems include encapsulation of antimicrobials into edible substances (Calo et al., 2015), thus forming micro or nanoparticles (Gyawalli and Ibrahim, 2014). Encapsulation of antimicrobial compounds can improve their distribution in the food matrix and stability, thus increasing their effectivity, resulting in an increase in the food's shelf life (Burgain et al., 2011). Furthermore, the encapsulation material can modulate the release or diffusion of the bioactive compounds into the food (Bhargava et al., 2015). Encapsulation has been highly recommended for obtaining homogeneous distribution of water-insoluble or low water soluble material such as EOs or nisin (El Asbahani et al., 2015; Zimet et al., 2018). Because of the hydrophobic nature of many antimicrobial compounds, they might not be homogenously distributed in the food matrix or would be concentrated in the lipid fraction when added to food. Meanwhile, it must be present in the hydrophilic fraction where microbial activity takes place.

The procedure of encapsulation consists of the preparation of small particles composed of an edible wall that covers the compound of interest so that it can be incorporated into the food as a suspension (Weiss et al., 2009). According to the size of the capsule the particles are considered to be micro or nanocapsules. For the preparation of microcapsules, different methods are used, which include spray drying, emulsification, extrusion, solvent precipitation, and coacervation, among other methods. The covering material for encapsulating needs to be food-grade quality, that is, made from materials authorized for human consumption. Moreover, the selection of the covering material should take into account the nature of the material to be encapsulated. Among the most common materials used are alginate, gellan and xanthan gum, carrageenan, starch, gelatin, and milk proteins for hydrophilic material to be encapsulated (Riaz and Masud, 2013). In the case of hydrophobic antimicrobials, emulsifying o surfactant agents, mainly phospholipids, are used to form micelles or liposomes (Weiss et al., 2009). Further information on the preparation of encapsulated material can be found in recent reviews, including Burgain et al. (2011), which describes the methodologies for producing encapsulated probiotic microorganisms and their incorporation in food systems, and El Asbahani et al. (2015), which describes the production of liposomes for dispersion of EOs added to food.

The antimicrobial effect of encapsulated bioactive compounds has been tested against foodborne-pathogens inoculated into foods (Bhargava et al., 2015; Zimet et al., 2018), or evaluated for shelf life of food products (Fernandes et al., 2018; Noori et al., 2018). Studies show that the effect of the antimicrobial was either not affected by the encapsulation or increased, and highlight that some encapsulating materials such as chitosan can strengthen their effectivity due to the mild antimicrobial effect of chitosan. Table 1 includes examples on the use of recent technologies for the application of natural antimicrobials in food including micro and nanoemulsion techniques as well as the delivery of natural antimicrobials in food systems from either edible coating or active packaging, which will be commented on further below (see Sections 3.3.2 and 3.3.3).

## 3.3.2 Edible films and coatings

A recent approach for the delivery of natural antimicrobials on to food surfaces is the incorporation of bioactive ingredients into edible films or coatings. These are thin layers of edible material. An edible film is a preformed sheet of edible material that is applied to the food surface, and an edible coating is a protective layer formed when food is dipped in the coating solution or the solution is sprayed or spread on the food

**Table 1** Delivery strategies of natural antimicrobials into food systems

| Natural antimicrobial | Delivery system | Carrying material | Microorganisms tested | Food system | References |
|---|---|---|---|---|---|
| Prickly pear fruit extract | Edible coating | Prickly pear fruit extract | Total mesophilic count | Packed sliced beef | Palmeri et al. (2018) |
| Satureja essential oil free and nanoencapsulated | Edible coating | Chitosan | Total viable counts, Lactic acid bacteria, *Pseudomonas* spp. | Refrigerated lamb meat | Pabast et al. (2018) |
| Oregano essential oil | Microencapsulation | Whey protein isolate | Mold and yeast | Grated Parmesan cheese | Fernandes et al. (2018) |
| Ginger essential oil | Nanoemulsion | Sodium caseinate | Psychrophilic bacteria, molds, and yeasts | Chicken breast fillet | Noori et al. (2018) |
| Allyl isothiocyanate | Edible coating | Washing with para acetic acid and coating with chitosan and corn biofiber gum | *Escherichia coli* O157:H7, Salmonella, natural microbiota | Fresh strawberries | Guo et al. (2018) |
| Nysin | Nanoencapsulation | Alginate and chitosan | *L. monocytogenes* | Vacuum sealed refrigerated lean beef | Zimet et al. (2018) |
| Thymol and carvacrol mixture | Active packing film | Low-density polyethylene with halloysite nanotubes | *E. coli* (ATCC 8739) | Hummus Dip | Krepker et al. (2017) |
| Nisin and polylysine | Edible coating | Chitosan | Total mesophilic, coliforms, Mold and yeasts. *S. aureus*, *Pseudomonas* spp. | Fresh cut carrots | Song et al. (2017) |

*Continued*

**Table 1** Continued

| Natural antimicrobial | Delivery system | Carrying material | Microorganisms tested | Food system | References |
|---|---|---|---|---|---|
| Natamycin | Edible coating | Chitosan | Total mesophilic count, mold and yeast, psychrotrophic, lactic acid bacteria, coliforms | Fresh "Phyllo" | Tsiraki et al. (2017) |
| Lactoperoxidase and α-tocopherol | Edible coating | Whey protein | Total viable counts, Psychrotrophic bacteria, *Shewanella putrefaciens*, *Pseudomonas fluorescens* | Pike-Perch fillets | Shokri and Ehsani (2017) |
| Oregano essential oil | Nanoemulsion | Tween 80 | *L. monocytogenes* ATCC 19115, *Salmonella typhimurium* ATCC 19585 and *E. coli* O157:H7 ATCC 700927 | Fresh lettuce | Bhargava et al. (2015) |
| Chitosan-lactoperoxidase | Edible coating | Chitosan and glicerol | Mold (*Colletotrichum gloeosporioides*, *Phomopsis* sp. RP257, *Pestalotiopsis* sp. and *Lasiodiplodia theobromae* ngr 05 A) | Mangoes | Cissé et al. (2015) |

surface and the material forms a cover once it is dried (Falguera et al., 2011). Both of them serve as physical barriers that protect the food surface from the surrounding environment by prevention of water loss, lowering the access of atmospheric oxygen, and as a protective cover against contamination by spoilage or pathogenic microorganisms. Additionally, the barrier can interfere with the transfer of volatile compounds and can contribute to the formation of a small space with a modified atmosphere on the adjacent area next to the food surface. When films or coatings contain bioactive compounds they provide a continuous and controlled release of the active compound on to the food surface during storage (Campos et al., 2011).

Many reports on the preparation of edible films and coatings have been published. Proteins, polysaccharides, and lipids, or a combination of them (composite materials) obtained from different sources of biomass are used as structural matrix material for the formation of edible films and coatings (Valdés et al., 2017). Bioactive antimicrobials incorporated in edible films and coatings include EOs, bacteriocins, and enzymes including lactoperoxidase, or natural extracts (Pabast et al., 2018; Song et al., 2017; Tsiraki et al., 2017; Valdés et al., 2017). In the preparation of the filming/coating solution, additional components can be added as plasticizers (sorbitol, glycerol) or cross-linking agents (enzymes such as transglutaminase). A novel approach to improve the effect of the antimicrobials incorporated in edible coatings is the development of nanoencapsulating, and multilayered/nanolaminate delivery systems (Aloui and Khwaldia, 2016).

The use of edible coatings, for example, sodium alginate, chitosan, or hydroxypropylmethylcellulose, containing antimicrobials has proved to be effective for solid and semisolid fresh perishable food products such as fruits, vegetables, meat, fish, wheat-based products, and dairy products, by controlling the development of spoilage and pathogenic microorganisms at the food surface (Aloui and Khwaldia, 2016; Guo et al., 2018; Palmeri et al., 2018; Song et al., 2017; Tsiraki et al., 2017; Valdés et al., 2017). Recent reports on the preparation of edible coatings intended to be used in food packaging have suggested chitosan as an optimum matrix material (Pabast et al., 2018; Song et al., 2017; Tsiraki et al., 2017).

Falguera et al. (2011) presented a review on the challenges of the use of edible films and coatings in processed food, as well as the importance of developing the corresponding regulatory issues. The use of agroindustrial by-products as a source of structural matrix material to be used in the preparation of edible films and coatings is also one of the current challenges in sustainable food production (Shit and Shah, 2014). An example is the case of prickly pear fruit extract, which was used as natural antimicrobial as well as structural matrix material for the preparation of an edible coating for preservation of packed sliced beef (Palmeri et al., 2018).

### 3.3.3 *Food packaging*

Traditional petroleum-based films where natural antimicrobials are incorporated can also be used as delivery systems (release systems). Antimicrobial packaging thus combines preservation with protection. Antimicrobials cannot only be incorporated into packaging formulation but they can also be immersed into the package headspace or included in a sachet inside the package, which allows the volatile antimicrobials to diffuse to the headspace. Furthermore, some antimicrobials can also be chemically bound to the packaging inner surface, thus immobilizing it (nonrelease systems; Galotto et al., 2015). All these packaging alternatives, which are called active packaging, have become the focus of recent attention regarding the control of microbial growth in the food surface during storage (Valdés et al. 2017).

In the release systems, packaging materials allow a steady diffusion of antimicrobial agents into the food surface or evaporation to the headspace, and consequently, can sustain a determinate concentration of antimicrobial agents on the surface of the food product for a long time (Aloui and Khwaldia, 2016). The amount of the antimicrobial in the packaging and its diffusion or evaporation rate, however, should be controlled so that the antimicrobial agents maintain their concentrations at the critical level of effectiveness. Diffusion or evaporation rate in the food surface or headspace depends on the packaging material, antimicrobial, and food composition (Galotto et al., 2015). Volatile antimicrobials in packaging have been proposed for food with a high surface/weight ratio and packaged under modified atmosphere, for example, salads. On the other hand, water-soluble or lipid-soluble antimicrobials have been recommended for vacuum packaged high moisture or high lipid containing food, respectively (Galotto et al., 2015; Krepker et al. 2017). As an example of recent research, the aforementioned authors prepared an active packaging film of low-density polyethylene with halloysite nanotubes added with thymol and/or carvacrol for a hummus dip with the film used as a cover.

# 4 Challenges and current use of natural antimicrobials in the food industry

## 4.1 Limitations and challenges in the application of natural antimicrobials to foods

The extensive use of natural preservatives in food processing is promising and is, therefore, deserving of extensive research. The reason for this could be due, at least partially, to consumers' increased preference towards chemical additive-free foods (Venskutonis, 2004) and also the food industry's interest in increasing food shelf life and safety at retail (Pisoschi et al., 2018). However, there are still many technological unknowns and challenges regarding the use of natural antimicrobials in food products. At present, the practical use of natural antimicrobials in food processing is somewhat limited by different causes regarding technological, sensory, safety, and regulatory issues, that is, unstandardized efficiency or lack of reproducibility of their activity, decreased activity due to incompatibility or interaction with packaging or food components, antimicrobial degradation during processing or storage, development of microbial resistance, negative impact on the flavor of food, safety concerns regarding the use of antimicrobials, or the prohibition or difficulty of being approved as a food ingredient by safety authorities (Aloui and Khwaldia, 2016; Burt, 2004; Galotto et al.,

2015; García et al., 2010; Ribeiro-Santos et al., 2017). Furthermore, the use of natural antimicrobials should have a reasonable cost for usage (David et al., 2013; Davidson, 2017). An effective use of antimicrobial compounds should take into account the characteristics of the target microorganisms, the interfering food product components, physicochemical factors such as pH, the way the antimicrobial is incorporated into the food, the processing and storage conditions, and the expected results in controlling the microbial growth (Davidson et al., 2015).

In order to identify or evaluate novel natural antimicrobials to be potentially used in foods, Davidson et al. (2015) suggested performing in vitro testing and application and challenging studies in actual food matrices. As for plant extracts and EOs in particular, evaluation of bioactive components, chemical interactions within the food matrix, toxicity or efficacy, and standardization and quality control procedures seem to be needed (Ribeiro-Santos et al., 2017).

Studies should not only be limited to antimicrobials and the food matrix but also to coatings and films where they could be incorporated. For example, some of the reported limitations of antimicrobial-containing edible coatings are a weak adhesion of coating materials to hydrophilic food surfaces, an eventual degradation of the antimicrobial agent, or a quick desorption through coating materials, among others (Aloui and Khwaldia, 2016). With regard to active packaging, one of the key challenges is to control the release of the antimicrobial agent from the film to the food (Singh et al., 2011). Furthermore, few commercial antimicrobial packages are available commercially, mainly because of their high cost (Silva et al., 2018).

Food flavor can be significantly affected due to the use of plant derived natural antimicrobials, such as EOs or polyphenol extracts. EOs are characterized by having an intense flavor and their addition to food most probably results in a high impact on the flavor. According to Burt (2004), the microbial critical inhibitory levels of EOs when they are mixed with food, which is within the range of 0.5–2 g per 100 g of food, would strongly affect food flavor. Thus, the amount of EOs added to food should allow for antimicrobial activity without depreciating the sensory characteristics of food products. Dussault et al. (2014) proposed a limit of 0.5 g of EO per 100 g of food to avoid sensory rejection. The combined use of different EOs and the use of EOs with other preservation technologies might fulfill the food safety requirements for food products without depreciation of sensory acceptation (Matan et al., 2006). The incorporation of EOs in active packaging appears to be an interesting alternative regarding the negative impact of EOs on flavor (Lorenzo et al., 2014; Ribeiro-Santos et al., 2017). Polyphenols are less intense in flavor than EOs, although, despite being odorless, they can result in a bitter taste or astringency (Ortega-Ramirez et al., 2014).

Moreover, the use of natural antimicrobials as constituents of processed food should follow a risk management decision in order to ensure the production of safe food. It is known that active compounds from EOs and polyphenols present oral toxicity at relatively low doses—the toxicity of carvacrol or quercetin 810 and 161 mg/kg body weight in rats (Burt 2004)—or may be related with food allergies (Panda and Rath, 2012; Ribeiro-Santos et al., 2017). On the other hand, EOs in food could exert beneficial medicinal effects (Davidson, 2017).

## 4.2 Legislatory aspects and current applications of natural antimicrobials in food

The use of antimicrobials by the food industry requires regulation, that is, their use has to be authorized and regulated by national or international regulations as proposed by the corresponding safety agencies. In the European Union, as an example, the use of natural antimicrobials can be subjected to the food additive (including flavorings and enzymes) regulations (Regulation (EC) No 1333/2008; Regulation (EC) No 1334/2008, Regulation (EC) 1332/2008), and to active packaging regulation (Commission Regulation (EC) 1935/2004 and Regulation (EC) 450/2009) and novel food regulations (Regulation (EC) 178/2002 and Regulation (EU) 2283/2015). Regulations do not allow the addition of certain natural undesirable substances, establish lists of permitted compounds and the maximum levels for certain substances, or request rigorous studies before a substance can be added to food.

The natural active substances, either directly incorporated into the food or released from the packaging to the food, that perform a preservative function should be authorized as food additives, including the enzymes. Some EOs that are authorized as flavors could also exert an antimicrobial effect, although the levels used for flavoring can be lower than those required for inhibiting microbial growth (Davidson, 2017). Regarding active packaging, substances interacting and enhancing the effect of the antimicrobials are subject to authorization under the active packaging legislation. Furthermore, when a potentially edible material rich in antimicrobials originating from plants, animals, microorganisms, cell cultures, minerals, etc., or formed by the result of production processes and practices or technologies that were not used or produced before 1997, is intended for use as a food ingredient, the food containing the edible material would require authorization under the Novel Food Regulation.

At present, apart from organic acids such as acetic, lactic, propionic, or citric acids and their salts, which are widely used in food products, only a relatively small number of other natural antimicrobials are allowed (Galotto et al., 2015). Nisin and natamycin are the only bacteriocins allowed, with the latter being used only to prevent mold growth on the surface of ripened cheeses and sausages (Pisoschi et al., 2018). Lysozyme is being used in cheese to prevent late blowing caused by *Clostridium tyrobutyricum*, and in other foods in accordance with good manufacturing practices (Lucera et al., 2012; Raybaudi-Massilia et al., 2009). The use of lactoperoxidase at limited amounts is permitted in food by Australian and US authorities (Raybaudi-Massilia et al., 2009). Lactoferrin has been used as an antimicrobial in a range of foods, for example, meat products (Hugo and Hugo, 2015). Furthermore, the application of a pediocin-producing strain of *P. acidilactici* or a spray with phages were approved in the United States for use respectively in reduced nitrite bacon (Smid and Gorris, 2007) and meat and poultry products (Silva et al., 2018).

EOs, which are being increasingly considered as antimicrobials to be incorporated in coatings and packaging films, are generally recognized as safe (GRAS; e.g., vanillin or methyl jasmonate; Raybaudi-Massilia et al., 2009). In the EU, some EOs or compounds obtained from them have been approved with a legal limit in foods such as flavorings, for example, camphor or myrtenol. Additionally, the use of polyphenol-rich plant-derived extracts such as grape seed extract (GRAS in United States) or rosemary extract (allowed additive in the UE) is also permitted.

Chitosan, commercially produced from crab and shrimp shell wastes, is also used in the food industry. It presents high biodegradability and antimicrobial properties (Lucera et al., 2012) and is considered to be nonantigenic and nontoxic (Raybaudi-Massilia et al., 2009). However, it has been indicated that its consumption should be avoided by people with a shellfish allergy, or who are pregnant or nursing.

## 5 Conclusions

Food products can be preserved by using a wide range of natural bioactive compounds from plants, animals, bacteria, algae, or mushrooms. The interest in their use has increased in recent years, in part in response to the consumer demands for food free of chemical additives. Several natural antimicrobials are currently being used combined with other antimicrobial hurdles for the extension of their shelf life or the control of pathogenic bacteria. However, there are still many unexplored possibilities in the use of novel antimicrobials that, as well as inactivates microorganisms, could also enhance the nutritional properties of food.

On the other hand, despite the promising insights concerning the potential use of natural antimicrobials in foods, they remain limited in their commercial adoption due to different problems such as poor standardization of natural antimicrobial ingredients, diminution of activity caused by interactions with food components or during processing, giving undesirable flavors, or regulatory burdens. Further research seems to be needed to improve the methodologies for the isolation of antimicrobials, determine their optimum levels, develop effective bio-based polymeric and active antimicrobial packaging materials acting as nanocarriers or delivery systems, optimizing their cost-in-use, and the assessment of safety concerns and health risks.

## References

Agourram, A., Ghirardello, D., Rantsiou, K., Zeppa, G., Belviso, S., Romane, A., et al., 2013. Phenolic content, antioxidant potential, and antimicrobial activities of fruit and vegetable by-product extracts. Int. J. Food Prop. 16, 1092–1104.

Ahmad, V., Khan, M.S., Jamal, Q.M.S., Alzohairy, M.A., Karaawi, M.A.A., Siddiqui, M.U., 2017. Antimicrobial potential of bacteriocins: in therapy, agriculture and food preservation. Int. J. Antimicrob. Agents 49 (1), 1–11.

Aloui, H., Khwaldia, K., 2016. Natural antimicrobial edible coatings for microbial safety and food quality enhancement. Compr. Rev. Food Sci. Food Saf. 15 (6), 1080–1103.

Aprodu, I., Stănciuc, N., Dumitraşcu, L., Râpeanu, G., 2014. Investigations towards understanding the thermal denaturation of lactoperoxidase. Int. Dairy J. 38 (1), 47–54.

Atarés, L., Chiralt, A., 2016. Essential oils as additives in biodegradable films and coatings for active food packaging. Trends Food Sci. Technol. 48, 51–62.

Ayala-Zavala, J.F., Vega-Vega, V., Rosas-Domínguez, C., Palafox-Carlos, H., Villa-Rodriguez, J.A., Siddiqui, M.W., et al., 2011. Agro-industrial potential of exotic fruit byproducts as a source of food additives. Food Res. Int. 44 (7), 1866–1874.

Ayala-Zavala, J.F., Silva-Espinoza, B., Cruz-Valenzuela, M., Leyva, J., Ortega-Ramírez, L., Carrazco-Lugo, D., et al., 2013. Pectin-cinnamon leaf oil coatings add antioxidant and antibacterial properties to fresh-cut peach. Flavour Fragance J. 28 (1), 39–45.

Balciunas, E.M., Castillo, M.F.A., Todorov, S.D., Franco, B.D.G.M., Converti, A., Oliveira, P.R., 2013. Novel biotechnological applications of bacteriocins: a review. Food Control 32 (1), 134–142.

Barbiroli, A., Bonomi, F., Capretti, G., Iametti, S., Manzoni, M., Piergiovanni, L., et al., 2012. Antimicrobial activity of lysozyme and lactoferrin incorporated in cellulose-based food packaging. Food Control 26 (2), 387–392.

Bhargava, K., Conti, D.S., da Rocha, S.R., Zhang, Y., 2015. Application of an oregano oil nanoemulsion to the control of foodborne bacteria on fresh lettuce. Food Microbiol. 47, 69–73.

Bintsis, T., 2018. Lactic acid bacteria: their applications in foods. J. Bacteriol. Mycol. 6 (2), 89–94.

Blaszyk, M., Holley, R.A., 1998. Interaction of monolaurin, eugenol and sodium citrate on growth of common meat spoilage and pathogenic organisms. Int. J. Food Microbiol. 39 (3), 175–183.

Boukroufa, M., Boutekedjiret, C., Petigny, L., Rakotomanomana, N., Chemat, F., 2015. Biorefinery of orange peels waste: a new concept based on integrated green and solvent free extraction processes using ultrasound and microwave techniques to obtain essential oil, polyphenols and pectin. Ultrason. Sonochem. 24, 72–79.

Burgain, J., Gaiani, C., Linder, M., Scher, J., 2011. Encapsulation of probiotic living cells: from laboratory scale to industrial applications. J. Food Eng. 104 (4), 467–483.

Burrowes, O.J., Hadjicharalambous, C., Diamond, G., Lee, T.C., 2004. Evaluation of antimicrobial spectrum and cytotoxic activity of *Pleurocidin* for food applications. J. Food Sci. 69 (3), FMS66–FMS71.

Burt, S., 2004. Essential oils: their antibacterial properties and potential applications in foods—a review. Int. J. Food Microbiol. 94 (3), 223–253.

Calo, J.R., Crandall, P.G., O'Bryan, C.A., Ricke, S.C., 2015. Essential oils as antimicrobials in food systems—a review. Food Control 54, 111–119.

Campos, C.A., Gerschenson, L.N., Flores, S.K., 2011. Development of edible films and coatings with antimicrobial activity. Food Bioprocess Technol. 4 (6), 849–875.

Carocho, M., Morales, P., Ferreira, I.C.F.R., 2015. Natural food additives: *Quo vadis*? Trends Food Sci. Technol. 45 (2), 284–295.

Chang, Y., McLandsborough, L., McClements, D.J., 2014. Interaction of cationic antimicrobial (ε-polylysine) with food-grade biopolymers: dextran, chitosan, carrageenan, alginate, and pectin. Food Res. Int. 64, 396–401.

Cissé, M., Polidori, J., Montet, D., Loiseau, G., Ducamp-Collin, M.N., 2015. Preservation of mango quality by using functional chitosan-lactoperoxidase systems coatings. Postharvest Biol. Technol. 101, 10–14.

Commission Regulation (EC) No 450/2009, 2009, Commission Regulation (EC) No 450/2009 of 29 May 2009 on active and intelligent materials and articles intended to come into contact with food.

Coss, E., Kealey, C., Brady, D., Walsh, P., 2018. A laboratory investigation of the antimicrobial activity of a selection of western phytomedicinal tinctures. Eur. J. Intern. Med. 19, 80–83.

Cruz-Chamorro, L., Puertollano, M.A., de Cienfuegos, G.A., de Pablo, M.A., 2006. In vitro biological activities of magainin alone or in combination with nisin. Peptides 27 (6), 1201–1209.

Cushnie, T.P.T., Lamb, A., 2011. Recent advances in understanding the antibacterial properties of flavonoids. Int. J. Antimicrob. Agents 38 (2), 99–107.

Daglia, M., 2012. Polyphenols as antimicrobial agents. Curr. Opin. Biotechnol. 23 (2), 174–181.

David, J.R.D., Steenson, L.R., Davidson, P.M., 2013. Expectations and applications of natural antimicrobials to foods: a guidance document for users, suppliers, research and development, and regulatory agencies. Food Prot. Trends 33 (4), 238–247.

Davidson, P.M., 1997. Chemical preservatives and natural antimicrobial compounds. In: Doyle, M.P., Beuchat, L.R., Montville, T.J. (Eds.), Food Microbiology: Fundamentals and Frontiers. ASM Press, Washington D.C.

Davidson, P.M., 2017. Natural antimicrobials from herbs and spices. In: Jen, J.J., Chen, J. (Eds.), Food Safety in China: Science, Technology, Management and Regulation. John Wiley & Sons Ltd, New Jersey.

Davidson, P.M., Bozkurt Cekmer, H., Monu, E.A., Techathuvanan, C., 2015. The use of natural antimicrobials in food: an overview. In: Taylor, M. (Ed.), Handbook of Natural Antimicrobials for Food Safety and Quality. Woodhead Publishing, Cambridge, pp. 1–19.

de Souza, C.E., Silva, C.M.H., Amaro, S.D., Cançado, V.J., Gomes, O.J.N., Miranda, V., et al., 2014. The use of versatile plant antimicrobial peptides in agribusiness and human health. Peptides 55, 65–78.

Duchateau, A.L.L., van Scheppingen, W.B., 2018. Stability study of a nisin/natamycin blend by LC-MS. Food Chem. 266, 240–244.

Dussault, D., Vu, K.D., Lacroix, M., 2014. In vitro evaluation of antimicrobial activities of various commercial essential oils, oleoresin and pure compounds against food pathogens and application in ham. Meat Sci. 96 (1), 514–520.

Ekengren, S., Hultmark, D., 1999. *Drosophila* cecropin as an antifungal agent. Insect Biochem. Mol. Biol. 29 (11), 965–972.

El Asbahani, A., Miladi, K., Badri, W., Sala, M., Addi, E.A., Casabianca, H., et al., 2015. Essential oils: from extraction to encapsulation. Int. J. Pharm. 483 (1–2), 220–243.

Elgayyar, M., Draughon, F.A., Golden, D.A., Mount, J.R., 2001. Antimicrobial activity of essential oils from plants against selected pathogenic and saprophytic microorganisms. J. Food Prot. 64 (7), 1019–1024.

Falguera, V., Quintero, J.P., Jiménez, A., Muñoz, J.A., Ibarz, A., 2011. Edible films and coatings: structures, active functions and trends in their use. Trends Food Sci. Technol. 22 (6), 292–303.

Faruck, M.O., Yusof, F., Chowdhury, S., 2016. An overview of antifungal peptides derived from insect. Peptides 80, 80–88.

Fernandes, R.V.B., Botrel, D.A., Monteiro, P.S., Borges, S.V., Souza, A.U., Mendes, L.E.S., 2018. Microencapsulated oregano essential oil in grated parmesan cheese conservation. Int. Food Res. J. 25 (2), 661–669.

Fitriyanti, M., Narsimhan, G., 2018. Synergistic effect of low power ultrasonification on antimicrobial activity of cecropin P1 against *E. coli* in food systems. LWT- Food Sci. Technol. 96, 175–181.

Fu, Y., Sarkar, P., Bhunia, A.K., Yao, Y., 2016. Delivery systems of antimicrobial compounds to food. Trends Food Sci. Technol. 57 (part A), 165–177.

Galanakis, C.M., 2012. Recovery of high added-value components from food wastes: conventional, emerging technologies and commercialized applications. Trends Food Sci. Technol. 26 (2), 68–87.

Galanakis, C.M., 2016. High value-added compounds from food waste. In: Galanakis, C.M. (Ed.), Reference Module in Food Sciences. Elsevier, Amsterdam, pp. 1–8.

Galotto, M.J., Guarda, A., López de Dicastillo, C., 2015. Antimicrobial active polymers in food packaging. In: Cirillo, G., Spirizzi, U.G., Iemma, F. (Eds.), Functional Polymers in Food Science. Scrivener Publishing Editorial, New Jersey, pp. 323–351.

García, P., Rodríguez, L., Rodríguez, A., Martínez, B., 2010. Food biopreservation: promising strategies using bacteriocins, bacteriophages and endolysins. Trends Food Sci. Technol. 21 (8), 373–382.

García-Salas, P., Morales-Soto, A., Sagura-Carretero, A., Fernández-Gutiérrez, A., 2010. Phenolic-compound-extraction systems for fruit and vegetable samples. Molecules 15 (12), 8813–8826.

Garde, S., Gómez-Torres, N., Delgado, D., Gaya, P., Ávila, M., 2016. Influence of reuterin-producing *Lactobacillus reuteri* coupled with glycerol on biochemical, physical and sensory properties of semi-hard ewe milk cheese. Food Res. Int. 90, 177–185.

Gavahian, M., Farahnaky, A., Farhoosh, R., Javidnia, K., Shahidi, F., 2015. Extraction of essential oils from Mentha piperita using advanced techniques: microwave versus ohmic assisted hydrodistillation. Food Bioprod. Process. 94, 50–58.

Gill, T.A., Singer, D.S., Thompson, J.W., 2006. Purification and analysis of protamine. Process Biochem. 41 (8), 1875–1882.

Gómez-Torres, N., Ávila, M., Delgado, D., Garde, S., 2016. Effect of reuterin-producing lactobacillus reuteri coupled with glycerol on the volatile fraction, odour and aroma of semi-hard ewe milk cheese. Int. J. Food Microbiol. 232, 103–110.

Gould, G.W., 1989. Mechanisms of Action of Food Preservation Procedures. Elsevier Applied Science, London.

Gould, G.W., 1996. Industry perspectives on the use of natural antimicrobials and inhibitors for food applications. J. Food Prot. 59 (13), 82–86.

Gram, L., Ravn, L., Rasch, M., Bruhn, J.B., Christensen, A.B., Givskov, M., 2002. Food spoilage—interactions between food spoilage bacteria. Int. J. Food Microbiol. 78 (1–2), 79–97.

Guil-Guerrero, J.L., Ramos, L., Moreno, C., Zúñiga-Paredes, J.C., Carlosama-Yepez, M., Ruales, P., 2016. Antimicrobial activity of plant-food by-products: a review focusing on the tropics. Livest. Sci. 189, 32–49.

Guo, M., Jin, T.Z., Gurtler, J.B., Fan, X., Yadav, M.P., 2018. Inactivation of *Escherichia coli* O157:H7 and salmonella and native microbiota on fresh strawberries by antimicrobial washing and coating. J. Food Prot. 81 (8), 1227–1235.

Gupta, S., Abu-Ghannam, N., 2011. Recent developments in the application of seaweeds or seaweed extracts as a means for enhancing the safety and quality attributes of foods. Innov. Food Sci. Emerg. Technol. 12 (4), 600–609.

Gutiérrez-del-Río, I., Fernández, J., Lombó, F., 2018. Plant nutraceuticals as antimicrobial agents in food preservation: terpenoids, polyphenols and thiols. Int. J. Antimicrob. Agents 52 (3), 309–315.

Gutiérrez-Larraínzar, M., Rúa, J., Caro, I., de Castro, C., de Arriaga, D., García-Armesto, M.R., del Valle, P., 2012. Evaluation of antimicrobial and antioxidant activities of natural phenolic compounds against foodborne pathogens and spoilage bacteria. Food Control 26 (2), 555–563.

Gyawalli, R., Ibrahim, S.A., 2014. Natural products as antimicrobial agents. Food Control 46, 412–429.

Hamed, I., Özogul, F., Regenstein, J.M., 2016. Industrial applications of crustacean by-products (chitin, chitosan, and chitooligosaccharides): a review. Trends Food Sci. Technol. 48, 40–50.

Hintz, T., Matthews, K.K., Di, R., 2015. The use of plant antimicrobial compounds for food preservation. Biomed. Res. Int. 2015, 1–12.

Hirsch, D.B., Baieli, M.F., Urtasun, N., Lázaro-Martínez, J.M., Glisoni, R.J., Miranda, M.V., et al., 2018. Sulfanilic acid-modified chitosan mini-spheres and their application for lysozyme purification from egg white. Biotechnol. Prog. 34 (2), 387–396.

Hoffmann, J.A., Hetru, C., 1992. Insect defensins: inducible antibacterial peptides. Immunol. Today 13 (10), 411–415.

Hondrodimou, O., Kourkoutas, Y., Panagou, E.Z., 2011. Efficacy of natamycin to control fungal growth in natural black olive fermentation. Food Microbiol. 28 (3), 621–627.

Hosseinnejad, M., Jafari, S.M., 2016. Evaluation of different factors affecting antimicrobial properties of chitosan. Int. J. Biol. Macromol. 85, 467–475.

Hugo, C.J., Hugo, A., 2015. Current trends in natural preservatives for fresh sausage products. Trends Food Sci. Technol. 45 (1), 12–23.

Jamaluddin, N., Stuckey, D.C., Ariff, A.B., Faizal Wong, F.W., 2017. Novel approaches to purifying bacteriocin: a review. Crit. Rev. Food Sci. Nutr. 13, 1–13.

Khan, M.K., Ahmad, K., Hassan, S., Imran, M., Ahmad, N., Xu, C., 2018. Effect of novel technologies on polyphenols during food processing. Innov. Food Sci. Emerg. Technol. 45, 361–381.

Khorshidian, N., Yousefi, M., Khanniri, E., Mortazavian, A.M., 2018. Potential application of essential oils as antimicrobial preservatives in cheese. Innov. Food Sci. Emerg. Technol. 45, 62–72.

Kowalska, H., Czajkowska, K., Cichowska, J., Lenart, A., 2017. What's new in biopotential of fruit and vegetable by-products applied in the food processing industry. Trends Food Sci. Technol. 67, 150–159.

Krepker, M., Shemesh, R., Poleg, Y.D., Kashi, Y., Vaxman, A., Segal, E., 2017. Active food packaging films with synergistic antimicrobial activity. Food Control 76, 117–126.

Li, Y., Xiang, Q., Zhang, Q., Huang, Y., Su, Z., 2012. Overview on the recent study of antimicrobial peptides: origins, functions, relative mechanisms and application. Peptides 37 (2), 207–215.

Li, Y.Q., Han, Q., Feng, J.L., Tian, W.L., Mo, H.Z., 2014. Antibacterial characteristics and mechanisms of ε-poly-lysine against *Escherichia coli* and *Staphylococcus aureus*. Food Control 43, 22–27.

Liao, H., Jiang, L., Cheng, Y., Liao, X., Zhang, R., 2018. Application of nisin-assisted thermosonication processing for preservation and quality retention of fresh apple juice. Ultrason. Sonochem. 42, 244–249.

Liu, Q., Meng, X., Li, Y., Zhao, C.-N., Tang, G.-Y., Li, H.-B., 2017. Antibacterial and antifungal activities of spices. Int. J. Mol. Sci. 18 (1283), 1–62.

Liu, W., Lu, J., Ye, A., Xu, Q., Tian, M., Kong, Y., et al., 2018. Comparative performances of lactoferrin-loaded liposomes under *in vitro* adult and infant digestion models. Food Chem. 258, 366–373.

Lopes, N.A., Brandelli, A., 2017. Nanostructures for delivery of natural antimicrobials in food. Crit. Rev. Food Sci. Nutr. 10, 1–11.

López, E.I., Recio, I., 2006. Antibacterial activity of peptides and folding variants from milk proteins. Int. Dairy J. 16 (11), 1294–1305.

Lopez-Pena, C.L., McClements, D.J., 2015. Impact of a food-grade cationic biopolymer (ε-polylysine) on the digestion of emulsified lipids: in vitro study. Food Res. Int. 75, 34–40.

Lorenzo, J.M., Batlle, R., Gómez, M., 2014. Extension of the shelf-life of foal meat with two antioxidant active packaging systems. LWT- Food Sci. Technol. 59 (1), 181–188.

Lucchesi, M.E., Chemat, F., Smadja, J., 2004. Solvent-free microwave extraction of essential oil from aromatic herbs: comparison with conventional hydro-distillation. J. Chromatogr. A 1043 (2), 323–327.

Lucera, A., Costa, C., Conte, A., Del Nobile, M.A., 2012. Food applications of natural antimicrobial compounds. Front. Microbiol. 3, 287. 1–13.

Martinez-Hernández, G.B., Amodio, M.L., Valeria de Chiara, M.L., Russo, P., Colelli, G., 2017. Microbial inactivations with hydrolysed lactoferrin and other natural antimicrobials in fresh-cut fennel. LWT- Food Sci. Technol. 84, 353–358.

Matan, N., Rimkeeree, H., Mawson, A.J., Chompreeda, P., Haruthaithanasan, V., Parker, M., 2006. Antimicrobial activity of cinnamon and clove oils under modified atmosphere conditions. Int. J. Food Microbiol. 107 (2), 180–185.

Meira, S.M.M., Zehetmeyer, G., Jardim, A.I., Scheibel, J.M., Oliveira, R.V.B., Brandelli, A., 2014. Polypropylene/montmorillonite nanocomposites containing nisin as antimicrobial food packaging. Food Bioprocess Technol. 7 (11), 3349–3357.

Mirabella, N., Castellani, V., Sala, S., 2014. Current options for the valorization of food manufacturing waste: a review. J. Clean. Prod. 65, 28–41.

Modugno, C., Loupiac, C., Bernard, A., Jossier, A., Neiers, F., Perrier-Cornet, J.M., et al., 2018. Effect of high pressure on the microbial activity and secondary structure of the bacteriocin nisin. Innov. Food Sci. Emerg. Technol. 47, 9–15.

Mohanty, D.P., Mohapatra, S., Misra, S., Sahu, P.S., 2016. Milk derived bioactive peptides and their impact on human health—a review. Saudi J. Biol. Sci. 23 (5), 577–583.

Montiel, R., Martín-Cabrejas, I., Peirotén, Á., Medina, M., 2016. Reuterin, lactoperoxidase, lactoferrin and high hydrostatic pressure treatments on the characteristics of cooked ham. Innov. Food Sci. Emerg. Technol. 35, 111–118.

Nazzaro, F., Fratianni, F., Neve, O.M., d'Acierno, A., Coppola, R., 2018. Recovery of biomolecules of high benefit from food waste. Curr. Opin. Food Sci. 22, 43–54.

Noori, S., Zeynali, F., Almasi, H., 2018. Antimicrobial and antioxidant efficiency of nanoemulsion-based edible coating containing ginger (*Zingiber officinale*) essential oil and its effect on safety and quality attributes of chicken breast fillets. Food Control 84, 312–320.

O'Connor, P.M., Ross, R.P., Hill, C., Cotter, P.D., 2015. Antimicrobial antagonists against food pathogens: a bacteriocin perspective. Curr. Opin. Food Sci. 2, 51–57.

Ortega-Ramirez, L.A., Rodriguez-Garcia, I., Leyva, J.M., Cruz-Valenzuela, M.R., Silva-Espinoza, B.A., Gonzalez-Aguilar, G.A., et al., 2014. Potential of medicinal plants as antimicrobial and antioxidant agents in food industry: a hypothesis. J. Food Sci. 79 (2), R129–R137.

Pabast, M., Shariatifar, N., Beikzadeh, S., Jahed, G., 2018. Effects of chitosan coatings incorporating with free or nano-encapsulated *Satureja* plant essential oil on quality characteristics of lamb meat. Food Control 91, 185–192.

Palmeri, R., Parafati, L., Restuccia, C., Fallico, B., 2018. Application of prickly pear fruit extract to improve domestic shelf life, quality and microbial safety of sliced beef. Food Chem. Toxicol. 118, 355–360.

Pan, M., Shen, S., Chen, L., Dai, B., Xu, L., Yun, J., et al., 2015. Separation of lactoperoxidase from bovine whey milk by cation exchange composite cryogel embedded macroporous cellulose beads. Sep. Purif. Technol. 147, 132–138.

Panda, S.K., Rath, C.C., 2012. Phytochemicals as natural antimicrobials: prospects and challenges. In: Gupta, V.K. (Ed.), Bioactive Phytochemicals-Perspectives for Modern Medicine 1. Daya Publishing House, New Delhi.

Pandey, K.B., Rizci, S.I., 2009. Plant polyphenols as dietary antioxidants in human health and disease. Oxidative Med. Cell. Longev. 2 (5), 270–278.

Parada, J.L., Caron, C.R., Medeiros, A.B.P., Soccol, R., 2007. Bacteriocins from lactic acid bacteria: purification, properties and use as biopreservatives. Braz. Arch. Biol. Technol. 50 (3), 521–542.

Patrignani, F., Siroli, L., Serrazanetti, D.I., Gardini, F., Lanciotti, R., 2015. Innovative strategies based on the use of essential oils and their components to improve safety, shelf-life and quality of minimally processed fruits and vegetables. Trends Food Sci. Technol. 46 (2, part B), 311–319.

Perinelli, D.R., Fagioli, L., Campana, R., Lam, J.K.W., Baffone, W., Palmieri, G.F., et al., 2018. Chitosan-based nanosystems and their exploited antimicrobial activity. Eur. J. Pharm. Sci. 117, 8–20.

Pina-Pérez, M.C., Ferrús, P.M.A., 2018. Antimicrobial potential of legume extracts against foodborne pathogens: a review. Trends Food Sci. Technol. 72, 114–124.

Pina-Pérez, M.C., Rivas, A., Martínez, A., Rodrigo, A., 2017. Antimicrobial potential of macro and microalgae against pathogenic and spoilage microorganisms in food. Food Chem. 235, 34–44.

Pisoschi, A.M., Pop, A., Georgescu, C., Turcuş, V., Olah, N.K., Mathe, E., 2018. An overview of natural antimicrobials role in food. Eur. J. Med. Chem. 143, 922–935.

Potter, R., Hansen, L.T., Gill, T.A., 2005. Inhibition of foodborne bacteria by native and modified protamine: importance of electrostatic interactions. Int. J. Food Microbiol. 103 (1), 23–34.

Poverenov, E., Arnon-Rips, H., Zaitsev, Y., Bar, V., Danay, O., Horev, B., et al., 2018. Potential of chitosan from mushroom waste to enhance quality and storability of fresh-cut melons. Food Chem. 268, 233–241.

Qin, Y.Y., Zhang, Z.H., Li, L., Xiong, W., Shi, J.Y., Zhao, T.R., et al., 2013. Antioxidant effect of pomegranate rind powder extract, pomegranate juice, and pomegranate seed powder extract as antioxidants in raw ground pork meat. Food Sci. Biotechnol. 22 (4), 1063–1069.

Rai, M., Pandit, R., Gaikwad, S., Kövics, G., 2016. Antimicrobial peptides as natural biopreservative to enhance the shelf-life of food. J. Food Sci. Technol. 53 (9), 3381–3394.

Raybaudi-Massilia, R.M., Mosqueda-Melgar, J., Soliva-Fortuny, R., Martín-Belloso, O., 2009. Control of pathogenic and spoilage microorganisms in fresh-cut fruits and fruit juices by traditional and alternative natural antimicrobials. Compr. Rev. Food Sci. Food Saf. 8 (3), 157–180.

Regulation (EC) No 1332/2008, 2008, Regulation (EC) No 1332/2008 of the European Parliament and of the Council of 16 December 2008 on food enzymes and amending Council Directive 83/417/EEC, Council Regulation (EC) No 1493/1999, Directive 2000/13/EC, Council Directive 2001/112/EC and Regulation (EC) No 258/97.

Regulation (EC) No 1333/2008, 2008, Regulation (EC) No 1333/2008 of the European Parliament and of the Council of 16 December 2008 on food additives. Official Journal of the European Union, L354/16, 31.12.2008.

Regulation (EC) No 1334/2008, 2008, Regulation (EC) No 1334/2008 of the European Parliament and of the Council of 16 December 2008 on flavourings and certain food ingredients with flavouring properties for use in and on foods and amending Council Regulation (EEC) No 1601/91, Regulations (EC) No 2232/96 and (EC) No 110/2008 and Directive 2000/13/EC.

Regulation (EC) No 178/2002, 2002, Regulation (EC) No 178/2002 of the European Parliament and of the Council of 28 January 2002 laying down the general principles and requirements of food law, establishing the European Food Safety Authority and laying down procedures in matters of food safety.

Regulation (EC) No 1935/2004, 2004, Regulation (EC) No 1935/2004 of the European Parliament and of the Council of 27 October 2004 on materials and articles intended to come into contact with food and repealing Directives 80/590/EEC and 89/109/EEC

Regulation (EU) 2015/2283, 2015, Regulation (EU) 2015/2283 of the European Parliament and of the Council of 25 November 2015 on novel foods, amending Regulation (EU) No 1169/2011 of the European Parliament and of the Council and repealing Regulation (EC) No 258/97 of the European Parliament and of the Council and Commission Regulation (EC) No 1852/2001.

Rhim, J.W., Park, H.M., Ha, C.S., 2013. Bio-nanocomposites for food packaging applications. Prog. Polym. Sci. 38 (10–11), 1629–1652.

Riaz, Q.U., Masud, T., 2013. Recent trends and applications of encapsulating materials for probiotic stability. Crit. Rev. Food Sci. Nutr. 53 (3), 231–244.

Riaz, A., Lei, S., Akhtar, H.M.S., Wan, P., Chen, D., Jabbar, S., et al., 2018. Preparation and characterization of chitosan-based antimicrobial active food packaging film incorporated with apple peel polyphenols. Int. J. Biol. Macromol. 114, 547–555.

Ribeiro-Santos, R., Andrade, M., Ramos de Melo, N., Sanches-Silva, A., 2017. Use of essential oils in active food packaging: recent advances and future trends. Trends Food Sci. Technol. 61, 132–140.

Said, L.B., Fliss, I., Offret, C., Beaulieu, L., 2018. Antimicrobial Peptides: The New Generation of Food Additives. Reference Module in Food Science. Elsevier, Amsterdam.

Shah, B., Davidson, P.M., Zhong, Q., 2012. Nanocapsular dispersion of thymol for enhanced dispersibility and increased antimicrobial effectiveness against *Escherichia coli* O157:H7 and *Listeria monocytogenes* in model food systems. Appl. Environ. Microbiol. 78 (23), 8448–8453.

Shen, H.S., Shao, S., Chen, J.C., Zhou, T., 2017. Antimicrobials from mushrooms for assuring food safety. Compr. Rev. Food Sci. Food Saf. 16 (2), 316–329.

Shit, S.C., Shah, P.M., 2014. Edible polymers: challenges and opportunities. J. Polym. 2014, 427259. 1–13.

Shokri, S., Ehsani, A., 2017. Efficacy of whey protein coating incorporated with lactoperoxidase and α-tocopherol in shelf life extension of pike-perch fillets during refrigeration. LWT- Food Sci. Technol. 85 (part A), 225–231.

Shukla, S.C., Singh, A., Pandey, A.K., Mishra, A., 2012. Review on production and medical applications of ε-polylysine. Biochem. Eng. J. 65, 70–81.

Sibel, A.A., 2014. Dairy-derived antimicrobial peptides: action, mechanisms, pharmaceutical uses and production proposals. Trends Food Sci. Technol. 36, 79–95.

Silva, L.M., Hill, L.E., Figueiredo, E., Gomes, C.L., 2014. Delivery of phytochemicals of tropical fruit by-products using poly (DL-lactide-co-glycolide) (PLGA) nanoparticles: synthesis, characterization, and antimicrobial activity. Food Chem. 165, 362–370.

Silva, F., Domingues, F.C., Nerín, C., 2018. Trends in microbial control techniques for poultry products. Crit. Rev. Food Sci. Nutr. 58 (4), 591–609.

Silvan, J.M., Martinez-Rodriguez, A.J., 2018. Food by-Products as Natural Source of Bioactive Compounds against Campylobacter. Reference Module in Food Science. Elsevier, Amsterdam.

Singh, P., Wani, A.A., Saengerlaub, S., 2011. Active packaging of food products: recent trends. Nutr. Food Sci. 41 (4), 249–260.

Smid, E.J., Gorris, L.G.M., 2007. Natural antimicrobials for food preservation. In: Rahman, S.R. (Ed.), Handbook of Food Preservation. CRC Press, Florida, pp. 216–221.

Sokovic, M., Glamoclija, J., Marin, P.D., Brkic, D., Griensven, L.J., 2010. Antibacterial effects of the essential oils of commonly consumed medicinal herbs using an in vitro model. Molecules 15 (11), 7532–7546.

Song, Z., Li, F., Guan, H., Xu, Y., Fu, Q., Li, D., 2017. Combination of nisin and ε-polylysine with chitosan coating inhibits the white blush of fresh-cut carrots. Food Control 74, 34–44.

Tajkarimi, M.M., Ibrahim, S.A., Cliver, D.O., 2010. Antimicrobial herb and spice compounds in food. Food Control 21 (9), 1199–1218.

Tang, S., Prodhan, Z.H., Biswas, S.K., Le, C., Sekaran, S.D., 2018. Antimicrobial peptides from different plant sources: isolation, characterization, and purification. Phytochemistry 154, 94–105.

Taveira, M., Silva, L.R., Vale-Silva, L.A., Pinto, E., Valentão, P., Ferreres, F., et al., 2010. *Lycopersicon esculentum* seeds: an industrial byproduct as an antimicrobial agent. J. Agric. Food Chem. 58, 9529–9536.

Tiwari, B.J., Valdramidis, V.P., O'Donell, C.P., Muthukumarappan, K., Bourke, P., Cullen, P.J., 2009. Application of natural antimicrobials for food preservation. J. Agric. Food Chem. 57 (14), 5987–6000.

Tsiraki, M.I., Karam, L., Abiad, M.G., Yehia, H.M., Savvaidis, I.N., 2017. Use of natural antimicrobials to improve the quality characteristics of fresh "phyllo"–A dough-based wheat product–shelf life assessment. Food Microbiol. 62, 153–159.

Turgis, M., Vu, K.D., Dupont, C., Lacroix, M., 2012. Combined antimicrobial effect of essential oils and bacteriocins against foodborne pathogens and food spoilage bacteria. Food Res. Int. 48 (2), 696–702.

Tylewicz, U., Nowacka, M., Martin-García, B., Wiktor, A., Gómez-Caravaca, A.M., 2018. Target sources of polyphenols in different food products and their processing by-products. In: Galanakis, C.M. (Ed.), Polyphenols: Properties, Recovery, and Applications. Woodhead Publishing, Duxtor, pp. 135–175.

Valdés, A., Ramos, M., Beltrán, A., Jiménez, A., Garrigós, M.C., 2017. State of the art of antimicrobial edible coatings for food packaging applications. Coatings 7 (4), 56. 1–23.

Venskutonis, P.R., 2004. Food additives: the dilemma of synthetic or natural. Acta Aliment. 33 (1), 1–5.

Verlee, A., Mincke, S., Stevens, C.V., 2017. Recent developments in antibacterial and antifungal chitosan and its derivatives. Carbohydr. Polym. 164, 268–283.

Wang, X., Yue, T., Lee, T., 2015. Development of Pleurocidin-poly(vinyl alcohol) electrospun antimicrobial nanofibers to retain antimicrobial activity in food system application. Food Control 54, 150–157.

Weiss, J., Gaysinsky, S., Davidson, M., McClements, J., 2009. Nanostructured encapsulation systems: food antimicrobials. In: Barbosa-Cánovas, G., Mortimer, A., Lineback, D., Spiess, W., Buckle, K., Cononna, P. (Eds.), Global Issues in Food Science and Technology. Academic Press, Massachusetts, pp. 425–479.

Wenquiang, G., Shufen, L., Ruixiang, Y., Shaokun, T., Can, Q., 2007. Comparison of essential oils of clove buds extracted with supercritical carbon dioxide and other three traditional extraction methods. Food Chem. 101 (4), 1558–1564.

Woraprayote, W., Malila, Y., Sorapukdee, S., Swetwiwathana, A., Benjakul, S., Visessanguan, W., 2016. Bacteriocins from lactic acid bacteria and their applications in meat and meat products. Meat Sci. 120, 118–132.

Zheng, L., Bae, Y.M., Jung, K.S., Heu, S., Lee, S.Y., 2013. Antimicrobial activity of natural antimicrobial substances against spoilage bacteria isolated from fresh produce. Food Control 32 (2), 665–672.

Zimet, P., Mombrú, Á.W., Faccio, R., Brugnini, G., Miraballes, I., Rufo, C., et al., 2018. Optimization and characterization of nisin-loaded alginate-chitosan nanoparticles with antimicrobial activity in lean beef. LWT- Food Sci. Technol. 91, 107–116.

Zou, Y., Lee, H.Y., Seo, Y.C., Ahn, J., 2012. Enhanced antimicrobial activity of nisin-loaded liposomal nanoparticles against foodborne pathogens. J. Food Sci. 77 (3), M165–M170.

# Improving the shelf-life of food products by nano/micro-encapsulated ingredients

5

*Hamed Hosseini**, *Zhila Tajiani*†, *Seid Mahdi Jafari*‡
*Food Additives Department, Food Science and Technology Research Institute, Research Center for Iranian Academic Center for Education, Culture and Research (ACECR), Khorasan Razavi Branch, Mashhad, Iran, †Agriculture Faculty, Ferdowsi University of Mashhad, Mashhad, Iran, ‡Faculty of Food Science and Technology, Gorgan University of Agricultural Sciences and Natural Resources, Gorgan, Iran

## Chapter outline



Food Quality and Shelf Life. https://doi.org/10.1016/B978-0-12-817190-5.00005-7

# 1 Introduction

Among about 1,250,000 references found in the field of food science and engineering by 2017, more than 52,000 references are related to food preservation, and based on the number of publications, the United States has the largest share (15.1%), followed by China, Spain, India, and Japan with 9.8%, 8.1%, 5.9%, and 5.7% of the publications, respectively (Konur and Grumezescu, 2017). Based on FAO statistics released in 2015 and 2016, 794 (10.8%) and 815 (11.0%) million people, respectively, have suffered from malnutrition throughout the world and these numbers are growing in spite of all the efforts made; as the World Bank reported in 2017, 83 million of people from 45 nations had nothing to eat (Prosekov and Ivanova, 2018). These statistics reveal that improvement in the production and protection of foods in both fresh and processed forms against various damages, namely chemical, physical, and microbial is a predominant concern for food scientists and societies. Besides microbial spoilage, the occurrence of some deteriorative reactions, including oxidation of lipids and proteins (chemical damage), enzymatic/nonenzymatic browning (biochemical reactions), moisture absorption, loss, or migration, and phase separation in emulsion systems (physical instability) are major factors contributing to many adverse changes in food products, resulting in rancidity and off-flavor, variation in appearance, texture, and aromatic properties of food, as well as loss of bioactive and functional ingredients (Kong and Singh, 2016).

Packaging is an important step in improving food stability (Hoseinnejad et al., 2018). Synthetic polymers with sufficient flexibility, transparency, and light weight are widely used for food packaging, while they are associated with serious environmental harms due to their nonbiodegradable properties (Siracusa et al., 2008). Microencapsulation of food ingredients to protect them against adverse process and storage circumstances is another significant issue being increasingly studied, such as microencapsulation of blueberry powder with whey protein isolate (WPI) as wall material by spray drying, resulting in effective alleviation of anthocyanin degradation and prolonging the blueberry extract shelf life (Flores et al., 2014).

Incorporating the natural bioactive compounds extracted from herbs and plants into fresh and processed products is another approach commonly used in order to extend the shelf life of food commodities (Ganje et al., 2016; Jafari et al., 2017). From the earliest times (about 1550 BC), various plant sources, such as vegetables, fruits, cereals, roots, pulses, etc., as well as their extracts have exhibited important roles in foods, medicines, and cosmetics. Today, from a world trade perspective, there are two regions for growing spices and herbs: the tropical zone, which is the source of capsicums, pepper, cardamom, nutmeg/mace, allspice/pimento, cloves, vanilla, ginger, cassia, cinnamon, and turmeric; and the nontropical area, the source of thyme, cumin, mustard, coriander, sesame seeds, oregano, bay, and the mints (Embuscado and Shahidi, 2015). These sources contain high-added value bioactive compounds, including the antioxidants (terpenoids, flavonoids, lignans, polyphenolics, sulfides, saponins, carotenoids, phthalides, curcumins, plant sterols, and coumarins), essential oils (EOs), and volatile compounds with remarkable antimicrobial, nutritive, and medicinal features, which made them suitable to formulate and produce functional foods (Granato et al., 2017; Murcia et al., 2004), and to extend their shelf life.

However, most of these compounds have some disadvantages like off-flavors/odors, low solubility in water and/or oil, degradation by oxygen, light, moisture, temperature, and presence of unsaturated bonds within their molecular structures, and low stability during food processing, formulation, and storage (Abd El-Salam et al., 2016; Shishir et al., 2018; Ozkan et al., 2019). Therefore, solving these issues by considering the retention of health features along with hiding the unpleasant sensory properties of the ingredients without any negative impact on the carrier food quality is a critical and challenging topic to be investigated by food professionals (Abd El-Salam et al., 2016). In this regard, micro/nanoencapsulation techniques have been introduced by many related studies as a reasonable and practical alternative (Jafari et al., 2008; Akhavan Mahdavi et al., 2014). Micro- and nanoencapsulation are the two foremost processes in encapsulation technology. Each of them presents a particular ability in promoting the functionality of bioactive ingredients (Shishir et al., 2018).

Encapsulation is a well-defined and developing process in which one substance (target material) could be surrounded by external (wall) materials (capsule form), or dispersed in surrounding material layers (matrix form), or a combination of these methods could be used (Mahdavee Khazaei et al., 2014). The formed particles will have a diameter varying from nanometer (nanoencapsulation) to millimeter scale (Ray et al., 2016). Encapsulation technology can be applied to produce bioactive ingredients effectively protected against light, oxygen, free radicals, etc. Many studies executed regarding the phytochemicals focused on their chemical classification, biological properties, and dietary sources, as well as the application of the nano- and micro-encapsulation approaches for protecting the phytochemicals against degradation and enhancing their bioavailability, solubility, and functionality (Lee and Wong, 2014).

This chapter aims to overview the influence of commonly used micro- and nano-encapsulation techniques by various food-grade biopolymers as encapsulants (mainly proteins, polysaccharides, and their blends) on the efficiency of bioactive ingredients for protection of food products against the most important chemical and microbial spoilage phenomena occurring throughout the food preparation chain.

# 2 Chemical spoilage of foods

## *2.1 Lipid oxidation*

According to the water activity ($a_w$) values and total moisture content (MC) of foods, they are usually classified into three main categories, namely low- ($0.00 < a_w < 0.60$; MC < 25%), intermediate- ($0.60 < a_w < 0.85$; 15 < MC < 50%), and high-moisture ($a_w > 0.85$; MC > 50%) (Knechtges, 2012). Milk and egg powders ($a_w \sim 0.20$ and 0.30, respectively), noodles ($a_w \sim 0.50$), crackers ($a_w \sim 0.10$), and dry kibbles ($a_w \sim 0.35$–0.55) are among food products in the first group. Wheat flour, raw peanuts, and dried fruits as intermediate-moisture foods show $a_w$ values of 0.70, 0.60–0.70, and 0.70, respectively, while most fresh foods are located in the high-moisture category, containing 60%–95% moisture, presenting $a_w$ levels close to 1.00 (Nelson and Labuza, 1992). Microbial- or enzymatic-induced deterioration is the predominant pathway responsible for quality loss in high-moisture products, while auto-oxidation is the major cause of deterioration in low-moisture products (dry foods) (Hu, 2016) such as nuts (Hosseini et al., 2014; Ghorbani and Hosseini, 2017).

Lipids are known as the main macrocomponents in foods with minimum stability compared to the other constituents, and hence, the shelf life of fatty foods (fried foods, margarines, nuts, and mayonnaise), particularly those containing high levels of polyunsaturated fatty acids like seafood, meat, and poultry, is strongly affected by oxidative rancidity (Taghvaei and Jafari, 2015). Among other products susceptible to oxidative rancidity are cookies, biscuits, prepared mixes, dried whole milk and fruits, ice cream powders, and coffee (Kong and Singh, 2016). As an example, since seafood, poultry, and meat have moderately high amounts of unsaturated fatty acids, metal catalysts, heme pigments, and several other oxidizing agents, they are prone to oxidative deterioration (Akhavan Mahdavi et al., 2014; Pourashouri et al., 2014a,b; Hosseini et al., 2019).

Progression of lipid oxidation in foods leads to development in off-flavors like "warmed-over flavors" in meat (Rahman et al., 2009), as well as degradation of proteins (Hosseini et al., 2014; Ghorbani and Hosseini, 2017), alteration in color, loss of vitamins, and generation of toxic substances (Singh and Cadwallader, 2004; Singh and Anderson, 2004), such as reactive aldehydes and free radicals. In fact, the double bonds present in fatty acids undergo autoxidation as they are confronted with hydrogen, oxygen, and enzymes, which is followed by a three-steps reaction (initiation, propagation, and termination), and catalyzed by enzymes, photosensitizers, and transition metals. The unstable hydroperoxides produced at earlier stages of lipid oxidation are subsequently degraded into secondary oxidation products, namely ketones, aldehydes, and alcohols; and these volatile products develop the off-flavor in food products (Rahman et al., 2009). Lipid oxidation occurrences in foods having both protein and lipid fractions like fish and meat, is associated with oxidation of the proteins that later leads to a considerable influence on the texture of the product. Therefore, lipid oxidation will lead to important changes in the sensory characteristics of the products including flavor, odor, texture and color, which are easily detected by the consumer, and hence, this reaction significantly limits the product shelf life (Jacobsen and Decker, 2010; Hosseini et al., 2014, 2019; Ghorbani and Hosseini, 2017).

The adverse influences associated with intake of free radicals/reactive oxygen species (ROS such as superoxide radicals, hydroxyl radicals, singlet oxygen, and hydrogen peroxide) on both food and biological ecosystems have been recognized. These molecules are chemically composed of one or more unpaired electrons; therefore, they are greatly unstable and could damage other biomolecules (Prakash et al., 2015). Despite some biologically positive roles defined for ROS (Halliwell, 1997), they frequently catalyze a number of unfavorable reactions, including lipid oxidation, decline in membrane fluidity, membrane damage, and DNA mutation that could result in cancer occurrences (Pietta, 2000). Today, most people, either consumers or food suppliers, are focusing on herbal-based antioxidants having low or no side effects (Hosseini et al., 2018b), because it is claimed that synthetic antioxidants like butylated hydroxyanisole (BHA), tertiary butylhydroquinone (THBQ), and butylated hydroxytoluene (BHT) have carcinogenic activity in mammalian systems (Shahidi and Zhong, 2010; Prakash et al., 2015; Rafiee et al., 2012).

## 2.2 Aroma and flavor loss

The smell/olfaction is referred to as an essential sense for identifying and evaluating foodstuffs as it can provide a pleasant feeling or warns about occurrences of deteriorative reactions in foods. The smell sense has always been related to aroma, perfume, or flavor (Ohloff, 1994). Flavor is a complex sensation involving the compounds attributed to odor and taste, as well as mouth and visual feeling. Representation of some molecular characteristics in aroma compounds is a necessary factor for their smell perception; relatively low water solubility (low polarity), sufficient vapor pressure, and high lipophilicity are among them. Generally, the aroma compounds have a molar mass lower than 400 Da (Kfoury et al., 2016). There are two perception limits for odorants, named Weber-Fechner's law: the lowest level of an odorant required to be detected, called the limit or threshold of detection; and the lowest concentration of an aroma compound required for its recognition, which is considered the limit or threshold of recognition (Bauer et al., 2001). Bauer et al. (2001) stated that the foodstuffs overall flavor is strongly affected by two principal components, namely the key odorants of the product providing the predominant aroma of the food, and the compounds developing off-flavors, which degrade the flavor substances in food.

# 3 The role of bioactive ingredients in preventing the chemical spoilage

Natural antioxidants are not only associated with food quality but are also promoting the healthiness of consumers. Most of these compounds represent radical scavenging capacity via donating a hydrogen atom to the active radicals present in the product, as can be seen for phenolic compounds, where they transfer a hydrogen atom to active radicals for delaying the propagation phase of the lipid oxidation reaction; while, the other types render such capacity through an electron transfer mechanism, as is seen for carotenoids having $^{\bullet}NO_2$ and $^{\bullet}CCl_3OO$ groups, and catechin analogs containing the peroxyl radicals ($^{\bullet}OH$) within their structures. These compounds are called free radical scavengers (FRS) and their efficiency is also influenced by the energy of the scavenging-induced antioxidant radicals, as the energy is low (such as phenolics, α-tocopherol, and catechol). The possibility of contributing a radical in the oxidation of other substrates is decreased and thus, the rate of oxidation reaction drops. On the other hand, reactions (electron transfer) between antioxidant radicals may result in formation of nonradical species, reducing the concentration of free radicals in the system. Phenolic compounds existing in plants could be categorized as derivatives of hydroxycinnamic acid (especially chlorgenic acid, and rosmarinic acid), phenolic acids (e.g., caffeic, gallic, and chlorogenic acids), and flavonoids (e.g., anthocyanins, catechins, quercetin, and rutin) (Martinez-Gracia et al., 2015; Chen and Xu, 2018; Lee and Wong, 2014). Furthermore, citric acid, α-tocopherol, and vitamin C are able to limit or prevent oxidation progression via scavenging the radical oxygen in the product (Kong and Singh, 2016).

## 3.1 Phenolipids

The antioxidant capacity of phenolic substances comes from their chemical structure and reducing properties, which, as they become able to neutralize the free radicals, chelate the transitional metals, as well as quenching the singlet oxygen (Martinez-Gracia et al., 2015; Hosseini et al., 2018b).

Lipophenols or phenolipids naturally found in marine organisms are a combination of phenolic compounds and lipids; they are FRSs soluble in oil, possessing many health benefits highly similar to common phenolics, including antiallergic, antiinflammatory, antiviral, antimicrobial, and anticarcinogenic. Despite similar chemical properties, physical properties of the phenolipids are different from phenolics, as phenolipids are capable of reacting with polar and nonpolar compounds (Chen and Xu, 2018).

## 3.2 Anthocyanins

Anthocyanins are secondary metabolites in many plants (e.g., blueberries), which are widely applied in the food industry as water-soluble natural colorants with antioxidant properties. Instability against many parameters, including pH, temperature, oxygen, light, metallic ions, and enzymes during storage and processing is the principal drawback associated with these pigments (Patras et al., 2010), which can be covered using a protective layer like encapsulation (Akhavan Mahdavi et al., 2016a,b). In this regard, a number of authors used different carrier agents, such as WPI (Flores et al., 2015), a mixture of maltodextrin and gum Arabic (Tatar Turan et al., 2015), β-cyclodextrin (Wilkowsk et al., 2015), and a combination of these compounds (Tao et al., 2017) for protecting the bioactive ingredients of blueberry juice. Based on their results, the encapsulation performance using a properly combination of dissimilar wall materials was significantly higher than single wall materials due to formation of new complexes with amphiphilic and interfacial characteristics, as was confirmed by several studies (Davidov-Pardo et al., 2013; Klein et al., 2010); for example, Davidov-Pardo et al. (2013) used a combination of maltodextrin, zein, and mesquite gum to encapsulate grape seed extract.

## 3.3 Catechins

Catechins are another important group of antioxidants and FRSs; they are flavonoids (classified as flavanols) soluble in water with a bitter taste, and are commonly found in red wine, tea, fruits (black grapes, apricots, strawberries), cocoa, broad beans, and green beans (Fan et al., 2017; Andersen and Markham, 2006). Epigallocatechin-3-gallate (EGCG) is the main fraction of tea catechins postulated as a powerful antioxidant, and its abilities in free radical scavenging, reduction, and chelating the metals has been demonstrated by in vitro experiments (Paximada et al., 2017). Catechin also exhibits antibacterial influence (Fan et al., 2017) as this impact on *Escherichia coli*, *Staphylococcus aureus*, and *Pseudomonas aeruginosa* confirmed by Riso et al. (2002). However, this ingredient is mostly sensitive to oxygen and light, followed by oxidation, alkaline pH, and elevated relative humidity that limits its application in large-scale by the food industry (Okabe et al., 1999; Ananingsih et al., 2013). Considering these limitations, a plausible method to overcome the drawbacks of catechins (e.g., low lipophilicity of EGCG, short half-life, restricting it to be incorporated in lipid-based matrices) is its encapsulation (Paximada et al., 2017).

## 3.4 Carotenoids

Carotenoids are more than 600 polyene compounds presenting diverse colors ranging between yellow to red containing the 3–13 conjugates double bonds that are or are not associated with the hydroxylated ring structures (six carbons) at one or two bottoms of the molecule (Chen and Xu, 2018). Their strong antioxidant characteristics are attributed to presence of conjugated double bonds, keto groups, hydroxyl groups, and cyclohexane rings within their molecular structure (Lee and Wong, 2014). Their importance in the food field is attributed to their function at inactivation of singlet oxygen (Chen and Xu, 2018; Tan et al., 2014). Singlet oxygen has two electrons with opposed spin directions in the outer orbitals, and hence, it shows an electrophilic nature that will allow it to attach to the double bonds in unsaturated fatty acids, resulting in the hydroperoxide formation. Carotenoids may directly react with singlet oxygen, leading to inactivation of both of them (chemical quenching). The other antioxidative mechanism revealed for carotenoids is physical quenching, by which carotenoids attract the extra energy (22.4 and 37.5 kcal) from singlet oxygen, resulting in formation of excited carotenoids and triplet oxygen (ground state). Finally, the excited carotenoids safely transfer their additional energy to the adjacent medium trough rotational and vibrational mechanisms. In addition, the carotenoids return the photo-activated molecules, such as excited forms of chlorophyll, heme-containing proteins, and riboflavin, to their ground state, which prevents them from contributing in the generation rout of the singlet oxygen (see above). The efficiency of the second mechanism is enhanced when a carotenoid contains more than eight conjugated double bonds, as well as the oxygenated ring-like terminals (Chen and Xu, 2018). However, carotenoids are prone to oxidation, isomerization, and degradation owing to their highly unsaturated structure, and hence, their protection against adverse environmental factors is an essential subject to be considered for food applications; for instance, by using encapsulation technique. The most common carotenoids applied in foods are β-carotene, astaxanthine, lutein, and lycopene, of which the last two are commonly used as functional ingredients and colorants in food products (Tan et al., 2014).

β-Carotene is a pigment with strong red-orange color. It is usually found in vegetables and fruits and is known as the precursor of vitamin A, as well as an antioxidant and free radical scavenging compound (Franceschi et al., 2010). However, its use in the food industry is associated with some limitations, namely low water solubility and poor stability (Jain et al., 2015).

Astaxanthin pigment is a ketocarotenoid derived from various sources, namely plants (Adonis sp.), bacteria (*Agrobacterium aurantiacum*, *Bacillus circulans*), yeasts (*Phaffia rhodozyma*), crustaceans (shrimp), algae (*Haematococcus pluvialis*), and fish (especially, salmon) (Cunningham and Gantt, 2011). The antioxidant capacity of this pigment is considered as an important technological advantage, representing 10-fold higher than β-carotene and lutein or 100-times higher than α-tocopherol (Rao et al., 2014). However, astaxanthin is naturally known as an unstable compound against some processing variables, such as high temperatures, oxygen, and light, especially when it is released from a biological matrix. In fact, the astaxanthin structure contains a large number of conjugated bonds, which considerably inhibits its application in the food industry (Chen et al., 2007). Taking this into account, different encapsulation procedures have been developed for protecting the astaxanthin in food systems (Chen et al., 2007; Gomez-Estaca et al., 2016; Montero et al., 2016).

### 3.5 Curcumin

Several studies have confirmed the antioxidant abilities of curcumin (Menon and Sudheer, 2007). However, this ingredient exhibits some disadvantages, including reduced water solubility and chemical stability, resulting in its limited biological activity (Strimpakos and Sharma, 2008). Encapsulation techniques are a versatile and efficient method to overcome this problem (Rafiee et al., 2018), and many studies have focused on encapsulation procedures for improving the stability of curcumin, such as electrospinning (Blanco-Padilla et al., 2015) and supercritical antisolvent (Arango-Ruiz et al., 2018).

### 3.6 Ascorbate

Ascorbic acid/vitamin C is another FRS, which acts as a water-soluble component in both plant and animal tissues (Chen and Xu, 2018). It can be used as a preservative ingredient or additive in foods due to its antioxidant feature (Comunian et al., 2013). Similar to phenolic compounds, reduction potential of ascorbate is lower than peroxyl radicals and thus, their interaction generates an inactivated peroxyl radical. Ascorbic acid is also able to renew the oxidized α-tocopherol due to having a lower reduction potential in comparison to the α-tocopherol radical. Ascorbate is found widely in plants including citrus fruits, green vegetables, tomatoes, potatoes, and berries. The degradation and loss in the ascorbic acid content of foods may happen due to heat processing, prolonged storage, transition metals, and contact with air (Chen and Xu, 2018; Comunian et al., 2013). Its degradation is also related to a color loss, both in the absence and presence of amines (Comunian et al., 2013). To solve this problem, different microencapsulation procedures, including spray drying (Pierucci et al., 2006), melt dispersion, solvent dispersion (Uddin et al., 2001), liposome entrapment (Farhang et al., 2012), etc., can be used for protection of ascorbic acid against various influences interfering with its stability, resulting in masking its acidic unsavory taste and controlled release (Comunian et al., 2013).

On the other hand, Troise and Fogliano (2013) revealed that ascorbic acid also acts as a pro-oxidant when it is associated with iron or other metals cations, resulting in the generation of hydrogen peroxide and dehydroascorbate as oxidized form of ascorbic acid. It has been claimed that the presence of ascorbic acid in a food system containing transition metals and proteins (e.g., infant formula), enhances the formation of Maillard reaction products via contributing in degradation and polymerization of amino acids and protein in the product formula (Dittrich et al., 2006). Leclere et al. (2002) reported a threefold increase in tryptophan degradation rate and considerable increase in *N*-carboxymethyllysine formation (lysine blockage) by preparing a whey-lactose model trial containing iron-ascorbate system. So, encapsulation technique can also be applied to prevent the detrimental features of ascorbic acid (Jang and Lee, 2008) and metal cations (Caillard et al., 2010) in relevant food products.

### 3.7 α-Tocopherol (vitamin E)

α-Tocopherol is widely accepted as an important lipid-soluble antioxidant applicable in the food industry. Its antioxidant capacity stems from its ability to stabilize or quench free radicals (Cervantes and Ulatowski, 2017). However, vitamin E rapidly undergoes degradation in the presence of free-radicals and oxygen. This problem, as well as low solubility of vitamin E in aqueous environments, can be properly overcome using microencapsulation methods, resulting in improvement of its physicochemical stability, as well as a sustainable release on prolonged storage (Parthasarathi and Anandharamakrishnan, 2016). These α-tocopherol loaded microencapsules may be applied as antioxidants in high fat bakery products for limiting the auto oxidation reaction, resulting in a prolonged shelf life (Singh et al., 2018).

### 3.8 Chelating agents (chelators)

In general, chelators can be used in order to control the lipid oxidation. In this regard, phosphates are able to chelate free metals, and hence, they are known as an effective factor limiting the oxidation reactions. However, this ingredient is sensitive to heat, as hydrolysis of polyphosphates and changes in their functionality may occur during thermal processing (Kuda et al., 2004). For example, cooking meat causes an 80% loss in added polyphosphate (0.5% legal limit) (Decker and Mei, 1996). Moreover, Li et al. (1993) observed a 100% loss in sodium tripolyphosphate incorporated into raw turkey after 1 day. The presence of phosphatases in raw meat is another factor negatively impacting the phosphate during storage and the early stages of cooking. Fortunately, it has been revealed that encapsulation of phosphates appropriately allows their protection from the above-mentioned damages (Table 1) (Sickler et al., 2013).

## 4 Microbial spoilage

According to consumer survey results, the three most important measures for food purchase from the consumers' viewpoint were, respectively, price (66%), fresh/not spoiled (37%), and quality (33%). In fact, today's consumers expect a wide range of issues, mainly more competitive prices combined with guaranteed safety, less processed or preserved foods, strong sensorial quality, and promoted nutritional and functional properties, as well as highly simple preparation and a prolonged shelf life for the product (Rohr et al., 2005). Based on the above-mentioned information, food spoilage

**Table 1** Some encapsulated ingredients applied as antioxidants for extending the shelf life of foods

| Ingredient | Effective component | Wall material | Wall material ratio | Wall/core | Encapsulation procedure—Preparation | EE[a] (%) | Particle size | Food system | Results | References |
|---|---|---|---|---|---|---|---|---|---|---|
| Quercetin- Canola oil | Quercetin | Alginate and novation 2300 starch (1.4% amylose) | 2:1 | – | Co-extrusion | – | | Canola oil | The lowest levels for PV[b] (8.47 meq/kg oil), p-AV[c] (2.16), TOTOX[d] (19.1), and FFA[e] (0.19%) | Sun-Waterhouse et al. (2014) |
| Ethanolic extract of Terminalia arjuna | Phenolics (19.49%), tannins (7.44%) and flavonoids (2.65%) | Maltodextrin and gum Arabic | 4:1 | 5:1 | Oven drying | | | Vanilla chocolate dairy drink | Improving the shelf-life (based on sedimentation stability, pH, TBARs[f], and lightness value) up to 120 days, better than free extract | Sawale et al. (2017) |
| Ethanolic extract of ginger | Phenolics (8.81 mg GAE/g) | Oil, lecithin and glycerol | | | Nanoliposome | | 164.5 nm | Sunflower oil | The higher antioxidant activities than free extract and BHT based on all tests (FRAP[g], DPPH, PV, TBARs) | Ganji and Sayyed-Alangi (2017) |
| (1) Collagen hydrolysate, (2) pomegranate extract, (3) shrimp lipid extract | (1) Antioxidant peptides, (2) polyphenols (mainly as ellagitannins, ellagic acid and anthocyanins), (3) Astaxanthin and α-tocopherol | Phosphatidylcholine and glycerol | 56:42 | 5:0.1 | Liposome followed by freeze-drying—Ultrasound | (1, 3) >90 (2) 63 | (1) 198.9, (2) 274.2, (3) 282.9 nm | Surimi gels | Slight decrease in gel strength and contributed to stability during long-term frozen storage, unable to prevent the lipid oxidation | Marin et al. (2018) |
| | Epigallocatechin-3-gallate | Whey protein isolate (WPI, 20%), bacterial cellulose (BC, 8%) and Tween 20 (5%) | 5:2 | | Electrospraying—Ultrasound | 80 | 268 nm | Emulsion containing 10% olive oil | Enhancing the shelf life and solubility of hydrophilic EGCG in oily phase | Paximada et al. (2017) |
| | Quercetin | Soybean lecithin (as liposome) and chitosan (as liposome coating) | | | Liposome followed by rotary evaporation—Ultrasound | 71.14 | 600 nm | Recommended to use in the food industry | Better storage stability than free quercetin at both 4 and 25°C in dark or natural light | Hao et al. (2017) |
| Grape seeds and skins extracts | Anthocyanins and phenolic compounds | Glycerol monooleate, soy lecithin, oil (liquid: sunflower oil or solid: palm oil) | 0.1:0.5:9 | 10:0.1 | Nanoemulsion—High pressure homogenization | | 176 nm | Emulsion | The sunflower oil-based nanoemulsions with no degradation of the encapsulated compounds at 4–55°C, improving the antioxidant activity due to controlled release | Sessa et al. (2013) |

| | | | | | | | | | | |
|---|---|---|---|---|---|---|---|---|---|---|
| | Limonene | β-CD | | 8:1 | Inclusion complex followed by Spray-drying—Homogenization | 66 | 1–3 μm | Nonalcoholic beverage | Enhancement of flavor and improved shelf-life compared to free ingredient | Saldanha do Carmo et al. (2017) |
| | Chlorogenic acid | β-Cyclodextrin (β-CD) | | 10:1 | Inclusion complex followed by freeze-drying—Magnetic stirrer | | | | Improving the storage stability of chlorogenic acid, without significant difference for antimicrobial and antioxidant activities of complex and free forms | Zhao et al. (2010) |
| | Lutein, β-carotene, lycopene, and canthaxanthin | Egg yolk phosphatidylcholine and Tween 80 | | 1:0.72 | Liposome followed by vacuum-dried—Ultrasound | | >80 | Lipid | Enhancing the antioxidant activity based on DPPH and FRAP (lutein >β-carotene >lycopene >canthaxanthin), and oxidative stability (specially, lutein and β-carotene), as compared with free forms | Tan et al. (2014) |
| | β-Carotene | WPI, gum acacia | 2:1 | 3:0.025 | Coacervation—Homogenization and magnetic stirrer | 77.30 | 140.2 μm | Soybean oil | Improving the stability, antioxidant activity and controlled release during 3 months storage, spherical shape with continuous surface | Jain et al. (2015) |
| Astaxanthin-containing lipid extract from shrimp | Astaxanthin | Gum Arabic and phosphatidylcholine | | 10:1 | Spray drying—Ultra Turrax and ultrasound | >90 | 6 μm | Recommended for food industry | The higher antioxidant activity (18-fold) and shelf life (110 days at 5°C) than free ingredient | Montero et al. (2016) |
| Green tea | Polyphenols, flavonoids (caffeine and catechins) | Cashew gum and maltodextrin | 1:2 | 4:1 | Spray drying | 33 | | Has been recommended | Increasing the effectiveness of bioactive ingredients | Silva et al. (2018a,b) |
| Rice bran oil containing 0.5% β-carotene | β-Carotene | Casein and gum tragacanth | 2:1 | 2:1 | Complex coacervation | 79.36 | 159.71 μm | | Improving the stability, controlled release and prolonged antioxidant activity | Jain et al. (2016) |
| | Curcumin | Amaranth protein isolate and pullulan | 1:1 | | Electrospinning | 93 | 224.50–248.60 nm | In vitro digestion process | Improving the thermal stability and retention of antioxidant activity | Blanco-Padilla et al. (2015) |

*Continued*

**Table 1** Continued

| Ingredient | Effective component | Wall material | Wall material ratio | Wall/core | Encapsulation procedure—Preparation | EE[a] (%) | Particle size | Food system | Results | References |
|---|---|---|---|---|---|---|---|---|---|---|
| | Anthocyanin | Maltodextrin (MD), β-CD, WPI and gum Arabic | (0.46–1.51):(0.06–0.85):(7.90–8.60):(0.10–0.49) | 4:1 | Freeze drying | 86.70–90.50 | 86.69–97.75 μm | | Improving the thermal stability | (Tao et al., 2017) |
| | Ascorbic acid | Gelatin, gum Arabic | 1:1 | 2:0.75 | Complex coacervation followed by freeze drying—Ultra-Turrax | 99.57 | 65.98 μm | Corn oil, usable in hydrophobic matrices | Increasing the stability, improving the controlled release | Comunian et al. (2013) |
| | Catechin | β-CD | | 1:1 | Inclusion complex | | | Oil (0%–80%), sugar (0%–80%) or protein (0%–40%): models to mimic different dairy products, beverages and confectioneries | Improving the stability of ingredient in alkaline pH and high relative humidity | Ho et al. (2017) |
| | α-Tocopherol | Sodium alginate (1.5% w/v) and pectin (2.0% w/v) | 1.5:2 | 3.5:1 | Air atomization technique using an encapsulator—Homogenization | 52.91 | 0.49 mm | Recommended as antioxidant in fat based bakery products | Sustainable release and improved stability | Singh et al. (2018) |
| | α-Tocopherol | Dipalmitoyl phosphatidylcholine | | | Liposome—Ultrasound | | 82.40–107.20 nm | DHA and EPA oil | Improving the oxidative stability of encapsulated DHA and EPA | Sahari et al. (2017) |
| | Sodium tripolyphosphate | Hydrogenated vegetable oil | | 1:1 | Provided by Rhodia Co. (Cranbury, NJ) | | | Ground turkey | Dramatically improving (80%) the oxidative stability of meat, as compared with free ingredient | Sickler et al. (2013) |
| Sour cherry pomace extracts | Polyphenols, especially anthocyanins | Whey or soy proteins | | 1:2 | Freeze drying—Homogenization | | | Cookies | Retention of polyphenols, anthocyanins and antioxidant activity, and hence, enhancing the storage stability during 4 months storage | Tumbas Šaponjac et al. (2016) |

| | | | | | | | | | | |
|---|---|---|---|---|---|---|---|---|---|---|
| | L-Ascorbic acid | Xyloglucan polysaccharide | | 1:0.04 | Spray drying—magnetic stirring | 96.34 | 4.40–34.0 μm | Tilapia fish burgers | More retention of ascorbic acid, thus, more desirable organoleptic properties after baking, as compared with free ingredient | Farias et al. (2018) |
| Cinnamon essential oil (EO) | | Chitosan | | 1.46:1 | Ionic gelification | 20.65 | 527 nm | Pork meat and meat products | Decrease of microbial growth, pH, PV, and TBARs | Hu et al. (2015) |
| | Astaxanthin | β-CD | | 200:1 | Inclusion complex followed by freeze drying—magnetic stirring | | 500 μm | Recommended for food industry applications | Enhancing the stability of astaxanthin against heat (100°C), light, and oxidation by 7–9-folds | Kim et al. (2010) |
| Holy basil EO | Phenolic and terpenoid compounds, mainly as methyl eugenol and beta-caryophyllene | Gelatin, followed by coating with carboxymethyl cellulose (CMC) solution containing palmitic acid 2 % (2 st.) | | | Coacervation and coating—magnetic stirring | | | Recommended for food industry applications | More retention of antioxidant and antimicrobial activities during 3 months storage period, as compared with free ingredient | Ngamakeue and Chitprasert (2016) |
| Apple skin extract | Polyphenols, mainly as rutin and quercetin | Sodium alginate | | | Extrusion—Encapsulator | >96 | 423–486 μm | *L. acidophilus* within milk and acidic water (pH = 2) | Great increase in survival of probiotic *L. acidophilus* in milk (stored at 4°C for 50 days) or in acidic water (pH = 2) | Shinde et al. (2014) |
| Olive leave extract | Phenolic compounds, mainly as oleuropein and its derivatives such as hydroxytyrosol and tyrosol | (1) WPC or complex of (2) WPC-pectin, span 80 | 40:1 | | W/O/W nano-emulsions, rotor-stator homogenizer | | (1) 675 nm and (2) 1443 nm | Soybean oil (SBO) | More promoting the oxidative stability (based on PV and TBARs) of SBO compared to free extract as a result of increased solubility and controlled release of olive leaf phenolic compounds through their nanoencapsulation, a high antioxidant activity comparable with the capacity of synthetic antioxidant (TBHQ) | Mohammadi et al. (2016) |

[a] EE: Encapsulation efficiency.
[b] PV: Peroxide value.
[c] p-AV: Para-anisidine value.
[d] TOTOX: p-AV + 2PV.
[e] FFA: Free fatty acids.
[f] TBARs: 2-Thiobarbituric acid reactive substances.
[g] FRAP: Ferric reducing antioxidant power assay.

is a critical issue that needs to be accurately managed and monitored. Food spoilage can be considered as a situation in which consuming a product is associated with safety hazards. The biochemical activity of definite microorganisms, namely molds, yeasts, or bacteria in foods and beverages, which is influenced by numerous factors, including extrinsic, intrinsic, implicit, and processing, will finally make them dominate in products, resulting in food spoilage (Lianou et al., 2016).

Products with a low pH or $a_w$ are mainly influenced by molds and yeasts. As can be seen for many foods like cheese and meat, the spoilage induced by yeasts and molds could be detected by their visible surface growth. *Aspergillus*, *Mucor*, *Penicillium*, and *Rhizopus* species are common fungi contributing to spoilage of both foods and beverages, particularly fruits and fruit-based products. Yeast species, including *Candida*, *Saccharomyces*, *Lachancea*, *Zygosaccharomyces*, and *Torulaspora* primarily contribute to the spoilage of products with a low $a_w$ and pH and high concentrations of sugar or salt, such as soft syrups, drinks, olives, and salad dressings. The growth of yeasts may also be associated with spoilage in animal-originated foods like dairy products and meat, although their population is much lower than that of bacteria (Lianou et al., 2016).

Bacterial species are commonly important in the spoilage of many foods and beverages, such as alcoholic beverages, bakery products, chocolate, dairy products, eggs, fruit juices, meat and poultry, seafood, vegetables, etc. The storage conditions, especially extrinsic factors like temperature and packaging, exhibit a strong impact on the growth of these bacterial species. For example, Pseudomonas spp. is commonly dominant in most fresh foods stored in aerobic conditions at different temperatures, while lactic acid bacteria (LAB), mainly Lactobacillus and Leuconostoc species, along with *Brochothrix thermosphacta*, are able to extend the spoilage and to constitute the dominant microflora in foods (chilled meat, dairy, fish, and freshly cut vegetable products) kept in modified atmosphere packaging (MAP) or vacuum conditions (Holzapfel, 1998). Any food ecosystem encounters five types of ecological determining factors:

**(1)** intrinsic factors: inherent features of any food, namely pH and $a_w$ levels, the redox potential, the nutrient amount, physical barriers against microbial invasion (e.g., skin, shell, or membrane in various food products), naturally antimicrobial elements or those which are formed or added during food processing;
**(2)** extrinsic determinants: environmental conditions, especially atmosphere gas type, relative humidity, and temperature applied during the production chain of food;
**(3)** implicit factors: the microbial population present in food product itself;
**(4)** processing parameters used in food preparations; and
**(5)** emergent influence.

These elements determine the microbial situation of food products, regarding both dominant population and propagation rate for microbes, either existing in or incorporated into food matrix (Hamad, 2012; Nychas et al., 2008). The term "implicit factors" illustrates the intrinsic biotic factors of foods, namely interactions between different microorganism populations contributing to food contamination, or between these microbes and the food matrix. Interactions among different microbial species existing in the food matrix occur in two main ways: they can show a synergistic effect together promoting their growth or exhibit an antagonistic phenomenon inhibiting their multiplication (Hamad, 2012). The term "antagonistic interactions" usually refers to the competition between various microbial species for receiving the nutrients and/or making the changes in the matrix, as the normal growth conditions for a specie is limited or prevented by another specie, as an example, via releasing some metabolites, including hydrogen peroxide, bacteriocins, and organic acids. If meat is stored under vacuum or MAP conditions, LAB and *B. thermosphacta* will dominate in the matrix as a result of antagonistic impact on growth of Gram-positive bacteria, so that they will be responsible for souring the sample, while in aerobic conditions, in turn, Gram-negative bacteria are well established, resulting in product putrefaction (Nychas et al., 1998).

During the storage period, transportation, and subsequent processing of food products, they are often contaminated by molds and the toxins produced by certain species of them. Based on the global statistics of the Food and Agriculture Organization (FAO) for this type of contamination, around 1000 million metric tons of stored food is annually spoiled owing to mycotoxins generated by molds (Martinez-Gracia et al., 2015). Moreover, oxidative deterioration is another parameter governing the agriproducts during the above-mentioned stages that has a synergistic influence on the extent of aflatoxin biosynthesis (Kim et al., 2008).

# 5 The importance of natural antimicrobials as food preservatives

The term "natural antimicrobials" refers to ingredients inherently found and directly obtained from biological materials without any manipulation on a laboratory scale. Plants, animals, algae, bacteria, and fungi are introduced as different resources for this type ingredient. The advantage of plant extracts as antimicrobial agent stems from their ability to control the natural spoilage pathways, called food preservation, and to inhibit/reduce the growth of pathogenic microorganisms, named food safety (Tajkarimi et al., 2010; Jafari et al., 2018); meanwhile, their main benefit is that they are generally recognized as safe (GRAS) without any residues, providing a perfect justification for them to be used instead of chemical additives (Martinez-Gracia et al., 2015). Two commonly used sources for natural antimicrobials are herbs and spices, introducing the materials obtained from the green parts (the stems, leaves, and buds) and other structures (flowers, seeds, rhizomes, roots, fruits, etc.) of a plant (Hossain and Shahidi, 2018). The most important natural ingredients used in food products as preservatives have been detailed below.

## 5.1 Essential oils

EOs from higher plants are among natural compounds consisting of a complex mixture of several bioactive components (>70 compounds), mostly nonpolar phenolics, terpenoids, terpenes, and phenylpropenes (Voon et al., 2012; Tajkarimi et al., 2010), which are broadly applied by people as food additives since they are proportionally highly volatile and are naturally biodegradable. Generally, among all constituents of EOs, two or three substances show a higher concentration (20%–70%) than the other components, determining the biological characteristics of the EOs (Pandit et al., 2016). For instance, carvacrol (30%) and thymol (27%) are consisted the predominant fractions in the Origanum EO (Bilia et al., 2014). EOs are mainly synthesized in aromatic plant species (>17,000) belonging to angiospermic families Lamiaceae, Myrtaceae, Rutaceae, Asteraceae, and Zingiberaceae (Regnault-Roger et al., 2012); the most common EOs having antimicrobial activity are thymol (thyme and ajowan), carvacrol (oregano), eugenol (clove), limonene (dill seed and citrus), cinnamic aldehyde (cinnamon), piperine (black pepper), anethole (aniseed), curcumin (turmeric), allicin (garlic), capsaicin (capsicum), gingerol (ginger), estragole (fennel seed), linalool (coriander), sabinene (curry leaf), and safranal (saffron) (Tajkarimi et al., 2010; Hossain and Shahidi, 2018; Kamimura et al., 2014; Hu et al., 2015; Khoshakhlagh et al., 2017).

Application of EOs instead of synthetic preservatives for improving the quality and shelf life of commodities produced by various industries, namely sanitary, pharmaceutical, cosmetic, food, and agricultural is considered a common strategy now, since EOs exhibit strong antimicrobial and antioxidant influences (Hossain and Shahidi, 2018). It has been revealed that some herbal extracts and EOs have great inhibiting impact on molds, pathogenic bacteria, viruses, microbial toxins synthesis, and insects, as well as limiting oxidative deterioration development (Prakash et al., 2015; Hossain and Shahidi, 2018). The studies performed by diverse toxicity analysis methods for detecting the safety extent of some EOs and their main components have shown that most of the EOs had high $LD_{50}$ values and more toxicity for nonmammalian than mammalian as many EO-bearing plants, such as thyme, nutmeg, clove, cinnamon, lemon grass, oregano, basil, etc. are considered as safe in "GRAS" category by United State Federal Regulations Code (21CFR part 182.20), and hence, are suitable as preservative ingredients in food commodities (U.S. Code of federal regulations, 2013).

However, in the case of EOs, their low solubility in water or the possibility of binding the EOs to fat and protein existing in the food matrix should be considered. As an example, the antimicrobial efficiency of thymol and eugenol EOs may be reduced due to binding with the constituents of complex food systems (Bor et al., 2016). Therefore, the sensitivity of bacteria to the EOs is mainly influenced by food constituents, namely protein, fat, water, preservatives, salt content, antioxidants, and pH, as well as the extrinsic elements such as the type of atmosphere (air, gas, vacuum), temperature, and microorganism properties. Therefore, the worse the microbial growth conditions, the higher the antimicrobial effect of EOs. In this regard, there is a problem, where high concentrations of EOs are needed for inhibiting or inactivating the microorganisms due to the above-mentioned issues, resulting in a negative change in the sensory properties of the food product and a decrease in consumers' acceptance when extra amounts of highly aromatic and volatile EOs are added into food matrix (Weiss et al., 2009). Therefore, an encapsulation approach could be used to overcome these concerns associated with the hydrophobic and free forms of the EOs (Gutierrez et al., 2008).

## 5.2 Phenolic compounds

The lipophilic property of some phenolics is known as a crucial factor reinforcing the antimicrobial effect of these ingredients. In fact, lipophilic fractions negatively affect two vital functions of the microbial cell, namely cell osmotic balance and membrane permeability, resulting in damage to the structural and functional properties of the microorganism, which is associated with loss in various substances such as ATP, ions, amino acids, and nucleic acids (Martinez-Gracia et al., 2015). The phenolic compounds naturally found in EOs exhibit a key role relating to antimicrobial activity of EOs against foodborne pathogenic bacteria. These ingredients are capable of disrupting the natural functions of the microbial cells, including alteration of cell membrane permeability, resulting in leakage of the internal contents of the cell (Bajpai et al., 2012), interfering with the ATP generation system, and disrupting the force driving the proton across the membrane, eventually leading to microbial cell death (Burt, 2004). Gram-negative bacteria (e.g., *E. coli* and *Salmonella Enteritidis*) are less susceptible to the phenolic compounds as antimicrobial agents than Gram-positive ones (e.g., *S. aureus*, *Listeria monocytogenes*, and *Bacillus cereus*) due to their lipopolysaccharide outer membrane limiting the diffusion rate across the cell membrane (Martinez-Gracia et al., 2015).

### 5.3 Curcumin

The main fraction found in *Curcuma longa* rhizome extract is curcumin, which is widely used by food and medicine industries due to its numerous biological activities and antimicrobial activity. Based on scientific research, the antimicrobial activity of curcumin is directly attributed to the methoxy and hydroxyl groups involved within its structure (Han and Yang, 2005). Comprehensive reviews about antimicrobial activity of curcumin and its encapsulation has already been released. Amongst microorganisms inhibited by curcumin, reference can be made to the Gram-positive bacteria, including *Bacillus subtilis*, *B. cereus*, *S. aureus*, *Staphylococcus epidermidis*, *Streptococcus mutans*, and Gram-negative ones, such as *P. aeruginosa*, *E. coli*, *Shigella dysenteriae*, and *Yersinia enterocolitica*, as well as *Helicobacter pylori* and fungi such as *Candida albicans*, *Aspergillus niger*, and *Penicillium notatum*. The MIC (minimum inhibitory concentration) index of curcumin for bacteria is higher than molds and yeasts (Silva et al., 2018a,b). However, the industrial interest in curcumin is under question due to its low water solubility and susceptibility against oxygen, basic pH, light, heat, metal ions, ascorbic acid, and enzymes. In this regard, micro/nano encapsulation of curcumin is a significant way to enhance its stability in the case of above-mentioned issues (Shlar et al., 2015; Wang et al., 2011) as it has been confirmed that this technique, especially nanoencapsulation, appropriately improves the antimicrobial activity of the curcumin. Despite many studies proving the positive role of encapsulation regarding curcumin antimicrobial activity, there is limited information comparing the performance of free form and encapsulated curcumin against foodborne pathogens existing in a processed complex food system or in vivo conditions; these issues need to be investigated (Silva et al., 2018a,b).

## 6 Enhancing the performance of bioactive ingredients by encapsulation

Encapsulation protects bioactive components from various impressing factors, mainly heat, water, oxygen, pH, shear, light, or other extreme conditions during processing, storage, and transportation (Devi et al., 2017; Akhavan et al., 2018; Assadpour and Jafari, 2018; Faridi Esfanjani et al., 2018; Rafiee and Jafari, 2018). This technique enables retention of versatile features of the food ingredients, namely solubility, functionality, nutritional value, and bioavailability, as well as masking off-flavors/odors, controlled releasing at the right time and right place, and increased targeting precision, and thus, maintaining and promoting their unique abilities (Shishir et al., 2018; Ray et al., 2016; Valdes et al., 2015).

The type of encapsulation procedure is the first issue needing to be specified before encapsulation of bioactive ingredients. For example, based on researches executed, the encapsulation techniques suitable for protecting the curcumin by lipids, zein, and gelatin were emulsification (Aditya et al., 2015), antisolvent precipitation (Dai et al., 2017), and electrospinning (Deng et al., 2017), respectively. Therefore, the most important subject generally required to be clear before initiation of process, is choice of wall materials and encapsulation technique (Shishir et al., 2018).

## 6.1 Common wall materials for encapsulation of bioactive ingredients

In this regard, the wall material type is a key issue contributing to the delivery mechanism of the surrounded material that is selected based on several parameters, such as the encapsulation procedure, the encapsulation purpose, encapsulation efficiency (EE), type of release, and representing restricted chemical reactivity with both the core component and food matrix, as well as the cost-effectiveness of the material (Nedovic et al., 2011; Wandrey et al., 2010). Moreover, the selected material must finally meet the safety standards issued by governmental organizations, such as the Food and Drug Administration (FDA, United States) or European Food Safety Authority (EFSA) (Wandrey et al., 2010). Bioactive components exhibit different properties in polarity, molecular weight, solubility, etc. and are therefore encapsulated by different approaches to achieve the definite molecular and physicochemical requirements. This explains why there is not a conclusive procedure for encapsulating the bioactive components (Ray et al., 2016). The different wall materials and techniques used for encapsulation of bioactive ingredients have previously been reviewed in the literature (Zuidam and Nedovic, 2010; Vinceković et al., 2017; Lee and Wong, 2014; Fang and Bhandari, 2010; Shishir et al., 2018; Ozkan et al., 2019; Abd El-Salam et al., 2016; Faridi Esfanjani and Jafari, 2016; Katouzian and Jafari, 2016), and here it is briefly reviewed with an emphasis on those applied for antioxidant and preservative ingredients as listed in Tables 1 and 2.

Ballesteros et al. (2017) mentioned that the type of both encapsulation process (spray-drying or freeze-drying) and the wall material (gum Arabic, maltodextrin, or a blend of these components) were the most effective factors for the encapsulation of antioxidant phenolics obtained from coffee grounds. Although gum Arabic exhibited a higher thermal stability, maltodextrin, especially using freeze-drying method, was the best wall material for retaining the phenolic compounds and flavonoids within the capsules (EE of 62% and 73%, respectively), as well as showing the highest antioxidant activity (86% of the original extract) as compared with other samples.

It is believed that natural biopolymers, particularly proteins and polysaccharides, are the best materials to encapsulate the bioactive ingredients incorporated into the food products (Abd El-Salam et al., 2016; Faridi Esfanjani and Jafari, 2016; Katouzian and Jafari, 2016). Proteins are known as perfect materials for this purpose, because they are amphipathic compounds that could well be linked to both the entrapped bioactive ingredients and the surrounded medium as an emulsifier. Micro/nanoparticles produced by natural proteins are eco-friendly, as they are biologically metabolized in the environment. Moreover, proteins are naturally occurring in food, and are available from ample renewable sources (Mohammadi et al., 2016; Abd El-Salam et al., 2016). Categories of proteins, including globular proteins (whey proteins, ovalbumin, pea proteins, and soy proteins), phosphoproteins (bovine milk caseins), and prolamines (cereal storage proteins) have been introduced as suitable wall materials for encapsulating the bioactive ingredients (Abd El-Salam et al., 2016).

Polysaccharides are another group of natural polymers that consists of monosaccharaide subunits with differences in the number, type, distribution, and bonding of these subunits throughout the chain. The enormous molecular structure and self-assembly ability of the polysaccharides makes them suitable for delivering bioactive ingredients. The most common polysaccharides used in the encapsulation operation of bioactive ingredients can be charge-based, classified into three main groups, namely anionic (gum Arabic, carrageenans, pectins, alginates, and xanthan), neutral (dextran, guar gum), and cationic (chitosan) (Abd El-Salam et al., 2016). da Rosa et al. (2014) used some polymers, namely β-cyclodextrin, xanthan gum, chitosan, and xanthan-chitosan mixture as the wall material for microencapsulating the phenolic compounds extracted from blackberry, using lyophilization approach. Based on their results, EE showed a significant dependence on the type of both bioactive ingredient and the wall polymer, as epicatechin and gallic acid were highly retained within the microcapsules when xanthan and β-cyclodextrin were used as coating agents. The highest antioxidant activity, corresponding with the higher retained phenolic compounds, recorded for microcapsules were provided by xanthan (90.75%) and β-cyclodextrin (84.43%). In addition, they claimed that the controlled release of phenolic compounds, as an important measure assuring the sustainable influence of a preservative during long storage period, was affected by solvent, polymer, and medium pH values.

Belščak-Cvitanović et al. (2015) investigated the efficiency of various natural biopolymers for entrapping green tea bioactive compounds using a spray drying technique. Regarding EE of different biopolymers, it ranged between 23.08% and 82.16%, a descending order reported as: inulin > modified starch > alginate > pectin > whey proteins > oligofructose > pea proteins > carrageenan > acacia gum > guar gum > xanthan. Based on their results, the lowest particle size (<5 μm), highest contents of epigallocatechin gallate (EGCG) and chlorophylls, as well as the most promising release kinetics was obtained using gums (acacia gum and xanthan) as the encapsulants. Finally, they claimed that a combination of these biopolymers can potentially provide functional ingredients, retain color, and improve sensory characteristics, and hence, the microencapsulated powders are appropriate to be incorporated into food systems, such as jelly candies (pectin) and chocolates, or cocoa drink powder mixtures (inulin or oligosaccharide).

**Table 2** Some encapsulated ingredients applied as antimicrobial agents for extending the shelf life of foods

| Ingredient | Effective component | Wall material | Wall material ratio | Wall/core | Encapsulation procedure—Preparation | EE[a] (%) | Particle size | Food system | Results | References |
|---|---|---|---|---|---|---|---|---|---|---|
| Cinnamon essential oil (EO) | | Chitosan | | 1.46:1 | Ionic gelification | 20.65 | 527 nm | Pork meat and meat products | Decrease of microbial growth, pH, PV[b], and TBARs[c] | Hu et al. (2015) |
| | Carvacrol | Hydroxypropyl-β-cyclodextrin (HPβ-CD) | | 1:1 | Inclusion complexes followed by freeze drying—magnetic stirring | 83.74 | 0.377 μm | Recommended for food industry applications | Decreasing the carvacrol concentration for inhibiting the *E. coli* and *S. Typhimurium*, and reducing the antioxidant activity, as compared with free carvacrol | Kamimura et al. (2014) |
| Citrus EO | D-limonene | *Alyssum homolocarpum* seed gum | | 5:1 | Electrospraying | 87.38 | 90 nm | Recommended for food industry applications | Increase in thermal stability of D-limonene | Khoshakhlagh et al. (2017) |
| Onion scales extract | Phenolic compounds, mainely as quercetin | Alginate | | 1:3 | Coacervation—magnetic stirring | 54.80 | 2.05 mm | Chicken muscle | Increasing the oxidative stability (TBARs), reducing the microbial (*Staphylococcus aureus, Pseudomonas fluorescens*) growth by 2 log cycle | Kanatt et al. (2018) |
| Thyme EO | Thymol, carvacrol, phenolic compounds | Gelatin and gum Arabic | 1:1 | 2:1 | Complex coacervation—Ultra- Turrax | 85 | | Bakery product (cake) | Enhancing the antimicrobial activity (10 times) against 9 microorganisms, especially *Staphylococcus aureus*, and *Aspergillus niger*) compared with the free EO—shelf life of 30 days—increasing induction period | Gonçalves et al. (2017) |
| *Mentha piperita* EO | | Chitosan and cinnamic acid nanogel | | | Coacervation—Ultrasound | | <100 nm | Fruit | Improving the antimicrobial activity (*Aspergillus flavus*) during the 1 month storage | Beyki et al. (2014) |
| *Origanum dictamnus L.* EO | Mainly as carvacrol (42.9%–51.7%) | Egg phosphatidyl choline and cholesterol | 5:1 | | Liposome—Ultrasound | 4.16 | | Recommended for food industry | Improvement of antioxidant and antimicrobial activities against Gram-positive and -negative bacteria, and pathogenic fungi, as well as *Listeria monocytogenes* | Liolios et al. (2009) |

| | | | | | | | | | | |
|---|---|---|---|---|---|---|---|---|---|---|
| *Satureja montana* EO | Mainly as carvacrol | β-CD | | 4:1 | Inclusion complexes—magnetic stirrer | 94.07 | | Recommended for food industry | Increasing the antibacterial, antifungal and antioxidant properties | Haloci et al. (2014) |
| | Allyl isothiocyanate | Gelatin and gum Arabic | 1:1 | 2:1 | Complex coacervation—freeze drying | >90 | | Mature green tomato | Sustained release of the ingredient and increasing the shelf life | Wu et al. (2015) |
| *Lippia turbinate* EO | | Gelatin and gum Arabic | 1:1 | ~2.5:1 | Complex coacervation followed by freeze drying—Magnetic stirring | 99.80 | | Peanut kernels | Sustained release and retention of antifungal power against *Penicillium* and *Aspergillus* during 78 days of storage | Girardi et al. (2017) |
| *Thymus capitatus* EO (70%) and soybean oil (30%) | Mainly as carvacrol (76.1%) | 1% aqueous SDS | | 9:1 | Nanoemulsion—High pressure micro-fluidizer | | 110 nm | Recommended for food industry | Enhancement of EO antibacterial activity against Gram-positive and -negative bacteria compared to free EO, and decreasing its antioxidant activity | Ben Jemaa et al. (2018) |
| Anise oil and medium-chain triacylglycerol (3:1) | Mainly as anetholes (97%) | Aqueous solution of soy lecithin | | | Emulsion delivery based system—Ultra Turrax | | 117.20 nm | Recommended for food industry | Controlled release, better physicochemical stability and antimicrobial activity (*Listeria monocytogenes* and *Escherichia coli*) than bulk anise oil, as well as improving the solubility in aqueous food systems | Topuz et al. (2016) |
| | Nisin | Solid lipid (hereon, Imwitor 900), poloxamer 188 and sodium deoxycholate | 10:5:0.125 | 15:2 | Solid lipid nanoparticles (SLNs)—Ultra-Turrax and high pressure homogenization | 73.6 | 159–167 nm | Recommended as a preservative in heat processed and low pH foods | Sustained release during 25 days period and antibacterial activity of nisin-loaded SLNs against *Listeria monocytogenes* and *Lactobacillus plantarum* for up to 20 and 15 days, respectively that was better than free form (1–3 days) | Prombutara et al. (2012) |
| Green tea extract | Polyphenols and catechins, mainly as epigallocatechin gallate (63.73%) | Chitosan and sodium tripolyphosphate | 5:1 | 1.2:0.5 | Microemulsion—Ultrasound | | 7.59 μm | Hamburger patties | More oxidation stability (TBARs) and antibacterial activity (*Coliform*, yeast and mold) than free extract form during 8 days storage at 4°C | Özvural et al. (2016) |

*Continued*

**Table 2** Continued

| Ingredient | Effective component | Wall material | Wall material ratio | Wall/ core | Encapsulation procedure—Preparation | EE[a] (%) | Particle size | Food system | Results | References |
|---|---|---|---|---|---|---|---|---|---|---|
| *Clove* oil | Lipophilic monoterpenes, mainly as Eugenol (83.73%) and Eugenyl acetate (11.37%) | Soy lecithin and cholesterol | 5:1 | | Liposome followed by vacuum oven—Cell ultra-fine grinding | 20.41 | 149.20 nm | Tofu | Great inhibition (99.99%) of *E. coli* and *S. aureus* via controlled release of *Clove* oil from liposome by secreting α-hemolysin (pore-forming cytotoxin) from *S. aureus*; increasing the food shelf life | Cui et al. (2015) |
| *Thymus capitatus* EO (70%) and soybean oil (30%) | Mainly as carvacrol (76.1%) | 1% aqueous SDS | | 9:1 | Nanoemulsion—high pressure micro-fluidizer | | 110 nm | Semi skimmed UHT milk | Improving the physicochemical and microbial (*S. aureus*) quality, and antioxidant capacity of milk, as compared with bulk EO, no difference between oxidative stability of free and encapsulated forms | Ben Jemaa et al. (2017) |
| *Peumus boldus* EO | | Gelatin and gum Arabic | | 1.25:2 | Complex coacervation followed by freeze drying | | 1000 μm | Peanut seed | Preventing the peanut deterioration by food spoilage microorganisms and insects during the 5 months storage | Girardi et al. (2018) |
| Oregano EO and sunflower oil (1:1) | Mainly as carvacrol | Deionized water, Cremophor RH 40 and Span 80 | 70:12:8 | 100:5 | Nanoemulsions by phase inversion temperature method—magnetic stirring-heating/cooling cycles | | 35 nm | Chicken pate | Preventing the microbial growth (*S. aureus* and *E. coli*) during 90 days storage and extending the shelf life better than free ingredient | Moraes-Lovison et al. (2017) |

| | | | | | | | | | | |
|---|---|---|---|---|---|---|---|---|---|---|
| *Illicium verum* Hook. f. EO (IvEO) | Mainly as anethole (89.12%) | Chitosan | | 1:0.8 | Freeze drying | 20.93 | 1000–2000 nm | Pistachio | Enhancing the antiaflatoxigenic inhibiting capacity twofold more than free IvEO, extending the nut shelf life | Dwivedy et al. (2018) |
| Black pepper EO (1) | Mainly as β-caryophyllene (2) and limonene | HPβ-CD | | 10:1 | Inclusion complex followed by freeze drying | (1) 50.55, (2) 85.30 | | | Enhancing the antibacterial activity against *S. aureus* and *E. coli* by fourfold, and slightly lower antioxidant activity than free EO | Rakmai et al. (2017) |
| *Satureja* plant EO | Phenolic compounds, especially carvacrol | Lecithin and cholesterol | 50:10 | | Nanoliposome—Ultra-Turrax homogenizer and ultrasound | 69.05 | 93 | Lamb meat | Improvement of microbiological safety, oxidative stability, and extension of shelf-life during 20 days chilled storage in comparison to bulk EO | Pabast et al. (2018) |
| Olive leaf extract | Phenolic compounds | Maltodextrin | | 4:1 | Homogenization | 62.86 | | Tomato paste | Improvement in physicochemical and microbial properties (*A. flavus*) of tomato paste over the storage period (20 days) at 25°C and 30°C. and nonencapsulated extract was more successful than microencapsulated one | Jafari et al. (2017) |

[a] EE: Encapsulation efficiency.
[b] PV: Peroxide value.
[c] TBARs: 2-Thiobarbituric acid reactive substances.

Application of cowpea and pea protein isolates in combination with maltodextrin has been suggested for microencapsulating ascorbic acid (7 and 8 μm particle size and 69% retention of ingredient), as well as the other bioactive ingredients (Pereira et al., 2009). Amaranth protein is a safe ingredient for celiac patients due to its low content of prolamins (Gorinstein et al., 2000). Aceituno-Medina et al. (2013) claimed that a combination of amaranth protein isolate and pullulan, as spinnable carbohydrate polymers, was suitable to provide electrospun microfibers, enabling the protection of biological activities in many bioactive ingredients when they are subjected to processing and gastrointestinal conditions or to the intrinsic factors of the food (e.g., redox potential, pH, enzyme content, water activity, etc.).

## 6.2 Encapsulation techniques commonly used for protecting the bioactive ingredients

Many encapsulation techniques classified as physical methods (e.g., spray drying, spray-bed-drying, fluid-bed coating, freeze drying, spray-freeze drying, extrusion, and supercritical fluids based techniques), physicochemical methods (e.g., complex coacervation, liposomes, and ionic gelation), chemical methods (e.g., interfacial polymerization, and molecular inclusion complexes) have been commonly used to encapsulate preservative ingredients (Ozkan et al., 2019; Zuidam and Heinrich, 2009). Some advantages and disadvantages have been mentioned for each of them, which has been reviewed in the literature in detail (Ozkan et al., 2019; Garcia-Moreno et al., 2018; Ray et al., 2016; Nedovic et al., 2011; Shishir et al., 2018).

### 6.2.1 Spray drying

Spray drying is known as the oldest approach; it is a continuous process, flexible, and cost-effective, which has the widest application in food industries, providing a great EE, solubility, stability, and controlled ingredient release for spherical-shape powders (Zuidam and Shimoni, 2009; Akhavan Mahdavi et al., 2014; Rajabi et al., 2015). A wide range of encapsulating materials, mostly including lipids (mono and di glycerides stearic acid), polysaccharides (gum Arabic, corn syrups, maltodextrins, and starches) and protein resources (gelatin, whey, milk, wheat, and soy) could be used for spray drying encapsulation (Ray et al., 2016). However, some disadvantages are attributed to this technique such as uneven conditions of the dryer interior space and complexity in controlling the particle size (ranging between 10 μm and 100 mm) (Fang and Bhandari, 2010) or in equipment (Nedovic et al., 2011). Additionally, some other disadvantages are:

**(1)** the bioactive ingredients, such as β-carotene, lycopene, anthocyanins, ascorbic acid, flavors, and colors undergo heat-induced damage due to relatively high temperatures during spray drying;
**(2)** reduction in product yield through loss of dry particles onto the wall of the drying chamber; and
**(3)** problems with carrier agents rich in sugar since they have a low glass transition temperature ($T_g$) and exhibit stickiness behavior (Ozkan et al., 2019; Fang and Bhandari, 2010).

Therefore, many factors should be considered before performing this method (Ozkan et al., 2019). Tatar Turan et al. (2015) introduced ultrasonic nozzle with spray-drying as a new atomization technology for improving the stability of blueberry's phenolic compounds incorporated in ice cream and cake, as the higher retention of blueberry's anthocyanin (79.35%) in cake recorded by this procedure, compared with those observed by conventional spray drying and freeze drying. Moreover, there was no significant difference between total phenolic content and antioxidant activity of blueberry extract microspheres prepared by ultrasonic nozzle and freeze-drying. They suggested application of this procedure in the food industry is for improving the stability and availability of bioactive compounds used as preservative or functional ingredients in food products.

### 6.2.2 Freeze drying

Lyophilisation, or freeze drying or sublimation drying, is a multistage developed technology particularly recommended for encapsulating heat-sensitive substances, which effectively ensures the retention of original texture, organoleptic properties, and biological function of the ingredients (Ceballos et al., 2012; Hosseini et al., 2018a). Gum Arabic, maltodextrin, modified starches, whey protein, etc. are the most common wall materials used in this procedure for encapsulating bioactive ingredients. The main drawback of the freeze-drying approach is related to its high time and energy use, as well as generating a porous structure surrounding bioactive substrate, which offers reduced protection over prolonged storage (Hosseini et al., 2018a,b; Ray et al., 2016; Gadkari and Balaraman, 2015).

### 6.2.3 Fluidized bed coating

Another widely used encapsulation procedure is fluidized bed coating (batch or continuous), by which powder particles are fluidized through a chamber, where encapsulating material (mostly gums, proteins, and starches) is spraying by a nozzle to cover the fluidized particles. The EE at this method could be effectively improved via optimizing the process variables, such as atomization pressure of nozzle, powder flow rate, feed rate of encapsulating agent, and process temperature at different zones. This technique is industrially applicable for many food ingredients and additives using three main procedures, namely top-, bottom-, and tangential-spray (Ray et al., 2016). The main advantages attributed to this technique include taste covering, prolonged shelf life, ease of handling, and controlled release (Gadkari and Balaraman, 2015).

### 6.2.4 Encapsulation with lipid vesicular carriers

Liposomes are the most commonly used lipid vesicular carriers in pharmaceutical and food production (Akhavan et al., 2018; Ghorbanzade et al., 2017; Rafiee and Jafari, 2018). The most popular ways for producing these systems are sonication, microfluidisation, high pressure homogenization, supercritical carbon dioxide, and electro spraying (Shishir et al., 2018). In fact, liposomes are phospholipid vesicles involving an interior aqueous phase and exterior continuous water phase that are separated from each other by a lipid bilayer structured through hydrophilic-hydrophobic interactions. The lipid bilayer consists of a polar head and a nonpolar tail, which gives it the semi-permeable membrane. Therefore, liposomes are very appropriate for encapsulating the bioactive ingredients with hydrophilic, hydrophobic, or amphiphilic characteristics. A major benefit of this system is the ability to control the release rate of incorporated materials and deliver them to the right place at the right time (Fang and Bhandari, 2010). Although soy or egg phospholipids are mostly used for preparing the liposomes, currently there is increasing attention on phospholipids derived from milk fat globule membrane as commercially available ingredients with functional properties and health benefits (Farhang et al., 2012).

Some successful works on the application of liposomes in encapsulating lutein, quercetin, anthocyanin, and vitamins have been published (Zhao et al., 2017a,b; Toniazzo et al., 2017). Ganji and Sayyed-Alangi (2017), after encapsulating the ethanolic extract of ginger in a liposome system consisted of sunflower oil (35%), lecithin (5%), glycerol (3%), and ginger ethanolic extract (GE, 2%), recommended that this system could be used instead of synthetic preservatives for food and biological matrices. Their results also showed that nanoliposomal GE was more effective in delaying the oxidation of oil in comparison to the control and free GE (Table 1). Similarly, Tan et al. (2014), and Aisha et al. (2014) confirmed that nanoliposomal carotenoids and *Orthosiphon stamineus* ethanolic extract had a higher radical scavenging activity than free forms of the extracts as a result of controlled release of bioactive compounds and an increasing contact between these antioxidants and preoxidants.

In contrast, a reverse trend was observed by Marin et al. (2018), where they used freeze-dried phosphatidylcholine liposomes for encapsulating collagen hydrolysate, shrimp lipid extract, and pomegranate peel extract from natural waste followed by incorporating them into surimi gels as antioxidant ingredients. They claimed that this event was a result of reduced availability for the encapsulated ingredients, as the oxidative index for pomegranate peel extract-liposome-incorporated gel with an entrapment efficiency of 63% was lower than other samples with EE values more than 90%. To justify this contradiction, reference can be made to the study of Tan et al. (2014), as mentioned in Table 1. They stated that the antioxidant properties of various carotenoids incorporated into the liposomes were dissimilar to those in solution, and their antioxidant activities correlated with the trial conditions including liposome preparation method, lipid type, concentration, and oxygen pressure. Their results showed that either lutein or β-carotene incorporated into the liposomes had a strong inhibition capacity against lipid peroxidation, as well as self-protective features in presence of the pro-oxidation elements, while in turn, lycopene- and canthaxanthin-loaded liposomes were weak, associated with pro-oxidation effects. Accordingly, they claimed that the antioxidant activity of carotenoids entrapped into liposome systems is related to both their chemical reactivity, and the amount of the ingredient incorporated into liposomal membrane, influencing the membrane characteristics.

On the other hand, it is believed that commercial application of liposomes is under question, owing to their low physicochemical stability, expensive operation, ingredient leakage, and fast releasing. Meanwhile, several studies demonstrated that these limitations could be overcame and improved using supplementary procedures, including chitosan coating, protein coating, micronized sucrose coating, pro-liposome hydration method, and ultrasound- or microfluidization-assisted encapsulation (Shishir et al., 2018; Farhang et al., 2012).

### 6.2.5 Coacervation

Coacervation, especially complex coacervation (de Souza Simões et al., 2017), is recognized as a simple and useful technique in both micro and nano encapsulation processes in pharmaceutical and food fields, which is performed over three main steps (Ghasemi et al., 2017, 2018):

**(1)** mixing the continuous liquid phase with an immiscible disperse phase, namely core fraction and wall material containing either single biopolymers such as gelatin (simple coacervation) or more than one biopolymer like gelatin and gum acacia (complex coacervation) (Fang and Bhandari, 2010);
**(2)** adjusting some parameters, including ionic strength, pH, wall fraction concentration, solution temperature, and molecular weight to establish a protective layer around the bioactive agent; and
**(3)** applying the cross-linking agents or heat to achieve solid capsules (Bakry et al., 2016).

Therefore, phase separation in complex coacervation technique occurs when polyelectrolytes with opposite charges are electrically balanced in an aqueous media. This event is influenced by a set of above-mentioned parameters (Jain et al., 2016). Complex coacervation is considered as a very good procedure because of the adequate protection on thermolabile bioactive ingredients (e.g., astaxanthin) (Gomez-Estaca et al., 2016), high EE and core trapping capacity, as well as superior controlled release (Jain et al., 2016), and mild temperatures compared with spray drying (de Souza Simões et al., 2017). However, high production cost, high dependence of stability on a narrow range of the aforementioned parameters, instability in aqueous medium, particles accumulation, and uneven particles size distribution are shortcomings, preventing this method from being used widely (Jia et al., 2016). As an example, Jain et al. (2015) applied a complex coacervation approach for improving the stability and release behavior of the β-carotene used in food systems over a 3-month storage period. They concluded that the best results, as listed in Table 1, only achieved by microencapsulating the ingredients using a defined condition, including the wall materials ratio of 2.0 (WPI):1.0 (gum acacia) and pH=4.2.

### 6.2.6 Co-extrusion

Co-extrusion is a procedure producing microcapsules consisted of a barrier coating (shell) that physically protects a liquid dispersion (core) against environmental damage (Sun-Waterhouse et al., 2014). Sun-Waterhouse et al. (2014) successfully used a combination of antioxidants (quercetin or vitamin E) and co-extrusion encapsulation with alginate and alginate-starch (1.4% or 89.8% amylose) as encapsulants to increase the oxidative stability of the unsaturated canola oil stored at 20°C and 38°C for 60 days storage period. The best combination from their results is presented in Table 1.

## 6.2.7 Cyclodextrins

Cyclodextrins (CDs) are cyclic oligosaccharides with 6, 7, or 8 glucopyranose units linked by a (1–4) glycosidic bonds, which result in a truncated cone structure with a hydrophobic cavity and a surface hydrophilic part (Gharibzahedi and Jafari, 2017; Rezaei et al., 2019). CD structure by having an external polar surface and inner apolar surface is, therefore, a suitable carrier for incorporating the nonpolar guest ingredients in order to improve their water solubility (Pinho et al., 2014) and bioavailability (Saldanha do Carmo et al., 2017), as well as protection of them from detrimental environmental conditions (e.g., elevated temperatures, oxidation, adverse pH values, and light) (Pinho et al., 2014; Saldanha do Carmo et al., 2017).

Beta-cyclodextrin (β-CD) with seven glucopyranose residues is fairly soluble in water (Ho et al., 2017) and is the most commonly involved CD in encapsulation techniques for its lowest cost (Pinho et al., 2014). Modified CDs contained a number of alkylated hydroxyl groups (e.g., accidentally methylated-β-cyclodextrin, Mβ-CD, and Hydroxypropyl-β-CD, HP β-CD), which are considerably more soluble in water (reversely depend on temperature) and less toxic compared to the parent β-CD (Ho et al., 2017). Several studies have examined applying the β-CD and its derivatives as encapsulating agents (Table 1). Jullian et al. (2007) successfully reported the stabilization of catechin by β-CD, DMβ-CD, and HPβ-CD, as was confirmed by Zhao et al. (2010), where they used β-CD to entrap chlorogenic acid, as a highly antioxidant and antimicrobial ingredient, extracted from tobacco leaf and purified by high performance liquid chromatography.

Ho et al. (2017) investigated the influence of alkaline pH values, high relative humidity (RH >65%), and several food component models, namely oil, protein, and sugar, on the stability of the catechin as a polyphenol antioxidant, before and after entrapping by β-CD, HPβ-CD, and Mβ-CD. They observed 61% retention in the antioxidant activity of the catechin by CD inclusion complexes. Flavors and aroma are commonly expensive, and it is therefore necessary to prevent them being lost or degraded (Kfoury et al., 2016). It has been established that CDs keep phenolic ingredients away from molecular oxygen, enzymatic oxidation and, therefore, extend the shelf life of foods (Çelik et al., 2011). In the case of EOs, it has been stated that their encapsulation in CDs provides a 16-fold increase in their aqueous solubility, as well as reducing their photodegradation extent up to 44-fold (Kfoury et al., 2019). CDs can properly be used to cover aroma and flavors, especially where other techniques are useless (Kfoury et al., 2016). As a protective barrier, CDs reduce the reaction between flavors and other ingredients existing in the food matrix, as well as volatility and degradation of flavors during storage, and enhance mechanical or thermal processes, resulting in a prolonged shelf life of food (Ho et al., 2017). Since this procedure is used for flavor covering, it is important to meet some benefits without addition of antioxidants, including preservation of original aromatic profile, inhibiting off-flavor development, minimizing the flavors-packaging interactions, and enhancing the photostability (Kfoury et al., 2016).

### 6.2.8 Solvent evaporation

Solvent evaporation technique consists of four main stages:

**(1)** preparing an emulsion, a suspension, or a solution by dissolving both coating polymer and bioactive ingredient in an organic solvent;
**(2)** mixing the organic phase, as dispersed phase, in an aqueous phase, as continuous phase, to form an emulsion;
**(3)** removing the solvent from the emulsion aiming liquid extraction or evaporation; and
**(4)** separation of particles by centrifugation or filtration followed by drying.

This method has been considered less commonly in encapsulation of food preservative ingredients in spite of its potential in production of nanoparticles (Ozkan et al., 2019).

### 6.2.9 Emerging encapsulation techniques

Todays, some emerging encapsulation techniques, such as electrohydrodynamic processes, including electrospraying and electrospinning, are recommended as alternative procedures having a high-performance, resulting in nano-microstructures with a large ratio of surface-to-volume, controlled release features, and high EE (Blanco-Padilla et al., 2015; Jafari, 2017a,b). Moreover, these low cost and versatile procedures are extremely suitable for encapsulating heat-sensitive ingredients, since the drying step is removed in these techniques (Shishir et al., 2018). Paximada et al. (2017) successfully used this procedure for improving the solubility of hydrophilic EGCG in an oily phase, as well as protection of EGCG from heating, moisture, and dissolution conditions, leading to enhanced EGCG shelf life when incorporated within different food items. To prevent the degradation of heat-sensitive ingredients by high temperatures used to melt the polymer, the bioactive compounds are commonly incorporated into the original polymer solutions. In this technique, a high-voltage potential is applied to charge the external layer of a polymer solution in a way that the polymer solution is driven through a conductive needle (spinning), the released thin polymer fibers are then crushed by an electrical potential applied between the needle and a collector. The electrostatic repulsion forces generated on the surface of droplets resulted in elongation of the droplet, converting from a semispherical form to a conical shape called a Taylor cone. As electric potential is increased, the repulsion forces are strengthened, resulting in the reduction of the droplet surface tension. More information about the mechanism of these techniques, as well as variables influencing the operation, is available in the literature (Garcia-Moreno et al., 2018; Shishir et al., 2018).

According to the role of supercritical fluid (e.g., carbon dioxide, nitrogen, propane, water, etc.), three procedures have been commonly developed:

**(1)** a process in which both bioactive compound and polymer, as solutes, are dissolved in a supercritical fluid, as a solvent, followed by passing through a small nozzle directed into a zone with a lower pressure, resulting in the co-precipitation of solutes caused by a considerable decrease in solvent power of supercritical fluids (Ozkan et al., 2019);
**(2)** supercritical fluid is used as an antisolvent in the most used procedure, in which a solution containing organic solvent and solutes is injected through a nozzle into a pressurized vessel

(Sosa et al., 2011) where it comes in contact with antisolvent, leading to a supersaturation state for solute in the atomized solution that is followed by nucleation, formation of micro- or nano-particles, and then, removing the organic solvent from the particles (Visentin et al., 2012). In fact, this procedure is used when the solubility of the solute in the supercritical fluid is poor, as can be seen for quercetin, as solute, and carbon dioxide, as antisolvent (Fernandez-Ponce et al., 2015); and

(3) formation of solid particles in two main stages, including the saturation of solutes (bioactive ingredient and wall material) by a supercritical fluid and expansion of this gas-saturated solution when it is passed from an atomization nozzle (Mattea et al., 2009).

Therefore, using supercritical fluids-assisted techniques, it will simply be possible to meet many advantages missed by conventional methods that include sufficient control on the morphology and size of the particles, retention of biological activity of thermo-sensitive substances, improvement in EE, and yield and purity of product. Meanwhile, determining the suitable procedure for supercritical processes according to the solubility of the core and wall materials in the supercritical fluid is the only restrictive factor to be considered (Ozkan et al., 2019).

## 7 Conclusion and further remarks

Food products are subject to various deteriorative routs, including microbial growth and physicochemical instability (e.g., oxidation), leading to a loss in quality, nutritional value, and safety of the food items, as well as reduction in consumer acceptance. Different techniques, such as addition of preservatives, could be used to reduce or prevent these adverse influences. In this regard, natural preservatives obtained from versatile resources, especially herbs, are the best choice due to their known and unique features. The antimicrobial and antioxidant activity of the extracts, pigments, and EOs from herbs and spices, as well as some vitamins (C and E) and chelating agents has been proved mainly in vitro. However, there are some limitations for them to be routinely used in the food industries, as they are mainly sensitive to changes in environmental conditions (e.g., pH, temperature, $a_w$, light, etc.) or to extreme conditions applied during processes (e.g., high temperatures and pressures). In addition, the activity of these natural preservatives may significantly decline during storage or through their interactions with food constituents, requiring a higher amount of these ingredients to be effective enough, which may undesirably influence the sensory features of the food products.

Considering these issues, micro/nano encapsulation, as an efficient delivery system, using different protein and carbohydrate polymers or their blends could be applied to stabilize/enhance the antimicrobial and antioxidative activities and controlled release of preservative ingredients within food matrices, promoting the quality and shelf life of the products. Meanwhile, the type of encapsulants, encapsulation method, and releasing mechanism of preservative ingredients are very important to meet a favored result. In this regard, optimizing the encapsulation procedure for each ingredient is a crucial factor to be considered. An optimized procedure assures protection of the ingredient in acute circumstances, as well as achieving a gradual release into the food matrix, and hence, encapsulation can significantly improve the performance of the ingredients compared to their free form. Higher encapsulation efficiencies can be achieved by nanoencapsulation approaches in comparison to the other scales. Finally, incorporating the micro/nanocapsule-loaded with bioactive ingredients into the food matrix is a promising and cost-effective technique to extend the shelf life of food on an industrial scale in accordance with today's demands of consumers who are interested in natural preservatives as alternatives to synthetic ones.

## References


Abd El-Salam, M.H., El-Shibiny, S., Grumezescu, A.M., 2016. Natural biopolymers as nanocarriers for bioactive ingredients used in food industries. In: Grumezescu, A.M. (Ed.), Encapsulations. Academic Press, pp. 793–829.

Aceituno-Medina, M., Lopez-Rubio, A., Mendoza, S., Lagaron, J.M., 2013. Development and characterization of food-grade electrospun fibers from amaranth protein pullulan blends. Food Res. Int. 54, 667–674.

Aditya, N.P., Aditya, S., Yang, H., Kim, H.W., Park, S.O., Ko, S., 2015. Co-delivery of hydrophobic curcumin and hydrophilic catechin by a water-in-oil-in-water double emulsion. Food Chem. 173, 7–13.

Aisha, A.F., Majid, A.M.S.A., Ismail, Z., 2014. Preparation and characterization of nano liposomes of *Orthosiphon stamineus ethanolic* extract in soybean phospholipids. BMC Biotechnol. 14 (1), 23.

Akhavan Mahdavi, S., Jafari, S.M., Ghorbani, M., Assadpoor, E., 2014. Spray-drying microencapsulation of anthocyanins by natural biopolymers: a review. Drying Technol. 32 (5), 509–518.

Akhavan Mahdavi, S., Jafari, S.M., Assadpoor, E., Dehnad, D., 2016a. Microencapsulation optimization of natural anthocyanins with maltodextrin, gum Arabic and gelatin. Int. J. Biol. Macromol. 85, 379–385.

Akhavan Mahdavi, S., Jafari, S.M., Assadpour, E., Ghorbani, M., 2016b. Storage stability of encapsulated barberry's anthocyanin and its application in jelly formulation. J. Food Eng. 181, 59–66.

Akhavan, S., Assadpour, E., Katouzian, I., Jafari, S.M., 2018. Lipid nano scale cargos for the protection and delivery of food bioactive ingredients and nutraceuticals. Trends Food Sci. Technol. 74, 132–146.

Ananingsih, V.K., Sharma, A., Zhou, W., 2013. Green tea catechins during food processing and storage: a review on stability and detection. Food Res. Int. 50, 469–479.

Andersen, Y.M., Markham, K.R., 2006. Flavonoids: Chemistry, Biochemistry, and Applications. CRC Press, Boca Raton.

Arango-Ruiz, A., Martin, A., Cosero, M.J., Jimenez, C., Londono, J., 2018. Encapsulation of curcumin using supercritical antisolvent (SAS) technology to improve its stability and solubility in water. Food Chem. 258, 156–163.

Assadpour, E., Jafari, S.M., 2018. A systematic review on nanoencapsulation of food bioactive ingredients and nutraceuticals by various nanocarriers. Crit. Rev. Food Sci. Nutr. 1–47.

Bajpai, V.K., Baek, K.H., Kang, S.C., 2012. Control of Salmonella in foods by using essential oils: a review. Food Res. Int. 45, 722–734.

Bakry, A.M., Abbas, S., Ali, B., Majeed, H., Abouelwafa, M.Y., Mousa, A., Liang, L., 2016. Microencapsulation of oils: a comprehensive review of benefits, techniques, and applications. Compr. Rev. Food Sci. Food Saf. 15, 143–182.

Ballesteros, L.F., Ramirez, M.J., Orrego, C.E., Teixeira, J.A., Mussatto, S.I., 2017. Encapsulation of antioxidant phenolic compounds extracted from spent coffee grounds by freeze-drying and spray-drying using different coating materials. Food Chem. 237, 623–631.

Bauer, K., Garbe, D., Surburg, H., 2001. Common Fragrance and Flavor Materials: Preparation, Properties and Uses. Wiley-VCH, Weinheim.

Belščak-Cvitanović, A., Levic, S., Kalušević, A., Špoljarić, I., Đorđević, V., Komes, D., Mršić, G., Nedović, V., 2015. Efficiency assessment of natural biopolymers as encapsulants of green tea (Camellia sinensis L.) bioactive compounds by spray drying. Food Bioproc. Tech. 8 (12), 2444–2460.

Ben Jemaa, M., Falleh, H., Neves, M.A., Isoda, H., Nakajima, M., Ksouri, R., 2017. Quality preservation of deliberately contaminated milk using thyme free and nanoemulsified essential oils. Food Chem. 217, 726–734.

Ben Jemaa, M.B., Falleh, H., Serairi, R., Neves, M.A., Snoussi, M., Isoda, H., Nakajima, M., Ksouri, R., 2018. Nanoencapsulated Thymus capitatus essential oil as natural preservative. Innov. Food Sci. Emerg. Technol. 45, 92–97.

Beyki, M., Zhaveh, S., Khalili, S.T., Rahmani-Cherati, T., Abollahi, A., Bayat, M., Tabatabaei, M., Mohsenifar, A., 2014. Encapsulation of Mentha piperita essential oils in chitosan–cinnamic acid nanogel with enhanced antimicrobial activity against *Aspergillus flavus*. Ind. Crop. Prod. 54, 310–319.

Bilia, A.R., Guccione, C., Isacchi, B., Righeschi, C., Firenzuoli, F., Bergonzi, M.C., 2014. Essential oils loaded in nanosystems: a developing strategy for a successful therapeutic approach. Evid. Based Complement. Alternat. Med. 2014, 651593.

Blanco-Padilla, A., Lopez-Rubio, A., Loarca-Pina, G., Gomez-Mascaraque, L.G., Mendoza, S., 2015. Characterization, release and antioxidant activity of curcumin-loaded amaranth-pullulan electrospun fibers. LWT- Food Sci. Technol. 63 (2), 1137–1144.

Bor, T., Aljaloud, S.O., Gyawali, R., Ibrahim, S.A., 2016. Antimicrobials from herbs, spices, and plants. In: Grumezescu, A.M. (Ed.), Encapsulations. Academic Press, pp. 269–293.

Burt, S., 2004. Essential oils: their antibacterial properties and potential applications in foods—a review. Int. J. Food Microbiol. 94, 223–253.

Caillard, R., Mateescu, M.A., Subirade, M., 2010. Maillard-type cross-linked soy protein hydrogels as devices for the release of ionic compounds an in vitro study. Food Res. Int. 43, 2349–2355.

Ceballos, A.M., Giraldo, G.I., Orrego, C.E., 2012. Effect of freezing rate on quality parameters of freeze dried soursop fruit pulp. J. Food Eng. 111, 360–365.

Çelik, S.E., Özyürek, M., Tufan, A.N., Güçlü, K., Apak, R., 2011. Spectroscopic study and antioxidant properties of the inclusion complexes of rosmarinic acid with natural and derivative cyclodextrins. Spectrochim. Acta A Mol. Biomol. Spectrosc. 78 (5), 1615–1624.

Cervantes, B., Ulatowski, L., 2017. Vitamin E and Alzheimer's disease—is it time for personalized medicine? Antioxidants 6 (3), 45.

Chen, B., Xu, M., 2018. Natural antioxidants in foods. In: Smithers, G.W. (Ed.), Reference Module in Food Science. Elsevier.

Chen, X., Chen, R., Guo, Z., Li, C., Li, P., 2007. The preparation and stability of the inclusion complex of astaxanthin with β-cyclodextrin. Food Chem. 101 (4), 1580–1584.

Comunian, T.A., Thomazini, M., Alves, A.J.G., de Matos Junior, F.E., de Carvalho Balieiro, J., Favaro-Trindade, C.S., 2013. Microencapsulation of ascorbic acid by complex coacervation: protection and controlled release. Food Res. Int. 52 (1), 373–379.

Cui, H., Zhao, C., Lin, L., 2015. The specific antibacterial activity of liposome-encapsulated Clove oil and its application in tofu. Food Control 56, 128–134.

Cunningham, F.X., Gantt, E., 2011. Elucidation of the pathway to astaxanthin in the flowers of *Adonis aestivalis*. Plant Cell 23 (8), 3055–3069.

da Rosa, C.G., Borges, C.D., Zambiazi, R.C., Rutz, J.K., da Luz, S.R., Krumreich, F.D., Benvenutti, E.V., Nunes, M.R., 2014. Encapsulation of the phenolic compounds of the blackberry (*Rubus fruticosus*). LWT- Food Sci. Technol. 58 (2), 527–533.

Dai, L., Sun, C., Li, R., Mao, L., Liu, F., Gao, Y., 2017. Structural characterization, formation mechanism and stability of curcumin in zein-lecithin composite nanoparticles fabricated by antisolvent co-precipitation. Food Chem. 237, 1163–1171.

Davidov-Pardo, G., Arozarena, I., Marín-Arroyo, M.R., 2013. Optimization of a wall material formulation to microencapsulate a grape seed extract using a mixture design of experiments. Food Bioproc. Tech. 6, 941–951.

de Souza Simões, L., Madalena, D.A., Pinheiro, A.C., Teixeira, J.A., Vicente, A.A., Ramos, Ó.L., 2017. Micro- and nano bio-based delivery systems for food applications: in vitro behavior. Adv. Colloid Interface Sci. 243, 23–45.

Decker, E.A., Mei, L., 1996. Oxidative processes in meat: antioxidant mechanisms and applications in muscle foods. In: Proceedings From the 49th Reciprocal Meat Conference. American Meat Science Assoc, Chicago, pp. 64–72.

Deng, L., Kang, X., Liu, Y., Feng, F., Zhang, H., 2017. Effects of surfactants on the formation of gelatin nanofibres for controlled release of curcumin. Food Chem. 231, 70–77.

Devi, N., Sarmah, M., Khatun, B., Maji, T.K., 2017. Encapsulation of active ingredients in polysaccharide–protein complex coacervates. Adv. Colloid Interface Sci. 239, 136–145.

Dittrich, R., Hoffmann, I., Stahl, P., Muller, A., Beckmann, M.W., Pischetsrieder, M., 2006. Concentrations of N-epsilon-carboxymethyllysine in human breast milk, infant formulas, and urine of infants. J. Agric. Food Chem. 54, 6924–6928.

Dwivedy, A.K., Singh, V.K., Prakash, B., Dubey, N.K., 2018. Nanoencapsulated Illicium verum Hook.f. essential oil as an effective novel plant-based preservative against aflatoxin B1 production and free radical generation. Food Chem. Toxicol. 111, 102–113.

Embuscado, M.E., Shahidi, F., 2015. Herbs and spices as antioxidants for food preservation. In: Shahidi, F. (Ed.), Handbook of Antioxidants for Food Preservation. Woodhead Publishing, pp. 251–283.

Fan, F.Y., Sang, L.X., Jiang, M., 2017. Catechins and their therapeutic benefits to inflammatory bowel disease. Molecules 22 (484), 1–29.

Fang, Z., Bhandari, B., 2010. Encapsulation of polyphenols—a review. Trends Food Sci. Technol. 21 (10), 510–523.

Farhang, B., Kakuda, Y., Corredig, M., 2012. Encapsulation of ascorbic acid in liposomes prepared with milk fat globule membrane-derived phospholipids. Dairy Sci. Technol. 92 (4), 353–366.

Farias, M.D.P., Albuquerque, P.B.S., Soares, P.A.G., de Sá, D.M.A.T., Vicente, A.A., Carneiro-da-Cunha, M.G., 2018. Xyloglucan from Hymenaea courbaril var. courbaril seeds as encapsulating agent of L-ascorbic acid. Int. J. Biol. Macromol. 107, 1559–1566.

Faridi Esfanjani, A., Jafari, S.M., 2016. Biopolymer nano-particles and natural nano-carriers for nano-encapsulation of phenolic compounds. Colloids Surf. B: Biointerfaces 146, 532–543.

Faridi Esfanjani, A., Assadpour, E., Jafari, S.M., 2018. Improving the bioavailability of phenolic compounds by loading them within lipid-based nanocarriers. Trends Food Sci. Technol. 76, 56–66.

Fernandez-Ponce, M.T., Masmoudi, Y., Djerafi, R., Casas, L., Mantell, C., Ossa, E.M.d.l., Badens, E., 2015. Particle design applied to quercetin using supercritical anti-solvent techniques. J. Supercrit. Fluids 105, 119–127.

Flores, F.P., Singh, R.K., Kong, F., 2014. Physical and storage properties of spray-dried blueberry pomace extract with whey protein isolate as wall material. J. Food Eng. 137, 1–6.
Flores, F.P., Singh, R.K., Kerr, W.L., Philips, D.R., Kong, F., 2015. In vitro release properties of encapsulated blueberry (*Vacciniu mashei*) extracts. Food Chem. 168, 225–232.
Franceschi, E., Cezaro, A.D., Ferreira, S.R.S., Kunita, M.H., Edvani, C., Rubira, A.F., Oliveira, J.V., 2010. Co-precipitation of β-carotene and bio-polymer using supercritical carbon dioxide as antisolvent. Open Chem. Eng. J. 4, 11–20.
Gadkari, P.V., Balaraman, M., 2015. Catechins: sources, extraction and encapsulation: a review. Food Bioprod. Process. 93, 122–138.
Ganje, M., Jafari, S.M., Dusti, A., Dehnad, D., Amanjani, M., Ghanbari, V., 2016. Modeling quality changes in tomato paste containing microencapsulated olive leaf extract by accelerated shelf life testing. Food Bioprod. Process. 97, 12–19.
Ganji, S., Sayyed-Alangi, S.Z., 2017. Encapsulation of ginger ethanolic extract in nanoliposome and evaluation of its antioxidant activity on sunflower oil. Chem. Pap. 71 (9), 1781–1789.
Garcia-Moreno, P.J., Mendes, A.C., Jacobsen, C., Chronakis, I.S., 2018. Biopolymers for the nano-microencapsulation of bioactive ingredients by electrohydrodynamic processing. In: Gutierrez, T.J. (Ed.), Polymers for Food Applications. Springer International Publishing, Cham, pp. 447–479.
Gharibzahedi, S.M.T., Jafari, S.M., 2017. 7—Nanocapsule formation by cyclodextrins. In: Nanoencapsulation Technologies for the Food and Nutraceutical Industries. Academic Press, pp. 187–261.
Ghasemi, S., Jafari, S.M., Assadpour, E., Khomeiri, M., 2017. Production of pectin-whey protein nano-complexes as carriers of orange peel oil. Carbohydr. Polym. 177, 369–377. Supplement C.
Ghasemi, S., Jafari, S.M., Assadpour, E., Khomeiri, M., 2018. Nanoencapsulation of d-limonene within nanocarriers produced by pectin-whey protein complexes. Food Hydrocoll. 77, 152–162.
Ghorbani, M., Hosseini, H., 2017. Determination of water activity corresponding to the proper quality of peanut kernels. J. Food Process. Preserv. 41 (6), e13260.
Ghorbanzade, T., Jafari, S.M., Akhavan, S., Hadavi, R., 2017. Nano-encapsulation of fish oil in nano-liposomes and its application in fortification of yogurt. Food Chem. 216, 146–152.
Girardi, N.S., García, D., Passone, M.A., Nesci, A., Etcheverry, M., 2017. Microencapsulation of *Lippia turbinata* essential oil and its impact on peanut seed quality preservation. Int. Biodeter. Biodegr. 116, 227–233.
Girardi, N.S., Passone, M.A., Garcia, D., Nesci, A., Etcheverry, M., 2018. Microencapsulation of *Peumus boldus* essential oil and its impact on peanut seed quality preservation. Ind. Crop. Prod. 114, 108–114.
Gomez-Estaca, J., Comunian, T.A., Montero, P., Ferro-Furtado, R., Favaro-Trindade, C.S., 2016. Encapsulation of an astaxanthin-containing lipid extract from shrimp waste by complex coacervation using a novel gelatin-cashew gum complex. Food Hydrocoll. 61, 155–162.
Gonçalves, N.D., Pena, F.d.L., Sartoratto, A., Derlamelina, C., Duarte, M.C.T., Antunes, A.E.C., Prata, A.S., 2017. Encapsulated thyme (*Thymus vulgaris*) essential oil used as a natural preservative in bakery product. Food Res. Int. 96, 154–160.
Gorinstein, S., Pawelzik, E., Delgado-Licon, E., Haruenkit, R., Weisz, M., Trakhtenberg, S., 2000. Characterisation of pseudocereal and cereal proteins by protein and amino acid analyses. J. Food Eng. 73, 151–161.
Granato, D., Nunes, D.S., Barba, F.J., 2017. An integrated strategy between food chemistry, biology, nutrition, pharmacology, and statistics in the development of functional foods: a proposal. Trends Food Sci. Technol. 62, 13–22.

Gutierrez, J., Barry-Ryan, C., Bourke, P., 2008. The antimicrobial efficacy of plant essential oil combinations and interactions with food ingredients. Int. J. Food Microbiol. 124, 91–97.

Halliwell, B., 1997. Antioxidants and human disease: a general introduction. Nutr. Rev. 55, 44–49.

Haloci, E., Toska, V., Shkreli, R., Goci, E., Vertuani, S., Manfredini, S., 2014. Encapsulation of *Satureja montana* essential oil in β-cyclodextrin. J. Incl. Phenom. Macrocycl. Chem. 80 (1), 147–153.

Hamad, S.H., 2012. Factors affecting the growth of microorganisms in food. In: Bhat, R., Karim Alias, A., Paliyath, G. (Eds.), Progress in Food Preservation. Wiley-Blackwell, Oxford, pp. 405–427.

Han, S., Yang, Y., 2005. Antimicrobial activity of wool fabric treated with curcumin. Dyes Pigm. 64 (2), 157–161.

Hao, J., Guo, B., Yu, S., Zhang, W., Zhang, D., Wang, J., Wang, Y., 2017. Encapsulation of the flavonoid quercetin with chitosan-coated nano-liposomes. LWT- Food Sci. Technol. 85, 37–44.

Ho, S., Thoo, Y.Y., Young, D.J., Siow, L.F., 2017. Cyclodextrin encapsulated catechin: effect of pH, relative humidity and various food models on antioxidant stability. LWT Food Sci. Technol. 85, 232–239.

Holzapfel, W.H., 1998. The Gram-positive bacteria associated with meat and meat products. In: Board, R.G., Davies, A.R. (Eds.), The Microbiology of Meat and Poultry. Blackie Academic and Professional, London, pp. 35–84.

Hoseinnejad, M., Jafari, S.M., Katouzian, I., 2018. Inorganic and metal nanoparticles and their antimicrobial activity in food packaging applications. Crit. Rev. Microbiol. 44 (2), 161–181.

Hossain, A., Shahidi, F., 2018. Bioactives in spices, and spice oleoresins: phytochemicals and their beneficial effects in food preservation and health promotion. J. Food Bioactives 3, 8–75.

Hosseini, H., Ghorbani, M., Sadeghi Mahoonak, A., Maghsoudlou, Y., 2014. Monitoring hydroperoxides formation as a measure of predicting walnut oxidative stability. Acta Aliment. 43, 412–418.

Hosseini, H., Ghorbani, M., Jafari, S.M., Sadeghi Mahoonak, A., 2018a. Investigating the effect of lipase from Candida rugosa on the production of EPA and DHA concentrates from *Kilka* fish (*Clupeonella cultiventris caspia*). LWT- Food Sci. Technol. 93, 534–541.

Hosseini, H., Bolourian, S., Yaghoubi Hamgini, E., Ghanuni Mahababadi, E., 2018b. Optimization of heat- and ultrasound-assisted extraction of polyphenols from dried rosemary leaves using response surface methodology. J. Food Process. Preserv. 42 (11), e13778.

Hosseini, H., Ghorbani, M., Jafari, S.M., Mahoonak, S.A., 2019. Encapsulation of EPA and DHA concentrate from Kilka fish oil by milk proteins and evaluation of its oxidative stability. J. Food Sci. Technol. 56 (1), 59–70.

Hu, M., 2016. Oxidative stability and shelf life of low-moisture foods. In: Hu, M., Jacobsen, C. (Eds.), Oxidative Stability and Shelf Life of Foods Containing Oils and Fats. AOCS Press, pp. 313–371.

Hu, J., Wang, X., Xiao, Z., Bi, W., 2015. Effect of chitosan nanoparticles loaded with cinnamon essential oil on the quality of chilled pork. LWT- Food Sci. Technol. 63 (1), 519–526.

Jacobsen, C., Decker, E.A., 2010. Understanding and reducing oxidative flavour deterioration in foods. In: Decker, E.A. (Ed.), Oxidation in Foods and Beverages and Antioxidant Applications. Woodhead Publishing, pp. 122–142.

Jafari, S.M., 2017a. 1—an overview of nanoencapsulation techniques and their classification. In: Nanoencapsulation Technologies for the Food and Nutraceutical Industries. Academic Press, pp. 1–34.

Jafari, S.M., 2017b. Chapter 1—an introduction to nanoencapsulation techniques for the food bioactive ingredients. In: Nanoencapsulation of Food Bioactive Ingredients. Academic Press, pp. 1–62.

Jafari, S.M., Assadpoor, E., He, Y., Bhandari, B., 2008. Encapsulation efficiency of food flavours and oils during spray drying. Drying Technol. 26 (7), 816–835.

Jafari, S.M., Ganje, M., Dehnad, D., Ghanbari, V., Hajitabar, J., 2017. Arrhenius equation modeling for the shelf life prediction of tomato paste containing a natural preservative. J. Sci. Food Agric. 97 (15), 5216–5222.

Jafari, S.M., Ghanbari, V., Dehnad, D., Ganje, M., 2018. Neural networks modeling of *Aspergillus flavus* growth in tomato paste containing microencapsulated olive leaf extract. J. Food Saf. 38 (1). e12396-n/a.

Jain, A., Thakur, D., Ghoshal, G., Katare, O.P., Shivhare, U.S., 2015. Microencapsulation by complex coacervation using whey protein isolates and gum acacia: an approach to preserve the functionality and controlled release of β-carotene. Food Bioproc. Tech. 8 (8), 1635–1644.

Jain, A., Thakur, D., Ghoshal, G., Katare, O.P., Shivhare, U.S., 2016. Characterization of microcapsulated β-carotene formed by complex coacervation using casein and gum tragacanth. Int. J. Biol. Macromol. 87, 101–113.

Jang, K.I., Lee, H.G., 2008. Stability of chitosan nanoparticles for l-ascorbic acid during heat treatment in aqueous solution. J. Agric. Food Chem. 56, 1936–1941.

Jia, Z., Dumont, M.J., Orsat, V., 2016. Encapsulation of phenolic compounds present in plants using protein matrices. Food Biosci. 15, 87–104.

Jullian, C., Moyano, L., Yanez, C., Olea-Azar, C., 2007. Complexation of quercetin with three kinds of cyclodextrins: an antioxidant study. Spectrochim. Acta A Mol. Biomol. Spectrosc. 67 (1), 230–234.

Kamimura, J.A., Santos, E.H., Hill, L.E., Gomes, C.L., 2014. Antimicrobial and antioxidant activities of carvacrol microencapsulated in hydroxypropyl-β-cyclodextrin. LWT- Food Sci. Technol. 57 (2), 701–709.

Kanatt, S.R., Tari, S., Chawla, S.P., 2018. Encapsulation of extract prepared from irradiated onion scales in alginate beads: a potential functional food ingredient. J. Food Meas. Charact. 12 (2), 848–858.

Katouzian, I., Jafari, S.M., 2016. Nano-encapsulation as a promising approach for targeted delivery and controlled release of vitamins. Trends Food Sci. Technol. 53, 34–48.

Kfoury, M., Hadaruga, N.G., Hadaruga, D.I., Fourmentin, S., 2016. Cyclodextrins as encapsulation material for flavors and aroma. In: Grumezescu, A.M. (Ed.), Encapsulations. Academic Press, pp. 127–192.

Kfoury, M., Auezova, L., Greige Gerges, H., Fourmentin, S., 2019. Encapsulation in cyclodextrins to widen the applications of essential oils. Environ. Chem. Lett. 17 (1), 129–143.

Khoshakhlagh, K., Koocheki, A., Mohebbi, M., Allafchian, A., 2017. Development and characterization of electrosprayed *Alyssum homolocarpum* seed gum nanoparticles for encapsulation of d-limonene. J. Colloid Interface Sci. 490, 562–575.

Kim, J.H., Yu, J., Mahoney, N., Chan, K.L., Molyneux, R.J., Varga, J., et al., 2008. Elucidation of the functional genomics of antioxidant-based inhibition of aflatoxin biosynthesis. Int. J. Food Microbiol. 22, 49–60.

Kim, S., Cho, E., Yoo, J., Cho, E., Choi, S.J., Son, S.M., Lee, J.M., In, M.-J., Kim, D.C., Kim, J.-H., Chae, H.J., 2010. β-CD-Mediated encapsulation enhanced stability and solubility of astaxanthin. J. Korean Soc. Appl. Biol. Chem. 53 (5), 559–565.

Klein, M., Aserin, A., Ishai, P.B., Garti, N., 2010. Interactions between whey protein isolate and gum Arabic. Colloids Surf. B Biointerfaces 79 (2), 377–383.

Knechtges, P.L., 2012. Food Safety: Theory and Practice. Jones & Bartlett Learning, Burlington, MA246.

Kong, F., Singh, R.P., 2016. Chemical deterioration and physical instability of foods and beverages. In: Anese, M., et al. (Eds.), The Stability and Shelf Life of Food. Second ed.. Woodhead Publishing, pp. 43–76.

Konur, O., Grumezescu, A.M., 2017. Scientometric overview in food nanopreservation. In: Food Preservation. Academic Press, pp. 703–729.

Kuda, T., Tsuda, N., Yano, T., 2004. Thermal inactivation characteristics of acid and alkaline phosphatase in fish and shellfish. Food Chem. 88 (4), 543–548.

Leclere, J., Birlouez-Aragon, I., Meli, M., 2002. Fortification of milk with iron-ascorbate promotes lysine glycation and tryptophan oxidation. Food Chem. 76, 491–499.

Lee, S.J., Wong, M., 2014. Nano- and microencapsulation of phytochemicals. In: Kwak, H. (Ed.), Nano- and Microencapsulation for Foods. John Wiley & Sons, Ltd, pp. 120–165.

Li, W., Bowers, J.A., Craig, J.A., Perng, S.K., 1993. Sodium tripolyphosphate stability and effect in ground turkey meat. J. Food Sci. 58 (3), 501–504.

Lianou, A., Panagou, E.Z., Nychas, G.J.E., 2016. Microbiological spoilage of foods and beverages. In: Subramaniam, P. (Ed.), The Stability and Shelf Life of Food. second ed.. Woodhead Publishing, pp. 3–42.

Liolios, C.C., Gortzi, O., Lalas, S., Tsaknis, J., Chinou, I., 2009. Liposomal incorporation of carvacrol and thymol isolated from the essential oil of *Origanum dictamnus* L. and in vitro antimicrobial activity. Food Chem. 112 (1), 77–83.

Mahdavee Khazaei, K., Jafari, S.M., Ghorbani, M., Hemmati Kakhki, A., 2014. Application of maltodextrin and gum Arabic in microencapsulation of saffron petal's anthocyanins and evaluating their storage stability and color. Carbohydr. Polym. 105 (1), 57–62.

Marin, D., Aleman, A., Sanchez-Faure, A., Montero, P., Gomez-Guillen, M.C., 2018. Freeze-dried phosphatidylcholine liposomes encapsulating various antioxidant extracts from natural waste as functional ingredients in surimi gels. Food Chem. 245, 525–535.

Martinez-Gracia, C., Gonzalez-Bermudez, C.A., Cabellero-Valcarcel, A.M., Santaella-Pascual, M., Frontela-Saseta, C., 2015. Use of herbs and spices for food preservation: advantages and limitations. Curr. Opin. Food Sci. 6, 38–43.

Mattea, F., Martín, Á., Cocero, M.J., 2009. Carotenoid processing with supercritical fluids. J. Food Eng. 93 (3), 255–265.

Menon, V., Sudheer, A.-R., 2007. The molecular targets and therapeutic uses of curcumin in health and disease. In: Aggarwal, B., Surh, Y.-J., Shishodia, S. (Eds.), Antioxidant and Anti-Inflammatory Properties of Curcumin. Springer, New York, pp. 105–125.

Mohammadi, A., Jafari, S.M., Esfanjani, A.F., Akhavan, S., 2016. Application of nanoencapsulated olive leaf extract in controlling the oxidative stability of soybean oil. Food Chem. 190, 513–519.

Montero, P., Calvo, M.M., Gomez-Guillen, M.C., Gomez-Estaca, J., 2016. Microcapsules containing astaxanthin from shrimp waste as potential food coloring and functional ingredient: characterization, stability, and bioaccessibility. LWT- Food Sci. Technol. 70, 229–236.

Moraes-Lovison, M., Marostegan, L.F.P., Peres, M.S., Menezes, I.F., Ghiraldi, M., Rodrigues, R.A.F., Fernandes, A.M., Pinho, S.C., 2017. Nanoemulsions encapsulating oregano essential oil: production, stability, antibacterial activity and incorporation in chicken pâté. LWT- Food Sci. Technol. 77, 233–240.

Murcia, M.A., Egea, I., Romojaro, F., Parras, P., Jimenez, A.M., Martinez-Tome, M., 2004. Antioxidant evaluation in dessert spices compared with common food additives. Influence of irradiation procedure. J. Agric. Food Chem. 52 (7), 1872–1881.

Nedovic, V., Kalusevic, A., Manojlovic, V., Levic, S., Bugarski, B., 2011. An overview of encapsulation technologies for food applications. Procedia Food Sci. 1, 1806–1815.

Nelson, K.A., Labuza, T.P., 1992. Relationship between water and lipid oxidation rates: water activity and glass transition theory. In: St Angelo, A.J. (Ed.), Lipid Oxidation in Food. ACS, Washington, DC, pp. 93–103.

Ngamakeue, N., Chitprasert, P., 2016. Encapsulation of holy basil essential oil in gelatin: effects of palmitic acid in carboxymethyl cellulose emulsion coating on antioxidant and antimicrobial activities. Food Bioproc. Tech. 9 (10), 1735–1745.

Nychas, G.-J.E., Drosinos, E.H., Board, R.G., 1998. Chemical changes in stored meat. In: Board, R.G., Davies, A.R. (Eds.), The Microbiology of Meat and Poultry. Blackie Academic and Professional, London, pp. 288–326.

Nychas, G.J.E., Skandamis, P.N., Tassou, C.C., Koutsoumanis, K.P., 2008. Meat spoilage during distribution. Meat Sci. 78, 77–89.

Ohloff, G., 1994. Scent and Fragrances: The Fascination of Odors and Their Chemical Perspectives. Springer-Verlag, Berlin.

Okabe, S., Ochiai, Y., Aida, M., Park, K., Kim, S., Nomura, T., Suganuma, M., Fujiki, H., 1999. Mechanistic aspects of green tea as a cancer preventive: effect of components on human stomach cancer cell lines. Jpn. J. Cancer Res. 90, 733–739.

Ozkan, G., Franco, P., De Marco, I., Xiao, J., Capanoglu, E., 2019. A review of microencapsulation methods for food antioxidants: principles, advantages, drawbacks and applications. Food Chem. 272, 494–506.

Özvural, E.B., Huang, Q., Chikindas, M.L., 2016. The comparison of quality and microbiological characteristic of hamburger patties enriched with green tea extract using three techniques: direct addition, edible coating and encapsulation. LWT- Food Sci. Technol. 68, 385–390.

Pabast, M., Shariatifar, N., Beikzadeh, S., Jahed, G., 2018. Effects of chitosan coatings incorporating with free or nano-encapsulated Satureja plant essential oil on quality characteristics of lamb meat. Food Control 91, 185–192.

Pandit, J., Aqil, M., Sultana, Y., 2016. Nanoencapsulation technology to control release and enhance bioactivity of essential oils. In: Grumezescu, A.M. (Ed.), Encapsulations. Academic Press, pp. 597–640.

Parthasarathi, S., Anandharamakrishnan, C., 2016. Enhancement of oral bioavailability of vitamin E by spray-freeze drying of whey protein microcapsules. Food Bioprod. Process. 100, 469–476.

Patras, A., Brunton, N.P., O'Donnell, C., Tiwari, B.K., 2010. Effect of thermal processing on anthocyanin stability in foods; mechanisms and kinetics of degradation. Trends Food Sci. Technol. 21 (1), 3–11.

Paximada, P., Echegoyen, Y., Koutinas, A.A., Mandala, I.G., Lagaron, J.M., 2017. Encapsulation of hydrophilic and lipophilized catechin into nanoparticles through emulsion electrospraying. Food Hydrocoll. 64, 123–132.

Pereira, H.V.R., Saraiva, K.P., Carvalho, L.M.J., Andrade, L.R., Pedrosa, C., Pierucci, A.P.T.R., 2009. Legumes seeds protein isolates in the production of ascorbic acid microparticles. Food Res. Int. 42 (1), 115–121.

Pierucci, A.P.T.R., Andrade, L.R., Baptista, E.B., Volpato, N.M., Rocha-Leão, M.H.M., 2006. New microencapsulation system for ascorbic acid using pea protein concentrate as coat protector. J. Microencapsul. 23 (6), 654–662.

Pietta, P.G., 2000. Flavonoids as antioxidants. J. Nat. Prod. 63, 1035–1042.

Pinho, E., Grootveld, M., Soares, G., Henriques, M., 2014. Cyclodextrins as encapsulation agents for plant bioactive compounds. Carbohydr. Polym. 101, 121–135.

Pourashouri, P., Shabanpour, B., Razavi, S.H., Jafari, S.M., Shabani, A., Aubourg, S.P., 2014a. Impact of wall materials on physicochemical properties of microencapsulated fish oil by spray drying. Food Bioproc. Tech. 7 (8), 2354–2365.

Pourashouri, P., Shabanpour, B., Razavi, S.H., Jafari, S.M., Shabani, A., Aubourg, S.P., 2014b. Oxidative stability of spray-dried microencapsulated fish oils with different wall materials. J. Aquat. Food Prod. Technol. 23 (6), 567–578.

Prakash, B., Kedia, A., Mishra, P.K., Dubey, N.K., 2015. Plant essential oils as food preservatives to control moulds, mycotoxin contamination and oxidative deterioration of agri-food commodities—Potentials and challenges. Food Control 47, 381–391.

Prombutara, P., Kulwatthanasal, Y., Supaka, N., Sramala, I., Chareonpornwattana, S., 2012. Production of nisin-loaded solid lipid nanoparticles for sustained antimicrobial activity. Food Control 24 (1), 184–190.

Prosekov, A.Y., Ivanova, S.A., 2018. Food security: the challenge of the present. Geoforum 91, 73–77.

Rafiee, Z., Jafari, S.M., 2018. Application of lipid nanocarriers for the food industry. In: Mérillon, J.-M., Ramawat, K.G. (Eds.), Bioactive Molecules in Food. Springer International Publishing, Cham, pp. 1–43.

Rafiee, Z., Jafari, S.M., Alami, M., Khomeiri, M., 2012. Antioxidant effect of microwave-assisted extracts of olive leaves on sunflower oil. J. Agric. Sci. Technol. 14 (Suppl), 1497–1509.

Rafiee, Z., Nejatian, M., Daeihamed, M., Jafari, S.M., 2018. Application of different nanocarriers for encapsulation of curcumin. Crit. Rev. Food Sci. Nutr., 1–77.

Rahman, M.S., Al-Belushi, R.M., Guizani, N., Al-Saidi, G.S., Soussi, B., 2009. Fat oxidation in freeze-dried grouper during storage at different temperatures and moisture contents. Food Chem. 114 (4), 1257–1264.

Rajabi, H., Ghorbani, M., Jafari, S.M., Sadeghi Mahoonak, A., Rajabzadeh, G., 2015. Retention of saffron bioactive components by spray drying encapsulation using maltodextrin, gum Arabic and gelatin as wall materials. Food Hydrocoll. 51, 327–337.

Rakmai, J., Cheirsilp, B., Mejuto, J.C., Torrado-Agrasar, A., Simal-Gándara, J., 2017. Physico-chemical characterization and evaluation of bio-efficacies of black pepper essential oil encapsulated in hydroxypropyl-beta-cyclodextrin. Food Hydrocoll. 65, 157–164.

Rao, A.R., Phang, S.M., Ravi, S., Ravishankar, G.A., 2014. Astaxanthin: sources, extraction, stability, biological activities and its commercial applications—a review. Mar. Drugs 12 (1), 128–152.

Ray, S., Raychaudhuri, U., Chakraborty, R., 2016. An overview of encapsulation of active compounds used in food products by drying technology. Food Biosci. 13, 76–83.

Regnault-Roger, C., Vincent, C., Arnason, J.T., 2012. Essential oils in insect control: low-risk products in a high-stakes world. Annu. Rev. Entomol. 57, 405–424.

Rezaei, A., Fathi, M., Jafari, S.M., 2019. Nanoencapsulation of hydrophobic and low-soluble food bioactive compounds within different nanocarriers. Food Hydrocoll. 88, 146–162.

Riso, P., Erba, D., Criscuoli, F., Testolin, G., 2002. Effect of green tea extract on DNA repair and oxidative damage due to H2O2 in jurkat T cells. Nutr. Res. 22, 1143–1150.

Rohr, A., Luddecke, K., Drusch, S., Muller, M.J., Alvensleben, R.V., 2005. Food quality and safety e consumer perception and public health concern. Food Control 16, 649–655.

Sahari, M.A., Moghimi, H.R., Hadian, Z., Barzegar, M., Mohammadi, A., 2017. Physicochemical properties and antioxidant activity of α-tocopherol loaded nanoliposome's containing DHA and EPA. Food Chem. 215, 157–164.

Saldanha do Carmo, C., Pais, R., Simplicio, A.L., Mateus, M., Duarte, C.M.M., 2017. Improvement of aroma and shelf-life of non-alcoholic beverages through cyclodextrins-limonene inclusion complexes. Food Bioproc. Tech. 10 (7), 1297–1309.

Sawale, P.D., Patil, G.R., Hussain, S.A., Singh, A.K., Singh, R.R.B., 2017. Effect of incorporation of encapsulated and free Arjuna herb on storage stability of chocolate vanilla dairy drink. Food Biosci. 19, 142–148.

Sessa, M., Casazza, A.A., Perego, P., Tsao, R., Ferrari, G., Donsi, F., 2013. Exploitation of polyphenolic extracts from grape marc as natural antioxidants by encapsulation in lipid-based nanodelivery systems. Food Bioproc. Tech. 6 (10), 2609–2620.

Shahidi, F., Zhong, Y., 2010. Novel antioxidants in food preservation and health promotion. Eur. J. Lipid Sci. Technol. 112, 930–940.

Shinde, T., Sun-Waterhouse, D., Brooks, J., 2014. Co-extrusion encapsulation of probiotic lactobacillus acidophilus alone or together with apple skin polyphenols: an aqueous and value-added delivery system using alginate. Food Bioproc. Tech. 7 (6), 1581–1596.

Shishir, M.R.I., Xie, L., Sun, C., Zheng, X., Chen, W., 2018. Advances in micro and nano-encapsulation of bioactive compounds using biopolymer and lipid-based transporters. Trends Food Sci. Technol. 78, 34–60.

Shlar, I., Poverenov, E., Vinokur, Y., Horev, B., Droby, S., Rodov, V., 2015. High throughput screening of nanoparticle-stabilizing ligands: application to preparing antimicrobial curcumin nanoparticles by antisolvent precipitation. Nanomicro Lett. 7 (1), 68–79.

Sickler, M.L., Claus, J.R., Marriott, N.G., Eigel, W.N., Wang, H., 2013. Reduction in lipid oxidation by incorporation of encapsulated sodium tripolyphosphate in ground turkey. Meat Sci. 95 (2), 376–380.

Silva, A.C.d., Santos, P.D.d.F., Silva, J.T.d.P., Leimann, F.V.r., Bracht, L., Goncalves, O.H., 2018a. Impact of curcumin nanoformulation on its antimicrobial activity. Trends Food Sci. Technol. 72, 74–82.

Silva, F., Torres, L., Silva, L., Figueiredo, R., Garruti, D., Araújo, T., Duarte, A., Brito, D., Ricardo, N., 2018b. Cashew gum and maltrodextrin particles for green tea (Camellia sinensis var Assamica) extract encapsulation. Food Chem. 261, 169–175.

Singh, R.P., Anderson, B.A., 2004. The major types of food spoilage: an overview. In: Steele, R. (Ed.), Understanding and Measuring the Shelf-Life of Food. Woodhead Publishing, Cambridge, pp. 1–23.

Singh, T.K., Cadwallader, K.R., 2004. Ways of measuring shelf-life and spoilage. In: Steele, R. (Ed.), Understanding and Measuring the Shelf-Life of Food. Woodhead Publishing, Cambridge, pp. 165–183.

Singh, J., Kaur, K., Kumar, P., 2018. Optimizing microencapsulation of α-tocopherol with pectin and sodium alginate. J. Food Sci. Technol. 55 (9), 3625–3631.

Siracusa, V., Rocculi, P., Romani, S., Rosa, M.D., 2008. Biodegradable polymers for food packaging: a review. Trends Food Sci. Technol. 19 (12), 634–643.

Sosa, M., Rodríguez-Rojo, S., Mattea, F., Cismondi, M., Cocero, M., 2011. Green tea encapsulation by means of high pressure antisolvent coprecipitation. J. Supercrit. Fluids 56 (3), 304–311.

Strimpakos, A., Sharma, R., 2008. Curcumin: preventive and therapeutic properties in laboratory studies and clinical trials. Antioxid. Redox Signal. 10, 511–537.

Sun-Waterhouse, D., Wang, W., Waterhouse, G.I.N., 2014. Canola oil encapsulated by alginate and its combinations with starches of low and high amylose content: effect of quercetin on oil stability. Food Bioproc. Tech. 7 (8), 2159–2177.

Taghvaei, M., Jafari, S.M., 2015. Application and stability of natural antioxidants in edible oils in order to substitute synthetic additives. J. Food Sci. Technol. 52 (3), 1272–1282.

Tajkarimi, M.M., Ibrahim, S.A., Cliver, D.O., 2010. Antimicrobial herb and spice compounds in food. Food Control 21 (9), 1199–1218.

Tan, C., Xue, J., Abbas, S., Feng, B., Zhang, X., Xia, S., 2014. Liposome as a delivery system for carotenoids: comparative antioxidant activity of carotenoids as measured by ferric reducing antioxidant power, DPPH assay and lipid peroxidation. J. Agric. Food Chem. 62 (28), 6726–6735.

Tao, Y., Wang, P., Wang, J., Wu, Y., Han, Y., Zhou, J., 2017. Combining various wall materials for encapsulation of blueberry anthocyanin extracts: optimization by artificial neural network and genetic algorithm and a comprehensive analysis of anthocyanin powder properties. Powder Technol. 311, 77–87.

Tatar Turan, F., Cengiz, A., Kahyaoglu, T., 2015. Evaluation of ultrasonic nozzle with spray-drying as a novel method for the microencapsulation of blueberry's bioactive compounds. Innov. Food Sci. Emerg. Technol. 32, 136–145.

Toniazzo, T., Peres, M.S., Ramos, A.P., Pinho, S.C., 2017. Encapsulation of quercetin in liposomes by ethanol injection and physicochemical characterization of dispersions and lyophilized vesicles. Food Biosci. 19, 17–25.

Topuz, O.K., Özvural, E.B., Zhao, Q., Huang, Q., Chikindas, M., Gölükçü, M., 2016. Physical and antimicrobial properties of anise oil loaded nanoemulsions on the survival of foodborne pathogens. Food Chem. 203, 117–123.

Troise, A.D., Fogliano, V., 2013. Reactants encapsulation and Maillard reaction. Trends Food Sci. Technol. 33 (1), 63–74.

Tumbas Šaponjac, V., Ćetković, G., Čanadanović-Brunet, J., Pajin, B., Djilas, S., Petrović, J., Lončarević, I., Stajčić, S., Vulić, J., 2016. Sour cherry pomace extract encapsulated in whey and soy proteins: incorporation in cookies. Food Chem. 207, 27–33.

U.S. Code of federal regulations (2013). Code of federal regulations. Title 21, Part 182, Section 182.20.

Uddin, M.S., Hawlader, M.N.A., Zhu, H.J., 2001. Microencapsulation of ascorbic acid: effect of process variables on product characteristics. J. Microencapsul. 18 (2), 199–209.

Valdes, A., Mellinas, A.C., Ramos, M., Burgos, N., Jimenez, A., Garrigos, M.C., 2015. Use of herbs, spices and their bioactive compounds in active food packaging. RSC Adv. 5 (50), 40324–40335.

Vinceković, M., Viskić, M., Jurić, S., Giacometti, J., Bursać Kovačević, D., Putnik, P., Donsì, F., Barba, F.J., Režek Jambrak, A., 2017. Innovative technologies for encapsulation of Mediterranean plants extracts. Trends Food Sci. Technol. 69, 1–12.

Visentin, A., Rodríguez-Rojo, S., Navarrete, A., Maestri, D., Cocero, M.J., 2012. Precipitation and encapsulation of rosemary antioxidants by supercritical antisolvent process. J. Food Eng. 109 (1), 9–15.

Voon, C.H., Bhat, R., Rusul, G., 2012. Flower extracts and their essential oils as potential antimicrobial agents for food uses and pharmaceutical applications. Compr. Rev. Food Sci. Food Saf. 11, 34–55.

Wandrey, C., Bartkowiak, A., Harding, S.A., 2010. Materials for encapsulation. In: Zuidam, N.J., Nedovic, V. (Eds.), Encapsulation Technologies for Active Food Ingredients and Food Processing. Springer, Dordrecht, Netherlands, pp. 31–100.

Wang, S., Tan, M., Zhong, Z., Chen, M., Wang, Y., 2011. Nanotechnologies for curcumin: an ancient puzzler meets modern solutions. J. Nanomater. 1, 1–8.

Weiss, J., Gaysinsky, S., Davidson, M., McClements, J., 2009. Nanostructured encapsulation systems: food antimicrobials. In: Gustavo, B.-C., Alan, M., David, L., Walter, S., Ken, B., Paul, C. (Eds.), Global Issues in Food Science and Technology. Academic Press, San Diego, CA. Chapter 24.

Wilkowsk, A., Ambroziak, W., CzyżOwska, A., Adamiec, J., 2015. Effect of microencapsulation by spray-drying and freeze-drying technique on the antioxidant properties of blueberry (Vaccinium myrtillus) juice polyphenolic compounds. Polish J. Food Nutr Sci. 66, 11–16.

Wu, H., Xue, N., Hou, C.-l., Feng, J.-t., Zhang, X., 2015. Microcapsule preparation of allyl isothiocyanate and its application on mature green tomato preservation. Food Chem. 175, 344–349.

Zhao, M., Wang, H., Yang, B., Tao, H., 2010. Identification of cyclodextrin inclusion complex of chlorogenic acid and its antimicrobial activity. Food Chem. 120 (4), 1138–1142.

Zhao, L., Temelli, F., Chen, L., 2017a. Encapsulation of anthocyanin in liposomes using supercritical carbon dioxide: effects of anthocyanin and sterol concentrations. J. Funct. Foods 34, 159–167.

Zhao, L., Temelli, F., Curtis, J.M., Chen, L., 2017b. Encapsulation of lutein in liposomes using supercritical carbon dioxide. Food Res. Int. 100, 168–179.

Zuidam, N.J., Heinrich, J., 2009. Encapsulation of aroma. In: Zuidam, N.J., Nedovic, V.A. (Eds.), Encapsulation Technologies for Food Active Ingredients and Food Processing. Springer, Dordrecht, Netherlands, pp. 127–160.

Zuidam, N.J., Nedovic, V., 2010. Encapsulation Technologies for Active Food Ingredients and Food Processing. Springer, New York, Dordrecht, Heidelberg402.

Zuidam, N.J., Shimoni, E., 2009. Overview of microencapsulates for use in food products or processes and methods to make them. In: Zuidam, N.J., Nedovic, V.A. (Eds.), Encapsulation Technologies for Food Active Ingredients and Food Processing. Springer, Dordrecht, Netherlands, pp. 3–31.

# Survival of probiotics in functional foods during shelf life

**6**

*Nayil Dinkçi, Vildan Akdeniz, A. Sibel Akalin*
Department of Dairy Technology Faculty of Agriculture, Ege University, Bornova-İzmir, Turkey

**Chapter outline**



## 1 Introduction

The role of healthiness in food preference is continuously increasing in the world as the scientific progress in understanding the connection between nutrition and health has an increasingly great effect on consumers' approach to nutrition. Nowadays, consumers believe that foods are directly related to their health and consumer demands on food production methods have changed considerably. Therefore, foods are not considered to only satisfy hunger and to supply nutrients for the human body, but to develop physical and mental health and to prevent nutrition related diseases of the consumers. For this reason, functional foods are in greater demand by health-conscious consumers, especially depending on the increasing life expectancy and healthcare cost of the aging population (Siro et al., 2008; Sarkar, 2010; Bhat and Bhat, 2011).

The term "functional food" is generally applied to food products that are claimed to have special beneficial physiological effects in the human body. According to EU concerted action, "a food can be regarded as functional if it has been satisfactorily demonstrated to affect beneficially one or more target functions in the body, beyond adequate nutritional effects, in a way that is relevant to either an improved state of health and well-being and/or a reduction of risk of disease." Functional foods must demonstrate their health effects as a part of diet as distinct from pills or capsules (Shiby and Mishra, 2013; Anadon et al., 2016). There is no unitary accepted definition for functional foods so far. In addition, there is no legislative definition of this term and a certain differentiation between conventional and functional foods in most countries (Bhat and Bhat, 2011).

Food Quality and Shelf Life. https://doi.org/10.1016/B978-0-12-817190-5.00006-9

There are several ways to manufacture functional foods based on the modification of processing methods or fortification with different substances. In this regard, the functionality of a product that is based on bioactive components can be targeted to a special disease or to improve overall wellbeing (Figueroa-Gonzalez et al., 2011; Shiby and Mishra, 2013). Functional food ingredients are generally used to optimize beneficial effects arising from the bioactive components in the product. These ingredients include mainly probiotics, prebiotics, soluble fiber, omega-3-polyunsaturated fatty acids, conjugated linoleic acid, some proteins, peptides and amino acids, antioxidants, vitamins, minerals, and phospholipids (Bhat and Bhat, 2011). Probiotics and prebiotics are main ingredients of fermented milk products, which account for the most important part of the functional food market (Figueroa-Gonzalez et al., 2011).

According to the most widely accepted FAO/WHO definition, probiotics are "live microorganisms which, when administrated in adequate amounts, confer a health benefit on the host" (FAO/WHO, 2002). This implies that probiotic cells must be alive when consumed. Therefore, the selection of adequate food matrix and also processing and storage conditions to provide probiotic delivery is critical (Capozzi et al., 2016). The most popular approach to consume probiotic cells is through food products and these products represent a significant part of functional foods (Tripathi and Giri, 2014).

In addition, probiotic foods are the fastest-growing area of the functional food sector as consumers are increasingly demanding products fortified with probiotic bacteria and the probiotic cultures are successfully applied in different types of food matrices (Bakr, 2015, 2016). >500 probiotic food products have been introduced to the global market during the last couple of decades (Tripathi and Giri, 2014).

A great number of food products including dairy products, and also meat, beverages, cereals, vegetables and fruits, and bread products have been produced to deliver probiotics (Bakr, 2015). Today, there has been a strong increase in the consumption of probiotic bacteria, especially through probiotic dairy products including fermented milks, ice cream, cheeses, baby foods, dairy desserts, whey-based beverages, sour cream, butter milk, liquid milk, and concentrated milk (Mohammadi et al., 2011; Tripathi and Giri, 2014). Growing research interest has also focused on the incorporation of probiotic bacteria into cultured dairy products to improve the nutritional quality of these products (Karimi et al., 2011). On the other hand, nondairy products such as vegetable-based and cereal-based products, fruit juices, and confectionary products have also been developed to deliver probiotics for people who are allergic to milk proteins or have severe lactose intolerance (Tripathi and Giri, 2014; Bakr, 2016).

It has been estimated that probiotic foods comprise between 60% and 70% of the total functional food market (Tripathi and Giri, 2014). Probiotic foods in the United States, European, and Japanese markets account for over 90% of the total functional foods worldwide, most of which comprise functional dairy products. These products constitute nearly 43% of the market, which is mostly based on fermented dairy products. However, the quality of most of the commercial probiotic foods are poor for viable probiotic count and do not meet their label claim on strain type (Sarkar, 2018).

Probiotic microorganisms exhibit a health benefit for the host when they are ingested, although the health benefits are genus- or strain-specific. A number of these benefits attributed to probiotics are related to the maintenance of normal intestinal flora, showing the ability to survive through the upper gastrointestinal system, and to be capable of surviving and growing in the intestinal region (Anadon et al., 2016).

Documented health benefits associated with probiotics include maintenance of intestinal microflora balance, protection against gastrointestinal pathogens, stimulation of immune system, reduction of serum cholesterol level and blood pressure, anticarcinogenic effects, alleviation of lactose intolerance symptoms, and nutritional enhancements. Probiotics can also be therapeutically applied to prevent infantile diarrhea, urogenital diseases, osteoporosis, food allergies, and atopic diseases, to reduce antibody-induced diarrhea, to alleviate constipation and high cholesterol level, to control inflammatory bowel diseases, and to protect from colon and bladder cancer. It is considered that these health benefits may result from the action of viable probiotics of cultured foods or the growth and the action of certain species of probiotics in the intestinal tract (Tripathi and Giri, 2014).

In fact, a wide variety of genera and species of microorganisms are reported as potential probiotics. However, lactic acid bacteria, mainly Lactobacillus and Bifidobacterium genera are the most widely used bacteria within probiotics in the food market. They are also normal inhabitants of the human intestine and have a long tradition of safe application within the food industry (Siro et al., 2008; Bakr, 2016). Other genera including the Enterococcus, Streptococcus, Leuconostoc genera, and others are also utilized (Abdollahi et al., 2016).

In Table 1, the potentially used probiotic species are given (Anadon et al., 2016; Sendra et al., 2016). Bacillus (*B. subtilis*, *B. cereus* var. *toyoi*) and yeasts that are not usual components of the gut microflora are also used as a human probiotic. The yeast *Saccharomyces boulardii* is used in capsule or powder form rather than in food preparation (Anadon et al., 2016). Primarily, a testing process including strain testing, identification by genotype and phenotype, functionalized characterization, safety assessment testing, and double-blind, placebo-controlled human trials to confirm their health benefits must be applied to organisms before being classified as probiotics. In addition, the guidelines for the evaluation of probiotics in food must be followed (FAO/WHO, 2002; Anadon et al., 2016).

Usually, the safety of novel strains has been revealed from the common presence of the species, either in foods or as normal commensals in the human gut. Lactobacillus and Bifidobacterium species, dominant inhabitants of the human intestine (Lactobacillus in the small intestine and Bifidobacterium in the large intestine) and also Lactococci and yeasts are considered as GRAS (generally recognized as safe). However, many probiotics are not categorized as GRAS and may be suspicious in terms of safety, such as enterobacteria or some strains of enterococci. For example, *Enterococcus faecium* and *Enterococcus faecalis* that have appeared as opportunistic pathogens in hospital environments, cause some infections of endocarditis and bacteraemia, as well as intra-abdominal, urinary tract, and central nervous system infections. Some enterococcus or bacillus strains may be problematic due to the presence of antibiotic resistance strains (e.g., vancomycin-resistant-Enterococcus strain) or the *B. cereus* group that is known to produce enterotoxins (Abdollahi et al., 2016; Anadon et al., 2016).

The first requirement for developing a probiotic food product is the utilization of suitable probiotic strains in adequate dose. The main criteria for selecting suitable

**Table 1** Probiotic cultures potentially used in foods or food supplements

| Genera | Species |
|---|---|
| *Lactobacillus* | *acidophilus* |
| | *casei* |
| | *plantarum* |
| | *reuteri* |
| | *rhamnosus* |
| | *salivarius* |
| | *paracasei* |
| | *salivarius* |
| | *delbrueckii subs. bulgaricus* |
| | *fermentum* |
| | *lactis* |
| | *brevis* |
| | *crispatus* |
| | *delbrueckii* subsp. *bulgaricus* |
| *Bifidobacterium* | *animalis/lactis* |
| | *bifidum* |
| | *breve* |
| | *longum* |
| | *adolescentis* |
| | *infantis* |
| | *essensis* |
| *Propionibacterium* | *freudenreichii* |
| *Bacillus* | *subtilis* |
| | *cereus* var. *toyoi* |
| | *clausii* |
| *Pediococcus* | *acidilacti* |
| *Enterococcus* | *faecalis* |
| | *faecium* |
| *Escherichia* | *coli* strain Nissle |
| *Saccharomyces* | *boulardii* |
| *Streptococcus* | *thermophilus* |

strains of probiotic bacteria are their viability during food processing and storage conditions, survival during intestinal transit, and potential health benefits to consumers. The survival of bacteria in the food matrix against different unsuitable parameters during product development, processing, and storage is strain specific (Tripathi and Giri, 2014).

Numerous criteria consisting of safety, technological, functional, and physiological characteristics of probiotic organisms have been recognized and suggested for selection of suitable strains. The selection criteria includes: safety criteria; origin, pathogeny, and infectivity properties; virulence factors; technological criteria; genetic stability and phage resistance; desired viability during processing and storage; good sensory and structural properties, and large scale production; functional criteria; resistance to acid, bile, and pancreatic enzymes, and adhesion to mucosal surface;

and for physiological criteria, documented health effects such as lactose metabolism, immunomodulation, antagonistic activity, anticholesterolemic effect, antimutagenic, and anticarcinogenic properties (Morelli, 2007; Tripathi and Giri, 2014; Anadon et al., 2016).

The efficacy of probiotic bacteria in food products depend on their viability during shelf life. In order to have beneficial health effects, probiotic foods should contain the required minimum viable and active cell count per gram or milliliter of food product at the time of consumption (Calinoiu et al., 2016). Up to now, there is no recommended standard for a specific count of live probiotic bacteria in the product during consumption except for the Codex standard on fermented milks, where a minimum of $10^6$ cfu $g^{-1}$ of product has been recommended (FAO/WHO, 2010). In fact, it is actually not possible to claim a specific level of probiotics for all beneficial effects (Raeisi et al., 2013).

However, the minimum necessary concentration of probiotic bacteria has been generally accepted as $10^6$ cfu $g^{-1}$ when the product is consumed. This recommendation is based on the assumption of daily consumption of 100 g of probiotic product. Because many authors propose that ingestion of $10^8$–$10^9$ viable cells per day is needed, which corresponds to 100 g or mL of probiotic food, to achieve probiotic action in the human organism. It has also been reported that probiotic foods should be consumed regularly as approximately 100 g per day in order to deliver $10^9$ viable cells into the intestine and to achieve health benefits for person (Karimi et al., 2011; Tripathi and Giri, 2014; Calinoiu et al., 2016).

In this context, the efficacy of probiotic foods can be provided by the addition of suitable strain to foods and beverages in sufficient counts and the protection of the strain survival during the shelf life of the product. Preliminarily, the production conditions and parameters of probiotic culture will be important as they refer to subsequent viability of the cultures. Thus, more cells are able to survive in food processing conditions and storage, and/or in the human gastrointestinal system (Farnworth and Champagne, 2016).

The most important parameter is probably the strain selection. The abilities of lactic cultures to grow on food matrices as well as to survive in process and storage conditions are very different even within a given species. Previously, the probiotic strains incorporated into food were chosen mainly according to their technological properties. The beneficial health effects of probiotics have gained importance over time, being a main parameter of strain selection for food applications. Therefore, the production parameters of probiotic cultures must be adapted to prevent lethal or sublethal damages to cells. Firstly, attention should be given to sublethal damages that can be seen on the cell wall or membranes and/or by the denaturation of internal cell components, such as enzymes, due to processing parameters. In addition, controlled stress conditions can be applied to increase survival ability of cultures to subsequent hard (troublesome) conditions such as heat treatment or freeze-drying. On the other hand, growth medium can contain some ingredients that improve the subsequent ability of cells to survive in the human intestine. For example, different sugars added to growth medium can affect cultures' sensitivity to bile by modifying their bile salt hydrolase activity (Ziar et al., 2014; Farnworth and Champagne, 2016).

Probiotic cultures are typically delivered as food supplements in caplets or capsules or by processed foods. Storage conditions, temperature, oxygen, and relative humidity are the main factors for the viability of probiotics delivered as food supplements. However, foods are increasingly seen as useful delivery matrices for probiotics. Food producers are seeking ways to incorporate probiotics into a wide range of foods and beverages. Yogurt and fermented milks are the most important foods to deliver probiotics. Probiotic cheese and other foods and beverages that carry probiotics are also gaining interest in the market. Many research projects have also been carried out that suggest the incorporation of probiotics in chocolate, sausages, cereal products, meat products, dried products, and vegetables. Therefore, the potential of delivery of probiotic bacteria by foods is enormous (Farnworth and Champagne, 2016).

The viability of probiotics in food is an essential property to obtain some beneficial health effects. The ability of probiotic cultures to provide benefits to the health of the host are usually only quantifiable by animal or clinical studies. Therefore, viability is the practical quality assurance test for probiotics. In clinical studies, health benefits have usually been attributed to counts of probiotics in excess of $10^8$–$10^9$ viable cells per day. Therefore, it is accepted that probiotic foods need to have $>10^6$–$10^7$ cfu $g^{-1}$ viable cells at the time of consumption. However, the viable counts generally decline during the storage of food products. In manufacture, higher numbers of probiotics (called overage) can be introduced to achieve an acceptable viable count. However, it can be expensive in practice, and overage may cause organoleptic problems. Therefore, the viability of probiotics should be maintained in foods during production and storage (Lee et al., 2008).

In fact, the stages of food manufacture, storage, and inoculation, including the production and storage of dried probiotic culture, as well as the food matrix itself, may be sources of considerable stress. Thus, food-related stress conditions including environmental factors should be well designed for the efficacy of probiotic bacteria (Marco and Tachon, 2013; Capozzi et al., 2016).

In addition, an ideal food matrix should preserve microorganisms from the hostile gut environment during digestion procedure, thus delivering high counts of probiotic bacteria to the main target organ such as the large intestine. Therefore, incorporating live probiotic cells into foods and then keeping them alive throughout shelf life is a significant task for food technologists (Lee et al., 2008; Capozzi et al., 2016).

Firstly, probiotic cultures are grown to high numbers on an industrial scale, using suitable food-grade culture media to provide a high load of probiotic cells in the final product following inoculation into the carrier food (Capozzi et al., 2016). The main strategies applied for incorporating probiotics into foods are as follows (Farnworth and Champagne, 2016).

Firstly, selection and production of suitable strain that is adaptable to the food matrix and then to the intestine of the host is important. Research findings report variability between strains with regard to survival ability in foods (Champagne and Gardner, 2008; Akalın et al., 2018) or in the human gastrointestinal system (Mainville et al., 2005).

Nowadays, strains are selected based on their recognized clinical effects and stability in food. Therefore, contemporary producers consider both technological and functional attributes in their strain selection process. Unfortunately, many strains offering beneficial health effects might be sensitive to production and storage conditions, and require technological adaptations to be incorporated in foods. Collaboration between the food producer and the probiotic supplier can lead to specially designed cultures that can be adapted to cope with heat, cold, acid, oxidative, high pressure, or osmotic stresses during food processing.

Secondly, the method of culture incorporation into food is important. Generally, probiotic cultures are added directly, in what is called "direct to the vat inoculation" (DVI). This adding procedure is preferred due to greater flexibility and better standardization for culture delivery. DVI can be carried out by simply adding the frozen or dried culture to the food matrix. However, it can cause substantial loses in viability if it is done inappropriately. In respect of frozen cultures, some thawing parameters, including temperature, need to be specifically adjusted. Although it is much easier to shop and store the freeze-dried cultures, the use of them in food processing plants is more difficult than frozen ones. When freeze dried cultures are used, many factors including plating medium, composition, and solids level of rehydration medium, as well as rehydration temperature and time, influence cell counts following addition of the culture to a food matrix.

Thirdly, processing steps in food manufacture are very important for probiotic culture survival. Many technological steps can be detrimental to the survival ability of probiotic bacteria and significant losses in probiotic viability can be seen following processing stages. Two main approaches have been emphasized to prevent these losses during processing:

**(1)** modifying the food matrix; and
**(2)** modifying food processing.

To modify the food matrix, appropriate pH conditions (neutral pH preferable), addition of antioxidants and growth factors such as prebiotics, plant or yeast extracts, and selection of nontoxic ingredients, for example, flavors and preservatives have been recommended.

To modify the food processing steps, lowering temperatures, including vacuum or nitrogen flushing, modifying the fermentation parameters, and adapting cells by applying sublethal stresses have been recommended (Farnworth and Champagne, 2016).

In addition, the best time to incorporate the probiotic culture should be assessed in the processing. For example, inoculation of milk before its renneting results in higher counts in Cheddar cheese production when compared to adding during the cheddaring or salting steps (Fortin et al., 2011). In ice cream production, the cultures are generally added to the ice cream mix before freezing in most studies (Akalın et al., 2017).

In fact, it is not easy to adapt the food matrix and processing and storage conditions for probiotic viability. As an example, numerous parameters should be considered to develop a new fermented milk containing probiotics. These parameters includes the properties of milk base (the protein and other nutrient content, the presence of ingredients and/or additives, the process parameters applied, such as heating, etc.), fermentation conditions (the kind of starter and probiotic culture, the form of cultures, the inoculation moment and level of the cultures, the fermentation time and temperature, etc.), and storage conditions (pH level, activity of starter culture except probiotics, redox level and antioxidants presence, encapsulation, etc.).

The storage stage is also important, as viability losses occur in that period depending on the temperature, moisture, pH degree, oxygen level, starter culture nature, redox level, and type of packaging. Therefore, appropriate packaging and environmental conditions should be supplied for adequate count of probiotic survival throughout storage (Farnworth and Champagne, 2016).

Probiotic cell counts in the product must be enumerated accurately to evaluate the given number of cells by the producer or the enumerated number of cells to regulatory compliance. Therefore, enumeration methods must be developed to enumerate probiotic cultures. The enumeration technique is carried out easily by traditional plating method for products containing only one strain, such as Yakult® with *Lactobacillus casei* Shirota. However, there can be numerous strains that will grow on petri plates for probiotic yogurt and cheese. Thus, selective media have been developed for these products (Shah, 2000; Karimi et al., 2012). Nevertheless, highly selective strain-related methods need to be developed due to variability in efficacy. Today, viable counts of specific probiotic strains can be assessed by PCR (Desfossés-Foucault et al., 2012). Recent developments in flow cytometry, where species-specific antibodies can be marked, provide the selective enumeration of bifidobacteria in dairy products (Geng et al., 2014).

On the other hand, microencapsulation, which is increasingly used to provide probiotic viability in food matrices, can lead to problems on cell enumeration by plating, PCR, or flow cytometry. Methodologies must be developed and applied to provide correct plating procedure by dissolving the microencapsulated particles (Champagne et al., 2011).

Sensory properties have a key position for consumer acceptability. It is thought that when the load of probiotics does not pass $10^7$ cfu g$^{-1}$, there is a negligible effect on sensory properties (Farnworth and Champagne, 2016). Sometimes probiotics can negatively affect the flavor. For example, bifidobacteria release acetic acid, which can be undesirable above a certain level (Mohammadi et al., 2012). Therefore, manufacturers must primarily establish the inoculation level of probiotics to maintain good sensory characteristics.

Finally, the functionality of probiotics in the human intestine strongly influences consumer attitude. The expression "probiotics enable health benefits" is one of the important challenges in the development of functional foods for manufacturers if regulatory approval of health claims is sought (Farnworth and Champagne, 2016).

## 2 Factors affecting the survival of probiotics during processing and storage of food

Since health effects of probiotic food products depend on the number of viable and active probiotic cells per gram or milliliter of the products at the time of consumption, it is crucial to have the minimum recommended level of probiotics in the final product and maintain this probiotic number over its shelf life (da Cruz et al., 2010). Probiotic foods require high technological demands to retain viability of probiotics in all production steps and during storage (Saarela, 2007). Hence, minimizing the loss of probiotic viability is significant for food technologists developing foods containing probiotics (Lee et al., 2008).

The number of viable probiotic bacteria generally declines 10–100-fold or more during production or storage (Mortazavian et al., 2006). Long-term survival of probiotics in foods depends on various factors related to food matrix, processing, and storage including temperature, heating, oxygen content, acidity, moisture content, water activity, osmotic stress, and packaging material (Boylston et al., 2004; Champagne et al., 2011). It is important to stabilize the viability of probiotics during processing and storage (Calinoiu et al., 2016).

## 2.1 The effect of probiotic strain selection on probiotic survival

Probiotics are typically members of Lactobacillus and Bifidobacterium species. The selection of appropriate probiotic strain is essential for the production of probiotic food because of the requirement for a high number of living cells during consumption (Talwalkar and Kailasapathy, 2004; Rouhi et al., 2013). Good sensory properties, phage resistance, viability during processing, and stability in the food during storage are important criteria in probiotic strain selection (Mattila-Sandholm et al., 2002). Probiotics are species and strain specific, therefore their technological robustness is also important besides their health-benefiting properties (Saarela, 2007; Douglas and Sanders, 2008). Lactobacilli such as *L. acidophilus*, *L. johnsonii*, *L. rhamnosus*, *L. casei*, *L. paracasei*, *L. fermentum*, *L. reuteri*, and *L. plantarum* are more robust and more suitable for food processes than bifidobacteria. Lactobacilli are resistant to low pH and have good adaption to food components in probiotic food formulation (Tripathi and Giri, 2014). Bifidobacterium species such as *B. longum* are not as acid-tolerant as Lactobacillus species, especially *L. acidophilus*, and require low oxidation reduction potential and specific growth factors (Shah et al., 1995).

In addition, the growth characteristics of probiotic strains should be known so that an appropriate probiotic strain for the processing technology could be selected or, if possible, processing conditions could be adjusted to optimize probiotic survival (Boylston et al., 2004). In addition, interactions between probiotic strains and other bacteria strains or starter cultures in food affect the numbers of probiotic bacteria during manufacture and storage (Rouhi et al., 2013). For example, during yogurt manufacture *Lactobacillus delbrueckii* subsp. *bulgaricus,* which is the yogurt starter culture, affects the viability of *L. acidophilus*, while bifidobacteria shows better stability against it (Dave and Shah, 1997). Other microorganisms in food could produce metabolic products that influence the viability of probiotic bacteria (Lourens-Hattingh and Viljoen, 2001). As probiotics are dependent on species and strain, the selection of the probiotic strain according to the requirements of the process and storage conditions is an important factor for their survival.

## 2.2 The effect of food matrix on probiotic survival

Ice cream, cheese, fermented milk, fermented meats, candy, chocolate, chewing gum, oat or soy-enriched milk, and infant formula are used as source of probiotics. Fermented milks are the most popular probiotic foods (Pennachia et al., 2006). Foods containing probiotic bacteria act as their vehicles and the delivery of probiotics to the human body is closely related to foods. The effect of food matrix has great importance on viability of probiotic bacteria (Vinderola et al., 2011). Different food compositions could represent a great challenge for probiotics survival (Shori, 2016). Functional and technological properties of same probiotic strains could vary due to different food ingredients. Solid content, fat and protein content, type and concentration of proteins and sugars, availability of nutrients, flavoring agents, thickeners, sweeteners, stabilizers, bioactive components, and growth promoters and inhibitors in the food matrix could affect the survival of probiotics during processing and storage (Lourens-Hattingh and Viljoen, 2001; Ranadheera et al., 2010).

### *2.2.1 Food ingredients and additives*

The compatibility of probiotics with the ingredients in food plays a major role in the growth and viability of probiotic bacteria. The ingredients could be protective, neutral, or detrimental for probiotics (Mattila-Sandholm et al., 2002). In addition, food additives such as salts (NaCl and KCl), sugars (sucrose and lactose), sweeteners (acesulfame and aspartame), aroma compounds (diacetyl, acetaldehyde, and acetoin), natural or artificial coloring and flavoring agents, nisin (a polypeptide-type antibiotic), natamycin, lysozyme, and nitrite could significantly affect the growth and viability of probiotic bacteria (Vinderola et al., 2002; Tripathi and Giri, 2014). High levels of certain additives, organic acids, curing agents such as sodium nitrite, especially in meat fermentation, antimicrobial preservatives in the food matrix, and some starter cultures producing bacteriocin could inhibit the growth of probiotics during fermentation, processing, and storage (Lee et al., 2008; Tripathi and Giri, 2014).

Although probiotic bacteria are more resistant to additives than lactic acid starter bacteria in dairy products, some additives could affect the probiotic bacterial growth depending on their concentrations. Vinderola et al. (2002) have found that aspartame used as a sweetener in fermented dairy beverages had no inhibitory effect at a concentration of 0.03%, whereas it had an inhibitory effect for some probiotic strains at a concentration of 0.12%. While natural colorings such as carmine, curcuma/bixin, and bixin are not inhibitory for the growth of probiotic bacteria at widely used concentrations in fermented milk and dairy products, some of the flavoring-coloring commercial mixtures have an important inhibitory effect even at concentrations recommended by suppliers (Vinderola et al., 2002).

Food ingredients and additives could also interact with probiotics and affect their growth and viability (Ranadheera et al., 2010). Some studies have reported that disaccharides and sorbitol supported cell survival by preventing cell membrane damage due to this interaction and stabilizing the cell membrane during storage (Yoo and Lee, 1993; Önneby et al., 2013). Food ingredients such as prebiotics, growth factors/promoters, and addition of fruit and plant products enhance the survival of probiotics in foods (do Espırito-Santo et al., 2011; Mohammadi et al., 2011; Rouhi et al., 2013).

#### Prebiotics

Prebiotics are nondigestible food ingredients that selectively stimulate growth and/or activity of probiotic bacteria (Akalın and Erisir, 2008). Probiotics generally grow poorly or do not grow in foods except fermented milk products due to the lack of proteolytic and glycolytic activities and the high requirement for some nutrients, such as nonprotein nitrogen and B-group vitamins (Champagne et al., 2005; Mohammadi and Mortazavian, 2011). Different prebiotics including inulin, inulin-type fructans, fructooligosaccharides (FOS), galactooligosaccharides (GOS), isomaltooligosaccharides, lactulose, high-amylose corn starch (hi-maize), β-glucan, lactitol, raffinose, maltodextrin, and verbascose are used (Rouhi et al., 2013) to encourage the growth and activity of probiotics in foods (Akalın and Erisir, 2008; Vinderola et al., 2011).

Prebiotic compounds could affect the viability of probiotics during fermentation and storage (Bruno et al., 2002; Akalın et al., 2004; Donkor et al., 2007; Rouhi et al., 2013). Although the efficiency of prebiotics on probiotic viability depends on various factors including the type of probiotic strains, the type of prebiotics, the prebiotic purity, the concentration of prebiotics, the formulation specifications of products, and the storage conditions, they generally enhance probiotic viability (Mohammadi and Mortazavian, 2011). Since fermented milks are the most common presentation of probiotics (Pennachia et al., 2006), the studies regarding the effect of prebiotics added into food matrix on the survival of probiotic is focused on dairy products. Table 2 represents some of these. Most of the studies showed that prebiotics, especially the fructan-type prebiotics inulin and oligofructose, have a significant increase in survival of bifidobacteria including *B. animalis*, *B. infantis*, *B. longum*, *B. pseudolongum* (Shin et al., 2000; Bruno et al., 2002; Akalın et al., 2004; Akalın and Erisir, 2008; Cardarelli et al., 2008), and *lactobacilli including L. acidophilus*, *L. casei*, *L. paracasei*, and *L. rhamnosus* (Desai et al., 2004; Donkor et al., 2007; Akalın and Erisir, 2008; Cardarelli et al., 2008; Hekmat et al., 2009; Koh et al., 2013). It seems prebiotic ingredients will expand depending on ingredient technology developments (Douglas and Sanders, 2008).

## Growth factors and growth promoters

Growth factors are used directly by probiotics as nutrients and growth promoters are used to enhance growth and/or activity of probiotic cells without direct use as nutrients (Mohammadi et al., 2011). Foods are fortified with different growth factors/growth promoters such as glucose, vitamins, minerals, casein, whey protein hydrolysates, L-cysteine, yeast extract, and antioxidant to increase the growth of probiotic bacteria. These supplements can significantly increase the survival of probiotic bacteria in food products, especially during storage (Mohammadi et al., 2011; Tripathi and Giri, 2014). Dave and Shah (1998) have reported that the addition of cysteine, whey protein concentrate, acid casein hydrolysates, or tryptone improved the viability of bifidobacteria during refrigerated storage for 35 days by providing growth factors as probiotic bacteria lack proteolytic activity. Akalın et al. (2007) found that supplementation with 1.5% whey protein concentrate in reduced-fat yogurt increased the viable counts of *Streptococcus salivarius* subsp. *thermophilus*, *L. delbrueckii* subsp. *bulgaricus*, and *Bacillus animalis* by o log cycle in the first week of storage when compared to control sample. Ramchandran and Shah (2008) have examined that the influence of protein-based fat replacer (1%–2%) on the growth and metabolic activities of yogurt starters (*S. salivarius* subsp. *thermophilus* and *L. delbrueckii* subsp. *bulgaricus*) and probiotics (*L. casei*, *L. acidophilus*, and *B. longum*). The results showed that the addition of protein-based fat replacer significantly improved growth of *S. salivarius* subsp. *thermophilus* and *B. longum*, but inhibited *L.*

**Table 2** Selected publications on the effects of prebiotics on survival of probiotics

| Aim of the study | Probiotic bacteria | Prebiotic | Survival of probiotics | References |
|---|---|---|---|---|
| Investigation of the effects of three types prebiotics at different concentration on viability of Bifidobacterium species during refrigerated storage at 4°C for 4 weeks | Bifidobacterium spp. (Bf-1 and Bf-6) | Fructooligosaccharide (FOS), galactooligosaccharide (GOS), and inulin | Increased viability was observed depending on prebiotic type and dose. Best retention of viability was observed when cultures were grown in the presence of 5% FOS or GOS | Shin et al. (2000) |
| Investigation of the effects of four types prebiotics on growth, activity, and viability of some Bifidobacterium species during fermentation and after refrigerated storage (4 weeks) | *Bifidobacterium infantis* (Bb-1), *B. longum* (Bb-2), *B. longum* (Bb-3), *B. pseudolongum* (Bb-4) and *B. animalis* (Bb-5) | Hi-maize, inulin, raftilose, and lactulose | Retention of viability during the 4 weeks storage was significantly higher ($P<.05$) in the presence of prebiotics then control. The highest viability of 75.3% was retained by strains *B. longum* (Bb-3) and *B. animalis* (Bb-5) using hi-maize, while inulin was least effective in retaining the viability | Bruno et al. (2002) |
| Investigation of the effects of FOS on the viability of yogurt bacteria and two commercial strains of bifidobacteria during 28 days storage at 4°C | *Bifidobacterium animalis* (Bb-12) and *B. longum* (Bb-46) | FOS | The recommended level was exceeded for *B. animalis* throughout 28 days storage (3.59–2.25 $10^7$ cfu $g^{-1}$) and viability of *B. longum* in yogurt containing FOS remained above $10^6$ cfu $g^{-1}$ for up to 21 days, whereas this level was maintained for only 7 days for control yogurt | Akalın et al. (2004) |
| Investigation of the effects of different prebiotics on the viability of Lactobacillus strains after 4 weeks storage at 4°C | Lactobacillus strains (*L. casei, L. paracasei, L. rhamnosus, L. zeae*) | Hi-maize, lactulose, inulin, or raftilose | Survival of lactobacilli in the presence of prebiotics was strain specific, but the viability was in many cases significantly higher ($P<.05$) than control. The best viability was observed with inulin | Desai et al. (2004) |

| | | | | |
|---|---|---|---|---|
| Investigation of the effects of prebiotic substances on the viability of *Lactobacillus rhamnosus* GG in reconstituted skim milk | *Lactobacillus rhamnosus* GG | Oligofructose-based or polydextrose-based prebiotic substances | Partial substitution of the solids content of the reconstituted skim milk by prebiotic substances resulted in a high level of probiotic survival | Ananta et al. (2005) |
| Investigation of the effects of inulin and lactulose on survival of *Lactobacillus acidophilus* LA-5 and *Bifidobacterium bifidum* BB-02 in Acidophilus-Bifidus yogurt | *Lactobacillus acidophilus* LA-5 and *Bifidobacterium bifidum* BB-02 | Inulin and lactulose | Inulin and lactulose stimulated the growth of *B. bifidum* BB-02 and lactulose was more effective. Only lactulose stimulated the growth of *L. acidophilus* LA-5 | Özer et al. (2005) |
| Investigation of the effects of inulin with various chain lengths on the count of *Lactobacillus acidophilus* in fat-free plain yoğurt | *Lactobacillus acidophilus* | Inulin with short, medium, and long chain | *L. acidophilus* counts were higher with the use of prebiotics compared to control | Aryana et al. (2007) |
| Investigation of the effects of inulin and hi-maize on the viability of *L. acidophilus* L10 and *L. casei* L26 in yogurt during cold storage | *L. acidophilus* L10 and *L. casei* L26 | Inulin and hi-maize | The supplementation of yogurt with prebiotics improved retention of viability of *L. acidophilus* L10 and *L. casei* L26 in yogurt during cold storage. Inulin was a better growth stimulant of probiotics than hi-maize | Donkor et al. (2007) |
| Investigation of the effects of β-glucan addition on growth and metabolic activity of *Bifidobacterium animalis* subsp. *lactis* | *Bifidobacterium animalis* subsp. *lactis* | β-Glucan from two different cereal sources (oat and barley) | The supplementation of yogurt with selected prebiotics improved viability and stability of *Bifidobacterium animalis* subsp. *lactis* in yogurt during 4 weeks cold storage | Vasiljevic et al. (2007) |

*Continued*

**Table 2** Continued

| Aim of the study | Probiotic bacteria | Prebiotic | Survival of probiotics | References |
|---|---|---|---|---|
| Investigation of the effects of oligofructose or inulin on the survival of *Lactobacillus acidophilus* La-5 and *Bifidobacterium animalis* Bb-12 in low-fat ice cream stored at −18°C for 90 days | *Lactobacillus acidophilus* La-5 and *Bifidobacterium animalis* Bb-12 | Oligofructose and inulin | Oligofructose significantly improved the viability of *L. acidophilus* La-5 and *B. animalis* Bb-12 in ice cream mix and the minimum level of $10^6$ cfu $g^{-1}$ was maintained for *B. animalis* Bb-12 in ice cream with oligofructose during storage | Akalın and Erisir (2008) |
| Investigation of the effects of inulin, oligofructose and oligosaccharides from honey on the viability of *Bifidobacterium animalis* subsp. *lactis* and *Lactobacillus acidophilus* in synbiotic petit-suisse cheeses at refrigerated storage for 28 days | *Bifidobacterium animalis* subsp. *lactis* and *Lactobacillus acidophilus* | Inulin, oligofructose and oligosaccharides from honey which serve as prebiotic agents | Probiotic populations varied from 7.20 up to 7.69 log cfu $g^{-1}$ for *Bifidobacterium animalis* subsp. *lactis* and from 6.08 up to 6.99 log cfu $g^{-1}$ for *Lactobacillus acidophilus*. The highest fructan contents were achieved by the cheese trials containing oligofructose and inulin combined | Cardarelli et al. (2008) |
| Investigation of the effects of yeast extract and combination of yeast extract and inulin on the viability of *L. rhamnosus* GR-1 in yogurt during refrigerated storage for 28 days | *Lactobacillus rhamnosus* GR-1 | Yeast extract or combination of yeast extract and inulin | The addition of yeast extract or combination of yeast extract and inulin positively affected the survival of L. reuteri RC-14 in yoğurt | Hekmat et al. (2009) |

| | | | | |
|---|---|---|---|---|
| Investigation of the effects of lactulose and hi-maize on the survival of *Lactobacillus acidophilus* La-5 and *Bifidobacterium lactis* Bb-12 in freshly made Iranian fermented milk drink (Doogh) | *Lactobacillus acidophilus* La-5 and *Bifidobacterium lactis* Bb-12 | Lactulose and hi-maize (1% and 3% *w*/w) | The highest viability was in treatment with 3% of hi-maize for bifidobacteria and in control treatment for *L. acidophilus*. The lowest viability was in control treatment for bifidobacteria and in treatment with 3% of hi-maize for *L. acidophilus* | Noubakhti et al. (2009) |
| Investigation of the effects of inulin on the production of probiotic fiber-enriched fermented milk stored at 4°C for 24 h and 7 days | *Lactobacillus acidophilus*, *Lactobacillus rhamnosus*, and *Bifidobacterium animalis* subsp. *lactis* | Inulin | Inulin had a stimulating effect on the growth of all probiotics. Probiotic counts were higher than 7.37 log cfu $mL^{-1}$. It was more effective for *B. lactis* | de Souza Oliveira et al. (2009) |
| Investigation of the effects of tagatose on the viability of *Lactobacillus casei* 01 and *Lactobacillus rhamnosus* CG | *Lactobacillus casei* 01 and *Lactobacillus rhamnosus* CG | Tagatose which is a low calorie monosaccharide epimer of fructose | Tagatose enhanced the growth *of L. casei* 01 and *L. rhamnosus* GG and their probiotic activities | Koh et al. (2013) |

*casei*, *L. acidophilus*, and *L. delbrueckii* subsp. *bulgaricus*. Nualkaekul et al. (2011) have confirmed the significant protective effect of protein on viability of *B. longum* in various types of juices. Protein derivatives enhance probiotic survival due to their nutritional value for the cells, their ability to reduce the redox potential of the medium and to increase the medium's buffering capacity resulting in a smaller decrease in pH (Dave and Shah, 1998; Mohammadi et al., 2011).

### Addition of fruit and plant products

Studies have shown that some fruit ingredients enhance probiotic survival. Buriti et al. (2014) found that tropical fruit pulps maintained the viability of *Bifidobacterium animalis* subsp. *lactis* BB-12 and *L. rhamnosus* Lr-32 over 7 log cfu $mL^{-1}$ during 21 days. Dos Santos Cruxen et al. (2017) have reported that the addition of fruit with bioactive properties such as Butia odorata provided viable *Bifidobacterium lactis* (Bl-04) above $10^6$ cfu $g^{-1}$ in ice cream during storage. Sah et al. (2015) evaluated the effects of pineapple peel powder addition on the viability and activity of *L. acidophilus* (ATCC 4356), *L. casei* (ATCC 393), and *L. paracasei* subsp. *paracasei* (ATCC BAA52) in yogurts throughout storage at 4°C for 28 days. The results showed that pineapple peel powder addition improved growth and retention of viability of probiotics and the probiotic counts in supplemented yogurts during storage ranged from 7.68 to 8.03 log cfu $g^{-1}$, which is higher 1 log cycle than control yogurt. do Espırito-Santo et al. (2011) have reviewed the influence of fruity food matrices on probiotic viability for dairy and nondairy products. They have reported that fruits such as mango, berries, banana, strawberry, passion fruit, and lemon maintained or enhanced the viability of probiotics when added into dairy products and the viability of some probiotics reduced only when fruits promoted a high release of lactic and acetic acids. Daneshi et al. (2013) studied the viability of four probiotic strains (*L. acidophilus* LA5, *B. lactis* BB12, *L. rhamnosus*, and *L. plantarum*) in nonfermented milk and carrot juice mix drink during refrigerated storage at $4 \pm 2$°C for up to 20 days. They have found that all strains had good viability in milk/carrot juice drink (88%–98%) and *L. acidophilus* LA5 was more stable than the three other strains. Green tea extract could be a potential food matrix for the survival of some probiotics, such as *L. paracasei* and *Bifidobacterium animalis*, most probably depending on the glycosidase activity of bacteria that could consume sugars from glycosylated flavonoids as an energy source. However, it may also depend on anaerobic conditions provided by the antioxidant and oxygen-scavenging properties of green tea (López de Lacey et al., 2014; Shori, 2016). Citrus fiber enriched fermented milks improved bacterial growth and survival of *L. casei* and *L. acidophilus*, whereas it affected *Bifidobacterium bifidum* irregularly (Sendra et al., 2008).

Fruit and plant extracts affect probiotic viability depending on phenolic compounds in the extract, the probiotic strain, the presence of co-cultures, and the concentration of fruit/plant extract added to the food (Sendra et al., 2016).

## 2.3 The effect of processing, fermentation and storage conditions on probiotic survival

The health effects of probiotic food products depend on maintenance of probiotic bacteria viability at the recommended amount in the food until consumption. Therefore, it is necessary to provide a high survival rate of probiotics throughout the product shelf life, including processing, fermentation, and storage, in order to ensure the benefits of consumer health (do Espırito-Santo et al., 2011; Tripathi and Giri, 2014). It is important that their viability and metabolic activity is retained during all the steps of the food processing operations and storage, from manufacture up to ingestion by the consumer (da Cruz et al., 2010). It is necessary to have $10^6$–$10^7$ cfu $g^{-1}$ probiotic bacteria in the final processed foods for therapeutic properties and a daily ingestion of $10^8$–$10^9$ cfu $g^{-1}$ probiotic bacteria for beneficial health effects (da Cruz et al., 2010). In terms of the survival of probiotics, food processing steps and food storage have hurdles such as temperature, heating, oxygen content, acidity, moisture content, water activity, osmotic stress, and packaging (Sendra et al., 2016).

### *2.3.1 Temperature and heating effect*

In food applications, especially in fermentation, the temperature at which probiotic bacteria could grow is one of the most important factors affecting probiotic viability. The optimum growth temperature of most probiotics ranges between 37 and 43°C. Most of the probiotic bacteria could not grow at temperatures below 15–20°C or above 45–50°C (Boylston et al., 2004; Calinoiu et al., 2016).

Heating is applied for two main purposes in food processing. High temperatures over 65°C are applied to destroy spoilage and pathogenic microorganism and lower temperatures are applied for technological purposes (Champagne et al., 2005). Food processing steps including high heat treatments are extremely destructive to probiotics because heating above 45–50°C has detrimental effects on probiotic survival and destroy at least one fraction of the population depending on the temperature and the strain (Calinoiu et al., 2016). During pasteurization, 4 log destruction of *L. acidophilus*, which is quite sensitive to heating, occurs (Champagne et al., 2005). Therefore, probiotics should be added after processes such as heating, cooking, and pasteurization in food manufacture (Tripathi and Giri, 2014).

The incubation temperature also has a major role on the viability of probiotics such as *L. acidophilus* and *Bifidobacterium* Bb-12. Mortazavian et al. (2006) reported that incubation at 37°C provided the highest viable counts, while incubation at 44°C provided the lowest count of probiotics in the yogurt. At higher incubation temperatures such as 45°C, *L. delbrueckii* subsp. *bulgaricus* becomes the dominant species in yogurt by rapid growth and enhances the antagonistic effect against probiotic bacteria. It is considered that this effect is due to production of a large amount of acid, hydrogen peroxide, and possibly bacteriocins, which cause the suppression of probiotic bacteria growth (Mortazavian et al., 2006).

Reale et al. (2015) have tested the growth capability of *L. paracasei*, *L. casei*, and *L. rhamnosus* at high temperatures and low temperatures. They have reported that none of the strains were able to grow at 50°C, some *L. casei* strains grew at 48–49°C, while most of the *L. paracasei* strains did not grow above 45°C. The results also showed that most of the strains grew at 10°C. Elevated temperature is a critical factor affecting stability of probiotics even during transport and storage. Some probiotic foods are stored at refrigerated temperature and some of them at ambient temperature (Lee et al., 2008). Randazzo et al. (2013) compared the survival of six probiotic wild strains of *L. rhamnosus* during 78 days of storage at 25 and 5°C in peach jam. All strains exhibited better performances during storage at 5°C, resulting in the number of probiotices higher than 7 log cfu $g^{-1}$ up to 78 days. In addition, all strains remained above the critical value of 6 log cfu $g^{-1}$ in samples stored at 25°C up to 45 days.

Lactobacilli are more tolerant to low temperatures than bifidobacteria; in general, low storage temperatures support the growth and survival of probiotics. In addition, limited postacidification depending on low temperatures enhances the viability of probiotics (Lourens-Hattingh and Viljoen, 2001). Nebesny et al. (2007) examined the survival of *L. casei* and *L. paracasei* in chocolate during storage for 12 months at different storage temperatures (4°C, 18°C, and 30°C). After 12 months, the number of probiotic bacteria in products stored at 4°C and 18°C was above required functional level ($10^6$–$10^7$ cfu $g^{-1}$) but the number of the probiotic bacteria was not sufficient ($10^5$ cfu $g^{-1)}$ in the products stored at 30°C. Mortazavian et al. (2007) studied the effect of refrigerated storage temperature (at 2°C, 5°C, and 8°C) on the viability of *L. acidophilus* and *Bifidobacterium animalis* subsp. *lactis* BB-12 in yogurt. They have found that the highest viability was at 2°C for *L. acidophilus* and at 8°C for *B. lactis* after 20 days' storage. Although it is preferable to cool the product for better probiotic survival, there are other factors interacting with temperature, such as oxygen content, acidity, and water activity (Lee et al., 2008).

### 2.3.2 Oxygen content

Dissolved oxygen has a negative effect on survival and growth of probiotic bacteria, as most of the species are strictly anaerobic (de Vuyst, 2000). Oxygen affects the probiotics in different ways. One of the ways is direct toxicity of dissolved oxygen to probiotic cells, which are very sensitive to oxygen, during processing and storage (Champagne et al., 2005). This is considered as the main reason of the reduction in the number of probiotics during storage under acidic conditions, most probably due to the intracellular production of hydrogen peroxide (Calinoiu et al., 2016). The other way is indirect effect of oxygen to probiotic cells. In the presence of oxygen, certain cultures such as *L. delbrueckii* strains, excrete peroxide in the medium (Champagne et al., 2005). While the aerobic bacteria completely reduce the oxygen to the water, the oxygen-scavenging system in the probiotic bacteria is either weak or completely absent. The absence of an electron transport chain in probiotic bacteria causes the reduction of oxygen to hydrogen peroxide (Talwalkar and Kailasapathy, 2004). Furthermore, probiotic bacteria do not have catalase enzymes, which decompose hydrogen peroxide, and the intracellular levels of $H_2O_2$ block the essential enzyme fructose 6 phosphofructoketolase, which is required for sugar metabolism.

Scavenging of $H_2O_2$ is significant for the survival of probiotic bacteria (Sarkar, 2010), because oxygen exposure causes cell death due to formation and accumulation of toxic oxygenic metabolites such as hydrogen peroxide ($H_2O_2$), superoxide anion ($O^{2-}$), and hydroxyl radical ($OH^-$) in the cell. This is known as oxygen toxicity (Talwalkar and Kailasapathy, 2004). In addition to $H_2O_2$ production, generation of free oxygen radicals due to oxidation of membrane lipids, which induce DNA damage in bacteria, has destructive effects on survival of probiotics during storage (Önneby et al., 2013). Therefore, the combination of antioxidant presence, storage under vacuum, and controlled water activity could be effective to minimize oxidation and maximize probiotic viability during storage (Tripathi and Giri, 2014).

Among these factors, oxygen toxicity is highly important, and it is crucial to lower oxygen levels during manufacture and storage. Oxygen can reach probiotic foods in many process steps, such as milk homogenization, cooling after pasteurization, and mixing after addition of starter cultures during fermented milk production (Ladero and Sanchez, 2017). Different methods have been attempted to reduce the oxygen content of probiotic foods and to prevent oxygen-derived toxicity. These include the elimination of peroxide-producing strains, fermentation under vacuum, the addition of strains that act as oxygen scavengers (such as *S. salivarius* subsp. *thermophilus*), the addition of antioxidants (such as ascorbic acid and catechins), and vacuum packaging. The packaging material also affects the survival of probiotics due to oxygen permeability (Tripathi and Giri, 2014; Sendra et al., 2016). Another way to prevent oxygen toxicity could be the selection of an aerotolerant Bifidobacterium spp. containing high levels of NADH oxidase and NADH peroxidase, which are important enzymes for scavenging oxygen from the intercellular medium (Sarkar, 2010). Furthermore, Mori et al. (1997) found that extracellular bifidogenic growth stimulators produced by *Propionibacterium freudenreichii* effectively suppress the production of peroxide under anaerobic conditions.

Although probiotics are anaerobes, their degree of oxygen sensitivity exhibits wide variation according to strain and species (Kawasaki et al., 2006). In general, bifidobacteria are more sensitive to oxygen than lactobacilli (Tripathi and Giri, 2014). Kawasaki et al. (2006) reported that most of the Bifidobacterium species were not oxygen tolerant except *Bifidobacterium boum* and *Bifidobacterium thermophilum*. While they grew well even under 20% oxygen content without $H_2O_2$ accumulation, the growth of *B. bifidum* and *B. longum* was inhibited under high oxygen content due to the production and accumulation of $H_2O_2$. Dave and Shah (1997) informed that reduction of dissolved oxygen enhanced the viability of both Lactobacillus and Bifidobacterium species.

### 2.3.3 Acidity and pH

Acidity and pH are important factors that restrict the viability of probiotic bacteria in foods. The optimum growth pH is between 5.5 and 6.0 for *L. acidophilus* and between 6.0 and 7.0 for bifidobacteria (de Vuyst, 2000). Lactobacilli, which could grow and survive in pH values between 3.7 and 4.3 (Boylston et al., 2004), is generally more tolerant to acidic conditions than bifidobacteria. Although the tolerance of Bifidobacterium spp. to acidic conditions is strain specific, most strains are sensitive to pH values below 4.6 and growth of bifidobacteria is generally retarded or inhibited below pH 5.0 or above 8.0 (Lourens-Hattingh and Viljoen, 2001; Karimi et al., 2011).

Low pH (<4.5) affects the probiotic bacteria cells negatively due to high energy requirement to maintain the intracellular pH. This leads to an insufficient quantity of ATP, which is necessary for other critical functions, and thus causes cell death (Nualkaekul et al., 2011). Low pH also increases the concentration of undissociated organic acids in the products resulting in high bactericidal effect of these acids (Tripathi and Giri, 2014). Nualkaekul et al. (2011) studied the effect of pH, citric acid, protein, and dietary fiber on the survival of *B. longum* NCIMB 8809 during refrigerated storage for 6 weeks in various types of fruit juices, including orange, grapefruit, blackcurrant, pineapple, pomegranate, and strawberry. They found that high levels of pH, citric acid, protein, and dietary fiber enhanced cell survival during refrigerated storage in the fruit juices, except in pomegranate and strawberry juices. The cell viability declined rapidly in these two juices, most likely due to the very high levels of phenolic compounds that they have. Nematollahi et al. (2016) have reported that low pH and presence of inhibitor phenolic compounds of cornelian cherry juice had a negative effect on viability of probiotics, especially industrial strains during refrigerated storage. In many fermented dairy-products, the loss of probiotic viability is related to fermentation process because of the acidity, which decreased to pH 4–5, and the accumulation of organic acids (Dave and Shah, 1997; Rivera-Espinoza and Gallardo-Navarro, 2010).

### 2.3.4 Moisture content and water activity

Probiotics are usually added to fresh foods with high water activity and short shelf life—a period of weeks—or to dry foods with low water activity and relatively long shelf life—a period of months. Water plays an important role in the cell survival mechanisms. Cell viability is particularly affected by high water activity ($a_w > 0{,}3$) and moisture level in food (Mugnier and Jung, 1985; Weinbreck et al., 2010). Exposure to moisture during storage has a detrimental effect on bacterial survival (Önneby et al., 2013). If moisture levels and water activity increase in food, the survival rate of probiotics decreases. Low water activity ($a_w < 0.3$) is necessary for good survival of probiotics during storage. For the most lactobacillus species, ideal water activity is between 0.11 and 0.23. High water activity leads to probiotic cell death by increasing the growth of other microorganisms and fungi, and undesirable chemical reactions (Anekella and Orsat, 2014). However, the amount of water remaining in the food after drying affects the viability of bacteria and the rate of viability loss during storage. Zayed and Roos (2004) found the optimum moisture content as between 2.8% and 5.6% for storage of freeze-dried *Lactobacillus salivarius* subsp. *salivarius*. Weinbreck et al. (2010) investigated the viability of *L. rhamnosus* GG in different water activities (between 0.15 and 0.7) at 37°C and results showed that high water activity (0.7) caused >10 log loss in cell viability with storage for 2 weeks. During storage, the relative humidity of the environment is also important and besides the relative humidity of the environment, moisture uptake properties and molecular mobility of the food matrix are effective on probiotic viability (Ying et al., 2010; Hoobin et al., 2013). In addition, the interaction between temperature and water activity is significant for bacterial survival: if the storage temperature increases, the detrimental effect of also moisture increases (Mugnier and Jung, 1985).

In spite of the positive effects of low water activity on probiotic viability, it is not possible to dry foods to very low water activity in many cases, because of the high energy cost of drying, negative effects on the sensory properties of foods, and difficulties in wetting and dispersing powders. Moisture barrier packaging, microencapsulation, and incorporation of probiotics into fat phases of foods could assist in preventing the detrimental effects of moisture (Lee et al., 2008).

### 2.3.5 Osmotic stress

Osmotic stress that effects the probiotic bacteria could occur in food applications if there are high amounts of salts or sugars in the medium (Reale et al., 2015). It is one of the inhibitory factors for bacterial growth (Ge et al., 2011). NaCl is widely used in the food industry as a preservative agent, but the salt content of foods could threaten the viability of the probiotic bacteria (Vinderola et al., 2002). Reale et al. (2015) investigated the capability of *L. paracasei*, *L. casei*, and *L. rhamnosus* to withstand osmotic stress in food processing. They found that the growth capability of probiotics reduced with increasing salt content, and most of the strains were extremely sensitive to a concentration of 4% NaCl and some strains were not able to grow in presence of 6% NaCl. Ge et al. (2011) reported that the growth of *L. casei* decreased with high osmotic stress caused by glucose in fermentation. However, the resistance of probiotic bacteria to osmotic stress decreases due to the increase in fermentation temperature (Ge et al., 2011).

The effects of osmotic pressure on the viability of probiotics are strain specific. Different strains have different osmotic resistance because of the differences in their membrane phospholipid composition. It is considered that certain strains such as *L. plantarum* are more resistant. Besides strain, concentrations of salt or sugar are very important criteria for survival, and low concentrations could be protective against osmotic stress. Wasko et al. (2013) reported that 0.2 M and 0.5 M NaCl in the growth medium significantly enhanced the survival of *L. rhamnosus* OXY after lyophilization, while 0.8 M NaCl, which is a higher salt concentration, in the growth medium did not improve cell viability (Wasko et al., 2013).

### 2.3.6 Packaging aspects

The packaging materials and the storage conditions play a significant role in maintaining the quality and probiotic survival in foods (Mattila-Sandholm et al., 2002). The oxygen permeability of the packaging material is important since oxygen affects the viability of probiotics during storage (Rouhi et al., 2013). As plastics used in the food industry have different porosity and permeability, plastic packaging could cause diffusion of oxygen into the food matrix depending on thickness of container (Ladero and Sanchez, 2017). For packaging of fermented milk products, usually high-impact polystyrene containers are used. However, this packaging material is ineffective as an oxygen barrier and allows oxygen permission, resulting in an unsuitable environment for probiotic survival (Miller et al., 2002). Studies on the packaging of probiotic foods are generally related to active packaging (oxygen absorbers) and plastic film packaging with high oxygen barrier properties (da Cruz et al., 2007). The use of impermeable containers or glass containers (Dave and Shah, 1997), oxygen barrier packaging (Miller et al., 2002), higher thickness plastic packaging with low oxygen permeability, multilayer packaging with selective permeability (Miller et al., 2003; da Cruz et al., 2007), and active packaging with incorporated oxygen scavengers in the packaging polymers (Miller et al., 2003) enhance the survival of probiotics due to low levels of oxygen.

### 2.3.7 Freezing and thawing

Freezing of a product has significant decreasing effects on the amount of probiotic microorganisms. The vital metabolic activities of the probiotic microorganisms are reduced or interrupted according to the conditions. The most important reasons for this are the mechanical stresses of the ice-crystals, which are formed in the external medium or inside the cells and can damage the cell membrane and cause to condensation of solutes in the extracellular/intracellular medium, or dehydration of the cells (Tripathi and Giri, 2014). The size of the ice crystals is important to the survival of the cells. Larger intracellular ice crystals, which are formed by slow freezing, cause greater damage to the cells. Therefore, rapid freezing is recommended for better maintenance of the probiotic microorganisms (Mohammadi et al., 2011).

Like freezing, the thawing process is also important for cell survivability. The cells are exposed to osmotic effects and also to high concentrations of harmful factors of organic acids, hydrogen ions, oxygen, and other poisonous components. It has been reported that the pH of the product also has a crucial role on the survivability of the cells: the lower the pH, the higher the detrimental effects on the microorganisms (Mohammadi et al., 2011). Also repeated freezing–thawing processes decrease viability and should be avoided.

Once the product is frozen, probiotic microorganisms can survive over a long shelf life. It is also possible to use protectants for cultures that have to be frozen or dried to prevent, or at least mitigate, loss of viability. The most commonly used protectants on an industrial scale are lactose/sucrose, monosodium glutamate, milk powder, and ascorbate (Akalın et al., 2017; Crittenden, 2009).

### 2.3.8 Drying

Probiotic cultures are generally dried for their ease of use in functional food applications, handling, storage, and transport, and mainly to increase their shelf life; however, drying also causes losses in the viability of probiotics (Santivarangka et al., 2006). Several techniques are used for drying foods. The most conventional ones are sun drying, hot air drying, spray drying, freeze drying, and vacuum drying. Freeze drying is the best method for the viability of the probiotics but is also an expensive method. Spray drying is the cheapest and most used method for drying of liquid foods but it causes losses of the probiotic microorganism according to the high temperatures and osmotic pressures (Tripathi and Giri, 2014). Fluidized bed drying and radiant energy vacuum drying techniques have been successfully used by Nag and Das (2013) and Noorbakhsh et al. (2013). The fluidized bed drying technique showed a higher viability of probiotics ($2.5 \log \mathrm{CFU\,g^{-1}}$) compared to the freeze drying technique.

#### Spray drying

During the spray drying technique, the liquid food is injected in atomized form into the hot drying medium at up to 200°C, where the probiotic microorganisms are exposed to very high temperatures that affecting the viability of the microorganisms (Brennan et al., 1986). During the spray drying technique, the cell wall, cytoplasmic membranes, DNA, and RNA are usually affected by the stresses, which lead to an increased cell permeability with loss of intracellular components from the cell into the environment (Crowe et al., 1988; Teixeira et al., 1997).

The outlet and inlet temperatures are important parameters for the survivability of the microorganisms. It is reported that the viable count of different species, such as *S. salivarius* subsp. *thermophiles*, *L. delbrueckii* subsp. *bulgaricus*, *L. paracasei* NFBC 338, and *L. salivarius* UCC 118 decreases with increasing outlet/inlet temperatures and the atomizing pressure (Kim and Bhowmik, 1990). In agreement with Kim and Bhowmik (1990), Gardiner et al. (2002) reported a survival rate of >80% at an outlet temperature of 85–90°C for the resistant variety of *L. paracasei* NFBC 338 in reconstituted skim milk. A survival rate of >60% was reported by Ananta and Knorr (2003) for *L. rhamnosus* GG under similar outlet conditions. A preheat treatment of 52°C for 15 min before the spray drying procedure is also recommended to enhance the survivability during drying and storage (Paéz et al., 2012). Besides these outlet/inlet temperatures, the tolerance to different stresses during spray drying caries also from species to species. Therefore, the selection of the appropriate strain is important. It is reported by Gardiner et al. (2000) that *L. paracasei* NFBC 338 showed a higher survivability than *L. salivarius* UCC 118 under similar spray drying conditions, which could be attributed to the greater thermal tolerance of the strain. It is also known that the thicker cell walls of Gram-positive cells; like Lactobacillus show a better survivability during spray drying; in addition, the cells in the early stationery phase survive better during spray drying and storage than the cells in the mid log phase (Pispan et al., 2013). According to the studies, it could be stated that the survival of probiotic cultures during spray drying depends on the species and strain of the used probiotics, the outlet/inlet temperature, and type of atomization (Tripathi and Giri, 2014). The viability of the probiotic cultures can be improved by reducing the outlet/inlet temperatures, but it should be noted that the final moisture content of the final product and its quality will also be influenced by the chosen outlet/inlet temperatures. It is stated that the moisture content for shelf-stable products should be not >3.5% (Zayed and Roos, 2004).

## Freeze drying

The freeze drying technique is carried out in three steps: freezing, primary drying, and secondary drying. In the first step, the cells are frozen at temperatures as low as −190°C and then dried in two steps under vacuum by sublimination. This technique has been used for decades for the manufacture of probiotic powders. Since the processing conditions are milder than other methods, like spray drying or hot air drying, high survival rates are being achieved (Wang et al., 2004). The first freezing phase is the most important step for the survivability of the probiotic microorganisms. Most of the cellular inactivation occurs at this step. It is known from further literature that the higher the surface area of the cell, the higher the membrane damage during the extracellular ice crystal formation at the freezing phase. It is also known that small, spherical cells like enterococci are more resistant to freezing than larger, rod shaped cells like lactobacilli (Tsvetkov and Brankova, 1983; Fonseca et al., 2000). The surface proteins, cell wall, and cell membrane of the bacterial cells are damaged during the removal of the bound water from bacterial cells during the drying phase. The lipid fraction of the cell membrane is the most sensitive area against drying, since lipid peroxidation may occur at this part. As the freeze drying parameters, the physicochemical formulation of the food is also critical for bacterial survivability (Brennan et al., 1986).

### 2.3.9 Rehydration

The rehydration step at the reuse of the dried probiotic products is carried out in four steps; wetting, submersion, dispersion, and dissolving (Freudig et al., 1999). The rehydration conditions such as temperature, volume of the rehydrating media, and rehydration time, as well as the physical properties of the material to be rehydrated and properties like osmolarity, pH, and nutritional energy of the rehydration solution are important factors affecting the viability of the probiotic bacteria (Carvalho et al., 2004).

#### Rehydration time

It is recommended to use slow rehydration procedures for optimal viability results. Poirier et al. (1999) reported increased cell recovery of *Saccharomyces cerevisiae* under controlled conditions and slow rehydration time (7–16 days) rather than immediate rehydration.

#### Temperature

The rehydration temperature of the spray-dried or freeze-dried product or probiotic cells is also very important for the viability of the probiotic bacteria. The rehydration temperature has to be chosen according to the species and strains of bacteria. There is no optimal rehydration degree for all bacteria. Temperatures between 30°C and 37°C should be chosen for thermophilic bacteria, and between 22°C and 30°C for mesophilic bacteria, while the rehydration temperature should not be >40°C in any case (Mille et al., 2004; Sinha et al., 1982).

#### The composition of the liquid medium

The composition of the liquid medium for rehydration is also an important factor on the viability of the probiotic bacteria. The use of the cryopreservation solution again as the rehydration medium is recommended, since increased viable counts of bacteria were counted by Abadias et al. (2001). It is thought that the high osmotic pressure that is provided by these solutions is controlling the rate of hydration. Costa et al. (2000) also reported that the use of a complex medium containing RSM, peptone/tryptone, and meat extract resulted in significantly higher bacterial cell viability than a phosphate buffer, sodium glutamate, water medium.

#### The ratio of dried powder to liquid medium

As well as the composition of the rehydration medium, the ratio of dried powder to liquid medium is an important factor on the viable bacterial counts. It is reported that 4–10 times higher viable counts of different probiotic cultures were found when the powder was added to a small amount of water—a ratio of 1:3 instead of 1:50 powder to liquid (De Valdez et al., 1985). The viability of the probiotic bacteria during rehydration depends also on the selected species and strains; therefore, it is important to standardize the rehydration procedure for each strain and product.

## 2.4 Microencapsulation

Microencapsulation is a useful tool for improving the delivery of probiotics in foods, providing higher stability, and easier handling and storage (Champagne and Fustier, 2007; Picot and Lacroix, 2003). Microencapsulation of probiotic cells is defined as the technology of packaging cells in miniature sealed capsules in order to segregate the cells from the surrounding environment. The capsules release their contents in controlled rates over a prolonged time under the influence of certain processing and environmental conditions in the intestine medium (Madene et al., 2006). It seems to be the best way for bacterial protection (Krasaekoopt et al., 2003). But microencapsulation of probiotics is not a simple technique. Many demands should be taken in account for a successful probiotic microencapsulant:

- The microencapsulation technique should be inexpensive, simple, and should not reduce the probiotic viability.
- The used materials have to be food grade and compatible with the food.
- The encapsulation efficiency should be as high as possible (~100% of the bacteria should be encapsulated).
- The microcapsules should contain a high amount of the probiotics.
- The microcapsules should not affect the sensory properties and texture of the foods.
- The microcapsules have to protect the probiotics against a range of environmental stresses during manufacture and storage.
- The microcapsules have to protect the probiotics during the gastrointestinal transit and release them in the gut at the required site of action.

Several techniques for the microencapsulation of probiotics or functional ingredients have been developed. Most of these methods are based on the entrapment of the probiotics in polymers such as alginate, carrageenan, and starch; coated in emulsions or fat; or dry impacting of prebiotics and enteric coats (Anal and Singh, 2007; Kailasapathy, 2002; Krasaekoopt et al., 2004). It is also possible to get commercial systems for the microencapsulation of probiotics, which are based on fat-coating, emulsion based, symbiotic-coating, and biopolymer systems (Crittenden, 2009).

It is known from further research that microencapsulation preserves probiotic cells from detrimental factors during processing and storage, such as low pH and high acidity (Wenrong and Griffiths, 2000), bile salts (Lee and Heo, 2000), heat and cold shocks caused by spray drying and freezing respectively (Shah and Ravula, 2004), in the case of anaerobic microorganisms from molecular oxygen (Sunohara et al., 1995), against bacteriophages (Steenson et al., 1987), and chemical antimicrobial agents (Sultana et al., 2000). Microencapsulation also has advantages on the improvement and stabilization of the sensory properties (Gomes and Malcata, 1999) and the immobilization of the cells for their homogeneous distribution through the product (Krasaekoopt et al., 2003).

Various factors influence the efficacy of microencapsulation. The method of microencapsulation, for example, spray drying, as well as the inlet and outlet temperatures, and the chosen probiotic strains are also very important for viability. The type of the wall material is also an important factor, since it has to protect the probiotics against oxygen, heat, and other environmental stresses during drying, processing, storage, low pH, and protease in the gastric tract (Critenden et al., 2006). Protein carbohydrate oil emulsions are preferred for microencapsulation, since they induce smaller beads and do not influence the sensory properties of the product in comparison to cellulose acetate phthalate, starch alginate, gellan, and xanthan gums (Sarkar, 2010).

# 3 Conclusion

It is an important fact that probiotics in food must retain their viability throughout shelf life. For this purpose, it is important to select a suitable strain-food matrix combination, to apply favorable food-processing conditions, and to provide convenient packaging and environmental conditions. The characteristics of the selected strains, the food matrix, food production stages, and storage conditions should be compatible for the survival of probiotics at the recommended level for health benefits. Selection of oxygen-tolerant, acid-tolerant, and bile-resistant strains, addition of prebiotics, growth factors/promoters, fruit and plant products to the food matrix, providing favorable processing, fermentation, and storage conditions are important factors to enhance the survival of probiotics. During processing and storage, providing an appropriate environment including optimum temperature and pH, low oxygen and water content, low osmotic stress, and suitable packaging is also crucial. Among these various applications, microencapsulation has emerged as a good alternative to overcome the problem of poor stability of probiotic microorganisms during processing, in the food matrix, during storage, and also in the gastrointestinal tract. Therefore, applications of microencapsulation are suggested for the food industry to enhance their prophylactic activities.

# References

Abadias, M., Teixido, N., Usall, J., Benabarre, A., Vinas, I., 2001. Viability, efficacy and storage stability of freeze dried biocontrol agent *Candida sake* using different protective and rehydration media. J. Food Prot. 64, 856–861.

Abdollahi, M., Abdolghaffari, A.H., Gooshe, M., Ghasemi-Niri, F., 2016. Safety of probiotic bacteria. In: Watson, R.R., Preedy, V.R. (Eds.), Probiotics, Prebiotics and Synbiotics. Bioactive Foods in Health Promotion. Elsevier, pp. 227–243.

Akalın, A.S., Erisir, D., 2008. Effects of inulin and oligofructose on the rheological characteristics and probiotic culture survival in low-fat probiotic ice cream. J. Food Sci. 73 (4), 184–188.

Akalın, A.S., Fenderya, S., Akbulut, N., 2004. Viability and activity of bifidobacteria in yogurt containing fructooligosaccharide during refrigerated storage. J. Food Sci. Technol. 39, 613–621.

Akalın, A.S., Gönç, S., Ünal, G., Fenderya, S., 2007. Effects of fructooligosaccharide and whey protein concentrate on the viability of starter culture in reduced-fat probiotic yogurt during storage. J. Food Sci. 72 (7), 222–227.

Akalın, A.S., Kesenkas, H., Dinkçi, N., Unal, G., Ozer, E., Kınık, O., 2017. Enrichment of probiotic ice cream with different dietary fibers: structural characteristics and culture viability. J. Dairy Sci. 101, 37–46.

Akalın, A.S., Unal, G., Dinkçi, N., 2018. Angiotensin-converting enzyme inhibitory and starter culture activities in probiotic yoghurt: effect of sodium–calcium caseinate and whey protein concentrate. Int. J. Dairy Technol. 71, 185–194.

Anadon, A., Martinez-Larranaga, M.R., Ares, I., Martinez, M.A., 2016. Probiotics and prebiotics: an assessment of their safety and health benefits. In: Watson, R.R., Preedy, V.R. (Eds.), Probiotics, Prebiotics and Synbiotics. Bioactive Foods in Health Promotion. Elsevier, pp. 3–23.

Anal, A.K., Singh, H., 2007. Recent advances in microencapsulation of probiotics for industrial applications and targeted delivery. Trends Food Sci. Technol. 18 (5), 240–251.

Ananta, E., Knorr, D., 2003. Pressure-induced thermotolerance of *Lactobacillus rhamnosus* GG. Food Res. Int. 36 (9), 991–997.

Ananta, E., Volkert, M., Knorr, D., 2005. Cellular injuries and storage stability of spray-dried *Lactobacillus rhamnosus* GG. Int. Dairy J. 15, 399–409.

Anekella, K., Orsat, V., 2014. Shelf life stability of lactobacilli encapsulated in raspberry powder: insights into non-dairy probiotics. Int. J. Food Sci. Nutr. 65 (4), 411–418.

Aryana, K.J., Plauche, S., Rao, R.M., McGrew, P., Shah, N.P., 2007. Fat-free plain yogurt manufactured with inulins of various chain lengths and *Lactobacilus acidophilus*. J. Food Sci. 72, 79–84.

Bakr, S.A., 2015. The potential applications of probiotics on dairy and non-dairy foods focusing on viability during storage. Biocatal. Agric. Biotechnol. 4, 423–431.

Bakr, S.A., 2016. Influence of food matrix on the viability of probiotic bacteria: a review based on dairy and non-dairy beverages. Food Biosci. 13, 1–8.

Bhat, Z.F., Bhat, H., 2011. Milk and dairy products as functional foods: a review. Int. J. Dairy Sci. 6, 1–12.

Boylston, T.D., Vinderola, C.G., Ghoddusi, H.B., Reinheimer, J.A., 2004. Incorporation of bifidobacteria into cheeses: challenges and rewards. Int. Dairy J. 14 (5), 375–387.

Brennan, M., Wanismail, B., Johnson, M.C., Ray, B., 1986. Cellular damage in dried *Lactobacillus acidophilus*. J. Food Prot. 49, 47–53.

Bruno, F.A., Lankaputhra, W.E.V., Shah, N.P., 2002. Growth, viability and activity of Bifidobacterium spp. in skim milk containing prebiotics. J. Food Sci. 67, 2740–2744.

Buriti, F.C.A., Freitas, S.C., Egito, A.S., dos Santos, K.M.O., 2014. Effects of tropical fruit pulps and partially hydrolysed galactomannan from *Caesalpinia pulcherrima* seeds on the dietary fibre content, probiotic viability, texture and sensory features of goat dairy beverages. LWT Food Sci. Technol. 59, 196–203.

Calinoiu, L.V., Vodnar, D.C., Precup, G., 2016. The probiotic bacteria viability under different conditions. Bull. UASVM Food Sci. Technol. 73 (2), 55–60.

Capozzi, V., Arena, M.P., Russo, P., Spano, G., Fiocco, D., 2016. Stressors and food environment: toward strategies to improve robustness and stress tolerance in probiotics. In: Watson, R.R., Preedy, V.R. (Eds.), Probiotics, Prebiotics and Synbiotics. Bioactive Foods in Health Promotion. Elsevier, pp. 245–256.

Cardarelli, H.R., Buriti, F.C.A., Castro, I.A., Saad, S.M.I., 2008. Inulin and oligofructose improve sensory quality and increase the probiotic viable count in potentially synbiotic petit-suisse cheese. LWT-Food Sci. Technol. 41 (6), 1037–1046.

Carvalho, A.S., Silva, J., Ho, P., Teixeira, P., Malcata, F.X., Gibbs, P., 2004. Relevant factors for the preparation of freeze-dried lactic acid bacteria. Int. Dairy J. 14, 835–847.

Champagne, C.P., Fustier, P., 2007. Microencapsulation for the improved delivery of bioactive compounds into foods. Curr. Opin. Biotechnol. 18, 184–190.

Champagne, C.P., Gardner, N.J., 2008. Effect of storage in a fruit drink on subsequent survival of probiotic lactobacilli to gastro-intestinal stresses. Food Res. Int. 41, 539–543.

Champagne, C.P., Gardner, N.J., Roy, D., 2005. Challenges in the addition of probiotic cultures to foods. Crit. Rev. Food Sci. Nutr. 45 (1), 61–84.

Champagne, C.P., Ross, R.P., Saarela, M., Hansen, K.F., Charalampopoulos, D., 2011. Recommendations for the viability assessment of probiotics as concentrated cultures and in food matrices. Int. J. Food Microbiol. 149, 185–193.

Costa, E., Usall, J., Teixido, N., Garcia, N., Vinas, I., 2000. Effect of protective agents, rehydration media and initial cell concentration on viability of *Pantoea agglomerans* strain CPA-2 subjected to freeze drying. J. Appl. Microbiol. 89, 793–800.

Critenden, R., Weerakkody, R., Sanguansri, L., Augustin, M., 2006. Synbiotic microcapsules that enhance microbial viability during non-refrigerated storage and gastrointestinal transit. Appl. Environ. Microbiol. 72, 2280–2282.

Crittenden, R., 2009. Incorporating probiotics into foods. In: Lee, Y.K., Salminen, S. (Eds.), Handbook of Probiotics and Prebiotics. Copyright© John Whiley & Sons, Inc, pp. 58–75.

Crowe, J.H., Crowe, L.M., Carpenter, J.F., Rudolph, A.S., Wistrom, C.A., Spargo, B.J., 1988. Interactions of sugarswith membranes. Biochim. Biophys. Acta 947, 367–384.

da Cruz, A.G., Faria, J.D.A.F., Van Dender, A.G.F., 2007. Packaging system and probiotic dairy foods. Food Res. Int. 40 (8), 951–956.

da Cruz, A.G., Faria, J.D.A.F., Cadena, R.S., Granato, D., Mortazavian, A.M., Bolini, M.A., 2010. Sensory evaluation of probiotic, prebiotic and symbiotic foods: relevance for product development. Compr. Rev. Food Sci. Food Saf. 9, 358–373.

Daneshi, M., Ehsani, M.R., Razavi, S.H., Labbafi, M., 2013. Effect of refrigerated storage on the probiotic survival and sensory properties of milk/carrot juice mix drink. Electron. J. Biotechnol. 16 (5), 1–12.

Dave, R.I., Shah, N.P., 1997. Viability of yoghurt and probiotic bacteria in yoghurt made from commercial starter cultures. Int. Dairy J. 7, 31–41.

Dave, R.I., Shah, N.P., 1998. Ingredient supplementation effects on viability of probiotic bacteria in yogurt. J. Dairy Sci. 81 (11), 2804–2816.

de Souza Oliveira, R.P.S., Perego, P., Converti, A., de Oliveira, M.N., 2009. The effect of inulin as a prebiotic on the production of probiotic fibre-enriched fermented milk. Int. J. Dairy Technol. 62, 195–203.

de Valdez, G.F., de Giori, G.S., de Ruiz Holgado, A.P., Oliver, G., 1985. Effect of drying medium on residual moisture content and viability of freeze dried lactic acid bacteria. Appl. Environ. Microbiol. 49, 413–415.

de Vuyst, L., 2000. Technology aspects related to the application of functional starter cultures. Food Technol. Biotechnol. 382 (2), 105–112.

Desai, A.R., Powell, I.B., Shah, N.P., 2004. Survival and activity of probiotic Lactobacilli in skim milk containing prebiotics. J. Food Sci. 69 (3), 57–60.

Desfossés-Foucault, E., Dussault-Lepage, V., Le Boucher, C., Savard, P., LaPointe, G., Roy, D., 2012. Assessment of probiotic viability during Cheddar cheese manufacture and ripening using propidium monoazide-PCR quantification. Front. Microbiol. 3, 1–11.

do Espırito-Santo, A.P., Perego, P., Converti, A., Oliveira, M.N., 2011. Influence of food matrices on probiotic viability: a review focusing on the fruity bases. Trends Food Sci. Technol. 22, 377–385.

Donkor, O.N., Nilmini, S.L.I., Stolic, P., Vasiljevic, T., Shah, N.P., 2007. Survival and activity of selected probiotic organisms in set-type yogurt during cold storage. Int. Dairy J. 17, 657–665.

Dos Santos Cruxen, C.E., Hoffmann, J.F., Zandona, G.P., Fiorentini, A.M., Rombaldi, C.V., Chaves, F.C., 2017. Probiotic butia (Butia odorata) ice cream: development, characterization, stability of bioactive compounds, and viability of *Bifidobacterium lactis* during storage. LWT- Food Sci. Technol. 75, 379–385.

Douglas, L.C., Sanders, M.E., 2008. Probiotics and prebiotics in dietetics practice. J. Am. Diet. Assoc. 108, 510–521.

FAO/WHO, 2002. Guidelines for the evaluation of probiotics in food. In: Food and Agriculture Organization of the United Nations and World Health Organization Working Group Report.

FAO/WHO, 2010. Codex standard for fermented milks. In: Codex Stan (Eds.), second ed. Food and Agriculture Organization, United Nations, Rome, pp. 1–5.

Farnworth, E.R., Champagne, C.P., 2016. Production of probiotic cultures and their incorporation into foods. In: Watson, R.R., Preedy, V.R. (Eds.), Probiotics, Prebiotics and Synbiotics. Bioactive Foods in Health Promotion. Elsevier, pp. 303–318.

Figueroa-Gonzalez, I., Quijano, G., Ramirez, G., Cruz-Guerrero, A., 2011. Probiotics and prebiotics-perspectives and challenges. J. Sci. Food Agric. 91, 1341–1348.

Fonseca, F., Beal, C., Corrieu, G., 2000. Method of quantifying the loss of acidification activity of lactic acid starters during freezing and frozen storage. J. Dairy Res. 67, 83–90.

Fortin, M.H., Champagne, C.P., St-Gelais, D., Britten, M., Fustier, P., Lacroix, M., 2011. Viability of *Bifidobacterium longum* in cheddar cheese curd during manufacture and storage: effect of microencapsulation and point of inoculation. Dairy Sci. Technol. 91, 599–614.

Freudig, B., Hogekamp, S., Schubert, H., 1999. Dispersion of powders in liquids in a stirred vessel. Chem. Eng. Prog. 38, 525–535.

Gardiner, G.E., O'Sullivan, E., Kelly, J., Auty, M.A., Fitzgerald, G.F., Collins, J.K., Ross, R.P., Stanton, C., 2000. Comparative survival rates of human-derived probiotic *Lactobacillus paracasei* and *L. salivarius* strainsduring heat treatment and spray drying. Appl. Environ. Microbiol. 66, 2605–2612.

Gardiner, G.E., Bouchier, P., O'Sullivan, E., Kelly, J., Kevin Collins, J., Fitzgerald, G., 2002. A spray-dried culture for probiotic Cheddar cheese manufacture. Int. Dairy J. 12 (9), 749–756.

Ge, X.Y., Yuan, J., Qin, H., Zhang, W.G., 2011. Improvement of L-lactic acid production by osmotic-tolerant mutant of *Lactobacillus casei* at high temperature. Appl. Microbiol. Biotechnol. 89, 73–78.

Geng, J., Chiron, C., Combrisson, J., 2014. Rapid and specific enumeration of viable bifidobacteria in dairy products based on flow cytometry technology: a proof of concept study. Int. Dairy J. 37, 1–4.

Gomes, A.M.P., Malcata, F.X., 1999. Bifidobacterium spp. and Lactobacikllus acidophilus: biological, biochemical, technological and therapeutical properties relevant for use as probiotics. Trends Food Sci. Technol. 10, 139–145.

Hekmat, S., Soltani, H., Reid, G., 2009. Growth and survival of *Lactobacillus reuteri* RC-14 and *Lactobacillus rhamnosus* GR-1 in yogurt for use as a functional food. Innov. Food Sci. Emerg. Technol. 10 (2), 293–296.

Hoobin, P., Burgar, I., Zhu, S.C., Ying, D.Y., Sanguansri, L., Augustin, M.A., 2013. Water sorption properties, molecular mobility and probiotic survival in freeze dried protein-carbohydrate matrices. Food Funct. 4, 1376–1386.

Kailasapathy, K., 2002. Microencapsulation of probiotic bacteria: technology and potential applications. Curr. Issues Intest. Microbiol. 3, 39–48.

Karimi, R., Mortazavian, A.M., da Cruz, A.G., 2011. Viability of probiotic microorganisms in cheese during production and storage: a review. Dairy Sci. Technol. 91, 283–308.

Karimi, R., Mortazavian, A.M., Amiri-Rigi, A., 2012. Selective enumeration of probiotic microorganisms in cheese. Food Microbiol. 29, 1–9.

Kawasaki, S., Mimura, T., Satoh, T., Takeda, K., Niimura, Y., 2006. Response of the microaerophilic Bifidobacterium species, *B. boum* and *B. thermophilum*, to oxygen. Appl. Environ. Microbiol. 72, 6854–6858.

Kim, S.S., Bhowmik, S.R., 1990. Survival of lactic acid bacteria during spray drying of plain yoghurt. J. Food Sci. 55, 1010–2008.

Koh, J.H., Choi, S.H., Park, S.W., Choi, N.J., Kim, Y., Kim, S.H., 2013. Synbiotic impact of tagatose on viability of *Lactobacillus rhamnosus* strain GG mediated by the phosphotransferase system (PTS). Food Microbiol. 36, 7–13.

Krasaekoopt, W., Bhandari, B., Deeth, H., 2003. Evaluation of encapsulation techniques of probiotics for yoghurt. Int. Dairy J. 13, 3–13.

Krasaekoopt, W., Bhandari, B., Deeth, H., 2004. The influence of coating materials on some properties of alginate beads and survivability of microencapsulated probiotic bacteria. Int. Dairy J. 737, 43.

Ladero, V., Sanchez, B., 2017. Molecular and technological insights into the aerotolerance of anaerobic probiotics: examples from bifidobacteria. Curr. Opin. Food Sci. 14, 110–115.

Lee, K.Y., Heo, T.R., 2000. Survival of *Bifidobacterium longum* immobilized in calcium alginate beads in simulated gastric juices and bile salts solution. Appl. Environ. Microbiol. 66, 869–873.

Lee, Y.K., Margolles, A., Mayo, B., Ruas-Madiedo, P., Gueimonde, M., de los Reyes-Gavilán, C., Chapman, T.A., Chin, J.J., Crittenden, R., Donohue, D., Tien Hung Chang, W., 2008. Probiotic microorganisms. In: Lee, Y.K., Salminen, S. (Eds.), Handbook of Probiotics and Prebiotics, second ed. John Wiley and Sons Inc., New Jersey, pp. 1–176. Chapter 1.

López de Lacey, A.M., Pérez-Santín, E., López-Caballero, M.E., Montero, P., 2014. Survival and metabolic activity of probiotic bacteria in green tea. LWT- Food Sci. Technol. 55, 314–322.

Lourens-Hattingh, A., Viljoen, B.C., 2001. Yogurt as probiotic carrier food. Int. Dairy J. 11, 1–17.

Madene, A., Jacquot, M., Scher, J., Desobry, S., 2006. Flavour encapsulation and controlled release—a review. Int. J. Food Sci. Technol. 41, 1–21.

Mainville, I., Arcand, Y., Farnworth, E.R., 2005. A dynamic model that simulates the human upper gastrointestinal tract for the study of probiotics. Int. J. Food Microbiol. 99, 287–296.

Marco, M.L., Tachon, S., 2013. Environmental factors influencing the efficacy of probiotic bacteria. Curr. Opin. Biotechnol. 24, 207–213.

Mattila-Sandholm, T., Myllarinen, P.M., Crittenden, R., Mogensen, G., Fonden, R., Saarela, M., 2002. Technological challenges for future probiotic foods. Int. Dairy J. 12, 173–182.

Mille, Y., Obert, J.P., Beney, L., Gervais, P., 2004. New drying process for lactic bacteria based on their rehydration behavior in liquid media. Biotechnol. Bioeng. 88, 71–76.

Miller, C.W., Nguyen, M.H., Rooney, M., Kailasapthy, K., 2002. The influence of packaging materials on the dissolved oxygen content of probiotic yogurt. Packag. Technol. Sci. 15, 133–138.

Miller, C.W., Nguyen, M.H., Rooney, M., Kailasapthy, K., 2003. The control of dissolved oxygen content in probiotic yogurts by alternative packing materials. Packag. Technol. Sci. 16, 61–67.

Mohammadi, R., Mortazavian, A.M., 2011. Technological aspects of prebiotics in probiotic fermented milks. Food Rev. Int. 27, 192–212.

Mohammadi, R., Mortazavian, A.M., Khosrokhavar, R., da Cruz, A.G., 2011. Probiotic ice cream: viability of probiotic bacteria and sensory properties. Ann. Microbiol. 61, 411–424.

Mohammadi, R., Sohrabvandi, S., Mohammad Mortazavian, A., 2012. The starter culture characteristics of probiotic microorganisms in fermented milks. Eng. Life Sci. 12, 399–409.

Morelli, L., 2007. In vitro assessment of probiotic bacteria: from survival to functionality. Int. Dairy J. 17, 1278–1283.

Mori, H., Sato, Y., Taketomo, N., Kamiyama, T., Yoshiyama, Y., Meguro, S., Sato, H., Kaneko, T., 1997. Isolation and structural identification of bifidogenic growth stimulator produced by *Propionibacterium freudenreichii*. J. Dairy Sci. 80 (9), 1959–1964.

Mortazavian, A.M., Ehsani, M.R., Mousavi, S.M., Reinheimer, J.A., Emamdjomeh, Z., Sohrabvandi, S., Rezaei, K., 2006. Preliminary investigation of the combined effect of heat treatment and incubation temperature on the viability of the probiotic micro-organisms in freshly made yogurt. Int. J. Dairy Technol. 59, 8–11.

Mortazavian, A.M., Ehsani, M.R., Mousavi, S.M., Rezaei, K., Sohrabvandi, S., Reinheimer, J.A., 2007. Effect of refrigerated storage temperature on the viability of probiotic microorganisms in yogurt. Int. J. Dairy Technol. 60 (2), 123–127.

Mugnier, J., Jung, G., 1985. Survival of bacteria and fungi in relation towater activity and the solvent properties of water in biopolymer gels. Appl. Environ. Microbiol. 50, 108–114.

Nag, A., Das, S., 2013. Improving ambient temperature stability of probiotics with stress adaptation and fluidized bed drying. J. Funct. Foods 5, 170–177.

Nebesny, E., Zyzelewicz, D., Motyl, I., Libudzisz, Z., 2007. Dark chocolates supplemented with *Lactobacillus* strains. Eur. Food Res. Technol. 225 (1), 33–42.

Nematollahi, A., Sohrabvandi, S., Mortazavian, A.M., Jazaeri, S., 2016. Viability of probiotic bacteria and some chemical and sensory characteristics in cornelian cherry juice during cold storage. Electron. J. Biotechnol. 21, 49–53.

Noorbakhsh, R., Yaghmaee, P., Durance, T., 2013. Radiant energy under vacuum (REV) technology: a novel approach for producing probiotic enriched apple snacks. J. Funct. Foods 5, 1049–1056.

Noubakhti, A.R., Ehsani, M.R., Mousavi, S.M., Mortazavian, A.M., 2009. Influence of laculose and Hi-maize addition on viability of probiotic microorganisms in freshly made symbiotic fermented milk drink. Milchwissenschaft 64, 191–193.

Nualkaekul, S., Salmeron, I., Charalampopoulos, D., 2011. Investigation of the factors influencing the survival of *Bifidobacterium longum* in model acidic solutions and fruit juices. Food Chem. 129, 1037–1044.

Önneby, K., Pizzul, L., Bjerketorp, J., Mahlin, D., Hakansson, S., Wessman, P., 2013. Effects of di- and polysaccharide formulations and storage conditions on survival of freezedried *Sphingobium* sp. World J. Microbiol. Biotechnol. 29 (8), 1399–1408.

Özer, D., Akin, S., Özer, B., 2005. Effect of inulin and lactulose on survival of *Lactobacillus acidophilus* LA-5 and *Bifidobacterium bifidum* BB-02 in Acidophilus-Bifidus yogurt. Food Sci. Technol. Int. 11, 19–24.

Paéz, R., Lavari, L., Vinderola, G., Audero, G., Cuatrin, A., Zaritzky, N., Reinheimer, J., 2012. Effect of heat treatment on spray drying on Lactobacilli viability and resistance to simulated gastrointestinal digestion. Food Res. Int. 48, 748–754.

Pennachia, C., Vaughan, E.E., Villani, F., 2006. Potential probiotic *Lactobacillus* strains from fermented sausages: further investigations on their probiotic properties. Meat Sci. 73, 90–101.

Picot, A., Lacroix, C., 2003. Optimization of dynamicloop mixer operating conditions for production of o/w emulsion for cell microencapsulation. Lait 83, 237–250.

Pispan, S., Hewitt, C.J., Stapley, A.G.F., 2013. Comparison of Cell survival rates of *E. coli* K12 and *L. acidophilus* undergoing spray-drying. Food Bioprod. Process. 91, 362–369.

Poirier, I., Marechal, P.A., Richard, S., Gervais, P., 1999. *Saccharomyces cerevisiae* viability is strongly dependant on rehydration kinetics and the temperature of dried cells. J. Appl. Microbiol. 86, 87–92.

Raeisi, S.N., Ouoba, L.I.I., Farahmand, N., Sutherland, J., Ghoddus, H.B., 2013. Variation, viability, and validity of bifidobacteria in fermented milk products. Food Control 34, 691–697.

Ramchandran, L., Shah, N.P., 2008. Effect of Versagel® on the growth and metabolic activities of selected Lactic acid bacteria. J. Food Sci. 73 (1), 21–26.

Ranadheera, R.D.C.S., Baines, S.K., Adams, M.C., 2010. Importance of food in probiotic efficacy. Food Res. Int. 43, 1–7.

Randazzo, C.L., Pitino, I., Licciardello, F., Muratore, G., Caggia, C., 2013. Survival of *Lactobacillus rhamnosus* probiotic strains in peach jam during storage at different temperatures. Food Sci. Technol. 33 (4), 652–659.

Reale, A., Renzo, T.D., Rossi, F., Zotta, T., Iacumin, L., Preziuso, M., Parente, E., Sorrentino, E., Coppola, R., 2015. Tolerance of *Lactobacillus casei, Lactobacillus paracasei* and *Lactobacillus rhamnosus* strains to stress factors encountered in food processing and in the gastro-intestinal tract. LWT- Food Sci. Technol. 60, 721–728.

Rivera-Espinoza, Y., Gallardo-Navarro, Y., 2010. Non-dairy probiotic products. Food Microbiol. 27, 1–11.

Rouhi, M., Sohrabvandi, S., Mortazavian, A.M., 2013. Probiotic fermented sausage: viability of probiotic microorganisms and sensory characteristics. Crit. Rev. Food Sci. Nutr. 53 (4), 331–348.

Saarela, M., 2007. Methods to improve the viability and stability of probiotics. In: Saarela, M. (Ed.), Functional Dairy Products. vol. 2. Woodhead Publishing, Abingdon, pp. 391–403. Chapter 18.

Sah, B.N.P., Vasiljevic, T., McKechnie, S., Donkor, O.N., 2015. Effect of refrigerated storage on probiotic viability and the production and stability of antimutagenic and antioxidant peptides in yogurt supplemented with pineapple peel. J. Dairy Sci. 98, 5905–5916.

Santivarangka, C., Kulozik, U., Foerst, P., 2006. Effect of carbohydrates on the survival of Lactobacillus helveticus during vacuum drying. Lett. Appl. Microbiol. 42, 271–276.

Sarkar, S., 2010. Approaches for enhancing the viability of probiotics: a review. Br. Food J. 112, 329–342.

Sarkar, S., 2018. Whether viable and dead probiotic equally efficacious? Nutr. Food Sci. 48, 285–300.

Sendra, E., Fayos, P., Lario, Y., Fernández-López, J., Sayas-Barberá, E., Pérez-Alvarez, J.A., 2008. Incorporation of citrus fibers in fermented milk containing probiotic bacteria. Food Microbiol. 25, 13–21.

Sendra, E., Sayas-Barberá, M.E., Fernández-López, J., Pérez-Alvarez, J.A., 2016. Effect of food composition on probiotic bacteria viability. In: Watson, R.R., Preedy, V.R. (Eds.), Probiotics, Prebiotics and Synbiotics. Bioactive Foods in Health Promotion. Elsevier, pp. 257–269.

Shah, N.P., 2000. Probiotic bacteria: selective enumeration and survival in dairy foods. J. Dairy Sci. 83, 894–907.

Shah, N.P., Ravula, R., 2004. Selling the cells in desserts. Dairy Ind. Int. 69, 31–32.

Shah, N.P., Lankaputhra, W.E.V., Britz, M., Kyle, W.S.A., 1995. Survival of *Lactobacillus acidophilus* and *Bifidobacterium bifidum* in commercial yoghurt during refrigerated storage. Int. Dairy J. 5, 515–521.

Shiby, V.K., Mishra, H.N., 2013. Fermented milks and milk products as functional foods—a review. Crit. Rev. Food Sci. 53, 482–496.

Shin, H.S., Lee, J.H., Pestka, J.J., Ustunol, Z., 2000. Growth and viability of commercial Bifidobacterium spp. in skim milk containing oligosaccharides and inulin. J. Food Sci. 65, 884–887.

Shori, A.B., 2016. Influence of food matrix on the viability of probiotic bacteria: a review based on dairy and non-dairy beverages. Food Biosci. 13, 1–8.

Sinha, R.N., Shukla, A.K., Lal, M., Ranganathan, B., 1982. Rehydration of freeze dried cultures of lactic streptococci. J. Food Sci. 47, 668–669.

Siro, I., Kapolna, E., Kapolna, B., Lugasi, A., 2008. Functional food. Product development, marketing and consumer acceptance-a review. Appetite 51, 456–467.

Steenson, L.R., Klaenhammer, T.R., Swaisgood, H.E., 1987. Calcium alginate immobilized cultures of lactic streptococci are protected from attack by lytic bacteriophage. J. Dairy Sci. 70, 1121–1127.

Sultana, K., Godward, G., Reynolds, N., Arumugaswamy, R., Peiris, P., Kailasapathy, K., 2000. Encapsulation of probiotic bacteria with alginate-starch and evaluation of survival in simulated gastrointestinal conditions and in yoghurt. Int. J. Food Microbiol. 62, 47–55.

Sunohara, H., Ohno, T., Shibata, N., Seki, K., 1995. Process for producing capsule and capsule obtained thereby. US Patent 5,478,570.

Talwalkar, A., Kailasapathy, K., 2004. A review of oxygen toxicity in probiotic yogurts: influence on the survival of probiotic bacteria and protective techniques. Compr. Rev. Food Sci. Food Saf. 3, 117–124.

Teixeira, P., Castro, H., Mohacsi-Farkas, C., Kirby, R., 1997. Identification of sites of injury in *Lactobacillus bulgaricus* during heat stress. J. Appl. Microbiol. 83, 219–226.

Tripathi, M.K., Giri, S.K., 2014. Probiotic functional foods: survival of probiotics during processing and storage. J. Funct. Foods 9, 225–241.

Tsvetkov, T., Brankova, R., 1983. Viability of micrococci and Lactobacilli upon freezing and freeze-drying in the presence of different cryoprotectants. Cryobiology 20, 318–323.

Vasiljevic, T., Kealy, T., Mishra, V.K., 2007. Effects of β-glucan addition to a probiotic containing yogurt. Food Chem. Toxicol. 72, 405–411.

Vinderola, C.G., Costa, G.A., Regenhardt, S., Reinheimer, J.A., 2002. Influence of compounds associated with fermented dairy products on the growth of lactic acid starter and probiotic bacteria. Int. Dairy J. 12, 579–589.

Vinderola, G., Binetti, A., Burns, P., Reinheimer, J., 2011. Cell viability and functionality of probiotic bacteria in dairy products. Front. Microbiol. 2 (70), 1–6.

Wang, Y.C., Yu, R.C., Chou, C.C., 2004. Viability of lactic acid bacteria and Bifidobacteria in fermented soymilk after drying, subsequent rehydration and storage. Int. J. Food Microbiol. 93, 209–217.

Wasko, A., Polak-Berecka, M., Gustaw, W., 2013. Increased viability of probiotic *Lactobacillus rhamnosus* after osmotic stress. Acta Aliment. 42 (4), 520–528.

Weinbreck, F., Bodnar, I., Marco, M.L., 2010. Can encapsulation lengthen the shelf-life of probiotic bacteria in dry products? Int. J. Food Microbiol. 136, 364–367.

Wenrong, S., Griffiths, M.W., 2000. Survival of bifidobacteria in yogurt and simulated gastric juice following immobilization in gellan-xanthan beads. Int. J. Food Microbiol. 61, 17–26.

Ying, D.Y., Phoon, M.C., Sanguansri, L., Weerakkody, R., Burgar, I., Augustin, M.A., 2010. Microencapsulated *Lactobacillus rhamnosus* GG powders: relationship of powder physical properties to probiotic survival during storage. J. Food Sci. 75 (9), 588–595.

Yoo, B., Lee, C.M., 1993. Thermoprotective effect of sorbitol on proteins during dehydration. J. Agric. Food Chem. 41, 190–192.

Zayed, G., Roos, Y.H., 2004. Influence of trehalose and moisture content on survival of *Lactobacillus salivarius* subjected to freeze drying and storage. Process Biochem. 39, 1081–1086.

Ziar, H., Gérard, P., Riazi, A., 2014. Effect of prebiotic carbohydrates on growth, bile survival and cholesterol uptake abilities of dairy-related bacteria. J. Sci. Food Agric. 94, 1184–1190.

# Modified atmosphere packaging for food preservation

**7**

*Umezuruike Linus Opara*, Oluwafemi J. Caleb[†], Zinash A. Belay**
*Postharvest Technology Research Laboratory, South African Research Chair in Postharvest Technology, Department of Horticultural Science, Faculty of AgriSciences, Stellenbosch University, Stellenbosch, South Africa [†]Post-harvest and Agro-Processing Technologies (PHATs), Agricultural Research Council (ARC), Infruitec-Nietvoorbij, Stellenbosch, South Africa

**Chapter outline**



## 1 Introduction

Global production and marketing of fresh produce has continued to rise over the past few decades in response to growing consumer demand. In addition to contributing to food and nutrition security, the production and marketing of agricultural, horticultural, and aquatic produce contributes to livelihoods through income generation. Packaging plays a crucial role by facilitating the containment, transportation, and logistics of fresh and processed commodities (Opara, 2009; Opara et al., 2007). Modified atmosphere packaging (MAP) of fresh and minimally processed commodities refers to the technique of sealing the products in a packaging system that alters normal air composition and provide optimal gas composition around the product. Under conditions of reduced $O_2$ and high $CO_2$, the metabolic processes of the horticultural product and microbial

**Food Quality and Shelf Life. https://doi.org/10.1016/B978-0-12-817190-5.00007-0**

activity are slowed down, which contributes to extended shelf life (Antmann et al., 2008; Ares et al., 2007). In contrast, when MAP is used for non-respiring products, such as meat and aquatic food products, the aim is to retain the introduced optimal atmosphere inside the package during storage. Therefore, high barrier films composed of different layers of materials are used.

Several gases have been used in MAP systems, each one having a different role in the preservation of food products. The main gases often used are oxygen, carbon dioxide, and nitrogen ($O_2$, $CO_2$, and $N_2$). MAP richer in $CO_2$ and poorer in $O_2$ than air can potentially reduce physiological and biochemical processes and retards senescence for fruit and vegetables (Antmann et al., 2008). Carbon dioxide is known for its fungistatic and bacteriostatic properties in food systems and is, therefore, the most important gas used in MAP systems. The reduction of $O_2$ slows down lipid oxidation and the development of rancidity as well as inhibiting the growth of aerobic microorganisms in fish, meat, and derived products. For these products, simple flushing with nitrogen is used as an alternative to vacuum packaging to replace $O_2$ in the package. Furthermore, nitrogen gas is used in a gas mixture for MAP as a filler (e.g., $CO_2$ and $N_2$) because of its low solubility in water and fat, which prevents packaging collapsing. Recently, particular attention has been paid to the use of some inactive or noble gases, such as argon (Ar), helium (He), and nitrous oxide ($N_2O$) (González-Buesa et al., 2014). These gases do not directly affect metabolism through modification of enzymes, however, they may increase the diffusivity of $O_2$, $C_2H_4$, and $CO_2$ from plant tissues because of their higher density than nitrogen, or inhibit respiration by affecting cytochrome oxidase activity in the mitochondria (Ma et al., 2017).

Furthermore, it is well established that package permeability and the breathable film area play an important role in MAP system design and successful application (Caleb et al., 2012). Therefore, the choice of film is a key factor in order to obtain optimum modification of the atmosphere and relative humidity (RH) to avoid extremely low concentrations of gases and accumulation of water. Most polymeric materials used in fresh produce packaging have lower water vapor transmission (Table 1). Therefore, most water molecules evaporated from the produce do not escape through the film and remain within the package, creating high in-package RH. Therefore, recent studies reported various approaches to control in-package RH during MAP. These include using perforated packaging films (Hussein et al., 2015; Bovi et al., 2016), creating humidity regulating packaging materials via direct incorporation of active substances into the package (Rux et al., 2015; Bovi et al., 2018), and designing humidity packaging systems incorporating two or more different packaging materials in one design (Belay et al., 2018; Rux et al., 2015).

The most dominant external factor influencing MAP systems is storage temperature. Temperature influences both gas exchange of the produce and the permeability of the film for $O_2$, $CO_2$, and water vapor (Jacxsens et al., 2000). Most of the physical, biochemical, microbiological and physiological reactions contributing to deterioration of produce quality are largely dependent on temperature (Tano et al., 2007). Metabolic processes including respiration, transpiration, and ripening are particularly temperature-dependent (Beaudry et al., 1992). MAP in combination with cold storage temperatures could be used to inhibit microbial growth (Caleb et al., 2013).

The market for MAP is categorized into health and personal care, food and beverages, pharmaceuticals, and other applications. This chapter focuses on MAP applied

**Table 1** Barrier and functional properties of packaging film for modified atmosphere packaging

| Types of film | Barrier properties | Functional properties |
|---|---|---|
| *Polyethylene*:<br>Low-density polyethylene (LDPE)<br>Linear low-density Polyethylene (LLDPE)<br>High-density polyethylene (HDPE) | LDPE has low permeability to water vapor and high permeability to gases<br>HDPE has better gas-barrier properties than LLDPE but poor clarity | LDPE has high sealing qualities, can be laminated, extrusion coated, or coextruded<br>LLDPE has better impact strength, tear resistance, higher tensile strength and elongation, greater resistance to environmental stress cracking, and better puncture resistance<br>HDPE has a higher softening point than LDPE and is not suitable as a sealant layer |
| Polypropylene (PP) | | PP can be extruded or coextruded to provide a sealant layer |
| Oriented polypropylene (OPP)<br>Coextruded-oriented polypropylene (COPP) | OPP is a high-water vapor and gas barrier<br>COPP is a good water vapor barrier and gas barrier properties | OPP is grease resistance<br>COPP can be laminated as a part of the lidding material and can be used as breathable packaging |
| *Styrene polymers*:<br>Polystyrene<br>Polyethylene terephthalate (PET) | Polystyrene is a poor barrier to gases and water vapor | Polystyrene is brittle<br>PET has high clarity and can be used as a low-gauge-oriented film. It can be used in crystalline or amorphous phase |
| *Vinyl polymers*:<br>Vinyl acetate Copolymer (VAC)<br>Ethylene vinyl alcohol (EVOH)<br>Polyvinyl chloride (PVC)<br>Polyvinylidene chloride (PVDC) copolymer | VAC has high permeability to gas and water vapor<br>EVOH is a high gases barrier but is moisture sensitive<br>PVDC is low permeability to water vapor and gases<br>PVC is good gas barrier and moderate barrier to water vapor | VAC can be used as a sealant layer, highly flexible<br>EVOH can be used between the main formable and sealant layer as a laminated sandwich<br>PVDC used as a coating on polyester and OPP for lidding films<br>PVC can be used as a thermoformable base |
| Cellulose based film<br>Cellulose acetate, cellulose butyrate, cellulose propionate | | |

to preserve food products, including fresh and fresh-cut fruit and vegetables, meat, poultry, seafood, and other food and beverages. It highlights the principles of MAP, discusses the latest and emerging technologies, and summarizes the applications of MAP in the fresh and minimally processed food industry.

# 2 Principles of MAP

Modified atmosphere packaging (MAP) of fresh produce relies on change of the atmosphere inside the package (Fig. 1). This could be achieved by the natural interplay between two processes, the respiration of the product and the transfer of gases through the packaging, which leads to an increase in $CO_2$ and decline in $O_2$ concentrations. This atmosphere can potentially reduce respiration rate, ethylene sensitivity, and physiological changes. MAP generally involves the packaging of a whole or fresh-cut product in plastic film bags, and can be either passive or active.

The success of atmospheric modification inside a package depends on factors summarized in Table 2. Under optimal storage temperatures, MAP systems have been successfully applied to preserve various fruit and vegetables, and non-respiring products such as meat and aquatic food products. Since responses differ for each type of fresh produce when packaged under MAP conditions, it is necessary to quantify the effect of the applied atmosphere for the individual product. Based on the product requirement, there are two approaches (passive and active MAP), their main differences are summarized in Table 3, and are discussed further in the next two sub-sections.

## 2.1 Passive MAP

Passive MAP can be created by using natural air composition and relying on produce respiration to attain the desired gas mixture. Effective atmosphere modification is derived from RR of products and gas permeability of the packaging film, which induces a passively established steady-state gas composition after a long transient period (Mahajan et al., 2008). After a given time, gas composition in the package of a fresh product reaches a definite balance between RR and permeability of packaging

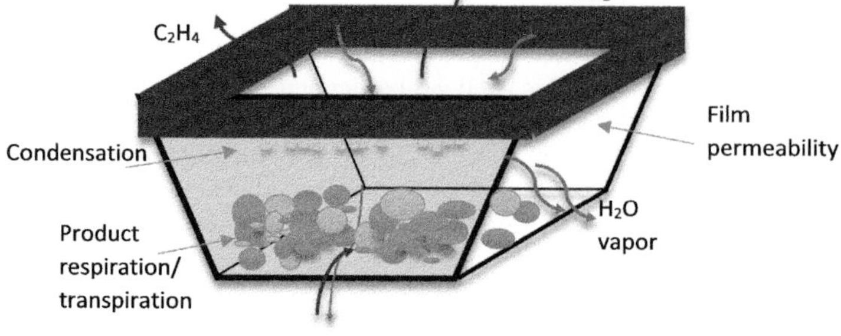


**Fig. 1** The dynamics of atmosphere inside modified atmosphere packaging.

**Table 2** Factors influencing optimal MAP for fresh or minimally processed fruit and vegetables

| Factor(s) | Variables |
|---|---|
| Produce | Produce weight |
| | Produce density |
| | Respiration rate |
| | Transpiration rate |
| | Ripeness stage |
| | Initial microbial load |
| Extrinsic | Gas ration and concentration |
| | Storage temperature |
| | Relative humidity |
| | Storage duration |
| Packaging | Volume |
| | Thickness of the film |
| | Film surface area for gas flux |
| | Gas permeability |
| Perforation | Number of perforations |
| | Radius of perforations |
| | Number of tubes |
| | Length of tubes |
| | Diameter of tubes and porosity of the tube packing |

From Fonseca, S.C., Oliveira, F.A., and Brecht, J.K., 2002. Modelling respiration rate of fresh fruits and vegetables for modified atmosphere packages: a review. J. Food Eng. 52(2), 99–119; Mahajan, P.V., Oliveira, F.A.R., and Macedo, I., 2008. Effect of temperature and humidity on the transpiration rate of the whole mushrooms. J. Food Eng. 84(2), 281–288.

**Table 3** Type of modified atmosphere packaging (MAP) for fresh-cut produce

| | Passive | Active |
|---|---|---|
| Definition | Modification of the gas composition inside a package due to interplay between the product respiration and the package permeability | Modification of the gas composition inside the package by flushing the package headspace with desired gas mixture |
| Equilibrium time | 1–2 days to 10–12 days | 1–2 h |
| Products suitable for | Mushrooms, carrots, strawberry, spinach | Cut apples, minimally processed leafy green vegetables, pomegranate arils |
| Cost | No extra cost involved if the package is properly designed | Extra investment is required for special machinery, i.e., gas mixer, gas flushing and packaging machine |
| Labeling requirements | No | Yes |

film. In this state of equilibrium, the total amounts of $CO_2$ produced and $O_2$ consumed by respiration are the same as that permeated through membrane exchange (Fonseca et al., 2002). The gas equilibrium state is thus influenced by RR of fresh produce, storage environmental factors (such as temperature and RH), and the permeability of the packing materials (Zhang et al., 2015a).

Passive MAP can effectively maintain the quality attributes and shelf life of horticultural commodities at lower storage temperature than unpacked commodities. Passive MAP in amide-PE bags improve the firmness and retain good fresh quality of fresh-cut kohlrabi (Escalona et al., 2007), control dehydration, and reduce mesophilic and psychrotrophic bacteria yeast growth. Passive MAP extends the shelf life of fresh-cut cantaloupe by sealing it with Cryovac LDX-5406 film for 9 days at 5°C (Bai et al., 2003). However, there are disadvantages of using passive MAP, including the protracted transient period required to achieve a desired atmosphere inside the packages and the risk of anoxia (Horev et al., 2012).

## 2.2 Active MAP

Active MAP involves gas flushing or scavenging, or the use of emitters. Active gas flushing relies on displacement of air with desired gas mixture to accelerate the establishment of the optimum gas composition and avoid exposure to unsuitable gas compositions for extending shelf life or improving safety, while maintaining quality (Fonseca et al., 2005). Numerous studies report the benefit of using replacement of normal air by low $O_2$ and enriched $CO_2$ atmosphere (Oms-Oliu et al., 2008; Costa et al., 2011), aiming to reduce the RR by providing a favorable atmosphere inside the package. On the other hand, the use of super atmospheric $O_2$ (>70%) is mainly aimed at reducing loss of firmness, reducing microbial growth, and maintaining antioxidants (Escalona et al., 2007). In the case of a gas-scavenging or emitting system, the most commercially important form of active packaging are small sachets of oxidizable iron based compounds used as $O_2$ scavengers (Kartal et al., 2012), which can prevent fruit discoloration and minimize chilling injuries. The $O_2$ scavengers have proven to be especially effective for reducing spoilage, delaying senescence, and reducing browning and mold incidence in fresh produce (Lee et al., 2014). On the other hand, $CO_2$ absorbers can prevent a build-up of $CO_2$ to injurious levels (Brody et al., 2001). Similarly, ethylene absorbers can help delay the climacteric rise in respiration and associated ripening for some fruit by scrubbing ethylene inside the package. The application of scavengers have been proven to be effective for kiwifruit, bananas, avocados, mushrooms, and persimmons (Ozdemir and Floros, 2004).

However, the use of scavenging sachets suffers from inadequate consumer acceptance and they are not appropriate for liquid foods, as direct contact of the liquid with the sachets usually causes spoilage of sachet contents (Ozdemir and Floros, 2004). Munro et al. (2009) reported that these sachets may be accidentally consumed with food or may be ingested by children. Synergistic effects of using oxygen scavenger sachets/pads utilizes the process of rusting or oxidation of iron compound in the presence of oxygen and water (Suppakul et al., 2003). Concerns are also expressed about oxygen scavengers allowing potential over growth of anaerobic pathogenic organisms (Munro et al., 2009). Thus, understanding the dynamics of MAP and individual food product characteristics and selecting the type of MAP according to the objective to be achieved is essential to benefit from MAP application.

# 3 Emerging MAP systems

## 3.1 High $O_2/CO_2$

Successful application of MAP requires selection of optimum gas as well as appropriate packaging film with gas permeability and water vapor transmission rate (WVTR) to balance moisture loss and water vapor condensation. Perforated (micro or macro) packaging films are commonly used to enhance $O_2$ and $CO_2$ gas permeability and control moisture around FFV. Such packaging films have the advantage of avoiding accumulation of $CO_2$ and maintaining higher concentration of $O_2$ (Hussein et al., 2015). Perforation of polymeric film is based on a compromise principle, since perforations affect the film's permeability to $O_2$ and $CO_2$ to a higher extent than water vapor transmission. Super atmospheric $O_2$ atmospheres ($O_2 > 70\%$) have been investigated as alternatives to low $O_2$ in MAP for fresh products.

Jacxsens et al. (2001) reported on the effects of super atmospheric $O_2$ on microbial and sensory properties of fresh-cut vegetables and mushrooms. The authors showed that super atmospheric $O_2$ atmospheres (80% and 92%) inhibited the growth of *A. flavus* and *B. cinerea*, prevented anaerobic fermentation, and inhibit enzymatic discoloration. Ayhan and Eştürk (2009) reported increases in antioxidant activity and bacteria counts for minimally processed pomegranate arils (cv. Hicaznar) stored under super atmospheric $O_2$ gas composition (70%). Other studies have explored the use of high or absolute concentration 100% $N_2$ atmospheres in MAP systems, because of the gas' ability to maintain produce quality (Ayhan and Eştürk, 2009; Ahmed et al., 2011). For instance, quality attributes of persimmon fruit packaged in 100% $N_2$ were maintained and shelf life extended, at 0°C for 90 days (Ahmed et al., 2011). Fresh-cut cabbage and lettuce MA-packaged with 100% $N_2$ stored at 1°C and 5°C, respectively, maintained their quality and appearance by the end of storage day 5 (Koseki and Itoh, 2002). Generally, $N_2$ is used to displace $O_2$; thereby, it helps to delay oxidative processes inside the package (Caleb et al., 2013).

## 3.2 High humidity

Conventional polymeric films used in fresh produce packaging often have lower water vapor transmission rates in comparison to the transpiration rates of fresh products. High in-package relative humidity due to water vapor released by horticultural products, if not adequately managed, can result in water condensation on the packaging material and products. This represents a risk to product quality and safety (Bovi et al., 2016; Linke and Geyer, 2013). In addition, excess moisture in packages can have a detrimental impact on powdered/flour products, which could lead to caking and/or softening. Water vapor condensation will occur on the surface of any product and packaging material that is at or below the dew point temperature of the surrounding air (Bovi et al., 2016). There is a critical challenge balancing between low (below required limits) and excessively high humidity. Humidity below required limits could result in excessive water loss and shriveling of fresh commodities; on the other hand, excessively high humidity is a favorable condition for accelerated microbial growth and decay. Hence, the effective management of high humidity in MAP systems for food application is essential.

Various strategies for managing high humidity inside packaged fresh produce have been reported in literature. For instance, micro-perforations are commonly used in fresh produce packaging to enhance mass transfer across the packaging material (Ben-Yehoshua et al., 1998; Almenar et al., 2007; Hussein et al., 2015). The application of moisture absorbers presents another innovative and versatile approach for managing high humidity. The absorbers can be classified into two main application approaches, based on:

**(i)** the use of contact moisture absorbers (Mahajan et al., 2008; Song et al., 2001); and
**(ii)** the use of non-contact moisture absorbers (Bovi et al., 2018; Rux et al., 2015).

Generally, water contact absorbers are commercially used for packaging of meat products, such as fish, beef, and pork (Fang et al., 2017). The different forms of contact absorbers used to regulate moisture in fish and meat foods include pads, superabsorbent polymeric laminate films, and sachets (Ozdemir and Floros, 2004). The use of non-contact moisture absorbers is most suitable for fresh produce application as these products actively respire and transpire, releasing water vapor inside the package headspace in the process (Bovi et al., 2018). In addition, a packaging material with high water vapor permeability or a combination of different polymeric films is used to optimize the in-package humidity (Caleb et al., 2016; Belay et al., 2018; Volpe et al., 2018).

## 3.3 Intelligent and smart MAP

Intelligent and smart MAP is a packaging system that is capable of carrying out intelligent functions (such as detecting, sensing, recording, tracing, communicating, and applying scientific logic) to facilitate decision making to extend shelf life, enhance safety, improve quality, provide information, and warn about possible problems (Yam et al., 2005; Sandhya, 2010). The conceptual framework describing the flow of information in an intelligent and smart system consists of four components: smart package devices, data layers, data processing, and information highway (wire or wireless communication networks) in the food supply chain (Ahvenainen, 2003; Urmila et al. (2015).

Smart package devices are inexpensive labels or tags that are attached onto primary packaging (for example, pouches, trays, and bottles) (Caleb et al., 2012). There are two basic types of smart package devices: data carriers (such as barcode labels and radio frequency identification [RFID] tags) that are used to store and transmit data, and package indicators (such as time–temperature indicators, gas indicators, biosensors) that are used to monitor the external environment and, whenever appropriate, issue warnings (Arvanitoyannis and Stratakos, 2012; Salinas et al., 2012).

In a MAP product, the respiration of the fresh produce, gas flux through the packaging film, gas generation by spoilage microbes, or leakage, may cause a change in gaseous composition within the package (Caleb et al., 2012). Gas indicators in the form of labels or print on the packaging films can help to monitor the safety and quality of the packed produce (Yam et al., 2005). Oxygen indicators are the most frequently used gas indicator, this is due to the ability of oxygen to cause oxidative color change and enhance microbial spoilage. Smiddy et al. (2002) suggested the use of a non-destructive oxygen sensor as a leakage indicator. These sensors were able to detect very low oxygen levels (0.07%), and their function was based on the measurement of phosphorescence intensity and phase shift. On the other hand, a carbon dioxide indicator can be used to detect early spoilage as well as to monitor the levels of carbon dioxide within modified atmosphere packages in transit and within storage facilities (Hong and Park, 2000; Neethirajan et al., 2009; Poyatos-Racionero et al., 2018).

Kuswandi et al. (2013) developed a bromophenol blue based dye to monitor guava (*Psidium guajava* L.) freshness. The dye was immobilized onto a cellulose membrane by absorption, and changes from blue to green were observed when the pH decreased as a consequence of the over-ripening of the guava due to the production of volatile organic compounds. For meat, colorimetric indicators have most often been applied in chicken and pork. Chen et al. (2014) and Urmila et al. (2015) developed an opto-electronic nose composed by pH indicators and metalloporphyrins. The authors developed a system by printing chemically responsive dyes (9 metalloporphyrins and 3 pH indicators (bromocresol green, bromocresol purple, and neutral red)) on a silica-gel flat plate. Salinas et al. (2012) developed an array composed of 16 sensing materials using the combination of different dyes for chicken meat packed in MAP (30% $CO_2$ and 70% $N_2$). Color changes of the array were characteristics of chicken meat aging in MAP. In general, the function of most of these indicators is based on color alteration as a result of a chemical or enzymatic reaction (Ahvenainen, 2003).

## 4 MAP of fresh and minimally processed fruit

Minimally processed fruit are different from intact fruit and vegetables, since processing results in tissue disruption and compromised cell integrity, with a concomitant increase in metabolic and microbiological activity (Chonhenchob et al., 2007). Furthermore, cutting could induce elevated ethylene production rates that could stimulate respiration rate and promote ripening of climacteric fruits, and consequently accelerate microbial growth (Brecht 1995). Studies conducted by Iqbal et al. (2009) demonstrated that fresh-cut fruit and vegetables being sliced, finely shredded, coarsely shredded, cubed, and grated could affect respiration rate and senescence. Minimal processing of fresh produce involves sorting, cleaning, washing, trimming/peeling/deseeding/coring, and cutting (such as chopping, slicing, shredding, chunking, and dicing).

MAP has been applied more extensively to whole fresh fruit and less to fresh, minimally processed fruit. This includes fresh produce such as strawberries (Sanz et al., 1999; Sanz et al., 2000; Almenar et al., 2007), loquat fruit (Amoros et al., 2008), mandarin (Del-Valle et al., 2009), and many other high and medium-respiring produce. However, researchers have recommended various optimal gas compositions for fresh-cut produce (Fig. 2). Cliff et al. (2010) reported on the quality of fresh-cut Gala apple

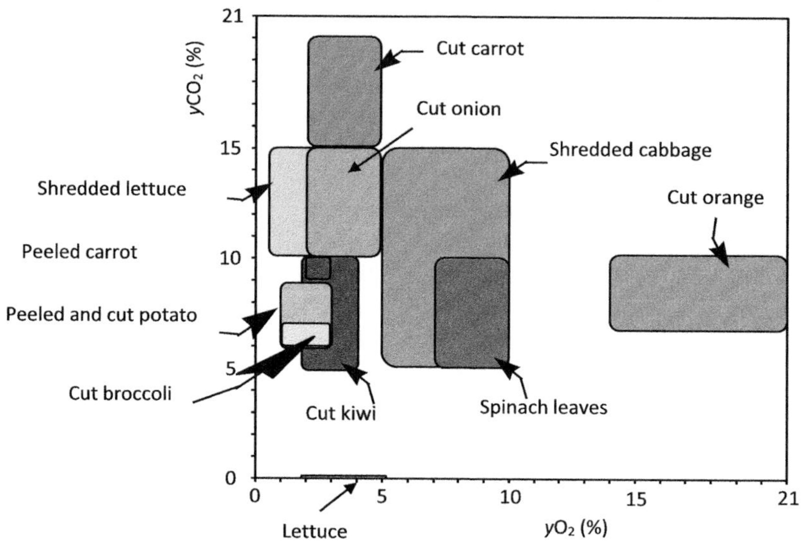


**Fig. 2** Optimal gas composition for fresh-cut produce.

slices stored in micro-perforated MAP systems. They found that packages with a high $O_2$ and high $CO_2$ atmosphere had better fruitiness, sweetness, firmness, and higher fruitiness-by-mouth quality, as compared to the standard solid film (non-perforated) with low $O_2$ and high $CO_2$ atmosphere. Quality attributes of fresh-cut Gala apple slices were best maintained under MAP composition (14% $O_2$ and 7% $CO_2$).

Successful use of MAP to maintain quality and extend shelf life of fresh produce must be accompanied by appropriate storage temperature, use of good quality of produce with minimal physiological damage, and the application of appropriate treatments to reduce microbial spoilage (Krasnova et al., 2012). Various measures can be taken to reduce deterioration of fresh produce, including good agricultural and processing practices (such as harvesting produce at optimum maturity stage and minimizing mechanical injuries), proper sanitation procedures, adherence to HACCP principles, as well as the application of the optimal postharvest treatment (Artés et al., 2009; Weerakkody et al., 2010; Mahajan et al., 2014). This would help to minimize quality deterioration and the risk of microbial contamination in perforated modified atmosphere packages (Boonruang et al., 2012; Oliveira et al., 2012).

## 5 MAP of fresh and minimally processed vegetables

Fresh and minimally processed vegetables are living, respiring, and the edible tissues continue to be metabolically active after harvest. These metabolic and biochemical processes can be influenced by changing the environmental conditions during handling and storage. The specific environmental conditions vary widely, not only between the different types and cultivars, but also within the same cultivar grown in different seasons or locations, as well as between the fresh and processed forms (Jayas and Jeyamkondan,

2002). Therefore, delaying the metabolic processes delays the postharvest quality loss, which is desirable during distribution and for short-term storage before marketing.

Unlike fresh vegetables, minimally processed vegetables are exposed to one or more processes such as trimming, peeling, washing, or slicing. The minimal processing steps induce or accelerate many physiological changes in vegetables, due to the disruption of plant cells and damage to membranes. This disruption increases the availability of nutrients, which in turn accelerates microbial growth, degrades product quality, and limits shelf life. One of the major challenges facing the production and marketing of fresh and minimally processed produce is rapid quality deterioration and reduced shelf life resulting from physiological disorders and presence of mechanical injuries (Hussein et al., 2015). Practical experience has demonstrated that tissues with high respiratory rates and/or low energy reserves have shorter postharvest lives (Rico et al., 2007).

MAP for fresh and minimally processed vegetables, flushed with low $O_2$ (1%–5%) and/or elevated $CO_2$ (5%–10%) levels, maintains quality and consequently extend shelf life (Ghidelli et al., 2018; Zhang et al., 2015a). As presented in Table 4, passive MAP shows a beneficial effect for fresh-cut Athena cantaloupe cubes, retaining saleable quality for 9 days at 5°C, whereas, active MAP (4% $O_2$ and 10% $CO_2$) maintained quality better than passive MAP with better color retention and reduced translucency, RR, and microbial populations. For Parthenon broccoli, low $O_2$ MAP containing 10% $O_2$ and 5% $CO_2$, reduced loss of quality parameters (overall appearance, odor, weight loss and color) and the decrease of functional compounds contents (chlorophyll and carotenoid pigments, vitamin C, total phenol content, and intact glucosinolates), compared to storage in air, at the end of storage (12 days) (Fernández-león et al., 2013). However, storage of some vegetables, such as spinach (Tudela et al., 2013) and fresh cut egg plants (Ghidelli et al., 2018) at low $O_2$ concentration and high $CO_2$ (1%–5% and 11%–15%, respectively) accelerated decay and accumulation of fermentative metabolites. For most vegetables, increasing $CO_2$ concentration significantly increases

**Table 4** Recommended gas mixture for various fresh and minimally processed vegetables during modified atmosphere packaging

| Commodity | Storage atmosphere ($O_2$:$CO_2$) (%) | Storage temperature (°C) |
|---|---|---|
| Broccoli | 2–3: 6–7 | 0–5 |
| Shredded cabbage | 5–7.5:15 | 0–5 |
| Sliced carrots | 2–5:15–20 | 0–5 |
| Chopped butterhead lettuce | 1–3:5–10 | 0–5 |
| Chopped green leaf lettuce | 0.5–3:5–10 | 0–5 |
| Chopped red leaf lettuce | 0.5–3:10–15 | 0–5 |
| Chopped romaine lettuce | 0.5–3:5–10 | 0–5 |
| Sliced red onion | 2–5:10–15 | 0–5 |
| Diced peppers | 3:5–10 | 0–5 |
| Sliced or whole peeled potato | 1–3:6–9 | 0–5 |
| Sliced zucchini | 0.25–1:0 | 5 |

Adapted from Oliveira, F., Sousa-Gallagher, M.J., Mahajan, P.V., and Teixeira, J.A., 2012. Development of shelf-life kinetic model for modified atmosphere packaging of fresh sliced mushrooms. J. Food Eng. 111(2), 466–473.

**Table 5** Types of gases in modified atmosphere packaging and their effect on fresh and minimally processed vegetables

| Types of gases | | General effects | References |
|---|---|---|---|
| Effect of carbon dioxide | Absence of $CO_2$ | *Advantages:* Reduce degradation of chlorophyll, which is related to greenness of harvest vegetables | Tudela et al. (2013) |
| | High $CO_2$ | *Advantages:* bacteriostatic effect, results inhibition of Gram-negative bacteria, such as *Pseudomonas spp.*, or *Enterobacteriaceae* | Scifò et al. (2009) and Goulas et al. (2005) |
| Effect of oxygen | Low $O_2$ | *Advantages:* with or without $CO_2$ can reduce the number of psychrophiles and Pseudomonas microorganisms $CO_2$ *Disadvantages:* off odor development | Tudela et al. (2013) |
| | Super atmospheric $O_2$ | *Advantages:* - inhibits microbial growth and decay - reduce deterioration of fresh processed vegetables and proliferation of aerobic spoilage microorganisms | Belay et al. (2018) |

tissue damage, ammonia production, and darkening of tissues, and decreases protein content (La Zazzera et al., 2010; Albornoz and Cantwell, 2015).

Recently, the application of super atmospheric $O_2$ (>70%) has shown beneficial effects, particularly for controlling anaerobic microorganisms (Table 5), preventing anaerobic fermentation and controlling enzymatic browning. Super atmospheric $O_2$ MAP (>85% $O_2$) showed a beneficial effect for fresh-cut red chard baby leaves by lowering the natural microflora growth throughout 7 days at 5°C (Tomás-callejas et al., 2011). The effect of $O_2$ concentrations reducing microbial growth can be related to the accumulation of reactive oxygen species that damage vital cell components, affecting cellular antioxidant protection systems of cell metabolism (Kader and Ben-Yehoshua, 2000).

Another advantage of MAP is providing desired RH inside the package when appropriate permeable packaging film is selected; this helps to prevent weight loss and subsequent change in appearance. MAP significantly controls the weight loss and wilting of broccoli for 28 days by using micro-perforated polypropylene and non-perforated polypropylene films (Serrano et al., 2006). Similarly, MAP techniques (7% and 9% $CO_2$) using OPP as amide-PE allowed extending the shelf life of kohlrabi sticks to 14 days at 0°C by reducing the RR and $C_2H_4$ production, whereas, sticks in perforated bags with a poor appearance reduced the shelf life (Escalona et al., 2007). Weight loss of pepper was reduced by 20% using perforated polyethylene packages at 8°C, 14°C, and 20°C, compared to unpacked pepper (Lownds et al., 1994).

# 6 MAP of fresh and minimally processed mushroom

Mushroom is a fruiting body, mostly above ground, of higher fungi formed from spacious underground mycelia (hyphae) by the process of fructification (Kalač, 2009). Mushrooms are a rich source of nutrients, particularly proteins and minerals, as well as vitamins B, C, and D, and contain bioactive constituents such as phenolic compounds, terpenes, steroids, and polysaccharides, which have various biological activities (Lin et al., 2017). Mushrooms are also known to exhibit antifungal, anti-inflammatory, antitumor, antiviral, antibacterial, hepatoprotective, anti-diabetic, hypolipedemic, antithrombotic, and hypotensive activities (Rathore et al., 2017). However, they are highly perishable, with a shelf life of 1–3 days at ambient temperature. Due to high metabolic activity and RR (Ares et al., 2007), overall structure (Oliveira et al., 2012), high moisture content, and high TR (Mahajan et al., 2008), causing difficulties in their distribution and marketing as fresh produce (Antmann et al., 2008). Loss of quality for mushrooms include browning, softening, cap development, off flavor, and secondary mold growth (Kim et al., 2006).

MAP is the simplest, most economical, and effective way of extending the shelf life of fresh mushrooms (Kim et al., 2006). As summarized in Table 6, various studies report MAP containing low $O_2$ and high $CO_2$ for different mushroom cultivars, and the

**Table 6** Recommended modified atmosphere packaging conditions for different mushroom cultivars

| Mushrooms cultivars | Storage temperature (°C) | Storage atmosphere ($O_2$:$CO_2$) (%) | References |
|---|---|---|---|
| Oyster mushrooms (*Pleurotus ostreatus*) | 4 | 15:5 | Villaescusa and Gil (2003) |
| White mushrooms | 4 | 5:4 | Tao et al. (2006) |
| Kohlrabi Mushroom | 5 | 21:0 | Escalona et al. (2007) |
| Shiitake mushrooms (*Lentinus edodes*) | 4 | 21:0 | Li et al., 2014 |
| Shiitake mushrooms (*Lentinus edodes*) | 4 | 2:10–13 | Ye et al. (2012) |
| Button mushrooms (*Agaricus bisporus*) | 0 and 5 | 3–5:<12 | Oliveira et al. (2012) |

recommended MAP condition for mushroom includes 3%–5% $O_2$ and <12% $CO_2$ (Ares et al., 2007; Oliveira et al., 2012). On the other hand, the deterioration of sensory characteristics of shiitake mushroom stored under both passive and active (5% $O_2$ and 2.5% $CO_2$) atmospheres using polyethylene and polypropylene at 5°C has been reported by Ares et al. (2007). For button mushroom, high $CO_2$ (95%–100%) treatment significantly reduced browning index and increased antioxidant ability, maintained flavor, quality, and consumer acceptance during postharvest storage (Lin et al., 2017). Using super $O_2$ atmospheric (>70%) concentration effectively inhibited enzymatic browning and preserving microbial quality of shredded Agaricus mushrooms has been reported by Jacxsens et al. (2001). Similarly, MAP with 50% and 100% $O_2$ concentration inhibited ethanol production, whereas active MAP with 3% $O_2$ and 5% $CO_2$ and passive MAP induced anaerobic fermentation for shitake mushroom stored at 10°C (Li et al., 2014).

One major effect of MAP for mushrooms is reducing weight loss by reducing water transpiration rate and $CO_2$ loss during respiration (Jiang et al., 2011). The weight loss of shiitake mushrooms significantly reduced by MAP compared to unpacked mushrooms (Ye et al., 2012). Even if MAP has been used for maintaining quality of mushrooms, one challenge of implementing MAP for mushrooms has been reported as the accumulation of moisture inside the package. Mushrooms are very sensitive to humidity levels; they have very high TR—over >90% of the weight at harvest is water—and they lack a skin as a barrier to diffusion (Mahajan et al., 2008). Therefore, the use of perforated packaging or moisture absorbers has been recommended in order to avoid quality deterioration (Antmann et al., 2008).

Rux et al. (2015) reported that mushrooms stored at 100% RH lost moisture at the rate of 0.03–0.22 $mg\,kg^{-1}\,s^{-1}$. Thus, considering the nature of mushrooms, proper selection and maintenance of temperatures, proper internal humidity, and optimum atmosphere are the most important parameters for extending the shelf life of mushrooms. Optimum MAP design by using perforated packaging material has been reported for button mushrooms by Oliveira et al. (2012) and for Shiitake mushroom by Antmann et al. (2008). Oliveira et al. (2012) recommended 1 (58 perforations per $m^2$) at 0°C, 1 (58 perforations per $m^2$) at 5°C, 3 (174 perforations per $m^2$) at 10°C and 6 (349 perforations per $m^2$) at 15°C in order to avoid exceeding the recommended limit for $CO_2$ concentration. Comparing the effect of active and passive MAP, active MAP (15% and 25% $O_2$) led to a smaller reduction of shiitake mushroom deterioration than passive MAP in films which have 17 perforations/$m^2$, 0.1 $mm^2$ surface than $9.0 \times 103$ perforations/$m^2$, 0.1 $mm^2$ surface film at 5°C (Antmann et al., 2008). The use of MAP accompanied with low temperature storage can effectively retard the quality changes and extend shelf life of fresh-cut mushrooms (Kim et al., 2006; Mahajan et al., 2008).

## 7 MAP of meat and fishery products

For meat and aquatic food products, elimination of $O_2$ from the MAP and enriching it with high $CO_2$ concentration helps to reduce microbial growth and oxidation of fat and retain color (Arvanitoyannis et al., 2011; Łopacka et al., 2017). The efficacy of

MAP in prolonging the shelf life of packaged meat and aquatic products mainly relies on the antimicrobial properties of $CO_2$ present inside the package. The presence of $CO_2$ in the headspace of the packages inhibits microbial growth and causes a change in the microbial content to bacteria with lower spoilage capacity (McMillin, 2008; Fernández et al., 2009; Sivertsvik et al., 2002). In some cases, additional gases are used in combination with the above-mentioned gases, such as nitrous, nitric oxides, ethane, chlorine, carbon monoxide, and sulfur dioxide, to inhibit microbial growth and prevent oxidation. Moreover, helium, argon, xenon, and neon have also been used under MAP because they are very inert and serve well as filler gases. Studies on muscle meat showed that a $CO_2$ range of 10% to 80% in a package enhances its bacteriostatic effect (Kerry et al., 2006). Therefore, the inclusion of $CO_2$ discharging sachets is beneficial in such systems.

Aquatic products are highly perishable, because of their high water activity, neutral pH, and presence of autolytic enzymes, and their deterioration is primarily because of bacterial action. Typical shelf life under icing and refrigerated storage conditions ranges from 2 to 14 days (Sivertsvik et al., 2002). The spoilage of fresh aquatic products is usually microbial; however, in some cases chemical changes, such as auto-oxidation or enzymatic hydrolysis of the lipid fraction, may result in off odors and flavors (Fernández et al., 2010). Moreover, fish and seafoods normally have a particularly heavy microbial load, owing to the method of capture and transport to shore, slaughtering methods, evisceration, and the retention of skin in the retail portions. Microbial activity causes the breakdown of fish protein and trimethylamine oxide (TMAO), with a resulting release of undesirable fishy odors (DeWitt and Oliveira, 2016).

Various active MAPs containing low $O_2$ and high $CO_2$ for aquatic food products have been shown to inhibit the normal spoilage flora and increase shelf life significantly (Table 7). Under MAP conditions, the growth of spoilage microorganisms is inhibited and the shelf life of fish can be extended by 1.5 to 2-fold compared with chilled storage in normal air (Fernández et al., 2010; Zhang et al., 2015b). However, the shelf life of fish products in MAP can be extended depending on raw material, temperature, gas mixtures, and packaging materials (Goulas et al., 2005). To avoid growth of anaerobes, the combination of $CO_2/N_2$ for fat fish products, and $CO_2/N_2/O_2$ for low-fat fish, is used (Zhang et al., 2015b). Most significantly, $CO_2$ is the most important gas used in MAP of aquatic products, because of its bacteriostatic and fungistatic properties. It inhibits growth of many spoilage bacteria and the inhibition is increased with increased $CO_2$ concentration in the atmosphere (Sivertsvik et al., 2002). Its inhibitory effect is further related to its high solubility in both water-phase and lipids of muscle foods, which is further affected by temperature, where the lower the temperature, the higher the solubility. The sensitivity of bacteria in aquatic products to $CO_2$ vary (Fernández et al., 2010; Özogul et al., 2002).

In addition, MAP has been shown to maintain the texture, odor, and overall sensory quality and appearance of aquatic products. For fish and seafood, most gas mixtures do not include $O_2$ (Table 4). This may be explained by the high rate of perishability of seafood, which results from psychrophilic spoilage bacteria growth, combined with muscle degradation by endogenous enzymes, and oxidative deterioration of the lipids (Torrieri et al., 2011). On the contrary, high $O_2$ concentrations, in combination with

**Table 7** Modified atmosphere packaging conditions in aquatic products

| Aquatic food products | Storage Temperature (°C) | Storage atmosphere ($O_2$:$CO_2$) (%) | Shelf life (days) |
|---|---|---|---|
| Salmon fillets | 4 | 0:60 | 14 |
| Sea bream | 4 | 30:40 | 27–28 |
| Chinese shrimp | 2±1 | 30:40 | 17 |
| Fresh sea bass fillets | 4 | 0: 60 | 14 |
| Atlantic salmon | 2±2 | 0: 75 | >28 |
| Striped red mullet | 1 | 0: 50 | 9–10 |
| Tuna | 2 | 60: 40 | >14 |
| Chub mackerel | 4 | 5: 50 | 12 |
| Cod loins | −2 | 5: 70 | 21 |
| Scallops | 0 | 10: 60 | 14 |
| Rainbow trout fillets | 4 | 2.5:7.5 | 14 |
| Cod loins | 1.5 | 5:50 | 14 |
| Abalone | 2±1 | 30:40 | 15 |
| Sardines | 2±2 | 0: 60 | 14 |
| Sea bass slices | 4 | 0: 80–100 | 20 |
| Mediterranean Swordfish | 4 | 0: 80–100 | 18 |
| Nile tilapia | 1 | 0: 50 | 23 |
| Cod Atlantic | 0 | 50: 50C | 11 |
| Cod | 2 | 0: 60 | 17 |
| Bass (gutted) | 3 | 30:50 | 9 |
| Dolphinfish | 1 | 5:45 | 18 |
| Halibut (fillets) | 4 | 0:50 | 23 |
| Mullet, red striped | 1 | 0:50 | 24 |
| Shrimp, brown | 4 | 0:40 | 7 |

From Zhang et al., 2015a; Fernández et al., 2009; Özogul et al., 2002, Goulas et al., 2005; DeWitt and Oliveira, 2016.

$CO_2$, are reported to preserve the red color of haem pigments (DeWitt and Oliveira, 2016). Similarly, Sivertsvik (2007) reported significant improvement in the overall quality of MAP pre-rigor farmed Atlantic cod fillets by the use of high oxygen content in a gas mixture of $CO_2$ and $O_2$. MAP for aquatic products also contributes to reduction of histamine formation than normal air packages (Özogul et al., 2002). The importance of estimating the concentration of histamine in fish and fish products is related to its impact on human health and food quality. Fish containing relatively high concentrations of histamine can cause poisoning or allergic reactions when consumed by some individuals (DeWitt and Oliveira, 2016).

Another important factor in MAP of aquatic food products is storage temperature, since rate of deterioration is temperature-dependent and can be inhibited by the use of low storage temperature (e.g., fish stored on ice). However, the single most important concern with the use of MAP products is the potential for the outgrowth and toxin production by the non-proteolytic, *Clostridium botulinum* type E, which can grow at low temperatures (Özogul et al., 2002). Therefore, MAP can be combined with super chilling processes to extend further the shelf life and safety of fresh fish and seafood. In this technique, also known as partial freezing, the temperature is reduced to 1–2°C below the initial freezing point without allowing partial freezing (Zhang et al., 2015b). As can be seen in Table 5, most of the studies in the literature recommended low temperature (<5°C) storage of various aquatic products coupled with MAP for self-life extension. However, only high quality fish and seafood should be used to benefit from the MAP application. The desirable quality and shelf life of these products under MAP conditions depends on the species, fat content, initial microbial load, gas mixture applied, and storage temperature. Therefore, for ensuring the safety and extending shelf life of fish and seafood, maintenance of recommended chilling temperature and atmosphere with good hygiene practice during the supply chain is essential.

## 8 Microbiological and other consequences of MAP

Microorganisms such as *Pseudomonas* spp., *Erwinia herbicola, Flavobacterium, Xanthomonas, Enterobacter agglomerans, Lactobacillus* spp., *Leuconostoc mesenteroides*, molds and yeasts are largely associated with spoilage of horticultural commodities (Caleb et al., 2013). The effectiveness of MAP to maintain safety of fresh and fresh-cut produce is mostly combined with optimal cold chain. Studies have shown that MAP could be used effectively to maintain the quality of fresh and fresh-cut fruit and vegetables (Kader and Watkins, 2000; Yahia, 2006). In contrast, the effect of MAP on spoilage microorganisms can vary depending on the type of produce packaged, the initial microbial load, and other extrinsic factors (Caleb et al., 2013).

For instance, Francis and O'Beirne, 1998suggested that MAP conditions of 3% $O_2$ and 5%–20% $CO_2$ might increase the growth rate of *L. monocytogenes*. However, they observed that at higher $CO_2$ concentrations, the growth of lactic acid bacteria increased and inhibited the growth of *L. monocytogenes*. Similarly, Bourke and O'beirne (2004) investigated the effects of packaging type, atmospheric composition, and storage temperatures on survival and growth of *Listeria* spp. in shredded dry coleslaw and its components. The authors observed a decline in *Listeria* spp. in shredded dry coleslaw at 3°C between days 10 and 12, while growth remained at initial inoculation levels under 8°C. The inhibition of *Listeria* spp. was attributed to the optimal gas composition between days 10 and 12, and the presence of competitive microflora. Artin et al. (2008) reported that $CO_2$ concentration had a significant impact on neurotoxin gene expression and neurotoxin formation in non-proteolytic *C. botulinum* type E. The authors showed that the expression of cntE mRNA and the formation of extracellular neurotoxin were 2-fold higher with a headspace $CO_2$ concentration of 70% (vol/vol) in comparison to headspace of 10% (vol/vol). A follow up study by Artin et al. (2010) reported that non-proteolytic *C. botulinum* differ from proteolytic *C. botulinum* type A1 ATCC 3502. They observed that $CO_2$ concentration had little effect on the expression of cntE mRNA and the formation of extracellular neurotoxin, and the expression of

*C. botulinum* type A1 ATCC 3502 neurotoxin cluster genes was dependent on growth phase rather than $CO_2$ concentration.

Generally, $CO_2$ is the main gas component used in MAP systems that confers a significant measure of antimicrobial effect against spoilage/pathogenic microbes. Various theories to explain the mechanism of its antimicrobial have been suggested over the years; these include direct inhibition of the enzymatic reactions; alteration of cell membrane function such as uptake and absorption of nutrient; $CO_2$ dissolving into bacterial membranes leading to decrease in intracellular pH; and direct changes in the physical and chemical properties of proteins. The inhibitory effect of $CO_2$ is overall dependent on the types of natural microbial population present and the characteristics of the fresh produce (such as topology, texture, and macro/micro pores). Furthermore, food-associated pathogens such as *C. perfringens*, *C. botulinum*, and *L. monocytogenes* are minimally affected by $CO_2$ concentration below 50% (Charles et al., 2003; Farber et al., 2003), which is detrimental to fruit tissue. Hence, the use of $CO_2$ is most effective on produce with aerobic and psychrotrophic Gram-negative bacteria. Adequate background knowledge of natural microflora for each MA-packaged produce is essential for a successful MAP design.

Extensive literature has shown MAP to control produce respiration rate and ethylene production and maintain high quality fresh fruit and vegetables. However, there are critical concerns on the management of microorganisms, which may be stimulated with moderate $CO_2$ concentration inside packages. Especially under inconsistent storage or transit temperature conditions, packaging material that performs optimally at a given temperature may default when respiration rate of produce does not correlate with packaging film permeability at higher temperature. Furthermore, it has been established that erratic temperature contributes to fresh produce accelerated deterioration, increased metabolic rate, which can enhance microbial proliferation in fresh or minimally processed MAP. Hence, the importance of maintaining a strict cold chain along the whole distribution continuum cannot be over emphasized.

# 9 Future prospects for MAP in food preservation

The global market for year-round supply of fresh, fresh-cut, and minimally processed produce has continued to rise, exceeding the financial value of the market for cereals and other agricultural products since the late the 1980s. Changes in consumer preference for fresh and safe foods has led to innovations in packaging technologies to deliver value to consumers. This trend assures the demand for modified atmosphere packaging (MAP) and related technologies for the handling and marketing of these commodities.

Modified atmosphere packaging (MAP) has been applied in the food industry for over 90 years to extend shelf life, and maintain quality and safety of fresh and fresh-cut produce. Recently, MAP has experienced a rapid development in both scientific research and industrial applications as a leading tool for packaging high value fresh and fresh-cut produce. MAP technologies have been applied to manage the postharvest quality and marketing of a wide range of commodities including fresh and fresh-cut fruits, vegetables, mushrooms, meat, and aquatic products. With foundations in passive and active modified atmospheres, more recent MAP systems include high-oxygen MAP, controlled MAP, and intelligent MAP. Passive MAP relies on applying knowledge of the produce respiration and permeability of the packaging material to determine the equilibrium atmosphere inside the MAP. On the other hand, active MAP refers to the incorporation of additives into the package with the aim of maintaining or extending the product quality and shelf life. Intelligent MAP systems incorporate technologies that monitor the condition of packaged food to give information regarding the quality during handling, transport, and storage.

The combination of advances in technological innovations and changing consumer demand have created push-pull factors that propel the demand for MAP in fresh, fresh-cut and, minimally processed foods. Developments in information and communication technologies for real time data and image capture, analysis, and transmission provide new possibilities, with better interaction between the consumer, product, and supply chains. The incorporation of biosensors into MAP for rapid detection and diagnosis of biological hazards offer new prospects for just-in-time management of food safety risks, including contamination. Drip loss continues to be a problem affecting fresh-cut MA packaged produce; in combination with MAP, this calls for the application of hurdle technologies such as edible coating to control the problem. Further research is warranted to improve our understanding of the impact of short-term breaks in the cold chain on the quality and safety of MA packaged produce.

## Acknowledgments

This work is based on the research supported wholly by the National Research Foundation of South Africa (Grant Number: 64813). The opinions, findings, and conclusions or recommendations expressed are those of the author(s) alone, and the NRF accepts no liability whatsoever in this regard.

## References

Ahmed, M., Yousef, A.R., Sarrwy, S.M.A., 2011. Modified atmosphere packaging for maintaining quality and shelf life extension of persimmon fruits. Asian J. Agric. Sci. 3 (4), 308–316.

Ahvenainen, R., 2003. Active and intelligent packaging: an introduction. In: Ahvenainen, R. (Ed.), Novel Food Packaging Techniques. Woodhead, Cambridge, pp. 5–21.

Albornoz, K., Cantwell, M.I., 2015. Fresh-cut kale quality and shelf-life in relation to leaf maturity and storage temperature. In: Proceedings of III International Conference on Fresh-Cut Produce: Maintaining Quality and Safety. Vol. 1141. pp. 109–116.

Almenar, E., Del-Valle, V., Hernández-Muñoz, P., Lagarón, J.M., Catalá, R., Gavara, R., 2007. Equilibrium modified atmosphere packaging of wild strawberries. J. Sci. Food Agric. 87 (10), 1931–1939.

Amoros, A., Pretel, M.T., Zapata, P.J., Botella, M.A., Romojaro, F., Serrano, M., 2008. Use of modified atmosphere packaging with microperforated polypropylene films to maintain postharvest loquat fruit quality. Food Sci. Technol. Int. 14 (1), 95–103.

Antmann, G., Lema, P., Lareo, C., 2008. Influence of modified atmosphere packaging on sensory quality of shiitake mushrooms. Postharvest Biol. Technol. 49, 164–170.

Ares, G., Lareo, C., Lema, P., 2007. Modified atmosphere packaging for postharvest storage of mushrooms. A Review. Fresh Produce 1 (1), 32–40.

Artés, F., Gómez, P., Aguayo, E., Escalona, V., Artés-Hernández, F., 2009. Sustainable sanitation techniques for keeping quality and safety of fresh-cut plant commodities. Postharvest Biol. Technol. 51 (3), 287–296.

Artin, I., Carter, A.T., Holst, E., Lövenklev, M., Mason, D.R., Peck, M.W., Rådström, P., 2008. Effects of carbon dioxide on neurotoxin gene expression in nonproteolytic *Clostridium botulinum* type E. Appl. Environ. Microbiol. 74 (8), 2391–2397.

Artin, I., Mason, D.R., Pin, C., Schelin, J., Peck, M.W., Holst, E., Carter, A.T., 2010. Effects of carbon dioxide on growth of proteolytic *Clostridium botulinum*, its ability to produce neurotoxin, and its transcriptome. Appl. Environ. Microbiol. 76 (4), 1168–1172.

Arvanitoyannis, I.S., Bouletis, A.D., Papa, E.A., Gkagtzis, D.C., Hadjichristodoulou, C., Papaloucas, C., 2011. Anaerobe The effect of addition of olive oil and "Aceto balsamico di Modena" wine vinegar in conjunction with active atmosphere packaging on the microbial and sensory quality of "Lollo Verde" lettuce and rocket salad. Anaerobe 17 (6), 303–306.

Arvanitoyannis, I.S., Stratakos, A.C., 2012. Application of modified atmosphere packaging and active/smart technologies to red meat and poultry: a review. Food Bioproc. Tech. 5 (5), 1423–1446.

Ayhan, Z., Eştürk, O., 2009. Overall quality and shelf life of minimally processed and modified atmosphere packaged "ready-to-eat" pomegranate arils. J. Food Sci. 74 (5), 399–405.

Bai, J., Saftner, R.A., Watada, A.E., 2003. Characteristics of fresh-cut honeydew (*Cucumis xmelo* L.) available to processors in winter and summer and its quality maintenance by modified atmosphere packaging. Postharvest Biol. Technol. 28, 349–359.

Beaudry, R.M., Cameron, A.C., Shirazi, A., Dostal-Lange, D.L., 1992. Modified-atmosphere packaging of blueberry fruit: effect of temperature on package $O_2$ and $CO_2$. J. Am. Soc. Hortic Sci. 117 (3), 436–441.

Belay, Z.A., Caleb, O.J., Mahajan, P.V., Opara, U.L., 2018. Design of active modified atmosphere and humidity packaging (MAHP) for 'wonderful' pomegranate arils. Food Bioproc. Tech. 1–17.

Ben-Yehoshua, S., Rodov, V., Peretz, J., 1998. Constitutive and induced resistance of citrus fruit against pathogens. In: Disease Resistance in Fruit: Proceedings of an International Workshop held at Chiang Mai, Thailand. Australian Centre for International Agriculture Research, Canberra. 18–21 May 1997. 1998.

Boonruang, K., Chonhenchob, V., Singh, S.P., Chinsirikul, W., Fuongfuchat, A., 2012. Comparison of various packaging films for mango export. Pack. Technol. Sci. 25 (2), 107–118.

Bovi, G.G., Caleb, O.J., Ilte, K., Rauh, C., Mahajan, P.V., 2018. Impact of modified atmosphere and humidity packaging on the quality, off-odour development and volatiles of 'Elsanta' strawberries. Food Pack. Shelf Life 16, 204–210.

Bovi, G.G., Caleb, O.J., Linke, M., Rauh, C., Mahajan, P.V., 2016. Transpiration and moisture evolution in packaged fresh horticultural produce and the role of integrated mathematical models: a review. Biosyst. Eng. 150, 24–39.

Bourke, P., O'beirne, D., 2004. Effects of packaging type, gas atmosphere and storage temperature on survival and growth of *Listeria spp.* in shredded dry coleslaw and its components. Int. J. Food Sci. Technol. 39 (5), 509–523.

Brecht, J., 1995. Physiology of lightly processed fruits and vegetables. HortScience 30 (1), 18–22.

Brody, A.L., Strupinsky, E.P., Kline, L.R., 2001. Active Packaging for Food Applications. CRC press.

Caleb, O.J., Opara, U.L., Witthuhn, C.R., 2012. Modified atmosphere packaging of pomegranate fruit and arils: a review. Food Bioproc. Tech. 5 (1), 15–30.

Caleb, O.J., Mahajan, P.V., Al-Said, F.A.J., Opara, U.L., 2013. Modified atmosphere packaging technology of fresh and fresh-cut produce and the microbial consequences—a review. Food Bioproc. Tech. 6 (2), 303–329.

Caleb, O.J., Ilte, K., Fröhling, A., Geyer, M., Mahajan, P.V., 2016. Integrated modified atmosphere and humidity package design for minimally processed Broccoli (*Brassica oleracea* L. var. italica). Postharvest Biol. Technol. 121, 87–100.

Charles, F., Sanchez, J., Gontard, N., 2003. Active modified atmosphere packaging of fresh fruits and vegetables: modeling with tomatoes and oxygen absorber. J. Food Sci. 68 (5), 1736–1742.

Chen, Q., Hui, Z., Zhao, J., Ouyang, Q., 2014. Evaluation of chicken freshness using a low-cost colorimetric sensor array with AdaBoost–OLDA classification algorithm. LWT-Food Sci. Technol. 57 (2), 502–507.

Chonhenchob, V., Chantarasomboon, Y., Singh, S.P., 2007. Quality changes of treated fresh-cut tropical fruits in rigid MAP containers. Packag. Technol. Sci. 20 (1), 27–37.

Cliff, M.A., Toivonen, P.M., Forney, C.F., Lu, C., 2010. Quality of fresh-cut apple slices stored in solid and micro-perforated film packages having contrasting $O_2$ headspace atmospheres. Postharvest Biol. Technol. 58 (3), 254–261.

Costa, C., Lucera, A., Conte, A., Mastromatteo, M., Speranza, B., Antonacci, A., Del Nobile, M.A., 2011. Effects of passive and active modified atmosphere packaging conditions on ready-to-eat table grape. J. Food Eng. 102 (2), 115–121.

Del-Valle, V., Hernández-Muñoz, P., Catalá, R., Gavara, R., 2009. Optimization of an equilibrium modified atmosphere packaging (EMAP) for minimally processed mandarin segments. J. Food Eng. 91 (3), 474–481.

DeWitt, C., Oliveira, A., 2016. Modified atmosphere systems and shelf life extension of fish and fishery products. Foods 5 (3), 48.

Escalona, V.H., Aguayo, E., Artés, F., 2007. Quality Changes of fresh-cut kohlrabi sticks under modified atmosphere packaging. J. Food Sci. 72 (5), S303–S307.

Fang, Z., Zhao, Y., Warner, R.D., Johnson, S.K., 2017. Active and intelligent packaging in meat industry. Trends Food Sci. Technol. 61, 60–71.

Farber, J.N., Harris, L.J., Parish, M.E., Beuchat, L.R., Suslow, T.V., Gorney, J.R., Busta, F.F., 2003. Microbiological safety of controlled and modified atmosphere packaging of fresh and fresh-cut produce. Compr. Rev. Food Sci. Food Saf. 2, 142–160.

Fernández-león, M.F., Fernández-león, A.M., Lozano, M., Ayuso, M.C., Amodio, M.L., Colelli, G., 2013. Retention of quality and functional values of broccoli 'Parthenon' stored in modi fi ed atmosphere packaging. Food Control 31 (2), 302–313.

Fernández, K., Aspe, E., Roeckel, M., 2009. Shelf-life extension on fillets of Atlantic Salmon (*Salmo salar*) using natural additives, superchilling and modified atmosphere packaging. Food Control 20 (11), 1036–1042.

Fernández, K., Aspé, E., Roeckel, M., 2010. Scaling up parameters for shelf-life extension of Atlantic Salmon (*Salmo salar*) fillets using superchilling and modified atmosphere packaging. Food Control 21 (6), 857–862.

Fonseca, S.C., Oliveira, F.A., Brecht, J.K., 2002. Modelling respiration rate of fresh fruits and vegetables for modified atmosphere packages: a review. J. Food Eng. 52 (2), 99–119.

Fonseca, S.C., Oliveira, F.A., Brecht, J.K., Chau, K.V., 2005. Influence of low oxygen and high carbon dioxide on shredded Galega kale quality for development of modified atmosphere packages. Postharvest Biol. Technol. 35 (3), 279–292.

Francis, G.A., O'Beirne, D., 1998. Effects of the indigenous microflora of minimally processed lettuce on the survival and growth of Listeria innocua. Int. J. Food Sci. Technol. 33 (5), 477–488.

Ghidelli, C., Pérez-gago, M.B., Ghidelli, C., 2018. Recent advances in modified atmosphere packaging and edible coatings to maintain quality of fresh-cut fruits and vegetables quality of fresh-cut fruits and vegetables. Crit. Rev. Food Sci. Nutr. 58 (4), 662–679.

González-Buesa, J., Page, N., Kaminski, C., Ryser, E.T., Beaudry, R., Almenar, E., 2014. Effect of non-conventional atmospheres and bio-based packaging on the quality and safety of Listeria monocytogenes-inoculated fresh-cut celery (*Apium graveolens* L.) during storage. Postharvest Biol. Technol. 93, 29–37.

Goulas, A.E., Chouliara, I., Nessi, E., Kontominas, M.G., Savvaidis, I.N., 2005. Microbiological, biochemical and sensory assessment of mussels (*Mytilus galloprovincialis*) stored under modified atmosphere packaging. J. Appl. Microbiol. 98 (3), 752–760.

Hong, S.-I., Park, W.-S., 2000. Use of colour indicators as an active packaging system for evaluating kimchi fermentation. J. Food Eng. 46 (1), 67–72.

Horev, B., Sela, S., Vinokur, Y., Gorbatsevich, E., Pinto, R., Rodov, V., 2012. The effects of active and passive modified atmosphere packaging on the survival of *Salmonella enterica* serotype Typhimurium on washed romaine lettuce leaves. Food Res. Int. 45 (2), 1129–1132.

Hussein, Z., Caleb, O.J., Linus, U., 2015. Perforation-mediated modified atmosphere packaging of fresh and minimally processed produce—a review. Food Pack. Shelf Life 6, 7–20.

Iqbal, T., Rodrigues, F.A.S., Mahajan, P.V., Kerry, J.P., 2009. Mathematical modelling of the influence of temperature and gas composition on the respiration rate of shredded carrots. J. Food Eng. 91, 325–332.

Jacxsens, L., Devlieghere, F., De Rudder, T., Debevere, J., 2000. Designing equilibrium modified atmosphere packages for fresh-cut vegetables subjected to changes in temperature. LWT-Food Sci. Technol. 33 (3), 178–187.

Jacxsens, L., Devlieghere, F., Van der Steen, C., Debevere, J., 2001. Effect of high oxygen modified atmosphere packaging on microbial growth and sensorial qualities of fresh-cut produce. Int. J. Food Microbiol. 71 (2–3), 197–210.

Jayas, D.S., Jeyamkondan, S., 2002. PH—postharvest technology: modified atmosphere storage of grains meats fruits and vegetables. Biosyst. Eng. 82 (3), 235–251.

Jiang, T., Zheng, X., Li, J., Jing, G., Cai, L., Ying, T., 2011. Integrated application of nitric oxide and modified atmosphere packaging to improve quality retention of button mushroom (*Agaricus bisporus*). Food Chem. 126 (4), 1693–1699.

Kader, A.A., Ben-Yehoshua, S., 2000. Effects of super atmospheric oxygen levels on postharvest physiology and quality of fresh fruits and vegetables. Postharvest Biol. Technol. 20 (1), 1–13.

Kader, A.A., Watkins, C.B., 2000. Modified atmosphere packaging—toward 2000 and beyond. HortTechnology 10 (3), 483–486.

Kalač, P., 2009. Chemical composition and nutritional value of European species of wild growing mushrooms: a review. Food Chem. 113 (1), 9–16.

Kartal, S., Aday, M.S., Caner, C., 2012. Postharvest biology and technology use of microperforated films and oxygen scavengers to maintain storage stability of fresh strawberries. Postharvest Biol. Technol. 71, 32–40.

Kerry, J.P., O'grady, M.N., Hogan, S.A., 2006. Past, current and potential utilisation of active and intelligent packaging systems for meat and muscle-based products: a review. Meat Sci. 74 (1), 113–130.

Kim, K.M., Ko, J.A., Lee, J.S., Park, H.J., Hanna, M.A., 2006. Effect of modified atmosphere packaging on the shelf-life of coated, whole and sliced mushrooms. LWT-Food Sci. Technol. 39 (4), 365–372.

Koseki, S., Itoh, K., 2002. Effect of nitrogen gas packaging on the quality and microbial growth of fresh-cut vegetables under low temperatures. J. Food Prot. 65 (2), 326–332.
Krasnova, I., Dukalska, L., Seglina, D., Juhnevica, K., Sne, E., Karklina, D., 2012. Effect of passive modified atmosphere in different packaging materials on fresh-cut mixed fruit salad quality during storage. World Acad. Sci. Eng. Technol. 67, 1095–1104.
Kuswandi, B., Maryska, C., Abdullah, A., Heng, L.Y., 2013. Real time on-package freshness indicator for guavas packaging. J. Food Meas. Characterization 7 (1), 29–39.
La Zazzera, M., Rinaldi, R., Amodio, M.L., Colelli, G., 2010. Influence of high $CO_2$ atmosphere composition on fresh-cut artichoke quality attributes. Acta Hortic. 934, 633–640.
Lee, J., Rudell, D.R., Watkins, C.B., 2014. Metabolic changes in 1-methylcyclopropene (1-MCP)-treated 'Empire' apple at different storage temperatures. Acta Hortic. 1048, 113–120.
Li, Y., Ishikawa, Y., Satake, T., Kitazawa, H., Qiu, X., Rungchang, S., 2014. Effect of active modified atmosphere packaging with different initial gas compositions on nutritional compounds of shiitake mushrooms (*Lentinus edodes*). Postharvest Biol. Technol. 92, 107–113.
Lin, Q., Lu, Y., Zhang, J., Liu, W., Guan, W., Wang, Z., 2017. Effects of high $CO_2$ in-package treatment on flavor, quality and antioxidant activity of button mushroom (*Agaricus bisporus*) during postharvest storage. Postharvest Biol. Technol. 123, 112–118.
Linke, M., Geyer, M., 2013. Condensation dynamics in plastic film packaging of fruit and vegetables. J. Food Eng. 116 (1), 144–154.
Łopacka, J., Półtorak, A., Wierzbicka, A., 2017. Effect of reduction of oxygen concentration in modified atmosphere packaging on bovine *M. longissimus* lumborum and *M. gluteus* medius quality traits. Meat Sci. 124, 1–8.
Lownds, N.K., Banaras, M., Bosland, P.W., 1994. Postharvest water loss and storage quality of nine pepper (Capsicum) cultivars. HortScience 29 (3), 191–193.
Ma, L., Zhang, M., Bhandari, B., Gao, Z., 2017. Recent developments in novel shelf life extension technologies of fresh-cut fruits and vegetables. Trends Food Sci. Technol. 64, 23–38.
Mahajan, P.V., Oliveira, F.A.R., Macedo, I., 2008. Effect of temperature and humidity on the transpiration rate of the whole mushrooms. J. Food Eng. 84 (2), 281–288.
Mahajan, P.V., Caleb, O.J., Singh, Z., Watkins, C.B., Geyer, M., 2014. Postharvest treatments of fresh produce. Philos. Trans. R. Soc. A 372 (2017), 20130309.
McMillin, K.W., 2008. Where is MAP going? A review and future potential of modified atmosphere packaging for meat. Meat Sci. 80 (1), 43–65.
Munro, I.C., Haighton, L.A., Lynch, B.S., Tafazoli, S., 2009. Technological challenges of addressing new and more complex migrating products from novel food packaging materials. Food Addit. Contam. 26 (12), 1534–1546.
Neethirajan, S., Jaya, D.S., Sadistap, S., 2009. Carbon dioxide sensors for the agric-food industry—a review. Food Bioproc. Tech. 2, 115–121.
Oliveira, F., Sousa-Gallagher, M.J., Mahajan, P.V., Teixeira, J.A., 2012. Development of shelf-life kinetic model for modified atmosphere packaging of fresh sliced mushrooms. J. Food Eng. 111 (2), 466–473.
Oms-Oliu, G., Soliva-Fortuny, R., Martín-Belloso, O., 2008. Edible coatings with antibrowning agents to maintain sensory quality and antioxidant properties of fresh-cut pears. Postharvest Biol. Technol. 50 (1), 87–94.
Opara, U.L., 2009. Quality management: an industrial approach to produce handling. Chapter 8 In: Florkowski, W.J., Shewfelt, R.L., Brueckner, B., Prussia, S.E. (Eds.), Postharvest Handling: A Systems Approach. second ed.. Elsevier, pp. 154–204.
Opara, L.U., Al-Jufaili, S.M., Rahman, M.S., 2007. Postharvest handling and preservation of fresh fish and seafoods. Chapter 6 In: Rahman, M.S. (Ed.), Handbook of Food Preservation. second ed.. CRC Press, Boca Raton, FL, pp. 151–172.

Ozdemir, M., Floros, J.D., 2004. Active food packaging technologies. Crit. Rev. Food Sci. Nutr. 44 (3), 185–193.

Özogul, F., Taylor, K.D.A., Quantick, P., Özogul, Y., 2002. Biogenic amines formation in Atlantic herring (*Clupea harengus*) stored under modified atmosphere packaging using a rapid HPLC method. Int. J. Food Sci. Technol. 37 (5), 515–522.

Poyatos-Racionero, E., Ros-Lis, J.V., Vivancos, J.L., Martinez-Manez, R., 2018. Recent advances on intelligent packaging as tools to reduce food waste. J. Clean. Prod. 172, 3398–3409.

Rathore, H., Prasad, S., Sharma, S., 2017. PharmaNutrition mushroom nutraceuticals for improved nutrition and better human health: a review. Biochem. Pharmacol. 5 (2), 35–46.

Rico, D., Martin-Diana, A.B., Barat, J.M., Barry-Ryan, C., 2007. Extending and measuring the quality of fresh-cut fruit and vegetables: a review. Trends Food Sci. Technol. 18 (7), 373–386.

Rux, G., Mahajan, P.V., Geyer, M., Linke, M., Pant, A., Saengerlaub, S., Caleb, O.J., 2015. Postharvest biology and technology application of humidity-regulating tray for packaging of mushrooms. Postharvest Biol. Technol. 108, 102–110.

Salinas, Y., Ros-Lis, J.V., Vivancos, J.L., Martínez-Máñez, R., Marcos, M.D., Aucejo, S., Lorente, I., 2012. Monitoring of chicken meat freshness by means of a colorimetric sensor array. Analyst 137 (16), 3635–3643.

Sandhya, 2010. Modified atmosphere packaging of fresh produce: current status and future needs. Food Sci. Technol. 43, 381–392.

Sanz, C., Perez, A.G., Olias, R., Olias, J.M., 1999. Quality of strawberries packed with perforated polypropylene. J. Food Sci. 64 (4), 748–752.

Sanz, C., Perez, A.G., Olias, R., Olias, J.M., 2000. Modified atmosphere packaging of strawberry fruit: effect of package perforation on oxygen and carbon dioxide/Envasado de fresas en atmósfera modificada: efecto de la perforación del envase en el contenido de oxígeno y dióxido de carbono. Food Sci. Technol. Int. 6 (1), 33–38.

Scifò, G.O., Randazzo, C.L., Restuccia, C., Fava, G., Caggia, C., 2009. Listeria innocua growth in fresh cut mixed leafy salads packaged in modified atmosphere. Food Control 20 (7), 611–617.

Serrano, M., Martinez-Romero, D., Guillen, F., Castillo, S., Valero, D., 2006. Maintenance of broccoli quality and functional properties during cold storage as affected by modified atmosphere packaging. Postharvest Biol. Technol. 39 (1), 61–68.

Sivertsvik, M., Jeksrud, W.K., Rosnes, J.T., 2002. A review of modified atmosphere packaging of fish and fishery products–significance of microbial growth, activities and safety. Int. J. Food Sci. Technol. 37 (2), 107–127.

Sivertsvik, M., 2007. The optimized modified atmosphere for packaging of pre-rigor filleted farmed cod (*Gadus morhua*) is 63 ml/100 ml oxygen and 37 ml/100 ml carbon dioxide. LWT-Food Sci. Technol. 40 (3), 430–438.

Smiddy, M., Papkovskaia, N., Papkovsky, D.B., Kerry, J.D., 2002. Use of oxygen sensors for the non-destructive measurement of the oxygen content in modified atmosphere and vacuum packs of cooked chicken patties; impact of oxygen content on lipid oxidation. Food Res. Int. 35, 577–584.

Song, Y., Lee, D.S., Yam, K.L., 2001. Predicting relative humidity in modified atmosphere packaging system containing blueberry and moisture absorbent. J. Food Process. Preserv. 25 (1), 49–70.

Suppakul, P., Miltz, J., Sonneveld, K., Bigger, S.W., 2003. Active packaging technologies with an emphasis on antimicrobial packaging and its applications. J. Food Sci. 68 (2), 408–420.

Tano, K., Oulé, M.K., Doyon, G., Lencki, R.W., Arul, J., 2007. Comparative evaluation of the effect of storage temperature fluctuation on modified atmosphere packages of selected fruit and vegetables. Postharvest Biol. Technol. 46 (3), 212–221.

Tao, F., Zhang, M., Hangqing, Y., Jincai, S., 2006. Effects of different storage conditions on chemical and physical properties of white mushrooms after vacuum cooling. J. Food Eng. 77 (3), 545–549.

Tomás-callejas, A., Boluda, M., Robles, P.A., Artés, F., Artés-hernández, F., 2011. LWT—food science and technology innovative active modified atmosphere packaging improves overall quality of fresh-cut red chard baby leaves. LWT- Food Sci. Technol. 44 (6), 1422–1428.

Torrieri, E., Antimo, P., Cavella, S., Fogliano, V., Attianese, I., Giuliana, G., Masi, P., 2011. Effect of modified atmosphere and active packaging on the shelf-life of fresh bluefin tuna fillets. J. Food Eng. 105 (3), 429–435.

Tudela, J.A., Marín, A., Garrido, Y., Cantwell, M., Medina-martínez, M.S., Gil, M.I., 2013. Postharvest biology and technology off-odour development in modified atmosphere packaged baby spinach is an unresolved problem. Postharvest Biol. Technol. 75, 75–85.

Urmila, K., Li, H., Chen, Q., Hui, Z., Zhao, J., 2015. Quantifying of total volatile basic nitrogen (TVB-N) content in chicken using a colorimetric sensor array and nonlinear regression tool. Anal. Methods 7 (13), 5682–5688.

Villaescusa, R., Gil, M.I., 2003. Quality improvement of *Pleurotus* mushrooms by modified atmosphere packaging and moisture absorbers. Postharvest Biol. Technol. 28 (1), 169–179.

Volpe, S., Mahajan, P.V., Rux, G., Gavella, S., Torrieri, E., 2018. Condensation and moisture regulation in packaged fresh-cut iceberg lettuce. J. Food Eng. 2016, 132–137.

Weerakkody, P., Jobling, J., Infante, M.M.V., Rogers, G., 2010. The effect of maturity, sunburn and the application of sunscreens on the internal and external qualities of pomegranate fruit grown in Australia. Sci. Hortic. 124 (1), 57–61.

Yahia, E.M., 2006. Modified and controlled atmospheres for tropical fruits. Stewart Postharvest Rev. 5 (6), 1–10.

Yam, K.L., Takhistov, P.T., Miltz, J., 2005. Intelligent packaging: concepts and applications. J. Food Sci. 70 (1), R1–R10.

Ye, J.J., Li, J.R., Han, X.X., Zhang, L., Jiang, T.J., Miao, X.I.A., 2012. Effects of active modified atmosphere packaging on postharvest quality of shiitake mushrooms (*Lentinula edodes*) stored at cold storage. J. Integr. Agric. 11 (3), 474–482.

Zhang, M., Meng, X., Bhandari, B., Fang, Z., 2015a. Recent application of modified atmosphere packaging (MAP) in fresh and fresh-cut foods. Food Rev. Int. 31 (2), 172–193.

Zhang, B., Ma, L., Deng, S., Xie, C., Qiu, X., 2015b. Shelf-life of pacific white shrimp (*Litopenaeus vannamei*) as affected by weakly acidic electrolyzed water ice-glazing and modified atmosphere packaging. Food Control 51, 114–121.

## Further reading

Soccol, M.C.H., Oetterer, M., 2003. Use of modified atmosphere in seafood preservation. Braz. Arch. Biol. Technol. 46 (4), 569–580.

Soliva-Fortuny, R.C., Martín-Belloso, O., 2003. New advances in extending the shelf-life of fresh-cut fruits: a review. Trends Food Sci. Technol. 14 (9), 341–353.

# Intelligent packaging to improve shelf life

8

*Valentina Siracusa**, *Nadia Lotti*[†]
*Department of Chemical Science (DSC), University of Catania, Catania, Italy, [†]Department of Civil, Chemical, Environmental and Materials Engineering, University of Bologna, Bologna, Italy

## Chapter outline



## 1 Introduction

It is well known that the main purpose of a food packaging system is to preserve the food from external contamination, maintaining its freshness and quality and, if possible, extending its shelf life. As reported by several authors, a traditional food packaging has four basic functions, shown in Fig. 1.

Traditional packaging has provided all those basic functions together with the traditional methods used to preserve food such as drying, refrigeration, freezing, curing, and sugaring (Lee et al., 2015); however, in recent years, due to the increasing demand for innovative and creative packaging to guarantee food safety, quality, and traceability, these methods have become insufficient. The requirement for new technologies, integrated in the food packaging, has become even more pressing. This new packaging must extend or maintain the shelf life and quality in a large range of fresh products such as vegetables, fruit, meat, and fish. Food producers, food processors, logistic operators, retailers, and consumers are the subjects of this innovation. But these innovations, before becoming commercially viable and adopted by all, must respect the regulatory requirements and overcome the possible major expenses for the newly adopted technology. Further, in order to take into account the environmental aspects, food packaging innovation is called to consider a broad range of sustainability issues,

Food Quality and Shelf Life. https://doi.org/10.1016/B978-0-12-817190-5.00008-2

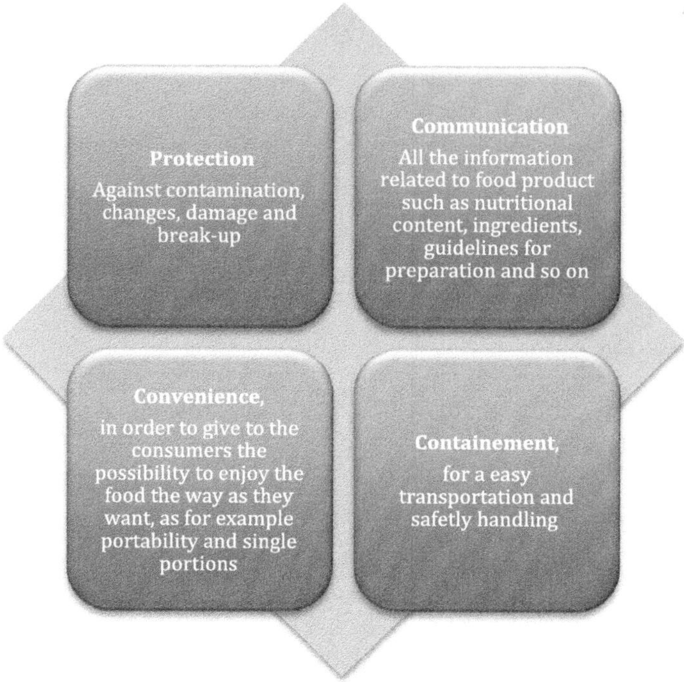


**Fig. 1** The four basic functions of traditional food packaging.

such as waste preservation, process optimization, recycling, reuse, correct use of resources, and so on. In the end, the innovation must be discussed, not only considering the four basic functions of traditional food packaging, but also taking into account the contribution to the environment. Consequently, a more sustainable solution is required in order to reduce the impact of both packaging waste and food loss.

Within the new generation of food packaging, active and intelligent packaging has gained even more interest as an innovative system (Realini and Marcos, 2014; Han, 2005). Active packaging technology has been extensively studied and reported in several studies (Lee et al., 2015). In this case, compounds are incorporated into the packaging system to extend shelf life and maintain and/or enhance food quality. It is a system where product, package and environment interact with each other in a positive way in order to extend food shelf life, improving food quality and safety. The type of active agents that could be used are diverse and they deliberately alter the inside packaging environment or the kind of interaction between the food and the packaging material. In general, as reported by Lee et al. (2015), active packaging application can be classified into four categories:

**a)** absorbing and scavenging of chemicals such as oxygen and carbon dioxide, moisture, ethylene, UV light, contaminants, and flavors;
**b)** removal of food chemicals such as, for example, lactose and cholesterol;
**c)** temperature control, self-heating or self-cooling; and
**d)** microbial control and quality control.

An example is the modified atmosphere packaging technology (MAP), used to extend the shelf life of fresh food, delay the ripening, reduce loss, and improve appearance. Temperature and different gas formulation are the main variables that must be considered, related to different product types and microorganisms. Other examples of active packaging are the use of antimicrobial packaging materials, gaseous agents, or antimicrobials (organic acid, pediocins, enzymes, essential oils) in the package headspace, and the use of oxygen scavengers, carbon dioxide absorbers and generators, ethylene scavenging, and moisture absorbers. Current active packaging systems, together with an indication of substance used, advantages, and applications, are reported in Fig. 2.

From the beginning of this millennium, research activity on food packaging innovation has greatly increased, devoted principally to the development of intelligent packaging. In this case, the function is to record, detect, sense, trace, and communicate between the packed food and the environment and vice versa, in order to extend the shelf life, provide information, improve the food quality, and advise of internal or external problems that reflect on the packed food products (Biji et al., 2015; Yucel, 2016).

The principal systems of intelligent packaging, together with the indication of the type, advantages, and applications, are shown in Fig. 3.

These systems differ from each other not only for composition but also in the type and quantity of data that can be obtained, and how it can be obtained and analyzed (Heising et al., 2014). As reported by Vanderroost et al., research on the three major technologies has increased since 2002, largely on sensor and indicator systems. They also report a list of research projects related to intelligent packaging, reflecting the trend of interest in the development, implementation, commercialization, and standardization of those materials and technologies.

According to the standard definition (EC 450-2009, 2009; Vanderroost et al., 2014), intelligent packaging permits continuous control of food package condition and the environment surrounding the food during storage and transport. Also, the package integrity is controlled during the whole lifecycle of the food/packaging system. In this manner, users are able to collect all the information concerning the packed products and its environment, thanks to the ability of the intelligent packaging to sense, detect, and record any changes.

In the last 20 years, the numbers of publications has increased considerably, especially from 2009. In fact, considering the financial crisis of 2007, the interest in new, efficient, and economic business processes, as well as in the reduction of losses (in terms of money, resources, and food), has driven research into these innovative types of packaging.

It is important to note that this packaging provides continuous information on food condition and on packaging integrity, which are important not only for the final consumers but also to detect any abuse or problems occurring in the entire supply chain, from the farm to the table.

Thanks to that, it is possible to reduce food loss, to reduce waste, and to control and eventually reduce unnecessary transport and storage steps.

Intelligent packaging is also considered a valid help in the Hazard Analysis and Critical Control Points (HACCP) and Quality Analysis and Critical Control Points

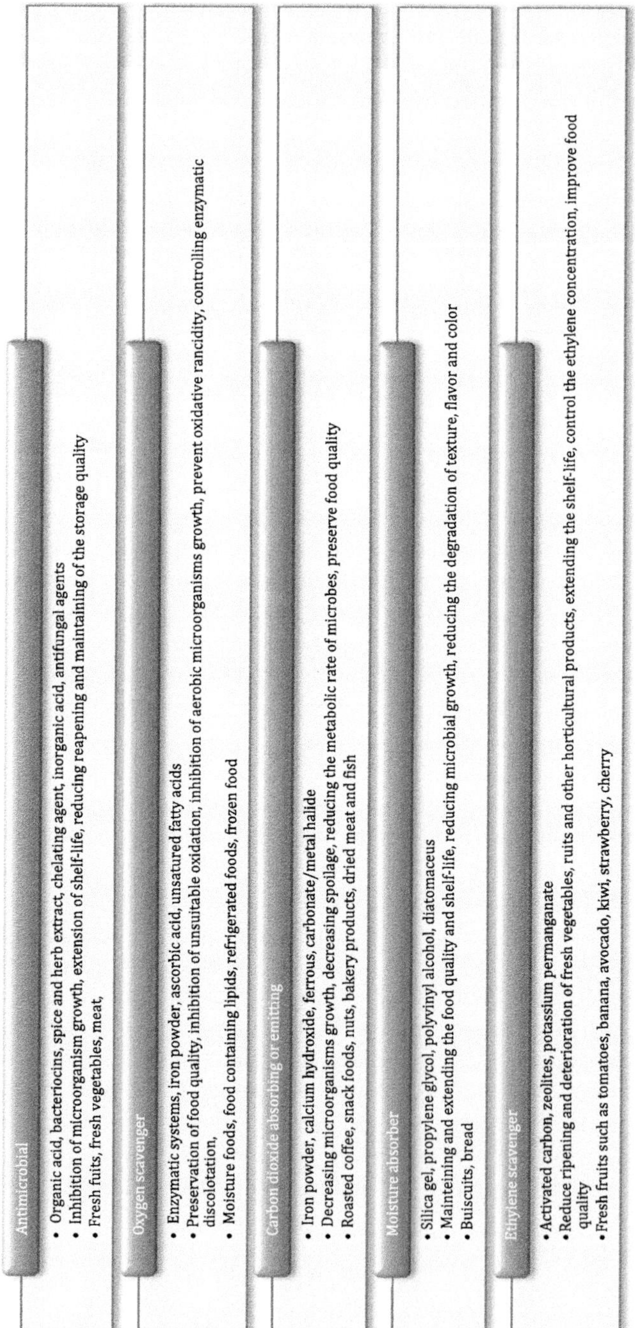


**Fig. 2** Active packaging systems.

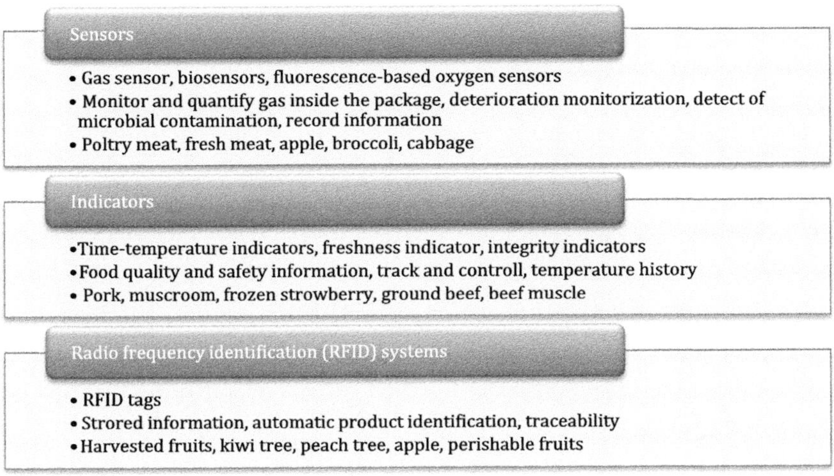


**Fig. 3** Intelligent packaging systems (Lee et al., 2015).

(QACCP) methodologies, used to control, detect, prevent, reduce, and eliminate any possible criticisms that could affect the final food quality. EC 2009, Intelligent packaging and active packaging can be used together in the realization of so-called "smart packaging". The product and the environment can be monitored continuously by the use of intelligent packaging and the changes that inevitably occur can be controlled by the use of active packaging.

The purpose of this chapter is to give an overview of recent scientific research results on recent and emerging technologies related to intelligent packaging systems that could be easily integrated with traditional packaging materials. The aim is to discuss the major technologies such as sensor, indicator, and radio frequency identification systems, considered the most promising technologies for the next generation of intelligent packaging systems.

Subsequently, market share and perspectives of intelligent packaging systems will be given, together with an indication of the approach that could be followed to design, produce, and apply those intelligent food packaging systems in a sustainable manner.

## 2 Intelligent packaging systems

Intelligent packaging represents a large and important step in avoiding food waste, while improving food logistics and traceability.

Indicators, sensors, and different devices are incorporated to detect, sense, record, and transmit all the recorded information from the package system and its environment, in order to monitor continuously the quality and the state of the product throughout the whole supply chain (Realini and Marcos, 2014; Yam et al., 2005).

## 2.1 Sensors

A sensor is an analytical device used to detect and quantify chemical or physical change in a food product inside a package. Sensors can provide quantitative results about, for example, concentrations, temperature, pH, and other quantitative data, and can store the data recorded. The positive and negative aspects of sensor systems are summarized in Fig. 4.

Chemical sensors are gaining much attention thanks to the fact that the receptor, the sensing part of the chemical sensor, is chemically capable of detecting the presence, activity, composition, and concentration of a specific chemical molecule or gas type through a surface adsorption mechanism. The result is a change of a coating property that is converted proportionally into a signal by the transducer (the measuring part of the sensor). The transducer is called "active" if it requires external power for the measurements, otherwise it is called "passive." The most important feature is that chemical sensors have to be small and flexible. They can be used to detect organic volatile compounds (VOCs) and small gas molecules such as $H_2$, CO, $CO_2$, $NO_2$, $CH_4$, $NH_3$, $H_2S$, $O_2$, and so on, which could be produced by food deterioration, or by the loss of package integrity. In general, these sensors could be used to substitute more binding measures performed with fixed or portable instruments such as gas chromatography mass spectrometer (GC-MS) or gas analyzers that require breakage of package, not real time measurements, not on-line control, and small scale usage. They can provide high sensitivity measures, but the chemical sensors commercially available still require high power (as for example, high operating temperature), are too large, not selective, too rigid, and not small enough to be inserted or integrated directly into the food package. As reported by Vanderroost et al. (2014), technologists are now devoting their interest to the development of a new generation of chemical sensors.

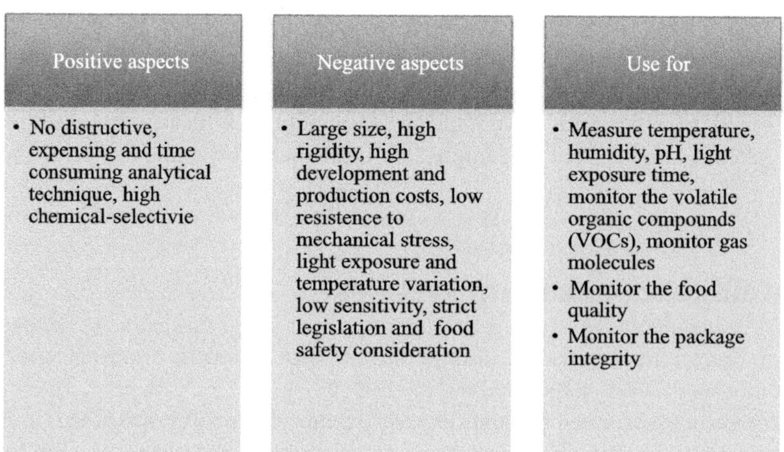


**Fig. 4** Positive and negative aspect of sensors.

### 2.1.1 Biosensors

Those intelligent systems are able to detect, record, and transmit target metabolites produced by biochemical reactions occurring in the food degradation process. According to the type of food packed, they can be properly tailored in order to detect the formation of specific degradation products. They are formed by two components: a bioreceptor that detects the chemical compounds and a transducer that converts the chemical signals into a quantitative response. Realini and Marcos (2014) reported some of those commercial biosensors, but not currently commercialized. The great interest is to insert those sensors directly in the food packaging. An interesting result was achieved by molecular imprinting technology where a polymer matrix is used as substrate for the sensor (Ghaani et al., 2016).

### 2.1.2 Printed electronics technology

Printed electronic devices could be considered the future of intelligent packaging technology. They are made by the combination of electrically functional inks on flexible substrates of film, such as polymers (polyimide, PEEK, PET, polyesters), steel, and paper. The production process is simple, with less material waste, and is less time consuming than traditional silicon based production processes. Inkjet printing is used for research and development studies, while gravure printing could be applied for mass production of printed electronic devices (Vanderroost et al., 2014). In addition, they present several advantages such as being lightweight, rollable, portable, foldable, bendable, thin, and, considering that the substrate could be different, they can be produced with shape and characteristics well tailored for each specific application.

Flexible printed chemical sensors are the most attractive intelligent packaging devices. The research in this field is still active in order to reach major challenges. In particular, the research is devoted to optimizing electronic ink formulation, in order to improve sensitivity and selectivity across the all kinds of environmental conditions that could occur during the transport and storage of food products. The minimization of power requirements for those sensors are also an important requirement as well as the optimization of processes to parameters such as printing speed, drying, ink viscosity, ink deposition and smoothness, wettability, reactivity, and roughness of the substrate (Monero et al., 2011).

### 2.1.3 Carbon nanotechnology

Carbon nanomaterials (CNs) such as carbon black, fullerenes, grapheme, graphite nanofibers, and nanotubes have been an object of great interest from researchers. Thanks to the fact that numerical simulation can be easily realized to predict the physicochemical behavior of CNs, the interest in such materials has increased their potential application. Those materials could be applied as chemical sensors (receptor and/or transduce) thanks to their excellent electrical properties and mechanical performance such as light-weight and high flexibility, even at low temperatures.

Carbon nanotubes (CNTs) and graphene are candidates for gas sensors because they show change in resistivity connected to change in gas concentration (Schedin et al., 2007). Their sensitivity is very high with detection limits at ppm levels of gas molecules concentration. In particular, thanks to its two dimensional shape, every atom of graphene could be considered as a surface atom. In this manner, every atom site could be involved in gas interaction, detection capability increasing enormously, with the detection of a single molecule.

Fullerenes and fullerene-based film are interesting materials because they can resist great pressure (over 3000 atm) (Grynko et al., 2009).

Carbon nanofibers (CNFs) are extremely pure materials, with high mechanical strength and high geometrical surface, that can easily be functionalized and surface modified, making them suitable to be used as chemical sensors (receptors).

Despite the great interest in CNs materials, there are still some problems concerning their application as intelligent packaging materials. First of all, the presence of contaminants on the surface must be absolutely avoided. This problem could be solved by functionalization of the CN surface with specific chemical or biological molecules. Further, the cost must be lowered and the production methods must be industrial scaled.

### *2.1.4 Silicon technology*

Due to the increasing performance of silicon photonics, research on transducer devices generating an optical signal is gaining even more interest in the scientific community as well as at industry level. This sensor does not need an electrical power supply but can be powered by using UV, visible, and IR light. The optical circuit is integrated in silicon semiconductor material. Actually, they were developed for very selective gas concentration detection in the headspace of food packaging. The major problems concerning this technology are noise sensitivity, low detection sensitivity, and the high development and operational costs of infrared lasers and detectors needed for sensor read-outs (Baets et al., 2013; Puyol et al., 1999). The advantages are low cost of production and the possibility to produce on a large scale, using the same methodologies and infrastructure used for the production of conventional silicon semiconductors. Thanks to the cost, size, scale, and energy employed, this technology could become a commercially alternative to the traditional technologies and applied in the near future for large-scale food-related applications.

### *2.1.5 Biotechnology*

Biotechnology has enabled the possibility of isolating and purifying living organisms such as cells, antibodies, and enzymes to be used in the realization of biosensors. In respect to chemical sensors, these are able to detect chemical analytes through biological receptors and are applied to identify allergens and analytes such as sugars, alcohols, lipids, nucleotides, and so on (Vanderroost et al., 2014). Most of the applications of these sensors are in the medicine field, but they can be applied in food and process control. Their integration in food packaging is still at the research stage. The most important need is to control the possibility of hazards due to the migration of the biological components of the food. As reported by Vanderroost et al. (2014), one application of a biosensor for the detection of pathogens in food grain, fruits, and wine was presented by Flex Alert—Scheelite Technologies LLC.

Biosensors could be used also to detect the presence of food spoilage associated with particular odors released by food ripening, fermentation, cooking, and so on. They are highly sensitive but limited in use due to their high selectivity for specific compounds.

### 2.1.6 Nose system

These systems are used to mimic the human nose to smell and taste flavor, odors, and savors associated with food. Sensors used for this purpose must have the same sensitivity as the human nose, a high selectivity to different compounds, small dimension, high stability, reproducibility, reliability of the data recorded, and short response and recovery time. Research is devoted to the production of small and cheap nose systems, that could be integrated directly in food packaging and that could be considered as a substitute for the traditional forms used for research purposes and control in food industry production (Gardner and Bartlett, 1993; Ouellette, 1999).

## 2.2 Indicators

In contrast to sensors, indicators can be used to obtain direct visual information about the food packed by evaluation of color change, increasing/decreasing of color intensity, diffusion of a color along a path (Kerry et al., 2006; Resem Brizio, 2016). The data are qualitative and the process irreversible. It means that the color diffusion or intensity change must be irreversible in order to avoid false information. The most commercially available indicators for food packaging application are shown in Fig. 5, together with specific information about their applications and some commercialized trade name indicators.

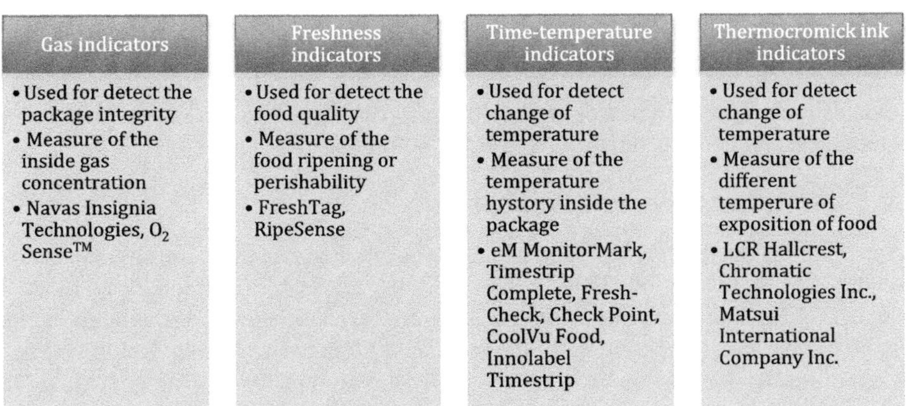


**Fig. 5** Indicators for food packaging applications.

### 2.2.1 Gas indicators

These indicators are principally used to detect package integrity, which is an essential requirement, especially when food is stored in a controlled internal atmosphere (MAP) (Vu and Won, 2013). It is a noninvasive approach that provides qualitative or semiquantitative data on alteration of internal gas concentration, performed by a visual color change. This uses a special pigment for plastic packaging that change clearly its color intensity, giving the retailer or manufacturers the possibility to remove the product from the supply chain before it reaches the supermarket or the final consumer. In addition, it alerts on the integrity of the package to avoid the use of unsafe products (Vanderroost et al., 2014). The most common gas indicator used for MAP application technology is the oxygen indicator, which detects the $O_2$ level inside the package (Yam et al., 2005). When it reaches a nonpermitted level, a color change is observed, due to a colorimetric oxidation of a dye-based indicator. If the oxygen level is reduced, the color returns to the original one. This reversibility is not always desired because the oxygen could be used by microbial growth inside the package (Hurme, 2003). Generally, $O_2$ gas sensor indicators are used together with $O_2$ scavengers or adsorbers (Kuswandi et al., 2011). Some example are the Ageless Eye® tablets of the Mitsubishi Gas Chemical Company Inc., the Tell-Tab $O_2$ indicator tablets of IMPAK supplier, the EMCO indicators, and the $O_2$Sense™ of the FreshPoint Lab supplier.

### 2.2.2 Freshness indicators

These indicators provide information about the quality of the food packed. By this kind of indicators, the microbial growth and/or the chemical change of food are detected. In this case, the reaction between the indicator and the microbial growth metabolites is recorded. The metabolites produced during storage, such as glucose, organic acids, ethanol, amines, nitrogen compounds, and so on, could be detected and their changes in concentration associated with the loss of food freshness. It could be used to detect, for example, volatile amines produced by the deterioration process of seafood, producing the characteristic fishy odor. These volatiles react with the dye indicator, causing a color change that indicates the loss of food freshness. Another example is the detection of the aromas released by fresh fruits ripening. The label indicator changes its color from red to orange to yellow, indicating the rate of freshness loss (Kuswandi et al., 2013; Vanderroost et al., 2014). Freshness indicators could be based not only on the detection by color indicator change but also by using biosensors to detect target metabolites, and could be incorporated into the food packaging.

### 2.2.3 Time-temperature indicators

Temperature is one of the most important parameters that must be carefully controlled in order to avoid loss in food quality and safety, especially for fresh food such as meat and fish. These indicators are useful because they are able to record the temperature history during the distribution and storage of food products, especially for chilled and frozen products (Pavelkova, 2012, 2013). Over a period of time, or due to change in temperature, a self-adhesive indicator composed of active zones changes its color irreversibly, faster at higher temperatures or slower at lower temperatures (Galagan and Su, 2008). Time-temperature indicators (TTIs) can be used to have information about the cold chain break. Producers and retailers, as well as consumers, are consequently informed on the shelf life of fresh products, in order to avoid food consumption over the data codes.

The TTIs available commercially are those based on physical, chemical, enzymatic, and biological systems (Kerry et al., 2006). Realini and Marcos (2014) reported a list of such indicators. All are related to color change. As an example, the TTI indicator 3M Monitor Mark® of the 3M Company, containing a fatty acid ester, changes its color due to the melting of the ester and its diffusion, when the temperature exceeds a critical value. The range of temperature detection is between −15 and 26°C, and the response indicator depends on the concentration and type of fatty acid ester (Kuswandi et al., 2011). The Keep-it® indicator of Keep-it Technologies is based on a chemical reaction between an immobilized chemical reactant and a mobile chemical reactant, stored in a separate compartment. When the sealing between the two compartments is removed, the two reactants come into contact in a time-temperature dependent manner, giving a visual signal. The Fresh-Check® indicator of Temptime Corporation is based on a polymerization reaction with the production of a colored polymer (Kerry et al., 2006). VITSAB® indicator of VITSAB International AB is based on an enzymatic reaction between an aqueous solution of enzymes and a substrate, contained in separate compartments. When the separation is removed, the contents are mixed and, by an enzymatic hydrolysis reaction of the substrate, a color change is detected (Galagan and Su, 2008). OnVu™ detector, supplied by Ciba and Freshpoint™, is an indicator based on a photochemical reaction. By UV exposure, a pigment changes its color gradually in a time-temperature dependent manner. The characteristic of this indicator is that it can be applied on the packaging as a label or it can be printed directly on the packaging surface (O'Grady and Kerry, 2008). The TopCryo™ indicator of the TRACEO supplier is based on a microbiological system, with microorganism, indicator, and nutritive medium closed in a plastic sachet attached in to the food package. Finally, FreshCode™ of Varcode Ltd. and Tempix® of Tempix AB are indicators in the form of labels, where the bare code is printed with fading inks that disappear with time-temperature increment (Realini and Marcos, 2014).

### 2.2.4 Thermochromic inks detector

These indicators are based on the use of thermo-sensitive ink that can be printed on the package. The color of these indicators could change reversibly or irreversibly depending on the temperature exposure. Irreversible inks are invisible until a certain temperature is not reached. Once the color appears, it remains constant if no change of temperature is recorded. If temperature increases, the color changes, leaving an indication of the temperature change history. Reversible inks change their color if heated but they return to original when the temperature decreases (Vanderroost et al., 2014). As reported by Ghaani et al. (2016), the color change could be accompanied by the indication of a "ready to serve" message.

## 2.3 Radio frequency identification (RFID) tags

RFID was originally used to control food products during distribution and storage. It represents a separate technology, different from those related to sensors and detectors. In general, this technology is not used to provide qualitative or quantitative information about the food quality status or shelf life. It is generally applied for the identification, automatization, theft prevention, and counterfeit protection of food products.

RFID is considered an automatic identification technology (Auto ID), and includes barcodes, QR-codes, magnetic ink, and so on.

Actually, this technology is used, not only to track the movement of raw materials and products, but also to monitor storage conditions and food quality during the entire supply chain (large production networks).

Depending on the type of power supply, RFID tags can be classified into three categories:

- passive RFID tags, with no battery, where the data between the tag and the reader are transferred by induction or backscatter coupling; these tags are small, light and can be read in a distance of a few meters;
- semipassive RFID tags, with a small battery, that records data in the tag; and
- active RFID tags, with an internal battery used to run the microchips and to transmit the signal to the reader; these are more expensive and can be read in a range of 100 m.

The communication between the RFID tags and the reader can be at low, high, and ultra-high frequency (LF, HF, UHF).

Despite this technology being used for many years in sectors such as ticketing, sports events timing, shipping, and so on, in the field of intelligent food packaging its use is still limited due to the fact that some technical adjustments, process and security issue must be controlled. In addition, in order to provide historical information of food distribution and storage, to monitor the package integrity, and to detect the quality status of the food, many parameters have to be controlled. Temperature, relative humidity, pH, pressure, light exposure, VOCs, and gas molecules concentration are some of them. Today, the combination of sensors with RFID tags is being studied by several researchers for future applications in the food supply chains. One or more sensors could be connected to an RFID tag, in order to record and collect continuously all the data measured by the sensors and memorized in the tag. Fruits, vegetables, meat, and fish, the most perishable and sensitive foods, could be easily controlled during their transport and storage, to avoid loss in freshness and quality, minimizing the amount of food spoiled before reaching retailers and consumers. There are several RFID suppliers that work in collaboration with food industries (Realini and Marcos, 2014).

Currently, research is being devoted to find how to integrate the sensors in the RFID tag or how to integrate the sensors-RFID tag directly in the packaging materials. Vanderroost et al. (2014) reported and described several research projects involved in the development of new sensors-enabled RFID tags for perishable foods such as fish, fresh fruits, and vegetables. Further, they reported the indication of some already widely commercially available sensors-RFID tags that can be inserted on the food packaging or in the box containing the food for transportation.

# 3 Global market and perspectives of intelligent food packaging

According to Realini and Marcos (2014), the market for active, intelligent, and advanced packaging is increasing at a compound annual growth rate (CAGR) of about 5.8% with about $5.3 billion of intelligent packaging sales in 2017. Unfortunately, the applications of such materials are still limited, probably due to the high price and limited integration with other packaging materials. The advances in biotechnology, chemistry, microelectronics, and material science can contribute to the development of new and low cost intelligent packaging solutions. An interesting solution could be the integration of RFID systems with sensors and biosensors, to be used to collect information in real time about the state of the packed food, leading to improved safety, quality, and management, while reducing food losses. Advancement in printed electronics for developing electrical devices by deposition of conductive inks on different substrates could help to integrate intelligent systems into packaging materials. Printed devices show the advantages of low price and easy disposability. Graphene technology, with the use of graphene inks with high mechanical flexibility, high electrical conductivity, and chemical stability, could represent the next generation of intelligent devices, as well as the use of CNTs (Realini and Marcos, 2014).

Organic photonics and electronic technologies are evaluating the possibility of integrating optical and electrical circuit devices with superior mechanical properties and similar, if not better, optical and electrical properties than silicon, into organic materials such as polymers. This technology is still in a fundamental stage and is not commercialized.

CNs technology could be developed in order to take advantage, not only of the superior electrical and mechanical properties, but also their unique optical properties that could be applied for the realization of optical sensors.

The adoption of this technology in the near future requires still research and development and, especially, cost reduction (Mlalila et al., 2016; Fuertes et al., 2016).

# 4 Life cycle assessment evaluation of intelligent food packaging

Intelligent food packaging is required to improve the overall food supply chain but also to contribute to the global reduction of food loss and food waste (Poyatos-Racionero et al., 2018; Ingrao et al., 2018a, b). For this purpose, intelligent packaging, as well as all kinds of packaging materials, has to be sustainable in terms of application but also in terms of design and production.

Packaging materials and intelligent devices such as tags, RFID, sensors, and so on, are not designed and conceptualized in respect of a closed-loop recyclability mechanism. Packaging has to be as cheap as possible and as functional as possible in order to improve food shelf life, as demanded by the industry, the market, and consumers; never mind if the packaging is difficult or even impossible to reuse or recycle. In addition, the use of different materials results in a complex and highly expensive recycling process and the demand for cheap and single use intelligent devices is more common thanks to the lower production process, ignoring the possibility of a future reuse.

According to the Sustainable Packaging Coalition, sustainable packaging must be characterized by different criteria. As reported from the Institute of Food Technologists (Brody et al., 2008), the criteria are:

1. sustainable packaging must be beneficial, safe, and healthy for consumers throughout its life cycle;
2. must meet market requirements (performance, cost);
3. sourced, manufactured, transported, and recycled by using renewable energy;
4. the use of renewable or recycled resources must be maximized for its production;
5. produced by clean technologies and best practices;
6. produced from healthy materials;
7. the production process must be optimized in terms of materials and energy; and
8. recovered and used in biological and/or industrial cradle-to-cradle cycles.

The production of any good, as well as goods produced for food packaging application, involves the individuation and definition of all life cycle phases, as well as the technologies and the materials used for production, in order to control and determine point by point all the possible impacts than can adversely affect environmental quality. In this context, Life Cycle Assessment (LCA) methodology can be used, together with the technical solutions that can be adopted in the industrial production processes. LCA, environmental impact, and sustainable development are key concepts for sustainable evolution.

The LCA methodology involves the determination and quantification of the environmental impacts associated with each of the production and developing phases, characterizing the life cycle of a given product or process. Through the individuation of the environmental impacts it is possible to evaluate possible solutions in order to respect the environmental sustainability of a product throughout its life cycle (Siracusa et al., 2014). The key factor for sustainable development is to achieve the perfect balance between technological innovation and environmental protection. Packaging systems are designed to protect and maintain the quality and safety of food after processing, while food is traveling from the origin point to the consumption point. The impacts linked to packaging production, transportation, use, and final disposal must be taken into consideration because all those phases affect environmental sustainability. The LCA methodology can be considered a valuable tool as a design-support for highlighting environmental criticalities related to packaging production, use, and final disposal. Improvement solutions as well as the promotion of more eco-friendly products must be part of decision-making processes.

According to Meneses et al. (2012), several packaging alternatives can be proposed to consumers, in terms of materials used, forms, and sizes. But, given that several factors must be considered in the food packaging field, such as food quality preservation, freshness conservation, pleasant image, good marketing appeal, correct product identification, storage and distribution information, and so on, choosing the best packaging solution while considering the environmental aspect is not always easy. For this purpose, LCA methodology is useful to have information about the main environmental hotspot related to the life cycle of each phase for a given packaging system (Ingrao et al., 2015, 2017).

It is well known that there is a massive consumption of materials in the packaging sectors, especially polymers materials, due to the combination of favorable factors but this is of course accompanied by consistent waste generation with consequent environmental pollution problems.

In addition, growing environmental awareness imposes the eco-friendly attributes to packaging products and processes. The development and use of materials with biodegradability and/or compostability attributes could be a good alternative to be followed in order to reduce environmental pollution as well as municipal solid waste (Peelman et al., 2013; Siracusa et al., 2008). Considering that the most important feature of packaging is to guarantee food conservation and preservation for long periods, reducing at the same time waste of food and materials and utilization of preservatives, to date, not many biopolymers have been fully employed for food packaging application. In addition, as shown by the results reported by Tabone et al. (2010), the comparison of the environmental impacts of 12 polymers, 7 obtained from fossil fuels, 4 from renewable resources, and 1 from both, showed that the employment of renewable resources instead of fossil ones does not necessarily guarantee reduction of the related environmental impacts.

## 5 Legal and safety aspects

In contrast to the United States, Australia, and Japan where intelligent packaging systems are widespread, in Europe the legislative regulations have hindered the diffusion of new packaging solutions into the market. The first law related to materials intended to come into contact with food appeared in 2004 (European Commission, 2004). In particular, this law intended to underline the safety of the packaging materials versus human health, reporting that any component present in the packaging shall not transfer to food in forbidden quantity, bringing unacceptable change in the food composition or organoleptic deterioration of food characteristics. In addition, intelligent materials used in food packaging shall not give misleading information to the consumers and their presence, as well as their suitability to food contact and appropriate use, must be fully detailed.

Dainelli et al. (2008) reported an interesting interpretation of such laws. The Regulation EC 2023/2006 implemented the previous law, taking into consideration the migration process—that is, the transfer of chemical or biological components into food in unacceptable quantities (European Commission, 2006). In 2009 a specific regulation on active and intelligent packaging appeared (European Commission, 2009), with specific indications for such materials intended to come into contact with food, and with a list of the authorized substances that can be used for their manufacture. Further, this law states that when active or intelligent devices are used for food packaging applications, it is mandatory to write the words "DO NOT EAT" and, if possible, put a specific symbol on the package. The risk for releasing substances from intelligent packaging occurs mainly when they are put inside the primary packaging. In fact, this is the main cause of reluctance among consumers to these materials. The main risk could be the leaching or the swallowing of active components (Mills, 2005; Han et al., 2005; Restuccia et al., 2010). In addition, when separate nonedible parts are included into the primary packaging, the reluctance of consumers increases, compared to parts attached to the package such as, for example, labels. In conclusion, in order to increase the acceptance of consumers toward this technology, it is necessary to minimize any potential risk of food contamination (Song and Hepp, 2005; Lee et al., 2008).

# 6 Conclusions

In the last 20 years, new packaging technologies have been developed in order to respond to ever more particular and sophisticated consumer demand. Preserved, fresh, tasty, and convenient food products with longer shelf life and controlled quality are objects of great interest to both consumers and industry. Market globalization has influenced the retailing process, with longer food distribution chains. In addition, consumer lifestyles have changed and, consequently, less time is spent shopping for fresh food and cooking. All these aspects have become driving forces for the development of a new package concept, created to extend the shelf life of food while maintaining and monitoring safety and quality. In this context, intelligent packaging is considered the newest technology in the food packaging field. It is seen as being able to improve the safety, quality, convenience, management, and traceability of food products, for suppliers, retailers, and final consumers. Despite the great interest in this technology, the documented results, and the interesting potentiality, it is still growing and not fully commercialized. There is still a gap between such technology and real market application. These systems are not fully commercially viable and not employed for everyday packaging commodities. Currently, intelligent packaging costs about 50%–100% of the whole cost attributed to the final packaging, while the commercial request is at most 10% of the total cost of the good available to the consumer. Despite the higher added benefits, which have been fully demonstrated to outweigh the higher costs of this technology, their commercial application is still limited.

Further advancement of this technology in any case is needed. The gap between the results obtained from laboratory tests and the real food applications must be closed. The great variability of the food is the heaviest parameter to be controlled and fully analyzed.

In addition, the possibility to integrate several functions within only one device is another technical goal for the future. For example, the possibility to detect and record several types of information at the same time could be of great relevance.

Even education of the final consumers of the benefits resulting from the use of intelligent packaging must be considered. Information about the correct use of the devices, what purpose it serves, how it works, that it is not dangerous, and so on must be well communicated to final consumers. The safety of the final packaging in respect to human health, food quality and safety, and possible migration of contaminants must be fully detailed.

Finally, but no less important, the sustainability of intelligent systems must be carefully checked, according to the universal concept of sustainable packaging. The interest versus reusable, reversible, and multiuse devices is the first challenge for an environmentally sustainable technology.

## References


Baets, R., Subramanian, A.Z., Dhakal, A., Selvaraja, S.K., Komorowska, K., Peyskens, F., 2013. Stectroscopy-on-chip applications of silicon photonics. In: Broquin, J.E. (Ed.), Integrated Optics: Devices, Materials, and Technologies XVII.

Biji, K.B., Ravishankar, C.N., Mohan, C.O., Srinivasa Gopal, T.K., 2015. Smart packaging systems for food application: a review. J. Food Sci. Technol. 52 (10), 6125–6135.

Brody, A.L., Bugusu, B., Han, J.H., Sand, C.K., McHugh, T.H., 2008. Innovative food packaging solutions. J. Food Sci. 73 (8), R107–R116.

Dainelli, D., Gontard, N., Spyropoulos, D., Zondervan-van den Beuken, E., Tobback, P., 2008. Active and intelligent food packaging: legal aspects and safety concerns. Trends Food Sci. Technol. 19, S103–S112.

EC 450-2009, 2009. EU Guidance to the Commission Regulation (EC No. 450–2009) on Active and Intelligent Materials and Articles Intended to Come into Contact with Food. European Commission.

European Commission, 2004. Commission Regulation (EC) No. 1935/2004 of 27 October 2004 on materials and articles intended to come into contact with food and repealing Directive 80/590/EEC. Official Journal of the European Union, OJ L 338, 4–17.

European Commission, 2006. Commission Regulation (EC) No. 2023/2006 of 22 December 2006 on good manufacturing practice for materials and articles intended to come into contact with food. Official Journal of the European Union 384, 75–78.

European Commission, 2009. Commission Regulation (EC) No 450/2009 of 29 May 2009 on active and intelligent materials and articles intended to come into contact with food. Official Journal of the European Union, OJ L 135, 3–11.

Fuertes, G., Soto, I., Carrasco, R., Vargas, M., Sabattin, J., Lagos, C., 2016. Intelligent packaging systems: sensors and nanosensors to monitor food quality and safety. J. Sens, 1–8. Article ID 4046061.

Galagan, Y., Su, W.F., 2008. Fadable ink for time-temperature control of food freshness: novel new time-temperature indicator. Food Res. Int. 41 (6), 653–657.

Gardner, J.W., Bartlett, P.N., 1993. A brief history of electronic noses. Sensors Actuators B 18, 211–220.

Ghaani, M., Cozzolino, C.A., Castelli, G., Farris, S., 2016. An overview of the intelligent packaging technologies in the food sector. Trends Food Sci. Technol. 51, 1–11.

Grynko, D., Burlachenko, J., Kukla, O., Kruglenko, I., Belyaev, O., 2009. Fullerene and fullerene-aluminum nanostructured films as sensitive layer for gas sensors. Semicond. Phys., Quantum Electron. Optoelectron. 12, 287–289.

Han, J.H., 2005. New technologies in food packaging: overview. In: Han, J.H. (Ed.), Innovation in Food Packaging, first ed. Elsevier Academic Press, Amsterdam, pp. 3–11. Chapter 1.

Han, J.H., Ho, C.H.L., Rodrigues, E.T., 2005. Intelligent Packaging. In: Han, J.H. (Ed.), Innovation in Food Packaging, first ed. Elsevier Academic Press, Amsterdam, pp. 138–155. Chapter 9.

Heising, J.K., Dekker, M., Bartels, P.V., Van Boekel, M.A., 2014. Monitoring the quality of perishable foods: opportunities for intelligent packaging. Crit. Rev. Food Sci. Nutr. 54 (5), 645–654.

Hurme, E., 2003. Detecting leaks in modified atmosphere packaging. In: Ahvenainen, R. (Ed.), Novel Food Packaging Techniques. Woodhead Publishing Ltd, Cambrige, UK, pp. 276–286.

Ingrao, C., Lo Giudice, A., Becenetti, J., Khaneghah, A.M., Sant'Ana, A.S., Rana, R., Siracusa, V., 2015. Foamy polystyrene trays for fresh-meat packaging: life-cycle inventory data collection and environmental impact assessment. Food Res. Int. 76, 418–426.

Ingrao, C., Gigli, M., Siracusa, V., 2017. An attributional life cycle assessment application experience to highlight environmental hotspots in the production of foamy polylactic acid trays for fresh-food packaging usage. J. Clean. Prod. 150, 93–103.

Ingrao, C., Becenetti, J., Bezama, A., Block, V., Goglio, P., Koukios, E.G., Lindner, M., Nemecek, T., Siracusa, V., Zabaniotou, A., Huisingh, D., 2018a. The potential roles of bio-economy in the transition to equitable, sustainable, post fossil-carbon societies: findings from this virtual special issue. J. Clean. Prod. 204, 471–488.

Ingrao, C., Faccilongo, N., Di Gioia, L., Messineo, A., 2018b. Food waste recovery into energy in a circular economy perspective: a comprehensive review of aspects related to plant operation and environmental assessment. J. Clean. Prod. 184, 869–892.

Kerry, J.P., O'Grady, M.N., Hogan, S.A., 2006. Past, current and potential utilization of active and intelligent packaging systems for meat and muscle-based products: a review. Meat Sci. 74, 113–130.

Kuswandi, B., Wicaksono, Y., Abdullah, A., Heng, L.Y., Ahmad, M., 2011. Smart packaging: sensors for monitoring of food quality and safety. Sens. & Instrumen. Food Qual. 5, 137–146.

Kuswandi, B., Maryska, C., Jayus, H., Abdullah, A., Heng, L.Y., 2013. Real time on-package freshness indicator for guavas packing. J. Food Meas. Charact. 7, 29–39.

Lee, S.Y., Lee, S.J., Choi, D.S., Hur, S.J., 2015. Current topic in active and intelligent packaging for preservation of fresh foods. J. Sci. Food Agric. 95, 2799–2810.

Lee, D.S., Yam, K.L., Piergiovanni, L., 2008. Food Packaging Science and Technology. CRC Press, Boca Raton, FL, p. 470.

Meneses, M., Pasqualino, J., Catells, F., 2012. Environmental assessment of the milk life cycle: the effect of packaging selection and the variability of milk production data. J. Environ. Manag. 107, 76–83.

Mills, A., 2005. Oxygen indicators and intelligent inks for packaging food. Chem. Soc. Rev. 34, 1003–1011.

Mlalila, N., Kadam, D.M., Swai, H., Hilonga, A., 2016. Transformation of food packaging from passive to innovative via nanotechnology: concepts and critiques. J. Food Sci. Technol. 53 (9), 3395–3407.

Monero, O., Boix, M., Claramunt, S., Prades, J.D., Cornet, A., Cirera, A., et al., 2011. Advanced performances in gas sensors: stretchable, flexible, wireless, wearable. Procedia Eng. 25, 1425–1428.

O'Grady, M.N., Kerry, J.P., 2008. Smart packaging technologies and their application in conventional meat packaging systems. In: Toldrá, F. (Ed.), Meat Biotechnology. Springer Science and Business Media, New York, USA, pp. 425–451.

Ouellette, J., 1999. Electronic noses sniff out new markets. Ind. Phys. 5, 1–26.

Pavelkova, A., 2012. Intelligent packaging as device for monitoring of risk factors in food. J. Microbiol. Biotechnol. Food Sci. 2, 282–292.

Pavelkova, A., 2013. Time temperature indicators as devices intelligent packaging. Acta Univ. Agric. Silvic. Mendelianae Brun. 56, 245–252.

Peelman, N., Rageart, P., De Meulenauer, B., Adons, D., Peeters, R., Cardon, L., Van Impe, F., Devlieghere, F., 2013. Application of bioplastics for food packaging. Trends Food Sci. Technol. 32, 128–141.

Poyatos-Racionero, E., Ros-Lis, J.V., Vivancos, J.-L., Martínez-Máñez, R., 2018. Recent advances on intelligent packaging as tools to reduce food waste. J. Clean. Prod. 172, 3398–3409.

Puyol, M., Del Valle, M., Garcés, I., Villuendas, F., Dominguez, C., Alonso, J., 1999. Integrated waveguide absorbance optode for chemical sensing. Anal. Chem. 1, 5037–5044.

Realini, C.E., Marcos, B., 2014. Active and intelligent packaging systems for a modern society. Meat Sci. 98, 404–419.

Resem Brizio, A.D.D., 2016. Use of indicators in intelligent packaging. Ref. Module Food Sci. https://doi.org/10.1016/B978-0-08-100596-5.03214-5. Elsevier Inc.

Restuccia, D., Spizzirri, U.G., Parisi, O.I., Cirillo, G., Curcio, M., Iemma, F., Puoci, F., Vinci, G., Picci, N., 2010. New EU regulation aspects and global market of active and intelligent packaging for food industry applications. Food Control 21, 1425–1435.

Schedin, K., Geim, A.K., Morozov, S.V., Hill, E.W., Blake, P., 2007. Detection of individual gas molecules adsorbed on grapheme. Nat. Mater. 6, 652–655.

Siracusa, V., Ingrao, C., Lo Giudice, A., Mbohwa, C., Dalla Rosa, M., 2014. Environmental assessment of a multilayer polymer bag for food packaging and preservation: an LCA approach. Food Res. Int. 62, 151–161.

Siracusa, V., Rocculi, P., Romani, S., Dalla Rosa, M., 2008. Biodegradable polymers for food packaging: a review. Trends Food Sci. Technol. 19, 634–643.

Song, Y.S., Hepp, M.A., 2005. US food and drug administration approach to regulating intelligent and active packaging components. In: Han, J.H. (Ed.), Innovation in Food Packaging, first ed. Elsevier Academic Press, Amsterdam, pp. 475–481. Chapter 26.

Tabone, M.D., Gregg, J.J., Beckam, E.J., Landis, A.E., 2010. Sustainability metrics: life cycle assessment and green design in polymers. Environ. Sci. Technol. 44, 8264–8826.

Vanderroost, M., Ragaert, P., Devlieghere, F., De Meulenaer, B., 2014. Intelligent food packaging: the next generation. Trends Food Sci. Technol. 38, 47–62. and related references.

Vu, C., Won, K., 2013. Novel water-resistant UV-activated oxygen indicator for intelligent food packaging. Food Chem. 140, 52–56.

Yam, K.L., Takhistov, P.T., Miltz, J., 2005. Intelligent packaging: concepts and applications. J. Food Sci. 70 (1), R1–R10.

Yucel, U., 2016. Intelligent packaging. Ref. Module Food Sci. https://doi.org/10.1016/B978-0-08-100596-5.03212-1.

# Food packaging and migration

9

*Rui M.S. Cruz*[*,†], *Bruna P.M. Rico*[*], *Margarida C. Vieira*[*,†]
[*]Department of Food Engineering, Institute of Engineering, University of Algarve, Faro, Portugal [†]Centre for Mediterranean Bioresources and Food (MeditBio), Faculty of Sciences and Technology, University of Algarve, Faro, Portugal

## Chapter outline



## 1 Introduction

Food safety and quality are the main concerns for food producers, industries, and consumers. The deterioration of food is caused by physical, chemical, and biological factors (Corrales et al., 2014). A number of prevention measures have been put in place to reduce food spoilage and extend its shelf life. Hazard analysis critical control points and regular inspections of food processing are currently applied to produce high quality food. Packaging is also used in a wide variety of foods with the objective of maintaining the safety and quality of end products. Food packaging is the largest growth sector in packaging and is an integral part of processing. Its main objective is to maintain quality from the time of packaging to the consumer. Another important function of packaging is to protect foods against physical, chemical, and biological hazards (oxygen, moisture, light, microbial, insect, etc.) during storage

**Food Quality and Shelf Life. https://doi.org/10.1016/B978-0-12-817190-5.00009-4**

and distribution. Other functions of the packing include containment, convenience, providing of food information, and to be aesthetically pleasing (Majeed et al., 2013). There are several materials used for food packaging; plastics being the one most used by different types of food industries.

Plastics are very versatile materials that have transformed our everyday lives for more than 60 years, providing many functionalities. The world production of plastics, mainly from fossil fuel raw materials, soared from 1.5 million tons in 1950 to 299 million tons in 2013 (Velis, 2014; Plastics Europe, 2014). Moreover, in 2015, almost 20 million metric tons of plastic packaging was used just in Europe, generating approximately 30 kg of plastic packaging waste per inhabitant per year (Plastics Europe, 2016; Eurostat, 2017).

Packaging includes more than 30 different plastics, the most common being the polyolefins, polyvinyls, and polyesters. New types of packaging materials are likely to be developed; some made from renewable biomass and with biodegradable properties due to the negative environmental impact of plastics and limited petroleum reserves (Raheem, 2012).

Plastic materials are moldable, heat sealable, and printable since they are composed of large, organic (carbon-containing) molecules that can be formed into a variety of useful products. These materials can also be included in production processes where the package is formed, filled, and sealed (Marsh and Bugusu, 2007). However, permeability to light, gases, vapors, and low molecular weight molecules and the possibility of chemical contaminants migrating into the food are some of the disadvantages of the use of these materials (Bhunia et al., 2013).

Different types of plastic materials, for example, polyethylene terephthalate (PET), polypropylene (PP), high (HDPE) and low (LDPE) density polyethylene, polystyrene (PS), polyvinylidene chloride (PVDC), and polycarbonate (PC) (Table 1), or combinations of several types of plastic layers, known as plastic multilayers, can be used in food packaging. Most of these polymers belong to the group of thermoplastics that can undergo mechanical recycling. In contrast, plastic multilayers are currently not recycled into new food packaging (LiferPack2, 2018).

**Table 1** Density, tensile strength and Young's modulus of some polymers used in food packaging (Robertson, 2006)

| Polymer | Density (kg/m$^3$) | Tensile strength (MPa) | Young's modulus (MPa) |
|---|---|---|---|
| LDPE | 910–940 | 8–12 | 200–400 |
| HDPE | 941–965 | 10–60 | 600–1400 |
| PET | 1400 | 55 | 1700 |
| PP | 900 | 30–38 | 1100–1500 |
| PC | 1200 | 62 | 2380 |
| PS | 1050 | 26–48 | 3200–4200 |
| PVDC | 1600–1700 | 73 | 483 |

# 2 Plastic packaging materials

## 2.1 Polyethylene (PE)

PE is produced by addition polymerization of ethylene gas in a high temperature pressure reactor, and is structurally the simplest plastic. Depending on the temperature, pressure, and catalyst of polymerization, a range of low, medium, and high-density resins are produced. The processing conditions control the degree of branching in the polymer chain and, therefore, the density and other properties of films and other types of packaging. Polyethylene is readily heat sealable and can be made into tough films, with a good barrier to moisture and water vapor. Nevertheless, they are not a particularly high barrier to oils and fats or gases, compared with other plastics (Kirwan et al., 2011). Low-density polyethylene (LDPE) and high-density polyethylene (HDPE) are two of the most widely used polyethylenes in the food industry. LDPE is heat sealable, inert, odor free, and shrinks when heated. It is a good moisture barrier but is relatively permeable to oxygen and is a poor odor barrier. It is less expensive than most films and is therefore widely used for bags, for coating papers or boards, and as a component in laminates. LDPE is also used for shrink or stretch-wrapping. HDPE is stronger, thicker, less flexible, and more brittle than LDPE and a better barrier to gases and moisture. Packages made from HDPE are waterproof, have high tear and puncture resistance, and also good seal strength (Fellows and Axtell, 2003).

## 2.2 Polyethylene terephthalate (PET)

PET is a form of thermoplastic polyester that is produced by condensation polymerization upon completion of an esterification reaction using mono-ethylene glycol and terephthalic acid or dimethyl terephthalate (Kim and Lee, 2012). Several beverages, including mineral water, soft drinks, juice, beer, and milk are packed in different types of PET bottles. Moreover, it can also be used in other food packaging applications such as trays for vegetables and other fresh foods. It can also be used for microwave production and conventional thermal processing due to its resistance to high temperatures (up to 220°C). However, analyses show that PET contains small levels of low molecular weight oligomers, from dimers to pentamers. The main volatile substance found in PET is acetaldehyde, which is highly significant because of its effects on odor quality, especially in cola drinks. These cyclic chemicals were detected at levels ranging from 0.06% to 1.0%, depending on the type of PET (Lau and Wong, 2000; Arvanitoyannis and Kotsanopoulos, 2014).

## 2.3 Polypropylene (PP)

PP is a clear glossy film with a moderate barrier to moisture, gases and odors, which is not affected by changes in humidity. It stretches, although less than polyethylene, and has high strength and puncture resistance. It is used in similar applications to LDPE. Oriented polypropylene is a clear glossy film with good optical properties and high tensile strength and puncture resistance. It is widely used to pack biscuits, snack foods,

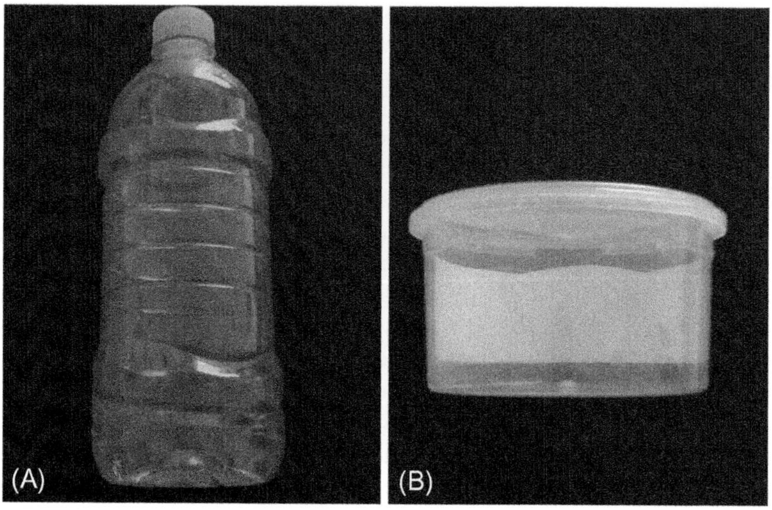


**Fig. 1** Examples of PET (A) and PP (B) packages.

and dried foods (Hirsch, 1991). PP, with its good resistance to oils and fats, is also the principal plastic type used for margarine tubs (Ilsi, 2002) (Fig. 1).

## *2.4 Polycarbonate (PC)*

Polycarbonate (PC) is a linear polyester containing carbonate groups in its structure and is produced by the polymerization of bisphenol A with carbonyl chloride [$COCl_2$]. PC is glass clear, heat resistant, and very tough. For these reasons, it is used in containers for hot filling or processing (Brown, 1992). The polymer is also used for reusable beverage containers, and oven trays for frozen food and prepared meals, since the low-temperature impact strength enhances the durability of the packaging and its toughness. Other applications of PC include boil-in-bag packs, retort pouches, and MW-oven cookware due to their stability at high temperatures (Robertson, 2006; Crompton, 2007). PC was also used for baby feeding bottles. However, a recent controversy over the safety of BPA has led to a market removal of PC infant feeding bottles, and the reduction and removal of BPA from food containers and from surface treatments for plastic, glass, and metallic containers that are thermally processed (Kirwan et al., 2011).

## *2.5 Polystyrene (PS)*

PS is made through the addition polymerization of styrene. Styrene is produced through the catalytic dehydrogenation of ethylbenzene. PS films are hard and transparent, and tend to be very brittle. PS has a high permeability to gases and vapors, making it suitable for food products with short shelf-lives. The main areas of application of polystyrene (PS) are foods such as coffee, ice cream, yogurt, and fruit juices.

PS is also used to make meat trays, biscuit trays, egg cartons, and boxes for fresh produce (Crompton, 2007; Piringer and Baner, 2000; Tawfik and Huyghebaert, 1998). The average detected levels of styrene monomer in food packages ranges from 100 to 3000 ppm (Tang et al., 2000).

### *2.6 Polyvinylidene chloride (PVDC)*

PVDC is a copolymer of vinyl chloride and vinylidene chloride (formed when two hydrogen atoms in ethylene are replaced by chlorine atoms). PVDC is heat sealable and heat shrinkable, and is a high barrier to moisture, aromas, and gases, and to fatty and oily products. As a result of the high gas and odor barrier, it is used to protect flavor and aroma sensitive foods from both loss of flavor and entrance of volatile contaminants. It is very strong and is therefore used in thin films in flexible packaging. However, it has a brown tint, which limits its use in some applications (Kirwan et al., 2011; Piringer and Baner, 2000).

## 3 Migration

In addition to the above-mentioned properties and functions of plastic packaging materials, there are secondary issues of increasing importance that packaging should consider: energy costs and raw materials, recycling, sustainability, and compliance with strict pollutant standards. Indeed, there should be a balance between the leading role and the postuse issues of packaging (Majeed et al., 2013). Moreover, the migration of chemical compounds from the package into the food is another issue regarding the use of plastic packages.

Depending on the food system, migration can be categorized as nonmigrating, volatile, and leaching systems. In the first system, there is an insignificant migration regarding high molecular weight polymeric materials, a small number of inorganic substances, or even pigments. In the volatile system the type of migration is related to dry solid foodstuff with low direct contact potential with the packaging material. The volatile compounds migrate following three stages: diffusion or evaporation of migrant, desorption, and adsorption onto product. In the third system the contact between packaging and food is essential (Lee et al., 2008).

The migration process can be divided into four main steps:

1) diffusion of chemical compounds through the polymers;
2) desorption of the diffused molecules from the polymer surface;
3) sorption of the compounds at the plastic–food interface; and
4) desorption of the compounds in the food (Ferrara et al., 2001).

The most current technique to obtain migration data is to measure the concentration of migrants directly in the foodstuffs, or in food simulants, after being in contact with a packaging material. The determination of migrating compounds from packaging materials to foods under real conditions is extreme complex due to the complexity of food matrices, especially when using fatty foods or oils. For that reason, the simulation of migration conditions can be performed using specific food simulants (Sanches Silva et al., 2006; Stoffers, 2005).

Using food simulants for testing the migration of chemical compounds from food packaging includes two steps. The first one is the exposure of the polymer packaging to the food simulant(s), which then allows the compounds from the packaging material to migrate into the simulant(s). The second is to quantify the migrants transferred to a food simulant in terms of overall migration or specific migration (Castle, 2007). The overall migration limit establishes the maximum amount of substances that the material can release to foods. The specific migration limit is applied to individual substances and is generally based on the toxicological nature of the concerning substance (EU, 2011).

Four food simulants are most commonly used to simplify testing for regulatory compliance purposes, since the simulants are less chemically complex than foods. Water is used to simulate aqueous foods ($pH > 4.5$), 3% aqueous acetic acid to simulate acidic aqueous foods ($pH < 4.5$), 10% aqueous ethanol to simulate alcoholic food products, and olive oil as fatty foods simulant (Grob, 2008). Moreover, to help to guarantee human health, the US Food and Drug Administration and the European Union established the type of food and recommended food simulants for food-contact materials to determine their probable migration into food (Tables 2 and 3).

**Table 2** FDA classification of food types and their description, and recommended food simulants for food-contact materials (US FDA, 2007)

| Type | Classification | Description | Recommended simulant |
|---|---|---|---|
| 1 | Aqueous | Nonacid, aqueous products; may contain salt, sugar, or both ($pH > 5$) | 10% ethanol |
| 2 | Acidic | Acid, aqueous products; may contain salt, sugar, or both, and including oil-in-water emulsions of low- or high-fat content | 10% ethanol |
| 3 | Fatty | Aqueous, acid or nonacid products containing free oil or fat; may contain salt, and including water-in-oil emulsions of low- or high-fat content | Food oil, HB307, Miglyol 812 |
| 4 | | Dairy products and modifications | |
| | Fatty | A. Water-in-oil emulsions, high or low fat | Food oil, HB307, Miglyol 812 |
| | Aqueous | B. Oil-in-water emulsions, high or low fat | 10% ethanol |
| 5 | Fatty | Low-moisture fats and oils | Food oil, HB307, Miglyol 812 |

**Table 2** FDA classification of food types and their description, and recommended food simulants for food-contact materials (US FDA, 2007)—cont'd

| Type | Classification | Description | Recommended simulant |
|---|---|---|---|
| 6 | | Beverages | |
| | Low alcohol | A. Containing up to 8% alcohol | 10% ethanol |
| | Aqueous | B. Nonalcoholic | 10% ethanol |
| | High alcohol | C. Containing more than 8% alcohol | 50% ethanol |
| 7 | | Bakery products (other than those under types VIII or IX) | |
| | Fatty | A. Moist bakery products with surface containing free fat or oil | Food oil, HB307, Miglyol 812 |
| | Aqueous | B. Moist bakery products with surface containing no free fat or oil | 10% ethanol |
| 8 | Dry | Dry solids with the surface containing no free fat or oil | Tenax |
| 9 | Fatty | Dry solids with the surface containing free fat or oil | Food oil, HB307, Miglyol 812 |

**Table 3** EU list of food simulants (EU Regulation 10/2011, 2011)

| Reference | Simulant | Food Type |
|---|---|---|
| A | 10% ethanol | Aqueous food |
| B | 3% acetic acid | Foods that have a hydrophilic character and are able to extract hydrophilic substances and which have a pH below 4.5 |
| C | 20% ethanol | Foods that have hydrophilic character and are able to extract hydrophilic substances, alcoholic foods with alcohol content of up to 20% and foods containing a relevant amount of organic ingredients that render the food more lipophilic |
| D1 | 50% ethanol | Alcoholic foods with an alcohol content of above 20% and dairy products |
| D2 | Vegetable oil | Fatty food and foods which contain free fats at the surface |
| E | Poly(2,6-diphenyl-phenylene oxide), particle size 60–80 mesh, pore size 200 nm | Dry foods |

The overall migration determination is an European Union (EU) regulatory requirement that establishes the migration limits for substances from food-contact materials. Chemical substances migration is dependent on the processing and storage conditions, the nature of the packaging material, and its own composition, as well as the chemical properties of the food (Bradley et al., 2009). According to EU 10/2011 (2011), the overall migration represents the total amount of nonvolatile substances transferred from the food-contact plastic to the food. An overall migration limit of 10mg/dm$^2$ on a contact area basis or 60mg/kg in the simulant or food (for plastics) is mandatory. Moreover, the EU also stablished the testing conditions for overall migration (Table 4).

The use of mathematical diffusion models, supported by scientific evidence, to predict the migration process is an alternative to direct measurement of migration using analytical testing. For compliance testing, these models are calibrated to provide a migration value close to the real one, but in the majority of cases, this is overestimated. Thus, there is a safety margin between predicted and real migration. Nevertheless, because the models are designed to be conservative, if any non-compliance is indicated through mathematical migration modeling, the results must be confirmed by traditional analytical tests (Petersen et al., 2005).

## 3.1 Factors affecting migration

The rate and speed of migration from packaging materials to food can be affected by several conditions: the type of contact (direct or indirect) of packaging materials with food, the packaging material, and its properties, including, among others, thickness and gas permeabilities, the initial concentration of migrant in the packaging material, its structure, molecular size, and polarity, the nature of the food, and the ratio of

**Table 4** Testing conditions for overall migration (EU Regulation 10/2011, 2011)

| Contact time and temperature | Food-contact condition |
|---|---|
| 10days at 20°C | Food at frozen and refrigerated conditions |
| 10days at 40°C | Long-term storage at or below room temperature, including 15min of heating up to 100°C or 70°C for up to 2h |
| 2h at 70°C | Any food heated up to 70°C for up to 2h, or up to 100°C for up to 15min, not followed by long-term room or refrigerated temperature storage |
| 1h at 100°C | High-temperature application up to 100°C for all food stimulants |
| 2h at 100°C or at reflux or alternatively 1h at 121°C | High-temperature applications up to 121°C |
| 4h at 100°C or at reflux | Any food-contact conditions with food simulants A, B or C, at temperature exceeding 40°C |
| 2h at 175°C | High-temperature applications with fatty foods exceeding the conditions heating up to 121°C |

surface area of the packaging to volume of food product (Anderson and Castle, 2003; Barnes et al., 2007; Poças and Hogg, 2007; Triantafyllou et al., 2007). Moreover, other factors, such as contact time and temperature need also to be controlled. On the one hand, migration increases with increasing temperature; on the other hand, a food packaging may be unsuitable at the end of shelf life (Barnes et al., 2007). Arvanitoyannis and Stratakos (2011) reported that concentration of the migrating compound is directly proportional to the square root of the contact time.

In a study reported by Fasano et al. (2012), the migration of phthalates (PAEs), alkylphenols, bisphenol A (BPA), and di(2-ethylhexyl) adipate (DEHA) from a wide range of food-packaging materials to food simulants (3% acetic acid, distilled water, and 15% ethanol) after 10 days of storage at 40°C was studied. The results showed higher amounts of plasticizers released from PE bread-bag compared to PE film. On the other hand, low level of PAEs and DEHA migrated from tetra pack packaging materials. Yogurt packed in PS showed very little dimethyl phthalate migration, but higher amounts of DEHA.

Xu et al. (2010) evaluated the migration of eight PAE compounds from plastics to cooking oil and mineral water under different storage conditions. The temperatures used were 20°C, 40°C, and 60°C, and a "dynamic" state 20°C (food simulant was treated at a frequency of 50 times/min for 5 min daily) and stored up to 2 months. The cooking oil (fatty food 1% to 14%) showed higher PAE content than mineral water (aqueous food <0.35%). The dynamic process also facilitated the migration of the compounds compared with the static state.

Several studies report the migration of styrene into food showing a daily exposure of 18.2–55.2 μg for individuals, with an annual exposure of 6.7–20.2 mg (Lickly et al., 1995; Tawfik and Huyghebaert, 1998; Jin et al., 2005; Khaksar and Ghazi-Khansari, 2009).

Lickly et al. (1995) studied the migration of styrene from different food-contact PS foam materials (meat trays, cups, plates) to oil (mixture of canola, sunflower, and other vegetable oil) and 8% ethanol at 21°C for 10 days, 49°C and 65.5°C for 1, 4, and 10 days. Migration increased from 1 to 10 days, and was found to be proportional to the square root of the increase in time at a specific temperature for all items except for drink cups. This behavior followed Fickian diffusion models where equilibrium partitioning was not a controlling factor. Khaksar and Ghazi-Khansari (2009) also showed that styrene migration in hot drinks such as tea, milk, and cocoa in milk was highly dependent upon temperature of drinks and fat content. Moreover, hot cocoa with milk showed the highest level of migration.

Bisphenol A has received a lot of attention from the food industry, since this compound migrates from PC baby bottles into food. Several studies have investigated BPA release from PC baby bottles (Kubwabo et al., 2009; Nam et al., 2010). Kubwabo et al. (2009) studied the migration of BPA into water, and 10% and 50% ethanol, using polycarbonate and other plastic containers (PC baby bottles, non-PC baby bottles, baby bottle liners, and reusable PC drinking bottles). The study indicated that higher temperatures and longer treatment periods resulted in higher BPA migration from PC bottles. The average concentration of residual BPA in 50% ethanol was higher (2.39 μg/L) compared to water (1.88 μg/L) at 40°C after 240 h.

Cwiek-Ludwicka et al. (2004) studied, in accordance with EU methods, the overall migration from food plastic packaging to water food simulants (distilled water, 3% acetic acid). Time and temperature testing conditions reflected the normal use of tested food packaging. The overall migration studies showed that the migration rate was very low, far below the allowed limit (10 $mg/dm^2$). The higher overall migration into 3% acetic acid (average 250.2 $mg/dm^2$) significantly exceeded the allowed limit, and was found in the case of a multilayer film. The tested multilayer film does not comply with the migration limit and it cannot be used as a food packaging for sour foodstuffs of pH below 4.5. Differences between the magnitude of overall migration into distilled water (0.5–1.1 $mg/dm^2$) and 3% acetic acid are probably due to the presence of easy washable compounds into the acidic medium. The study concluded that the application of such food packaging materials must be limited.

## 3.2 Release of new active compounds

Changes in the lifestyle of consumers, who are looking for environmental friendly and safe products, presents a major challenge for the packaging industry. Food acts as a driving force for the development of new concepts and packaging, where the useful interaction between packaging, environment, and food occurs. For this purpose, the Regulation of the European Union 1935/2004 allowed, for the first time, the opportunity for active packaging to be used in Europe, allowing the application of materials with agents that could migrate to the food (Restuccia et al., 2010).

According to the European Directive 94/62/EC Management of Packaging and Packaging Waste, recycling and reduction of packaging should reduce disposal of waste. Active packaging, that is, films, can help solve the problem of excessive use of plastic packaging, replacing them partially. These can also act as a barrier to the outside (in particular water vapor and oxygen) and as a carrier for functional compounds, at the same time, reducing physical damage and breakage and, therefore, preserve/improve the integrity of food (Rubilar et al., 2013). In order to solve the environmental problems arising from the excessive use of plastics, there has been an increasing interest in developing new biodegradable materials. These novel materials can be obtained from various natural sources, such as proteins, polysaccharides, and lipids, or natural polyesters produced by various microorganisms. The total replacement of petrochemical polymers by biopolymers is not yet possible due to its physico-chemical characteristics and higher production costs. Therefore, it is necessary to continue the studies to improve the properties of the materials produced with these natural polymers so that they can be used commercially. The use of natural polymers has emerged as an alternative to deal with the problem of plastic packaging and as part of the basis of films (Abugoch et al., 2011).

## 3.3 Microencapsulation

Control of the release of compounds, active agents, from films to food is a difficult point to treat. For many years, the microencapsulation was used in the pharmaceutical industry for the release and stability of formulations and to mask strong flavors and odors. Microencapsulation is a useful technique to protect conditions to increase the shelf life of nutraceutical distribution systems, through the incorporation of bioactive compounds in food systems (Nori et al., 2011). The technology associated with the modification of the release of active principles, such as pharmaceuticals, pesticides, colorants, flavorings, or natural compounds, is vast. In between these technologies, polymer matrix systems are widely applied in the form of microparticles. Microparticles are subdivided into microspheres and microcapsules, according to their structure. Microspheres are particles consisting of a polymer network in which the active substance is distributed in its solid or molecular state. Microcapsules are particles consisting of an inner core containing the active agent covered by a polymer layer of variable thickness (Suave et al., 2006).

The concept of microcapsules arose from the idealization of the cellular model. In this, a membrane that surrounds and protects the cytoplasm and the other components exerts other functions such as controlling the input and output of the cell. Similarly, the microcapsule consists of a layer of an encapsulating agent, generally a polymeric material, which acts as a protective film, isolating the active substance and avoiding the effect of its inadequate exposure. The membrane undergoes specific stimulation, releasing the substance at the ideal moment (Suave et al., 2006). Microencapsulation has been studied since 1929 in the textile sector and, more recently, also in the food sector, for the encapsulation of compounds to prevent oxidation and loss of volatile substances (Butstraen and Salaün, 2014). The application of this technology extends to the incorporation of dyes, seasonings, acidulants, vitamins, and minerals. The microencapsulation technique protects these ingredients against nutritional losses and preserves or masks color and taste (inhibiting the reaction with other materials), and increases shelf life and incorporates release control of certain components.

The release of the active agents incorporated into the films (directly or through microencapsulation) can occur through mechanical rupture, through the action of temperature and pH, by means of biodegradation, solubility in the mediums and also by diffusion. The diffusion of the active agents in the polymer matrix is defined as a process of mass transfer of individual molecules from a substance by means of a random molecular motion and associated with a concentration gradient (Suave et al., 2006). Controlled release technology can be used to increase the efficiency in the production of many ingredients. It was initially used by the pharmaceutical industry, with further expansion in other areas, such as controlled release of agrochemicals, fertilizers, veterinary drugs, and in assets for processed foods. The controlled release may be considered as a method whereby one or more encapsulated ingredients are available at specific time periods. One of the advantages of liberation is that the active ingredient is released during prolonged and controlled periods of time (Parize, 2009).

In general, the release of the encapsulated ingredient depends on the type of geometry of the particle and the encapsulating agent employed. These factors dictate the release mechanism, which may be based on solvent effects, diffusion, degradation, or fracture. Any type of stimulus can be used for the release of the encapsulated ingredient, such as change in pH, mechanical stress, temperature, enzyme activity, and time. From Fig. 2, it is observed that the release of the active agent occurs in two steps. In the first step, the swelling of the polymer matrix, rapid solubilization in the medium,

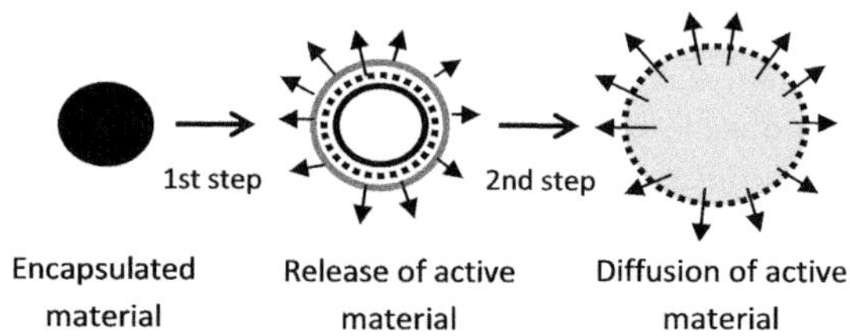


**Fig. 2** Release of active compounds from microcapsules.
Adapted from Matté, G. M., Rosa, S. (2013). Microencapsulation of chitosan microspheres. *Revista Iberoamericana de Polímeros*, 14, 206–218.

and active agent present on the surface of the encapsulant. The second step is characterized by the relaxation and swelling of the polymer matrix and the occurrence of diffusion of the active agent present in the microcapsules (Matté and Rosa, 2013; Parize, 2009).

An important factor to be considered in the diffusion process of an active agent is its solubility in the polymer matrix. When an active agent is dispersed in the matrix, diffusion occurs as its solubilization occurs in the polymer. Already in a system in which the active agent is solubilized in the matrix polymer, this step will be suppressed, allowing faster diffusion. The ability of a polymer to swell also directly influences the diffusion of the active agent. When the microencapsulation system contacts with water, for example, hydration of the material may occur and also a progressive gelation of the polymer chains, forming a high viscosity layer in the water-polymer interface. This viscous layer increases in thickness as the hydration or swelling progresses. Consequently, the diffusion of the active agent is determined by the rate of swelling of the polymer. Since there is not a single type of active agent release curve that satisfies all needs, four theoretical models of release can be defined. The first considers a trigger mechanism, which initiates the release. The second mechanism assumes that the wall of the microcapsule acts as a reservoir, assuming that the rate of release is constant. The third assumes the passage through the wall of the microcapsule. The fourth model considers the wall as a semipermeable and selective membrane of different molecular weights (Matté and Rosa, 2013; Parize, 2009; Suave et al., 2006).

## 3.4 Quantification of active agents

Laboratory methods, including high performance liquid chromatography (HPLC), capillary electrophoresis, ultraviolet-visible spectrophotometry, spectrometry, and electrochemical detectors have been used to analyze the total phenolic compounds and flavonoids in various substances (Šeruga et al., 2011). Phenolic compounds are widely distributed in the plant kingdom. They are defined as substances having aromatic rings, having one or more hydroxyl substituents, including functional groups thereof. Among the fruits, the grape is one of the major sources of phenolic compounds. The main phenolics present in grapes are flavonoids (anthocyanins and flavonols), stilbenes (resveratrol), phenolic acids (derived from cinnamic and benzoic acids), and a wide variety of tannins. Anthocyanins are flavonoids that are distributed in nature and are responsible for most of the blue and violet colors, and all the shades of red that appear in flowers, fruits, some leaves, stems, and roots of plants. In vines, they accumulate in the leaves during senescence and are responsible for the coloring of the skin of red grapes, also found in the pulp and grains of some varieties of grapes (Malacrida and Motta, 2005).

## 3.5 Release kinetics

The idea of controlled release of biologically active substances has been used since the 1950s with the development of materials of lipid, inorganic, and polymeric nature (Witt, 2013).

Whenever an encapsulated material is produced, it is necessary to predict how it behaves in terms of release. It is necessary to ensure the release of a suitable form of the encapsulated active substance, as well as to predict how the release behaves over time. For the accomplishment of these studies, it is common to resort to mathematical models that express the behavior of the system under study. These models become important in the development of a new formulation and experimental verification of the release process. The experimental data obtained can be analyzed by means of kinetic equations that relate the amount of active substance released as a function of time (Moreira, 2010).

Controlled release formulations are intended to produce increasingly effective products. For this purpose, the use of mathematical modeling becomes very useful as it allows the prediction of the kinetic release before the release systems are executed. More often, this allows the measurement of some important physical parameters, such as the diffusion coefficient of the active substance, using the experimental release data adjustment model. Mathematical modeling, whose development requires the understanding of all phenomena that affect the release kinetics of active substances, has a very important value for process optimization (Dash et al., 2010).

The model can simply be thought of as a mathematical metaphor for some aspects of reality, which, in this case, is identified with the set of phenomena that govern the kinetics of liberation. Thus, mathematical modeling is widely used in different areas such as genetics, medicine, psychology, biology, economics, engineering, and technology. The methods dependent on the model are based on different mathematical functions, which describe the dissolution profile. Once a suitable function has been selected, the dissolution profiles are evaluated according to the parameters of the derived model. To determine the release kinetics of a substance, the model should describe its dissolution profile (Dash et al., 2010).

The Korsmeyer-Peppas, Weibull, Higuchi, and Fick models are considered the most widely applied models in the study of the controlled release of compounds (Moreira, 2010).

### 3.5.1 Korsmeyer-Peppas model

In 1983, Korsmeyer described the release of substances from a polymer system. To find out the release mechanism, the Korsmeyer-Peppas model was developed:

$$\frac{M_t}{M_\infty} = k.t^n \tag{1}$$

where $Mt/M_\infty$ is the fraction of active agent released over time $t$, $k$ is the rate constant, and $n$ is the release exponent.

The value of $n$ is used to characterize different release mechanisms and provides information on the mechanism of release of active substances from the polymer matrix, as can be observed in Table 5 (Dash et al., 2010; Parize, 2009; Shoaib et al., 2006).

The Arrhenius model was developed to evaluate the influence of temperature on the effective diffusion coefficient of a given active substance. Thus, a model can be transformed by inserting a reference temperature, Tref. This transformation not only confers stability to numerical integration and parameter estimation, but also allows the use of the reference velocity constant, kref (Portella et al., 2014; Goneli et al., 2014a, b)

$$k(t) = k_{ref}.e^{-\frac{E_a}{R}\left(\frac{1}{T}-\frac{1}{T_{ref}}\right)} \tag{2}$$

where $k(t)$ is the reaction rate constant, $E_a$ is the activation energy of the reaction, $T$ is the temperature, and $R$ is the ideal gas constant ($R=8.314$ J/mol K).

Thus, the Korsmeyer-Peppas model with temperature dependence can still be written in another way, using as reference the data of Eqs. (1), (2):

$$\frac{M_t}{M_\infty} = k_{ref}.e^{-\frac{E_a}{R}\left(\frac{1}{T}-\frac{1}{T_{ref}}\right)}.t^n \tag{3}$$

### 3.5.2 Weibull model

This model has been described by different dissolution processes according to the equation:

$$\frac{M_t}{M_\infty} = 1 - e^{-k(t-t')^b} \tag{4}$$

where $M_t/M_\infty$ is the fraction of the active agent released over time $t$, $k$ is the rate constant, $t'$ is the so-called latency time, the elapsed time between the initial time and

**Table 5** Diffusion exponent and mechanism of release of the solute for a spherical form (Parize, 2009).

| n | Release mechanism |
|---|---|
| ≤0.43 | Fickian diffusion Case I: Diffusion of compound through the matrix layers |
| $0.43 < n < 0.85$ | Non Fickian: Diffusion and swelling occur simultaneously (anomalous diffusion) |
| 0.85 | Fickian diffusion Case II: swelling/relaxation of the matrix |
| > 0.85 | Super Fickian diffusion Case II: Diffusion, swelling, relaxation and erosion of the matrix |

the beginning of the release of the active substance, and b is a parameter of form that characterizes the curves as exponential ($b=1$), S-shaped ($b>1$), or parabolic ($b<1$) (Dash et al., 2010; Kalam et al., 2007; Oms-Oliu et al., 2009).

This model can be transformed if we introduce the temperature dependence, according to the Arrhenius model:

$$\frac{M_t}{M_\infty} = 1 - e^{-k_{ref}.e^{-\frac{E_a}{R}\left(\frac{1}{T}-\frac{1}{T_{ref}}\right)}.(t-t')^b} \tag{5}$$

### 3.5.3 Higuchi model

Another example of a mathematical model to describe the release of a substance from a matrix system was proposed by Higuchi in 1961 (Dash et al., 2010). In general the Higuchi model is applied through the equation:

$$\frac{M_t}{M_\infty} = k.\sqrt{t} \tag{6}$$

where $M_t/M_\infty$ is the active agent fraction released over time $t$, and $k$ is the velocity constant of the Higuchi model (Moreira, 2010).

This model can be transformed if we introduce the temperature dependence, according to the Arrhenius model:

$$\frac{M_t}{M_\infty} = k_{ref}.e^{-\frac{E_a}{R}\left(\frac{1}{T}-\frac{1}{T_{ref}}\right)}.\sqrt{t} \tag{7}$$

### 3.5.4 Fick model

Diffusion is the process by which matter is transported from one place to another within the system itself and results from random molecular movements occurring at small distances. Adolf Fick, in 1855, was the first to attempt to quantify the diffusion process (Manadas et al., 2002). The Fick model is based on diffusion fluxes, where the diffusion of the bioactive compound from a region with high concentration to a low concentration occurs (Penim, 2015).

The diffusion transfer rate of a substance can be expressed by Fick's first law:

$$F = -D\frac{dC}{dx} \tag{8}$$

where $F$ is the transfer rate per unit area, $D$ is the effective diffusion coefficient, $C$ is the concentration of the diffusing substance, and $x$ is the distance in the diffusion direction (Pires, 2011).

The second law of Fick is widely used when it is interesting to know the diffusion of a fluid, due to its concentration, and the diffusion of the material; this law is especially useful when the fluid in question is water (Santos et al., 2014):

$$\frac{dC}{dt} = D\left(\frac{d^2C}{dx^2}\right) \tag{9}$$

Crank (1975), based on Fick's second law for a flat plate in contact with a solution, arrived at the simplified equation of the Fick model:

$$\frac{M_t}{M_\infty} = \left(\frac{2}{L}\right)\left(\frac{D.t}{\pi}\right)^{0.5} \tag{10}$$

where $M_t/M_\infty$ is the active agent fraction released over time $t$, $D$ is the effective diffusion coefficient, and $L$ is the film thickness (El-Aquar and Murr, 2003).

In order to evaluate the influence of temperature on the effective diffusion coefficient, the Arrhenius equation was used, described as follows (Lisboa et al., 2015):

$$D = D_{ref}.e^{-\frac{E_a}{R}\left(\frac{1}{T}-\frac{1}{T_{ref}}\right)} \tag{11}$$

Thus, the Fick model can be transformed if we introduce the temperature dependence, according to the Arrhenius model:

$$\frac{M_t}{M_\infty} = \left(\frac{2}{L}\right)\left(\frac{D_{ref}.e^{-\frac{E_a}{R}\left(\frac{1}{T}-\frac{1}{T_{ref}}\right)}.t}{\pi}\right)^{0.5} \tag{12}$$

## 3.6 Migration of compounds in new packaging systems

In a study with tripolyphosphate-crosslinked chitosan particles, Hamamura et al. (2010) found that some factors influenced the stability of chitosan particles, such as the effect of the contact time between chitosan and solutions, the ionic strength of the crosslinking between chitosan and tripolyphosphate, and the effect of pH. In this study it was proved that chitosan and tripolyphosphate particles undergo dissolution at pH 1.2 and at pH 7.4 only appear to suffer from swelling. Hamamura et al. (2010) reported that if the pH of a solution decreases, the higher the concentration of H+ ions that neutralize tripolyphosphate and the lower the degree of chitosan-tripolyphosphate crosslinking. If the pH is too low the ionic dissociation and dissolution of the chitosan-tripolyphosphate network occurs. Otherwise, if the pH increases, protonation of the chitosan amino groups also increases, which induces a decrease in crosslinking density with tripolyphosphate, allowing swelling due to increased chitosan chain distance, and rapid release of the chitosan to the surrounding solution.

Rodriguez et al. (2007) studied the addition of essential oils to a wax coating in order to develop an antimicrobial active packaging. The results showed the ability of these systems to preserve strawberries from microorganism contamination by the release of antimicrobials from the coating. During this study, there was no direct contact between the essential oils and the food product. Therefore, the natural volatile compounds (eugenol, carvacrol, thymol) present in the headspace packaging were considered the responsible compounds for the inhibition of the pathogen growth.

Torres-Arreola et al. (2007), in a study with fresh sierra fish fillets, reported a delay in lipid oxidation and protein denaturation incorporating BHT into LDPE packaging. Compared with the control films, samples packed in BHT-LDPE films demonstrated lower lipid oxidation, expressed as thiobarbituric acid-reactive substances (TBARS) values ($4.20 \pm 0.52$ mg compared to $11.95 \pm 1.06$ mg malonaldehyde (MDA)/kg), peroxide index values ($7.20 \pm 1.38$ meq/kg compared with $15.15 \pm 1.48$ meq/kg), and free fatty acid contents ($7.98 \pm 0.43$ compared with $11.83 \pm 1.26$ of oleic acid). Samples packed in BHT-LDPE films also exhibited less tissue damage and retained firmness better than fillets packed in LDPE.

Graciano-Verdugo et al. (2010) showed that sealable LDPE films containing 1.9% and 3% of α-tocopherol maintained the oxidation stability of corn oil for 16 weeks at 30°C, compared to the control bag without the antioxidant (12 weeks).

Manzanarez-López et al. (2011) reported that poly(lactic acid) films containing 2.58% of α-tocopherol were also able to delay the induction of oxidation, measured as peroxide value, of soybean oil at 20°C (max. values of 9.9 meq/kg compared with 19.5 meq/kg), 30°C (max. values $<10$ meq/kg compared with 27.5 meq/kg), and 40°C (max. values of 13.5 meq/kg compared with 33.9 meq/kg).

Another study, by Carrizo et al. (2016), in which green tea extract was added to a laminating adhesive and thus not in direct contact with the packaged food (peanuts and cereals covered with chocolate), showed that the package was able to protect food against oxidation during a long-term period of 16 months.

Bolumar et al. (2016) tested LDPE films coated with rosemary extract to protect pork patties from high-pressure-processing-induced lipid oxidation and, consequently, extend the shelf life. The new package delayed oxidation up to 25 days, demonstrated by lower peroxide values ($7.2 \pm 1.38$ meq/kg compared with $15.15 \pm 1.48$ meq/kg), FFA ($7.98 \pm 0.43$% compared with $11.83 \pm 1.26$% oleic acid), and TBARS ($4.20 \pm 0.52$ mg MDA/kg compared with $11.95 \pm 1.06$ mg MDA/kg), while the lipid oxidation of chicken breast patties submitted to high-pressure treatment and, stored at 5°C,was higher in the surface part of samples (Bolumar et al., 2011).

## 4 Conclusion

The use of plastic packaging materials and the migration of chemical compounds into food must be continuously controlled by all stakeholders to warrant the use of maximum legal limits and consequently the production of safer food products. Moreover, the search for new techniques that may contribute to the safety assessment for all potential migrants is also mandatory. Nevertheless, the continuous demand for healthier food processes and food products, and the reduction of the use of plastic materials, creates new opportunities for the production and use of more natural and sustainable food packaging materials, and thus, contributing to the environment and human health.

## References

Abugoch, L.E., Tapia, C., Villamán, M.C., Yazdani-Pedram, M., Díaz-Dosque, M., 2011. Characterization of quinoa protein and chitosan blend edible films. Food Hydrocoll. 25, 879–886.

Anderson, W.A.C., Castle, L., 2003. Benzophenone in cartonboard packaging materials and the factors that influence its migration into food. Food Addit. Contam. 20, 607–618.

Arvanitoyannis, I.S., Kotsanopoulos, K.V., 2014. Migration phenomenon in food packaging. Food—package interactions, mechanisms, types of migrants, testing and relative legislation—a review. Food Bioprocess Technol. 7, 21–36.

Arvanitoyannis, I.S., Stratakos, A.C., 2011. Migration from Food Packaging Materials in Technologies of Preservation of Food and Food Packaging. University Studio Press, Thessaloniki, Hellas (Greece).

Barnes, K.A., Sinclair, C.R., Watson, D.H., 2007. Chemical Migration and Food Contact Materials. Woodhead, Cambridge1–13.

Bhunia, K., Sablani, S.S., Tang, J., Rasco, B., 2013. Migration of chemical compounds from packaging polymers during microwave, conventional heat treatment, and storage. Compr. Rev. Food Sci. Food Saf. 12, 523–545.

Bolumar, T., Andersen, M.L., Orlien, V., 2011. Antioxidant active packaging for chicken meat processed by high pressure treatment. Food Chem. 129, 1406–1412.

Bolumar, T., La Peña, D., Skibsted, L.H., Orlien, V., 2016. Rosemary and oxygen scavenger in active packaging for prevention of high-pressure induced lipid oxidation in pork patties. Food Packing Shelf Life 7, 26–33.

Bradley, E.L., Castle, L., Jickells, S.M., Mountfort, K.A., Reada, W.A., 2009. Use of overall migration methodology to test for food-contact substances with specific migration limits. Food Addit. Contam. 26, 574–582.

Brown, W.E., 1992. Primary food packaging materials. Plastics in Food Packaging. Marcel Dekker, Inc.

Butstraen, C., Salaün, F., 2014. Preparation of microcapsules by complex coacervation of gum Arabic and chitosan. Carbohydr. Polym. 99, 608–616.

Carrizo, D., Taborda, G., Nerín, C., Bosetti, O., 2016. Extension of shelf life of two fatty foods using a new antioxidant multilayer packaging containing green tea extract. Innov. Food Sci. Emerg. Technol. 33, 534–541.

Castle, L., 2007. Chemical migration into food: an overview. In: Barnes, K.A., Sinclair, C.R., Watson, D.H. (Eds.), Chemical Migration and Food Contact Materials. Woodhead.

CEC, 2011. Regulation 10/2011 of 4 January 2011 on plastics materials and articles intended to come into contact with food. Off. J. Eur. Union L12, 1–89.

Corrales, M., Fernández, A., Han, J.H., 2014. Antimicrobial packaging systems. In: Han, J. (Ed.), Innovations in Food Packaging. Amsterdam, Elsevier, pp. 133–170.

Crank, J., 1975. The Mathematics of Diffusion, second ed. Clarendon Press, Oxford.

Crompton, T.R., 2007. Additive Migration From Plastics into Food, first ed. Smithers Rapra Press, Shawbury, UK.

Cwiek-Ludwicka, K., Stelmach, A., Mazańska, M., Jurkiewicz, M., Półtorak, H., 2004. Determination of the overall migration of chemicals from the plastic packaging into the aqueous models of food using the EU methods. Ann. Natl. Inst. Hyg. 55, 1–8.

Dash, S., Murthy, P.N., Nath, L., Chowdhury, P., 2010. Kinetic modeling on drug release from controlled drug delivery systems. Acta Pol. Pharm. Drug Res. 67, 217–223.

El-Aquar, Â.A., Murr, F.E.X., 2003. Kinetics and modeling on osmotic dehydration of papaya (*Carica papaya* L.). Food Sci. Technol. 23, 69–75.

EU, 2011. Commission Regulation (EU) No. 10/2011 of 14 January 2011 on plastic materials and articles intended to come into contact with food. Off. J. Eur. Union L12, 1–89.

Eurostat, 2017. Packaging Waste Statistics. http://ec.europa.eu/eurostat.

Fasano, E., Bobo-Blay, F., Cirillo, T., Montuori, P., Lacorte, S., 2012. Migrationof phthalates, alkylphenols, bisphenol A and di(2-ethylhexyl)adipate fromfood packaging. Food Control 27, 132–138.

Fellows, P.J., Axtell, B.L., 2003. Appropriate Food Packaging: Materials and Methods for Small Businesses, second ed. Practical Action Publishing.
Ferrara, G., Bertolodo, M., Scoponi, M., Ciardelli, F., 2001. Diffusion coefficient and activation energy of Irganox 1010 inpoly(propylene-*co*-ethylene) copolymers. Polym. Degrad. Stab. 73, 411–416.
Goneli, A.L.D., Nasu, A.K., Gancedo, R., Araúlo, W.D., Sarath, K.L.L., 2014a. Drying kinetics of *Cordia verbenacea* DC. Leaves. Revista Brasileira de Plantas Medicinais 16, 434–443.
Goneli, A.L.D., Vieira, M.C., Vilhasanti, H.C.B., Gonçalves, A.A., 2014b. Mathematical modeling and effective diffusion of *Schinus terebinthifolius* leaves during drying. Pesquisa Agropecuária Tropical 44, 56–64.
Graciano-Verdugo, A.Z., Soto-Valdez, H., Peralta, E., Cruz-Zárate, P., Islas-Rubio, A.R., Sánchez-Valdes, S., Sáchez-Escalante, A., González-Mández, N., González-Ríos, H., 2010. Migration of α -tocopherol from LDPE films to corn oil and its effect on the oxidative stability. Food Res. Int. 43, 1073–1078.
Grob, K., 2008. The future of simulants in compliance testing regarding the migration from food contact materials. Food Control 19, 263–268.
Hamamura, N.R., Negrón, A.V., Cavero, H.A., Milla, A.L., 2010. Preparation of chitosan particles cross-linked with tripolyphosphate and modified with polyethylene glycol. Rev. Soc. Quim. Peru 76, 336–354.
Hirsch, A., 1991. Flexible Food Packaging. Van Nostrand Reinhold, New York.
ILSI Europe Packaging Material Task Force, 2002. Packaging materials: 3. Polypropylene as a packaging material for foods and beverages. International Life Sciences Institute.
Jin, O.C., Jitsunari, F., Askawa, F., Dong, S.L., 2005. Migration of styrene monomer, dimmer and trimers from polystyrene to food simulants. Food Addit. Contam. 22, 693–699.
Kalam, M.A., Humayun, M., Parvez, N., Yadav, S., Garg, A., Amin, S., Sultana, Y., Ali, A., 2007. Release Kinetics of modified pharmaceutical dosage forms: a review. Continental J. Pharm. Sci. 1, 30–35.
Khaksar, M.R., Ghazi-Khansari, M., 2009. Determination of migration monomer styrene from GPPS (general purpose polystyrene) and HIPS (high-impact polystyrene) cups to hot drinks. Toxicol. Mech. Methods 19, 257–261.
Kim, D.J., Lee, K.T., 2012. Determination of monomers and oligomers in polyethylene terephthalate trays and bottles for food use by using high performance liquid chromatography -electrospray ionization-mass spectrometry. Polym. Test. 31, 490–499.
Kirwan, M.J., Plant, S., Strawbridge, J.W., 2011. Plastics in food packaging. In: Coles, R., Kirwan, M.J. (Eds.), Food and Beverage Packaging Technology. John Wiley and Sons.
Kubwabo, C., Kosarac, I., Stewart, B., Gauthier, B.R., Lalonde, K., Lalonde, P.J., 2009. Migration of bisphenol a from plastic baby bottles, baby bottle liners and reusable polycarbonate drinking bottles. Food Addit. Contam. 26, 928–937.
Lau, O., Wong, S., 2000. Contamination in food from packaging material. J. Chromatogr. A 882, 255–270.
Lee, B.S., Yam, K.L., Piergiovanni, L., 2008. Migration and food package. In: Food Packaging Science and Technology. CRC Press, USA.
Lickly, T.D., Lehr, K.M., Welsh, G.C., 1995. Migration of styrene from polystyrene foam food-contact article. Food Chem. Toxicol. 33, 475–481.
LiferPack2L, 2018. Recycling Packs to Life. http://www.liferpack2l.eu/en.
Lisboa, J.F., Silva, J.N., Cavalcanti, M.T., Silva, E.M.C., Gonçalves, M.C., 2015. Analysis of hydration of grains of birdseed. Revista Brasileira de Engenharia Agrícola e Ambiental 19, 218–223.
Majeed, K., Jawaid, M., Hassan, A., Bakar, A.A., Khalil, H.P.S.A., Salema, A.A., Inuwa, I., 2013. Potential materials for food packaging from nanoclay/natural fibres filled hybrid composites. Mater. Des. 46, 391–410.

Malacrida, C.R., Motta, S., 2005. Total phenolics and anthocyanins in grape juice. Food Sci. Technol. 25, 659–664.

Manadas, R., Pina, M.E., Veiga, F., 2002. Dissolution studies in vitro as a prognostic tool for oral absorption of modified release pharmaceutical dosage forms. Revista Brasileira de Ciências Farmacêuticas 38, 375–399.

Manzanarez-López, F., Soto-Valdez, H., Auras, R., Peralta, E., 2011. Release of α -tocopherol from poly(lactic acid) films, and its effect on the oxidative stability of soybean oil. J. Food Eng. 104, 508–517.

Marsh, K., Bugusu, B., 2007. Food packaging–roles, materials and environmental issues. J. Food Sci. 72, 39–55.

Matté, G.M., Rosa, S., 2013. Microencapsulation of chitosan microspheres. Revista Iberoamericana de Polímeros 14, 206–218.

Moreira, J.S., 2010. Estudo de síntese e cinética de libertação controlada em bionanocompósitos. Dissertação de Mestrado em Engenharia Química. Universidade de Aveiro, Departamento de Química.

Nam, S.H., Seo, Y.M., Kim, M.G., 2010. Bisphenol a migration from polycarbonate baby bottle with repeated use. Chemosphere 79, 949–952.

Nori, M.P., Favaro-Trindade, C.S., Alencar, S.M., Thomazini, M., Balieiro, J.C.C., Castillo, C.J.C., 2011. Microencapsulation of propolis extract by complex coacervation. LWT Food Sci. Technol. 44, 429–435.

Oms-Oliu, G., Odriozola-Serrano, I., Soliva-Fortuny, R., Martín-Belloso, O., 2009. Use of Weibull distribution for describing kinetics of antioxidant potential changes in fresh-cut watermelon. J. Food Eng. 95, 99–105.

Parize, A.L., 2009. Desenvolvimento de sistemas microparticulados e de filmes à base de quitosana e corante natural cúrcuma. Dissertação de Doutoramento em Química, Universidade Federal de Santa Catarina, Centro de Ciências Físicas e Matemáticas, Programa de Pós-Graduação em Química, Florianópolis.

Penim, D.F.L., 2015. Microencapsulação de própolis em matrizes de polissacáridos e estudos de libertação controlada. Dissertação de Mestrado em Engenharia Alimentar. Universidade de Lisboa, Instituto Superior de Agronomia.

Petersen, J.H., Trier, X.T., Fabech, B., 2005. Mathematical modelling of migration: a suitable tool for the enforcement authorities? Food Addit. Contam. 22, 938–944.

Pires, R.O., 2011. Modelagem matemática de perfis de liberação de fármacos a partir de nanocarregadores. Dissertação de Mestrado em Nanociências, Centro Universitário Franciscano de Santa Maria.

Piringer, O.G., Baner, A.L., 2000. Plastic Packaging Materials for Food. Wiley-VCH.

Plastics Europe, 2014. Association of Plastics Manufactures, Plastics—the Facts. An analysis of European plastics production, demand and waste data. Available from: www.plasticseurope.org.

Plastics Europe, 2016. Plastics—the Facts, 2016. http://www.plasticseurope.org.

Poças, M.F., Hogg, T., 2007. Exposure assessment of chemicals from packaging materials in foods: a review. Trends Food Sci. Technol. 18, 219–230.

Portella, A.C.F., Munaro, M., Ascêncio, S.D., Siqueira, C.A., Ferreira, T.P.S., Aguiar, R.W.S., 2014. Physical and chemical characterization of essential oil of *Siparuna guianensis* Aublet. Quim. Nova 37, 844–849.

Raheem, D., 2012. Application of plastics and paper as food packaging materials—an overview. Emirates J. Food Agric. 25, 177–188.

Restuccia, D., Spizzirri, U.G., Parisi, O.I., Cirillo, G., Curcio, M., Iemma, F., Puoci, F., Vinci, G., Picci, N., 2010. New EU regulation aspects and global market of active and intelligent packaging for food industry applications. Food Control 21, 1425–1435.

Robertson, G.L., 2006. Food Packaging Principles and Practice, second ed. CRC Press, Taylor and Francis Group, Boca Raton, F.L.
Rodriguez, A., Batlle, R., Nerin, C., 2007. The use of natural essential oils as antimicrobial solutions in paper packaging: part II. Prog. Org. Coat. 60, 33–38.
Rubilar, J.F., Cruz, R.M.S., Silva, H.D., Vicente, A.A., Khmelinskii, I., Vieira, M.C., 2013. Physico-mechanical properties of chitosan films with carvacrol and grape seed extract. J. Food Eng. 115, 466–474.
Sanches Silva, A., Sendón García, R., Cooper, I., Franz, R., Paseiro Losada, P., 2006. Compilation of analytical methods and guidelines for the determination of selected model migrants from plastic packaging. Trends Food Sci. Technol. 17, 535–546.
Santos, J.A., Lima, D.R., Cardoso, A.O., Reis, A.B., 2014. Comparativo entre diferentes solubilizadores usados na produção de filmes de quitosana. XX Congresso brasileiro de Engenharia Química, Florianópolis.
Šeruga, M., Novak, I., Jakobek, L., 2011. Determination of polyphenols content and antioxidant activity of some red wines by differential pulse voltammetry, HPLC and spectrophotometric methods. Food Chem. 124, 1208–1216.
Shoaib, M.H., Tazeen, J., Merchant, H.A., Yousuf, R.I., 2006. Evaluation of drug release kinetics from ibuprofen matrix tablets using HPMC. Pak. J. Pharm. Sci. 19, 119–124.
Stoffers, N.H., 2005. Certified reference materials for food packaging specific migration tests: development, validation and modeling. http://www.library.wur.nl/WebQuery/edepot/121626.
Suave, J., Dall'agnol, E.C., Pezzin, A.P.T., Silva, D.A.K., Meier, M.M., Soldi, V., 2006. Microencapsulação: Inovação em diferentes áreas. Health Environ. J. 7, 12–20.
Tang, W., Hemm, I., Eisenbrand, G., 2000. Estimation of human exposure to styrene and ethylbenzene. Toxicology 144, 39–50.
Tawfik, M.S., Huyghebaert, A., 1998. Polystyrene cups and containers: styrene migration. Food Addit. Contam., Part A 15, 592–599.
Torres-Arreola, W., Soto-Valdez, H., Peralta, E., Cardenas-López, J.L., Ezquerra-Brauer, J.M., 2007. Effect of a low-density polyethylene film containing butylated hydroxytoluene on lipid oxidation and protein quality of sierra fish (*Scomberomorus sierra*) muscle during frozen storage. J. Agric. Food Chem. 55, 6140–6146.
Triantafyllou, V.I., Akrida-Demertzi, K., Demertzi, P.G., 2007. A study on the migration of organic pollutants from recycled paper board packaging materials to solid food matrices. Food Chem. 101, 1759–1768.
US FDA, 2007. Guidance for Industry: Preparation of Premarket Submissions for Food Contact Substances: Chemistry Recommendations. Ingredients, Additives, GRAS & Packaging. Office of Food Additive Safety.
Velis, C., 2014. Global Recycling Markets—Plastic Waste: A Story for One Player—China. Report Prepared by FUELogy and Formatted by D-Waste on Behalf of International Solid Waste. Association—Globalization and Waste Management Task Force, ISWA, Vienna.
Witt, E.P., 2013. Produção de blendas poliméricas obtidas a partir de fécula de mandioca e álcool polivinílico—PVA para estudo cinético de liberação controlada de fármaco. Dissertação de Bacharelato em Química. Universidade Tecnológica Federal do Paraná, Campus Pato Branco.
Xu, Q., Yin, X., Wang, M., Wang, H., Zhang, N., Shen, Y., Xu, S., Zhang, L., Gu, Z., 2010. Analysis of phthalate migration from plastic containers to packaged cooking oil and mineral water. J. Agric. Food Chem. 58, 11311–11317.

# Assessment of fresh fruit and vegetable quality with non-destructive methods

10

*Hulya Cakmak*
Department of Food Engineering, Hitit University, Corum, Turkey

**Chapter Outline**



## 1 Introduction

Harvesting maturity is the primary indicator of the postharvest shelf life and final quality of fruit and vegetables (Lamikanra, 2002). Unripe or overripe fruits and vegetables have limited shelf life with respect to their vulnerability to mechanical damage, softening, and physiological and microbiological disorders; and eventually, these cause great economic loss. Therefore, monitoring quality especially with non-destructive methods, will create a big difference for limiting these avoidable food wastes. As it is shown in Fig. 1, what we understand from quality of food is a combination of several aspects. Every single branch of this complex structure correlates with all others, but they also provide a different perspective of quality individually. Most of the food losses are associated with the internal and external defect formation due to the expiry of shelf life. However, contamination of food with foreign materials (soil, dirt, feces, chemical residues, worms, pathogens, etc.), adulteration with similar but lower quality duplicates, cracks, cuts, scars, scabs, and punctures occurring during handling, and nonstandard size and shapes also affect these losses (Lamikanra, 2002; Lu and Lu, 2017; Su et al., 2017).

Quality attributes of foods are divided into two major categories: internal and external quality. The internal quality involves maturity, texture (firmness, mealiness, etc.), nutrient composition (total soluble solid or moisture, pH, acidity, sugar, chlorophyll, anthocyanin, etc.), internal defects or diseases (browning, chilling injury, water core, hollow, pit, rot, microbial spoilage, etc.), chemical contaminants (pesticides, antibiotics,

**Food Quality and Shelf Life. https://doi.org/10.1016/B978-0-12-817190-5.00010-0**

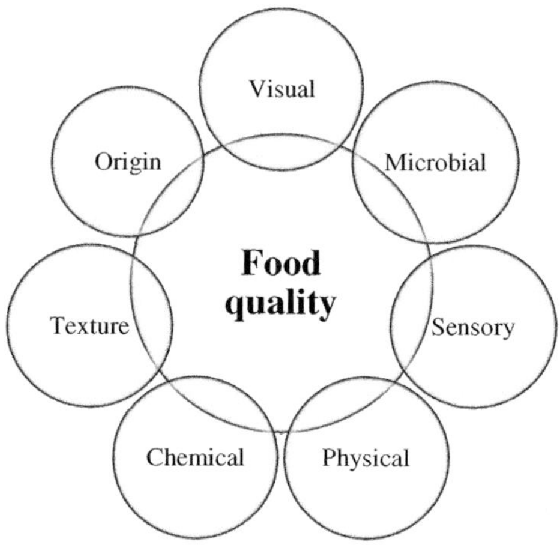


**Fig. 1** Different aspects of food quality.

mycotoxins, etc.) taste, and aroma. External quality involves size, shape, color (visual color, glossiness, sunburn, etc.), and defects and bruises (puncture, cut, scald, cracks, splits, sprouts, etc.), as well as foreign materials (soil, feces, pesticides, insects, etc.) (Studman, 2001; Nicolaï et al., 2014; Ali et al., 2017; Su et al., 2017; Gao et al., 2018; Li et al., 2018; Su and Sun, 2018). In Fig. 2, some internal and external defects and disorders are shown. The external and internal parameters are mostly monitored and analyzed with manual visual inspection assays like sensory tests, or with destructive methods. However, due to the nature of the analysis method, the food sample is either partly or completely destroyed. Especially for foods with high volume, such as watermelon, melon or pumpkin, the destructive and sensory methods are often time-consuming, subjective, and expensive (Sun et al., 2010; Ali et al., 2017).

The development of non-destructive quality evaluation methods has become a necessity for cost reduction, and improving the accuracy and repeatability of analysis, along with reduction of the analysis time (Irudayaraj and Reh, 2008). Non-destructive methods allow real-time, on-line, and in-field applications, also the automated monitoring of food quality can be performed with minimal or no preparation of sample for these analyses. Fruits and vegetables are classified with respect to the authenticity of variety, size, shape, color, maturity, internal and external defects, diseases, or presence of contamination, by non-destructive methods. There is a vast amount of literature

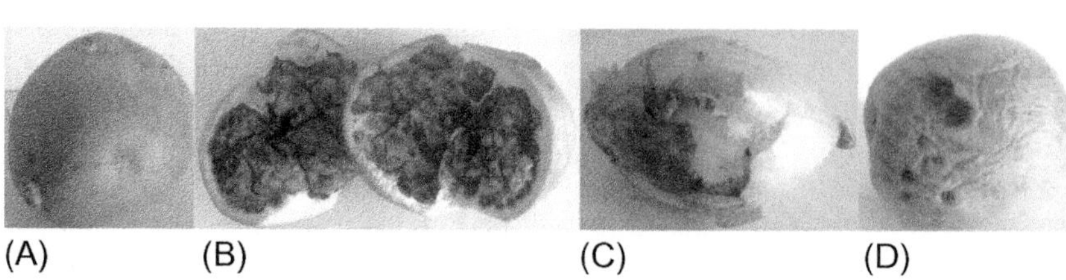


**Fig. 2** (A) Sprout on potato, (B) mold formation in pomegranate, (C) black mold on onion, (D) sunburn and brown spots on apple.

about non-destructive analysis of fruits and vegetables. However, most of these studies focus on employing these methods for classification according to maturity and shelf life quality. Although the measure varies significantly with food type, maturity is determined according to the water content and its mobility, total soluble solid content (TSS), titratable acidity (TA), TSS/TA, sugar content and its distribution, starch, acidity, color, chlorophyll, anthocyanin, firmness, and astringency (Chen and Sun, 1991; Studman, 2001; Butz et al., 2005; Sun, 2010; Nicolaï et al., 2014; Mahesh et al., 2015; Patel et al., 2015; Ali et al., 2017; Arendse et al., 2018; Li et al., 2018; Srivastava and Sadistap, 2018).

This chapter presents non-destructive methods for internal and external quality evaluation of fruits and vegetables. Fundamentals and theoretical considerations of each technique are briefly summarized, and their recent commercial or laboratory-scale applications are presented. The details of image acquisition, pre- and postprocessing are discussed under the title of hyperspectral and multispectral imaging (MSI). The limitations and the advantages of each non-destructive method are also shared for building a common knowledge about these methods.

# 2 Non-destructive methods for food quality evaluation

## 2.1 Spectroscopic methods

### 2.1.1 Nuclear magnetic resonance (NMR) and magnetic resonance imaging (MRI) spectroscopy

Magnetic resonance is the interaction between atomic particles (electrons and the nucleus) and the external magnetic field ($B_0$) (McCarthy, 2012). If the particle that is interacting with the magnetic field is the nucleus, then it is called nuclear magnetic resonance (McCarthy, 2012). For example, the nuclei of $^{1}H$, $^{13}C$, $^{15}N$, $^{19}F$, $^{23}Na$, and $^{31}P$ have magnetic moments, and when placed in a strong magnetic field, they absorb the resonance energy in the radiofrequency range of spectrum (1 cm–10 m) (Du and Sun, 2004; Butz et al., 2005; Sun, 2009; McCarthy, 2012; Reusch, 2013; Arendse et al., 2018). Without any external magnetic field, the protons are randomly aligned, but with application of high magnetic field the protons align their spin in the same direction as the magnetic field (Patel et al., 2015; Kirtil and Oztop, 2016). A nucleus containing an odd number of nucleons, such as $^{1}H$ and $^{13}C$, is required to obtain net magnetization moment (Kirtil and Oztop, 2016). $^{1}H$ (hydrogen nuclei) produces the greatest NMR signal and gives excellent contrast between different structures; therefore most of the studies related with the solid food quality evaluation adapt $^{1}H$ NMR, due to presence of abundant hydrogen atoms (Clark et al., 1997; Kuperman, 2000; Kirtil and Oztop, 2016).

Monitoring proton motion provides information about the spatial distribution of proton density ($\rho$), relaxation times ($T_1$ and $T_2$), and diffusion rates of biological materials (Kuperman, 2000; Nicolaï et al., 2014; Suchanek et al., 2017). Bloch proposed an equation about time evolution of the nuclear magnetic moments under external magnetic field (McCarthy, 2012). With the removal of net magnetic moment (M) from equilibrium, the system recovers with an exponential behavior and returns to the original state (Kuperman, 2000; McCarthy, 2012). $T_1$ is spin-lattice (longitudinal) relaxation time, and is a constant, which demonstrates the period of growing of longitudinal magnetization ($M_z$) and returning to the initial maximum, while $T_2$ is spin-spin (transverse) relaxation time, which is a constant that refers to the period of transverse magnetization ($M_{tr}$ or $M_{x,y}$) decaying or dephasing (Kuperman, 2000; Hills, 1998; McCarthy, 2012; Kirtil and Oztop, 2016). Spin-echo or gradient-echo pulse sequences are produced with variations in echo time (TE), repetition time (TR) of pulse, number of pulses (single or consecutive), repetition, and order of these parameters (Hills, 1998; McCarthy, 2012). Upon the specified sequence procedure, the weight of images ($T_1$, $T_2$ or proton density-weighted) is modified (Hills, 1998; Kuperman, 2000; McCarthy, 2012; Kirtil and Oztop, 2016). Cell size and structure, chemical composition, viscosity of cellular content, and magnetic susceptibility influence the proton relaxation (Clark et al., 1997; Patel et al., 2015). Therefore, variations in the internal quality such as ripeness, firmness, decay formation, bruising, water core, browning, mealiness, wooly breakdown, or distribution of water within the structure, as well as chemical composition (water, sugar, oil, pectin, starch, etc.), of fruits and vegetables can easily be monitored with NMR (Chen and Sun, 1991; Sonego et al., 1995; Butz et al., 2005; Hills, 2006; Bizzani et al., 2017; Lu and Lu, 2017; Arendse et al., 2018). Damaged or affected structures within the fruit or vegetable are characterized from the darker areas, due to their low proton density (Sonego et al., 1995; Clark et al., 1997; Lammertyn et al., 2003; Patel et al., 2015).

Various magnetic field strengths between 0.1 T and 9.5 T are employed for internal quality evaluation of fruits and vegetables (Clark et al., 1998; Clark and MacFall, 2003; Lammertyn et al., 2003; Andaur et al., 2004; Thybo et al., 2004; Marigheto and Hills, 2005; Cho et al., 2008; Musse et al., 2009; Kotwaliwale et al., 2012a; Zhang and McCarthy, 2012; Zhang and McCarthy, 2013; Tao et al., 2014; Mazhar et al., 2015; Suchanek et al., 2017). The exposure to subzero temperatures during storage may cause formation of an internal injury, called chilling injury or cold damage, in agricultural products. However these injuries can be detected with MRI either with diffusion weighted or $T_1$ and $T_2$-weighted imaging sequences (Kotwaliwale et al., 2012a; Tao et al., 2014). Chilling injury in tomatoes is associated with water mobility due to membrane disintegration; however, the signal intensity acquired from pericarp and core tissues have rather different responses in MRI (Tao et al., 2014). Similarly, cold injury has occurred in cucumbers unevenly and resulted in different proton density responses in rind and seed sections (Kotwaliwale et al., 2012a). It was not possible to classify the control and freeze-damaged cucumbers with $T_1$-weighted images, though $T_2$-weighted images have been more sensitive to detect the freeze-damaged samples, especially if the rind section and whole section was freeze-damaged.

In addition to these injuries, spraing disease and internal bruising in potatoes have been identified with low signal in proton-weighted images and the presence of mobile water from the high signal in $T_2$-weighted images obtained with a 7 T MRI system (Thybo et al., 2004). When mealiness develops in apples, the diffusion of water between cells gradually decreases, and it was stated in the study of Marigheto and

Hills (2005) that the sound apples had lower average spin-spin relation time (267 ms) compared to the mealy apples (348 ms). In addition, average $T_1$ values of sound apples (579 ms) were rather lower than the mealy apples (917 ms) (Marigheto and Hills, 2005). Soft fruits such as apple, avocado, kiwifruit, and pear are more susceptible to mechanical damage, and further softening, core breakdown, and browning of the internal tissues might occur (Clark et al., 1998; Lammertyn et al., 2003; Taglienti et al., 2009; Mazhar et al., 2015; Suchanek et al., 2017). After core breakdown of pears, core breakdown disorder symptoms have occurred, the moisture diffused towards the skin, and affected brown tissues have been distinguished from the healthy tissues with their comparably lower pixel intensities (Lammertyn et al., 2003). Small cavities have been formed in further steps of browning in pear, and they have been identified from black areas found in $T_1$-weighted images of low-field MRI (Suchanek et al., 2017).

In kiwifruit, softening of the tissue has been characterized from the movement of water, such as the areas that have had low mobility appearing dark in $T_2$-weighted images, while the bright areas corresponded to high water mobility (Taglienti et al., 2009). Moreover, little black spots that have appeared in the mesocarp have been associated with gases such as ethylene and carbon dioxide, released due to the ripening process. Avocados that have been exposed to impact have also been recognized from image contrast of $T_2$-weighted $^1$H-MR images (Mazhar et al., 2015). The area exposed to impact had higher signal intensity as a result of released intracellular material, thus supplying protons.

Magnetic resonance imaging creates 2D or 3D high quality images of an object nondestructively (Butz et al., 2005; Patel et al., 2015). Strong magnetic field is required for high signal-to-noise ratio, better image contrast, and higher spatial resolution, so instrumentation costs are rather expensive, depending on the type of magnet and radiofrequency coil used (Kuperman, 2000; Butz et al., 2005; McCarthy, 2012; Patel et al., 2015). Poor signal-to-noise ratio can be improved by involving many image scans, particularly for low field MRI (Suchanek et al., 2017). However, low field MRI is capable of simultaneous identification of several compounds (Su et al., 2017). In addition, chemical shift can be converted for detection of protons linked with the presence of water, sugar, and oil only (Patel et al., 2015). An online MRI measurement system was not very effective for distinguishing the pits attached to cherries with respect to the additional peak appearance, noise, and artifacts (Hills, 1998). In addition, it takes longer for image acquisition and obtained data needs to be resolved with complex algorithms (McCarthy, 2012; Patel et al., 2015).

### 2.1.2 Raman spectroscopy

Raman spectroscopy is based on the changes of rotational, vibrational, or low-frequency modes of molecules that are irradiated by a monochromatic light beam (Qin et al., 2016; Su et al., 2017; Arendse et al., 2018). Due to the irradiation of the sample, a small portion of photons is inelastically scattered (Lu, 2016; Arendse et al., 2018). The energy transfer from or to photons defines the Stokes or anti-Stokes scattering status of energy, but both of them comprise Raman scattering (Qin et al., 2016; Lu and Lu, 2017; Arendse et al., 2018).

Raman spectrum is unique for each molecule, like a fingerprint, which provides spatial and spectral information about the chemical composition, the structure, and their interactions (Dhakal et al., 2014; Ropodi et al., 2016; Lu and Lu, 2017; Su et al., 2017; Hussain et al., 2018). The equipment consists of a light (excitation) source at a wavelength of visible or infrared region, a wavelength separator, and a detector such as a charge-coupled device (CCD) (Dhakal et al., 2014; Qin et al., 2016). Raman spectroscopy signals are week due to a very low number of photons belonging to the Raman scattering (Liu et al., 2013; Zheng and He, 2014; Qin et al., 2016). These weak signals are improved by employing metal nanostructures such as gold or silver nanoparticles that are attached to the surface of sample; this technique is called Surface-enhanced Raman Spectroscopy (SERS) (Liu et al., 2013; Zheng and He, 2014; Qin et al., 2016).

Raman and SERS methods have been used for analyzing chemical properties of various food and agricultural products; however, most of the studies have been implemented for food safety evaluation, such as pesticide residue detection in fruits, mycotoxins, and chemical adulterants (Liu et al., 2013; Dhakal et al., 2014; Müller et al., 2014; Zheng and He, 2014; Qin et al., 2016). Dhakal et al. (2014) studied the detectable limit of chlorpyrifos (organophosphorous pesticide) residues on apples by a prototype Raman spectroscopy design with a 785 nm excitation laser source. Raman peaks at 341, 631, and 677 $cm^{-1}$ were attributed to the fingerprint peaks of chlorpyrifos, and they concluded that the system was able to detect a minimum of 6.69 ppm pesticide residue on apples.

Similarly, Liu et al. (2013) studied azinphos-methyl and phosmet (organophosphorous) and carbaryl (carbamate) pesticide residues on apple and tomato samples with SERS including gold nanoparticles. The SERS was equipped with a 785 nm NIR diode laser source, and as a result, the detection limits of these pesticides varied between 4.51 and 6.66 ppm for apples, and between 2.91 and 5.35 ppm for tomato samples. Although the prediction capacity of the proposed partial least squares (PLS) model was lower for carbaryl on tomato ($R$=0.82), and phosmet and azinphos-methyl ($R$=0.87 and 0.88) (Liu et al., 2013). Thiabendazole presence in banana and citrus fruits (orange, grapefruit, lemon, and bio lemon) has been investigated with SERS with Ag colloidal nanoparticles and Raman at 532 nm excitation laser line (Müller et al., 2014). Raman was able to detect the thiabendazole $10^{-2}$ moL/L level in banana and citrus fruits, while SERS was able to detect at $10^{-11}$ mol/L that was supporting the enhancement of signals employing Ag nanoparticles.

Raman and SERS are rapid methods, require no sample preparation, and allow detection of very small material amounts. In addition, these methods are very sensitive to detect low levels of chemicals (pesticides, fungicides, mycotoxins, adulterants); especially if the signal intensity is improved with incorporation of metallic nanoparticles in SERS measurements.

### 2.1.3 Visible and near infrared spectroscopy

The visible region of the electromagnetic spectrum is located between the UV and infrared region, from 380 to 780 nm. The infrared spectrum is divided into three sections: near-infrared (NIR) 780–2500 nm, mid-infrared (MIR) 2500–25,000 nm, and far-infrared (FIR) 25,000–300,000 nm (Su and Sun, 2018). Molecules vibrate at their specified energy level, but when they are exposed to radiation with a light source, they absorb the light at particular wavelengths (Sun, 2016). The absorption bands observed in the infrared region, while their overtones and combination bands of O–H, N–H, C–H, and S–H stretching vibrations and stretching-bending combinations, appear in the NIR spectral range (Ozaki et al., 2006; Sun, 2010). These overtones and their combinations provide subtle data about the chemical composition of food (Chen and Sun, 1991; Magwaza et al., 2012a; Porep et al., 2015; Arendse et al., 2018). O–H bonds are associated with the absorption bands related mostly with the water content, while C–H bonds are attributed to the organic compounds like fatty acids, proteins, and carbohydrates (Ozaki et al., 2006; Sun, 2010; Park and Lu, 2015). Water has a strong absorption in the NIR region (Magwaza et al., 2012a; Park and Lu, 2015; Porep et al., 2015), although some of spectral bands attributed to water may have overlapping peaks with other constituents like carbohydrates (Sun, 2009; Nicolaï et al., 2014; Park and Lu, 2015; Porep et al., 2015).

Measurement modes of NIR spectroscopy are transmittance, transflectance, interactance, diffuse transmittance, and diffuse reflectance (Huang et al., 2008; Sun, 2016). Transmittance mode in the NIR region provides information about internal characteristics, while reflectance mode is more appropriate for determining surface or subsurface characteristics, due to low penetration (Chen and Sun, 1991; Schaare and Fraser, 2000; Ozaki et al., 2006; Magwaza et al., 2012a). Therefore, reflectance mode is preferred in fruits or vegetables with thin peel (apricot, strawberry), and transmittance is more applicable to the fruits with thicker peel such as orange (Ozaki et al., 2006; Camps and Christen, 2009; Sánchez et al., 2012). Diffuse transmittance mode is used for solid samples, having a thickness from 10 to 20 mm, and is measured between 800 and 1000 nm. However, for thicker samples, the diffuse reflectance mode is preferred, in the 1100–2500 nm range, because transmittance is insignificant (Huang et al., 2008; Sun, 2016).

An NIR spectrometer mainly consists of an infrared source (lamp), monochromator (wavelength isolator), detector, and data processor (Ozaki et al., 2006; Cen and He, 2007; Sun, 2009; Porep et al., 2015). Radiation sources are grouped into continuum and line sources (Sun, 2009). Continuum sources, such as tungsten, halogen, tungsten-halogen, and quartz-halogen lamps, are widely used in food related applications (Cen and He, 2007; Sun, 2009). Line sources, such as light emitting diodes (LEDs), lasers, and laser diodes, are also used, but they emit light at certain discrete wavelengths (Ozaki et al., 2006; Cen and He, 2007; Sun, 2009) and, therefore, are not practical in full spectrum analysis (Sun, 2009). Wavelength selection devices (wavelength isolators) may include filters, monochromator, prisms, and grating (Ozaki et al., 2006; Sun, 2009). Optical interference filters (narrow-band interference filter, wedge-interference filter) and electronically tunable filters (liquid crystal tunable filter, acousto-optical tunable filter) are two major groups that help to select the wavelength (Ozaki et al., 2006; Sun, 2009). The detectors are used for measuring the intensity of dispersed light striking them, and convert the radiation into electrical signals (Sun, 2009; Sun, 2010; Porep et al., 2015). Raw spectral data is collected in a computer and special software programs are employed for processing the data (Sun, 2009). However, raw data may include light scattering, artifacts, instrumental noise, and baseline shift, which can be eliminated through suitable data preprocessing and further multivariate statistical methods (Ozaki et al., 2006; Sun, 2009; ElMasry and Nakauchi, 2016; Malegori et al., 2017). Calibration and prediction models are built upon multivariate data analysis, and the number of effective wavelengths on evaluated classification features is then reduced.

Total soluble solid content (TSS), titratable acidity (TA), maturity index (TSS/TA), pH, sugar content, starch content, color (L*, a*, b*, C*, hue), and firmness are the main quality parameters that help to classify fruits and vegetables according to their maturity stage by using NIR spectroscopy. NIR is highly penetrative and no sample preparation is needed before the analysis (Srivastava and Sadistap, 2018). Intense research has been carried out with NIR spectroscopy for maturity classification depending on the given quality parameters of tropical fruits such as acerola, jujube, kiwi, mango, passion fruit, pomelo, or fruits and vegetables that have high economic value due to mass consumption rates such as apple, carrot, cucumber, grapes, orange, pumpkin, strawberries, tomatoes, etc. (Jamshidi et al., 2012; Sánchez et al., 2012; Sirisomboon et al., 2012; Jha et al., 2014; Liu et al., 2015; Chen et al., 2016; Guo et al., 2016; Rungpichayapichet et al., 2016; Malegori et al., 2017; Ncama et al., 2017; Beghi et al., 2018; Cavaco et al., 2018; Oliveira-Folador et al., 2018; Xiao et al., 2018).

TSS content prediction with NIR spectroscopy may be influenced by the sampling position, specifically for the fruits or vegetables with a thick rind (Schaare and Fraser, 2000; Maniwara et al., 2014; Lu et al., 2015; Muenmanee et al., 2016). Lu et al. (2015) determined that TSS models obtained from NIR spectra (500–1010nm) greatly varied with respect to the sampling point. The TSS content of melons was better predicted from the stylar-end position compared to the equatorial sampling position. Similarly, the total soluble solid content of Newhall navel oranges was found dependent on the spectral data obtained from different positions (Liu et al., 2015). Not only TSS, but titratable acidity, maturity index, and vitamin C content were better predicted in the equatorial spectra rather than distal surface. The astringency of intact persimmon fruit was also evaluated with respect to six different spectral measurement points with Vis-NIR and NIR spectra (Cortes et al., 2017). However, the authors stated that combination of all measurement points had higher prediction capacity than four measuring points that were found below the equatorial axis. NIR spectra is also employed for external disorder evaluation, such as rind break-down in mandarin, bruise formation in apple, scalding in pomegranate, and chilling injury and rot formation in kiwi fruit (Ozaki et al., 2006; Magwaza et al., 2012b; Arendse et al., 2018).

In addition to the aforementioned classification parameters, Dall'Acqua et al. (2015) studied the discrimination of Euterpe species (açaí and juçara fruits) with diffuse reflectance NIR spectra in the 1000–2500nm range. Following data preprocessing, key wavelengths, of 1438nm (first overtone of O–H bonds) and 1928nm (combination bonds of O–H vibrations), had the highest correlation with respect to the principal component analysis (PCA).

NIR spectroscopy is also applicable for on-tree and in-field quality evaluation of fruits by handheld or portable devices (Abasi et al., 2018). These devices are used for ripeness evaluation, and oil and free fatty acid content determination in olive and palm fruits for optimum harvest time (Makky and Soni, 2014; Fernández-Espinosa, 2016), measurement of TSS, moisture content, titratable acidity, sugar content, and firmness for evaluation of maturity in apricot, date, grape, mango, orange, sugar beet, and red and green tomatoes (Camps and Christen, 2009; Tiwari et al., 2013; Pan et al., 2015; Marques et al., 2016; Urraca et al., 2016; Alhamdan and Atia, 2017; Nordey et al., 2017; Xiao et al., 2017; Cavaco et al., 2018; Huang et al., 2018). NIR measurement with portable devices is practical, especially in-field applications; however the number of samples analyzed is important for prediction accuracy of total soluble solids (Urraca et al., 2016). In addition, the classification of fruits and vegetables according to firmness, color, and titratable acidity parameters had rather lower prediction capacity compared to the soluble solid determinations (Camps and Christen, 2009; Pan et al., 2015; Marques et al., 2016; Nordey et al., 2017; Cavaco et al., 2018; Huang et al., 2018).

In the study of Makky and Soni (2014), ripeness of palm fruit bunches was evaluated with respect to the reflectance data in the visible range, and the color of the fruit was associated with the chlorophyll and carotenoid concentrations. They concluded that the reflectance spectra between 580 and 680 nm had significant effect on prediction of ripeness from the proposed models, due to the higher absorption values of chlorophyll-a and chlorophyll-b. However, the oil content was better predicted in the 400 and 500 nm spectral range, which was associated with the carotene concentrations of fruit. In contrast to the palm fruit ripeness evaluation, Fernández-Espinosa (2016) studied in the NIR range (1000–2300 nm) for on-line ripeness evaluation of intact olives. Oil absorption bands have been observed at 1210 and 1245 nm (stretching vibrations of C–H bands of alkyl groups and alkene), 1145 and 1160 nm (stretching of C–O bonds of aliphatic esters), 1175, 1185, 1260, and 1275 nm (alkyl groups and alkenes). While absorptions at 1745 and 1693 nm were associated with triglycerides, and at 1728 nm absorption was linked with oleic acid. Overall, the proposed models were successful for prediction of moisture, oil content, and free acidity during ripening of different olive cultivars with high correlation coefficients ($R^2 > 0.90$).

Fourier transform infrared spectrometers (FTIR) are a special type of infrared spectroscopy, which include interferometers instead of monochromators (Sun, 2009; Smith, 2011). An interferometer consists of radiation source, beam splitter, a fixed and a moving mirror, and this optical device creates an interferogram of the sample, which is then Fourier transformed into a spectrum (Ozaki et al., 2006; Sun, 2009). Recent MIR spectrometers are an FTIR using interferometer; therefore, the term FTIR spectrometer might be used instead of MIR spectrometer (Sun, 2009). This equipment has several advantages over regular infrared spectrometers: high signal-to noise ratio, speed of analysis, precise measurement, throughput advantage, sensitivity, and cost (Ozaki et al., 2006; Sun, 2009; Smith, 2011). Likewise, it is an effective method for quantitative analysis of total soluble solids, pectin, starch, and sugars (glucose, fructose, sucrose) (Sun, 2009; Magwaza et al., 2012a; Xiao et al., 2017; Oliveira-Folador et al., 2018).

### 2.1.4 Hyperspectral imaging and multispectral imaging

Hyperspectral and multispectral imaging are novel methods that unite imaging and spectroscopy techniques for improvement of the spatial and spectral data they provide alone (Park and Lu, 2015). These data are combined to create three-dimensional data for the evaluated object (Sun, 2010; Magwaza et al., 2012a; Su et al., 2017; Arendse et al., 2018). Hyperspectral images consist of two spatial dimensions ($x$ and $y$) and one spectral dimension ($\lambda$), therefore, acquired data from these hyperspectral images are known as hypercube or data cube (Sun, 2010; Nicolaï et al., 2014; Giovenzana et al., 2015; Su et al., 2017; Arendse et al., 2018). Spectral information is extracted from contagious spectral bands, which generally fall into the Vis/NIR or NIR range of the electromagnetic spectrum (Sun, 2010; Magwaza et al., 2012a; Nicolaï et al., 2014; Su et al., 2017); however, the spectrum is narrowed down for the improvement of signal-to-noise ratio (SNR) and precision, as well as required time for processing obtained data (Adão et al., 2017; Sun et al., 2017).

Signal-to-noise ratio is a feature of image quality that measures the desired signal over the level of background noise (Sun, 2010; Park and Lu, 2015). Unlike hyperspectral imaging (HSI), MSI deals with discrete wavebands imaging (Park and Lu, 2015; Su and Sun, 2018). HSI employs hundreds of spectral bands, while in MSI, the number of wavelengths is very limited, reaching up to a maximum of 20 (Su et al., 2017; Su and Sun, 2018). MSI is generally applicable for real-time imaging of samples with its comparably fast image acquisition and short image processing period compared to the HSI system (Qin et al., 2013a; Giovenzana et al., 2015; Lu and Lu, 2017). Higher spectral and spatial resolution of hyperspectral images creates voluminous data and this makes HSI unavailable for real-time applications (Arendse et al., 2018; Su and Sun, 2018). However, selecting key wavelengths for HSI reduces the amount of data to be processed, and increases commercial application potential of HSI (Islam et al., 2018).

HSI systems acquire the spectral image from three different scanning techniques, that is, point scanning (whisk-broom method), line scanning (push-broom method), and area scanning (staring imaging/tunable filter) (Sun, 2010; Qin et al., 2013a; Park and Lu, 2015; ElMasry and Nakauchi, 2016; Li et al., 2018). The line scanning method is suitable for online food quality monitoring and inspections, especially for scanning of a line of samples with an array detector moving on a conveyor belt (Sun, 2010; Qin et al., 2013a; ElMasry and Nakauchi, 2016). The scanning method choice varies depending on the purpose of spectral imaging, sample and detector positioning (fixed/moving), length of image acquisition period, and spatial resolution or SNR of images, etc.

In spectral measurements, reflectance, transmittance, and interactance sensing modes are used; however they have rather different approaches for image acquisition (Park and Lu, 2015; Lu, 2016). For example, there is no contact between the detector and measured sample in reflectance mode; but measurement is affected by surface properties of the sample (Schaare and Fraser, 2000; Lu, 2016); whereas, in transmission mode, the measurements are less susceptible to the surface characteristics of the sample, and this mode is more applicable for detecting internal defects (Schaare and Fraser, 2000; Park and Lu, 2015; Lu, 2016).

Hyperspectral images are processed through consecutive steps. First, wavelength calibration, spatial calibration, and radiometric calibrations (for reflectance, transmittance, and normalization) are employed (Sun, 2010; Park and Lu, 2015). Following the calibration steps, noise reduction or removal is needed for enhancing the image quality; moreover, some methods are employed like standard normal variate (SNV), median filtering, Savitzky-Golay filtering (SG), normalization (N), first and second order derivative (1st and 2nd), multiplicative scatter correction (MSC), vector normalization (VN), mean centering (MC), and baseline correction (Sun, 2010; Park and Lu, 2015; Ropodi et al., 2016; Rahman et al., 2017; Sun et al., 2017). Although calibration and smoothing of images are required for further information extraction, care must be taken for appropriate filter selection and noise reduction method, as some of the extracted data might be lost due to false processing applications (Park and Lu, 2015; Keresztes et al., 2017). The next step of hyperspectral data evaluation is image classification with linear or nonlinear optimal wavelength selection or, in other words, a feature extraction step, which reduces the dimensions of the data and specifies the representatives of original features (Sun, 2010; Park and Lu, 2015; ElMasry and Nakauchi, 2016). The classification methods involve PCA, PLS regression, linear discriminant analysis (LDA), Fishers discriminant analysis (FDA), multiple linear regression, artificial neural network (ANN), and successive projections algorithm (SPA) (Gowen et al., 2007; Sun, 2010; Mahesh et al., 2015; Park and Lu, 2015; ElMasry and Nakauchi, 2016).

The HSI system has various uses in foods such as variety detection, maturity/ripening classification, internal and external defect/disease detection, as well as evaluation of quality parameters, such as TSS, pH, firmness, phenolic compounds, sugar content and distribution, and dietary fiber, etc. (Sun, 2010; Mahesh et al., 2015; Park and Lu, 2015; Sun et al., 2017; Yan et al., 2017). An overview of some recent HSI applications for non-destructive evaluation of fruit and vegetables is presented in Table 1. TSS content and acidity determination is a general approach for predicting the quality and ripening stage of fruits. However, in traditional methods, destructive analyses are performed for measuring these parameters. HSI technology has been used in several studies for the prediction of soluble solid content of fruits, even the distribution of soluble solid of apples and kiwifruit were successfully determined with proposed regression models (Mo et al., 2017b; Zhu et al., 2017). Another quality parameter that also reveals the maturity stage is firmness. In contrast to the soluble solid and acidity classification, ripeness evaluation with respect to firmness is rather challenging. High local variations of firmness in an individual fruit may result in lower prediction capacity of HSI (Khodabakhshian and Emadi, 2017). Besides given applications, internal or external defects, injuries, bruise or insect damage can be detected in various fruits and vegetables with HSI. Internal damages such as hollow heart in radish, black spot in potatoes, bruise on apple surface, sprouting of potato, slug and worm contamination on lettuce leaves, fecal contamination on spinach leaves, skin injury, and decay have been detected with HSI technology (Everard et al., 2014; Li et al., 2016; López-Maestresalas et al., 2016; Keresztes et al., 2017; Mo et al., 2017a; Yu et al., 2017; Gao et al., 2018).

MSI similar to HSI has been used for determination of TSS and sugar content, acidity, firmness, and ripening stage, and mechanical damage and injuries have been detected with this method (Lu, 2004; Liu et al., 2014; Huang et al., 2015; Khodabakhshian et al., 2017; Hashim et al., 2018; Islam et al., 2018; Tang et al., 2018). Huang et al. (2015) studied the detection of bruise on apples with MSI; however, hyperspectral image acquisition was first performed in the wavelength range of 325–1100 nm in

**Table 1** Hyperspectral imaging of various fruits and vegetables

| Food | Quality parameter | Wavelength (nm) | Light source[a] | Calibration/ preprocessing | Regression | References |
|---|---|---|---|---|---|---|
| Apple | Bruise detection | 1000–2500 | 20 W HL (6) | 1st, MC, S, AN, PN | PLS-DA | Keresztes et al. (2017) |
| Apple | TSS | 400–1000 | 150 W QTHL (2) | MN, MxN, RN, S, 1st, 2nd, SNV, BC, MSC | PLS-DA | Mo et al. (2017b) |
| Banana | MC, TSS, firmness | 400–1000 | 50 W HL (2) | Dark-white reference | PCA, PLS-DA | Rajkumar et al. (2012) |
| Blueberry | TSS, firmness | 360–2000 | 20 W HTL | TS | PLS-DA | Leiva-Valenzuela et al. (2013) |
| Celery | SDF, IDF | 378–1008 | HTL | Dark-white reference, SNV | PLS-DA | Yan et al. (2017) |
| Cherry | Pit detection | 400–1000 | 100 W HL(2) | SG | BNN, PCA-BNN | Siedliska et al. (2017) |
| Kiwi | TSS, pH, firmness | Vis/NIR: 450–1000 NIR: 951–1670 | 150 W QTHL (2) | Dark-white reference | PLS-DA, MLR, LSSVM | Zhu et al. (2017) |
| Lettuce | Slug and worm detection | 400–1000 | HTL (2) | Dark-white reference | one way-ANOVA | Mo et al. (2017a) |
| Lettuce | Decay detection | 380–1012 | HL | – | PLS-DA | Simko et al. (2015) |
| Lime | TSS, TA, TSS/TA | 897–1752 | HL (2) | SG, 1st, 2nd, MSC,SNV | PLS-DA | Teerachaichayut and Ho (2017) |
| Melon | TSS, TA, firmness | 400–1000 | 150 W HL (2) | SNV, MSC, VN, MC, SG, 1st | GA, SPA, CARS | Sun et al. (2017) |
| Nectarine | Cultivar detection | 450–1040 | 37 W HL (12) | K-means clustering | PLS-DA | Munera et al. (2018) |
| Onion | Sorting (sound/ diseased) | 1016–1742 | HL (2) | SG, SNV, MSC | PCA, PLS-DA | Islam et al. (2018) |
| Peach | Sorting (sound/ defected) | 325–1100 | 500 W QHL (2) | Dark-white reference | PCA | Li et al. (2016) |

| | | | | | | |
|---|---|---|---|---|---|---|
| Pear | Ripeness classification | 325–1100 | 500W QLA (2) | Dark-white reference, TS | SPA | Khodabakhshian and Emadi (2017) |
| Persimmon | Firmness, ripeness, astringency | 450–1020 | HL (12) | Dark-white reference | LDA, QDA, SVM, PLS-DA | Munera et al. (2017) |
| Potato | Black spot detection | Vis/NIR: 400–1000 SWIR: 1000–2500 | 20W HL (6) | SG, 1st, 2nd | PLS-DA | López-Maestresalas et al. (2016) |
| Potato | Sprouting eye | 400–1000 | 150W HL (2) | SG | FDA, LSSVM | Gao et al. (2018) |
| Radish | Hollow heart | 400–1000 | – | MSC, SNV, MN | SVM | Yu et al. (2017) |
| Raspberry | Maturity (phenolics) | 884–1717 | 70W TIH (2) | MSC, SNV, detrending | MPLS | Rodríguez-Pulido et al. (2017) |
| Spinach | Fecal contamination | UV: 464–800 V: 464–800 Vis/NIR:456–950 | 10W LEDs (4) | 3-point average | PLS-DA | Everard et al. (2014) |
| Spinach | Nitrate distribution | 913–2519 | HL | Dark-white reference, 2D-FT | PLS-DA | Yang et al. (2017) |
| Strawberry | MC, TSS, pH | 400–1000 | 50W HL (2) | Dark-white reference | PLS-DA | ElMasry et al. (2007) |
| Tomatoes | MC, TSS, pH | 948–2494 | 100W HTL (4) | S, SG, MSC, SNV, 1st, 2nd | PLS-DA | Rahman et al. (2017) |

IDF: Insoluble dietary fiber, MC: moisture content, SDF: soluble dietary fiber, TSS: total soluble solid content, TA: titratable acidity.

HL: Halogen lamp, HTL: halogen tungsten lamp, LEDs: light emitting diodes, QHL: quartz halogen light area source, QLA: quartz light area source, QTHL: quartz tungsten halogen lamp, TIH: tungsten iodine halogen lamp.

AN: Area normalization, BC: baseline correction, MC: mean centering, MN: mean normalization, MSC: multiplicative scatter correction, MxN: maximum normalization, PN: peak normalization, RN: range normalization, S: Smoothing, SG: Savitzky-Golay smoothing, SNV: standard normal variate, SVM: support vector machine, TS: threshold segmentation, VN: vector normalization, 1st: first derivative, 2nd: second derivative, 2D-FT: 2D discrete Fourier transform.

BNN: Backpropagation neural networks, CARS: competitive adaptive reweighted sampling, FDA: Fisher discriminant analysis, GA: genetic algorithm, LDA: linear discriminant analysis, LSSVM: least square support vector machine, MLR: multiple linear regression, MPLS: modified partial least squares regression, PCA-BNN: principal component analysis backpropagation neural networks, PLS-DA: partial least square discriminant analysis, QDA: quadratic discriminant analysis, SPA: successive projections algorithm.

[a] **Number of light sources given in brackets.**

order to determine optimum bands for MSI. Static and online bruise measurements were carried out at optimum wavelengths of 780, 850, and 960nm, and both imaging methods correctly identified the healthy apples with 100% accuracy. However, static measurement performed better for detection of bruised apples (91.5%) compared to online measurement (74.6%).

Cold injury in mango was also evaluated in with MSI at 5 different wavelengths (450, 545, 630, 750, and 915nm) and PCA was carried out for finding the important features (Hashim et al., 2018). Color (L* and a*), moisture content, TSS, pH, and firmness have been measured destructively for evaluating the parameters of mangoes having cold injury. L* and moisture content were the key parameters for detection of this damage, and PCA with least square support vector machine (LS-SVM) model resulted in higher prediction accuracy (83.3%) for classification of the fruits. In addition to these studies, several other discrete spectral bands were adopted for determination of TSS, pH, titratable acidity, firmness, and ripeness. For example, Khodabakhshian et al. (2017) employed 6 discrete bands (450, 521, 630, 780, 853, and 950nm) in order to predict the TSS and titratable acidity of pomegranate, while Liu et al. (2014) studied with 19 discrete bands between 400 and 1000nm for evaluation of TSS, firmness, and ripeness stage of strawberries. Similarly, healthy and diseased onion bulbs have been classified using MSI at 18 discrete bands between 400 and 1000nm (Islam et al., 2018). Sugar content in Fuji apples has been predicted with comparably lower number of spectral bands, but the number of optimum wavelengths has been narrowed down from a broad image acquisition range (350–1200nm) (Tang et al., 2018). The authors concluded that multiple linear regression model employing four optimal wavelengths at the same time (461, 469, 947, and 1049nm) performed better for prediction of sugar content ($r$=0.8861).

## 2.2 X-ray computed tomography (CT)

X-rays correspond to the wavelength range of 0.01 to 10nm; however, lower wavelengths up to 1nm are known as hard X-rays, while over this limit they have low energy and less penetration and are known as soft X-rays (Renu and Chidanand, 2013; Kotwaliwale et al., 2014; Sun, 2016). Due to lower penetration power, soft X-rays are more applicable for agricultural materials (Renu and Chidanand, 2013; Kotwaliwale et al., 2014). X-ray images are created by penetration of high-energy photons with energies in the range of 120eV to 120keV, and attenuation of X-ray radiation generates the transmission images (Kotwaliwale et al., 2014; Sun, 2016; Verboven et al., 2018).

When X-rays pass through an object, the energy of the X-rays exponentially decrease and the weakening of the X-ray beam is expressed as attenuation (Kotwaliwale et al., 2014; Schoeman et al., 2016; Verboven et al., 2018). The magnitude of attenuation or absorption of X-ray beam is measured with Haunsfield unit or CT number (Kotwaliwale et al., 2012b; Schoeman et al., 2016). This number varies with the incident energy level, material composition (moisture, acidity, and pH), density, sample thickness, and its absorption coefficient (Sonego et al., 1995; Butz et al., 2005; Kotwaliwale et al., 2012b; Muziri et al., 2016; Schoeman et al., 2016; Verboven et al., 2018). X-ray CT equipment consists of an X-ray tube (produces the polychromatic beam), beam collimator (narrows the beam of waves), and a detector (Kotwaliwale et al., 2012b; Schoeman et al., 2016; Sun, 2016; Verboven et al., 2018). Following the production of polychromatic divergent X-ray beam with an X-ray source rotating around the sample, the collimated sharp beam directed towards the sample and detector measures the remaining attenuated radiation (Schoeman et al., 2016; Sun, 2016; Verboven et al., 2018).

The lack of uniformities in the microstructures of foods and inclusion of air or other gases may cause some problems regarding the visualization of internal structure, although X-ray CT has a great advantage that attenuation of X-rays from food matrix can easily be discriminated from the air within the structure (Verboven et al., 2018). In fact, the quantitative determination of voids or air spaces present in a pomegranate has successfully been predicted by X-ray CT (Magwaza and Opara, 2014). Bruise and Braeburn (browning) disorder in apples, mealiness in pear, wooly breakdown in nectarine, tough fibrous tissue formation in asparagus and carrot, ripening in mango, and internal structure of cherry tomato have been evaluated with X-ray CT and X-ray micro-CT (Sonego et al., 1995; Kotwaliwale et al., 2012b; Herremans et al., 2013; Cantre et al., 2014; Magwaza and Opara, 2014; Donis-González et al., 2016a, b; Muziri et al., 2016; Diels et al., 2017; Wang et al., 2017b). Wooly structure in nectarines has been visualized by dark appearance in X-ray images, which corresponded to the gas enclosed in the structure that had no X-ray absorption (Sonego et al., 1995). Also, bruising of apples due to impact has been observed according to their different intensities (Diels et al., 2017). Compressed pores beyond the impact regions have had higher intensities, while the bruised area had lower intensity compared to the healthy tissues. In case of mealy pear, the highest porosity was observed at the neck of the fruit; especially big cavities were formed around large individual cells (Muziri et al., 2016). The mealiness in pear has been characterized by the rate of high density dark voxel present in the neck region.

Ripeness degree evaluations in mango have been carried out by evaluation of 3D microstructure from the outer and inner mesocarp excised from the equator region with X-ray micro-CT (Cantre et al., 2014). Defect and void detection in Conference pears have been also implemented with 3D construction of X-ray CT images, and the authors stated that internal defects bigger than 3.5 mm size had been detected with 100% accuracy when the detector pixel resolution had been arranged to 0.5 mm (Van Dael et al., 2017). It has been observed that, ripe fruit cells had highly irregular shape with low sphericity, and in addition, pore size had decreased and pore connection had disappeared due to ripening. Herremans et al. (2013) carried out a systematic study for Braeburn browning disorder formation in Braeburn apples with respect to radial anisotropy. They concluded that the outer cortex tissues had more isotropic configuration than the inner and middle cortex, which had higher anisotropy. In addition, the pores and tissues had not been perceived as different after excessive disorder formation during long storage period (Herremans et al., 2013).

Deposition of lignin inside asparagus and carrot develops undesirable tough (stringy) tissue, which creates difficulty in eating; therefore these samples have been classified according to their internal structure by in-line X-ray CT systems (Donis-González et al., 2016a, 2016b). The X-ray images of both asparagus and carrot have been processed with Gabor filter, which is widely used for texture classification (Donis-González et al., 2016a, 2016b; Sun, 2016). The fibrous tissue in carrot has lower intensity than the healthy tissue; however, the undesirable fibrous tissue had high variations in pixel surrounded by healthy tissue (Donis-González et al., 2016a). High pixel range and contrast values were obtained from X-ray images, even though the carrots that had healthy tissues had been classified with 87.9% accuracy with a total of 4 s had been used for image acquisition and processing for each carrot. Similarly, the fibrous tissue in asparagus was identified from the low intensity regions in the X-ray images and healthy tissues were classified with 91.2% accuracy rate (Donis-González et al., 2016b). The image acquisition and processing time was 2 s per asparagus in total, so X-ray CT measurement was successful for use as an in-line classifications equipment (Donis-González et al., 2016b).

As an example, X-ray micro-CT images of sound mandarin, mushroom, cardamom, and clove samples (Fig. 3) were kindly supplied by the Centre for Rapid and Sustainable Product Development (CDRSP) of Polytechnic Institute of Leiria, and were recorded with Bruker Skyscan 1174 micro-CT (Belgium). The source parameters are held constant at 50 kV and 800 μA. The rotation step of samples was 1.2, 0.8, 0.7 and 0.7°, respectively. And total image reconstruction time of mandarin was 107 s (376 slice), mushroom was 404 s (971 slice), cardamom was 249 s (963 slice), and clove was 285 s (998 s), respectively. The porous and fibrous microstructures, air holes, and voids are clearly visible on these X-ray images, depending on the differences in image intensities.

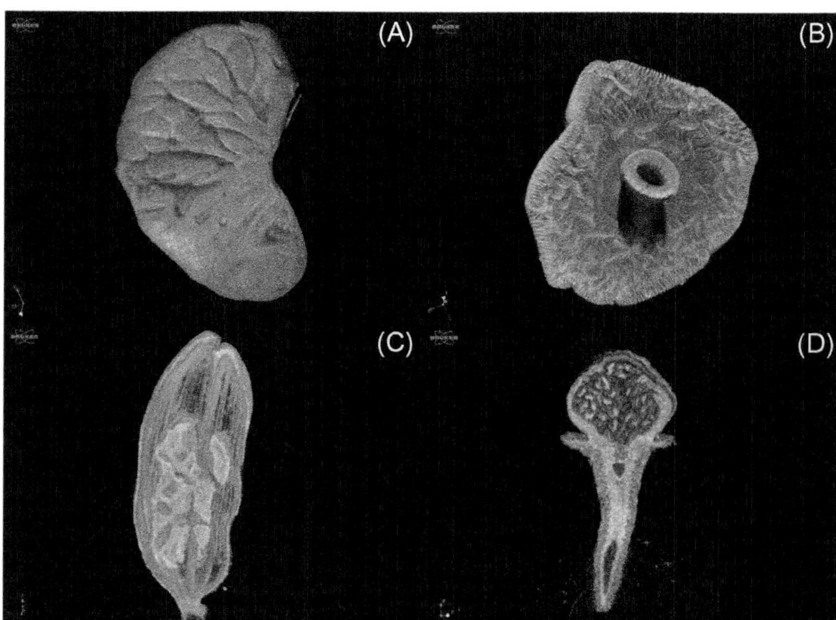


**Fig. 3** X-ray CT images of (A) mandarin and (B) mushroom, (C) cardamom, (D) clove. Figure courtesy of Margarida Cabrita Franco, Polytechnic Institute of Leiria, CDRSP.

X-ray CT and X-ray μ-CT is highly capable of evaluation of microstructure of fruits and vegetables, allowing 2D and 3D model reconstruction from stacking of the image slices. Although the image acquisition and processing is very time-consuming, recent studies also confirm that this technique is also applicable for inline visualization of microstructure with high accuracy levels. The primary concerns about using X-rays are that the exposure to X-rays may damage living tissues; however, an appropriate shielding of the equipment can eliminate these concerns (Renu and Chidanand, 2013; Kotwaliwale et al., 2014).

## 2.3 Mechanical methods

Non-destructive firmness measurement is sensitive to transformations at structural level, whereas destructive methods are not very capable in detecting small variations in food texture (Kilcast, 2004). Mechanical methods, such as acoustic, impact, ultrasonic, and vibration, are common methods in non-destructive firmness (texture) determination of foods (Kilcast, 2004; Nicolaï et al., 2014; Mireei et al., 2015).

Acoustic analysis involves the scattering and reflection of sound waves. The sound signal is generated with a slight tap by small rod or pendulum, and the sound wave is transmitted through the food sample (Studman, 2001; Kilcast, 2004; Sun et al., 2010; Aboonajmi et al., 2015). The sound output (response) is captured with a microphone that is attached to the sample by contact or noncontact method (Muramatsu et al., 1997; Studman, 2001; Kilcast, 2004; Elbatawi, 2008; Mao et al., 2016). The response is then transferred to the computer to analyze the signals, and a fast Fourier Transform (FFT) algorithm converts the data from time-domain to the frequency domain (Studman, 2001; Kilcast, 2004; Verlinden et al., 2004; Elbatawi, 2008; Zhang et al., 2018). The geometry, mass, and modulus of elasticity of the sample affect the obtained resonant frequency (Studman, 2001). Hollow formation in white asparagus and potatoes, and the firmness of apple, kiwi, pear, tomato, and watermelon have been analyzed with acoustic-based methods (Muramatsu et al., 1997; Schotte et al., 1999; Elbatawi, 2008; Mendoza et al., 2012; Foerster et al., 2013; Li et al., 2016a; Mao et al., 2016; Zhang et al., 2018).

Geometry-based acoustic responses of pear has been evaluated with a sampling rate of 51.2 kHz for 0.16 s in order to evaluate the firmness of pears with different shapes (Zhang et al., 2018). It was stated that the measurements from equator and calyx region of pear had overlapped, and the classification accuracy of pears with respect to the acoustic firmness values had increased when taking into account both sampling regions. The effect of hitting balls of various on the acoustic responses that create the sound wave has been studied by Mao et al. (2016). The authors have tested rubber, glass, and stainless steel balls, and the lowest standard deviation had been obtained with the stainless steel ball. In further analysis, they investigated the firmness of watermelon and determined that the mass of sample negatively correlated with resonance frequency (Mao et al., 2016). In addition, the index of first moment (acoustic parameter used as firmness indices) had higher prediction ability of watermelon firmness with the highest number of nodes in the middle layer used in ANN model. Kiwifruit firmness has been evaluated both destructive and nondestructively, based on compression, impact, and acoustic impulses (Muramatsu et al., 1997; Ragni et al., 2010; Li et al., 2016a). Ragni et al. (2010) studied the parameters of impact, such as drop height, and the speed of conveyor belt that carries the kiwifruit. It has been concluded that the drop height was not as effective as belt speed, and the prediction capacity involving the peak impact force-based ANN back-propagation prediction models were better compared to the model involving the impact duration.

Laser Doppler vibrometry (LDV) is a noncontact acoustic vibration measuring method, and the response of vibration signals are recorded following excitation of the sample (Terasaki et al., 2006; Abbaszadeh et al., 2013; Ali et al., 2017). In addition to firmness and ripeness evaluation, this method is also used for prediction of TSS, TA, and TSS/TA (Terasaki et al., 2006; Taniwaki et al., 2009a, 2009b; Abbaszadeh et al., 2013; Abbaszadeh et al., 2015). The analyzed samples were excited using swept sine waves with a frequency range between 0 and 3000 Hz, and the vibrational response of the sample was detected with LDV. Then the acquired signals (frequency response) were analyzed with FFT algorithm for extracting the vibrational spectrum of the sample (Terasaki et al., 2006; Taniwaki et al., 2009a, b; Abbaszadeh et al., 2013). A second resonant frequency ($f_2$) obtained from the spectrum was used to calculate the elasticity index (EI) of the samples as a measure of fruit firmness (Terasaki et al., 2006; Taniwaki et al., 2009a, b). Taniwaki et al. (2009a, b) stated in their study that the EI values of persimmon samples were found to be highly correlated with the sensory tests especially at the lower frequency range. Similarly, EI values of pear have been better correlated with the sensory test at the lower frequencies (Terasaki et al., 2006). Watermelon ripeness evaluation was carried out using TSS, TA, and TSS/TA evaluation and the responses of LDV results were correlated with these parameters (Abbaszadeh et al., 2013). The prediction of step-wise multiple linear regression and PLS models were compared, and the higher correlation has observed in multiple linear regression model for each quality parameter ($R^2 > 0.95$).

Ultrasonic measurements are often used for the evaluation of firmness and internal disorders of fruits such as chilling injury (Verlinden et al., 2004; Aboudaoud et al., 2012; Morrison and Abeyratne, 2014; Hitchman et al., 2016). The ultrasonic system mainly involves a transmitting transducer, which transfers the ultrasonic pulse to the sample, and a receiving transducer, which gathers the transmitted ultrasonic waves and converts them into electrical signals (Kilcast, 2004; Verlinden et al., 2004). Although this method is used as a non-destructive measurement method, chilling injury in tomatoes had been evaluated destructively by using ultrasonic transducers (Verlinden et al., 2004). The authors have determined that the time-of-flight of the ultrasonic waves has increased depending on the softening of tomatoes, and the apparent attenuation has also increased due to the chilling injury development (Verlinden et al., 2004). The maturity of Braeburn apple has been monitored with laser ultrasound and during the aging of apple, the attenuation of waves has increased likewise (Hitchman et al., 2016). In a recent study, a novel portable single-transducer ultrasonic device has been employed in firmness evaluation of oranges; upon the obtained data, it has been concluded that the reflected ultrasonic energy was highly correlated ($R = 0.99$) with the firmness of the oranges (Morrison and Abeyratne, 2014).

Non-destructive acoustic firmness measurements have compatible results compared to the sensory measurements, with respect to its accuracy and repeatability in the analysis (Aboonajmi et al., 2015). Similarly, LDV is advantageous with its accurate and fast measurements, which increase the potential use in commercial-scale. The impact measurements are not influenced by the shape of the sample, although deformation after small impact is low for the samples that have a high firmness value (Nicolaï et al., 2014). Therefore the short contact time produce low signal-to-noise ratio, which may limit the accuracy of the measurement (Nicolaï et al., 2014).

## 2.4 Miscellaneous methods

Terahertz (THz) radiation, or THz spectroscopic imaging method, corresponds to the region between the infrared and microwave region of electromagnetic spectrum, and, depending on the development in radiation sources and detectors, its potential use in food quality evaluation is increased (Butz et al., 2005; Gowen et al., 2012; Qin et al., 2013b; Wang et al., 2017a). Continuous wave radiation (CW-THz) terahertz imaging has been introduced for improving the low speed required for imaging of time-domain spectroscopy (THz-TDS) and simplification of imaging and image processing (Gowen et al., 2012; Qin et al., 2013b). Some THz spectroscopy applications include; determination of TSS in apple, classification of oils with their authenticity, evaluation of adulteration in honey, identification of transgenic (genetically modified) sugar beets, evaluation of insect infestation in wheat, identification of pesticide residues in sweet potatoes and antibiotic residues in milk and egg powder, and foreign material detection in chocolate and noodles (Gowen et al., 2012; Qin et al., 2013b; Wang et al., 2017a). Although having a range of applications, this method has limitations such as difficulty in analysis of high-moisture foods, high instrumental costs, and low imaging speed. However, portable and affordable CW-THz imaging, with its higher speed of imaging and higher resolution, has been introduced for commercial applications (Gowen et al., 2012; Qin et al., 2013b).

When a coherent light such as laser illuminates a sample, back-scattered rays form a granular pattern that includes randomly distributed bright and dark spots (Szymanska-Chargot et al., 2012; Arefi et al., 2017; Lu and Lu, 2017). The distribution pattern is called a biospeckle pattern if the biological sample is illuminated. This method is applicable for disease and defect detection in fruit and vegetables. Ripeness evaluation (TSS, TA, and starch), mealiness (firmness) formation, and monitoring of bull's eye rot in apples have been evaluated with respect to the biospeckle activity (Szymanska-Chargot et al., 2012; Arefi et al., 2017; Lu and Lu, 2017; Abasi et al., 2018). Fast imaging in biospeckle method is not recommended for accurate classifications with respect to the quality parameters, since it requires enough time for monitoring the numerous frames (Arefi et al., 2017; Lu and Lu, 2017).

Fluorescence imaging or, more specifically, chlorophyll fluorescence imaging (CFI), is a simple non-destructive method for monitoring of photosynthetic electron transfer (Kalaji et al., 2014). Photosynthetic activity in plant tissues is linked to water, temperature, UV intensity, chemical, and air pollution related stresses (Kalaji et al., 2014; Gorbe and Calatayud, 2012; Lu and Lu, 2017). Minimal processing of leafy vegetables also creates a stress and CFI has been used as an indicator of monitoring the stress in mini romaine lettuce and endive (Hägele et al., 2016). The authors stated that the maximum quantum efficiency had decreased significantly due to the stress of minimal processing applications. In addition, freeze injury in lettuce, rind injury in lemon, scald formation in apple, senescence of pear, nutrient deficiency in tomato and maize during plant growing, and apple and tomato maturity assessment can be listed as other CFI applications (Gorbe and Calatayud, 2012; Kalaji et al., 2014; Simko et al., 2015). The fruit or vegetable tissues without any chlorophyll or tissues with decay may emit some of the signals that are not related with chlorophyll; therefore, it may produce a false signal or noise in the acquired images (Gorbe and Calatayud, 2012; Simko et al., 2015).

# 3 Conclusions

In this chapter, non-destructive spectroscopic, computed tomography, mechanical, and other miscellaneous methods were discussed based on fruit and vegetable applications. The implementation cost of equipment, the length of data processing, as well as the amount of data to be extracted, push ahead the technological improvements to build portable, practical, on-line, real-time, in-field, and inexpensive methods. Precise, accurate, and repeatable measurements are required for correct classification of foods; however, in most cases the amount of data is voluminous, and employing appropriate preprocessing methods and algorithms may increase the accuracy of results and decrease the complexity of data to be processed. Several recent non-destructive applications have been discussed, and each of them has some limitations or advantages over other methods. The continuous developments in equipment and software technology, especially introduction of artificial intelligence, for in-line measurements provides fast and accurate prediction of food quality, thus avoidable food losses will gradually be decreased the with these improvements.

# References

Abasi, S., Minaei, S., Jamshidi, B., Fathi, D., 2018. Dedicated non-destructive devices for food quality measurement: a review. Trends Food Sci. Technol. 78, 197–205.

Abbaszadeh, R., Rajabipour, A., Ahmadi, H., Mahjoob, M.J., Delshad, M., 2013. Prediction of watermelon quality based on vibration spectrum. Postharvest Biol. Technol. 86, 291–293.

Abbaszadeh, R., Rajabipour, A., Ying, Y., Delshad, M., Mahjoob, M.J., Ahmadi, H., 2015. Nondestructive determination of watermelon flesh firmness by frequency response. LWT-Food Sci. Technol. 60 (1), 637–640.

Aboonajmi, M., Jahangiri, M., Hassan-Beygi, S.R., 2015. A review on application of acoustic analysis in quality evaluation of agro-food products. J. Food Process. Preserv. 39 (6), 3175–3188.

Aboudaoud, I., Faiz, B., Aassif, E., Moudden, A., Izbaim, D., Abassi, D., et al., 2012. The maturity characterization of orange fruit by using high frequency ultrasonic echo pulse method. In: IOP Conference Series: Materials Science and Engineering. vol. 42. IOP Publishing, p. 012038. 1.

Adão, T., Hruška, J., Pádua, L., Bessa, J., Peres, E., Morais, R., Sousa, J.J., 2017. Hyperspectral imaging: a review on UAV-based sensors, data processing and applications for agriculture and forestry. Remote Sens. (Basel) 9 (11), 1110.

Alhamdan, A.M., Atia, A., 2017. Non-destructive method to predict Barhi dates quality at different stages of maturity utilising near-infrared (NIR) spectroscopy. Int. J. Food Prop. 20 (suppl. 3), S2950–S2959.

Ali, M.M., Hashim, N., Bejo, S.K., Shamsudin, R., 2017. Rapid and nondestructive techniques for internal and external quality evaluation of watermelons: a review. Sci. Hortic. 225, 689–699.

Andaur, J.E., Guesalaga, A.R., Agosin, E.E., Guarini, M.W., Irarrázaval, P., 2004. Magnetic resonance imaging for nondestructive analysis of wine grapes. J. Agric. Food Chem. 52 (2), 165–170.

Arefi, A., Ahmadi Moghaddam, P., Modarres Motlagh, A., Hassanpour, A., 2017. Towards real-time speckle image processing for mealiness assessment in apple fruit. Int. J. Food Prop. 20 (suppl. 3), S3135–S3148.

Arendse, E., Fawole, O.A., Magwaza, L.S., Opara, U.L., 2018. Non-destructive prediction of internal and external quality attributes of fruit with thick rind: a review. J. Food Eng. 217, 11–23.

Beghi, R., Giovenzana, V., Tugnolo, A., Guidetti, R., 2018. Application of visible/near infrared spectroscopy to quality control of fresh fruits and vegetables in large-scale mass distribution channels: a preliminary test on carrots and tomatoes. J. Sci. Food Agric. 98 (7), 2729–2734.

Bizzani, M., Flores, D.W.M., Colnago, L.A., Ferreira, M.D., 2017. Non-invasive spectroscopic methods to estimate orange firmness, peel thickness, and total pectin content. Microchem. J. 133, 168–174.

Butz, P., Hofmann, C., Tauscher, B., 2005. Recent developments in noninvasive techniques for fresh fruit and vegetable internal quality analysis. J. Food Sci. 70 (9), R131–R141.

Camps, C., Christen, D., 2009. Non-destructive assessment of apricot fruit quality by portable visible-near infrared spectroscopy. LWT- Food Sci. Technol. 42 (6), 1125–1131.

Cantre, D., Herremans, E., Verboven, P., Ampofo-Asiama, J., Nicolaï, B., 2014. Characterization of the 3-D microstructure of mango (*Mangifera indica* L. cv. Carabao) during ripening using X-ray computed microtomography. Innov. Food Sci. Emerg. Technol. 24, 28–39.

Cavaco, A.M., Pires, R., Antunes, M.D., Panagopoulos, T., Brázio, A., Afonso, A.M., et al., 2018. Validation of short wave near infrared calibration models for the quality and ripening of 'Newhall'orange on tree across years and orchards. Postharvest Biol. Technol. 141, 86–97.

Cen, H., He, Y., 2007. Theory and application of near infrared reflectance spectroscopy in determination of food quality. Trends Food Sci. Technol. 18 (2), 72–83.

Chen, P., Sun, Z., 1991. A review of non-destructive methods for quality evaluation and sorting of agricultural products. J. Agric. Eng. Res. 49, 85–98.

Chen, H.Z., Xu, L.L., Tang, G.Q., Song, Q.Q., Feng, Q.X., 2016. Rapid detection of surface color of shatian pomelo using Vis-NIR spectrometry for the identification of maturity. Food Anal. Methods 9 (1), 192–201.

Cho, B.K., Chayaprasert, W., Stroshine, R.L., 2008. Effects of internal browning and watercore on low field (5.4 MHz) proton magnetic resonance measurements of T2 values of whole apples. Postharvest Biol. Technol. 47 (1), 81–89.

Clark, C.J., MacFall, J.S., 2003. Quantitative magnetic resonance imaging of 'Fuyu' persimmon fruit during development and ripening. Magn. Reson. Imaging 21 (6), 679–685.

Clark, C.J., Hockings, P.D., Joyce, D.C., Mazucco, R.A., 1997. Application of magnetic resonance imaging to pre-and post-harvest studies of fruits and vegetables. Postharvest Biol. Technol. 11 (1), 1–21.

Clark, C.J., MacFall, J.S., Bieleski, R.L., 1998. Loss of watercore from Fuji' apple observed by magnetic resonance imaging. Sci. Hortic. 73 (4), 213–227.

Cortes, V., Rodriguez, A., Blasco, J., Rey, B., Besada, C., Cubero, S., … Aleixos, N., 2017. Prediction of the level of astringency in persimmon using visible and near-infrared spectroscopy. J. Food Eng. 204, 27–37.

Dall'Acqua, Y.G., Cunha Júnior, L.C., Nardini, V., Lopes, V.G., Pessoa, J.D.D.C., de Almeida Teixeira, G.H., 2015. Discrimination of *Euterpe oleracea* Mart.(Açaí) and Euterpe edulis Mart.(Juçara) intact fruit using near-infrared (NIR) spectroscopy and linear discriminant analysis. J. Food Process. Preserv. 39 (6), 2856–2865.

Dhakal, S., Li, Y., Peng, Y., Chao, K., Qin, J., Guo, L., 2014. Prototype instrument development for non-destructive detection of pesticide residue in apple surface using Raman technology. J. Food Eng. 123, 94–103.

Diels, E., van Dael, M., Keresztes, J., Vanmaercke, S., Verboven, P., Nicolai, B., … Smeets, B., 2017. Assessment of bruise volumes in apples using X-ray computed tomography. Postharvest Biol. Technol. 128, 24–32.

Donis-González, I.R., Guyer, D.E., Pease, A., 2016a. Postharvest noninvasive assessment of undesirable fibrous tissue in fresh processing carrots using computer tomography images. J. Food Eng. 190, 154–166.

Donis-González, I.R., Guyer, D.E., Pease, A., 2016b. Postharvest noninvasive classification of tough-fibrous asparagus using computed tomography images. Postharvest Biol. Technol. 121, 27–35.

Du, C.J., Sun, D.W., 2004. Recent developments in the applications of image processing techniques for food quality evaluation. Trends Food Sci. Technol. 15 (5), 230–249.

Elbatawi, I.E., 2008. An acoustic impact method to detect hollow heart of potato tubers. Biosyst. Eng. 100 (2), 206–213.

ElMasry, G.M., Nakauchi, S., 2016. Image analysis operations applied to hyperspectral images for non-invasive sensing of food quality—a comprehensive review. Biosyst. Eng. 142, 53–82.

ElMasry, G., Wang, N., ElSayed, A., Ngadi, M., 2007. Hyperspectral imaging for nondestructive determination of some quality attributes for strawberry. J. Food Eng. 81 (1), 98–107.

Everard, C.D., Kim, M.S., Lee, H., 2014. A comparison of hyperspectral reflectance and fluorescence imaging techniques for detection of contaminants on spinach leaves. J. Food Eng. 143, 139–145.

Fernández-Espinosa, A.J., 2016. Combining PLS regression with portable NIR spectroscopy to on-line monitor quality parameters in intact olives for determining optimal harvesting time. Talanta 148, 216–228.

Foerster, J., Truppel, I., Bochow-Neß, O., Geyer, M., 2013. Comparison of acoustic sensor systems for quality analysis of asparagus using scanning laser vibrometry for visualization. Comput. Electron. Agric. 91, 10–18.

Gao, Y., Li, Q., Rao, X., Ying, Y., 2018. Precautionary analysis of sprouting potato eyes using hyperspectral imaging technology. Int. J. Agric. Biol. Eng. 11 (2), 153–157.

Giovenzana, V., Beghi, R., Civelli, R., Guidetti, R., 2015. Optical techniques for rapid quality monitoring along minimally processed fruit and vegetable chain. Trends Food Sci. Technol. 46 (2), 331–338.

Gorbe, E., Calatayud, A., 2012. Applications of chlorophyll fluorescence imaging technique in horticultural research: a review. Sci. Hortic. 138, 24–35.

Gowen, A.A., O'Donnell, C., Cullen, P.J., Downey, G., Frias, J.M., 2007. Hyperspectral imaging—an emerging process analytical tool for food quality and safety control. Trends Food Sci. Technol. 18 (12), 590–598.

Gowen, A.A., O'Sullivan, C., O'Donnell, C.P., 2012. Terahertz time domain spectroscopy and imaging: emerging techniques for food process monitoring and quality control. Trends Food Sci. Technol. 25 (1), 40–46.

Guo, Y., Ni, Y., Kokot, S., 2016. Evaluation of chemical components and properties of the jujube fruit using near infrared spectroscopy and chemometrics. Spectrochim. Acta A Mol. Biomol. Spectrosc. 153, 79–86.

Hägele, F., Baur, S., Menegat, A., Gerhards, R., Carle, R., Schweiggert, R.M., 2016. Chlorophyll fluorescence imaging for monitoring the effects of minimal processing and warm water treatments on physiological properties and quality attributes of fresh-cut salads. Food Bioproc. Tech. 9 (4), 650–663.

Hashim, N., Onwude, D.I., Osman, M.S., 2018. Evaluation of chilling injury in mangoes using multispectral imaging. J. Food Sci. 83 (5), 1271–1279.

Herremans, E., Verboven, P., Bongaers, E., Estrade, P., Verlinden, B.E., Wevers, M., … Nicolai, B.M., 2013. Characterisation of 'Braeburn' browning disorder by means of X-ray micro-CT. Postharvest Biol. Technol. 75, 114–124.

Hills, B., 1998. Magnetic Resonance Imaging in Food Science. John Wiley & Sons.

Hills, B.P., 2006. Applications of low-field NMR to food science. In: Annual Reports on NMR Spectroscopy. vol. 58. Academic Press, pp. 177–230.

Hitchman, S., van Wijk, K., Davidson, Z., 2016. Monitoring attenuation and the elastic properties of an apple with laser ultrasound. Postharvest Biol. Technol. 121, 71–77.

Huang, H., Yu, H., Xu, H., Ying, Y., 2008. Near infrared spectroscopy for on/in-line monitoring of quality in foods and beverages: a review. J. Food Eng. 87 (3), 303–313.

Huang, W., Li, J., Wang, Q., Chen, L., 2015. Development of a multispectral imaging system for online detection of bruises on apples. J. Food Eng. 146, 62–71.

Huang, Y., Lu, R., Chen, K., 2018. Prediction of firmness parameters of tomatoes by portable visible and near-infrared spectroscopy. J. Food Eng. 222, 185–198.

Hussain, A., Pu, H., Sun, D.W., 2018. Innovative nondestructive imaging techniques for ripening and maturity of fruits—a review of recent applications. Trends Food Sci. Technol. 72, 144–152.

Irudayaraj, J., Reh, C. (Eds.), 2008. Nondestructive Testing of Food Quality. vol. 18. John Wiley & Sons.

Islam, M.N., Nielsen, G., Stærke, S., Kjær, A., Jørgensen, B., Edelenbos, M., 2018. Novel non-destructive quality assessment techniques of onion bulbs: a comparative study. J. Food Sci. Technol. 55 (8), 3314–3324.

Jamshidi, B., Minaei, S., Mohajerani, E., Ghassemian, H., 2012. Reflectance Vis/NIR spectroscopy for nondestructive taste characterization of Valencia oranges. Comput. Electron. Agric. 85, 64–69.

Jha, S.N., Narsaiah, K., Jaiswal, P., Bhardwaj, R., Gupta, M., Kumar, R., Sharma, R., 2014. Nondestructive prediction of maturity of mango using near infrared spectroscopy. J. Food Eng. 124, 152–157.

Kalaji, H.M., Oukarroum, A., Alexandrov, V., Kouzmanova, M., Brestic, M., Zivcak, M., … Goltsev, V., 2014. Identification of nutrient deficiency in maize and tomato plants by in vivo chlorophyll a fluorescence measurements. Plant Physiol. Biochem. 81, 16–25.

Keresztes, J.C., Diels, E., Goodarzi, M., Nguyen-Do-Trong, N., Goos, P., Nicolai, B., Saeys, W., 2017. Glare based apple sorting and iterative algorithm for bruise region detection using shortwave infrared hyperspectral imaging. Postharvest Biol. Technol. 130, 103–115.

Khodabakhshian, R., Emadi, B., 2017. Application of Vis/SNIR hyperspectral imaging in ripeness classification of pear. Int. J. Food Prop. 20 (suppl. 3), S3149–S3163.

Khodabakhshian, R., Emadi, B., Khojastehpour, M., Golzarian, M.R., Sazgarnia, A., 2017. Development of a multispectral imaging system for online quality assessment of pomegranate fruit. Int. J. Food Prop. 20 (1), 107–118.

Kilcast, D. (Ed.), 2004. Texture in Food: Solid Foods. Elsevier.

Kirtil, E., Oztop, M.H., 2016. 1 H nuclear magnetic resonance relaxometry and magnetic resonance imaging and applications in food science and processing. Food Eng. Rev. 8 (1), 1–22.

Kotwaliwale, N., Curtis, E., Othman, S., Naganathan, G.K., Subbiah, J., 2012a. Magnetic resonance imaging and relaxometry to visualize internal freeze damage to pickling cucumber. Postharvest Biol. Technol. 68, 22–31.

Kotwaliwale, N., Kalne, A., Singh, K., 2012b. Monitoring of mango (*Mangifera indica* L.)(Cv.: Chousa) ripening using X-ray computed tomography. In: Sensing Technology (ICST), 2012 Sixth International Conference on (pp. 326–330). IEEE.

Kotwaliwale, N., Singh, K., Kalne, A., Jha, S.N., Seth, N., Kar, A., 2014. X-ray imaging methods for internal quality evaluation of agricultural produce. J. Food Sci. Technol. 51 (1), 1–15.

Kuperman, V., 2000. Magnetic Resonance Imaging: Physical Principles and Applications. Elsevier.

Lamikanra, O. (Ed.), 2002. Fresh-Cut Fruits and Vegetables: Science, Technology, and Market. CRC Press.

Lammertyn, J., Dresselaers, T., Van Hecke, P., Jancsók, P., Wevers, M., Nicolaı, B.M., 2003. Analysis of the time course of core breakdown in 'Conference' pears by means of MRI and X-ray CT. Postharvest Biol. Technol. 29 (1), 19–28.

Leiva-Valenzuela, G.A., Lu, R., Aguilera, J.M., 2013. Prediction of firmness and soluble solids content of blueberries using hyperspectral reflectance imaging. J. Food Eng. 115 (1), 91–98.

Li, H., Pidakala, P., Billing, D., Burdon, J., 2016a. Kiwifruit firmness: measurement by penetrometer and non-destructive devices. Postharvest Biol. Technol. 120, 127–137.

Li, J., Chen, L., Huang, W., Wang, Q., Zhang, B., Tian, X., Fan, S., Li, B., 2016. Multispectral detection of skin defects of bi-colored peaches based on vis–NIR hyperspectral imaging. Postharvest Biol. Technol. 112, 121–133.

Li, B., Lecourt, J., Bishop, G., 2018. Advances in non-destructive early assessment of fruit ripeness towards defining optimal time of harvest and yield prediction—a review. Plants 7 (1), 3.

Liu, B., Zhou, P., Liu, X., Sun, X., Li, H., Lin, M., 2013. Detection of pesticides in fruits by surface-enhanced Raman spectroscopy coupled with gold nanostructures. Food Bioproc. Tech. 6 (3), 710–718.

Liu, C., Liu, W., Lu, X., Ma, F., Chen, W., Yang, J., Zheng, L., 2014. Application of multispectral imaging to determine quality attributes and ripeness stage in strawberry fruit. PLoS One 9 (2), e87818.

Liu, C., Yang, S.X., Deng, L., 2015. Determination of internal qualities of Newhall navel oranges based on NIR spectroscopy using machine learning. J. Food Eng. 161, 16–23.

López-Maestresalas, A., Keresztes, J.C., Goodarzi, M., Arazuri, S., Jarén, C., Saeys, W., 2016. Non-destructive detection of blackspot in potatoes by Vis-NIR and SWIR hyperspectral imaging. Food Control 70, 229–241.

Lu, R., 2004. Multispectral imaging for predicting firmness and soluble solids content of apple fruit. Postharvest Biol. Technol. 31 (2), 147–157.

Lu, R., 2016. Light Scattering Technology for Food Property, Quality and Safety Assessment. CRC Press.

Lu, Y., Lu, R., 2017. Non-destructive defect detection of apples by spectroscopic and imaging technologies: a review. Trans. ASABE 60 (5), 1765–1790.

Lu, J., Qi, S., Liu, R., Zhou, E., Li, W., Song, S., Han, D., 2015. Nondestructive determination of soluble solids and firmness in mix-cultivar melon using near-infrared CCD spectroscopy. J. Innov. Opt. Health Sci. 8 (06), 1550032.

Magwaza, L.S., Opara, U.L., 2014. Investigating non-destructive quantification and characterization of pomegranate fruit internal structure using X-ray computed tomography. Postharvest Biol. Technol. 95, 1–6.

Magwaza, L.S., Opara, U.L., Nieuwoudt, H., Cronje, P.J., Saeys, W., Nicolaï, B., 2012a. NIR spectroscopy applications for internal and external quality analysis of citrus fruit—a review. Food Bioproc. Tech. 5 (2), 425–444.

Magwaza, L.S., Opara, U.L., Terry, L.A., Landahl, S., Cronje, P.J., Nieuwoudt, H., … Nicolaï, B.M., 2012b. Prediction of 'Nules Clementine' mandarin susceptibility to rind breakdown disorder using Vis/NIR spectroscopy. Postharvest Biol. Technol. 74, 1–10.

Mahesh, S., Jayas, D.S., Paliwal, J., White, N.D.G., 2015. Hyperspectral imaging to classify and monitor quality of agricultural materials. J. Stored Prod. Res. 61, 17–26.

Makky, M., Soni, P., 2014. In situ quality assessment of intact oil palm fresh fruit bunches using rapid portable non-contact and non-destructive approach. J. Food Eng. 120, 248–259.

Malegori, C., Marques, E.J.N., de Freitas, S.T., Pimentel, M.F., Pasquini, C., Casiraghi, E., 2017. Comparing the analytical performances of Micro-NIR and FT-NIR spectrometers in the evaluation of acerola fruit quality, using PLS and SVM regression algorithms. Talanta 165, 112–116.

Maniwara, P., Nakano, K., Boonyakiat, D., Ohashi, S., Hiroi, M., Tohyama, T., 2014. The use of visible and near infrared spectroscopy for evaluating passion fruit postharvest quality. J. Food Eng. 143, 33–43.

Mao, J., Yu, Y., Rao, X., Wang, J., 2016. Firmness prediction and modeling by optimizing acoustic device for watermelons. J. Food Eng. 168, 1–6.

Marigheto, N., Hills, B., 2005. MRI as a potential on-line sensor of apple quality. In: s. pp. 455–464.

Marques, E.J.N., de Freitas, S.T., Pimentel, M.F., Pasquini, C., 2016. Rapid and non-destructive determination of quality parameters in the 'Tommy Atkins' mango using a novel handheld near infrared spectrometer. Food Chem. 197, 1207–1214.

Mazhar, M., Joyce, D., Cowin, G., Brereton, I., Hofman, P., Collins, R., Gupta, M., 2015. Non-destructive 1H-MRI assessment of flesh bruising in avocado (*Persea americana* M.) cv. Hass. Postharvest Biol. Technol. 100, 33–40.

McCarthy, M.J., 2012. Magnetic Resonance Imaging in Foods. Springer Science & Business Media.

Mendoza, F., Lu, R., Cen, H., 2012. Comparison and fusion of four nondestructive sensors for predicting apple fruit firmness and soluble solids content. Postharvest Biol. Technol. 73, 89–98.

Mireei, S.A., Sadeghi, M., Heidari, A., Hemmat, A., 2015. On-line firmness sensing of dates using a non-destructive impact testing device. Biosyst. Eng. 129, 288–297.

Mo, C., Kim, G., Kim, M.S., Lim, J., Lee, K., Lee, W.H., Cho, B.K., 2017a. On-line fresh-cut lettuce quality measurement system using hyperspectral imaging. Biosyst. Eng. 156, 38–50.

Mo, C., Kim, M.S., Kim, G., Lim, J., Delwiche, S.R., Chao, K., Lee, H., Cho, B.K., 2017b. Spatial assessment of soluble solid contents on apple slices using hyperspectral imaging. Biosyst. Eng. 159, 10–21.

Morrison, D.S., Abeyratne, U.R., 2014. Ultrasonic technique for non-destructive quality evaluation of oranges. J. Food Eng. 141, 107–112.

Muenmanee, N., Joomwong, A., Natwichai, J., Boonyakiat, D., 2016. Assessment of internal quality of Japanese pumpkin (*Cucurbita maxima*) using near infrared spectroscopy. Pak. J. Biotechnol. 13 (1), 55–65.

Müller, C., David, L., Chiş, V., Pînzaru, S.C., 2014. Detection of thiabendazole applied on citrus fruits and bananas using surface enhanced Raman scattering. Food Chem. 145, 814–820.

Munera, S., Besada, C., Aleixos, N., Talens, P., Salvador, A., Sun, D.W., Cubero, S., Blasco, J., 2017. Non-destructive assessment of the internal quality of intact persimmon using colour and VIS/NIR hyperspectral imaging. LWT- Food Sci. Technol. 77, 241–248.

Munera, S., Amigo, J.M., Aleixos, N., Talens, P., Cubero, S., Blasco, J., 2018. Potential of VIS-NIR hyperspectral imaging and chemometric methods to identify similar cultivars of nectarine. Food Control 86, 1–10.

Muramatsu, N., Sakurai, N., Yamamoto, R., Nevins, D.J., Takahara, T., Ogata, T., 1997. Comparison of a non-destructive acoustic method with an intrusive method for firmness measurement of kiwifruit. Postharvest Biol. Technol. 12 (3), 221–228.

Musse, M., Quellec, S., Cambert, M., Devaux, M.F., Lahaye, M., Mariette, F., 2009. Monitoring the postharvest ripening of tomato fruit using quantitative MRI and NMR relaxometry. Postharvest Biol. Technol. 53 (1–2), 22–35.

Muziri, T., Theron, K.I., Cantre, D., Wang, Z., Verboven, P., Nicolai, B.M., Crouch, E.M., 2016. Microstructure analysis and detection of mealiness in 'Forelle' pear (*Pyrus communis* L.) by means of X-ray computed tomography. Postharvest Biol. Technol. 120, 145–156.

Ncama, K., Opara, U.L., Tesfay, S.Z., Fawole, O.A., Magwaza, L.S., 2017. Application of Vis/NIR spectroscopy for predicting sweetness and flavour parameters of 'Valencia'orange (*Citrus sinensis*) and 'Star Ruby'grapefruit (Citrus x paradisi Macfad). J. Food Eng. 193, 86–94.

Nicolaï, B.M., Defraeye, T., De Ketelaere, B., Herremans, E., Hertog, M.L., Saeys, W., Toricelli, A., Vandendriessche, T., Verboven, P., 2014. Nondestructive measurement of fruit and vegetable quality. Annu. Rev. Food Sci. Technol. 5, 285–312.

Nordey, T., Joas, J., Davrieux, F., Chillet, M., Léchaudel, M., 2017. Robust NIRS models for non-destructive prediction of mango internal quality. Sci. Hortic. 216, 51–57.

Oliveira-Folador, G., de Oliveira Bicudo, M., de Andrade, E.F., Renard, C.M.G.C., Bureau, S., de Castilhos, F., 2018. Quality traits prediction of the passion fruit pulp using NIR and MIR spectroscopy. LWT–Food Sci. Technol. 95, 172–178.

Ozaki, Y., McClure, W.F., Christy, A.A. (Eds.), 2006. Near-Infrared Spectroscopy in Food Science and Technology. John Wiley & Sons.

Pan, L., Lu, R., Zhu, Q., McGrath, J.M., Tu, K., 2015. Measurement of moisture, soluble solids, sucrose content and mechanical properties in sugar beet using portable visible and near-infrared spectroscopy. Postharvest Biol. Technol. 102, 42–50.

Park, B., Lu, R. (Eds.), 2015. Hyperspectral Imaging Technology in Food and Agriculture. Springer, New York.

Patel, K.K., Khan, M.A., Kar, A., 2015. Recent developments in applications of MRI techniques for foods and agricultural produce—an overview. J. Food Sci. Technol. 52 (1), 1–26.

Porep, J.U., Kammerer, D.R., Carle, R., 2015. On-line application of near infrared (NIR) spectroscopy in food production. Trends Food Sci. Technol. 46 (2), 211–230.

Qin, J., Chao, K., Kim, M.S., Lu, R., Burks, T.F., 2013a. Hyperspectral and multispectral imaging for evaluating food safety and quality. J. Food Eng. 118 (2), 157–171.

Qin, J., Ying, Y., Xie, L., 2013b. The detection of agricultural products and food using terahertz spectroscopy: a review. Appl. Spectrosc. Rev. 48 (6), 439–457.

Qin, J., Chao, K., Kim, M.S., 2016. Raman scattering for food quality and safety assessment. In: Light Scattering Technology for Food Property, Quality and Safety Assessment. CRC Press, pp. 399–440.

Ragni, L., Berardinelli, A., Guarnieri, A., 2010. Impact device for measuring the flesh firmness of kiwifruits. J. Food Eng. 96 (4), 591–597.

Rahman, A., Kandpal, L.M., Lohumi, S., Kim, M.S., Lee, H., Mo, C., Cho, B.K., 2017. Nondestructive estimation of moisture content, pH and soluble solid contents in intact tomatoes using hyperspectral imaging. Appl. Sci. 7 (1), 109.

Rajkumar, P., Wang, N., Elmasry, G., Raghavan, G.S.V., Gariepy, Y., 2012. Studies on banana fruit quality and maturity stages using hyperspectral imaging. J. Food Eng. 108 (1), 194–200.

Renu, R., Chidanand, D.V., 2013. Internal quality classification of agricultural produce using non-destructive image processing technologies (soft x-ray). Int. J. Latest Trends Eng. Technol. 2 (4), 535–543.

Reusch, W., 2013. Nuclear Magnetic Resonance Spectroscopy. https://www2.chemistry.msu.edu/faculty/reusch/virttxtjml/spectrpy/nmr/nmr1.htm. Access date: 13, October, 2018.

Rodríguez-Pulido, F.J., Gil-Vicente, M., Gordillo, B., Heredia, F.J., González-Miret, M.L., 2017. Measurement of ripening of raspberries (*Rubus idaeus* L) by near infrared and colorimetric imaging techniques. J. Food Sci. Technol. 54 (9), 2797–2803.

Ropodi, A.I., Panagou, E.Z., Nychas, G.J., 2016. Data mining derived from food analyses using non-invasive/non-destructive analytical techniques; determination of food authenticity, quality & safety in tandem with computer science disciplines. Trends Food Sci. Technol. 50, 11–25.

Rungpichayapichet, P., Mahayothee, B., Nagle, M., Khuwijitjaru, P., Müller, J., 2016. Robust NIRS models for non-destructive prediction of postharvest fruit ripeness and quality in mango. Postharvest Biol. Technol. 111, 31–40.

Sánchez, M.T., De la Haba, M.J., Benítez-López, M., Fernández-Novales, J., Garrido-Varo, A., Pérez-Marín, D., 2012. Non-destructive characterization and quality control of intact strawberries based on NIR spectral data. J. Food Eng. 110 (1), 102–108.

Schaare, P.N., Fraser, D.G., 2000. Comparison of reflectance, interactance and transmission modes of visible-near infrared spectroscopy for measuring internal properties of kiwifruit (*Actinidia chinensis*). Postharvest Biol. Technol. 20 (2), 175–184.

Schoeman, L., Williams, P., du Plessis, A., Manley, M., 2016. X-ray micro-computed tomography (μCT) for non-destructive characterisation of food microstructure. Trends Food Sci. Technol. 47, 10–24.

Schotte, S., De Belie, N., De Baerdemaeker, J., 1999. Acoustic impulse-response technique for evaluation and modelling of firmness of tomato fruit. Postharvest Biol. Technol. 17 (2), 105–115.

Siedliska, A., Zubik, M., Baranowski, P., Mazurek, W., 2017. Algorithms for detecting cherry pits on the basis of transmittance mode hyperspectral data. Int. Agrophys. 31 (4), 539–549.

Simko, I., Jimenez-Berni, J.A., Furbank, R.T., 2015. Detection of decay in fresh-cut lettuce using hyperspectral imaging and chlorophyll fluorescence imaging. Postharvest Biol. Technol. 106, 44–52.

Sirisomboon, P., Tanaka, M., Kojima, T., Williams, P., 2012. Nondestructive estimation of maturity and textural properties on tomato 'Momotaro'by near infrared spectroscopy. J. Food Eng. 112 (3), 218–226.

Smith, B.C., 2011. Fundamentals of Fourier Transform Infrared Spectroscopy. CRC Press.

Sonego, L., Ben-Arie, R., Raynal, J., Pech, J.C., 1995. Biochemical and physical evaluation of textural characteristics of nectarines exhibiting woolly breakdown: NMR imaging, X-ray computed tomography and pectin composition. Postharvest Biol. Technol. 5 (3), 187–198.

Srivastava, S., Sadistap, S., 2018. Non-destructive sensing methods for quality assessment of on-tree fruits: a review. J. Food Meas. Charact. 12 (1), 497–526.

Studman, C.J., 2001. Computers and electronics in postharvest technology—a review. Comput. Electron. Agric. 30, 109–124.

Su, W.H., Sun, D.W., 2018. Multispectral imaging for plant food quality analysis and visualization. Compr. Rev. Food Sci. Food Saf. 17, 220–239.

Su, W.H., He, H.J., Sun, D.W., 2017. Non-destructive and rapid evaluation of staple foods quality by using spectroscopic techniques: a review. Crit. Rev. Food Sci. Nutr. 57 (5), 1039–1051.

Suchanek, M., Kordulska, M., Olejniczak, Z., Figiel, H., Turek, K., 2017. Application of low-field MRI for quality assessment of 'Conference'pears stored under controlled atmosphere conditions. Postharvest Biol. Technol. 124, 100–106.

Sun, D.W. (Ed.), 2009. Infrared Spectroscopy for Food Quality Analysis and Control. Academic Press.

Sun, D.W. (Ed.), 2010. Hyperspectral Imaging for Food Quality Analysis and Control. Elsevier.

Sun, D.W. (Ed.), 2016. Computer Vision Technology for Food Quality Evaluation. Academic Press.

Sun, T., Huang, K., Xu, H., Ying, Y., 2010. Research advances in nondestructive determination of internal quality in watermelon/melon: a review. J. Food Eng. 100 (4), 569–577.

Sun, J., Ma, B., Dong, J., Zhu, R., Zhang, R., Jiang, W., 2017. Detection of internal qualities of hami melons using hyperspectral imaging technology based on variable selection algorithms. J. Food Process Eng. 40 (3), e12496.

Szymanska-Chargot, M., Adamiak, A., Zdunek, A., 2012. Pre-harvest monitoring of apple fruits development with the use of biospeckle method. Sci. Hortic. 145, 23–28.

Taglienti, A., Massantini, R., Botondi, R., Mencarelli, F., Valentini, M., 2009. Postharvest structural changes of Hayward kiwifruit by means of magnetic resonance imaging spectroscopy. Food Chem. 114 (4), 1583–1589.

Tang, C., He, H., Li, E., Li, H., 2018. Multispectral imaging for predicting sugar content of 'Fuji'apples. Opt. Laser Technol. 106, 280–285.

Taniwaki, M., Hanada, T., Sakurai, N., 2009a. Postharvest quality evaluation of "Fuyu" and "Taishuu" persimmons using a nondestructive vibrational method and an acoustic vibration technique. Postharvest Biol. Technol. 51 (1), 80–85.

Taniwaki, M., Takahashi, M., Sakurai, N., 2009b. Determination of optimum ripeness for edibility of postharvest melons using nondestructive vibration. Food Res. Int. 42 (1), 137–141.

Tao, F., Zhang, L., McCarthy, M.J., Beckles, D.M., Saltveit, M., 2014. Magnetic resonance imaging provides spatial resolution of chilling injury in micro-tom tomato (*Solanum lycopersicum* L.) fruit. Postharvest Biol. Technol. 97, 62–67.

Teerachaichayut, S., Ho, H.T., 2017. Non-destructive prediction of total soluble solids, titratable acidity and maturity index of limes by near infrared hyperspectral imaging. Postharvest Biol. Technol. 133, 20–25.

Terasaki, S., Sakurai, N., Zebrowski, J., Murayama, H., Yamamoto, R., Nevins, D.J., 2006. Laser Doppler vibrometer analysis of changes in elastic properties of ripening 'La France'pears after postharvest storage. Postharvest Biol. Technol. 42 (2), 198–207.

Thybo, A.K., Jespersen, S.N., Lærke, P.E., Stødkilde-Jørgensen, H.J., 2004. Nondestructive detection of internal bruise and spraing disease symptoms in potatoes using magnetic resonance imaging. Magn. Reson. Imaging 22 (9), 1311–1317.

Tiwari, G., Slaughter, D.C., Cantwell, M., 2013. Nondestructive maturity determination in green tomatoes using a handheld visible and near infrared instrument. Postharvest Biol. Technol. 86, 221–229.

Urraca, R., Sanz-Garcia, A., Tardaguila, J., Diago, M.P., 2016. Estimation of total soluble solids in grape berries using a hand-held NIR spectrometer under field conditions. J. Sci. Food Agric. 96 (9), 3007–3016.

Van Dael, M., Verboven, P., Dhaene, J., Van Hoorebeke, L., Sijbers, J., Nicolai, B., 2017. Multisensor X-ray inspection of internal defects in horticultural products. Postharvest Biol. Technol. 128, 33–43.

Verboven, P., Defraeye, T., Nicolai, B., 2018. Measurement and visualization of food microstructure: fundamentals and recent advances. In: Food Microstructure and Its Relationship with Quality and Stability. Woodhead Publishing, pp. 3–28.

Verlinden, B.E., De Smedt, V., Nicolaï, B.M., 2004. Evaluation of ultrasonic wave propagation to measure chilling injury in tomatoes. Postharvest Biol. Technol. 32 (1), 109–113.

Wang, K., Sun, D.W., Pu, H., 2017a. Emerging non-destructive terahertz spectroscopic imaging technique: principle and applications in the agri-food industry. Trends Food Sci. Technol. 67, 93–105.

Wang, S.X., Hu, R.F., Gao, K., Wali, F., Zan, G.B., Wang, D.J., … Wei, S.Q., 2017b. Non-destructive study of fruits using grating-based X-ray imaging. Nucl. Sci. Tech. 28 (2), 24.

Xiao, H., Sun, K., Sun, Y., Wei, K., Tu, K., Pan, L., 2017. Comparison of benchtop Fourier-transform (FT) and portable grating scanning spectrometers for determination of total soluble solid contents in single grape berry (*Vitis vinifera* L.) and calibration transfer. Sensors 17 (11), 2693.

Xiao, H., Li, A., Li, M., Sun, Y., Tu, K., Wang, S., Pan, L., 2018. Quality assessment and discrimination of intact white and red grapes from *Vitis vinifera* L. at five ripening stages by visible and near-infrared spectroscopy. Sci. Hortic. 233, 99–107.

Yan, L., Xiong, C., Qu, H., Liu, C., Chen, W., Zheng, L., 2017. Non-destructive determination and visualisation of insoluble and soluble dietary fibre contents in fresh-cut celeries during storage periods using hyperspectral imaging technique. Food Chem. 228, 249–256.

Yang, H.Y., Inagaki, T., Ma, T., Tsuchikawa, S., 2017. High-resolution and non-destructive evaluation of the spatial distribution of nitrate and its dynamics in spinach (*Spinacia oleracea* L.) leaves by near-infrared hyperspectral imaging. Front. Plant Sci. 8, 1937.

Yu, X.L., He, Y., Feng, X., Feng, S., 2017. Non-destructive detection of local radishes' hollow hearts by hyperspectral imaging combined with support vector machine. In: 2017 ASABE Annual International Meeting. American Society of Agricultural and Biological Engineers, pp. 1.

Zhang, L., McCarthy, M.J., 2012. Measurement and evaluation of tomato maturity using magnetic resonance imaging. Postharvest Biol. Technol. 67, 37–43.

Zhang, L., McCarthy, M.J., 2013. Assessment of pomegranate postharvest quality using nuclear magnetic resonance. Postharvest Biol. Technol. 77, 59–66.

Zhang, H., Wu, J., Zhao, Z., Wang, Z., 2018. Nondestructive firmness measurement of differently shaped pears with a dual-frequency index based on acoustic vibration. Postharvest Biol. Technol. 138, 11–18.

Zheng, J., He, L., 2014. Surface-enhanced Raman spectroscopy for the chemical analysis of food. Compr. Rev. Food Sci. Food Saf. 13 (3), 317–328.

Zhu, H., Chu, B., Fan, Y., Tao, X., Yin, W., He, Y., 2017. Hyperspectral imaging for predicting the internal quality of kiwifruits based on variable selection algorithms and chemometric models. Sci. Rep. 7 (1), 7845.

# Sensory shelf life estimation

**11**

*Ana Giménez, Gastón Ares*
Sensometrics & Consumer Science, Instituto Polo Tecnológico de Pando, Universidad de la República, Canelones, Uruguay

**Chapter outline**



## 1 Introduction

Shelf life of a food product can be defined as the time period acceptable eating quality is retained, from a safety, nutritional, and sensory standpoint (IFST, 1993). During this time period, the product should provide consumers its intended sensory experience, performance and benefits (ASTM E2454-05, 2011). Shelf life dates are compulsorily included on the labels of prepackaged food products in most countries as a means to provide consumers with an anchor point that provides guarantees on their quality and safety (Department for Environment, Food and Rural Affairs, 2011). Two typologies of shelf life dates exist: *use by* and *best before*. According to the EU label system, *use by* date is determined by safety aspects, whereas *best before* date is related to sensory and nutritional quality (European Commission, 2000). Food manufacturers are responsible for determining the type of shelf life date appropriate for a specific food product (European Commission, 2002). Therefore, accurately estimating shelf life dates poses a challenge to food scientists, since manufacturers, retailers, and consumers rely on this information along the food production and consumption chain.

Manufacturers need to assure products meet food safety criteria, along with quality thresholds that ensure consumer satisfaction throughout the labeled shelf life. Brand and product loyalty might be compromised if consumer expectations are not met (Harcar and Karakaya, 2005). In an attempt to reduce this risk, manufacturers tend to become more conservative when establishing the maximum period of time a product can be stored, which leads to an excessively short shelf life (Newsome et al., 2014). In addition, in many instances, the need to introduce new products to the market within a

Food Quality and Shelf Life. https://doi.org/10.1016/B978-0-12-817190-5.00011-2

short time frame leads companies to establish shelf life based on prior experience with similar products. These practices might lead to an early retrieval of a food product from the marketplace that is still acceptable for consumption, from sensory, nutritional, and safety perspectives. In fact, empirical data shows that a considerable amount of products are discarded from the market if they are not sold within their shelf life (Spada et al., 2018). This implies both economic and environmental consequences, particularly given the need to reduce food waste to achieve a sustainable transformation to keep human development within safe boundaries (Rockström et al., 2009).

Consumers look for shelf life dates on food labels motivated by their interest for fresh, natural, safe, and superior quality food products. Shelf life dates are a proxy for healthiness, nutritional value, freshness, and safety during their decision making (Alongi et al., 2018; Giménez et al., 2012; Milne, 2013; Newsome et al., 2014; Wansink and Wright, 2006; Wilson et al., 2017; Ragaert et al., 2004). In addition, shelf life dating can influence acceptability, taste perceptions, and perceived quality of food products (Priefer et al., 2016; Wansink and Wright, 2006). Research has shown that shelf life labeling is one of the reasons underlying food waste at the household level, as a proportion of consumers tend to discard products when they are past their shelf life, without even trying them (Gaiani et al., 2018).

Strategic efforts to prolong shelf life and improve consumer understanding of shelf life dating typologies have been highlighted as potential contributors to reducing food waste worldwide and improving the sustainability of the food sector (Spada et al., 2018; Gaiani et al., 2018). In this sense, recent research has shown that shelf life dating focused on maximizing consumer acceptance and minimizing changes with respect to the fresh product are usually too conservative, which can lead to unnecessary food waste (Man, 2016).

During storage, food products change as a consequence of biological, enzymatic, and physicochemical reactions that take place (Labuza and Szybist, 2001). These changes might impact the nutritional, microbiological, or sensory quality of a food product, limiting the time period it should be consumed. However, sensory changes are the ones limiting the shelf life of most food products (Hough, 2010). According to Hough and Garitta (2012), "once the sanitary and nutritional hurdles have been overcome, the remaining barrier depends on the sensory properties of the product". For this reason, *best before* dates based on sensory shelf life are the most common in packaged food products. Although some sensory changes may go unnoticed by consumers, others would affect food quality perception. Thus, information on how the characteristics of the products change throughout storage is needed to determine when a food product no longer has the intended sensory characteristics and may be rejected by consumers. In the upcoming sections, methodologies for shelf life estimation are presented and the methodological considerations needed to obtain accurate results are discussed.

## 2 Sensory shelf life estimation

Sensory shelf life can be defined as the period of time a product retains its intended sensory characteristics under specific storage conditions. Therefore, sensory shelf life estimation aims at determining the specific point in time when changes in the sensory characteristics are no longer acceptable. A typical shelf life study involves evaluating sensory or hedonic characteristics of food samples stored for different periods of time. In order to design this kind of study adequately, a number of issues should be carefully addressed in advance.

The total length of the study should surpass the expected shelf life of the product. This is necessary to identify the time point at which the product reaches the minimum acceptable sensory quality or failure criterion. The frequency of sampling should be such that it allows estimates to have the desired level of precision. For this reason, a minimum of six sampling times is generally advised (Hough, 2010; Kilcast, 2011). Sampling can be equally distributed along the total storage period or an increased sampling frequency can be used as the product approaches the end of its expected shelf life. This last approach can provide more useful information than equally spaced sampling times in the case of products with long shelf lives or that deteriorate following an exponential kinetics (Gacula and Singh, 1984). As Robertson (2010) pointed out, a key concept when designing a shelf life study is to minimize the cost and time of the testing, yet provide reliable and statistically valid data.

## 2.1 Experimental designs for sensory shelf life experiments

The samples included in the study should be selected taking into account production variability and study duration. Two different experimental designs can be used for obtaining samples with different storage times when conducting a sensory shelf life study: basic and reversed design.

### 2.1.1 Basic design

The basic storage design requires a single batch of product to be stored under certain pre-established storage conditions. Samples are extracted at each of the selected sampling times and analyzed from a sensory and/or hedonic perspective, using the methodologies described later in the chapter. This design is particularly useful when dealing with products that largely vary across batches (e.g., fresh products). Examples of application include minimally processed vegetables (Ares et al., 2009; Paulsen et al., 2018), apple juice (Ferrario and Guerrero, 2016), yogurt (Salvador et al., 2005), and cod (Østli et al., 2013).

Although basic design is the most widely used for sensory shelf life estimation, it is not efficient regarding the use of time and resources (Lawless and Heymann, 2010). Basic design can be particularly cumbersome in the case of sensory methods, as they require gathering a panel of trained assessors or consumers at each of the storage times. In addition, as stated by Hough (2010), participants might realize they are participating in a sensory shelf life study and provide biased results or change their criteria, particularly if long storage times are required or sampling periods are infrequent. These difficulties could lead to expectation errors, and inconsistent use of scales and learning effects. An alternative to minimize these biases is to include one or more fresh samples in each evaluation.

### 2.1.2 Reversed design

This kind of testing design allows all samples with different storage times to be evaluated in one single instance, reducing the cost and resources needed for the sensory measurements. This can be achieved in three different ways:

1. Drawing samples from different production batches and storing them under the desired conditions at the preestablished sampling times so that all samples with different storage times are evaluated on the same day (Fig. 1A). In this case, consistent production quality is usually sought for assuring minimal batch variability.
2. Storing samples from one single batch under conditions that stop all deterioration processes and removing samples at each selected sampling time to store them under normal conditions until their evaluation (Fig. 1B).
3. Storing samples from one single batch under the desired conditions and once the elapsed time for sampling has been reached, keeping the sample under conditions where no further deterioration occurs (Fig. 1C).

In the first case, differences among samples are expected to be due to deterioration processes undergone during storage and differences between batches assumed to be small. Gámbaro et al. (2005) followed this approach for estimating the sensory shelf life of "alfajor," a chocolate-coated cake (individually wrapped), working with different industrial batches stored at 20.0±0.5°C for 0, 28, 46, 60, 65, 70, 75, and 80 days. Fresh samples were stored at the different selected times, so that samples with eight different storage times were evaluated on the same day.

The other two alternatives are only applicable to some product categories, for which deterioration processes can be stopped without modifying the sensory characteristics of the products. For some categories, freezing or storing at very low refrigeration temperatures might be a feasible option for stopping deterioration. The storage conditions selected should guarantee not only that product aging stops, but also that the freezing-defrosting cycle does not introduce sensory changes. An example of the application of a reversed design was reported by Jacobo-Velázquez and Hernández-Brenes (2011), when estimating the shelf life of high hydrostatic pressure processed avocado pulp at 4°C. These authors stored processed avocado pulp under refrigerated conditions for 45 days, collecting samples every 5 days and storing them at −80°C until they were all analyzed at the end of the storage period (45 days).

The reversed design overcomes the main drawbacks of the basic storage design. All samples are tested on the same day, reducing time and resources necessary to conduct the testing. It is particularly useful when consumer studies are used to estimate sensory shelf life of food products. However, selecting storage conditions that halt all deterioration processes might not always be possible. Products such as fresh or minimally processed fruits and vegetables cannot be frozen to stop their aging process.

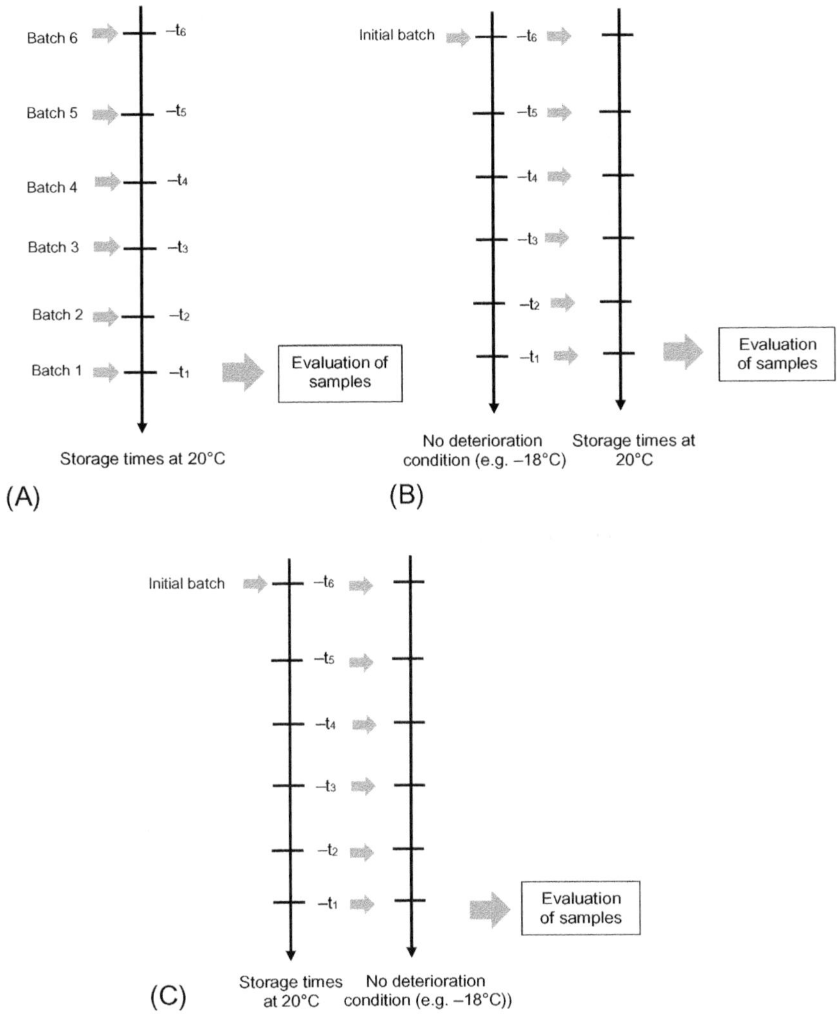


**Fig. 1** Shelf life test reversed designs: (A) samples from different batches are removed at each selected sampling time and stored until all samples are gathered for evaluation; (B) samples from one single batch are stored under a nondeterioration condition and subsets are removed to selected storage condition when sampling time is reached; (C) samples from one single batch are stored in selected conditions until each of the sampling times is reached and each subset is removed to a nondeterioration condition until evaluation.

## 3 Methodologies for sensory shelf life estimation

Sensory evaluation offers a toolbox from which to draw when carrying out sensory shelf life testing. Sensory evaluation can be defined as a scientific discipline that "evokes, measures, analyzes, and interprets responses to the characteristics of products as perceived by the senses" (Stone and Sidel, 2004). This discipline can be divided into two areas: analytical tests that provide objective information about the sensory characteristics of products, and hedonic tests, which evaluate consumers' hedonic reaction towards products (Ares and Varela, 2017). Sensory shelf life estimation has been based on both analytical and hedonic tests, as will be presented later in this section. Regardless of the methodological approach, standard sensory evaluation practices should be followed in order to gather reliable data from which accurate and reproducible sensory shelf life estimations can be drawn (Lawless and Heymann, 2010). Readers not familiar with sensory evaluation techniques should refer to textbooks for detailed information about sensory methodologies and recommendations for best practice (e.g., Stone and Sidel, 2004; Lawless and Heymann, 2010).

Sensory shelf life estimation requires establishing a sensory endpoint or a failure criterion, that is, the maximum tolerable deterioration in sensory quality. Each methodology involves a different failure criterion, which means a different sensory quality at the end of a product's shelf life. This has economic and environmental implications that researchers should be aware of. In the following sections, the implications of each methodology are discussed, together with their advantages and disadvantages.

## 3.1 Sensory shelf life estimation based on analytic tests

These methods are based on the objective evaluation of changes over time in the sensory quality of products or in the intensity of the main sensory characteristics responsible for product deterioration. Analytic tests should be conducted with panels of trained assessors, who should be screened and selected based on their sensory acuity for basic sensory characteristics and their ability to detect differences among samples (Stone and Sidel, 2004). Then, assessors should be extensively trained to evaluate the sensory characteristics of products using specific methodological approaches. In the case of sensory shelf life estimation, three methodologies have been extensively used: difference from control, rating overall quality, and quantifying attribute intensity.

### 3.1.1 Difference from control test

The difference from control test has been traditionally applied to sensory quality control of food products (Meilgaard et al., 1991; Yantis, 1992). An overall degree of difference from a control product is rated by a trained assessor panel, generally using a 10 cm scale, and a criterion for product acceptance/rejection is agreed. In the case of sensory shelf life studies, the degree of difference between stored samples and a fresh control sample is measured. In order to apply this methodology, several issues are to be taken into account.

First is the need to keep a fresh sample to serve as control as storage time progresses. Depending on the kind of product under study, this might be a relatively simple task; in some product categories, storage conditions can be easily selected to guarantee no deterioration takes place. The selected condition to keep a control sample should be previously tested in order to assure no significant sensory changes occur. For example, Man (2015) suggested storing a control product at −18°C when studying the shelf life of a cooked chilled potato dish, whereas Chouliara et al. (2007) kept a control sample at −30°C when studying the effect of modified atmosphere packaging and oregano essential oil addition on the shelf life of refrigerated chicken breast meat. However, it should be stressed that keeping the same fresh control over the whole study period might not be possible for other product categories. In those cases, an alternative would be to have the control sample replaced periodically, but the replacement should be validated using discriminative tests to assure that no differences in the sensory characteristics of the control samples exist. An example of this approach can be found in Hough et al. (1999). These authors kept a control sample of ricotta cheese at 2°C–3°C for 7 days without significant sensory changes when conducting a sensory shelf life study of this product category. In order to replace the control sample, every 7 days a triangle test with a trained panel was performed and a new fresh sample replaced the previous control sample. However, this approach would be difficult to implement in the case of natural fresh products, such as fruit and vegetables.

Second, assessors should undergo an adequately designed training program prior to engaging in this kind of task. It is crucial that they become familiar, not only with the characteristics of the control sample, but also with the sensory characteristics of samples representing different points along the scale. During the training sessions, samples with different storage times are evaluated and through open discussion with the panel leader, assessors agree on the scores to be assigned to the different samples when compared to the control sample. A blind control sample is to be included in every test session as part of the test set, serving as a baseline. During the shelf life study, assessors receive the control sample labeled as K, and the stored samples and the blind control sample labeled with 3-digit numbers. They are asked to rate the samples as to the degree of difference from the control sample, generally using a 10 cm line scale. Additional verbal descriptions of different degrees of difference are sometimes included along the scale. Data from this test can be analyzed by analysis of variance and stored samples significantly different from the control sample can be identified. However, these data do not provide any information on the reasons for the difference or whether those differences are relevant to the consumer. Additional questions exploring attributes perceived as different—even providing a scale to rate attribute difference—may be added.

A cut-off score that defines the maximum tolerable difference between a stored sample and the fresh control should be set in advance to establish the end of the product's shelf life. The average difference between the stored samples and the fresh sample could be regressed over time and the sensory shelf life estimated as the time the stored product reaches the cut-off point (Hough et al., 1999; Freitas and Costa, 2006). Practitioners are advised to avoid arbitrarily selection of this failure criterion and to rely on information about consumer perception for its definition.

Other methodologies can be used to estimate differences between stored samples and the fresh control. In particular, discriminative tests such as paired comparisons, triangle, tetrad, or duo-trio are useful alternative approaches. These methodologies are frequently used in industrial environments, being easier to implement than difference from control scales (Kilcast, 2011; Bouillé and Beeren, 2016). An example of the application of this approach is the estimation of "high-quality life" (HQL) for chilled and frozen food products. HQL is defined as "the time elapsed between freezing of an initially high-quality product and the moment when, by sensory assessment, a statistically significant difference ($p < .01$) from the initial quality can be established" (IIR, 1999; Evans, 2016).

In closing this section, it should be emphasized that sensory limits established when working with this kind of methodology usually lead to conservative estimations that might impact on an early withdrawal of food products from the shelf, leading to undesired financial and environmental consequences.

### 3.1.2 Overall quality scoring

This method is based on direct measurement of sensory quality throughout storage. In order to provide valid and reliable information, researchers should define quality specifications and set tolerance limits in advance. The trained assessors' task is to rate sensory quality by weighting different sensory characteristics of the products and integrating them into a single score. This requires trained assessors to have a mental reference of the sensory characteristics of the fresh product and to know which weight to assign to the sensory changes the product undergoes during storage when evaluating overall sensory quality. This task is particularly complex considering that an increase or decrease of characteristic sensory attributes or the onset of sensory defects can affect overall quality differently. Extensive training and a minimum number of experts or trained assessors are required (8 to 20) to get accurate and reliable results (Lawless and Heymann, 2010).

Quality scores are averaged across assessors and regressed over time. Sensory shelf life can be estimated as the time point when sensory quality reaches a specific quality score or failure criterion, which corresponds to the minimum sensory quality of a commercial product. The failure criterion is usually arbitrarily determined (e.g., the mid-point of the quality scale), which is not a recommended practice.

Quality-based methods have been used to estimate the sensory shelf life of a variety of products such as fresh fruits and vegetables, dairy, meat, and sea products. Allende et al. (2008) evaluated the sensory quality of fresh cut escarole and iceberg lettuce during storage with a trained panel of eight assessors, measuring the overall visual quality with a 10-point scale (0 = very bad, not characteristic, 10 = very good, very characteristic of the product). The end of the sensory shelf life was reached when the overall visual quality of the products reached a score of 5. Similarly, Medina et al. (2012) used this approach for sensory shelf life estimation of minimally processed baby spinach.

One of the most popular applications of quality-based methods for sensory shelf life estimation is the Quality Index Method (QIM), widely used in the fish industry. This method measures the degree of change in selected sensory attributes using a scoring system that ranges from 0 to 3 (the lower the score, the fresher the fish). A qualitative description of each score for each sensory attribute is provided to simplify and standardize the evaluation. Demerit scores are summed up to a total to provide an overall evaluation. A QIM scheme has to be developed for each species, since it does not take into account the differences between species. Considering that this method does not take into account the differences between species, QIMs have been developed for a large number of commercially important species (Hyldig et al., 2010) and they continue to be developed as commercial interest in other species grows (e.g., Borges et al., 2013; Baixas-Nogueras et al., 2003; Ritter et al., 2016; Sant'Ana et al., 2011; Fogaça et al., 2017). Ritter et al. (2016) developed a QIM protocol to study the sensory shelf life of tambatinga fish, working with 12 panelists. As a first training step, panelists were asked to observe and describe the sensory changes perceived in fish samples with six different storage times for 12 sessions. A second training stage was carried out, adjusting the preliminary QIM protocol, working with 18 fish samples during eight sessions. This stage was followed by a validation stage where assessors had to evaluate nine samples with different storage times using the 13 quality parameters they had previously defined (Table 1).

Although quality-based methods have been frequently applied for sensory shelf life estimations of different product categories, a large number of published studies do not comply with recommendations for best practice in sensory evaluation. This reduces the validity of the sensory shelf life estimation and signals the need for advancements in the application of standard practices when dealing with sensory trained panels, particularly for the evaluation of sensory quality. It is not unusual to find published research where three to five assessors have performed the quality scoring task with little or no training (Gómez-López et al., 2007; Lopez-Galvez et al., 2013; Medina et al., 2012; Mendes et al., 2011) and, in some cases, hedonic information gathered from trained assessors has been regarded as an indication of product quality (Costa et al., 2012; Gómez-López et al., 2007; Siripatrawan and Noipha, 2012). Such drawbacks could limit the validity of the data and the accuracy of sensory shelf life estimations.

### 3.1.3 Attribute intensity measurement

Descriptive methodologies aim at providing a complete qualitative and quantitative description of the sensory characteristics of products (Lawless and Heymann, 2010). In the specific case of sensory shelf life estimation, descriptive methodologies can be used to obtained a detailed description of the sensory changes a product undergoes during storage. These methods provide intensity ratings for individual sensory attributes, enabling the estimation of sensory shelf life, as the storage time when the intensity of a critical attribute reaches a preestablished value (Giménez et al., 2012). In order to do so, specifications must be set identifying the sensory attributes considered critical for sensory quality and establishing intensity limits, that is, the maximum or minimum intensity of sensory attributes of a product that is still acceptable for consumption.

The use of intensity scales demands 8–12 assessors to be selected and extensively trained in attribute identification and scaling. First, a set of descriptors are developed by the trained assessors, working with a wide array of samples with different storage times to identify those sensory characteristics that best describe differences among samples (ISO, 2016). Assessors individually select the descriptors they consider appropriate to describe the differences among samples and then, through an open discussion with the panel leader, agree on the terms to be used. Once descriptors are selected, a training stage should follow where the panelists are exposed to reference standards that anchor descriptors' intensity. Unstructured or structured line scales

**Table 1** Quality Index Method (QIM) protocol to study sensory shelf life of gutted ice-stored tambatinga (***Colassoma macrosporum*** x ***Piaractus brachypomum***) (Ritter et al., 2016)

| Quality attributes | Parameters | Description | Points |
|---|---|---|---|
| Overall aspect | Brightness | Intense | 0 |
| | | Bright | 1 |
| | | Opaque | 2 |
| | Godet test | Firm/elastic | 0 |
| | | Softened | 1 |
| Eyes | Shape | Convex | 0 |
| | | Flat/irregular | 1 |
| | Cornea | Crystalline | 0 |
| | | Cloudy | 1 |
| | | Opaque | 2 |
| | Pupil | Black/well-defined | 0 |
| | | Reddish/well-defined | 1 |
| | | Black/not-defined | 2 |
| Gill | Color | Red blood | 0 |
| | | Brownish red | 1 |
| | | Brownish | 2 |
| | Odor | Fresh/neutral | 0 |
| | | Rancid/sour | 1 |
| | | Putrid | 2 |
| Abdomen | Color | Whitish | 0 |
| | | Light salmon | 1 |
| | | Grayish pink | 2 |
| | Odor | Fresh/neutral | 0 |
| | | Rancid/sour | 1 |
| | | Putrid | 2 |
| Fins | Dorsal/Ventral/ Caudal | Full/wet/with elasticity | 0 |
| | | Full/resected tips/with elasticity | 1 |
| | | Failed/resected tips/no elasticity | 2 |
| Total QIM | | Sum of all the scores | 0–18 |

(10 or 15 cm) are generally used to rate attribute intensity. Panelists are trained to the use of the intensity scales for the selected attributes using intensity references and selected samples. Only when panel performance is considered reliable and consistent are the samples under study evaluated.

Once the panel is trained, the average intensity of the sensory attributes at each sampling time is calculated and regressed over time (Fig. 2). Sensory shelf life can be determined as the time point when the intensity of a sensory attribute reaches the predetermined criterion. Researchers are strongly advised to avoid the selection of arbitrary failure criterion for attribute intensity.

Several authors have used this methodology to estimate sensory shelf life of different food products (Cardenas Bonilla et al., 2007; Ares et al., 2009; Garitta et al.,

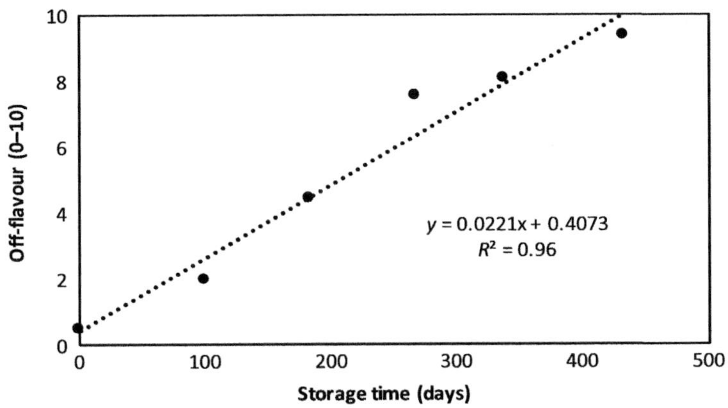


**Fig. 2** Average off-flavor intensity of orange juice with different storage time, evaluated by a trained assessors' panel using 10cm intensity scales. A reversed design based on the evaluation of different batches (c.f. Fig. 1A) was used.

2004). Jacobo-Velázquez and Hernández-Brenes (2011) followed this procedure when estimating shelf life of high hydrostatic pressure processed avocado paste. A trained panel developed a list of six descriptors related to changes during storage, working with samples with different storage times. A training phase followed, where assessors were trained to recognize those descriptors and their references in several sessions, using a 10cm scale (Table 2). In a third phase, through open discussion with the panel leader, two of the descriptors, sour and rancid, were agreed to be critical. Assessors were then trained with samples that covered the whole range of intensities of the two

**Table 2** Reference samples to train assessors on avocado descriptors (Jacobo-Velázquez et al., 2010)

| Descriptor | Preparation of reference sample |
|---|---|
| Green color | Freshly HHP[a]-processed avocado paste (100g) was allowed to oxidize at room temperature until the paste changed to brown color (approximately 30min). Browning was stopped by adding citric acid (1.2g) to the avocado paste. Fresh and brown avocado pastes were used as reference samples and were visually evaluated by the trained panel |
| Bitter | Caffeine (3000ppm) was mixed into freshly HHP-processed avocado paste |
| Sour | Citric acid (12,500ppm) was mixed into freshly HHP-processed avocado paste |
| Rancid | Hexanal (200ppm) was mixed into freshly HHP-processed avocado paste |
| Sweet | Sucrose (50,000ppm) was mixed into freshly HHP-processed avocado paste |
| Oily | Canola oil (50,000ppm) was mixed into freshly HHP-processed avocado paste |

[a] HHP: High hydrostatic pressure.

critical descriptors, until they provided reproducible and consistent results. Finally, the trained panel was requested to evaluate the samples under study. The failure criterion for sensory shelf life estimation was selected based on consumer data, following the approach that will be discussed later in the chapter.

The focus when using this methodology is on the sensory profile of the test samples and the insights it provides regarding a product's deterioration processes. Unlike other applications of descriptive analysis, when applied to shelf life studies, the number of descriptors is usually limited to those that are critical, generally ranging in number from 3 to 6 (Hough, 2010). A relatively small number of descriptors helps reduce the time and resources needed to maintain a well-trained and calibrated panel. The reader could refer to other sensory methodologies when time or resources hinder the proper development of a highly trained panel.

Although the method provides rich, valid, and reliable information, several published studies do not follow recommendations for best practice in sensory science. It is not uncommon to find published research where a reduced number of assessors is used (Medina et al., 2012; Costa et al., 2011, 2012) or where hedonic expressions are used to quantify attribute intensities (Chouliara et al., 2007; Costa et al., 2011). These practices compromise the accuracy and reliability of sensory shelf life estimations. In addition, several studies have asked trained assessors for hedonic information—asking them to decide whether a product was acceptable or not for consumption—and established a shelf life limit based on the panel's average acceptability (Allende et al., 2008; Chouliara et al., 2007; Costa et al., 2017; El-Obeid et al., 2018; Lopez-Galvez et al., 2013; Vandekinderen et al., 2009).

## 3.2 Sensory shelf life estimation based on hedonic tests

Hedonic tests gather information directly from the consumer: how much a product is liked or disliked, how it is perceived, whether it is accepted or rejected for consumption, whether it would be bought or not. The consumer perspective becomes an invaluable asset when estimating how long a food product should remain on the shelf (retail and home). The sensory changes that take place during storage do not necessarily modify how much a product is liked. In addition, trained assessors are expected to be more sensitive than consumers (Lawless and Heymann, 2010). Therefore, the sensory changes a trained panel detects and quantifies could go unnoticed to consumers or even if noticed, consumers might not care about them. For this reason, establishing a shelf life limit based solely on differences perceived by a trained panel could become too conservative and reduce the commercial life of a food product, which has negative economic and environmental consequences.

One of the important concepts that researchers should acknowledge is that sensory shelf life does not depend only on how the sensory characteristics of the product change over time, but rather on how consumers react to these changes, as they are the ones who decide whether a product after a certain storage time is still acceptable or not (Hough et al., 2003). Consumers should determine whether a product is still acceptable for consumption from a sensory perspective or how much it is liked after a certain period of storage time has elapsed. Consumer perception has been recognized as a key input for the selection of failure criteria for shelf life estimation, as they are "the ultimate arbiters of food quality" (Robertson, 2010). Two consumer methodologies can be used for sensory shelf life estimation: acceptability limit and survival analysis. Both methodologies require evaluations performed by large groups of naïve consumers, who should evaluate samples based on their subjective perception.

### 3.2.1 Acceptability limit

Consumers' hedonic reaction towards a food product can be measured by asking their degree of liking using a hedonic scale (Lawless and Heymann, 2010). The most widely used hedonic scale is the traditional 9-point category scale, with phrases ranging from dislike extremely to like extremely and a neutral point at the center, developed at the U.S. Army Food and Container Institute (Peryam and Girardot, 1952). Alternatively, an anchored 9-box scale (Fig. 3) could also be used (Curia et al., 2001; Giménez et al., 2007). Consumers should be chosen to represent the target consumer population. A minimum of 100 consumers is generally recommended to obtain reliable information (Hough et al., 2006).

Participants are presented with a set of samples with different storage times and are asked to score their overall liking using the above-mentioned scale. Depending on the sensory shelf life design, the product, and the time frame, it might be possible to have the same consumers taste all test samples. Consumers may also be asked to score their liking of different attributes, or even provide additional information on how different sensory characteristics are perceived, by means of additional tasks, for example, answering a check-all-that-apply (CATA) question. Overall acceptability scores for each sample are averaged, and a linear or nonlinear regression of overall liking against time is usually carried out (Giménez et al., 2007; Montes Villanueva and Trindade, 2010). The shelf life of the product is defined as the storage time at which the product's overall acceptability falls below a previously set value.

In order to set this overall acceptability limit, authors have used different criteria. Some authors have used an acceptability value of 6.5 on the 9-point scale, a limit initially proposed by Muñoz et al. (1992) for quality control specifications, whereas a number of authors have considered an overall liking score below 6 (like slightly) (Tomac et al., 2017; Gámbaro et al., 2005; Giménez et al., 2008). Montes Villanueva and Trindade (2010) and Latou et al. (2014) chose an average score of 5.0 (neither like nor dislike) as the acceptability limit. Fig. 4 provides an example of shelf life estimation using the acceptability limit methodology.

The selection of a single critical acceptability limit as cut-off point across product categories has certain drawbacks. For some food categories, overall acceptability of the fresh product might be in a range close to the cutoff point, which in turn could lead to too conservative shelf life estimations that limit their commercial life. On the other

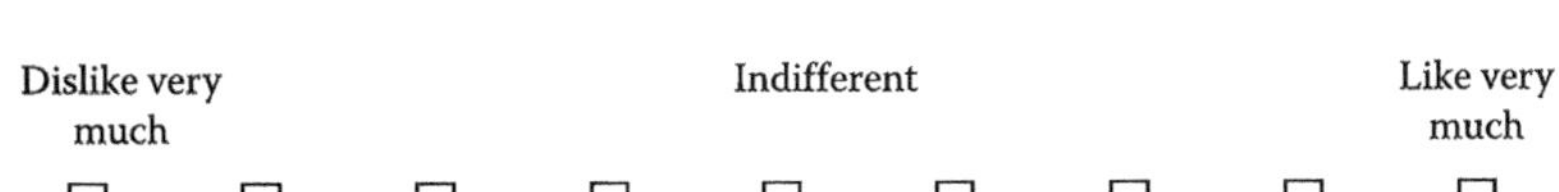


**Fig. 3** Anchored hedonic box scale for determining consumer overall liking of food products.

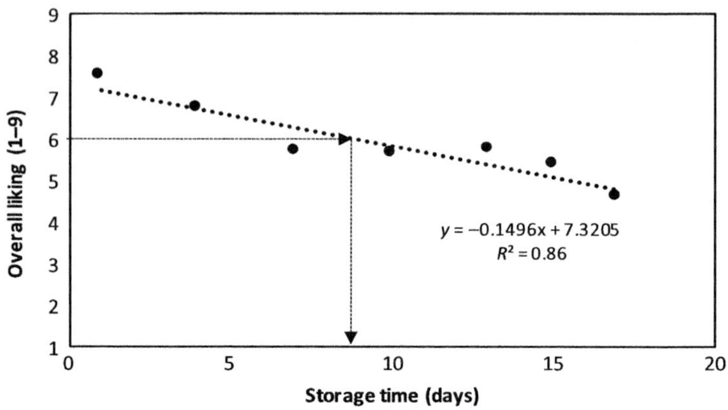


**Fig. 4** Average overall liking scores of toast bread samples with different storage time (following a basic design), evaluated by consumers using 9-point hedonic scores. Sensory shelf life was estimated as the storage time when overall liking reached a score of 6 (9 days).

hand, consumers would probably reject products that are usually well-liked and have high overall liking scores when fresh before a low acceptability score is reached.

Hough et al. (2002) proposed a different criterion to establish an overall acceptability limit. The first significant difference in overall acceptability from that of the fresh product is calculated and the sensory shelf life is defined as the time at which this acceptability score is reached. The acceptability limit (S) is calculated using the following equation:

$$S = F - Z_{\alpha}\sqrt{\frac{2.\mathrm{MSE}}{n}} \tag{1}$$

where $S$ is the minimum tolerable acceptability of stored sample; $F$ is acceptability of fresh sample; $Z\alpha$ is one-tailed coordinate of the normal curve for $\alpha$ significance level; MSE is the mean square of the error derived from the analysis of variance of the consumer data; and $n$ is the number of consumers.

This criterion reflects the time consumers perceive a difference in the stored product with respect to the fresh one that relates to changes in its sensory characteristics during storage.

It should be noted that considering this criterion can lead to overly conservative estimations. A slight decrease in acceptability does not necessarily relate to consumers rejecting or discarding the product for consumption. Giménez et al. (2007) compared different failure criteria to establish the shelf life of brown pan bread and concluded that by the time the acceptability limit was reached a relatively low percentage of consumers rejected the product, providing an estimation that was not compatible with commercial practices. Other authors have also reported similar conclusions (Garitta et al., 2015) and recommended other bases to calculate cut-off points, as will be discussed later.

### 3.2.2 Survival analysis

Overall liking scores provide no insight as to what a consumer would do when confronting a food product that has been stored for some time, whether he/she would decide to eat it or discard it. Alternatively, sensory shelf life can be estimated considering consumers' rejection to consume or purchase a product. Hough et al. (2003) were the first to apply survival analysis statistics to sensory shelf life, shifting the focus from the product deteriorating to the consumer decision of accepting or rejecting the product for consumption.

Survival analysis statistics, extensively used in clinical studies, epidemiology, biology, sociology, and reliability studies (Klein and Moeschberger, 1997; Meeker and Escobar, 1998) could be applied to sensory shelf life estimations based on consumers' acceptance or rejection of samples with different storage times. Participants are asked whether they would accept or reject the different samples for consumption. They are usually instructed to answer considering they have already bought the product or the product is being served at home (Hough et al., 2003).

Early publications using survival analysis to estimate sensory shelf life of food products worked with 50–80 consumers. However, Hough et al. (2007) performed simulation studies on the ideal number of consumers to recruit for this kind of study, and they concluded a number close to 120 consumers is recommended. In those cases where a reversed storage design is used, each consumer evaluates the whole sample set in a single session. If a basic storage design is followed, then the consumer panel should be assembled on each sampling occasion. Another alternative is to have different consumers evaluate a single sample at each of the storage times. Libertino et al. (2011) performed simulation studies to estimate the number of consumers to obtain valid estimations and recommended a minimum of 50 for each of storage time, considering a total of six storage times. This approach would be also useful in those studies where a fresh control cannot be kept (Araneda et al., 2008) or for certain types of sample sets where a significant carry-over effect could be expected.

Survival analysis has been extensively used in the literature to estimate sensory shelf life of a wide variety of food products such as dairy (Garitta et al., 2004; Curia et al., 2005; Salvador et al., 2005), bakery (Giménez et al., 2008, 2007, 2017; Gámbaro et al., 2005), minimally processed fruits and vegetables (Lareo et al., 2009; Ares et al., 2009; Jacobo-Velázquez and Hernández-Brenes, 2011; Garitta et al., 2018), and fish (Østli et al., 2013). The test setup is simple, participants are presented with a randomized set of samples with different storage times and asked to answer the question "would you normally consume this product" with a yes or no answer.

The data gathered from these studies are summarized in a table like Table 3, where the acceptance/rejection data for each consumer is included in each row (acceptance=Yes, rejection=No). The exact moment at which a consumer rejects a product is not actually observed, considering a consumer evaluates a fixed number of samples with different storage times providing, as a result, censored data (Hough et al., 2003). Data collected is classified in three categories: left, interval, and right censored. If a consumer rejects the sample at the first storage time considered, then rejection was not observed and happened before that first observation time ($T \leq t_1$), and that data is classified as left censored. If a consumer accepts a sample at $t_2$ and rejects a sample at $t_3$, the exact time at which he/she rejects the product occurs between $t_2$ and $t_3$ ($t_2 < T \leq t_3$) and his/her data is interval censored. If rejection happens after the last storage time ($T > t_6$), then the data is classified as right censored (Table 3).

**Table 3** Typical data matrix obtained from survival analysis

| Consumer | Storage time | | | | | | Censoring |
|---|---|---|---|---|---|---|---|
| | $t_1$ | $t_2$ | $t_3$ | $t_4$ | $t_5$ | $t_6$ | |
| 1 | No | No | No | No | No | No | Left |
| 2 | Yes | Yes | No | No | No | No | Interval |
| 3 | Yes | Yes | No | Yes | No | No | Interval |
| … | … | … | … | … | … | … | … |
| n | Yes | Yes | Yes | Yes | Yes | Yes | Right |

Acceptance = Yes; rejection = No.

Survival analysis allows estimation of the rejection function, and modeling of the proportion of consumers that reject a food product as a function of storage time. As storage time increases, the proportion of consumers that reject the product increases as well. If a random variable $T$ is defined as the storage time a consumer rejects the sample, the survival function $S(t)$ can be defined as the probability of a consumer accepting a product stored for a time period longer than $t$, that is $S(t)=P(T>t)$. To estimate the parameters of the survival function, the likelihood function is used. The likelihood function describes the joint probability of obtaining the given observations for the $n$ consumers (Klein and Moeschberger, 1997):

$$L=\prod_{i\in R} S(ri)\cdot\prod_{i\in L}\left(1-S(li)\right)\cdot\prod_{i\in I}\left(S(li)-S(ri)\right) \tag{2}$$

where: $R$ is the set of right-censored observations; $L$ is the set of left-censored observations; and $I$ is the set of interval-censored observations.

Each type of censoring makes a different contribution to the likelihood function.

The distribution of the survival times could be assumed to follow a Weibull or log-normal distribution and estimates of the of the survival function $S(t)$ could be obtained (Klein and Moeschberger, 1997). If the Weibull distribution is chosen, the survival function is:

$$S(t)=S_{sev}\left(\frac{\ln(t)-\mu}{\sigma}\right) \tag{3}$$

where: $S_{sev}$ (•) is the survival function of the smallest extreme value distribution: $S_{sev}(w)=\exp(-e^{w})$; and $\mu$ and $\sigma$ are the model's parameters.

When the lognormal distribution is chosen for $T$, then the survival function is:

$$S(t)=1-\Phi\left(\frac{\ln(t)-\mu}{\sigma}\right) \tag{4}$$

where: $\Phi(.)$ is the standard normal cumulative distribution function; and $\mu$ (location parameter) and $\sigma$ (shape parameter) are the model's parameters.

For a given set of experimental data, the model's parameters are estimated by maximizing the likelihood function and substituting $S(t)$ in Eq. (2) by the expression

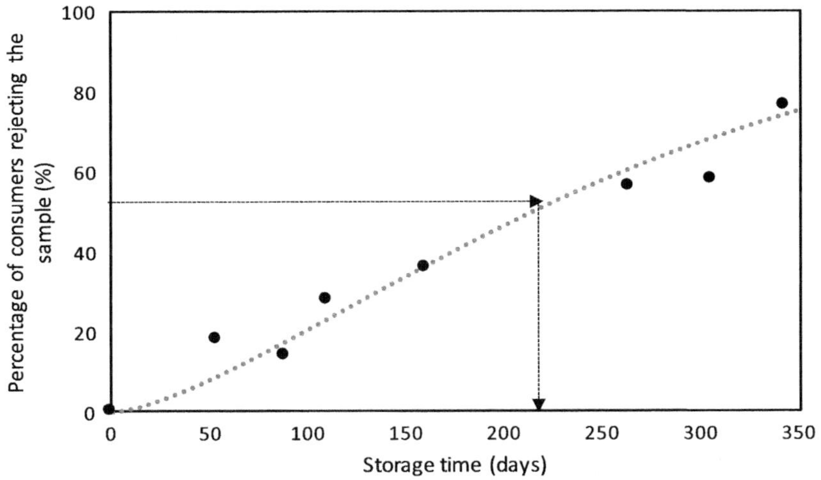


**Fig. 5** Consumers' rejection to consume as a function of storage time for samples of dulce de leche in plastic packages (basic design).

given in Eqs. (3) or (4), depending on the model chosen, using specialized software. Calculations using R software have been provided by Hough (2010).

These parameters are used to graph the proportion of consumers rejecting the product as a function of storage time, as shown in Fig. 5. In order to estimate the shelf life of the product, a probability level of consumers' rejection should be set. A shelf life value considering 50% rejection has been recommended in the literature (Gacula and Singh, 1984; Cardelli and Labuza, 2001; Hough et al., 2003). Other authors (Ares et al., 2006; Giménez et al., 2007; Jacobo-Velázquez et al., 2010; Giménez et al., 2017) have used a 25% rejection to make shelf life estimations; a more conservative criterion, but still considered reasonable from a practical point of view.

## 3.3 Sensory shelf life estimation based on the combination of analytic and hedonic tests

This approach builds on the combination of the methodologies described above. It aims at obtaining less arbitrary failure criterion for sensory attributes or sensory quality using consumer perception as input.

Sensory changes during storage can be evaluated with a panel of assessors, provided that attributes that limit the product's shelf life are identified. The intensity of these critical attributes may increase, as would the occurrence of off-flavors or off-odors, or decrease during storage, as would the loss of aroma or desirable texture attributes. Hough and Garitta (2012) proposed, as a first step, to obtain samples with different intensity levels of the critical descriptors. These samples are to be evaluated by a trained panel for attribute intensity and by a consumer panel for acceptability. Analytical or instrumental measurements may be gathered as well, since this kind of data is simpler and easier to collect. However, it is not always possible to rely on analytical and instrumental measurements to monitor changes during storage. A regression of the acceptability data versus the trained sensory panel data is performed and the sensory cut-off point is obtained by introducing a failure criterion for overall liking (6 or the minimum difference in overall liking scores) in the regression equation (Fig. 6A). This cut-off point indicates a limit on the intensity of a sensory or instrumental parameter above which (or below, for a positive characteristic) there is a significant decrease in consumer acceptability. Finally, a regression of the sensory or instrumental data versus storage time allows prediction of the sensory shelf life as the storage time necessary for the product to reach that cut-off point intensity of the measured parameter (Fig. 6B).

A limitation of this approach is that the estimation would depend on the critical descriptor selected. Considering all the sensory characteristics of the product change simultaneously, and that their impact altogether on consumers' reaction can be different

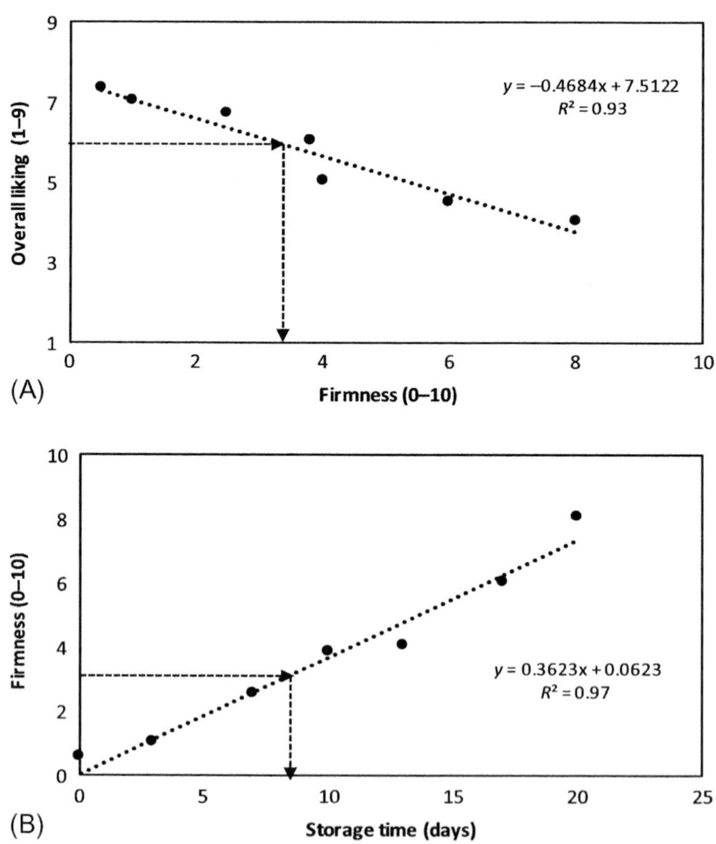


**Fig. 6** (A) Correlation between overall liking and firmness of hamburger buns with different storage time (reversed design based on frozen samples, Fig. 1B). The cut-off corresponding to an overall liking score of 6 is shown; and (B) Firmness of hamburger bans with different storage time. The shelf life corresponding to the firmness score determined using the cut-off point is shown.

from the impact of one high intensity single sensory defect, selecting the most relevant attribute might be cumbersome. A single measurement of the sensory deterioration of products could provide a single failure criterion. Ares et al. (2009) developed a sensory quality index for strawberries based on the correlation between trained assessors' data and consumer acceptability scores. Instead of using one single critical descriptor to establish a sensory cut-off limit, the authors performed a principal component analysis (PCA) on the mean scores of the trained assessors' data followed by a multiple regression analysis considering acceptability as a dependent variable, and sample loadings for the first two principal components of the sensory data as explanatory variables. In this case, the authors selected a score of 6 as acceptability limit and were able to predict a sensory quality index value that could, in turn, be correlated to storage conditions to predict the product's shelf life. Similarly, Derossi et al. (2016) applied PCA to sensory data to define a cut off criteria for shelf life estimation of fresh cut lettuce.

This same approach can be used by combining other analytic and hedonic tests. For example, survival analysis based on consumer acceptance/rejection data can be used to estimate cut-off points for sensory attributes (Garitta et al., 2015). Similarly, in order to relate sensory quality data as perceived by trained assessors to consumer perception, some authors have proposed to integrate consumer data with quality indexes. For instance, Djekic et al. (2018) developed a total quality index (TQI) to evaluate the quality of sliced apples dried with different technologies during storage. Their approach took into account instrumental and sensory parameters, selecting quality parameters according to consumers' previous ranking of attribute importance to the sensory quality of the product. A panel of eight trained assessors used a 5-level quality scale to score the four sensory parameters selected (appearance, odor, oral texture, and flavor). A mathematical model was developed for calculating a single TQI using quality scores. Earlier, Ares et al. (2009) developed a sensory quality index based for strawberries correlating consumer and trained panel data.

## 4 Recommendations and challenges

For most food products, shelf life is determined by changes in their sensory characteristics throughout storage. Sensory shelf life estimation has traditionally focused on assuring that minimum changes in the sensory characteristics of products occur during storage. This approach can lead to an unnecessarily high amount of food waste, since consumers may still consume a stored product that is different from the fresh one. In this sense, shelf life extension has been identified as a potential strategy to reduce food waste worldwide, thus contributing to an improvement of the sustainability of the food sector (Spada et al., 2018; Gaiani et al., 2018). Therefore, sensory shelf life estimation should aim at maximizing the storage period, while assuring that consumers would not find an unacceptable product within its stated shelf life. For this reason, practitioners are advised to use consumer input to determine the maximum tolerable deterioration for consumers. In this sense, consumer-based methodologies or methodologies that combine analytic and hedonic tests are recommended over those based exclusively on analytic tests.

One of the main challenges of consumer-based methods for sensory shelf life estimation is the definition of the failure criterion, that is, the minimum overall liking score or the maximum percentage of consumers rejecting the sample. Research is necessary to understand better the implications of setting different failure criteria from both an economic and environmental perspective. In this sense, information about the amount of food discarded at the retailer and household levels for different failure criteria is necessary to refine the methodologies used for estimating sensory shelf life.

In addition, another challenge for improving the accuracy of sensory-shelf life estimations is to apply more ecologically valid approaches to measure consumer perception of products with different degrees of deterioration. This requires acknowledging that a wide range of variables beyond sensory characteristics influence consumer perception and choice, including personal characteristics (e.g., socio-economic level, gender, age), product variables (e.g., brand, price, packaging), and context (e.g., where, when, and how the product is consumed). In line with that, virtual reality and evoked context studies attempt to provide industry a better prediction of consumers' behavior (Galiñanes Plaza et al., 2018). The recent use of evoked contexts in sensory shelf life studies (Giménez et al., 2015) has shown that providing context to currently used methodologies could lead to more reliable estimations.

Finally, it is important to highlight that accurate sensory shelf life estimation requires the application of methodologies that comply with recommendations and best practices in sensory and consumer science. However, there are a large number of examples of studies published in peer-reviewed journals reporting invalid estimations due to methodological flaws. Improving the methodological approaches used for sensory shelf life estimation will bring benefits at the academic and industrial level, as well as for society as a whole.

## References

Allende, A., Selma, M.V., Lopez-Galvez, F., Villaescusa, R., Gil, M.I., 2008. Role of commercial sanitizers and washing systems on epiphytic microorganisms and sensory quality of fresh-cut escarole and lettuce. Postharvest Biol. Technol. 49, 155–163.

Alongi, M., Sillani, S., Lagazio, C., Manzocco, L., 2018. Effect of expiry date communication on acceptability and waste of fresh-cut lettuce during storage at different temperatures. Food Res. Int. https://doi.org/10.1016/j.foodres.2018.09.056. in press.

Araneda, M., Hough, G., Wittig de Penna, E., 2008. Current-status survival analysis methodology applied to estimating sensory shelf life of ready-to-eat lettuce (Lactuca sativa). J. Sens. Stud. 23, 162–170.

Ares, G., Varela, P., 2017. Trained vs. consumer panels for analytical testing: fueling a long lasting debate in the field. Food Qual. Prefer. 61, 79–86.

Ares, G., Parentelli, C., Gámbaro, A., Lareo, C., Lema, P., 2006. Sensory shelf life of shiitake mushrooms stored under passive modified atmosphere. Postharvest Biol. Technol. 41, 191–197.

Ares, G., Barrios, S., Lareo, C., Lema, P., 2009. Development of a sensory quality index for strawberries based on correlation between sensory data and consumer perception. Postharvest Biol. Technol. 52, 97–102.

ASTM E2454–05, 2011. Standard Guide for Sensory Evaluation Methods to Determine the Sensory Shelf Life of Consumer Products. ASTM International, West Conshohocken, PA. 2011, http://www.astm.org.

Baixas-Nogueras, S., Bover-Cid, S., Veciana-Nogués, T., Nunes, M.L., Vidal-Carou, M.C., 2003. Development of a quality index method to evaluate freshness in Mediterranean hake (*Merluccius merluccius*). J. Food Sci. 68, 1067–1071.

Borges, A., Conte-Júnior, C.A., Franco, R.M., Freitas, M.Q., 2013. Quality index method (QIM) developed for pacu Piaractus mesopotamicus and determination of its shelf life. Food Res. Int. 54, 311–317.

Bouillé, A.G., Beeren, C.J.M., 2016. Sensory evaluation methods for food and beverage shelf life assessment. In: Subramaniam, P. (Ed.), The Stability and Shelf Life of Food. second ed. Woodhead Publishing Series in Food Science, Technology and Nutrition: Number 297, Elsevier Ltd., Kidlington, UK, pp. 199–224.

Cardelli, C., Labuza, T.P., 2001. Application of Weibull hazard analysis to the determination of the shelf life of roasted and ground coffee. Lebensm.-Wiss. Technol. 34, 273–278.

Cardenas Bonilla, A.C., Sveinsdottir, K., Martinsdottir, E., 2007. Development of Quality Index Method (QIM) scheme for fresh cod (*Gadus morhua*) fillets and application in shelf life study. Food Control 18, 352–358.

Chouliara, E., Karatapanis, A., Savvaidis, I., Kontominas, M.G., 2007. Combined effect of oregano oil and modified atmosphere packaging on shelf-life extension of fresh chicken breast meat, stored at 4°C. Food Microbiol. 24 (6), 607–617.

Costa, C., Lucera, A., Conte, A., Mastromatteo, M., Speranza, B., Antonacci, A., Del Nobile, M.A., 2011. Effects of passive and active modified atmosphere packaging conditions on ready-to-eat table grape. J. Food Eng. 102, 115–121.

Costa, C., Conte, A., Buonocore, G.G., Lavorgna, M., Del Nobile, M.A., 2012. Calcium-alginate coating loaded with silver–montmorillonite nanoparticles to prolong the shelf-life of fresh-cut carrots. Food Res. Int. 48, 164–169.

Costa, C., Lucera, A., Licciardello, F., Conte, A., Del Nobile, M.A., 2017. Application of preservation strategies to improve the shelf life of spreadable cheese. Food Packaging Shelf Life 11, 16–20.

Curia, A.V., Hough, G., Martínez, M.C., Margalef, M.I., 2001. How argentine consumers understand the Spanish translation of the 9-point hedonic scale. Food Qual. Prefer. 12, 217–221.

Curia, A., Aguerrido, M., Langohr, K., Hough, G., 2005. Survival analysis applied to sensory shelf life of yogurts I: argentine formulations. J. Food Sci. 70, s442–s445.

Department for Environment, Food and Rural Affairs, 2011. Guidance on the Application of Date Labels to Food. DEFRA, London, pp. 1–26.

Derossi, A., Mastrandrea, L., Amodio, M.L., de Chiara, M.L.V., Colelli, G., 2016. Application of multivariate accelerated test for the shelf life estimation of fresh-cut lettuce. J. Food Eng. 169, 122–130.

Djekic, I., Tomic, N., Bourdoux, S., Spilimbergo, S., Smigic, N., Udovicki, B., Hofland, G., Devlieghere, F., Rajkovic, A., 2018. Comparison of three types of drying (supercritical $CO_2$, air and freeze) on the quality of dried apple—quality index approach. LWT- Food Sci. Technol. 94, 64–72.

El-Obeid, T., Yehia, H., Sakkas, H., Lambrianidi, L., Tsiraki, M., Savvaidis, I., 2018. Shelf-life of smoked eel fillets treated with chitosan or thyme oil. Int. J. Biol. Macromol. 114, 578–583.

European Commission, 2000. European Commission. Directive 2000/13/EC of the European Parliament and of the council of 20 march 2000 on the approximation of the laws of the member states relating to the labelling, presentation and advertising of foodstuffs. Off. J. Eur. Union 109, 29–41.

European Commission, 2002. Commission Regulation (EC) No 178/2002 of 28 January 2002 laying down the general principles and requirements of food law, establishing the European Food Safety Authority and laying down procedures in matters of food safety. Off. J. Eur. Communities 31, 1–24.

Evans, J., 2016. Effects of food and beverage storage, distribution, display, and consumer handling on shelf life. In: Subramaniam, P. (Ed.), The Stability and Shelf Life of Foods, second ed. Woodhead Publishing Series in Food Science, Technology and Nutrition: Number 297, Elsevier ltd, Kidlington, UK, pp. 107–157.

Ferrario, M., Guerrero, S., 2016. Effect of a continuous flow-through pulsed light system combined with ultrasound on microbial survivability, color and sensory shelf life of apple juice. Innov. Food Sci. Emerg. Technol. 34, 210–224.

Fogaça, F.H.D.S., Gonzaga Junior, M.A., Vieira, S.G.A., Farias, E., Ferreira-Bravo, I., Silva, T., Calvet, R., Pereira, A.L.M., Prentice-Hernández, C., 2017. Appraising the shelf life of farmed cobia, Rachycentron canadum, by application of a quality index method. J. World Aquacult. Soc. 48 (1), 70–82.

Freitas, M.A., Costa, J.C., 2006. Shelf life determination using sensory evaluation scores: a general Weibull modeling approach. Comput. Ind. Eng. 51, 652–670.

Gacula, M.C., Singh, J., 1984. Statistical Methods in Food and Consumer Research. Academic Press, New York, NY.

Gaiani, S., Caldeira, S., Adorno, V., Segrè, A., Vittuari, M., 2018. Food wasters: profiling consumers' attitude to waste food in Italy. Waste Manag. 72, 17–24.

Galiñanes Plaza, A., Delarue, J., Saulais, L., 2018. The pursuit of ecological validity through contextual methodologies. Food Qual. Prefer. https://doi.org/10.1016/j.foodqual.2018.11.004. in press.

Gámbaro, A., Garitta, L., Giménez, A., Varela, P., Hough, G., 2005. Sensory shelf-life estimation of *Alfajor* by survival analysis. J. Sens. Stud. 19, 500–509.

Garitta, L., Hough, L., Sánchez, R., 2004. Sensory shelf life of dulce de leche. J. Dairy Sci. 87, 1601–1607.

Garitta, L., Langhor, K., Gómez, G., Hough, G., Beeren, C., 2015. Sensory cut-off point obtained from survival analysis statistics. Food Qual. Prefer. 43, 135–140.

Garitta, L., Langhor, K., Elizagoyen, E., Ottaviano, F., Gómez, G., Hough, G., 2018. Survival analysis model to estimate sensory shelf life with temperature and illumination as accelerating factors. Food Qual. Prefer. 68, 371–376.

Giménez, A., Varela, P., Salvador, A., Ares, G., Fiszman, S., Garitta, L., 2007. Shelf life estimation of brown pan bread: a consumer approach. Food Qual. Prefer. 18, 196–204.

Giménez, A., Ares, G., Gámbaro, A., 2008. Survival analysis to estimate sensory shelf life using acceptability scores. J. Sens. Stud. 23, 571–582.

Giménez, A., Ares, F., Ares, G., 2012. Sensory shelf-life estimation: a review of current methodological approaches. Food Res. Int. 49, 311–325.

Giménez, A., Gagliardi, A., Ares, G., 2015. Influence of evoked contexts on consumers' rejection of two products: implications for shelf life estimations. Food Res. Int. 76 (3), 527–531.

Giménez, A., Gagliardi, A., Ares, G., 2017. Estimation of failure criteria in multivariate sensory shelf life testing using survival analysis. Food Res. Int. 99, 542–549.

Gómez-López, V., Ragaert, P., Debevere, J., Devlieghere, F., 2007. Pulsed light for food decontamination: a review. Trends Food Sci. Technol. 18, 464–473.

Harcar, T., Karakaya, F., 2005. A cross-cultural exploration of attitudes toward product expiration dates. Psychol. Mark. 22, 353–371.

Hough, G., 2010. Sensory Shelf Life Estimation of Food Products. Taylor & Francis Group, Boca Raton, FL.

Hough, G., Garitta, L., 2012. Methodology for sensory shelf-life estimation: a review. J. Sens. Stud. 27, 137–147.
Hough, G., Puglieso, M.L., Sánchez, R., Mendes Da Silva, O., 1999. Sensory and microbiological shelf-life of a commercial ricotta cheese. J. Dairy Sci. 82, 454–459.
Hough, G., Sánchez, R.H., Garbarini De Pablo, G., Sánchez, R.G., Calderón Villaplana, S., et al., 2002. Consumer acceptability versus trained sensory panel scores of powdered milk shelf-life defects. J. Dairy Sci. 85, 1–6.
Hough, G., Langohr, K., Gómez, G., Curia, A., 2003. Survival analysis applied to sensory shelf life of foods. J. Food Sci. 68, 359–362.
Hough, G., Wakeling, I., Mucci, A., Chambers IV, E., Méndez-Gallardo, I., Alves, L., 2006. Number of consumers for sensory acceptability tests. Food Qual. Prefer. 17, 522–526.
Hough, G., Calle, M.L., Serrat, C., Curia, A., 2007. Number of consumers necessary for shelf life estimations based on survival analysis statistics. Food Qual. Prefer. 18, 771–775.
Hyldig, G., Martinsdóttir, E., Sveinsdóttir, K., Schelvis, R., Bremner, A., 2010. Quality index methods. In: Nollet, L.M.L., Toldrá, F. (Eds.), Handbook of Seafood and Seafood Products Analysis. CRC Press, New Jersey, pp. 463–481.
IFST, 1993. Shelf Life of Foods: Guidelines for Its Determination and Prediction. Institute of Food Science and Technology, London.
IIR, 1999. Recommendations for the Processing and Handling of Frozen Foods. IIF-IIR, France.
ISO, 2016. Sensory Analysis—Methodology—General Guidance for Establishing a Sensory Profile. European committee for standardization, Brussels, Belgium. ISO 13299: 2016.
Jacobo-Velázquez, D.A., Hernández-Brenes, C., 2011. Sensory shelf-life limiting factor of high hydrostatic pressure processed avocado paste. J. Food Sci. 76, s388–s395.
Jacobo-Velázquez, D.A., Ramos-Parra, P.A., Hernández-Brenes, C., 2010. Survival analysis applied to the sensory shelf-life dating of high hydrostatic pressure processed avocado and mango pulps. J. Food Sci. 75, s286–s291.
Kilcast, D., 2011. Sensory evaluation methods for food shelf life assessment. In: Kilcast, D., Subramaniam, P. (Eds.), Food and Beverage Stability and Shelf Life. Woodhead Publishing, pp. 350–380.
Klein, J.P., Moeschberger, M.L., 1997. Survival Analysis, a Self-Learning Text. Springer-Verlag, New York.
Labuza, T.P., Szybist, L.M., 2001. Open Dating of Foods. Food & Nutrition Press, Trumbull, CT.
Lareo, C., Ares, G., Ferrando, L., Lema, P., Gámbaro, A., Soubes, M., 2009. Influence of temperature on shelf life of butterhead lettuce leaves under passive modified atmosphere packaging. J. Food Qual. 32, 240–261.
Latou, E., Mexis, S.F., Badeka, A.V., Kontakos, S., Kontominas, M.G., 2014. Combined effect of chitosan and mofdified atmosphere packaging for shelf life extension of chicken breast fillets. LWT- Food Sci. Technol. 55, 263–268.
Lawless, H.T., Heymann, H., 2010. Sensory Evaluation of Food. Principles and Practices, second ed. Springer, New York.
Libertino, L.M., López Osornio, M.M., Hough, G., 2011. Number of consumers necessary for survival analysis estimations based on each consumer evaluating a single sample. Food Qual. Prefer. 22, 24–30.
Lopez-Galvez, F., Ragaert, P., Palermo, L., Eriksson, M., Devlieghere, F., 2013. Effect of new sanitizing formulations on quality of fresh-cut iceberg lettuce. Postharvest Biol. Technol. 85, 102–108.
Man, D., 2015. Shelf Life, second ed. John Wiley & Sons, Ltd, West Sussex, UK, p. 81.
Man, D., 2016. Food storage trials. In: Subramaniam, P. (Ed.), The Stability and Shelf Life of Food, second ed. Woodhead Publishing Series in Food Science, Technology and Nutrition: Number 297, Elsevier Ltd., Kidlington, UK, pp. 171–198.

Medina, M.S., Tudela, J.A., Marín, A., Allende, A., Gil, M.I., 2012. Short postharvest storage under low relative humidity improves quality and shelf life of minimally processed baby spinach (Spinacia oleracea L.). Postharvest Biol. Technol. 67, 1–9.

Meeker, W.Q., Escobar, L.A., 1998. Statistical Methods for Reliability Data. John Wiley & Sons, New York.

Meilgaard, M., Civille, G.V., Carr, B.T., 1991. Sensory Evaluation Techniques. CRC Press, Inc., Boca Raton, Florida.

Mendes, R., Alves Silva, H., Anacleto, P., Cardoso, C., 2011. Effect of $CO_2$ dissolution on the shelf life of ready-to-eat Octopus vulgaris. Innov. Food Sci. Emerg. Technol. 12, 551–561.

Milne, R., 2013. Arbiters of waste. Date labels, the consumer and knowing good, safe food. Sociol. Rev. 60 (2), 84–101.

Montes Villanueva, N.D., Trindade, M.A., 2010. Estimating sensory shelf life of chocolate and carrot cupcakes using acceptance tests. J. Sens. Stud. 25, 260–279.

Muñoz, A.M., Civille, G.V., Carr, B.T., 1992. Sensory Evaluation in Quality Control. Van Nostrand Reinhold, New York.

Newsome, R., Balestrini, C.G., Baum, M.D., Corby, J., Fisher, W., Goodburn, K., et al., 2014. Applications and perceptions of date labeling of food. Compr. Rev. Food Sci. Food Saf. 13 (4), 745–769.

Østli, J., Esaissen, M., Garitta, L., Nøstvold, B., Hough, G., 2013. How fresh is fresh? Perceptions and experience when buying and consuming fresh cod fillets. Food Qual. Prefer. 27 (1), 26–34.

Paulsen, E., Barrios, S., Baenas, N., Morenoc, D., Heinzen, H., Lema, P., 2018. Effect of temperature on glucosinolate content and shelf life of ready-to-eat broccoli florets packaged in passive modified atmosphere. Postharvest Biol. Technol. 138, 125–133.

Peryam, D.R., Girardot, N.F., 1952. Advanced taste-test method. Food Eng. 24 (58–61), 194.

Priefer, C., Jörissen, J., Bräutigam, K.R., 2016. Food waste prevention in Europe—a cause-driven approach to identify the most relevant leverage points for action. Resour. Conserv. Recycl. 109, 155–165.

Ragaert, P., Verbeke, W., Devlieghere, F., Debevere, J., 2004. Consumer perception and choice of minimally processed vegetables and packaged fruits. Food Qual. Prefer. 15, 259–270.

Ritter, D.O., Lanzarin, M., Novaes, S.F., Monteiro, M.L.G., Almeida Filho, E.S., Mársico, E.T., Franco, R.M., Conte-Junior, C.A., Freitas, M.Q., 2016. Quality index method (QIM) for gutted ice-stored hybrid tambatinga (Colossoma macropomum Piaractus brachypomum) and study of shelf life. LWT Food Sci. Technol. 67, 55–61.

Robertson, G., 2010. Food Packaging and Shelf Life: A Practical Guide. CRC Press, Inc., Boca Raton, Florida.

Rockström, J., Steffen, W., Noone, K., Persson, A., Chapin, F.S., Lambin, E.F., … Foley, J.A., 2009. A safe operating space for humanity. Nature 461 (7263), 472–475.

Salvador, A., Fiszman, S., Curia, A., Hough, G., 2005. Survival analysis applied to sensory shelf life of yogurts-II: spanish formulations. J. Food Sci. 70 (7), s446–s449.

Sant'Ana, L.S., Soares, S., Vaz-Pires, P., 2011. Development of a quality index method (QIM) sensory schee and study of shelf-life of ice-stored blackspot seabream (Pangellus bogaraveo). LWT- Food Sci. Technol. 44, 2253–2259.

Siripatrawan, U., Noipha, S., 2012. Active film from chitosan incorporating green tea extract for shelf life extension of pork sausages. Food Hydrocoll. 27, 102–108.

Spada, A., Conte, A., Del Nobile, M.A., 2018. The influence of shelf life on food waste: a model-based approach by empirical market evidence. J. Clean. Prod. 172, 3410–3414.

Stone, H., Sidel, J.L., 2004. Sensory Evaluation Practices. Elsevier Academic Press, London.

Tomac, A., Cova, M., Narvaiz, P., Yeannes, M., 2017. Sensory acceptability of squid rings gamma irradiated for shelf-life extension. Radiat. Phys. Chem. 130, 359–361.

Vandekinderen, I., Devlieghere, F., Van Camp, J., Denon, Q., Sanchez Alarcon, S., Ragaert, P., De Meulenaer, B., 2009. Impact of decontamination step with peroxyacetic acid on the shelf life, sensory quality and nutrient content of grated carrots packed under equilibrium modified atmosphere and stored at 7°C. Postharvest Biol. Technol. 54, 141–152.

Wansink, B., Wright, A.O., 2006. 'Best if used by …' how freshness dating influences food acceptance. J. Food Sci. 71 (4), s354–s357.

Wilson, N., Rickard, B., Saputo, R., Ho, S.T., 2017. Food waste. The role of date labels, package size, and product category. Food Qual. Prefer. 55, 35–44.

Yantis, J., 1992. The Role of Sensory Analysis in Quality Control. ASTM Manual Series: MNL 14, Baltimore, USA.

## Further reading

Aschemann-Witzel, J., Jensen, J.H., Jensen, M.H., Kulikovskaja, V., 2017. Consumer behaviour towards price-reduced suboptimal foods in the supermarket and the relation to food waste in households. Appetite 116, 246–258.

Cruz, A.G., Walter, E.H.M., Silva Cadena, R., Faria, J.A.F., Bolini, H.M.A., Pinheiro, H.P., et al., 2010. Survival analysis methodology to predict the shelf-life of probiotic flavored yogurt. Food Res. Int. 43, 1444–1448.

Dethmers, A.E., 1979. Utilizing sensory evaluation to determine product shelf life. Food Technol. 33 (9), 40–43.

Djekic, I., Vunduk, J., Tomasevíc, I., Kozarski, M., Petrovic, P., Niksic, M., Pudja, P., Klaus, A., 2017. Application of quality function deployment on shelf-life analysis of Agaricus bisporus Portobello. LWT-Food Sci. Technol. 78, 82–89.

Fernández-León, M.F., Fernández-León, A.M., Lozano, M., Ayuso, M.C., Amodio, M.L., Colelli, G., 2013. Retention of quality and functional values of broccoli Parthenon stored in modified atmosphere packaging. Food Control 31 (2), 302–313.

Gacula, M.C., Kubala, J.J., 1975. Statistical models for shelf life failures. J. Food Sci. 40, 404–409.

Heldman, D.R., Hartel, R.W., 1997. Principles of Food Processing. Aspen Publishers, Inc, Gaithersburg, Maryland.

Kilcast, D., 2000. Sensory evaluation methods for shelf-life assessment. In: Kilcast, D., Subramaniam, P. (Eds.), The Stability and Shelf-Life of Food. CRC/Woodhead, Boca Raton, LF, pp. 79–105.

Meiselman, H., 2013. The future in sensory/consumer research: …evolving to a better science. Food Qual. Prefer. 27, 208–214.

Mexis, S.F., Chouliara, E., Kontominas, M.G., 2012. Shelf life extensión of ground chicken meat using an oxygen absorber and a citrus extract. LWT- Food Sci. Technol. 49, 21–27.

Pedro, A.M.K., Ferreira, M.M.C., 2006. Multivariate accelerated shelf life testing: a novel approach for determining the shelf life of foods. J. Chemom. 20, 76–83. NO VA.

# Accelerated shelf life testing

# 12

*Sonia Calligaris, Lara Manzocco, Monica Anese,*
*Maria Cristina Nicoli*
Department of Agriculture, Food, Environmental and Animal Sciences,
University of Udine, Udine, Italy

**Chapter outline**



## 1 Introduction

Shelf life (SL) is an important property of any food. Each food has its own shelf life and all stakeholders of the supply chain should be aware of it. From a theoretical point of view, shelf life has been defined as a finite length of time after production (in some cases after maturation or aging), during which the food product retains a required level of quality under well-defined storage conditions (Nicoli, 2012). Even if the rationalization of the shelf life concept has required the effort of many researchers over more than four decades, the assessment of shelf life still represents an exciting challenge for food scientists. On the other hand, it is increasingly recognized as a vital process to comply with current regulation, maintain company brand reputation, and, eventually, increase market share.

Nowadays food companies are required by law to attribute a shelf life to their products under storage conditions declared on the label. Open dating is compulsory in Europe, in many South American and Arabic countries, Israel, and Taiwan (Robertson, 2009). In the United States, dating is not compulsory, with exception of infant formula and baby foods, but, at present, 20 US States require dating of some food categories. Considering EU law,

**Food Quality and Shelf Life. https://doi.org/10.1016/B978-0-12-817190-5.00012-4**

Regulation 1169/2011 (EC, 2011) defines the date of minimum durability of a food as the date until which the food retains its specific properties when properly stored. In the case of highly perishable foods undergoing microbial spoilage during storage with consequent safety implications, the minimum durability date is replaced by a "use by" date (art. 10, EU 1169/2011). In any case, the responsibility of food dating lies with the food producer and no further official indications are provided about the methodologies to obtain a reliable estimation of food dating. Food dating is generally applied to packed foods. However, there is a variety of situations, both at consumer and industrial level, in which the consumption of food, raw ingredient, or semi manufactured product is spread over a certain length of time after pack opening. For instance, the use of flour or bulk fat in the production line of a bakery food could be extended over a significant period of time after pack or container opening. When this happens, quality alteration might proceed faster and sometimes follow reaction pathways other than those occurring in the originally packed food. For this reason, the secondary SL concept (SSL) has been introduced (Nicoli and Calligaris, 2018). The latter is the period after pack opening during which a food product maintains an acceptable quality level. To this regard, EU regulation (Article 25—1169/2011) (EC, 2011) requires the indication of appropriate storage or use conditions after opening of the package as well as the time limit for product consumption, where appropriate. Information on shelf life after opening should be communicated, such as "best within xx days of opening."

Fig. 1 schematically shows the different life stages of a food product from manufacturing to pack opening (if present) and storage on the shelves. It should be pointed out that shelf life definition is always a quality issue not related to any safety risk for the consumer. For this reason, the product safe life is in any case longer than the product shelf life. It is noteworthy that secondary shelf life is generally shorter than shelf life and both are shorter than the time associated with possible hygienic and safety issues for consumers (safe life of the product).

Food shelf life is traditionally determined by monitoring changes in the quality of the food products, packed in their original package, under environmental conditions (e.g., temperature, light exposure, relative humidity) that mimic those actually experienced by the product on the market shelf. This approach, which is called real-time shelf life testing, might be ideally exploited for any food category, but it is actually advantageous only for perishable foods, which undergo quick alterative events. If no considerable differences between primary and secondary shelf life are expected after pack opening, the SL indications applied for the originally packaged food could be considered valid also in the case of the opened product. A similar consideration would also hold for ingredients and semi processed foods whose use in the production line or in the kitchen is so rapid that product spoilage after package opening can be regarded as negligible. The remaining products, whose use is parceled over a significant length of time and that can be affected by the changes of the environmental conditions after package opening, should be considered for SSL assessment. Real-time testing, used to define shelf life of foods, can also be adopted to assess secondary shelf life, since alterative events are expected to proceed fast, and secondary shelf life to be shorter (Fig. 1). However, as reported in literature, appropriate modifications of the methodology might be required (Cappuccio et al., 2001; Nicoli and Calligaris, 2018).

In contrast, when alterative events occur slowly, it might be much more useful to accelerate shelf life tests by monitoring food quality depletion under environmental conditions

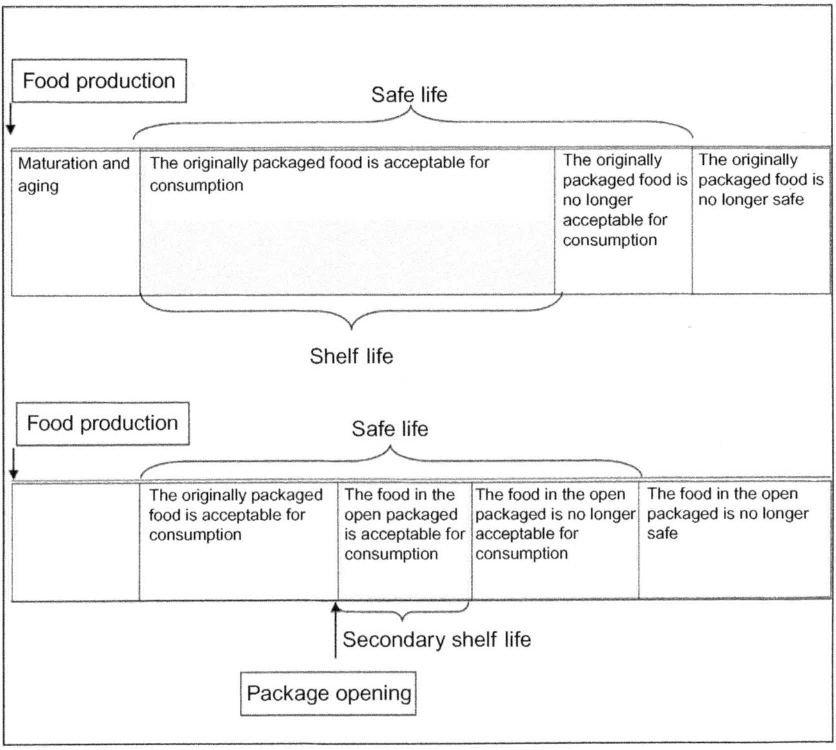


**Fig. 1** Life stages of a food product from manufacturing to maturation and aging, to storage in the original package or after package opening. Shelf life, secondary shelf life, and safe life are indicated.

that allow deteriorative events to proceed faster. If properly applied, this procedure, known as accelerated shelf life testing (ASLT), allows estimation of the shelf life data at storage conditions usually experienced by the product on the market by using data acquired at accelerated storage conditions (Labuza and Schmidl, 1985; Corradini and Peleg, 2007; Mizrahi, 2011). This approach is particularly advantageous for those products that are characterized by a long shelf life, such as ambient-stable and frozen foods. In these cases, the producers are required to generate shelf life data in times that fit with industrial and market needs. An exemplary situation is the need of attributing an expiry date to novel shelf stable products (e.g., shelf life reasonably longer than 10–12 months) while, according to the common timing of industrial food development, the time available to turn a novel idea into a commercialized food is often shorter than 6 months.

## 2 Fundamental phases of shelf life assessment process

The shelf life assessment process could be divided in three fundamental phases. These phases should be carefully tackled to generate a reliable shelf life dating by considering

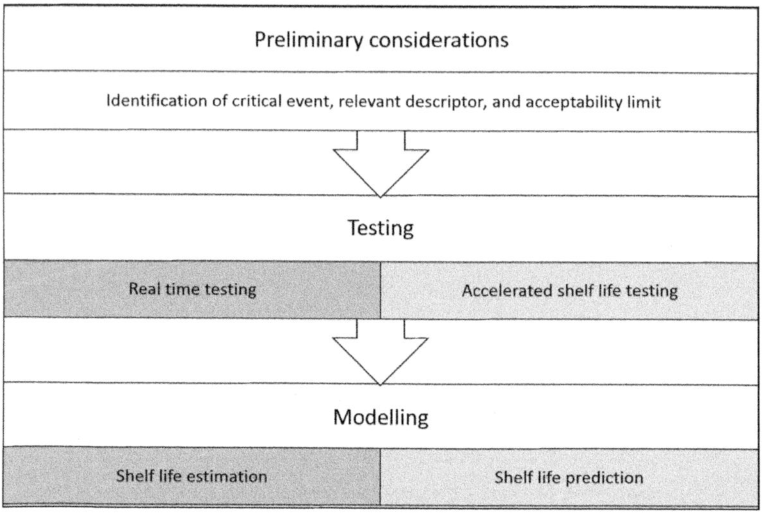


**Fig. 2** Schematic representation of the shelf life testing protocol.

both real time and ASLT protocols (Nicoli, 2012). A schematic representation of these phases is shown in Fig. 2.

## 2.1 *Preliminary considerations*

The initial phase of the shelf life assessment protocol requires the clear identification of the event associated with the most critical impact on food quality. As widely reported in literature, depending on food features and storage conditions, various microbiological, enzymatic, chemical, or physical phenomena could take place simultaneously during storage (Table 1). In the case of chilled foods, the growth of alterative microorganisms and the occurrence of enzymatic reactions impairing the sensory properties of the food are generally regarded as the main events responsible for food quality decay. On the contrary, the shelf life of frozen foods is critically limited by the development of enzymatic or chemical reactions, since microbial growth does not represent an issue at these storage temperatures. Finally, the events leading to the quality loss of shelf stable products stored at ambient temperature are generally chemical, such as oxidation of lipids, flavors, and pigments, and nonenzymatic browning, as well as physical phenomena including starch retrogradation and structural collapse.

Defining, among all biological, chemical, and physical events possibly affecting food quality, the one limiting shelf life might be an arduous task. An easy criterion for the choice of the quality depletion indicator is based on its earliness. In other words, the first indicator that changes during storage is often the one selected for monitoring food quality evolution in shelf life assessment.

However, this important choice becomes even more complex considering that it is inevitably connected to the selection of the acceptability limit. The latter represents the level of the quality change that discriminates acceptable products from those no

**Table 1** Food products and main expected deteriorative events

| Food | Main alterative events |
|---|---|
| Chilled | Microbial growth<br>Enzymatic reactions<br>Senescence<br>Tissue mechanical damage |
| Frozen | Oxidative reactions<br>Enzymatic reactions<br>Surface drying<br>*Re*-crystallization |
| Ambient | Oxidative reactions<br>Non enzymatic browning<br>Caramelization<br>Structural collapse<br>Starch retrogradation<br>Packaging contaminants migration |

longer acceptable (Manzocco, 2016). It should be stressed that the definition of the acceptability limit is the aspect that converts a stability test into a shelf life test. In fact, the aim of a stability test is to predict the deteriorative reaction kinetics as a function of different variables. By contrast, a shelf life study has the objective to estimate correctly the time needed to reach the acceptability limit that discriminates acceptable from unacceptable products (Nicoli, 2012).

A number of different limit values (e.g., pathogen microbes, contaminants migrating from the packaging into food, and toxic or potentially toxic compounds formed during storage) are indicated by regulatory bodies for commercialization and consumption of food under safe conditions. Shelf life operators are often intrigued by the possibility of using these safety limits as possible acceptability limits for shelf life studies. However, unacceptability due to food intake under unsafe conditions is exceptionally critical for food producers. For this reason, the acceptability limit for shelf life assessment should never be related to safety issues. Any product exceeding a safety limit during storage should be regarded as out of basic standards required for consumption because it has reached the end of its safe life (Fig. 1). Thus, the acceptability limit should be chosen as the value allowing estimation of a shelf life, included in the safety time interval, during which the product retains acceptable quality characteristics and no risk for consumer health is present (Manzocco, 2016).

As previously noted, the decision maker of food SL is the producer itself, who is completely responsible for consumer satisfaction and legal requirement adherence. This role is covered by means of the choice of the acceptability limit, through the definition of a set of requirements for the commercialization of its own product. Compulsory acceptability limits can be derived from producer voluntary label claims. For instance, when the concentration of a bioactive compound, added to increase food functionality, is declared on the product label, its value must be compulsory guaranteed until shelf life. In this case, the producer could choose the bioactive concentration declared on the label as a possible acceptability limit. However, given the low number of acceptability limits derived from voluntary claims, in the vast majority of cases the acceptability limit can be freely chosen according to the quality policy of the producer. The latter can actually select a more restrictive indicator, based on a certain level of sensory dissatisfaction of consumers, who might accidentally consume the food at shelf life. In this case, the producer voluntary defines the tolerable level of sensory dissatisfaction of the target consumers. This choice is the result of considerations merging the different needs of consumers, production, delivery structure, and business. It is basically an issue of rational evaluation of potential negative consequences deriving from the risk of consumer dissatisfaction. Marketing need of increasing/maintaining market share can actually exert a critical role on the final decision of the acceptability limit. Business aspects might become so critical as to dominate the final choice. For instance, the shelf life of canned food mainly results from the need of increasing product turnover on the shelves rather than from a sensible decrease in product quality and consumer satisfaction.

## 2.2 Testing

After the preliminary considerations reported above, the shelf life assessing protocol (Fig. 2) requires accurate planning of the testing phase. When a real-time shelf life test is performed, the evolution of the selected quality indicator is monitored under storage conditions mimicking the foreseeable storage conditions. It is useful to remember that the rate of deteriorative events in a food product is a function of intrinsic food characteristics ($C_i$), packaging related factors ($P_i$), and environmental factors ($E_i$) (Nicoli, 2012):

$$k = f\left(C_i, P_i, E_i\right) \tag{1}$$

The observed reaction rate, and thus the product shelf life, is actually the result of the complex interactions among these factors.

For this reason, the basic premise to correctly perform a real-time shelf life test is that all the factors critically affecting reaction rate are defined and maintained constant during the test. For instance, it is recommended that shelf stable products, such as bakery, canned, and dried foods, be stored at 20–25°C; chilled products at 3–4°C, and frozen foods at −18°C (EU conditions) or −12°C (US conditions). However, since temperature fluctuations are inevitably frequent during food storage, shelf life tests could be profitably carried out under the worst situation that the product might be expected to experience during storage. Based on this consideration, shelf life tests of chilled products are often performed at 8–10°C instead of 3–5°C. Similarly, for ambient stable products, 30–35°C could represent a reasonable room temperature simulating typical storage conditions during summer time. Such a temperature range is also applied when the product is expected to be transported from producer country to market country with limited possibility of temperature control during shipping. In addition, if the food is packaged in a see-through container, light exposure could become a critical factor in determining its shelf life, often showing a higher importance than all other environmental factors (Calligaris and Manzocco, 2012). In these cases, the effect of this factor during shelf life testing should be not neglected, nor underestimated.

Conversely, when performing ASLT, all compositional and packaging related variables in Eq. (1) must be kept constant, whilst a proper environmental factor is chosen and exploited to accelerate the reaction rate responsible for food quality depletion. As previously pointed out, the main objective of ASLT is to produce a reliable shelf life value while significantly reducing the testing time required for its generation.

In an attempt to define the ASLT conditions, a number of different environmental parameters could be potentially applied as accelerating factors, including temperature, relative humidity, light intensity, or gas partial pressure. Independently of the accelerating factor chosen, it is essential to perform ASLT only when the quality depletion rate increases as a function of the intensity of the selected accelerating factor. To this aim, the selected quality indicator should be monitored during storage under different levels of intensity of the accelerating factor.

## 2.3 Modeling

Data acquired during the testing phase are used to describe the evolution of the selected quality indicator as a function of storage time. To this aim, they are generally submitted to modeling according to the fundamental kinetic principles (Van Boekel, 2008). Accordingly, the rate of changes of a quality indicator ($I$) as a function of storage time ($t$) can be calculated by integration of the general kinetic equation:

$$\int_{I_o}^{I} \frac{dI}{I^n} = \int_{o}^{t} k dt \tag{2}$$

where $k$ is the rate constant and $n$ the reaction order.

Since the reaction order $n$ does not give any indication of the actual mechanisms of the reaction involved in food quality depletion, $k$ is considered to be an "apparent" rate constant. The choice of the reaction order $n$ of the rate law is generally performed based on the selection of the model best fitting the experimental values. Apparent zero, first, and second orders are the reaction orders most frequently used to describe food quality depletion. In these cases, the rate constant is calculated by linear regression analysis of $I$, $\ln I$ or $1/I$ as a function of storage time. The choice of the reaction order is generally supported by statistical analysis based on the comparison of the values of the determination coefficient of the linear regressions ($R^2$), the $P$ value, as well as on the visual analysis of the regression residues (Van Boekel, 2008).

Once the reaction rate constant ($k$) has been estimated by linear regression, the shelf life under the storage conditions selected to perform the test can be estimated by solving the integrated form of Eq. (2) as a function of time:

$$\mathrm{SL} = \frac{1}{k} \int_{I_0}^{I_{\lim}} \frac{dI}{I^n} \tag{3}$$

where $I_0$ is the value of the critical indicator just after food production and $I_{\lim}$ is the critical indicator value corresponding to the previously defined acceptability limit.

Fig. 3 shows a schematic representation of quality indicator changes following zero, first and second reaction orders as well as relevant shelf life equations.

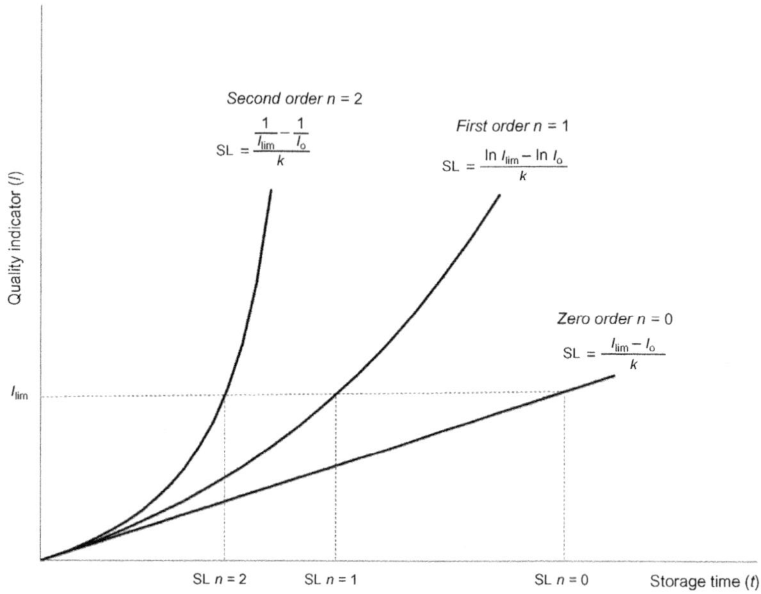


**Fig. 3** Schematic representation of zero, first, and second orders reactions. Relevant shelf life equations are also reported.

It should be noted that complex alterative phenomena, based on reaction pathways characterized by the occurrence of concomitant and consecutive reactions, might lead to the inapplicability of the classic kinetics approach described above. A typical example in which these models fail is the evolution of phenomena showing sigmoidal evolution. These events are characterized by an induction period (lag phase) followed by the progressive increase of the shelf life indicator that finally reaches a maximum level (Fig. 4). Different examples of this behavior can be found in literature monitoring the development of oxidative reactions (Corradini and Peleg, 2007; Calligaris et al., 2008b; Odriozola-Serrano et al., 2009), nonenzymatic browning (Vaikousi et al., 2009), and enzymatic activity (Terefe et al., 2004). Also the evolution of sensory consumer acceptability during food storage might present similar evolution, which does not allow conventional modeling according to the fundamental kinetic principles (Hough et al., 2006; Guerra et al., 2008; Buvè et al., 2017). In these cases, empirical descriptive models other than the classic ones are identified to describe properly the evolution of the shelf life indicator.

When ASLT is carried out, the same procedure of reaction rate ($k$) estimation of the deteriorative event has to be repeated at different levels of intensity of selected accelerating factor.

Only in this way can the relationship between the reaction rate and the acceleration factor be estimated. In other words, it is essential that the dependence of the reaction rate on the selected accelerating factor is known. In fact, the requirement to apply the ASLT procedure is the knowledge of the mathematical model allowing estimation of the reaction rate as a function of the intensity of the accelerating factor. Only if this

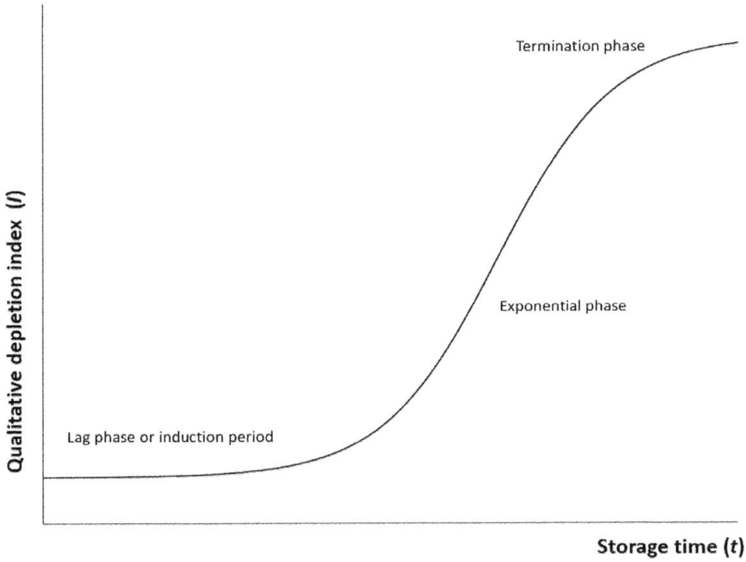


**Fig. 4** Schematic representation of the changes in the quality depletion indicator of a complex reaction following a sigmoidal evolution.

requisite is complied with is the extrapolation of reaction rate at actual storage conditions possible, and a shelf life value predictable.

## *2.4 Basic procedure to perform an ASLT*

Table 2 shows the basic procedure to be applied when performing an ASLT. It obviously includes the preliminary phase (step 1), described above (Fig. 2), which is common to both real-time and ASLT. Following this preliminary step, it is necessary to define the acceleration factor and choose at least three relevant accelerating conditions (step 2). In the attempt to plan an ASLT (step 3), that is, a time-costing and labor-consuming process, the careful planning of experiments must be developed, anticipating as far as possible the eventual occurrence of critical issues that could potentially undermine test success (Calligaris and Manzocco, 2012).

The design of a detailed experimental plan in which sample size, sampling frequency, analytical tools, resources, and costs are clearly defined appears essential at this point. In fact, it would be disastrous to be, for instance, out of samples before the product spoils simply due to the application of the wrong sampling plan. The latter, of course, is not identical at each acceleration level but requires modulation according to the expected acceleration of the alterative event. Monitoring data while progressively acquiring them during storage, together with a certain level of flexibility in the management of time intervals, might certainly improve ASLT performance. Regarding resources, the confirmation of the availability of resources is essential before running the test. Resources include personnel, equipment, and other facilities needed to store the samples during the test and perform the chemical, physical, or sensory analysis at

**Table 2** Basic procedure of an ASLT

| Procedure |
|---|
| (1) Definition the SL critical event, relevant descriptor and acceptability limit |
| (2) Definition of at least three levels of intensity of the accelerating factor to be used during the test |
| (3) Definition of the number of products packages and relevant batches to be stored under each accelerated condition |
| (4) Start the test by collecting and analyzing the samples at defined time intervals |
| (5) Estimation of reaction rate by statistical elaboration of data acquired during the test (classic kinetic approach or other appropriate models) |
| (6) Modeling of the dependence of the deteriorative event rate ($k$) on the accelerating factor |
| (7) Computation of shelf life data—if no deviations are observed—by using the mathematical model generated in the step 6 and acceptability limit defined in step 1 |

the expected storage times. The design of experiments is certainly an easy task when enough information on product stability and its dependence on the accelerating factor is available. For instance, such information may be inferred from prior stability tests performed within the company or from literature data. On the other hand, if this information is not exhaustive, or only partial, a straightforward experimental plan cannot be defined at the beginning of the ASLT and online rearrangements are necessary based on the results progressively acquired during the testing phase. Thus, only after steps 1–3 have been carefully defined (Table 2), the test could be developed by monitoring the changes of the selected index during storage under accelerated conditions (step 4).

Afterwards, data must be modeled by using, when possible, the classic kinetic approach above described to compute the reaction rate ($k$) (step 5). Fig. 5A schematically shows a hypothetical graph obtained when the critical indicator increases linearly with the storage time under different accelerated conditions. In the case here considered, the $k$ values can be computed from quality depletion data by applying a zero order reaction model. The estimated $k$ values can now be used to estimate the relation between the accelerating factor (A) and the reaction rate ($k$) (step 6):

$$k = f(A) \tag{4}$$

It should be noted that a minimum of three different acceleration conditions are needed to proceed with the step 6.

However, to improve the statistical reliability of the regression analysis or when the $A$ dependence of $k$ is complex, more data points are absolutely necessary to improve the model accuracy. Fig. 5B represents the general and simplest case of a dependence of $k$ versus the acceleration factor according to the equation of a straight line. In this case, the equation generated by the regression analysis can be regarded as shelf life prediction model:

$$k = mA + k_0 \tag{5}$$

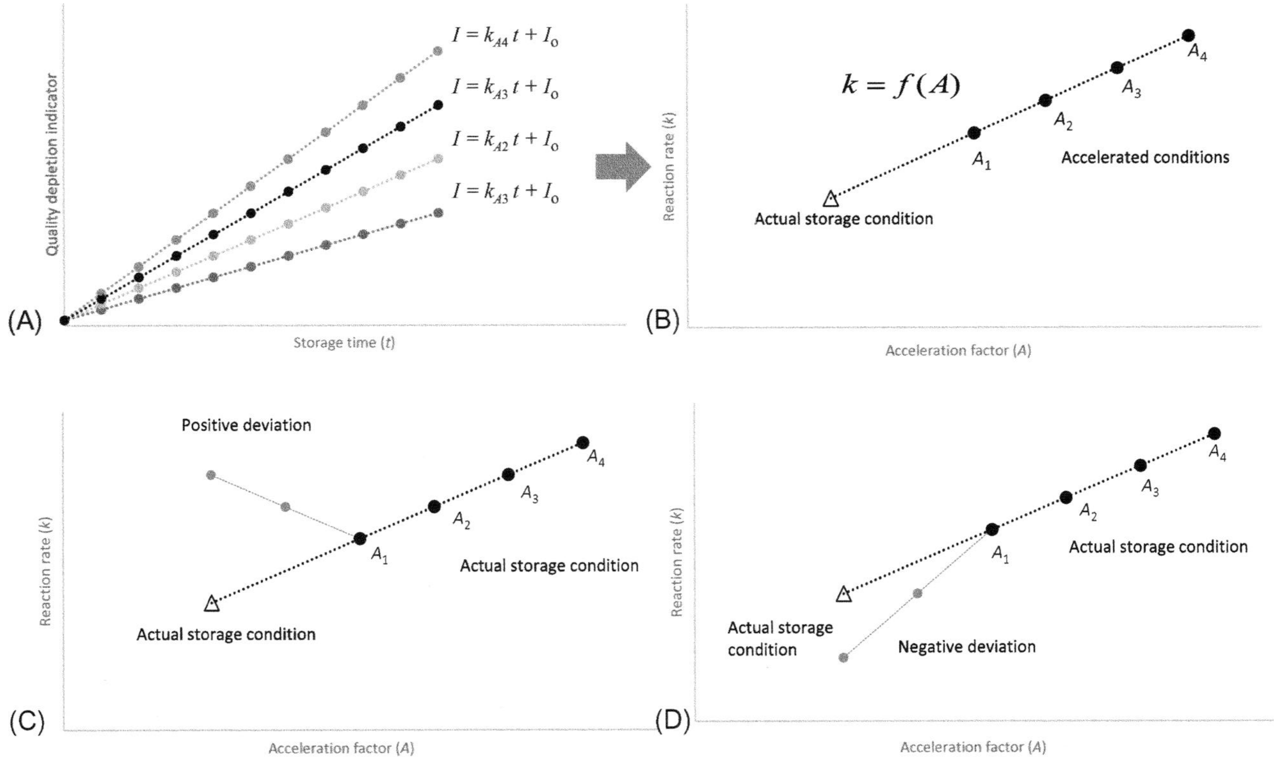


**Fig. 5** Art PDF of replacement images for CE edit Schematic representation of steps 5 and 6 described in Table 2: (A) estimation of reaction rate by statistical elaboration; (B) modeling of the dependence of the deteriorative event rate on the accelerating factor; (C) positive deviation from expected behavior; (D) negative deviation from expected behavior.

where $m$ represents the slope of the linear equation and $k_0$ the value of $k$ at $A=0$.

This equation can be used as a predictive tool to estimate the reaction rate at the desired actual storage condition ($A_{actual}$) and finally to produce shelf life data:

$$\mathrm{SL} = \frac{I - I_0}{k_{actual}} = \frac{I - I_0}{mA_{actual} + k_0} \tag{6}$$

where $k_{actual}$ and $A_{actual}$ are the reaction rate and the value of the accelerating factor at the conditions at which the shelf life has to be computed (actual storage condition), respectively. By using this equation, it is possible to estimate the shelf life of the product stored under the conditions experienced by the product on the shelf.

This procedure could appear quite simple. However, different pitfalls could arise depending on the nature of the acceleration factor applied in the test. In fact, the successful application of this approach requires that food is able to withstand the accelerating factor increase without inducing the development of phenomena different from the deteriorative event selected as SL index. When this happens, positive or negative deviations from the values predicted by the application of the model that describes the $A$ dependence of $k$ are often observed (Fig. 5C and D). This means that the extrapolation of shelf life data from accelerated data shall be performed only within the accelerated condition range experimentally proven to conform to the model describing the accelerating factor dependence of the reaction rate. In other words, the generated model needs to be adapted to the specific circumstances of the product being considered.

# 3 Accelerating factors in ASLT

As aforementioned, different environmental factors, such as temperature, oxygen, light, and relative humidity, could be potentially exploited as accelerating factors (AF) in ASLT. Theoretically, the procedure proposed in Table 2 can be used independently of the nature of the environmental factor selected to speed up the deteriorative reactions. Among all environmental factors that might be potentially exploited in ASLT, temperature is certainly the most frequently used. This is not only due to the fact that temperature is one of the most critical factors affecting food reaction kinetics, but also for the availability of a universally recognized mathematical model able to describe the temperature sensitivity of the rates of food quality loss; that is, the Arrhenius model (see Section 3.1). Beside temperature, light has also been used in literature to accelerate quality depletion, as a single factor or in combination with temperature. Finally, sporadic scientific works on the application of oxygen and relative humidity in ASLT can be found in literature. Unfortunately, a unique accepted model is not available for light or for other AF. Table 3 reports some relevant references on the use of temperature, light, oxygen concentration, and relative humidity in ASLT. Due to the huge application of temperature in ASLT, references reported in the table are only some recent examples of the use of this methodology and their list is far from exhaustive.

**Table 3** Examples of literature data relevant to the application of different accelerating factors in ASLT

| Accelerating factor | Product | References |
|---|---|---|
| Temperature | Biscuits | Wang et al. (2015) |
| | Extra virgin olive oil | Li and Wang (2018) |
| | Strawberry juice | Buvè et al. (2017) |
| | Ascorbic acid degradation | Peleg et al. (2018) |
| | Tomato paste | Jafari et al. (2017) |
| Light | Beverage containing crocin | Manzocco et al. (2008) |
| | Lemon flavored drink | Garitta et al. (2018) |
| | Sunflower oil and soybean oil | Manzocco et al. (2012) |
| Oxygen | Powder infant formula | An et al. (2018) |
| Relative humidity | Dehydrated cabbage | Mizrahi et al. (1970) |
| | Egg white powder | Rao et al. (2012) |

In the following sections, the use of temperature and light as accelerating factors in ASLT is discussed with reference to the mathematical models reported in the literature to predict the rate of alterative reaction in food.

## 3.1 Temperature

The extensive use of temperature in ASLT is basically due to the fact that Arrhenius Eq. (7) (Arrhenius, 1901), developed theoretically on the molecular basis for reversible chemical reactions, has been shown to hold empirically for estimating the rate of a wide range of complex chemical, physical, and sensory changes occurring in foods (Labuza and Riboh, 1982; Peleg et al., 2012):

$$k = k_0 \cdot e^{-\frac{E_a}{RT}} \tag{7}$$

where:

$k$ is the reaction rate constant;

$R$ is the molar gas constant ($8.31\,\mathrm{J\,K^{-1}\,mol^{-1}}$);

$T$ is the absolute temperature (K);

$E_a$ is the apparent activation energy ($\mathrm{J\,mol^{-1}}$); and.

$k_0$ is the so called preexponential factor.

The possibility to apply the Arrhenius equation to describe the temperature dependence of quality depletion rate can be easily assessed by plotting $\ln k$ as a function of the reciprocal of temperature (reported in Kelvin degree) and rewriting the equation in its linearized form as follows:

$$\ln k = \ln k_0 - \frac{E_a}{RT} \tag{8}$$

When the Arrhenius behavior is fulfilled, the relation between these two variables is visually observed as a straight line. Hence, the temperature dependence of the reaction rate can be calculated performing a linear regression analysis and estimating the parameters of Eq. (8). The latter, with relevant parameters, can be used to estimate the reaction rate at the desired temperature.

From the statistical point of view, the estimation of the Arrhenius equation parameters raises some critical issues and a better estimation can be generated by applying a reparametrization by inserting a reference temperature $T_{ref}$, corresponding to the average of the temperatures used in the ASLT (Van Boekel, 2009). The reparametrized Arrhenius equation is reported in Eq. (9) in the linearized form:

$$\ln k = \ln k_{ref} - \frac{E_a}{R}\left(\frac{1}{T} - \frac{1}{T_{ref}}\right) \tag{9}$$

The reparametrization process is performed to enhance the reliability of parameter estimates; indeed, it reduces the correlation between the reaction rate constant and the activation energy parameter estimation.

The temperature dependence of the shelf life indicator can be integrated into the shelf life equation and a shelf life predictive model can be generated:

$$\mathrm{SL} = \frac{\int_{I_0}^{I_{lim}} \frac{dI}{I^n}}{k_{ref} \exp\left[-\frac{E_a}{R}\left(\frac{1}{T} - \frac{1}{T_{ref}}\right)\right]} \tag{10}$$

A number of data can be found in the literature with reference to the application of the Arrhenius equation to describe the temperature dependence of alterative reactions in foods (see Table 3 with some recent literature examples). The focus of most of these studies is the understanding of the temperature dependence of food alteration. The latter is actually described by the reaction activation energy ($E_a$), which is a powerful tool to predict reaction rate at the desired temperature.

Unfortunately, activation energy does not assume a constant value for each alterative event. It changes according to the food characteristics, processing and storage conditions. Fig. 6 visually compares the indicative range of magnitude of $E_a$ for some alterative events occurring during food storage. It can be observed that the range of the $E_a$ values is definitely wide. For instance, in the case of lipid oxidation, the range of $E_a$ reported in literature varies from about 20 to 150kJ/mol (Calligaris et al., 2004; Orlien et al., 2006; Dermesonlouoglou et al., 2007; Huang and Sathivel, 2008; Topuz, 2008; Tazi et al., 2009). Similar considerations can be made for the development of Maillard reaction during food storage. In this case the range of $E_a$ values moves from 50 to 250kJ/mol (Burdurlu et al., 2006; Burdurlu and Karadeniz, 2003; Sithole et al., 2004; Thomson et al., 2005; Pereyra Gonzales et al., 2010; Schmitz-Schug et al., 2014). By contrast, photochemical oxidation and enzymatic catalyzed reactions generally show low values of $E_a < 50$ kJ/mol (Kristensen et al., 2001; Manzocco et al., 2008; Sothornvit and Kiatchanapaibul, 2009); whereas microbial inactivation and protein denaturation are characterized by high $E_a$ values. The wide range of activation energies reported for the same alterative event greatly depends on the different compositional, processing,

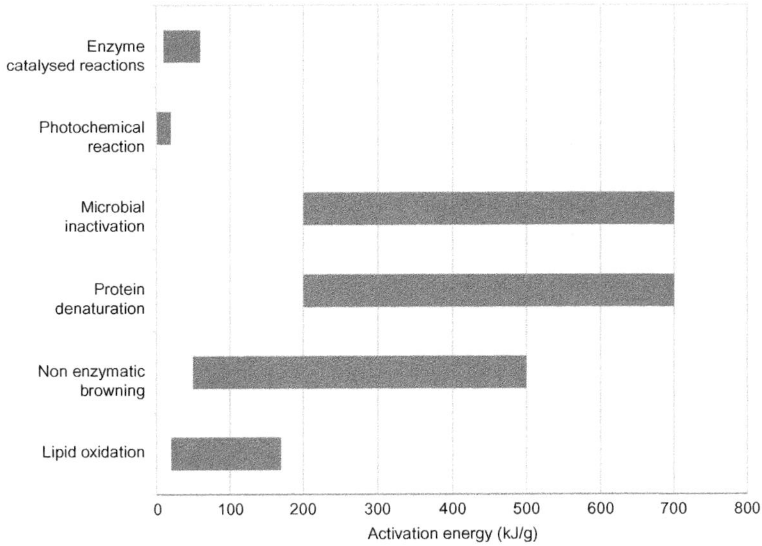


**Fig. 6** Indicative range of magnitude of activation energies ($E_a$) for some typical reactions in foods.

and packaging variables, which are specific to the considered food product. This means that for each food product, the activation energy value should be experimentally calculated since the use of an inadequate activation energy could lead to dramatic prediction errors in the SL data.

The knowledge of the activation energy of the alterative reaction could be very profitable in the attempt to plan an ASLT. In fact, the existence of different magnitudes of $E_a$ indicates that quality depletion phenomena are differently accelerated by the temperature increase. This means that the time saved when performing an ASLT may be considerably different. For instance, the time saved in SL determination by a 10°C temperature increase is about 50% when the $E_a$ of the alterative event value is circa 50 kJ/mol and only 20% for deteriorative events with $E_a$ lower than 20 kJ/mol. Thus, for critical indicators characterized by $E_a$ values lower than 50 kJ/mol, the acceleration obtained increasing temperature could be too low to be of interest in the attempt to save time during shelf life assessment (Calligaris et al., 2012).

In the temperature range where the Arrhenius model is fulfilled, a direct and simple index of how the reaction rate or shelf life change with temperature can be computed. In particular, a shelf life factor can be calculated as the ratio between the shelf life value at a given temperature $T$ and the shelf life at a reference temperature $T_{\text{ref}}$ (Anese et al., 2012):

$$SL_{\text{factor}} = \frac{SL_T}{SL_{T_{\text{ref}}}} \tag{11}$$

where $SL_T$ is the shelf life at a certain temperature and $SL_{\text{ref}}$ is the SL at a reference temperature.

This index could give an indication of the decrease of shelf life as a consequence of a given temperature increase. When $T_{ref}$ is 10°C higher than a certain temperature, this factor is generally called $Q_{10}$ value (Labuza and Schmidl, 1985):

$$Q_{10} = \frac{SL_T}{SL_{T+10}} \tag{12}$$

where $SL_{T+10}$ is the shelf life at a temperature of 10°C above.

Both indices are very easily understandable and applicable in common practice. However, it should be borne in mind that these indexes can be used only in the temperature range in which the Arrhenius behavior is fulfilled. In other words, preliminary data modeling according to the Arrhenius equation is needed to set up an experimental plan correctly for the assessment of the $SL_{factor}$ or $Q_{10}$ values. This aspect should be carefully considered because, although the Arrhenius equation is frequently applied in SL studies, many pitfalls could arise in practice leading to deviations from the Arrhenius equation with consequent errors in shelf life prediction (Fig. 5C and D). As previously noted, the successful application of the Arrhenius model requires that food is able to withstand the increase in temperature without developing phenomena other than the event monitored by the selected shelf life indicator. In this regard, the choice of the temperature interval for running the ASLT is crucial to avoid positive or negative deviations from linearity of the expected Arrhenius equation causing errors in shelf life prediction. Table 4 summarizes the main phenomena leading to deviations from the Arrhenius behavior during ASLT performed by increasing food storage temperature. The temperature range in which deviations are customary observed, along with some key-literature, is also reported.

For instance, the application in ASLT of an excessive temperature could induce changes in the reaction pathway. In other words, when complex reactions, such as oxidation and nonenzymatic browning, develop during storage, deviations from the Arrhenius behavior (Fig. 5C and D) are highly probable, due to changes of the reaction pathway as a consequence of temperature increase. Thus, the prediction based on high temperature behavior could underestimate or overestimate food stability. This is the case for fats and oils undergoing oxidation (Calligaris et al., 2016). In fact, the development of excessive oxidation levels as a consequence of temperature increase is reported to be critical in performing ASLT because reaction pathways dominating at lower temperatures might became negligible at higher ones (Frankel, 2005). Moreover, at such a high temperature, intermediate oxidative products (e.g., peroxides), which are often monitored as critical indicators for shelf life assessment, could be degraded fast, accounting for an underestimation of the final shelf life. Not only the degradation of alterative reaction products, but also the thermal degradation of compounds able to steer reaction rate (e.g., pro-oxidant or antioxidant molecules) could affect the overall alteration kinetics, modifying data adherence to the expected Arrhenius behavior. These phenomena are more intense in complex food products, undergoing parallel and consequential alterative events. For instance, in vegetable derivatives containing a lipid fraction, the oxidation of the lipid phase occurs often in concomitance with the depletion of natural occurring antioxidants (polyphenols) and the development of newly formed ones (nonenzymatic reaction products) (Zamora and Hidalgo, 2005; Echavarria et al., 2012).

**Table 4** Possible events occurring as a consequence of temperature increase and leading to deviations from the Arrhenius behavior and relevant literature evidence

| Product | Deteriorative event | Phenomena leading to deviations from Arrhenius equation | Expected deviation temperature range (°C) | References |
|---|---|---|---|---|
| Oil and fats | Lipid oxidation | Formation or degradation of reaction reactants | $T > 60$ | Frankel (2005), Zamora and Hidalgo (2005), and Echavarria et al. (2012) |
| Tomato derivatives | Carotenoid oxidation | Change of the prevalent deteriorative event | $T > 0$ | Manzocco et al. (2006) |
| Frozen tomato derivatives<br>Frozen emulsions | Carotenoid oxidation<br>Lipid oxidation | Increase in oxygen concentration | $T > 0$ | Manzocco et al. (2006)<br>Calligaris et al. (2007b) |
| Fats and oils<br>Bakery products | Lipid oxidation | Lipid crystal melting | $T > T_{melting}$ | Calligaris et al. (2004, 2006, 2007a, 2008b), Manzocco et al. (2006) |
| Frozen foods | Diffusion limited reactions (e.g., EB) | Transition from glass to rubber or viscous liquid structure | $T > T_g$ | Parker and Ring (1995), Champion et al. (1997) |
| Dried food | Diffusion limited reactions (e.g., EB, NEB, caramelization) and physical phenomena (caking, collapse, shrinkage, flavor release) | Transition from glass to rubber or viscous liquid structure | $T > T_g$ | Roos (1995) |

In some cases, the choice of the ASLT range could be crucial to avoid the shift of the prevalent alterative events upon the increase in temperature. An example is that of tomato derivatives stored at 0°C (Manzocco et al., 2006). In these products, carotenoid oxidation is expected to be the prevalent event leading to the end of product shelf life, causing food color depletion. However, pigment bleaching could be masked, as temperature increases, by the concomitant formation of brown compounds due to the development of nonenzymatic browning reactions.

Another factor, seldom considered in ASLT studies, is the decrease of gas solubility upon temperature increase. In this regard, it is noteworthy that the solubility of oxygen in aqueous media is reduced by almost 50% upon a 10°C increase in the typical temperature range of ASLT for frozen foods (Fig. 7).

Being oxygen the critical reactant of oxidative reactions in frozen foods, a clear deviation from Arrhenius behavior might be observed when the alteration is accelerated by increasing temperature. This is the case with frozen tomato sauces undergoing oxidative bleaching of carotenoids during storage (Manzocco et al., 2006). When tests are accelerated without taking into account oxygen solubility changes, a critical overestimation of shelf life could be observed.

More intriguing is the possible effect of physical structure modifications upon temperature changes. Different authors reported deviations from the Arrhenius behavior when phase transitions occur in the temperature range considered for ASLT (Parker and Ring, 1995; Fennema 1996; Champion et al., 1997; Calligaris et al., 2004, 2006, 2007a,b, 2008b; Manzocco et al., 2006). A representative case is that of ASLT of foods containing a crystalline or partially crystalline phase (water, lipid, starch) showing a solid–liquid transition upon temperature increase. Following crystal melting, a wide number of compositional characteristic could change in the liquid phases surrounding crystals. In particular, as crystals melt, the reactant concentration in the remaining liquid phase could significant decrease. For instance, in the case of melting of lipid crystals, the concentration of unsaturated triacylglycerols, $O_2$, antioxidants, and pro-oxidants in the liquid phase is bound to change. Beside reactant concentration, other changes in physico-chemical properties (i.e., reactant solubility, pH, ionic

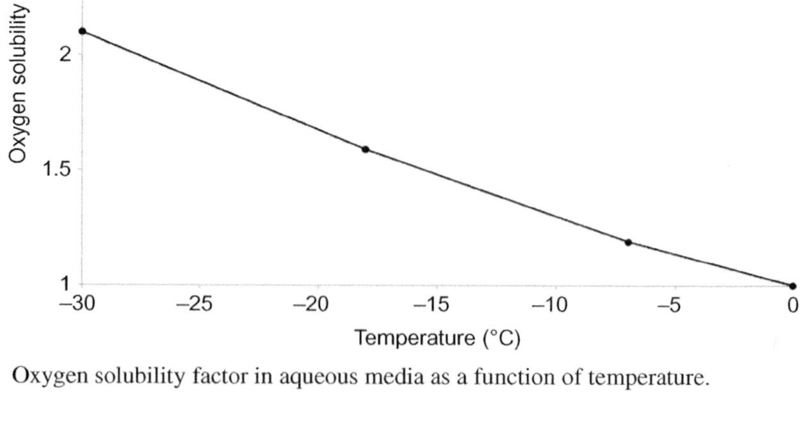


**Fig. 7** Oxygen solubility factor in aqueous media as a function of temperature.

strength, water activity, viscosity) that could affect food stability might occur (Parker and Ring 1995; Champion et al., 1997; Calligaris et al., 2004). These compositional modifications could counterbalance and/or even oppose the direct effect of temperature on the reaction rate, giving reason to the observed Arrhenius deviations (Calligaris et al., 2012). It should be noted that the phase transition of water prevalently affects frozen food shelf life; whereas lipid phase transition could affect Arrhenius behavior in a wide temperature range associated with the phase transition temperature of the lipid matrix present in the product. Table 5 shows some examples of phase transition range for some lipid sources widely used in foods.

Based on data shown in Table 5, it is likely that deviations from Arrhenius behavior are highly probable when the testing temperature is increased above the melting temperature of the fat material used in the food formulation.

Regarding fat crystallization, the effect of composition changes on Arrhenius deviations has been experimentally proven, considering different case studies by our research group. A modified Arrhenius equation has been proposed by inserting in the conventional equation a corrective factor ($\Delta k$) taking into account the main expected changes affecting reaction rate (Calligaris et al., 2004):

$$k = k_0 \cdot \Delta k \cdot e^{-\frac{E_a}{RT}} \tag{13}$$

COMP: in the equation above I cannot see Δ but when I open the formula, the formula is correct. Please checkwhere $\Delta k$ is a corrective factor introduced in the Arrhenius equation.

Since at a given temperature the rate at which any reaction develops can be considered as the result of the ratio between driving forces and resistances, $\Delta k$ can be defined by the identification of the proper forces and resistances responsible for the Arrhenius deviation. The application of this approach requires a deep understanding of the complex phenomena involved and the analytical possibility to quantify the changes occurring in the matrix as a consequence of temperature changes. In any case, this approach has been efficaciously applied to correct the curvature in the Arrhenius plot for different food matrixes (Calligaris et al., 2004, 2006, 2007a,b, 2008a; Manzocco et al., 2006).

**Table 5** Indicative melting temperature obtained by differential scanning calorimetric analysis of selected lipid sources

| Lipid source | Melting temperature (°C) | References |
|---|---|---|
| Palm kernel oil | 28 | Tan and Che Man (2002) |
| Palm stearin | 50 | Toro Vasquez et al. (2000) |
| Coconut oil | 23 | Tan and Che Man (2002) |
| Milk fat | 40 | Breitschuh and Windhab (1998) |
| Extra virgin olive oil | 10 | Barba et al. (2013) |
| Sunflower oil | 0 | Calligaris et al. (2008b) |
| Peanut oil | 8 | Tan and Che Man (2000) |

Moreover, deviations from Arrhenius behavior are well known also to be observed when the glass transition temperature of the food system is overcome. In these cases, the transition from the glassy state to the viscous or rubbery one causes the rate of any diffusion-limited phenomena (EB, NEB, caking, collapse, shrinkage, flavor release) to proceed at a rate much higher than that expected based on the Arrhenius equation. When this happens, the Williams-Landel-Ferry (WLF) model (1955), which shows a higher curvature than the Arrhenius one, can be successfully exploited to estimate reaction rate:

$$\ln\frac{k_{\mathrm{ref}}}{k}=\frac{C_1\left(T-T_{\mathrm{ref}}\right)}{C_2+\left(T-T_{\mathrm{ref}}\right)} \tag{14}$$

where:

$C_1$ and $C_2$ are system dependent constants:
$T$ is the temperature and $T_{\mathrm{ref}}$ is a reference temperature ($T_{\mathrm{ref}} > T_g$);
$k$ is the rate constant; and
$k_{\mathrm{ref}}$ is the rate constant $T_{\mathrm{ref}}$.

Beside these models, other simply descriptive equations have been proposed when Arrhenius behavior is not fulfilled. One example is the model proposed by Waterman and Adami (2005) accounting for generic nonlinear Arrhenius behaviors:

$$k = AT^n e^{-\frac{E_a}{RT}} \tag{15}$$

where $A$, $n$, and $E_a$ are parameters determined using nonlinear fitting programs.

The use of this equation allows for a better mathematical description of some level of curvature in Arrhenius plot, but parameters are pure corrective factors with no specific relation to phenomena causing deviation.

Based on these considerations, it should be said that, due to the huge number of pitfalls potentially arising during ASLT, the use of predictive models to extrapolate deteriorative event rates at usual storage temperatures from accelerated data shall be performed only within the temperature range experimentally proven to conform to the considered model. In other words, modeling methodology requires being adapted to the specific circumstances of the product being examined.

## 3.2 Light

Beside temperature, light can be regarded as an alternative acceleration factor in ASLT. In fact, light exposure could strongly accelerate the development of light-induced reactions, mainly oxidative ones. As a consequence, photosensitive foods are generally packed in opaque materials, able to protect them from both UV and visible light. However, several products, such as fats and oils, soft-drinks, beer, and wine, which equally suffer light-induced oxidation, are often packed in see-through materials and displayed on highly lit shelves (600–800 lx) in order to attract consumers. It is noteworthy that shelf life testing of these photo-sensitive foods is rarely performed by assessing quality changes during storage under different light conditions. ASLT tests of these products are conventionally performed by increasing temperature and largely ignoring the effect of light. Such an approach may obviously lead to dangerous overestimations of product shelf life, since light-induced reactions generally show a slight temperature dependence with very low activation energy values (Fig. 6). In these cases, the estimation of shelf life based on light rather than on temperature could be much more effective.

In the attempt to perform ASLT exploiting light, the basic procedure described in Table 2 is applied. Samples are exposed to light at increasing intensity, selecting at least three or more lighting conditions. However, as previously stressed, the basic requirement for the exploitation of light as an acceleration factor in ASLT is the existence of a mathematical model accounting for the effect of light on the reaction rate. Differently from temperature, there are no universally accepted mathematical models predicting the effect of light on the deteriorative event rate. For this reason, the light dependence of food degradation kinetics should be tested case by case. To our knowledge, few examples on the exploitation of light as acceleration factors in ASLT can be found in literature and the light dependence models generated resulted different depending on the product under study.

Fig. 8 shows a schematic representation of the light dependence of different alterative phenomena, namely bleaching of carotenoids in beverages and oxidation of bulk oils (Manzocco et al., 2008, 2012). As shown in the figure, light dependence of carotenoid bleaching has been mainly described by the equation of a straight line where the intercept represents reaction rate under dark ($k_d$) and the slope represents the electromagnetic

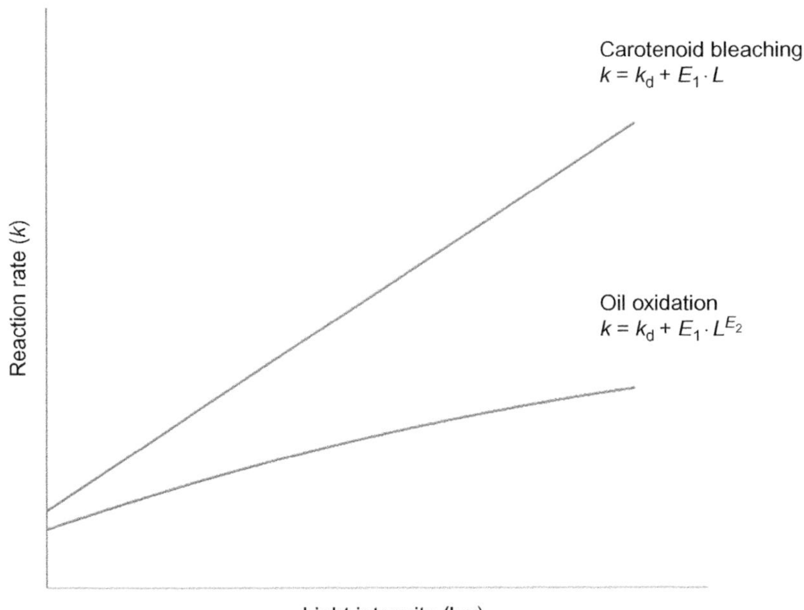


**Fig. 8** Light dependence of the rate ($k$) of carotenoid bleaching in beverages and oxidation of oils. Relevant mathematical models reported in the literature to describe the light dependence are also shown (Manzocco et al., 2008; Manzocco et al., 2010).

energy required to activate the reaction ($E_1$) (Manzocco et al., 2008). More complex dependence has been observed in the case of oxidation of bulk oils where a power law equation was found to well fit experimental data (Manzocco et al., 2012).

These equations actually allow prediction of reaction rate as a function of the light intensity at a given temperature. However, to our knowledge, little effort has been made in the attempt to rationalize the use of light as accelerating factor in shelf life assessment. This is surprising considering the huge number of food products that suffer light-induced alteration during storage on the shelf.

## 3.3 Temperature and light

The synergic exploitation of both temperature and light as accelerating factors in ASLT has also been proposed. This approach was applied to estimate the shelf life of bulk oils and colored beverages, which are generally stored in the market on well-lit shelves. In this case, the products are stored in incubators at increasing temperature under different lighting levels and data are elaborated to develop mathematical models predicting the product shelf life as a function of the changes of both accelerating factors. For instance, by applying this method, an accurate prediction of oil shelf life at 20°C under 600 lx light intensity (which are the typical conditions experienced by the product on the market shelves) would take circa 2–3 weeks. By contrast, performing the test at 30°C under 5000 lx, the same result could be obtained in less than 2 days (Manzocco et al., 2012). This result is, however, dependent on the availability of a robust mathematical model predicting the combined effects of both accelerating factors. The latter can be produced based on data generated by monitoring the rate of the alterative event in a wide range of storage conditions. As an example, Table 6 shows the rate constants of peroxide formation in sunflower oil stored at 10°C, 20°C, or 30°C under light intensity between 0 and 8000 lx.

The effect of temperature ($T$) on the light ($L$) dependence of oxidation rate ($k$) in sunflower oil was evaluated by best fitting analysis.

$$k = \underbrace{k_{\text{ref}} e^{-E_a\left(\frac{1}{T}-\frac{1}{T_{\text{ref}}}\right)}}_{\text{Temperature dependence of reaction rate}} + \underbrace{\left(a_1 \cdot T + b_1\right) L^{\left(a_2 \cdot T + b_2\right)}}_{\text{Light dependence of oxidation rate}} \tag{16}$$

where $T_{\text{ref}}$ is chosen in the middle of the temperature range considered in the experimental plan (293.15 K, corresponding to 20°C), and $k_{\text{ref}}$, $E_a$, $a_1$, $a_2$, $b_1$, and $b_2$ are experimental parameters, as shown in Table 7.

Such a model, which was validated on external data, is particularly useful and versatile since it allows the prediction of oxidation rate of sunflower oil in different experimental conditions. In fact, if the oil is stored in the dark or in opaque packaging material ($L$=0 lx) at increasing temperatures, the model is brought back to the Arrhenius equation. By contrast, if the oil is stored at room temperature under increasing light intensity, the model describes the light dependence of oxidation rate at constant storage temperature.

**Table 6** Rate constants of peroxide formation in sunflower oil stored at 10°C, 20°C, or 30°C under light intensity between 0 and 8000 lx

| Temperature (°C) | Light intensity (lx) | Oxidation rate (meq $O_2$/$kg_{oil}$/days) |
|---|---|---|
| 10 | 0 | 0.31 |
| | 1000 | 6.00 |
| | 2700 | 5.21 |
| | 5300 | 7.68 |
| | 8000 | 8.65 |
| 20 | 0 | 0.72 |
| | 600 | 5.78 |
| | 1400 | 8.40 |
| | 2000 | 8.75 |
| | 2700 | 10.02 |
| | 5300 | 11.76 |
| | 8000 | 14.22 |
| 30 | 0 | 1.97 |
| | 1000 | 9.50 |
| | 2700 | 13.23 |
| | 5300 | 19.96 |
| | 8000 | 23.94 |

Data from Manzocco, L., Panozzo, A., Calligaris, S., 2012. Accelerated shelf life testing (ASLT) of oils by light and temperature exploitation. J. Am. Oil Chem. Soc. 89(4), 577–583.

**Table 7** Experimental parameters of the temperature and light dependence of oxidation rate in sunflower oil

| | Experimental parameter |
|---|---|
| $k_{ref}$ | 0.80 |
| $E_a$ | 64.76 |
| $a_1$ | −0.04 |
| $a_2$ | 0.016 |
| $b_1$ | 12.44 |
| $b_2$ | −4.27 |

Data elaborated from Manzocco, L., Kravina, G., Calligaris, S., Nicoli, M.C. 2008. Shelf life modelling of photosensitive food: the case of coloured beverages. J. Agric. Food Chem. 56, 5158–5164.

In order to achieve a shelf life model accounting for both the effect of light and temperature on oil oxidation, Eq. (17) was substituted in Eq. (16). The general shelf life model results as follows:

$$\mathrm{SL} = \frac{PV_{\lim} - PV_o}{k_{\mathrm{ref}} \cdot e^{-E_a\left(\frac{1}{T}-\frac{1}{T_{\mathrm{ref}}}\right)} + \left(a_1 \cdot T + b_1\right) L^{\left(a_2 \cdot T + b_2\right)}} \tag{17}$$

This example relevant to sunflower oil shows that consideration of both temperature and lightning conditions in the shelf life model allows to concomitantly reach two apparently incompatible objectives: accuracy and rapidity of prediction.

In this context, the analysis of the magnitude of thermal and electromagnetic activation energy of photosensitive foods could represent a valid tool to support the operator in selecting the accelerating factor providing the utmost advantage in terms of time saved during the shelf life test. This is certainly an open research field, which will require more attention from food scientists and researchers in the future.

# 4 Examples of the application of the basic procedure of ASLT

The application of the basic procedure to obtain shelf life data by the application of the Arrhenius equation will be illustrated step by step in this section. Data reported are original figures obtained in our shelf life testing laboratory.

## 4.1 Example 1

The challenge was to develop a mathematical model allowing estimation of the shelf life at 20°C of a yellow beverage by exploiting data relevant to its alteration kinetics under accelerated conditions. The beverage had a low pH, around 3, and was ambient stable. The beverage color was due to the presence in the formulation of β-carotene, which is known to undergo oxidative reactions during storage, leading to fading of the typical yellow color of the product.

In order to perform the shelf life test under accelerated conditions, the beverage was stored at 30°C, 40°C, 50°C, and 60°C for up to 8 months. Table 8 reports data relevant to the color changes of the beverage during storage at the selected temperatures. Color modifications were acquired by measuring the hunter color parameter $b^*$ and were expressed as percentage decrease as compared to the color of the just prepared beverage. First of all, experimental data may be plotted by reporting the changes of the indicator ($I$) (Fig. 9A) and of the natural logarithm of the indicator ($\ln I$) (Fig. 9B) as a function of storage time. These two graphical representations allow visually evaluation

**Table 8** Dataset of % of color changes as a function of storage time at different temperatures (original data)

| | Color bleaching (%) | | | |
|---|---|---|---|---|
| **Time (months)** | **30°C** | **40°C** | **50°C** | **60°C** |
| 0.0 | 100 | 100 | 100 | 100 |
| 0.2 | 97 | 98 | 95 | 95 |
| 0.6 | 95 | 90 | 88 | 82 |
| 2.0 | 90 | 84 | 77 | 63 |
| 3.0 | 82 | 73 | 65 | 39 |
| 4.0 | 77 | 66 | 51 | 26 |
| 6.0 | 72 | 60 | 41 | 17 |
| 8.0 | 66 | 52 | 34 | 10 |

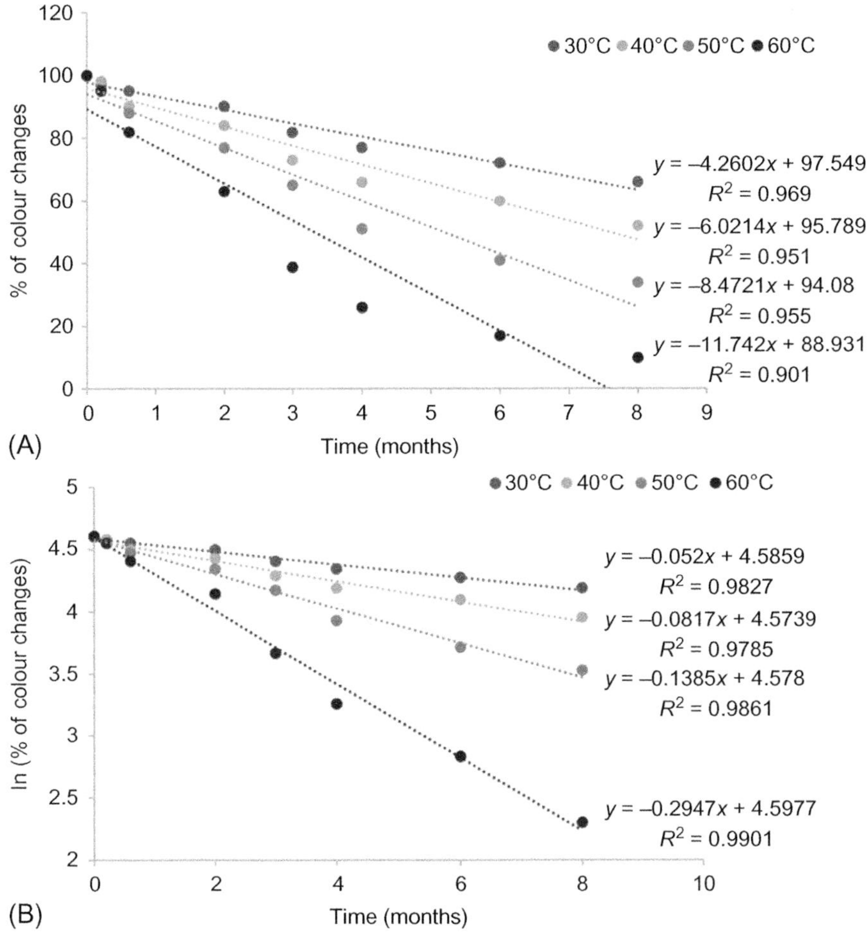


**Fig. 9** Beverage color bleaching as a function of storage time considering a zero order reaction kinetics (A) and first order reaction kinetics (B).

of experimental data and indicatively attributing the development of the deteriorative event to a zero or first reaction order. The reaction order should be then confirmed by applying linear regression analysis on reported data. The results of the linear regression analysis and goodness of fit are also reported in the Fig. 9. The rate constants ($k$) considering the zero and first order reaction are represented by the value of the slope of the linear equations. To define the best model to be used in the description of the evolution of beverage color during storage, the value of the determination coefficient ($R^2$) can be used to discriminate between the models. In the experimental case here considered, independently of temperature, the first order model was the one allowing reduction of estimation errors, as highlighted by the higher value of $R^2$ in comparison to those corresponding to the zero order model. Thus, first order reaction rates were further applied in the application of the Arrhenius equation.

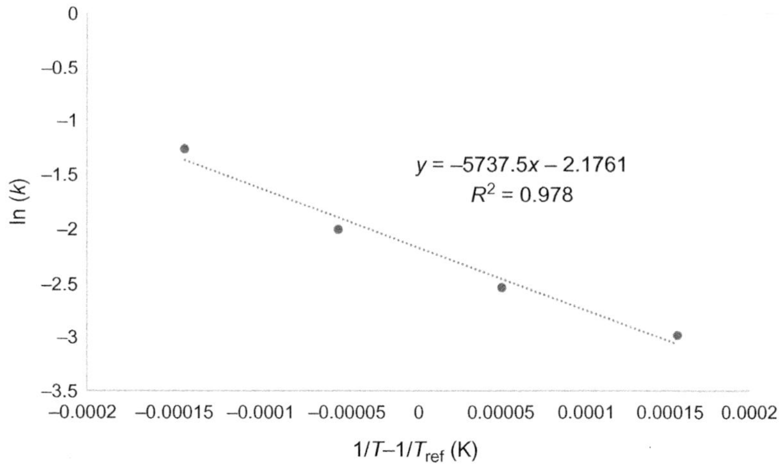


**Fig. 10** Rate constant, expressed in logarithmic value, as a function of the reciprocal of temperature. The relevant Arrhenius equation is also reported.

The estimated values of the first order rate constants may now be used to compute the parameters of the Arrhenius equation. To do this, a plot of the logarithmic values of reaction rate ($\ln k$, measured in month$^{-1}$) versus the reciprocal value of the absolute temperature ($1/T$, measured in K) (Fig. 10) is needed to evaluate if the Arrhenius equation is appropriate to describe the relation between the rate constant and temperature.

To apply the methodology of reparametrization (Eq. 9), temperature data should be converted in the following way:

$$\left(\frac{1}{T}-\frac{1}{T_{ref}}\right) \tag{18}$$

where $T_{\text{ref}}$ is set equal to the mean value of the temperatures tested (in our case, 45°C). Then, parameters of the Arrhenius equation may be estimated with linear regression analysis (Fig. 10).

The mathematical model predicting reaction rate as a function of temperature according to Arrhenius equations results as follows:

$$k = k_{\text{ref}} \exp\left[-\frac{E_a}{R}\left(\frac{1}{T^*}-\frac{1}{T_{\text{ref}}}\right)\right] \tag{19}$$

where $k_{\text{ref}}$ and $E_a$ must be substituted by the corresponding estimates and $T^*$ is the temperature at which to predict shelf life.

By applying the obtained Arrhenius equation (Fig. 10), it is possible to estimate the values of $k$ at the temperature of interest: in the example 20°C, thus 293 K. The $k_{20°\text{C}}$ resulted −0.024 months$^{-1}$. From the equation it can be also estimated the $E_a$ of the reaction that resulted 47.67 kJ/mol.

**Table 9** Shelf life values estimated by using reaction rates reported in Fig. 9 at different temperatures

| Temperature (°C) | Estimated shelf life (months) |
|---|---|
| 20 | 28.7 |
| 30 | 13.0 |
| 40 | 8.1 |
| 50 | 4.4 |
| 60 | 1.7 |

Given that a first order kinetic equation has been used, shelf life at 20°C may be estimated accordingly to the following equation:

$$\mathrm{SL} = \frac{\ln I_0 - I_{\mathrm{lim}}}{k_{\mathrm{ref}} \exp\left[-\frac{E_a}{R}\left(\frac{1}{T^*} - \frac{1}{T_{\mathrm{ref}}}\right)\right]} \tag{20}$$

where $I_0$ is the experimental value of $I$ at time zero.

Considering the shelf life equation reported in Fig. 3 for a first order reaction model, and assuming as acceptability limit a color bleaching equal to 50%, the shelf life can be estimated as follows:

$$\mathrm{SL} = \frac{\ln\left(I_0 - I_{\mathrm{lim}}\right)}{k_{20^\circ\mathrm{C}}} = \frac{\ln(100) - \ln(50)}{0.024} = 28.7\,\mathrm{months} \tag{21}$$

It should be considered that different sources of uncertainty affect the final shelf life value, begetting a very wide confidence interval. Its computation may be very demanding from the mathematical point of view and is generally not performed in common practice.

By using the kinetics data reported in Fig. 9 and Eq. 21, it is also possible to calculate the shelf life at the different temperatures considered in the stability test (Table 9). These data can be further used to generate a shelf life model describing the product shelf life as a function of temperature (Fu and Labuza, 1993). Fig. 11 shows the so called shelf life plot. If Arrhenius is fulfilled, a linear equation describing the temperature dependence of SL can be obtained:

$$\ln \mathrm{SL} = -bT + c \tag{22}$$

where SL is the shelf life, $T$ is the temperature expressed in °C, and $b$ and $c$ are experimental parameters corresponding to the slope and the intercept of the regression line, respectively.

At the end of the described process to generate shelf life data, a validation at 20°C is highly recommended to avoid possible errors associated with the specific criticism of the product.

Although the time required to set up the model was rather long (about 8 months), the equations developed can be practically used on a routine basis as simple tools to

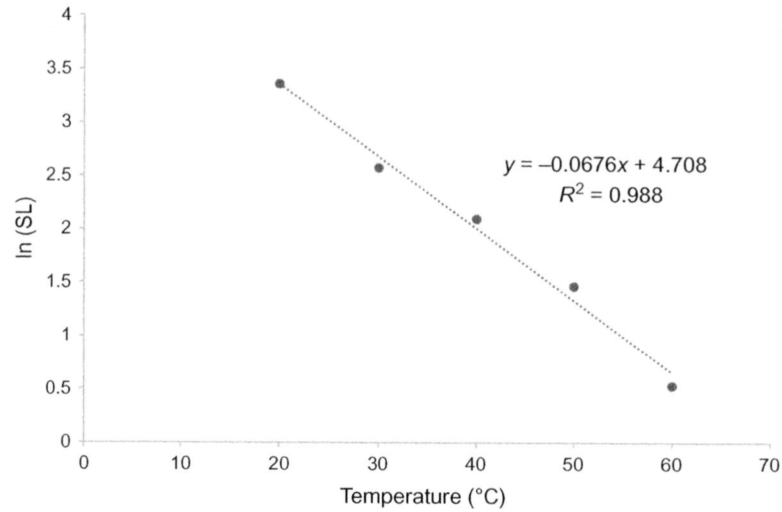


**Fig. 11** Shelf life plot of the yellow beverage.

estimate the shelf life of analogous products in the entire temperature range considered. In other words, once set up, the model only requires periodic validation to verify that minor changes in formulation or process conditions do not modify the reaction order or the activation energy of the alterative phenomena. By contrast, the reliability of such a model might be dramatically impaired when major changes in the beverage production process are carried out. In such cases, the model structure and parameter estimates should be carefully reconsidered by performing further accelerated tests on the product prepared according to the novel processing conditions.

## 4.2 Example 2

The challenge was to estimate the shelf life at 20°C of glucose syrup, intended as an ingredient for different formulations of sweet products. In this case, the shelf life limiting factor was the occurrence of crystallization phenomena: when sugar crystallization occurs, the product is no longer considered acceptable.

As is well known, crystallization phenomena of sugars can be favored by decreasing storage temperature. Based on this consideration, the acceleration of the phenomena leading to the quality depletion of the glucose syrup was obtained by reducing storage temperature. Specifically, the product was stored at 5°C, 10°C, and 15°C for up to 150days. The extent of crystallization was assessed by evaluating the crystal formation by microscopy analysis. Table 10 reports the percentage of crystallization during storage of the glucose syrup at the selected temperatures. The graphical representation of these data is shown in Fig. 12. The plot of the percentage of crystallization as a function of storage time clearly shows that the changes in the selected shelf life indicator evolved according to a complex behavior. In this case, the time required for crystallization to be initially observed (i.e., the induction period before crystallization)

**Table 10** Percentage of crystallization of glucose syrup during storage at different temperatures

| Storage time (days) | Percentage of crystallization (%) | | |
|---|---|---|---|
| | 5°C | 10°C | 15°C |
| 0 | 0 | 0 | 0 |
| 5 | 0 | 0 | 0 |
| 10 | 0 | 0 | 0 |
| 15 | 0 | 0 | 0 |
| 20 | 0 | 0 | 0 |
| 25 | 0 | 0 | 0 |
| 30 | 2 | 0 | 0 |
| 40 | 10 | 0 | 0 |
| 50 | 20 | 0 | 0 |
| 60 | 30 | 3 | 0 |
| 70 | 40 | 9 | 0 |
| 80 | 50 | 20 | 0 |
| 90 | 60 | 30 | 0 |
| 100 | 70 | 40 | 0 |
| 110 | 80 | 50 | 0 |
| 120 | 90 | 60 | 5 |
| 150 | 100 | 70 | 10 |

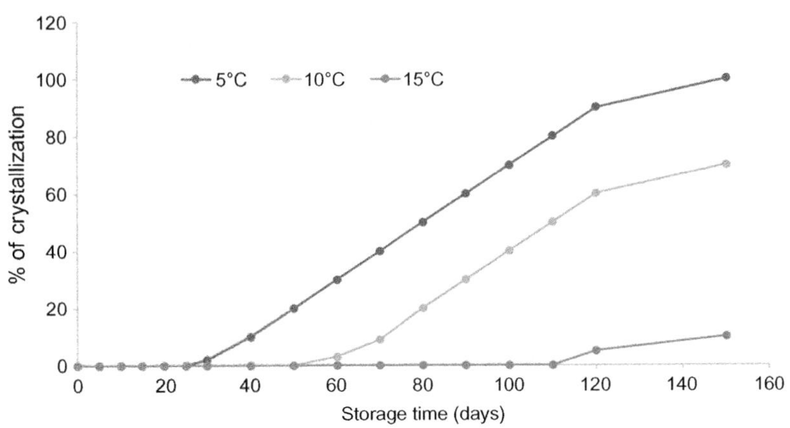


**Fig. 12** Percentage of crystallization of a syrup as a function of storage time at 5°C, 10°C, and 15°C.

is the shelf life determining factor. Thus, instead of studying the crystallization kinetics in terms of rate constant, the attention was focused on the extent of the lag phase before glucose crystallization (Table 11). In particular, the lag phase of crystallization (expressed in days) was computed, and these data were used to estimate the temperature dependence of glucose syrup crystallization.

**Table 11** Lag phase of crystallization of glucose syrup stored at 5°C, 10°C, and 15°C

| Temperature (°C) | Lag phase (days) |
|---|---|
| 5 | 25 |
| 10 | 50 |
| 15 | 110 |

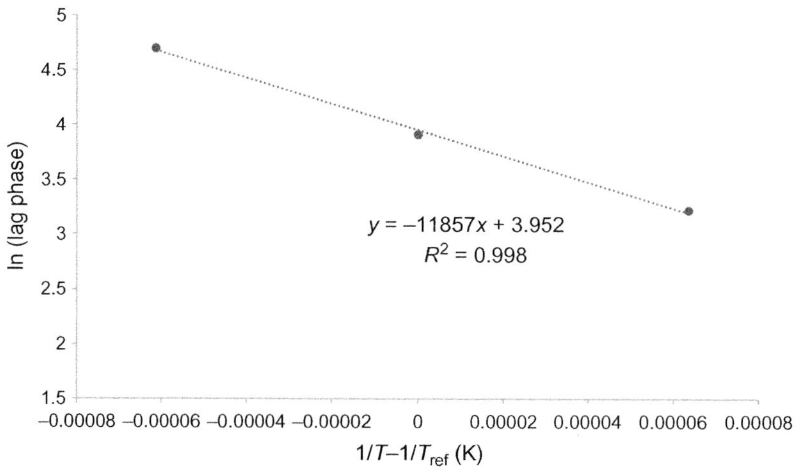


**Fig. 13** Lag phase, expressed in logarithmic value, as a function of the reciprocal of temperature. The relevant Arrhenius equation is also reported.

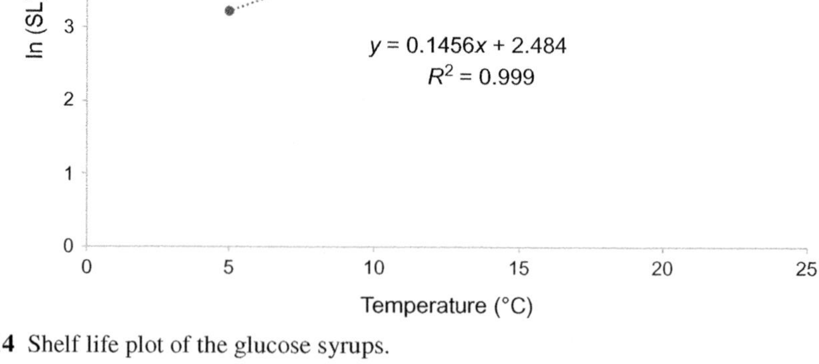


**Fig. 14** Shelf life plot of the glucose syrups.

To this aim, the logarithmic values of the lag phase of crystallization were graphically represented as a function of the reciprocal value of temperature (Fig. 13). It can be noted that a linear relationship between lag phase and absolute temperature was observed. Regression analysis of this data was thus performed according to the reparametrized Arrhenius equation, with the reference temperature set in correspondence at 10°C. This analysis produced the parameter estimates reported in Fig. 13. From the equation, it can be also estimated that the $E_a$ of glucose crystallization in the syrup resulted 95.5 kJ/mol. Finally, lag phase at 20°C, and thus, product shelf life can be estimated by using the Arrhenius equation, resulting in 217 days. Fig. 14 reports the shelf life plot relevant to the glucose syrup case study.

# 5 Conclusions and future need

Actual literature information provides limited help to solve the problem of shelf life determination under accelerated conditions. Although many environmental factors can be virtually exploited to accelerate alterative phenomena, temperature is still the only one that is extensively used for this purpose. Company managers dealing with the shelf life issue covet an accurate and easily applicable tool allowing shelf life prediction based on sample storage at temperatures higher than those experienced on the shelves. Such a tool requires the exact knowledge of the activation energy of the alterative phenomena. As discussed in this present chapter, this information is rarely available and more effort should be made to produce a comprehensive database of activation energy data for classes of alterative events and food products. Even more research efforts are needed to turn the potentiality of exploitation of accelerating factors other than temperature into real ASLT testing protocols.

The generation of predictive models to assess shelf life is certainly an arduous task but provides valuable advantages allowing the reduction of time, cost, and labor required for estimation of shelf life of long-lasting foods.

# References

An, D.S., Lee, J.H., Lee, D.S., 2018. Shelf life model of powdered infant formula as a function of temperature and oxygen concentration. Food Packag. Shelf Life 15, 130–133.

Anese, M., Lanciotti, R., Gardini, F., Lagazio, C., 2012. Case studies. In: Nicoli, M.C. (Ed.), Shelf Life Assessment of Food. CRC Press, Boca Raton, pp. 247–294.

Arrhenius, S.A., 1901. Larobok I Teoretisk Elektrokeni. Quando&Handel, Leipzig.

Barba, L., Arrighetti, G., Calligaris, S., 2013. Crystallization and melting properties of extra virgin olive oil studied by synchrotron XRD and DSC. Eur. J. Lipid Sci. Technol. 115, 322–329.

Breitschuh, B., Windhab, E.J., 1998. Parameters influencing co-crystallization and polymorphism in milk fat. J. Am. Oil Chem. Soc. 75 (8), 897–904.

Burdurlu, H.S., Karadeniz, F., 2003. Effect of storage on nonenzymatic browning of apple juice concentrates. Food Chem. 80 (1), 91–97.

Burdurlu, H.S., Koca, N., Karadeniz, F., 2006. Degradation of vitamin C in citrus juice concentrates during storage. J. Food Eng. 74, 211–216.

Buvè, C., van Betds, T., Haenen, A., Kebede, B., Braekers, R., Hendrickx, M., Van Loey, A., Grauwet, T., 2017. Shelf life dating of shelf stable strawberry juice based on survival analysis of consumer acceptance information. J. Sci. Food Agric. 98, 3437–3445.

Calligaris, S., Manzocco, L., 2012. In: Nicoli, M.C. (Ed.), Critical Indicators in Shelf Life Assessment. CRC Press, Boca Raton, pp. 61–73.

Calligaris, S., Manzocco, L., Conte, L.S., Nicoli, M.C., 2004. Application of a modified Arrhenius equation for the evaluation of oxidation rate of sunflower oil at subzero temperatures. J. Food Sci. 69, 361–366.

Calligaris, S., Sovrano, S., Manzocco, L., Nicoli, M.C., 2006. Influence of crystallisation on the oxidative stability of extra virgin olive oil. J. Agric. Food Chem. 54 (2), 529–535.

Calligaris, S., Manzocco, L., Kravina, G., Nicoli, M.C., 2007a. Shelf-life modeling of bakery products by using oxidation indexes. J. Agric. Food Chem. 55, 2004–2009.

Calligaris, S., Manzocco, L., Nicoli, M.C., 2007b. Modelling the temperature dependence of oxidation rate in water-in-oil emulsions stored at sub-zero temperatures. Food Chem. 101, 1019–1024.

Calligaris, S., Da Pieve, S., Kravina, G., Manzocco, L., Nicoli, M.C., 2008a. Shelf-life prediction of bread sticks by using oxidation indices: a validation study. J. Food Sci. 73 (2), E51–E56.

Calligaris, S., Barba, L., Arrighetti, G., Nicoli, M.C., 2008b. Phase transition of sunflower oil as affected by the oxidation level. J. Am. Oil Chem. Soc. 85, 591–598.

Calligaris, S., Manzocco, L., Lagazio, C., 2012. Modeling chemical, physical and sensory indicators. In: Nicoli, M.C. (Ed.), Shelf Life Assessment of Food. CRC Press, Boca Raton, pp. 75–126.

Calligaris, S., Manzocco, L., Anese, M., Nicoli, M.C., 2016. Shelf-life assessment of food undergoing oxidation—a review. Crit. Rev. Food Sci. Nutr. 56, 1903–1912.

Cappuccio, R., Full, G., Lonzarich, V., Savonitti, O., 2001. Staling of roasted and ground coffee at different temperature. Combining sensory and GC analysis. In: Proceeding of the 19th International Scientific Colloquium on Coffee. CD ROM, Trieste, Italy.

Champion, D., Blond, G., Simatos, D., 1997. Reaction rates at sub-zero temperatures in frozen sucrose solution: a diffusion controlled reaction. Cryo-Letters 18, 251–260.

Corradini, M.G., Peleg, M., 2007. Shelf life estimation from accelerated data. Trends Food Sci. Technol. 18, 37–47.

Dermesonlouoglou, E.K., Giannakourou, M.C., Taoukis, P.S., 2007. Kinetic modelling of the degradation of quality of osmo-dehydrofrozen tomatoes during storage. Food Chem. 103, 985–993.

EC (European Commission), 2011. Regulation no. 1169/2011 of the European Parliament and of the council of 28 January 2002. Off. J. Eur. Communities, L304/18, 22/11/2011. 1–46.

Echavarria, A.P., Pagan, J., Ibarz, A., 2012. Melanoidins formed by Maillard reaction in food and their biological activity. Food Eng. Rev. 4, 203–223.

Fennema, O., 1996. Freezing preservation. In: Fennema, O. (Ed.), Physical Principles of Food Preservation. Marcel Dekker Inc, New York, pp. 173–215.

Frankel, E.N., 2005. Stability methods. In: Frankel, E.N. (Ed.), Lipid Oxidation. The Oil Press, Davis, pp. 165–186.

Fu, B., Labuza, T.P., 1993. Shelf-life prediction: theory and application. Food Control 4 (3), 125–133.

Garitta, L., Langohr, K., Elizagoyen, E., Ottaviano, F.G., Gomez, G., Hough, G., 2018. Survival analysis model to estimate sensory shelf life with temperature and illumination as accelerating factors. Food Qual. Prefer. 68, 371–376.

Guerra, S., Lagazio, C., Manzocco, L., Barnabà, M., Cappuccio, R., 2008. Risks and pitfalls of sensory data analysis for shelf life prediction: data simulation applied to the case of coffee. LWT Food Sci. Technol. 41, 2070–2078.

Hough, G., Garitta, L., Gómez, G., 2006. Sensory shelf-life predictions by survival analysis accelerated storage models. Food Qual. Prefer. 17, 468–473.

Huang, J., Sathivel, S., 2008. Thermal and rheological properties and effects of temperature on the viscosity and oxidation rate of unpurified salmon oil. J. Food Eng. 89, 105–111.

Jafari, S.M., Ganje, M., Dehnad, D., Ghanbari, V., Hajitabar, J., 2017. Arrhenius equation modelling for shelf life prediction of tomato paste containing a natural preservative. J. Sci. Food Agric. 97, 5216–5222.

Kristensen, D., Hansen, E., Arndal, A., Tinderup, R.A., Skibsted, L.H., 2001. Influence of light and temperature on the colour and oxidative stability of processed cheese. Int. Dairy J. 11, 837–843.

Labuza, T.P., Riboh, D., 1982. Theory and application of Arrhenius kinetics to the prediction of nutrient loss in food. Food Technol. 10, 66–74.

Labuza, T.P., Schmidl, M.K., 1985. Accelerated shelf life testing of foods. Food Technol. 39, 57–64.

Li, X., Wang, S.C., 2018. Shelf life of extra virgin olive oil and its prediction models. J. Food Qual. 2018, 1639260.

Manzocco, L., 2016. The acceptability limit in food shelf life studies. Crit. Rev. Food Sci. Nutr. 56 (10), 1640–1646.

Manzocco, L., Calligaris, S., Nicoli, M.C., 2006. Modeling bleaching of tomato derivatives at subzero temperatures. J. Agric. Food Chem. 54 (4), 1302–1308.

Manzocco, L., Kravina, G., Calligaris, S., Nicoli, M.C., 2008. Shelf life modelling of photosensitive food: the case of coloured beverages. J. Agric. Food Chem. 56, 5158–5164.

Manzocco, L., Panozzo, A., Calligaris, S., 2012. Accelerated shelf life testing (ASLT) of oils by light and temperature exploitation. J. Am. Oil Chem. Soc. 89 (4), 577–583.

Mizrahi, S., 2011. Accelerated shelf-life tests. In: Kilcast, D., Subramaniam, P. (Eds.), Food and Beverage Stability and Shelf Life. Woodhead Publishing Limited, Cambridge, pp. 482–507.

Mizrahi, S., Labuza, T.P., Karel, M., 1970. Feasibility of accelreated tests for browning in dehydrated cabbage. J. Food Sci. 35, 804–807.

Nicoli, M.C., 2012. The shelf life assessment process. In: Nicoli, M.C. (Ed.), Shelf Life Assessment of Food. CRC Press, Boca Raton, pp. 17–36.

Nicoli, M.C., Calligaris, S., 2018. Secondary shelf life: an underestimated issue. Food Eng. Rev. 10 (2), 57–65.

Odriozola-Serrano, I., Soliva-Fortuny, R., Martin-Belloso, O., 2009. Influence of storage temperature on the kinetics of the changes in anthocyanins, vitamin C, and antioxidant capacity in fresh-cut strawberries stored under high-oxygen atmospheres. J. Food Sci. 74, C184–C191.

Orlien, V., Risbo, J., Rantanen, H., Skibsted, L.H., 2006. Temperature-dependence of rate of oxidation of rapeseed oil encapsulated in a glassy matrix. Food Chem. 94, 37–46.

Parker, R., Ring, S.G., 1995. A theoretical analysis of diffusion-controlled reactions in frozen solution. Cryo-Letters 16, 197–208.

Peleg, M., Normand, M.D., Corradini, M.G., 2012. The Arrhenius equation revisited. Crit. Rev. Food. Sci. Nutr. 52, 830–851.

Peleg, M., Normand, M.D., Dixon, W.R., Goulette, T.R., 2018. Modelling the degradation kinetics of ascorbic acid. Crit. Rev. Food. Sci. Nutr. 9, 1478–1494.

Pereyra Gonzales, A.S., Naranjo, G.B., Leiva, G.E., Malec, L.S., 2010. Maillard reaction kinetics in milk powder: effect of water activity and mild temperatures. Int. Dairy J. 20, 40–45.

Rao, Q., Rocca-Smith, J.R., Schoenfuss, T.C., Labuza, T.P., 2012. Accelerated shelf life testing of quality loss for commercial hydrolysed hen egg with powder. Food Chem. 135, 464–472.
Robertson, G.L., 2009. Food packaging and shelf life. In: Robertson, G.L. (Ed.), Food Packaging and Shelf Life: A Practical Guide. CRC Press, Taylor & Francis Group, Boca Raton, pp. 1–16.
Roos, Y.H., 1995. Phase Transition in Foods. Academic Press, San Diego.
Schmitz-Schug, I., Kulozik, U., Foerst, P., 2014. Reaction kinetics of lysine loss in a model dairy formulation as related to the physical state. Food Bioprocess Technol. 7, 877–886.
Sithole, R., McDaniel, M.R., Meunier Goddik, L., 2004. Rate of Maillard reaction in sweet whey powder. J. Dairy Sci. 88, 1636–1645.
Sothornvit, R., Kiatchanapaibul, P., 2009. Quality and shelf-life of washed fresh-cut asparagus in modified atmosphere packaging. LWT- Food Sci. Technol. 42, 1484–1490.
Tan, C.P., Che Man, Y.B., 2000. Differential scanning calorimetric analysis of edible oils: comparison of thermal properties and chemical compostion. J. Am. Oil Chem. Soc. 77 (2), 143–155.
Tan, C.P., Che Man, Y.B., 2002. Differential scanning calorimetric analysis of palm oil, palm oil based products and coconut oil: effect of scanning rate variation. Food Chem. 76, 89–102.
Tazi, S., Plantevin, F., Di Falco, C., Puigserver, A., Ajandouz, E.H., 2009. Effect of light, temperature and water activity on kinetics of lipoxidation in almond-based products. Food Chem. 15, 958–964.
Terefe, N.S., van Loey, A., Hendrickx, M., 2004. Modelling the kinetics of enzyme-catalysed reactions in frozen systems: the alkaline phosphatase catalysed hydrolysis of di-sodium-p-nitrophenyl phosphate. Innov. Food Sci. Emerg. Technol. 5, 335–344.
Thomson, M.K., Lauridsen, L., Skibsten, L.H., Risbo, J., 2005. Temperature effect on lactose crystallization, maillard reaction and lipid oxidation in whole milk powder. J. Agric. Food Chem. 53, 7082–7090.
Topuz, A., 2008. A novel approach for color degradation kinetics of paprika as a function of water activity. LWT Food Sci. Technol. 41, 1672–1677.
Toro Vasquez, J.F., Briceno-montelongo, M., Dibildox-Alvarado, E., Charò-Alonso, M., Reyes-hernadez, J., 2000. Crystallization kinetics of palm stearin in blends with sesame oil. J. Am. Oil Chem. Soc. 77 (3), 297–309.
Vaikousi, H., Koutsoumanis, K., Biliaderis, C.G., 2009. Kinetic modelling of non-enzymatic browning in honey and diluted honey systems subjected to isothermal and dynamic heating protocols. J. Food Eng. 95, 541–550.
Van Boekel, M.A.J.S., 2008. Kinetic modelling of food quality: a critical review. Compr. Rev. Food Sci. Food Saf. 7, 144–158.
Van Boekel, M.A.J.S., 2009. Kinetic Modelling of Reactions in Foods. Taylor and Francis, Boca Raton.
Wang, B., Xiao, L., Jiang, L., Li, B., Quian, P., 2015. Evaluation of accelerated test factors through the development of predictive models in vacuum-packed compressed biscuits. Food Anal. Methods 8, 1618–1628.
Waterman, K.C., Adami, R.C., 2005. Accelerating testing: prediction of chemical stability of pharmaceuticals. Int. J. Pharm. 293, 101–125.
Zamora, R., Hidalgo, F., 2005. Coordinate contribution of lipid oxidation and Maillard reaction to the nonenzymatic food browning. Crit. Rev. Food Sci. Nutr. 45, 49–59.

## Further reading

Williams, M.L., Landel, R.F., Ferry, J.D., 1955. The temperature dependence of relaxation mechanisms in amorphous polymers and other glass-forming liquids. J. Am. Oil Chem. Soc. 77, 3701–3706.

# Index

Note: Page numbers followed by *f* indicate figures and *t* indicate tables.

## Q



## R

CPI Antony Rowe
Eastbourne, UK
November 20, 2019

to map chemical variability in bare soils in elemental concentrations of soils amended with biosolids, and in to use LANDSAT TM data to map these elemental concentrations of the soils when they are not covered by vegetation. 4) In **paperless solution for the collection, management and dissemination of marine data** were addressed many of the problems related to the complexity of the logical linkages between the different types of data in fisher management, Olrac (www.olrac.com) a South African company, that has developed a data collection and management system it has named Olfish (www.olfish.com) for the specific use of operators and managers in the marine environment with a special focus on the commercial fishing industry. 5) In **Vegetation mapping and management** was studied the vegetation types in the semiarid to arid region of Mond Protected Area, south-west Iran, based on unsupervised classification of the Spot XS bands and then produced updated maps. Based on field observation and supported by satellite maps, three major habitat zones in the study area, namely, the coastal, riverine and inland zones, were recognized. Twelve vegetation types were recognized in the field that showed a good compatibility with the satellite image. 6) **Earth to space link** describes how to calculate link-power budget in order to relate to two quantities, the transmitted power and the received power, and show in detail how the difference between these two powers is accounted. To arrive at an accurate answer, factors such as the uplink power amplifier gain and noise factors, transmit antenna gain, slant angles and corresponding atmospheric loss over distance and climatic attenuation factors must be taken into account.

**In the fourth issue** Optimization problems and develop of new technologies are presented with ten chapters: 1) **Guidelines for satellite tracking** raise the need to know where to look for the artificial and natural satellites in the evening sky has been the obsession for many. So that mathematical models and positioning mechanisms were explained in this chaper to paint the picture of satellite tracking and give the brief insight in when to use what mathematical model to pinpoint the object in sky, even if the final goal is just a view through a telescope. 2) **Interference studies in cellular satellite systems** accounts to reduce the array antenna aperture on board the satellite for multiple-spot-beam cellular coverage. The reduction in the antenna aperture with this approach translates into significant reductions in number of array elements, RF components, and A/D and D/A converters. Analysis has shown that in spite of the smaller aperture and the broader beams of the sub-beams, the co-channel interference between sub-beams using the same frequency segment is not adversely affected. 3) In **Ultra-life cycle utilization of GEO** have proposed and verified practical development plan for satellites navigation based on GEO satellites. Particularly was introduced one kind of satellite communication techniques with micro-terminal low-information rate developed in this section. This technique can satisfy some communications and positioning requirements such as unattended measurement and short-message emergency communications. 4) **Planar antennas for satellite communications** was devoted to planar antennas not only for that already in use but proposing other kind that could be used for satellite applications. In this way was described actual planar antennas used in satellite communications systems and finished with a proposal of new developments of planar antennas that authors think could be used in the near future. 5) **Power and spectral efficient multiuser broadband wireless communication system** focuses on different aspects of the communication system, namely PAPR reduction and power control in transmitter, channel estimation for design of adaptive and optimized system, multiuser detection at the receiver for increase in user capacity. *The newly proposed CI/MC-CDMA system discussed in this chapter may become an efficient multiple access scheme for*

*Satellite communication used for long distance broadband signal transmission in conjunction with cellular system.* 6) In **Quantum based information transfer** was introduced the quantum communication to help to establish a secure communication link, and present solutions with zero redundancy error correction. 7) **Satellite coverage optimization problems with shaped reflector antennas** was devoted to optimization formulations of satellite coverage problems and their solution which are specially useful in satellite broadcasting applications, and where the information goes from one to many. 8) In **Satellite Laser Communication** is used a cloud attenuation model derived from key Italsat results to find laser cloud attenuation for satellite-ground links with gaseous attenuation included for 2-10 micron attenuation. Link availability is raised with suitable Northern Latitude satellites, such as Earth Observation Satellites. Soviet cloud correlation results indicate that link availability would be raised to acceptable levels with ground sites separated by 100-290 km. 9) In **Satellite motion** is described a little history on the satellite motion and its mathematical formulation. This includes the dimensions of orbit and the position of satellite on its orbits, and what velocity for a satellite is requested.
10) **System aspects of active phased arrays** concluded this book by reviewing early developments of phased arrays for multiple-beam satellite communications applications. A key component in these developments is the modular monolithic microwave integrated circuit (MMIC) beam-forming matrices that generate a number of simultaneous and independently digitally controlled beams.

Editor

**Nazzareno Diodato**
*MetEROBS – Met European Research Observatory, GEWEX-CEOP Network,*
*World Climate Research Programme, via Monte Pino snc, 82100 Benevento*
*Italy*

1

# About QoS in DVB-S2/RCS Systems

Baptiste Jacquemin[1,2], Pascal Berthou[1,2],
Thierry Gayraud[1,2] and Lionel Bertaux[1,2]
[1]*CNRS; LAAS; 7 avenue du Colonel Roche, F-31077 Toulouse, France*
[2]*Université de Toulouse; UPS, INSA, INP, ISAE; LAAS; F-31077 Toulouse, France*

## 1. Introduction

The standardization of a Return Channel via Satellite and the satellite community efforts in term of interoperability over the last few years stands for major milestones in the development of reliable, efficient and low cost satellite equipments. It leads to quite a positive outcome: geostationary satellite networks are expected to play a decisive role in bridging the existing digital divide through providing broadband access to multimedia services in low terrestrial infrastructure areas.
However, unlike cable or 3GPP access networks, a lot of work on IP over satellite has been needed, especially about Quality of Service (QoS). The QoS architecture takes benefits from DVB-RCS dynamic allocation schemes and IP QoS architecture to cope with the satellite delay and the scarce uplink resources. This chapter deals with design and evaluation of Quality of Service Architecture to be implemented in DVB-S2/RCS systems.

The first section of this chapter aims at introducing DVB-S2/RCS Systems.
The long term efforts to optimize the DVB-S standard to lower the price of satellite access networks led to a new evolution of the standard: DVB-S2. A better encapsulation mechanism of IP packets and a new adaptative transmission scheme are the main concerns for the QoS architecture.
The encapsulation of IP packets in DVB-S has always been a complex problem. This section presents the evolution of the standard from the Multiprotocol Encapsulation (MPE) and the Ultra Lightweight Protocol (ULE) to the Generic Stream Encapsulation (GSE).
The Adaptive Coding and Modulation (ACM) technique that increase the network efficiency according to the weather conditions is a major evolution. The variable transmission rate impacts the QoS management and offers new perspectives for future system evolution.

DVB-S Satellite Terminals can only receive frames from the satellite. The need for a return link rapidly becomes essential so as to support emerging Internet services via satellite.
The return link access scheme in DVB-S/RCS systems is MF-TDMA. The return link is segmented into portions of time and frequency ("superframes. A Network Control Center (NCC) performs the entire satellite system control, especially Satellite Terminals synchronization and resource allocation. It periodically broadcasts a signaling frame, the TBTP (Terminal Burst Time Plan), which updates the timeslot allocation within a

superframe between every competing ST. This allocation can be dynamically modified on STs demand thanks to a bandwidth on demand protocol called Demand Assignment Multiple Access (DAMA). This system is presented here.

The next section rapidly overviews the concepts and mechanisms of Quality of Service management in basic architectures such as IETF Intserv and IETF / Diffserv. Others mechanisms such as Traffic shaping / conditioning, SLA, Scheduling and Admission control that have a main impact on the QoS are also described.

The next part aims at describing what means QoS in satellite networks thanks to the DVB-S2/RCS QoS Architecture example.
From the very first system only based on MPLS, a first architecture based on Diffserv was proposed. It was then enhanced to better fit to the DVB-RCS system in the IST project Satsix.

The next part will answer a main question related to the satellite networking systems that is: How to develop new services with Satellite Systems?
Based on our research work and results in the field, we'll explain how to use Simulation (using NS-2 or NS-3), Basic Emulation (using Linux TC/Simnet) and Advanced Emulation testbed like the one that was developed in various projects we were involved in. And we'll conclude that part with our skill on Real Deployed Systems.

The last part deals with Performance Evaluation of the described proposals. We first evaluated DVB-S/RCS NS-2 emulation model with QoS. The next way used to evaluate the proposed architecture was done through the PLATINE emulation testbed coming as the main result of the Satsix project. Our last experiment was done in the OURSES project, labeled in the Aerospace Valley research center with the following main devices from Astra (satellite), Thales Alenia Space (Gateway) and Advantech (Satellite Terminal).

To conclude this chapter, results summary and lessons learned will be given and future work will be described.

## 2. DVB-S2/RCS main features

### 2.1 First overview of DVB-S/RCS systems

Started in 1993, the international European DVB Project published, in the end-nineties, a family of digital transmission specifications, based upon MPEG-2 (Motion Picture Expert Group) video compression and transmission techniques. Data are transported within MPEG-2 transport streams (MPEG2-TS) which are identified through DVB Service Information Tables. Adapted for satellite systems, DVB-S defines one of the most widespread formats used for Digital TV over the last years and still nowadays. However, DVB-S Satellite Terminals can only receive frames from the satellite. The need for a return link rapidly becomes essential so as to support emerging Internet services via satellite, leading to 3 solutions:

- UDLR (UniDirectional Link Routing) which emulates a cheap bidirectional solution through a terrestrial return link,
- DVB-S system with low speed terrestrial return link,

– DVB-RCS, which provides a full bidirectional satellite architecture [Fig. 1].

A good overview of DVB-S/RCS satellite networks architecture is given in Fig. 1, compliant with the architecture adopted within the ETSI BSM [3] group and the DVB-RCS standards. It consists in a geostationary satellite network with Ka MF-TDMA (Multiple Frequency Time Division Multiple Access) uplinks and Ku TDM (Time Division Multiplexed) downlinks.

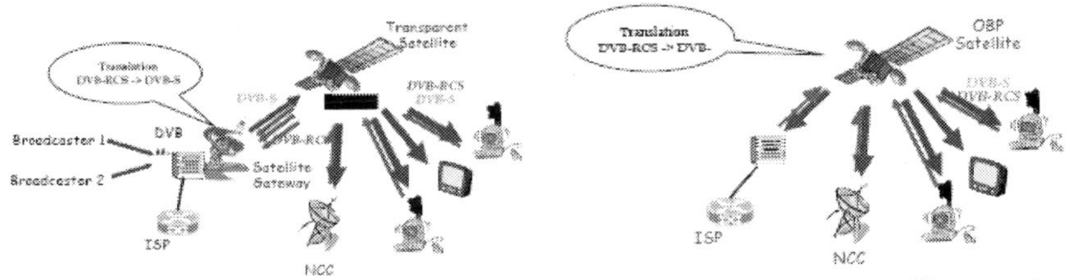


a) transparent satellite b) regenerative satellite

Fig. 1. DVB-S/RCS architecture

Satellite Terminals (RCST) provide single PC or LANs with the access to the network, while Gateways (GWs) allow the connection with Internet core networks. The uplink access from each RCST is managed through DVB-RCS interfaces.

On the 2 topologies, the end-user side of the platform is on the right. On the left is shown the provider/enterprise/Internet side of the platform. It can be distinguished also between the satellite network side (in the middle) and the IP network sides (on left and right ends), interconnected by RCSTs. So, the 3 main components in the satellite network side (middle) are the Satellite, the Return Channel by Satellite Terminals (RCST) and the Network Control Center (NCC).

In Fig. 1-a, the architecture relies on a transparent satellite offering a star topology. All the forward links from GW to the RCSTs are DVB-S links and all the return links are DVB-RCS links. This allows the satellite payload to work in a simple transparent way without any computation to be made on the received frames before resending. Such a payload is easier to design and was the first implemented in such GEO satellites. But the main constraints of such architecture are due to the mandatory double hop to be done to go from one RCST to another one as it is needed to go through the GW to access to an RCST.

On the opposite, this problem is solved in the second kind of architecture shown in Fig. 1-b. In this topology, the uplinks (to the satellite) are DVB-RCS links only and the downlinks (to the RCST) are DVB-S. The complexity of this solution is located in the satellite where the payload has to be regenerative to translate incoming frames in DVB-RCS in outgoing frames in DVB-S. More complex to implement, the regenerative payload was designed later than the transparent one.

It has to be noticed that it is now time to implement hybrid payload including two parts one transparent and the other one regenerative inducing more complexity of the payload, but nothing new in the architecture components where the two kinds of network components coexist, but in separated configurations.

## 2.2 Specific DVB-RCS features

DVB-RCS systems involve lots of specific techniques, but only a few of them impact the QoS of such a satellite network. So this section is dedicated to the 2 main ones that are DAMA and Encapsulation.

### 2.2.1 DVB-RCS Demand Assignment Multiple Access (DAMA)

Furthermore, DVB-RCS requires a Medium Access Control (MAC) protocol because Satellite Terminals (ST) is able to simultaneously access the return channel capacity. The standard method relies on a Multi-Frequency Time Division Multiple Access (MF-TDMA). It basically relies on the availability of several TDMA channels (corresponding to different carrier frequencies), each subdivided into frames and further into timeslots of fixed length (bursts) during which the STs are able to transmit data through MPEG2-TS or ATM traffic bursts.

The entity responsible for this timeslot allocation within the Superframe shared by competing STs is the **NCC (Network Control Center)** that centralizes the satellite resources management. Thus it periodically broadcasts a signaling frame, the TBTP (Terminal Burst Time Plan) that contains the information on which STs relies to know when to transmit their bursts. This allocation can be dynamically modified by STs requests so as to prevent from wasting satellite resources that would be otherwise statically allocated. The implementation of such a mechanism is generally known as bandwidth on demand (BoD) algorithm.

In order to dynamically manage the bandwidth allocation, a bandwidth on demand protocol called Demand Assignment Multiple Access (DAMA) is defined. It relies on the STs ability to request frequently "capacities" to the NCC which enables a regular update of the TBTP to fit to the STs respective traffic load [Fig. 2]. The latter provides signaling schemes as well as MAC QoS Classes and their mapping on capacity types.

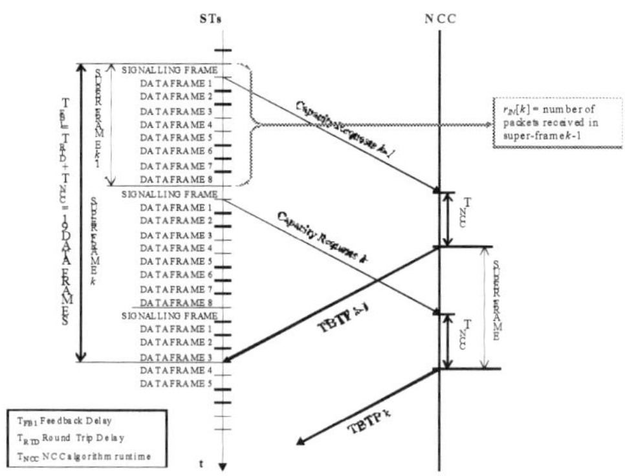


Fig. 2. DAMA algorithm: TBTP computed from RCST requests

Thus, the norm defines 5 Capacity Categories to fit the applications needs that will be detailed further in this paper. Capacity types are vital to return path QoS support at MAC layer; therefore they will be described in more details in the following. Any given ST can be

assigned one or a mix of the four capacity types. Generally, higher priority classes of service are associated with guaranteed capacity (CRA, RBDC), while lower priority classes are predominantly given best effort capacity (VBDC, FCA).
The DVB-RCS standard has left many issues open, e.g. how capacity requests are triggered, how and when certain parameters are negotiated (CRA), and if they can be re-negotiated, etc. It defines that when the NCC assigns timeslots to a certain RCST through the TBTP table, it can indicate a "channel" to which the timeslots are assigned.
It is obvious that this DAMA mechanism has great impact on what we will discuss later in this paper.

### 2.2.2 Encapsulation: from MPE to ULE

The multiprotocol encapsulation (MPE) provides a mechanism for transporting data networks on top of the MPEG2 TS in DVB networks. It has been adapted for carriage of IP packets, both IPv4 and IPv6. The encapsulation shall be done in accordance with the "Multiprotocol Encapsulation" technique described in the ETSI/DVB standard EN 302 192 and TR 101 202]. MPE includes methods for addressing the receivers of the data in the MPE header, which is necessary when many users have access to the same data channel. This feature allows several logical networks to be established without assigning PID values to each service.
IP datagrams are encapsulated in "datagram_sections" as defined in ISO 13818-6. The section_number and last_section _number must be "0" when carrying the IP protocol. The section format provides a format for mapping the datagram to the MPEG2 TS, and support filtering of datagram based on the MAC address using hardware or software demultiplexers.
The mapping of the datagram_section into MPEG2-TS is defined in ISO 13818-1. The sections are inserted into the payload of the MPEG2 packets (only first packet). The MPEG2 header is added to each packet. The resulting stream is the output of the data encapsulator/multiplexer, which is fed to the DVB modulator and satellite uplink equipment.
Many network operators and manufacturers of electronic equipment have adapted the MPE standard. That means that the standard is already in use and working well. Even though it is not the most efficient scheme for IP data transport.

An alternative encapsulation method has been defined by the IETF, RFC 4259. This directly places packet data into a Stream. This is called the Uni-Directional Link Encapsulation (ULE) defined in RFC 4326. The design of ULE simplifies processing, by reducing the number of header bytes and by significantly reducing the number of protocol fields that a receiver needs to parse. ULE also uses a Type field that resembles that adopted by the IEEE Ethernet standard, permitting easy interfacing to a wide range of network service (including IEEE 802.1pQ; MPLS; IPv4; IPv6).

ULE allows transmission of SNDUs up to 32 KB (compared to a maximum of 4KB in MPE). ULE also provides an extension header format (with an associated IANA registry), which allows future addition of new protocol fields to an encapsulated PDU, while providing backwards compatibility with existing implementations. This method is used to provide an efficient bridging method, but in future could also be used for encryption, compression, etc.

ULE is still old fashioned and solutions better fitted to Internet communications for instance have led to other proposal. The most promising called GSE will be presented later in this paper.

### 2.3 DVB-S enhancement: DVB-S2 standard and its new mechanisms

This section deals with the presentation of the new standard DVB-S2 and will be dedicated to the presentation of the main new features of such a satelite network, that are the DRA/ACM and GSE encapsulation scheme.

#### 2.3.1 DRA/ACM

##### 2.3.1.1 Return link

For the return link, different physical layers for individual terminals and for collective terminals have to be considered. The combination of adaptive coding, adaptive modulation and variable symbol rate can lead to different trade-offs depending on the type of scenario and of terminal.

The return link physical layer is based on the DVB-RCS standard with an adaptive waveform. DRA is considered, this means it is possible to change the coding rate, the modulation scheme and/or the transmission symbol rate. DRA is not included in the current DVB-RCS standard, but can be implemented using the standard capabilities without any additional information or changes (except for adaptive modulation using 8PSK and BPSK, which are not currently available in the DVB-RCS standard).

A DRA scheme is defined as an association of coding rate, modulation scheme and carrier symbol rate. The coding rate may be chosen among the set of DVB-RCS coding rates. Modulation can be either QPSK or 8PSK (with BPSK if required by link budgets). There is no specific constraint on the transmission symbol rate. This must be adapted to the terminal requirements in terms of the peak data rate range. However, due to equipment constraints (demodulation in particular), frequency plan constraints (in particular if the frequency plan needs to be reconfigured dynamically), the only transmitted symbol rates that are considered are multiples of each other.
A total of 70 combinations are thus possible.

The choice of DRA schemes to be retained is a trade-off between:

- the spectral efficiency (and thus the system capacity, the total number of users in the system and the average bit rate per user),
- the useful peak data rate (maximum data rate seen by a user)
- and the required SNR for each DRA scheme (the increase in required SNR should be reasonable with respect to the increase in peak rate or spectral efficiency).

When going from lowest DRA schemes to higher ones (i.e. from low SNR to a better SNR), we should:

- increase the required C/N0 with at least 2 or 3 dB steps between schemes to ensure the stability of the control loop,

- increase the spectral efficiency or the symbol peak data rate or both (that is increase the useful data rate).

When considering all possible combinations, we have 14 possible combinations of coding rate and modulation scheme, and 70 possible DRA schemes (combination of coding rate, modulation scheme and symbol rate). To select the set of DRA schemes required for the system, we consider some additional constraints resulting from implementation issues, as well as some specificity from the scenarios. We target a residential scenario with individual terminals, where we will try to optimise the spectral efficiency (to get a higher system capacity and a large number of users) rather than the peak data rate (that will be more a priority for collective terminals). However a reasonable peak data rate should still be offered to remain competitive with terrestrial solutions.
At the end, the selection process leads operational real systems to focus on around ten schemes only.

#### 2.3.1.2 Forward link

The forward link physical layer is based on the DVB-S2 standard that supports an adaptive physical layer thanks to ACM (Adaptive Coding and Modulation). It addresses different kinds of terminals and different link conditions by allowing a large set of possible MODCOD schemes. With ACM, the coding rate and Modulation scheme can be chosen depending on the link quality within the set of available MODCOD (MODulation and CODing) schemes. The link can also be adapted dynamically by the implementation of a control loop for each station and dynamic measurement of the channel quality based on the SNIR estimation (with pilot symbols transmitted within the DVB-S2 frame).

For the forward link, there is no specific trade-off in the physical layer definition for the individual versus the collective terminals. They will all use a subset of the overall DVB-S2 possible MODCOD schemes. The MODCOD schemes retained for such systems only depend on the performances of the MODCODs proposed by the DVB-S2 standard. We select the best schemes in terms of spectral efficiency versus required Es/N0. Then the MODCODs that are actually used by individual or collective terminals will only depend on the Es/N0 range that they can reach.

This means the same DVB-S2 downlink carrier could be used to address both individual and collective terminals, the only constraint being that all the MODCOD schemes shall be supported at the same time. However, for the computations hereafter we will separate the MODCOD schemes distribution computation for individual and collective terminals.
And, as before on the return link, the selection process leads operational real systems to focus on around ten schemes only.

### 2.3.2 Encapsulation in DVB-S2: GSE

S2 introduces a new physical layer that supports a set of transmission waveforms that use a combination of higher-order modulation and powerful FEC coding. Although backwards-compatibility with existing DVB-S is supported, the main advantage arises when S2 is used with a range of terminal capabilities, particularly when the waveform is dynamically chosen to match the prevailing conditions of the receivers. A transmission frame consists of a 90-bit physical layer header providing a preamble and identifying the ModCod used. The payload of a physical layer frame is known as a baseband frame (BBframe) and includes a 10 byte signalling header, which is followed by the BBframe payload. The size of this payload depends on the ModCod that was selected and can be up to 8 KB, significantly larger than an MPEG-2 TS Packet. The BBframe payload may carry a sequence of TS Packets, as in other MPEG-2 networks. Therefore, PDUs can be sent by encapsulating them using MPE or ULE. In S2, the transmission is optimised by omitting any Null TS Packets from the BBframe, and re-inserting these at the Receiver (preserving end-to-end timing).

The requirements for GSE differ to those of MPEG-2 TS, in that the GS uses a strong FEC code, and a much larger frame compared to that of a TS Packet. The requirement to check the integrity of a received SNDU therefore differs from that for the MPEG-2 TS. The current proposal is not to check every SNDU, but instead to place a 32-bit CRC at the end of each BBframe. Fig. 3 shows a general picture of the different encapsulation mechanisms involved in IP communication over DVB-S2 through GSE.

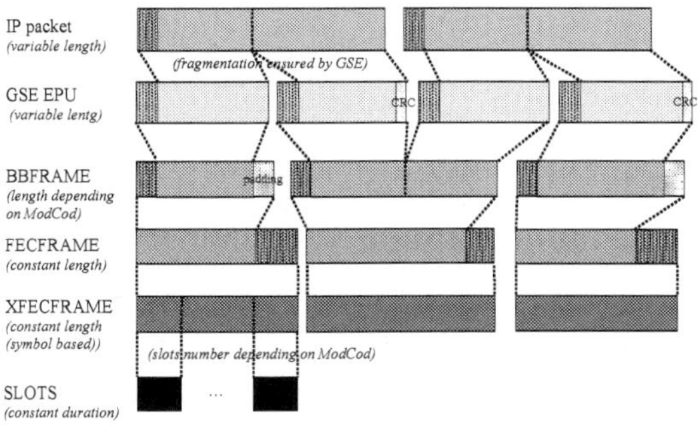


Fig. 3. GSE encapsulation in DVB-S2

The BBFRAME is the container that adapts GSE to fit the DVB-S2 properties to a data stream. DVB-S2 allows the use of a different ModCod (Modulation and Coding scheme) for each BBFRAME. This permits to take into account (while sending DVB-S2 data) the reception conditions, if available (e.g. through DVB-RCS).
A scheduling algorithm needs to take such variations into account.

In the DVB-S2 architecture, both modulation and coding can change from one BBFRAME to another, leading to variable length and variable duration BBFRAMEs. These two parameters are, however, tackled independently with the help of FECFRAMEs and XFECFRAMEs.
However, once the ModCod (more precisely the coding scheme) has been chosen, the BBFRAME length is implicitly fixed so that a padding mechanism is needed.

The FECFRAME is a constant size container that is used to encapsulate the BBFRAME with its associated coding. There are two possible values for a FECFRAME length (16200 bits or 64 800 bits) only one of which is implemented in a given system.

We will not describe more detail about FECFRAMEs, as the only pertinent information regarding a scheduler implementation is the constant length of the FECFRAME and the "variable" length of its payload.

The GSE protocol uses the BBframe header to determine the frame size (and presence of padding). The encapsulated PDUs also known as EPUs (Encapsulated Payload Units) are carried in DVB-S2 BBFRAMEs.

Not all receivers necessarily support the same set of ModCods (e.g. implementation of different sets of ModCods, different locations within the satellite footprint, local propagation conditions, or other reasons). Care therefore needs to be taken to ensure that all SNDUs or SNDU fragments are sent using GSUs with a ModCod that can be received by the intended recipient(s). Sending a SNDU or a SNDU fragment with a higher ModCod than required consumes unnecessary satellite capacity, while sending with a lower ModCod could result in loss.

At the time of writing, issues remain open such as flow security, support for signalling/control, the effect of small packets (concatenation) and these are the overall performance of the different features when used in an environment where traffic and terminal reception conditions are both varying.

## 3. QoS in terrestrial networks

The concept of quality of service (QoS) has gradually emerged to meet the various requirements demanded by certain types of applications, primarily in the world of interactive. Indeed, initially, the majority of Internet traffic consisted of textual data having no specific requirements but progressively, tools involving simultaneously files transfer, instant messaging, audio and video appeared. Consequently, guarantees on bandwidth, delay or jitter should be provided to users to ensure proper functioning.

However, the architecture of the Internet, based on TCP/IP, is not designed to differentiate the types of traffic and is currently dominated by a single model of service: best effort. This architecture can only guarantee correct functioning of all types of applications by offering the oversizing of the network that consists in configuring a network with a capacity that far exceeds the requirements. But this approach only postpones the problem and can hardly be applied to wireless access technologies with limited bandwidth.

Thus, effective management of resources is necessary to provide to Internet users QoS guarantees adapted to their needs. In this part, we will therefore firstly introduce the fundamentals of QoS. Then, the precursor models of QoS, IntServ and DiffServ will be detailed and finally we will show how the two signaling protocols, COPS and SIP, can participate in the implementation of QoS.

### 3.1 The basis of QoS

According to the standard ISO 8402 (ISO, 2000), quality of service is defined as "the totality of characteristics of an entity that bear on its ability to satisfy stated and implied needs". The E800 recommendation ITU-T (ITU-T, 1993) defines it as "the collective effect of service performance which determines the degree of satisfaction of a user of the service".

Thus, different types of QoS can be defined from these definitions and can be separated into three categories as defined in (Hardy, 2001):

- The intrinsic quality of service that is provided directly by the network itself and described by objective parameters such as, for example, the delay or the losses. It is the primary IETF focus.
- The perceived quality of service that matches the quality perceived by the user (also called QoE, Quality of Exeprience). It depends heavily on network performance but is measured by an average of users' opinions. The most common method is the MOS (Mean Opinion Score) in which a group of users separately evaluate perceived quality of an application between 1 and 5, an average score being then performed. The MOS is typically used for audio or video quality of an application, but may also involve perceived QoS connection time, the perceived security of user, service availability, etc... In addition, there is not necessarily a correspondence between intrinsic QoS and perceived QoS, the latter being very subjective. ETSI and ITU primarily use the term QoS as perceived QoS and prefer the term network performance to describe the technical part of QoS.
- The assessed quality of service that refers to the willingness of a user to continue using a particular service. It depends on the QoS perceived but also the price, service support offered by the supplier and other commercial aspects.

The main parameters that describe the intrinsic QoS in IP networks are:

- The transfer delay of packets, in milliseconds. It is usually measured from end to end but may be calculated on a portion of the network.
- The jitter or delay variation of packet forwarding, expressed in milliseconds.
- The bit rate, expressed in bits per second (bit/s or bps) or bytes per second.
- The rate of packet loss, defined as the percentage of packets lost in relation to the total number of packets transmitted.

### 3.2 The main existing models to ensure QoS

Two major proposals were made by the IETF to ensure QoS and proper functioning of IP services, real time or not: IntServ (Integrated Services) and DiffServ (Differenciated Services) each associated with a working group of the same name.

#### 3.2.1 The IntServ model

The work of the IntServ working group led in 1994 to the definition of an integrated services architecture (Braden et al., 1994) composed essentially of two parts: an extended service model, called the IS model and a reference implementation framework for the development of this model.

This architecture, allowing to support QoS without changing the IP, is based on a per flow resource reservation. Each router must maintain the state of the flows going through it, which fundamentally modifies the principle of the Internet, which, on the contrary, was till now based on a conservation of the state of flows at the user terminals. That's why routers are then equipped with four additional features:

- The packet scheduler, responsible for the delivery of packets streams, using a set of queues as well as other mechanisms such as timers.
- The classifier that realizes the correspondence between an incoming packet and the class of service to which it is associated. The level of QoS provided by each service class is programmable for each stream.

- The admission control that implements the decision algorithm used by the router to determine whether a new flow may or may not obtain the requested QoS without degrading the guarantees offered previously.
- The reservation establishment protocol, needed to create and maintain the state of flow on routers. The protocol chosen for this function is RSVP (ReSerVation Protocol), defined later as Resource reSerVation Protocol in (Braden et al., 1997).

Two new classes of service are then defined in addition to best-effort that receives no special treatment at routers:

- The Guaranteed Service (GS) (Shenker et al., 1997) provides guarantees in terms of bandwidth and maximum transfer delay of packets, expressible quantitatively. If the stream respects the reserved parameters, this service ensures that all packets will arrive with a maximum delay and that they will not be lost in the queues in case of congestion. This service is suitable for real-time applications with strong time constraints such as videoconferencing or VoIP. However, no average delay is guaranteed, so it's the application itself which has to manage the delay variations at the receiver side by using buffering mechanisms.
- The Controlled-Load service (LC) (Wroclawski, 1997) is a service expressible qualitatively in terms of bandwidth, which ensures the user that its data stream will be transmitted with a QoS close to the one obtained in a network not overloaded (not congested).

The guarantees are obtained from end-to-end by the concatenation of such assurances, offered separately by each router crossed along the path. Furthermore, as stated previously, the protocol used to configure these routers is RSVP and we will now explain its functioning.

Initially, a PATH message is propagated from the transmitter (sending application) to the receiver. This message contains the traffic specification (TSPEC) that will be generated by the application. This specification can not be modified along the path but other information can be added through additional specification (AdSpecs) to precise specific resources constraints. Once the message arrived, the receiver responds with a RESV message that contains the description of traffic flow to which the resource reservation should apply (TSPEC Receiver) and the parameters demanded to implement the required service (RSpec). These descriptions may also change along the path.

RESV messages must follow the reverse path of PATH messages and trigger an effective reservation of resources (state have to be stored at each router) if the admission control mechanisms of each router validate the request. If a router validates the request, it creates and maintains a state corresponding to this flow. However, the reservations lifespan is limited and PATH/RESV messages must be exchanged periodically so that the reservation remains valid. It allows the system to be robust to changes in routing for exemple.

Once QoS is configured, when a reserved flow is going through a router, the classifier identifies it by its 5-tuple (Source IP address, destination IP address, Protocol, TCP/UDP source port, TCP/UDP destination port). The scheduler then handles the queue management.

RSVP is so a signaling protocol that allows to reserve dynamically bandwidth and ensure a maximum delay from end-to-end. This reservation, initiated by the receiver, can prevent that some applications monopolize resources unnecessarily and allows in the case of multicast communication to differentiate the reservation (and billing) for each receiver.

Moreover, its dynamic functioning can adapt to changing communications (changes in the number of participants, route changes, etc…). Finally, this protocol also has the advantages of being adaptable to both IPv4 and IPv6 and seamlessly passing non-RSVP routers.
However, RSVP requires maintaining state information for each flow at each node or router along a path connecting a transmitter to its receiver. So, when the number of users and flows increases, the number of states becomes significant and the traffic is all the more saturated as refreshments between RSVP routers become important and create overhead. The main shortcoming of the IntServ model and the associated RSVP protocol remains their lack of adaptation to the scale factor, especially as the reservation in RSVP is unidirectional. So, for a bidirectional application requiring QoS in both directions, the amount of messages is twice as high. The IntServ/RSVP model is therefore more adapted to small networks such as LAN.
It is one of the reasons that explain that another model of architecture has been proposed: the DiffServ architecture.

### 3.2.2 The DiffServ model

To solve the problem of scalability posed by the per flow management of QoS in the core network routers, the DiffServ working group has therefore proposed ((Blake et al., 1998) and (Grossman, 2002)) to separate traffic by classes of service. Thus, per flow QoS treatment is realized only at edge routers that aggregate flows by traffic class. The number of state maintained by core routers is so reduced to the number of classes and not anymore to the number of flows, which greatly reduces the complexity.
Each service class is identified by a value encoded in an existing field of IP header, redefined by the DiffServ group and named DSCP (DiffServ Code Point), which presents the advantage of not requiring the use of an additional signaling protocol. This is the TOS (Type Of Service) for IPv4 and TC (Traffic Class) for IPv6.

The DiffServ architecture is primarily based on the concept of DiffServ domain that consists in the grouping of one or more networks under a single administrative authority. This domain is composed of nodes or core routers that are only connected to nodes within the same DiffServ domain and edge routers that interconnect the DiffServ domain to other domains, DiffServ or not. The edge routers can also acts as well an incoming router when the traffic enters into a DiffServ domain as an outgoing router when the traffic leaves the domain.
A client of a DiffServ domain (which can be either a user or another domain DiffServ) must negotiate a contract with the service provider responsible for this area that specifies the terms and conditions of use of the concerned services: this contract is called an SLA (Service Level Agreement) and the technical part is specified by different SLS (Service Level Specification). An SLA contains the following information:

- The traffic that the customer is likely to generate in terms of data volume, rate, number of users, etc...
- The QoS that the service provider has to provide the customer in terms of availability, security, reliability or performance (delay, bandwidth, etc…).
- The policy followed by the service provider in case of overflow traffic (rejected, accepted but surcharge, etc...).

Finally, (Nichols, 1999) defines an entity called bandwidth broker with knowledge of the resources availability and policies of the associated domain. One of its main tasks is the admission control. In addition, to allow an end-to-end allocation of resources across all the domains taking into account the different SLAs negotiated between them, this entity must communicate with the bandwidth broker of the neighbouring domains.

To manage a DiffServ domain, the service provider in charge of it first starts by sizing its network according to the contracted SLAs with all of its customers.

The processing of packets entering the DiffServ domain is then realized at the edge routers in charge of flows classification by class of service, and traffic conditioning. To do this, they are composed of a classifier, a meter, a marker, a regulator and finally a dropper of non-conforming traffic. The classifier identifies a flow from either the DSCP field only, or a combination of one or more fields such as source IP address, destination IP address, DSCP, protocol ID, source and destination ports or other information as the input interface. Then, mechanisms for profiling and traffic measurement allow on the one hand marking or re-marking the packets and, on the other hand, to shape flows or drop them totally or partially to meet the negotiated traffic profiles.

During the marking, the DSCP field is updated using one of the different classes of service or PHB (Per Hop Behaviour). In addition to the Best Effort class which is not subject to any special treatment, two PHBs have been defined by IETF:

- Expedited Forwarding (EF) (Jacobson et al., 1999) which corresponds to the highest priority and ensures the transfer of flows with high temporal constraints (eg VoIP, videoconferencing) guaranteeing a certain bandwidth and low delays, jitter and loss rate.
- Assured Forwarding (AF) (Heinanen et al., 1999) which defines four levels of priority (AF1, AF2, AF3, AF4) on the delivery of certain packets in case of congestion.

Once packets have been marked by the edge routers, they are treated within the DiffServ domain by the core routers depending on the PHB deducted from the DSCP field which describes the forwarding behavior of these routers. The priorities are then performed by scheduling algorithms such as PQ (Priority Queuing), WFQ (Weighted Fair Queuing) or CBQ (Class Based Queuing).

The DiffServ model therefore provides a QoS management more adapted and more realistic than the IntServ one. The per class management (or per aggregate) can indeed be much more resistant to the scale factor. Moreover, Diffserv does not require signaling protocol like RSVP by adapting the header of IP packets, which saves bandwidth.

However, the sizing of a DiffServ domain taking into account the SLAs contracted with neighboring domains and users is a heavy and complex task to implement that does not dynamically adapt to rapid changes in traffic. Similarly, not using a signaling protocol at the application level implies that the user is unable to dynamically change the resources according to its needs.

### 3.3 QoS Signaling protocol

The previous paragraphs show that, although IntServ is unsuitable for scaling up, the use of a resource reservation protocol (RSVP) allows a more precise (adapted to each stream) and dynamic QoS management. The idea of using one or more signaling protocols to negotiate and establish end-to-end QoS has consequently been the subject of a number of research among the scientific community.

The objective of this part is then to present some of these protocols that enable the implementation of QoS, the other concepts (security, supervision, etc.) being not addressed.

### 3.3.1 COPS and the notion of QoS management by policy

The RAP (Resource Allocation Protocol) working group of the IETF defined in 2000 an architecture based on the notion of policy to improve the admission control mechanisms of a network. A policy is defined as a set of rules for the administration, management and access control to network resources. Each rule is then associated with a set of conditions which corresponds to a series of actions to do in case of these conditions are met. To enable the exchange of such policies on a client/server model, a protocol is also defined by that working group, the protocol COPS (Common Open Policy Service) (Durham et al., 2000).

This model of by policy management is composed of two core elements: PEP (Policy Enforcement Point), responsible for implementing policy decisions and the PDP (Policy Decision Point), in charge of making decisions based on defined policy. These two components communicate via the COPS protocol. To take its decisions, the PDP communicates with a database of policies through the LDAP protocol (Leightweight Directory Access Protocol) and can query other entities such as an authentication server or a bandwidth broker using SNMP (Simple Network Management Protocol) for example.

This generic model of network control by policy allows for two distinct modes of control: outsourcing and configuration (also known as provisioning).

In the outsourcing model, a router (including PEP) which has to make a decision on acceptance of a reservation sends a COPS request to the PDP and the latter responds with the decision taken in accordance with the policy rules. It can especially be used in relation with the IntServ model when a router receives a RESV message and must decide to accept or decline the resource reservation. Indeed, that decision may require other information than the resources locally available at the router and the use of a PDP can thus be judicious.

In the configuration model, when external events require a change in the configuration of routers (and thus PEPs), the PDP may communicate to them new rules to apply through the COPS protocol. They no longer have to seek the PDP before making a decision. This model can overcome the main weakness of DiffServ model in which classes of service configuration is static. Indeed, a network administrator can define, for example, two types of policies, one appropriate to the day where many VoIP calls are held and one for the night which suits rather backup's servers or data downloads. In this case, edge routers act as PEPs, while the bandwidth broker plays the role of PDP.

### 3.3.2 SIP: the session control and the QoS

SIP (Session Initiation Protocol) is a signaling protocol standardized by the IETF (Rosenberg et al., 2002) designed to establish, modify and terminate sessions with one or more participants. Its main use is now the session management of voice over IP (VoIP), for which it is currently the most common open standard, but the fact that it is independent of the type of data transmitted and the type of protocol used also allows him to develop many other applications such as instant messaging, video conferencing, distance learning, video games or virtual reality. The parameters of these sessions are described by the Session Description Protocol or SDP (Handley et al., 2006) and, in this part, we will see how these parameters can be helpful for the dynamic reservation and release of resources.

These parameters are negotiated during the establishment of the session but also during the session modifications. This negotiation can be conducted directly by the SIP clients located on users' terminals but, although they are aware of some important parameters, such as codecs they can use, the latter have no real means to know the status of available resources along the path their communications will cross. Moreover, it would imply the integration of QoS mechanisms such as RSVP or COPS on SIP clients, which has the double disadvantage of making the SIP clients more complex and not allowing SIP clients which are unaware of QoS to use this option.

That's why these new features must be rather implemented at the SIP proxy: the intermediate entities where SIP clients are registered and that intercept SIP messages. Indeed, they may allow operators to know the duration of the session, the number of involved media, the codecs used and their associated characteristics (bandwidth, etc..). From this information, automatic management of the resources reservation/release and of admission control according to the policy adopted by the operator becomes feasible and no changes are required at the SIP clients.

Two modes of session establishment with QoS reservation can then be distinguished:

- The "enabled" mode in which the session establishment and resource reservation are performed in parallel but do not depend on each other. The session will start regardless of the outcome of the reservation. For example in the case of DiffServ networks, QoS will be BE if the reservation has failed and EF if it worked.
- The "assured" mode in which the establishment of the session is done only if the QoS reservation has been realized.

A standard (Camarillo et al., 2002) detailing the operation of these two modes has been proposed. It is based on new messages such as Session Progress, UPDATE or PRACK and a set of preconditions added to the session descriptors. The SIP clients can then specify whether the implementation of QoS is "mandatory" ("assured" mode) or "optional" ("enabled" mode) for each direction of communication (reception, transmission) and if the QoS must be e2e (end-to-end) or local (at the access network). However, RFC 3312 (Camarillo et al., 2002) considers that QoS is implemented by the participants of the SIP session while we will consider that it is implemented by the SIP proxy. The Fig. 4 illustrates an example of a SIP session integrating resource reservation/release in "assured" mode.

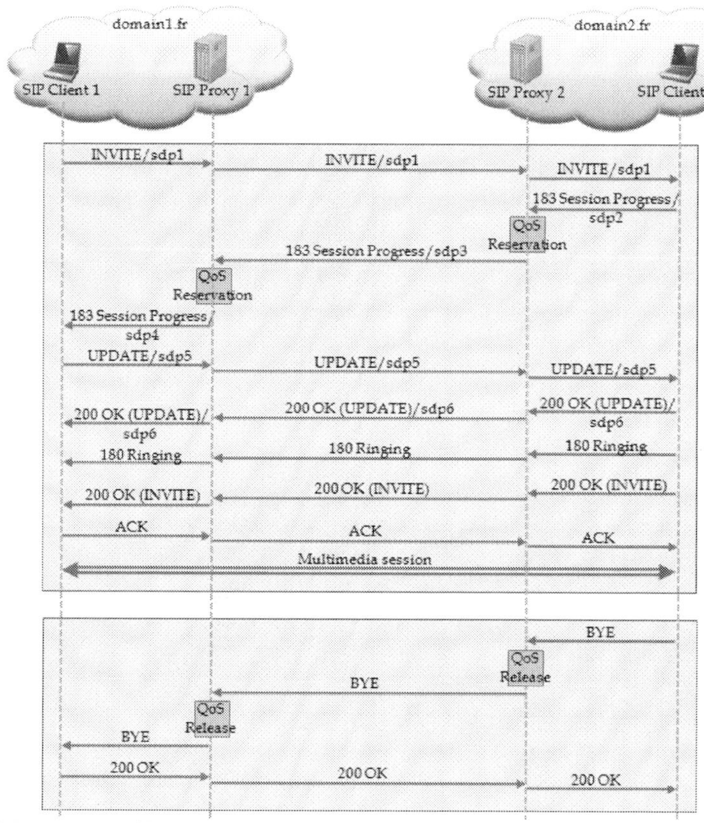


Fig. 4. SIP session with resources reservation/release

PRACK and 200 OK (PRACK) messages are not presented to make the figure more comprehensible but they are normally exchanged at the reception of a 1xx message to make it reliable. It can also be noticed that, in this example, all messages are exchanged via all the SIP proxies to manage the QoS along the whole path. Finally, QoS reservation/release in Fig. 4 may for example represent the exchange of messages (COPS, RSVP, etc..) allowing SIP proxies to communicate with the entities responsible for resources management within the concerned domains. For example, if we consider a DiffServ domain between domain1.fr and domaine.fr, the SIP proxies will be able to exchange COPS messages with the Bandwidth Broker in charge of the DiffServ domain. The latter can then configure the edge routers of this domain to prioritize the flows of the future SIP session.

Moreover, whatever the session establishment mode is, if a change of session is triggered by a re-INVITE message, SIP proxies can analyze the new parameters and automatically warn the entities responsible for resource management. The SIP signaling thus allows much more dynamic management of QoS for the applications it controls.

SIP, although it was designed originally to allow the session control, can be used very efficiently in the dynamic management of QoS.

## 4. QoS in satellite networks: DVB-S2/RCS QoS Architecture

The quality of service in DVB-S2/RCS networks is basically managed at layer 2. The first section gives an overview of classical layer 2 QoS architecture. The QoS offer should satisfy the application QoS requirements then this relationship is covered by the upper layers. Three QoS management strategies that can be considered as three different architecture maturity degrees are presented in this section. The first one, corresponding to a short term technique to introduce QoS in satellite system is based on MPLS over satellite systems. Then a fully IP based solution is presented as a medium term solution and the complete integration of satellite networks in Next Generation Networks (NGN) consists in the long term evolution of theses architectures.

### 4.1 Layer 2 QoS management

- In star architectures, the gateway (GW) centralizes the traffic and the signaling path and uses all of the offered bandwidth. Then, the quality of service management could be simply ensured by a correct scheduling algorithm, according to the packets destination and their service classes.

  But, when considering meshed topologies, the return link is shared among multiples satellite terminals. As one of the first things that impact the quality of service in the satellite network is the network access, packets should be sent on the air interface as soon as possible or even compliantly with the delay or the bandwidth required by the application. Then, in theses architectures, the quality of service management is essentially assured for the return link, i.e the DVB-RCS part.

  As said previously, the DAMA request/assignment cycle exhibits a non negligible latency and additional delays that cannot always match interactivity requirements of multimedia services. In order to both maximize satellite resource use and meet multimedia requirements, the DVB-RCS norm discriminates RCST capacity requests into 5 categories:
- Continuous Rate Assignment (CRA): Fixed slots are assigned in each MF-TDMA frame for the whole duration of a RCST connection
- Rate-Based Dynamic Capacity (RBDC): a dynamic rate capacity (in slots/frame) granted in response to explicit RCST requests
- Volume-Based Dynamic Capacity (VBDC & AVBDC): a dynamic cumulative volume capacity (in slots), granted in response to explicit RCST requests
- Free Capacity Allocation (FCA), which is assigned to STs on an "as available" basis from unused capacity

The standard defines separate MAC traffic priority queues and suggests a requesting strategy for each of them, that is to say a relevant mapping between traffic and request categories. Any given RCST can be assigned one or a mix of the four capacity types. In general, higher priority classes of service are associated with guaranteed capacity (CRA, RBDC), while lower priority classes are predominantly given best effort capacity (VBDC, FCA).

The generally recommended MAC queues are the following:

- Real Time (RT) queue used by temporally constrained applications as VoIP or visioconference. A CRA allocation mechanism is often used to feed it.
- Critical Data (CD) queue used by critical applications. It could be file transfer applications as well as low importance visioconference. RBDC requests are generally used.
- Best Effort (BE) queue used by non critical applications, as emailing or web browsing. RBDC, VBDC, or AVBDC are recommended.
- Network Management (NM) queue is recommended to offer a guaranteed path to signaling protocols as those presented in the following sections.

The CRA capacity allocated to the RCST should be consumed by the Real Time queue, but could be shared with other classes if unused.

However those suggestions are not yet sufficient to seamlessly integrate satellite networks into an end-to-end NGN QoS Architecture. The next sections propose three solutions, from the simple one to the NGN compatible one.

### 4.2 QoS with MPLS based architectures

One of the challenges the Internet is facing since the middle of 1990's is the exponential increased level of the IP traffic, sometime beyond what routers are able to handle. One solution to overcome the router bottlenecks it to delegate more of the IP forwarding to the layer 2. This allows the router to shortcut the heavy layer 3 processing based on routing table and implemented in software, while using instead a lightweight layer 2 switching table implemented in dedicated ASICs in the forwarding process.

MPLS is being defined by the Internet Engineering Task Force (IETF) as a standards-based approach to apply label-switching technology to large-scale networks. Although MPLS, as its name suggests, can conceptually support multiple protocols, the current work is focused on the integration of IPv4 and IPv6 with ATM, Ethernet, Wavelength Division Multiplexing (WDM) and Frame Relay. Thus, MPLS could be seen as the simplest solution for the seamless IPv4 to IPv6 migration of a heterogeneous network.

MPLS forwards data using labels that are attached to each data packet. Intermediate MPLS nodes do not need to look at the content of the data in each packet. In particular the destination IP addresses in the packets are not examined, which enables MPLS to offer an efficient encapsulation mechanism for private data traffic traversing the SP backbone. Service providers are therefore offering MPLS as a VPN technology.

#### 4.2.1 Intserv / Diffserv QoS support in MPLS

MPLS acts as a level 2/3 protocol, its role is to transport a native network level traffic. IP includes in its header ToS field in IPv4 or Traffic Class field in IPv6, which provides information on the kind of traffic to transport. The traffic level quality information that should be borne by MPLS is fully dependent on its upper layer protocol or the resources reservation provided by RSVP. Both IntServ and DiffServ classes are supported over the MPLS architecture. IntServ and DiffServ classes are mapped onto DVB-S/RCS classes. Bandwidth allocations and flow isolation is efficiently supported by means of the MPLS VPN mechanisms.

The **IntServ** service architecture incorporates one main component to support the IntServ QoS model: RSVP-TE (Awduche, 2001). This IntServ Signalling protocol for MPLS is an extension of RSVP for MPLS. It allows traffic engineering via label distribution. RSVP-TE works on the Downstream-on-Demand principle, i.e. the upstream router requests a label from the downstream router. This allows an ingress router to specify a route for a given flow. Resources reservation is optional with RSVP-TE; this allows the label distribution even if the paths do not need reserved network resources. Unlike RSVP, RSVP-TE works only between ingress and egress routers, and not between the sources and the destinations of the flows. RSVP-TE also allows the traffic policing working with th‹e LSP instead of the destination IP address.

Unlike IntServ, **DiffServ** does not require the reservation of resources end-to-end. It actually runs on a Per-Hop-Behaviour (PHB) principle. This means that each router filters the incoming packets. In MPLS, LSRs do not read the IP headers, thus they do not have any direct information regarding the PHB for arriving packets. MPLS encapsulates IP packets with a shim header that includes the label and the Exp field. This last one is composed of 3 bits intended to support marking of packets for DiffServ, which on its side defines its classes of service with the 6-bit DSCP field. The problem is that DiffServ can have 64 possible DSCPs whereas MPLS/Exp bits can only address up to 8 possible types of Per-Hob Behaviors (PHBs). In order to support for instance 1 EF class, 4 AF classes with 3 dropping priorities and a default Best Effort class larger field containing QoS control information is needed. Two possible labeling configurations exists (Le Faucheur, 2002):

- A configuration called E-LSP (Exp LSP) is able to treat most of the QoS requirements. E-LSP allows one label, thus 8 different PHBs. This is enough to support a minimal configuration: EF, BE, and four AF classes with two drops precedence level.
- If a network has more than 8 PHBs, then 3 Exp bits will not be able to convey all the PHBs to LSPs. One way is to use the label itself to convey PHBs, therefore, an LSR uses the label to determine the PHB. For L-LSP PHB is determined from label and AF drop precedence is determined from Exp bits.

Both E-LSP and L-LSP can be used, but it depends on which kind of traffic is going to be forwarded. E-LSP is simpler to implement and is close to DiffServ. However, L-LSP is more sophisticated and could offer more possibilities when the QoS over MPLS is more mature. It depends on the requirements.

### 4.2.2 IntServ/Diffserv Mapping over DVB-S/RCS

Performances of MPLS are also determined by the capabilities of the satellite architecture lower layers, based on DVB-S/RCS technologies, especially, in terms of bandwidth allocation capabilities. MAC/DAMA concerns the satellite segment procedures for bandwidth assignments, which, on one hand, aim at an efficient use of the valuable satellite capacity while, on the other hand, should allow the respect of QoS requirements for IP traffic streams.

Two algorithms of bandwidth allocation exist:

- A Connection Admission Control (CAC) provides the RCST with a statically assigned capacity, which is always available to the ST.
- While a Bandwidth-on-Demand (BoD) mechanism provides the RCST with an additional amount of capacity which dynamically adapts to the current RCST traffic conditions.

Due to the long propagation delay associated with the satellite links, the BoD schemes implicitly introduce an additional delay. This additional delay, which is given by the queuing delay in the RCST buffers, is in the range of 0.5-1 sec., depending on the BoD control policy.

Taking into consideration the **CoS contract** characterizing the MPLS traffic classes, the crucial point is to supply a proper mapping; possibly, the traffic classes might be supported by a mix of the capacity request categories.

The **CoS requirements** can be classified in bandwidth, delay and jitter constraints:

- Bandwidth constraints might impose that a defined amount of capacity is granted to the connection for the connection lifetime;
- Delay constraints enforce an upper limit on the queuing delay in the buffers;
- Jitter constraints require a limited delay variation in the transmission of the packets.

Due to the additional queuing delay introduced by BoD mechanisms, MPLS connections with stringent delay and jitter requirements - i.e., real time (RT) connections - cannot rely on dynamic access to the network: the proper amount of capacity, computed on the basis of the bandwidth requirements, must be statically reserved via Connection Admission Control (CAC) procedures.

On the other hand, Best Effort (BE) connections can avail of dynamic capacity assignments (via the BoD mechanism), since no particularly stringent delay requirements have to be satisfied. Furthermore, since no bandwidth requirements are defined, also the CAC procedure is not triggered; this means that BE connections are the most subject to congestions.

Finally, the connections with no jitter requirements, no (or loose) delay requirements but with bandwidth constraints - i.e., Critical Data (CD)- must be subject to CAC admission decisions but they may access a portion of reserved capacity via the BoD mechanism. In this way, these connections are not subject to congestions but, when the reserved capacity is not fully requested, the leftover part can be distributed among the best effort connections.

For each identified MPLS traffic category, the following table shows the CoS requirements, the suitable DVB priority class, the QoS guarantees and the functional modules, which allow the QoS requirements to be met.

| *IP classes* | *CoS requirements* | *DVB priority class* | *QoS guarantees* | *CAC* | *BoD* |
|---|---|---|---|---|---|
| GS - IntServ,<br>EF - DiffServ | Bandwidth<br>Delay<br>Jitter | RT | Bandwidth<br>Delay<br>Jitter | yes | no |
| CS-IntServ,<br>AF-DiffServ | Bandwidth | CD | Bandwidth | yes | optional |
| Best Effort | - | BE | - | no | yes |

Table 1. IntServ/DiffServ QoS classes mapping onto DVB-S/RCS

### 4.3 IP oriented architecture

Motivated by the need of more efficient and simplier to manage architectures, the next evolution of DVB-S2/RCS satellite network resides in IP over DVB, or at least as closed as it can. In fact, several encapsulation layers are still needed (MPE, ULE, GSE...) according to the referred architectures, but MPLS could be avoided to save its encapsulation overhead.

The satellite terminal becomes an **IP/Diffserv edge router**, and embeds all the mechanisms required to manage several PHB. The QoS is managed in three steps (Fig. 5).

- The first step take place at Medium Access Control level where the DAMA allocates the bandwidth on a fixed basis for the real time queue and on an on-demand basis for other queues (critical data and best efforts queues).
- The second step is at IP layer as described above. A dedicated IP level module implements a queue management system aiming at providing a differentiated service with regards to several service classes.
- The third step is at user layer level. The user classifies its own flows and provides the classification to a QoS server, located in the RCST, that configures its QoS architecture. The goal is to exploit the capabilities offered by the IP-level QoS.

In other words each step involves a service offering to the above.

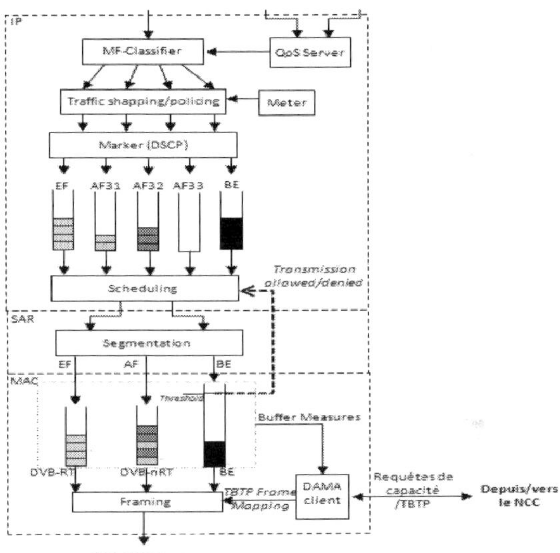


Fig. 5. The QoS Architecture

### 4.3.1 RCST QoS architecture components

A *Service Class* (SC) is given by the underlying applications behaviour that uses the service. In our example there exist five Service Classes for IP layer and three for MAC layer. Service classes for MAC layer are: Real-Time, Critical Data and Best Effort. Service classes for IP layer are: Real-Time (EF), three Assured Forwarding level (AF1, AF2, AF3) and one Best-Effort (BE).

- Real time (RT) service class: this is the service class for non-jitter, low delay flows typically filled with small packets. The underlying satellite bandwidth allocation mechanism (the DAMA at MAC level) should have reserved sufficient static bandwidth. This could be done at the terminal logon.
- AF service classes: these service classes deserve a special treatment with regards to best effort but with more priority. For exemple, signalling could use the AF1 queue, streaming the AF2, and web browsing the AF3.

- Best effort (BE) service Class: all the traffic that does not belong to real-time or critical data service class belongs to best effort service class.

A *scheduler* is used to differentiate IP queues and to map IP queues over MAC queues. This is a crucial component of the architecture. The EF class is generally mapped with the DVB-RT class and the AF queues are mapped over the CD queue that could use RBDC and VBDC allocation schemes. For this purpose, the traffic categories could be served by a scheduler using a simple priority queuing (PQ) discipline or a Weight Fair Queuing (WFQ). In fact, MAC queues should be as small as possible (only the necessary space to save the fragment of several IP packets) to reduce the end-to-end delay. To avoid MAC queues overflooding, interaction mechanisms (cross layering) between IP and MAC layer should be setup. For instance, when the BE queue (at MAC level) exceed a given threshold, a "backpressure" algorithm stops the scheduler to feed the MAC queue. This threshold could be set up according to the DAMA algorithm.

A *Traffic Flow* (TF) is a given data flow generated by protocol entities of a given application. A TF is uniquely recognized by a TF ID designing the originating network address, the destination network address, the originating transport service access point and the destination transport service access point. Classically in (TCP, UDP)/IP networks, this is given by the 4-tuple <IP-source, IP-destination, port source, port destination>. The *MF-Classifier* sets up the DSCP packet field according to the QoS policy.

A *shaper* is a specific algorithmic that enforces the emission profile of a flow according to its specifications (rate, delay, etc). A dual leaky bucket (DLB) is a simple solution to implement it but a Hierarchical Dual Leacky Bucket (HDLB) offers better flexibility in the queue management.

A *dropper* is a more stringent version of a shaper. In place of attempting to shape the traffic, it simply drops the non-conforming traffic. It does this by dropping each packet of the flow until the traffic goes back to a normal state. This can be implemented by reducing the input queue size of a dual leaky bucket to one.

#### 4.3.2 Network components

The *Connection Admission Control* (CAC) algorithm in the NCC performs preventive congestion control by deciding the admission of new connections in the network. CAC decisions are taken on the basis of the QoS requirements and guarantee the availability of bandwidth and/or the fulfillment of delay requirements.

The elements in this QoS architecture that perform or can perform CAC are the following: the QoS server on the RCST side, which is informed of the user QoS requests; the ARC, Access Resource Controller (in IP-oriented architecture), which is a separate component located at the NCC side that takes the final decision according to the SLA, the network capacity, etc. To ensure the CAC procedure, a dedicated protocol has to be set up between RCST, NCC and ARC.

These components are mandatory to set up a Diffserv compliant QoS architecture in a DVB-S2/RCS system. However, this is not sufficient to have a perfect integration of satellite systems in NGN networks. The section 0 reviews the lack components.

### 4.3.3 Satlabs

To reach a better interoperability of DVB-RCS products, after several year of extensive work and discussions, both "Quality of Service" and "Management and Control Harmonization" Satlabs Working Groups have proposed recommendations (Combes & Pirio, 2008). With the incorporation of the recommendations into the SatLabs certification program, DVB-RCS certified that terminals will be able to interoperate in different systems requiring only changes to their configuration and to provide real-time services efficiently.

In line with the latest revision of the DVB-RCS standard, these recommendations detail how the following enhanced features operate in a DVB-RCS system:

- Harmonized support for IP Classes of Service according to the IETF DiffServ recommendations, and their mapping on DVB-RCS specific MAC layer Capacity Categories and Dynamic Requests,
- Standardized management procedures based on the IETF SNMP recommendations,
- A Common SNMP Management Information Base (MIB) for all DVB-RCS terminals,
- Standardized procedures and protocols for terminal's Configuration file and log files management,
- A Common protocol for terminal's software updates based on Multicast distribution.

As part of the short term QoS harmonization, SatLabs defines that the RCST shall support at least the following PHBs:

- Expedited Forwarding (EF)
- At least one Assured Forwarding PHB Class (AF3), as defined in RFC2597 , with at least two drop precedences
- Best effost PHB, as defined in RFC 2474.

## 4.4 Next Generation Satellite Networks

The following definition is an excerpt of ETSI's NGN-Starter Group conclusions: « The term NGN is commonly used to give a name to the changes to the service provision infrastructures that have already started in the telecom and IT industry. As such it is not a term that can be precisely defined but is rather an umbrella term to describe developments following PSTN/ISDN/GSM phase 2+ era.

One of the main characteristics of NGN is the *uncoupling of services and networks*, allowing them to be offered separately and to evolve independently. Therefore, in the NGN architectures proposals, there is a clear separation between the services functions and the transport functions. An *open interface* is provided between both. NGN allow the provisioning of both existing and new services independently of the network and the access type used. In NGN the *functional entities* controlling policy, sessions, media, resources, service delivery, security, etc. may be distributed over the infrastructure, including both existing and new networks. When they are physically distributed, they communicate over open interfaces. That is why the NGN architectures proposed in standards bodies and fora consist of layers and planes and show a lot of reference points. New protocols are being standardised to provide the communication between those functional entities. *Interworking* between NGN and existing networks such as PSTN, ISDN and GSM is provided by means of Gateways.

The following deals with the IP Multimedia Subsystem (IMS) architecture that could be considered as the most NGN mature architecture. The integration of DVB-RCS system in an IMS global network is presented.

### 4.4.1 IMS architecture

The QoS management in IMS architecture relies on several functional components:

- The Call Session Control Function (CSCF) is the standardized access point for an IMS user. It is in charge of the service policy management. According to the network architecture it could be a proxy (P-CSCF), a server (S-CSCF) or an edge entity (I-CSCF).
- The Policy Decision Function (PDF) is in charge of the service policy decisions.
- The Policy Enforcement Function (PEF) is an IP packet conditioner (traffic shaping) and a classifier.
- The Application Functions (AF) offers the services to the user.

IMS doesn't implement directly the QoS management but relies on the underlying networks. IMS standardizes the interface for the QoS management of these networks.

When considering access networks, the interface between PDF and the access link is ensured by COPS and the interface between the user terminal and the P-CSCF is based on SIP protocol.

### 4.4.2 Resource provisioning using SIP in DVB-RCS networks

In the IP oriented satellite architecture, the QoS functions are triggered in the RCST by the QoS Server as explain before, but this could directly be done by application servers to avoid user configuration. A SIP Proxy (Alphand, 2005) with extended functionalities has been developed to automate QoS reservations with multimedia applications. SIP (Session Initiation Protocol) should be used by multimedia applications for the session management and codec negotiation. The main functionality of a SIP proxy is to route SIP messages (session establishment for instance) from a SIP entity (an application) to another SIP entity. To remain the QoS reservation transparent for the application, the enhanced SIP proxy has to extract SDP (Session Description Protocol) descriptions from SIP messages exchanged between caller and callee, translate the parameters of each media in the session in QoS characteristics and send QoS reservations with the QoS Server on behalf of the application. The QoS-Aware SIP proxy could be centralized in the operator network part, but it could be also implemented in each RCST to reduce the connection setup time.

### 4.4.3 Integration of DVB-S2/RCS networks

The IP oriented architecture presented before is compatible with the implementation of the QoS in the IMS architecture (Baudoin, 2009) according to several modifications.

Compared to this medium term architecture, the P-CSCF functions are managed by a QoS-Aware SIP proxy collocated with the Gateway. The ARC should act as a PDF. It has "enforcement" functionalities when it configures at the same time the DAMA. The QoS Server has the role of QoS enforcement function (PEF) and could be triggered by the ARC using the COPS protocol. The SIP protocol is used continuously during the QoS configuration.

However, the IMS concept is well suited to a deployment in mobile networks, where end-users are directly linked across the wireless link. Conversely, in the DVB-RCS model, end users and satellite networks are separate entities, which is the major obstacle for IMS integration in DVB-based networks. DNS or DHCP could partially provide solutions. Moreover, accounting in satellite systems is generally private and centralized. Efforts have to be done to adapt it to a larger and compatible accounting architecture.

## 5. Evaluation of a DVB-S/RCS satellite system

### 5.1 Methodologies

In this section, we will describe the different methodologies allowing evaluating the performance of a DVB-S/RCS satellite system.

#### 5.1.1 Simulation: NS-2

Development and implementation of new services on satellite networks take time and money, so before being implemented on a real system, an implementation on a test-bed has to be done. The downside of emulation is the time needed to build and configure it, moreover it needs consequent resources and it is not easily scalable. Simulations allow overcoming some of these issues by providing a scalable and effortlessly reconfigurable network at lower cost, the major disadvantage being the realness of the results obtained since they are ruled by the model accuracy.

Network Simulator 2 (NS-2) is a discrete event simulator for networks research. Supported by DARPA and NSF, its modularity made it one of the most famous network simulators: researchers use it all around the world to implement new components or to check theoretical works. NS-2 provides protocols and communication mediums that can be settled depending on simulations needs. Updates from the development team and patches from research groups allow NS-2 to simulate current networks and latest protocols. NS-2 handles simulations of satellite networks but several components are missing, dynamic bandwidth allocation is provided by patch and there is neither QoS architecture nor encapsulation scheme. The TDMA-DAMA mechanism has been implemented and tested in (Gotta et al., 2006); the MAC layer introduced is used by our model to represent the dynamic bandwidth allocation (see section 0).

#### 5.1.2 Advanced Emulation testbed: PLATINE

Simulation and emulation both provide the opportunity to evaluate performances, at low cost, on more or less realistic systems. When simulation needs a complete modeling of the systems from applications to physical network and operates in virtual time, emulation is more demonstrative since real applications can be deployed over the model describing transfer characteristics, delay and error behavior for instance.

The (IST SATSIX, 2004) project relies on these different methods of evaluation and validation, with the design and development of simulations, emulations and trials. The emulation testbed relies on the satellite emulation platform (PLATINE) (Baudoin et al., 2007) formerly developed within the frame of the (IST SATIP6, 2001) project. It aims to demonstrate the network and application services integration on next generation satellite systems and the possibility to interoperate with terrestrial networks. With regards to the previous release, PLATINE includes a DVB-RCS and DVB-S2 emulation, with ULE/MPEG2-TS and AAL5/ATM stacks together with the adaptive physical layer simulation and the associated radio resources management (RRM). A complete QoS architecture mixes ehanced SIP proxies (as described in 0), IP/MAC scheduling and cross layer techniques is available.

### 5.1.3 Real experiments: OURSES

During the (OURSES, 2006) project, we had he opportunity to use a DVB-S2/RCS system. A platform compliant with the IP oriented architecture was setup during the project. The gateway and the terminals are compliant with the Satlabs recommendations. The four (VoIP, ViC, Critical Data, Best Effort) Diffserv class of service are offered on the STM satlink 1000 terminals we used. A Service Level Agreement (SLA) is setup on the gateway side (Thales A9780 model) for each customer. It fixes the limits in terms of bandwidth with each MAC service classes. The tests were done using a satellite channel emulator and the Ka band.

## 5.2 Performance evaluations

### 5.2.1 DVB-S/RCS NS-2 simulation model with QoS

This section briefly describes the DVB-S/RCS NS-2 simulation model with QoS architecture that have been developed at LAAS/CNRS, further details on implementation and simulations can be found in (Gayraud et al., 2009). Such model can be used to simulate new protocols or to compare results with measurements obtained through emulation or experimentations done on a real link. To be efficient, architecture and behaviour have to be as closed as possible from the chosen satellite network (the one from OURSES project in our case). The model is using a TDMA-DAMA MAC layer above the physical layer defined by NS-2. The simulation made on the model without our contribution shows a really efficient behaviour with a dynamic bandwidth allocation and a fast establishment of connections. However, to be closer from the real system and to improve performances, some features need to be added:

- Dynamic encapsulation of IP packets,
- Substitution of the single queue at MAC layer by two distinct queues,
- Addition of queues at IP layer (inspired from DiffServ architecture).

Since packets fragmentation is not possible with Network Simulator 2, the MAC layer adjusts the sending time to the available bandwidth based on the assigned slots. The dynamic encapsulation (from IP to ATM frames) doesn't fragment the packets either, but their size is settled according to AAL5 protocol, resulting in a consistent overhead (around 10 percent).

QoS architecture is implemented by duplicating the queue at MAC layer and by adding buffers at IP layer: flows are aggregated, differentiated and stocked according to the DIFFSERV architecture from terrestrial network; management is done by a packet scheduler below the queues. To study the model behaviour, two kind of traffic with specific constraints were generated:

- Constant Bit Rate (CBR) needing low delay and jitter, associated to Real Time flows (RT),
- File Transfer Protocol (FTP) needing large bandwidth regardless to delay, associated to non Real Time flows (n-RT).

The chosen transport protocols are respectively UDP and TCP, the most commons for such flows. During experiments, the available bandwidth is settled to 128kbps per slot (this ratio depending on weather conditions and chosen coding scheme), since one satellite terminal can have at most two slots (256kbps), CBR rate has been settled to 128 kbps (without encapsulation overhead). Flows will compete with each other, the main point of simulations being to show the efficiency of the QoS architecture added to the model: CBR flows should get the lowest delay possible while FTP flows would still be able to establish communication and transfer data.

The rest of this section will focus on the QoS architecture by taking a look at the model's behavior. To illustrate the competition between the two types of flows, delay suffered by the communications and the throughput they can achieved are shown on Fig. 6.

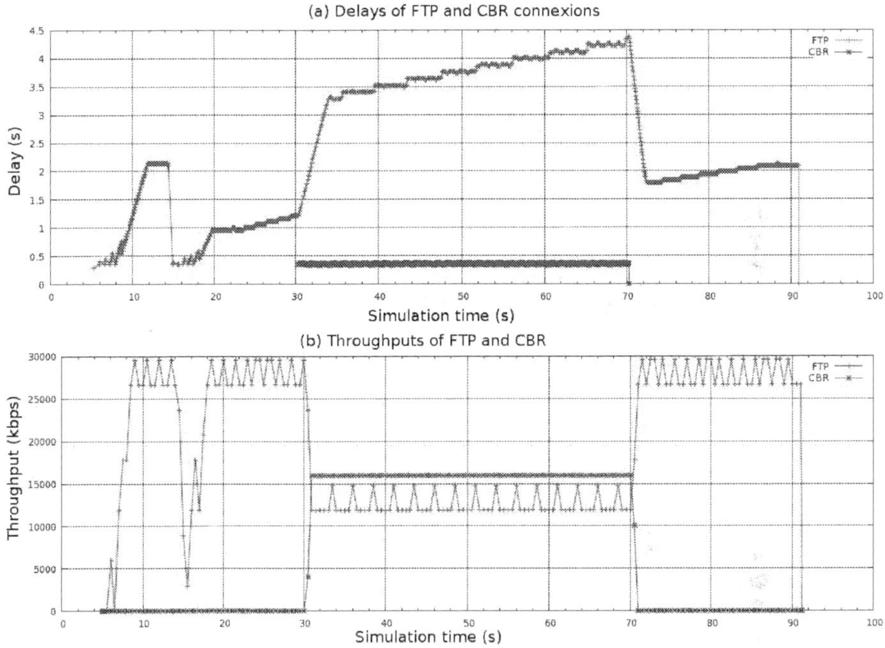


Fig. 6. a) Delay suffered by connections. b) Throughput of connections.

Differences between RT and n-RT flows are clearly visible: delay suffered by CBR is stable and below 500ms while FTP delay fluctuates and is above 3s (Fig. 6a). On Fig. 6.b, it is noticeable that FTP throughput is restricted by bandwidth taken by CBR. These two results illustrate the model's behavior by showing the differentiation done on those flows; the one with more constraints is getting the lower delay possible and enough bandwidth so no loss occurs. For n-RT flow, the throughput and the delay are fluctuating depending on network load and bandwidth allocated to the satellite terminal.

The model behaves properly and reacts as we expect: indeed; it provides an efficient QoS architecture to the basic satellite network from NS-2. But some improvements can still be done on the model: using a more efficient manager below the MAC buffers and providing a thinner encapsulation mechanism. There are also some features needing to be tested: using RED instead of DropTail policy in satellite terminal buffers or using a more realistic error model (already implemented but not used during simulations). The latest experiments were done to study SCTP (Stream Control Transport Protocol) behavior on a QoS satellite network and compare it with TCP; results can be found in (Bertaux et al., 2010).

### 5.2.2 PLATINE performances evaluations

In the following parts, we will show two exemples of QoS management in DVB-S2/RCS satellite systems, using the VisioSIP client (a SIP videoconferencing tool) and a QoS-aware SIP proxy (located behind each ST and based on the NIST-SIP Proxy) that send reservation or release messages to a QoS Server, located on RCSTs and able to reconfigure DiffServ queues to prioritize flows with strong time-constraints (VoIP, videoconferencing, etc...). Moreover, we consider that each ST has a total bandwidth of 1000kbps.

#### 5.2.2.1 Impact of the queue management: BE vs EF

We consider here that a SIP videoconferencing session is initiated between two SIP clients located behind two separate STs. The SIP session starts at t=t0 + 10s and then 3 concurrent UDP flows (500 kbps) start respectively at t=t0+60s, t=t0+120s and t=t0+180s and terminate at t=t0+240s. Finally the SIP session ends at t=t0+300s. Moreover, 150 kbps of CRA is allocated to the studied ST to support, in terms of bandwidth, the video and audio flows.
We will make the analysis on the audio delays graphs presented on Fig. 7, but the same analysis will apply to the video delays graphs that are similar.
These two series of delays' graphs show a real benefit of the IPv6 QoS usage and a fair separation of the classes of service can be observed on the first graphs. Detail analysis of those graphs is now provided.
First, concerning the comparison of the graphs with and without QoS, it can be observed a real improvement when the QoS architecture is running especially when background traffic is high: The "moving average delay" graphs show that when two or three concurrent flows are running (between 120 and 240 ms) a very high increase of average delay is experienced by the audio flow when the QoS is not set (above 4 seconds delay) while the average delay remains below 360 ms when the QoS is set, which is compatible with audio conference requirements. In the case of high load on the satellite return link, the impact of the QoS architecture is clearly shown here.

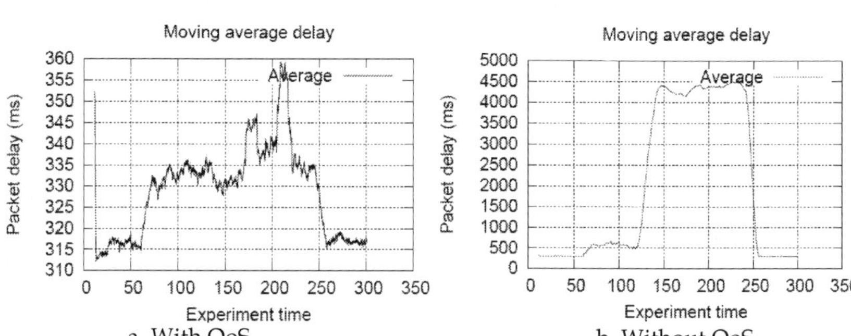


Fig. 7. Moving average delay for the audio flow

When no concurrent flows are running, delay for the audio flow is around 300ms in both cases (with and without QoS), cf. graphs between 0 and 60 seconds. This can be explained by the fact that all CRA resources, in this case, are used by the multimedia flows and no on-demand capacity is needed. When just one concurrent UDP flow of 500kbps is running the delay of

VoIP application is increasing in both cases but very slightly when QoS is set (from 315 ms up to 330 ms on average) while it's increasing up to 500 ms on average in the case no QoS is set. The capacity of the channel should be enough for both flows but the CRA capacity is not enough and on-demand RBDC bandwidth is required. So the audio flow experiences more delay when QoS is not set; this is due to the fact that all flows (audio, video and best-effort flows) are using the same MAC buffer and PVC, and so the same delay is experienced by all packets in this buffer, implied by the capacity allocation scheme. When the QoS is set, a different MAC buffer and PVC is used for high priority traffic (audio and video packets) and is served first compared to the low priority MAC buffer. Consequently, the audio flow is protected and the delay is increasing very slightly: it's experiencing an end-to-end delay compatible with audio conference application requirements (under 400ms).

Secondly, concerning the classes of service separation, we can notice on the first graph (a) with QoS that the impact on high priority classes of service of concurrent flows is rather low, and does not degrade the overall quality for end-to-end users: the delay remains below 400ms which is acceptable for interactive audio conference applications. The delay increases from 315 ms up to 360 ms, which can be explained by the sending time for large low priority packets.

#### 5.2.2.2 Impact of the RBDC mechanism on interactive applications

The following experiments show the impact of the DAMA algorithm on interactive applications. The teleconferencing application takes the EF service class and the background traffic the Best Effort service class, but, unlike the previous experiment, there is no CRA allocated to this Satellite Terminal, all the capacity is given with RBDC requests. The teleconferencing application is first started, then 3 concurrent UDP flows of 400 kbps start and terminate at the same time than the previous experiment.

On Fig. 8.a., the delay experienced by the audio stream is less than 700 ms (this is the same values for video stream) and decreases to very low delay. The first noticeable thing is that the DAMA algorithm works fine with audio and video streams. The delay stays stable, around 650 ms, even with the throughput variation. The second noticeable thing is the delay diminution that occurs during the experiments. This can be explained by the fact that the teleconferencing application takes benefit from the RBDC requests made for background traffic as this traffic has a better priority.

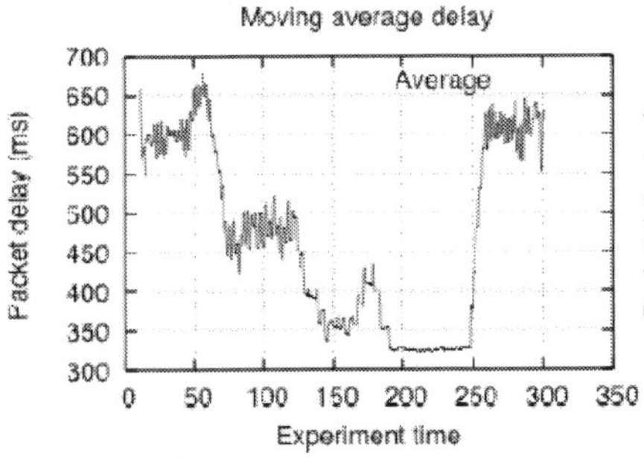


a. Audio stream delay

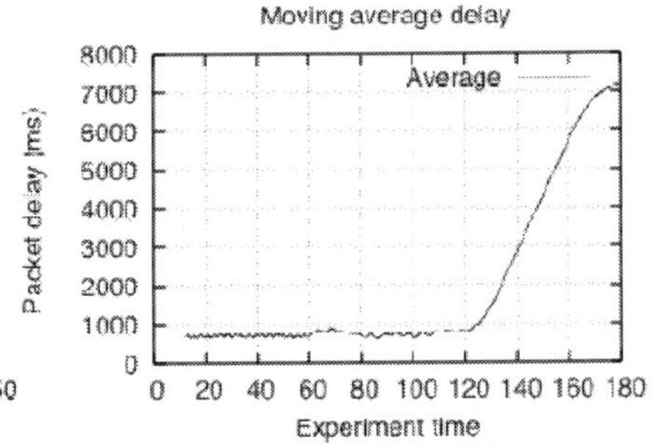


b. First UDP flow delay

Fig. 8. Moving average delay for the audio flow

On Fig. 8.b, as the link capacity is not reached, the packet delay is stable, below one second but, of course, when there is no more capacity, the delay increase, but only for the Best Effort Class.

The main problem, in this case, is that the delay of audio and video flows is often higher than what is advised in ITU-T recommandations (ITU-T, 2001), namely a value inferior to 400 ms. Consequently, to provide a solution to lower the delay given with the RBDC mechanism when no CRA (or no sufficient CRA) are allocated to a specific ST, a new extension of the SIP Proxy has been proposed to allow it to communicate with an entity located at the NCC side: the Access Resource Controller (ARC). When a SIP session is initiated, the SIP proxy can intercept the SDP, deduct the codec bitrate and ask to the ARC to increase the quantity of CRA allocated to the concerned ST corresponding to the sum of codec bitrates. The ARC checks if the SIP clients are authorized to use this service and decides to accept or reject the resource reservation.

## 6. Conclusion

This chapter has explained the way that can be used to provide such a satellite network client with the QoS he requested. It was proven that these QoS archtectures are feasible, that their performances are good enough by several actions like simulation, emulation, and real systems.

The work on QoS architecture is still ongoing and heterogeneous access networks mixing satellite and other radio techniques such as Wimax, and wireless systems in general. This work will lead in the very next future to the implementation of some of ours. It seems that the first network ensuring QoS may be the satellite systems that were described, designed and evaluated in the work as described in this paper.

## 7. References

Baudoin, C.; Dervin, M.; Berthou, P.; Gayraud, T.; Nivor, F.; Jacquemin, B.; Barvaux, D. & Nicol, J. (2007). PLATINE: DVB-S2/RCS enhanced testbed for next generation satellite networks. *Proceedings of International Workshop on IP Networking over Next-generation Satellite Systems (INNSS'07)*, pp. 251-267, ISBN: 978-0-387-75427-7, Budapest, July 2007, Springer New-York.

Bertaux, L.; Gayraud, T. & Berthou, P. (2010). How is SCTP Able to Compete with TCP on QoS Satellite Networks ? *The Second International Conference on Advances in Satellite and Space Communications (SPACOMM'10)*, Greece, June 2010.

Blake, S.; Black, D.; Carlson, M.; Davies, E.; Wang, Z. & Weiss, W. (1998). *An Architecture for Differentiated Service*, IETF RFC 2475.

Braden, R.; Clark, D. & Shenker, S. (1994). *Integrated Services in the Internet Architecture : an Overview*, IETF RFC 1633.

Braden, R.; Zhang, L.; Berson, S.; Herzog, S. & Jamin, S. (1997). *Resource ReSerVation Protocol (RSVP) – Version 1 Functional Specification*, IETF RFC 2205.

Camarillo, G.; Marshall, W. & Rosenberg, J. (2002). *Integration of Resource Management and Session Initiation Protocol (SIP)*, IETF RFC 3312.

Durham, D.; Boyle, J.; Cohen, R.; Herzog, S.; Rajan, R. & Sastry, A. (2000). *The COPS (Common Open Policy Service) Protocol*, IETF RFC 2748.

Gayraud, T.; Bertaux, L. & Berthou, P. (2009). A NS-2 Simulation model of DVB-S2/RCS Satellite network. *Proceedings of the 15th Ka and Broadband Communications - KaBand'09)*, pp.663-670, Italia, September 2009.

Gotta, A.; Potorti, F. & Secchi, R (2006). Simulating Dynamic Bandwidth Allocation on Satellite Links. *Proceeding from the 2006 workshop on ns-2: the IP network simulator (WNS2)*, ISBN:1-59593-508-8, Italia, October 2006, ACM New York.

Grossman, D. (2002). *New Terminology and Clarifications for DiffServ*, IETF RFC 3260.

Handley, M.; Jacobson, V. & Perkins, C. (2006). *SDP : Session Description Protocol*, IETF RFC 4566.

Hardy, W. C. (2001). *QoS Measurements and Evaluation of Telecommunications Quality of Service*, ISBN : 0-471-49957-9, Wiley.

Heinanen, J.; Baker, F.; Weiss, W. & Wroclawski, J. (1999). *Assured Forwarding PHB Group*, IETF RFC 2597.

ISO8402 (2000). *Quality Management and Quality Assurance Vocabulary*. Technical Report, International Organization for Standardization.

ITU-T-Rec. E.800 (1993). *Terms and Definitions Related to Quality of Service and Network Performance Including Dependability*, Technical Report, International Telecommunication Union.

ITU-T-Rec. G.1010 (2001). *End-user Multimedia QoS Categories*, Technical Report, International Telecommunication Union.

Jacobson, V.; Nichols, K. & Poduri, K. (1999). *An Expedited Forwarding PHB*, IETF RFC 2598.

Nichols, K.; Jacobson, V. & Zhang, L. (1999). *A Two-bit Differentiated Services Architecture for the Internet*, IETF RFC 2638.

Rosenberg, J.; Schulzrinne, H.; Camarillo, G.; Johnston, A.; Peterson, J.; Sparks, R.; Handley, M. & Schooler, E. (2002). *SIP: Session Initiation Protocol*, IETF RFC 3261.

Shenker, S.; Partridge, C. & Guerin, R. (1997). Specification of Guaranteed Quality of Service, IETF RFC 2212.

Wroclawski, J. (1997). Specification of the Controlled-Load Network Element Service, IETF RFC 2211.

D. Awduche and al., (2001), RFC 3209: RSVP-TE: Extensions to RSVP for LSP Tunnels.

F. Le Faucheur and al. (2002), RFC 3270: Multi-Protocol Label Switching (MPLS) Support of Differentiated Services.

S. Combes, S. Pirio, (2008), ESA/ESTEC, SatLabs System Recommendations - Quality of Service Specifications.

C. Baudoin and al., (2009), On DVB Satellite Network Integration in IMS, IWSSC, Sienna, Italy.

O. Alphand, and al, (2005), QoS Architecture over DVB-RCS satellite networks in a NGN framework, Globecom, St Louis, United States.

IST SATIP6 Project, (2001), (Contract IST-2001-34344)

IST SATSIX Project (2004), (Contract IST-2004-26950)

OURSES project, (2006), http://www.ourses-project.fr

2

# Antenna System for Land Mobile Satellite Communications

Basari, Kazuyuki Saito, Masaharu Takahashi and Koichi Ito
*Chiba University*
*Japan*

## 1. Introduction

Personal wireless communications is a true success story and has become part of people's everyday lives around the world. Whereas in the early days of mobile communications Quality of Service (QoS) was often poor, nowadays it is assumed the service will be ubiquitous, of high speech quality and the ability to watch and share streaming video or even broadcast television programs for example is driving operators to offer even higher uplink and downlink data-rates, while maintaining appropriate QoS.

Terrestrial mobile communications infrastructure has made deep inroads around the world. Even rural areas are obtaining good coverage in many countries. However, there are still geographically remote and isolated areas without good coverage, and several countries do not yet have coverage in towns and cities. On the other hand, satellite mobile communications offers the benefits of true global coverage, reaching into remote areas as well as populated areas. This has made them popular for niche markets like news reporting, marine, military and disaster relief services. However, until now there has been no wide-ranging adoption of mobile satellite communications to the mass market.

Current terrestrial mobile communication systems are inefficient in the delivery of multicast and broadcast traffic, due to network resource duplication (i.e. multiple base stations transmitting the same traffic). Satellite based mobile communications offers great advantages in delivering multicast and broadcast traffic because of their intrinsic broadcast nature. The utilization of satellites to complement terrestrial mobile communications for bringing this type of traffic to the mass market is gaining increasing support in the standards groups, as it may well be the cheapest and most efficient method of doing so.

In order to challenge the great advantages of mobile satellite communications, the Japan Aerospace Exploration Agency (JAXA) has developed and launched the largest geostationary S-band satellite called Engineering Test Satellite-VIII (ETS-VIII) to meet future requirements of mobile communications. The ETS-VIII conducted various orbital experiments in Japan and surrounding areas to verify mobile satellite communications functions, making use of a small satellite handset similar to a mobile phone. The mobile communication technologies adopted by ETS-VIII are expected to benefit our daily life in the field of communications, broadcasting, and global positioning. Quick and accurate directions for example, can be given to emergency vehicles by means of traffic control information via satellite in the event of a disaster (JAXA, 2003).

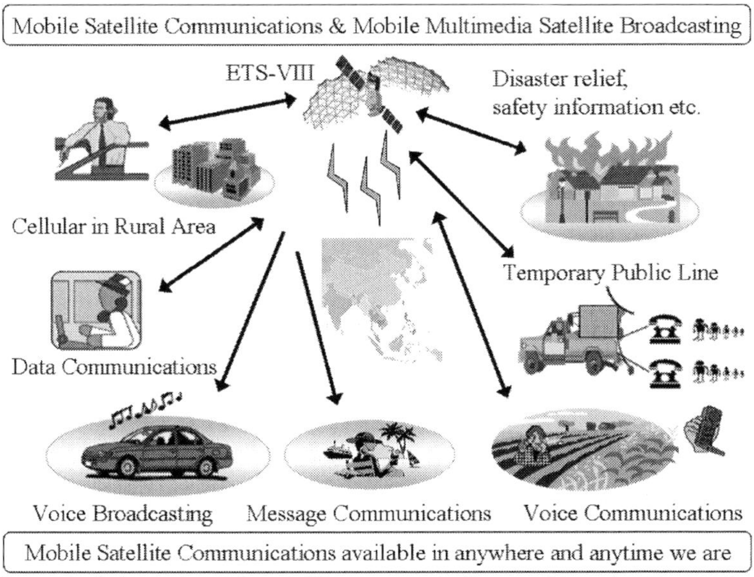


Fig. 1. Conceptual chart of mobile satellite communications and broadcasting system (JAXA & i-Space, 2003)

Figure 1 shows some of services made possible through the technological developments with the ETS-VIII. The mission of ETS-VIII is not only to improve the environment for mobile-phone based communications, but also to contribute to the development of technologies for a satellite-based multimedia broadcasting system for mobile devices. It will play as important role in the provision of services and information, such as the transmission of CD-quality audio and video; more reliable voice and data communications; global positioning of moving objects such as cars, broadcasting; faster disaster relief, etc (JAXA & i-Space, 2003).

In addition, nowadays as can be seen with the spreading of the GPS or the Electronic Toll Collection (ETC), the vehicular communications systematization is remarkable. From this phenomenon, in the near future, system for mobile satellite communications using the Internet environment will be generalized and the demand for on-board mobile satellite communications system as well as antenna is expected to increase. So far, we are enrolled in the experimental use of ETS-VIII and develop an onboard antenna system for mobile satellite communications, in particular for land vehicle applications.

In this chapter, we will figure out realization of an antenna system and establishing a mobile communication through a geostationary satellite by designing smaller and more compact antenna, developing a satellite-tracking program which utilizes Global Positioning System (GPS) receiver or gyroscope sensor, and data acquisition program which utilizes spectrum analyzer for outdoor measurement using the signal from the satellite. First, in order to minimize the bulky antenna system, a new structure of active integrated patch array antenna is proposed and developed without phase shifter circuit, to realize a light and low profile antenna system with more in reliability and high-speed beam scanning possibility. Then, the antenna system is built by the proposed antenna which its beam-tracking characteristics is determined by the control unit as the vehicle's bearing from a navigation system (either gyroscope or GPS receiver). Here, the antenna system will be installed in a vehicle and communicate with the satellite by tracking it during travelling as a concept of the antenna system.

This chapter will be divided by several sections from the research background, antenna design, numerical results, chamber measurement verification, realization on overall antenna system design, and finally antenna system verification by conducting measurement campaign using the satellite.

This chapter is organized as follows. Section 2 will provide review of mobile satellite communications systems in particular its design parameters. An example of a link budget for a mobile satellite application is given. Section 3 describes designing issues on vehicle antennas for mobile satellite system from their mechanical and electrical requirements, and also their tracking functions. In this section, we also describe our proposed antenna system, especially aimed at ETS-VIII applications. Section 4 will focus on the planar antenna design for compactness and integrated construction. It provides details about the measurement results of some basic antenna performances, such as $S_{11}$, axial ratio and radiation pattern characteristics that compared with the numerical results which are calculated by use of moment method. Section 5 will describe about verification of all antenna system in laboratory test and experimentally confirm in outdoor immobile-state measurement to verify the satellite-tracking performances using gyro sensor system under pre-test for field measurement campaign. The effect of radome and ground plate also will be discussed. Section 6 will show various field experiments results by utilizing the satellite to verify the validity of our developed antenna system. Overall system is tested for its performance validity not only propagation characteristics but also bit error rate performance. Finally, the last section draws conclusions on the work, and provides scope and direction for promotion in the future applications.

## 2. Mobile Satellite System Communication

### 2.1 Mobile Satellite System Architecture

Figure 2 describes a typical design for mobile satellite communication system. Three basic segments: satellite, fixed and mobile earth station are included. A propagation path is added as another fourth segment owing to its importance factor that mainly affects the channel quality of the communication system. In land mobile satellite system, the most serious propagation problem is the effect of blocking caused by buildings and surroundings objects, which cause losing the satellite signal completely. The second problem is shadowing caused by tree and foliage, resulting the signal attenuation. The other is multipath fading, which is mainly caused by surrounding buildings, poles and trees. However, such an effect can usually be ignored when the directional antenna is used since less reflected signals approach the receiver.

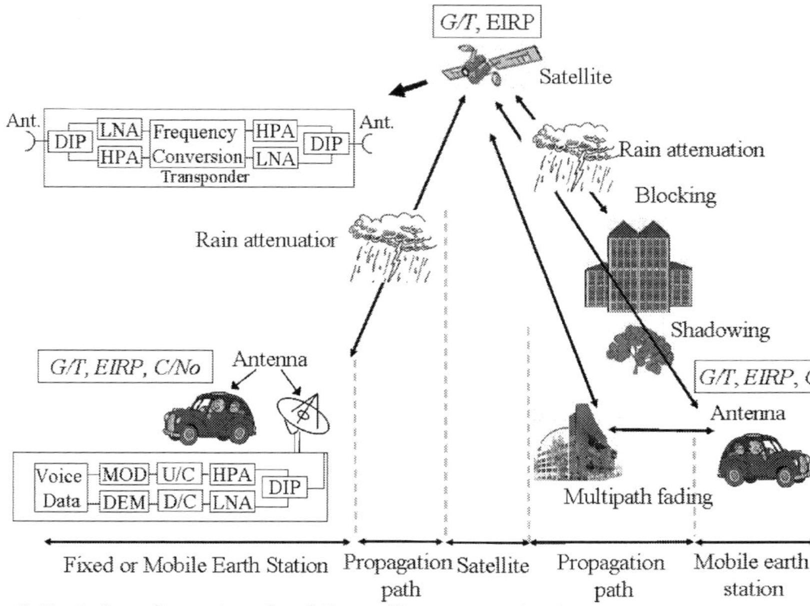


Fig. 2. Typical configuration of mobile satellite communications

Fixed or mobile earth station system consists of antenna, diplexer (DIP), up-converter and down-converter (U/C and D/C), high power amplifier (HPA) and low noise amplifier (LNA), as well as modulator (MOD) and demodulator (DEM). The satellite system is almost similar either for fixed or mobile earth station which can be constructed by antenna and up-converter and down-converter, called a transponder. Most of commercial satellites do not have modulator and demodulator. They only transmit a signal after converting its frequency and amplify the received weak signals, or usually called a bent pipe transponder or a transparent transponder.

### 2.2 Mobile Satellite System Link Parameters

Performance of mobile satellite system is characterized by three main parameters for link budget. Those parameters indicate the performance of three segments -namely satellite, fixed and mobile earth station- are *G/T* (ratio of antenna gain to system noise temperature or usually called figure of merit), effective isotropically radiated power (*EIRP*) and $C/N_0$ (ratio of carrier power to noise power density). The *G/T* and *EIRP* denote the receiving and transmitting capabilities, respectively, of satellite, fixed earth station and mobile terminal. The $C/N_0$ indicates the quality of the communication channel.

The $G/T$ is calculated from a value of $G$ which means system gain at the input port to the Low Noise Amplifier (LNA). Consequently, the ratio of antenna gain to noise temperature at the input port to the LNA can be written as:

$$\frac{G}{T} = \frac{G_R}{T_a + T_0(L_f - 1) + T_R L_f} \quad \text{(dBK)} \tag{1}$$

where $G/T$: figure of merit, $G_R$: gain of receiving antenna, $T_a$: antenna noise temperature, $T_0$: physical temperature when the circuit immersed, $T_R$: receiver noise temperature, $L_f$: total loss of feed lines and components such as diplexers, cables, and phase shifters (if used). As for the transmitter, the *EIRP* is one of the important parameters to describe the capabilities of transmission. The *EIRP* can be expressed as:

$$\begin{aligned} EIRP &= G_T \cdot P_T \quad \text{(watts)} \\ \text{or} \quad [EIRP] &= [G_T] + [P_T] \quad \text{(dBW)} \end{aligned} \tag{2}$$

where $G_T$: gain of transmitting antenna and $P_T$: transmitted power.

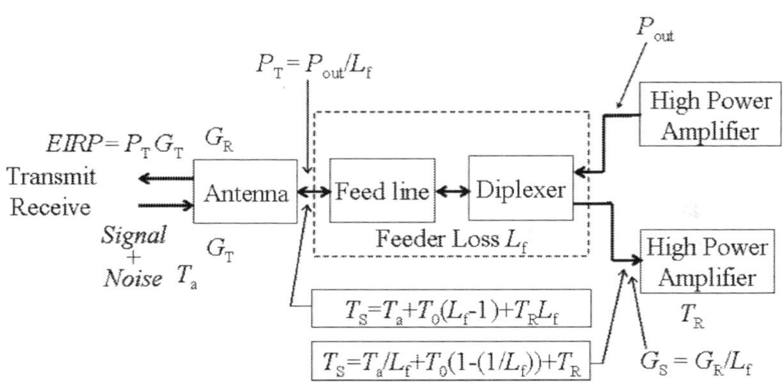

Fig. 3. Typical RF stage at earth station and satellite

In general, the radio frequency stages of earth station and satellite consist of antenna, feed line, diplexer, high power amplifier (HPA), and low noise amplifier (LNA), as shown in Fig. 3. From the figure, the ratio of input signal power ($C$) to noise power density ($N_0$) or simply called carrier to noise density ratio ($C/N_0$) at the input point to the antenna can be written as follows:

$$\begin{aligned} \frac{C}{N_0} &= \frac{EIRP}{L_P}\left(\frac{G_R}{T_S}\right)\frac{1}{\kappa} \\ \left[\frac{C}{N_0}\right] &= [EIRP] - [L_P] + \left[\frac{G_R}{T_S}\right] + 228.6 \end{aligned} \tag{3}$$

where $L_P$: free space propagation, $G_R/T_S$: figure of merit, and $K$ : Boltzman's constant ($1.38 \times 10^{-23}$ watt/sec/K).

The total channel quality in the system is calculated by including the uplink and downlink channels, given by:

$$\left(\frac{C}{N_0}\right)_{Total} = \left(\frac{1}{\left(\frac{C}{N_0}\right)_{Uplink}} + \frac{1}{\left(\frac{C}{N_0}\right)_{Downlink}}\right)^{-1} \quad \text{(dBHz)} \tag{4}$$

In land mobile satellite communications, the gain of mobile station is quite smaller than the satellite has, allowing the total quality is dominated by the poor uplink and the total channel quality will never exceed the uplink quality no matter how much downlink quality is increased.

Once the system channel quality is calculated, the next is what kind of modulation scheme is suitable for communication. Mobile satellite communication system currently uses digital modulation schemes, such as π/4-QPSK, OQPSK, or MSK (Lodge, 1991); low-bit rate digital voice encoder-decoder of about 4.8 to 6.7 kbps, such as vector sum excited linear prediction (VSELP), low-delay code excited linear prediction (LD-CELP), adaptive differential pulse code modulation (ADPCM), regular pulse excited linear prediction code with long-term prediction (RPE-LTP); and powerful forward error correction (FEC) technologies.

Next, another most important performance parameter for mobile satellite modulation schemes is efficiency, which includes both power and bandwidth efficiency, since mobile satellite communication systems usually have limited availability of both power and bandwidth. The power efficiency is defined as the ratio of required signal energy per bit to noise density ($E_b/N_0$) required to achieve a given bit error rate (BER) over an additive white Gaussian noise (AWGN) channel, although in fact, mobile satellite communication channel is Ricean fading channel. However, let we deal with performance over AWGN for simplicity in design. The bandwidth efficiency is defined as ratio of information rate $R$ [bit/s] and the required channel bandwidth $B$.

For analog signals (passband signals) $C/N_0$ is used in the same way as $E_b/N_0$ for digital (baseband signals). $C$ and $E_b$ are related by the bit rate by:

$$C = E_b \cdot R_b \tag{5}$$

So, $C/N_0$ is:

$$C / N_0 = \left(E_b / N_0\right) \cdot R_b \tag{6}$$

Due $C/N_0$ does not relate with bit rate, as increasing bit rate does not affect $C/N_0$ value, but $E_b/N_0$ get decreasing.

Some parameters besides efficiency as described earlier, such as immunity to nonlinearity, simplicity for implementation are also be considered for designing a simple and small mobile satellite modem. Table 1 shows comparison of several modulation schemes in terms of bandwidth, $E_b/N_0$ for $BER$ $10^{-5}$, non linearity immunity, and implementation simplicity for BPSK, QPSK, OQPSK, π/4-PSK, and MSK.

* in normalized frequency offset from the center frequency

** A is highest value

| Modulation scheme | Half-power Bandwidth* | Noise Bandwidth* | $E_b/N_0$ *for* $BER = 10^{-5}$ | Nonlinearity Immunity** | Implementation Simplicity** |
|---|---|---|---|---|---|
| BPSK | 0.88 | 1.00 | 9.6 dB | D | A |
| QPSK | 0.44 | 0.50 | 9.6 dB | C | B |
| OQPSK and π/4-PSK | 0.44 | 0.50 | 9.6 dB | B | C |
| MSK | 0.59 | 0.62 | 9.6 dB | A | D |

Table 1. Several digital modulation schemes for mobile satellite communications (Xiong, 1994)

Finally, we calculate the above mentioned parameters for link budget. Here, we design vehicle to vehicle communication via satellite for low rate data and voice communications. We assume that the data rate is 8 kbps by using a convolutional code with viterbi decoder, and 5.6 kbps PSI-CELP voice codec are suitable for conveying low data rate and voice signal from ground vehicle to another vehicle through the geostationary satellite. An example of link budget calculation is depicted in Table 2 below. From this example, by use of a large satellite antenna (i.e. very high gain antenna), we can design a small mobile earth station to enable development of a compact land vehicle communication.

| Link parameter | Forward link | |
|---|---|---|
| Uplink | | |
| Uplink frequency (GHz) | | 2.6575 |
| $T_x$ power (Watt) | | 1.00 |
| Feed loss (dB) | | 1.70 |
| Antenna gain (dBi) | | 5.00 |
| $T_x$ EIRP (dBW) | Vehicle | 3.20 |
| Tracking loss (dB) | | 3.00 |
| Propagation loss (dB) | | 192.35 |
| Received level (dBW) | | - 190.25 |

| | | |
|---|---|---|
| Satellite antenna gain (dBi)<br>Feed loss (dB)<br>Satellite $G/T$ (dB/K)<br>System noise temperature (K)<br>Uplink $C$ (dBW)<br>$N_0$ (dB/Hz)<br>Uplink $C/N_0$ (dBHz) | Satellite | 43.80<br>2.60<br>14.04<br>520.00<br>- 151.95<br>- 201.44<br>49.49 |
| Downlink | | |
| Downlink frequency (GHz)<br>$T_x$ power (Watt)<br>Feed loss (dB)<br>Satellite gain (dBi)<br>Pointing loss (dB)<br>$T_x$ EIRP (dBW)<br>Propagation loss (dB)<br>Received level (dBW) | Satellite | 2.5025<br>40.00<br>2.60<br>43.80<br>3.00<br>54.22<br>191.83<br>- 138.71 |
| Antenna gain (dBi)<br>Feed loss (dB)<br>Tracking loss (dB)<br>Satellite $G/T$ (dB/K)<br>System noise temperature (K)<br>Downlink $C$ (dBW)<br>$N_0$ (dB/Hz)<br>Downlink $C/N_0$ (dBHz) | Vehicle | 5.00<br>1.70<br>3.00<br>- 22.92<br>418.60<br>- 138.41<br>- 202.38<br>63.97 |
| Calculation Results | | |
| Total $C/N_0$ (dBHz)<br>Bit rate (kbps)<br>$E_b/N_0$ (dB)<br>Coding gain (Convolutional code $R$=1/2, $K$=5, with Viterbi decoder and without interleaver) for BER=$10^{-5}$<br>Required $C/N_0$ (dBHz)<br>Margin (dB) | | 49.34<br>8.00<br>10.31<br>5.00<br>43.63<br>5.71 |

Table 2. Link budget calculation for land mobile satellite communication

## 3. Antenna System Design for Vehicle Application

### 3.1 Vehicle Antennas

In order to develop an antenna system for land vehicle application in mobile satellite communication system, considering the requirements of antenna properties both in mechanical and electrical characteristics is required. Thus, their characteristics are briefly explained in the following subsection.

### 3.1.1 Mechanical Characteristics

#### 3.1.1.1 Compactness and Lightweight

Design of mobile antennas is required as compact and lightweight as possible to minimize the space and easy installation (Ohmori et al., 1998 & Rabinovich et al., 2010). However, a compact antenna has two major disadvantages in electrical characteristics such as low gain and wide beamwidth. Due to its low gain and limited electric power supply, it is quite difficult for mobile antennas to have enough receiving capability (i.e. $G/T$) and transmission power (i.e. *EIRP*). Nonetheless, such disadvantages of mobile terminals can be compensated by providing a satellite has a large antenna and huge power amplifier with enough electric power.

The second demerit is that a wide beam antenna is likely to transmit undesired signals to, and receive them from, undesired directions, which will cause interference in and from other systems. The wide beam is also suffered from multipath fading in land mobile satellite communication. Therefore, a directive sufficient-gain antenna is expected to prevent fading and interference.

#### 3.1.1.2 Installation

Easy installation and appropriate physical shape are worthwhile requirements besides compactness and lightweight (Ohmori et al., 1998 & Rabinovich et al., 2010). The requirements antennas for cars or aircraft are different from shipborne which has enough space for antenna system installation. In case of cars, low profile and lightweight equipment is required. Aircraft antenna is required more stringent to satisfy aerodynamics standard such as low air drag (Ohmori et al., 1998). Our research concern on designing and developing a compact, lightweight and easy installation.

### 3.1.2 Electrical Characteristics

#### 3.1.2.1 Frequency and Bandwidth

The Radio Regulations of International Telecommunication Union (ITU) regulates the satellite services including allocated frequency according to each region (three regions, i.e. Region 1: Europe, Russia, & Africa; Region 2: North & South America and Region 3: Asia). The typical frequencies allocated to mobile satellite communications are the L (1.6/1.5 GHz) and S (2.6/2.4) bands which being operated in the present, Ka (30/20 GHz) band and millimeter wave for future systems (ITU-R Radio Regulation, 2004).

The required frequency bandwidth is about 7% for L band, 10% for S band, and 40% for Ka band. This chapter provides an antenna for S band application with wide frequency bandwidth and will be discussed in the next subsection.

#### 3.1.2.2 Polarization and Axial Ratio

In mobile satellite communications, circular polarized waves are used to avoid polarization tracking and Faraday rotation. When both satellite and mobile earth stations use linearly (vertical or horizontal) polarized waves, the mobile earth stations have to keep the antenna coinciding with the polarization. If the direction of the mobile antenna rotates 90°, the antenna cannot receive signals from the satellite. Even if circular polarization waves are used, the polarization mismatch loss caused by the axial ratio has to be taken into account to link budget. Generally, we design a circular polarized antenna below 3 dB axial ratio (Sri Sumantyo et al., 2005).

### 3.1.2.3 Gain and Beam Coverage

Required antenna gain is determined by a link budget, which is calculated by taking into account the satellite capability and the required channel quality. The channel quality ($C/N_0$) depends on the $G/T$ and the EIRP values of the satellite and mobile earth stations. Typical gains are shown in Table 3 according to their application at L band satellite communications.

| Antenna | Typical Gain (dBi) | Typical $G/T$ (dBK) | Typical antenna (dimension) | Typical service |
|---|---|---|---|---|
| Directional | 20–24 | -4 | Dish (1mφ) | Voice, high speed data |
| | 17–20 | -8 to -6 | Dish (0.8mφ) | Ship (Inmarsat-A,B) |
| Semi directional | 8–16 | -18 to -10 | SBF (0.4mφ) | Voice/high speed data |
| | | | Phased array | Aircraft (Inmarsat-Aero) |
| | 4–8 | -23 to -18 | Array (2-4 elements) | Ship (Inmarsat-M) |
| | | | Helical, patch | Land mobile |
| Omni directional | 0–4 | -27 to -23 | Quadrifilar, | Low speed data (message) |
| | | | Drooping- | Ship (Inmarsat-C) |
| | | | dipole | Aircraft |
| | | | Patch | Land mobile |

Table 3. Typical gain for L band satellite communications (Ohmori et al., 1998 & Ilcev, 2005)

The beams of mobile antennas are required to cover the upper hemisphere independent of mobile motions. Low gain antennas have advantages in terms of establishing communication channel without tracking the satellite because of their omnidirectional beam patterns. In contrary, high gain antennas have to track satellites owing to their narrow directional beam patterns. We design a medium gain antenna owing to the use of large dimension and high gain satellite antenna in this application.

### 3.1.2.4 Satellite Tracking

Unlike omnidirectional antennas, medium and high gain antennas need a tracking function. Tracking capabilities depend on the beamwidth of the antennas and the speed of mobile motions, where the directional antennas with narrow beams have to track the satellite both in elevation and azimuth directions. In general, the required accuracy of tracking is considered to be within 1 dB (Ohmori et al., 1998), which is an angular accuracy within about a half of half power beam width (HPBW). However, directional antennas with relatively narrow beams should track the satellite only in the azimuth directions because the elevation angles to the satellite are almost constant, especially in land mobile satellite communications.

Figure 4 classifies satellite-tracking functions. Tracking function divide into two function groups, namely beam steering and tracking control method. There are two types of satellite tracking systems: mechanical and electrical. A mechanical tracking system uses mechanical structures to keep the antenna in the satellite direction by utilizing a motor or mechanical drive system. An electrical tracking system tracks the satellite by electrical beam scanning.

There are two tracking algorithm, namely an opened-loop method and closed-loop method. The difference between them is whether the satellite signal is considered or not. The opened-loop uses information of mobile position and its bearing from one or several sensors regardless the satellite signal. In contrary, the closed-loop method utilizes the satellite signal to track it. To use this method, received signals from the satellite must be stable without severe fading. It is adopted in aeronautical and maritime mobile communications but quite difficult to apply in land mobile satellite communication due to its stability is predominantly affected by shadowing and fading.

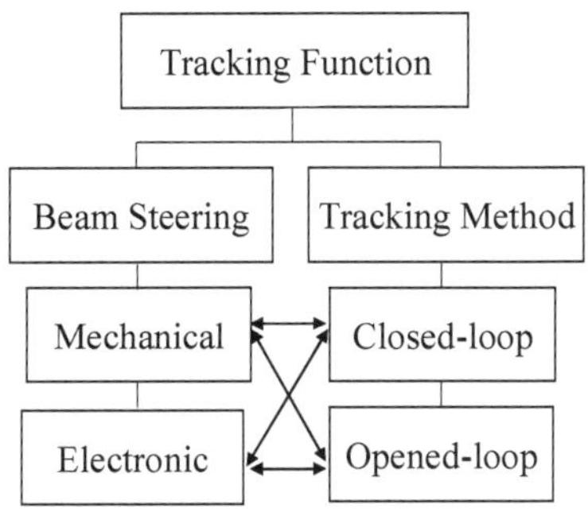


Fig. 4. Classification of satellite tracking function

### 3.2 Design of Vehicle-Mounted Antenna System

In mobile satellite communications, an antenna model is expected to be able to respond to changes in the direction of a mobile object. Several antennas were able to meet mobile satellite antenna requirements have been extensively investigated, are widely available in the literature include the conical beam antennas by using wire antennas such as quadrifilar or bifilar helix (Kilgus, 1975, Terada & Kagoshima, 1991, Nakano et al., 1991, Yamaguchi & Ebine, 1997), drooping dipole (Gatti & Nybakken, 1990) or even patch antenna in higher mode operation (Nakano et al., 1990, & Ohmine et al., 1996) and the satellite-tracking antennas (Ito et al., 1988). As described in the previous subsections, an attractive feature of the former antenna design is that, as the radiation is omnidirectional in the conical-cut direction and also their beam is broad in the elevation plane, satellite-tracking is not necessary. However, such antennas offer typical gain about 0 - 4 dBi (Ohmori et al., 1998, Ilcev, 2005, Fujimoto & James, 2008) because of their isotropic energy in the conical-cut direction. Further, owing to our application target for ETS-VIII, which is described by a link budget calculation in Table 2, the gain is designed by more than 5 dBi in the overall azimuth coverage area at specified elevation angles (Table 4). Therefore, in case of omnidirectional antenna, their typical gain will not satisfy the specification. Therefore, a beam-tracking antenna is selected to suit the target. By utilizing a beam-tracking antenna, owing to its directional beam property, the beam can be deflected towards the satellite direction when the vehicle moves. Although such antenna type needs a tracking function, owing to the generation of directional beam and smaller fading effects from surrounding terrain, higher transmission rate are possible.

Most recent decades of the developed antenna system for vehicle-based applications are impractical since their design, based on mechanical steering that makes them extremely bulky. This type of antenna system is heavyweight and high power consumption as well as low tracking-speed owing to the use of electric motors responsible for mechanical steering (Kuramoto et al., 1988, Huang & Densmore, 1991, Jongejans et al., 1993, Strickland, 1995). An alternative solution is a planar phased array antenna which performs beam steering by electronic means (Nishikawa et al., 1989, Ohmori et al., 1990, Alonso et al., 1996, Konishi, 2003). However, the use of phase shifters for beam forming is quite expensive owing to their large quantities requirement. Such phase shifters, need to be properly designed in order to avoid the beam squinting in which the beam direction may considerably differ at receive and transmit frequencies. Moreover, nonlinear effects from electronic phase shifter and switches generate the noise problem in phased array antenna (Ohmori, 1999). Design of antenna system requires as compact and lightweight as possible to minimize the space and easy installation. In the following subsection, we describe specifications, targets and structure of our vehicle antenna system.

### 3.2.1 Specifications and Targets

The specifications and targets of the antenna are shown in Table 4. The ETS-VIII is providing voice/data communications with satellite mobile terminals in the S-band frequency (2.5025 GHz and 2.6575 GHz for reception and transmission, respectively). The polarization is left-handed circular (LHCP) for both transmission and reception units. As this antenna is assumed to be used in Tokyo and its vicinity, the targeted elevation angle is set to 48°. In the system, the antenna beam is expected to be steered towards the satellite and cover the whole azimuth space by more than 5 dBic and less than 3 dB for the gain and the axial ratio, respectively.

| Specifications | | |
|---|---|---|
| Frequency bands | Transmission ($T_x$) | 2655.5–2658.0 MHz |
| | Reception ($R_x$) | 2500.5–2503.0 MHz |
| Polarization | Left-handed circular polarization for both $T_x$ and $R_x$ | |
| **Targets** | | |
| Angular ranges | Elevation angle (*El*) | 48° (Tokyo) ±10° |
| | Azimuth angle (*Az*) | 0° to 360° |
| Minimum gain | 5 dBic | |
| Maximum axial ratio | 3 dB | |

Table 4. Specifications and targets for ETS-VIII applications

### 3.2.2 Antenna System Architecture

Figure 5 depicts a satellite-tracking system built with the beam switching method. As shown in this figure, the localization of the satellite is determined, based on the location and travelling direction of the mobile station by use of currently available car navigation systems and gyroscope, and the appropriate beam direction is selected. Then the signal emitted by the tracking unit is received and by appropriately controlling the activation of the feedings of each element, through the switching circuit used to control the feeding of the antenna, the beam is switched in three directions in the azimuth plane and the satellite can be followed. Since the antenna system utilizes the gyroscope, the satellite-tracking can be kept as the GPS satellite is out of sight.

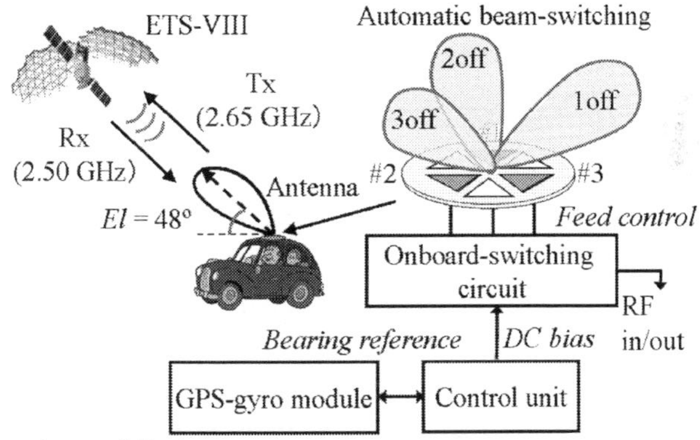


Fig. 5. Antenna system architecture

The array antenna configuration has a beam switching capability, where in principle, $n$ circularly polarized elements are ($360°/n$) sequentially rotated (Teshirogi et al., 1983 & Hall et al., 1989) and set with an equal distance between each elements following a circular path. Our antenna is a 120° sequentially physical rotated and set with an equal distance between each element following a circular path. With such alignment, each element is fed in-phase allowing their relative phase is physically shifted. The feeding of each antenna element is successively turned off by controlling the onboard-switching circuit and thus the whole azimuth range can be scanned by step of 120°. Three beams can be generated to cover all of the azimuth angles. The beam is generated in the azimuth plane at -90° from the element that is turned off. As a result, if each element i.e. element no. 1, 2 and 3 is turned off, the beam is generated in the direction $Az$ = 0, 120° and 240° (Sri Sumantyo et al., 2005), respectively as shown in Fig. 1. In addition, the satellite-tracking is conducted in the azimuth plane regardless considering the elevation direction owing to the antenna gain is predicted quite enough to communicate with the geostationary satellite as earlier listed in Table 2, Section 2.2.

The antenna system is operated by a control unit allowing the antenna beam is automatically steered. The beam-forming of array antenna is generated by providing bias voltages to switch on and off the P-I-N diodes on the onboard-switching circuit and thus two elements of the array can be correctly fed and afterwards a desired beam is created among three selectable-beams. As the ETS VIII satellite lies at southern from the Japanese archipelago, by considering position of array antenna installation on car's roof, the antenna beam can be invariably controlled in south direction. By such a way, antenna tracks the satellite in good reliability and high-speed beam scanning.

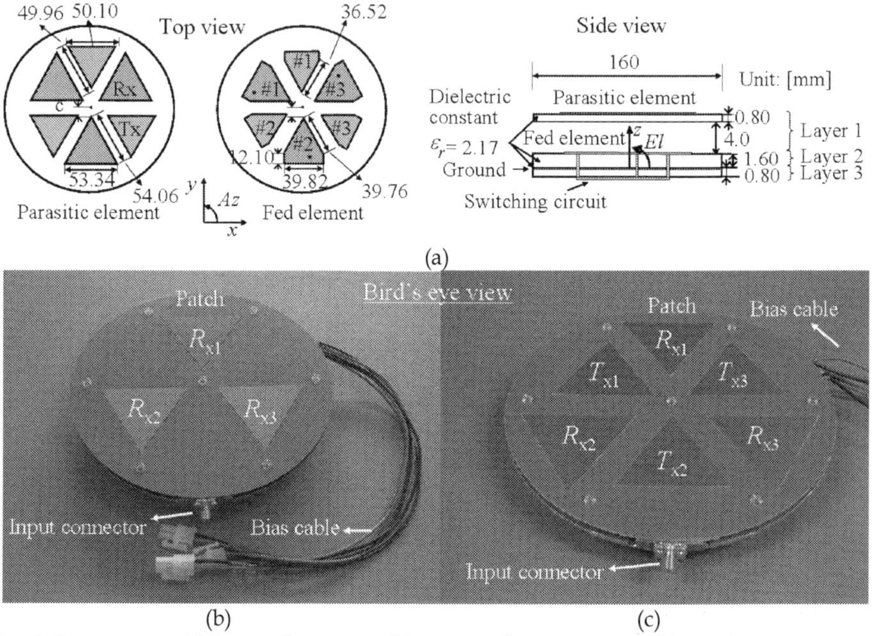


Fig. 6. Structure and fabricated antenna: (a) proposed construction, (b) receive-use antenna, (c) receive and transmit-use antenna

### 3.2.3 Array Antenna Structure

We basically design an antenna for both of transmit and receive use by arranging them on the same layer for achieving a compactness. We develop the antenna started from a single element, a receive-use antenna up to a final structure namely a receive–transmit-use antenna. Examples of antenna structures and their developed antenna are depicted in Fig. 6. The antenna is composed of three layers, i.e. parasitic elements with air gap (layer 1), fed elements (layer 2) and switching circuit (layer 3). The fed elements are three pentagonal patch antennas that directly excited from the feeding line on layer 3. In the top of the construction is put three isosceles triangular patches as parasitic elements to enhance bandwidth and gain of the antenna. In order to match on 50-ohm feeding, air gap is inserted in the layer between fed and parasitic elements. This design excites two near-degenerate orthogonal modes of equal amplitudes and 90° phase difference for left-handed circular polarization (LHCP) operation. Good axial ratio performance can be obtained by adjusting position of feeding point, air gap height, and parasitic patch size.

The antenna structure is integrated with a power divider and a switching circuit, which mounted on the backside of the structure. The circuit is functioned as a feeding control of the antenna. The mounted circuit composed of a simple power divider and a Double Pole Triple Throw (DP3T) switching circuit developed by (Kaneko et al., 2006). This integrated antenna allows its compactness and low loss because no additional cable required.

## 4. Laboratory Test Performance

In earlier section, configuration of array antenna structure is proposed for receive-use and receive-transmit-use as well. Nevertheless, due to satellite problem (NICT, 2007), we utilize the antenna for receive-use only in our measurements. Basically, the transmit-use antenna performance and its characteristics are similar to the receive-use one. Once we fabricate the antenna, we test in laboratory for some basic antenna measurements. Following is the measurement results such as $S_{11}$, axial ratio, and radiation pattern characteristics.

### 4.1 Input Characteristics

Owing to the stacked-parasitic patch structure, two generated-resonant modes enhance impedance bandwidth as described in Fig. 7(a). Impedance bandwidth of the antenna i.e. ($|S_{11}|$ < -10 dB) is about 8.50%, quite enough for ETS-VIII applications. Moreover, the measured antenna gives a good input matching at the target frequency 2.5025 GHz by (50.28 - j24.86) ohm.

### 4.2 Axial Ratio Characteristics

The array antenna gives good performance at $El$ = 48° in the target frequency. Good axial ratio is required to eliminate polarization tracking because of circular polarization. The measured result shows the axial ratio is 1.0 dB at center frequency 2.5025 GHz for each of three generated-beams. In addition, the 3 dB axial ratio bandwidth gives about 1.8% as shown in Fig. 7(b).

### 4.3 Radiation Characteristics

#### 4.3.1 Radiation Pattern in Elevation-Cut Plane

Figure 7(c) shows the radiation characteristics of the array antenna in the elevation-cut plane when element no. 1 is switched off. The antenna main beam is generated at $Az$ = 0 which is shown at the right side of the figure. Same manner is obtained in case of no. 2 is off and no.3 is off where each beam occurs at $Az$ = 120° and $Az$ = 240°, respectively. Noted that the gain more than 5.2 dBic and the axial ratio less than 1.7 dB meets the requirements for elevation angle $El$ = 38°-58° where this is an approximation of elevation range of the satellite by considering the Japanese archipelago from northern to southern. We confirmed that the gain is 6.6 dBic and the axial ratio 1.2 dB at $El$ = 48°. The measurement is taken at center frequency 2.5025 GHz. These results surely allow us to track the satellite regardless its elevation with regard to vehicle.

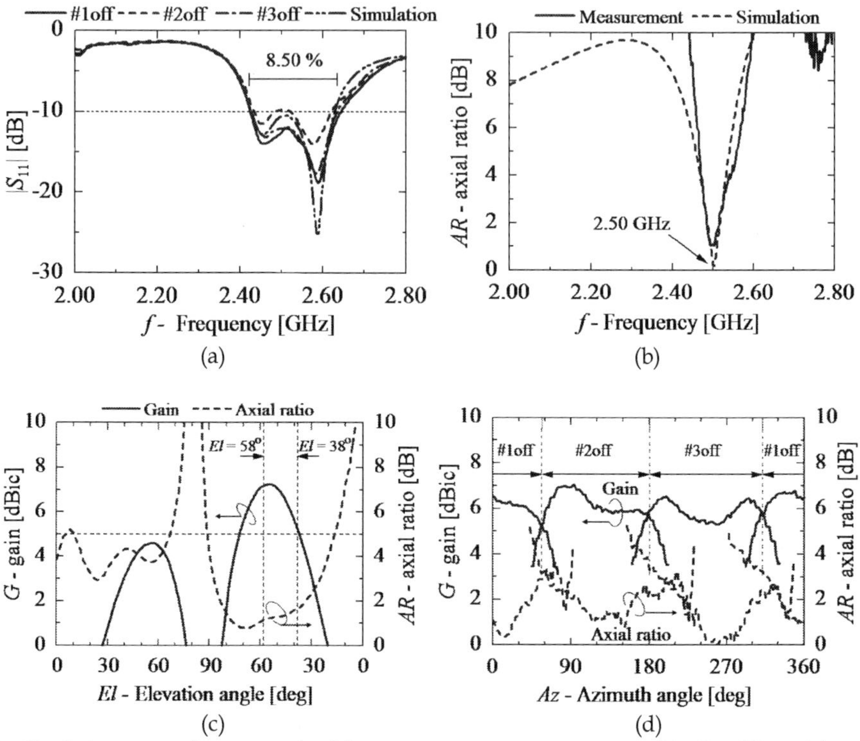


Fig. 7. Antenna performance in laboratory test measurement: (a) $S_{11}$ (b) axial ratio characteristics, (c) radiation pattern in *El*-plane and (d) radiation pattern in *Az*-plane

#### 4.3.2 Radiation Pattern for Beam-Switching

Figure 7(d) shows the measurement results of gain and axial ratio for each of three antenna beams in the azimuth plane at $El = 48°$. Small difference among each antenna beam-shape is observed. Such unsymmetrical property is considered due to the phase difference effect of the switching circuit and dimensionally discrepancy antenna fabrication. However, the 5 dBic-coverage in the 360° of the conical direction is satisfied enough. In addition, the beam is possibly switched at minimum gain 5.2 dBic and the axial ratio below 3 dB is possibly to be obtained for automatic satellite-tracking.

## 5. Verification of Antenna System

Following all the required components of the system are developed and individually examined, testing of the completed array antenna is performed in anechoic chamber. Here, we test the tracking system performance of the antenna with regard to the conical-cut radiation pattern at specified elevation angle. Figure 8(a) illustrates an antenna system measurement set up where a transmitting (*T*x) antenna is employed as though a satellite and the array antenna as a receiving antenna. The antenna is covered by a radome and put on a ground plane because in mobile measurement campaign the antenna will be located on car roof as well as to avoid mechanical restriction or weather hindrance like snow, wind, and rainfall. The antenna is put 8 mm upper from the ground plate to avoid the switching circuit of the antenna on the rear-side touches the ground plane which may cause a shorted-circuit due to the bias cables connections. This measurement evaluates capability of the array antenna allowing the beam automatically switched pursuing the transmitting antenna.

In order to realize such a measurement we developed a simple control program on PC to control the antenna beam by use of a gyro sensor inside of GPS module. The measurement is performed without and with ground plane–radome. Such measurement is carried out to grasp the influence of ground plane and radome on the antenna performance specifically the gain and the axial ratio in azimuth scanning. In fact, we confirmed that a hemisphere radome is less effect on gain and axial ratio performance as well.

The beam of the antenna is generated by a mechanism that consists of switching off one of the radiating elements as earlier mentioned in Section 3.2.2 where the tested results is depicted in Fig. 7(d). Fig. 7(d) represents the beam-switching characteristics in manual operation that each beam is separately measured. Having performed a manual beam measurement in Fig. 7(d), we decide at which point we should to switch the beam automatically with regard to the gain value at the coincide point of each beam. Since the beam is possibly to be switched at minimum gain more than 5 dBic, then the antenna beam is switched at specific azimuth angle (this measurement is at $Az$ = 56°, 180° and 312°). With such decision, we confirmed that the gain can be switched automatically at the given azimuth angles and the axial ratio for each beam satisfies below 3 dB to cover 360° conical-plane. This tracking result is shown in Fig. 8(b). In this case, the beam is switched by itself as the azimuth angle changes to track the transmitting antenna.

In order to figure out the differences of the antenna characteristics caused by ground plane and radome installed, Fig. 8(c) depicts the automatic tracking when ground plane and radome installed. It is shown that the characteristics of each beam does not change drastically when the radome and ground plate are employed (Fig. 8(c)) compared with the only antenna structure (Fig. 8(b)), except the axial ratio owing to the scattering from the ground plane affects the incident phase onto the antenna. The effect of radome is considerably neglected since its hemisphere shape gives minimum scattering and its thin material provided less loss. In general, our antenna system is able to automatically control and correctly select the beam in the azimuth direction whose gain more than 5 dBic and axial ratio less than 3 dB, although when the ground plane is mounted, its axial ratio rises by 0.5 dB at the coincide point.

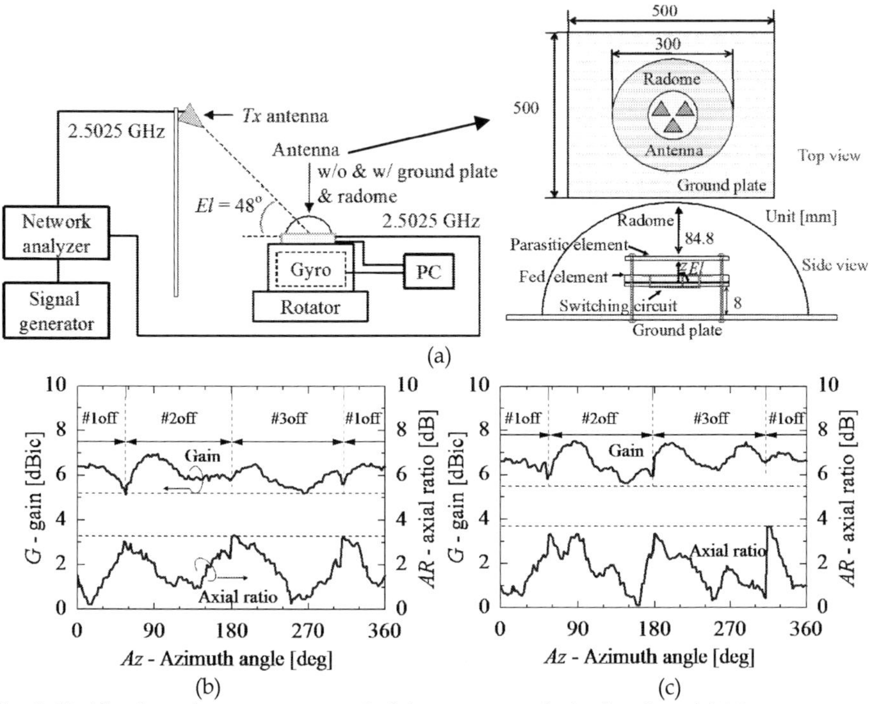


Fig. 8. Verification of antenna system in laboratory anechoic chamber: (a) Measurement set up (b) antenna performance without radome and ground, (c) antenna performance with radome and ground

Additionally, before we perform a measurement campaign, we also confirm the validity of the overall antenna system by carrying out an outdoor measurement using signal from the ETS-VIII satellite. The antenna system is tested by a fixed-state testing rig in Chiba Prefecture area ($El$ = 48°) without obstacle present (direct signal) as illustrated in Fig. 9(a). This measurement utilizes a spectrum analyzer (Agilent E4403B) to measure the received signal power from the satellite. In order to compensate the weak satellite signal, an amplifier (Agilent 83017A) is associated with the antenna and thus the signal level has enough $C/N_0$. Measured result showed that $C/N_0$ is about 47.30 dBHz with link margin 1.45 dB, sufficiently to make a satellite-tracking measurement.

The measurement of satellite-tracking for array antenna is performed as same as the antenna system test in the anechoic chamber. For this purpose, we take the received signal power while the antenna is rotating. A gyro sensor inside of GPS module is put on the beneath of the antenna and connected to PC for automatic satellite-tracking. As a result, the satellite-tracking is well operated with good level as depicted in Fig. 9(b). Moreover, three selectable beams are smoothly switched to cover all the azimuth direction.

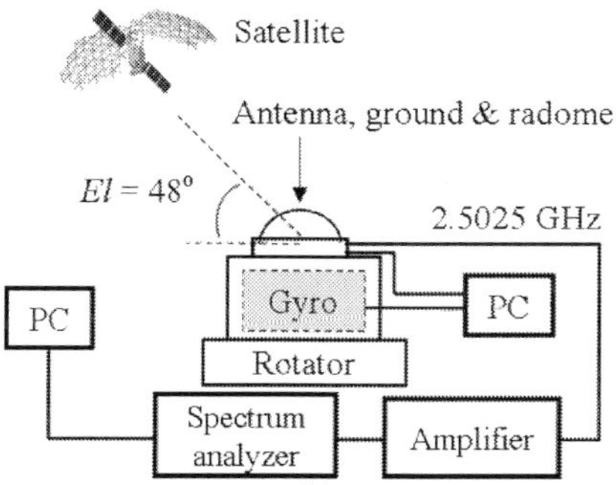


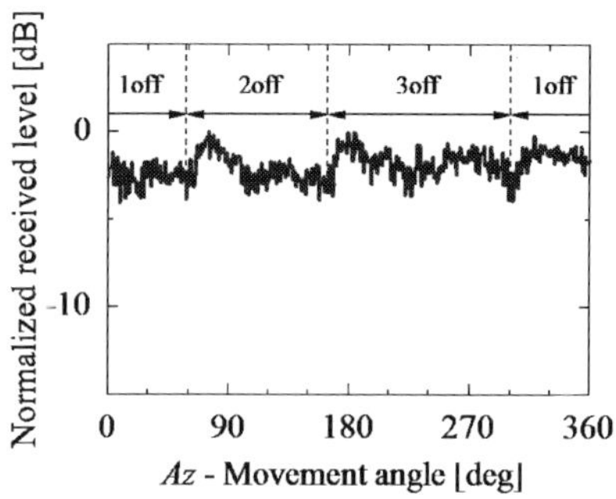


Fig. 9. Verification of antenna system in outdoor test: (a) Measurement set up (b) satellite tracking performance

## 6. Measurement Campaign using Geostationary Satellite

We have validated the antenna system performance by immobile-state measurements in Section 5. Afterwards, we carry out a measurement campaign to verify possibility for application in real environment. Reported by (NICT, 2007), however, large deployable reflector (LDR) antenna that installed on ETS-VIII satellite cannot be used because of improper situation at power supply of low noise amplifier (PS-LNA), thus the measurement campaign is conducted by using high accuracy clock (HAC) receiving antenna with lower gain 25 dBi instead of 43.80 dBi of the LDR antenna. Because of such circumstance, the measurement is also assigned only for forward link namely from ground fixed-station (transmitter) to vehicle (receiver) through the ETS-VIII satellite (Basari et al., 2009). In this case, we modify the calculated-link budget instead of listed-parameters in Table 2, to deal with the situation, for 8 kbps of transmission rate and BER = $1.0\times10^{-4}$ . We use a parabolic antenna (gain 22.4 dBi) at the transmitter to boost a transmitted signal. Afterwards, we measure received signal power and average bit error rate (BER) at the receiver. The received signal power is retrieved from intermediate frequency (IF) signal of handset terminal. Configuration of measurement is represented by a block diagram in Fig. 10(a) and an example of measurement is viewed in Fig. 10(b).

### 6.1 Verification of Satellite-Tracking

At first, verifying a single antenna beam is performed in line of sight area without obstacles present. We utilize a spectrum analyzer (Agilent E4403B) to retrieve the IF signal amplitude. By setting it at zero-span in specified frequency, received signal power can be recorded. One of three selectable beams of the antenna is viewed in Fig. 11(a). The experiment result tends to agree with the calculation even though small difference is observed. We employ an aluminium ground plane on the antenna under which is mounted (Fig. 10(b)). Such plate affects the antenna performance especially its beam. Small difference exists, particularly the beam-shape and the beam-width. However, such differences do not significantly worsen the received signal of the antenna.

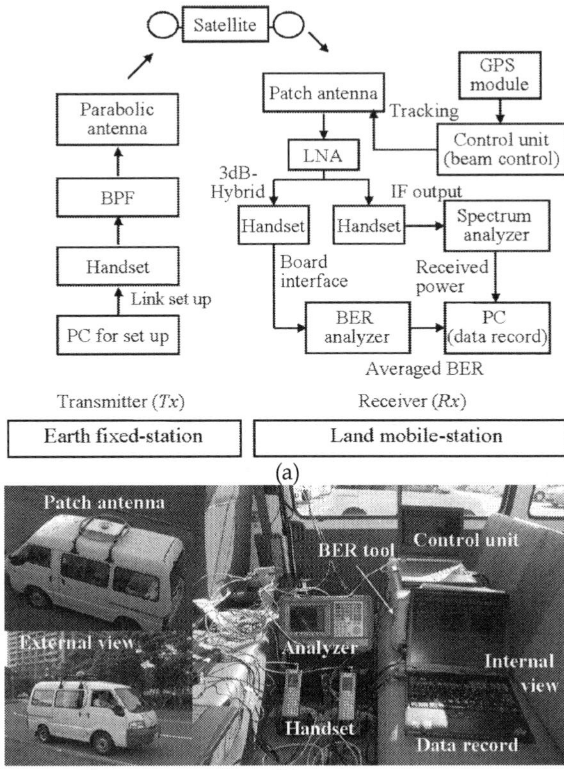

(b)

Fig. 10. Experimental set-up for measurement campaign: (a) block diagram of configuration (b) measurement view

The measurement also verifies the satellite-tracking of antenna system. While vehicle moves around at the rotary, the antenna beam is electronically steered pursuing the satellite associated with vehicle's orientation. Three selectable beams are smoothly switched towards the satellite in azimuth direction as described in Fig. 11(b). In addition, due to we use spectrum analyzer for measuring signal power, the $C/N_0$ can be calculated by the following formula (Hranac & Currivan, 2006 and Agilent, 2003):

$$C/N_0 = \text{(signal power)} - \text{(noise floor)} + \text{(10 log RBW)} \tag{7}$$

where, RBW is resolution bandwidth of measurement. By considering the received signal power is averaged by -60 dBm and noise floor -83 dBm, the $C/N_0$ is approximately 47.77 dBHz and link margin 1.94 dB is obtained.

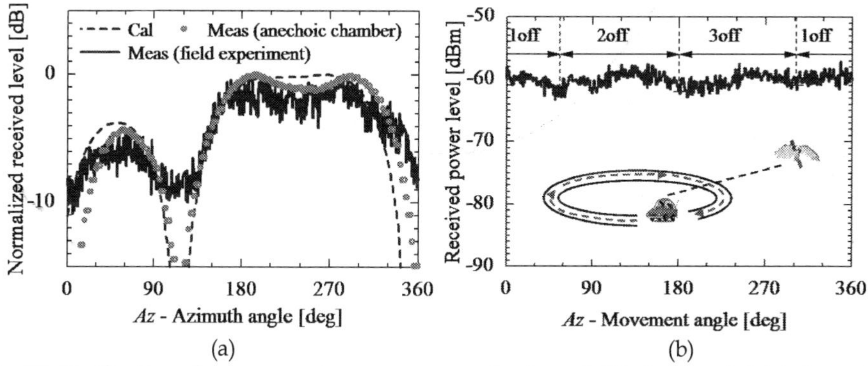


Fig. 11. Verification of antenna system measurement using satellite in line of sight places: (a) Single-beam verification (b) automatic satellite tracking verification

### 6.2 Evaluation in Blockage Areas

In this measurement, we thoroughly evaluate the effect of obstacle objects namely buildings, foliages of trees, utility poles such as electricity power poles, and highway overpasses or bridge roof, to the received signal qualities of the antenna. Fig. 12(a) shows an experiment at buildings area. There are two high buildings on the test field which expected to block the signal from the satellite. While vehicle moves on straight path, the signal from the satellite is blocked becoming same level with the noise floor and thus it cannot be detected well as a received signal. If it happens for too long period, communication link will be disconnected.

We also examine the blockage due to roadside-trees. The measurement is carried out in summer season since dense foliages exist and will affect the performance of the received signal power. As shown in Fig. 12(b), the measurement is performed on straight path where the satellite is situated at the left side of the measurement path. Unlike the blockage due to buildings, roadside-trees are less effect to the received signal even though they barred by an average 3-5 dB, decreasing well-received signal. It is obvious shown in Fig. 12(b) in comparison with the direct wave signal. In addition, almost 75% of the satellite signal is approximately attenuated by 2-6 dB at this dense foliages area.

In order to confirm the antenna system usability in common land mobile applications, we test the system in expressway and common road. The field experiment of expressway is carried out in Chiba prefecture area ($El = 48^{\circ}$) between Kisarazu-kita and Soga interchange by speed 70-80 km/h. With almost no obstacles present, well received signals are obtained and yet the beam-switch happens at the beginning and the last measurement path. We confirmed that the signal drastically dropped in very short time (0.2-0.4 s) when the vehicle passes through overpasses or bridges. This result is shown in Fig. 13(a).

Our antenna has radiation characteristics in the elevation direction, so thus it is required to confirm with real environment for vehicle application. For that purpose, we test the antenna system at inclined-road. The vehicle passes on the $5^{\circ}$ inclined-road with speed of motion is constantly kept by 40 km/h without any beam-switch happens. Fig. 13(b) shows an example of the measured satellite signal for downward-motion. The received signal tends to increase in 0.5 dB. In this case, downward-motion means increasing elevation allowing increasing the antenna gain.

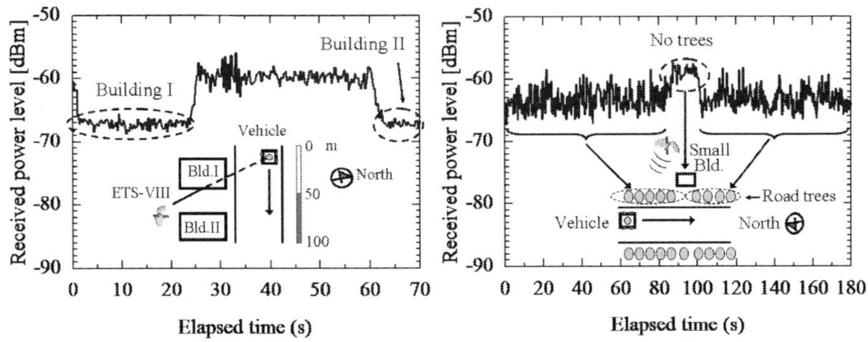


(a) (b)

Fig. 12. Evaluation on blockage signal measurements: (a) due to buildings (b) due to roadside trees

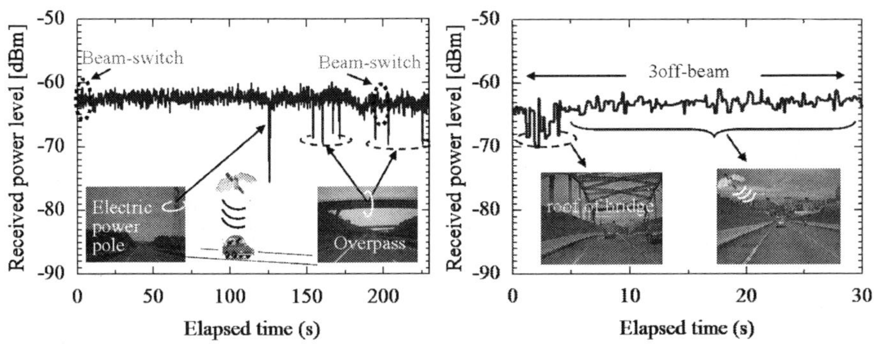


(a) (b)

Fig. 13. Evaluation on blockage signal measurements: (a) due to overpasses and pole (b) due to roof of bridge and inclined-road constraint

### 6.3 Bit Error Rate Verification

As we earlier designed the antenna system, BER is set by $1.0 \times 10^{-4}$. Therefore, we confirm it by performing a measurement on BER characteristics besides the propagation measurements. Here, the measurement is carried out at rotary in line of sight area to verify the satellite-tracking. As previously reported, binary-phase shift keying (BPSK)-modulated pseudo-noise sequences (PN-code) is transmitted from the ground-station through the satellite and received at our vehicle-station. We use a BER analyzer (Anritsu MD6420A) to retrieve the BER value that measured from interface board to which connected with the analyzer from handset terminal. The modulation rate is set to be 8 kbps. The measurement circumstance is similar to measurement path in Subsection 6.1 for satellite-tracking. As shown in Fig. 14, the BER performance is kept stable in range $5\text{-}7\times10^{-4}$ while beam-switching occurs. Thus, validity of our antenna system design in the beginning Section is confirmed.

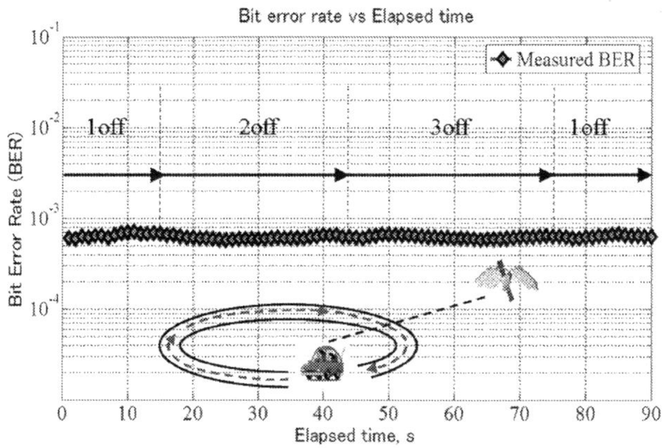


Fig. 14. Bit error rate characteristics on satellite-tracking

## 7. Conclusion

Simple selectable-beam antenna system for vehicle-mounted mobile satellite communication applications was presented. We have discussed it from design, laboratory test as well as outdoor test and verification in measurement campaign. System components discussed include array antenna includes its switching system to activate the array, and satellite-tracking function as well. The measurement in anechoic chamber was thoroughly examined with satisfactory performances, well suited with the calculations. The basic measurement included $S_{11}$, axial ratio, and radiation pattern characteristics. Overall antenna system performances in anechoic chamber test gave satisfactory results. Further, the antenna system was also confirmed by using signal from satellite for immobile-state outdoor measurement on the testing-rig. Without any obstacles present, the system was able to correctly select the beam for tracking the satellite with regard to the rotation of the antenna.

Following the satisfactory performances in anechoic chamber measurement and fixed-state outdoor measurement, the antenna system was also examined in measurement campaign using the satellite signal to verify the validity of the developed system for possibility use in land mobile satellite applications. We built a connection with the satellite and tested the antenna system in particular its tracking capability. As a result, the system was correctly track the satellite while vehicle was moving by considering its position and bearing information that retrieved from GPS module.

In addition, we also considered and tested the antenna system under environment constraints that affecting its received signal power, such as buildings, roadside-trees, utility poles, highway overpasses and inclined-road present. The results showed us the blockage and shadowing happened and significantly attenuated the received signal owing to its dependence on direct wave signal. Moreover, the BER performance of the system has been verified with satisfactory result by about $10^{-4}$.

In addition, this vehicle-mounted antenna system design and its measurement results will help engineers and practitioners to design a costly-effective land-mobile satellite communication system. Ultimately, our developed antenna system is simple, compact and promising in low cost, for contribution in the future mobile satellite communication applications.

## 8. References

Agilent Technologies (2003). Spectrum Analyzer Measurements and Noise: Measuring Noise and Noise like Digital Communications Signals with a Spectrum Analyzer. *Application Note 1303.*

Alonso, J.I.; Blas, J.M.; Garcia, L.E.; Ramos, J.; de Pablos, J.; Grajal, J.; Gentili, G.G.; Gismero, J. & Perez, F. (1996). Low cost electronically steered antenna and receiver system for mobile satellite communications. *IEEE Trans. on Microwave Theory and Techniques,* Vol. 44, No. 12, (Dec. 1996) pp. 2438–2449, ISSN: 0018-9480

Basari; Purnomo, M.F.E; Saito, K.; Takahashi, M. & Ito, K. (2009). Realization of simple antenna system using ETS-VIII satellite for land vehicle communications. *IEICE Trans. Communications,* Vol. E92-B, No. 11, (Nov. 2009) pp. 3375–3383, ISSN: 0916-8516

Basari; Purnomo, M.F.E; Saito, K.; Takahashi, M. & Ito, K. (2009). Simple switched-beam array antenna system for mobile satellite communications. *IEICE Trans. Communications,* Vol. E92-B, No. 12, (Dec. 2009) pp. 3861–3868, ISSN: 0916-8516

Fujimoto, K. & James, J.R. (2008). *Mobile Antenna Systems Handbook,* Artech House, ISBN: 978-1-59693-126-8, Norwood, MA, USA

Gatti, M.S. & Nybakken, D.J. (1990). A circularly polarized crossed drooping dipole antenna, *Proceedings of IEEE Antennas and Propagation Society Intl. Symposium,* pp. 254-257, ISBN: 978-0780370708, Dallas, TX, USA, May 1990, IEEE, Piscataway, NJ

Hall, P.S.; Dahele, J.S. & James, J.R. (1989). Design principles of sequentially fed, wide bandwidth, circularly polarised microstrip antennas. *IEE Proceedings H Microwaves Antennas Propagation,* Vol. 136, No. 5, (Oct. 1989) pp. 381– 389, ISSN: 0950-107X

Hranac, R. & Currivan, B. (2006). Digital Transmission: Carrier-to-Noise Ratio, Signal-to-Noise Ratio, and Modulation Error Ratio. *Broadcom Corporation and Cisco Systems, Inc.,* white paper.

Huang, J. & Densmore, A.C. (1991). Microstrip yagi array antenna for mobile satellite vehicle application. *IEEE Trans. Antennas Propagation,* Vol. 39, No. 7, (July 1991) pp. 1024–1030, ISSN: 0018-926X

i-Space (2003). *Engineering Test Satellite (ETS-VIII).* http://i-space.jaxa.jp/experiments2002/01_ETS8ver6.pdf

Ilcev, S.D. (2005). Global Mobile Satellite Communications: For Maritime, Land and Aeronautical Applications. Springer, ISBN: 1-4020-7767-X, Dordrecht, the Netherlands

International Telecommunication Union Radiocommunication Sector (ITU-R) Radio Regulations (2004). ICT Regulation Toolkit, Module 5 Radio Spectrum Management

Ito, K.; Ohmaru, K. & Konishi, Y. (1988). Planar antennas for satellite reception. *IEEE Trans. Broadcasting,* Vol. 34, No. 4, (Dec. 1988) pp. 457–464, ISSN: 0018-9316

Japan Aerospace Exploration Agency-JAXA (2003). *Engineering Test Satellite (ETS-VIII).* http://www.jaxa.jp/projects/sat/ets8/index_e.html

Jongejans, A.W.; Rinous, P.J. & Roederer, A.G. (1993). Review of ESA mobile antenna developments for satellite communications, *Proceedings of IEEE Antennas and Propagation Society Intl. Symposium,* pp. 1838–1841, ISBN: 978-0780370708, Ann Arbor, MI, USA, June–July 1993, IEEE, Piscataway, NJ

Kaneko, K.; Tanaka, T.; Takahashi, M. & Ito, K. (2006). Electronic scanning characteristics of simple satellite-tracking array antenna. *IEICE Trans. Communications (Japanese Edition),* Vol. J89-B, No. 9, (Sept. 2006) pp. 1696–1704, ISSN: 0915-1877

Kilgus, C.C. (1975). Shaped-conical radiation pattern performance of the backfire quadrifilar helix. *IEEE Trans. Antennas Propagation,* Vol. 23, No. 3, (May 1975) pp. 392–397, ISSN: 0018-926X

Konishi, Y. (2003). Phased Array Antennas. *IEICE Trans. Communications,* Vol. E86-B, No. 3, (Mar. 2003) pp. 954–967, ISSN: 0916-8516

Kuramoto, T.; Yamane, T. & Endo, N. (1988). Mechanically steered tracking antenna for land mobile satellite communications, *Proceedings of IEEE Antennas and Propagation Society Intl. Symposium,* pp. 1314–1317, ISBN: 978-0780370708, Syracuse, NY, USA, June 1988, IEEE, Piscataway, NJ

Lodge, J.H. (1991). Mobile satellite communication systems: toward global personal communications. *IEEE Communications Magazine,* Vol. 29, No. 11, (Nov. 1991) pp. 24–30, ISSN: 0163-6804

Nakano, H.; Vichien, K.; Sugiura, T. & Yamauchi, J. (1990). A singly-fed patch antenna radiating a circularly polarised conical beam. *Electronic Letters,* Vol. 26, No. 10, (May 1990) pp. 638-640, ISSN: 0013-5194

Nakano, H.; Saura, Y.; Mimaki, H. & Yamauchi, J. (1991). Circularly polarized conical beam formation by backfire helical antennas. *IEICE Trans. Communications,* Vol. E74-B, No. 10, (Oct. 1991) pp. 3246–3252, ISSN: 0916-8516

National Institute of Information and Communications Technology-NICT (2007). *Engineering Test Satellite (ETS-VIII) Project.* http://www2.nict.go.jp/p/p463/ETS8/ETS8.html

Nishikawa, K.; Sato, K. & Fujimoto, M. (1989). Phased array antenna for land mobile satellite communications. *IEICE Trans. Communications (Japanese Edition),* Vol. J72-B, No. 7, (July 1989) pp. 323–329, ISSN: 0915-1877

Ohmine, H.; Sunahara, Y. & Matsunaga, M. (1996). A TM21 mode annular ring microstrip antenna for personal satellite communication use. *IEICE Trans. Communications,* Vol. E79-B, No. 9, (Sept. 1996) pp. 1227–1233, ISSN: 0916-8516

Ohmori, S.; Wakana, H. & Kawase, S. (1998). *Mobile Satellite Communications,* Artech House, ISBN: 0-89006-843-7, Norwood, MA, USA

Ohmori, S.; Tanaka, K.; Yamamoto, S. & Tsuchiya, M. (1990). A phased array tracking antenna for vehicles, *Technical Report of the IEICE (Japanese Edition),* AP90-75, pp. 33–40, ISSN: 09135685, Oct. 1990, IEICE, Tokyo

Ohmori, S. (1999). Phased array antennas for mobile communications. *Springer Journal Annals of Telecommunications,* Vol. 54, No. 1–2, (Jan. 1999) pp. 92–102, ISSN: 1958-9395

Rabinovich, V.; Alexandrov, N. & Alkhateeb, B. (2010). *Automotive Antenna Design and Applications,* CRC-Press, ISBN: 978-1-4398-0407-0, Boca Raton, FL, USA

Sri Sumantyo, J.T.; Ito, K. & Takahashi, M. (2005). Dual band circularly polarized equilateral triangular patch array antenna for mobile satellite communications. *IEEE Trans. Antennas Propagation,* Vol. 39, No. 7, (July 1991) pp. 1024–1030, ISSN: 0018-926X

Strickland, P.C. (1995). Planar arrays for MSAT and INMARSAT land mobile satellite communications, *Proceedings of IEEE Antennas and Propagation Society Intl. Symposium,* pp. 1838–1841, ISBN: 978-0780370708, Newport Beach, CA, USA, June 1995, IEEE, Piscataway, NJ

Terada, N. & Kagoshima, K. (1991). Compatible mobile antenna for mobile satellite and cellular communication systems, *Technical Report of the IEICE (Japanese Edition),* AP 91-65, pp. 29–34, ISSN: 09135685, Sept. 1991, IEICE, Tokyo

Teshirogi, T.; Chujo, W.; Tanaka, M.; Itoh, T. & Komuro, H. (1983). Circular polarization array antenna with sequential rotation and phase shifts, *Technical Report of the IEICE (Japanese Edition),* AP83-57, pp. 49–54, ISSN: 09135685, Aug. 1983, IEICE, Tokyo

Xiong, F. (1994). Modem technique in satellite communications. *IEEE Communications Magazine,* Vol. 32, No. 8, (Aug. 1994) pp. 84–98, ISSN: 0163-6804

Yamaguchi, R. & Ebine, Y. (1997). Design of vehicle antenna using helical array for mobile satellite communication system, *Technical Report of the IEICE (Japanese Edition),* AP 97-64, pp. 97–64, ISSN: 09135685, July 1997, IEICE, Tokyo

# 3

# Cooperative Strategies for Satellite Access

Luca Simone Ronga and Rosalba Suffritti
*CNIT*
*Italy*

Enrico Del Re
*University of Florence*
*Italy*

## 1. Introduction

Satellite communications have become an important node of the global telecommunication infrastructure. Satellite capacity request is growing quickly, driven not only by broadcast applications but, mainly, by broadband services, in particular by the expectation of "always-on" broadband services available everywhere. Thus, new "killer" applications such as HDTV (High Definition Digital Television) and broadband Internet access, provided through satellites, can help to face the growth of capacity demand foreseen in the near future. Moreover, in addition to the provision of satellite multimedia services to fixed terminals, there is an increasing demand for broadband communications on the move (i.e. on ships, trains, aircrafts, vans, cars).

Analysing such an increasing demand of satellite communications, the work reported in this chapter is focused on the study of different techniques which allow the improvement of the performance of satellite users displaced in severe environments. The analysis of this context, in fact, has revealed the need to adopt adequate advanced techniques to achieve a sufficient quality in satellite links, especially in those scenarios where the link budget is tighter, such as, for example, the mobile satellite one. These considerations have motivated the study of *cooperative strategies* which allow the mitigation of the deleterious effects of fading. This is obtained thanks to a new form of spatial diversity in which the diversity gain can be achieved through the cooperation of different users which generate a virtual MIMO (Multiple-Input Multiple-Output) system. The adoption of these methodologies can be very helpful in those scenarios characterised by continuous occurrences of NLOS (Non-line-of-sight) and LOS (Line-of-sight) channel conditions and, therefore, it is interesting to assess their implementation in critical satellite contexts. Considering such a context, the chapter will investigate the adoption of different cooperative techniques in some satellite access scenarios, pointing out its advantages and drawbacks. The chapter is organised as follows. Starting from the identification of critical issues in different satellite access scenarios, reported in Section 2, a general overview on cooperative strategies and, in particular, on the selected cooperative approaches, is provided in Section 3. Then, Section 4 and Section 5 report some different satellite case studies in order to show the advantages of using this kind of approach in uplink and in downlink satellite access, respectively. Finally, Section 6 provides some concluding remarks.

## 2. Satellite Access: scenarios and critical issues

Satellite communications have developed a global success in the field of digital audio/TV broadcasting because they offer a wide coverage area and, therefore, they are suitable for the distribution of multimedia contents to a large number of potential users, also in rural environments. Moreover, they allow the extension of the coverage area of terrestrial, fixed and mobile, networks. One of the most interesting example concerning this capability, is provided by Inmarsat which has developed a broadband global area network service for mobile terminals on land, at sea and in the air. Users can send and receive voice and data services nearly everywhere on Earth. In particular, in some specific cases as the transoceanic maritime and aeronautical communications, satellites are the only practical solution to telecommunications requirements.

Broadband satellite systems can also help to bridge the digital divide because they can provide a rapid deployment compared with other terrestrial infrastructures, without gigantic investments. For example, continents (e.g. Africa) and large countries which, currently, lack in infrastructures could satisfy their needs (mobile phones, Internet access, etc.) and create new opportunities for human development. Applications like telemedicine, e-learning or simply an easy access to information can allow economic activities to grow and develop.

Satellite systems can allow a multitude of valuable services and applications to emerge. Besides for commercial services such as broadcasting, multimedia transmission and broadband services, the use of satellite for telecommunication is also considered for other application scenarios such as public services, emergency services, data relay services, etc. For example, the monitoring and the protection of critical infrastructures such as pipelines and oil platforms, depend on data transmission via satellite. And also coastal and maritime security has increased thanks to the use of new satellite technologies suitable for tracking the position and the state of goods transported by sea. In fact, vessels are required to carry satellite terminals that transmit their identity and position. The benefits of satellite communications are well visible also in emergency applications wherein the world-wide Civil Protection is involved in order to guarantee safety to population. In case of floods, earthquakes, volcanic eruptions and other major disasters, terrestrial communication networks could be damaged and not be able anymore to provide the services required by first responder teams, such as, for example, a robust voice communication system. Rescue teams terminals should be also compatible with other different kinds of terminals if the disaster involves more than one country and so multinational rescue operations are needed. In such a situation, satellites can flexibly connect different first responder team clusters over large distance across incompatible standards. In fact, for large disasters, only satellites are actually able to cover the whole scene and provide broadband services. A satellite communication component is considered in the Air Traffic Management scenario, as well. Also in this application, the main satellite communication strengths are the large coverage area and the rapid deployment. Thanks to the use of satellites, a seamless service between air traffic controllers and pilots could be provided in Europe, including not only areas of dense traffic but also remote areas such as Mediterranean sea, transatlantic routes, deserts, etc.

However, analysing all these scenarios, some critical issues in the use of satellite systems, common to many contexts, can be highlighted. In particular, the presence of link impairments and fading conditions (multipath, long periods of shadowing and blockage) or the mobility effects (occurrence of visibility and not visibility conditions) require the adoption of solutions in order not to reduce system performance and capabilities. Moreover, power constraints have to be taken into account, as well, especially in case mobile terminals are considered.

## 3. Overview on Cooperative Communications

Some years ago, a new class of techniques, called *cooperative communications*, has been proposed as a valuable alternative to the spatial diversity techniques which require the deployment of additional antennas in order to mitigate the fading effects.

Cooperative communications are based on the concept that a group of mobile terminals can share their single antennas in order to generate a "virtual" multiple antenna, obtaining the same effects than a MIMO system, (Nosratinia et al., 2004; Ribeiro & Giannakis, 2006). This approach can be seen as a new form of spatial diversity in which, however, the diversity gain can be achieved through the cooperation of different users, opportunely grouped in clusters, which can assume the double role of *active user*, i.e. the user which transmits its own information data and *cooperator*, i.e. the user which "helps" the active user in its transmission, (Sendonaris et al., 2003a;b).

The key concept is that each user sees an independent fading process and that spatial diversity can be generated by transmitting each user's data through different paths, as shown in Fig. 1.

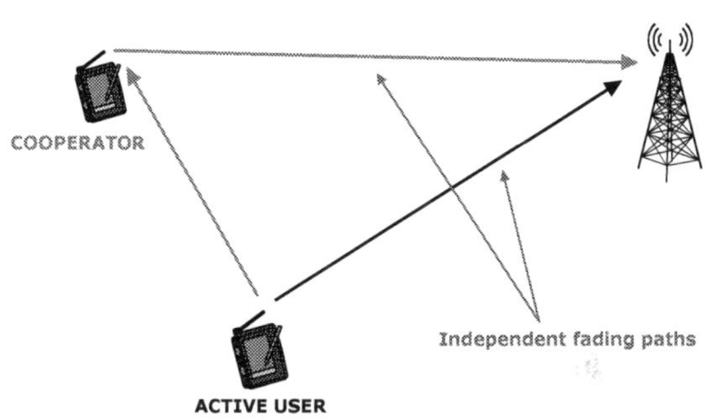


Fig. 1. Example of cooperative communications

An effective way to mitigate fading is to supply the receiver with multiple replicas of the same information-bearing signal transmitted over independent channels. Because of this independence, the probability that all the considered signals are simultaneously vanishing due to fading, is considerably reduced.

If $p$, $(0 \leq p \leq 1)$, is the probability that any signal is faded below a threshold value, the probability that all $L$ independent fading channels, containing the same signal, are faded below the threshold value, is given by:

$$p_{tot} = \prod_{i=1}^{L} p = p^L \tag{1}$$

and, therefore, it is lower than $p$, (Lee & Chugg, 2006).

The cooperative approach turns to be useful for mobile terminals which, because of their size constraints, cannot support multiple antennas and it allows them to increase their performance in terms of Bit Error Rate, Packet Error Rate and Outage probability.

The scenarios wherein the idea of cooperation has been applied so far are, mainly, the cellular networks, the wireless sensor networks and the ad hoc networks, but it can be very interesting to consider the adoption of such strategies also in mobile satellite scenarios which are characterised by the continuous occurrence of LOS and NLOS conditions.

There are several cooperative methods which have been proposed in literature (Nosratinia et al., 2004; Ribeiro & Giannakis, 2006; Sendonaris et al., 2003a;b). However, the main cooperative strategies can be summarised in:

- *Amplify and Forward* (*AF*)
- *Decode and Forward* (*DF*)
- *Selective Forwarding* (*SF*)
- *Coded-Cooperation*

### 3.1 Amplify and Forward

The *Amplify and Forward* is the simplest cooperative method. In this scheme cooperators receive a noisy version of the signal transmitted by active users which, then, amplify and re-transmit towards the final destination. Thus, in this case, also the noise component is amplified and retransmitted by cooperators.
Considering the case of one active user and one cooperator, the amplification factor $A$ can be written as follows, (Darmawan et al., 2007; Ribeiro & Giannakis, 2006):

$$A^2 = \frac{P_c}{P_u|h(u,c)|^2 + N} \tag{2}$$

being $P_c$ the power of the signal transmitted by the cooperator, $P_u$ the power of the signal transmitted by the active user, $|h(u,c)|^2$ is the coefficient of the channel between active user and cooperator, and $N$ is the noise power.
The *Amplify and Forward* strategy requires minimal processing at cooperator terminals but needs a consistent storage capability of the received signal consuming, therefore, memory resources. This method is particularly efficient when the cooperator is close to final destination, as shown in Fig. 2, so that the link from the cooperator to the destination, $d_2$, is characterized by high signal-to-noise ratios and, hence, the link between the active user and the cooperator, $d_1$, becomes comparable to the link between the active user and the destination, $d_3$.

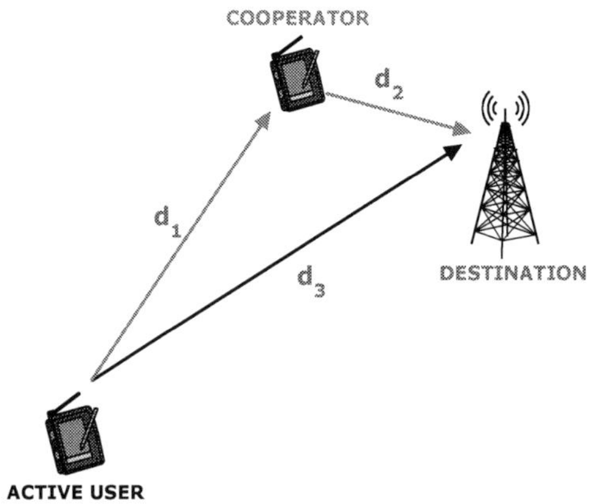


Fig. 2. *Amplify and Forward*: efficient terminals displacement

### 3.2 Decode and Forward

In the traditional *Decode and Forward* scheme, instead, each cooperator always decodes signal coming from the active users, $u(i)$ (with $i = 1 \ldots N_u$, where $N_u$ is total of active users), obtaining an estimate of transmitted signal, $\hat{u}(i)$. Then, it retransmits the signal, $c(i)$:

$$c(i) = \hat{u}(i) \qquad i = 1 \ldots N_u \tag{3}$$

after a re-encoding generally with a repetition-coded scheme.

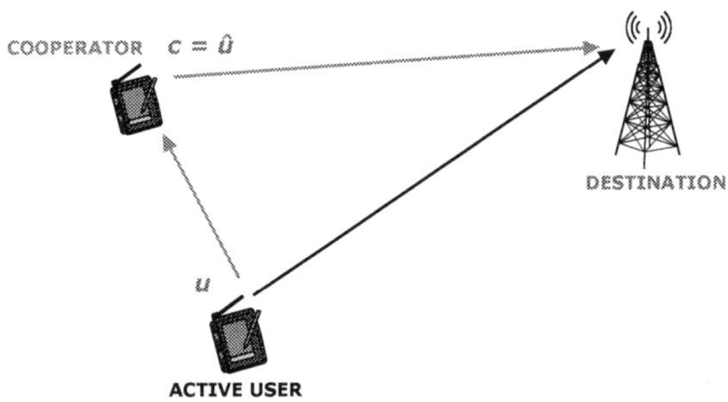


Fig. 3. *Decode and Forward* scheme

Although it has the advantage to be a simple scheme, this cooperative method does not achieve diversity gain. In fact, considering the case of one active user and one cooperator, it is proven that the diversity order is only one, because the overall error probability over two links is dominated by the error probability in the link between the active user and the cooperator, (Laneman et al., 2004; Ribeiro & Giannakis, 2006).

### 3.3 Selective Forwarding Cooperation

The *Selective Forwarding* strategy derives from the Decode and Forward technique and it is based on the concept that cooperators repeat active users' packets by transmitting them through different channel paths with the condition that only the successfully decoded packets received from active users, are sent toward the final destination.

This strategy is more complex than the Decode and Forward method, (Nosratinia et al., 2004; Ribeiro & Giannakis, 2006), because it requires FEC (Forward Error Correction) decoding followed by a CRC (Cyclic Redundancy Check) check to detect possible errors in the packets sent from the active users to the cooperators, but it has some important advantages.

First of all, *Selective Forwarding* is the simplest cooperative method from the perspective of the destination even though it overworks the digital processor at cooperating terminals. Moreover, differently from the Decode and Forward, it allows to achieve diversity and, therefore, to increase the diversity order. Assuming that wireless links between active users and cooperators ($d_1$), are much better than links between active users and their final destinations, ($d_3$), as shown in Fig. 4, and that all users in the considered cluster see uncorrelated channels, the diversity order can be considered equal to the number of users involved in a transmission (active user and its cooperators), (Alamouti, 1998). In this case, *Selective Forwarding* turns to be the best choice for implementing a cooperation process.

Since, for example, in a return link satellite scenario the previous assumptions can be considered valid, the *Selective Forwarding* scheme can be selected as a right cooperative strategy to be implemented in such kind of environments.

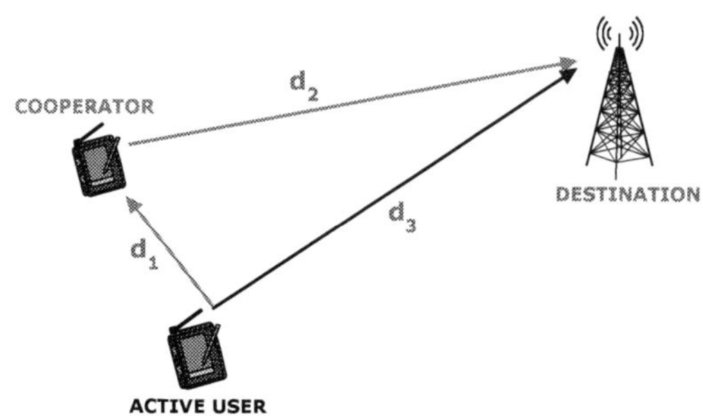


Fig. 4. *Selective Forwarding*: best implementation scenario

### 3.4 Coded-Cooperation

In the *Coded-Cooperation*, the cooperative strategy is integrated with channel coding techniques. In this case, instead of producing more replicas of the active user's signal, as it happens in other cooperative methods, the codewords produced by each user belonging to a determined cluster, are divided in different portions which are transmitted through different independent fading channels, by the considered user and by a selected group of users, called *partners*, which are involved in the cooperation process, (Hunter & Nosratinia, 2002; 2006; Janani et al., 2004).

The basic idea is that each user tries to transmit an incremental redundancy of its partners data, besides its own data. Considering, for example, the case of two users, they cooperate by dividing their own codewords of length $N$, in two successive segments, as shown in Fig. 5. In the first segment, each user transmits a codeword of length $N_1$ containing its own data,

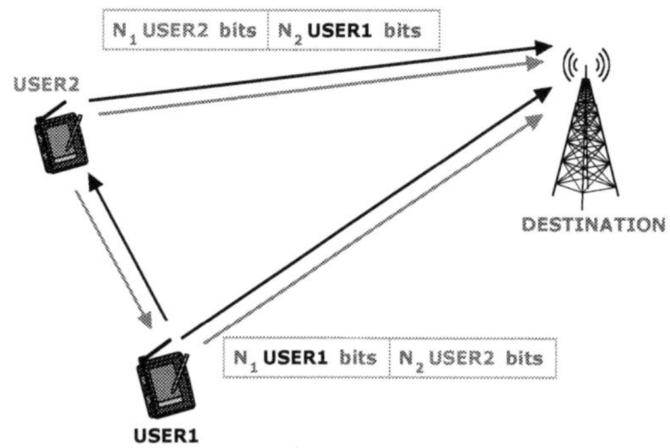


Fig. 5. *Coded-Cooperation* scheme

obtained by its original codeword. Then, each user receives and decodes its partner's first segment. If this is correctly decoded, each user can compute the additional parity bits of the partner's data and transmit the new codeword of length $N_2$ containing the partner's data, in the second segment. If the partner's info cannot be correctly decoded, the user reverts to the non-cooperative mode and it transmits its own data. In fact, if a certain terminal is unable to cooperate, because of the wrong reception of the partner's data, it can always use the available capacity to transmit its own data.

The idea of *Coded-Cooperation* is to use the same overall code rate and power for transmission as in a comparable non-cooperative system, i.e. the same system resources are used. Moreover, this cooperation methodology can provide a higher degree of flexibility with respect to other cooperation methods and a higher adaptability to channel conditions, by allowing the use of different channel coding and partitions schemes. For example, the overall code can be a block code or a convolutional code or a combination of both and, then, coded bits to put into the different segments, can be selected through puncturing, product codes, etc., (Hunter & Nosratinia, 2006).

## 4. Cooperation Techniques for Uplink Satellite Access

Considering what said above, the *Selective Forwarding* and the *Coded-Cooperation* turn to be two cooperative strategies which are suitable to be used in critical satellite scenarios, in particular in the return link suffering from a tighter link budget especially if the involved users are mobile terminals. Therefore, in the following, a specific uplink satellite scenario which presents some tricky issues, is proposed as "case study", in order to show the advantages deriving from the adoption of such cooperative strategies.
The considered model is composed of a set of $N_u$ vehicular users which are interconnected through reliable wireless links and connected to a terrestrial gateway through a geostationary satellite, as shown in Fig. 6.

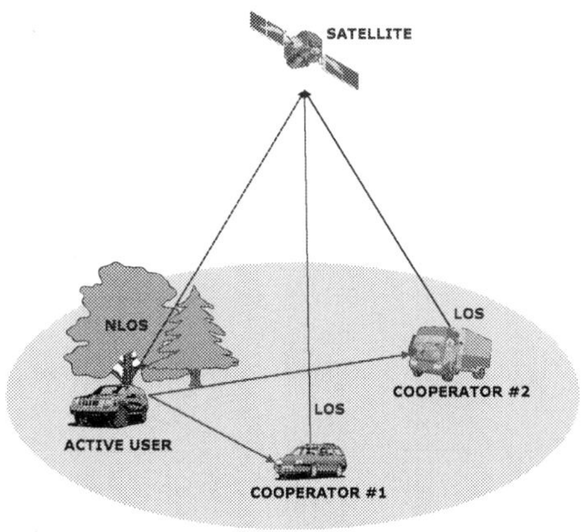


Fig. 6. Satellite cooperative scenario

The forward link is based on the DVB-S2 (Digital Video Broadcasting - Satellite second generation) standard, (DVB-S2 standard, 2009), while the return link (on which this analysis is focused) is based on DVB-RCS (Digital Video Broadcasting - Return Channel Satellite), (DVB-RCS standard, 2005). According to the MF-TDMA (Multi Frequency - Time Division Multiple Access) scheme employed by such a standard, a certain number of frequency/time slots are assigned to users within a superframe depending on their specific demand. The adopted propagation satellite channel model is mainly taken from (Ernst et al., 2008), and it is summarised here for the sake of completeness. The model considers a frequency non-selective

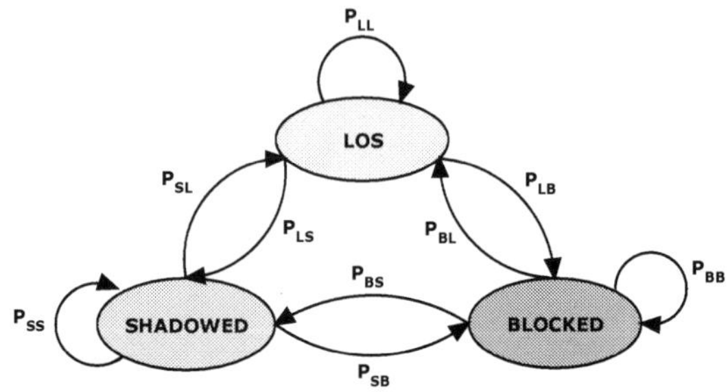

Fig. 7. 3-states channel model

channel at *Ku* band. In these conditions, a generic passband received signal, $r(t)$, can be written as:

$$r(t) = \mathrm{Re}\{A(t) \cdot \tilde{s}(t - t_0)e^{j2\pi f_0 t}\} + n(t) \tag{4}$$

where $A(t)$ is the multiplicative time-varying channel coefficient, $\tilde{s}(t)$ the complex-envelope of the transmitted signal, $t_0$ the propagation delay, $f_0$ the carrier frequency and $n(t)$ the additive thermal noise.

The channel coefficient is a complex term and, therefore, it can be expressed through its absolute value (also called modulus), $|A(t)|$, and its phase $\phi(t)$:

$$A(t) = |A(t)|e^{\phi(t)} \tag{5}$$

The amplitude of the channel coefficient, $|A(t)|$, represents the amplitude of the fading term which, according to this class of models, can be divided into fast and slow fading. Slow fading events, commonly referred to as shadowing, model the attenuation caused by the orography and large obstacles, such as hills, buildings, trees, etc., through absorption and diffraction mechanisms, and they are normally modelled as a finite state machine. Fast fading events, instead, due to the irregularity of the obstacles (e.g. vegetative shadowing) and to the multipath propagation phenomena caused by reflections over surrounding surfaces, can be additionally modelled as superimposed random variations that follow a given Probability Density Function (PDF) for each state.

At an arbitrary time instant $t$ and assuming that the transmitted signal $\tilde{s}(t)$ has unitary amplitude[1], the overall PDF describing the received signal amplitude, called below $R(t)$, can be written as:

$$p_R(r) = \sum_{k=1}^{N} P_k \cdot p_{R,k}(r) \tag{6}$$

being $N$ the number of states, $P_k$ the absolute probability of being in the state $k$ (that can be easily obtained from the *State Transition Matrix* $S = [p_{ij}]$, containing in each element the probability of transition from the state $i$ to the state $j$) and $p_{R,k}(r)$ the PDF associated to the fast fading within state $k$.

Following this approach, a three states (LOS, Shadowed and Blocked) Markov-chain based model is assumed for the fading process, as shown in Fig. 7.

[1] Under this hypothesis, the received signal amplitude, $R(t)$ corresponds to the amplitude of the fading term, i.e. $R(t) = |A(t)|$.

The *LOS* state is characterised by a Rician PDF of the following form:

$$p_R(r) = \frac{r}{\sigma^2} \cdot \exp\left(-\frac{r^2+z^2}{2\sigma^2}\right) \cdot I_0\left(\frac{r \cdot z}{\sigma^2}\right), \qquad r \geq 0 \tag{7}$$

being $I_0$ the zero-order modified Bessel function of the first kind, $z$ the amplitude of the line-of-sight component and $\sigma^2$ the power of the real part or the imaginary part of the scattered component.

The *Shadowed* state is characterised by a Suzuki PDF, (Suzuki, 1977). The Suzuki process is a product process of a Rayleigh process and a Lognormal (LN) process, (Finn & Flemming, 1977; Pätzold, 2002). The slow signal fading is, in this case, modelled by the Lognormal process taking the slow time variation of the average local received power into account. The Rayleigh process models, instead, the fast fading. The Suzuki PDF can be expressed as follows, (Lin et al., 2005):

$$\begin{aligned} p_R(r) &= \int_0^{+\infty} \left[\frac{r}{\sigma_{ray}^2 L^2} \cdot \exp\left(-\frac{r^2}{2\sigma_{ray}^2 L^2}\right)\right] \cdot \\ &\left[\frac{1}{\sqrt{2\pi}\phi\sigma_{ln} L} \cdot \exp\left\{-\frac{1}{2}\left(\frac{\ln(L) - \phi\mu_{ln}}{\phi\sigma_{ln}}\right)^2\right\}\right] dL \end{aligned} \tag{8}$$

wherein the first term represents the conditional joint Lognormal and Rayleigh PDF while the second term is the Lognormal PDF which characterises the random variable $L$. Moreover, $\phi = \ln 10/20$ while $\mu_{ln}$ and $\sigma_{ln}$ are the mean and standard deviation, respectively, of the associated Gaussian distribution in dB unit.

Finally, the *Blocked* state is characterised by no signal availability. The set of considered parameters is provided in Table 1 for the environment considered next, namely *highway*. The average state transition period is equal to 0.0417 s, corresponding to blocks of 1000 samples at the sampling frequency of 24 kHz. The above mentioned state duration refers to average speed $v$ of 100 Km/h.

| *Environment* | *State Transition Matrix* | *P (LOS, SH, BL)* | *Rice z* | *Rice $\sigma$* | *Rice Factor* | $\sigma_{ln}$ | $\mu_{ln}$ |
|---|---|---|---|---|---|---|---|
| Highway | 0.9862 0.0138 0.0000 | 0.8922 | 0.9892 | 0.0947 | 17 dB | 1.5 dB | -8 dB |
| | 0.1499 0.8378 0.0123 | 0.0823 | | | | | |
| | 0.0008 0.0396 0.9596 | 0.0255 | | | | | |

Table 1. *Ku*-band land-vehicular channel parameters

Doppler Spectrum is estimated as proposed in (Dubey & Wee Teck Ng, 2002; Law et al., 2001), taking into account a realistic antenna beamwidth and the angle between satellite position and terminal direction by means of the following equation:

$$S(f) = \begin{cases} \dfrac{A}{f_d\sqrt{1-\left(\dfrac{f}{f_d}\right)^2}} & \text{if} \quad f_d\cos(\phi+\alpha) < f < f_d\cos(\phi-\alpha) \\ 0 & \text{otherwise} \end{cases} \tag{9}$$

The following values have been considered:

- $\alpha = \pi/2$

- $f_d = v \cdot f_0 / c$
- $2\phi = \theta_{3\,dB} = 70\lambda / D$
- $D = 65$ cm

being $D$ the antenna diameter, $v$ the terminal speed defined above and $f_0 = c/\lambda$, the carrier frequency at $Ku$ band equal to 14 GHz.

### 4.1 Selective Forwarding Cooperation for Critical Satellite Scenarios

The analysis considers the adoption, in the scenario described above, of a cooperative strategy which allows the users to share the uplink effort according to the *Selective Forwarding* cooperation scheme. Fig. 8 shows an example of the used procedure which describes how the resources are allocated and managed in the TDMA scheme. Groups of timeslots, named *frames*, are assigned to active users and cooperators in order that they can transmit their *traffic bursts* (in the following named simply "packets").

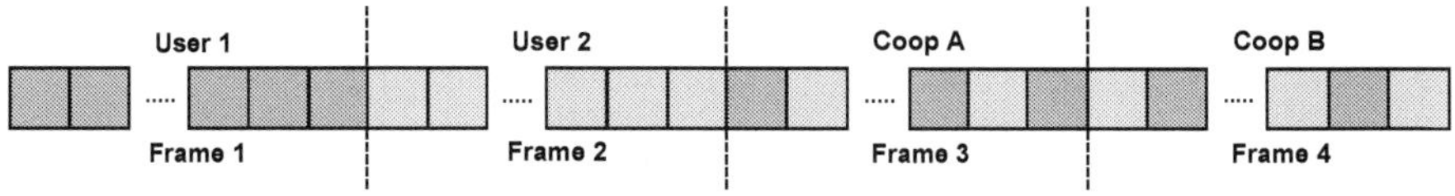


Fig. 8. Example of timeslot assignation in a superframe: 2 active users and 2 cooperators

Within each superframe, the active users (*User1* and *User2*) convey their informative packets while the cooperators (*Coop A* and *Coop B*) repeat each one half *User1*'s packets and half *User2*'s packets in an alternate way. In particular, *Coop A* retransmits before a *User1*'s packet and then a *User2*'s packet, whereas, vice versa, *Coop B* starts repeating before a *User2*'s packet and then a *User1*'s packet. Hence, in this case, two replicas of the same packet for each active user are sent through the satellite and the receiver can apply a CRC mechanism in order to detect the correct packets among those received. Such a method can be simply extended to a different number of active users and cooperators.

The benefits of this procedure can be assessed observing Fig. 9 wherein the received signal power of each active user and its cooperators, is reported. In some time portions, in fact, the cooperators can experiment better satellite channel conditions than the active users and their retransmission of packets becomes fundamental in order to not to lose some pieces of information sent by the active users. The receiver can process differently corrupted replicas of the same packet and the probability to detect packets successfully increases considerably.

In the model, the terrestrial wireless links between active users and cooperators, used to share packets, are characterized by error-free conditions in order to evaluate the efficiency of the cooperative strategy in the satellite land-vehicular scenario.

In the following, some results achieved through computer simulations are presented. First of all, it is shown how the number of involved cooperators affects the system performance. In particular, in Fig. 10, the performance comparison in terms of average PER (Packet Error Rate) between the no cooperation and cooperation (with 2 cooperators and 4 cooperators) cases in the highway environment is reported. The number of active users is considered equal to 2 in all simulated cases. Focusing mainly on this Figure, it can be seen that as the number of cooperators increases, the PER values decrease considerably for fixed $E_b/N_0$ values and, in particular, it can be noted that, the case considering 4 cooperators has a PER floor at about $2 \cdot 10^{-3}$ for $E_b/N_0$ values starting from 2 dB with respect to the no cooperation case which

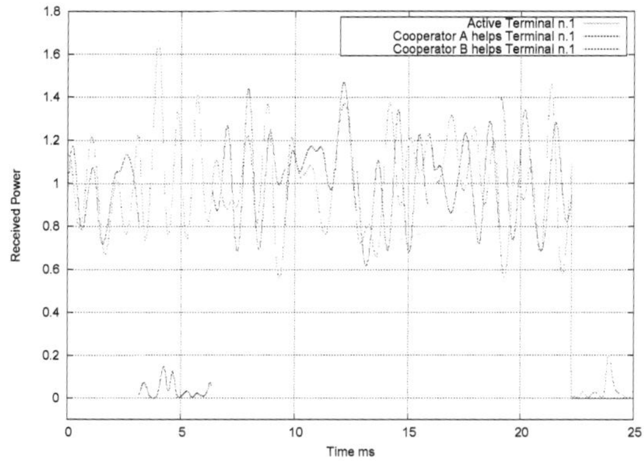


(a) *Active user: User1*

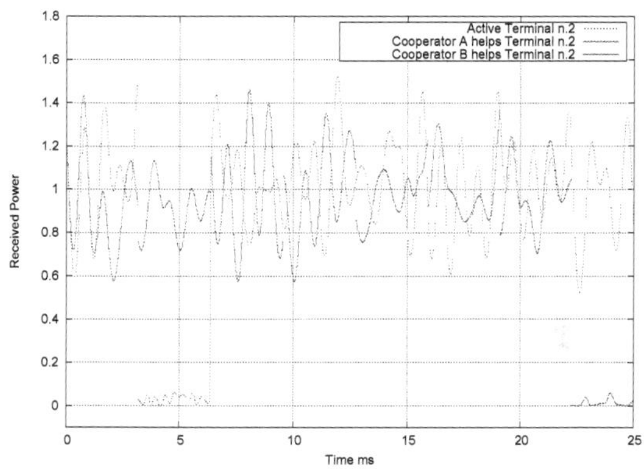


(b) *Active user: User2*

Fig. 9. Received signal power of *Active user*, *Cooperator A* and *Cooperator B*

has, instead, a PER floor at $1.1 \cdot 10^{-1}$. The presence of PER floors is due to the occurrence, with the given probabilities already shown in Table 1, of Shadowed and Blocked state channel conditions. However, the context taken into account for satellite broadband communications is, mainly, that of elastic IP traffic generated by applications like e-mail, web browsing, FTP and TELNET services, which are not completely compromised by a delay, loss or bandwidth limitations, due also to the occurrence of NLOS channel conditions. For these reasons, it is worth analysing how the cooperation strategy affects the system performance when the satellite channel is only in LOS or in NLOS conditions in order to evaluate the realistic behaviour of the system which works for the most part of the time in LOS conditions. The LOS state is, as a matter of facts, the state with the highest absolute probability (89.22% in the considered highway environment).

Fig. 11 shows, therefore, a comparison in terms of PER between no cooperation and cooperation (4 cooperators) cases considering the satellite channel being only in the LOS state or

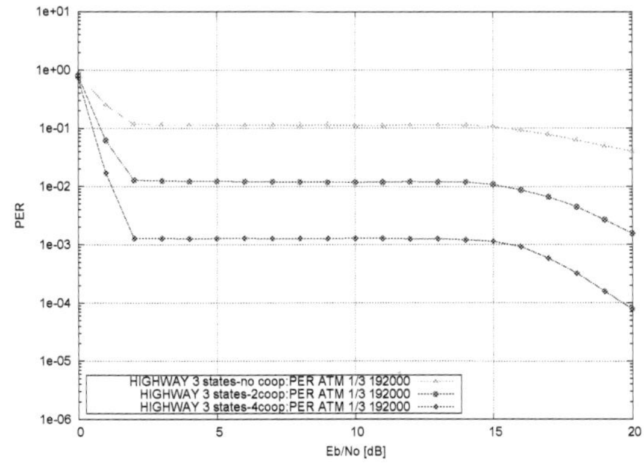


Fig. 10. PER performance for ATM cell, code rate 1/3, data rate 192 kbit/s, HIGHWAY environment: 3 states - *Ideal case* 4 cooperators, 2 cooperators and no cooperation cases

only in the Shadowed state. The Blocked state, as already said, is characterised by no signal availability so the achieved BER (Bit Error Rate) values are equal to 0.5.
The results concerning the LOS state are encouraging because they show that the adoption of the cooperation (4 cooperators) allows improving the system performance achieving the PER value $10^{-6}$ with a gain equal to 1.4 dB with respect to the case of absence of cooperation.

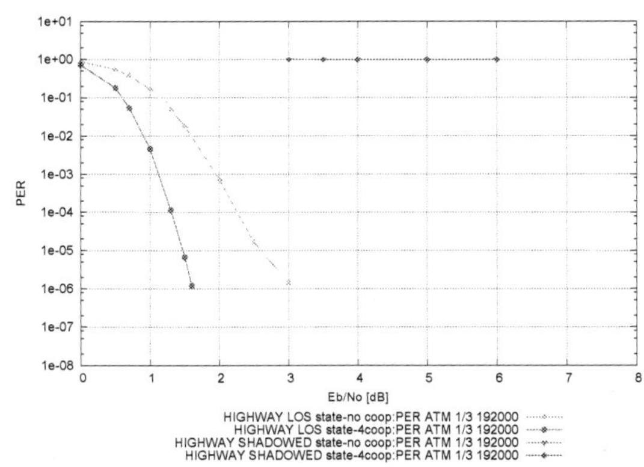


Fig. 11. PER performance for ATM cell, code rate 1/3, data rate 192 kbit/s, HIGHWAY environment: LOS state and Shadowed state - *Ideal case* 4 cooperators and no cooperation cases

### 4.2 Coded-Cooperation in Mobile Satellite Systems

In the following, the adoption of *Coded-Cooperation* in the same return link scenario previously described, is taken into account. In this case, the analysis starts considering the $i$-th user (with $i = 1 \ldots N_u$) which aims at transmitting a message of size $k$ bits. The message is first encoded by the physical layer encoder, obtaining the codeword $\mathbf{c(i)}$ of size $n$ bits. Once all

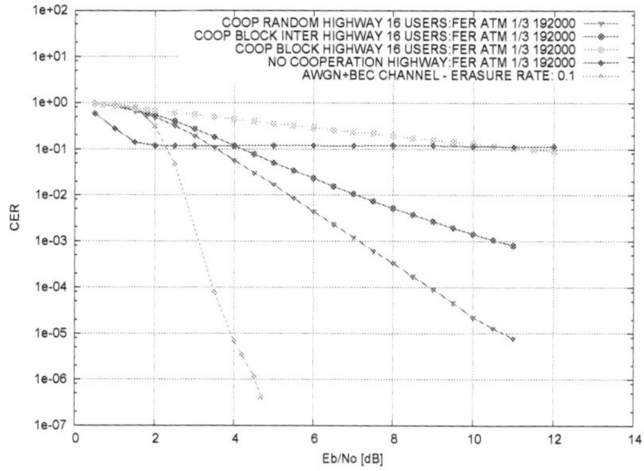


Fig. 12. Performance comparison in terms of CER between cooperative (16 users) and non-cooperative schemes for ATM cell, code rate 1/3, data rate 192 kbit/s: HIGHWAY environment

codewords $\mathbf{c}(\mathbf{i})$ are ready, they are exchanged through terrestrial links among the $N_u$ users. At each user $i$, each generic message $\mathbf{c}(\mathbf{j})$ coming from the other users, is divided in $N_u$ sub-blocks, $\mathbf{c}(\mathbf{j}) = [c_1(j), c_2(j), \ldots, c_{N_u}(j)]$. A new vector bit $\mathbf{x}(\mathbf{i})$, hereafter referred to as *combined codeword*[2], is then produced by the generic $i$-th user by combining $N_u$ sub-blocks belonging to different users' codewords. The vector $\mathbf{x}(\mathbf{i})$ is, then, sent by the $i$-th user through the satellite link. The selection of the sub-blocks involved in the combined codewords can be based on predefined or random patterns depending on the considered *Coded-Cooperation* scheme, under the constraint that all the sub-blocks of a codeword $\mathbf{c}(\mathbf{i})$ are sent through different combined codewords.

Some results which prove the effectiveness of such a procedure are presented in the following. Performance has been analysed in terms of CER (Codeword Error Rate) vs. $E_b/N_0$ at the output of the FEC decoder in the gateway. In the plot in Fig. 12, a comparison among three different coded-cooperative schemes considering sixteen users, and the non-cooperative case is reported. In the first two schemes, named *cooperation block* and *cooperation block inter*, the codeword of the i-th user, constituted by a systematic part and a parity part, is divided in as many portions as the number of cooperative users and each of them transmits a combined codeword, as previously explained. The difference between these two schemes is in the rule that assigns each portion of the original codeword to each user. In the first scheme, a simple rule is used: the first user transmits the first portion of the systematic part and the first portion of the parity part of all codewords, the second one transmits the second portion of both parts and so on for all users. In the second scheme, instead, the portions sent by each user are assigned pseudo-randomly bearing however in mind that all sub-blocks of each codeword $\mathbf{c}(\mathbf{i})$ shall be transmitted. So, for instance, the first user transmits the first portion of systematic part but not the first one of the parity part. In the third scheme, named *cooperation random*, the partitioning of the codeword between systematic part and parity part is not considered

[2] Note that a combined codeword does not belong to a specific code book, i.e. it is not a result of an encoding procedure. It represents a concatenation of portions belonging to different actual codewords.

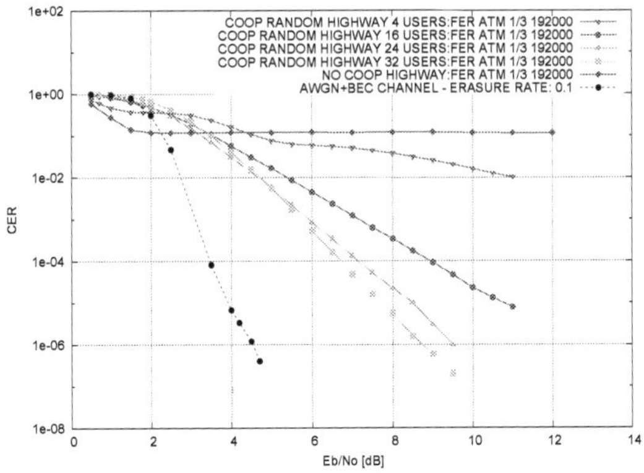


Fig. 13. Performance in terms of CER of the cooperation random scheme for different number of users, for ATM cell, code rate 1/3, data rate 192 kbit/s: HIGHWAY environment

anymore. In this case, the codeword portions composing the combined codeword are constituted by the bits of the original codeword of each user, which are assigned to each user using a random rule. Thus, the i-th user can transmit a portion composed by as many systematic bits as parity bits depending on the distribution of the bits that the random rule has generated. Using this last scheme the highest randomization level is guaranteed and, as it can be seen in Fig. 12, the deleterious effects of fading can be more effectively counteracted. Also the performance over the AWGN (Additive White Gaussian Noise) channel with erasures, in the following named *AWGN+BEC*, is reported. This curve represents a reasonable reference which, for high $E_b/N_0$ values, could be taken as an acceptable lower bound to the system performance: under the assumption that only the LOS state can be successfully decoded, and in case the diversity introduced by cooperation could break any channel correlation effect, each codeword would in fact virtually face an uncorrelated channel with an erasure rate equal to the NLOS share, given by the sum of $P_{SHADOWED}$ and $P_{BLOCKED}$.
In Fig. 13, the *cooperation random* scheme is further investigated and it is shown how the number of users affects the system performance. It can be seen how, as the number of users increases, the CER values decrease for a fixed $E_b/N_0$ value. The performance improvement is more remarkable for increasing $E_b/N_0$ values. Using this scheme it is possible to achieve CER values performing a feasible system which does not present anymore a high floor value as it is, instead, for the non-cooperative case which has a CER floor at $10^{-1}$. In particular, it can be noted that the CER value $10^{-5}$ is achieved for $E_b/N_0$ equal to 7.7 dB. This results is encouraging also because, if the channel state information were introduced in the simulation model, the achieved improving could be more relevant.

## 5. Cooperation Techniques for Downlink Satellite Access

Generally, in a downlink scenario, the link from the satellite to the active terminal is comparable with the links from the satellite to cooperating devices and, therefore, the *Amplify and Forward* strategy can be particularly efficient in this kind of scenarios. For this reason, a particular downlink satellite scenario is taken into account in order to show how the use of such a strategy can led to improvements in the system performance.

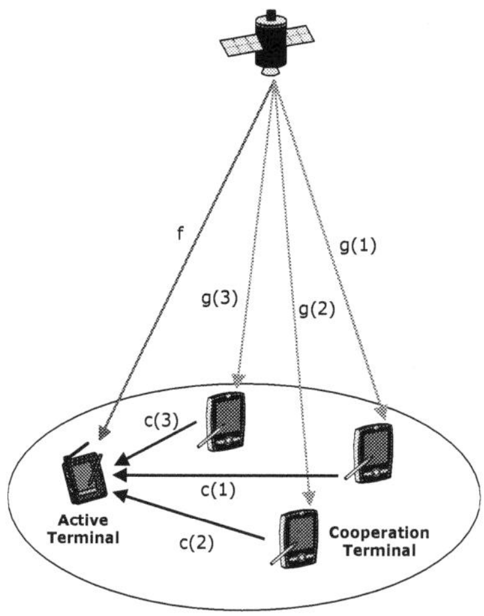


Fig. 14. Downlink Satellite Cooperation Scenario

| | | | |
|---|---|---|---|
| $d_{sat}$ | 36000 | [Km] | satellite terminal distance |
| $d_{coop}$ | 10 | [Km] | cooperative terminal |
| $L_{sat}$ | -205.34 | [dB] | satellite terminal path loss |
| $L_{coop}$ | -118.5 | [dB] | cooperative terminal path loss |
| $B_{sat}$ | 36 | [MHz] | transpoder bandwidth |
| $P_{sat}$ | 70 | [dBW] | satellite power |
| $P_{max}$ | 250 | [mW] | cooperative terminal maximum power |
| $G/T_{Rx}$ | -24 | [dB/K] | handheld receiver $G/T$ |
| $T_{sys}$ | 290 | [K] | system temperature |
| $F_c$ | 2000 | [MHz] | cooperation channel frequency |
| $F_d$ | 11750 | [MHz] | downlink channel frequency |

Table 2. Main operational parameters

The adopted downlink cooperation scenario is depicted in Fig. 14. A DVB-S2 hub processes and sends digital signals to some users grouped in a cluster, through the satellite. A potential mobile DVB-S2 receiver (the active terminal) combines the signals coming from the satellite and from several mobile cooperators belonging to the same cluster. The satellite-to-earth link is modelled with a Corazza-Vatalaro process, (Corazza & Vatalaro, 1994), while the cooperator-to-active user link is represented only by an AWGN channel. The Corazza-Vatalaro channel model is a combination of a Rice and a Log-normal factors, with shadowing affecting both direct and diffused components. The cooperative path-loss value of 118 dB, reported in Table 2, derives from the choice of a cooperation frequency $F_c = 2$ GHz and a cooperator distance $d_{coop} = 10$ Km.
The fading effect on the cooperative links is not considered, as expected in environments characterized by limited distances (within 10 Km) and good visibility among terminals. The model considers a time resolution equal to:

$$\frac{1}{2B_{sgn}} = \frac{1}{14.8}\,\mu s \tag{10}$$

being $B_{sgn}$ the bandwidth of the modulated QPSK signal ($FEC = 1/2$) considering an useful data rate of 7.2 Mbaud.

### 5.1 Amplify and Forward Cooperation for Mobile Satellite Terminals

The basic idea of *Amplify and Forward* strategy is that around a given terminal, there can be other single-antenna terminals which can be used to enhance diversity by forming a virtual (or distributed) multiantenna system where the satellite signal is received from the active terminal and a number of cooperating relays. Cooperating terminals retransmit the received signal after amplification. As said before, the AF strategy is particularly efficient when cooperating terminals are located close to the active one so that the cooperative links ($c(1)$,$c(2)$,$c(3)$ in Fig. 14) are characterized by high signal-to-noise ratios and the link from the satellite to the active terminal ($f$) is comparable with the links from the satellite to cooperating devices ($g(1)$,$g(2)$,$g(3)$ in Fig. 14). Starting from Eq. (2), the considered amplification factor $A$ is given by:

$$A_i^2 = \frac{P_{max}}{P_{sat}|g(i)|^2 + N} \qquad (11)$$

where $P_{sat}$ is the satellite downlink power and $P_{max}$ the cooperative terminal maximum power, $g(i)$ the $i$-th link pathloss and $N = KT_{sys}B_{sat}$ the noise power at the earth terminals.
With this choice, the resulting $C/N$ on the active terminal is given by the following expression, assuming that all of the cooperating terminals, $M$, have the same characteristics and the cooperative channels, $c$, are similar:

$$\frac{C}{N} = \frac{P_{sat}|f|^2}{N}(1 + M\frac{A^2|c|^2}{1 + A^2|c|^2}) \qquad (12)$$

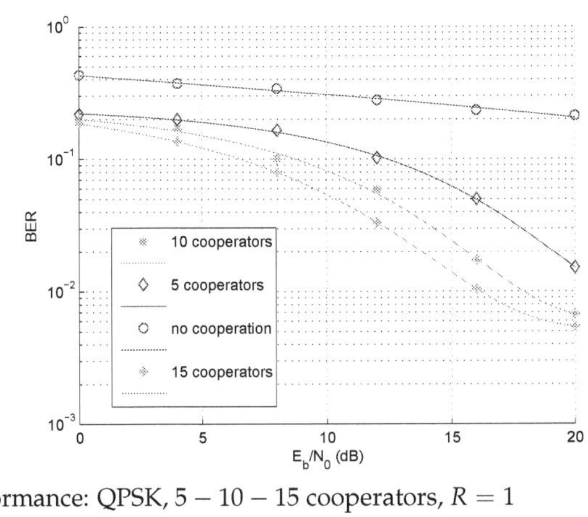


Fig. 15. BER performance: QPSK, 5 – 10 – 15 cooperators, $R = 1$

System performance has been analysed in terms of BER and the resulting BER versus $E_b/N_0$ curves for different configurations have been plotted. The curves of Fig. 15 show the advantages deriving from the use of the cooperation AF with a QPSK modulation for various number of cooperators (5, 10 and 15). All the handsets share the same Rice factor $R = 1$ (medium

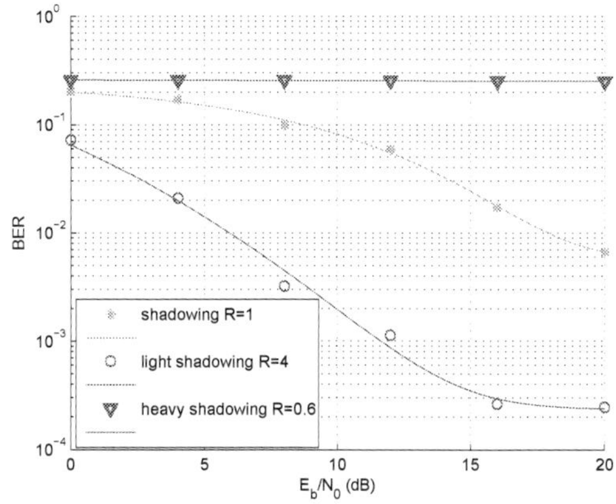


Fig. 16. BER performance: QPSK for variable Rice Factor $R = 0.6 - 1 - 4$ and 10 cooperators

shadowing), modeling the situation where the consumers cooperators all work under homogeneous operational conditions. Fig. 16 shows QPSK performances obtained by varying the Rice factor $R$. The case of heavy shadowing ($R = 0.6$), medium shadowing ($R = 1$) and light shadowing ($R = 4$) are compared. For $R = 4$ the performance is close to the target ($BER = 10^{-4}$), while for $R = 0.6$ the BER values are higher than target, resulting unacceptable for the DVB-S2 system.

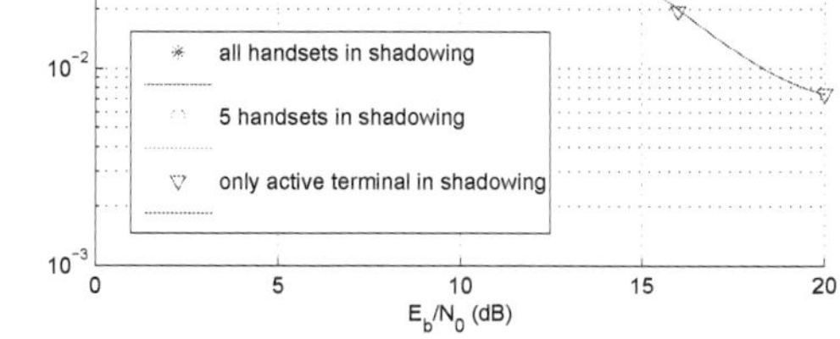


Fig. 17. BER performance: QPSK, varying handset number in heavy shadowing for $R = 0.6$

Finally, Fig. 17 shows the BER performance in the case a varying number of handsets are in heavy shadowing ($R = 0.6$) while the remaining ones have $R = 1$. By considering such a less critical situation, where only a subset of cooperating terminals are subject to heavy shadowing, it can be seen that the system performance improves.

## 6. Conclusion

This chapter has presented the possible adoption of cooperation strategies in satellite access, focusing on two case studies showing an uplink and downlink mobile satellite scenario. The use of these different techniques and methodologies in various applications scenarios, can led to the achievement of improvement of the system performance in terms of Bit Error Rate and Packet Error Rate.

In particular, in the uplink scenario, the introduction of the *Coded-Cooperation* for DVB-RCS terminals working in a land vehicular scenario, allows improving considerably, for increasing $E_b/N_0$ values, the system performance compared with the non-cooperative system, especially if a codeword partitioning scheme maximising the level of randomness in the distribution of the sub-blocks among different users is adopted. In the best simulated scenario, if it is considered a Codeword Error Rate value of $10^{-5}$, the system performance is, however, still roughly 3.8 dB away from the reference (AWGN with erasure rate equal to NLOS share) case, leaving significant room for further optimisation of the system. However, a trade-off between the number of cooperative users, the resulting system complexity and the achievable performance is necessary. Moreover, also the adoption of a *Selective Forwarding* cooperation in a DVB-RCS land-vehicular scenario, allows improving sensibly the system performance in the considered environments, depending on the number of users involved in the cooperation process. The simulation results have shown that, considering 4 cooperators which cooperate with 2 active users, a cooperation gain equal to 1.4 dB can be achieved with respect to the case of absence of cooperation.

As what concerns, instead, the downlink scenario the idea was to build a cooperation among a set of mobile terminals, in a way that the signal received by each single device is the result of the composition of more replicas of the same signal sent by other cooperating devices. Link cooperation, in this case, enables the reception of satellite services from handheld terminals when a cluster of cooperating users is present.

## 7. References

Alamouti, S. M. (1998). A Simple Transmit Diversity Technique for Wireless Communications, *IEEE Journal on Selected areas in Communications*, Vol. 16, pp. 1451-1458, October 1998.

Corazza, G. & Vatalaro, F. (1994). A Statistical Model for Land Mobile Satellite Channels and Its Applications to Nongeostationary Orbit Systems, *IEEE Transactions on Vehicular Technology*, vol. 43, pp. 738-741, August 1994.

Darmawan, A. & Kim, S.W. & Morikawa, H. (2007). Amplify-and-Forward Scheme in Cooperative Spatial Multiplexing, *16th IST Mobile and Wireless Communications Summit*, July 2007, Budapest, Hungary.

Dubey, V.K. & Wee Teck Ng, (2002). Comments on On the Doppler spectrum at the mobile unit employing a directional antenna, *IEEE Communication Letters*, Vol. 6, No. 11, pp. 472-474, November 2002.

Ernst, H. & Harles, G. & Scalise, S. (2008). Measurement and Modelling of the Land Mobile Satellite Channel at Ku-Band, *IEEE Transactions on Vehicular Technology*, Vol. 57, No. 2, pp. 693-703, March 2008.

ETSI EN 301 790 v 1.4.1 (2005). Digital Video Broadcasting (DVB): Interaction channel for satellite distribution systems, September 2005.

ETSI EN 302 307 v 1.2.1 (2009). Digital Video Broadcasting (DVB): Second generation framing structure, channel coding and modulation system for Broadcasting, Interactive Services, News Gathering and other broadband satellite applications (DVB-S2), August 2009.

Finn, M.I. & Flemming, H. (1977). Mobile Fading-Rayleigh and Lognormal Superimposed, *IEEE Transactions on Vehicular Technology*, Vol. 26, No. 4, pp. 332-335, November 1977.

Hunter, T.E. & Nosratinia, A. (2002). Cooperation Diversity through Coding, *IEEE International Symposium on Information Theory (ISIT)*, July 2002, Lausanne, Switzerland.

Hunter, T.E. & Nosratinia, A. (2006). Diversity through Coded Cooperation, *IEEE Transaction on Wireless Communications*, Vol. 5, No. 2, pp. 283-289, February 2006.

Janani, M. & Hedayat, A. & Hunter, T.E. & Nosratinia, A. (2004). Coded Cooperation in Wireless Communications: Space-Time Transmission and Iterative Decoding, *IEEE Transaction on Signal Processing*, Vol. 52, No. 2, pp. 362-371, February 2004.

Laneman, J.N. & Tse, D.N.C. & Wornell, G.W. (2004). Cooperative Diversity in Wireless Networks: Efficient Protocols and Outage Behavior, *IEEE Transaction on Information Theory*, Vol. 50, No. 12, pp. 3062-3080, December 2004.

Law, C.L. & Yoshida, S. & Xu, C.Q. (2001). On the Doppler power spectrum at the mobile unit employing a directional antenna, *IEEE Communication Letters*, Vol. 5, No. 1, pp. 13-15, January 2001.

Lee, D.K. & Chugg, K.M. (2006). A Pragmatic Approach to Cooperative Communication, *IEEE Military Communications Conference (MILCOM)*, October 2006, Washington, D.C.

Lin, D.B. & Lin, H.P. & Tseng, M.C. (2005). Performance analysis of M-ary PSK adaptive modulation system over Rayleigh-lognormal fading channel, *IEEE Vehicular Technology Conference Spring (VTC2005Spring)*, May 2005, Stockholm, Sweden.

Nosratinia, A. & Hunter, T.E. & Hedayat, A. (2004). Cooperative Communication in Wireless Networks, *IEEE Communications Magazine*, Vol. 42, No. 10, pp. 74-80, October 2004.

Pätzold, M. (2002). *Mobile Fading Channels*, Wiley, January 2002.

Ribeiro, A. & Giannakis, G.B. (2006). Fixed and Random Access Cooperative Networks, *EURASIP Newsletter*, pp. 3-24, March 2006.

Sendonaris, A. & Erkip, E. & Aazhang, B. (2003). User cooperation diversity - part I: System Description, *IEEE Transactions on Communications*, Vol. 51, No. 11, pp. 1927-1938, November 2003.

Sendonaris, A. & Erkip, E. & Aazhang, B. (2003). User cooperation diversity - part II: Implementation Aspects and Performance Analysis, *IEEE Transactions on Communications*, Vol. 51, No. 11, pp. 1939-1948, November 2003.

Suzuki, H. (1977). A Statistical Model for Urban Radio Propagation, *IEEE Transactions on Communications*, Vol. 25, No. 7, pp. 673-680, July 1977.

4

# MIMO Channel Models for Satellite Communications

Abbas Mohammed and Asad Mehmood
*Blekinge Institute of Technology*
*Sweden*

## 1. Introduction

In recent years considerable attention has been drawn to the systems with multiple element transmitter and receiver arrays. It has been demonstrated both in theory and practice that multiple-input multiple-output (MIMO) systems offer the promise of increased capacity, high spectral efficiency and high gains by exploiting space-time processing techniques in particular space-time coding (STC) techniques under different propagation environments. This superior performance is achieved by the decomposition of the radio channel into a set of independent spatially uncorrelated channels. The basic method to achieve diverse channels, in addition to the exploitation of time and frequency dimensions, is to use the additional dimension of space, i.e., antennas are spatially separated from each other resulting in space-time processing. The advantages obtained by MIMO channels are highly dependent on the orientations of scatterers and the correlations among signal carriers which limit the performance of MIMO systems. The antenna separation, in terms of wavelength of the operating frequency, has significant impact on the spatial correlation. To achieve uncorrelated fading paths adequate antenna spacing along with rich scattering environment is necessary. An alternative solution to achieve low fading correlation is to use antenna arrays with cross polarizations, i.e., antenna arrays with polarizations in orthogonal or near orthogonal orientations without increasing the bandwidth and in particular, the concept of three-dimensional (3D) polarization has a significant role in achieving diversity by polarization. It has been shown in recent research that it is possible to attain even more channels by using benefits of the combined spatial and polarization diversity in rich scattering environments.

Satellite communication systems are not immune from this wave of innovation. However, due to difference in the propagation conditions in satellite and terrestrial links, the applicability and designs of MIMO systems are different as well. Due to very large path lengths, transmit and/or receive antennas must be placed at appropriate distances from each other to realize diverse paths. To achieve this, the possible diversity sources, i.e., satellite diversity and site diversity can be exploited in forming the MIMO channels for satellites. In the case of satellite diversity, the satellites are far apart from each other to achieve diversity and as a result the path lengths and the time of arrival of signals can vary vastly between the satellites and the ground terminals resulting in synchronization problem. These issues can be dealt with by employing cooperative satellite diversity concept or the use of compact antennas in which the problem of synchronization does not exist. The frame work for the most recent developments in satellite communications includes satellite land mobile and fixed communication systems, satellite navigation systems, Earth Observation systems and the state of art propagation models and evaluation tools for these systems. The influence of radio channel is a critical issue for the design, performance assessment and real-time operation of these highly reconfigurable hybrid (satellite and terrestrial) radio networks providing voice, text and multimedia services operating at RF frequencies ranging from 100 MHz to 100 GHz and optical frequencies.

The organization of the chapter is as follows: in section 2, a brief introduction to directional channel modelling including MIMO channel modelling is presented. In section 3 we present an overview of MIMO channel models for satellite communications. Section 4 describes MIMO channel models for satellites based on polarization concept. Finally, conclusions and suggestions are given in section 5.

## 2. Directional Channel Modelling

The difficulties in modelling a radio channel are due to complex and varying propagation environments. In case of land mobile satellite (LMS) communications the signal travelling between a land mobile and a satellite suffers from different propagation impairments (chapter 1). The diverse nature of propagation media (e.g., ionosphere, troposphere and local effects) adds further a dimension of complexity in predicting the affects of propagation impairments on radio signals. It is important to note that the level of information obtained from a channel model about an environment is highly dependent on the type of system under assessment. In order to predict the performance of single sensor narrowband system it may be sufficient to consider the time varying amplitude distribution and the received signal power. Thus, classical channel models which provide information about signal power level distributions and Doppler shifts may be adequate for narrowband systems. Broadband communication systems build on the classical understanding of the received signal power distributions and Doppler spread also exploit spatial processing to operate in highly complex and diverse propagation environments. Thus, it is necessary to incorporate new concepts such as adaptive antenna arrays, angles of arrival (AoA), angles of departure (AoD), delay spread and multiple antennas at both the ends of a communication link, the so-called MIMO systems.

When investigating MIMO channels, the additional dimension that comes into play is space which needs to be modelled in a similar way as frequency and time variations have been modelled for the wideband single-input single-output (SISO) systems. In contrast to the systems which deal with only temporal spreading, the MIMO channels require the angular distribution of energy at both the ends of the communication link. The impulse response of the double directional channel between a transmitter positioned at $P_t$ and a receiver positioned at $P_r$ with $n$ paths in 3D space can be written as (Steinbauer et al., 2001):

$$h(P_t, P_r, \tau, \Omega_t, \Omega_r) = \sum_{i=1}^{n} h_i(p_t, p_r, \tau, \Omega_t, \Omega_r) \quad (1)$$

where $\tau$, $\Omega_t$ and $\Omega_r$ denote the delay, AoD and AoA, both in 3D space, respectively and $h_i$ is the contribution of the $i^{th}$ component, i.e.,

$$h_i(p_t, p_r, \tau, \Omega_t, \Omega_r) = \alpha_i e^{j\phi_i} \delta(\tau - \tau_i)\delta(\Omega_t - \Omega_{t,i})\delta(\Omega_r - \Omega_{r,i}) \quad (2)$$

where $\alpha_i$, $\varphi_i$, $\tau_i$, $\Omega_{t,i}$ and $\Omega_{r,i}$ represent the amplitude, phase, time delay, AoD and AoA of the $i^{th}$ multipath contribution, respectively.

These parameters are determined by the relative location of the transmitter and the receiver. When either the transmitter or the receiver moves, these variables become a function of time and can change drastically over large periods of time (long distances). Therefore a more compact representation of time variant double-directional channel impulse response is given by,

$$h(t, \tau, \Omega_t, \Omega_r) = \sum_{i=1}^{n} h_k(t, \tau, \Omega_t, \Omega_r) \quad (3)$$

In addition to these parameters, the double directional channel impulse response is also dependent on the antenna patterns and the modelling bandwidth.

### 2.1 The MIMO Propagation Channel

If multiple antennas are deployed at both the ends of a communication link, a MIMO system is obtained as shown in Fig. 1. The key idea underlying MIMO theory is that signals sampled in the spatial domain at both the ends are combined in such a way that multiple parallel channels are created. The double-direction description of the channel can be extended to MIMO channel by considering $n_t$ transmit and $n_r$ receive spatially separated antennas at both the ends. The corresponding MIMO channel matrix can be defined as:

$$H(t,\tau) = \begin{bmatrix} h_{11}(t,\tau) & h_{12}(t,\tau) & \cdots & h_{1m}(t,\tau) \\ h_{21}(t,\tau) & h_{22}(t,\tau) & \cdots & h_{2m}(t,\tau) \\ \vdots & \vdots & \ddots & \vdots \\ h_{n1}(t,\tau) & h_{n2}(t,\tau) & \cdots & h_{nm}(t,\tau) \end{bmatrix} \quad (4)$$

where $h_{nm}(.)$ denotes the narrowband channel (wide sense stationary uncorrelated scattering process) between the $m^{th}$ transmit and $n^{th}$ receive antenna. The elements of the channel matrix (i.e., individual narrowband SISO channels) are taken as independent circularly symmetric complex Gaussian variables with equal variance.

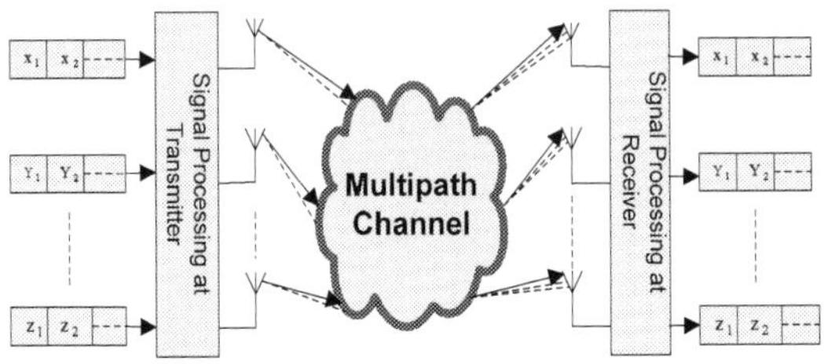


Fig. 1. The MIMO propagation channel.

The representation of a MIMO channel by individual SISO channels in not complete description of the multiple antenna systems. In order to get full benefit from MIMO systems (i.e., diversity gain, spatial multiplexing gain and array gain) certain trade-offs exist between these gains to achieve an adequate bit error rate at all times in an interference and noise limited system and at the same time maximizing throughput. The performance of MIMO techniques requires the exploitation of spatial correlation between all channel matrix elements. In (3) the elements of the channel matrix are assumed to be independent and therefore any two elements are uncorrelated. Practically, there is always some correlation between the channel matrix elements. These correlations are owing to small antenna array separation, antenna geometry and small amount of angular spread at the transmitter or the receiver side or both. Different studies have been found in literature to investigate the effects of correlation on the performance of systems employing multiple antenna techniques. In (Lokya, 2001) $n$ equal power and equal rate parallel sub-channels are considered with 'r' as correlation coefficient between any two sub-channels. The capacity of such a channel can be written as:

$$C(r) = n.\log_2\left[1+\frac{\rho}{n}(1-r)\right]+\log_2\left[1+\frac{n\rho r}{n+\rho(1-r)}\right] \tag{5}$$

where '$\rho$' represents the signal-to-noise ratio (SNR). When $r = 0$ the above equation reduces to the well-known formula for capacity:

$$C(r) = n.\log_2\left[1+\frac{\rho}{n}(1-r)\right] \quad \text{bps/Hz} \tag{6}$$

Comparison of (5) and (6) illustrate that the SNR decreases inversely with increase in the value of correlation coefficient (e.g., $r = 0.7$ results in 3 dB reduction in SNR). The MIMO channel capacity versus correlation coefficient for different values of '$n$' at SNR value of 30 dB is shown in Fig. 2.

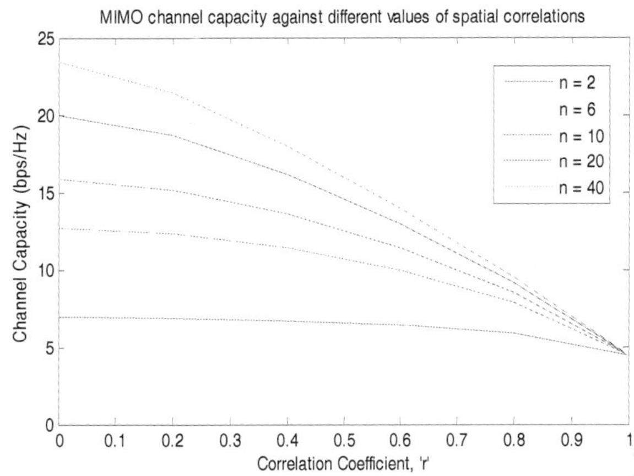


Fig. 2. MIMO channel capacity for 'n' channels against different values of correlation coefficient 'r'.

The other factors that limit the performance of MIMO channels are: number of scatterers, presence of line-of-sight (LOS) path, correlation between antenna elements at the transmitter or receiver side or both, keyholes, antenna patterns and geometry etc.

## 3. MIMO Channel Models for Satellite Communications

The wireless world has seen substantial increase in the demand of high quality broadband wireless services during recent years which results in the evolution of MIMO communication systems. The satellite communication systems are not immune to this change. In order to evaluate the performance of these systems employing MIMO technology, radio propagation models with high accuracy are required which can capture all those effects that affect different aspects of these systems whilst remaining computationally simple and require less simulation times.

Different approaches can be used to build a channel model with certain trade-offs (Almer et al., 2007). For example, physical or deterministic channel models based on ray-tracing algorithms can provide accurate results for a particular scenario, however due to global coverage of satellites this approach is not often used. On the other hand statistical models are built around measurement data and provide reasonable accuracy for the environments similar to that for which the model was built. However, they provide little insight into the propagation mechanisms and depend on the accuracy of measured data. An intermediate approach between these models is physical-statistical model. The physical-statistical modelling approach is the most appropriate in predicting the 'ON/OFF' nature and finding the small scale fading effects over large coverage areas applicable to LMS communication systems (Saunders et al., 2007). In this section some of the recently developed MIMO channel models based on different channel modelling approaches are presented.

### 3.1 The Physical-Statistical MIMO Channel Model

This physical-statistical MIMO channel (King et al., 2005) for LMS communications is based on 'clusters' concept and uses the same methodology given in (Correia, 2001; Molisch, 2004). The ray-tracing algorithm has been exercised to find different propagation effects like roof-top diffraction, specular reflection, shadowing caused by trees and foliage and blockage by buildings in LMS communication links in urban and high-way environments. The model can generate high resolution time series data and power delay profile for communication links between the satellite and mobile terminal antennas and can also predict the correlation between these links. In this model, obstacles (e.g., buildings, trees etc) are grouped into clusters of spherical shapes where clusters centers are randomly positioned. The building heights follow lognormal distribution. Twenty scatterers are placed randomly around the cluster centre each with dimension following Laplacian distribution (Correia, 2001). The building densities are assumed to be 90% in clusters representing urban environment, whereas trees are considered 90% of time in clusters for high-way environment. In order to validate the model, the parameters used are obtained from measurements data collected in Munich, Germany at L-band (1.54 GHz), for urban and high-way environments.

When signals reflected by distant clusters are blocked by buildings in the local clusters, these contributions are rejected. The signals from the satellite antennas to the mobile antennas are selected through appropriate clusters of scatters. Each scatterer in the cluster is assigned randomly the same reflection coefficient from a uniform magnitude distribution between 0 and 1 and phase 0 to $2\pi$. This channel model considers three paths between a satellite and moving mobile terminal: a line-of-sight (LOS) path, a blocked LOS path and an attenuated path by trees. The time series data '*I*' for a satellite or a high altitude platform (HAP) antenna *M* and each moving mobile antenna *N* can be written as:

$$\alpha_{M,N} = \begin{cases} P_{M,N} + b\sum_{i=1}^{n} T_i F_i P_{M,N,i} exp(jkd_{M,N,i}) & Clear\ \ Path \\ D_{M,N}P_{M,N} + b\sum_{i=1}^{n} T_i \Gamma_i P_{M,N,i} exp(jkd_{M,N,i}) & Blocked\ \ Path \\ T_{M,N}P_{M,N} + b\sum_{i=1}^{n} T_i \Gamma_i P_{M,N,i} exp(jkd_{M,N,i}) & LOS\ \ Path \end{cases} \tag{7}$$

where $P_{M,N}$, $D_{M,N}$ and $T_{M,N}$ denote the LOS path loss, the diffraction loss and the LOS trees loss, respectively, between satellite antenna *M* and moving mobile terminal antenna *N*. The term '*b*' is the clutter factor derived from measurements in each environment, $T_i$ is the tree attenuation from scatter *i*, $\Gamma_i$ is the reflection coefficient at scatterer *i*, $P_{M,N,i}$ is the path loss and $D_{M,N,i}$ is the distance between satellite antenna *M* and mobile terminal antenna *N* via scatterer *i*. The small scale fading parameters such as AoA distribution, shadowing depth and wideband parameters like root mean square (RMS) delay spread or coherence bandwidth can be approximated in each environment using the output time series and spatial power delay profile data of the model.

### 3.2 Analytical MIMO LMS channel model at Ku-Band and above

The physical-statistical channel model described earlier is designed for L (1-2 GHz) and S (2-4 GHz) frequency bands for LMS communication systems. The application of MIMO systems for satellite communications at Ku frequency band (12-18 GHz) and above has been discussed in (Liolis et al., 2007). In this model, two features of MIMO technology are presented: (i) a 2x2 MIMO spatial multiplexing system is used to achieve capacity improvements and a closed form expression for the outage capacity is derived (ii) MIMO spatial diversity scheme with receive antenna selection is applied in order to reduce interference in LMS communication links. In addition, an analytical closed form expression for interference mitigation on forward link of a satellite 2x2 MIMO diversity system with antenna selection is also obtained. In order to discuss the features of MIMO techniques, the model assumes propagation phenomena such as clear LOS, high antenna directivity, rain fading and rainfall spatial homogeneity. The propagation delay offset (synchronization problem) in LMS communications is also considered and a practical solution to this problem is found where matched filters are applied, first to the received signals for the detection of propagation delay offset and then the resulting signals are fed to the timing aligner. Subsequently, the delay offsets are eliminated by adjusting the timing of a signal serial-to-parallel converter.

The figure of merit for the analysis of MIMO fading channel is the outage capacity. A dual-satellite MIMO communication channel at Ku-band and above is shown in Fig. 1 in (Liolis et al., 2007). The terminal station is equipped with two highly directive collocated antennas to communicate with two satellites S1 and S2. The separation, $\Delta\theta$, between the antennas is kept large enough so that the spatial correlation due to rain along the relevant paths is as low as possible. Considering the clear LOS between the terminal station and each satellite and that each terminal station antenna is at bore-sight with the corresponding satellite, the total path loss along each link can be written as follows (Liolis et al., 2007):

$$A_i = FSL_i + A_{Ri} \qquad i = 1, 2 \tag{8}$$

where $FSL_i = 10\log_{10}(4\pi\, d_i f / c)^2$ is the free space loss along each link, $f$ is the operating frequency and $c$ is the speed of light. The term $A_{Ri}$ represents the rain induced attenuation. The convective raincell model using Crane's assumptions is employed for the description of vertical variation of the rain fall structure using the same approach as in (Panagopoulos & Kanellopoulis, 2002). Based on these suppositions if $\Delta\theta$ is sufficiently large, the spatial correlation between random variables $A_{Ri}$ representing channel coefficients is low and a decorrelated (ideal) MIMO satellite channel model is obtained.

To find out the capacity of dual satellite MIMO channel, it is assumed that the two satellites transmit different and independent data streams and the channel is perfectly known at the terminal side while the transmitting satellites have no information about the channel. Equal powers are allocated to the two satellites owing to distributive nature of the system in the absence of channel state information (CSI). The capacity of the dual satellite MIMO channel based on the standard MIMO theory, taking into the account above assumptions, can be written as follows:

$$C = \sum_{i=1}^{2} log_2(1 + 0.5SNR_{CS_i} 10^{-A_{Ri}/10}) \tag{9}$$

where $SNR_{CS_i}$ are the nominal SNR ratio values under clear sky conditions. The above equation gives the instantaneous capacity of deterministic 2x2 MIMO channel. However due to rain induced attenuation and stochastic behavior of the channel, the appropriate metric to characterize the resulting fading channel is the outage capacity which can be written in the following form:

$$(C \leq C_{out,q}) = \frac{1}{2} \int_{u_A}^{+\infty} du_1 fv_1(u_1) erfc\left( \frac{u_B \rho_{n12} u_1}{\sqrt{2(1 - \rho_{n12}^2)}} \right) = q \tag{10}$$

where $C_{out,\,q}$ is the information rate guaranteed for *(1-q) 100%* of channel realizations. The term $\rho_{n12}$ is logarithmic correlation coefficient between normal random variables $u_i$ and $u_A$, $u_B$. Complete mathematical details about these variables can be found in (Liolis et al., 2007) This model is applicable over a large range of SNR values and shows significant capacity gains of the MIMO systems over the SISO systems for moderate and high SNR levels.

The major factor limiting the capacity of LMS system is the adjacent satellite cochannel interference on its forward link. A satellite 2x2 MIMO diversity system based on receive antenna selection to mitigate the cochannel interference problem on the forward link due to differential rain attenuation from adjacent satellites is presented in (Liolis et al., 2007). In addition an analytical model is provided for interference mitigation based on satellite 2x2 MIMO diversity systems. To ease the complexity and cost associated with multiple RF chains only one RF chain is used at the receiver that performs antenna selection, i.e., it utilizes receive selection diversity to detect the signal related to the path with the highest SNR (Sanayei & Nosratinia, 2004). For optimal selection of a signal, the receiver scans the two antennas for a training signal transmitted along with data and the signal with the highest SNR is selected for the reception of the next data burst. This model uses a distinct approach for interference mitigation as compared to the conventional approaches found in MIMO theory. The effect of rainfall on interference analysis, differential rain attenuation related to an adjacent satellite and spatial inhomogeneity of rainfall medium are taken into account especially for congested urban areas where the increased demand for link capacity and radio coverage imposes the coexistence of many satellite radio links over the same geographical and spectral area. The satellite 2x2 MIMO diversity model using antenna selection and operating at Ku-band and above is illustrated in Fig. 4 in (Liolis et al., 2007). This model does not require channel knowledge at the transmitter side and only the information about wanted signals' channels at the receiver is desired. Mathematical derivation of this model and influence of various geometrical and operational system parameters on system performance can be found in (Liolis et al., 2007).

## 4. Application of Polarized MIMO Channels to Satellite Communications

The large MIMO gains can be achieved with low correlation between antenna elements at both ends of MIMO communication system. A fundamental way of achieving low antenna correlation is to use antenna elements with significant separation such that the relative phases of multipath contributions arriving at the receiving antennas are significantly different. However, owing to size restriction it is difficult to deploy multiple antennas with large separation at the mobile terminal. In addition, in communications where LOS is dominant (e.g., satellite communications), the MIMO systems offer reduced performance since LOS components overpower multipath components in the received signal. An alternative solution is to use polarized arrays for multiple antenna systems (Mohammed & Hult, 2008; Hult et al., 2010). With spatially separated cross-polarized antenna arrays, both the polarization diversity and polarization multiplexing can be achieved (Jiang et al., 2004; Moriatis et al., 2009) (e.g., two dual polarized spatially separated arrays form 4-antenna arrays).

In order to get benefit from polarization dimension, the cross-polar transmissions (e.g., transmission from vertically polarized antenna to horizontally polarized antenna) should be zero. However, in real scenarios there is always some polarization mismatch since linearly polarized antenna arrays have non-zero patterns for cross-polar fields (Hult & Mohammed, 2008). In addition, multipath effects (e.g., diffraction, scattering, reflection etc.) may change the plane of polarization of incident electromagnetic waves at the receiver.

### 4.1 Physical-Statistical Polarized MIMO Channel

In order to achieve additional diversity in satellite communication links, an extension of the physical-statistical MIMO model to 2x2 dual polarized MIMO model is presented in (Horvath et al., 2007) for a single satellite serving land mobile. A single satellite containing two antennas with right-hand circular polarization (RHCP) and left-hand circular polarization (LHCP), respectively, communicates with mobile terminal using the same antenna configuration. It is assumed in this model that LOS paths between co- and cross-polar channels are fully correlated and diffused multipaths are fully uncorrelated between co- and cross-polar components. The polarization characteristics are described by Stoke's theorem and the related concepts. The channel model construction is similar to (King, 2005) with additional polarization features are included as follows: In the case of unobstructed LOS signal, path loss is applied to the co-polar channels while cross-polar channels are neglected. When the LOS signal is blocked by buildings, rooftop diffraction loss is applied to both the co- and cross-polar channels. In the case of LOS path through vegetation, attenuation is applied depending upon the path length and using an attenuation factor of -1.3 dB. A mathematical representation of data generated by this model can be found in (Horvath et al, 2007). Extensive measurements were carried out along tree-lined/highway and urban environments using an artificial platform in order to optimize the model by fitting its parameters to the measured data. The model is capable of producing statistically accurate wideband channel first and second order characteristics and can be used to evaluate the capacity and diversity benefits of MIMO applications in LMS communication systems. The use of dual polarized system results in 2-fold increase in capacity and 4-fold boost in diversity gain (Horvath et al., 2007). The diversity gain can be further increased by employing the concept of compact 3D-polarized antennas (Horvath & Friyges, 2006). In the case of synchronization problem in multi-satellite to ground links, one solution can be that MIMO antennas have to be collocated onboard a single satellite. This case is similar but not identical to handheld devices where the available space is limited. Here the available space is significant and antennas with large aperture and high gain can be applied. The channel statistics and power delay profiles are dependent on size and positioning of scatterers and reflection coefficients. The details of the experiments conducted to fine-tune the model can be found in (King, 2007).

## 4.2 Empirical-Statistical MIMO Channel Models for Satellites

The empirical-statistical channel models are built statistically around field measurements and can produce appropriate channel statistics for similar environments. An empirical-statistical 2x2 dual polarized LMS-MIMO channel model based on measurement campaigns is presented in (King, 2007). This MIMO channel model characterizes two models for LMS communication systems: the narrowband and the wideband channels. In order to describe 'ON/OFF' phenomenon occurring in LMS communication links, Markov chain approach is used. In narrowband model, single or dual lognormal fading, which is correlated over MIMO domain, is applied across each Markov state. In the case of wideband model lognormal fading is correlated over the delay domain as well and Rician factor is dependent on the large scale fading level.

To design a narrowband channel model, large scale data (lognormal fading) for each of the four channels in each Markov state is generated using a Gaussian random number generator with zero mean and unit standard deviation. The memoryless data streams (data samples) are then filtered using first order recursive linear time invariant digital filter in order to obtain correct temporal fading. The cross-correlation between large scale fading channels can be obtained in the following way:

$$vec(Y) = (\chi)^{1/2} vec(Y) \tag{11}$$

where $\chi$ is the correlation matrix obtained from measurements. Two large scale time series vectors are defined to represent dual lognormal fading which produces 8 time series correlated vectors: four vectors for each Markov state (each state represents a MIMO channel). Markov chains are illustrated by their state frame lengths, state vector $W$ and transition matrix $P$. A minimum state frame of length 1 meter is found appropriate to capture the rapid changes in the level by observations. The probability of changing from state-$i$ to state-$j$ (both for co- and cross polar) can be calculated as follows:

$$P_{i,j} = \frac{M_{i,j}}{M_i} \tag{12}$$

where $M_{i,j}$ is the number of transitions from state-$i$ to state-$j$ and $M_i$ is the number of state frames corresponding to state-$i$. A Markov chain Monte Carlo method can create random walk through the states for co- and cross polar channels using the Metropolis-Hastings algorithm (Chib & Greenberg, 1995). Markov states are associated with each of the two channels obtained from 4 MIMO channels (2 co-polar and 2 cross-polar) resulting in 4 channels for each MIMO channel. Now Rician factor is defined using polynomial approach for large scale fading and Rician samples are generated by Rice's sum of sinusoids method (Pätzold et al., 1998). The average correlation coefficient value can be calculated on the basis of small scale MIMO channel correlation measurement data. The small scale and large scale fading time series data for each MIMO channel are added at each sampling instant to get the narrowband channel model.

A wideband channel model is constructed based on measurement campaigns and the schematic diagram of such a model is shown in Fig. 7.4 in (King, 2007). The 2x2 LMS MIMO channel matrix for this model can be written as:

$$Y_{R,L} = \begin{bmatrix} h(t,\tau)_{R,R} & h(t,\tau)_{R,L} \\ h(t,\tau)_{L,R} & h(t,\tau)_{L,L} \end{bmatrix} X_{R,L} + N_{R,L} \tag{13}$$

where $Y_{R,L}$, $X_{R,L}$ and $N_{R,L}$ are the received, transmitted and noise vectors, respectively. The subscripts, *L* and *R*, are used to denote the antennas polarizations at each end of the radio link. The amplitudes, propagation delays and phases of each path are randomly time varying parameters due to motion of the vehicle or the satellite. The phases of the paths are assumed to be mutually independent random variables with uniform distribution in 0 to 2π interval. All taps are considered to maintain frequency and delay domain resolution spaced *10 ns* and maximum resolvable delay is *400 ns* for the tree-line roads and urban environments and *200 ns* for the suburban environment. Large scale fading samples with correct autocorrelation and cross-correlation properties are generated for each tap using parameters obtained from the measurement data by the method similar to the narrowband model. In the case of wide band, the lognormal fading is correlated across both the MIMO domain and the delay domain. All taps in the wideband case are represented by the Markov state probability and state transition probability matrices with all taps in state 1 or in state 2 simultaneously. Each state in Markov chain is assigned a set of lognormal fading generators obtained from Gaussian random number generator with zero mean and unit standard deviation. Rician samples are generated in a similar way as for the narrowband case. The first tap of the delay line model is correlated with large scale fading level using polynomial fit. The value of K factor for remaining taps, derived from the measurement data, is lognormally distributed and independent of large scale fading level. The small scale fading generators are uncorrelated for all taps due to uncorrelated scattering effects. The small scale MIMO matrix samples are partially correlated for each delay tap. Now the small scale and large scale fading generators are combined by summing the outputs of the taps to obtain the channel impulse response time series generator.

The models described above can generate the narrowband and wideband channel models with well conserved first order, second order statistics and MIMO channel cross-correlations. However the parameters used to generate these models are based on specific measurement campaigns for particular environments. Thus, these models can be used to generate fading effects for similar types of environments which limit the application of these models for different environments.

## 5. Conclusion

This chapter presented an overview of standard MIMO channel models for LMS radio communication systems. Standard channel models play a vital role in the design and performance assessment of advanced transceivers techniques and smart antennas employed to establish reliable communication links in LMS communication systems.

The quality of service and spectral efficiency in a LMS communication system suffer from limited transmit power, high path loss, blockage, shadowing and high link delay. In order to overcome these effects by employing MIMO techniques to enhance LMS performance, physical-statistical MIMO channel models including 3D polarization concept are developed. The ray-tracing algorithm is used which can model small and large fading effects in an efficient way and can cover large coverage areas for satellites systems. The physical-statistical channel models with multiple satellites and/or dual polarized antenna configurations can be simulated under different propagation environments and satellite elevations and show significant improvements in capacity and link reliability. Analytical channel models are developed to investigate the effects of MIMO techniques on capacity improvement and interference mitigation in LMS systems. The empirical-statistical MIMO channel models (narrowband and wideband) are constructed around experimental data. These channel models can generate the narrowband and wideband channel models with well conserved first order, second order statistics and MIMO channel cross-correlations. However, these models have some limitations since they are built from experimental data obtained from specific measurement campaigns for particular environments, they can be used to generate fading effects for similar types of environments which limit the application of these models for all environments.

The MIMO channel models for LMS systems are built and validated using some available experimental data and the data obtained from measurement campaigns. These channel models need to be refined to discover realistic hybrid physical-statistical channel models and satellite to indoor channel models using the data acquired from experimental campaigns that have been done or still going on in urban, sub-urban, rural, forest or indoor areas at L, S and C band and also the data gathered from Earth Observation systems. The spatial characteristics in multi-antenna channel modelling including polarization effects (especially 3D polarization) are expected to be crucial in future LMS communications systems. Thus, new and improved hybrid physical-statistical channel models, satellite-to-indoor propagation models and propagation impairment mitigation techniques based on multiple antennas techniques are necessary to assess the parameters and performance of satellite communication systems. The MIMO channel models for LMS/HAPs and indoor propagation environments employing compact antennas including 3D polarization concept based on different measurement campaigns and using the multiple antenna features similar to terrestrial communications systems will be the topic of interest for future satellite communication systems.

## 6. References

Almers, P., Bonek, E., Burr, A., Czink, N., Debbah, M., Degli-Esposti, Hofstetter, H., Kyosti, P., Laurenson, D., Matz, G., Molisch, A., F., Oestages, C., & Ozcelik, H. (2007). Survey of Channel and Radio Propagation Models for Wireless MIMO Systems. EURASIP Journal of Wireless Communication and Networking, volume 2007.

Chib, S., & Greenberg. (1995). Understanding the Metropolis-Hastings Algorithm. The American Statistician, 49(4), 327-335.

Correia, L. M. (2001). Wireless Flexible Personalized Communications. Cost 259: European Co-operation in Mobile Radio Research, 148-200.

Horvath, P., & Friyges, I. (2006). Application of 3D-Polarization Concept in Satellite MIMO Systems. In proceedings of 49th Annual IEEE Global Telecommunications Conference, San Francisco, Calif, USA.

Horvath, P., Karagiannidis, K. G., King, P. R., Stavrou, S., & Frigyes, I. (2007). Investigation in MIMO Satellite Channel Modelling: Accent on Polarization. EURASIP Journal of Wireless Communications and Networking.

Hult, T., & Mohammed, A. (2008). Evaluation of Depolarization Effects on the Performance of High Altitude Platforms (HAPs). IEEE 67th Vehicular Technology Conference, VTC08-Spring, Singapore.

Hult, T., Mohammed, A. Yang, Z., & Grace, D. (2010). Performance of a Multiple HAP System Employing Multiple Polarization. Invited Paper, Special Issue, Springer Wireless Personal Communications Journal, 52(1), 105-117.

Jiang. L., Thiele, L., & Jungnickel, V. (2004). On the Modelling of Polarized MIMO Channel. Fraunhofer Institute for Telecommunications Heinrich-Hertz-Institute Einsteinufer 37, D-10587, Berlin, Germany.

King, P. R., Evans, B. G., & Stavrou, S. (2005). Physical-Statistical Model for the Land Mobile-satellite Channel Applied to Satellite/HAP MIMO. 11th European Wireless Conference.

King, P. R. (2007). Modelling and Measurement of the Land Mobile Satellite MIMO Radio Propagation Channel. Ph. D Thesis, Centre for Communication Systems Research, University of Surrey, Guildford, UK.

Liolis, P.K., Panagopoulos, D. A., & Cottis, G. P. (2007). Multi-Satellite MIMO Communications at Ku-Band and above: Investigation on Spatial Multiplexing for Capacity Improvement and Selection Diversity for Interference Mitigation. EURASIP Journal of Wireless Communication and Networking, volume 2007.

Lokya, L. S. (2001). Channel Capacity of MIMO Architecture Using the Exponential Correlation Matrix. IEEE Communication Letters, 5(9), 369-370.

Mohammed, A., & Hult, T. (2008). Performance Evolution of a MIMO Satellite Diversity System. European Space Agency 10th International Workshop on Signal Processing for Space Communications, (SPSC 2008), Rhodes Island, Greece.

Molisch, A. F. (2004). A Generic Model for MIMO Wireless Propagation Channels in Macro- and Microcells. IEEE Transactions on Signal Processing, 52(1), 61-71.

Moraitis, N., Horvath, P., Constantinou, P., & Friyges, I. (2009). On the Capacity Evaluation of a Land Mobile Satellite System using Multiple Antennas at the Receiver. 3rd European Conference on Antennas and Propagation, Berlin, Germany.

Panagopoulos, A. D., & Kanellopoulos, J. D. (2002). Prediction of Triple-Orbital Diversity Performance in Earth-Space Communications. International Journal of Satellite Communications, 20(3), 187-200.

Pätzold, M., Killat, U., & Laue, F. (1998). On the Statistical Properties of Deterministic Simulation Models for Mobile Fading Channels. IEEE Transactions on Vehicular Technology, 47(1), 254-269.

Sanayei, S., & Nosratinia, A. (2004). Antenna Selection in MIMO Systems. IEEE Communication Magazine, 42(10), 68-73.

Saunders, S., & Zavala, A. (2007). Antennas and Propagation for Wireless Communication Systems. John Wiley & Sons Ltd.

Steinbauer, M., Molisch, A. F., & Bonek, E. (2001). The Double Directional Radio Channel. IEEE Antennas and Propagation Magazine, 43(4), 51-63.

commissioned by the CIDAP to carry out a market research study for the hotel resort. This study, which was carried out by a consultant from Toulouse and financed in part by the regional council, concluded in particular that there was no regional market for industrial seminars that would justify the deployment of a videoconference system.
In La Mongie, all the people taking part in the experiment switched to ADSL as soon as that technology was deployed in the resort, well before the end of the experiment. ADSL is a system which seems to "kill" any alternative solution, irrespective of its quality.
The list of people taking part in the experiment is closed, because of the speed proposed and the need for sharing as well as the equipment to be financed, but several structures and individuals would like to participate in the experiment: a restaurant owner, as well as two people who have recently moved to the area: an English teacher who wants to organise online language courses and a person who wants to organise martial arts videoconferences. This situation creates a certain amount of jealousy among those who are "excluded" from the experiment.
Despite the failures experienced by the people taking part in the experiment, it appears that the Twister solution is unanimously perceived very positively and interests a growing number of people who have recently settled in the valley. The absence of ADSL and the fact that the experimental system is free encourages people who have recently arrived in the area (English people, neo-rural residents, etc.) to try and participate in the project. However, there is a problem with the quality of service of the system for users having large bandwidth needs.

### 5.2 The experiment in Tibiran-Jaunac

Tibiran-Jaunac-Las Aribes, a commune in the Hautes-Pyrénées, has 249 inhabitants, including a retirement home with 62 residents. Its territory is very vast: the subscriber node is 8 km from the nearest telephone subscriber and 14 km from the furthest subscriber. Tibiran is a perennial non-covered area, which, according to the mayor, does not interest the operators and whose network is not maintained by France Télécom.
The majority of the commune's inhabitants are elderly and are from the farming sector. Active inhabitants work mainly in Tarbes and Toulouse. The development plans include a communal estate.

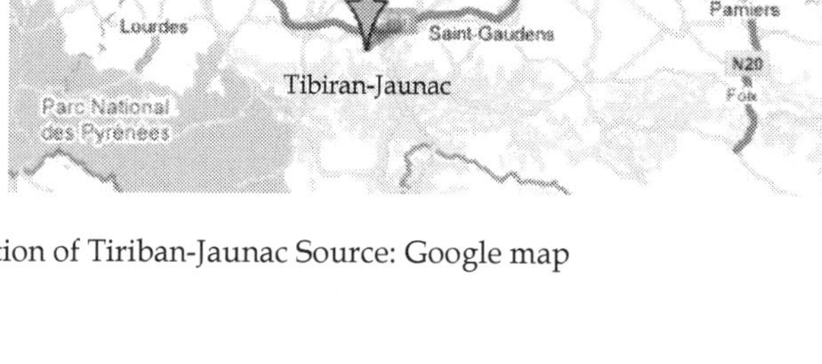


Fig. 4. Localization of Tiriban-Jaunac Source: Google map

### 5.2.1 The genesis of the project

Local demand, composed of recently retired people and young people of school age, is strong. At the end of 2004 the mayor contacted France Télécom to connect the commune to ADSL. The operator considered that the commune was not eligible, but indicated that it could consider making a commercial offer if the commune financed the installation, at an estimated cost of €300,000.

There is still a strong demand within the commune and the mayor would like to find another cheaper solution. The project was then included in the Twister project at the end of 2005.

The mayor's personality and his ICT knowledge play a fundamental role in the high-speed project for the commune. He is familiar with the technology involved since he is seconded from the Interior Ministry to the ICT department of the Prefecture of the Hautes Pyrénées. The mayor decided, as soon as Tibiran was accepted as a Twister site, to put in place a veritable infrastructure policy for his territory by investing in a radio network to connect all the inhabitants.

### 5.2.2 The needs and projects

High-speed access is seen as a possible way of delivering more neighbourhood services to the commune's inhabitants:

- at the current time the town hall is open only on Saturdays: therefore the town hall intends to propose uses linked to the dematerialisation of procedures. It also wants to put in place an Internet access point, with three stations, linked to the town hall (training, etc.);
- the retirement home that provides medical care has considerable needs. These include, in particular, being able to connect to the local management system in Tarbes via a high-speed link; personal services are also planned (remote transmission, telediagnosis, etc.);
- since June 2006, it has been necessary to provide notaries public or purchasers of property with information regarding the commune and/or the risks related to the plot of land. The idea is to provide direct online access to such information;
- finally, the farm/inn present in the commune wants to develop Internet tools, such as an Internet site with online reservations.

### 5.2.3 The use made of the experimental system and the problems encountered

The system was deployed in Tibiran-Jaunac at the end of 2005. It connects the 12 people taking part in the experiment. The participants were selected by the mayor, based on the first applications, while trying to achieve a mix of ages and types of usage. Some participants were used to high-speed access, while others were primarily motivated by the geographical distance separating them from their families.

The installation of the Wi-Fi system required some trees, some of which were a hundred year old, to be pruned or chopped down. The owners had to be "pushed" to accept this sacrifice. Health concerns regarding the effects of Hertzien waves were also an obstacle.

The town hall considered that the bandwidth of 1Mb allocated by EADS Astrium[14] to all the people taking part in the experiment was less than the speed announced initially. No *peer-to-peer* problems have been encountered. An intelligent bandwidth manager was installed to this end (the last usage allocated was *peer-to-peer*).
The system has been affected by numerous losses of connection, due in particular to the failure of Aramiska, whose role was taken over by Ouranos Networks and other service interruptions to the satellite network. In the case of persistent failures, EADS Astrium's *hotline* is contacted. This is the only means used. The Wi-Fi network functions correctly.

Mrs B.: a participant aged 70
*Mrs B. is a keen user of IT and the Internet. She has had a computer since 2001 and uses it intensively. She connects to the Internet for several hours every day. She uses it to make numerous searches and to pursue her leisure activities: information on embroidery and consulting forums on the subject, looking up train timetables, use of Google Earth to see how places she has visited have changed, genealogy searches, sending photos and looking more closely into information read in "Micro hebdo", etc. She also uses Skype every day (with headphones and Webcam) to call her daughter in Toulon. For her, high-speed access is indispensable.*
*She does not shop online (fears regarding bank card security) but looks up information, makes her choice and sends a shopping list to her daughter who then buys the items in question for her on eBay or other online shopping sites.*
*Thanks to the Internet and its search facilities, she avoids certain physical journeys and saves on the cost of telephone calls and postage stamps.*
*Finally, the only problem which she encounters is the English vocabulary specific to IT. Her dictionary is incomplete.*
Interview conducted on 7 December 2006

Mr C.: a regular Internet user
*Mr C. works at the University of Toulouse and commutes every day to and from Tibiran. He has used high-speed access for his professional activities since 1999. The Internet enables him to carry out a large number of searches on his passions in various fields: astronomy (in his view the photos on the Internet are better than direct viewing via an astronomical telescope), archaeology, the daily news, to supplement the radio and written press, local weather forecasts, etc.*
*The Internet is a means of putting daily information into perspective and making readers more responsible. "With a stone, you can crack a nut, or the neighbour's head."*
*Alongside the searches, which he also carries out at work, he also sends e-mails and photos to his friends. He neither uses Internet telephony services nor online shopping sites. The Internet enables him to avoid having to do certain tasks physically (buying train tickets) but, on the other hand, has had the opposite effect (discovering a museum online then physically visiting the museum).*

---

[14] At the end of 2005, Aramiska (then Ouranos) provided the Tibiran site with a free access service of 2Mbps/512kbps. From October 2006, the speed was reduced to 1Mbps/256kbps.

*He has sometimes carried out searches to help neighbours who did not want to take part in the experiment, but who now want to be connected.*
Interview conducted on 7 December 2006

### 5.2.4 Assessment of the results of the experiment

Certain positive elements have emerged. At the end of the experiment, Wi-Fi technology is seen as both reliable and perennial. Local demand has increased strongly. Over and above the hard core of 12 people taking part in the experiment, thanks to word of mouth 300 people are currently awaiting a high-speed connection. For the town hall, which has a project for a communal estate, high-speed access is important to attract and keep people in the commune. The neighbouring communes also want to be connected to Tibiran-Jaunac's high-speed network.

There are however certain negative points:

- frequent losses of connection;
- the latency time is seen as long: this is a user perception of the system's performance. The relation with the latency time caused by the satellite may seem hazardous without any further information;
- the current project (the Twister experiment with its time and financial limits) does not allow more people to be connected[15];
- the Internet satellite service available on the market does not support VoIP and there is strong demand for the voice-data coupling, "at the current time the network is too poor";
- Wi-Fi nuisances (waves, old trees, leaves, etc.) have been important;
- the continuity of the system after the experiment was not envisaged in the project. It depends to a large extent on the mayor and his investment in Wi-Fi. Twister has contributed the satellite part.

### 5.2.5 The future prospects

Since December 2006, the commune has been fully Wi-Fi covered, thanks to 6 relays and one satellite access point. The whole of this infrastructure has been financed by the commune (€50,000) and subsidised for the time being solely by the State. The Midi-Pyrénées region is planning a call for projects to which the commune must respond. To that end, it must establish that no tenders have been submitted in response to its call for tenders (in accordance with the LEN), but the context is experimental, which changes things somewhat. The departmental council has not yet been mobilised on the question but lobbying has started.

For the experiment to be transformed into a commercial offering at an affordable cost for inhabitants, an economic model must be found. There are two thresholds fixed by the satellite offering and the ADSL offering in cities (not more than €3 for 1Mb per subscriber). There is also a psychological appropriation threshold by the user (irrespective of the user's location and the technology used) of the Internet access service which is that of ADSL, in urban areas. Indication: "The 2Mbs satellite connection is €700 a month. For 30 users, the

[15] The satellite technical system could connect more users, in exchange for a financial contribution.

cost is acceptable, but it will not be feasible to amortize the equipment and it will be difficult to renew the technical equipment every four years."

The town hall is holding discussion with TDF for the installation of a 2Mbs SDSL link to replace the satellite access point (approximately €400 a month). The aim is to conclude a public service delegation contract on a leasing basis with an operator to operate the network. But the changeover is both difficult (the software and equipment used for the network are proprietary systems and it would therefore be difficult for an operator to use them) and urgent (the project ends at the beginning of February 2007).

"For the time being the transition from experimentation to a commercial operation would be counter-productive. There would be some dissatisfied people."

### 5.3 Conclusion

The provisional evaluation of the Twister projects is positive. The solutions meet the high-speed needs identified beforehand. The deployment of the experiment has stirred interest in the valley, with increasing demand (successful school courses, reflection on the creation of a public Internet space, etc.). As a result other people have expressed an interest in benefiting from this offering which is attractive because, on the one hand, the equipment and material are fully financed and installed and, on the other hand, the connection is offered free of charge throughout the experiment. Subsequently, the community of communes could finance part of the installations to help maintain the system on the territory on a permanent basis, confirming the existence on this territory of a true digital development project, through the deployment of a technical system.

Nevertheless, the deployment of ADSL is developing as well as its technology (Re ADSL, etc.), the progressive covering of the territory by standard technology is a threat to the experiment. The technical difficulties encountered by the people taking part in the experiment and Aramiska's bankruptcy have hindered the development of this technology.

Finally, a problem of "governance" (lack of information, the fact that the technical system and the types of usage and services proposed were not co-constructed, etc.) was identified. The people taking part in the experiment want the circulation of information to be improved.

## 6. Conclusion

The research problem revolves around one question: that of analysing the *process of adoption and the prospects of wider deployment of technical systems*. The question of the wider deployment of the technical systems concerns not only the analysis of learning mechanisms and the appropriation of technology, but also that of the impact/effects on the territories used in the experiment and of the interplay of the stakeholders.

The analysis of the couplings referred to below is based on the means chosen to achieve the objective. Four "variables" interact here (technical systems; applications; territories and users) and can be considered from the point of view of various couplings:

the "applications/territories" coupling, in order to identify, for example, the variables explaining the adoption and appropriation of an application by a territory and the conditions necessary to ensure its effective transposition to other territories;

the "applications/technical systems" coupling, in order to identify, for example, the attributes of the alternative technologies used, which might be an obstacle to the use of an application: problems of latency, sharing bandwidth in the case of use by communities, etc.);
the "systems and territories" coupling: where the application plays the role of an intermediary variable in analysing the success factors of a technology, etc.;
the "technical systems and users" coupling: in order to identify, for example, the process of the adoption of a system by users.
Our analysis has enabled us to confirm the relevance of certain couplings of variables.
For the "applications-territories" coupling, certain territorial stakeholders want to develop more particularly structuring applications, such as, for example, e-administration services (for citizens), telemedicine (for retirement homers, ski resorts, etc.). However these projects are still on the drawing-board and have not been implemented. It seems that the first phase is to use the system and assess its reliability, credibility, etc., which is a key stage in the process of taking things further and actually implementing projects. However, satellite connections do not support certain uses for which there is now strong demand (basic uses are no longer enough. The interaction of the "applications-systems" coupling is in that case weak. However, interest in high-speed Internet access is not called into question. The "systems-users" coupling (through the appropriation process) and the "systems-territories" coupling (through the action of the local authorities) are very strong, confirming the major attraction of high-speed access.
Our assessment of the results of the experiment reveals certain negative elements.
1) problems regarding the quality of service in relation to types of usages, which can consume bandwidth or a demand that tends to replicate urban demand (success of VoIP);
2) a question of territorial governance: how to manage the sharing of bandwidth? Self-management or a 'technical' solution (software but who decides what and according to which criteria?);
3) the choice of territories used in the experiment: depending on the strategies of operators and regulatory and technological developments, the number of so-called perennial non-covered is shrinking.
There seem to be two types of territories: the transitional non-covered territories where satellite technology coupled with Wi-Fi seems to be a good temporary solution; in such cases the operators are preparing the market for the arrival of France Télécom since the business model is not competitive in the case of the French market. This raises several questions: how to transform these transitional solutions into a financially feasible market for operators? How long can this transitional period last? What prices should be charged? Is it necessary for the public authorities to contribute funding? For the perennial non-covered territories, satellite technology can be a solution if it can solve the problems of quality of service (see above) and provide a more efficient response to user requirements (*peer-to-peer*, VoIP);
4) finally, the local stakeholders participating in the experiment have quickly identified a decisive element, which is a stumbling block to territorial development: support services. The communes involved are small territories, which have limited financial resources and have difficulty not only finding the necessary technical and human competences locally, but also recruiting a "support co-ordinator" for the development of ICT.
However, numerous positive elements have undoubtedly emerged from the experiment.
1) The deployment of technology on territories, which satisfies the objectives of the public authorities to bring rural territories into the information society.

In France, the prevailing view favours the digital development of the territory and the deployment of high-speed access. This is confirmed by the numerous policies implemented and aids provided at the level of the European Commission (Plan i2010), the State (Plan RE/SO 2007) and its calls for projects (DATAR), etc. This prescriptive approach helps to stimulate (while accentuating the feeling of exclusion in the territories concerned) the high-speed offering by the operators irrespective of their type. The LEN (2004) supports this movement;

2) the development of a market on these territories; are we not seeing the ADSL operators arrive after or during the experiment and capture users?

These high-speed projects can find themselves competing with and "abandoned" by the sudden arrival of standard technology on the territory. Indeed, numerous experiments were cancelled, moved or under-used, owing to the deployment of ADSL in the territories concerned;

3) an ongoing improvement in the technical offering by the operators (Wi-Fi or satellite) in interaction with the complex forms of demand (peer-to-peer or VoIP). Satellite technical solution are evolving (undoubtedly less quickly than terrestrial solutions), but the commercialisation of offerings using these technologies is to a large extent held back by the difficulty of creating a market that is significantly large market (all the non-covered areas in France) and sustainable (competition with ADSL)

Initially, the project did not provide for bandwidth management, but in a Wi-Fi network, the bandwidth is shared and heavy consumption by one or more network can considerably undermine the quality of the service offered. At the request of the co-ordinators of certain sites, an intelligent bandwidth manager was installed locally. This is based on the prioritisation of the protocols used: the last protocol allocated being peer-to-peer;

4) the wider territorial deployment of the technology beyond the first people taking part in the experiment (waiting lists) and its extension to neighbouring communes.

The experiment has produced in most cases, a surge in local interest in the project by non-participating citizens and residents of neighbouring communes. This interest is also boosted by word of mouth, especially as regards the advantageous conditions of the experiment (free, equipment, etc.);

5) improved anticipation by local authorities of the needs of their citizens not only as regards infrastructure needs (reflection on costs, project management, etc.), but also as regards new innovative services (combined with a real territorial strategy) and, finally, in the area of types of usage (taking account of training needs).

In general, these experiments have proved a way of meeting needs expressed and have also helped to mobilise and energise the stakeholders involved locally (public authorities, citizens wanting high-speed access, etc.). As a result, local authorities are considering providing financial support so as not to interrupt the existing connections to increasingly insistent and demanding citizens, at the end of the end of experiments. When one measures the financial effort made and envisaged by small local authorities, even when helped by public co-financing, it is clear that these experiments have satisfied to a large extent the expectations of the territorial stakeholders.

## 7. References

Bijker, W. E. (1995). Of Bicycles, Bakelites, and Bulbs Toward a Theory of Sociotechnical Change, MIT Press, Cambridge

Bolton P.; Harris C. (1999). Strategic Experimentation. *Econometrica*, vol.67, n°2, pp. 349-374

Grossmann, S.J.; Kihlstrom, R.E.; Mirman, L.J. (1977). A Bayesian Approach to the Production of Information and Learning by Doing, *Review of Economic Studies*, vol. 44, pp. 533-547

Leach, J.C.; Madhavan, A.N. (1993). Price Experimentation and Security Market Structure, *The Review of Financial Studies*, vol. 6, n°2, pp. 375-404

Orlikowski, W. (1992). The duality of technology: rethinking the concept of technology in organizations, *Organization Science*, vol 3, pp. 398-427

Swanson, E.; Ramiller, N.C. (1997). The Organizing Vision in Information System Innovation, *Organization Science*, vol 8, n°5, pp. 458-474

Yin, R.K. (2003). Case study research: Design and methods, 3rd edition, Applied social research methods series, Sage publications, vol. 5, London

# 6

# Design and Implementation of Satellite-Based Networks and Services for Ubiquitous Access to Healthcare

Georgi Graschew, Theo A. Roelofs, Stefan Rakowsky and Peter M. Schlag
*Surgical Research Unit OP 2000, Experimental and Clinical Research Centre ECRC, Max-Delbrück-Centre for Molecular Medicine and Charité – University Medicine Berlin, Charité Campus Berlin-Buch, Lindenberger Weg 80, D-13125 Berlin Germany*

## 1. Introduction and Background

Telemedicine describes the use of Information and Communication Technologies (ICT) for the delivery of medical services. It aims at equal access to medical expertise irrespective of the geographical location of the person in need. New developments in ICT have enabled the transmission of medical images in sufficiently high quality that allows for a reliable diagnosis to be determined by the expert at the receiving site (Pande et al., 2003; Lacroix et al., 2002). Through Telemedicine patients can get access to medical expertise that may not be available at the patients' site. Networks for Telemedicine enable the integration of distributed medical competence and contribute to the improvement of the quality of medical care, to the cost-effective use of medical resources and to quick and reliable decisions.

For optimal performance of telemedical applications, the networks and communication tools used must be optimised for medical applications, both with respect to the Quality-of-Service (QoS, a set of parameters characterising the performance of the communication channel per se, such as transmission bandwidth, delay, jitter, data loss, etc.) as well as to the Class-of-Service (CoS; a set of terms specifying the medical services offered in the network, like Telesurgery, Telepathology, Telesonography, Tele-Teaching, -Training & -Education, etc.).

The use of specifically designed networks for telemedicine (distributed medical intelligence) contributes to the continuous improvement of patient care. Experience over the last decade has shown that the goals of Telemedicine are not automatically reached by the introduction and use of singular new technologies per se, but rather require the implementation of integral services.

At the same time, however, these innovative developments in ICT over the last decade bear the risk of creating and amplifying a digital divide in the world, creating a disparity in the quality of life, e.g. between the northern and the southern Euro-Mediterranean area (Graschew et al., 2003a; Dario et al., 2005; Graschew et al., 2004a). In recent years different projects have demonstrated how the digital divide is only one part of a more complex problem: the need for integration. (Wootton et al., 2005; Rheuban & Sullivan, 2005; Graschew et al., 2003b; Graschew et al., 2002).

## 2. Methodology

During the last years OP 2000 (Operating Room of the future) has designed, developed and validated various modules for interactive telemedicine services (Schlag et al. 1999; Graschew et al. 2000). One of the key elements is the interactive telecommunication module WoTeSa / WinVicos: WoTeSa, a dedicated Workstation for Telemedical applications via Satellite that uses the communication software WinVicos (Wavelet-based interactive Video communication system).

WoTeSa is a PC with sufficient processing capacity (≥ 3 GHz Pentium IV, ≥ 512 MB RAM), one or more Osprey video capture boards (Osprey 100 or Osprey 500), a camera with composite and s-video outputs as live source (e. g. Canon VC-C4); a second camera as document camera for transmission of non-digital images; standard headset or microphone with small loudspeakers. The different video inputs of the Osprey video capture card can be used for direct connection to various medical video sources. WoTeSa thus serves quasi as a medical video hub. It is noteworthy that WoTeSa is a dedicated workstation and yet it can be realised with off-the-shelf components, thus making it readily available and at the same time very flexible and adaptive.

WinVicos is an all-software high-quality interactive video communication system, supplying real-time video, still-image and audio-transmission. WinVicos is especially designed for telemedical applications (e.g. telesurgery, teleradiology, telepathology) using a hybrid speed-optimised wavelet-codec that is based on the concepts of Partition, Aggregation and Conditional Coding (PACC; Patent DE 197 34 542 A1 from "Deutsche Telekom", Darmstadt, Germany). In contrast to most mainstream video coding systems that are mostly optimised for cinema and home entertainment, the PACC codec does not employ motion estimation but maximises the frame resolution to allow for maximal details to be visualised. WinVicos communicates IP-based and allows for online scaling of the transmission parameters (bit rate, frame rate and frame size from 128x96 up to 640x480 pixel). It supports both point-to-point and multipoint communication scenarios. Besides high quality live video transmission using moderate bandwidths (0.5–1 Mbit/s) it also allows for still-image transmission. WinVicos is very easy to use: there is a main user interface that is sufficient for the standard actions of the user. This includes calling the video conference partner via a telephone book, adjustment of both local and remote transmission parameters, as well as speaker and microphone volume control (see Fig. 1).

In both the video windows and the still-image windows WinVicos supports the use of common cursors shared by the conference partners. As WinVicos is an all-software system not only the implementation of other video codecs can be readily realised, but also a continuous performance improvement is supported to keep up with recent developments in the field. The most recent version of WinVicos also supports the transmission of video streams in full high definition (HD) resolution. Further readings on WoTeSa and WinVicos can be found in the literature (Graschew et al., 2001).

Other telemedicine systems are used e.g. for tele-ultrasound in rural areas where telementoring by live videoconferencing allowed to guide the ultrasound technician to record additional images of the patient (O'Neill et al., 2000) for clinical assessment of pediatric burns which showed a good agreement between the face-to-face consultation and seeing the patient via videoconference (Smith et al., 2004) and for home telecare services likely to improve quality of health services (Guillen et al., 2002). Other systems are described in Sable 2002; Latifi et al., 2004, Eadie et al., 2003.

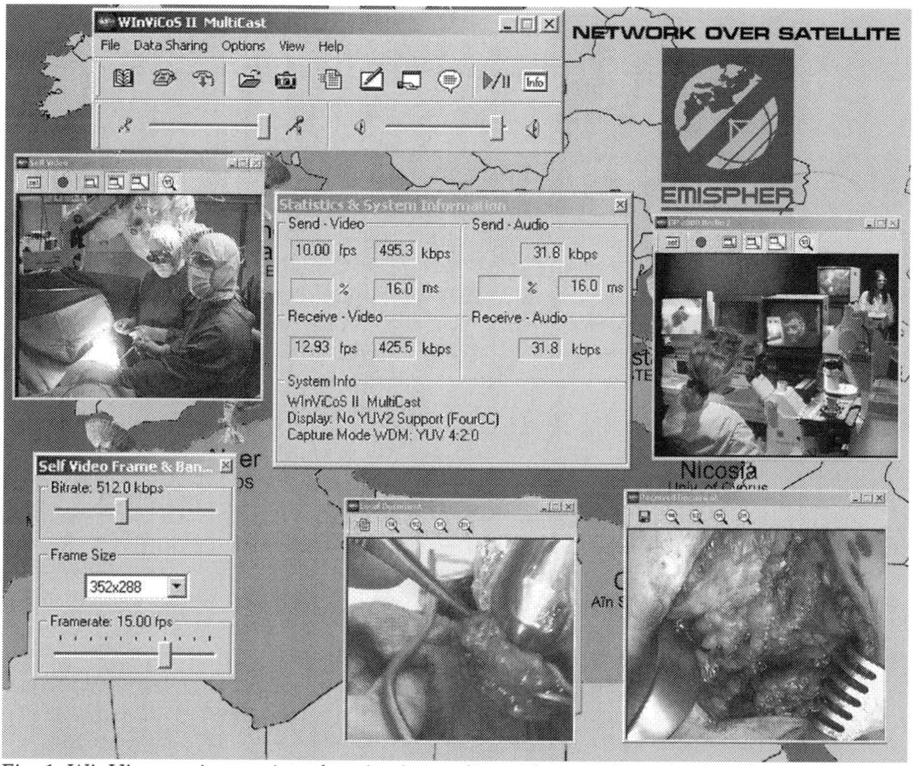


Fig. 1. WinVicos main user interface (top): two live video windows (top) and two still-image windows (bottom), online flexible adjustment of transmission parameters (bottom left), transmission performance can be monitored online (centre).

## 3. Satellite-based Networks and Services

OP 2000 has been a prominent partner in various telemedical networks using satellite-based technology for the deployment of services for interactive video communication and medical telepresence (Graschew et al., 2009). In the following, four projects will be described in which WoTeSa / WinVicos have been used as interactive tool for real-time telemedical application services and e-learning scenarios.

### 3.1 Generic Advance Low-cost trans-European Network Over Satellite - GALENOS

In cooperation with: Eutelsat (F), Nortel Dasa, (D), Alcatel Space Industries (F), Telespazio (I), MEDSAT (F), NCSR Demokritos (Gr); financially supported by the European Union (EU).

In the GALENOS project a European competence network via satellite, dedicated to telemedical applications, has been realised (Graschew et al., 2001). Owing to the participation of industrial partners in integrating the communication network and getting satellite transmission capacity available, several telemedical services (e.g. offline access to archived data, live consultation of experts, tele-teaching, etc.) have become available in a unified and low-cost technology. This possibility to get support from external experts, the improvement of the precision of the surgical treatment, and last but not least, online documentation and hence improved analysis of the available data on a patient, contributes to a continuous improvement in treatment and care of patients.

The trans-European communication network over satellite provides various collaborative services and applications, avoiding the use of specific technologies, thus promoting a wide availability. Furthermore, the network exhibits inter-operability with previously existing telecommunication networks. The network offers up to 2 Mbit/s interfaces with satellite link and covers a total of 14 clinics in Bulgaria, France, Germany, Greece, Italy and Tunisia. Access to the local area network (LAN) at each site is also provided for the exchange of patient data, medical images, videoconferencing, etc. Substantial reactivity and technical flexibility allow treatment protocols, adapted to the patient's pathology, in the shortest time.

The GALENOS network enables physicians not only to receive virtual "classical" hands-on medical training, but also trains them in using the state-of-the-art communication-, video- and computer-technologies needed for collaborative work in a network of distributed medical intelligence (interactive Teleteaching). Such distributed medical intelligence in the competence network enables e.g. local physicians or local hospitals that are confronted with unexpected test results to get online advice from the nearest academic hospital (intraoperative radiological Teleconsultation). It also becomes possible to perform intraoperative Telepathology where, after a biopsy has been taken and a slice has been prepared, this sample is put under a remotely-controllable, camera-equipped microscope and the diagnosis is then formulated by a remote expert. Or intraoperative Teleradiology, where anytime during the operation, the surgeon can access all the preoperative radiological data (like X-ray, CT, MRT, etc.) and discuss it with the radiologist in real-time.

### 3.2 Disaster Emergency Logistic Telemedicine Advanced Satellites System - DELTASS

The DELTASS Consortium consists of: Centre National D'Etudes Spatiales (F), MEDES (Institut de Médecine et de Physiologie Spatiales (F), SPACEBEL (B), Alcatel Space Industries (F), EADS - DORNIER (European Aeronautic Defence and Space Company (D), EADS - MS&I (European Aeronautic Defence and Space Company - Matra Systémes & Information (F), and Surgical Research Unit OP 2000 at Charité (D).

The DELTASS project was funded by the European Space Agency (ESA) as part of the ARTES-5 programme (Advanced Research in Telecommunications Systems).

Further general information on the DELTASS project can also be found at:
http://telecom.esa.int/telecom/www/object/index.cfm?fobjectid=750.

In the DELTASS-project (Graschew et al., 2008) a disaster scenario was analysed and an appropriate telecommunication system for effective rescue measures for the victims was set up and evaluated. Satellite-based systems are well suited for these circumstances, where generally ground infrastructures are partly or even totally destroyed. In such situations, even on a large geographic area or isolated area, space-based services can be easily and quickly deployed. DELTASS demonstrates operational performance of various services, covering the different aspects/phases of disaster emergency medicine. According to these phases, the DELTASS system is made up of the various corresponding subsystems (Fig. 2):

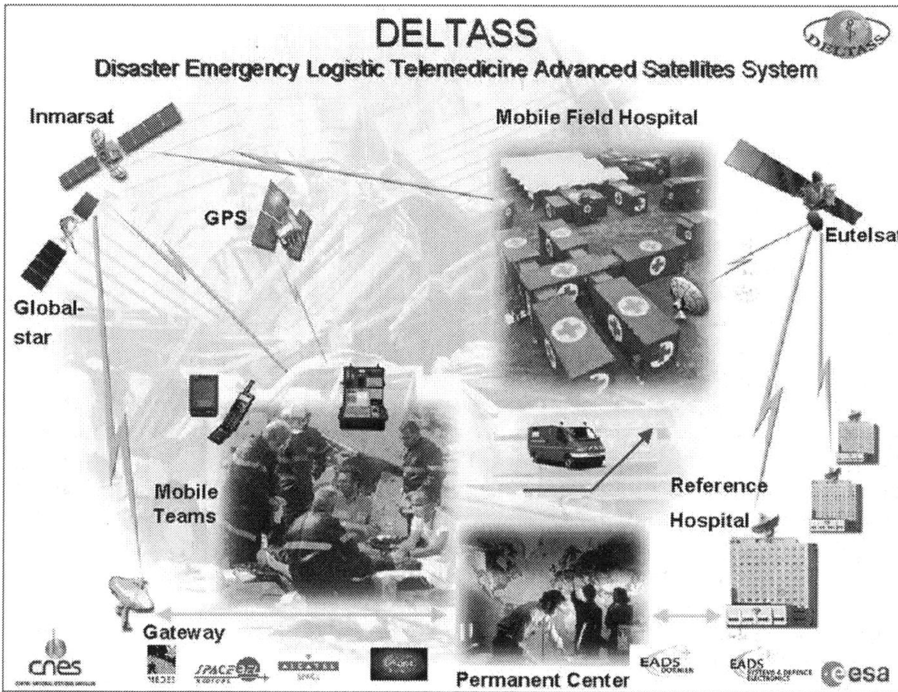


Fig. 2. DELTASS System Architecture: Mobile Teams, Permanent Centre (PC), Mobile Field Hospital (MFH) and Reference Hospital (RH) are interconnected via several satellite systems with different bandwidths. Additionally, terrestrial communication channels support the data exchange between the PC and the RH.

*Mobile Teams*

Mobile teams are deployed on the disaster site for search, identification, triage and evacuation of victims. They communicate with the coordination- and medical- teams located in Permanent Centre or Mobile Field Hospital via low-rate (Globalstar, 9.6 kbps) and medium-rate (Inmarsat, 64 kbps) satellite telecommunication systems. Their positions are tracked via the established Global Positioning System (GPS satellite system).

*Permanent Centre*
The Permanent Centre is located outside the disaster area. The Permanent Centre constitutes a new element in the architecture of support systems for disaster emergencies and is unique to the DELTASS system. In conventional set-ups the mobile teams at the disaster site are coordinated and supported by the staff of a Mobile Field Hospital deployed at or close to the disaster site. However, complete deployment of such a Mobile Field Hospital takes at least ~6 hours, usually ~12 hours, and consequently the activities of mobile teams in these first, highly critical hours, are ill-coordinated and far from optimal. To improve this bottleneck, DELTASS has a designated Permanent Centre that is in control of coordination and medical support to the mobile teams from time zero on. The Permanent Centre is equipped with terrestrial gateways to the Globalstar and Inmarsat satellite systems through which it receives all data from the mobile teams. It coordinates all actions of the mobile teams and manages all medical and logistic data, thus assuring efficient operation during the first critical phase. All data received at the Permanent Centre are processed, appropriate Reference Hospitals (RH; see below) are identified and the logistic and medical data are transferred to these RH via terrestrial telecommunication links.

*Mobile Field Hospital (MFH)*
A Mobile Field Hospital (MFH), which will be deployed at or close to the disaster site, provides all activities related to the co-ordination of the mobile teams on the disaster site, the victims' medical triage, reception, first aid treatment, conditioning for transportation, further medical expertise for some patients by teleconsultations between MFH and Reference Hospital(s).

*Reference Hospital (RH)*
The Reference Hospital(s) (RH), located outside of the disaster area, acts as an expert centre by providing telemedical services to the MFH using the high-bandwidth satellite link (VSAT, 2 Mbps). These services consist of off-line and on-line telediagnosis, access to external medical databases, as well as real-time interactive telemedical services such as live teleconsultations, live telesonography, intraoperative virtual reality simulation and interactive telemicrobiology (see Figs. 3 and 4).
Statistics show that in cases of disaster emergency medicine, approx. 40% more amputations are performed, as compared to normal situation. One of the aims of providing live second opinion by remote experts is to reduce this number of unneeded amputations, manipulations and subsequent complications substantially, by expert support during triage, diagnosis and medical treatment.
These interactive telemedical services between the MFH and the RH are realised using a dedicated WoTeSa (Workstation for Telemedical Applications via Satellite) with the communication software WinVicos (Wavelet-based interactive Video communication system). WoTeSa/WinVicos combines the user-friendliness and flexibility of IP-based communication protocols with the security and sufficiently-high quality of the live video transmission at a satellite bandwidth of only up to 2 Mbps. Medical experts at the RH support the medical treatments in the MFH and enable a quick and reliable decision concerning treatment and/or evacuation of the patient/victim. In this way, the quality of the provided medical service during and after disaster emergencies is strongly improved.

The performance of the DELTASS system has been shown during various full-size live demonstrations (http://telecom.esa.int/telecom/www/object/index.cfm?fobjectid=6324).

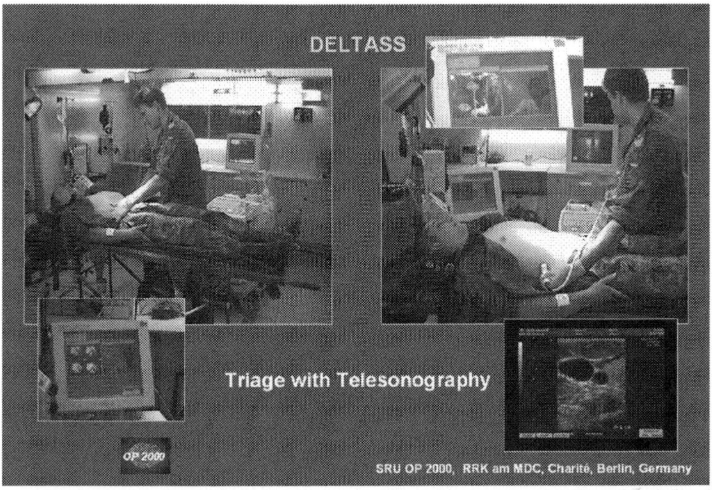


Fig. 3. Live Telesonography: The ultrasound video data stream (insert right bottom) is transmitted live via WoTeSa / WinVicos to the experts of the RH (insert right top). Additionally the signal of a room camera (e.g. for telementoring; insert right top) as well as further medical data (e.g. X-ray, CT, etc.; insert left bottom) can be transmitted live to the RH.

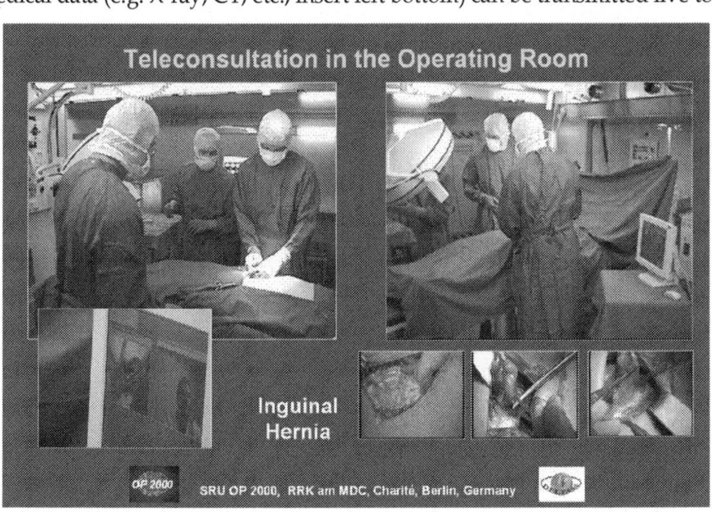


Fig. 4. Live Telesurgery of Inguinal Hernia: During the surgery the video data stream of a camera integrated in the centre of the surgical light can be transmitted live to the experts in the RH using WoTeSa / WinVicos (insert left bottom). In this way these experts receive a live video of the situs of surgery (inserts right bottom) and can advise the colleague in the MFH.

### 3.3 Medical Assistance for Ships - MEDASHIP

Cooperating partners: D'Appolonia (I), AVIENDA (UK), Eutelsat (F), NCSR Demokritos (GR); co-funded by the European Union (EU) under the eTEN programme.

The main objective of the service developed by the MEDASHIP project is to supply integrated solutions for medical consultations on-board of ships (Graschew et al., 2004b). The satellite-based telemedicine services address both passenger ships and merchant vessels and are intended to provide passengers and crew members with an effective medical assistance in cases of emergency and in all those cases where the on board medical staff requires second opinion. During the validation phase the service was tested on board of three ships with the possibility to have it connected to three land medical centres. (Fig. 5).

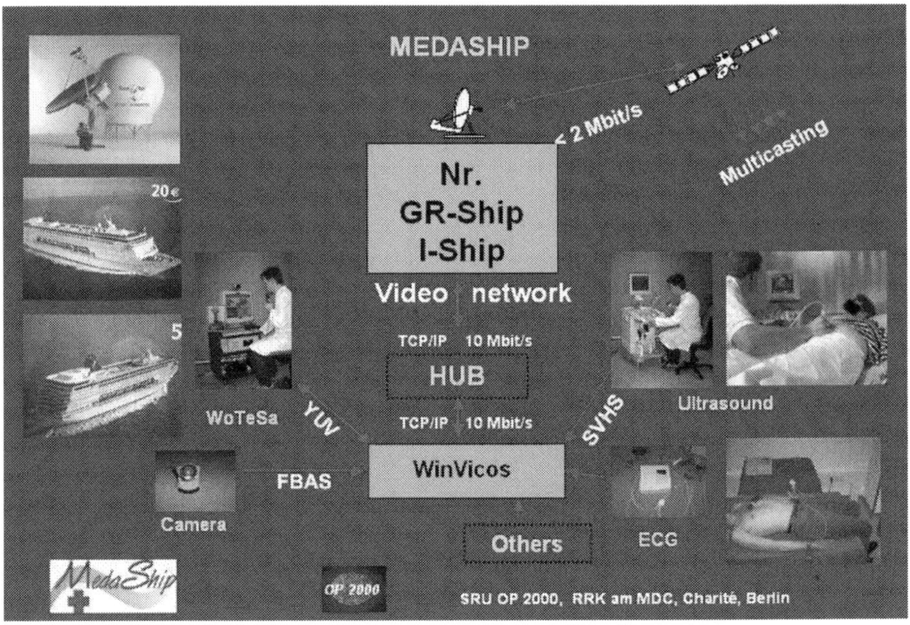


Fig. 5. Video network on-board of the ships: Video camera, ultrasound and electrocardiography equipment connected with the satellite modem via WoTeSa/WinVicos.

In addition to the standard medical equipment aboard the ships, two video cameras, an electrocardiograph (ECG) and an ultrasound (US) equipment are used. With this equipment the following telemedical services have been realised using satellite transmission at a bandwidth of 512 kbps up to 1 Mbps offering the required high quality of images and video transmission:

*Teleconsultation*

The live camera on-board of the ship can be used to transmit the image of the doctor who is leading the examination on-board of the ship or the image of the patient when being questioned by the land-based expert. It can also be used to show the land-based expert an injured part of the patient's body which he needs to see for his consultation. Thus a very realistic and effective live communication is possible.

*Electrocardiography*

The ECG system is connected to WoTeSa on board the ship and can be controlled by the physician from this workstation. Via application sharing software also the expert can control the ECG system from the land-based workstation. The main menu that includes all the functions of the ECG as well as the patient's ECG is transmitted to the expert. Thus the expert and the physician on board can jointly acquire and analyse the ECG report.

*Telesonography*

The S-video output of the US equipment is directly connected to the Osprey video capture board. Satellite transmission tests have shown that not only still images can be transferred but also live ultrasound investigations can be transmitted at 500-700 kbps (see Fig. 6).
With a document camera analogous patient data can be captured and digitised by WinVicos as a document. For example X-ray or CT-images can be captured from an illumination board and displayed locally and transmitted using this document camera function.

*Data Security*

Respect for patient confidentiality is a clear requirement in the MEDASHIP system and the level of privacy protection has been addressed throughout the MEDASHIP project. In MEDASHIP, from the telemedicine platform viewpoint, the following requirements should be taken into account to prevent abuse and to safeguard the privacy and confidentiality of managed clinical and personal information of the patients.

- *Authentication:* Monitoring and verifying all the accesses to the information. A control over the users should be carried out at each access to the system, by utilising the users' credentials (username and password) in order to verify that the user is who he/she claims to be.
- *Encryption:* Scrambling a sender's transmission according to an algorithm that the recipient then uses to unscramble and decipher the transmission. All the process is not visible to the user, the client software knows all it needs.
- *Access control:* Authorising access to specific and clearly identified resources to certain users based on their company responsibilities and the security classification of the resources.
- *Integrity:* Developing or utilising applications and data management software that is secure from unauthorised modification of their code.
- *Confidentiality:* Developing or utilising applications and data management software that is secure from disclosure to unauthorised persons or programs.
- *Auditing and accounting system:* Monitoring the system and maintaining records of system and user activity.
- *Security policy:* Establishing clear security policies with customers and end-users.

Unlike open network satellite equipment technologies Linkway uses a proprietary acquisition and synchronisation technique making signal interception and decoding virtually impossible. Data Security will also be assured by the coding algorithm of the WinVicos software. The transmitted data can only be decoded by this software. As one member owns the software licence this provides an initial level of data security through the licence distribution.

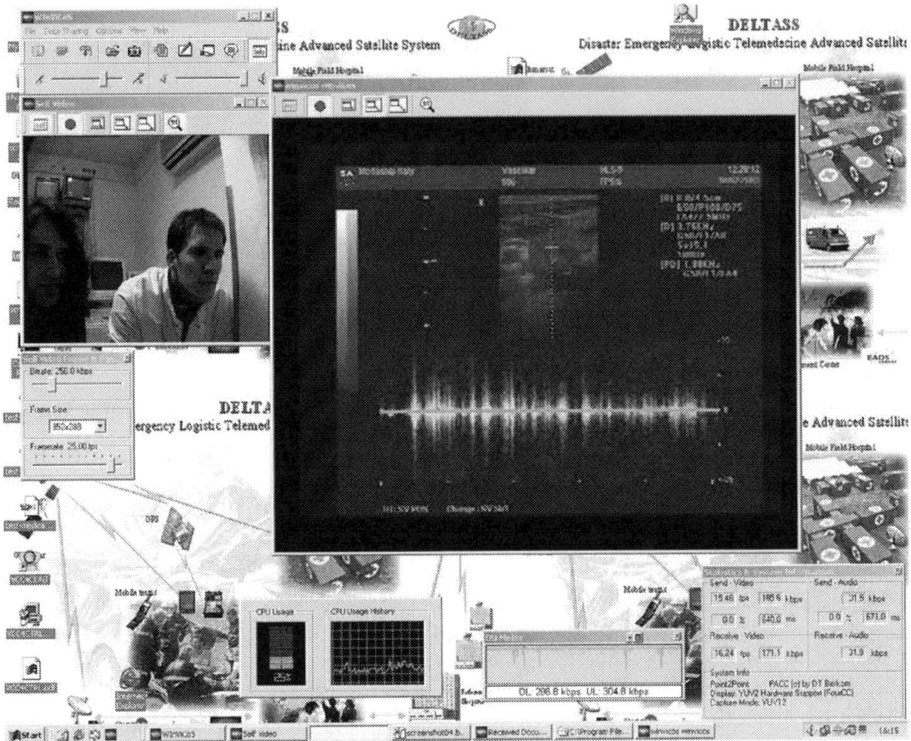

Fig. 6. Telesonography: the live signals of the ultrasound equipment on-board of the ship are transmitted to the reference hospital. A physician at Charité is consulting the ultrasound examination of a patient on board of the cruise ship.

*Reduction of cost*

The costs for emergency interventions for removing a passenger from the ship and hospitalisation abroad are not to be undervalued. The removal of a passenger in the Carribean can cost up to $ 11.000 and the cost for hospitalisation can range from 500-1000 € per day. Consequently, market trends force passenger shipping lines to offer services that help to improve the response to on-board clinical emergencies, improve the customer satisfaction and the companies' image.

### 3.4 Euro-Mediterranean Internet-Satellite Platform for Health, medical Education and Research - EMISPHER

In cooperation with: FMPC - Faculty of Medicine and Pharmacy of Casablanca, Morocco; ANDS - Agence National de Documentation de la Santé (Ministère de la Santé), Algiers, Algeria; Tunis - Faculty of Medicine of Tunis, Tunisia; ASU - Aïn Shams University, Cairo,

Egypt; NIFRT - Nasser Institute for Research and Treatment (Ministry of Health and Population, MOHP), Cairo, Egypt; UCY - University of Cyprus, Nicosia, Cyprus; ISTEM - Continuing Medical Education and Research Centre, University of Istanbul, Turkey; NCSR Demokritos, Athens, Greece; IsMeTT - Istituto Mediterraneo per i Trapianti e Terapie ad Alta Specializzazione, Palermo, Italy; CICE - Centre International de Chirurgie Endoscopique, Clermont-Ferrand, France; and Charité University Medicine Berlin, Germany; co-funded by the European Union (EU) under the EUMEDIS / MEDA programme.

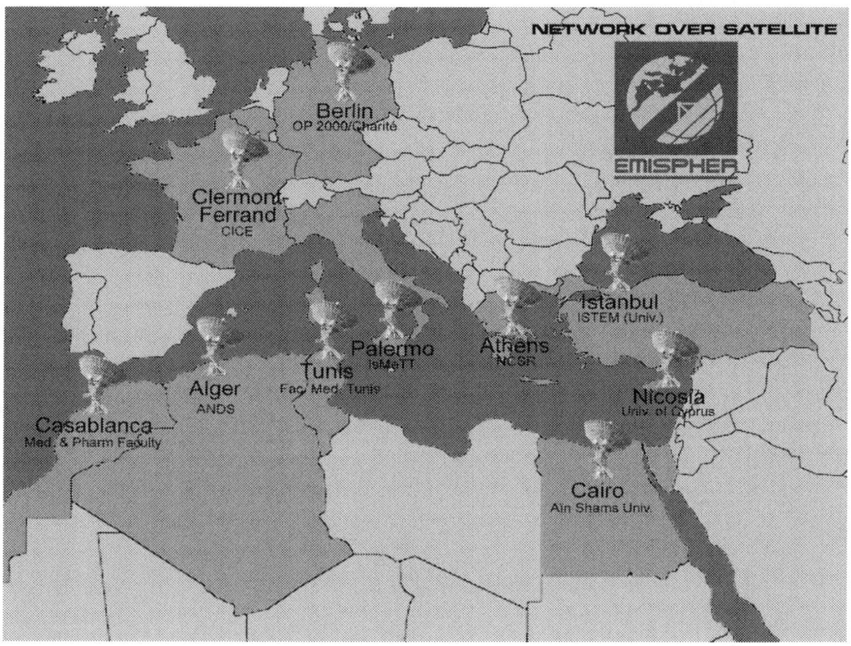


Fig. 7. Medical Centres in the EMISPHER Network

EMISPHER is dedicated to establish an equal access for most of the countries of the Euro-Mediterranean area to real-time and on-line services for healthcare in the required quality of service (see www.emispher.org). In the project an integrated Internet-Satellite platform has been set up on which three main areas of work have been realised: Virtual Medical University, Real-Time Telemedicine, and Medical Assistance (Graschew et al., 2005). The platform includes a bi-directional satellite network (up to 2 Mbps) between 10 Centres of Excellence in the Euro-Mediterranean region (Morocco, Algeria, Tunisia, Egypt, Cyprus, Turkey, Greece, Italy, France and Germany; see Fig. 7). For dissemination of the achieved results and for maximising its impact, EMISPHER has organised international conferences at each of the Mediterranean partner sites.

*The EMISPHER Virtual Medical University*

The formation and operation of the EMISPHER Virtual Medical University (EVMU) for e-learning (teleteaching) is one of the main efforts in the project. The EVMU uses real-time broadcast of lectures, live surgical operations and pre-recorded video sequences etc., as well as web-based e-learning applications. The target population of the EVMU is comprised of medical students (both undergraduate and postgraduate) hospital staff, general practitioners and specialists, health officers and citizens.

Each of the leading medical centres provides didactical material and modules for synchronous and asynchronous e-learning in their medical specialties. The central gateway to EVMU is the project's website: www.emispher.org.

Some of the multimedia teaching material needs to be presented in real-time. Live transmission of surgical operations from operating theatres, lectures, etc. from one site to one or several sites simultaneously (point-to-point or multipoint) are possible in the network between the 10 partners (Fig. 8).

Fig. 8. Interactive multipoint e-learning session with transmission of live ultrasound video data from Charité (Berlin) to Ain Shams University (Cairo), Agence Nationale de Sante (Algiers), Faculté de Medecine et de Pharmacie (Casablanca) and University of Cyprus (Nicosia).

*Real-Time Telemedicine*

EMISPHER has set up a satellite-based network using the combined WoTeSa and WinVicos modules for real-time telemedicine. In the field of real-time telemedicine the following categories of applications are offered: second opinion, teleteaching & teletraining (demonstration and spread of new techniques), telementoring (enhancement of staff qualification), and undergraduate teaching courses and optimisation of the learning curve. The leading medical centres in the project provide expertise in the following medical fields: open and minimally-invasive surgery, multi-organ transplantation, endoscopy, pathology, radiology, interventional imaging, neurology, infectious diseases, oncology, gynaecology and obstetrics, reproductive medicine, etc. These real-time telemedical applications contribute to improved quality of patient care and to accelerated qualification of medical doctors in their respective specialty. The main target audience are specialist doctors (Fig. 9).

Fig. 9. Interactive multipoint radiological teleconsultation in EMISPHER. Experts from different centres interactively discuss radiology results from a particular patient case.

*Medical Assistance*

The third field of service operated in EMISPHER is medical assistance. As tourism constitutes a substantial economical factor in the Mediterranean region and because of the increasing mobility of the population, continuity of care through improved medical assistance is of major importance for improved healthcare in the Euro-Mediterranean region. Introduction of standardised procedures, integration of the platform with the various local communication systems and training of the medical and non-medical staff involved in the medical assistance chain allow for shared management of files related to medical assistance (medical images, diagnosis, workflow, financial management, etc.) and thus for improved care for travellers and expatriates.

*User Evaluation*

Two types of evaluation were performed for the EMISPHER satellite-internet platform and the services delivered from the platform. Firstly, the evaluation of all technical aspects of the platform and connecting services (WinVicos and MEDSKY) and secondly the evaluation of the operational elements of the system meeting the medical user requirements.

Overall point-to-point connections were successful with normal delays (~650 ms). Realised maximum bandwidths for symmetric point-to-point connections are 512-640 kbps in each direction. Up to these bandwidths the video and audio streaming is smooth. When going beyond these settings frozen images and audio problems start occurring. For asymmetric video bandwidth allocation (for example for e-learning applications) the maximum bandwidth settings for good operation are 768 kbps in one direction and 192 kbps in the other direction. For multipoint connections between three sites the maximum bandwidth settings for good operation are 384, 384 and 512 kbps. If the results presented above are compared with the nominal bandwidths required for the various services, it is clear that the EMISPHER platform provides sufficient bandwidth for most of the telemedicine applications with smooth video and audio streams in sufficient resolution.

During the first months of full operation of the network continuous evaluations have been performed and user-feedback was collected. This resulted in the upgrade of the WinVicos telemedical communication software as well as various upgrades of the MEDSKY software platform. The valuable user feedbacks were taken into account and have lead to even improved functionalities of the system and thus improved quality of medical work.

This demonstrates that the satellite platform with WoTeSa/WinVicos telemedical stations is very suited for the envisioned variety of medical services and is easy to learn for a wide range of medical professionals from the various countries.

## 4. Barriers and their possible solutions

Although the market is promising and the technology ready (or nearly) to be used, the take-up and commercialisation of the telemedicine services is still uncertain due to a number of barriers:

*External*

- Decision-making in health is fragmented with respect to procurement policies, which hinders progression towards real integration.
- Establishing and building confidence with physicians takes time.
- It still seems that it will be a further 1-2 years before full deployment and therefore a critical mass can be built.
- How the services are packaged and attitudinal changes from within the health sector are also necessary.
- There are numerous organisational issues.

*Economic*

- It is difficult to predict the breakeven level of services and therefore to determine the minimum level of investment required for breakeven of the services.

- There is a lack of investment in health and therefore significant investment is still required.
- Considerable investment in improved management, training and education of personnel, re-designing of care and logistic processes is necessary.

*Legal*

- Ethical issues
- Regulatory framework issues
- Reimbursement issues
- Lack of standardisation

*Technical*

- Software development is still evolving.
- IP levels of connectivity are also required as a minimum.
- All the services need to be fully integrated into one platform.

In various pilot projects, the technology has been put in place, the necessary applications have been developed and it had been proved that it can be used successfully and meet the needs of its end users. Yet numerous trials and demonstrations carried out during the projects have also highlighted a certain number of issues that could potentially hinder the commercialisation process of e-services in the medical sector.
In fact, various market analysis reports prepared for the European Commission also insist on the fact that the telemedicine market growth will be dependent upon a number of vital conditions and enabling factors already noted above. It is generally accepted that four key action lines should be initiated in parallel: consolidation on the supply side, technical integration, investment on the demand side and "accompanying measures" that could be the enabling factors required to allow the health telematics market to achieve substantial and exponential growth.

*Suggested solutions to overcome barriers*

The solutions to such 'barriers' to commercialisation do not lie within the scope of the technology and appear to be generic to those 'up and running' tele-health activities worldwide. It is only in North America that new legislation has been introduced to respond to the particular needs of this practice and this legislation are not yet being applied on a federal level but rather on a state by state basis. However, the organisational and cultural aspects that must accompany any new form of practice need the input of other actors both on the governmental, legal and political scale. Tele-health is not a simple extension of current health systems and cannot be perceived as such.
Solutions to the "barriers" can be consolidated under the following three main categories:

- Awareness of telemedicine and tele-health as integral part of medical practice;
- The need for common standards and policies;
- The need for specific legislation.

## 5. Conclusions and Perspectives: Virtual Hospitals

Satellite-based networks and services are crucial to support ubiquitous access to healthcare. Appropriate enabling technologies must be deployed and interconnected via networks with an appropriate design. In this chapter we have presented WoTeSa/WinVicos as a flexible high-end module for real-time interactive telemedical services. Besides video communication in medically expedient quality, the provision of interactivity for the remote control of medical equipment is indispensable. Both video communication and interactivity require a (nearly) real-time mode of bi-directional interactions. Various examples have been given of particular networks and services that have been deployed, each to support medical telepresence in specific functional scenarios (GALENOS, DELTASS MEDASHIP and EMISPHER).

However, despite substantial improvements that have been realised, these developments bear the risk of creating and amplifying digital divides in the world. To avoid and counteract this risk and to fulfill the promise of Telemedicine, namely ubiquitous access to high-level healthcare for everyone, anytime, anywhere (so-called ubiquitous Healthcare or u-Health) a real integration of both the various platforms (providing the "Quality-of-Service", QoS) and the various services (providing the "Class-of-Service", CoS) is required (Graschew et al., 2002; Graschew et al., 2003b; Wootton et al., 2005; Rheuban & Sullivan 2005; Graschew et al., 2006a). A virtual combination of applications serves as the basic concept for the virtualisation of hospitals. Virtualisation of hospitals supports the creation of ubiquitous organisations for healthcare, which amplify the attributes of physical organisations by extending its power and reach. Instead of people having to come to the physical hospital for information and services the virtual hospital comes to them whenever they need it. The creation of Virtual Hospitals (VH) can bring us closer to the ultimate target of u-Health (Graschew et al., 2006b).

The methodologies of VH should be medical-needs-driven, rather than technology-driven. Moreover, they should also supply new management tools for virtual medical communities (e.g. to support trust-building in virtual communities). VH provide a modular architecture for integration of different telemedical solutions in one platform (see Fig. 10).

Due to the distributed character of VH, data, computing resources as well as the need for these are distributed over many sites in the Virtual Hospital. Therefore, Grid infrastructures and services become useful for successful deployment of services like acquisition and processing of medical images (3D patient models), data storage, archiving and retrieval, as well as data mining, especially for evidence-based medicine (Graschew et al., 2006c).

The possibility to get support from external experts, the improvement of the precision of the medical treatment by means of a real medical telepresence, as well as online documentation and hence improved analysis of the available data of a patient, all contribute to an improvement in treatment and care of patients in all circumstances, thus supporting our progress from e-Health and Telemedicine towards real u-Health.

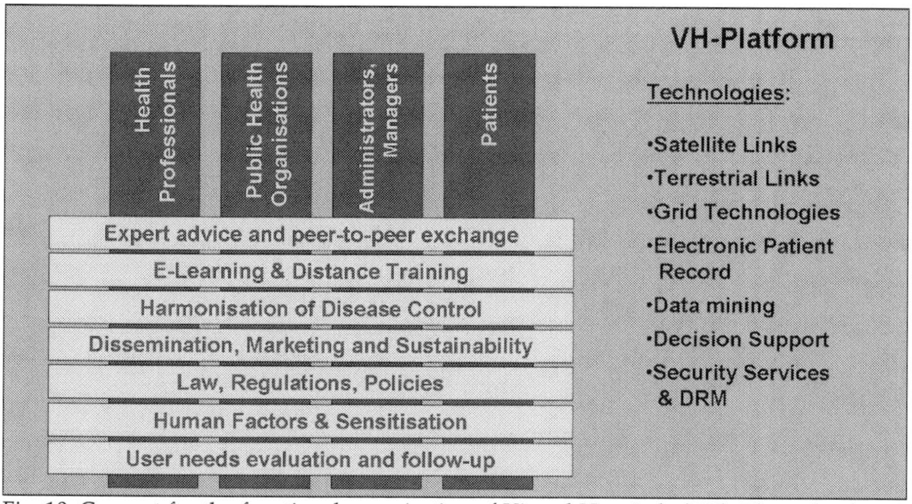


Fig. 10. Concept for the functional organisation of Virtual Hospitals (VH): The technologies of VH (providing the "Quality-of-Service", QoS) like satellite-terrestrial links, Grid technologies, etc. will be implemented as a transparent layer, so that the various user groups can access a variety of services (providing the "Class-of-Service", CoS) such as expert advice, e-learning, etc. on top of it, not bothering with the technological details and constraints.

## 6. References

Dario, C. et al. (2005). Opportunities and Challenges of eHealth and Telemedicine via Satellite. *Eur J. Med. Res., Vol. 10, Suppl I, Proceedings of ESRIN-Symposium, July 5, 2004, Frascati, Italy, (2005),* pp. 1-52.

Eadie, L.H. et al., (2003). Telemedicine in surgery. *Br. J. Surg.,* Vol. 90, pp. 647-58.

Graschew, G. et al. (2000). Interactive telemedicine in the operating theatre of the future. *J Telemedicine and Telecare* Vol. 6, Suppl. 2, pp. 20-24.

Graschew, G. et al. (2001). GALENOS as interactive telemedical network via satellite, In: *Optical Network Design and Management,* Proc. of SPIE, Vol 4584, pp. 202-205.

Graschew, G. et al. (2002). Broadband Networks for Interactive Telemedical Applications, *APOC 2002, Applications of Broadband Optical and Wireless Networks, Shanghai 16.-17.10.2002, Proceedings of SPIE,* Vol. 4912, pp. 1-6.

Graschew, G. et al. (2003a). Telemedicine as a Bridge to Avoid the Digital Divide World, *8. Fortbildungsveranstaltung und Arbeitstagung Telemed 2003, Berlin, 7.-8. November 2003, Tagungsband,* pp. 122-127.

Graschew, G. et al. (2003b). Telepresence over Satellite, *Proceedings of the 17th International Congress Computer Assisted Radiology and Surgery, London, 25.-28.6.2003,* International Congress Series, Vol. 1256, ed. H.U. Lemke et al., pp. 273-278.

Graschew, G. et al. (2004a). Interactive Telemedicine as a Tool to Avoid a Digital Divide of the World, In: *Medical Care and Compunetics 1,* L. Bos (Ed.), pp. 150-156, IOS Press, Amsterdam.

Graschew, G. et al., (2004b). MEDASHIP - Medizinische Assistenz an Bord von Schiffen, In: *Telemedizinführer Deutschland, ed. 2004,* A. Jäckel (Ed.), Deutsches Medizin Forum, Ober-Mörlen, Germany, pp. 45-50.

Graschew, G. et al., (2005). Überbrückung der digitalen Teilung in der Euro-Mediterranen Gesundheitsversorgung - das EMISPHER-Projekt, In: *Telemedizinführer Deutschland, ed. 2005,* A. Jäckel (Ed.), Ober-Mörlen, Germany, pp. 231-236.

Graschew, G. et al., (2006a). VEMH - Virtual Euro-Mediterranean Hospital für Evidenz-basierte Medizin in der Euro-Mediterranen Region, In: *Telemedizinführer Deutschland, Ausgabe 2006,* A. Jäckel (Ed.), Medizin Forum AG, Bad Nauheim, Germany, pp. 233-236.

Graschew, G. et al., (2006b). New Trends in the Virtualization of Hospitals - Tools for Global e-Health, In: *Medical and Care Compunetics* 3, L. Bos et al. (Eds.) Proceedings of ICMCC 2006, The Hague, 7-9 June 2006, IOS Press, Amsterdam, pp.168-175.

Graschew, G. et al., (2006c). Virtual Hospital and Digital Medicine - Why is the GRID needed?, In: *Challenges and Opportunities of HealthGrids,* V. Hernandez et al. (Eds.) Proceedings of HealthGrid 2006, Valencia, 7-9 June 2006, IOS Press, Amsterdam, pp.295-304.

Graschew, G. et al., (2008). DELTASS - Disaster Emergency Logistic Telemedicine Advanced Satellites System - Telemedical Services for Disaster Emergencies. *International Journal of Risk Assessment and Management* Vol. 9, pp. 351-366.

Graschew, G. et al., (2009). *New developments in network design for telemedicine.* Hospital IT Europe, Vol. 2 No. 2, pp. 15-18.

Guillen, S. et al., (2002). User satisfaction with home telecare based on broadband communication. *J. Telemed. Telecare,* Vol. 8, pp. 81-90.

Lacroix, L. et al., (2002). International concerted action on collaboration in telemedicine: recommendations of the G-8 Global Healthcare Applications Subproject-4. *Telemed. J. E-Health,* Vol. 8, pp. 149-157.

Latifi, R. et al., (2004). Telepresence and telemedicine in trauma and emergency care management. *Stud. Health Technol. Inform.,* Vol. 104, pp. 193-199.

O'Neill, S.K. et al., (2000). The design and implementation of an off-the-shelf, standards-based tele-ultrasound system. *J. Telemed. Telecare,* Vol. 6, suppl 2, pp. 52-53.

Pande, R.U. et al., (2003). The telecommunication revolution in the medical field: present applications and future perspective. *Curr. Surg.,* Vol. 60, pp. 636-640.

Rheuban, K.S. & Sullivan, E. (2005). The University of Virginia Telemedicine Program: traversing barriers beyond geography. *J. Long-Term Eff. Med. Implants,* Vol. 15, pp. 49-56.

Sable, C. (2002). Digital echocardiography and telemedicine applications in pediatric cardiology. *Pediatr-Cardiol.* Vol. 23, pp. 358-369.

Schlag, P.M. et al., (1999). Telemedicine - The New Must for Surgery. *Archives of Surgery* Vol. 134, pp. 1216-1221.

Smith, A.C. et al., (2004). Diagnostic accuracy of and patient satisfaction with telemedicine for the follow-up of paediatric burns patients. *J. Telemed. Telecare,* Vol. 10, pp. 193-198.

Wootton, R. et al., (2005). E-health and the Universitas 21 organization: 2. Telemedicine and underserved populations. *J. Telemed. Telecare,* Vol. 11, pp. 221-224.

# 7

# Characterisation and Channel Modelling for Satellite Communication Systems

Asad Mehmood and Abbas Mohammed
*Blekinge Institute of Technology*
*Sweden*

## 1. Introduction

The high quality of service, low cost and high spectral efficiency are of particular interest for wireless communication systems. Fundamental to these features has been much enhanced understanding of radio propagation channels for wireless communication systems. In order to provide global coverage of broadband multimedia and internet-based services with a high signal quality to diverse users, seamless integration of terrestrial and satellites networks are expected to play a vital role in the upcoming era of mobile communications. The diverse nature of propagation environments has great impact on the design, real-time operation and performance assessment of highly reconfigurable hybrid (satellite-terrestrial) radio systems providing voice, text and multimedia services operating at radio frequencies ranging from 100 MHz to 100 GHz and optical frequencies. Therefore, a perfect knowledge and modelling of the propagation channel is necessary for the performance assessment of these systems. The frame work for most of the recent developments in satellite communications includes satellite land mobile and fixed communications, satellite navigation and earth observation systems and the sate-of-art propagation models and evaluation tools for these systems.

The organization of the chapter is as follows: Section 2 describes the multipath propagation impairments in land mobile satellite (LMS) communications. In Section 3, the probability distributions that characterize different impairments on radio waves are discussed. Section 4 provides an overview of statistical channel models including single-state, multi-state and frequency selective channel models for LMS communications. The chapter ends with concluding remarks.

## 2. Propagation Impairments Effecting Satellite Communication Links

The use of satellite communication systems for modern broadband wireless services involves propagation environments for radio signals different from that in conventional terrestrial radio systems. The radio waves propagating between a satellite and an earth station experience different kinds of propagation impairments: the effects of the ionosphere, the troposphere and the local fading effects as shown in Fig. 1. The combined effect of these impairments on a satellite-earth link can cause random fluctuations in amplitude, phase, angles of arrivals, de-polarization of electromagnetic waves and shadowing which result in degradation of the signal quality and increase in the error rates of the communication links.

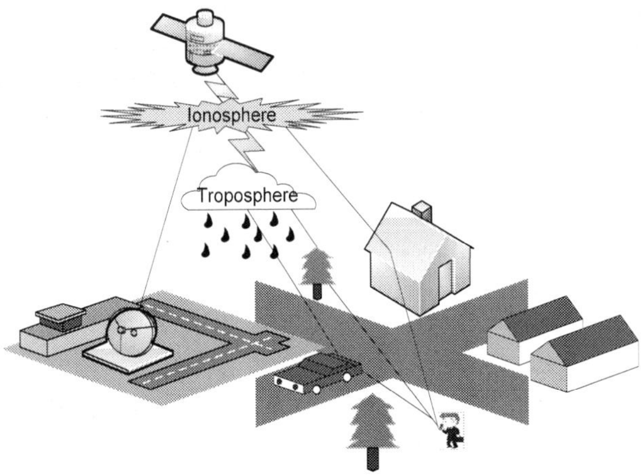


Fig. 1. The land-Mobile-Satellite Communication System

### 2.1 Ionospheric Effects

The effects of the ionosphere (an ionized section of the space extending from a height of 30 km to 1000 km) have adverse impact on the performance of earth-satellite radio propagation links. These effects cause various impairments phenomena such as scintillation, polarization rotation, refraction, group delays and dispersion etc, on the radio signals. The scintillation and polarization rotation effects are of foremost concern for satellite communications.

Ionospheric scintillations are variations in the amplitude level, phase and angle of arrival of the received radio waves. They are caused by the small irregularities in the refractive index of the atmosphere owing to rapid variations in the local electron density. The main effect of scintillation is fading that strongly depends on the irregularities or inhomogeneities of the ionosphere (Ratcliffe, 1973; Blaunstein, 1995; Saunders & Zavala, 2007). Scintillation effects are significant in two zones: at high altitudes (E and F layers of ionosphere) and the other is ±20° around the earth's magnetic equator. The effects of scintillation decrease with increase in operating frequency. It has been observed in various studies that at the operating frequency of 4 GHz ionospheric scintillations can result in fades of several dBs and duration between 1 to 10 seconds. The details about ionospheric scintillation can be found in International Telecommunication Union Recommendations (ITU-R, 2009a).

The orthogonal polarization (linear or circular) is used in satellite communication systems to increase the spectral efficiency without increasing the bandwidth requirements. This technique, however, has limitations due to depolarization of electromagnetic waves propagating through the atmosphere. When linearly polarized waves pass through the ionosphere, the free electrons present in the ionosphere due to ionization interact with these waves under the influence of the earth's magnetic field in a similar way as the magnetic field of a motor acts on a current carrying conductor. This results in rotation of the plane of polarization of electromagnetic waves, recognized as Faraday rotation. The magnitude of Faraday rotation is proportional to the length of the path through the ionosphere, the geomagnetic field strength and the electron density, and inversely proportional to the square of the operating frequency. The polarization rotation is significant for small percentages of time at frequencies 1 GHz or less. The effect of Faraday rotation is much lower at higher frequencies even in the regions of strong ionospheric impairments and low elevation angles, e.g., at frequency of 10 GHz, Faraday rotation remains below 1° and can be ignored (ITU, 2002). Cross-polarization can also be caused by the antenna systems at each side of the link. The effects of depolarization are investigated by two methods: cross-polarization discrimination (XPD) and polarization isolation. The details can be found in (Roddy, 2006; Saunders & Zavala, 2007).

### 2.2 Tropospheric Effects

The troposphere is the non-ionized lower portion of the earth's atmosphere covering altitudes from the ground surface up to a height of about 15 km of the atmosphere. The impairments of this region on radio propagations include hydrometeors, e.g., clouds, rain, snow, fog as well as moisture in atmosphere, gradient of temperature and sporadic structures of wind streams both in horizontal and vertical directions. The effects imparted by these impairments on radio signals are rain attenuation, depolarization, scintillation, refraction, absorption, etc. The radio waves are degraded by these effects to varying degrees as a function of geographic location, frequency and elevation angle with specific characteristics. The tropospheric effects in LMS communication links become significant when the operating frequency is greater than 1 GHz.

One of the major causes of attenuation for LMS communication links operating at frequency bands greater than 10 GHz (e.g., Ku-Band) is rain on the transmission paths in tropospheric region. The rain attenuation in the received signal amplitude is due to absorption and scattering of the radio waves energy by raindrops. The attenuation is measured as a function of rainfall rate and increases with increase in the operating frequency, rainfall rate and low elevation angles (Ippolito, 2008). The rainfall rate is the rate at which rain would accrue in a rain gauge placed in a specific region on the ground (e.g., at base station). The procedure to calculate attenuation statistics due to rainfall along a satellite-earth link for frequencies up to 30 GHz consists of estimating the attenuation that exceeds 0.001% of the time from the rainfall rate that exceeds at the same percentage of time and has been detailed in ITU-R recommendations (ITU-R, 2007).

The LMS channel utilization can be augmented without increasing the transmission bandwidth by the use of orthogonally polarized transmissions (linear or circular). The polarization of radio waves can be altered by raindrops or ice particles in the transmission path in such a way that power is transferred from the desired component to the undesired component, resulting in interference between two orthogonally polarized channels. The shape of small raindrops is spherical due to surface tension forces, but large raindrops adopt shape of spheroids (having flat base) produced by aerodynamic forces acting in upward direction on the raindrops. When a linearly polarized wave passes through raindrops of non-spherical structure, the vertical component of radio wave parallel to minor axis of raindrops experiences less attenuation than that the horizontal component. As a result, there will be a difference in the amount of attenuation and phase shift experienced by each of the wave components. These differences cause depolarization of radio waves in the LMS links and are illustrated as differential attenuation and differential phase shift. Rain and ice depolarization have significant impacts on satellite-earth radio links for frequency bands above 12 GHz, especially for systems employing independent dual orthogonally polarized channels in the same frequency band in order to increase the capacity. The method of predicting the long-term depolarization statistics has been described in ITU-R recommendations (ITU-R, 2007).

A radio wave propagating through satellite-earth communication link will experience reduction in the received signal's amplitude level due to attenuation by different gases (oxygen, nitrogen, hydrogen, etc.) present in the atmosphere. The amount of fading due to gases is characterized mainly by altitude above sea level, frequency, temperature, pressure and water vapour concentration. The principal cause of signal attenuation due to atmospheric gases is molecular absorption. The absorption of radio waves occurs due to conversion of radio wave energy to thermal energy at some specific resonant frequency of the particles (quantum-level change in the rotational energy of the gas molecules). Among different gases only water vapours and oxygen have resonant frequencies in the band of interest up to 100 GHz. The attenuation due to atmospheric gases is normally neglected at frequency bands below 10 GHz. A procedure to find out the effects of gaseous attenuation on LMS links has been discussed in ITU-R recommendations (ITU-R, 2009b).

Scintillations (rapid variations in the received signal level, phase and angle-of-arrival) occur due to inhomogeneities in the refractive index of atmosphere and influence low margin satellite systems. The tropospheric scintillations can be severe at low elevation angles and frequency bands above 10 GHz. Multipath effects can be observed for small percentages of time at very low elevation angles ($\leq 4^{o}$) due to large scale scintillation effects resulting in signal attenuation greater than 10 dB.

## 2.3 Local Effects

In addition to the ionospheric and the tropospheric attenuation effects, radio waves suffer from energy loss due to complex and varying propagation environments on the terrain. An earth station is surrounded by different obstacles (buildings, trees, vegetation etc) of varying heights, dimensions and of different densities. These obstructions cause different multipath propagation phenomena: diffraction due to bending of the signal around edges of buildings, dispersion or scattering by the interaction with objects of uneven shapes or surfaces, specular reflection of the waves from objects with dimensions greater than the wavelength of the radio waves, absorption through foliage etc. In addition, the movement of mobile station on earth over short distances on the order of few wavelengths or over short time durations on the order of few seconds results in rapid changes in the signal strength due to changes in phases (Doppler Effect). All these effects result in loss of the signal energy and degrade the performance and reliability of LMS communications links. A detailed discussion about local effects on LMS communication links can be found in (Goldhirsh & Vogel, 1998; Blaunstein & Christodoulou, 2007).

## 3. Probability Distribution Functions for Different Types of Fading

The performance of satellite-earth communication links depends on the operating frequency, geographical location, climate, elevation angle to the satellite etc. The link reliability of a satellite-based communication system decreases with the increase in operating frequency and at low elevation angles. In addition, the random and unpredictable nature of propagation environments increases complexity and uncertainty in the characterization of transmission impairments on the LMS communication links. Therefore, it is suitable to describe these phenomena in stochastic manner in order to assess the performance of LMS communication systems over fading channels. Various precise and elegant statistical distributions exist in the literature that can be used to characterize fading effects in different propagation environments (Simon & Alouini, 2000; Corraza, 2007). In general signal fading is decomposed as large scale path loss, a medium slowly varying component following lognormal distribution and small scale fading in terms of Rayleigh or Rice distributions depending on the existence of the LOS path between the transmitter and the receiver. In this section, we give a brief overview of standard statistical distributions used to model different fading effects on the LMS communication links.

### 3.1 Rayleigh Distribution

In case of heavily built-up areas (Urban Environments) the transmitted signal arrives at the receiver through different multipath propagation mechanisms (section 2.3). The resultant signal at the receiver is taken as the summation of diffuse multipath components characterized by time-varying attenuations, different delays and phase shifts. When the number of paths increase the sum approaches to complex Gaussian random variable having independent real and imaginary parts with zero mean and equal variance. The amplitude of the composite signal follows Rayleigh distribution and the phases of individual components are uniformly distributed in the interval 0 to $2\pi$ . The received signal (real part) can be written as:

$$R_{Ray} = \sum_{i=1}^{n} a_i(t)\cos(\omega_c t + \theta_i(t)) \qquad i = 0,1,2,\ldots,n \tag{1}$$

where $a_i(t)$ is the amplitude, $\theta_i(t)$ is the phase of the $i^{th}$ multipath component and $\omega_c$ represents the angular frequency of the carrier. The corresponding probability density function (pdf) of the received signal's envelope is expressed in the following mathematical form:

$$P_{Ray}(r) = \frac{r}{\sigma^2}\exp(-r^2/2\sigma^2) \qquad r \geq 0 \tag{2}$$

where $\sigma$ denotes the standard deviation and '$r$' represents envelop of the received signal.

### 3.2 Rician Distribution

In propagation scenarios when a LOS component is present between the transmitter and the receiver, the signal arriving at the receiver is expressed as the sum of one dominant vector and large number of independently fading uncorrelated multipath components with amplitudes of the order of magnitude and phases uniformly distributed in the interval (0,2$\pi$). The received signal is characterized by Rice distribution and is given as follows:

$$R_{Rice} = C + \sum_{i=1}^{n} a_i(t)\cos(\omega_c t + \theta_i(t)) \qquad i = 0,1,2,\dots,n \tag{3}$$

where constant 'C' represents the magnitude of the LOS signal between the transmitter and the receiver. Other parameters are the same as described for Rayleigh distribution. The pdf of the envelope of the received signal is illustrated in the following mathematical form:

$$P_{Rice}(r) = \frac{r}{\sigma^2}\exp\left[\frac{-(r^2+C^2)}{2\sigma^2}\right] I_0\left(rC/\sigma^2\right) \tag{4}$$

where $I_0$ represents the modified Bessel function of first kind and zero order, and $C^2/2$ is the mean power of the LOS component. If there is no direct path between the transmitter and the receiver (i.e., C = 0) the above equation reduces to Rayleigh distribution. The ratio of the average specular power (direct path) to the average fading power (multipath) over specular paths is known as Rician factor ($a^2/2\sigma^2$) and is expressed in dBs.

### 3.3 Log-Normal Distribution

In addition to signal power loss due to hindrance of objects of large dimensions (buildings, hills, etc), vegetation and foliage is another significant factor that cause scattering and absorption of radio waves by trees with irregular pattern of branches and leaves with different densities. As a result the power of the received signal varies about the mean power predicted by the path loss. This type of variation in the received signal power is called shadowing and is usually formulated as log-normally distributed over the ensemble of typical locals. Shadowing creates holes in coverage areas and results in poor coverage and poor carrier-to-interference ratio (CIR) in different places. The pdf of the received signal's envelope affected by shadowing follows lognormal distribution that can be written in the following mathematical form:

$$P_{\log normal}(r) = \frac{1}{\sqrt{2\pi}\sigma r}\exp\left[-\frac{1}{2}\left(\frac{(\ln r - \mu)^2}{2\sigma^2}\right)\right] \qquad r \geq 0 \tag{5}$$

where $\mu$ and $\sigma$ are mean and standard deviation of the shadowed component of the received signal, respectively.

### 3.4 Nakagami Distribution

As discussed in (Saunders & Zavala, 2007), the random fluctuations in the radio signal propagating through the LMS communication channels can be categorized into two types of fading: multipath fading and shadowing. The composite shadow fading (line-of-sight and multiplicative shadowing) can be modelled by lognormal distribution. The application of lognormal distribution to characterize shadowing effects results in complicated expressions for the first and second order statistics and also the performance evaluation of communication systems such as interference analysis and bit error rates become difficult. An alternative to lognormal distribution is Nakagami distribution which can produce simple statistical models with the same performance. The pdf of the received signal envelope following Nakagami distribution is given by,

$$P_r(r) = \frac{2}{\Gamma(m)}\left(\frac{m}{2\sigma^2}\right)^m r^{2m-1} \exp\left(\frac{-mr^2}{2\sigma^2}\right) \qquad r \geq 0 \qquad (6)$$

where $\Gamma(.)$ represents the Gamma function, $2\sigma^2 = E(r^2)$ is the average power of the LOS component and $m \geq \frac{1}{2}$ is the Nakagami-m parameter which varies between $\frac{1}{2}$ to $\infty$. When $m$ increases the number of Gaussian random variables contributions increases and the probability of deep fades in the corresponding pdf function decreases. Non-zero finite small and large values of $m$ correspond to urban and open areas, respectively. The intermediate values of $m$ correspond to rural and suburban areas.

### 3.5 Suzuki Distribution

The Suzuki process is characterized as the product of Rayleigh distributed process and lognormal process (Pätzold et al., 1998). Consider a Rayleigh distributed random variable $\alpha$ with pdf $P_\alpha(r)$ and another random variable $\beta$ following lognormal distribution with pdf $P_\beta(r)$. Let $\xi$ be a random variable defined by the product of these two independent variables ($\xi = \alpha.\beta$). The pdf $P_\xi(r)$ of $\xi$ can be written as follows:

$$P_\xi(r) = \frac{r}{\sqrt{2\pi\sigma_0^2}\sigma_\mu} \int_0^\infty \frac{1}{r^3}.\exp\left(\frac{-r^2}{2z^2\sigma_0^2}\right).\exp\left(-\frac{1}{2}\left(\frac{\ln r - m_\mu}{\sigma_\mu}\right)^2\right) \qquad r \geq 0 \qquad (7)$$

This type of distribution can be used to represent propagation scenarios when LOS component is absent in the received signal.

## 4. Statistical Channel Models for Land-Mobile-Satellite Communications

The influence of radio channel is a critical issue for the design, real-time operation and performance assessment of LMS communication systems providing voice, text and multimedia services operating at radio frequencies ranging from 100 MHz to 100 GHz and optical frequencies. Thus, a good and accurate understanding of radio propagation channel is of paramount significance in the design and implementation of satellite-based communication systems.

The radio propagation channels can be developed using different approaches, e.g., physical or deterministic techniques based on measured impulse responses and ray-tracing algorithms which are complex and time consuming and statistical approach in which input data and computational efforts are simple. The modelling of propagation effects on the LMS communication links becomes highly complex and unpredictable owing to diverse nature of radio propagation paths. Consequently statistical methods and analysis are generally the most favourable approaches for the characterization of transmission impairments and modelling of the LMS communication links.

The available statistical models for narrowband LMS channels can be characterized into two categories: single state and multi-state models (Abdi et al., 2003). The single state models are described by single statistical distributions and are valid for fixed satellite scenarios where the channel statistics remain constant over the areas of interest. The multi-state or mixture models are used to demonstrate non-stationary conditions where channel statistics vary significantly over large areas for particular time intervals in nonuniform environments. In this section, channel models developed for satellites based on statistical methods are discussed.

### 4.1 Single-State Models

**Loo Model:** The Loo model is one of the most primitive statistical LMS channel model with applications for rural environments specifically with shadowing due to roadside trees. In this model the shadowing attenuation affecting the LOS signal due to foliage is characterized by log-normal pdf and the diffuse multipath components are described by Rayleigh pdf. The model illustrates the statistics of the channel in terms of probability density and cumulative distribution functions under the assumption that foliage not only attenuates but also scatters radio waves as well. The resulting complex signal envelope is the sum of correlated lognormal and Rayleigh processes. The pdf of the received signal envelope is given by (Loo, 1985; Loo & Butterworth, 1998).

$$P(r) = \begin{cases} \frac{1}{r\sqrt{2\pi d_0}} \exp\left[-\frac{(\ln r-\mu)^2}{2d_0}\right] & \text{for } r >> \sqrt{b_0} \\ \frac{r}{b_0}\exp\left(-\frac{r^2}{2b_0}\right) & \text{for } r << \sqrt{b_0} \end{cases} \tag{8}$$

where μ and $\sqrt{d_0}$ are the mean and standard deviation, respectively. The parameter $b_0$ denotes the average scattered power due to multipath effects. Note that if attenuation due to shadowing (lognormal distribution) is kept constant then the pdf in (8) simply yields in Rician distribution. This model has been verified experimentally by conducting measurements in rural areas with elevation angles up to $30^\circ$ (Loo et al., 1998).

**Corraza-Vatalaro Model:** In this model, a combination of Rice and lognormal distribution is used to model effects of shadowing on both the LOS and diffuse components (Corazza & Vatalaro, 1994) The model is suitable for non-geostationary satellite channels such as medium-earth orbit (MEO) and low-earth orbit (LEO) channels and can be applied to different environments (e.g., urban, suburban, rural) by simply adjusting the model parameters. The pdf of the received signal envelop can be written as:

$$P_{(r)}(r) = \int_0^{\infty} p(r/S)\, p_S(s) dS \tag{9}$$

where $p(r/S)$ denotes conditional pdf following Rice distribution conditioned on shadowing $S$ (Corazza et al., 1994)

$$P(r/S) = 2(K+1)\frac{r}{S^2}\exp\left[-(K+1)\frac{r^2}{S^2} - K\right] I_0\left(2\frac{r}{S}\sqrt{K(K+1)}\right) \qquad r \geq 0 \tag{10}$$

where $K$ is Rician factor (section 3.2) and $I_0$ is zero order modified Bessel function of first kind. The pdf of lognormal of shadowing S, is given by:

$$P_S(S) = \frac{1}{\sqrt{2\pi} h\sigma S}\exp\left\{-\frac{1}{2}\left(\frac{\ln S - \mu}{h\sigma}\right)^2\right\} \qquad S \geq 0 \tag{11}$$

where $h = \ln(10)/20$, $\mu$ and $(h\sigma)^2$ are mean and variance of the associated normal variance, respectively. The received signal envelop can be interpreted as the product of two independent processes (lognormal and Rice) with cumulative distribution function in the following form (Corraza & Vatalaro, 1994):

$$P_r(r_0) = P(r < r_0) = \int_0^{r_0}\int_0^{\infty} \frac{P_S(S)}{S} P_r\left(\frac{r}{S}\right) dr dS = 1 - E_S\left\{Q\left(\sqrt{2K}, \frac{r_0}{S}\sqrt{2(K+1)}\right)\right\} \tag{12}$$

where $E(.)$ denotes the average with respect to $S$ and $Q$ is Marcum $Q-$ function.

The model is appropriate for different propagation conditions and has been verified using experimental data with wide range of elevation angles as compared to Loo's model.

**Extended-Suzuki Model:** A statistical channel model for terrestrial communications characterized by Rayleigh and lognormal process is known as Suzuki model (Suzuki, 1977). This model is suitable for modelling random variations of the signal in different types of urban environments. An extension to this model, for frequency non-selective satellite communication channels, is presented in (Pätzold et al., 1998) by considering that for most of the time a LOS component is present in the received signal. The extended Suzuki process is the product of Rice and lognormal probability distribution functions where inphase and quadrature components of Rice process are allowed to be mutually correlated and the LOS component is frequency shifted due to Doppler shift. The pdf of the extended Suzuki process can be written as (Pätzold et al., 1998):

$$P_{\xi}(r) = \int_{-\infty}^{\infty} \frac{1}{|y|} P_{\alpha\beta}(\tfrac{r}{y}, y)dy \qquad (13)$$

where $P_{\alpha\beta}(x, y)$ denotes the joint pdf of the independent Rician and lognormal processes $\alpha(t)$ and $\beta(t)$, and $x = r/y$ where y is variable of integration. The pdfs of Rice and lognormal processes can be used in (13) to obtain the following pdf:

$$P_{\xi}(r) = \frac{r}{\sqrt{2\pi}\psi_0 \sigma} \int_{0}^{\infty} \frac{1}{y^3} \exp\left\{-\left[{((\tfrac{r}{y})^2 + p^2)}\big/{2\psi_0}\right]\right\} . I_0({rp}/{y\psi_0}) . \exp\left[-({(\ln y - m)^2}/{2\sigma 2})\right] \quad r \geq 0 \qquad (14)$$

where $\psi_0$ is the mean value of random variable $x$, $m$ and $\mu$ are the mean and standard deviation of random variable y and $p$ denotes LOS component.

The model was verified experimentally with operating frequency of 870 MHz at an elevation angle $15^o$ in rural area with 35% trees coverage. Two scenarios were selected: a lightly shadowed scenario and a heavily shadowed scenario with dense trees coverage. The cumulative distribution functions of the measurement data were in good agreement with those obtained from analytical extended Suzuki model.

**Xie-Fang Model:** This model (Xie & Fang, 2000), based on propagation scattering theory, deals with the statistical modelling of propagation characteristics in LEO and MEO satellites communication systems. In these satellites communication systems a mobile user or a satellite can move during communication sessions and as a result the received signals may fluctuate from time to time. The quality-of-service (QoS) degrades owing to random fluctuations in the received signal level caused by different propagation impairments in the LMS communication links (section 2). In order to efficiently design a satellite communication system, these propagation effects need to be explored. This channel model deals with the statistical characterization of such propagation channels.

In satellite communications operating at low elevation angles, the use of small antennas as well as movement of the receiver or the transmitter introduces the probability of path blockage and multipath scattering components which result in random fluctuations in the received signal causing various fading phenomena. In this model fading is characterized as two independent random processes: short-term (small scale) fading and long-term fading. The long term fading is modelled by lognormal distribution and the small scale fading is characterized by a more general form of Rician distribution. It is assumed, based on scattering theory of electromagnetic waves, that the amplitudes and phases of the scattering components which cause small scale fading due to superposition are correlated. The total electric field is the sum of multipath signals arriving at the receiver (Beckman et al., 1987):

$$E_{tot} = E\exp(j\psi) = \sum_{i=1}^{n} A_i \exp(j\phi_i) \quad (15)$$

where $n$ denotes the number of paths, $A_i$ and $\phi_i$ represent the amplitude and phase of the $i^{th}$ path component, respectively. The pdf of the received signal envelope can be obtained as follows (Xie & Fang, 2000):

$$P_r(r) = \frac{r}{\sqrt{S_1 S_2}} \exp\left[-\frac{S_1 r^2 + S_2 \alpha^2 + S_1 \beta^2}{2 S_1 S_2}\right] \times \frac{1}{2\pi}\int_0^{2\pi} \exp\left[\frac{2 S_2 \alpha r \cos\theta + 2 S_1 \beta r \sin\theta + (S_1 - S_2) r^2 \cos^2\theta}{2 S_1 S_2}\right] d\theta \quad (16)$$

and the pdf of the received signal power envelope is given by:

$$P_p(W) = \frac{1}{2\sqrt{S_1 S_2}} \exp\left[-\frac{S_2 W + S_1 \alpha^2 + S_2 \beta^2}{2 S_1 S_2}\right] \times \frac{1}{2\pi}\int_0^{2\pi} \exp\left[\frac{2 S_2 \alpha \sqrt{W} \cos\theta + 2 S_1 \beta \sqrt{W} \sin\theta + (S_1 - S_2) W \cos^2\theta}{2 S_1 S_2}\right] d\theta \quad (17)$$

where the parameters $S_1$, $S_2$, $\alpha$, and $\beta$ denote the variances and means of the Gaussian distributed real and imaginary parts of the received signal envelope '$r$', respectively, and '$W$' represents the power of the received signal.

This statistical LMS channel model concludes that the received signal from a satellite can be expressed as the product of two independent random processes. The channel model is more general in the sense that it can provide a good fit to experimental data and better characterization of the propagation environments as compared to previously developed statistical channel models.

**Abdi Model:** This channel model (Abdi et al., 2003) is convenient for performance predictions of narrowband and wideband satellite communication systems. In this model the amplitude of the shadowed LOS signal is characterized by Nakagami distribution (section 3.4) and the multipath component of the total signal envelop is characterized by Rayleigh distribution. The advantage of this model is that it results in mathematically precise closed form expressions of the channel first order statistics such as signal envelop pdf, moment generating functions of the instantaneous power and the second order channel statistics such as average fade durations and level crossing rates (Abdi et al., 2003). According to this model the low pass equivalent of the shadowed Rician signal's complex envelope can as:

$$R(t) = A(t)\exp[j\phi(t)] + Z(t)\exp[j\varphi(t)] \quad (18)$$

where $A(t)$ and $Z(t)$ are independent stationary random processes representing the amplitudes of the scattered and LOS components, respectively. The independent stationary random process, $\phi(t)$, uniformly distributed over (0, $2\pi$) denotes the phase of scattered components and $\varphi(t)$ is the deterministic phase of LOS component. The pdf of the received signal envelop for the first order statistics of the model can be written as (Abdi et al., 2003):

$$P_r(r) = \left(\frac{2b_0 m}{2b_0 m + \Omega}\right)^m \cdot \frac{r}{b_0} \exp\left(-\frac{r^2}{2b_0}\right) \bullet {}_1F_1\left(m,\ 1,\ \frac{\Omega r^2}{2b_0(2b_0 m + \Omega)}\right) \quad r \geq 0 \quad (19)$$

where $2b_0$ is the average power of the multipath component, $\Omega$ is the average power of the LOS component and ${}_1F_1(.)$ is the confluent hypergeometric function.

The channel model's first order and second order statistics compared with different available data sets, demonstrate the appropriateness of the model in characterizing various channel conditions over satellite communication links. This model illustrates similar agreements with the experimental data as the Loo's model and is suitable for the numerical and analytical performance predictions of narrowband and wideband LMS communication systems with different types of encoded/decoded modulations.

### 4.2 Multi-state Models

In the case of nonstationary conditions when terminals (either satellite or mobile terminal) move in a large area of a nonuniform environment, the received signal statistics may change significantly over the observation interval. Therefore, propagation characteristics of such environments are appropriately characterized by the so-called multi-state models.

Markov models are very popular because they are computationally efficient, analytically tractable with well established theory and have been successfully applied to characterize fading channels, to evaluate capacity of fading channels and in the design of optimum error correcting coding techniques (Tranter et al., 2003). Markov models are characterized in terms of state probability and state probability transition matrices. In multi-state channel models, each state is characterized by an underlying Markov process in terms of one of the single state models discussed in the previous section.

**Lutz Model:** Lutz's model (Lutz et al., 1991) is two-state (good state and bad state) statistical model based on data obtained from measurement campaigns in different parts of Europe at elevation angles between 13° to 43° and is appropriate for the characterization of radio wave propagation in urban and suburban areas. The good state represents LOS condition in which the received signal follows Rician distribution with Rice factor $K$ which depends on the operating frequency and the satellite elevation angle. The bad state models the signal amplitude to be Rayleigh distributed with mean power $S_0 = \sigma^2$ which fluctuates with time. Another important parameter of this model is time share of shadowing '$A$'. Therefore, pdf of the received signal power can be written as follows (Lutz et al., 1991):

$$p(S) = (1-A).p_{Rice}(S) + A\int_0^\infty p_{Ray}(S/S_0)p_{LN}(S_0)dS_0 \tag{20}$$

The values of the parameters A, K, means, variances and the associated probabilities have been derived from measured data for different satellite elevations, antennas and environments using curve fitting procedures. The details can be found in (Lutz et al., 1991). Transitions between two states are described by first order Markov chain where transition from one state to the next depends only on the current state. For two-state Lutz' model, the probabilities $P_{ij}$ ($i, j = g, b$) represent transitions from sate $i$ to state $j$ according to good or bad state as shown in Fig. 2.

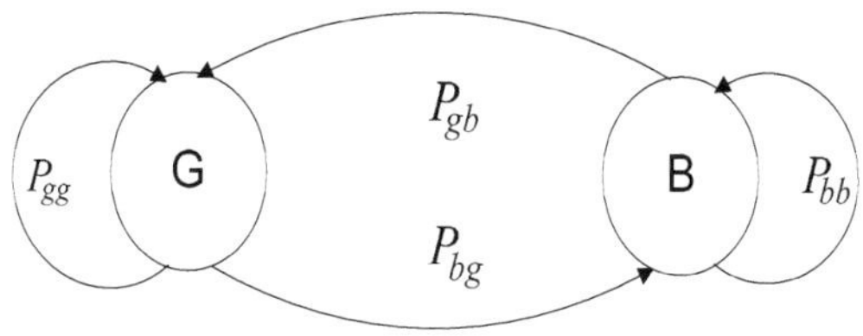


Fig. 2. Lutz's Two-state LMS channel model.

The transition probabilities can be determined in terms of the average distances $D_g$ and $D_b$ in meters over which the system remains in the good and bad states, respectively.

$$P_{gb} = {}^{vR}\!/_{D_g} \qquad\qquad P_{bg} = {}^{vR}\!/_{D_b} \tag{21}$$

where $v$ is the mobile speed in meters per second, R is the transmission data rate in bits per second. As the sum of probabilities in any state is equal to unity, thus $P_{gg} = 1 - P_{gb}$ and $P_{bb} = 1 - P_{bg}$. The time share of shadowing can be obtained as:

$$A = \frac{D_b}{D_b + D_g} \tag{22}$$

The parameter $A$ in this model is independent of data rate and mobile speed. For different channel models, the time share of shadowing is obtained according to available propagation conditions and parameters. For example in (Saunders & Evans, 1996) time share of shadowing is calculated by considering buildings height distributions and street width etc.

**Three-State Model:** This statistical channel model (Karasawa et al., 1997), based on three states, namely clear or LOS state, the shadowing state and the blocked state, provides the analysis of availability improvement in non-geostationary LMS communication systems. The clear state is characterized by Rice distribution, the shadowing state is described by

Loo's pdf and the blocked state is illustrated by Rayleigh fading as shown in Fig. 3(a), where $a_1$ denotes the LOS component, $a_2$ represents shadowing effects caused by trees and $a_3$ represents blockage (perfect shadowing). Similarly, multipath contributions in the form of coherently reflected waves from the ground are denoted by $b_1$ and incoherently scattered components from the land obstructions are represented by $b_2$. The pdf of the received signal envelop is weighted linear combination of these distributions:

$$P_R(r) = MP_{Rice}(r) + LP_{Loo}(r) + NP_{Rayleigh}(r) \tag{23}$$

where $M$, $L$, and $N$ are the time share of shadowing of Rice, Loo and Rayleigh distributions, respectively. The distribution parameters for the model were found by means of the data obtained from measurements using "INMARSAT" satellite and other available data sets. The model was validated by comparing the theoretical cumulative distributions with those obtained from measurement data. The state transitions characteristics of the model were obtained using Markov model as shown in Fig. 3(b). The state occurrence probability functions $P_A$, $P_B$ and $P_c$ (where $P_A + P_B + P_C = 1$) can be computed as follows (Karasawa et al., 1997):

$$P_A = (90 - \theta)^2 / a \tag{24}$$

where $\theta$ is the elevation angle of satellite ($10^{\circ} \le \theta \le 90^{\circ}$) and 'a' is a constant with values:

$$a = \begin{cases} 7.0 \times 10^{3} & \text{for urban areas} \\ 1.66 \times 10^{4} & \text{for suburban areas} \end{cases}$$

$$P_B = \begin{cases} P_C/4 & \text{for urban areas} \\ 4P_C & \text{for suburban areas} \end{cases} \tag{25}$$

In order to characterize the state duration statistics such as the average distances or time spans during which a particular state tends to persist, a model capable of providing time-variant features is essential. A Markov process suitable for this purpose is expressed as three-state model as shown in Fig. 3(b) (Karasawa et al., 1997). In this model short-term fluctuations in the received signal are represented by specific pdfs within the states and long-term fading is described by the transitions between the states. This model is also suitable for the performance assessment of satellite diversity.

A significant aspect of the LMS systems is that a single satellite is not adequate for achieving the desired coverage reliability with a high signal quality. Thus, it is desirable that different satellite constellations should be employed which can improve the system availability and signal quality by means of satellite diversity. If a link with one of the satellites is interrupted by shadowing, an alternative satellite should be available to help reduce the outage probability. This channel model also provides analysis for the improvement of the signal quality and service availability by means of satellite diversity where at least two satellites in LEO/MEO orbit, illuminate the coverage area simultaneously in urban and suburban environments.

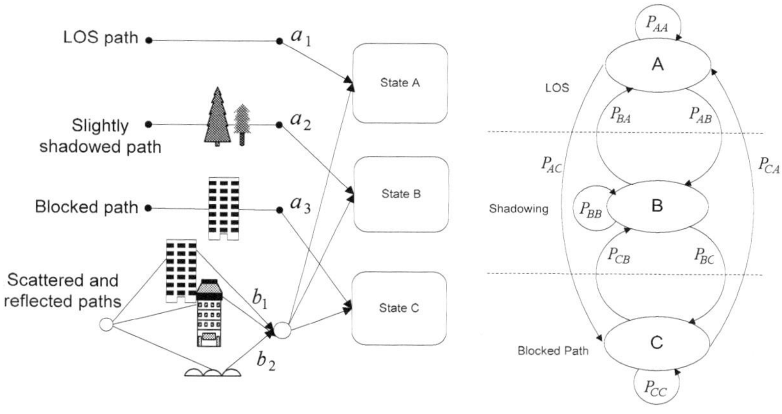


Fig. 3. Three-sate LMS channel model (a) Propagation impairments (b) Markov model.

**Five-State Model:** This channel model is based on Markov modelling approach in which two-state and three-state models are extended to five-state model under different time share of shadowing (Ming et al., 2008). The model is basically a composition of Gilbert-Elliot channel model and the three-state Markov channel model in which shadowing effects are split into three states: 'good' state represents low shadowing, 'not good not bad' state characterizes moderate shadowing and 'bad state' describes heavy or complete shadowing as shown in Fig. 4 (Ming et al., 2008). The 'good' state has two sub-states: clear LOS without shadowing and LOS state with low shadowing. Similarly, the 'bad' state has two sub-states: heavily shadowed areas or completely shadowed or blocked areas. A state transition can occur when the receiver is in low or high shadowing areas for a period of time. The transitions can take place from low and high shadowing conditions to moderate shadowing conditions but cannot occur directly between low and high shadowing environments.

For different shadowing effects, the statistical signal level characteristics in terms of the pdf are described as: low shadowing follows Rice distribution, moderate shadowing is represented by Loo's pdf and high shadowing conditions are described by Rayleigh-lognormal distribution. The pdf of the received signal power is a weighted linear combination of these distributions:

$$P(s) = X_1 P_{Rice1}(s) + X_2 P_{Rice2}(s) + X_3 P_{Loo}(s) + X_4 P_{Ray\_L1}(s) + X_5 P_{Ray\_L2}(s) \quad (26)$$

where $X_i \ (i = 1,...,5)$ are time share of shadowing of the states $S_i \ (i = 1,...,5)$, respectively.

The state probability and state transition probability matrices are determined using the time series of the measured data. The channel model has been validated using available measured data sets and different statistical parameters are obtained using curve fitting procedures. The channel statistics like the cumulative distribution function, the level crossing rate, the average fade duration, and the bit error rate are computed which show a good agreement with the statistics of the data obtained from measurements. The channel model is appropriate for urban and suburban areas.

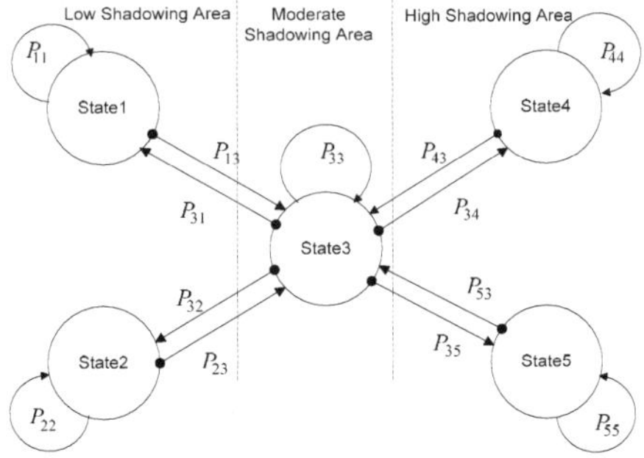


Fig. 4. Five-state Markov channel model for LMS communications.

**Modelling Frequency Selective LMS Channel:** The LMS propagation channel effects depend on the propagation impairments (section 2), geographical location, elevation angles and operating frequency band. Extensive measurements are needed for the characterization of LMS fading caused by different propagation impairments. When components of a signal travelling through different paths arrive at the receiver with delays significantly larger as compared to the bit or symbol duration, the signal will undergo significant amount of distortion across the information bandwidth, it results in frequency selective fading or wideband fading (e.g., in the case of broadband services or spread spectrum). The impulse response of a wideband channel model (also known as tapped-delay line model) under wide sense stationary uncorrelated scattering (WSSUS) assumption can be written as:

$$h(t,\tau)=\sum_{i=1}^{N} a_i(t)\delta[t-\tau_i(t)]\exp[j(2\pi f_{d,i}(t)+\theta_i(t))] \tag{27}$$

where $a_i(t)$, $\tau_i(t)$, $f_{d,i}$ and $\theta_i(t)$ are the amplitude, delay, Doppler shift and phase of the $i^{th}$ component of the received signal, respectively, and $\delta(t)$ denotes the Dirac delta function.

A tapped-delay line model that describes the wideband characteristics of LMS communication link has been given in (Jahn, 2001). The parameters for this model are extracted using extensive measurement data at L-band for different applications, scenarios and environments. In order to adopt the channel for LMS communications, the channel impulse response is divided into three components: the direct path, near echoes and far echoes as shown in Fig. 5 (Jahn, 2001). The delays $\tau_i$ $(i=1,2,\ldots,N)$ of the taps are taken with respect to the delay of the direct path. The power of all taps is normalized to the power of the direct path. The amplitude distributions of the echoes follow Rice or Rayleigh distribution (section 3) depending on the presence of LOS or non-LOS situations, respectively. The number $N^n$ of near echoes in the locality of the receiver follows Poisson distribution with parameter $\lambda$ (*i.e.*, $f_{Poisson}(N) = (\lambda^N/_{N!})e^{-\lambda}$) and the corresponding delays $\tau_i$ $(i = 1,2,...,N)$ characterizing near echoes follow exponential distribution with parameter b $\{i.e., f_{\exp}(\tau_i^n) = (e^{-\tau_i^n/b}/b)\}$. The power of the taps decay exponentially. The far echoes $N^f = N - N^n - 1$, which are few in numbers are characterized by Poisson distribution. The amplitude distributions of the far echoes are described by Rayleigh distribution. The description of different regions of the wideband LMS channel impulse response can be found in (Jahn, 2001). Another physical-statistical channel model that deals with the frequency selectivity of LMS channels is found in (Parks et al., 1996). This model consists of two cascaded processes. The first one deals with propagation effects from satellite to earth and the second process illustrates the terrestrial propagation impairments.

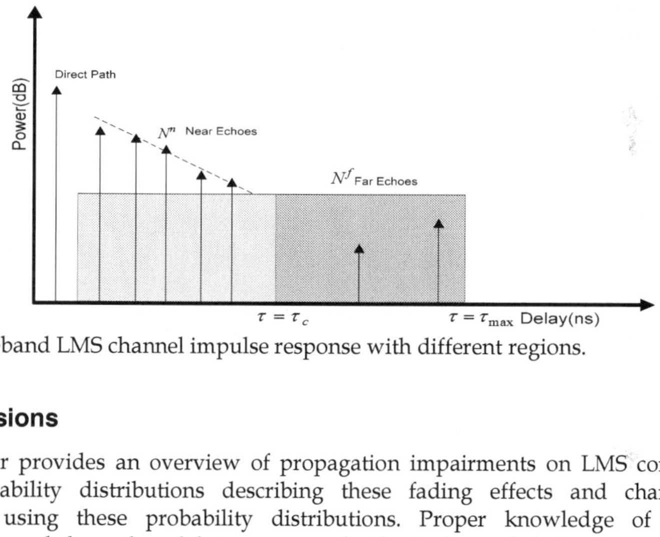


Fig. 5. Wideband LMS channel impulse response with different regions.

## 5. Conclusions

This chapter provides an overview of propagation impairments on LMS communication links, probability distributions describing these fading effects and channel models developed using these probability distributions. Proper knowledge of propagation impairments and channel models is necessary for the design and performance assessment of advanced transceiver techniques employed to establish reliable communication links in LMS communication systems. The main focus lies on highlighting which are the effects and the relevant propagation models need to be considered for LMS communication links in order to accurately estimate the propagation impairments. The performance of LMS communication systems depend on different factors including operating frequency, elevation angles, geographic location, climate etc. Different approaches can be used to find the effects of these factors on LMS communication links such as physical-statistical channel models which are more accurate but require long simulation times and are complex. On the other hand statistical methods are simple and require less computational efforts. In addition, due to diverse nature of propagation environments, it is appropriate to use stochastic approaches for the performance assessment of LMS communication links.

## 6. References

Abdi, A., Lau, C. W., Alouini, M., & Kaveh, M. (2003). A New Simple Model for Land Mobile Satellite Channels: First- and Second-Order Statistics. *IEEE Trans. Wireless Comm.*, 2(3), 519-528.

Blaunstein, N., & Christodoulou, C. G. (2007). Radio Propagation and Adaptive Antennas for Wireless Communication Links. John Wiley & Sons, Inc., Hoboken, New Jersey.

Corraza, G. E., & Vatalaro, F. (1994). A Statistical Channel Model for Land Mobile Satellite Channels and Its Application to Nongeostationary Orbit Systems. *IEEE Trans. Vehicular Technology*, 43(3), 738-742.

Corazza, G. E. (2007). Digital Satellite Communications. Springer Science plus Business Media, LLC, New York.

Goldhirsh, J., & Vogel, W. J. (1998). Handbook of Propagation Effects for Vehicular and Personal Mobile Satellite Systems, Over of Experimental and Modelling Results.

Ippolito, J. L., Jr. (2008). Satellite Communications Systems Engineering, Atmospheric Effects, Satellite Link Design and System Performance. John Wiley & Sons Ltd.

ITU. (2002). Handbook on Satellite Communications, Wiley-Interscience, 3rd Edition.

ITU-R. (2007). Ionospheric Propagation data and Prediction Methods Required for the Design of Satellite Services and Syatems. ITU-R P. 618-9.

ITU-R. (2009a). Ionospheric Propagation data and Prediction Methods Required for the Design of Satellite Services and Syatems. ITU-R P. 531-10.

ITU-R. (2009b). *Attenuation by Atmospheric Gases.* ITU-R P. 676-8.

Jahn, A. (2001). Propagation Considerations and Fading Countermeasures for Mobile Multimedia Services. *Int. Journal of Satellite Communications*, 19(3), 223-250.

Karasawa, Y., Kimura, K. & Minamisono, K. (1997). Analysis of Availability Improvement in LMSS by Means of Satellite DiversityBased on Three-State Propagation Channel Model. *IEEE Trans. Vehicular Technology*, 46(4), 1047-1056.

Loo, C. (1985). A Statistical Model for a Land Mobile Satellite Links. *IEEE Trans. Vehicular Technology*, Vol. 34, no. 3, pp. 122-127.

Loo, C., & Butterworth, J. S. (1998). Lan Mobile Satellite Measurements and Modelling. *IEEE Proc.*, 86(7), 1442-14462.

Lutz, E., Cygan, D., Dippold, M., Donalsky, F., & Papke, W. (1991). The Land Mobile Satellite Communication Channel- Rceording, Statistics and Channel Model. *IEEE Transactions on Vechicular Technology*, 40(2), 375-386.

Ming, H., Dongya, Y., Yanni, C., Jie, X., Dong, Y., Jie, C. & Anxian, L. (2008). A New Five-State Markov Model for Land Mobile Satellite Channels. *Int. Symposium, Antennas, Propagation and EM Theory*, 1512-1515.

Parks, M. A. N., Saunders, S. R., Evans, B. G. (1996). A wideband channel model applicable to Mobile Satellite Systems at L-band and S-band. *IEE Colloquim on Propagation Aspects of Future Mobile Systems*, 12, 1-6.

Pätzold, M., Killat, U., & Laue, F. (1998). An Extended Suzuki Model for Land Mobile Satellite Channels and Its Statistical Properties. *IEEE Trans. Vehicular Technology*, 47(2), 617-630.

Ratcliffe, J. A. (1973). Introduction in Physics of Ionosphere and Magnetosphere. Academic Press, New York.Blaunstein, N. (1995). Diffusion spreading of middle-latitude ionospheric plasma irregularities. Annales Geophasice, 13, 617-626.

Roddy, D. (2006). Satellite Communications, The McGraw Hill Companies, Inc, Fourth Edition.

Saunders, S. R., & Evans, B. G. (1996). Physical Model for Shadowing Probability for Land Mobile Satellite Propagation. *IEE Electronic Letters,* 32(17), 1248-1249.

Saunders, S. R., & Zavala, A. A. (2007). Antennas and Propagation for Wireless Communication Systems. J. Wiley & Sons, New York.

Simon, M., & Alouini, M. (2000). Digital Communication over Fading Channels: A Unified Approach to Performance Analysis. John Wileys & Sons, Inc, ISBN 0-471-31779-9.

Suzuki, H. (1977). A Statistical Model for Urban Radio Propagation. *IEEE Trans. Comm.,* 25(7), 673-680.

Tranter, W., Shanmugan, K., Rappaport, T., and Kosbar, K. (2004). Principles of Communication Systems Simulation with Wireless Applications. Pearson Education, Inc.

Xie, Y., & Fang, Y. (2000). A General Statistical Channel Model for Mobile Satelllite Systems. *IEEE Trans. Vehicular Technology,* 49(3), 744-752.

# 8

# Combining satellite and geospatial technologies for exploring rainstorm hazard over Mediterranean Central Area[1]

Nazzareno Diodato
*MetEROBS – Met European Research Observatory, GEWEX-CEOP Network, World Climate Research Programme, via Monte Pino snc, 82100 Benevento Italy*
*e-mail:* scodalabdiodato@gmail.com

## 1. Introduction

Modelling is not an alternative to observation but,
under certain circumstances, can be a powerful tool
in understanding observations and in developing and testing theory.

Mulligan M., and Wainwright J., 2004. Modelling and Model Building.
In: *Environmental Modelling*, Wiley, p. 2

Multiple Damaging Hydrological Events (MDHE, Petrucci & Polemio, 2003) are rapidly developing into deluges, flashfloods, floods, mudflows, accelerated erosion, and landslides (Kar & Hodgson, 2008; Younis et al., 2008), with tragic consequences on the viable habitat for humankind and ecosystems, and agriculture (Clarke & Rendell, 2005). In this context, MDHE could have more impact than the frequently cited hazard of global warming due to intensification of the hydrological cycle and the concentration of rainfall in sporadic- but more intense events (Allen & Ingram, 2002).
There is, in fact, evidence available from different parts of the world of a rising trend of natural disasters since 1993 (Sivakumar, 2005), included Medietarrean basin (Diodato & Bellocchi, 2010). For Southern Italy, in particular, the catstrophic events of Sarno in 1998 (Mazzarella & Diodato, 2002), with the more recent devastating deluges in Naples in 2001, 2003, 2004, 2006, and in southeastern of Sicily in 2009, were caused by extremes rain of 100-400 mm fallen in few hours over little areas. Therefore, global vision in remote sensing coverage and surveillance loop are important, since we do not know where an event might take place (Bacon et al., 2008). However, estimating rainfall from satellite imagery is rather complex (Ymeti, 2007), and due to limited success of deterministic rainstorm impact modelling techniques (Heneker et al., 2001).

[1] This chapter is a revision of the paper appeared on **The Open Environmental Engineering Journal, 2009, 2, 97-103.**

Also, while the literature on general model theory is vast, the aims of modellers usually consist of improving our understanding of a phenomenon and its process, and ultimately predicting the response of the landscape (Kelly et al., 2004; Diodato, 2005). In this context, data assimilation models, that combine ground measurements with remote sensing of rain-data, need to accommodate many specific aspects of the observations and models (Pan et al., 2008).

While surface data will always remain important cornerstones of reference for monitoring and modelling geospatial data, ground data suffers especially due to mutability of their patterns, even as the modeller is compelled to adapt frequently to maintain sufficient condition of temporal and spatial homogeneity, with time-series that are difficult to update.

The advent of Geographical Information Science (GISsci) can confer an innovative role on hazard modelling development, satellite data assimilation, model outputs uncertainty assessment, spatial data scaling, and mapping visualization. Although satellite data are regarded as indirect information and not as reliable as surface data, they can be of great help when used for scaling and assisting the modelling of a dynamic system (Su et al., 2008). However, the problem is that we have a significant increase in uncertainty when the measurements and forecasts move from the global to local scale, especially in their landscape response to change, such as downpours, heavy runoffs and flash-floods, deluges, sediment transport, and urban stormwater (after Beven, 2008). An interesting study for assessing rainfall impact was recently done by (Shoji & Kitaura, 2006) that analyzed precipitation with the parametric geostatistical approach in order to obtain information for predicting natural hazards caused by heavy rains.

In this paper, a different geostatistical criterion was applied - specifically a non-parametric approach - by transforming ground and satellite information into a continuous probabilistic response consistent with *soft* descriptions of hazards which is referred to in this study to mitigate the uncertainties in downscaling and geocomputational tracking (e.g., spatio-temporal non-homogeneity in the primary variable pattern, accuracy of the supplementary variables, errors involving sampling and hazard modelling). Processes operating to these multiple spatial and temporal scales, however, challenge the predictive capability of environmental models and integration or scaling of data from different sources (Allen et al., 2004). Non-parametric geostatistical multivariate analysis, via co-indicator coding criteria, is able to combine rainstorm indicators (which are recorded at sparse raingauge station-points) and supplementary satellite rain data (which are recorded across regular patterns). So that, the novelty of our approach lies in how methods and different tools might incorporate uncertainty associated with satellite data into a model of rainstorm hazard accounting, and to illustrate how model performs at sub-regional scale. In this way, the expansion of a Rainstorm Hazard Index (*RHI*) data from point to spatial information can be assessed with the Indicator CoKriging (ICK) technique, using Tropical Rainfall Mission Monitoring (TRMM-NASA) satellite rain data as covariate. Thus, spatial information is visualized with examples of probability estimations for different precipitation durations - ranging from 3 to 48 hours - and the quantification of hydrological hazard fields is done using probability maps of damaging rainstorms prone-areas.

## 2. Reference Data Sets and Methodology

### 2.1 Study area and problem setting

Heavy rainfall between 1951 and 2007 show Northern Mediterranean more affected than Southern one (Fig. 1a). Worldwide temporal pattern is also shown with a trend of hydrological disasters strongly increasing (Fig. 1b).

The rainstorms most perceived by the public are the large-scale damaging events; however, there is evidence that the most deadly floods are those with short lead times - flash floods - which in Mediterranean Europe have mostly a spatially limited character and can occur far away from major rivers (Lalsat et al., 2003).

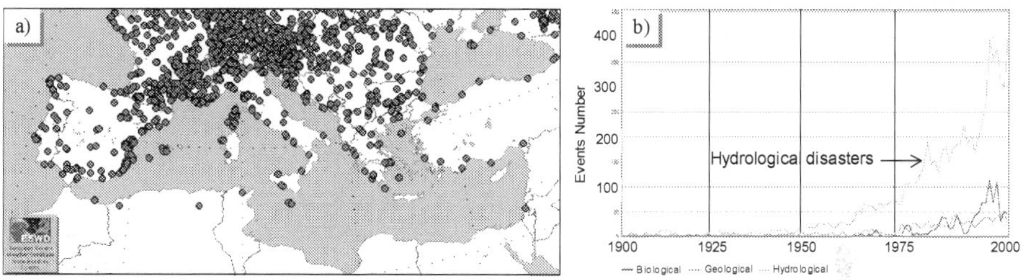


Fig. 1. (a): Occurrence of the heavy rain and hail during 1951-2007 period across Mediterranean lands (http://essl.org/cgi-bin/eswd/eswd.cgi); (**b**): Global natural disasters trends upon 1900-2005 period from EM-DAT (OFDA/CRED International Disaster Database, http://www.emdat.be).

In this respect, a test-area extending approximately 60000 km$^2$, was selected from Mediterranean central area (Fig. 2a corner). SCIA-APAT Database (www.apat.it/) was utilized for collecting rainfall ground data. However, ground data are not always updated and not all the networks uniformly coincide at all times with this database. Then satellite rain-data were also derived from the TRMM-NASA platform, algorithm 3B42 multi-satellite precipitation estimates (Huffman et al., 2007), that uses an optimal combination (HQ) of 2B-31, 2A-12, SSMI, AMSR, and AMSU precipitation estimates, with a resolution of 0.25x0.25 degree (about 25x25 km) grid boxes (http://disc.sci.gsfc.nasa.gov/).

In this way, a reference classification was constructed from RHI, driven by rainstorm events on 14 November 2004, 24 January 2003, and 4-5 May 1998. Data assimilation pattern in the region under study were obtained from 64 raingauges (Fig. 2a), and 143 supplementary satellite rain grid-data (Fig. 2b).

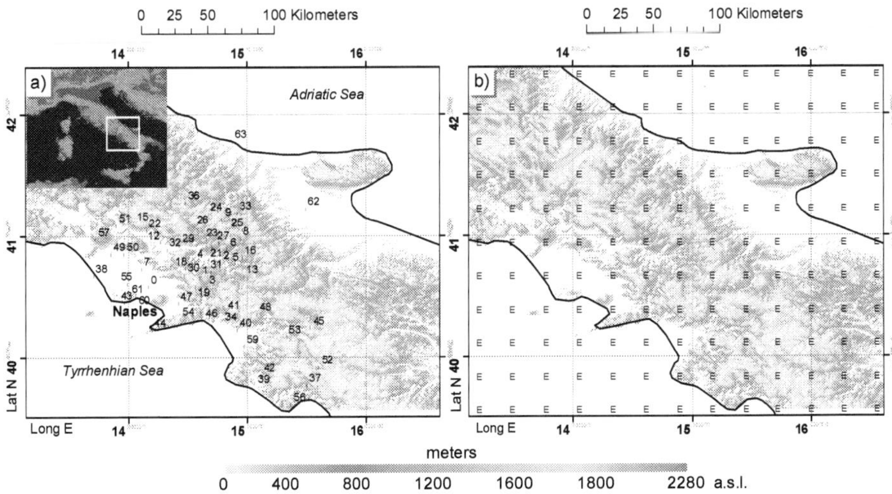


Fig. 2. (**a**): Geographical setting and data assimilation patterns from in-situ-raingauges with coded-station-points, and (**b**): TRMM-RS satellite rain data pixel centroid grid of 25 x 25 km, superimposed on elevation data of hillshade land derived from DEM (SRTM)-90 meters (http://srtm.csi.cgiar.org/).

## 2.2 Rainstorm hazard problem-solving logic process

Expert systems can be designed to model processes when carried out using the IF-THEN logic statement to impose an event contingent upon the condition (Moody & Katz, 2004). Problem-solving logic process frameworks include first an invariant spatial model recognizing critical-thresholds from the response ratios between the two following components of the landscape:

- pulsing force that disturbs the system, including current rainstorm depth, and;
- resistance force, including storm variability that occurred in the system's climate history.

As a more concrete application, we can incorporate, for each rainy step of duration $h$ at sampled location $\mathbf{s}_\alpha$, two processes into the rainstorm logic statement linking the the *RHI* to the following power equation (after Diodato, 2006; Diodato & Petrucci, 2009):

$$RHI_h(\mathrm{s}_\alpha) = \max : \left\{ \left[ \frac{1 + RSD_h}{f_{(Rclim)}} \right]^2_{h(\mathrm{s}_\alpha)} \right. \quad \forall \quad h = 1 \ldots 48 \quad \text{hours} \tag{1}$$

where $RSD_h$ is the Rain-Storm Depth (mm), that represents the **pulsing force** that disturbs the system during an event of duration $h$, and:

$$f_{(Rclim)} = \text{Med}\left(RSD_h\right) - \left(8 - \sqrt{h}\right) \cdot S_{wet} \quad (2)$$

is a function that represents the **system resistance state**, that is the intrinsic ability of the system to resist change because of its history (recent and past). Med($RSD_h$) -the threshold value - is the median of the annual maximum rainfall (mm) of duration $h$, and the term (8-$\sqrt{h}$)$S_{wet}$, is a function adjusting the threshold value with the current variation of the soil humidity. As proxies of the soil humidity, three coefficients was introduced as $S_{wet}$ equal to 0.5, 0, and 2 according to dry, humid, and very humid soil conditions before the event, respectively; these coefficients can be derived, in turn from remote sensing; the duration of rainstorm ($h$) under square root is to explain a major accommodation of the system for rainfall spanning over a longer period. Whereas, for each sampled location with a Rainstorm Hazard Index ($RHI$)$\rightarrow$0, no-rainstorm hazard occurs, and with $RHI$ value barely over 1, the probability of occurrence of a rainstorm hazard commences at 0.50.

### 2.3 Matching coding approach for decision–making under uncertainty

While the above RHI-model is utilized to arrive at conclusions at the puntual-scale, the use of geostatistics method may help to overcome the inherent difficulties in spatial scaling, when the above $RHI$ discrete data must accommodate a continuous spatial solution and data collection across sampled- and unsampled locations. Thus, the $RHI$-results are converted to a binary vector and matched to satellite rain-data under a GIS flow and supported by indicator cokriging technique (Fig. 3).
Consider the following information obtained over the study area:

- values of the random primary variable $Z$ ($RHI$), at m locations $\mathbf{s}_\alpha$, $z(\mathbf{s}_\alpha)$, $\alpha = 1,2 \ldots n_1$; and
- $y(\mathbf{s})$ TRMM satellite rain-data at supplementary grid locations $\mathbf{s}$ within the area.

Indicator approach of the primary variable requires that all data be coded as local prior probability values. Precise measurements of $z_k$ at hard data locations $\mathbf{s}_\alpha$ are then coded into a set of $K$ binary (hard) indicator data defined as:

with

$$i\left(\mathrm{s}_\alpha ; z_k\right) = \begin{cases} 1 \rightarrow z\left(\mathrm{s}_\alpha\right) > z_k \\ 0 \rightarrow \text{otherwise} \end{cases} \quad (3)$$

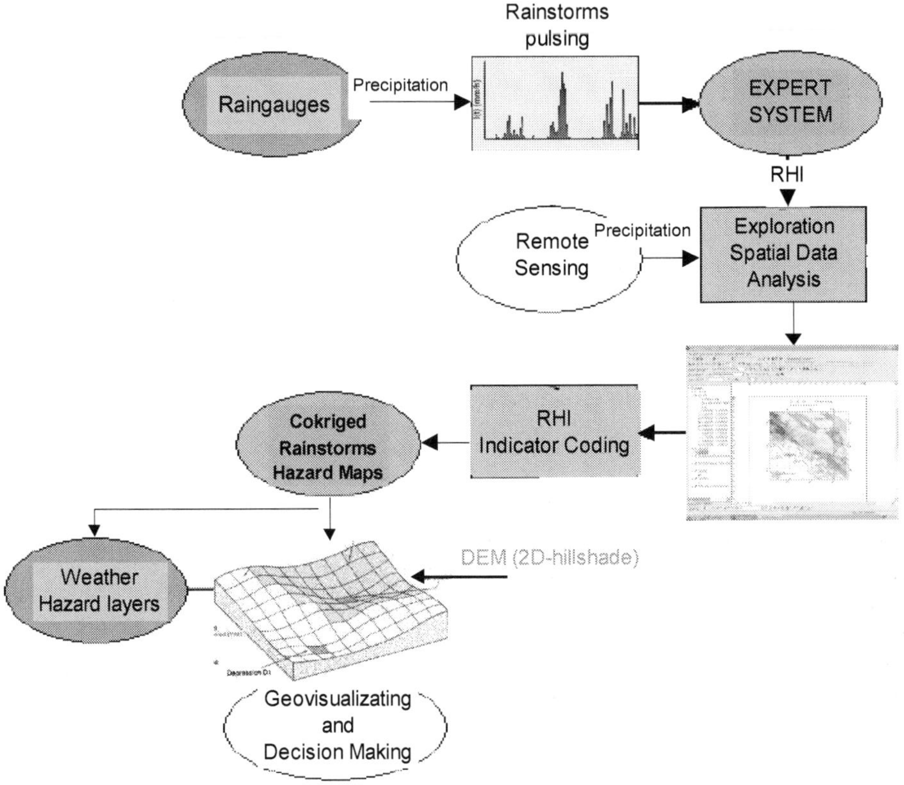


Fig. 3. Flow chart of process for estimation of rainstorm hazard mapping via GIS rules.

The $z$-values are *hard* in both senses: (1) they are directly derived from measurements of ground rainfall-data, and (2) are successively transformed into binary vector data. These measurements are often supplemented by a relatively large amount of indirect data, such as those conditioned on remotely sensed spectral response $y(\mathbf{s})$.
Each of these data provides only indirect information about the value of the variable $Z$. Using both *ground* and *satellite* information such as matching data, the approach is aimed at assessing the probability that the value of $z$ at any unsampled site $\mathbf{s}$ is greater than a given critical $z_k$ value. In this way, Indicator CoKriging (ICK) is able to take into account both the information to be processed together, and then used in the ordinary cokriging equations (Goovaerts, 1997; Johnston et al., 2001). To account for both categorial (*RHI*) and continuous (satellite data), we used standardized variables to produce composite indices compatible to indicator cokriging (Johnston et al., 2001; Hengle et al., 2004). So that, both covariance and cross-covariance functions were applied on the above standardized primary and auxiliary variables for incorporating exhaustively sampled satellite data using the indicator datum that is collocated with the location being estimated. Availability of coregionalization between indicator *ground* and *satellite* at critical values of *RHI* for each location $\mathbf{s}_o$ within the study area

allows a grid layer of: the hazard $\alpha(\mathbf{s}_\alpha)$ of declaring a location vulnerable to damage by rainstorms on the basis of the estimate $[I(\mathrm{s}; z_k)]_{\mathrm{oIOK}}^*$ when actually $Z(\mathbf{s}) > z_k = p_c$ (critical value = 1).

## 3. Results and Discussions

### 3.1 Mapping the rainstorms hazard prone-areas

Figure 4 (a,b,c) shows high-probability cokriged (p>50%) maps of areas prone to rainstorm hazards (dark grey zones), and superimposed by areas where multiple damaging hydrological events (MDHE) were observed. It was found that areas with high probability of predicted hazard matched the area actually subject to injurious phenomena, such as severe erosion, landslides, floods, and mudflows.

The severity of the damage suffered in these areas was not uniform for each rainstorm level, i.e., the damage observed depended not only on the amount of rainfall but also on the sensitivity of each specific landscape and on soil humidity (others topographical conditions were not considered in this work).

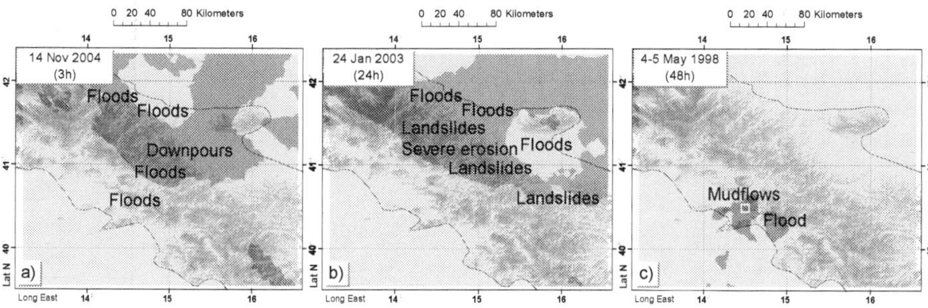


Fig. 4. High-probability (p>50%) cokriged maps of areas prone to rainstorm hazards (dark grey zone) for rainstorms of duration 3, 24, and 48 hours (a, b, and c, respectively). Note: the damaging hydrological events superimposed were almost all matched by the cokriged model.

The most extreme hydrogeomorphological processes occur over orographically complex terrain where vegetation is sparse (especially lands that are under autumn tilling, or after the rainy season), and where drainage systems may be obstructed by sediment erosion to contain large volumes of runoff. Prolonged rain usually occurs during the winter season only, when rainstorm prone-areas assume winding configurations, as for instance in Fig. 4b. On the contrary, MDHE are more spatially limited in the warm season (May-September), but more intensive, such as those which occurred on 14 November 2004 (Fig. 4a). Although the event of May 4-5, 1998 was expected to be of lower intensity, because of its long duration, the impact was catastrophic at the Sarno location (Campania region), where the several mudflows destroyed over one hundred people (Fig. 4c). This occurred because the meteorological perturbation originated from convective-clouds in larger systems, which are today more dominant in rain-producing mechanisms of high-impact over small areas (Dünkeloh & Jacobeit, 2003). This complexity is reflected in stormy pattern concentrate to the end of summer, as it is possible observe by positive anomalies of rain amount in

September occur during the recent decade (1999-2008), compared to the climatological period (1950-2000) (Fig. 5a).

The graph of Figure 5b also suggests that Southern Italy is subjected to a increasing precipitation, in term of intensity and of course of hazard, within September months spanning from 1948 to 2009. After 1996 in fact, rain rates were unusually very high, showing an abrupt change during the last decade. This can be an important advice when RHI modelling is used in an operative phase, where the climate information and human experiences becoming essential for successful completion an alert forecasts.

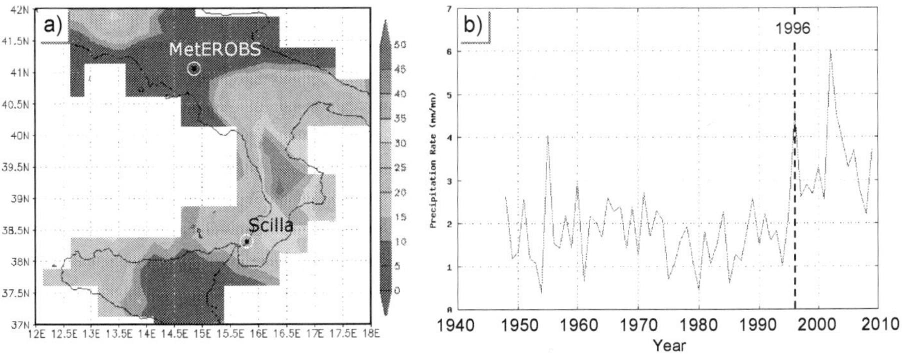


Fig. 5. (**a**): Spatial pattern of the September rain anomalies (mm) during recent decade (2000-2008) compared to climatological rain (1950-2000) upon Southern Italy (arranged from TRMM remote sensing by NASA Earth Science Data); (**b**): September rain rate evolution from 1948 to 2009 for the same region (from NOAA, ESRL Physical Sciences Division http://www.esrl.noaa.gov/ Acker & Leptoukh, 2007).

In this respect, the our results show that sub-regional rainstorm hazard modelling can provide probability maps for damaging events in Italy with a spatial variability resolution of approximately 20 km. Spatially finer estimates (e.g., at local-scale: < 10 km) can be ensured only with the availability of more accurate and detailed supplementary satellite-rain data, although, as noted by Anagnostou (2004), all satellite sensors are affected by errors originating from the non-unique, non-linear relationship of rainfall characteristics to observations and by sampling frequency and sensor resolution issues.

## 4. Conclusion

The model presented here provided the minimum but valuable set of data from which a rough tool for estimating early impacts soon after rainstorms can be derived. Damaging rainstorms collected for this retrospective experiment are documented in the category of localized events. Impact of the damage was determined by an optimum scaling critical value for predicting hazard prone-areas of three rainstorm types, although the *RHI*-model is capable of performing with data of storms of different intensities. These first results show that sub-regional rainstorm hazard modeling can provide probability maps for damaging events in Italy with a spatial variability resolution of approximately 20 km. Spatially finer estimates can be ensured only with the availability of more accurate- and detailed satellite-rain data, or during forecast stages, if real-time monitoring is implemented on an operational basis, where supplementary satellite information is then replaced by Quantitative Precipitation Forecasting.

## 5. References

Acker J.G. & Leptoukh, G. (2007). Online analysis enhances use of NASA earth science data. *EOS Trans.* AGU 88, 14–17.

Allen, M.R. & Ingram, W.J. (2002). Constraints on future changes in climate and the hydrologic cycle. *Nature* 419, 224-231.

Allen, T.R.; Walsh, S.J.; Cairns, D.M., Messina, J.P., Butler, D. & Malanson, G.P. (2004). Geostatistics and spatial analysis: Characterizing form and pattern at the Alpine treeline. In: Geographical Information Science and Mountain Geomorphology, Bishop, M.P. & J. F. Shroder Jr, (eds.). Springer-Praxis Publishing: Chichester, 190-218.

Anagnostou, E.N. (2004). Overview of overland satellite rainfall estimation for hydro-meteorological applications. *Surv. Geophys.*, 25, 511-537.

Bacon, D.P.; Ahmad, N.N.; Dunn, T.J.; Montheit, M.C. & Sarma, A. (2008). An operational multiscale system for hazards prediction, mapping, and response. *Nat. Hazardsn* 44, 317-327.

Beven, K. (2008). Models, Management and Uncertainty: The future of Hydrological Science. In: *Rivers Basins-From Hydrological Science to Water Management*, I. Tchiguirinskaia, S. Demuth, and P. Hubert (eds.), IAHS Publication No. 323 ISBN 978-1-901502-69-5, 154.

Clarke, M.L. & Rendell, H,M. (2005). Climate, extreme events and land degradation. In: *Extreme Weather Events and Public Health Responses*, W. Kirch, B. Menne, and R. Bertollini (eds.), Springer, 136-152.

Diodato, N. (2005). Geostatistical uncertainty modelling for the environmental hazard assessment during single erosive rainstorm events. Environ. Monit. Assess. 105, 25-42, 2005.

Diodato, N. (2006). Spatial uncertainty modeling of climate processes for extreme hydrogeomorphological events hazard monitoring. ASCE's *J. Environ. Eng.* 132, 1530-1538.

Diodato, N.; Petrucci, O. & Ceccarelli, M. (2009). Rainstorm hazard problem-solving spatial-time scale invariant process model designing. *Geophys. Res. Abstracts*, vol. 11, EGU2009-4107, 2009.

Diodato, N. & Bellocchi G. (2010). Storminess and Environmental Changes in the Mediterranean Central Area. Earth Interaction (in press).

Dünkeloh, A. & Jacobeit, J. (2003). Circulation dynamics of Mediterranean precipitation variability 1948-1998. *Int. J. Clim.* 23, 1843-1866.

Goovaerts, P. (1997). *Geostatistics for Natural Resources Evaluation*, Oxford: Oxford University Press.

Hengle, T.; Heuvelink, G.B.M. & Stein, A. (2004). A generic framework for spatial prediction of soil variables based on regression-kriging. *Geoderma* 120, 75–93.

Heneker, T.M.; Lambert, M.F. & Kuczera, G. (2001).A point rainfall model for risk-based design. *J. Hydrol.*, 247, 54-71.

Huffman, G.J.; Adler, R.F.; Bolvin, D.T.; Gu, G.; Nelkin, E.J.; Bowman, K.P.; Hong, Y.; Stocker, E.F. & Wolff, D.B. (2007). The TRMM Multi-satellite Precipitation Analysis: Quasi-Global, Multi-Year, Combined-Sensor Precipitation Estimates at Fine Scale", *J. Hydrometeorology* 8, 38-55.

Johnston, K.; Ver Hoef, J.M.; Krivoruchko, K. & Lucas, J. (2001). *Using ArcGis Geostatistical Analyst*, New York: ESRI.

Kar, B. & Hodgson, M.E. (2008).A GIS-Based Model to Determine Site Suitability of Emergency Evacuation Shelters. *Trans. in GIS* 12, 227–248.

Kelly, E.J.; Drake, N.A. & Barr, S.L. (2004). Spatial modelling of the terrestrial environment: the coupling of remote sensing with spatial models. In: *Spatial modelling of the terrestrial environment*, E.J. Kelly, N. A. Drake, & S.L. Barr (eds.), Wiley & Sons Ltd: Chichester, 1-6.

Llasat, M.C.; Rigo, T. & Barriendos, M. (2003). The 'Montserrat-2000' flash flood event: A comparison with the floods that have occurred in the northeast Iberian Peninsula since the 14th century. *Int. J. Climatol.* 23, 453–469.

Mazzarella, M. & Diodato, N. (2002). The alluvial events in the last two Century at Sarno, southern Italy: their classification and power-low time occurrence. *Theor. App. Climatol.* 72, 75-84.

Moody, A. & Katz, D.B., (2004). Artificial intelligence in the study of mountain landscapes", in *Geographical Information Science and Mountain Geomorphology*, M. P. Bishop, M.P., and J. F. Shroder Jr, Ed. Springer-Praxis Publishing: Chichester, 220-251, 2004.

Pan, M.; Wood, E.F.; Wojcik, R. & McCabe, M.F. (2008). Estimation of regional terrestrial water cycle using multi-sensor remote sensing observations and data assimilation. *Remote Sens. Environ.* 112, 1282-1294.

Petrucci, O. & Polemio, M. (2003). The use of historical data for the characterisation of multiple damaging hydrogeological events. *Nat. Hazards Earth Syst. Sci.*, 3, 17–30.

Shoji, T. & Kitaura, H. (2006). Statistical and geostatistical analysis of rainfall in central Japan. *Comput. Geosci.* 32, 1007-1024.

Sivakumar, M.V.K. (2005). Impact of natural disasters in agriculture, rangeland and forestry: an overvie. In: *Natural Disasters and Extreme Events in Agriculture*, M. V. K. Sivakumar, R. P. Motha, & H. P. Das (eds.), Springer-Verlag: Berlin, 1-22.

Su, H.; Wood, E.F.; Wang, H. & Pinker, R.T. (2008). *Spatial and Temporal Scaling Behavior of Surface Shortwave Downward Radiation Based on MODIS and In Situ Measurements. Geosci. Rem. Sens. Lett. IEEE* 5, 542-546.

Ymeti, I. (2007). Rainfall estimation by Remote Sensing for conceptual rainfall-runoff modelling in the Upper Blue Nile basin. M.S. Thesis, International Institute for Geo-information Science and Earth Observation, The Netherlands.

Younis, J.; Anquetin, S. & Thielen, J. (2008). The benefit of high-resolution operational weather forecasts for flash flood warning. *Hydrol. Earth Syst. Sci.* 12, 1039–1051.

# Design and Simulation of a DVB-S2-like Adaptive Air interface Designed for Low Bit Rate Emergency Communications Satellite Link in Ku/Ka/Q/V Bands

Ponia Pech[1], Marie Robert[2], Alban Duverdier[2] and Michel Bousquet[3]
[1]*TeSA*, [2]*CNES*, [3]*ISAE/SUPAERO*
*France*

## 1. Introduction

Access to information is of paramount importance in emergency situations when a disaster or a natural catastrophe (earthquakes, landslides, tsunamis, tidal wave, volcanic eruptions, floods, cyclones, other tornados or forest fires) occurs in a given place. The rescuers must have access to telecommunications means in order to be able to transmit critical information (such as the number of victims, of persons needing medical assistance or to be transported towards care centers, hospital locations, the nature of injuries, etc.) to the center that is in charge of coordinating the heavy logistical rescue operations. However, when an extensive area is affected (in developing countries in particular, but this holds true in first world countries also), it is most often impossible to use some telecommunications equipments, because of a destruction of the base stations or an unavailability of electric power. Thus, it is an absolute priority, in order to manage a situation of post-disaster, to restore the communications infrastructure. Three distinct and consecutive phases can be distinguished (Pech et al., 2008):

- **Phase 1**: This is the phase that immediately follows the disaster, where no means of communications is available, and where minimal emergency communications must be established as quickly as possible. This is in this phase that the very rapid deployment (less than some hours) of a satellite system seems most suited. The ground equipments that can be envisaged in this type of system are transportable or portable terminals, not necessarily mobile, that can be operational everywhere between 5 and 30 minutes without the assistance of a technical expert staff. Transmissions must be secured and have a strong availability.
- **Phase 2**: In a second stage, the restoration of a more elaborated communications network involving a transportable user terminal with a higher capacity also ensuring wireless connectivity (GSM, Wi-Fi, WiMax, etc.) is necessary to exchange more complex information. This local infrastructure can be linked to the national telecommunications

public network through a DVB-RCS-like professional satellite terminal, or an Inmarsat's BGAN (Global Broadband Area Network) device (Inmarsat).

- **Phase 3**: After several weeks, even several months, once the first necessities are met and the situation is being reestablished little by little, a restoration of the nominal communication infrastructure can be envisaged.

Emergency satellite systems intended to address the needs of the rescue teams both in phase 1 and phase 2 have been the subject of numerous recent studies and developments, such as for instance the FP6 WISECOM project (cf. Fig. 1) (WISECOM, 2007), diverse standardization works TISPAN (ETSI. EG 202 339, 2004), EMTEL (ETSI. TS 102 181, TS 102 182 and TR 102 410), and ITU-R WP-4B work on wideband spreading signals (IUT-R. IUT-R WP 4B, 2006), and research studies on UWB (*Ultra Wide Band*) satellite transmission carried out with the French Space Agency (CNES - *Centre Nationale d'Etudes Spatiales*) and Thales Alenia Space France (Leconte et al., 2006; Dervin, 2007). Another related research field which has drawn much attention these last years from academia and industries alike is telemedicine, for instance with the OURSES (French acronym meaning *Offer of Services Rural Use using Satellite*) project (Girault et al., 2008; Mailhes et al., 2008).

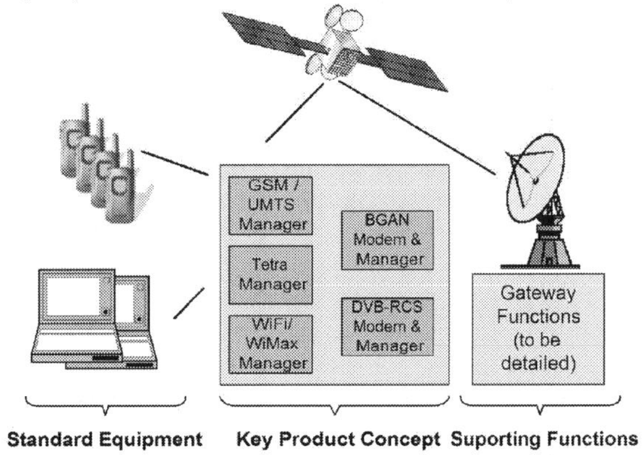


Fig. 1. The WISECOM architecture

The present study inserts itself only within the framework of the aforementioned phase 1. The purpose here is to propose and study a solution allowing to establish very quickly a minimal low bit rate satellite link (in the order of several tens of kbps) in an emergency situation, while using the available resources of the satellite, and a single-user ground terminal with a low transmission power, a small diameter antenna, and a dedicated Tx/Rx system, characterized by an electric consumption which shall be as low as possible (battery or solar panels). The bi-directional transmission link is modeled within the Juzzle [4] open source environment software with an emphasis on the return channel, deploying an adaptive strategy based on the DVB-S2 adaptive modulation and coding (ACM) scheme. This paper expounds the link budget dimensioning and a customized, enhanced DVB-S2 air interface proposed to support minimal emergency communications in a severely impaired channel environment in high frequencies; it also outlines the combined Excel/Matlab/Juzzle high-level transmission link software simulation platform that is being developed in order to assess the performance of the proposed transmission scheme. Focus is placed on the underlying theoretical models involved in the simulator which follows a cross-layer approach, integrating propagation, DVB-S2-based physical layer, and traffic components.

## 2. System architecture, scenario, and traffic characterization

### 2.1 System architecture and scenario

The proposed system architecture and scenario are as follows (Pech et al., 2008): the emergency mission signals are superimposed with those of the primary system characterized by a star topology, a classic multibeam, multicarrier, broadband bent-pipe satellite operating either in Ku/Ka or Q/V-bands, and with a 120-MHz transponder (cf. Fig. 2). Thus a dedicated channel for the emergency mission is not required. The gateway has at its disposal a bandwidth of 480 MHz, while the user links share a total bandwidth of 240 MHz, in four 120-MHz sub-bands and two polarizations, with a reuse frequency factor of 1 over 4. The frequencies used for the return link are enlisted in Table 1. Furthermore, all-IP (Internet Protocol) architecture is assumed.

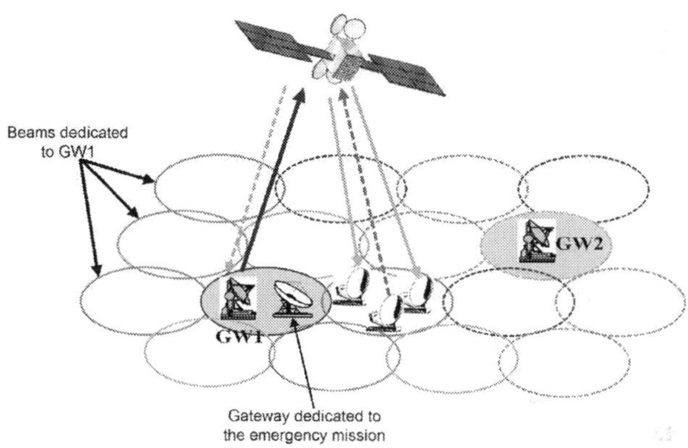


Fig. 2. System architecture

The main characteristics of the satellite are summarized in Table 2. For the emergency mission, a set of 4 different types of user terminals is considered: two mobile user terminals of very low transmission power (between 0.5 and 2 W), the first one (UTA) having a patch antenna, and the second one (UTB) an omnidirectional antenna; and two deployable or transportable user terminals: UTC is a rapidly deployable "mini-gateway" mounted on a van, transmitting at up to 50 W, while UTD can be transported by a human user and has a transmission power of up to 5 W.

### 2.2 Traffic characterization

In phase 1 of a post-disaster emergency situation, the services which are most required to be provided as a minimum requirement are voice services, and transfer of small text or video files (SMS - Short Message Service / MMS - Multimedia Messaging System). Special emphasis will be laid upon Voice Over IP (VoIP), and more generally IP applications, as the ever increasing widespread use of IP technologies will greatly contribute to enhancing mobility and nomadicity between private and corporate access. The data to be transmitted are characterized mainly in terms of bit rate, error rates (Bit Error Rate - BER, or Packet Error Rate - PER), delay, jitter, average and maximum packet sizes. The salient Quality of Service (QoS) features of the two types of applications which are envisaged are presented hereafter:

- Real-time, delay-sensitive but not or very little loss-sensitive applications: especially VoIP. A strictly minimum bit rate of 5.3 kbps (assuming ITU H.323 G.723.1 ACELP codec) is required (Nguyen et al., 2001). It has been shown however that optimal bandwidth occupation for VoIP over satellite is around 12 kbps. VoIP bit rate also varies depending on the codec, on whether RTP (Real-Time Protocol) is compressed, and on the overhead introduced by the header of the protocol suite (Ethernet, IP, UDP, RTP). For instance, with RTP, a total of 54 bytes are transmitted, out of which only 20 belong to the payload. The bit rate can thus be considered to range between 5.3 and 13 kbps. A minimum MOS (Mean Opinion Score) requirement of 3.5 as suggested in WISECOM (WISECOM, 2007) ensures a good voice quality. In this respect, ITU-T G.114 Recommendation (IUT-T, 2000) specifies a maximum latency value of 150 ms for one-way VoIP communications. As for the jitter, the mean packet inter-arrival time at the receive side must be roughly close to that at the transmit side, with a small standard deviation. Lastly, in terms of packet corruption and loss, some experiments have shown that the satellite link is quite robust to packet corruption in clear sky or moderately degraded channel environment with a BER of up to a $10^{-5}$ (that is, Frame Erasure Rate or FER of 2%) (Nguyen et al., 2001). The performance of VoIP over satellite as obtained by Nguyen et al. (Nguyen et al., 2001) from the COMSAT laboratories will be used as a reference test bench.

- Data-like loss-sensitive, but not or very little delay-sensitive applications, for instance SMS/MMS, email applications, file exchange and Internet browsing: the transmitted mean bit rate shall be at least 32 kbps for Web browsing and file exchange, and 200 kbps for email applications (Inmarsat). BER values of up to $10^{-6}$ can be supported (WISECOM, 2007). Moreover, the time interval between the sending of an SMS and its reception by the receiver must be between 6 and 8 s in average, given that actually 98% of sent SMSs are successfully delivered by a mobile user to a fixed network within a 5-s time period, according to some telecom operators [ETSI, February 2006; ETSI, October 2004]. Since the integrity of SMS messages is 100%, it is obvious that SMSs are well fitted to emergency communications.

## 3. Air interface, channel modelling and link budget

### 3.1 Enhanced DVB-S2 air interface

It is proposed to adopt the ETSI DVB-S2 ACM MODCODs (ETSI, EN 302 307, January 2004) for the return channel as it provides excellent performance close to Shannon's theoretical limit due to an advanced Forward Error Correction (FEC) scheme (concatenated BCH and LDPC codes), and allows an attractive waveform flexibility in presence of channel fading, with its inherent ACM capability. Incidentally, this adoption of DVB-S2 ACM schemes for the return channel was recently standardized within the DVB-RCS+M working group (ETSI, EN 301 790 V1.5.1, May 2009), but in the latter standard, very short (4 kbps) DVB-S2

PLFRAMEs were envisaged instead of normal and short lengths (64,800 bits for the normal frame, and 16,200 bits for the short frame). By contrast, here, standard-length DVB-S2 PLFRAMEs are assumed. In quasi-error free (QEF) environment (PER of $10^{-7}$), in an Additive White Gaussian Noise (AWGN) channel, DVB-S2 operates at ideal $E_s/N_0$ ranging from 16 dB down to -2.35 dB. The performance of the receiver in terms of signal acquisition/ reacquisition time, decoding thresholds, etc., all are are well known (ETSI, TR 102 376, February 2002). The present novelty is the use of DVB-S2 along with spread spectrum and other adaptive mechanisms for return channel interactive services while DVB-S2 was specifically designed for the forward link and broadcasting services. This thus raises some performance challenges to be coped with.

Since the considered applications are all assumed here to be IP-based ones, the DVB-S2 waveform is coupled with an efficient encapsulation mechanism, namely the GSE (*Generic Stream Encapsulation*) protocol at the Segmentation and Reassembly (SAR) layer, aimed at segmenting network layer IP datagrams (PDUs or *Payload Data Units*) into link layer DVB-S2 basic data units called BBFRAMEs (*base band frames*). The GSE mechanism was designed with the purpose of fully taking advantage of the innovative features of the DVB-S2, primarily in terms of reliability, flexibility, and enhanced capacity. MPE (*Multi-Protocol Encapsulation*) and ULE (*Unidirectional Lightweight Encapsulation*) have been the standard encapsulation techniques that were classically used in DVB-typed satellite systems, and as such they received abundant attention in the literature. Nonetheless, GSE constitutes a much more efficient encapsulation scheme fitted to the DVB-S2 standard in that it allows to fully exploit the adaptive ACM capability, implementing QoS scheduling decisions and flexible placement and enhanced fragmentation of PDUs in the flow (Cantillo, 2008).

In particular, DVB-S2 GS (*Generic Stream*) data flows may be packetized or continuous streams. The first ones are suited to carrying PDUs of constant size, whereas the latter category was designed to seamlessly adapt to input stream of any format, including continuous bit streams and variable-sized PDUs such as IP datagrams. GSE can avoid using MPEG2 packets as with MPE and ULE, which would be sub-optimal in the framework of DVB-S2. In effect, the GS flow is more suited to interactive services as it overcomes the inadequate MPEG2-TS constraints of constant bit rate and end-to-end delay. In addition, due to the large sizes of a BBFRAME payload (up to 40 times as long as an MPEG2 packet), datagram fragmentation occurs less often. Measures in the Internet network backbone show that the mean size of an IP datagram is about 500 bytes, which roughly amounts to 7000 \ 500 $\approx$ 14 IP datagrams carried in the longest available BBFRAME, against 2 or 3 fragmentations on the MPEG2-TS layer, and up to 10 in the case of ATM (Cantillo, 2008). GS streams are tailored into 21 possible BBFRAME frames, thus offering a variety of efficiency versus error protection compromises, and predefined sizes ranging from 384 to 1,779 bytes (short BBFRAMEs), or 2,001 to 7,274 bytes (for long BBFRAMEs). Consequently, all these characteristics of GSE result in IP datagrams being delivered more rapidly, efficiently and optimally in a cross-layer perspective, with reduced overhead and complexity (Cantillo, 2008).

A last technique deployed for the purpose of enhancing the DVB-S2 transmission performance is Spread Spectrum (SS) in its Direct Sequence (DS) variant. Besides its resistance to interference, jamming and multipath impairments, a quite powerful property of SS exploited in the framework of this study is its processing gain $G_P$, defined as the ratio of the spread bandwidth over the original bandwidth in dB. This processing gain can thus be added to the signal side of the SNR calculation (Ayala et al., 2004). This fruitful property is due to the power–bandwidth trade-off that exists in any radio communication system: using a spread spectrum signal enables the system to operate at negative signal to noise ratios, thus allowing to deploy smaller terminals with reduced transmission power with respect to the non-spread case. This consequently means improved battery lifetime in the case of portable terminals, as well as an easier and quicker deployment of the terminals. Therefore the adequacy of SS is straightforward for emergency communications, in heavy rain environment (Yoon et al., 2008).

It must be pointed out that spread spectrum should not be confused with the standard DVB-S2 scrambling process. Spreading must be applied to each symbol of the PLFRAME including the PLHEADER and the pilot symbols, and is followed by scrambling, which applies a scrambling code to the spread signal (ETSI, EN 302 307, April 2009). Although there is some similarity between the two processes since both multiply an original signal with a pseudo-random noise (PRN), SS enlarges the bandwidth of the signal whereas scrambling does not since it only randomizes the $(I+jQ)$ samples of the PLFRAME for energy dispersal (ETSI, EN 301 790 V1.5.1, May 2009). The Direct Sequence Spread Spectrum (DS-SS) block can only be inserted before the base-band filter and the modulator. In this technique, the PRN is directly applied to the data entering the carrier modulator. The modulator therefore sees a much higher bit rate, which corresponds to the chip rate of the PRN sequence. The purpose of modulating an RF carrier with such a code sequence is to produce a direct-sequence-modulated spread spectrum with $((\sin x)/x)^2$ frequency spectrum, centered at the carrier frequency. There is no changing the point in the system where the DS-SS must be placed otherwise it would be a quite different SS technique. As a result, that means that a standard DVB-S2 transmitter cannot be used as a black box, but that the transmission chain must undergo some design adaptations, so as to conform to the provision envisaged in clause 5.1 of the mobile version of the DVB-RCS standard (ETSI, EN 301 790 V1.5.1, May 2009).

### 3.2 Channel modelling and link budgets

On the basis of such an air interface, a thorough point-to-point link budget analysis has been carried out for the four selected frequency bands Ku, Ka, Q and V in order to determine, first how much admissible capacity on the return link is provided for each MODCOD for the different user terminals, UTA, UTB, UTC and UTD as referred to in section *§2.1*; and second, which DVB-S2 MODCOD is required when a specific attenuation threshold is crossed (or equivalently to meet a given system availability). Link budget calculations and analyses have been performed both with and without SS, and have carefully focused on interference calculations, since the ratio *C/I* is of paramount importance in the link budget. However, link budget computation over the whole coverage is not needed, since it is assumed that the geographical area affected by the disaster is quite limited in extent (less than 100 $km^2$ which is the size of a large city such as Paris in France) which allows to consider that the link budget parameters are roughly constant over the area.

For this purpose, static (that is to say, temporal and spatial variability is not taken into account) statistical ITU-R channel models are needed to calculate the total atmospheric attenuation to be introduced into the link budgets.

Actually, the effect of each attenuation component may strongly depend on the link frequency. From Ku band to higher frequency bands, propagation through the satellite

channel is mostly impaired by rain, gas (oxygen), clouds, water vapor, amplitude scintillation, depolarization, and a degradation of the receive terminal figure of merit (*G/T*) due to an increase of its noise temperature (Castanet et al., 2003).
For example, oxygen attenuation is neglectable at Ku/Ka bands, but is permanently present in the atmosphere at higher frequencies, and exhibits close lines of strong absorption between 50 and 70 GHz, making satellite communications impossible at 60 GHz. Rain attenuation impairs the propagation channel most, with fade depth of up to 20 dB already at Ka band, for 0.01 % of the time (cf. Fig. 3).

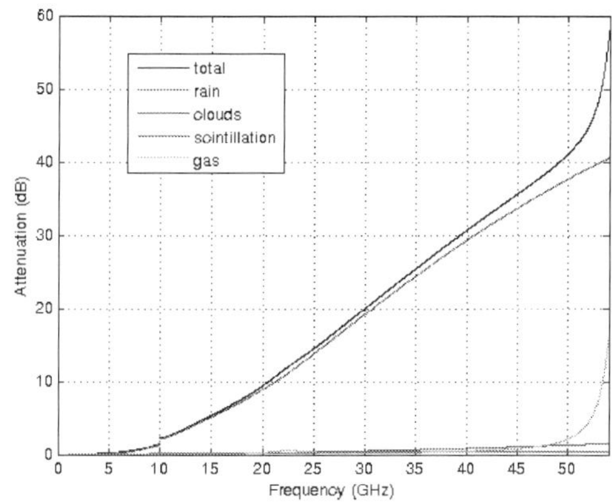


Fig. 3. Total attenuation vs the frequency at Louvain la Neuve, Belgium (Pech, 2003)

The acceptable maximum bit rates for each user terminal were determined according to the channel propagation state in Ku, Ka, Q and V bands, and the calculations also took into account the SS spreading factor $L$ ($L$=1 corresponds to the case where SS is unused), as well as the size of the DVB-S2 PLFRAME (64,800 bits for normal frame, or 16,200 bits for short frame). A sample of obtained results is given in Table 1 for an uplink attenuation corresponding to 0.01 % of the time in the case of normal PLFRAME, and assuming that all $M$ = 200 user terminals are of the same type. When there is only one user terminal transmitting, no spread spectrum interference is present. As a matter of fact, the link budget computations have to be refined by taking into account the effect of the mutual interferences due to spread spectrum. Such analysis was presented for instance by Viterbi in (Viterbi, 1985). It leads to considering the following signal energy per chip (of duration $T_s$ and bandwidth $W_s = 1/T_s$ ) over noise ratio as the actual operating point of the system:

$$\frac{E_s}{N_0^{'}} = \frac{E_b/N_0}{1+(M-1)(R_b/W_s)(E_b/N_0)} \tag{1}$$

where there are $M$ identical user terminals transmitting to a same gateway at the same power and with spread spectrum; $N_0$ is the background noise spectral density received by each user; $E_b$ is the received energy per information bit of duration $T_b$ and rate $R_b = 1/T_b$ bps. $N_0$ is the sum of thermal noise $N_0W_s$ and of the interference contribution of the other $M$-1 users. In other words, the total noise spectral density is related to $N_0$ through the equation:

$$N_0' = N_0 + (M-1)E_s \quad (2)$$

| Frequency band | Uplink atten. (dB) | Maximum bit rate (kbps) (L, MODCOD) | | | |
|---|---|---|---|---|---|
| | | User terminal | | | |
| | | *UTA* | *UTB* | *UTC* | *UTD* |
| **Ku (14 GHz)** | 6.5 | **889.7 (1,9)** | **889.7 (1,9)** | 889.7 (1,9) | 889.7 (1,9) |
| **Ka (30 GHz)** | 24.7 | 236 (32,5) | **102 (32,6)** | 3977 (8,28) | 3977 (8,28) |
| **Q (45 GHz)** | 43.5 | 0 | 0 | 3977 (16,28) | **591 (8,5)** |
| **V (54 GHz)** | 71.5 | 0 | 0 | 1770 (8,13) | **29 (32,1)** |

Table 1. Maximum allowed bit rates for an uplink attenuation corresponding to 0.01% of the time, with their required spread factors *L* and MODCODs. The 28 MODCODs of the DVB-S2 standard are involved, and referred to by their indices from 1 (QPSK 1/4) to 28 (32 APSK 9/10). By required MOCODs, it is meant the least robust, but robust enough MODCODs required by the terminals to establish a communication link.

From the results presented in Table 1, the following conclusions can be drawn:

- In Ku band, the portable/mobile terminals UTA and UTB yield quite sufficient bit rate performance, for a moderate channel attenuation level of 6.5 dB.
- In Ka band, these terminals can still transmit very reasonably: maximum bit rates of 236 kbps and more than 100 kbps are yielded by UTA and UTB terminals respectively.
- In Q band, the portable user terminals UTA and UTB can still transmit at low bit rates for an uplink attenuation of up to 30 dB, but require the activation of DS-SS. For $A_{UL}$ greater than 32 dB, increasing *L* is not sufficient enough to establish the communications link.
- In Q and V bands, the link cannot be established for these low transmit power terminals even though SS is used when the uplink attenuation exceeds 32 dB. Therefore the use of a more powerful terminal, at least UTD (man-transportable terminal, with a transmit power of 5W) is necessary. For the latter, a maximum bit rate of almost 600 kbps is attainable for an uplink attenuation of 43.5 dB (Q band).
- In V band, the link cannot be established anymore for the low to moderate transmit power terminals UTA, UTB and UTD even though SS is used. Nonetheless, minimum communications with VoIP could be established with an uplink attenuation of up to 60 dB, but at the cost of having a very high spread factor *L*, in the order of 5,000, which is not permitted by the bandwidth constraint. UTD is still usable with a maximum bit rate yielding 29 kbps. As the transmit power must be kept as low as possible, it is straightforward that the mini-gateway UTC does not need to be deployed to reach the goal of establishing a minimum emergency link.

## 4. Integrated simulation platform

### 4.1 General description

A software model of the enhanced DVB-S2 transmission link was developed in mixed standard C and Java languages within an open source software environment called Juzzle (SILICOM). The whole simulation platform is an integrated Excel/Matlab/Juzzle software package which also implements the C DISLIN (Michels, 2008) data plotting library. GSL GNU scientific library (Galassi, 2008) could also be incorporated in a future revision of the simulator. The former library is a set of C subroutines and functions that are used to display data graphically as post-processing functions within Juzzle, whereas the latter is a collection of routines written from scratch in C for numerical computing. These routines present a modern Applications Programming Interface (API) for C programmers, allowing wrappers to be written for very high level languages.

The philosophy of the integrated simulation platform tool is as follows:

*Step 1:* The user has the possibility of tuning the values of all the parameters of the point-to-point link budget in the Excel sheet, including the detailed characteristics of the terminals (location, EIRP, transmit power, etc.), the operating frequency band, and the DVB-S2 physical layer parameters (modem roll-off, type of PLFRAME, etc.). Once the system parameters have been validated, an Excel macro enables to create all the system and transmission parameters input files that feed both an off-line propagation pre-processing Matlab routine and the Juzzle core component.

*Step 2:* The offline pre-processing propagation module written in C must be run in order to generate the needed attenuation time series for the return link in Ka and EHF bands using the enhanced stochastic Maseng-Bakken model developed by Lacoste and Carrié (Lacoste, 2005; Lacoste et al., 2005; Castanet et al., 2008; Carrié et al.; Lemorton et al., 2007). The module uses the Juzzle C data-driven engine. An example of such an attenuation time-series is shown in Fig. 4 (taken from Pech et al., 2009). The synthesizer relies on the theoretical principles of the enhanced Maseng-Bakken model which was used by Lacoste (Lacoste et al., 2005; Castanet et al., 2008), and recently improved by Carrié (Carrié et al.). Actually, this new stochastic model enables to either stochastically interpolate initial samples to synthesize fast fluctuations of rain attenuation or generate "on-demand" rain attenuation events starting from the three parameters: duration of the event, maximum attenuation and position of the peak attenuation.
The model improvement lies in the characterization of the conditional probability $p(A(t)\,|\,A(t-Dt_1),\ A(t+Dt_2))$ which enables very fast interpolation or stochastic generation of rain events "on-demand". Three of the input parameters ($m$, $s$, $A_{offset}$) can be assessed for all link configurations using ITU-R recommended models. For the last parameter, $b$, a rough estimate equal to $10^{-4}$ $s^{-1}$ can be used independently of the sampling rate and whatever the link considered in North-western Europe for elevation angles between 25° and 38° and frequencies between 12 GHz and 50 GHz (Carrié et al.).
The "event-on-demand" time series synthesizer derived from this first model offers the possibility to configure the maximum level and the duration of the rain attenuation events to be generated. This model exhibits the physical soundness of the real rain impairments

phenomenon and enables generation of rain attenuation events databases representative of experimental databases (Carrié et al.).

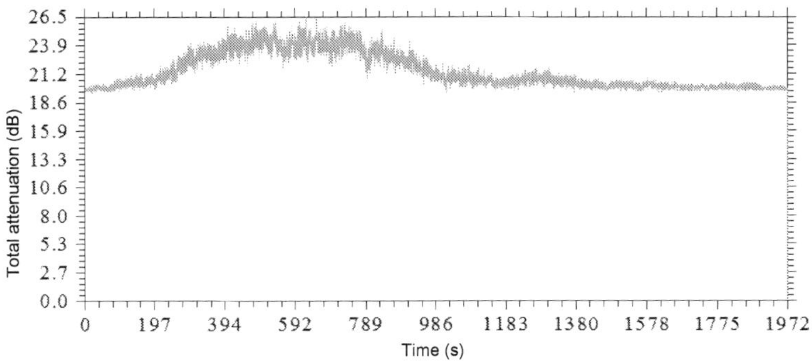


Fig. 4. Total attenuation time-series (in dB) generated by the offline module within the Juzzle simulator at 54 GHz

*Step 3:* The Juzzle on-line core component contains the complete high level system link model and enables to post-process and display a number of performance results.

*Step 4:* Last but not least, a Matlab routine is provided with the aforementioned simulator components, allowing to post-process the results file containing the collected performance parameters. It could be envisaged to replace this Matlab component in the future with a GSL-based module embedded in the Juzzle environment, as mentioned previously.

### 4.2 The Excel sheet link budget calculator

The Excel sheet link budget calculator was developed in order to help dimension the satellite emergency system to be simulated. It is a powerful tool allowing to:

- configure the whole system (satellite, air interface, characteristics of Earth terminals both from the emergency system and the primary system, etc.);
- compute the return link budgets both for the emergency system and the primary system.

The Excel sheet is composed of 12 window tabs. The tool was developed with a library of Visual Basic functions grouped into 10 distinct categories. It also calls an extern dynamic link library (DLL) named "propa.dll" (Lacoste, 2006; Lacoste, February 2006), which was developed by the CNES and comprises routines that enable to calculate the various propagation attenuation and scintillation components on an Earth-space link according to standardized ITU statistical prediction methods:

- Rec. ITU-R P.676 (ITU-R, 2005) due to dry air and water vapour;
- Rec. ITU-R P.840 (ITU-R, 1999): attenuation due to clouds;
- Rec. ITU-R P.618 (ITU-R, 2007): attenuation due to rain and melting layer, scintillation fade, total impairment.

### 4.3 The Juzzle simulation platform

The Juzzle simulation platform is composed of two separate and standalone Juzzle components (a propagation generator component and a processing core component), and an extern post-processing Matlab routine.

The architecture of the Juzzle link model is shown in Fig. 5 hereafter with all its components:

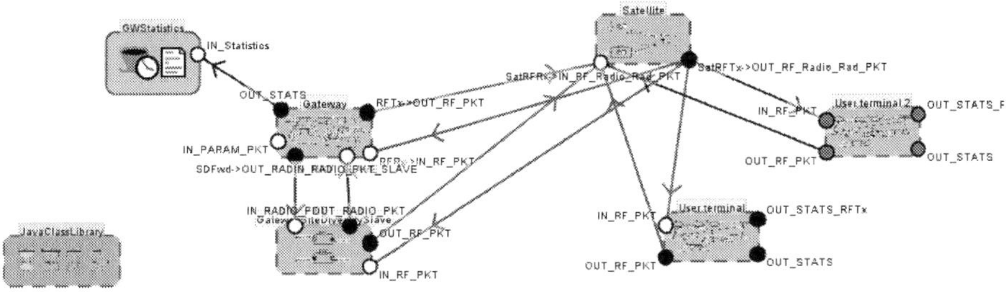


Fig. 5. Juzzle link model

The core processing module was developed in Java due to the portability, reusability, and object-oriented feature of the programming language. This module uses the Java event engine allowing to model the various delays within the system more easily.
The module models the emergency satellite system at high level as shown in Fig. 5, with the following nodes: two user terminals ("User terminal 1" and "User terminal 2"), the gateway, and the satellite, especially in terms of radio packet transmission, and its related link budget computation functions. As a result, the simulator is a powerful and complex integrated simulation platform handling more than 700 parameters, and being composed of 44 Java classes.
The component also possesses a statistics node whose tasks is to write down in an output ASCII text file a number of probe parameters allowing to compute performance statistics related to the link quality (BER, system margin, $C/(N_0+I_0)$, MODCOD, DS-SS spread factor, bit rate, etc.). As of today, exactly 30 parameters are monitored, but their number can be increased at will, with the restriction that the larger the number of parameters, the larger the results file. For instance, a simulation over a simulated time of 59,900 seconds produces a 1.428-GB output file.

The software simulation platform is described in more details in the next sections.

#### 4.3.1 Traffic modules

Four types of traffic are modelled: VoIP, Web browsing, and SMS for the return link, and Internet aggregate traffic as an example of continuous forward link traffic. The corresponding underlying theoretical models employed are given in the following:

<u>a) VoIP traffic</u>

This source is modelled by a classic ON/ OFF state machine (Pech, 2003). The model is represented by an ON/OFF source which operates as an alternating process of talk and silence periods which are distributed according to negative exponential laws with means $a^{-1}$ and $b^{-1}$ respectively.

**b) Web browsing traffic model**

Basically, the Web traffic model relies on an ON/OFF two-state model. The specificity of the model lies in its representing the traffic from a user behavioural point of view, by modeling a user's session, that is, a user Web request instead of focusing on the notion of a Web page. The overall traffic stream of a user is composed of a superimposition of the packet arrival processes of all TCP connections within a user's session. Moreover, it is a hierarchical model in which session, connection and IP levels are included (Choi & Limb, 1999).

**c) SMS traffic**

SMS traffic is simply modeled by a generator of SMS messages whose mode of operation is as follows (ETSI, 3GPP TS 23.040, October 2004; ETSI, 3GPP TS 23.038, September 2004): a random number N ranging from 1 to 10 (produced by a uniform random number generator) of concatenated 140-byte (amounting to 160 alphanumeric characters) SMS messages is generated every period $T_{SMS}$ where $T_{SMS}$ is distributed according to a negative exponential distribution. The number N is generated each time the SMS traffic source is invoked, that is, when the basic service is switched into. The scheme is repeated (ARQ: Automatic Repeat reQuest) every two minutes provided the basic service (i.e. a bit rate of less than 4 kbps) is still selected.

**d) IP aggregate traffic source**

The traffic source is modeled by a Fractional Gaussian Noise (FGN) using the Fast Fourier Transform (FFT). Paxson's FFT method (Paxson, 1997) consists in synthesizing a sample path having a same power spectral density (PSD) as a Fractional Gaussian Noise (FGN) process. This sample path can then be used in simulations as traces of real self-similar traffic. The algorithm is basically based on a fast approximation of the PSD of an FGN process using the FFT. The algorithm relies on an implementation of Paxon's self-similar generator written in ANSI C, and provided with in the form of a routine fft_fgn.c developed by Christian Schuler.

### 4.3.2 Adaptive strategy module

The adaptive strategy mechanism is at the heart of the simulator, and relies on a combination of several techniques aimed at ensuring high availability of the transmission links in spite of severe channel impairments in the selected frequency bands (*e.g.* about 20 dB in Ka band and more than 80 dB in Q band 0.01 % of the time). The techniques employed are: ACM, spread spectrum with varying DS-SS spread factor (*L*), gateway site diversity (SD) (to improve the downlink budget for the return channel when it is impaired by rain), and ARQ-like time diversity (TD). ARQ is only applied to SMSs, the user terminal automatically attempting at retransmitting the same SMS message at different times, when the channel conditions improve. Currently, retransmission is done on a pure deterministic basis, that is, periodically after a fixed time interval. Nevertheless, in the future, more elaborate strategies could be devised exploiting fade duration, fade slope, and inter-fade statistics so as to implement an efficient method able to predict the channel attenuation in the medium term (several minutes).

It is moreover assumed that all of the user terminals of the emergency mission which are deployed over the geographical area affected by the disaster are of the same type and undergo the same ACM mode. This makes sense since the area is assumed to be quite limited in extent, more precisely smaller than 100 $km^2$, which allows to consider that the link budget and transmission parameters are roughly constant over the whole area. Nevertheless, it must be kept in mind that this uniform ACM scheme over the whole area is also a consequence of a simplistic assumption concerning the channel propagation modelling, in which no spatial variability is taken into account. Only a temporal variability is modeled using Carrié's enhancement of Lacoste's CNES/ONERA rain fading time series stochastic generator based on Maseng-Bakken model (Lacoste, 2005; Lacoste et al., 2005; Castanet et al., 2008; Carrié et al.; Lemorton et al., 2007).

The tabulated results of the link budget analysis previously mentioned enabled to carefully construct appropriate and efficient channel-aware link adaptive strategies, specific to each frequency band, the purpose of which is to ensure very high availability of the link, and enough capacity to authorize at least SMS communications even during a severe rain event (extremely strong storms), and VoIP as a minimum guaranteed service the rest of the time. The adaptive strategy combines ACM with spread-factor-varying SS, the spread factor $L$ being adapted (increased) before ACM activation depending on the attenuation level with respect to predefined thresholds, and with three different service classes being defined:

1. *Premium service*: VoIP, and Web browsing/SMS with a bit rate of about 180 kbps;
2. *Gold service*: VoIP/SMS with a bit rate of 4 to 16 kbps;
3. *Basic service*: SMS only with a bit rate of less than 4 kbps.

For each frequency band, these services are switched from each other through the DS-SS spread factor $L$ (from 1 to up to several thousands). Within each service (or primary bit rate mode), the channel fluctuations are dynamically compensated for by full DVB-S2 ACM MODCOD switching.

Whenever the return link downlink channel is impaired by rain above a pre-determined attenuation threshold, and when the above compensation techniques still do not allow to close the link budget, SD is deployed on the feeder link. In other words, the downlink signal from the satellite is rerouted towards a slave gateway station which is located in a place not affected by rain. This brings an additional site diversity gain $G_{SD,DL}$ into the link budget, that can be calculated with the ITU-R P.618-9 Recommendation (ITU-R, P.618-9, January 2007).

It should now be noted that an optimal adaptive mechanism would obviously need to exploit, not only the static statistical characteristics of the channel fade components, but also their dynamic properties - in other words, an informed knowledge of fade / inter-fade duration, and fade slope ITU-R statistics and models (cf. Fig. 6). Furthermore, two additional considerations must be pondered when devising a relevant adaptive strategy to meet the system requirements for the targeted emergency mission: *(a)* the system total delay requires some short-term fade prediction to be made within a time horizon of several seconds; *(b)* and a higher layer MODCOD switching criterion other than merely the $E_s/N_0$ or the BER, such as the PER could be used so as to yield better performance at the application level. The former consideration is due to a necessity of correctly estimating at time *t* the channel attenuation $A(t+\Delta t)$ at time $t+\Delta t$, knowing that otherwise using $A(t)$ to decide at time *t* which MODCOD is to be used would be mostly inadequate since quite rapid variations may occur in the channel attenuation time evolution. To properly model a rain attenuation predictor, an accurate link delay budget must be established, including in particular the 500-ms satellite Round-Trip Delay (RTD), and the channel estimator calculation time (Aroumont et al., 2006) (using the DVB-S2 pilot symbols which are spread over different frames). The second consideration relies on some error modelling that enables to derive the PER from a physical layer-oriented QoS parameter like the BER (Pech et al., 2002).

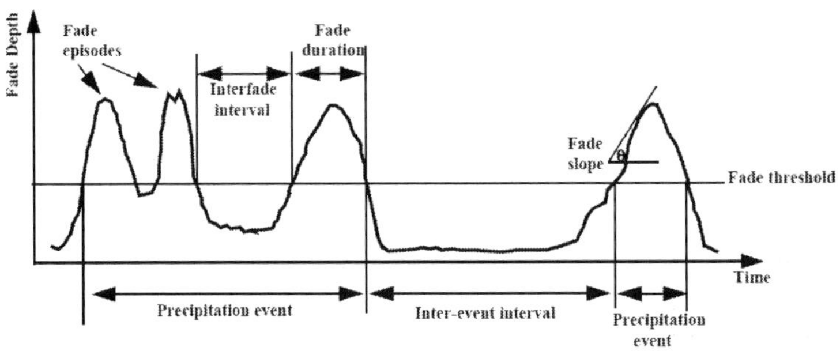


Fig. 6. Secondary statistics of fade

Currently, to obtain the attenuation to be applied to the channel, the simulator performs an attenuation prediction by linear interpolation or extrapolation, and takes into account the satellite geostationary Round Trip Propagation Delay (RTPD).

### 4.4 Extern Matlab post-processing module

This module is a post-processing module in charge of computing the transmission link performance mainly in terms of link availability, bit rate, error rates (BER or PER), delay, jitter, average and maximum packet sizes in order to characterize the application-level QoS, but also in terms of IP throughput, IP traffic characteristics (Tou et al., 2008) (inter-packet delay, IP delay variation, one way delay with maximum and minimum values, packet losses), and IP over frame efficiency over the DVB-S2 (Girault et al., 2008) (in connection with the encapsulation overhead and the MODCODs selected, and more generally the implemented adaptive strategy), as well as in terms of attenuation prediction errors. Some performance figures related to VoIP over DVB-S2 links have been made available in recent studies (Jegham et al., 2008), and could serve as a reference basis for the present project, in particular regarding a comparison between GSE and MPE/ULE encapsulation schemes. The module also allows to investigate the impacts of rain fading on the service performance. It provides a set of post-processing graphical plotting functions that will enable to optimize the performance of the adaptive strategy, tune its parameters (*e.g.* hysteresis and detection margins), and assess the ability of the proposed air interface to ensure a robust low bit rate satellite communication link for emergency mission.

## 5. Some results

The acceptable maximum bit rates for each user terminal were determined according to the channel propagation state in Ku, Ka, Q and V bands, and the calculations also took into account the SS spreading factor *L* (*L*=1 corresponds to the case where SS is unused), as well as the size of the DVB-S2 coded frame FECFRAME (64,800 bits for normal frame, or 16,200 bits for short frame). A sample of obtained results regarding to the admissible maximum bit rates and their associated MODCODs and spread factors was given in Table 1 which is copied into Table 2 hereafter for the sake of commodity:

| **Frequency band** | **Uplink atten. (dB)** | **Maximum bit rate (kbps) (L, MODCOD)** | | | |
|---|---|---|---|---|---|
| | | **User terminal** | | | |
| | | ***UTA*** | ***UTB*** | ***UTC*** | ***UTD*** |
| Ku (14 GHz) | 6.5 | **889.7 (1,9)** | **889.7 (1,9)** | 889.7 (1,9) | 889.7 (1,9) |
| Ka (30 GHz) | 24.7 | 236 (32,5) | **102 (32,6)** | 3977 (8,28) | 3977 (8,28) |
| Q (45 GHz) | 43.5 | 0 | 0 | 3977 (16,28) | **591 (8,5)** |
| V (54 GHz) | 71.5 | 0 | 0 | 1770 (8,13) | **29 (32,1)** |

Table 2. Maximum bit rates allowable for an uplink attenuation corresponding to 0.01% of the time, with their required spread factors *L* and MODCODs

The results obtained without SS show in particular that the portable/mobile terminals UTA and UTB yield quite sufficient bit rate performance in Ku band, and for a moderate channel attenuation level of 6.5 dB. These terminals can still transmit very reasonably in Ka band. But in Q and V bands, the link cannot be established for these low transmit power terminals even though SS is used. Therefore, the use of at least UTD is necessary. For the latter, a maximum bit rate of almost 600 kbps is attainable for an uplink attenuation of 43.5 dB (Q band). In V band, UTD still is usable with a maximum bit rate yielding 29 kbps. The behaviour of the adaptive strategy for UTD in function of the total uplink attenuation is shown in Fig. 7. As the transmit power must be kept as low as possible, it is straightforward that the mini-gateway UTC does not need to be deployed in order to establish a minimal emergency link.

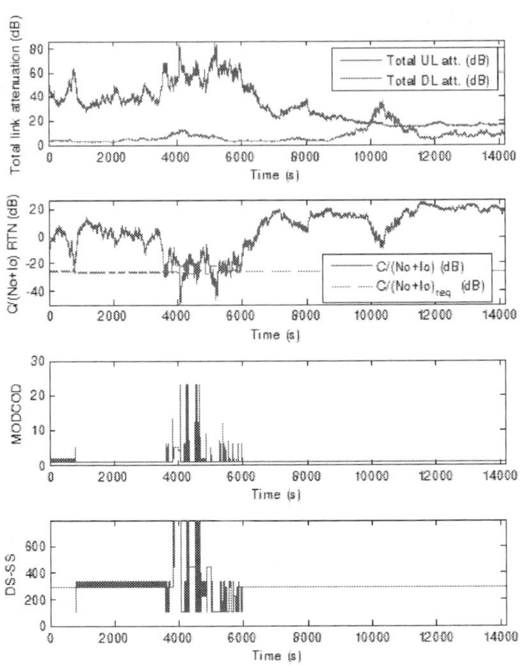


Fig. 7. Adaptive strategy behaviour for UTD in V band vs the total uplink attenuation

## 6. Conclusion

This chapter has presented the architecture of a multibeam, bent-pipe satellite system, used to rapidly establish a low bit rate link for emergency communications in Ku/Ka and Q/V bands. The characteristics of the proposed system have been described. An enhanced DVB-S2-like air interface has been proposed, involving DS-SS technique and other adaptive mechanisms such power control, and site diversity. Link budget analyses have shown that even though SS may be deactivated, very low transmit power user terminals UTA and UTB yield reasonable performance for the envisaged purpose up to the Q band, but for rather moderate channel attenuation values. This highlights the relevance and efficiency of a solution combining ACM, SS, bit reduction, and SD techniques in a new adaptive strategy in order to improve transmission performance. Such an adaptive strategy has been successfully implemented within the more general framework of an integrated Excel/Juzzle/Matlab software simulation platform designed to model the system in a high level cross-layer approach mixing propagation, physical layer, and higher layers components. Some selective results of the simulations carried out using this DVB-S2-like satellite link software simulator have been partially presented highlighting its ability to perform relevant analyses of the performance of the system both at the physical layer and the network level.

## 7. Acknowledgments

The work was carried out within the framework of an R&T project funded by the French Space Agency (CNES, Centre National d'Etudes Spatiales), Toulouse, France. The project developed in three phases: I. *Traffic and service characterization*, II. *Air interface and link budgets*, and III. *Computer link modelling and simulations*.

## 8. References

Ponia Pech, Puming Huang, Michel Bousquet. Low Bit Rate Satellite Link for Emergency Communications, *International Workshop on Satellite and Space Communications 2008* (IWSSC'2008), Toulouse, France, 1-3 October 2008.

Inmarsat. Inmarsat BGAN description: http://www.inmarsat.com/Services/Land/BGAN/default.aspx.

WISECOM. Deliverable 1.1-1; Survey of use cases, contract number 034673, 9 August 2007, available at: http://www.wisecom-fp6.eu/deliverables/D1.1-1_Survey_of_Use_Cases.pdf.

SILICOM. Official Juzzle Web site: http://www.juzzle.org/.

Thuan Nguyen, Ferit Yegenoglu, Agatino Sciuto, and Ravi Subbarayan. Voice over IP Service and performance in satellite, *IEEE Communications Magazine* **March 2001; 39**, pp. 164 - 171.

ITU-T G.114 Recommendation: Transmission systems and media, general recommendations on the transmission quality for the entire Internet telephone connection: One-way transmission time, number 114 in G. ITU-T, 2000.

ETSI TR 102 444 V1.1.1: Analysis of the short message service (SMS) and cell broadcast service (CBS) for emergency messaging applications; emergency messaging; SMS and CBS, February 2006.

ETSI. Digital cellular telecommunications system (Phase 2+); Universal mobile telecommunications system (UMTS); Technical realization of short message service (SMS) (3GPP TS 23.040 version 5.8.1 Release 5), October 2004.

ETSI. EN 302 307 V1.1.1: Digital video broadcasting (DVB); Second generation framing structure, channel coding and modulation systems for broadcasting, interactive services, news gathering and other broadband satellite applications, January 2004.

ETSI. ETSI TR 102 376 V1.1.1: Digital video broadcasting (DVB); User guidelines for the second generation system for broadcasting, interactive services, news gathering and other broadband satellite applications (DVB-S2), February 2002.

Juan Cantillo. Codage multi-couches pour systèmes de communication par satellites, Ph.D thesis report, Télécom Paris, Toulouse, France, 19 May 2008.

Mario Reyes Ayala, Edgar Alejandro Andrade Gonzáles, and José de Jesús Roa Franco. Performance in QPSK and BPSK synchronous sequence DS-CDMA satellite systems, *1st International Conference on Electrical and Electronics Engineering (ICEEE)*, Iaghouat, Algeria, 24-27 June 2004.

Ju-Hyun Yoon, Jae-Kwon Lee, Dae-Ig Chang, and Deock-Gil Oh. Spread spectrum technique for digital broadcasting system in rain environment, *10th International Conference on Advanced Communication Technology (ICACT)*, Phoenix Park, South Korea, 17-20 February 2008.

Laurent Castanet, Laurent Féral, and Frédéric Lacoste. Adaptive coding-modulation techniques for Ka/Q-band systems; Review of available Ka/Q-band propagation measurements and models, ESA contract confidential technical report, ESA/ESTEC/Contract n°16533/02/NL/EC. Ref.: IFMT-ACM_TR1, issue 2, 3 March 2003.

Ponia Pech. Incidence de la prise en compte des effets de techniques de mécanismes de lutte contre les affaiblissements (FMT) en bande Ka sur la gestion des ressources dans un système d'accès multimédia par satellite géostationnaire, Ph.D thesis report, Supaero, Toulouse, 19 December 2003.

A. J. Viterbi. When Not to Spread Spectrum - a Sequel, *IEEE Communications Magazine* **April 1985; 23(4)**.

Helmut Michels, DISLIN 9.4, A Data Plotting Library, 15 October 2008. Downloadable at the following address: http://www.mps.mpg.de/dislin/document.html.

Mark Galassi, Jim Davies, James Theiler, Brian Gough, Gerard Jungman, Michael Booth, and Fabrice Rossi. GNU Scientic Library, Reference Manual, Edition 1.11, for GSL Version 1.11, 5 February 2008. Downloadable at the following address: http://www.gnu.org/software/gsl/.

Frédéric Lacoste. Modelling of the dynamics of the Earth-space propagation channel at Ka and EHF bands, Ph.D thesis report, SUPAERO, Toulouse, France, September 2005.

Hyoung-Kee Choi, John O. Limb. A Behavioral Model of Web Traffic, *Proceedings of the 7th International Conference on Network Protocols (ICNP'99)*, Toronto, Ontario, Canada, October/November 1999.

Alexander Klemm, Christoph Lindemann, and Marco Lohmann. Traffic Modeling and Characterization for UMTS Networks. *Proceedings of the Globecom, Internet Performance Symposium*, San Antonio TX, United States, November 2001.

Vern Paxson. Fast, Approximate Synthesis of Fractional Gaussian Noise for Generating Self-Similar Network Traffic, *Computer Communication Review* **October 1997; 27(5)**, pp.5-18.

W. Willinger, M. S. Taqqu, R. Sherman, D. V. Wilson, Self-Similarity Through High-Variability: Statistical Analysis of Ethernet LAN Traffic at the Source Level, *IEEE/ACM Transactions on Networking* **1997; 5**, pp. 71-86.

Frédéric Lacoste, Michel Bousquet, Laurent Castanet, Frédéric Cornet, Joël Lemorton. Improvement of the ONERA-CNES rain attenuation time series synthesiser and validation of the dynamic characteristics of the generated fade events, *Space Communication Journal* **2005; 20(1-2)**.

Laurent Castanet (editor) et al. Influence of the variability of the propagation channel on mobile, fixed multimedia and optical satellite communications, *SatNEx JA-2310 book*, ISBN 978-3-8322-6904-3, Shaker, 2008.

Guillaume Carrié, Frédéric Lacoste, Laurent Castanet. New "on-demand" channel model to synthesise rain attenuation time series at Ku-, Ka- and Q/V-bands, submitted to *International Journal of Satellite Communications and Networking*, Special issue on Channel Modelling and Propagation Impairment Simulation.

Anbazhagan Aroumont, Ana Bolea Alamanac, Laurent Castanet, Michel Bousquet, Stefano Cioni, and Giovanni E. Corazza. Performance of channel quality estimation algorithms for fade mitigation techniques with Ka/Q/V band satellite systems, *9th International Workshop on Signal Processing for Space Communication*, Noordwijk, Netherlands, September 2006.

Ponia Pech, Laurent Castanet, and Michel Bousquet. A prediction model to convert propagation distributions in statistics of quality of service performance parameters, *1st COST 280 International Workshop on Propagation Impairment Mitigation for Millimetre Wave Radio Systems*, Malvern, United Kingdom, 1-3 July 2002.

Ihsane Tou, Mathieu Gineste, Thierry Gayraud, Pascal Berthou. Quality of Service Evaluation in Satellite Systems, *International Workshop on Satellite and Space Communications 2008* (IWSSC'2008), Toulouse, France, 1-3 October 2008.

Nicolas Girault, Nizar Jegham, Nicolas Lerouge, Cédric Le Guern, André-Luc Beylot. OURSES: Efficiency of IP Encapsulation over DVB-S2 Links, *International Workshop on Satellite and Space Communications 2008* (IWSSC'2008), Toulouse, France, 1-3 October 2008.

Nizar Jegham, Nicolas Girault, Cédric Le Guern, Gilles Roussel, Stéphane Lohier, André-Luc Beylot. VoIP over a DVB-S2 ACM link, *International Workshop on Satellite and Space Communications 2008* (IWSSC'2008), Toulouse, France, 1-3 October 2008.

Joël Lemorton, Laurent Castanet, Frédéric Lacoste, Carlo Riva, Emilio Matricciani, Uwe-Carsten Fiebig, Max Van de Kamp, and Antonio Martellucci, Development and validation of time-series synthesizers of rain attenuation for Ka-band and Q/V-band satellite communication systems, *International Journal of Satellite Communications and Networking*, 2007; 25:575–601.

International Telecommunication Union/ITU Radiocommunication Sector, Recommendation ITU-R P.618-9 - Propagation data and prediction methods required for the design of Earth-space telecommunication systems, 1 January 2007.

International Telecommunication Union/ITU Radiocommunication Sector, Recommendation ITU-R P.676-6 - Attenuation by atmospheric gases (Question ITU-R 201/3), 2005.

International Telecommunication Union/ITU Radiocommunication Sector, Recommendation ITU-R P.840-3 -Attenuation due to clouds and fog (Question ITU-R 201/3), 1999.

Ponia Pech, Puming Huang, Michel Bousquet, Marie Robert, and Alban Duverdier. Simulation of an Adaptive Strategy Designed for Low Bit Rate Emergency Satellite Communications Links in Ku/Ka/Q/V Bands. *International Workshop on Satellite and Space Communications 2009* (IWSSC 2009), 10th-11th September 2009, Siena-Tuscany, Italy.

ETSI, « Digital cellular telecommunications system (Phase 2+); Universal Mobile Telecommunications System (UMTS); Alphabets and language-specific information (3GPP TS 23.038 v6.1.0)", September 2004.

CNES freeware PROPAGATION Dynamic Link Library (DLL). http://logiciels.cnes.fr/PROPA/en/logiciel.htm.

Frédéric Lacoste. IUT-R propagation models software library. CNES, 13 February 2006. http://logiciels.cnes.fr/PROPA/en/usermanual.pdf.

ETSI EN 301 790 V1.5.1. Digital Video Broadcasting (DVB); Interaction channel for satellite distribution systems. May 2009.

Netdish VoIP over satellite. http://www.netdish.com/index.php?option=com_content&task=view&lang=en&id=37&Itemid=55.

ETSI. EG 202 339 v1.1.1: Telecommunications and Internet converged Services and Protocols for Advanced Networking (TISPAN); Definition of requirements on the functional architecture for supporting Emergency and Priority user services. September 2004.

ETSI. TS 102 181 V1.1.1: Emergency Communications (EMTEL); Requirements for communication between authorities/organizations during emergencies, December 2005.

ETSI. TS 102 182 V1.2.1: Emergency Communications (EMTEL); Requirements for communications from authorities/organizations to individuals, groups or the general public during emergencies. December 2006.

ETSI. TR 102 410 V1.1.1: Emergency Communications (EMTEL); Basis of requirements for communications between individuals and between individuals and authorities whilst emergencies are in progress. August 2007.

Katia Leconte, Martial Coulon et Marie-laure Boucheret. Analyse d'applicabilité de standards de télécom terrestres aux systèmes de télécommunication par satellite - Signal très large bande pour scénario d'urgence. Rapport d'étude du CNES, version 02, 1er septembre 2006.

Matthieu Dervin. Définition des missions et des scénarii de référence. Rapport d'étude du CNES, 1er septembre 2007.

IUT-R. IUT-R WP 4B: Draft new recommendation ITU-R: Characteristics of fixed-satellite service systems using wideband spreading signals, ref 4/92 (Rev.1)-E. 3 October 2006.

C. Mailhes, B. Comet, H. de Bernard, E. Campo, A. Prieto, and S. Bonhomme. Telemedicine Applications in OURSES Project, *International Workshop on Satellite and Space Communications 2008* (IWSSC'2008), Toulouse, France, 1-3 October 2008.

# 10

# Mapping and Estimation of Chemical Concentrations in Surface Soils Using LANDSAT TM Satellite Imagery

B.B. Maruthi Sridhar and Robert K. Vincent
*Department of Geology, Bowling Green State University, Bowling Green, OH 43403 United States*

## 1. Introduction

Application of fertilizers, manures and other chemicals to agricultural fields is a continuous practice that is required for the optimum crop production. The rate of fertilizer application to soils is often determined by farmers based on the results of periodic soil sampling and analysis that were conducted prior to the cropping season. However conventional methods of soil sampling and analysis for soil variability in chemical characteristics are too time-consuming and expensive for multi-seasonal monitoring over large-scale areas. Hence, remote sensing has been used as an alternative method for determining and mapping the physical and chemical characteristics of the soil.

High resolution aerial imagery were used to map the organic carbon (Chen et al., 2000), clay content (Sullivan et al., 2005), organic matter and Bray-1 phosphorus concentration (Varvel et al., 1999), total soil phosphorus, copper and sulfur (Sridhar et al., 2009a,b) in bare soils. Dematte et al. (2003) reported that chemical variations in soil resulting from fertilizer applications can be detected, based on the intensity of reflectance. Several studies showed the use of spectral reflectance to determine the soil color (Post et al., 2000), texture and particle size distribution (Chang et al., 2001), soil moisture (Lobell & Asner, 2002), iron oxides (Ji et al., 2002), carbonates (Ben-Dor & Banin, 1990), clay (Ben-Dor & Banin, 1995), organic carbon (Dalal & Henry, 1986; Morra et al., 1991; Reeves et al., 2002) organic matter (Henderson et al, 1992) and soil phosphorus (Bogreckci & Lee, 2005, 2007; Sridhar et al., 2009a ).

In this chapter, we are focuing on the use of LANDSAT TM data to map the changes in elemental concentrations of the soils that were applied with sewage sludge.

### 1.1. Land application of sewage sludge

Application of treated sewage sludges (biosolids) to agricultural land has become a prominent and acceptable method of waste disposal in recent years. Largely because of the 1991 ban on ocean dumping of sewage sludge and the higher costs of sludge incineration and land-filling, the application of sewage sludge to agricultural fields has gained momentum. Sewage sludge is the solid, semi-solid or liquid residue generated during the treatment of domestic sewage in waste water treatment plants (USEPA, 2002). Sewage sludge that is land applied has to meet the treatment criteria and the standard pollutant and pathogen limits set by USEPA under Part 503 rule (USEPA, 2002).

Biosolids are known to improve soil physical characteristics (Epstein et al., 1975; Wei et al., 1985), increase the organic matter and cation exchange capacity and supply the nutrients required for crop growth (Sommers, 1977; Singh & Agrawal, 2008). However, the potential for excess application of biosolids, resulting in a build up of nitrogen, phosphorus (Mantovi et al., 2005), zinc, copper, lead (Mantovi et al., 2005; Udom etal., 2004; Nyamangara & Mzezewa, 1999) and cadmium (Bergkvist et al., 2003) in the surface soils of agricultural fields continues to be an area of concern. Accumulation of phosphorus at high concentrations is a major environmental concern, as it affects the water quality of lakes and rivers in the event of runoff (Shober & Sims, 2003).

Hence, there is an increasing need to continuously monitor the extent of soil contamination in biosolid-applied fields. Even though conventional methods of soil sampling and testing are being used for this purpose, they are often expensive, time-consuming and unsuitable for mapping soil contamination over large areas. The addition of soil contaminants as a result of biosolid application tend to be concentrated in surface soil samples (Mantovi et al., 2005; Bergkvist et al., 2003; Udom etal., 2004; Nyamangara & Mzezewa, 1999).

## 2. Methodology

### 2.1. Soil sampling and chemical analysis

Two adjacent agricultural fields, F34 and F11, that received a cummulative amount of 34 ton acre$^{-1}$ (76 Mg hac$^{-1}$) and 11 ton acre$^{-1}$ (25 Mg hac$^{-1}$) of Class B biosolids on a dry weight basis during the period of 1985-2002 were selected for this study (Fig 1). Soil samples were collected at 0, 30, and 50 cm depths from each of the 70 sampling locations across the two fields. These fields were selected because they are representative of large areas of northwest Ohio where land application of biosolids has become an important agricultural practice. The soil samples were collected on May 19 of 2005, one day prior to LANDSAT over pass, and the sampling locations were marked using a Trimble GeoExplorer (Trimble Navigation Limited, CA, USA) global positioning system (GPS) receiver. The collected soil samples were dried and passed through a 2 mm sieve. The moisture content of the surface soil samples was measured using the gravimetric method. The source of sewage sludge for the agricultural fields in the study area was the Oregon Waste Water Treatment Plant (OWWTP). The basic composition of the sewage sludge of OWWTP is typical of the bio-solids produced in Ohio, which is regulated within the limits set by the U.S. Environmental Protection Agency (USEPA) under the part 503 rule (USEPA 2002).

Soil samples (approximately 0.5 g) were digested with concentrated $HNO_3$, according to USEPA method SW846-3051A (USEPA, 1998) using a Mars Xpress microwave digestion unit (CEM, Matthews, NC, USA). The digested solution was filtered and then analyzed for As, B, Be, Ca, Cd, Cr, Cu, Fe, K, Mg, Mn, Mo, Na, Ni, P, Pb, S, Se, Si and Zn concentrations using inductively coupled plasma-optical emission spectrometry (ICP-OES) (IRIS Intrepid II, Thermo Scientific, Waltham, MA, USA). Quantification was achieved using matrix matched high and low concentration standards for each element. Internal quality controls and blanks were run every ten samples in order to quantify cross-contamination and recovery efficiencies.

Analysis of variance (ANOVA) was used to compare the accumulation of each element in the F34 and F11 fields using SAS version 9.1 statistical software (SAS Institute Inc., Cary, NC, USA). An alpha level of 0.05 was used to determine the significance.

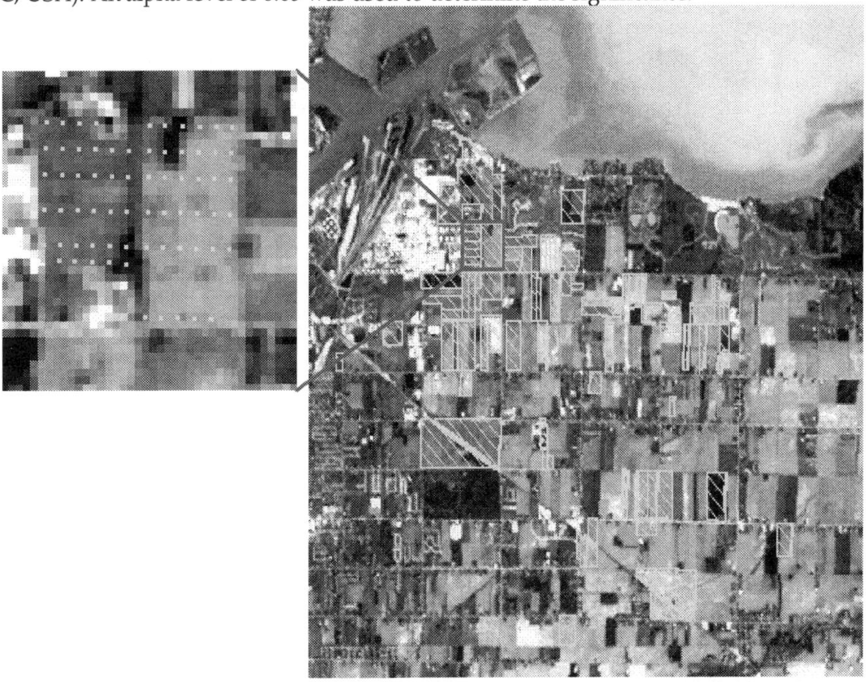

Fig. 1. The LANDSAT 5 TM natural color image (TM bands 1, 2, and 3 displayed as BGR, respectively) obtained on May 20, 2005 showing the eastern part of Lucas County in northwest Ohio; this area drains into Lake Erie, which is towards the northern side (top) of the image. The fields permitted for Class B biosolid application in the area are outlined in the image. The fields marked with red borders are the experimental fields used in this study. Soil sampling locations of the study area were shown as yellow dots in the insert image. (Reprinted from Science of Total Environment 47, Sridhar et al., 2009a, with permission from Elsevier)

### 2.2. LANDSAT data acquisition and analysis

The LANDSAT image frames of May 20 and June 5, 2005, covering the study area were downloaded soon after soil sampling. They were then processed with the ER Mapper image processing software, a commercial product of Earth Resources Mapping, Inc. The study area was located within the LANDSAT overpass region of Path 20, Row 31. The natural color image of the study area, overlaid with outlines of the fields permitted for Class B biosolid applications, is shown in Figure 1. The locations of all the 70 soil sampling points collected one day prior to LANDSAT 5 overpass were also shown separately in Figure 1 on the natural color image of the study area. The study site was dry, without any vegetation, implying that image spectral reflectances represent the spectral reflectance of soil. The procedure for developing the GIS database of the Class B biosolid permitted fields in Wood and Lucas counties of northwest Ohio was reported in detail by McNulty (2005).

Based on the locations of the 70 soil samples, the dark object subtracted (DOS) pixel values corresponding to the LANDSAT TM bands 1-5 and 7 were derived from the original May 20, 2005 image. The spectral range of these LANDSAT TM bands are as follows: Band 1: 450-520 nm; Band 2: 520-600 nm; Band 3: 630-690 nm; Band 4: 760-900 nm; Band 5: 1550-1750 nm; and Band 7: 2080-2350 nm. The dark object of each spectral band is defined as one value less than the minimum digital number found in all the pixels of the image (Vincent et al., 2004). The detailed procedure for DOS and its effects on removal of atmospheric haze was given in Vincent (1997) and Vincent et al. (2004). From the DOS-corrected digital number (DN) values of the six LANDSAT single bands, 15 non-reciprocal spectral ratios were calculated. These spectral ratios are: $R_{2,1}$; $R_{3,1}$; $R_{3,2}$; $R_{4,1}$; $R_{4,2}$; $R_{4,3}$; $R_{5,1}$; $R_{5,2}$; $R_{5,3}$; $R_{5,4}$; $R_{7,1}$; $R_{7,2}$; $R_{7,3}$; $R_{7,4}$; $R_{7,5}$ where R represents the ratio and the numbers represent the LANDSAT TM band numbers (Vincent, 1997). The spectral ratios were calculated using the MINITAB statistical software (MINITAB Inc., State College, PA, USA).

### 2.3. LANDSAT TM best spectral ratio model development and validation

The relationship between the chemical concentrations of the surface soil samples and the DOS DN values corresponding to the six single bands and the 15 non-reciprocal spectral ratios were developed by regression analysis. Using the MINITAB regression analysis component the best subsets regression was employed, and only the top two models with highest $R^2$ adjusted values were chosen to report for each number of variables. The best subsets procedure was used for sequentially entering independent variables one at a time to improve the regression equation's predictive ability. The reported models from the best subset regression output were tested for autocorrelation with a Durbin-Watson (DW) statistical test (Durbin & Watson, 1951). This tests for autocorrelation in the input parameters. Finally, the model which had the highest $R^2$ adjusted and that also passed the DW test was selected as the best model for given inputs. This procedure was reported in detail elsewhere by Vincent (2000) and Vincent et al. (2004). The identified best model was then applied to the same May 20, 2005, LANDSAT image, which was used in developing the model to map the elemental concentration of the surface soils. The model was also applied and validated using the June 5, 2005, LANDSAT image, which was obtained 17 days after the soil sampling. In the LANDSAT images that were applied with the best model, masks were created to limit the display to only bare soil fields.

## 3. Results and discussion

### 3.1. Soil chemical concentration

The chemical concentration of the soils at 0, 30 and 50 cm depths and the moisture content of the surface soils in both the F34 and F11 treated fields are shown in Table 1. Among all the chemicals that were analyzed, the accumulation of Ba, Cd, Cu, S and P were significantly ($p<0.05$) higher in the surface soils of F34, compared to F11 (Table 1). There was no significant difference in the chemical concentrations at 30 and 50 cm depths among the F34 and F11 soils. Also, the moisture content of the surface soils in both the fields was similar (Table 1). The soils were of the prevalent latty silty clay type with surface soils having 40-55% of clay and 3-5% of organic matter (Soil Survey Staff, 2007).

| Treatment | Soil Depth (cm) | Ba (mg/kg) | Cd (mg/kg) | Cu (mg/kg) | S (mg/kg) | P (mg/kg) | Moisture (%) |
|---|---|---|---|---|---|---|---|
| F34 | 0 | 161 (± 28.5) | 5.7 (± 0.5) | 55 (± 5) | 405 (± 57) | 2550 (± 625) | 9.7 (± 7.3) |
| | 30 | 103 (± 16.7) | 5.7 (± 0.46) | 37 (± 5.1) | 197 (± 55) | 796 (± 439) | |
| | 50 | 97 (± 14.4) | 5.6 (± 1.1) | 35 (± 6.1) | 154 (± 31) | 557 (± 103) | |
| F11 | 0 | 98 (± 19.1) | 3.6 (± 1.9) | 37 (± 6.6) | 265 (± 68) | 988 (± 303) | 5.5 (± 2.5) |
| | 30 | 100 (± 13.3) | 3.5 (± 0.9) | 31 (± 5.1) | 165 (± 49) | 588 (± 177) | |
| | 50 | 99 (± 13.5) | 3.7 (± 0.9) | 31 (± 4.9) | 132 (± 51) | 558 (± 149) | |

Table 1. Chemical concentration of soils applied with 34 ton acre$^{-1}$ (F34) and 11 ton acre$^{-1}$ (F11) of class B biosolids. The given values are means ± standard deviation of 35 replicates. (Reprinted from Science of Total Environment 47, Sridhar et al., 2009a, with permission from Elsevier)

### 3.2. LANDSAT spectral ratio model

Regression equations were established to determine the chemical concentrations of Ba, Cd, Cu, S and P, which are significantly ($p<0.05$) higher in the surface soils of F34 compared to F11, using the DOS-corrected six TM bands and the 15 non-reciprocal spectral ratios. The best spectral ratio input models that pass the DW test of significance along with their $R^2$ adjusted and standard error values are given in Table 2.

| Chemical | Best spectral ratio model | $R^2$ Adjusted (%) | SE (mg/kg) |
|---|---|---|---|
| Phosphorus | 4156 -1690 (R51) + 2257 (R73) | 67.9 | 531.2 |
| Copper | 75 -17.9 (R51) + 21.9 (R73) | 59 | 6.9 |
| Sulfur | 507 - 14.7 (R51) + 214 (R73) | 49.3 | 66.8 |

Table 2. Best spectral ratio input models for phosphorous, copper, and sulfur that pass the Durbin Watson test along with the values of $R^2$ adjusted and SE (standard error). Note: The models developed for Cd and Ba did not pass the Durbin Watson test at 5% level of significance. (Reprinted from Science of Total Environment 47, Sridhar et al., 2009a, with permission from Elsevier).

None of the single band models passed the DW test. Phosphorus had the highest $R^2$ adjusted value (67.9%) followed by Cu among the chemical attributes that passed the DW test (Table 2) and are considered for mapping with LANDSAT TM data. Hence, only the P and Cu results were shown in this paper. The P values obtained from chemical analysis of the 70 surface soil sampling locations versus the predicted values of P for the same locations obtained by applying the P spectral ratio model P(mg/kg) = 4156 -1690 (R51) + 2257 (R73) to the May 20, 2005, LANDSAT TM frame is given in Figure 3.

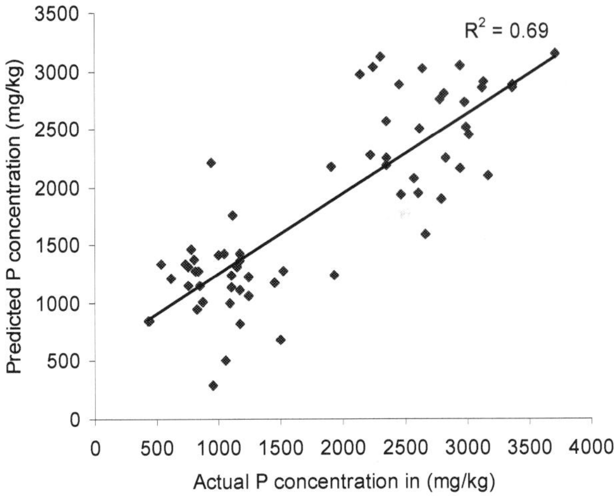


Fig. 3. Actual versus predicted P concentration (in mg/kg) of surface soil samples using the dark object subtracted best P spectral ratio model being applied to the LANDSAT 5 TM frame of May 20, 2005, which was also used for developing the model. (Reprinted from Science of Total Environment 47, Sridhar et al., 2009a, with permission from Elsevier)

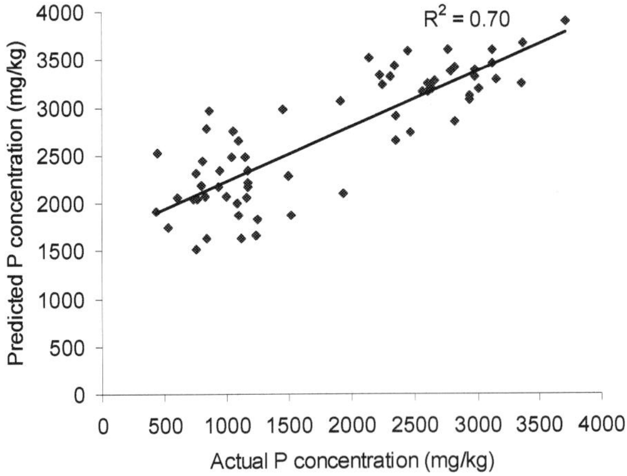


Fig. 4. Actual versus predicted P concentration (in mg/kg) of surface soil samples using the dark object subtracted best P spectral ratio model being applied to the LANDSAT 5 TM frame of June 5, 2005. (Reprinted from Science of Total Environment 47, Sridhar et al., 2009a, with permission from Elsevier).

The P spectral ratio model was also applied to the June 5, 2005 LANDSAT image frame and the predicted P values were plotted against the P values obtained by the soil analysis (Fig. 4). The model performed well in predicting the P concentrations of surface soil when applied to either of the LANDSAT TM images.

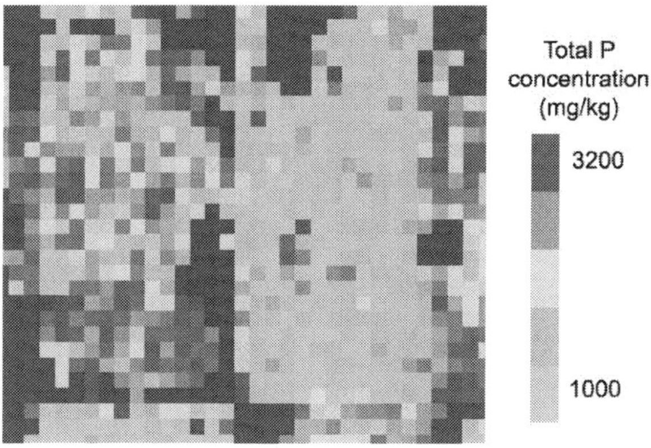


Fig. 5. Image showing the total P concentration (mg/kg) in surface soil samples of fields F34 (left side of the image) and F11 (right side of the image), displayed as red (high P content) to Turquoise (low P content), obtained by applying the best P spectral ratio model to the LANDSAT 5 TM frame of May 20, 2005 which was also used for developing the model. (Reprinted from Science of Total Environment 47, Sridhar et al., 2009a, with permission from Elsevier)

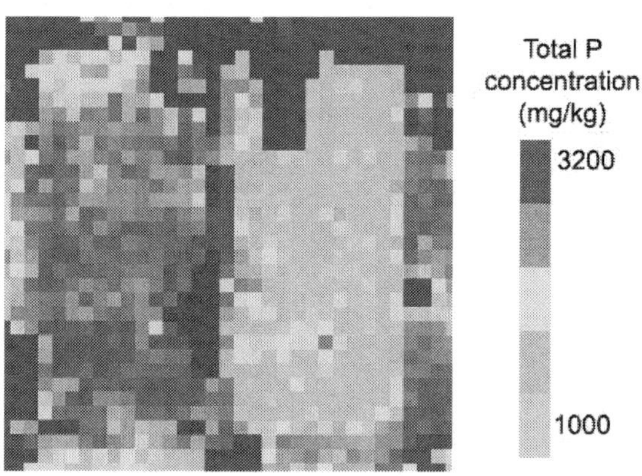


Fig. 6. Image showing the total P concentration (mg/kg) in surface soil samples of fields F34 (left side of the image) and F11 (right side of the image), displayed as red (high P content) to Turquoise (low P content), obtained by applying the P spectral ratio model to the LANDSAT 5 TM frame of June 5, 2005. (Reprinted from Science of Total Environment 47, Sridhar et al., 2009a, with permission from Elsevier)

The application of the best P spectral ratio model to the LANDSAT 5 TM frame of May 20, 2005, which was also used in developing the model, is shown in figure 5. The redder color in this image corresponds to higher amounts of P in surface soil. Figure 6 shows the image of the same spectral ratio model that was developed using the LANDSAT 5 frame of May 20, 2005, being applied to the LANDSAT 5 frame of June 5, 2005. Note that the P concentration in the F34 field is significantly higher than the F11 field in both the images (Fig 5 and 6). The application of the best P spectral ratio model to the May 20, 2005, LANDSAT TM image, showing the part of the watershed that drains into Lake Erie, is given in Figure 7. The fields outlined in this figure are permitted for Class B biosolid application.

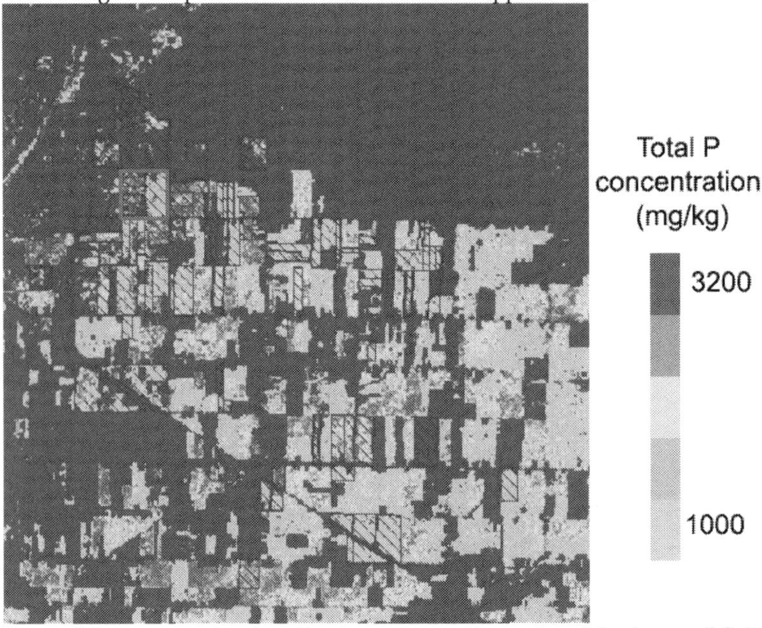


Fig. 7. The total P concentration (mg/kg) in surface soil samples of the bare soil fields in the eastern part of the Lucas County of northwest Ohio, which is a part of the drainage basin of Lake Erie, which is located at the northern side (top) of the image. The image is obtained by applying the P spectral ratio model to the LANDSAT 5 TM frame of May 20, 2005 which was used for developing the model. The fields permitted for Class B biosolid application in the vicinity are outlined in the image. The fields marked with a pink border are the experimental fields used in this study. Not all of the permitted fields in this image would have recently received sewage sludge, and some non-permitted fields could have recently received other types of fertilizers. (Reprinted from Science of Total Environment 47, Sridhar et al., 2009a, with permission from Elsevier)

The Cu values obtained from chemical analysis of the 70 surface soil sampling locations versus the predicted values of Cu for the same locations obtained by applying the Cu spectral ratio model Cu(mg/kg) = 75 -17.9 (R51) + 21.9 (R73) to the May 20, 2005, LANDSAT TM frame is given in Figure 8. The Cu spectral ratio model was also applied to the June 5,

2005 LANDSAT image frame and the predicted Cu values were plotted against the Cu values obtained by the soil analysis (Fig. 9). The model performed well in predicting the Cu concentrations of surface soil when applied to either of the LANDSAT TM images.

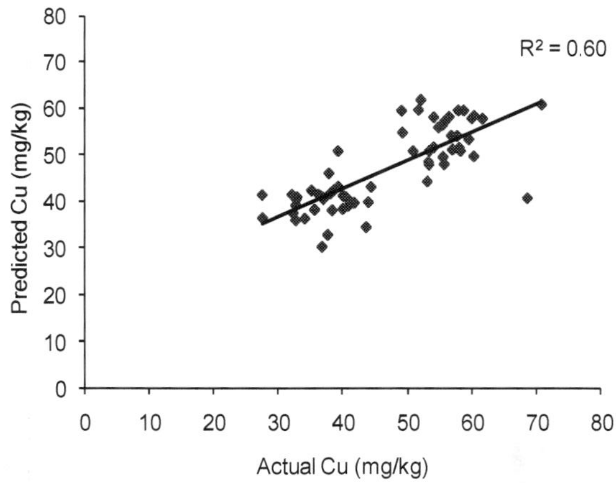


Fig. 8. Actual versus predicted Cu concentration (in mg/kg) of surface soil samples using the dark object subtracted best Cu spectral ratio model being applied to the LANDSAT 5 TM frame of May 20, 2005, which was also used for developing the model.

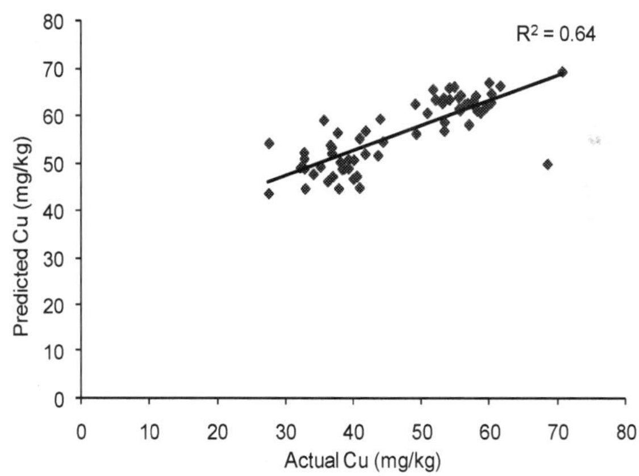


Fig. 9. Actual versus predicted Cu concentration (in mg/kg) of surface soil samples using the dark object subtracted best Cu spectral ratio model being applied to the LANDSAT 5 TM frame of June 5, 2005.

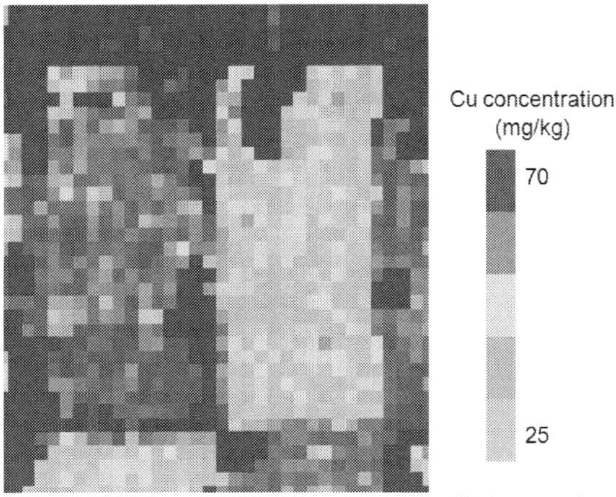


Fig. 10. Image showing the total Cu concentration (mg/kg) in surface soil samples of fields F34 (left side of the image) and F11 (right side of the image), displayed as red (high Cu content) to Turquoise (low Cu content), obtained by applying the best Cu spectral ratio model to the LANDSAT 5 TM frame of May 20, 2005 which was also used for developing the model.

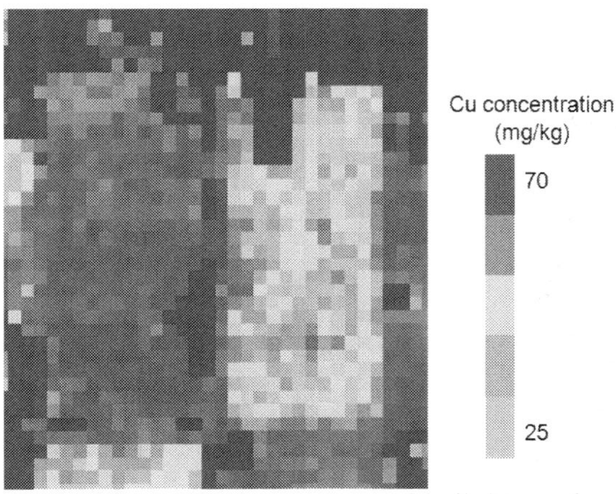


Fig. 11. Image showing the total Cu concentration (mg/kg) in surface soil samples of fields F34 (left side of the image) and F11 (right side of the image), displayed as red (high Cu content) to Turquoise (low Cu content), obtained by applying the P spectral ratio model to the LANDSAT 5 TM frame of June 5, 2005.

The application of the best Cu spectral ratio model to the LANDSAT 5 TM frame of May 20, 2005, which was also used in developing the model, is shown in figure 10. The redder color in this image corresponds to higher amounts of Cu in surface soil. Figure 11 shows the image of the same spectral ratio model that was developed using the LANDSAT 5 frame of May 20, 2005, being applied to the LANDSAT 5 frame of June 5, 2005. Note that the Cu concentration in the F34 field is significantly higher than the F11 field in both the images (Fig 10 and 11). The application of the best Cu spectral ratio model to the May 20, 2005, LANDSAT TM image, showing the part of the watershed that drains into Lake Erie, is given in Figure 12.

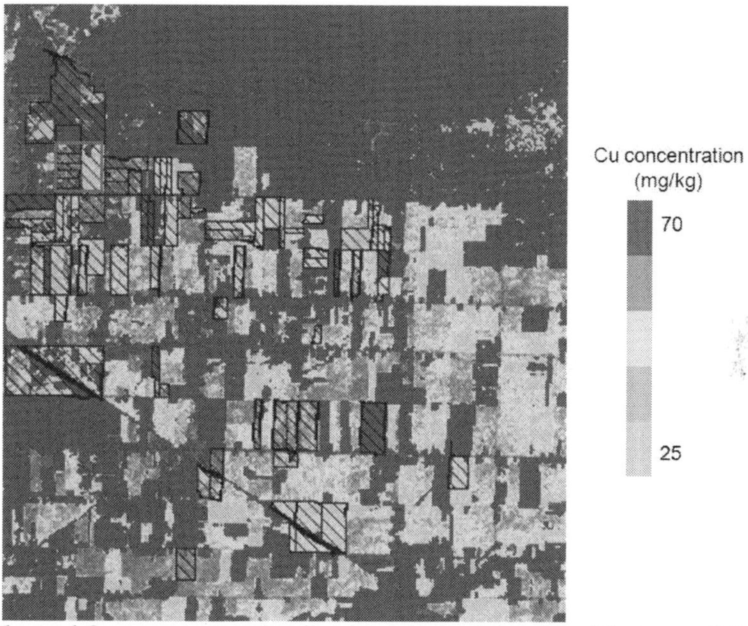


Fig. 12. The total Cu concentration (mg/kg) in surface soil samples of the bare soil fields in the eastern part of the Lucas County of northwest Ohio, which is a part of the drainage basin of Lake Erie, which is located at the northern side (top) of the image. The image is obtained by applying the Cu spectral ratio model to the LANDSAT 5 TM frame of May 20, 2005 which was used for developing the model. The fields permitted for Class B biosolid application in the vicinity are outlined in the image. The fields marked with a pink border are the experimental fields used in this study.

The analytical results showed that the accumulation of P and Cu in surface soil samples of F34 was about 2.6 and 1.5 times higher than for the F11 soils. This confirms the report of Chang et al. (1983) that five continuous years of biosolids application in two California soils at 0, 22.5, 45 and 90 ton per hectare increased the total P concentration of surface soil (0-15 cm) from 515-540 mg/kg to 1092-1312, 1657-2163 and 2617-3470 mg/kg, respectively. Similarly Maguire et al. (2000) reported that the concentration of total soil P in surface soils (0-20 cm) of biosolid amended soils was 738 mg/kg, or nearly double the values in unamended soils, where the total soil P was 403 mg/kg. High concentrations of Cd and Cu in the surface soils of F34 compared to F11 agree with the reports of Nyamangara and Mzezewa (1999), that the long-term application of biosolid increases the accumulation of Cd and Cu in the surface soils

## 4. Discussion

LANDSAT TM data can be used to estimate and map some chemical characteristics of soils, such as total phosphate content, as shown in this study. Our results enabled us to conclude that remotely sensed imagery of bare soil fields can be used to quantify and map the spatial variation of total P and Cu concentration in surface soils. The technology is simple enough to be applied to the entire watershed. The P and Cu spectral ratio models were more robust and reliable than the single band input models and can be applied to bare soil fields with low (<13%) soil moisture.

Nanni and Dematte (2006) have successfully employed LANDSAT TM data to estimate the sand, silt, clay, organic matter, cation exchange capacity (CEC) and sum of cations in Brazilian soils. They derived spectral reflectance values from the corrected LANDSAT image to develop multiple regression equations in order to estimate the different physical and chemical characteristics of the soils; however, no soil maps were presented in that study (Nanni & Dematte, 2006). Our study is significant because it represents the first successful effort in using LANDSAT TM data to estimate and map P concentration in surface soils. We also successfully validated the P spectral ratio model by applying to another LANDSAT image obtained on June 5, 2005.

Aerial imagery was used to map the organic carbon (Chen et al., 2000), clay content (Sullivan et al., 2005), organic matter and Bray-1 phosphorus concentration (Varvel et al., 1999) and LANDSAT TM imagery was used to estimate the physical and chemical properties (Nanni & Dematte, 2006) of surface soils in the previous studies. However, the algorithms developed in these studies were based on the reflected image intensity values of the soils, which required correction for atmospheric haze with atmospheric models before applying the algorithms to another date. The P spectral ratio model developed in our study is based on the DOS-corrrected spectral ratios and is more robust that any model that could be derived from a combination of single spectral bands. Vincent et al. (2004) showed that the DOS spectral ratio models were more robust than single band models and can be applied with reasonable accuracy to different times of data collection, though their subject was cyanobacteria blooms in lakes or streams, and the present study is about P concentrations in bare soils on dry land.

By applying our P and Cu spectral ratio model, we can identify and map the P and Cu concentration in surface soils as a result of biosolid application. Because P accumulation in soils can also result from the application of biosolids, animal manures, and man-made fertilizers, this research has significant implications in identifying the fields with high concentrations of surface soil P, thus helping implementation of P-based management practices on agricultural fields, with an aim toward reduction of P runoff into nearby surface water bodies.

Shober and Sims (2003) reported that twenty-four of the states and territories in the United States now have regulations to restrict the land application of biosolids, based on phosphorus concentration in soil. Thirteen of these 24 states have established actual numerical limits for soil test phosphorus (STP), with an aim to cease the application of biosolids once these limits are reached. As the total soil P and STP are linearly related to each other (Allen & Mallarino, 2006), our P spectral ratio model can be used to monitor P levels in surface soils.

The remainder of this chapter is devoted for the presentation of a sample application which demonstrates how quantitative remote sensing of soil P and Cu concentrations can be used to meet the variety of information needs. This particular case study shows the use of our remote sensing models for monitoring a recent environmental disaster.

### 4.1 Sample application

#### a. Coal fly ash spill

A vast amount of fly ash, a by-product of coal incineration, was spilled over a wide area on December 22, 2008, at approximately 0100 hours EST, when an earthen wall of a fly ash disposal pond broke at the Tennessee Valley Authority (TVA) Kingston Fossil Plant, located at Harriman, Roane County, Tennessee (TVA, 2009; Fig 13). This is the largest environmental disaster of its kind involving a coal fly ash spill in the US history (New York Times, 2008). Approximately 5.4 million cubic yards of fly ash were spilled over an area of 300 acres outside the ash storage ponds. Fly ash was also spilled into the Emory River and the tributaries that flow into the Emory River, which serves as a source of drinking water (TVA, 2009; New York Times, 2008; Fig 13).

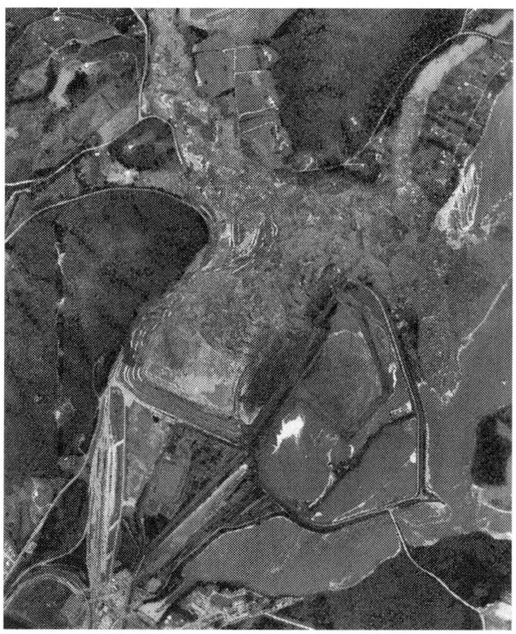

Fig 13. The aerial image of the Tennessee fly ash spill area acquired soon after the spill. (Image courtesy of Tennessee Valley Authority and can be viewed at http://www.tva.gov/kingston/photo_gallery/pages/123002.htm)

Accumulation of large amounts of fly ash as a result of coal combustion is becoming a major environmental concern in the United States. With the Nation's increasing demand for energy, the increase in coal combustion has required the disposal of large quantities of coal combustion residues. Approximately 131 million tons of coal combustion residues were generated within the United States in 2007, of which 36% was disposed in landfills, 21% in surface impoundments, and the remainder was reused for beneficial purposes. There are approximately 300 landfills and 300 surface impoundment facilities used by 440 coal fired plants across the United States (Fig 13).

The physical and chemical properties of the fly ash generated at a given coal-fired plant depends on the nature of the coal being burned, type of the combustion method, and the storage and handling methods involved. In general, fly ash is substantially rich in many elements (including heavy metals) and is usually stored in either landfills or artificial lagoons on open land. The general chemical composition of fly ash is given in Table 3 (Page *et al.*, 1979).

| Element | Fly ash concentration |
|---|---|
| *Major elements (%)* | |
| Al | 0.1-17.3 |
| Ca | 0.11-22.2 |
| Fe | 1-29 |
| Mg | 0.04-7.6 |
| Na | 0.01-2.03 |
| K | 0.15-3.5 |
| S | 0.002-0.004 |
| P | 0.04-0.8 |
| *Trace elements (ppm)* | |
| As | 2.3-6300 |
| B | 10-618 |
| Cd | 0.7-130 |
| Co | 7-520 |
| Cr | 10-1000 |
| Cu | 14-2800 |
| Hg | 0.02-1 |
| Mn | 58-3000 |
| Mo | 7-160 |
| Ni | 6.3-4300 |
| Pb | 3.1-5000 |
| Se | 0.2-134 |
| Zn | 10-3500 |

Table 3. General chemical characteristics of Fly ash. (After Page et al., 1979; Reprinted from Photogrammetric Engineering and Remote Sensing 75, Sridhar et al., 2009b)

Mapping the spatial distribution of chemical concentrations as a result of fly ash spill is important for determining its effect on human and environmental health and for conducting remediation and recovery efforts. Previous studies have shown that fly ash pollution can be monitored through several methods: conventional soil sampling and analysis, measurement of Sr-87 to Sr-86 ratios in plant samples growing in the vicinity, and measurement of ferromagnetic mineral concentrations present within the fly ash by magnetic mapping. However, all of these methods are based on point measurements at the ground-level, requiring intensive sampling of large-scale contaminated areas, which is expensive, time-consuming and sometimes not even possible as the areas may be sufficiently hazardous (e.g., poor slope stability) and inaccessible to scientists and engineers.

### b. LANDSAT TM Images

Remote sensing has been used as an alternative method for determining and mapping the physical and chemical characteristics of exposed soils. In this study, we have used LANDSAT Thematic Mapper (TM) sensor data to identify and map the TVA's fly ash spill in Tennessee. The Thematic Mapper and Enhanced Thematic Mapper plus (ETM+) sensors were aboard the LANDSAT 5 and 7 respectively, each of which has a 16-day repeat cycle with a pixel size of 30 m X 30 m for bands 1-5 and 7. Because the two satellites have an 8-day offset from one another, it is possible to obtain coverage of a given ground area every 8 days, though the TM sensor aboard LANDSAT 7 (called ETM+) has approximately a quarter of its pixels missing, due to the loss of its scan line converter since May 31, 2003. In order to monitor the before and after fly ash spill events of the Kingston fly ash plant, we downloaded LANDSAT 5 TM data acquired on November 20, 2008 (32 days before fly ash spill); December 22, 2008 (about 9 hours after fly ash spill) and February 1, 2009 (40 days after fly ash spill) for this study. The LANDSAT TM images were then processed with the ER Mapper image processing software, a commercial product of Earth Resources Mapping, Inc., as mentioned earlier for the sewage sludge study.

For the purpose of measuring the P and Cu concentrations in the exposed fly ash and soil in the spill vicinity, the spectral ratio models given in Table 2 were applied to each of the three LANDSAT TM images covering the fly ash spill study area.

### c. Results of Image Analysis

Figure 14 shows the natural color images, total P and Cu concentration images of the fly ash-spill vicinity for each of the following dates of LANDSAT TM overpass: November 20, 2008, December 22, 2008, and February 1, 2009, representing the periods of 32 days before the fly ash spill, 9 hours after, and 40 days after the fly ash spill, respectively. The fly ash is seen as a grey-colored area in the December 22, 2008 natural color imagery. Image interpretation of a standard natural color image (TM bands 1, 2, and 3 displayed as blue, green, and red, respectively) yields limited success for mapping the fly ash deposits because the contrast between fly ash and background soils is small in visible wavelength regions. The LANDSAT TM derived P concentrations range from 1,500 to 4,500 ppm and the Cu concentrations from 25 to 75 ppm in the November, 2008 and February, 2009 images. We have assigned turquoise, yellow, orange and red colors to increasing P and Cu concentrations as indicated by the respective color coding bars in Figure 1. The intense red color in the December 22, 2008 image clearly stands out due to larger areas of high P and Cu concentrations exposures at the surface, compared to the November 20, 2008 image acquired before the fly ash spill. The 25 to 75 ppm range of Cu concentrations derived from the December 22, 2008 image correlates well with the 29.9 to 69.4 ppm range of Cu values obtained through laboratory analysis from eight (8) ash sampling locations collected from

December 23, 2008 through January 5, 2009 in the fly ash spill vicinity (Tetra Tech, 2009). The P concentrations were not reported through soil analysis (Tetra Tech, 2009).

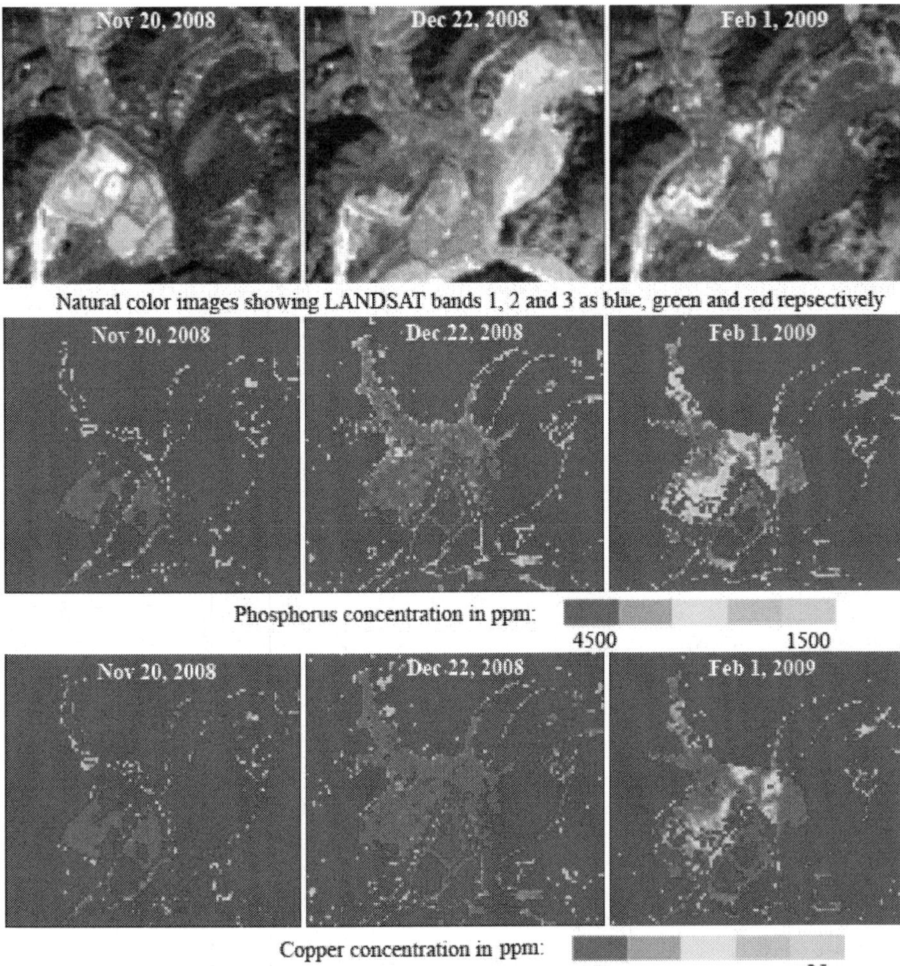


Fig. 14. The LANDSAT TM images of the Tennessee fly ash spill area acquired on November 20, 2008 (32 days before fly ash spill); December 22, 2008 (9 hours after fly ash spill) and February 1, 2009 (40 days after fly ash spill), respectively. The images in the rows 1-3 represent the natural color image, surface P concentration image, and surface Cu concentration image, respectively. (Reprinted from Photogrammetric Engineering and Remote Sensing 75, Sridhar et al., 2009b, with permission from the American Society for Photogrammetry & Remote Sensing)

In November, before the fly ash spill, the high concentrations of P and Cu were shown to be confined to the fly ash ponds. In December, a large extent of increased P and Cu concentrations is clearly evident towards the northern and north eastern side of the ash holding ponds as shown in Figure 14. These high P and Cu concentration features agree with the fly ash spill towards the northern and north eastern sides of the ash holding ponds (TVA, 2009). In February, the P and Cu concentrations mapped with the same LANDSAT TM algorithms were lower than those recorded in December. As an action item for dust suppression, TVA spread 85 tons of winter rye grass, 650 tons of straw, and other erosion-control mulch on the surface of the fly ash spill areas from January 3 through January 15, 2009 (TVA, 2009). For areas of where grass and straw were spread over, the February 1, 2009 image shows lower P and Cu concentrations, compared to the December 22, 2008 image. This series of images shows that LANDSAT TM data can be used to quantitatively monitor the remediation and recovery efforts of P and Cu contaminated sites.

## 5. Conclusions

In summary, the application of P and Cu mapping algorithms from LANDSAT TM images have allowed us to map and estimate the increase in surface concentrations of these two elements in surface soil samples. These results show the effective use of multispectral satellite image analysis in determining surface soil chemical concentrations. Compared to point measurements made by ground-based soil sampling and chemical analysis, the satellite-based measurements have a great advantage in mapping the spatial distribution and concentration of elements and chemical compositions over time. As LANDSAT TM has 30-m spatial resolution, there is a measurement by the satellite for every 1/5th of an acre (the area covered by one pixel) during each overpass, which would be cost prohibitive for point measurements that require manual soil sample collection and laboratory analysis of each sample. Satellite monitoring of surface soil elemental concentrations for environmental purposes can surpass point measurements on the ground and measured in the laboratory, at least for P and Cu. We have looked at several other elements for similar algorithms but found none as good as these two. The limitations of our spectral ratio models is that they were developed using bare soil fields that had a low soil moisture (<13%) and the soils tested were of the prevalent soil type (Latty Silty Clay) in this region. Also the P and Cu concentration that went into these models were in the ranges of 700-3000's and 30-60's respectively. Thus we have no evidence that these models will be accurate for soils with P and Cu concentrations outside these limits and with different soil types of higher soil moisture contents. Further research is being done to test the model to determine the P and Cu concentration in surface soils of other soil types in this region.

## 6. References

Allen, B.L. & Mallarino, A.P. (2006) Relationships between extractable soil phosphorus and phosphorus saturation after long-term fertilizer or manure application. *Soil Sci Soc Am J*, Vol.70, pp.454-463.

Ben-Dor, E. & Banin, A. (1990) Near infrared reflectance analysis of carbonate concentration in soils. *Appl. Spectrosc*, Vol.44, pp.1064-1069.

Ben-Dor, E. & Banin, A. (1995) Near infrared analysis (NIRA) as a method to simultaneously evaluate spectral featureless constituents in soils. *Soil Sci*, Vol.4, pp.259-270.

Bergkvist, P.; Jarvis, N.; Berggren, D. & Carlgren, K. (2003) Long-term effects of sewage sludge applications on soil properties, cadmium availability and distribution in arable soil. *Agric. Ecosyst. Environ*, Vol.97, pp.167-179.

Bogrekci, I. & Lee, W.S. (2005) Spectral soil signatures and sensing phosphorus. *Biosyst. Eng*, Vol.92, pp.527-533.

Bogrekci, I. & Lee, W.S. (2007) Comparison of ultraviolet, visible and near infrared sensing for soil phosphorus. *Biosyst. Eng*, Vol.96, pp.293-299.

Chang, A.C.; Page, A.L.; Sutherland, F.H. & Grgurevic, E. (1983) Fractionation of phosphorus in sludge affected soils. *J. Environ. Qual*, Vol.12, pp.286-290.

Chang, C.W.; Laird, D.A.; Mausbach, M.J. & Hurburgh, Jr C.R. (2001) Near-infrared reflectance spectroscopy - principal component regression analysis of soil properties. *Soil Sci. Soc. Am. J*, Vol.65, pp.480-490.

Chen, F.; Kissel, D.E.; West, L.T. & Adkins, W. (2000) Field-scale mapping of surface soil organic carbon using remotely sensed imagery. *Soil Sci. Soc. Am. J*, Vol.64, pp.746-753.

Dalal, R.C. & Henry, R.J. (1986) Simultaneous determination of moisture, organic carbon, and total nitrogen by near infrared reflectance spectrophotometry. *Soil Sci. Soc. Am. J*, Vol.50, pp.120-123.

Dematte, J.A.M.; Pereira, H.S.; Nanni, M.R.; Cooper, M. & Fiora, P.R. (2003) Soil chemical alterations promoted by fertilizer application assessed by spectral reflectance. *Soil Sci*, Vol.168, pp.730-747.

Durbin, J. & Watson, G.S. (1951) Testing for serial correlation in least squares regression: II. *Biometrica*, Vol.38, pp.159-178.

Epstein, E.; Taylor, J.M. & Chaney, R.L. (1975) Effects of sewage sludge on some soil physical properties. *J. Environ. Qual*, Vol.4, pp.139-142.

Henderson, T.L.; Baumgardner, M.F.; Franzmeieir, D.P.; Stott, D.E & Coster, D.C. (1992) High dimensional reflectance analysis of soil organic matter. *Soil Sci. Soc. Am. J*, Vol.56, pp.865-872.

Ji, J.F.; Balsam, W.L.; Chen, J. & Liu, L.W. (2002) Rapid and quantitative measurement of hematite and goethite in the Chinese Loess-Paleosol sequence by diffuse reflectance spectroscopy. *Clay Miner*, Vol.50, pp.208-216.

Lobell, D.B. & Asner, G.P. (2002) Moisture effects on soil reflectance. *Soil Sci. Soc. Am. J*, Vol.66, pp.722-727.

Maguire, R.O.; Sims, J.T. & Coale, F.J. (2000) Phosphorus fractionation in biosolids-amended soils: Relationship to soluble and desorbable phosphorus. *Soil Sci. Soc. Am. J*, Vol.64, pp.2018-2024.

Mantovi P, Baldoni G, Toderi G. Reuse of liquid, dewatered, and composted sewage sludge on agricultural land: effects of long-term application on soil and crop. Water Res. 2005; Vol.39, pp.289-296.

McNulty, W.S. (2005) The creation of a GIS database and the determination of sludge's spectral signature in an agricultural setting. M.S. Thesis. Department of Geology, Bowling Green State University, Bowling Green, OH, USA.

MINITAB Statistical Software Version 15. State College, PA: MINITAB Inc.; 2007-2008.

Morra, M.J.; Hall, M.H. & Freeborn, L.L. (1991) Carbon and nitrogen analysis of soil fractions using near infrared reflectance spectroscopy. *Soil Sci. Soc. Am. J*, Vol.55, pp.288-291.

Nanni, M.R. & Dematte, J.A.M. (2006) Spectral reflectance methodology in comparison to traditional soil analysis. *Soil Sci. Soc. Am. J*, Vol.70, pp.393-407.

New York Times (2008) Tennessee ash flood larger than initial estimate, Published on December 26, 2008, URL: *http://www.nytimes.com/2008/12/27/us/27sludge. html?_r=2&ref=us* (last date accessed: 1 July 2009).

Nyamangara, J. & Mzezewa, J. (1999) The effect of long-term sewage sludge application on Zn, Cu, Ni and Pb levels in a clay loam soil under pasture grass in Zimbabwe. *Agric. Ecosyst. Environ*, Vol.73, pp.199-204.

Page, A.L.; Elseewi, A.A. & Straughan, I.R. (1979) Physical and chemical properties of fly ash from coal-fired power plants with special reference to environmental impacts. *Residue Reviews*, Vol.71, pp.83-120.

Post, D.F.; Fimbres, A.; Matthias, A.D.; Sano, E.E.; Accioly, L.; Batchily, A.K. & Ferreira, L.G. (2000) Predicting soil albedo from soil color and spectral reflectance data. *Soil Sci. Soc. Am. J*, Vol.64, pp.1027-1034.

Reeves, III J.B.; McCarty, G.W.; Mimmo, T. (2002) The potential of diffuse reflectance spectroscopy for the determination of carbon inventories in soil. *Environ. Pollut.* Vol.116, pp.S264-S277.

SAS Institute. SAS Software Version 9.1. Cary, NC: SAS Institute, Inc.; 2002-2003.

Shober, A.L. & Sims, J.T. (2003) Phosphorus restrictions for land application of biosolids: current status and future trends. *J. Environ. Qual*, Vol.32, pp.1955-1964.

Singh, R.P. & Agrawal, M. (2008) Potentail benefits and risks of land application of sewage sludge. *Waste Management*, Vol.28, pp.347-358.

Soil Survey Staff, Natural Resources Conservation Service, United States Department of Agriculture. Web Soil Survey. http://websoilsurvey.nrcs.usda.gov. 2007.

Sommers, L.E. (1977) Chemical composition of sewage sludges and analysis of their potential use as fertilizers. *J. Environ. Qual*, Vol.6, pp.225-232.

Sridhar, B.B.M. & Vincent, R.K. (2009) Mapping and estimation of phosphorus and copper concentrations in fly ash spill area using LANDSAT TM Images. *Photogrammetric Engineering and Remote Sensing*, Vol.75, Nb.9, pp.1030-1033.

Sridhar, B.B.M.; Vincent, R.K.; Witter, J.D. & Spongberg, A.J. (2009) Mapping the total phosphorus concentration of biosolid amended surface soils using LANDSAT TM data. *Science of Total Environment*, Vol.47, pp.2894-2899.

Sullivan, D.G.; Shaw, J.N. & Rickman, D. (2005) IKONOS imagery to estimate surface soil property variability in two Alabama physiographies. *Soil Sci. Soc. Am. J*, Vol.69, pp.1789-1798.

Tennessee Valley Authority (2009) Corrective Action Plan for the TVA Kingston Fossil Plant Ash Release, URL: *http://www.tva.gov/kingston/cap/ TVA_ Corrective _ Action_ Plan_Draft _D5.pdf* (last date accessed: 1 July 2009).

Tetra Tech EM Inc. (2009) Final CERCLA emergency response report, Kingston fossil plant Fly ash response Harriman, Roane County, Tennessee Tetra Tech Inc. Soil and ash sampling results Kingston fossil fly ash response Harriman, Roane County, Tennessee, URL: *http://www.epaosc.org/sites/4642/files/ erfinal reporttvakingston.pdf* (last date accessed: 1 July 2009).

U. S. Environmental Protection Agency. Standards for the use or disposal of sewage sludge. Office of Water, Washington D. C, 2002.

U. S. Environmental Protection Agency. Test methods for evaluating solid waste. Office of Solid Waste and Emergency Response, Washington D. C, 1998.

Udom, B.E.; Mbagwu, J.S.C.; Adesodun, J.K. & Agbim, N.N. (2004) Distributions of zinc, copper, cadmium and lead in a tropical ultisol after long-term disposal of sewage sludge. *Environ. Int.* Vol.30, pp.467-470.

Varvel, G.E.; Schlemmer, M.R. & Schepers, J.S. (1999) Relationship between spectral data from an aerial image and soil organic matter and phosphorus levels. *Precision Agric*, Vol.1, pp.291-300.

Vincent, R.K. (1997) Fundametals of geological and environmental remote sensing. Prentice Hall, Upper Saddle River, NJ.

Vincent, R.K. (2000) Forecasts of monthly averaged daily temperature highs in Bowling Green, Ohio from monthly sea surface temperature anomalies in Eastern Pacific ocean during the previous year. *Photogramm. Eng. Remote Sens*, Vol.66, pp.1001-1009.

Vincent, R.K.; Qin, X.; McKay, R.M.L.; Miner, J.; Czajkowski, K.; Savino, J. & Bridgeman, T. (2004) Phycocyanin detection from LANDSAT TM data for mapping cyanobacterial blooms in Lake Erie. *Remote Sens. of Environ*, Vol.89, pp.381-392.

Wei, Q.F.; Lowery, B. & Peterson, A.E. (1985) Effect of sludge application on physical properties of a silty clay loam soil. *J. Environ. Qual*, Vol.14, pp.178-180.

11

# OLFISH - A complete, paperless solution for the collection, management and dissemination of marine data

Dr. Amos Barkai, Fatima Felaar, Karl Geggus, Zahrah Dantie and Arno Hayes
*Olrac (Ocean Land Resource Assessment Consultants)*
*South Africa*

## 1. Introduction

Fisheries management is continually frustrated by the lack, or poor quality, of critical data on fishing operations (catches, duration, gear, locations and relevant environmental conditions). While quantitative methods for managing fisheries have developed with considerable complexity, the quality of the available data remains an obstacle for meaningful advances in fisheries management. There are a number of aspects to the problem, not all of which are technical. A culture of protecting catch data and disinformation is common amongst fishers, fishing companies and even formal state-run offices, and significant education is needed in order to change this culture.

Another problem is the poor quality of historic data in many fisheries around the world. Much energy is wasted and important opportunities lost because of the uncertainty surrounding crucial historic data. For example, there are typically many factors related to catch-per-unit-effort data, a key index of trends in resource abundance, which are not recorded, and hence cannot be incorporated into statistical analyses. Frequently, these missing data are crucial to management decisions. For scientists, unreliable data leads to a poor basis for stock assessment models and management programs. For industry, the lack of sound data significantly reduces its fishing efficiency, since past performance cannot be studied properly. As a result, poor management decisions based on unreliable analyses are made, often with substantial cost and risk to fish resources and the fishing industry.

Although there is presently greater awareness amongst scientists and fisheries managers about the importance of collecting fishing data, there is still confusion about exactly which data are needed, and how to collect and store them. It is common for skippers to record scientific data on one form, for shore managers to use another for commercial purposes, and for skippers to keep separate fishing logbooks. These data are then transferred to different computer systems, often complex spreadsheets, or, sometimes, are left in paper format in large, inaccessible books and files. There is a degradation in the quality of data because of the multi-stage process of transcription from handwritten logbook sheets to paper forms and then to computer databases.

So even when good will is present, technically, the absence of a flexible and comprehensive system for capturing essential data during fishing operations is a major obstacle. A large amount of logistical and environmental data is lost simply because of the difficulty of recording this information easily in real time. This is despite the advent of a complex array of sensory equipment available in the bridge of modern fishing vessels. As a result, environmental patterns become part of a skipper's experience and seldom, if ever, become formally available to scientists or managers of fishing operations.

The most logical first point of data entry, through the fishing vessel skipper, should occur in digital format directly into a computer. One of the difficulties with fisheries data is the complexity of the logical linkages between the different types of data. Any reasonable approach to the problem requires the use of modern relational databases which are able to address the multidimensional complexity of the problem.

In order to address many of the problems described above, Olrac (www.olrac.com) a South African company, has developed a data collection and management system it has named Olfish (www.olfish.com) for the specific use of operators and managers in the marine environment with a special focus on the commercial fishing industry.

## 2. Electronic Logbooking

### 2.1 Benefits

An obvious approach to the "data crises" is to bring modern data and information technology (Elog) to the marine environment in general, and to the commercial and recreational fishing industry in particular. Providing fishers with accurate yet easy to use data logging tools could potentially transform the entire fishing fleet and the fishers community into the largest surveyors group of the marine environment in the world. The calibre of data produced through electronic logbooks has the potential to benefit all sectors of the fishing industry, from the fishers themselves to seafood consumers, resource managers, scientists and government enforcement agencies in between. In addition, the international shift towards a greater emphasis on output control measures, such as annual catch limits (ACL's) and total allowable catches (TAC's), requires the implementation of sophisticated catch monitoring tools in order to allow for a near-real-time auditing of catch versus TAC. However, it is important to note that the benefits of electronic data logging go beyond merely adhering to regulations. It is crucial for the industry to realise that it will ultimately be the greatest beneficiary of accumulated good quality data. A few obvious benefits derived from the collection of a large amount of accurate data in a near-real-time environment are:

#### 2.1.1 Better Stock Assessments

The accuracy and timely delivery of electronically recorded data will allow for more exact indications of catch in a current year. In the past, due to the delays of paper-based reporting, incomplete data from preceding years has been used to estimate the TAC of the following year. The uncertainty associated with such calculations has resulted in conservative stock assessments which lead to overly restrictive TAC's. This means that there is often an over-discard of fish which would otherwise be commercially viable. Electronic data logging would allow for up-to-date and accurate data to be used for TAC estimation, thus eliminating much uncertainty and adding weight and justification to the TAC's allocated.

### 2.1.2 Better Targeting and Gear Utilization

The security and verification features of electronic logbooks, (see Security and Data Integrity - 5.4.7 below), as well as multimedia photographic and video utilities (see Multinote Taker and Notebook - 6.2 below), can potentially replace the role of an observer onboard a vessel. This can then be adapted into an incentive scheme for improved gear and fishing-ground selectivity, thus reducing unintended bycatch. Capturing target species may also lead to a decrease in days at sea, which is often beneficial for the skipper.

### 2.1.3 Faster Transmission

Faster transmission of information from sea to shore allows for "on-time reaction", i.e. decisions made on a regulatory, managerial, commercial or environmental basis are relevant to what is actually happening at sea. Back-logged, non-electronic reporting means that any event at sea is only registered on shore sometimes up to several weeks after it has occurred. Responding to month-old information, particularly in an ever-dynamic ocean environment, is practically pointless. Faster transmission will have a substantially positive effect on, for example, quota management, conservation and even commercial decision-making.

### 2.1.4 Catch Prediction and Management

Built-in analytical tools available within electronic logbook software (see Olfish Explorer - 6.4 below) are able to harness historical information stored in their electronic databases to help fishers calculate and predict fish migration, fishing hotspots etc. This greatly increases efficiency in a number of fields, such as targeting areas and the selection of fishing grounds and techniques. Similarly, fishers will be able to avoid "dry" areas, maximizing their time at sea and ultimately reducing discarding rates.

### 2.1.5 Traceability

Traceability is the ability to locate the source and "journey" of a fish from ocean to supermarket shelf. Legal organizations, such as the Marine Stewardship Council, prohibit fish without certification logos from entering the market. Such logos are obtained through traceability, i.e. proving that the fish in question had been caught in a certified area under certified conditions. Electronic data logging makes traceability a simple and speedy process. Information from the vessel at sea can be efficiently transmitted to market authorities who can then clear the catch for sale. Furthermore, electronic data logging allows for a highly detailed recording of catch information. Thus, catch freshness can easily be proven, increasing its market value. Buyers then benefits from being able to accurately estimate the shelf-life of the product they have bought. None of this would be possible without verifiable and immediate traceability.

## 2.2 The Olfish System: A Short Overview

Olfish is a third-generation, data logging and data management, software tool which was initially developed for the commercial fishing industry, but now provides a complete solution for the collection, management and reporting of other vessel-based activities, such as commercial and recreational fishing trips, oceanographic surveys, marine inspections, cargo and service trips, surveillance missions, etc.

The present version of Olfish includes three basic components in order to cater for the entire data flow, from at-sea collection to the generation and dissemination of reports.
The onboard, data collection component named Olfish Dynamic Data Logger (Olfish-DDL) is a standalone data collection tool installed onboard the vessel's PC.
Olfish-DDL also has a shore component which is identical to the vessel version but allows data from many vessels to be stored and viewed on one user-interface. This component is available in two versions:

a. A Single Fleet unit that aggregates operational data from vessels of a single company or organisation.
b. A Meta-Shore unit, which can aggregate operational data received from many shore units. The Meta-Shore unit can be used by a government agency, fishing association or even a union of states to manage data from a number of countries/states.

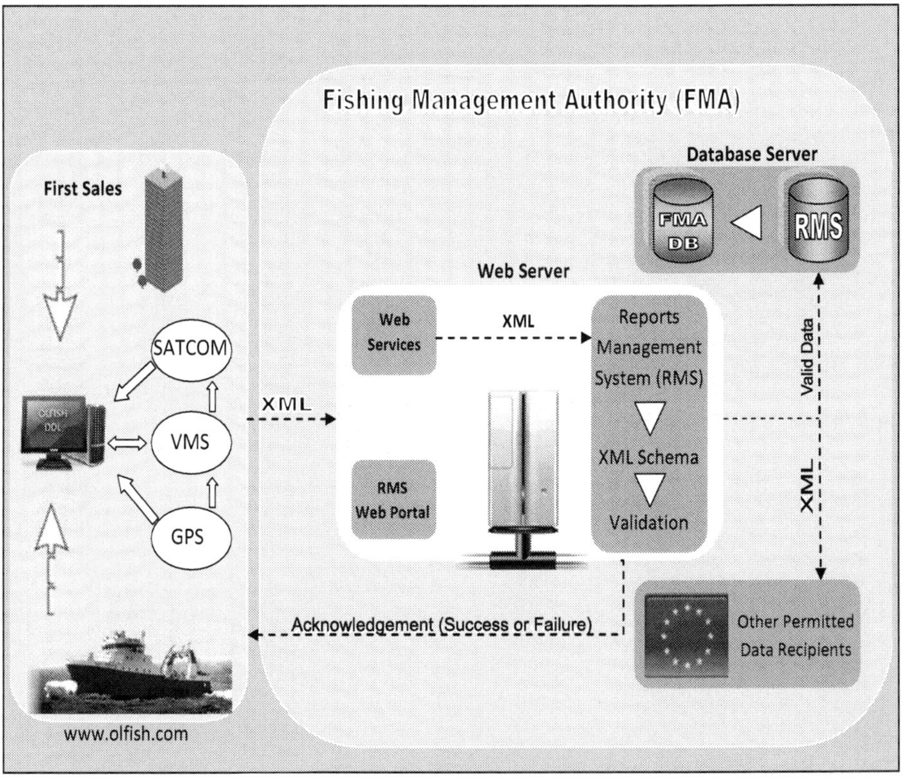


Fig. 1. Overall structure of the Olfish data collection and management system

The third component of Olfish is a web application named Olfish- Report Management System (Olfish-RMS$^{TM}$) and its main function is to receive, store and disseminate reports coming from Olfish-DDL (or, if necessary, other, third party, data logging systems). Olfish-RMS also allows for the direct entry of data via an internet interface for cases where the use of an onboard data-logger is not practical (cost or unsuitable working environment). With Olfish-RMS the entire fleet of vessels can be managed. It includes a vessel registry, a full quota management system and an elaborate administrative component which allows Olfish-RMS to be customised to satisfy many needs.

### 2.3 Olfish Dynamic Data Logger

#### 2.3.1 Basic Functionality

Olfish-DDL is a touch-screen-ready utility that captures data in real-time and/or after the fishing activity has taken place. Olfish-DDL can read GPS input via an additional GPS logging utility and it incorporates GIS capabilities for easy viewing of vessel movements and other operational fishing data. With Olfish-DDL, the user can collect any type of data in any form. These include images, video clips, numerical and alphanumeric fields, free text comments, date, time, location, etc. Olfish allows data to be inserted from guiding images ("infograph") to guide it through complex data entry needs. Each mode of data entry has its own unique data entry interface, specifically designed for the type of data recorded. Olfish-DDL is highly customisable and can be easily modified to address vastly different data recording and reporting needs.

Fig. 2. Olfish vessel unit on a tablet PC

### 2.3.2 Overall Structure

Olfish-DDL consists of the following:

- Configuration files defining levels, fields, parameters
- Database for working data
- Database for archived data
- User interface elements: Data Entry, Data Browser, Mapper, Data Centre, Mini Reporter, Explorer
- Input/output modules for the following types of data:
  a. Reports to specific agencies and third-parties
  b. Import / export of operational data
  c. Backup of the complete system
  d. Error / exception handling reports to Olfish Support

There are two main levels of configuration in Olfish-DDL:

User interface: This is a developer-level configuration which governs the way the command bar menu (Dynamic Commands Bar - DCB) functions, based on client specific needs.

Field and lookup values: As a business model, Olrac ships Olfish-DDL with as many predefined fields as possible. However, within Olfish-DDL, the user can:

- modify field parameters, such as: display names, maximum and minimum values, set mandatory and carry over fields, capture on start/on end, make visible etc. Olfish-DDL ensures that changes which could affect underlying data capture logic are not allowed.
- hide and show lookup table records.
- add, edit and delete lookup table records.
- add fields - these fields have as much functionality and legitimacy as any original predefined fields.

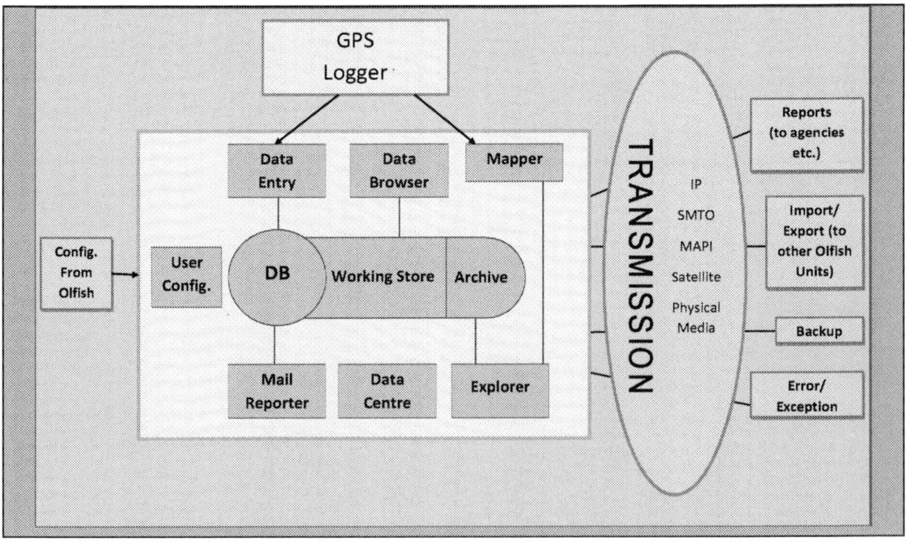


Fig. 3. Olfish-DDL basic structure

Olfish-DDL makes extensive use of dropdown lists whenever possible. The use of dropdown lists to enter data helps to maintain data integrity, thus minimising typos and saving time. However, new fields and values can be added by users if necessary.

While Olfish-DDL is normally shipped with the client's basic user configuration, it can be easily configured to fit the "taste" and needs of different users of the same basic configuration. With Olfish-DDL, the user can decide which fields should be visible, which are compulsory and which are remembered from previous entries. The user can also decide which data fields should be visible in which phase of the vessel operation. Examples are 'trip start', 'trip end' and, within a trip activity, 'start' and 'end'. Olfish-DDL also allows users to set up upper and lower limits for any numerical field in order to reduce the chance of typos.

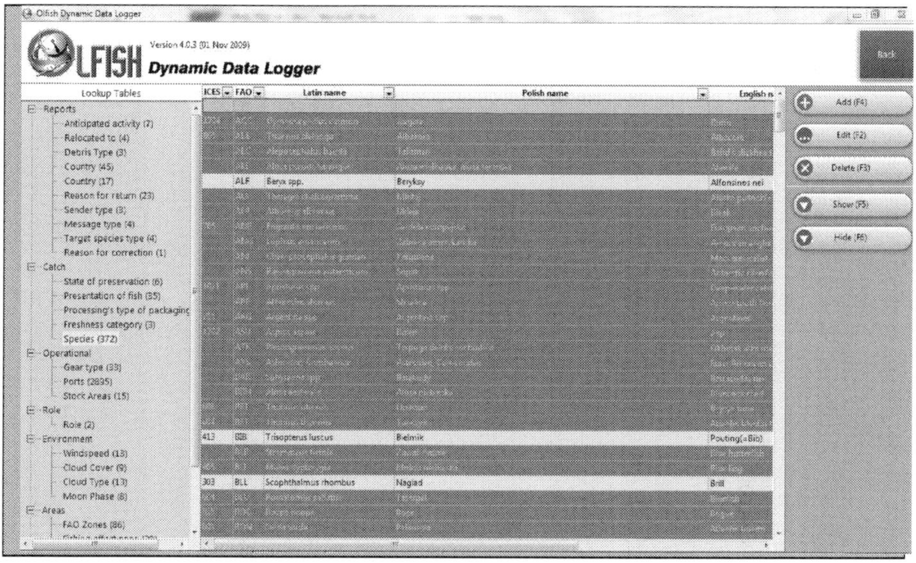


Fig. 4. Olfish-DDL lookup table customisation form

Another feature of Olfish-DDL is a Dynamic Commands Bar (DCB) which can be configured to "intelligently" guide the user during its data logging activities. The DCB can be configured to reflect data logging actions of vastly different activities. For example, the same basic underlying user interface can have different DCBs and can be used to collect data for totally different forms of fishing (trawl, longline, purse seine, traps, etc.) or other vessel activities (sea-farms maintenance, cargo delivery, coastal guard patrols, oceanographic surveys, etc.).

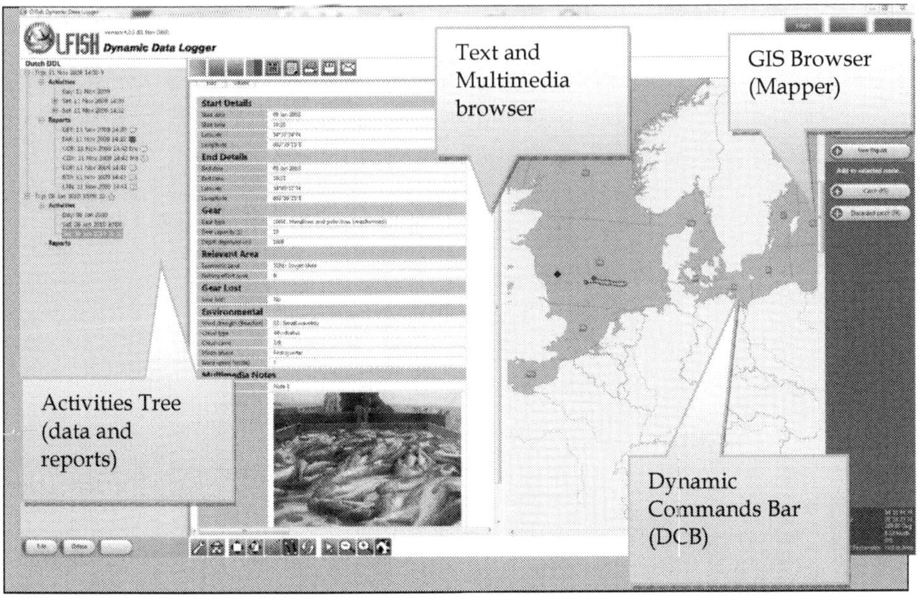


Fig. 5. Olfish-DDL main dashboard screen

Olfish-DDL has been designed to be a highly customisable data logging tool. However, under certain circumstances, it is undesirable to allow the user to change the basic configuration and customisation pre-setup. With Olfish-DDL, it is possible to prevent unwanted configuration changes in cases where data definition is strictly controlled by a higher level management body (examples are: company head offices, management agencies, scientific program managers, etc.). In such cases, it is possible to "hard" configure Olfish-DDL and constrain the user's ability to hide or ignore certain fields. This is mainly done in order to "force" uniformity and full data logging execution when Olfish-DDL is used for regulation-controlled data logging activities.

Data collected by Olfish-DDL can be used to generate any type of report in any format (XML, HTML, CSV, PDF, etc.). These reports can be saved and transferred to other databases (such as Olfish-DDL shore version, Olfish-RMS or other third party databases) either directly, using portable storage devices, or in real-time using the onboard VMS (Vessel Monitoring System) or other onboard satellite communication systems.

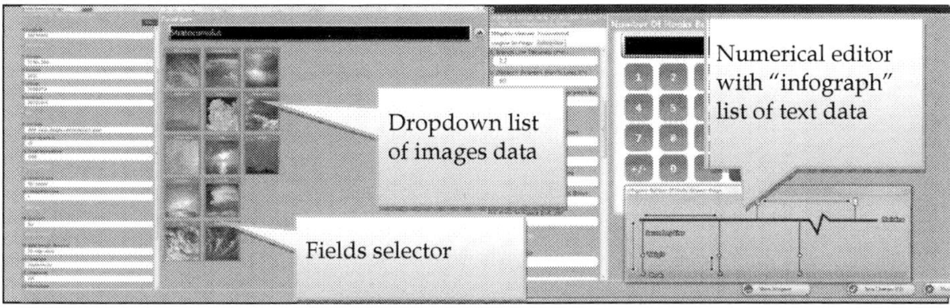


Fig. 6. Examples for Olfish-DDL Data Editor

### 2.3.3 GPS-Logger

Olfish-DDL can plot vessel movements and trips and set tracks, as well as automatically fill in date, time and location and other GPS related fields, if it has access to a GPS unit. The GPS unit can either be a VMS transponder or a standard GPS outputting NMEA strings on a serial or USB connection.

The Olfish-DDL application does not, in fact, talk directly to the GPS unit. Rather, a small, "light-weight", stand-alone application runs continually on the computer hosting Olfish-DDL. This application is the GPS-Logger, developed by Olrac. Olfish-DDL communicates with this GPS-Logger via a simple application programming interface, allowing all the low-level interfacing with the various GPS units to be handled exclusively by the GPS-Logger. This means that only the GPS-Logger application needs to be updated as Olfish develops support for new GPS or VMS units.

Another advantage to having the GPS-Logger run continuously is that it still logs GPS information even if Olfish-DDL is not running. Olfish-DDL can then use this logged information to provide marker values by means of a small "time machine" utility, when marker data are needed during non-real-time data recording activities. The data stored in the GPS-Logger can also be used to plot vessel tracks even if the user has not actively recorded GPS points.

The GPS-Logger can actually read any serial port information and can be extended to extract data from any set of NMEA 0183 sentences. This allows the GPS-Logger to record not only GPS information, but also any information outputted by devices conforming to the NMEA standard. These devices could include many analogue sensors such as echo sounders and anemometers amongst others.

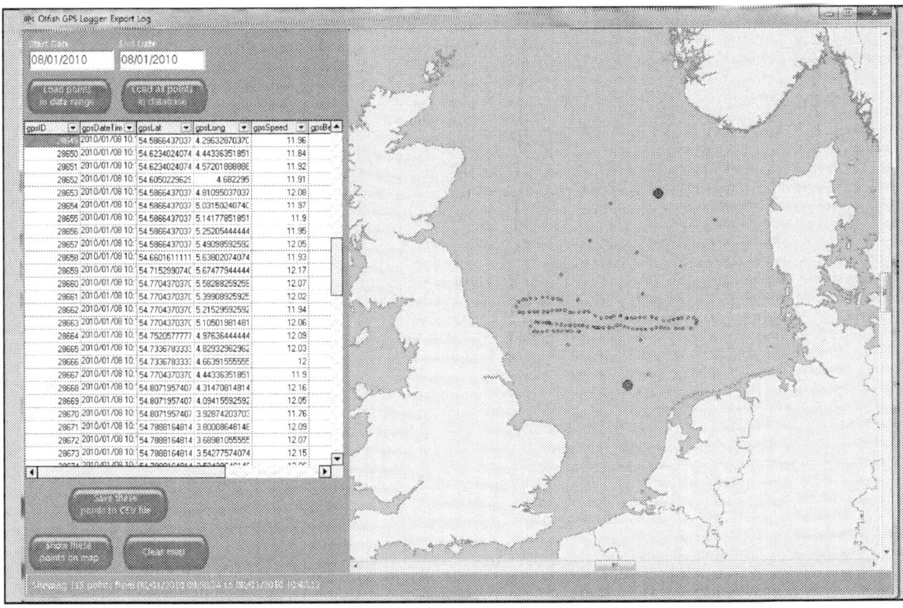


Fig. 7. GPS-Logger

## 3. Data Management Approach

### 3.1 Real-Time Activities and Events

Olfish-DDL data classes can be grouped into the following three kinds:

Activities - the data class encapsulates an activity, that is, something that has a start time and location as well as an end time and location. The combination of time and location properties is called a Marker. Examples include Trips, Shots, Sets, Hauls, etc.

Events - the data class encapsulates an event, that is, something that is of short (negligible) duration or has duration of no interest, and recorded as a single temporal point. Examples include placing a trap and retrieving catch from a long-line hook.

Non-spatial, non-temporal data - the data class encapsulates data without time or space markers. Examples include catch records and gear details.

The vessel unit of Olfish- DDL allows for the "real-time" data entry of activities and events, guiding the user in data entry to ensure a linear sequence of events (as they happen). For example, only once a trip has been started can a fishing operation be started, and only once a fishing operation has been started can catch events and catch be recorded. With Olfish-DDL it is possible, however, to switch off the "real time" mode and to enter data at "post event" mode when it might be more convenient for the user to do so. In non-real time mode the entire operation data can be captured on one form in any order desired by the user. In this mode it is assumed that the user will read his/her data from other data logging platform(s) such as a paper logsheet, spreadsheet, or even form memory. However, in order to ease data entry in non-real time mode, the user can access the entire GPS-Logger database and, for example, enter vessel location at any particular time by simply pointing at the relevant date/time point in the GPS-Logger database. This utility in Olfish-DDL is called the "Time Machine".

### 3.2 Data Storage

The initial Olfish configuration defines this hierarchical data structure of classes and fields, as well as operating parameters for each level and field. Olfish-DDL uses a hybrid approach to data storage. It uses XML files to store "active" trip data during vessel operations and relational databases to store historic data ("MS Access" on the onboard PC and MySQL on the shore unit PC). Active trip data can be archived into the storage database once the trip has ended and un-archived back, anytime, if any editing is needed (Olfish-DDL does not allow archived data to be edited directly and it is possible to block data from being un-archived if there is such a request). Regardless of which database is used, the database structure is always kept in synchronization with the configuration of the user-specific version of Olfish-DDL. This is necessary, as configuration may change (either by updates from the development team or configuration and customisation changes by the user). On start-up, Olfish-DDL compares the data store structures with the configuration and automatically reconfigures the structure, if necessary and without data loss, to bring it into alignment with the configuration.

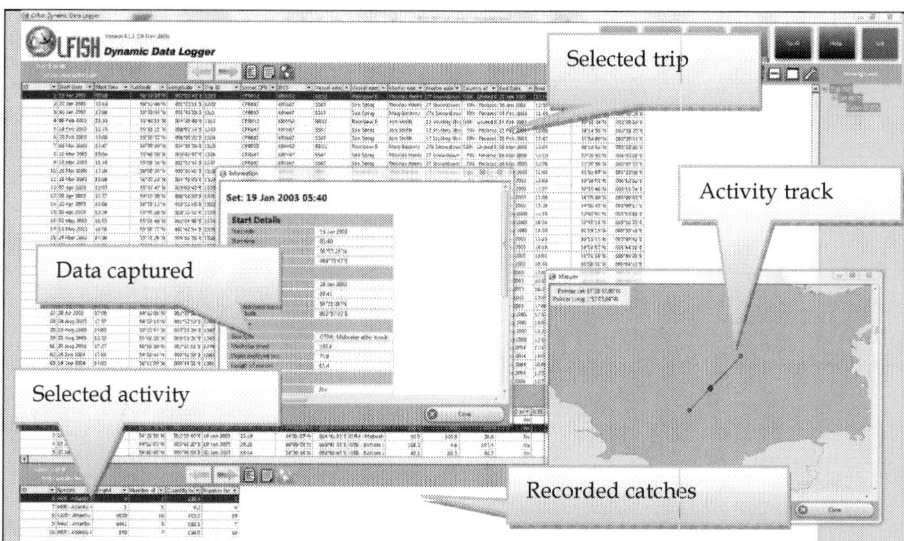


Fig. 8. Olfish-DDL main database interface (Data Centre)

All data captured by Olfish-DDL are organised into classes (or levels or tables). Each class has fields and a set of children classes. For example, one Olfish-DDL might let the user capture information for levels called Trips (a term often used to define the period between vessel departure and return) Sets (a defined activity within a trip), and Catches. A Trip class can have many Sets and a Set can have many Catches. Each class can contain any number of fields. Typical fields for a Trip could be, for example, Departure Date, Departure Time, Departure Port and Skipper Name. Possible Set fields could include Start Time, Start Latitude, Start Longitude, Gear Used, etc. Possible Catch fields include Species Caught, Weight, Products, etc.

## 4. System Configuration

### 4.1 The Need for Flexibility

An electronic logbook is not and cannot be a simple "off-the-shelf" product. Apart from multiple discrepancies within the fishing industry (differing regulations for different countries, different vessel-types, fishing methods, gear, targeted species etc.), the product must also cater for the data logging needs of conservation groups, observers and recreational fishers, to name a few. A rigid software configuration would be unable to adapt itself to the unique data recording requirements of each party. It is therefore necessary that any Elog tool be highly customizable, either at a developmental or user level. In this way, each company or organization will have a logbook configured specifically for its recording needs, with only the fields, input data and report types relevant to its operation appearing on the interface. Such configuration prevents clutter and confusion, and ensures maximum accuracy regarding data collection and transmission.

### 4.2 Editing Field Properties and Adding New Fields

The user can define new fields either on the vessel or on the shore units. These field definitions are then added to the configuration store, in the same manner as the Olfish-defined fields. These user fields are therefore then indistinguishable from fields defined by Olfish. Olfish-DDL automatically updates the database structure in accordance with these new field definitions.

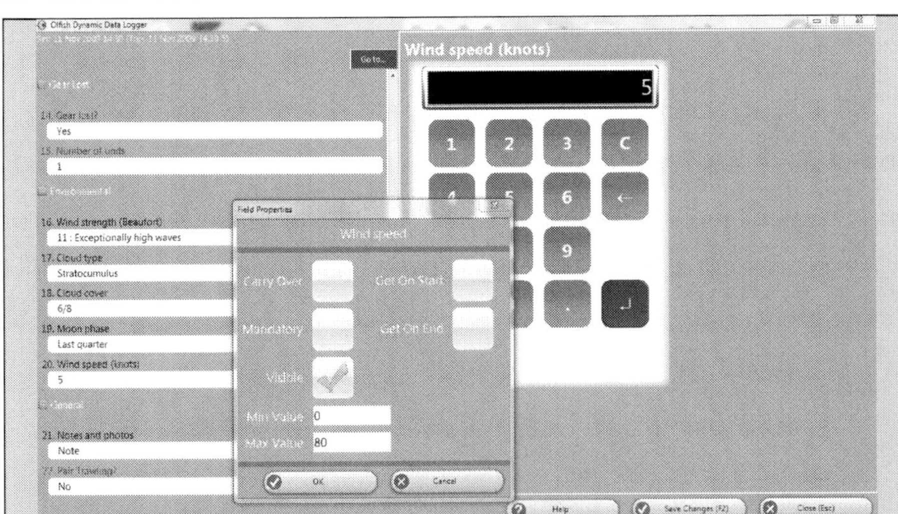


Fig. 9. Olfish-DDL user configuration form

When data is exported, new field definitions are included in the export file. Then, during import on the target system, Olfish-DDL adds the new fields to the configuration store and updates the database, providing a suitably configured data store for all the data in the imported file.

## 5. Reporting

Faster transmission of accurate data from vessel to land-based authorities can allow for near–real-time quota management. Current paper-based data-logging systems cause serious delays regarding quota calculations. In order for the quota allocation to ensue, data needs to be submitted, cross-checked and processed, a procedure which, using current data-logging techniques, may take up to a few months. This can substantially influence the last few months of a fishery, where delays in data reception can lead to some quota not being taken or, in the eyes of many, even worse: to be over- taken. Without updated knowledge as to how much quota has been thus far consumed, quota calculations have to be over-conservative and fishers often find themselves ultimately not reaching their entire quota. An electronic dat-logging system, which allows for near real time data entry and submission, and which can communicate via the web with all necessary authorities, reduces transmission time to days (see Web Based reporting, 7 below). Fishers will be able to use up their entire quota confidently, as it will be managed by near-real-time, good quality data. Similarly, management decisions will then also be made in near-real-time, decreasing the risk of over-fishing, as quota deficiency can also be identified and prevented in time.

### 5.1 An Ecosystem-Based Approach to Fisheries Management

A global move towards a more conservation-directed fisheries management system is underway. In the past, the only preventative measure which vessels were bound to take, regarding ecosystem conservation, involved using only approved gear types. There was very little surveillance and research done on the effect of the non-target-species by-catch on the ecosystem. This is changing. A paper from the Joint Nature Conservation Committee (Pope, J. G. & Symes, D. 2002) of the UK lists a few of the new regulations which are to be implemented in order to achieve this change. Amongst them:

- Scientists are required to provide ecosystem impact assessments alongside the advice on fisheries management options;
- Scientists are required to establish limit reference points for all target and non-target species within the ecosystem and managers should determine the best fishing practice to guarantee minimum levels of disturbance to the ecosystem;
- Fisheries managers are required to take all necessary actions to ensure that damage to the ecosystem (viz. excessive non-target species mortality and degradation of marine habitats) is not caused by intensities or forms of fishing activity beyond those required for rational and responsible exploitation of target species within commercial fisheries. In effect, this will require most - if not all - fishing effort to be reduced to levels commensurate with limit reference points.

Achieving such goals is dependent on the ability to collect data other than retained catches. These should include non-commercial and commercial by-catch, sea-bird and marine mammal interactions, impact on benthic species, detailed information on gear used and many others. For such a quantity and variety of data, paper logbooks are hopelessly inadequate.

### 5.2 Real-Time Management

Another advantage of electronic logbook near-real-time reporting is the ability to report trial fishing results (common in mixed species pelagic fisheries) almost immediately. The outcome of such trial fishing can determine if permission to fish is giving or not. Such permission, if not granted immediately (when supported by catch results), can compromise the ability of fishers to fish the discovered shoal of fish since pelagic shoal are often dispersed or move away in a very short time. Presently, it common that the time-lapse between the provision of catch results and the warranty of permission is too long and fish shoals disappear before the vessel is given an opportunity to fish them. Through near-real-time reporting, a sample-catch report can be sent, received, analyzed and responded to in time for the resource to be properly exploited.

### 5.3 Security

Finally, electronic reporting, able to harness the latest in encryption technology, can heighten report confidentiality and security to the maximum. Near real time reporting means that little time lapses between event and report delivery, thus minimizing any opportunity for tampering with the report by third-party "messengers". Furthermore, not only can reports themselves be encrypted, but access to the reports can be restricted through the use of unique codes and passwords.

### 5.4 Reporting Through Olfish

#### 5.4.1 Data Transmission Process

Data collected by Olfish-DDL can be transmitted to the shore in two basic forms:

- Raw, original data
- Summarized reports

In principle, Olfish-DDL transmits original data only to its shore mirror units while other data recipients receive data in a summarised form as a report.

With Olfish-DDL onboard, groups of vessels can "export" vessel data to different local companies/agencies (each running an Olfish shore unit) and the local companies can report in turn to a global company/agency running an Olfish meta-shore unit. Operational data can only flow from the vessel units to the shore units. Configuration data (lookup table records, field display names, new field definitions and field properties such as max, min values and mandatory, carryover status) can flow in either direction (from vessel to shore or from shore to vessel). This is illustrated in Figure 10 below.

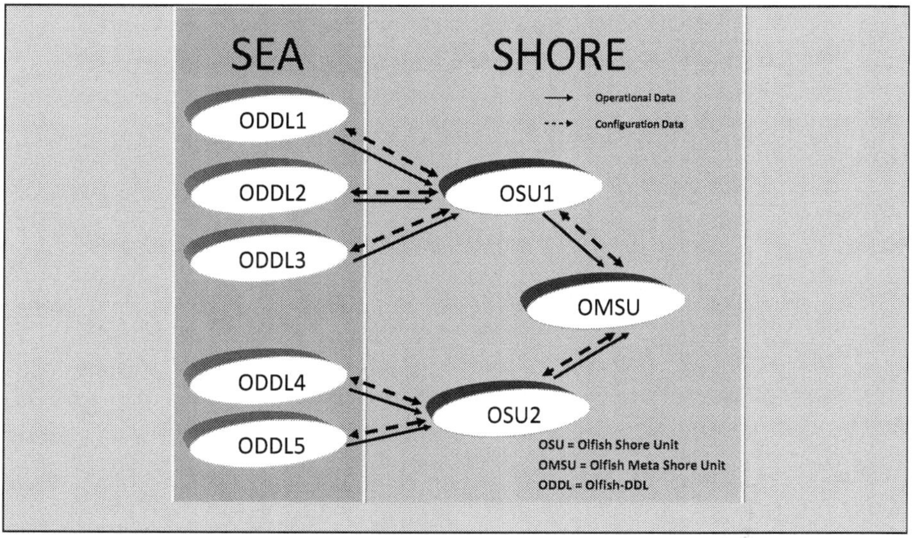


Fig. 10. Data flow between Olfish-DDL vessel units to Olfish-DDL shore units

### 5.4.2 Reports Transmission Process

Olfish-DDL can also generate specific reports as required by the regulatory bodies and compliance agencies. These specific reports use only a subset of the operational data recorded by the onboard unit. The recipient's of these reports can be either or both Olfish-RMS and other, third party, databases (European ERS for example).

The flow of reports amongst all these entities is configurable and can be customized to match different requirements. Implementations can include one or several of the following:

- Vessel units exporting all operational data to the shore unit. The shore unit then extracts and generates relevant reports for Olfish-RMS or other agencies.
- Vessel unit generates reports and sends them to Olfish-RMS. These reports can be accessed via the internet by permitted users directly or can be passed to other agencies' and/or countries' Reports Management Systems.
- Vessel unit generates reports and sends directly to other agencies' databases.
- Olfish-DDL can compile reports required by commercial clients for their own internal use. However, reports can also be used by any other authorised data recipients such as: compliance agencies, scientific and conservation groups and any other permitted third party commercial or non-commercial data recipients. Eligible third party data recipients can receive data selectively based on their needs and permission level.

Olfish DDL can provide output in a generic (native) XML or CSV format, or can comply with regulations and XSD schemas. Examples of management agencies whose regulations and schemas have been implemented by the Olfish-DDL are: EU regulations (EU 1077/2008), AFMA (Australian Fisheries Management Authorities) regulations etc.

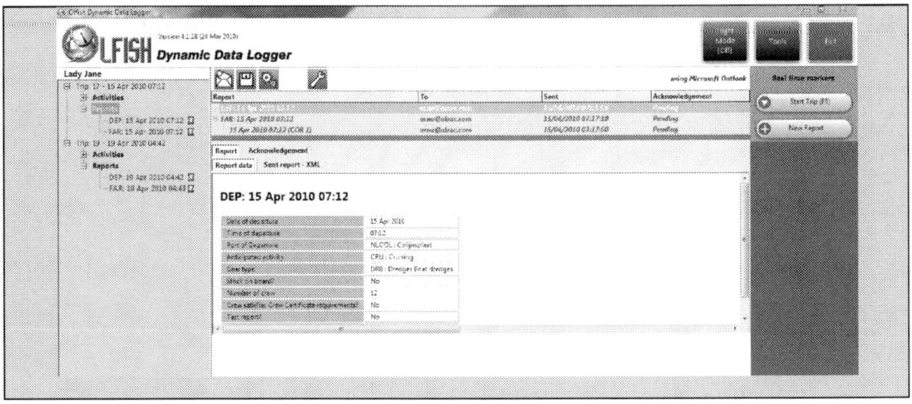


Fig. 11. Report Manager

In addition Olfish- DLL can generate output reports as PDF, HTML, BMP files or even as formal hardcopy logheets as may be required by the compliance authorities.

Reports can be generated automatically by predefined triggered events such as the crossing of certain geographical lines and date/time signals or manually by the user as required. Once generated, reports can be saved to a report queue, viewed, edited, sent and resent. The user can view the reports in legible HTML format or as XML or CSV before sending them out.

Reports can be compressed, authenticated and encrypted to fit any set of transmission requirements. For example, Olfish-DDL can implement secure, end-to-end transmission protocol that allows users to transmit secure XML reports, using X.509 digital certificates according the W3C XML security standards.

In addition to XML or CSV reports, Olfish-DDL can generate reports in the form of SQL scripts (with embedded data) that load the data into the third party database directly, without the need for a front end data loader tool.

Reports can be sent as email attachments, embedded text in the email message or as saved files on any mobile storage container. Olfish-DDL can also send data via a secured, dedicated FTP port if this is the preferred or required transfer mechanism.

Olfish-DDL includes an administrative module for the creation and mailing of predefined reports. These include agency compliance reports as defined by the EU regulations (EU 1077/2008), the Australian Fisheries Management Agency (AFMA) regulations and other management agencies (Poland, Holland, New Zealand, Canada and the USA amongst others).

Sending a report from Olfish-DDL to a third party on the shore can involve some or all of the following steps (Figure 12 below):

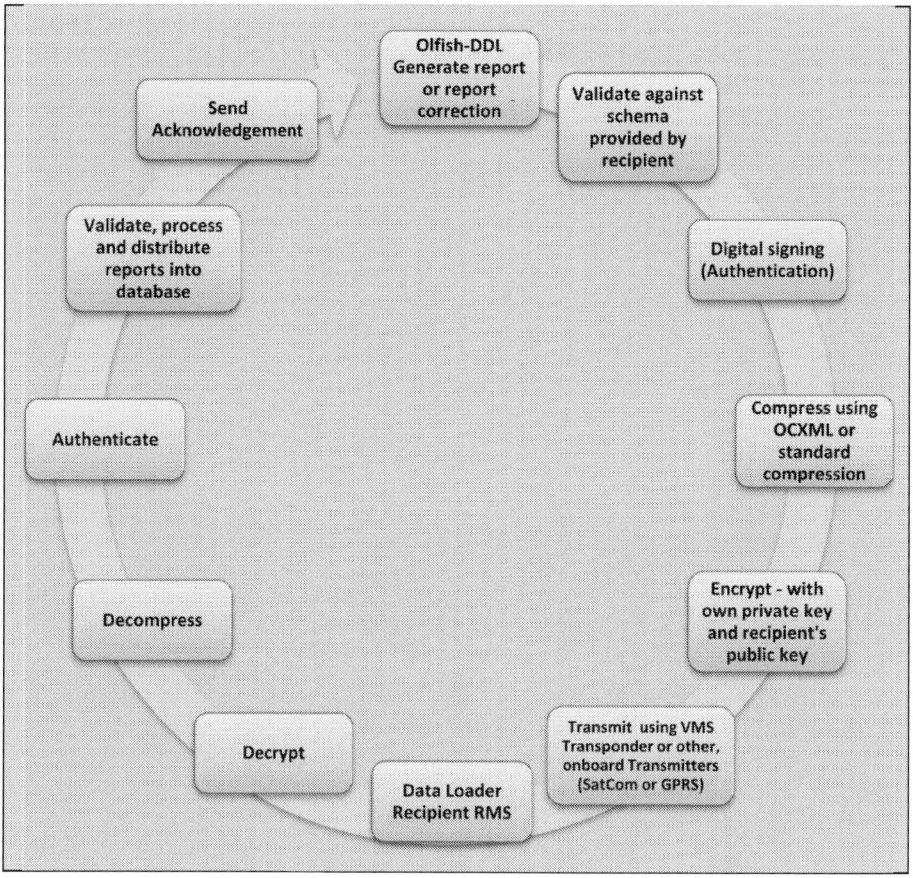


Fig. 12. Transmission process from Olfish-DDL to Shore-RMS

Olfish-DDL can, potentially, generate reports in any format which the client requires. Currently, formats catered for are XML and CSV. Olfish-DDL uses XSLT 2.0 to generate reports in XML format. In Poland, where the shore Reporting Management System was also developed by Olrac Olfish-DDL, EU and Polish specific reports are generated as XML files, which are then compressed as OCXML before being sent to the Polish RMS. On the shore these files are decompressed using ODCXML before validation and data integration take place. Once the reports are validated and processed successfully, a success message is sent back to the sender Olfish-DDL. If the report is not processed successfully, an error message is sent back to the sender, along with a reason for the process failure.

#### 5.4.3 Correction Reports

Olfish-DDL makes a very clear distinction between data and reports. Data is information such as catch, gear used, fishing time, etc. which is captured in a raw (original) form by

Olfish-DDL. Reports are extracts of specific data as required by different schemas. As such, reports are really short documents (XML, CSV, HTML, etc.) compiled from the original data. If data is changed after the report has been sent, it is a requirement in some cases to send a new, corrected report. Any change of data triggers the Olfish-DDL to automatically update reports which make use of this data. Thus, any change of data means a change in the corresponding report.

Olfish-DDL marks sent reports, which have been corrected in this way, visually on the interface. As such, it is very clear to the user that a report has been corrected and must be resent. When the report gets resent, Olfish-DDL adds additional elements to the XML report, indicating that the message is a correction, the reason for the correction, as well as which report is being corrected. Upon receipt of a correction message, the original message is discarded by the recipient authorities, and the new message is processed. Each management authority has a different set of regulations regarding the handling of corrections and updated reports. AFMA, for example, allows only ten corrections per fishing trip or non-fishing period.

### 5.4.4 XML File Transmission

Many reports sent from Olfish-DDL to Olfish-RMS or other third party RMS, such as enforcement/regulatory agencies, are in XML (Extensible Markup Language) format. XML is not an ideally compact data format, and transmission between vessels and shore using onboard satellite communication systems is often costly. It is therefore useful to compress the XML before sending it. Olrac has developed a method for compressing XML (OCXML) that achieves higher data compression than generally available compression utilities.

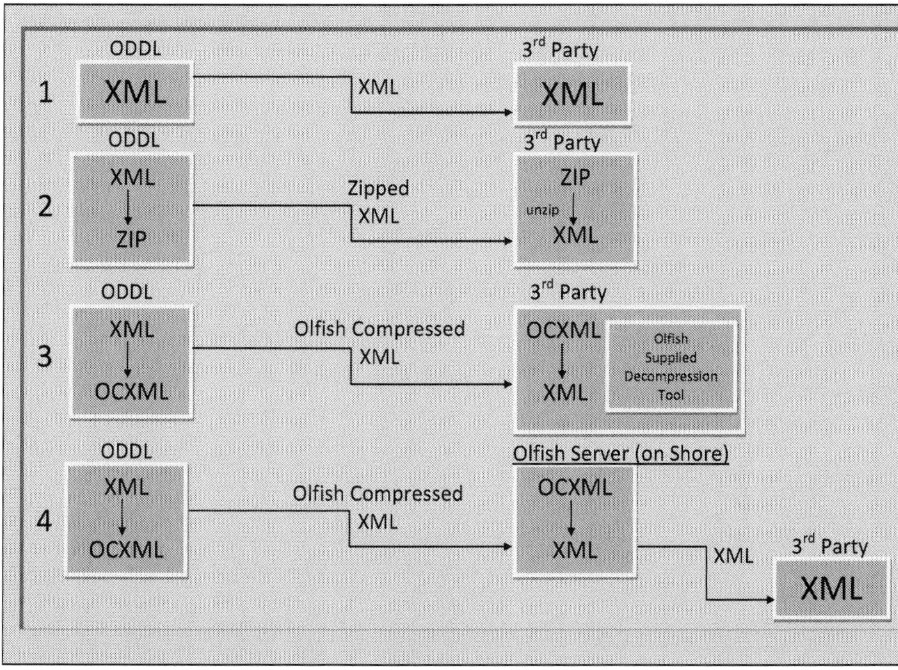


Fig. 13. Transmission of XML files from Olfish- DDL (vessel unit) to third-party (Olfish-DDL shore unit or RMS)

Figure 13 above shows four different ways in which the XML files could be sent:

- uncompressed
- compressed using a mutually agreed-upon compression standard (e.g. ZIP)
- compressed as OCXML, with a paired ODCXML decompression utility on the recipient RMS (either Olfish-RMS or third party RMS).
- compressed as OCXML, sending to a shore-based intermediate Olfish server, which decompresses the OCXML and forwards the standard uncompressed XML file on to the intended third party RMS.

### 5.4.5 Secure Transmission

A number of transmission methods have been developed by Olrac in order to address different client requirements. In the Netherlands, various reports, as determined by EU regulations 1077/2008 and Dutch regulations, are sent to authorities as XML documents. These XML documents conform to the XSD (XML Schema Definition) offered by Holland's VISHUB agency. The VISHUB agency provides an encryption add-on, by means of a user specific USB key. The type of encryption implemented in Holland is an asymmetric private/public key encryption.

After a report is generated, Olfish-DDL validates the report against the VISHUB's schema. If the report validates successfully, Olfish-DDL invokes the Dutch provided encryption add-on to compress, encrypt and sign reports generated aboard the vessel. The encryption add-on performs these functions in the following way:

- Compresses the report using g-zip.
- Generates random AES key.
- Encrypts the report with AES key.
- Encrypts AES key with recipient's public key.
- Signs the message with sender's private key - digital signature.
- Concatenates and encodes the contents of the report along with signature and encrypted AES key.

  Upon receipt of the report from the vessel, the VISHUB follows the following process:
- Separates message into report, signature content etc.
- Decrypt encrypted AES key with recipient's private key.
- Decrypt encrypted report with AES key.
- Decompress message.
- Verifies signature using the sender's public key.

The end result of this process is an unencrypted report, exactly as it was sent from the vessel. After the report has been verified, the VISHUB validates the report against its schema. If the report validates successfully, a success message is sent back to the sender. If the report does not validate, an error message is sent back to the sender, along with a reason for the validation failure. If the report validates successfully, the VISHUB forwards a portion of the message to the AID agency.

In Australia, where Olfish-DDL needs to conform with the Australian Fisheries Management Agency (AFMA) regulations, Trip and non-fishing period reports are sent as XML documents that conform to AFMA's published XML Schema. The reports are sent to AFMA's native RMS (eLogs) with a pre-allocated username and password attached to them. Before a report gets transmitted, Olfish-DDL validates the report against AFMA's XSD. If the report validates successfully, Olfish-DDL encrypts the sender's username and password using AFMA's public key. The encrypted username and password is added to the report. The report is compressed by Olfish-DDL using a standard Zip utility before transmission takes place.

Upon receipt of the report by eLogs, a locally developed data loader utility decrypts the username and password using its private key, thereby verifying that the report comes from a valid sender. Once authenticated, the report is validated by AFMA against its own XSD. If the data validates, AFMA imports the report data into its database.

### 5.4.6 Communication Techniques

Olfish-DDL allows data and reports to be sent out using a variety of transmission methods, depending on their availability on the vessel.

MAPI: Olfish-DDL can send email using any standard or maritime email software that supports the MAPI standard (the "simple" MAPI subset is sufficient). MAPI is the Messaging Application Programming Interface, a messaging architecture for Microsoft Windows. The email software must be:

installed on the same computer as Olfish-DDL

correctly configured to use some transmission mechanism (such as TCP/IP or Inmarsat-C) and

set up to be the default email client.
Examples of e-mail clients that can be used to send e-mail using MAPI:
Stratos AmosConnect
Vizada SkyFile
Outlook
Any other e-mail client that offers simple MAPI support
Direct interface to satellite transceiver on vessel: Olfish-DDL can send email directly using several different types of satellite transceivers. Olfish-DDL achieves this by using the Olfish GPS-Logger utility, which communicates directly via a serial port connection to the transceiver and lets the transceiver handle the transmission using its own proprietary technology to send emails and attached reports. Examples are:
Thrane & Thrane Inmarsat-C transceivers (Inmarsat)
BlueTraker (GPRS/Iridium hybrid)
Integrating with maritime e-mail solutions: Olfish-DDL can send mail using these programs in the following way: The report is generated by Olfish-DDL and then copied automatically to a special "outbox" folder. In some cases, the user then browses to this folder from within one of these third-party email applications, finds the report that has just been generated, and sends it. In other cases, the third-party software automatically "polls" the outbox folder, and sends the reports at set intervals. Sometimes reports cannot be sent as binary attachments by the third party software. In these cases, Olfish-DDL uuencodes the binary reports into text before they are placed in the outbox folder. The recipient e-mail client then uudecodes the text back into its binary report form.
Examples of third-party software with which Olfish-DDL interacts are:
Boatracs
SkyMate
Faria WatchDog
METOCEAN MetMailer
Thrane & Thrane EasyMail
Direct SMTP and POP: Olfish-DDL can send and receive email via its own internal emailing utility (Olfish-Mailer), using internet SMTP (Simple Mail Transfer Protocol) and POP (Post Office Protocol), if the vessel has access to the Internet (TCP/IP), either broadband or dialup. Olfish-DDL then sends email without any need for third party email software.
Other direct internet protocols: If the vessel has access to the internet (TCP/IP), then Olfish-DDL can be configured to make use of several different internet transfer protocols. These include SMTP and POP as described above (specifically for email) but also FTP (File Transfer Protocol) and HTTP/HTTPS (Hypertext Transfer Protocol/Secure Sockets layer), for text and binary files.

### 5.4.7 Security and Data Integrity

Olfish-DDL includes numerous security and validation features to ensure data security and integrity. Examples are:
User logins - the software requires a unique user ID and password to access each software installation. Each software installation can have a number of authorized users, for example an administrator, owner/operator and mate.

Edit/Delete control - although with the standard Olfish-DDL the user has the ability to delete and/or edit any recorded data, these features can be shut off or amended to meet the data standards of the customer.
Criteria monitoring - many fields in Olfish are valid or permitted only when certain criteria have occurred. With Olfish-DDL, only valid fields and values are visible depending on predefined combinations of criteria.
Data validation: Olfish-DDL allows the user to set up a minimum and maximum for any numerical data to prevent the entry of out-of-range typos.
GPS Puller - a feature located on a remote computer, which allows authorities to call for and receive location data directly from the Olfish GPS logger at any point during a vessel's trip.

## 6. Add-On Modules and Other Utilities

### 6.1 Change Control

In addition Olfish-DDL allows permitted users to track back any change that was made to data previously saved. This utility is password protected and can be activated only as part of an overall agreement between management authorities and the user. When in auditing mode, the user can see any sequences of changes made to previously saved data. These include the date and time of a change which was made, the username and the original value/s.

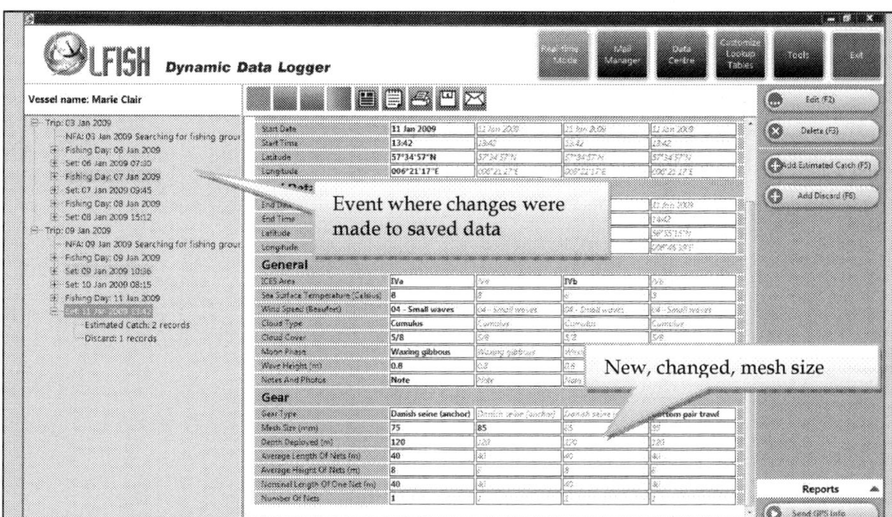


Fig. 14. Change-Control utility: pink markers on the activity tree show places where changes took place, bold figures in 2nd, 3rd and 4th columns were changed.

### 6.2 Multinote Taker and Notebook

Olfish DDL includes a utility named Multinote taker which allows users to enter multimedia data in the form of free text, pictures and video clips. These notes are part of the

Olfish-DDL database and the user can browse them anytime by pointing at the activity where the Multinote was taken or by using a specially built Notebook browser which allows users to browse through all their notes, pictures and videos in one location while keeping the note traced to its original location on the DLL activity tree.

Fig. 15. Multinote Taker

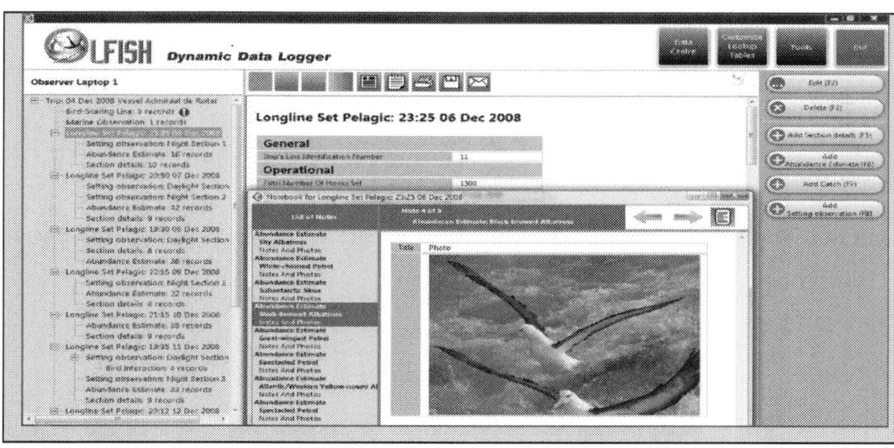


Fig. 16. Multimedia Notebook browser

### 6.3 Mini Reporter

The mini-report query utility (Mini-Reporter) is a relatively simple query tool which allows users to extract basic statistics of recorded data during an active trip before they are archived in simple tabulated form. The mini-reporter allows users to group data by different grouping variables.

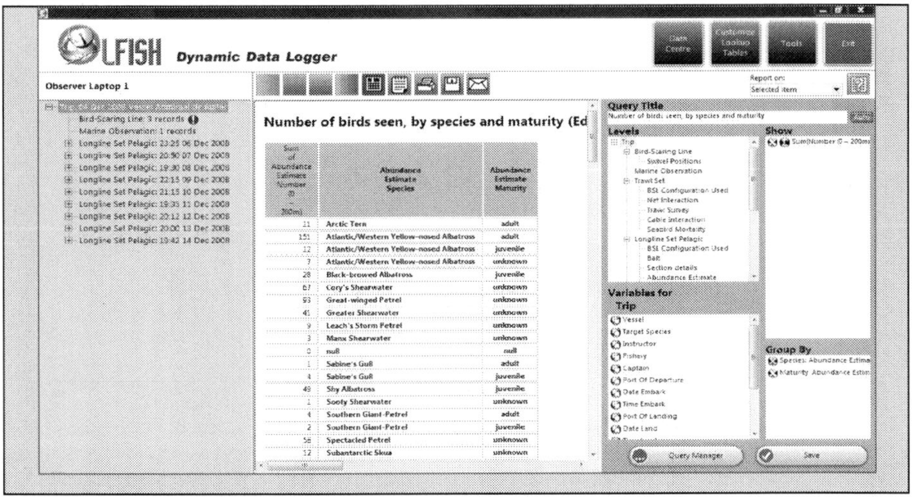


Fig. 17. Mini-Reporter: The HTML browser in the centre is changed to reflect the query output as was set up on the right.

## 6.4 Olfish Explorer

Olfish-Explorer is an add-on, optional, visual data analysis module specifically designed to work with the Olfish-DDL main database. It allows the user to analyse subsets of data captured by Olfish-DDL. The DDL lets the user capture data grouped into classes. Each class has fields. For example, one DDL might let the user capture information on Trips, Sets, and Catches. A trip has, for example, Departure Date, Departure Port and Skipper Name fields. Possible Set fields include Start Time, Start Latitude, Start Longitude, Gear Used, etc. Possible Catch record fields include Species, Weight, etc. From a selected set the user can create Subsets of data for a particular analysis or presentation. For example, graphs can be drawn showing CPUE (Catch per Unit Effort) as a function of time, moon phase, current strength etc. Also, spatial CPUE density distributions can be plotted on a map, which can then be filtered for different target species or to reflect differing environmental conditions. Subsets can be swapped in order to explore different scenarios and data relationships. For example, CPUE at new moon time compared to CPUE at full moon. The subset definitions, i.e. the list of classes and fields selected for the subset can be saved for re-use.

With Olfish-Explorer it is possible to add calculated fields and to use them just as pre-entered data. For example, the calculations of CPUE by dividing catch weight by fishing duration. The Explorer allows the user to create and set up any number of calculated fields and to edit and delete them.

The table created by the Olfish-Explorer query engine can be shown on a graph (such as a bar graph or a pie chart) or shown in a density distribution map (if the data contains longitude/latitude information for each row in the table).

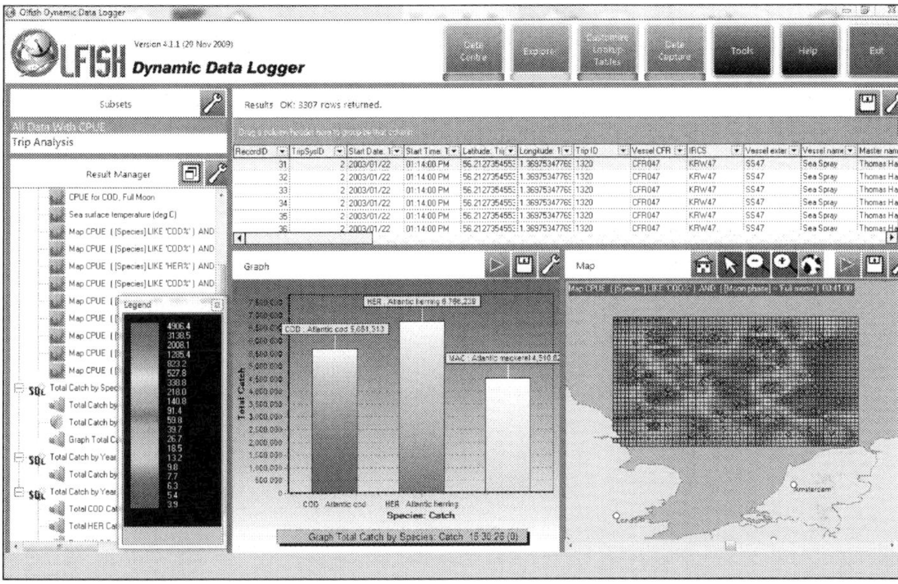


Fig. 18. Olfish-Explorer view mode

### 6.5 Fishing Consultant

The Fishing Consultant can predict the future catch rates of a particular species for specified fishing areas and environmental conditions. The consultant uses catch rate and any other information that is available about the conditions in which each recorded catch occurred to produce a density distribution, indicating which areas are most likely to have the densest population of fish.

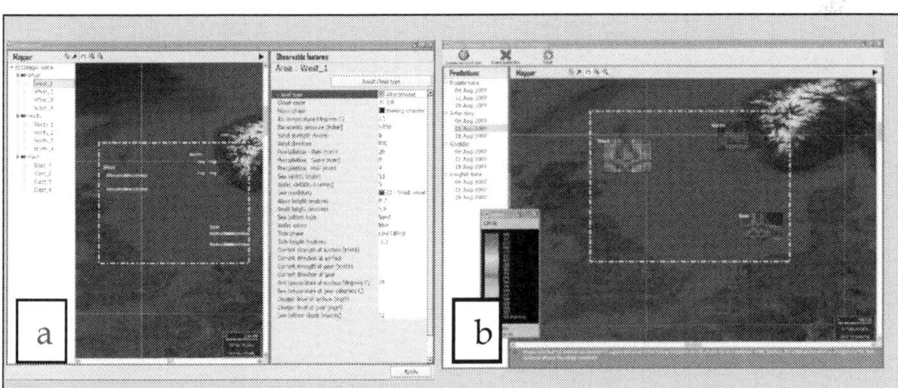


Fig. 19. Fishing Consultant. Setting up prediction parameters (a). Projected good fishing sites (b)

A mathematical prediction model is trained on the historical data of catch and all recorded environmental factors. The trained model can then predict future catch rates under conditions specified by the user. The predictive accuracy of the model depends on the quality and quantity of data available in the database. The predicted catch rates can be visualised on a GIS component that provides a colour-coded predicted catch rate grid of each fishing area specified by the user.

### 6.6 Fleet Activities Optimizer

OLFISH Fleet Activities Optimiser is a utility which has been designed to aid fishing managers in scheduling ship time optimally in order to meet certain objectives. The program can manage a fleet of numerous vessels, fishing in multiple areas, for several species of fish, one of which will usually be identified as the primary target species. The scheduling mechanism can be set to achieve any of the following tasks:

- Land a specified tonnage of the primary species at the minimum possible cost.
- Land a specified tonnage of the primary species in the shortest possible time
- Land the maximum possible catch of the primary species in a specified time period.
- Maximise the value of the catch (all species) in a specified time period.
- Land the primary species at minimum cost, in order to maintain stock for processing at a specified daily rate without exceeding specified stock limits.

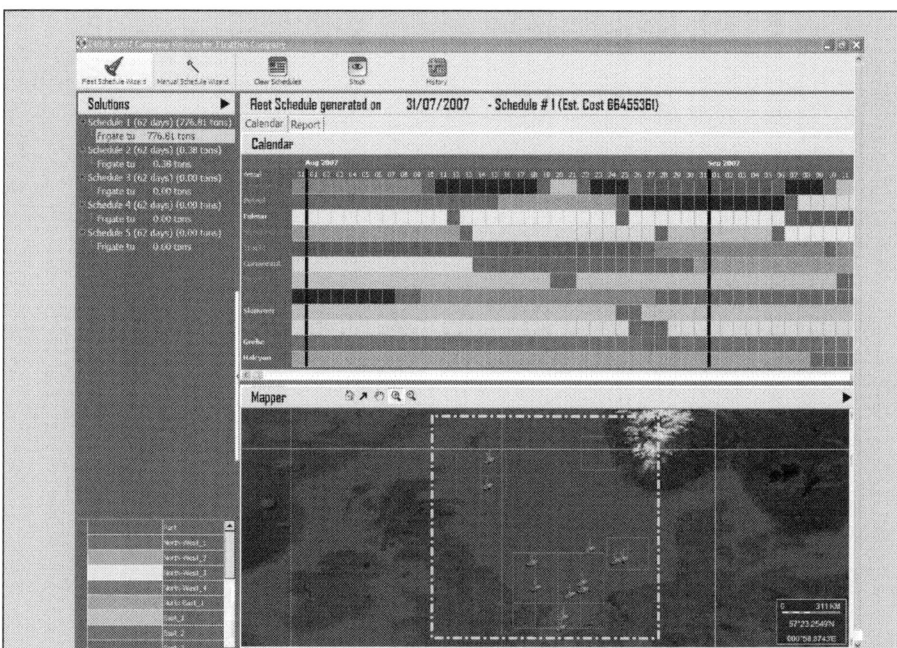


Fig. 20. Fishing fleet activity optimiser - vessel locations and proposed schedule

In determining a solution, the program uses input information with regard to vessel specifications, the distance from port to each area, the expected catch rate in each area for each species (which may be generated using the "OLFISH Fishing Consultant") and the specified constraints. The output of the program is a visual representation of a number of possible fleet schedules, each of which is close to the optimal solution, but which may differ with respect to bycatch landings etc., allowing the user to choose the most appropriate solution for his/her needs.

### 6.7 Form Maker

It is also envisaged that there will be a transition period, where a vessel may be using Olfish-DDL to log vessel activities and catches, but those vessel activities and catches will still need to be reported to agencies and government authorities using paper-based industry standard log sheets. In order to address such needs, Olfish-DDL allows any paper-based log-sheet to be scanned, incorporated into, and completed within Olfish-DDL, and then to be printed out and delivered to the relevant authority. The paper logsheet can be populated automatically by the form maker utility if data have been entered using Olfish-DDL.

The form maker utility is a developer utility and cannot be used by the user. It does, however, make the creation of hard-copy forms within Olfish-DDL an easy and fast process. It is our hope, nevertheless, that paper forms will soon be a thing of the past since the secure and efficient way of recording and transferring data offered by Olfish-DDL eliminates the need for paper forms of any kind for this process.

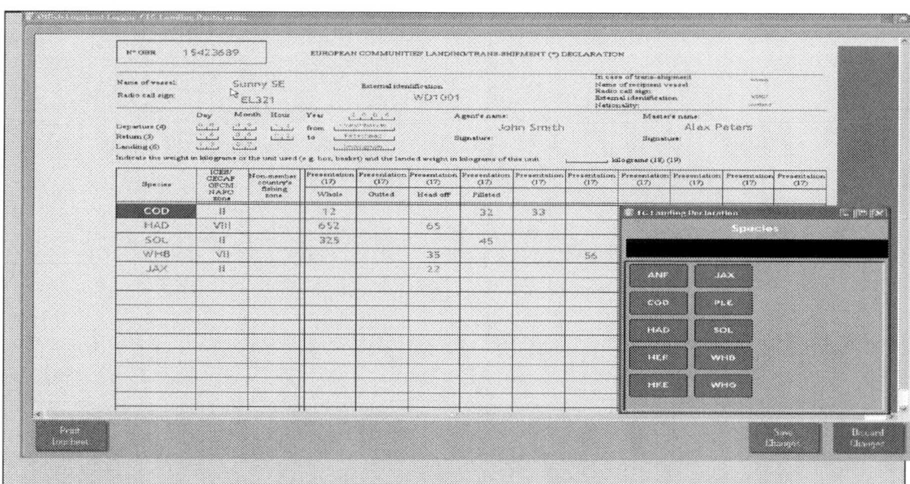


Fig. 21. Form Maker

### 6.8 Scales Data Logging Utility

The Scales data logging utility facilitates the incorporation of data obtained from mechanical scales (regarding the species, weight, grade and the processing state of catch). The data received from the scales is sent to a computer by either a TCP/IP network connection or a standard RS232 communications port. Olfish-DDL currently caters for four different industry standard scales. These scale types are NESCO, Pols, Scanvaegt and Marel scales. A scales code component allows a user to link data such as species, weight, grade and the processing state used in the software to the scales codes that are used by the scales. An individual scales code might indicate a particular species of a particular grade and processing state that is being weighed. The scale sends only the scales code and the weight of the fish currently being weighed to the software. The Scales Interface component uses the scales code to retrieve the species, grade and processing state and then ties these values to the received weight.

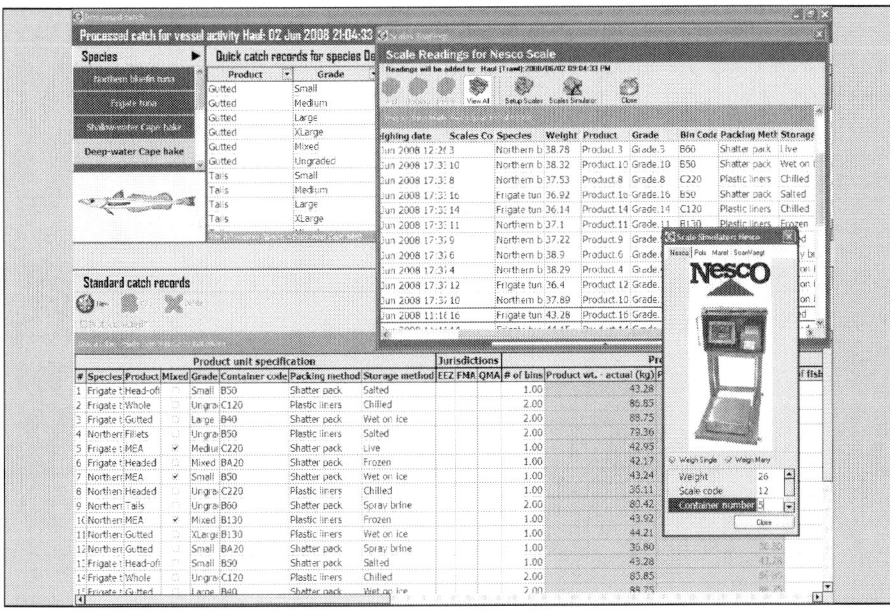


Fig. 22. Scale interface

## 6.9 Fishing Inspector

The Olfish-DDL inspector utility is a small program which can reside on the compliance inspector's USB key. This utility allows permitted inspectors to extract certain (predefined) information from the user's Olfish-DDL without the need to give them access to the full software.

Fig. 23. Olfish-DDL Inspector utility

## 7. Olfish-RMS: A Web-based Reports Management System

### 7.1 Functional Overview

In those installations for which a client does not have or does not wish to use a pre-existing database, Olfish has developed a shore management reporting system it has named Olfish-RMS. This system is an in-house implementation of the Electronic Reporting System (ERS) as requested by the European Union. Olfish manages reports generated by vessels and allows users to access their sent reports via the internet. The system allows the administrator to see, in real time, reports coming from different vessels into the RMS. Once reports are saved and processed, they can be viewed and data can be summarized (such as total catch by vessel and by species). The RMS also incorporates Google Maps to graphically visualize vessels and catch locations.

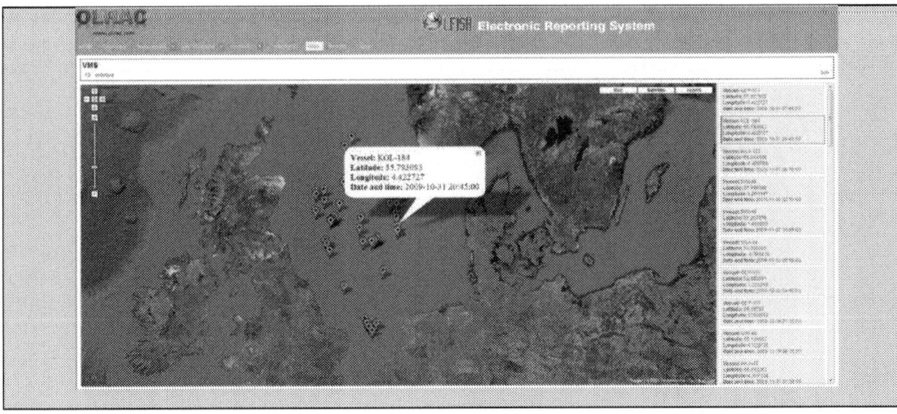


Fig. 24. Olfish-RMS WebViewer: Vessels location using Google Maps

Olfish-RMS is a complete system for the management and dissemination of reports (catch, landing, vessel movement, sales and transhipment) received from fishing vessels or entered directly when relevant. The Olfish-RMS server was constructed according to the EU regulation (EC) 1077/2008 as was adopted by the Commission in 2008. However, the Olfish-RMS server has been designed to allow its scope to broaden to include other national data and reporting needs. Olfish-RMS is deployed on an MS Windows Apache server running PHP version 5 and MySQL Version 5.

The Olfish-RMS consists of 2 components: A web-based interface (WebViewer) and a communication module (MailPuller).

The web-based interface is controlled by user access control. The levels of access are:

- Administrator - Has complete control to add/edit/delete data on the system as well as viewing of aggregate data.
- Inspector - Has access to add/edit/delete data related to EU reports and the viewing of lookup fields as well as the viewing of aggregate data.
- Vessel user - Allows the vessel owner/master to view all reports submitted by the vessel.
- Member state - Allows other member states to view landings when this occurs in their waters, as well as prior notification reports to their ports.

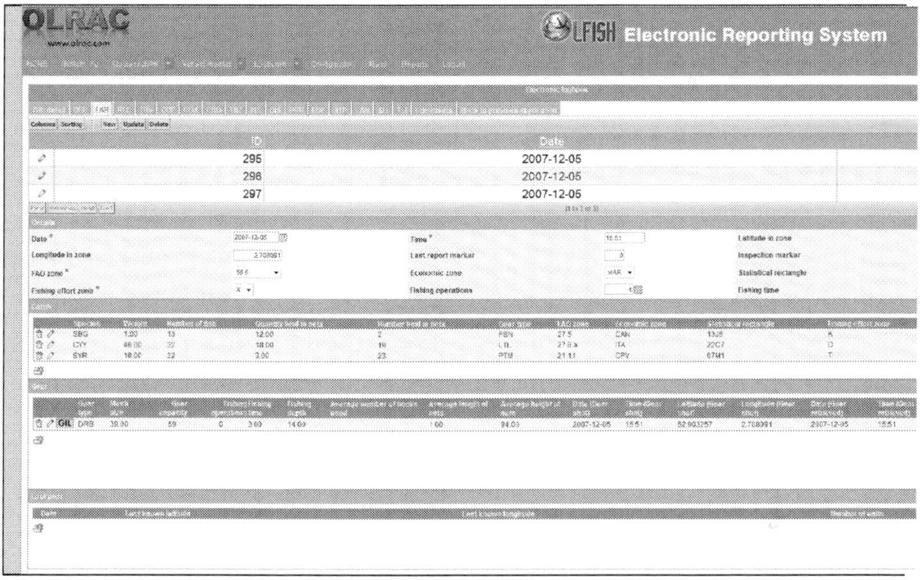


Fig. 25. Olfish-RMS WebViewer. Example listing - Vessel Registers

The communication module (MailPuller) is configurable to connect to any SMTP server account dedicated to the receipt of electronic reports from Vessels and First Sale units as well as other member states. The MailPuller also scans all incoming messages, and where relevant (according to regulation EC 1077/2008) forwards landing and prior notification reports to other member states. The configuration of which member states to send to is done from within the Olfish-RMS web interface. Once incoming reports have been validated, they are imported into the Olfish-RMS MySQL database. If validation fails, a failure report is sent back to the sender. If the report is imported, a success acknowledgement is sent back to the sender.

If required, inspectors can manually add report data. This may be necessary due to the failure of onboard systems or for vessels that have not yet adopted an onboard reporting system. Inspectors are also able to enter comments on any submitted report and reports are marked as pending until they are accepted by an inspector.

Olfish-RMS allows the user to view reports on an individual level (in their original format) as well as aggregated reports based on imported data. The system also allows for the import of corrected data as per EU regulation (EC) 1077/2008. Vessel location and summary data can also be viewed spatially, using Google Maps. Olfish-RMS facilitates data cross-checking, system status, data validation and the conditional dissemination of data and reports to other member states as dictated by EU regulations. Olfish-RMS is also preconfigured with all lookup values as specified by (EC) 1077/2008.

Olfish-RMS allows the user to swap between languages. The interface, as well as all lookup values, change based on the language selection. Olfish-RMS is able to export all lookup data and reports to Excel and Word formats.

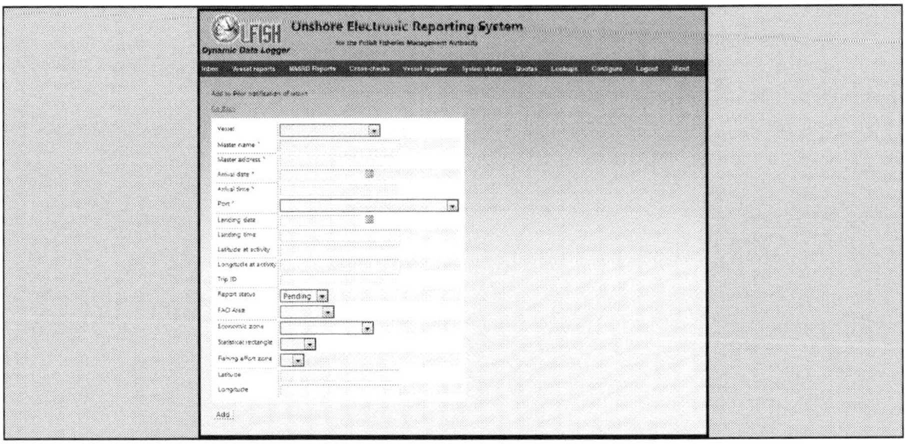


Fig. 26. Olfish-RMS WebViewer. Example of data entry screen for manual entry of report data

## 7.2 Technical Overview

### 7.2.1 Definitions

Olfish-RMS - The web-based Olfish Report Management System for the processing of electronic reports

Mail Processor - Windows-based application that connects to a mail account and downloads, processes and acknowledges electronic reports

MS - member states / other countries

Flag state - vessel country of registration

Windows service - always on Windows application running in the background

Web service - A standard method of integrating web-based applications over the Internet. All communication is done using XML

SOAP - a protocol for exchanging structured data in the implementation of web-services. It essentially provides the envelope for sending web-service messages.

Apache - a deployment platform on the server on which Olfish-RMS is installed which returns a web page to a web browser when the web browser requests it.

PHP - a widely-used general-purpose scripting language that is especially well-suited for Web development and can be embedded into HTML

MySQL - open source relational database

Ajax - a group of interrelated web development techniques used on the client-side to create interactive web applications

Delphi - the programming language used to write the data logging, data pulling and data pushing applications

### 7.2.2 The Basic Functional Infrastructure

Olfish-RMS is developed to run on a Microsoft Windows platform with an Apache distribution containing MySQL and PHP. The exact version requirements are:

Apache 2.2
MySQL 5.1
PHP 5.3.1

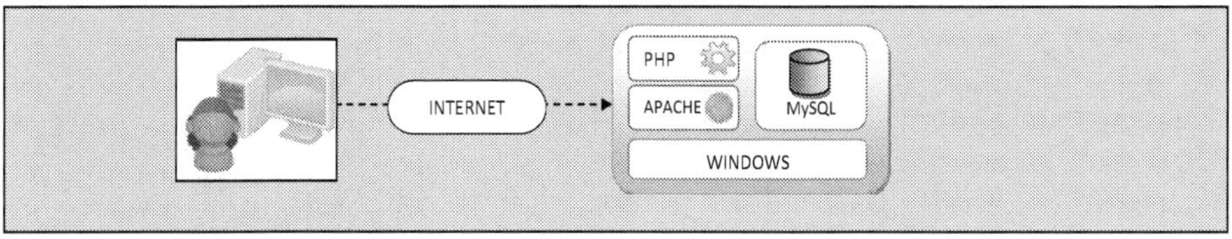


Fig. 27. Olfish-RMS infrastructure

### 7.3 System Components

Olfish-RMS consists of four components:

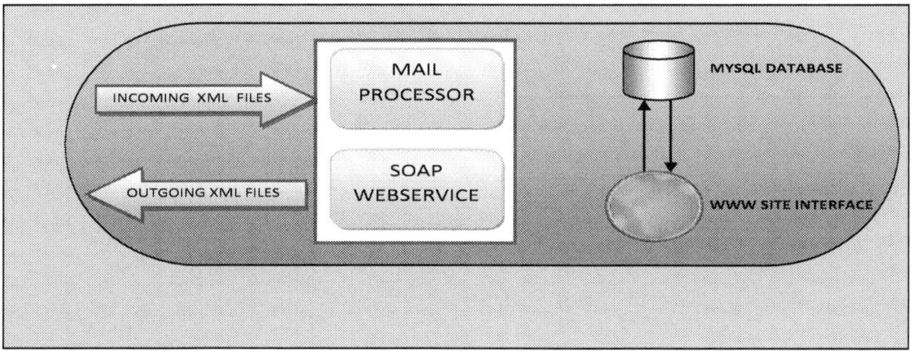


Fig. 28. Olfish-RMS components

#### 7.3.1 MySQL Database

Olfish-RMS uses a well-designed relational database that makes extensive use of indexes and stored procedures for the fast execution of reporting requirements on growing databases. The database implements the following modules:

Security - controls access to the interface

System logs - logs all automatic and manual processes on the database. This includes the receipt and sending of electronic reports

Electronic reports - stores received electronic reports in a relational way with links between report data. E.g. a departure report and a return to port report are linked to the same trip

"Paper logbook" reports - stores manually inputted data from the paper logsheets until these are to be phased out completely.

Monthly reports - stores manually inputted data from the monthly paper logsheets in use by smaller vessels

Quota management - stores yearly TAC allocations as well as the distribution of these allocations to vessels and vessel groups

Vessel registry - stores entire fleet details, movement into and out of the fleet as well as all changes to the vessels (physical or management related)
Lookups - stores all lookups in use throughout the system. All regulated lookups come pre-populated and are not modifiable

### 7.3.2 Web Site/ Interface

Olfish-RMS has a PHP written front-end utilising AJAX technology for ease of use and speed. All forms and grids are implemented uniformly for quick understanding of the user interface. All grids throughout the system have a header section with one or more of the following functionalities:
Search - the user can specify filtering for each field in the grid.
Columns - the user can select which columns to display in the grid
Sorting - the user can set the column sorting (ascending or descending)
XLS - export entire grid contents to Excel format
Print - prepare the grid for printing
Likewise, a footer section with one or more of the following functionalities:
First, Previous, Next, Last - Move page by page through the dataset in the grid
Go to - jump to a particular page
View - change the number of records displayed in a grid on each page.
The website provides an interface for the eight database modules listed above as well as five additional interfaces:
Reporting - Interrogates tables for user defined reports in Excel and third party data formats
Cross-checks - compares reported catch weights across different reports
Maps - visually displays vessel locations across different reports using Google Maps.
Tables - Tabulated presentation of data such as catch distribution, quota caught, etc.
Graphs - graphic presentation of similar data (above).

### 7.3.3 Mail Processor

Olfish-RMS utilizes a Delphi written application that registers three Windows services performing the following:
Monitors a POP3 account for incoming electronic messages from vessels. In order to configure the Mail Processor POP account, the user needs three pieces of information:
The name of the user ISP's mail server that holds the user email. Typically it's something like "mail.example.com".
The name of the account the user was assigned by the user ISP. This may or may not be the user email name, or something like it, or something completely unrelated.
The password to the user account.
Processes electronic messages in the following steps:
Decodes if required
Decompresses
Validates against schema
Saves to relational database
Prepares acknowledgements
Connects to SMTP server for the sending of electronic messages (acknowledgements) to vessels.

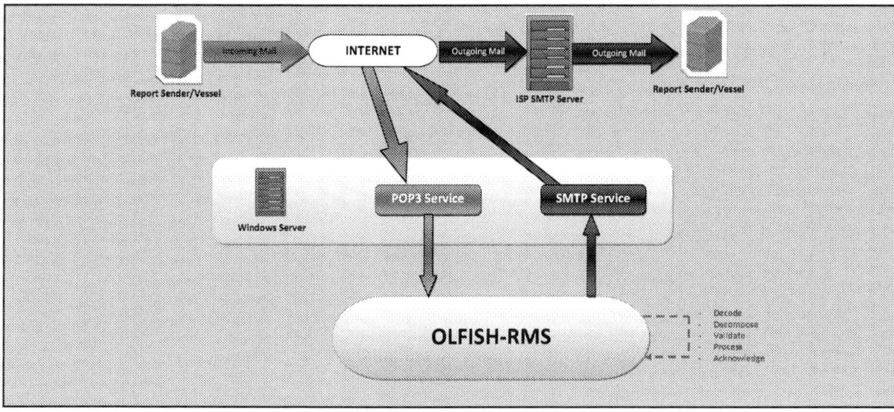


Fig. 29. Processor processes

### 7.3.4 SOAP Web-Service

Olfish-RMS registers a SOAP web service for interaction with other member states. This allows for the sending and request of reports between MS. The unique procedure of the webservice is called setERS - this is the only entry point of the webservice. MS only need to communicate the IP address and port of the server the webservice is running on. The procedure is called using the SOAP protocol and the parameter of the setERS procedure is an XML message containing an OPS element.

OPS elements have the following structure:

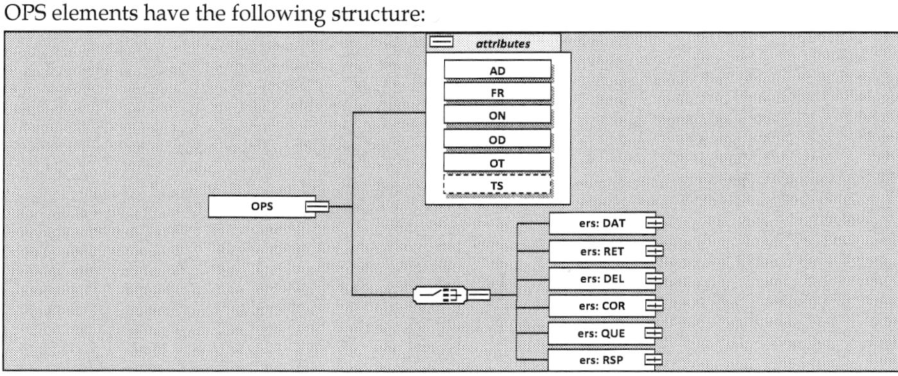


Fig. 30. Operations XSD

The attributes are:

AD: 2 letter country code of the recipient MS
FR: 2 letter country code of the sender MS
ON: Operation Number (AAA 99999999 999999)
OD: Operation date (Date the operation was initiated)
OT: Operation time (Time the operation was initiated)
TS: Test flag (if TS is present and free text is filled in)

Sub-elements (Operations) are:

DAT: Pushing of data to another MS. This happens under the following circumstances:

- When a vessel lands its catch in an MS other than the flag MS
- When a vessel intends to enter a port in an MS other than the flag MS
- When the first marketing takes place in an MS other than the flag MS
- When the first marketing does not take place in the MS where the fish was landed

RET: Acknowledgement of a previous operation
DEL: Deletion of previously sent data
COR: Correction to previously sent data
QUE: Query to pull data from another MS
RSP: Response to a pull query (QUE)

### 7.3.5 Web-Service Security

Data exchanges between MS are secured using SSL certificates. The web service uses a double hand-shake mechanism when establishing the SSL connection. The coastal state certificate is checked by the flag state and the coastal state checks the flag state certificate. No maximum downtime is set, best efforts are deployed to maintain the web-service availability 24/7.

## 8. References

022615 CEDER, 2008. Final Activity Report [online] available: https://ceder.jrc.ec.europa.eu/c/document_library/get_file?folderId=68984&name=DLFE-11544.pdf

SSP8-CT-2003-502153 SHEEL. 2004. Electronic logbook information to be exchanged. Report 1.2.Specific Targeted Research Project (STREP)

Barkai, A. & Bergh, M. Use and Abuse of data in fishery management. Deep Sea 2003: Conference on the Governance and Management of Deep-sea Fisheries. 27- 29 November 2003. Dunedin. Theme 4. Technology requirements.

FAO Fisheries Report. No. 761. Rome, FAO. 2005. 16p Data Formats and Procedures for Monitoring, Control and Surveillance. Report of the Expert Consultation on Bergen, Norway, 25-27 October 2004.

Henninger, H. 2009 Environmental Defense Fund Electronic Logbook Pilot (Phase 1) Final Report

Pope, J. G. & Symes, D. 2002. An ecosystem based approach to fisheries management, English Nature, ISSN, jncc.gov.uk [online] available: http://www.jncc.gov.uk/pdf/Achieving_2.pdf

Richard Banks Ltd. 2004. Evaluation of the NAFO OBSERVER SCHEME. Final Report. Fish 2002/03

12

# Vegetation Mapping of the Mond Protected Area of Bushehr Province (SW Iran)

Ahmadreza Mehrabian*[1], Alireza Naqinezhad[2], Abdolrassoul Salman Mahiny[3], Hossein Mostafavi[1], Homan Liaghati[4] and Mohsen Kouchekzadeh[5]

[1]*Department of Botany, Faculty of Biological Sciences, Shahid Beheshti University, Tehran, Iran*

[2]*Department of Biology, Faculty of Sciences, University of Mazandaran, Babolsar, Iran*

[3]*College of the Environment, Gorgan University of Agriculture and Natural Resources Sciences, Iran*

[4]*Department of environmental and resources economic, Environmental Sciences Research Institute, Shahid Beheshti University, Tehran, Iran*

[5]*Department of planning and design of the environment, Environmental Sciences Research Institute, Shahid Beheshti University, Tehran, Iran*

## 1. Introduction

Ecosystems dominated by natural and semi-natural vegetation, occupy large portions of the Earth's surface and provide important services that should be preserved (Balvanera et al., 2001). Vegetation mapping is one of the most important phases of vegetation conservation. Satellite data such as those produced by Landsat and Spot have became ever more available to the public (Mahiny 2004) and advances in the automatic classification of satellite data make this technique an important tool for vegetation mapping nowadays (Jenes, 1996). The main goal of traditional vegetation mapping has been the identification of plant communities which are defined as the repetitive combination of species, structural types, growth forms and other terrain attributes (e.g. McGraw and Tueller, 1983; Wallens et al., 2000; Calarck et al., 2001; Zak and Cabido, 2002; Tobler et al., 2003). The mixing of traditional and advance methods can be used for comprehensive studies in the vegetation mapping. Due to the vulnerability of arid regions, comprehensive vegetation studies are necessary in these areas. The arid regions of the world occupy 26-35 percent of the earth's land surface, much of this wide region lying latitudes of between 15 and 30 degrees Northern and reflects various types based on the climatic conditions (Archibold, 1995). There is a shortage of knowledge on the vegetation of the Middle East, but investigations have been carried out on the ecology of individual plants and their associations (Zohary, 1973). The coastlines in the Middle East can support a diverse range of flowering plants; some are tolerant to highly saline soil and inundation to various degrees, while others inhabit low salinity soil. Regional climatic, topographic and geographic conditions are assumed to be the main causes of vegetation forming in the desert and semi desert areas of Iran (Zohary, 1966-1986). In the hot southern parts of Iran with relatively high temperatures in both winter and summer and scant rainfall, a climatic regime governs which is similar to that of tropical northeast Africa, and the hot Sindian desert dominates, with occasionally more server temperature maxima and minima (Rechinger, 1963-1999; Zohary, 1966-1986 and Assadi, 1984). Iran is the classic country of great salines and Kavirs; Saline and alkaline soils are expanding in arid and semi-arid regions and cover 12.5% of the total land area of the country. These include Solenchak and Solontez soils, salt marsh soils, desert soils, Sierozem mixed with Solenchalk soils and saline alluvial soils (Dewan & Famouri, 1964). The elevation of the regions varies between -28 m on the shores of the Caspian Sea to about 1650 m in Kavire-Meyghan, Markazi Province (Akhani & Ghorbanli, 1993). Halophytic communities in Iran have been studied by many researchers. Zarinkafshe (1977) studied salty regions of Iran for flora, while Kunkel (1977) addressed the plants in the Hormoz, Qeshm and neighbouring islands. Moreover, some investigations have been carried out on the plants and vegetation of the Qeshm and Kish Island, the Persian Gulf region was also studied by Termeh & Moussavi (1982), Hamzehée (2001) and Attar et al. (2004).

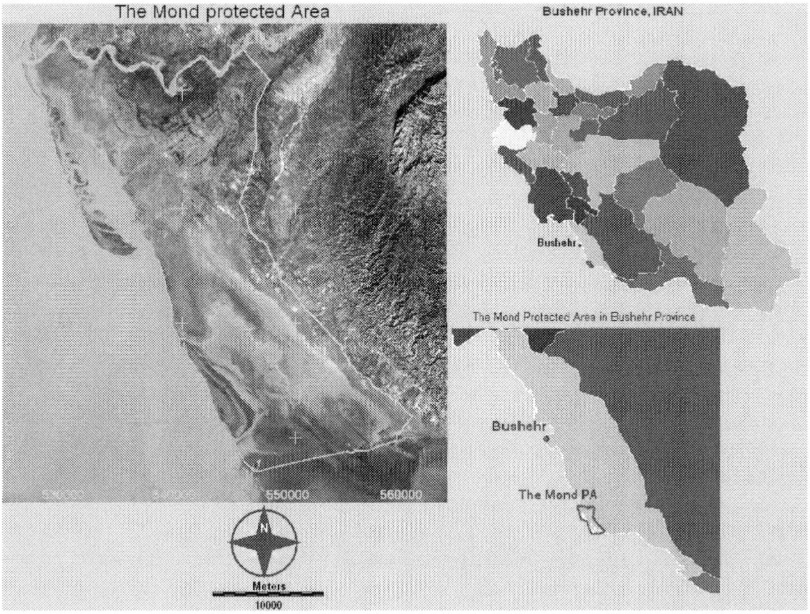


Fig. 1. Location of the Mond Protected Area in the coastal zone of Persian Gulf, Southern Iran.

The distribution of halophytic communities has been depicted cartographically by Mobayen & Tregubov (1970), Mobayen (1976), Freitag (1977), Carle and Frey (1977), Frey (1982), and Kramer (1984). Further physiognomic and ecological-geographic data on such plant communities have been provided by Kunkel (1977), Ghorbanli and Lambinon (1978), Breckle (1983), Assadi (1984), Frey and Probst (1986), and Akhani and Ghorbanli (1993).

Many researchers have been carried out on salt desert vegetation e.g., Zohary (1963, 1973), Termeh and Moussavi (1976), Leonard (1981-1988), Asri et al. (1995) and Asri and Ghorbanli (1997), Mehrabian et al., 2008). Due to the ecological and conservational values of Mond Protected area (Bushehr Province, Iran), this area was selected for vegetation mapping based on an integrative description of community structure and floristic attributes. The most important goals of this paper are 1) to provide a case of vegetation type mapping in the arid study area using field work, GIS and RS techniques, and 2) to compare these results with those other arid regions of the world.

## 2. Important

Mond Protected Area covers 53227 hectares and is located to the southwest of Bushehr between Northern latitude 27°15' to 28° 45' and Eastern longitude 51°15' to 51°35' (Fig. 1). The average yearly temperature is 14 0C and annual precipitation is 155 mm. The study area is very flat, with its highest altitude at only 12 m. There are three physiographical units in the area including alluvial and colluvial fans, river alluvial plains and lowlands. The soils consist of alluvial, regosols, saline alkaline soils, solonchak and solontez. Administratively, tree islands called Omolgorm, Tahmadoon and Nakhiloo have been included in the area. Soils of the islands belong to the saline-alkaline type with a sandy texture (Fig. 2). The Mond area can be phytogeographically classified within the Sahara-Sindian region (Leonard, 1981-1988). However, it can also be classified in the Sudanian region (sensu Zohary, 1973).

Vegetation sampling was carried out during 2005 to 2007 when the soils and vegetation map units were studied. We used all four bands of the Spot5 Satellite XS imagery acquired on 26 January 2005 to investigate the vegetation attributes. Image projection was WGS 84, and the zone number was 39n. Unsupervised classification was conducted and sampling units were chosen for the field work. Owing to the sparse vegetation of the area and based on a visual examination of the image, we found that a combined visual, unsupervised and supervised method should be used for vegetation mapping of the area. For visual assessment, we generated a pseudo-color composite image using bands 2, 3 and 4 of the Spot5 imagery. We also used bands 2 and 4 to produce a preliminary NDVI layer (Normalized Difference Vegetation Index) showing crude vegetation density for the area. This was used along with unsupervised map and other ancillary data to sample vegetation in the field. Vegetation sampling was conducted following Braun-Blanquet cover scale (Braun-Balnquet, 1964). We used 156 geographically positioned sampling points to assess vegetation. The size of samples varied between 4 m2 to 32 m2 based on the minimal area taken at each point. The field work and satellite images were mutually complementary. Dominant and companion species and their coverage were recorded in samples. Vegetation types for each area were recognized according to the occurrence of specific perennial species accompanied by some companion species. These dominant species were used for naming each vegetation type. Species emerging in each season were added to the plant list of each vegetation type during the investigations. Geo-positioning of sampling points made using with GPS The visual boundary of the map units was digitized and stored on GIS for future analysis. Using data gathered on the field, unsupervised classification of the Spot5 XS bands through is cluster module of the Erdas Imagine 8.4 software (Leica Geosystems Geospatial Imaging) and visual examination of the pseudo-color composite of the area, we distinguished different vegetation types (as map units on the GIS map) delineated them on the image and produced a final vegetation map. Vegetation map units are defined as areas where vegetation is relatively homogenous (Samira et al., 2001). A map unit is defined and named according to the taxonomic classification of the dominant community. Each map unit for the area comprised a vegetation type with the exception of those areas empty of vegetation. Water bodies and bare lands. Information about soils in the study area is based on previous soil studies in different parts of the area. Based on these studies, four major types of soil were recognized which can be subdivided to 13 detailed soil units (Fig. 2). Moreover, a classification of the habitats in the study area was provided according to fieldworks and complementary GIS methods which helped in vegetation type mapping.

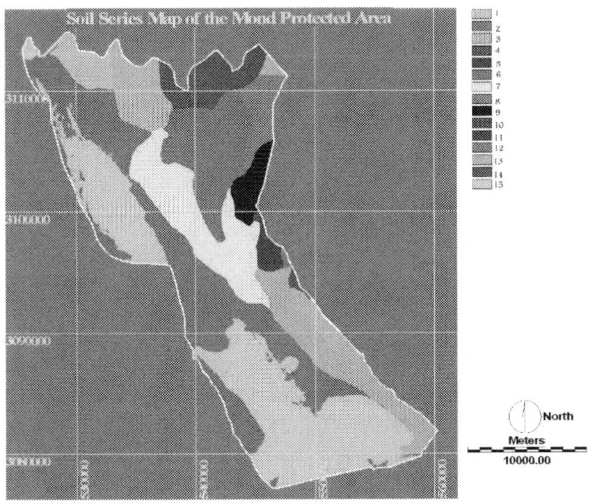


Fig. 2. Soil map of the Mond Protected Area. (1,3,4,13=Alluvial 2, 7, 8,10,11, 12 14=Solenchak 9=Regosoil, 15=Water. Different number represent gradient in each one.

Based on field observation and supported by satellite maps, three major habitat zones in the study area i.e. coastal zone, riverine zone and inland zone were recognized. These habitat zones are covered with three broad plant formations in the area. These are shrublands (northern parts), bushy grasslands (inland parts) and mangrove forests (southeastern parts of the coastal zone). In each formation, different vegetation types were recognized on the basis of field vegetation sampling guided by an unsupervised classification of the Spot XS data. Twelve vegetation types were recognized in the field that showed a good compatibility with the satellite image (map units). Moreover, large parts of the study area near the sea coast were bare lands or filled by sea water. These parts together with cultivated areas were defined as separate map units on the final map manipulated by GIS (Fig. 3). The vegetation types were variable in size and flora composition. Some vegetation types covered more than 20% of the area while others had coverage of less than 5% (Table 1). Some vegetation types, e.g. Halocnemum strobilaceum, Suaeda aegyptiaca, Lycium edgworthii are widely distributed, but Ephedra foliata (Nakhiloo island), Salsola drummundi (Eastern area), Atriplex leucoclada (Nakhiloo island), Salicornia europaea-Suaeda heterophylla (northwestern area) and Avicennia marina (south eastern area) are restricted to small habitats (Fig. 3). There are three vegetation types (Arthrocnemum macrostachyum, Ephedra foliata and Cyperus conglomerates-Halopyrum mucronatum) on Omolgorm Island, two vegetation types (Arthrocnemum macrastachyum and Cyperus conglomerates-Halopyrum mucronatum) on Tahmadoon Island and three vegetation types (Arthrocnemum macrostachyum, Cyperus conglomerates-Halopyrum mucronatum and Atriplex leucoclada on Nakhiloo Island (Fig. 3). The density of vegetation was presented as a map using bands 2 and 4 of the Spot XS data in NDVI calculation. The density indicated an increase in vegetation southward to northward and westward to eastward (Fig. 4).

## A-Shrubland formation (along the Mond River)

### 1-Tamarix leptopetala-Phragmites australis vegetation type (no. 16 in Fig. 3)

Tamarix leptopetala Bge. and Phragmites australis (Cav.) Trin. ex Steud are two dominant species of this vegetation type. Phragmites australis is a hygrophilous plant in rivers and saline marshes (Asri and Ghorbanli 1997). Tamarix leptopetala is one of the most characteristic genera in the Middle East (Zohary 1973). It comprises of about 35 species in the Middle East, many occurring in saline habitats, saline river beds and desert wadies, saline and sandy soils, estuaries of central depressions and vast areas of inland salines with a relatively high water table (Zohary 1973).

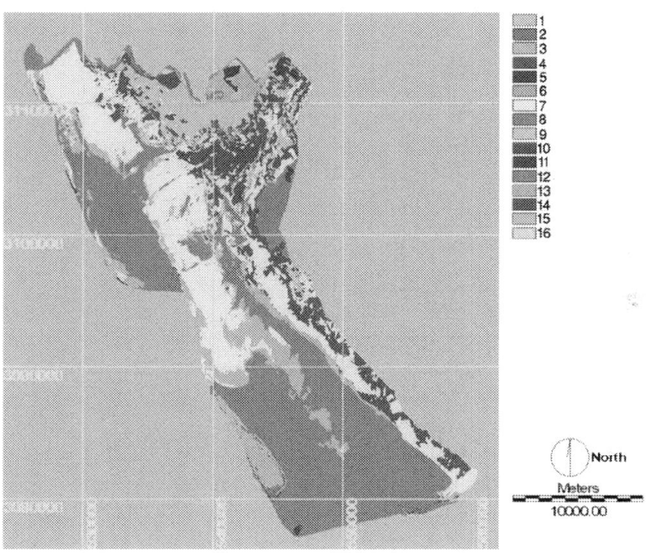


Fig. 3. Vegetation types and map units in the Mond Protected Area (Zohary, 1973): 1- Suaeda aegyptiaca, 2- Arthrocnemum macrastachyum, 3- Bare lands, 4- Halocnemum strobilaceum ((high density), 5- Farmlands, 6- Halocnemum strobilaceum (low density), 7- Water, 8- Avicennia marina, 9- Cyperus conglomerates-Halopyrum mucronatum, 10- Atriplex leucoclada, 11- Salsola drummondii, 12- Lycium edgworthii, 13- Suaeda fruticosa, 14- Salicornia europaeae- suaeda heterophylla, 15- Tamarix leptopetala

The Tamarix leptopetala- Phragmites australis vegetation type is situated in the banks of the Mond River. The first zone of this riverine vegetation belt comprises Phragmites australis and towards the inland Tamarix leptopetala replaces it and dominates over a wide area and is also dominant in many small dried rivulets and stream beds inside the area and this vegetation type is the most important vegetation type of the Iranian salt lands. This vegetation type shows a coverage of 50-75 % over the area. The most important companion species are Alhagi persarum Boiss. & Buhse., Artemisia scoparia Waldst., Cressa cretica L., Cyperus rotundus L. Spergula fallax (Lwe.) E. H. L. Krause and Suaeda aegyptiaca (Hasselq.) Zoh. This vegetation type is situated in alluvial soils.

#### 2-*Lycium edgworthii* vegetation type (no. 12 in Fig. 3)

Lycium edgworthii is distributed over certain parts of Iran. The vegetation type dominated by latter species is found on the margins of wet salty inland soils and also the external zone of the Mond River after the Tamarix-Phragmites vegetation type. Lycium edgworthii has a high density in some parts of the river margin. The coverage of this vegetation type varies between 60-70%. The most important companion species of this vegetation type are Aloina aloides (Schultz) Kindb., Anagallis arvensis L., Bromus rubens L., Calendula persica C. A. Mey., Centaurium pulchellum (Swartz.) Druce., Cuscuta chinensis Lam., Lophochlora phleoides (Vill.) Reichenb., Phlaris minor Retz. This vegetation type occupies alluvial soils in the study area.

#### 3-*Suaeda fruticosa* vegetation type (no. 13 in Fig. 3)

Suaeda fruticosa (L.) Forssk. is a dark green bushy plant which distributed across saline lands of the Sahara-Sindian region that in places penetrate into the Irano-Turanian region. This species is geographically distinct in central and southern saline (Zohary, 1973). The species is dominant in the vegetation type distributed over northern parts regions of the study area as well as Omolgorm, Tahmadoon and Nakhiloo Islands with alluvial soils. The coverage of this vegetation type varies between 75 and 100 %. The most important companion species of this type are Aeluropus lagopoides (L.) Trin. ex Thwaites, Cyperus rotundus L. Ephedra foliata Boiss and Kotschy., Lycium edgeworthi and Salsola drummondii Ulbrich.. Although this vegetation type shows some mixed situations with Lycium edgworthii vegetation type in some parts of the study area, there are many pure spots of this vegetation type dominated by Suaeda fruticosa in the area.

### B-bushy and grassland formations (vast inland area)

#### 4,7- *Halocnemum strobilaceum* vegetation type (no. 4 & 7 in Fig. 3)

Halocnemum strobilaceum is a dwarf shrub or richly branched perennial herb turning dark green as an adult. This species is an penetrative element to coast lines and inland marshes. It occupies broad belts on the fringe of salt lakes and Kavirs with relatively higher water table (Akhani & Ghorbanli, 1993). In Iranian inlands, it forms dense and almost pure stands for hundreds of miles around the smaller and large salt pans and also in «lost rivers» (Zohary, 1973). It also covers broad zones in the South and South west of Iran, extending far inwards from the seashores of the Persian Gulf, the Gulf of Oman and the Arabian Sea (Zohary, 1973). On the peripheries of most of the inland salines, it forms a pioneer halophytic community or the second phase after the Salicornia europaea vegetation (Zohary, 1973). The vegetation type dominated and characterized by Halocnemum strobilaceum is the largest vegetation type and distributed in almost all of inland parts of the study area with alluvial soils. The coverage of this type is 5-75%. Due to the intensively salty conditions of the habitats of this vegetation type, companion species are very poorly represented. The most important companion species are Aelurupus lagopoides (L.) Trin. Ex Thwaites., Asphodelus tenuifolius Cav., Gynandriris sisyrinchium (L.) Parl., Plantago amplexicaulis Cav., Plantago coronopus L., Plantago psyliium L., Plantago stocksii Boiss. & Decne., Psylliostachys spicata (Willd.), Sonchus tenerrimus L. and Suaeda heterophylla (Kar. et Kir.) Bge.

**6-*Salsola drummundi* vegetation type (no. 12 in Fig. 3)**

The genus Salsola comprises about 30 species in the Middle East. Except for a few annual and uncommon species, they are mostly dominant species in various plant communities (Zohary, 1973). Most Salsola species are xero-halophytes (Zohary, 1973). Salsola drummundi is the dominant species for the vegetation type distributed over the eastern parts of the study area with the Solonchak soils. The coverage of this vegetation type is 50-75 %. The most important companion species are Atriplex leucoclada (Boiss.) Aellen., Limonium iranicum (Bornm.) Lincz., Plantago psyliium L., Salsola cyclophylla Barker. and Suaeda aegyptiaca (Hasselq.) Zoh.

**7-*Arthrocnemum macrastachyum* vegetation type (no. 2 in Fig. 3)**

Arthrocnemum macrastachyum as a leafless, bushy succulent species with rather deep roots that is very common in the west over part the Middle East (Zohary, 1973). This species with the main distribution in Mediterranean region, occupies large stretches of littoral marshes (Akhani & Ghorbanli, 1993). It penetrates however deeply into desert areas such as the Dead Sea area, inner Anatolia, the Syrian Desert and Iraq. In the coastal marshes of the East Mediterranean, this species forms large pure stands along the salt-water bodies. Arthrocnemum macrostachyum is distributed in northwestern, southern (Omolgorm Island) and southwestern (Tahmadoon and Nakhiloo Islands) areas. The Arthrocnemum macrostachyum vegetation type is unique to high salty and wet soils on the margins of salt lakes, banks and estuaries of high saline rivers and streams and of littoral marshes of the Persian Gulf. In other localities it is less exclusive but still very abundant (Zohary, 1973). The coverage of this vegetation type is 75-100 % and it occurs on alluvial soils. Companion species are Atriplex leucoclada (Boiss.) Aellen, Cistanche tubulosa (Schrenk.) R. Wight., Halocnemum strobilaceum M. B., Limonium Iranicum (Bornm.) Lincz., Salicornia europaea L. and Suaeda heterophylla (Kar. et Kir.) Bge.

**8-Salicornia europaea-Suaeda heterophylla vegetation type (no. 14 in Fig. 3)**

Salicornia europaea and Suaeda heterophylla are two dominant annual species in this vegetation type distributed over the north west of the area. This vegetation type constitutes the first vegetation zone in salty habitats near maritime and estuary areas with Solenchak soils. The coverage of this type is 75-100%. Companion species of this vegetation type are Arthrocnemum macrastachyum and Halocnemum strobilaceum. This vegetation type was previously considered to be one of the obligatory hygro-halophtic communities in the classification presented by Akhani and Ghorbanli (1993).

**9-Cyperus conglomeratus-Halopyrum mucronatum vegetation type (no. 10 in Fig. 3)**
Cyperus conglomeratus and Halopyrum mucronatum are dominant species of this vegetation type which is located on the coastal shores to the south east of the study area with coverage of 25-50 %. The most important companion species are Atriplex leucoclada (Boiss.) Aellen., Cistanche tubulosa (Schrenk.) R. Wight., Heliotropium bacciferum Forssk., Salsola jordanicola Eig., Senecio vulgaris L. and Helianthemum lippi (L.) Pers. This vegetation type is found in alluvial or sandy soils.

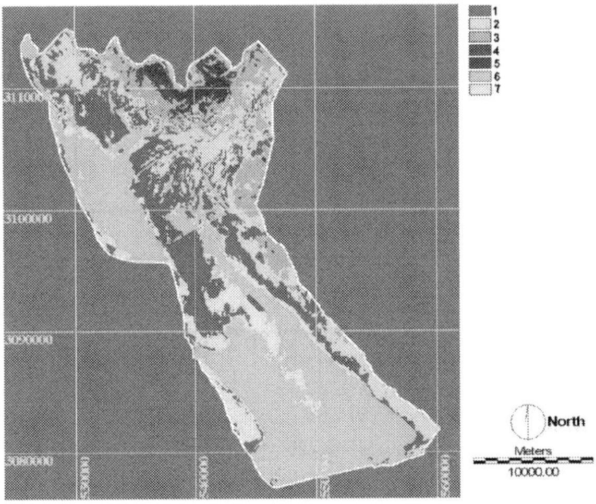


Fig. 4. Vegetation density map of the Mond Protected Area (1=very scanty, 2=scanty, 3=semi-scanty, 4=low dense, 5=dense, 6=water, 7=bare land)

**10- Atriplex leucoclada vegetation type (no. 11 in Fig. 3)**
The genus Atriplex is represented by the number of dominant species in various vegetation types of which some species (e.g. A. halimus and A. leucoclada) are distributed in salty habitats (Zohary, 1973). The Atriplex leucoclada vegetation type is distributed on Nakhiloo Island (south west part). The coverage of this vegetation type is 50-75 %. The most important companion species are Aizoon canarensis L., Cornulaca aucheri, Heliotropium bacciferum Forssk., Spergularia marina (L.) Griseb., Polycarpon tetraphyllum (L.) L. and Senecio vulgaris L. This vegetation type occupies sandy soils.

**11-Ephedra foliata vegetation type (no. 6 in Fig. 3)**
Ephedra is an excellent example of a world-wide genus represented in the Middle East by about ten species (Zohary, 1973). The Ephedra foliata is a Sudanian element in the Middle East and confined to the Gulf of Eilat, southern Arabia and southern Iran (Zohary, 1973). Ephedra foliata vegetation type is distributed only on Omolgorm Island, and over very small patches in northern parts of the area. The coverage of this type is 25-50 %. The companion species are Cyperus conglomerates Rottb., Senecio vulgaris L. and Suaeda fruticosa (L.) Forssk. This vegetation type occupies sandy soils in the study area.

## C-Mangrove forest formations (coastal vegetation)

### 12-*Avicennia marina* vegetation type (mangrove forest) (no. 9 in Fig. 3)

Mangrove forests are sensitive habitats due to their ecotonic (transitional) condition, i.e. they are affected by both marine and land ecosystems. These forests are confined to shores of the Persian Gulf (including its islands) and Oman Sea in the Middle East. The Avicennia marina vegetation type in Mond protected area is the furthest western range of Mangrove distribution in the world. Avicennia marina is the typical component of this vegetation, a species belonging to eastern mangroves that are distributed broadly in throughout world (Zohary, 1973). This vegetation type is somewhat mixed with other vegetation types, e.g. Arthrocnemum macrastachyum and Halocnemum strobilaceum vegetation types. The coverage of this type is 50-100%. It occupies alluvial and Solenchak soils in the area. Other companion species of this vegetation type are Arthrocnemum macrastachyum, Cistanche tubulosa (Schrenk) R. Wight and Salicornia europaea.

## D-vegetation affected by human activities (anthropogenic effects)

### 13-Suaeda aegyptiaca vegetation type in abandoned farmlands (no. 1 in Fig. 3)

There are about 22 species of this genus in the Middle East, of which only a few are important for their broad distribution. They can be divided into hydro- and xero-halophytes and occur in both littoral and inland saline habitats. Suaeda aegyptiaca is a ruderal species and therefore the vegetation type dominated by this species establishes in abandoned farmlands. This vegetation type with coverage of 50-75 % occurs mainly in northern parts of the area with alluvial soils. The companion species includes Artemisia scoparia Waldst. & Kit., Atriplex leucoclada, Calendula persica C. A. Mey., Chrozophora hierosolymitana Spreng., Senecio vulgaris L., Sonchus oleraceus L., Sporobolus arabicus Boiss. and Stellaria media (L.) Cyr.

### 14-Farmlands (anthropogenic vegetation) (no. 5 in Fig. 3)

One of the important categories in the vegetation map prepared is farmlands representing areas strongly affected by man and livestock. This map category is largely the result of human agricultural activity, and is comprised of three large units, namely cultivated, segetal and ruderal (Zohary, 1973). This map unit is distributed over the margins of the northeastern and eastern parts with alluvial and Solenchak soils.

E-Map units without vegetation

### 15-Bare lands (no. 3 in Fig. 3)

Bare lands constitute one of the largest units in the vegetation map of the study area located along the coastal line. Due to high salinity, vegetation density was as low as nearly 0. This vegetation type is established in Solenchak soils.

### 16-Water (no. 8 in Fig. 3)

The eastern and southern parts of the study area are fully occupied by water connected to the Sea (Fig. 3).

## 3. Conclusion

The current study is a new approach to vegetation mapping in Iran using remote sensing (RS) and the geographic information system (GIS). One of the most important problems in remote sensing of desert vegetation is that the reflectance from soil and rocks is often much greater than that of sparse vegetation and this makes it difficult to separate out the vegetation signal (Gates et al. 1965); and there is spectral variability within shrubs of the same species (Duncant et al., 1993). These properties hamper accurate classification of vegetation in these areas. This study proved the usefulness of the Spot XS imagery for vegetation mapping but also it showed that in arid regions, mapping can only be completed satisfactorily if it is accompanied by extensive field sampling, visual image interpretation and hybrid classification methods. Thus, hybrid approaches that include field work, GIS and RS are required in such circumstances.

In a study of the halophytic vegetation of the Middle East, Zohary (1973) discussed that the most halophytic communities of Iran belong to the phytosociological classes Halocnemum strobilacei irano-anatolica. Breckle (1983) classified the halophytic vegetation of Iran and Afghanistan as follows: (1) saline flats (very sparse vegetation, soil with very high salt content); (2) euhalophytic vegetation (halo-hammada on gravel-sandy but probably clay soil) and (3) mesohalophytic vegetation (with less salt in the soil profile). Frey and Probst (1986) provided a geographical classification for total halophyte vegetation including (1) salty pans of the central Iranian undrained basin and their peripheries; (2) shore zones of salt lakes; (3) areas the Persian Gulf and (4) southern Caspian coastal zone. Akhani and Ghorbanli (1993) with a geographical-ecological approach classified halophytic communities of Iran as (1) Halocnemum strobilaceum communities (on muddy salt flats); (2) obligatory hydro-halophytes communities; (3) Tamarix communities; (4) Hydrophilous euryhlophytic communities; (5) Mangrove communities; (6) Hydrohalophytic plant communities; (7) Halophytic shrub communities on salty and dry soils; (8) Herbaceous perennial and hemicryptophyte halophytic communities; (9) xeromorphytic communities with salt-tolerant xerophytes and (10) annual halophytic communities.

Mond Protected Area is one of the best indicators of halophytic vegetation in the arid lands of Iran. Salty river, salty pans, coastal and muddy salt flats induce habitat diversity supporting the establishment of different halophytic vegetation types (see Figs. 2, 3, 4). This study identified 15 map units of which 12 are concerned with vegetation types. These vegetation types reveal plant communities adapted to different habitats and environmental gradients in the area.

Most vegetation types identified in the current study were previousy recognized in the other arid or salty areas of Iran (Akhani and Ghorbanli, 1993; Asri & Ghorbanli, 1997; Ghahreman et al., 2000; Alaei et al. 2001). The occurrence of Halocnemum strobilaceum, Avicennia marina, Salicornia europaea-Suaeda heterophylla, Suaeda fruticosa, Tamarix leptopetala and Arthrocnemum macrostachyum vegetation types is identical with the results of Akhani and Ghorbanli (1993). The study area can be considered as the first and third vegetation zones defined in the classification offered by Breckle (1983). This research revealed that the establishment of the vegetation types is largely regulated by edaphic factors (texture, chemical composition and humidity). Some parts of the area are impacted by agricultural activity, grazing and the destruction of vegetation. These activities have led to major changes in habitats and ecosystems and have threatened wildlife. These ecosystems represent landscapes that can be restored and managed for uses such as ecotourism and recreation.

| Vegetation types | Surface (ha) | Percentage (%) |
|---|---|---|
| *Halocnemum stroboilaceum* | 19320.1 | 36.30 |
| *Suaeda aegyptica* | 3044.52 | 5.72 |
| *Salsola drummondii* | 1080.88 | 2.03 |
| *Lycium edgworthii* | 708.72 | 1.33 |
| *Suaeda fruticosa* | 263.96 | 0.50 |
| *Arthrocnemum macrastachyum* | 209.52 | 0.39 |
| *Tamarix leptopetala* | 189.04 | 0.36 |
| *Cyperus conglomerates-Halopyrum mucronatum* | 87.04 | 0.16 |
| *Atriplex leucoclada* | 18.92 | 0.04 |
| *Avicennia marina* | 14.56 | 0.03 |
| *Ephedra foliata* | 8.92 | 0.02 |
| *Salicorniaeuropaea-suaeda teterophylla* | 9 | 0.02 |
| Farmalnds | 364.52 | 0.68 |
| Water | 21052.2 | 39.55 |
| Bare lands | 6855.12 | 12.88 |
| Total | 53227 | 100.00 |

Table 1. Areal coverage of vegetation types in the Mond Protected Area

## 4. References

Akhani. H., Ghorbanli M. (1993). A contribution to the halophytic vegetation and flora of Iran, in H. Leith and A. Al- Masoom (eds.). Towards the Rational use of High Salinity Tolerant Plants, vol. 1, p. 35-44, Kluwer Academic publishers, Netherlands.

Archibold, O. W. (1995). Ecology of World vegetation. London. Chapman & Hall. 1-510.

Asri Y., Ghorbanli M. (1997). The halophilous vegetation of the Orumieh lake salt marshes, NW. Iran. Plant Ecology 132, 155-170.

Asri Y., Hamzehee B., Ghorbanli M. (1995). Etude Phytosociologique de la vegetation halophile de l'est du lac Orumieh (Nord Ouest de l'Iran). Doc. Phytosociology, 15: 299-308.

Assadi M. (1984). Studies on the autumn plants of Kavir, Iran. Iran. j. Bot. 2, 125-148.

Attar F., Hamzehée B, Ghahreman A (2004). A Contribution to flora of Qeshm Hsland, Iran. Iran. J. Bot. 10, 199-218.

Balvenera, P., Daily, G., Ehrlich, P., Ricketts, T., Bailey, S., Kark, S., Kermen, C. & Pereira, H (2001). Conserving biodiversity and ecosystem services. Sciences, 291, 2047.

Braun-Blanquet J. (1964). Pflanzensoziologie: Grundzüge der Vegetationskunde. 3. neu bearb. Aufl. - Springer-Verlag. Wien.
Breckle S.W. (1983). Temperate deserts and semi-deserts of Afghanistan and Iran. In: N. E. West (ed). Ecosystems of the World. 5 : Temperate Deserts and Semi-Deserts. Elsevier. Amsterdam. 271-319.
Calarck P.E., Seyfried M.S., Harris B. (2001). Intermountain plant community classification using Landsat TM and spot HRV data. Range Management 54, 152-160.
Carle R., Frey W. (1977). Die vegetation des Maharlu-Beckenis bei Shiraz (Iran) Unter besonderer beruccksichitigung der vegetation im bereich der suss-und salzawasserquellen am seeufer. Beih. TAVO, Reihe A (Naturwiss) 2, Dr. Ludwig Reuchert Verlag, Wiesbaden.
Dewan M., Famouri J. (1964). The Soils of Iran. FAO, Rome.
Duncant J., Stow D., Franklin J., Hope A. (1993). Assessing the relationship between spectral vegetation indices and shrub cover in the Jourdan Basin, New Mexico. International journal of Remote Sensing, 14, 3395-3416.
Frey W., Probst W. (1986). A synopsis of the vegetation of Iran. In : H Kürschner (ed). Contribution of the vegetation of southwest Asia. Beih. TAVO. Naturwiss, 24, 9-24.
Frey W. (1982). Maharlu-Becken bei Shiraz (Iran). Mittlerer Teil. Vegetation, 1: 100,000, Karte AVI 10.2 TAVO, Dr. Ludwig Reichert Verlag, Wiesbaden.
Freitag H. (1977). Turan Biospher Reserve, prelimnary vegetation map, pp. 86-89. im : Spooner, B. (ed.). Case Study on Desertification Iran : Turan. Department of Environment, Tehran.
Gates D.M., Keegan H.J., Schleter J.C., Weidner V.R. (1965). Spectral properties of plants. Applied Optics, 4, 11-20.
Ghahreman A., Hamzehée B., Attar F. (2000). Natural Vegetation Map of the Qeshm island, Publication of Qeshm Free Area.
Ghorbanli M., Lambinon J. (1978). Permier. Apereu de la zonation de la vegetation halo-gypsophile du lac Ghom. Lenjunia Rev. Bot. 92 :1-20.
Hamzehée B. (2001). Application of Anaphyto Software in Phytosociological data analysis (A Case Study: Eroded Terraces of the Queshm Island). - Research Institute of Forests and Rangelands. Tech. Publ. 237. Tehran. 379 pp. (In Persian).
Jensen, J. R. (1996). Introductory Digital Image Processing. A Remote Sensing Perspective (2nd ed). New Jersey : Prentice Hall.
Kramer W. (1984). Mittleter Zagros (Iran) Vegetation, 1:600,000, Karte AVI 6 TAVO, Dr. Ludwig Riechert Verlag, Wiesbaden.
Kunkel G. (1977). The vegetation of Hormoz, Qeshm and neighboring islands (Persian Gulf area). Flora et vegetatio Mundi 6. 186p.
Leica Geosystems Geospatial Imaging (2002). Erdas Imagine 8.4 software, USA.
Leonard J. (1981-1988). Contribution a l'étude de la flore et de la végétation des déserts d Iran. Fasc 1-9. Meise.
McGraw J.F., Tueller P.T. (1983). Landsat computer-aided analysis teqniques for range vegetation mapping. Journal of Range Managements, 36: 627-631.
Mehrabin A. R., Naqinezhad A., Mostafavi H., Kiabi B., Abdoli A. Flora and habitats of Mond Protected area. Mohit Shenasi., Vol 46(1) : 1-18
Mobayen S. (1976). Structure geobotanique du Lut. Acta Ecol. Iranica 1, 73-86.

Mobayen S., Tregubov V. (1970). Carte de la végétation naturelle de l,Iran, 1; 250,000. University Tehran, UNDP/FAO no.Ira 7.

Rechinger K.H. (ed.) (1963-1999). Flora Iranica. Vien, Graz : Akademische Druck.

Salman Mahiny A. ( 2004). A Modelling Approach to Cumulative Effects Assessment for Rehabilitation of Remnant Vegetation. Ph.D. Thesis, ANU, Australia.

Omar. Samir. A. S., Misak. R., King. P., Shahid. Shabbir. A., Abo-Rizk. Hanna, Grealish. G. and Roy. W. (2001) Mapping the vegetation of Kuwait through reconnaissance soil survey. Journal of Arid Environment, 48: 341-355.

Termeh F., Moussavi M. (1976). Contribution a l'étude de la végétation automnale du Dashte-Lut. Institut de Recherches Entomologiques et Phytopathologiques d Evine, Department of Botany. Tehran. No. 7.

Termeh F., Moussavi M. (1982). Plants of Kish Island, Iran. Wildenowia, 12, 253-286.

Tobler M.W., Cochard R., Edward P.J. (2003). The impact of cattle ranching on large-scale vegetation patterns in coastal Savanna in Tanzannia. journal of Applied ecology, 30, 430-444.

Wallens J., Milington A.C., Hichkin W., Arquepino R., Jones S. (2000). Mapping humid tropical vegetation in eastern Bolivia in R. Alexander and A. C. Millington (Eds) Vegetation Mapping. Chichester : John Willey and sons.

Zak M.R., Cabido M. (2002). Spatial pattern of the Chaco vegetation of central Argentina. Integration of remote sensing and phytosociology. Applied Vegetation Sciences, 5, 213 -226.

Zarrin-kafche M. (1977). Etude de la relation existent entre la compositon chemique des plants et celle des Sols du Sud de karadj (pressé de Tehran). Acta Ecology. Iran. 1, 60-69.

Zohary M. (1963). On the Geobotanical Structure of Iran. Bull. Res. Counc. Isr., Sect. d, Bot., Suppl. 1. 113p.

Zohary M. (1966-1986). Flora Palestine, Vol : 1-4. The Israel Academy of Sciences and Humanities.

Zohary M. (1973). Geobotanical Foundations of the Middle East. 2 vols. Gustav Fischer Verlag. Amesterdam.

13

# Earth to space link

Mandeep Jit Singh, Mardina Abdullah, Baharudin Yatim,
Mahamod Ismail and Wayan Suparta
*Malaysia*

## 1. Introduction

The measurements were taken continuously from 1st January 2006 to 31st December 2008. The measured rainfall data was analyzed and presented in this chapter. Three years averaged measured rain rate and rain attenuation were compared with the cumulative probability distributions existing prediction models. Some rainfall data from selected tropical climates sites were used to make a comparison of rain rate and rain attenuation prediction.

## 2. The Variation of Rainfall Amount

Malaysia experiences heavy rain throughout the year and the rainfall distribution is patterned by monsoon activities. The Northeast monsoon is from October to March and the Southwest monsoon is from May to September. The Northeast Monsoon is the major rainy season in the country. The Northeast Monsoon blows from South China Sea during the months of October to March, this affects high brought rainfall to the East Coast of Peninsular Malaysia and a few other locations at Peninsular Malaysia. It was recorded by Malaysian Meteorological Department that the Northeast Monsoon has give the highest rainfall in all location in Malaysia because the wind of this monsoon blows from South China Sea without obstruction by island, peninsular and mountain. Monsoon weather systems that develop in the conjunction with cold air outbreaks from Siberia produce heavy rains, which often cause severe floods along the east coast of Malaysia. The Southwest monsoon is comparatively drier throughout the country except for the state of Sabah. During this season, most states experience monthly minimum rainfall. This is attributed to relatively stable atmospheric conditions in the equatorial region. In particular, the dry condition in Peninsular Malaysia is accentuated by the rain shadow effect of the Sumatran mountain range. During the inter-monsoon periods, winds are light and variable. Morning skies are often clear and this favours thunderstorm development in the afternoon. In the states of west coast of Peninsular Malaysia, thunderstorms contribute to a mean monthly rainfall maximum in each of the two transition period.
Figure 1 shows the monthly variation of rainfall amount at USM. A relationship between the monsoon activities and monthly rainfall amount can be studied. Based on the measurement, the total rainfall for the year 2006 was 1835 mm, for the year 2007 was 2065 mm and for the

year 2008 was 2111 mm. The total average annual rainfall at USM for three years was 2004 mm. The total average rain accumulation for the Northeast Monsoon (October - March) and Southwest Monsoon (May - September) was 1192 mm (59% of the annual averaged) and 640 mm (32% of the annual averaged), respectively. The maximum rainfall amount was in November for the year 2006 that is 274 mm, in September for the year 2007 that is 287 mm and in December for the year 2008 that is 280 mm. The minimum rainfall amount was in July for year 2006 that is 66mm, in August that is 58 mm and in May that is 5 mm. From this figure, it indicates that the rainfall amount for Northeast Monsoon is higher than Southwest Monsoon.

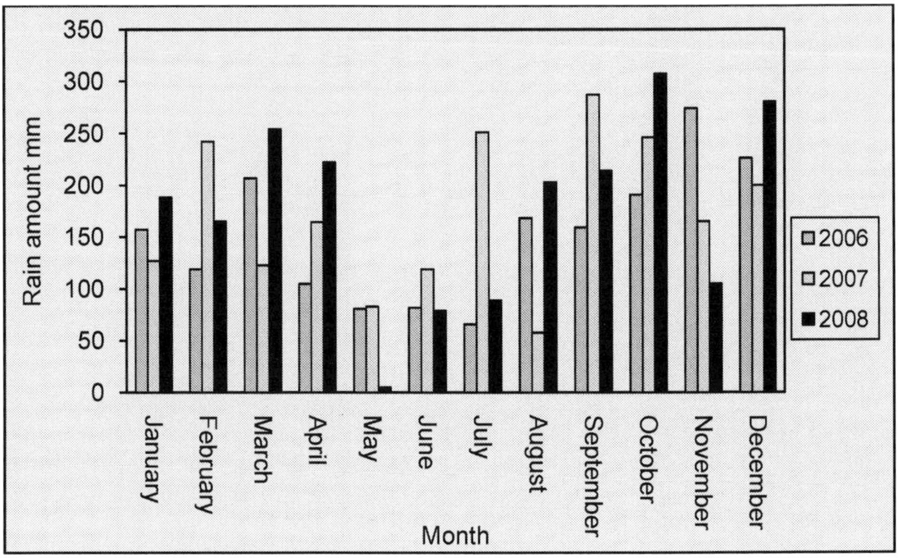


Fig. 1. Variation of monthly rainfall amount at USM for the year 2006, 2007 and 2008

## 3. Testing of Prediction Models

The achievement of high availability targets in advanced satellite link design requires a deep knowledge of the radio channel behavior. The effects due to different atmospheric causes, such as precipitation, clouds, atmospheric gases and tropospheric scintillation can be measured quite accurately by means of satellite beacon signals. Most prediction models of probability distribution function (pdf) of exceeding rain attenuation in slant path, based on rain rate pdf, when tested against concurrent satellite beacon measurements show large errors (Matricciani, *el at.*, 2008).

Two types of prediction tests used for this analysis:

1. The percentage error
2. The Real Mean Square (RMS) error

For certain percentage of time (from 0.001 to 1 percent of the year), for which data are available, percentage relative error, $E_{rel}$ (percent) between the predicted value and the measured value are calculated

$$E_{rel} = \frac{A_{predicted} - A_{measured}}{A_{measured}} \times 100\% \quad (1)$$

The mean error, $\mu_e$ and standard deviation, $\sigma_e$ are used to calculate the Root Mean Square, $D_e$ (RMS). The parameter is defined as follows

$$D_e = [(\mu_e)^2 + (\sigma_e)^2]^{1/2} \quad (2)$$

According to evaluation procedures adopted by the CCIR the preferred prediction method is the one producing the smallest RMS values (CCIR, 1983)

## 4. Statistical analysis rainfall

### 4.1 The correlation of Rain Rate and Rain Attenuation

Fig. 2 shows a typical variation of rain rate and rain attenuation measured on a rainy day (2nd March 2008). The rapid small-scale fluctuations of the signal level are because of ionospheric and tropospheric scintillations and slower variation correspond to rain events - fade. At certain times, the decrease in the relative signal strength (attenuation) and the increase in the rain rate are coincidental, however at other times, the decrease in the signal strength was leading the increased rain rate. It is probably the result of raindrops held aloft by updrafts in the cloud, and then their time of arrival at the ground would be considerably delayed. This delay time is variable and is as long as 4-8min (Ramachandran, *et. al.*, 2004, Mandeep, 2008). In the data presented in Fig. 2, delay (corrections) time of 8 min was accounted for when investigating the correlation between rain rate and rain attenuation.

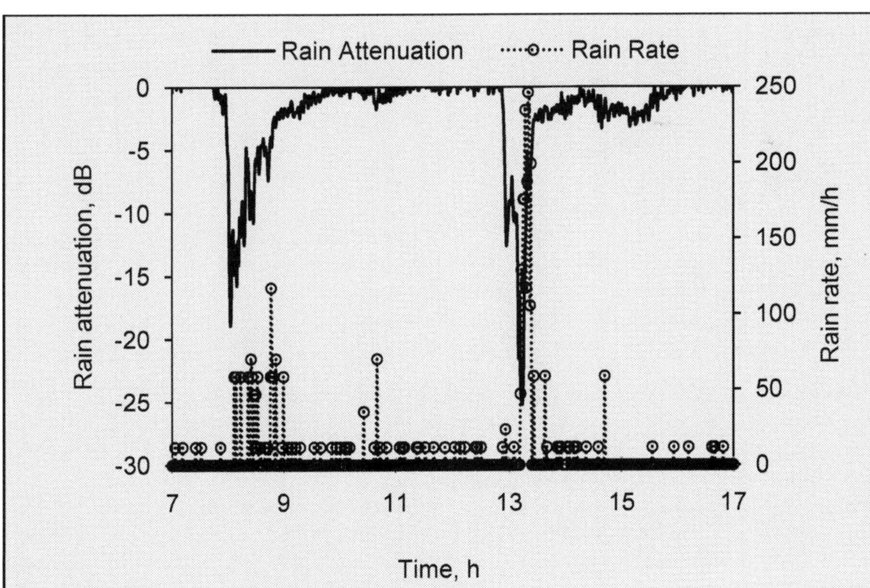


Fig. 2. Time series record of attenuation and rain rate

To determine the nature and strength of the relationship between rain rate and rain attenuation, correlation analysis has been done to indicate the strength of the relationship between rain rate and rain attenuation is high. Fig 3 shows the relation between rain rate and rain attenuation. Form that figure, the dots are obtained by straightforwardly plotting rain attenuation versus rain rate and the solid line is the best fit curve obtained by fitting rain attenuation versus rain rate. The R- square is 91.39%. It indicates that the correlation between rain rate and rain attenuation is high.
The attenuation for the site was a logarithmic function of rain rate and can be simplified as:

$$A\ (\mathrm{dB}) = 0.054R^{1.2315}$$

where A denotes as attenuation and R denotes as the site rain rate, mm/h. This empirical equation would be helpful to system designer to determine the link margin at the site. Calculating by using this empirical equation, a threshold of 19.6 dB is considered as the economical limits.

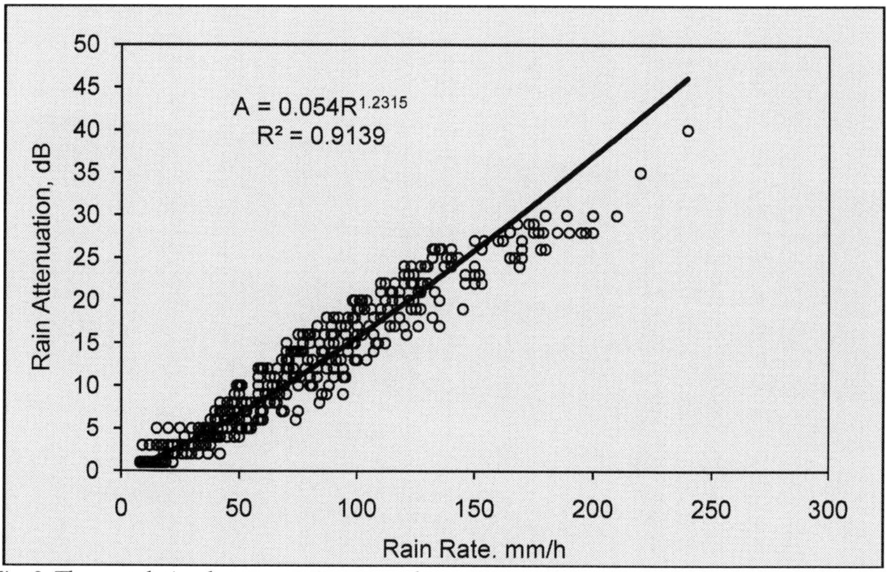

Fig. 3. The correlation between rain rate and rain attenuation

## 4.2 Statistical Confidence Level and Interval of the Measured Data

In statistics, a confidence interval (CI) is an interval estimate of a population parameter. Instead of estimating the parameter by a single value, an interval likely to include the parameter is given. Thus, confidence intervals are used to indicate the reliability of an estimate. How likely the interval is to contain the parameter is determined by the confidence level or confidence coefficient. The end points of the confidence interval are referred to as confidence limits. Increasing the desired confidence level will widen the confidence interval. For analysis purpose, a 95% confidence level was used because it is recommended by many statistical analysts. By using Minitab software, the confidence limits for 95%

confidence level were calculated. Fig 4, 5 and 6 show the confidence level of cumulative distribution of measured rain rate data in the year 2006, 2007 and 2008, respectively. In any statistical investigation, the variations of values in a measured data are due to random and systematic errors. The measured rain rate data for the year 2006, 2007 and 2008 lie in between the 95% confidence limits. The R-square is 99.7%, 99.2%, 97.8%, respectively. Therefore, these rain rate data are suitable to be used for comparison with models.

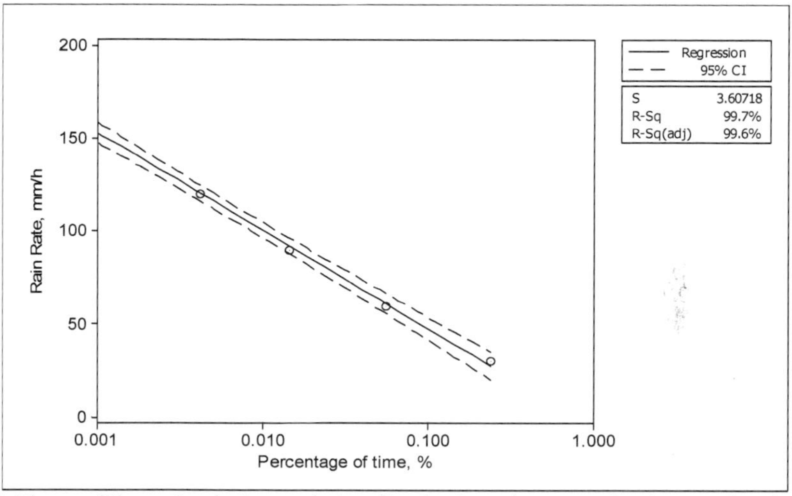


Fig. 4. The confidence level of cumulative distribution of measured rain rate data in year 2006

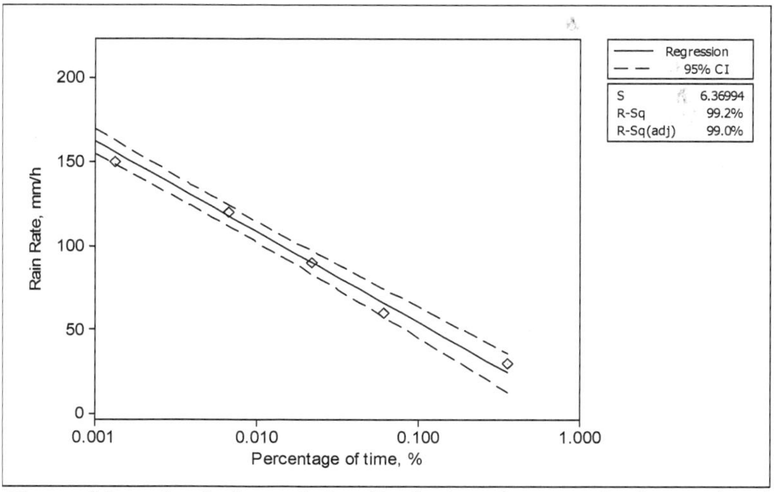


Fig. 5. The confidence level of cumulative distribution of measured rain rate data in year 2007

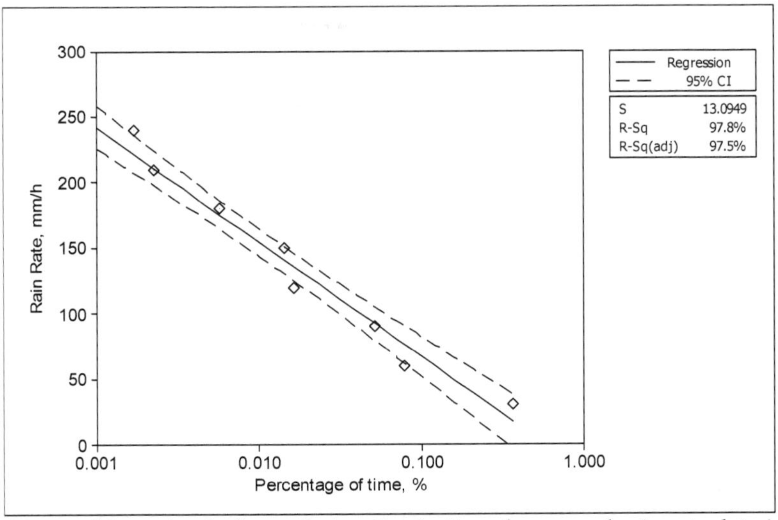


Fig. 6. The confidence level of cumulative distribution of measured rain rate data in year 2008

To calculate the confidence limit, having drawn a random sample and calculated the mean and standard deviation, the lower confidence limit (LL) is found by

$$LL = \bar{X} - z\left(\frac{\alpha}{2}\right) s_{\bar{x}} \tag{3}$$

In a similar way, the upper confidence limit (UL) is found by

$$UL = \bar{X} + z\left(\frac{\alpha}{2}\right) s_{\bar{x}} \tag{4}$$

where $a$ is the significance level. At 95% confidence level, $z\left(\frac{\alpha}{2}\right)$ is equal to 1.96 and $s_{\bar{x}} = \frac{s}{\sqrt{N}}$, where $s$ is the standard deviation and N is the sample size.
By using equation (3) and (4), the confidence limits were calculated for 95% confidence level. The confidence limits are calculated for the rain attenuation data in year 2006, 2007 and 2008 data are shown in Fig 7, 8 and 9. The measured rain attenuation data lies in between the 95% confidence limits for the entire measurement time. Hence, these rain attenuation data are suitable to be used for comparison with models.

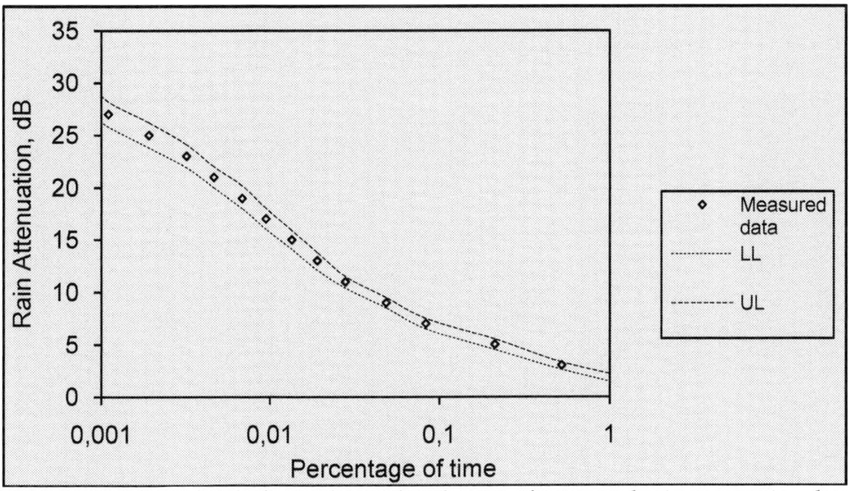


Fig. 7. The confidence level of cumulative distribution of measured rain attenuation data in year 2006

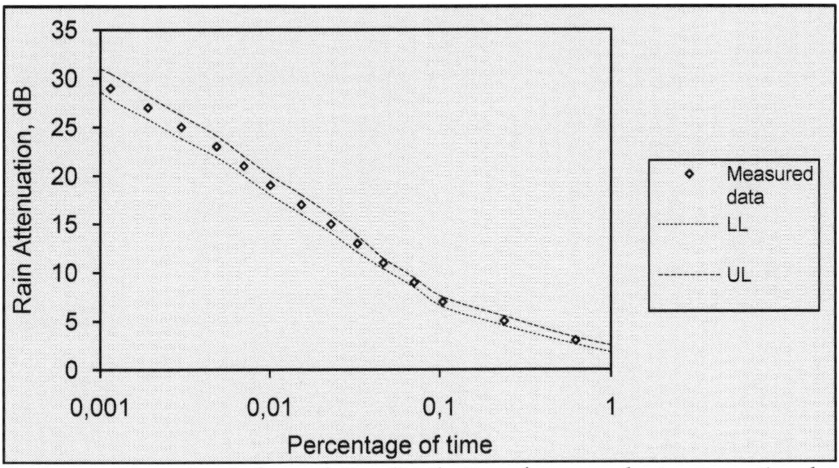


Fig. 8. The confidence level of cumulative distribution of measured rain attenuation data in year 2007

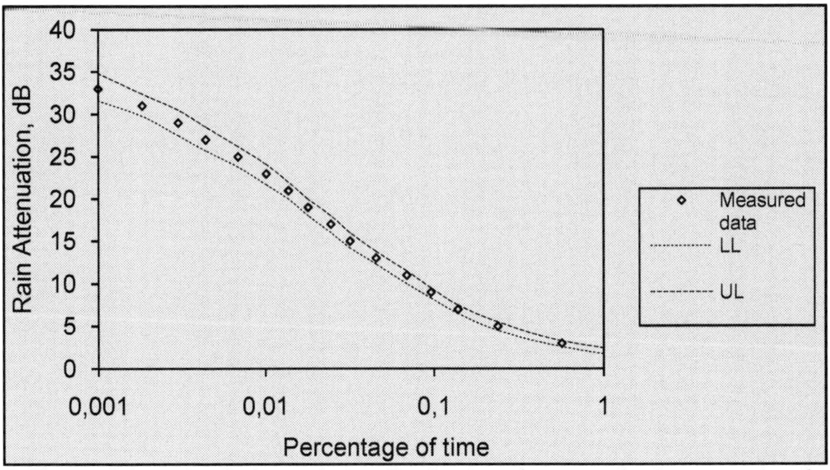

Fig. 9. The confidence level of cumulative distribution of measured rain attenuation data in year 2008

## 5. Rain Rate, Rain Attenuation Analysis

Communication systems operating at frequencies beyond 10 GHz in tropical and equatorial climates are subjected to many fade occurrences due to heavy rain. The information on rain rate and attenuation statistic is useful in link budget design. The considerable average worst month from year to year and within individual years is important in the planning of satellite earth-link design.

Rain rate is the volume of liquid water that falls through a unit area per unit time period. Rain attenuation is the depletion of electromagnetic energy during propagation through rain, caused by raindrop scattering and absorption. Worst month distribution of rain rate and attenuation is defined as the highest rain rate and attenuation, respectively within a set of individual monthly distributions of rain rate and attenuation for a year.

Best fit lines of the annual cumulative distribution of measured rain rate and rain attenuation for these three years have been used to make the comparison.

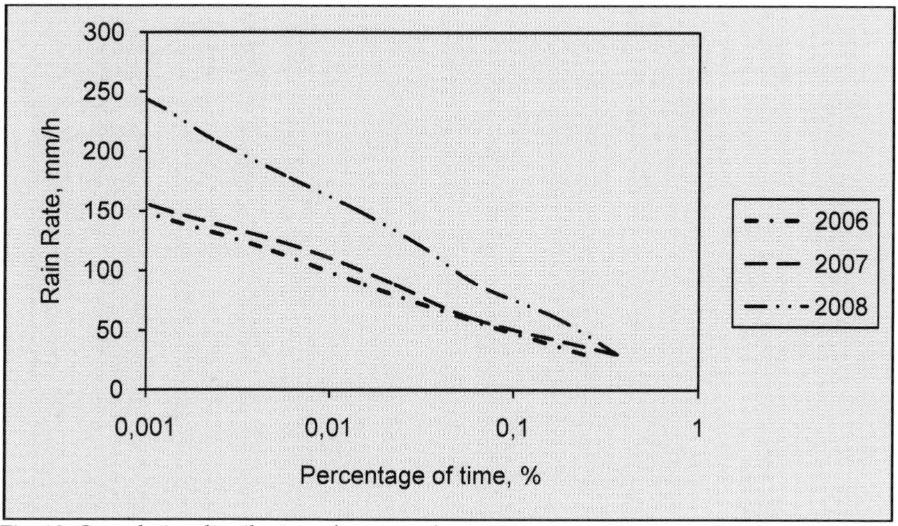


Fig. 10. Cumulative distribution of measured rain rate

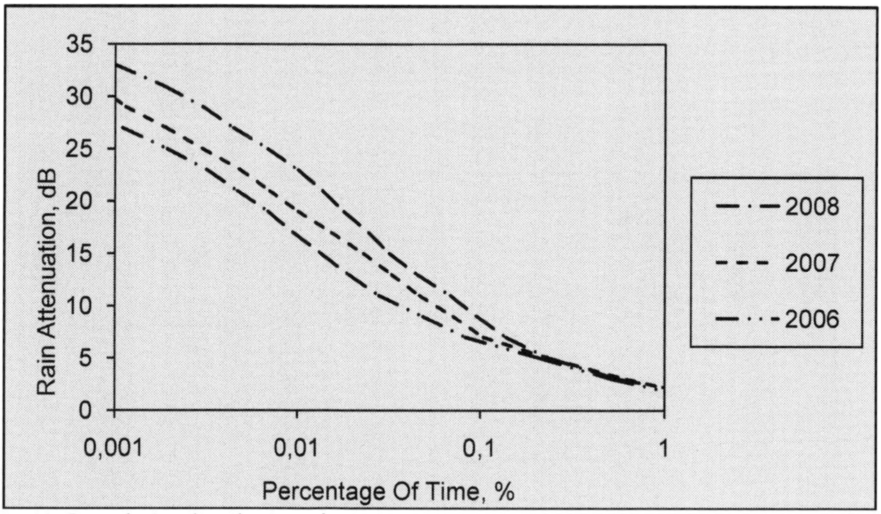


Fig. 11. Cumulative distribution of rain attenuation

Fig 10 and Fig 11 show the annual cumulative distribution of rain rate and rain attenuation for three years (2006-2008), respectively. Table 1 shows the record of rain rate and rain attenuation at point 0.01% of time and maximum point for the year 2006, 2007, 2008.

| Year | Rain Rate. mm/h | | Rain Attenuation, dB | |
|---|---|---|---|---|
| | At point 0.01% | Maximum | At point 0.01% | Maximum |
| 2006 | 100 | 180 | 17 | 29 |
| 2007 | 110 | 210 | 18 | 31 |
| 2008 | 160 | 270 | 28 | 31 |

Table 1. The record of rain rate and rain attenuation at point 0.01% of time and maximum point for the year 2006, 2007, 2008

## 6. Worst Month Statistics

Worst month distribution of rain attenuation is defined as the highest rain attenuation within a set of individual monthly distributions of rain attenuation for one year. Rain attenuation data for year 2006, 2007 and 2008 at USM was analyzed.

In planning of the design of reliable communication systems, the required statistics of propagation effects is relevant to the worst month reference. On the other hand, the reference statistics for several propagation prediction methods are the long term average annual distribution. Therefore, the conversion of the yearly statistic to the worst month is important. The annual and worst-month rain attenuation exceedance curve is shown in Fig.12, Fig 14 and Fig 16 for year 2006, 2007 and 2008, respectively. The worst month for 2006, 2007 and 2008 were November, October and March, respectively. The Q factor as a function of annual percentage of exceedance for Ku-band rain attenuation is shown in Fig.13, Fig 15 and Fig 17 for year 2006, 2007 and 2008, respectively.

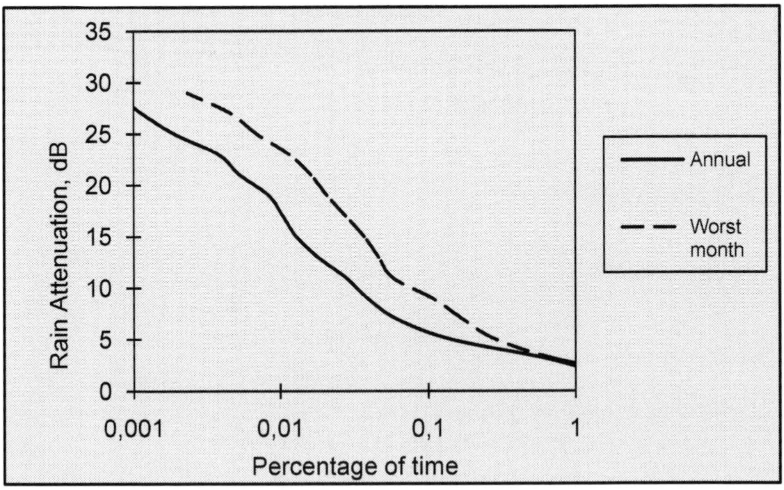


Fig. 12. The annual and worst-month rain attenuation exceedance curve at Ku-band for year 2006

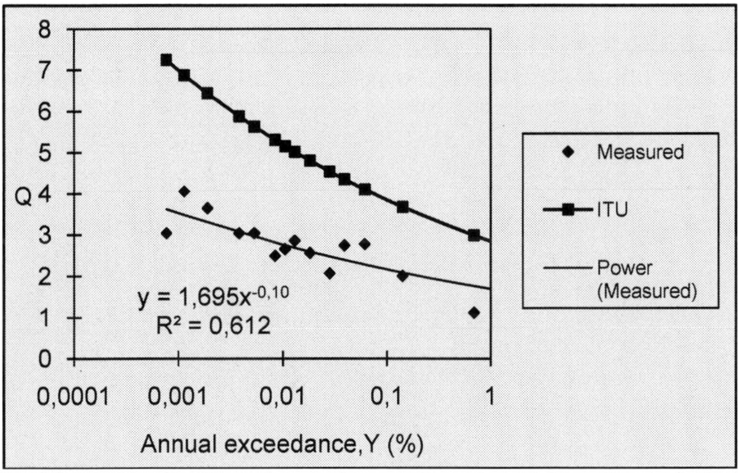


Fig. 13. The Q factor as a function of annual percentage of exceedance for rain attenuation at Ku-band for year 2006

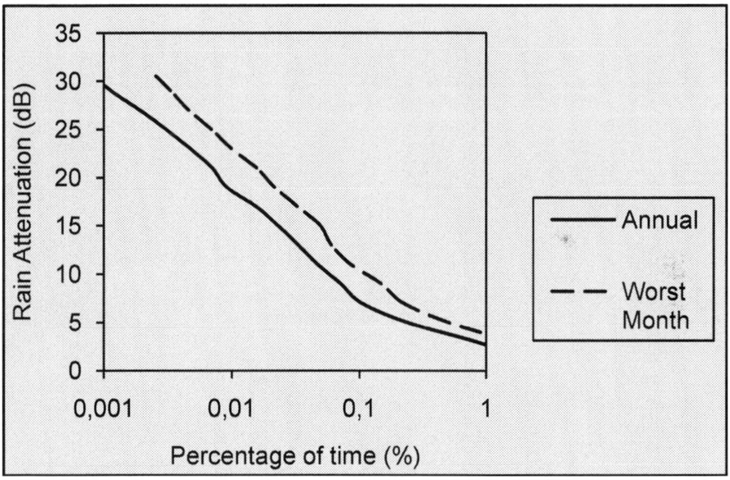


Fig. 14. The annual and worst-month rain attenuation exceedance curve at Ku-band for year 2007

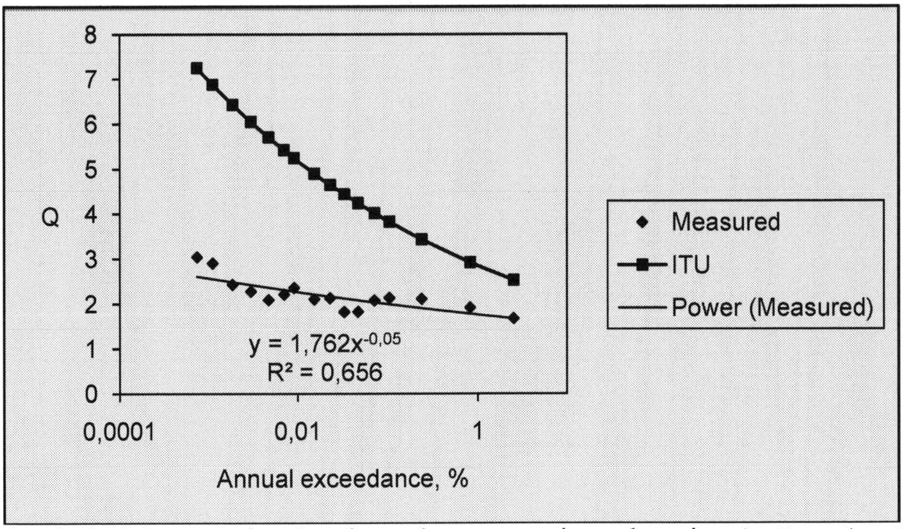


Fig. 15. The Q factor as a function of annual percentage of exceedance for rain attenuation at Ku-band for year 2007

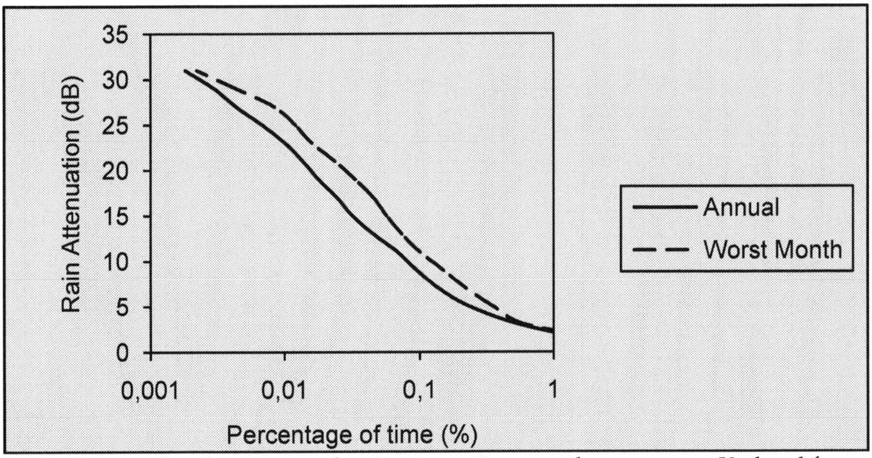


Fig. 16. The annual and worst-month rain attenuation exceedance curve at Ku-band for year 2008

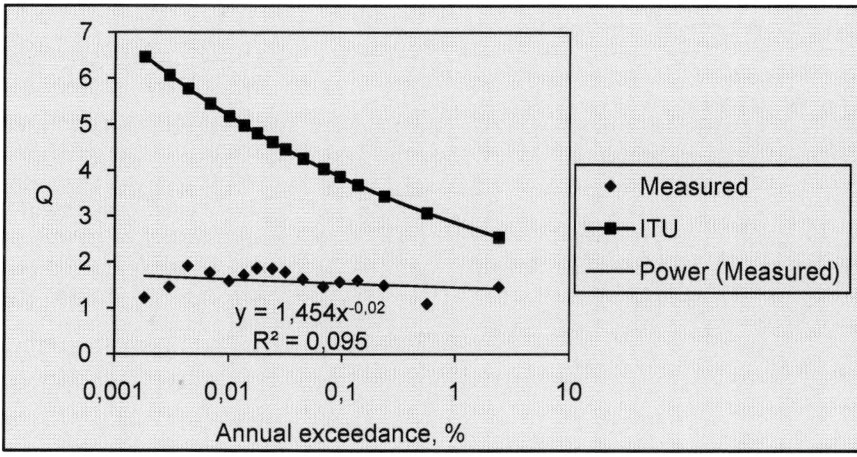


Fig. 17. The Q factor as a function of annual percentage of exceedance for rain attenuation at Ku-band for year 2008

The Q factor for rain attenuation was found to follow the power law of the form $Q=AY^{-\beta}$. Table 2 shows the parameter A and β for year 2006, 2007 and 2008.

| Year | A (Proposed by ITU-R = 2.82) | B (Proposed by ITU-R = 0.15) |
|---|---|---|
| 2006 | 1.6953 | 0.106 |
| 2007 | 1.7624 | 0.055 |
| 2008 | 1.4547 | 0.024 |

Table 2. The parameter A and β for year 2006, 2007 and 2008

For global rain rate applications, the ITU P.841-4 has recommended values of **A** = 2.82 and β = 0.15 for tropical, subtropical and temperate climate regions with frequent rain. The percentage of time for worst month distribution of rain attenuation is significantly higher than annual distribution of rain attenuation. There is a large difference between the A and β values obtained for tropical area with the ITU proposed. It shows that the ITU values are not suitable for use in worst-month analysis for tropical area. This indicates that the Q factors are climatic dependant. The worst month curve was strongly influenced by the Northeast Monsoon during which the highest levels of attenuation occurred. Worst month attenuation statistics are very important for the study of the performance of a communication system during periods of up to 31 days.

Ku-band TV services are affected by outages for time-critical transmission such as real-time news and sports broadcasting. Service providers need to consider the use of appropriate forward error correction codes, the choice of modulations, and the range of uplink power controls to use during severe rain fade periods in the overall design of their communications networks (Pan, *et al*, 2008). These techniques can be used to provide a significant improvement in both performance and availability for worst month attenuation.

## 7. Specific Attenuation Analysis

The measurement ran continuously from 1st January 2006 to 31st December 2008. Three years of cumulative distributions are obtained. The measurement ran continuously from 1st January 2006 to 31st December 2008. Three years of cumulative distributions are obtained. The relation between rain attenuation and rain rate in year 2006, 2007 and 2008 were shown in Fig. 18, 19 and 20. In these figures, the symbols × are obtained by directly plotting the rain attenuation against rain rate and the solid line is the best-fit curve obtained by fitting the rain attenuation against rain rate. The effective path length, $L_{eff}$, is a function of rain rate and has a direct correlation with the measured rain rate. The effective path length has been found by using equation 2.5 and 2.6 and shown in Fig. 21. The effective path length for the site is a power-fitting function of rain rate and can be simplified as

$$L_{eff}\,(\text{km}) = 13.367\ R^{-0.21} \tag{5}$$

The $L_{eff}$ is obtained from the equation 5 is used to divide the total attenuation to obtain specific attenuation, γ at different rain rate. Fig. 22, 23 and 24 show the relation between specific attenuation, γ and rain rate in the year 2006, 2007 and 2008.

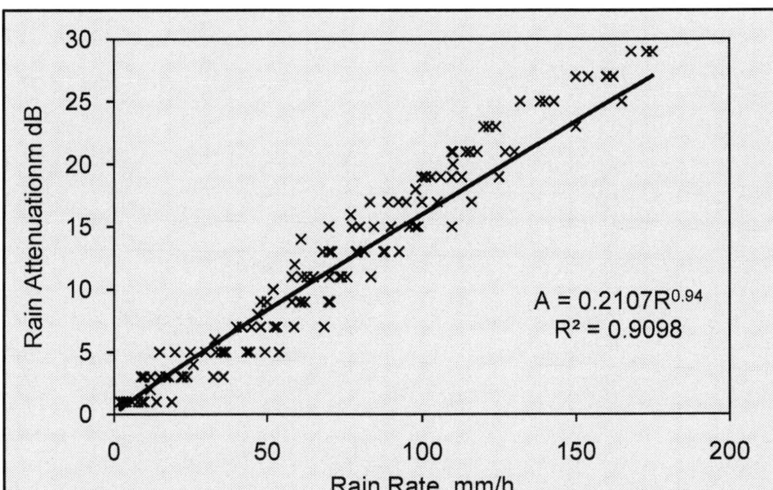


Fig. 18. The correlation between rain attenuation and rain rate in year 2006

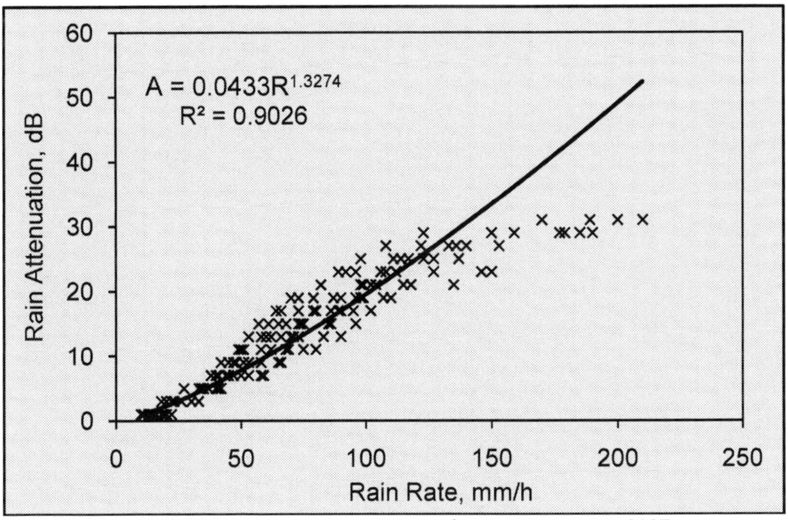


Fig. 19. The correlation between rain attenuation and rain rate in year 2007

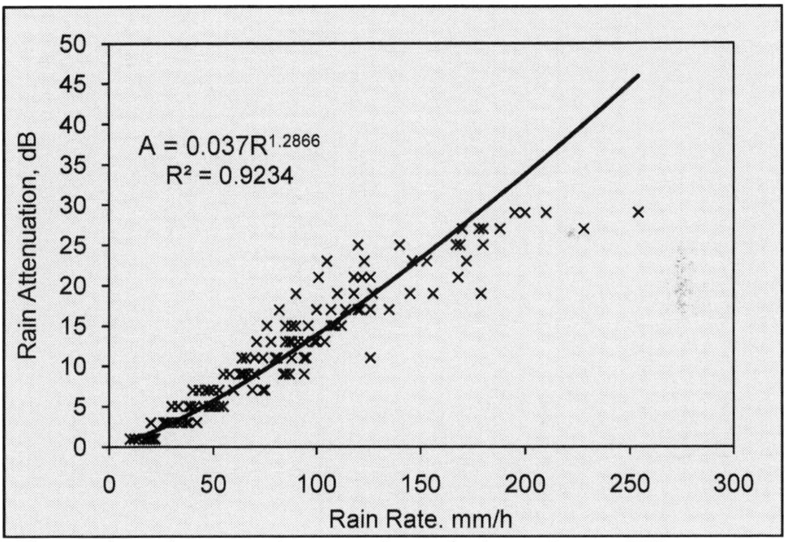


Fig. 20. The correlation between rain attenuation and rain rate in year 2008

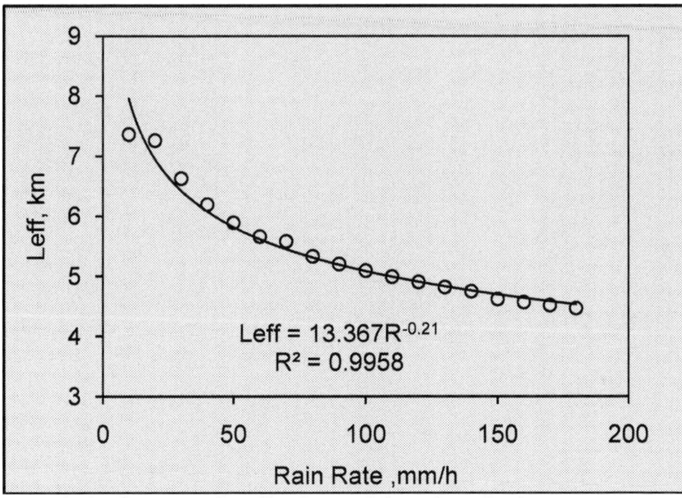


Fig. 21. The correlation between the effective path length and rain rate.

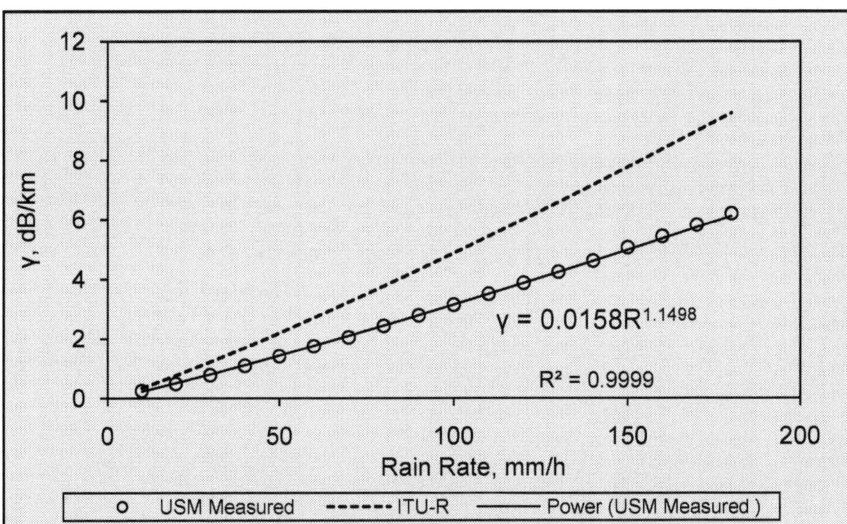


Fig. 22. Relationship between specific attenuation and rain rate compared with ITU-R in year 2006

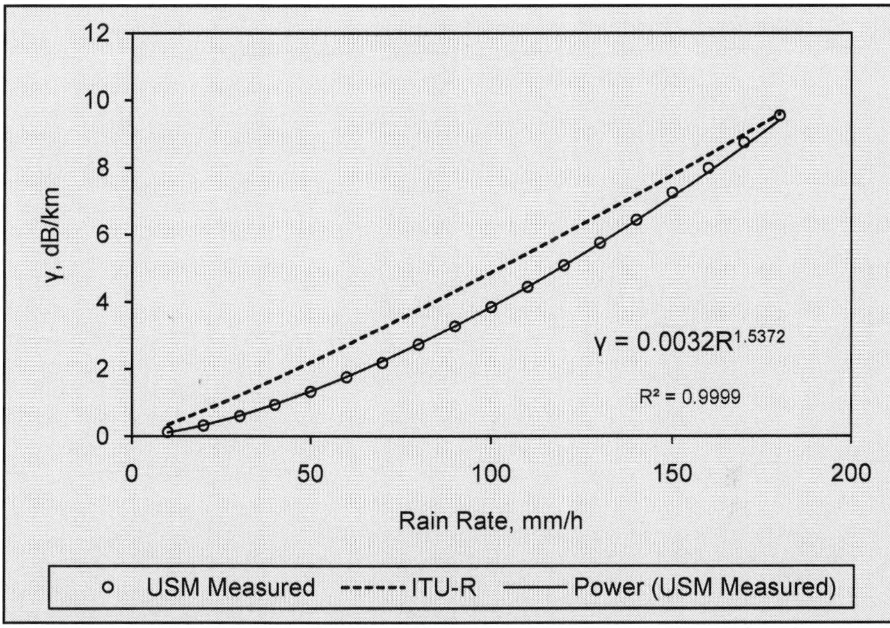


Fig. 23. Relationship between specific attenuation and rain rate compared with ITU-R in year 2007

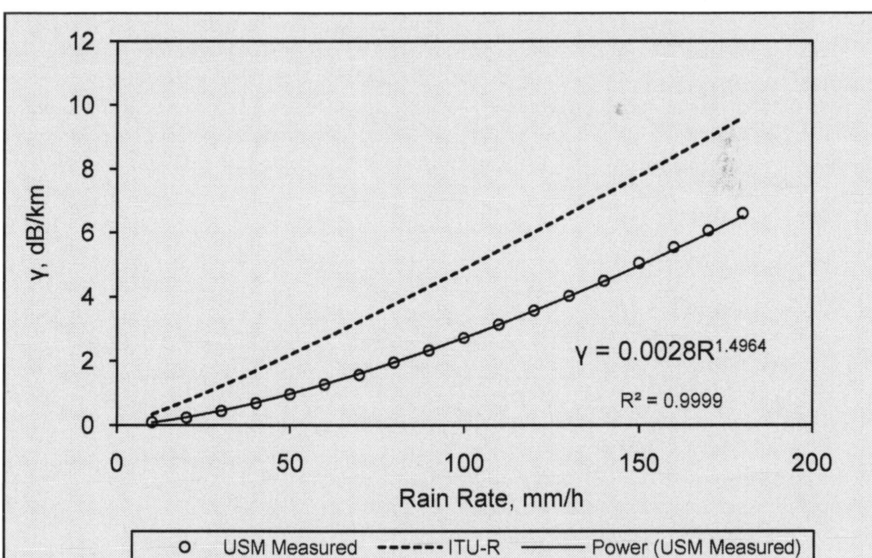


Fig. 24. Relationship between specific attenuation and rain rate compared with ITU-R in year 2008

Using equation 2.2 and 2.3, *k* and *a* that obtained are 0.0242 and 1.152 respectively. Table 3 shows the regression coefficients for k and α by using empirical procedure.

| Year | *k* | α |
|---|---|---|
| 2006 | 0.0158 | 1.1498 |
| 2007 | 0.0032 | 1.5372 |
| 2008 | 0.0028 | 1.4964 |

Table 3. Regression coefficients for k and α by using empirical procedure

Based on rain rate and rain attenuation measurements, the ITU-R has overestimated the specific rain attenuation due to tropical rainfall at least in the 3 years term view. The coefficients of k and α are found that can significantly vary and be considerably different from the ITU-R proposed for regression coefficients and it implies that the raindrop size distribution (DSD) in Malaysia's tropical region is quite different from that adopted by ITU-R, at least in our experiment period. There are many factors influencing the specific attenuation. This is considered due to the verity of the drop size from temperate regions to the tropical region. The availability and accuracy of measured data is the factor to influence the empirical value. Therefore, ITU-R recommendation for regression coefficients of rain specific attenuation is not suitable use in predicting rain attenuation for Malaysia.

## 8. Analysis of One-Minute Rain Rate Measured Data with Existing Models

The comparison of the measured one minute rain rate values with existing rain rate models is shown in this section. There are 5 tropical climates sites (e.g. USM, Bangkok, Bandung, Manila, Fiji) 2 years average (from the years 2002 to 2003) measured one-minute rain rate that used in comparison. The 2 years (from 1$^{st}$ January 2007 to 31$^{st}$ December 2008) average USM measured rain rate has been used in the comparison. The existing models that applied in the prediction one minute rain rate are Moupfouma model, ITU-R model, KIT simplified model, Rice & Holmberg model and Dutton & Dougherty model. The prediction rain rate depends on the annual rainfall values. The annual rainfall values for these tropical climates sites are shown below:

| Sites | Annual rainfall (mm) |
|---|---|
| USM | 2088.0 |
| Bangkok | 1565.0 |
| Bandung | 1956.0 |
| Manila | 2300.0 |
| Fiji | 3087.5 |

Table 4. The average annual rainfall

The comparison of one minute rain rate prediction models with measured data for the 6 tropical climates sites are shown in Fig 25, 26, 27, 28 and 29.

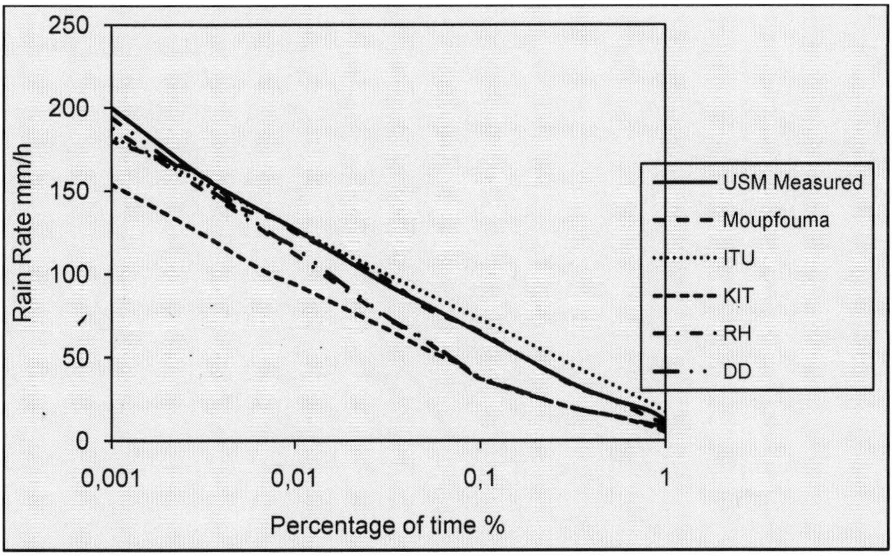


Fig. 25. Comparison of one minute rain rate prediction models with measured data for USM site.

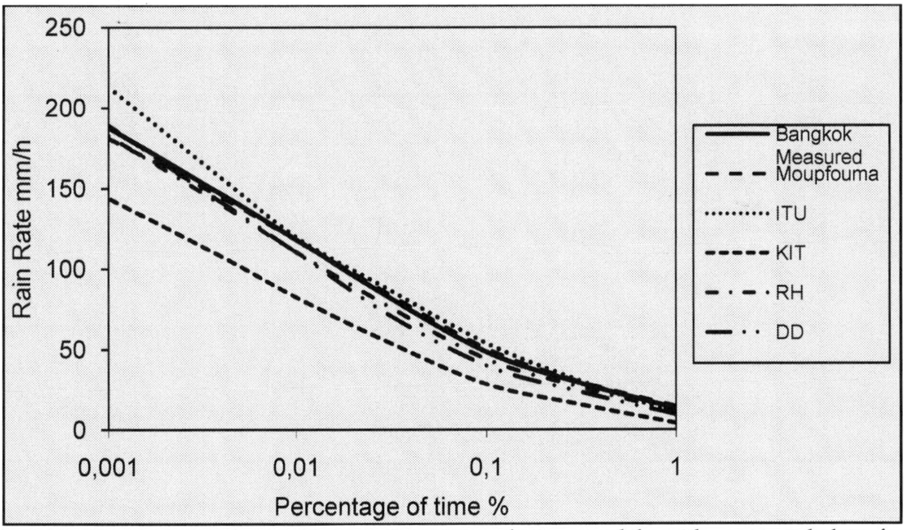


Fig. 26. Comparison of one minute rain rate prediction models with measured data for Bangkok site.

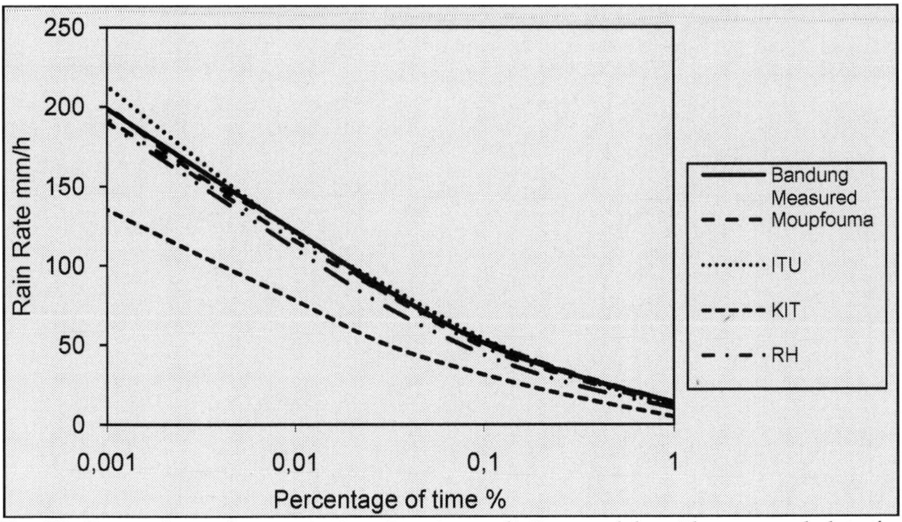


Fig. 27. Comparison of one minute rain rate prediction models with measured data for Bandung site.

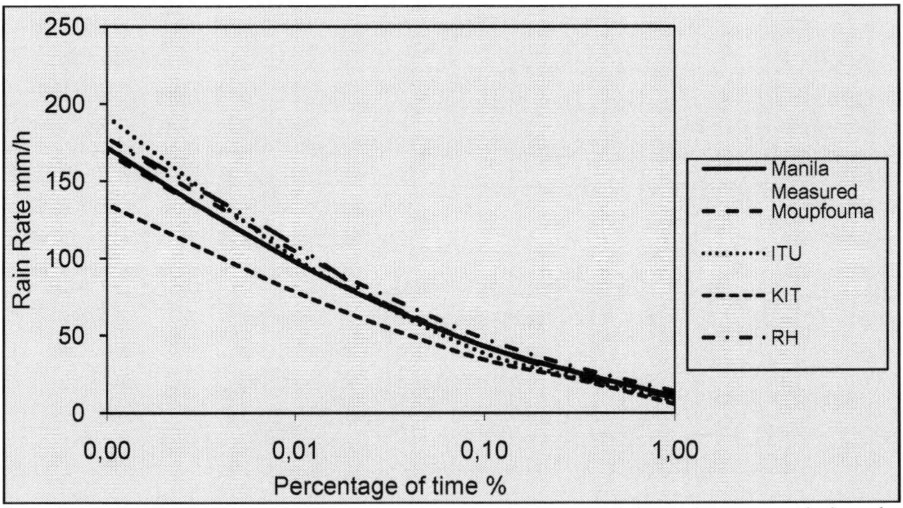


Fig. 28. Comparison of one minute rain rate prediction models with measured data for Manila site.

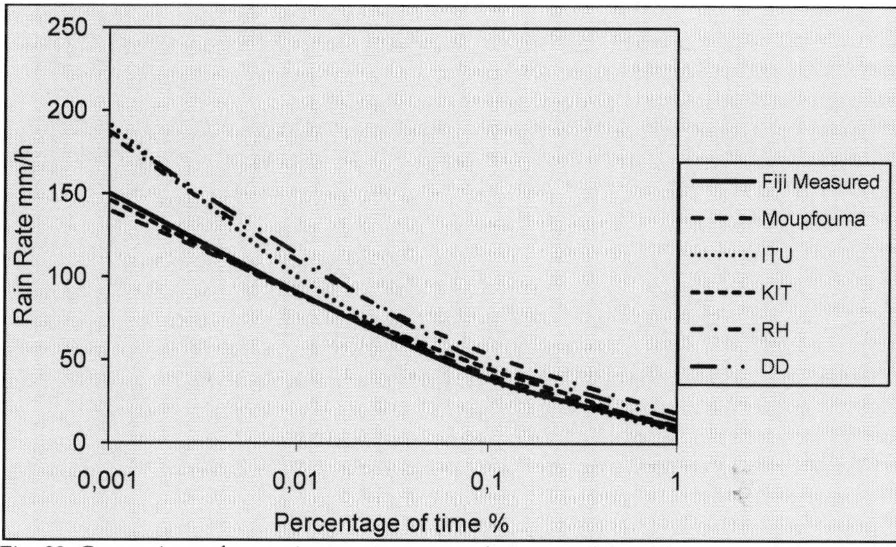


Fig. 29. Comparison of one minute rain rate prediction models with measured data for Fiji site.

The Moupfouma model overestimates the one minute rain rate from 0.01% to 1% of time and underestimates the rain rate from 0.001% to 0.01% at USM sites. The model gave a RMS value of 8.64% for USM. This is because the model has a probability law behavior that underlines the complexity of the rain rate distribution according to the climate of the zone of interest. For Bangkok, Bandung, Manila and Fiji sites, the model follows closely the measured rain rate values throughout the entire percentage of time that the rain rate is exceeded. The model gave a low RMS value for those tropical sites. The model's RMS values were 53% (Bangkok), 2.33% (Bandung), 1.69% (Manila) and 6.22% (Fiji). The coefficients of $\lambda$ and Y values from the slope of rain rate curve in equation 2.18 depend strongly on the measured rain rate data. For tropical and sub-tropical localities, $\lambda = 1.066$ and $Y = 0.214$ are used in calculation of rain rate cumulative distribution slopes.

The ITU-R model overestimates the one minute rain rate from 0.01% to 1% of time and underestimates the rain rate from 0.001% to 0.01% at USM sites. The model gave a RMS value of 20.72% for USM. For Bangkok, Bandung, Manila and Fiji sites, the model follows closely the measured rain rate values up to 0.01% of time that rain rate is exceeded before the model overestimates the measured values. The model gave a RMS value of 13.18% for Bangkok site, 11.75% for Bandung, 13.65% for Manila and 17.90% for Fiji. The ITU-R has the climate zones used in the equatorial region are subdivided further that includes region with the similar rain rate characteristics and a large number of measured rain rate database that are available for equatorial region.

For USM, Bangkok, Bandung and Manila, the Kitami Institute of Technology (KIT simplified) model underestimates the measured rain rate throughout the entire percentage of time that the rain rate is exceeded. The model's RMS value was 36.56% (USM), 41.59% (Bangkok),

42.72% (Bandung) and 25.57% (Manila). The model gave a high RMS value for these sites because the annual rainfall amount at these sites were not more than 2300 mm. The KIT model prediction at Fiji, gave a low RMS value of 15.62%. The model follows closely the measured rain rate values at the entire percentage of time that the rain rate is exceeded. This is because the annual rainfall at Fiji was above 3000 mm. The KIT model states that the accuracy of the model depends largely on the annual rainfall values, where the higher the annual rainfall values better the prediction gets.

The RH model underestimates the measured rain rate at USM, Bangkok and Bandung throughout the entire percentage of time that the rain rate is exceeded. The model gave a RMS value of 29.65% at USM, 8.59% at Bangkok and 7.58% at Bandung. The RH model overestimates the measured rain rate at Manila and Fiji throughout the entire percentage of time that the rain rate is exceeded. The model gave a RMS value of 149% at Manila and 42.16% at Fiji. The RH considered the convective rain activity and stratiform rain activity was neglected. The thunderstorm ratio, $\beta$ was based on thunderstorm rain but on the convective rain activity days to total rain days. The model gave a high RMS value at Fiji site because the $\beta$ value given by RH is 0.3, however the $\beta$ value calculated to be 0.75.

The Dutton and Dougherty (DD) model underestimates the measured rain rate at USM, Bangkok and Bandung and overestimates the measured rain rate at Manila and Fiji throughout the entire percentage of time that the rain rate is exceeded. The model gave a RMS value of 29.04% at USM, 16.03% at Bangkok, 186% at Bandung, 7.73% at Manila and 28.10% at Fiji. The M (average annual total rainfall depth, ,mm) values used to calculate the coefficient constant in Europe were below 1200mm per year, but the annual rainfall, M is above 1800mm per year in tropical climate.

A summary of the info is as show in Table 5 and a conclusion of the best and worst model is given. The comparison of rain rate was done between measured data and five pre-existing mathematical models. For the tropical region, it was found that Moupfouma model revealed a close fit to the measured data for low, medium and high rain rates. The Moupfouma model is judged suitable for use in predicting rates in tropical climates. The KIT simplified model exhibited poor performance in comparison.

| Site | Annual rainfall, mm | RMS value, % | | | | | Conclusion | |
|---|---|---|---|---|---|---|---|---|
| | | Moupfouma | ITU-R | KIT | RH | DD | Best Model | Worst Model |
| USM | 2088.00 | 8.64 | 20.72 | 36.56 | 29.65 | 29.04 | Moupfouma | KIT |
| Bangkok | 1565.00 | 35 | 13.18 | 41.59 | 8.59 | 16.03 | Moupfouma | KIT |
| Bandung | 1956.00 | 2.33 | 11.75 | 42.72 | 7.58 | 186 | Moupfouma | KIT |
| Manila | 2300.00 | 1.69 | 13.65 | 25.59 | 149 | 7.73 | Moupfouma | KIT |
| Fiji | 3087.50 | 6.22 | 17.90 | 15.62 | 42.16 | 28.10 | Moupfouma | RH |

Table 5. The summary of the comparison of rain rate prediction model

## 9. Analysis of Rain Attenuation Measured Data with Existing Models

The rain attenuation prediction models exposed in literature calculate the attenuation related to a given rain rate or else to a given percentage of time. For terrestrial as well as satellite microwave links, one of the fundamental needs for the link designer is to have at his disposal an effective model that predicts attenuation caused by rain on the propagation path with a good accuracy. Most of the models available are empirical or semi empirical and their accuracy are based on the accuracy of the measured rain rate cumulative distribution (Moupfouma, 2009).

The comparison of the measured rain attenuation values with existing rain attenuation models is shown in this section. There are 5 tropical climates sites that are USM, Bangkok, Bandung, Manila, Fiji measured rain attenuation that is used for the comparison. The 2 years (from 1st January 2007 until 31st December 2008) average USM measured rain attenuation has been used in the comparison. The existing models that applied in the prediction rain attenuation are ITU-R model, Ong model, Ramachandran and Kumar model, CETUC model, Leitao and Watson model, Garcia-Lopez model, SAM model and Assis-Einloft model. The comparison measured rain attenuation with existing predicted models at these 8 tropical climates sites are shown in Fig 30, 31, 32, 33 and 34.

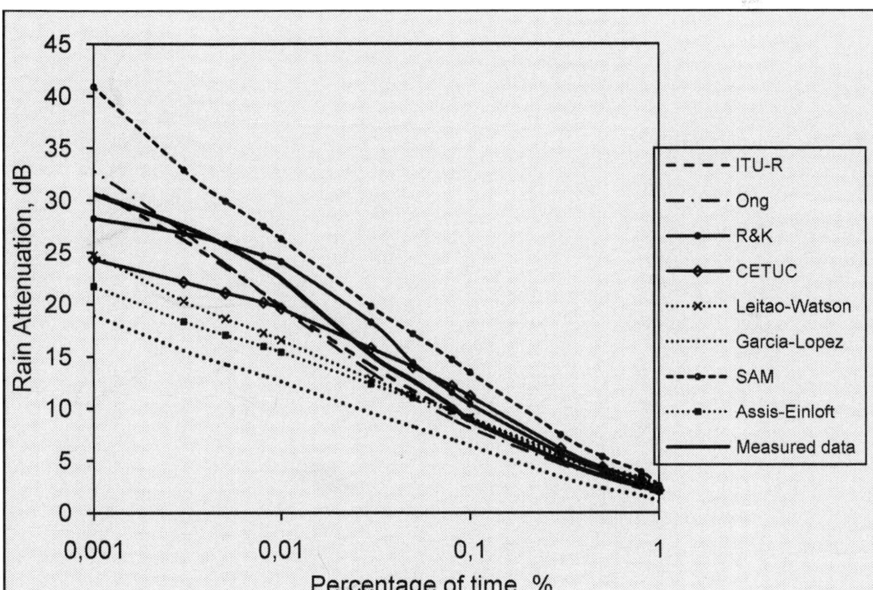


Fig. 30. The comparison measured rain attenuation with existing predicted models at USM

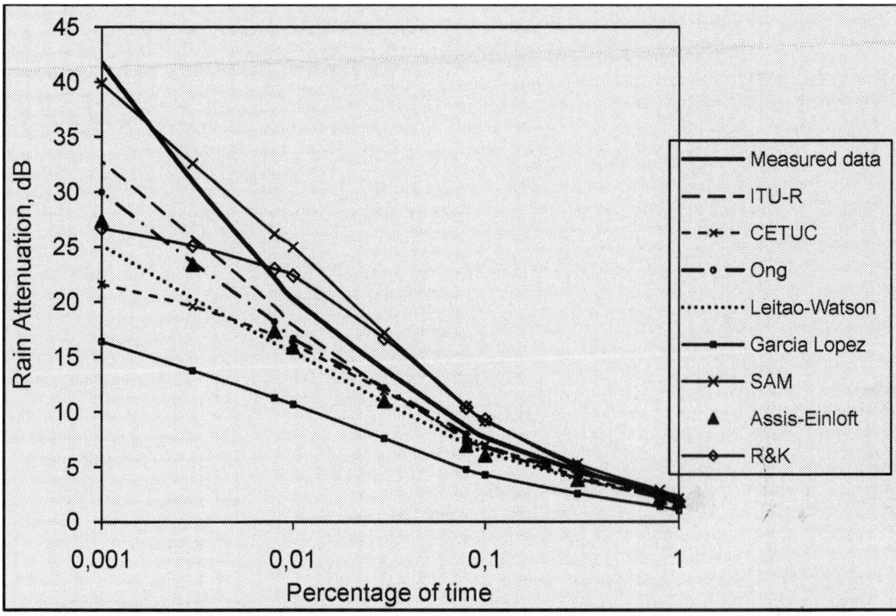


Fig. 31. The comparison measured rain attenuation with existing predicted models at Bangkok

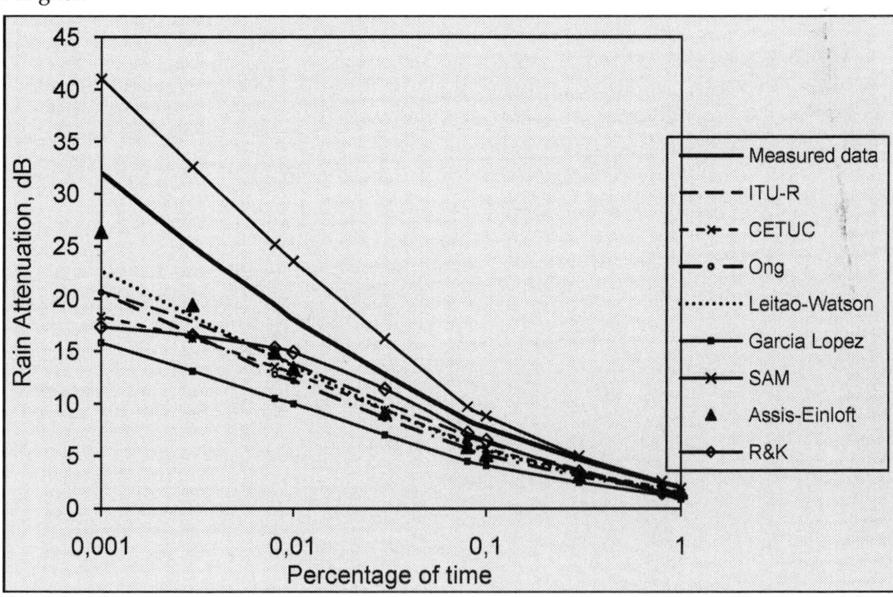


Fig. 32. The comparison measured rain attenuation with existing predicted models at Bandung

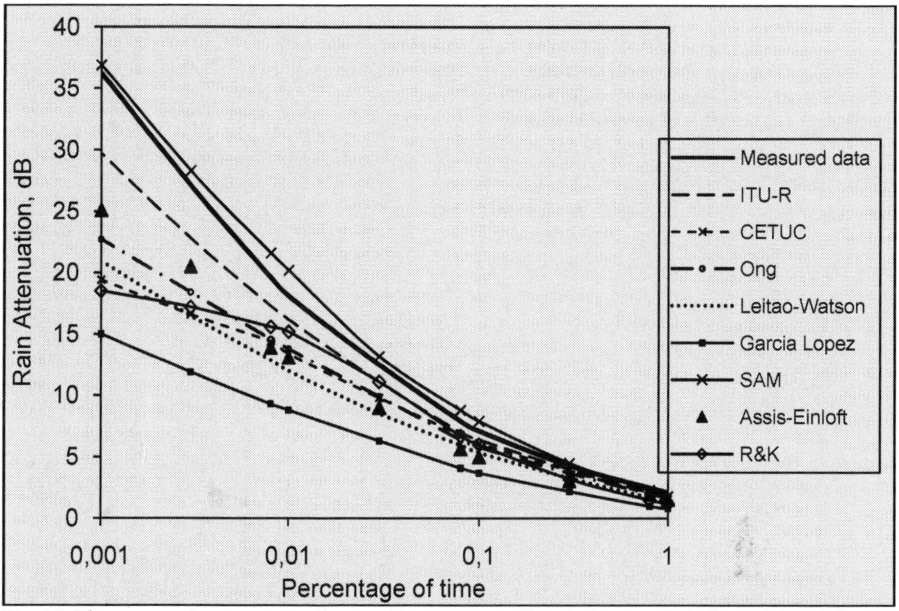

Fig. 33. The comparison measured rain attenuation with existing predicted models at Manila

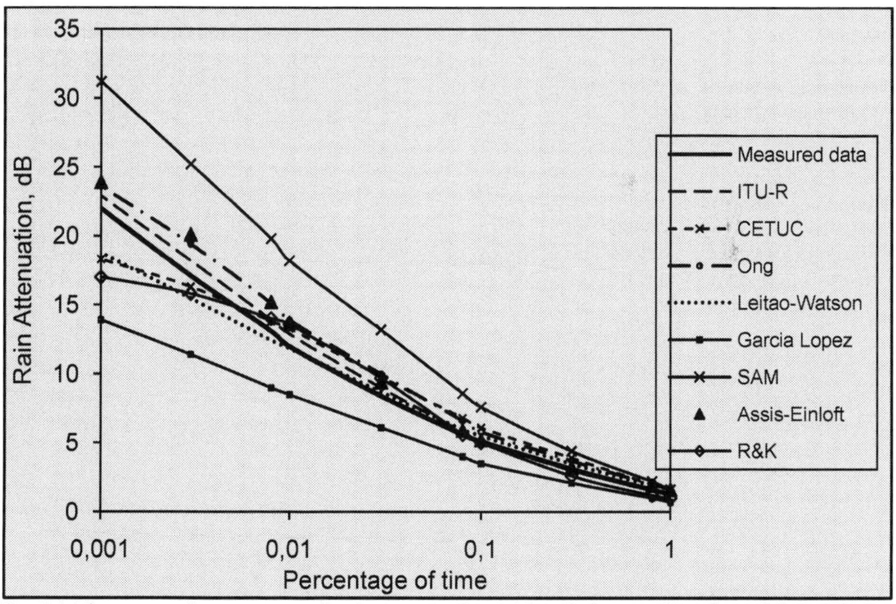

Fig. 34. The comparison measured rain attenuation with existing predicted models at Fiji

The ITU-R model underestimates all of these 5 tropical climates, except Fiji throughout the entire percentage of time. The model follows closely with the USM measured data from 0.05% - 1% of time. ITU-R underestimates the rain attenuation at the lower percentage of time because of the roll over effect, where as the rain rate increases, the attenuation reduce. This is because of the lack of high rain rate data from tropical climates. The rain column height is constant and maximum (10 km) when the rain reaches its saturation point, but the rain-cell diameter continues to decrease with increasing rain rate. Hence, the proportional increase of rain volume, which is a combination of rain-cell diameter, rain column height and rain rate would cause saturation (Ramachandran and Kumar, 2004). The vertical path reduction coefficient was used to minimize the prediction error. At Bangkok, Manila and Fiji, the ITU-R models gave a lower RMS value. At Fiji, the ITU-R model follows closely the measured rain attenuation throughout the entire percentage of time. The model gave a low RMS value because the rain rate of 90.7 mm/h was used for calculating the rain attenuation at 0.01% of time. This model was developed based on low rain rate of 85 mm/h at 0.01% of time from temperate climates. At Bangkok and Manila, the model gave a high RMS value because of the high rain rate vales at 0.01% of time have been used in calculating the rain attenuation. At Bandung, the model gave a high RMS values. At Bandung, the high elevation angle of above $60^0$ was applied in experimental. The station height above sea level that used was 700m, whereas this model was developed by station heights above mean sea level from 20m to 400m. The Ong model at USM underestimates rain attenuation at the entire measurement time. The model gave a percentage error of ±14% with a range RMS value of 9.62% at USM. This model was revised from ITU-R model. The model has a roll over effect at lower percentage of time, because it was developed for 4/6 GHz. At Fiji, the Ong model follows closely the measured rain attenuation for the entire measurement time. The station height above sea level at both of these sites is below 60m. The station height above sea level that was used to develop this model was below 60m. At Bangkok, the model agrees reasonably well with the measured rain attenuation down to an outage time of 0.03% and deviates considerably from the measured values from 0.03% to 0.001%. At lower percentage of time above 0.01%, the model relative error increases because of the model was developed for 4/6 GHz. When the higher operational frequency gets, the higher rain attenuation will be at lower percentage of time. The Ong model at Bandung and Manila give poor performance for the entire measurement time. The model gave a high RMS value at these sites. It is because the station heights above sea level are above 80m and the elevation angle of the measurement site was above 55°.

At USM, Bangkok and Fiji, Ramachandran and Kumar model (R&K) follows closely with the USM rain attenuation measured from 0.03% to 1% of time. However, the model underestimates the rain attenuation from 0.03% to 0.001% of time. For this model takes into account the effect of the breakpoint to predict the attenuation exceedance in the tropics. In the tropics when the rain rate increase and approach the breakpoint the rain structure gradually changes from stratiform to convective. If the breakpoint is reached at a lower rain rate, then the rain tends to saturate fast (Ramachandran and Kumar, 2004). Because of this reason, the model has a roll over effect at lower percentage of time. The model gave a RMS value of 19.91% at Bangkok and 16.41% at Fiji. For the $0.003 \leq p \leq 1$, the rain attenuation increase gradually with increasing rain rate. Beyond 0.003% of time, the rain attenuation tended to saturation finally leading to total outage. At Bandung and Manila, the model is rejected for prediction for the entire measurement time. The model gave high RMS values for these sites. This is because of the rain rate ($R_{AB}$) at the breakpoint is above 70mm/h at these measurement sites. The rain rate of 58mm/h at the breakpoint was used to determine the model coefficient.

The CETUC model is simple to apply and uses the full rain rate distribution to predict the attenuation distribution, avoiding extrapolations functions dependent on the percentage of time. The model keeps the concept of an equivalent rain cell. The attenuation dependence on frequency is completely described by the parameters k and α (ITU-R recommendation parameters that used in calculating specific attenuation). At Fiji, the CETUC model agrees reasonably well with the measured values from 0.008% -1% of time and deviates considerably from the measured values from 0.001% to 0.008% of outage time. It gave a percentage error of ±25% with a RMS value of 18.96%. The model gave a low RMS value because the elevation angle was 45.5$^0$ and the station height was 13m. The model was developed based on an average elevation angle 42$^0$ and the altitudes below 50m. At USM, the model gave a high RMS value of 19.34% at USM, because the parameters k and α that recommended are not suitable used in USM. The highest rain rate and rain attenuation values were 200 mm/h and 30 dB, respectively used to apply the model. At Bandung, Manila and Bangkok, the CETUC model deviated considerably the measured rain attenuation for the entire measurement time. At Bandung and Bangkok sites, the model gave a high percentage error with a RMS value because the rain height calculated by CETUC was 3.18 km. The height above sea level for Bandung station was 700m. However, the rain height used to develop the model was based on limited number of stations with height above sea level below 50m. At Manila, the model gave a high percentage error of ±38.7% with a RMS value of 26.6%. This is because the effective length of the rain cell was developed by the rain rate values from 12 mm/h at 1% to 150 mm/h at 0.001% of time and the rain height calculated by CETUC was 3.18 km.

At Fiji, the Leitao-Watson model appears to work well down to the entire measurement time. The model gave a lower RMS values. The model parameters *s, t, u, v* and *w* are suitable to be used at Fiji site. Besides Fiji, the Leitao-Watson model underestimates the rain attenuation for the entire measurement time at the other sites. The model gave high RMS value of 24% and above. The model developed according to radar observation of rainfall structure, proposed the same set of equations with different parameters for widespread and convective rain (discrimated by a rain rate threshold of 20 mm/h) (Capsoni, *et. al.,*2009). The model was developed by using Europe rainfall data. It will make the model cannot perform well and give a high RMS value in predicting rain attenuation in tropical countries.

The Garcia-Lopez gave a high RMS value of above 30% and above for all these measurement sites. The model is underestimates the rain attenuation values for the entire outage time of an average year at all these measurement sites. This is because the rain height of 4km was given by Garcia-Lopez. The rain heights in tropical countries are above 5km, which are given by the ITU-R map of rain height above mean sea level. The coefficient constant of the model was obtained based on low rain rate of 60 mm/h at 0.01% of time. The range of rain rate at 0.01% of time for tropical climates is from 100mm/h to 130mm/h of time depending on the geographical area. The station height used to determine coefficient constant was averaged to 200m above mean sea level.

The SAM model at USM and Fiji site overestimates and shows poor performance in predicting rain attenuation. The model gave a high RMS value. This is because the model of the parameter controlling the rate of decay of the horizontal profile (γ). The model would give a lower RMS value below 10% if γ parameter was optimized against the set of data obtained from the measurement sites (Stutzman & Yon, 1986). At Bangkok and Bandung, the model follows closely the measured rain attenuation from 0.08%-1% of time and overestimates the rain attenuation from 0.08% to 0.001%. At low percentage of times, the large errors are due to the fact that the predicted rain rates are less accurate in regions of high occurrence levels. The median values of the observations were estimated for each probability level in order to develop the model because long data sets were not available and the pooling of data from a number of locations was necessary to reduce the estimation error. The model appears to work well for the entire measurement time at Manila. The model gave a low RMS value of 9.8% because the parameter $\gamma$ value given by SAM was optimized against the measured data sets.

At Fiji, Assis-Einloft model agrees reasonably well with the measured values for the entire measurement time. This is because the development of the reduction factor for this model was based on the measurement done at temperate and tropical climates, whereby at tropical climates 80 data sets at antenna elevation angles from $40^0$ to $50^0$ were used in order to reduce the prediction error at high rainfall intensity regions. Assis-Einloft model is not show the good agreement for the entire measurement time at Bangkok, Bandung and Manila. This is because the antenna elevation angles that used at these measurement sites were above $50^0$. The path length that was considered by Assis was from 6km to 20.7km, but the path lengths at these measurement sites were below 6km. At USM, the model gave a high RMS value. A uniform rain rate was assumed for developing the model by introducing the concept of path length reduction factor. The path length at USM sites was below 6km.

The summary of the best model and worst model at the comparison sites was done and shown in Table 6. In this section, the comparison of rain attenuation was done for the measured data. For the tropical climate, it was found that no model revealed a close fit to the measured data for low, medium and high rain rates. The models do not predict rain attenuation at the lower percentage of time. The noticeable difference between the measured and the predicted variation is the existence of the breakpoint. The exceedance curves show that as the rain rate increases, the trend of the slope of the curve gradually decreases from large negative value, and then the trend that changes is referred to as the breakpoint in the exceedance curve (Ramachandran, *et. al.*, 2004, Mandeep, *et. al.*, 2008). The breakpoint exceedance curve usually occurs at high rain rate. When the rain structure is stratiform, the rainfall is widespread with low rain rates.

| Site | RMS value, % | | | | | | | | Conclusion | |
|---|---|---|---|---|---|---|---|---|---|---|
| | ITU-R | Ong | R&K | CETUC | Leitao | Garcia | SAM | Assis | Best Model | Worst Model |
| USM | 7.11 | 9.62 | 11.50 | 19.34 | 18.75 | 38.58 | 37.96 | 25.80 | ITU-R | Garcia |
| Bangkok | 17.36 | 15.82 | 19.91 | 221 | 251 | 49.98 | 16.45 | 22.35 | Ong | Garcia |
| Bandung | 28.78 | 32.79 | 29.43 | 29.29 | 27.35 | 47.86 | 22.36 | 26.70 | SAM | Garcia |
| Manila | 13.02 | 202 | 30.45 | 26.60 | 32.73 | 53.75 | 9.57 | 27.63 | SAM | Garcia |
| Fiji | 5.57 | 18.91 | 16.41 | 18.96 | 137 | 33.85 | 46.49 | 100 | ITU-R | SAM |

Table 6. The summary of the comparison of rain attenuation prediction models

The ITU-R model is judged suitable for use in predicting rain attenuation in these measurement tropical climates sites. The Garcia-Lopez model exhibited poor performance in comparison. The results are particularly important for the tropical and equatorial region because not much of research that has been done in these regions.

**Acknowledgment**

The authors are grateful thanks to Electrical and Electronic School of USM Engineering Campus for technical support and like to acknowledge Ministry of Science, Technology and Innovation (MOSTI) 01-01-02-SF0670 for financial support. The authors would like to express sincere thanks to the reviewers for their comments.

# 14

# Guidelines for Satellite Tracking

Dusan Vuckovic, Petar Rajkovic and Dragan Jankovic
*Faculty of Electronic Engineering, University of Nis*
*Serbia*

## 1. Looking at the sky

Whether you are willing to confess or not, the most useful instrument for observing the sky, by day or by night, is the naked eye. For the ones with less experience, it is important to become familiar with the general aspects of the sky. Wide field of view that human eye is naturally equipped with is ideal for this purpose. Binoculars and telescope are also indispensable tools, but they may be introduced, once the familiarity with the sky is established.

Since celestial bodies have a tendency of moving, it is very hard to predict their future position. It is the big mathematical engine that will help us pinpoint the object in the sky, so we would be able to look there, or point our telescope. Many objects in the sky look very similar. In NASA's[1] 1997 report it's being said that there are 25,000 man-made objects in space. 8681 are currently orbiting Earth, and 16,000 objects are in a state of decay. Course, situation is quite different today, considering that's been 13 years since that report was published.

Large portion of 16,000 objects are orbital debris: parts, such as nosecone shrouds, lens, hatch covers, rocket bodies, payloads that have disintegrated or exploded, and even objects that "escape" from manned spacecraft during operations. Most of these objects are small, but quite a few are large enough to be seen with unaided eye.

Additional difficulties come with the fact that many military or government satellites are painted in black, thus very hard to spot even with very sophisticated optical instruments.

Three things become immediately obvious when you take a look at the night sky. Not all the stars look the same, also stars twinkle, and finally, it's easy to arrange stars into recognizable patterns. Some stars are bright, some are less bright and some almost too dim to see with naked eye. Astronomers have had to devise some means of expression the variation precisely.

Ancient Greek astronomer Hipparchos was the first to produce the star catalogue, introducing the new value of importance, or magnitude, the star had. He thought that important stars should be of higher importance and termed them magnitude 1. Those somewhat dimmer he considered of second importance (magnitude 2), and so on. Altogether, he created six classes of star's magnitudes. At that time, he was not aware of the fact that his six divisions were based on the way the human eye recognizes a brightness

[1] NASA - US National Aeronautics and Space Administration

difference, where one object seems half as bright as another does. It was later discovered that Hipparchos' six magnitudes give a difference between magnitude 1 and 6 of 100 times. Although this system has been refined, most scientists use six degrees magnitude system to explain the stars, or other celestial body's brightness in the sky.
Predictions are necessary if you want to observe satellites or other objects in the sky. You have to know where to look and when to look. Time is very important, so a stopwatch with an accuracy of at least 0.5 seconds is essential. Set the watch by radio time signals, or by telephone system's speaking clock. More innovative method would include mobile phone with stopwatch that synchronizes the clock through network operator time, or over the network, with NTP[2] protocol.
Next step in simplified satellite tracking would include gauging the positions of a satellite by imaging a line drawn in the middle of two fairly close stars between which the satellite passes, or by imaging a vertical line drawn down from a particular star. The precise moment at which the satellite crosses this line is the most important one. With the help of a Star Atlas, satellite track can be indicated. Identification of a position is not too difficult, but the problem is accurate satellite's transit time. Reliable results require constant repetition and practice.
Artificial satellites may be divided in three categories, depending on their brightness: Bright, visible to the naked eye (magnitude < 3), dimmer (3 < magnitude < 9) and extremely faint ones. Satellites dimmer than magnitude 6 require at least binoculars to be able to spot them.
More sophisticated satellite tracking method would involve specialized computer software, and mathematical apparatus. Today, time synchronization, which is the requirement for better preciseness is simplified in a great manner with the introduction of NTP boxes, primarily used in network performance monitoring. With a set of relatively simple mathematical procedures and the selection of adequate, so called, "propagation model", we'll be able to spot, follow or wait for a satellite to pass above us in the evening sky.
Necessary steps needed to achieve this goal will be explained in detail in the following chapters.

## 2. Propagation Models

To be able to distinct satellite orbits we need to describe them. There are seven, and sometimes, eight numbers to tell us the orbit of each satellite. Those are orbital elements, or sometimes "Keplerian" [3] elements. The ellipse oriented about the Earth with the satellite on a specific position in time is described with these numbers. Keplearian model introduced orbits with constant shape and orientation, with the Earth at one focus of the ellipse, not the centre, unless the orbit itself is circular.
To increase the exactness of the tracking model, another value is introduced. This value represents the corrections of Keplerian model known as perturbations. Corrections are introduced due to lumpiness of the Earth's gravitational field, and the "drag" on the satellite due to atmosphere. That way, Drag becomes an optional eight orbital element.
Basic orbital elements are:

---

[2] Network Time Protocol, protocol for synchronizing the clocks of computer systems over packet-switched, variable-latency data networks.

[3] Johann Kepler [1571-1630] was a German mathematician, astronomer and astrologer, and key figure in the 17th century scientific revolution.

1. **Epoch**, or "Epoch Time" (T0) is a number that specifies the time at which the snapshot of other orbital elements was taken.

2. **Orbital inclination** (I0) indicates the angle between the equator and the orbit when looking from the centre of the Earth. It ranges from 0 to 180 degrees. (Fig. 1)

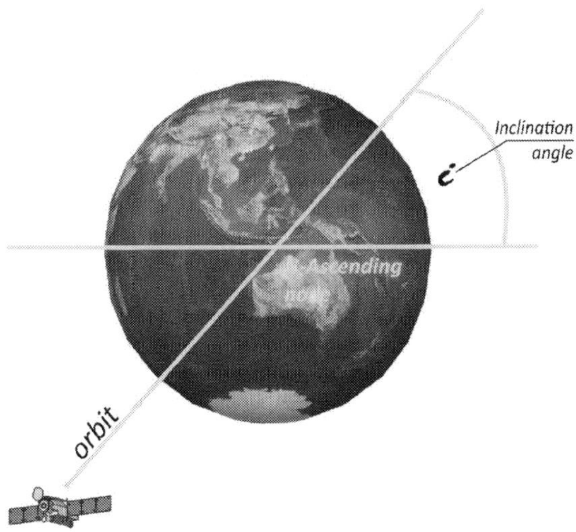


Fig. 1. Inclination angle

3. **Right Ascension of Ascending Node** (RAAN or O0) is the second number that orients the orbital plane in space. The first one is Inclination. Ascending node is the place where the satellite crosses the equator while going from the Southern Hemisphere to the Northern Hemisphere. Due to Earth's rotation, fixed object in space is necessary to measure Ascension. (Fig. 2.)
   Solution is to use Aries, which has the same location as vernal equinox, so the angle between the Aries and Ascending node is called Right Ascension of Ascending Node.

4. **The Eccentricity** defines the flatness of the orbit. If the orbit is a perfect circle, then eccentricity is 0, and 1 when very flat.

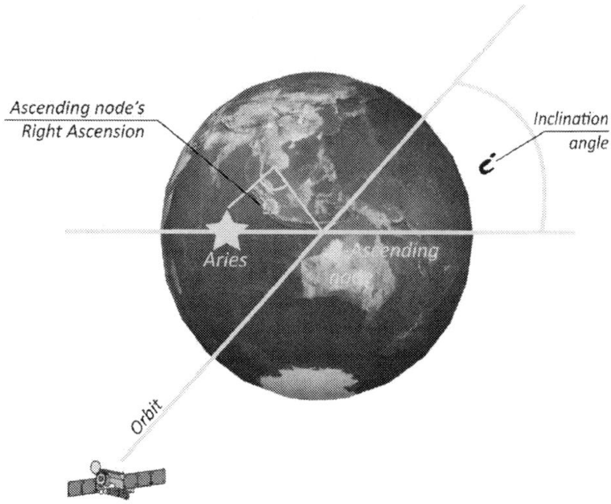


Fig 2. Ascending node's Right Ascension

5. Due to elliptical shape of the orbit, satellite will sometimes be closer and sometimes further from the Earth. The point where the satellite is closest to the Earth is called Perigee, and the point where it's furthest from the Earth is called Apogee. (Fig 3.)

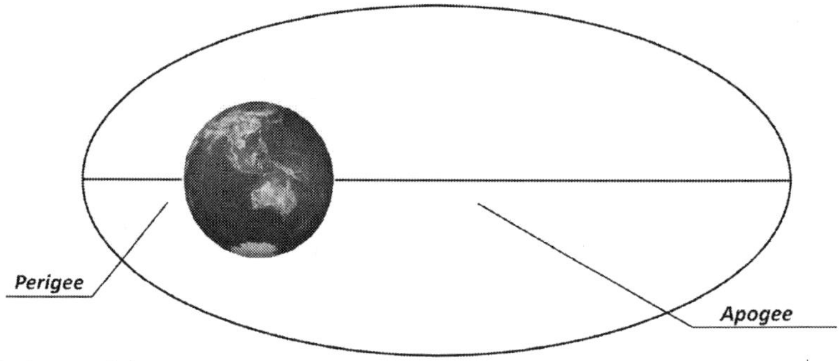


Fig. 3. Perigee and Apogee

The angle formed between the perigee and the ascending node is called **the Argument of Perigee**. The argument of perigee would have the value 0 if the perigee would occur at the ascending node. (Fig. 4.)

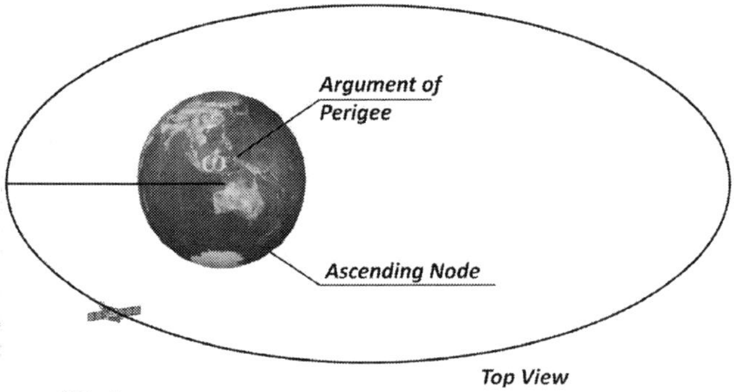


Fig. 4. Argument of Perigee

6. **The Mean Motion** is the value that illustrates how fast the satellite is going. According to Kepler's Law:

$$v^2=GM/r$$

*v = velocity of the satellite (m/s)*
*M = Earth's mass (5.98*10$^{24}$)*
*G = gravitational constant (6.672*10$^{-11}$ Nm$^2$/kg$^2$)*
*r = distance between the satellite and centre of the Earth (m)*

The closer the satellite gets to the Earth, more speed it gets. This value will help us obtain the satellite's altitude.

7. **The Mean Anomaly** ranges from 0 to 360 degrees, it's referenced to the perigee and represents satellite's position on orbital path. In the perigee itself, mean anomaly would be 0.

8. **The Drag** is final, optional value that encapsulates atmospheric drag as well as gravitational pull from stellar bodies such as Sun or the moon. Knowledge of the atmospheric density at the satellite's position is required to calculate the drag force on the satellite. It may also be required in terms of establishing a reliable linkto the satellite. Nevertheless, this may require more information than density alone. (King-Hele, 1983.)
Just for density calculations two models should be mentioned. The Harris-Priester model and the Jacchia model (Bindebrink et al., 1989.)

If only gravitational force is assumed by Newton's law is acting on the satellite the first five parameters are constant and the orbit is an ideal Keplerian orbit (Bunnell, P. 1981.)

Considering the tracking of Earth orbiting objects, institution that has the leading role in providing data related to all satellites in the orbit is NORAD[4]. NORAD periodically releases

[4] NORAD - North American Aerospace Defense Command

element sets that, in conjunction with certain models, can be used to track satellites and all other objects. These element sets are periodically refined so as to maintain a reasonable prediction capability.

It is very important to emphasize that not just any prediction model will do the job. The NORAD element sets are "mean" values obtained by removing periodic variations in a particular way. In order to obtain good predictions, periodic variations must be reconstructed. Reconstruction is done by prediction model in the exactly the same way they were removed by NORAD. To gain maximum prediction accuracy NORAD element sets must be used with one of the following models:

### 2.1 SGP/SDP propagation models

SGP propagation model, developed by Hilton and Kuhlman in 1966 is used for near-Earth satellites. This model simplifies drag effect on mean motion by taking it as linear in time.

SGP4 model was developed by Ken Cranford in 1970 and is also used for near-Earth satellites. Afterwards SGP4 model was extended with deep-space equations by Hujsak in 1979 so it incorporated gravitational effects of the moon and Sun, sectoral and tesseral Earth harmonics important for calculation of half-day and one-day period orbits. This model was named SDP4 model as for deep-space model.

Hoots created SGP8 model in 1980 for near-Earth satellites as a simplification of his own theory that uses same gravitational and atmospheric models as Lane and Cranford but with differential equations integrated in a different manner. As a follow up SDP8 model was formed for deep-space satellites.

As previously mentioned the first orbit model was developed in 1966 and used to provide orbital elements in the form of two-line elements. They contain the Keplerian elements (Montenbruck et al., 2000.) inclination $i$, eccentricity $e$, right ascension $\Omega$, argument of perigee $\omega$ and Mean anomaly $M_0$ at epoch time $t_0$. Instead of the semi-major axis a, anomalistic mean motion $n_0$ is used. Modified model SGP4 is used since 1970 for the production of two-line element sets for low-Earth satellites by NORAD. It accounts Earth gravitational field through zonal terms $J_2$, $J_3$, $J_4$ and atmospheric drag (assuming a non-rotating spherical atmosphere) through a power density function.

The first step in SGP4 calculation is constants calculation:

$$a_1 = \left( \frac{k_e}{n_o} \right)^{\frac{2}{3}} \tag{1}$$

whereas $k_e = \sqrt{GM}$ , and $G$ is Newton's universal gravitational constant.

$M$ = the mass of the Earth

$n_o$ = the SGP type "mean" motion at epoch

$$\delta_1 = \frac{3}{2} \frac{k_2}{a_1^2} \frac{\left(3\cos^2 i_o - 1\right)}{\left(1 - e_o^2\right)^{\frac{3}{2}}} \tag{2}$$

$e_o$ is the "mean" eccentricity at epoch

$i_o$ is the "mean" inclination at epoch

$$k_2 = \frac{1}{2} J_2 a_E^{\,2} \tag{3}$$

$J_2$ is , as previously mentioned, the second gravitational zonal harmonic of the Earth, and $a_E$ is the equatorial radius of the Earth.

$$a_o = a_1 \left( 1 - \frac{1}{3}\delta_1 - \delta_1^{\,2} - \frac{134}{81}\delta_1^{\,3} \right) \tag{4}$$

$$\delta_o = \frac{3}{2} \frac{k_2}{a_o^{\,2}} \frac{\left(3\cos^2 i_o - 1\right)}{\left(1 - e_o^{\,2}\right)^{\frac{3}{2}}} \tag{5}$$

The original mean motion $n_o''$ and semimajor axis $a_o''$ are first recovered from the input elements by the equations

$$n_o'' = \frac{n_o}{1 + \delta_o} \tag{6}$$

$$a_o'' = \frac{a_o}{1 - \delta_o} \tag{7}$$

If the perigee stands between 98 kilometers and 156 kilometers, the value of the constant $s$ used in SGP4 model will change to

$$s^* = a_o''\left(1 - e_o\right) - s + a_E \tag{8}$$

$s$ is parameter for the SGP4/SGP8 density function

For perigee below 98 kilometers, s becomes

$$s^* = 20/\text{XKMPER} + a_E \tag{9}$$

XKMPER = 6378.135kilometers/Earth radii.

If the value of s is changed, then the value of $(q_o - s)^4$ must be replaced by

$$\left(q_o - s^*\right)^4 = \left[ \left[ \left(q_o - s\right)^4 \right]^{\frac{1}{4}} + s - s^* \right]^4 \tag{10}$$

Whereas $q_o$ is parameter for the SGP4/SGP8 density function.

The next step is calculation of the constants with the use of appropriate values of $s$ and $\left(q_o - s\right)^4$.

$$\theta = \cos i_o \tag{11}$$

$$\xi = \frac{1}{a_o'' - s} \tag{12}$$

$$\beta_o = \left(1 - e_o^{\ 2}\right)^{\frac{1}{2}} \tag{13}$$

$$\eta = a_o'' e_o \xi \tag{14}$$

$$C_2 = \left(q_o - s\right)^4 \xi^4 n_o'' \left(1-\eta^2\right)^{-\frac{7}{2}} \left[ a_o'' \left(1 + \frac{3}{2}\eta^2 + 4e_o\eta + e_o\eta^3\right) + \frac{3}{2}\frac{k_2\xi}{\left(1-\eta^2\right)}\left(-\frac{1}{2} + \frac{3}{2}\theta^2\right)\left(8 + 24\eta^2 + 3\eta^4\right) \right] \tag{15}$$

$$C_1 = B^* C_2 \tag{16}$$

$B^*$ is the SGP4 type drag coefficient

$$C_3 = \frac{\left(q_o - s\right)^4 \xi^5 A_{3,0} n_o'' a_E \sin i_o}{k_2 e_o} \tag{17}$$

$$A_{3,0} = -J_3 a_E^{\ 3}, \tag{18}$$

Whereas $J_3$ is the third gravitational zonal harmonic of the Earth.

$$C_4 = 2n_o'' \left(q_o - s\right)^4 \xi^4 a_o'' \beta_o^{\ 2} \left(1-\eta^2\right)^{-\frac{7}{2}} \left( \left[ 2\eta\left(1 + e_o\eta\right) + \frac{1}{2}e_o + \frac{1}{2}\eta^3 \right] - \frac{2k_2\xi}{a_o''\left(1-\eta^2\right)} \times \right.$$
$$\left. \left[ 3\left(1-3\theta^2\right)\left(1 + \frac{3}{2}\eta^2 - 2e_o\eta - \frac{1}{2}e_o\eta^3\right) + \frac{3}{4}\left(1-\theta^2\right)\left(2\eta^2 - e_o\eta - e_o\eta^3\right)\cos 2\omega_o \right] \right) \tag{19}$$

$\omega_o$ is the "mean" argument of perigee at epoch.

$$C_5 = 2\left(q_o - s\right)^4 \xi^4 a_o'' \beta_o^{\ 2} \left(1-\eta^2\right)^{-\frac{7}{2}} \left[ 1 + \frac{11}{4}\eta\left(\eta + e_o\right) + e_o\eta^3 \right] \tag{20}$$

$$D_2 = 4a_o''\xi C_1^{\ 2} \tag{21}$$

$$D_3 = \frac{4}{3}a_o''\xi^2\left(17a_o'' + s\right)C_1^{\ 3} \tag{22}$$

$$D_4 = \frac{2}{3}a_o''\xi^3\left(221a_o'' + 31s\right)C_1^{\ 4} \tag{23}$$

The secular effects of atmospheric drag and gravitation are included through the equations

$$M_{DF} = M_o + \left[ 1 + \frac{3k_2\left(-1+3\theta^2\right)}{2a_o''^2\beta_o^{\ 3}} + \frac{3k_2^{\ 2}\left(13 - 78\theta^2 + 137\theta^4\right)}{16a_o''^4\beta_o^{\ 7}} \right] n_o''\left(t - t_o\right) \tag{24}$$

Where $M_o$ is the "mean" mean anomaly at epoch and $\left(t - t_o\right)$ is time since epoch

$$\omega_{DF} = \omega_o + \left[ -\frac{3k_2\left(1-5\theta^2\right)}{2a_o''^2\beta_o^4} + \frac{3k_2^2\left(7-114\theta^2+395\theta^4\right)}{16a_o''^4\beta_o^8} + \frac{5k_4\left(3-36\theta^2+49\theta^4\right)}{4a_o''^4\beta_o^8} \right] n_o''\left(t-t_o\right) \quad (25)$$

$$k_4 = -\frac{3}{8}J_4 a_E^4 \quad (26)$$

According to previous cases, $J_4$ is the fourth gravitational zonal harmonic of the Earth

$$\Omega_{DF} = \Omega_o + \left[ -\frac{3k_2\theta}{a_o''^2\beta_o^4} + \frac{3k_2^2\left(4\theta-19\theta^3\right)}{2a_o''^4\beta_o^8} + \frac{5k_4\theta\left(3-7\theta^2\right)}{2a_o''^4\beta_o^8} \right] n_o''\left(t-t_o\right) \quad (27)$$

And $\Omega_o$ is the "mean" longitude of ascending node at epoch.

$$\delta\omega = B^* C_3\left(\cos\omega_o\right)\left(t-t_o\right) \quad (28)$$

$$\delta M = -\frac{2}{3}\left(q_o-s\right)^4 B^* \xi^4 \frac{a_E}{e_o\eta}\left[\left(1+\eta\cos M_{DF}\right)^3 - \left(1+\eta\cos M_o\right)^3\right] \quad (29)$$

$$M_p = M_{DF} + \delta\omega + \delta M \quad (30)$$

$$\omega = \omega_{DF} - \delta\omega - \delta M \quad (31)$$

$$\Omega = \Omega_{DF} - \frac{21}{2}\frac{n_o'' k_2\theta}{a_o''^2\beta_o^2} C_1\left(t-t_o\right)^2 \quad (32)$$

$$e = e_o - B^* C_4\left(t-t_o\right) - B^* C_5\left(\sin M_p - \sin M_o\right) \quad (33)$$

$$a = a_o''\left[1 - C_1\left(t-t_o\right) - D_2\left(t-t_o\right)^2 - D_3\left(t-t_o\right)^3 - D_4\left(t-t_o\right)^4\right]^2 \quad (34)$$

$$IL = M_p + \omega + \Omega + n_o''\left[\frac{3}{2}C_1\left(t-t_o\right)^2 + \left(D_2+2C_1^2\right)\left(t-t_o\right)^3 + \frac{1}{4}\left(3D_3+12C_1D_2+10C_1^3\right)\left(t-t_o\right)^4 + \frac{1}{5}\left(3D_4+12C_1D_3+6D_2^2+30C_1^2D_2+15C_1^4\right)\left(t-t_o\right)^5\right] \quad (35)$$

$$\beta = \sqrt{\left(1-e^2\right)} \quad (36)$$

$$n = k_e / a^{3/2} \quad (37)$$

where $\left(t-t_o\right)$ is time since epoch.

In cases where perigee height is less than 220 kilometers, the equations for a and $IL$ should be truncated after the $C_1$ term, and all terms involving $C_5$, $\delta\omega$, and $\delta M$ should be dropped.
Add the long-period periodic terms

$$a_{xN} = e\cos\omega \tag{38}$$

$$IL_L = \frac{A_{3,0}\sin i_o}{8k_2 a\beta^2}(e\cos\omega)\left(\frac{3+5\theta}{1+\theta}\right) \tag{39}$$

$$a_{yNL} = \frac{A_{3,0}\sin i_o}{4k_2 a\beta^2} \tag{40}$$

$$IL_T = IL + IL_L \tag{41}$$

$$a_{yN} = e\sin\omega + a_{yNL} \tag{42}$$

Next step involves solving the Kepler's equation for $(E+\omega)$ by defining

$$U = IL_T - \Omega \tag{43}$$

$$(E+\omega)_{i+1} = (E+\omega)_i + \Delta(E+\omega)_i \tag{44}$$

$$\Delta(E+\omega)_i = \frac{U - a_{yN}\cos(E+\omega)_i + a_{xN}\sin(E+\omega)_i - (E+\omega)_i}{-a_{yN}\sin(E+\omega)_i - a_{xN}\cos(E+\omega)_i + 1} \tag{45}$$

$$(E+\omega)_1 = U \tag{46}$$

The following equations are used to calculate preliminary quantities needed for short-period periodics.

$$e\cos E = a_{xN}\cos(E+\omega) + a_{yN}\sin(E+\omega) \tag{47}$$

$$e\sin E = a_{xN}\sin(E+\omega) - a_{yN}\cos(E+\omega) \tag{48}$$

$$e_L = \left(a_{xN}{}^2 + a_{yN}{}^2\right)^{1/2} \tag{49}$$

$$p_L = a\left(1 - e_L{}^2\right) \tag{50}$$

$$r = a(1 - e\cos E) \tag{51}$$

$$\dot{r} = k_e\frac{\sqrt{a}}{r}e\sin E \tag{52}$$

$$r\dot{f} = k_e \frac{\sqrt{p_L}}{r} \tag{53}$$

$$\cos u = \frac{a}{r}\left[\cos(E+\omega) - a_{xN} + \frac{a_{yN}(e\sin E)}{1+\sqrt{1-{e_L}^2}}\right] \tag{54}$$

$$\sin u = \frac{a}{r}\left[\sin(E+\omega) - a_{yN} - \frac{a_{xN}(e\sin E)}{1+\sqrt{1-{e_L}^2}}\right] \tag{55}$$

$$u = \tan^{-1}\left(\frac{\sin u}{\cos u}\right) \tag{56}$$

$$\Delta r = \frac{k_2}{2p_L}\left(1-\theta^2\right)\cos 2u \tag{57}$$

$$\Delta u = -\frac{k_2}{4{p_L}^2}\left(7\theta^2-1\right)\sin 2u \tag{58}$$

$$\Delta\Omega = \frac{3k_2\theta}{2{p_L}^2}\sin 2u \tag{59}$$

$$\Delta i = \frac{3k_2\theta}{2{p_L}^2}\sin i_o \cos 2u \tag{60}$$

$$\Delta\dot{r} = -\frac{k_2 n}{p_L}\left(1-\theta^2\right)\sin 2u \tag{61}$$

$$\Delta r\dot{f} = \frac{k_2 n}{p_L}\left[\left(1-\theta^2\right)\cos 2u - \frac{3}{2}\left(1-3\theta^2\right)\right] \tag{62}$$

The osculating quantities are calculated with the addition of short-period periodics

$$r_k = r\left[1 - \frac{3}{2}k_2\frac{\sqrt{1-{e_L}^2}}{{p_L}^2}\left(3\theta^2-1\right)\right] + \Delta r \tag{63}$$

$$u_k = u + \Delta u \tag{64}$$

$$\Omega_k = \Omega + \Delta\Omega \tag{65}$$

$$i_k = i_o + \Delta i \tag{66}$$

$$\dot{r}_k = \dot{r} + \Delta\dot{r} \tag{67}$$

$$\dot{rf}_k = \dot{rf} + \Delta \dot{rf} \quad (68)$$

Unit orientation vectors are calculated by

$$\mathbf{U} = \mathbf{M} \sin u_k + \mathbf{N} \cos u_k \quad (69)$$

$$\mathbf{V} = \mathbf{M} \cos u_k - \mathbf{N} \sin u_k \quad (70)$$

where

$$\mathbf{M} = \begin{Bmatrix} M_x = -\sin \Omega_k \cos i_k \\ M_y = \cos \Omega_k \cos i_k \\ M_z = \sin i_k \end{Bmatrix} \quad (71)$$

$$\mathbf{N} = \begin{Bmatrix} N_x = \cos \Omega_k \\ N_y = \sin \Omega_k \\ N_z = 0 \end{Bmatrix} \quad (72)$$

Then, position and velocity are given by

$$\mathbf{r} = r_k \mathbf{U} \quad (73)$$

and

$$\dot{\mathbf{r}} = \dot{r}_k \mathbf{U} + \left(\dot{rf}\right)_k \mathbf{V} \quad (74)$$

### 2.2 Propagation models modifications

SGP propagation model was modified in time. Several minor points in the original SGP4 paper emerged where performance of SGP4 could be improved. To maximise the usefulness of all of these features, one should ideally use Two Line Elements formed with differential correction, using an identical model as well (Vallado, D. et al. 2006). Next chapter will shed some light on what Two Line Elements are.

## 3. Two Line Elements

Orbit tracking programs require information about the shape and orientation of satellite orbits. That information was available from different websites. One of most common quoted sources is CelesTrak.com website, maintained by the group of satellite tracking enthusiasts. As a result of legislation passed by the US Congress and signed into law on 2003 November 24 (Public Law 108-136, Section 913), which was updated in 2006 (US National Archives, 2006.), Air Force Space Command (AFSPC) has embarked on a three-year pilot program to provide space surveillance data—including NORAD two-line element sets (TLEs)—to non-US government entities (NUGE). Since US Public Law prohibits the redistribution of the data obtained from this new NUGE service "without the express approval of the Secretary of Defence"a lot of other sources were immediately shut down.

CelesTrak has received continuing authority to redistribute Space Track data from US government and that way become one of the most useful information sources for the community.
TLE's are redistributed in a form shown in Fig. 5. All relevant parameters are color-coded and explained in Table 1.

```
ISS
1 25544U98067A10102.85853206  .0002565400000-017456-309629
2 25544  51.6472205.9374 0004892 166.2878 293.9622 15.74716373653188
```

Fig. 5. Two Line elements set

| Element | Description |
|---|---|
| ISS | Satellite name |
| 1 25544U<br>2 25544 | Satellite number |
| 98067A | International designator |
| 51.6472 | Inclination |
| 10102.85853206 | Epoch Year & Day Fraction |
| 205.9374 0004892 | Right Ascension of the Ascending Node |
| .00025654 | First Derivative of Mean Motion |
| 00000-0 | Second Derivative of Mean Motion |
| 166.2878 | Eccentricity |
| 293.9622 | Argument of Perigee |
| 15.747163736 | Mean Anomaly |
| 53188 | Revolution Number at Epoch & Checksum |
| 17456-3 | Drag Term |
| 0 | Ephemeris Type |
| 9629 | Element number & Checksum |

Table 1. Two Line Elements Explained

There are several things to consider. The accuracy of the original TLEs is not known. Some TLE data propagates into future quite well, while, the next set of elements can depart dramatically after only a day or less. Methods to overcome this problem are explained in (Vallado, D. et al. 2006).

## 4. Integrating Mathematical Models

Our intention was to integrate all orbital propagation models into one C# program. Integral version of the program can be downloaded from http://medlab.elfak.ni.ac.rs/spacetrack/sgpsdp.rar. It is important to highlight main program methods used for satellite position calculations:

```
publicvoid SGP(int IFLAG, double TSINCE)
publicvoid SGP4(int IFLAG, double TSINCE)
publicvoid SDP4(int IFLAG, double TSINCE)
publicvoid SGP8(int IFLAG, double TSINCE)
publicvoid SDP8(int IFLAG, double TSINCE)
```

Previous FORTRAN IV code produced by T.S. Kelso in 1988 according to (Hoots, F. R et al. 1980) was not optimized and hard to execute on modern parallel (multi-core) architectures.
The FORTRAN implementation of the SGP4 and SDP4 model in respective methods is rudimentary for the propagation process. It was necessary to produce functions which would help us calculate position $\vec{r}$ and velocity $\dot{\vec{r}}$ of a satellite at any given time by using the TLE data. Models specified in (Hoots, F. R. and al., 1980) from the original FORTRAN IV code are ported to C# in respect to the corrections made during the years, especially in the SDP4 subroutine DEEP. C# code contains the same variable names and structures as in the original FORTRAN routines to ensure compatibility and expandability.
Additional encapsulation was done with the creation of ActiveX component ready to be integrated in any .NET project.

## 5. NAVSTAR Satellite Tracking Software

NAVSTAR satellite tracking software presented in this paper is also based on the mathematical SGP4/SDP4 model. Program uses two line elements set as an input to calculate and visualize satellite's position in Space. It can be used to navigate telescopes to space objects passing over certain point on Earth. The complete mathematical model is encapsulated in ActiveX control, so it acts like a black box. The data is provided from TLEs and on the other end viewport coordinates are calculated.
NAVSTAR has three basic functions:

- Graphical display of satellite positions in real-time, simulation, and manual modes;
- Tabular display of satellite information in the same modes;
- Generation of tables (ephemerides) of past or future satellite information for planning or analysis of satellite orbits.

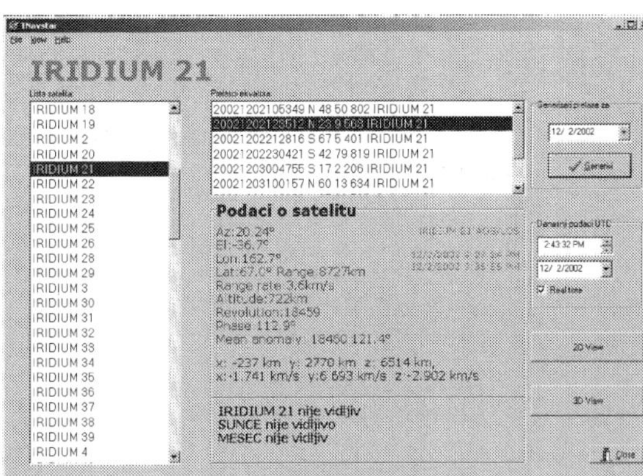


Fig. 6. Satellite selection dialog and Table Window

The principal feature of NAVSTAR is a series of Map Windows, which display the current position of satellites and observers on a simple world map, together with information such as bearing (azimuth), distance, and elevation above the observer's horizon. The maps may

be updated in real time, simulated time, or manually set to show the situation at any given moment of time, past or future.
An additional Table Window displays much more-detailed information about one or more satellites in a tabular form. The tabulated items can be selected and rearranged to fit the screen. This information can be updated in real-time, manual, or simulation modes as illustrated in Fig. 6.

Also, satellite 2D footprint (Fig.7) tracking is available, as well as a 3D view (Fig.8). Tracking algorithms SGP4 and SDP4 give considerable accuracy and opportunity of efficient computation of viewing opportunities. It's also possible in 3D view to make a prediction on satellites position in the future, or to see its position in the past. All is based on the information gathered from TLE's.
The preciseness of visualization depends on accuracy and age of gathered TLE data.

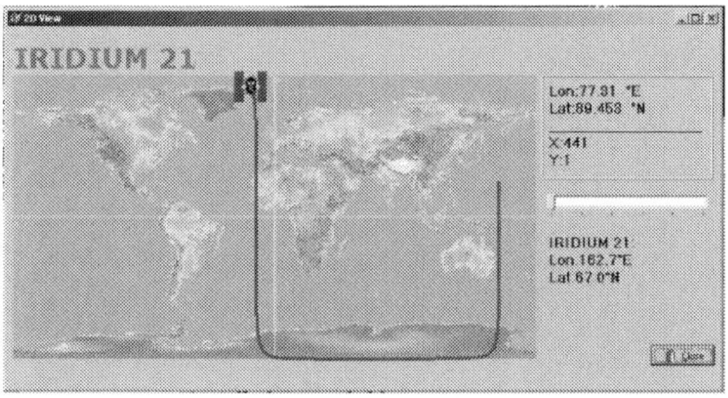


Fig. 7. 2D View

Regarding the 3D View (Fig.8), options for variable view angle, zoom and time increment are implemented. This gives a user the opportunity to view satellite from all angles and possibility to see its path (orbit), area on the Earth covered by its signal (in a form of beam) and real-time movement, as well as possible faster movement caused by a time speed up.

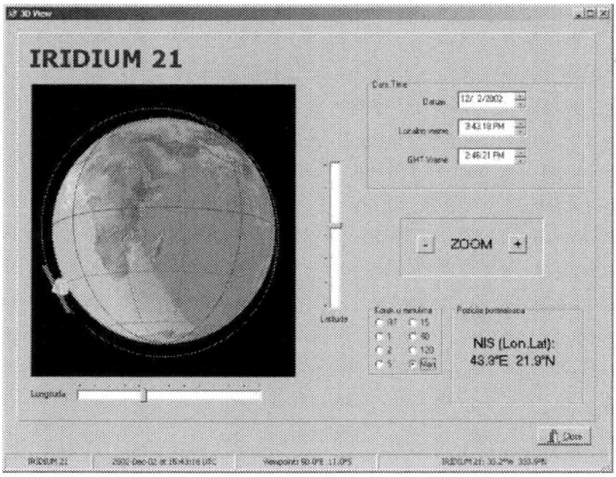


Fig. 8. 3D Satellite tracking View

The part of Earth not covered with the Sun light is dimmed on the globe, so the user can predict when it will become possible to see the satellite by a naked eye.
This software can be used to visualise the orbit trajectory of a satellite under different points of view. It gives the user the freedom of being able to study the satellite's ground conjunctions by tracking the satellite over the ground, or, with another approach, by calculating the elevation and azimuth angle of the satellite from a static ground station position.
Another very useful functionality, previously mentioned is satellite's footprint visualisation. These functionalities can be used, from an engineering point of view, to adjust ground station dishes in order to establish reliable links to the satellite by calculating it's precise position with the help of TLE data sets.
3D view can be used to simulate satellite's orbit around the Earth. Time lapse function enables user to see the orbit in the future, and exact satellite's position so we would be able to see which orbit parameters the satellite has to have in order to fulfil its task.

## 6. Conclusion

It is assumed that "space age" started with the first artificial satellite in the orbit around the planet. Nowadays, satellites are used for various different purposes: telecommunications, broadcasting, observation, imaging and even espionage. What they all have in common is the fact that they all must obey the rules of celestial mechanics. To be able to visualise the motion, software presented in this chapter was created.
For the satellite dynamics, the SGP4 and SDP4 models by NORAD were implemented. Both SGP4 and SDP4 are based on fundamental laws stated by Newton and Kepler. One of the biggest advantages of SGP4/SDP4 models is that they've been recognized and verified by NORAD thus providing a precise and manageable mathematical framework for the orbital calculations. But bear in mind that those are not perfect models. They work with mean values. NORAD has removed periodic variations in a particular way, and the models, in their present form do not contain numerical integration methods.
Future investigations and updates will improve propagation making it more precise. This process will certainly increase the complexity, so the balance between complexity and preciseness must be kept.

## 7. References

Binderink, A. L.; Radomski, M.S.; Samii M. V. (1989) Atmospheric drag model calibrations for spacecraft lifetime prediction; *In NASA, Goddard Space Flight Center, Flight Mechanics/ Estimation Theory Symposium*; 445-458

Bunnell, P. (1981) Tracking Satellites in Elliptical Orbits; *Ham Radio Magazine*; 46-50.

Hoots, F. R.; Roehrich R. L. (1980) Models for propagation of NORAD element sets; *Project Spacecraft Report No. 3*; Aerospace Defence Command, United States Air Force, 3-6

King-Hele, D.G (1983), Observing Earth Satellites, Macmillan

Montenbruck, O.; Gill, E.. (2000) Real-Time estimation of SGP4 Orbital Elements from GPS Navigation Data; *International Symposium SpaceFlight Dynamics, MS00/28*; 2-3

US National Archives and Records Administration.(2006) US Public Law, *Section 109–364*; Oct. 17, 2006, Stat. 2355.

Vallado A. D.; Crawford P. (2006) SGP4 Orbit Determination; *American Institute of Aeronautics and Astronautics publication*; 19-21.

# 15

# Interference in Cellular Satellite Systems

Ozlem Kilic[(1)] and Amir I. Zaghloul[(2,3)]
[(1)]*The Catholic University of America, Washington, DC, U.S.A.*
[(2)]*Virginia Polytechnic Institute and State University, VA, U.S.A*
[(3)]*US Army Research Laboratory, MD, U.S.A.*

## 1. Introduction

In cellular satellite communications systems, a given coverage area is typically filled with a network of contiguous spot beams, which carry concentrated radiation along preferred directions. The coverage regions for such applications are typically large areas, such as continents and many beams need to be generated.

Due to bandwidth limitations in cellular communications, the same bandwidth is allocated to beams which are isolated spatially. This is known as frequency reuse, and the beams operating at the same frequency are referred as co-channel beams. While this approach allows a large coverage area with limited bandwidth, the co-channel beams have the potential to interfere with each other. This is known as co-channel interference and its nature and how it could be reduced is the focus of this chapter.

The interference in multiple beam satellite communications systems will be investigated under two different approaches. First approach, which is the conventional way of defining beam coverage on earth, is discussed in Section 2. This will be referred to as spot beam coverage as explained in further detail later. The interference will be investigated for two cases; first is the uplink where interference at the satellite antenna is the main concern, and the second is the downlink where interference at the user terminal is calculated. Section 3 discusses a new method of defining beam coverage on earth, referred to as sub-beam coverage. The motivation is to keep the coverage on earth identical but reduce the satellite antenna size as much as 50% (Kilic & Zaghloul, 2009). The advantages and overall performance of the sub-beam approach in terms of interference is the subject of Section 3.

## 2. Interference in Cellular Satellite Systems

In multibeam satellite systems, the coverage area is divided into a number of beams often referred to as spot beams, which are much smaller in size and cover the area contiguously.

---

Since satellite systems are bandwidth limited, the sub-division of beams into smaller portions allow for frequency reuse to increase capacity. The available bandwidth is shared among these beams as depicted in Figure 1 for reuse factor of n.

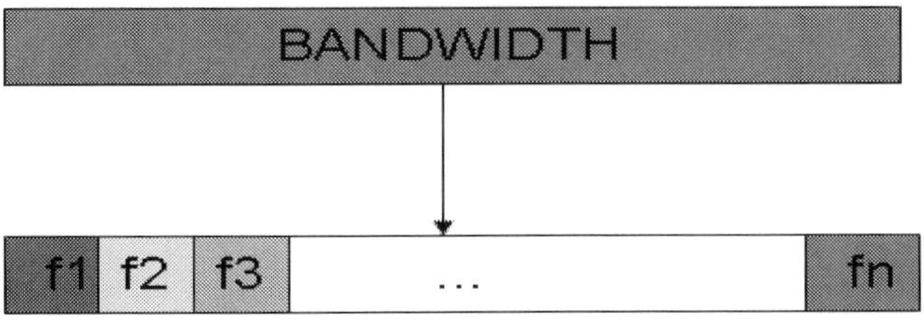


Fig. 1. Frequency reuse in multi-beam satellite communications

The size of the antenna that generates these beams on earth is related directly to the peak gain at the center of the spot beams and the smallest spot beam size. The spot beams are typically defined by the contours at 3 or 4 dB down from the peak power at the center of the beam.

## 2.1 Coverage on Earth

Achieving a contiguous coverage is important so that there are no regions without service in the coverage area. Since the beams are defined by the projection of antenna patterns on earth at a certain contour value, they tend to be close to circular shape. These circles on earth need to be structured so that they overlap with each other to avoid any gaps in the coverage area. In order to have a systematic approach, these can be represented by various geometric lattices that tessellate. A few such possibilities are shown in Figure 2. It is often the hexagon that is used in the system design as it closely represents a circle, i.e. for the same distance from the center to the edge, the hexagon area is closest to that of the circle that circumscribes it. Therefore the hexagon represents the beam which circumscribes it as shown in Figure 3. This assures that there are no gaps between beams. Then the system is designed based on this artificial hexagonal geometry on earth as depicted in Figure 4.

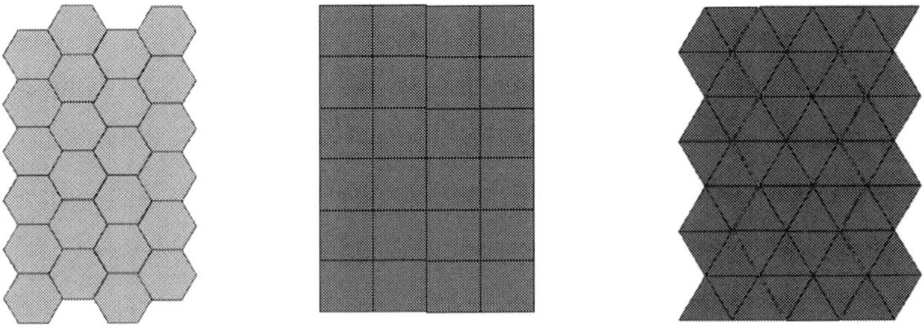

Fig. 2. Contiguous coverage on earth using tessalating shapes - hexagonal, square and triangular lattices

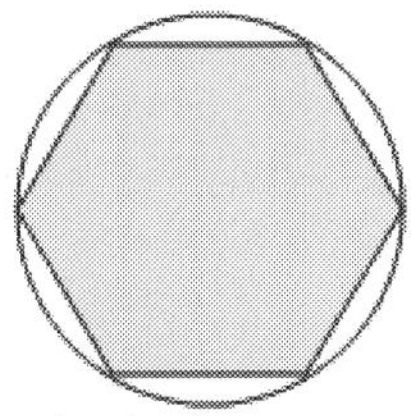

Fig. 3. Hexagonal representation of a circular beam

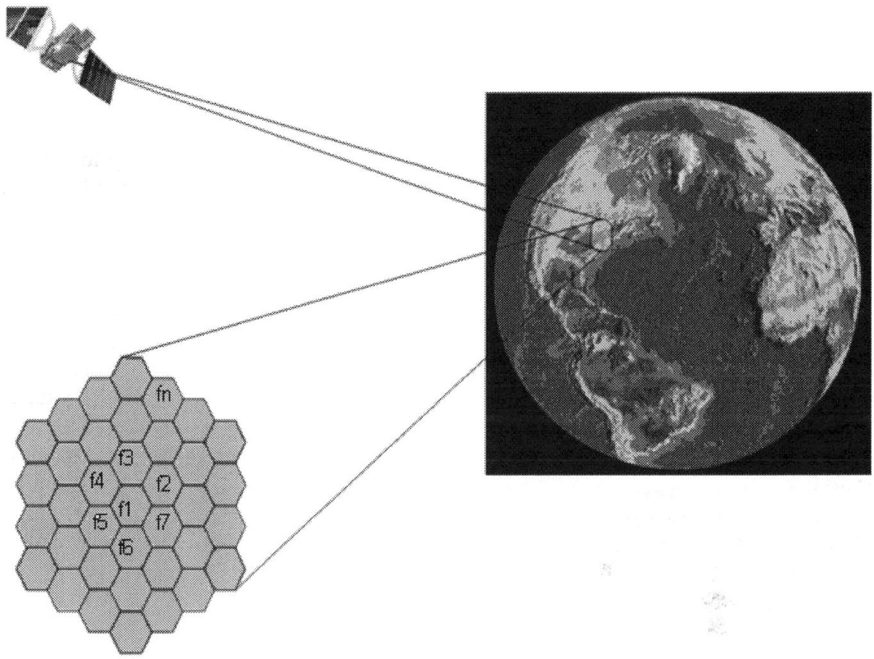


Fig. 4. Hexagonal Coverage on Earth

### 2.2 Frequency Reuse

Since the satellite systems typically serve large ares such as countries or continents, a large number of beams need to share the available bandwidth. Therefore, the available bandwidth within a beam becomes a very limited resource, as is implied in Figure 1 earlier. To circumvent this, a frequency reuse scheme is often utilized. This is based on reusing the same frequencies in spatially isolated beams. Therefore, the available bandwidth is divided into a smaller number of beams than the total number of beams in the coverage area. The set of contiguous beams that share the total available bandwidth is known as a cluster. The clusters are then repeated in the coverage area relying on the fact that the beams operating at the same bandwidth will be separated from each other sufficiently so that they do not interfere with each other.

There are only a discrete set of possible cluster sizes, N, to accommodate a contiguous coverage for a hexagonal geometry [Mehrotra, 1994]. The possible number of beams in a cluster which would form a tessalating shape is given by:

$$N = i^2 + j^2 + i \times j \tag{1}$$

where N, is the number of beams in the cluster; i.e. the number of beams that share the total bandwidth, and i, j are non-negative integer numbers. This results in possible cluster sizes of 3, 4, 7, 9, etc. An example of how these clusters are formed is shown in Figure 5.

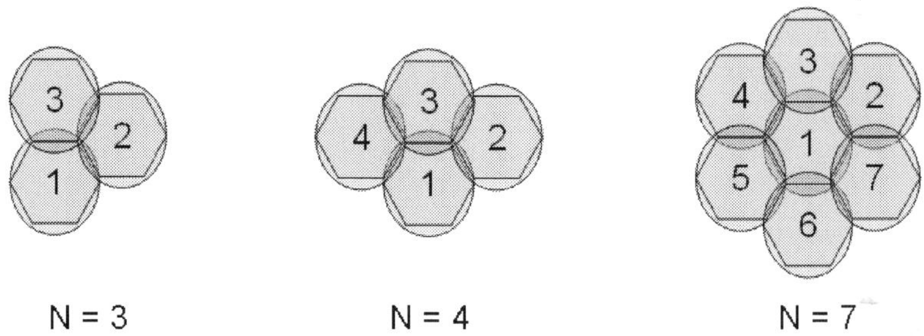


Fig. 5. Different cluster options for hexagonal lattice

### 2.3 Co-channel Beams and Tiers

The clusters as depicted in Figure 5 for reuse factors of 3, 4 and 7, are repeated to fill the required coverage area on earth. An example of how a cluster size of three (i.e. i=1, j=1) would be used to fill a given area is shown in Figure 6. A cluster size of three implies that the total available bandwidth is shared between three beams. The numbers 1, 2, 3 are used to identify the beams using the corresponding bandwidth segments. Therefore, beams with the same number imply that they use the same frequency.

Beams operating at the same frequency are known as co-channel beams. In Figure 6, the co-channel beams are shown in the same color and labeled with the same bandwidth segment number. It can be observed that the location of co-channel beams follow a pattern. They can be grouped by their distance with respect to a reference beam. The set of co-channel beams which have same distance from a reference beam are said to fall on a tier. Therefore tiers define a set of beams equidistant from a reference point. In a hexagonal geometry, each tier consists of six or twelve beams. Figure 7 shows the first four tiers for the frequency reuse of three.

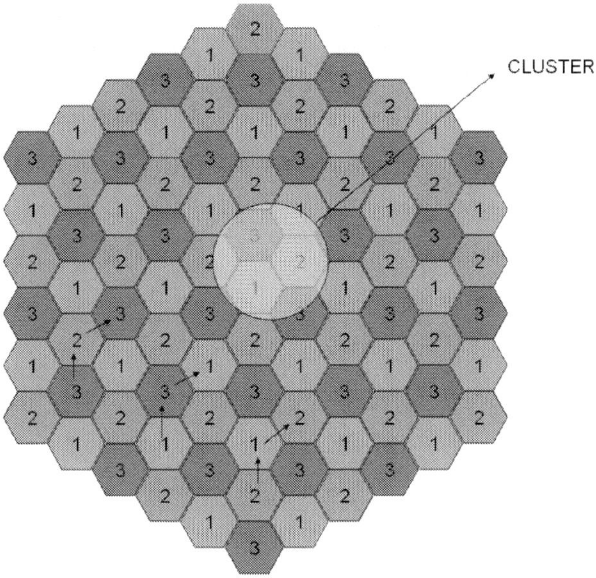


Fig. 6. Cluster coverage, N=3

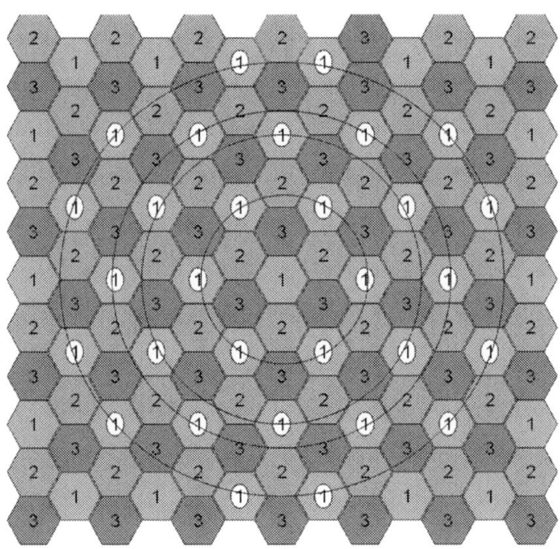

Fig. 7. Tiers in frequency reuse

The beams that lie on a tier are equidistant from the beam at the center of the tier, and for an azimuthally symmetric power distribution, beams on a tier would contribute the same amount of interference. As the tier's number increases, the diameter hence the distance from the center beam increases, reducing the contribution from the beams in that tier compared to the beams on closer tiers assuming the radiation decreases away steadily from the antenna peak.

As can be observed from Figures 6 and 7, higher frequency reuse numbers result in tiers with larger diameters, thereby increasing the distance between co-channel beams and reducing the total number of beams operating at the same frequency. However, this is done at the expense of reduced bandwidth within a beam, a trade off which needs to be decided by system engineers based on the requirements of a particular system.

### 2.4 Antenna Pattern and Spot Beam Generation

Due to their ability to generate multiple beams simultaneously, phased arrays are a natural choice for multi-beam satellite antennas. Each beam in the coverage is generated by electronically scanning the beam. A key parameter in satellite antenna design is the directivity of antenna, which defines how well the antenna focuses the power in the desired direction.

The radiation pattern of an array antenna depends on the array factor. For a MxN planar array, the array factor is given by (Balanis, 2005)

$$AF = \sum_{m=1}^{M}\sum_{n=1}^{N} I_{mn} e^{j(m-1)(kd_x \sin\theta\cos\phi+\beta_m)} e^{j(n-1)(kd_y \sin\theta\sin\phi+\beta_n)} \tag{2}$$

where $I_{mn}$ are the voltages that feed the mn^th^ element in the array. The array factor is related to the antenna directivity which measures how well the input power is focused along a given direction and can be computed as follows (Stutzman, 1998)

$$D_0 = \frac{4\pi[AF(\theta_0,\phi_0)][AF(\theta_0,\phi_0)^* |_{max}}{\int_0^{2\pi}\int_0^{\pi}[AF(\theta_0,\phi_0)][AF(\theta_0,\phi_0)]^* \sin\theta \cdot d\theta \cdot d\phi} \tag{3}$$

In the case of a large array with uniform excitation, this equation can be approximated by

$$D_0 = \frac{\pi^2}{\Omega_A} \tag{4}$$

where the denominator is the subtended beam angle defined by

$$\Omega_A = \frac{\theta_{x0}\theta_{y0}\sec\theta_0}{\left[\sin^2\varphi_0 + \frac{\theta_{y0}^2}{\theta_{x0}^2}\cos^2\varphi_0\right]^{\frac{1}{2}}\left[\sin^2\varphi_0 + \frac{\theta_{x0}^2}{\theta_{y0}^2}\cos^2\varphi_0\right]^{\frac{1}{2}}} \tag{5}$$

In equation (5) the angles $\varphi_0$ and $\theta_0$ represent the direction at which the directivity is computed, and $\theta_{x0}$ and $\theta_{y0}$ denote the half power beamwidths of the array along the directions x and y, the plane on which the array resides. In this chapter we will assume a 40x40 square planar array with isotropic elements separated by half wavelength as shown in Figure 8.

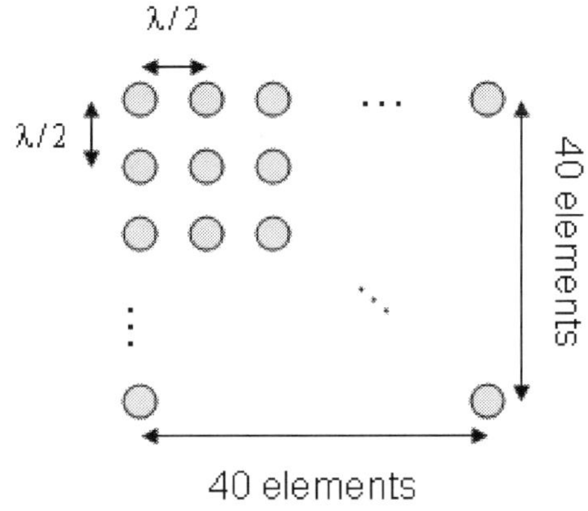


Fig. 8. A uniformly spaced 40x40 element planar array

The plot of the directivity in two different planes for this particular array is shown in Figures 9-10. The 40x40 element antenna is sized to have an edge gain of 36 dB defined at its -4dB contour relative to the peak. Thus, the peak gain of the antenna is 40 dB, with a beamwidth of 1.44 degrees at the - 4 dB down contour from the peak gain. The spot beam in the coverage area is defined by this contour. The contour plot for the same antenna is shown in Figure 11. The beam is generated at the center as denoted by the white circle.

As seen in the contour plot, the energy is radiated to the whole coverage area even though most of the power is focused in the spot beam. This "leak" into other regions is not a problem for beams operating at different frequencies as filtering will eliminate this energy. But for co-channel beams; i.e. beams operating at the same frequency, this leaking energy creates the interference which is the focus of this chapter.

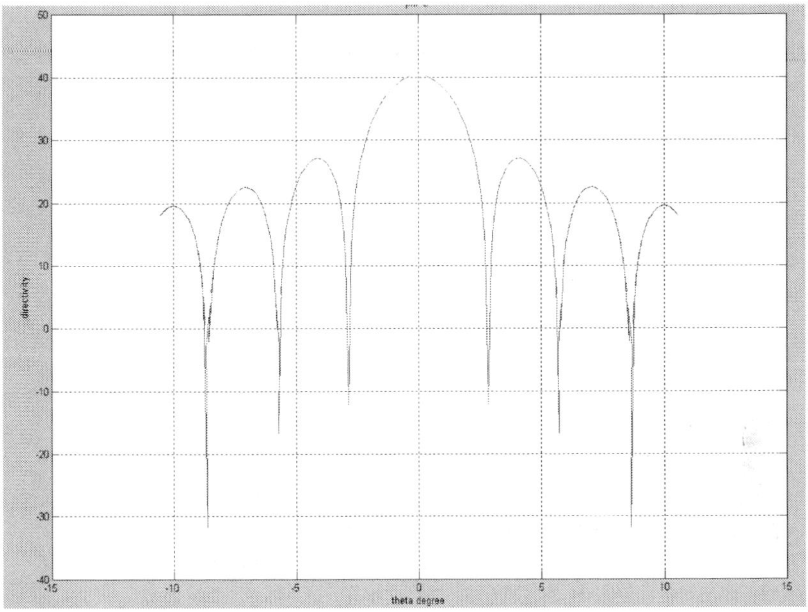


Fig. 9. Array factor for the 40x40 element array, φ = 0 plane

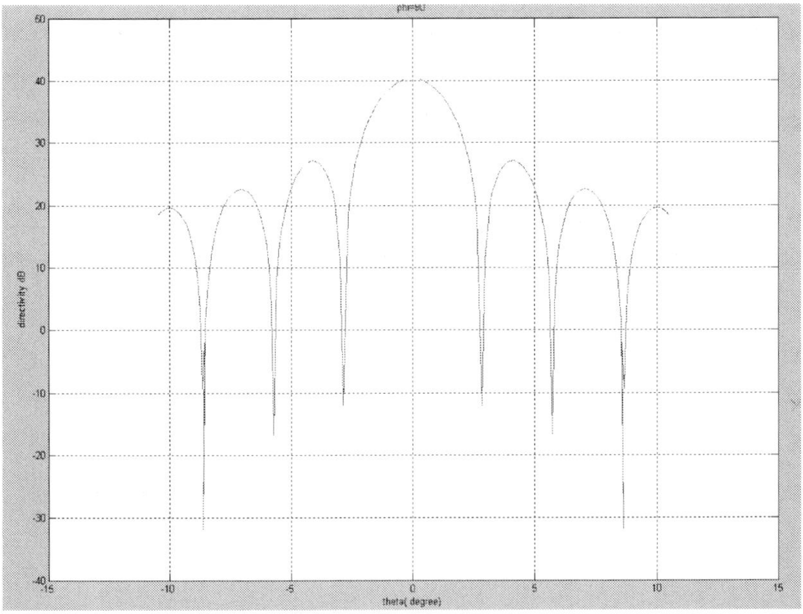


Fig. 10. Array factor for the 40x40 element array, φ = 90 plane

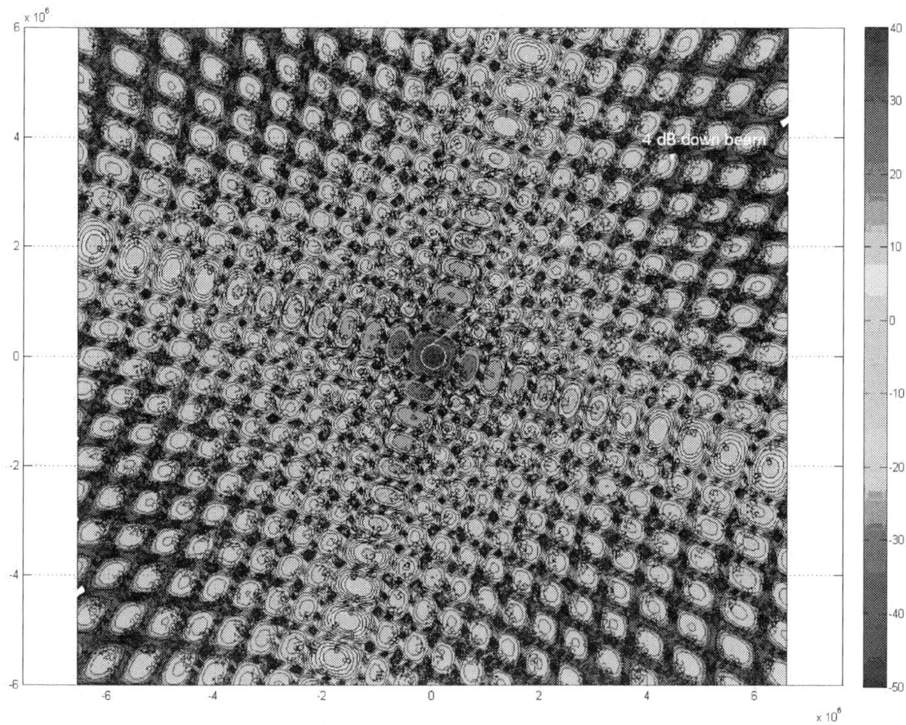


Fig. 11. Contour plot of directivity for 40x40 element array

### 2.5 Co-channel Interference

The energy leak from a beam into other beams operating at the same frequency causes co-channel interference. Figure 12 shows a simplistic diagram of the co-channel interferers in both the uplink and the downlink. The downlink interference is primarily a function of the reuse number and of the aggregate power due to the power in the side lobes of the interfering co-channel spot beams that is received in an earth station receiver. On the other hand, the uplink co-channel carrier-to-interference ratio is dependent upon the reuse number and the number of co-channel users transmitting simultaneously and received at the side lobes of the interfered beam. This section will focus solely on the downlink interference.

The farther the co-channel beams are separated from each other, the lower is the expected interference levels as the antenna pattern gradually tapers off with increasing angles away from the peak point. However, this gradual taper in the antenna pattern is oscillatory due to the side lobes. Therefore, the co-channel interference in a multiple beam system depends on the antenna pattern as well as the cluster configuration.

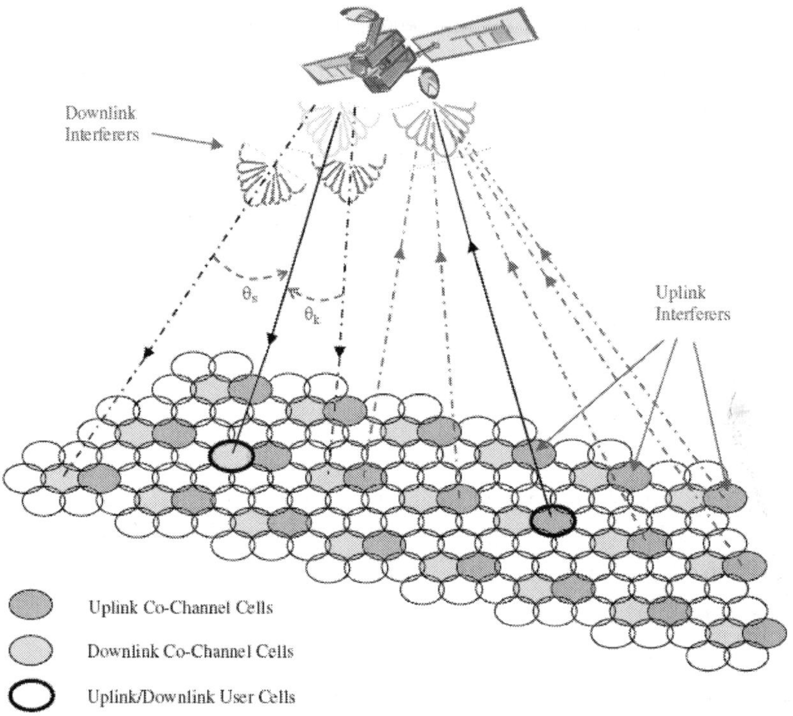


Fig. 12. Co-channel interference concept in multibeam satellite communication systems

A parameter for the system performance is the carrier to interference ratio, which defines how much better the intended signal is compared to the interfering signals:

$$\left.\frac{Carier}{Interference}\right|_{\text{at beam center}} = \left.\frac{C}{I}\right|_{\text{at beam center}} = \frac{D_{Max}}{D_{Interference}} \tag{6}$$

where the interference term is the sum of powers received from all co-channel beams. We can express equation (6) equivalently in dB as:

$$10\log(\frac{C}{I}) = 10\log(D_{\max} - \log D_{Interference}) \tag{7}$$

For an azimuthally symmetric antenna pattern, the co-channel beams lying on the same tier would result in equal interference levels as they are equidistant from the reference point. Therefore, categorizing all co-channel beams by their tier number helps simplify the interference calculations. Although the square array does not produce an azimuthally symmetric pattern, the power levels tend to decrease as the tier number increases, i.e. as the co-channel beams are farther from the reference point. This concept is shown in Figures 12 and 13 for frequency reuse factors of 3 and 7. We observe that for the larger size cluster (i.e. N=7), the diameter of the tiers are larger for the same tier number. This implies that, larger cluster sizes will likely have less co-channel interference.

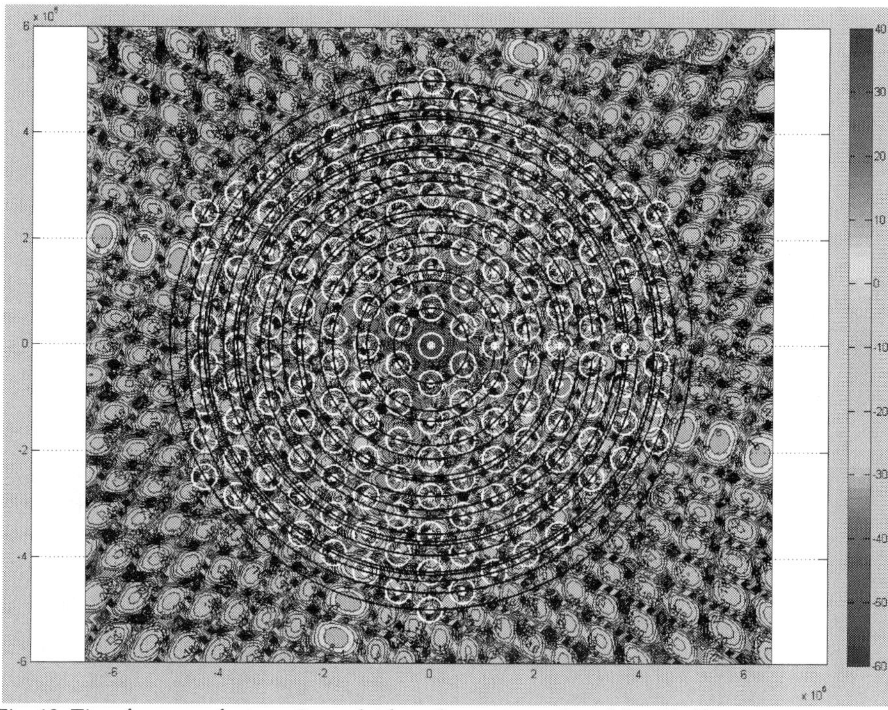

Fig. 13. Tiers for reuse factor = 3 overlaid on the contour plot of the 40x40 array

Fig. 14. Tiers for reuse factor = 7 overlaid on the contour plot of the 40x40 array

### 2.6 Example – LEO System Downlink Interference

In this section we calculate the total downlink interference in using a Low Earth Orbiting (LEO) satellite system. The relevant parameters of the system are as follows:

Altitude = 9400 km
Peak Gain = 40.0 dB
Spot Beam edge @ 4dB (36.0 dB)
Frequency = 1.9 GHz
Minimum elevation angle = 22°
Number of beams on earth = 500

The spot beams generated by the 40x40 element phased array antenna for frequency reuse of three are shown in Figure 15. The radii of all spot beams are assumed equal to that of the center spot beam, and are computed using the beam width at -4dB relative to the peak gain and the altitude as follows:

$$Radius = H(altitude) * \tan(\theta_{4dB\_down}) \tag{8}$$

For the 40x40 element array antenna, the beam width is 0.0252 radians (1.44 degrees), resulting in a radius of 236km. The beam coverage on earth for this system for reuse factor of 3 is shown in Figure 15. The red lines depict the different tiers for this reuse factor, and same numbers denote the same bandwidth of operation, hence co-channel beams.

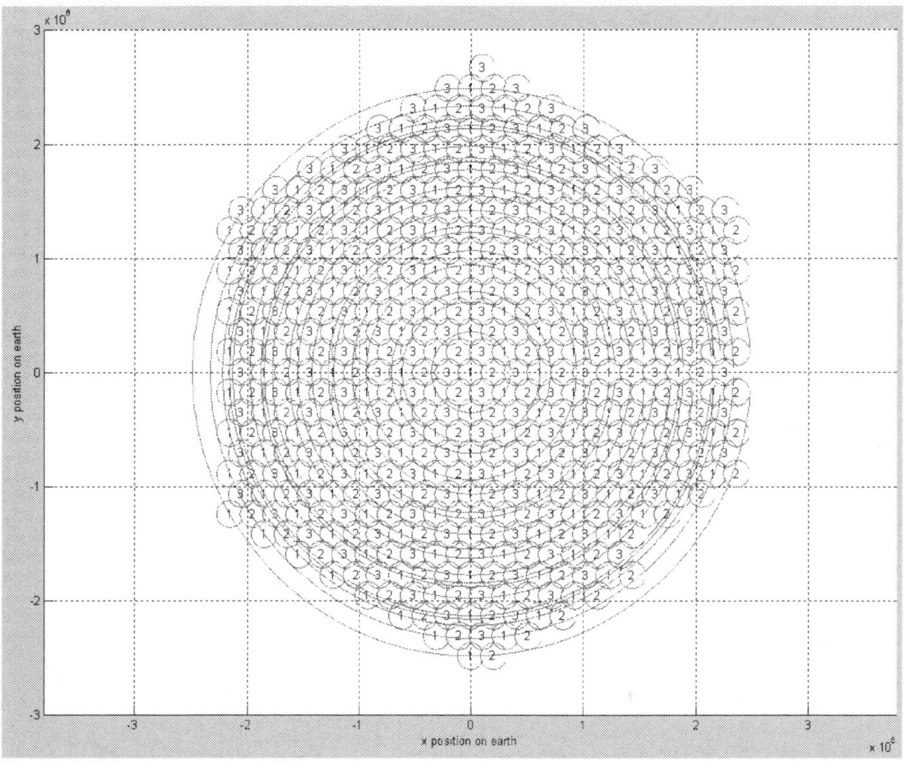


Fig. 15. LEO system, reuse of 3

The total interference from each tier for the center beam is shown for different reuse factors in Figure 16. We observe that there are fewer tiers for the same coverage when a higher frequency reuse factor is used. It is also observed that the interference levels are higher (hence the carrier to interference ratio is lower) from the tiers closest to the beam of interest. Since the square array is not azimuthally symmetric, we observe oscillations in the interference levels between tiers due to the side lobe level variations. We also observe that higher frequency reuse factor results in lower interference as there are fewer co-channel beams in the coverage area and they are spread farther apart.

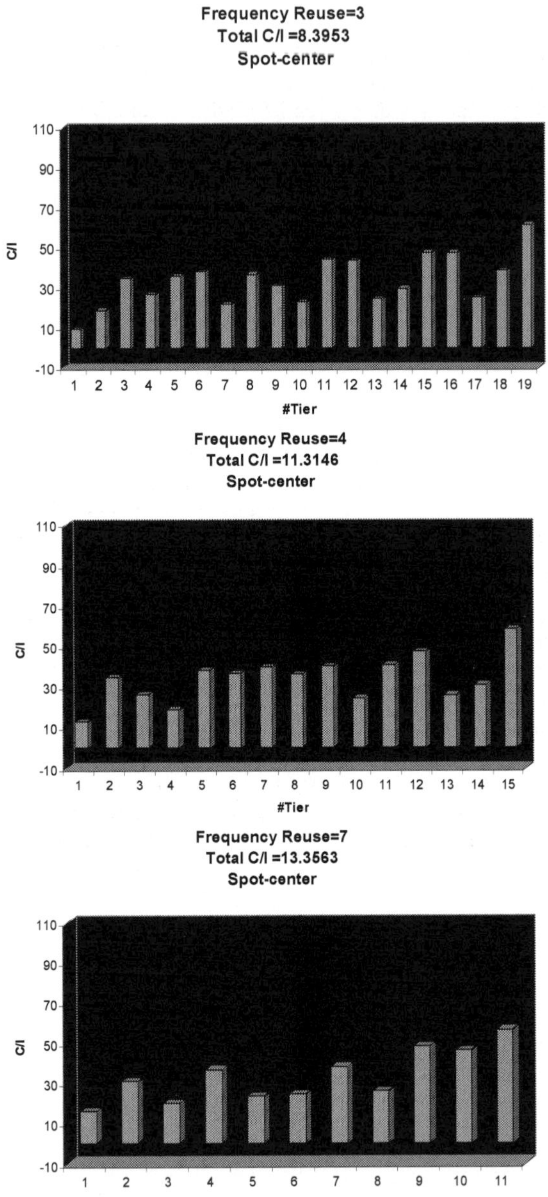


Fig. 16. Interference versus tier number (a) reuse = 3 (b) reuse = 4 (c) reuse = 7

## 3. Sub-Beam Concept

As we have seen so far, the coverage area in multibeam satellite systems is divided into a number of beams often referred to as spot beams, which are much smaller in size. As discussed before, since satellite systems are bandwidth limited, the sub-division of beams into smaller portions allow for frequency reuse to increase capacity. The size of antenna that generates these beams on earth is related directly to the peak gain at the center of the spot beam and the smallest spot beam size. The spot beams are typically defined by the contours at 3 or 4 dB down from the peak gain at the center of the beam. While it is the edge gain requirement which needs to satisfy the communication link, the satellite antennas have to be sized for the maximum gain to achieve the 3-4 dB drops around the perimeter of the spot beam. Therefore it can be argued that the satellite antenna is typically oversized as lower gain levels within the spot beam would satisfy the link requirements. Having a larger sized antenna leads to higher cost. To achieve 3 dB higher gain levels, one needs to double the aperture size, which directly translates to cost.

Recently a new method has been presented for reducing the cost of satellite antennas by sizing the antennas for lower power levels than the conventional approach, while maintaining the edge gain requirements, (Kilic and Zaghloul, 2009). This approach relies on the partitioning of the spot beam into a number of sub-beams. The basic idea behind the sub-beam partitioning approach is to replace the spot beams with a number of smaller sub-beams with lower peak gain and same edge gain values. This translates to an increase in the total number of beams to be generated by the antenna. However, since the sub-beams have lower peak gain, the antenna size can be reduced significantly. The increase in the number of beams results in a more complex beamforming structure, but this can be handled digitally without significant cost effects. Therefore the sub-beam partitioning results in significant reduction in the cost of the antenna compared to the spot beam approach.

Although the antenna cost is reduced in sub-beam approach a key issue which remains to be addressed is the interference effects in the new beam configuration. It is necessary to satisfy the carrier to interference ratio throughout the coverage area for this approach to be of practical value.

The objective of this section is to investigate interference performance of the sub-beam approach in comparison to the spot beam configuration. First the definition of the spot beam concept is presented and the sources of interference in a multiple beam communication system are identified. Then the sub-beam concept is introduced. Finally, the performance of the conventional spot beam approach with the sub-beam partitioning technique is compared in terms of co-channel beam interference.

### 3.1 Partitioning of spot beams into sub-beams

The sub-beam design reduces the peak gain requirements at the beam center while maintaining the edge gain requirements. In order to do this, each spot beam in the coverage area is divided into a number of smaller beams defined by contour levels less than the typical 3 or 4 dB, as depicted in Figure 17, where four sub-beams are used to replace a spot beam. For an edge gain requirement of 36 dB, it is observed that the sub-beam approach provides a more uniform gain distribution within a beam's coverage area (36-37 dB) compared to the spot beam approach (36-40 dB). A group of these smaller sub-beams, i.e. "sub-beam clusters," represent each spot beam.

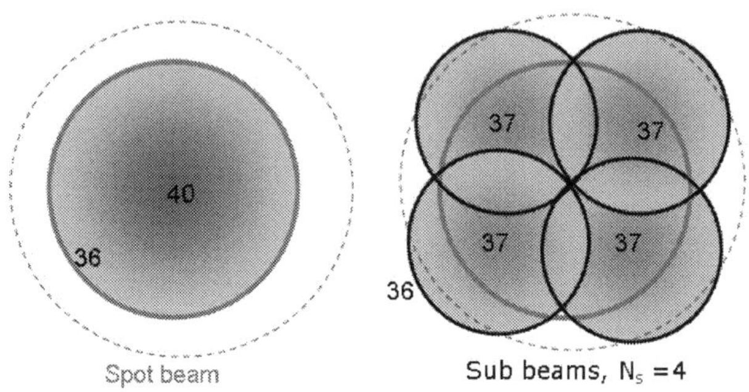


Fig. 17. Sub-beam representation of a spot beam

To satisfy the contiguity requirements of the coverage, the sub-beams intersect at the boundaries of the spot beam. Figure 18 demonstrates the sub-beam concept, where the hexagons correspond to the spot beams in the coverage area. The numbers at the center of each hexagon denote the frequency band assigned in each beam. A frequency reuse index (or spot beam cluster size) of seven is depicted in the figure. The large circles encompassing sets of seven hexagons represent the spot-beam clusters based on this reuse factor. The sub-beams are shown for the center spot beam of each cluster. A sub-beam cluster size of four is assumed in the demonstration.

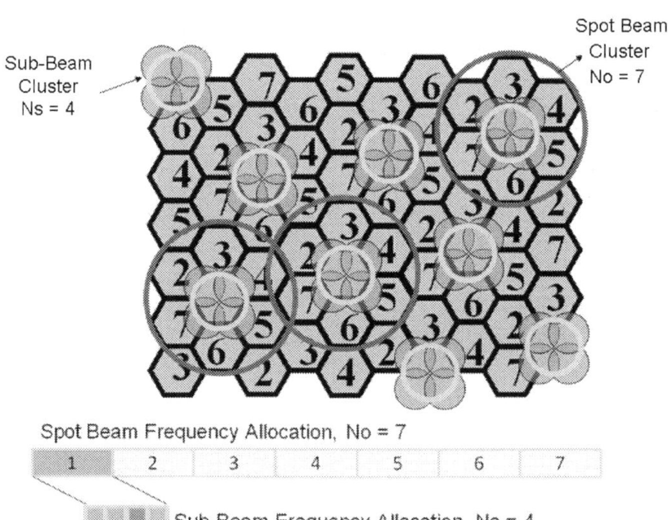


Fig. 18. Sub-beam partitioning for spot beams, reuse =7

As a result of this approach, the antenna is required to generate four times more beams than the conventional spot beam design. However, the sub-beams do not need to satisfy the peak gain requirements of the spot beam. Since only the edge gain needs to be satisfied, the sub-beam peak gain is lower and the overall antenna size is reduced as explained in the following sections.

### 3.2 Spectrum Subdivision with the Sub-beams

The frequency reuse configuration is conserved with the sub-beam approach. The allocated bandwidth per spot beam is shared among the sub-beams in a sub-beam cluster. As Figure 18 suggests, each sub-beam in a sub-beam cluster uses the bandwidth allocated to the spot beam they represent. While each sub-beam is designated a smaller portions of the bandwidth, the bandwidth allocation for the entire spot beam remains the same. Hence, the capacity per spot beam is conserved. Furthermore, this approach makes it possible to have a non-uniform distribution of the available bandwidth in a spot beam to address local high traffic areas. Depending on the sub-beam cluster size, some areas are served by more than one sub-beam from different spot beams, adding to the flexibility of traffic assignment.

### 3.3 Gain Reduction Concept of the Sub-beam Approach

The amount of reduction in the antenna size depends on the edge gain relative to the peak, and beam width requirements of the coverage. The sub-beam cluster generates an equivalent coverage to the spot beam by using a higher number of smaller beams. This enables using smaller antennas. The number of beams in a sub-beam cluster needs to be chosen such that a contiguous coverage is achieved. This results in values such as 3, 4, 7 ...etc.

The relationship between the relative edge gain of the spot beam and the relative edge gain of the sub-beams can be determined as a function of the number of sub-beams that make up a sub-beam cluster. In the derivations that follow, the subscripts 'o', and 's' denote the spot beam and sub-beams, respectively. The sub-beam edge contours are defined by $x_s$ dB down relative to the peak gain of $G_s$ dB. The spot beam contours are assumed to be $x_0$ dB lower than the peak gain, which is denoted by $G_0$ dB. Since the edge gain requirements are satisfied for both cases, the gain relationship can be written as follows:

$$G_o - x_o = G_s - x_s \tag{9}$$

This equation can be rewritten in terms of gain reduction as:

$$\Delta G \equiv G_o - G_s = x_o - x_s \tag{10}$$

It is observed that the reduction in gain is directly proportional to the difference of the beam contour levels of the spot beam and the sub-beams. While the typical contour level to define a spot beam is 3-4 dB, the sub-beam configuration is flexible, and the choice is based on increasing the gain reduction. As equation (10) suggests, defining the sub-beams at a low level would increase the reduction in gain. However, the sub-beam cluster size, Ns, constrains the beam width of the sub-beams, and the value of $x_s$ cannot be assigned arbitrarily. The choice will depend on the edge gain of the spot beam as well as the partition size, $N_S$.

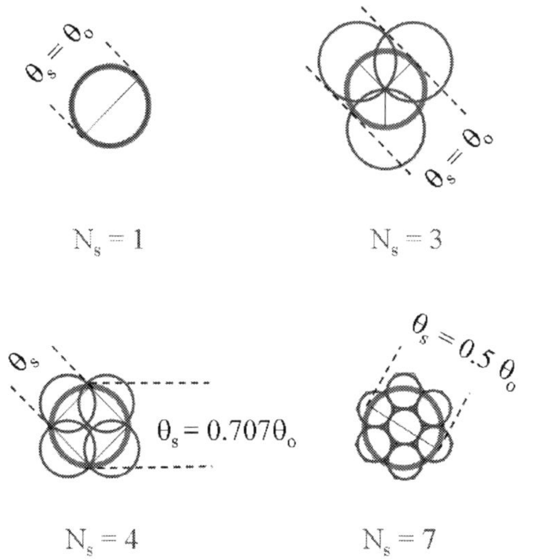


Fig. 19. Possible Sub-beam Partitions for Contiguous Coverage

### 3.4 Defining the Sub-beams to Achieve Desired Gain Reduction

A contiguous coverage of the sub-beam clusters can be obtained by satisfying Equation 1, in which case the number of sub-beams that define a spot beam are restricted to values such as 1, 3, 4, 7, ..., etc. Examples of the geometry for different cluster sizes are shown in Figure 19, where the circle at the center corresponds to the spot beam and the surrounding circles denote the sub-beam contours. The beam width ratio, $\theta_s/\theta_o$ depends on the sub-beam cluster geometry, and is equal to 1, 1, 0.707 and 0.5 for sub-beam cluster sizes of 1, 3, 4 and 7, respectively. The higher sub-beam clusters approximate the spot beam better and reduce the overlap of sub-beams from different clusters. Higher cluster sizes also reduce the sub-beam peak gain that is needed to satisfy the spot beam edge gain requirement.

An empirical relationship has been obtained to relate the beam width of a phased array at an arbitrary contour level, x to its HPBW by using the definition of the array factor as:

$$\frac{\theta_X}{\theta_3} = 0.59x^{0.4806} \tag{12}$$

where x is in dBs, (Zaghloul , et. al. 2000). This empirical relationship assumes isotropic radiating elements, which combine in space to form the directive pattern, and has been verified for large array sizes for broadside radiation. The expression is determined by fitting a curve to different size square arrays that range from 20x20 to 50x50 elements with half wavelength spacing between them. The results of the simulations and the agreement with the empirical relation above are shown in Figure 20.

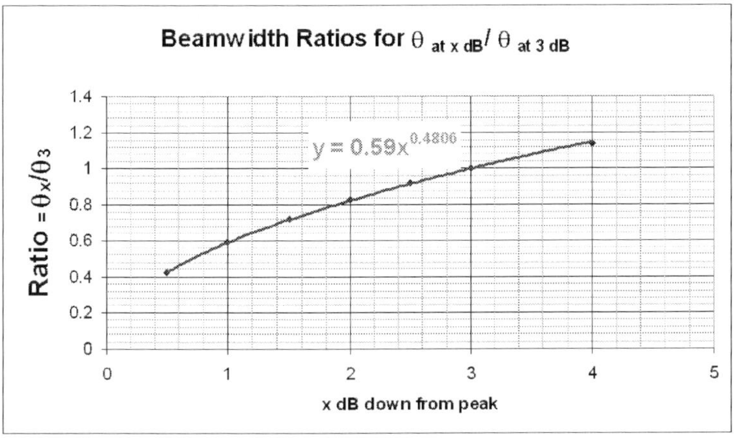


Fig. 20. Beam width at arbitrary contour levels for large arrays

Using equation (12) in (2), the contour levels of the spot and sub-beams can be related to their beam widths, and as follows:

$$9.612\log\left(\frac{x_s}{x_o}\right)+x_o-x_s=20\log\left(\frac{\theta_s}{\theta_o}\right) \tag{13}$$

Equation 13 can be rewritten in terms of gain reduction, using (10) as

$$9.612\log\left(\frac{x_s}{x_o}\right)+\Delta G=20\log\left(\frac{\theta_s}{\theta_o}\right) \tag{14}$$

where $\Delta G = G_o - G_s$ is the gain reduction achieved by using the sub-beam concept and the beam width ratio on the right is a constant determined by the sub-beam cluster configuration. Thus, for a given sub-beam configuration, the peak gain reduction and the ratio of the relative edge gains are uniquely related. In other words, if two of the parameters $\Delta G$, $x_c$ and $x_s$ are given, the third is uniquely determined for a specific sub-beam cluster configuration.

The relation between reduction in gain and beam contour levels is demonstrated in Figure 21. Two cases are plotted for comparison. The blue line corresponds to the sub-beam cluster size of 4, while the red line represents the sub-beam cluster size of 7. The dashed lines show the contour level for the sub-beams as a function of gain reduction, and the solid line describes the relation between spot-beam contour levels and the gain reduction. It is observed that, for a desired amount of gain reduction, the beam contour levels for both sub-beam and spot beam are uniquely defined for each sub-beam cluster size.

Since the sub-beam approach is offered as a solution to reduce the antenna aperture size, it is worth investigating the improvement by assuming that the spot beam contour levels and the peak gain are already defined for the system to be replaced. For instance, if a spot beam width at 4 dB contour levels is being replaced with sub-beams of lower contour levels, the graph in Figure 21 suggests that a gain reduction of 3dB can be achieved for Ns = 4 by defining the sub-beam contours at 1 dB down from the sub-beam peak gain. Similarly, for Ns = 7, a gain reduction of 3.6 dB is achieved by defining the sub-beams at 0.4 dB down from their peak gain.

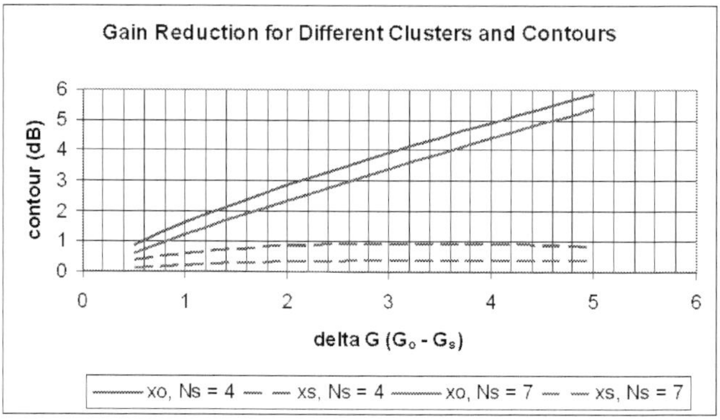


Fig. 21. Contour levels of spot beams and sub-beams as a function of gain reduction.

This example implies that for a given spot beam configuration, the larger sub-beam clusters can offer higher gain reduction, which is associated with a corresponding reduction in the antenna aperture size and cost. However, since the higher cluster sizes define beams at lower contour levels, a legitimate concern is to make sure the energy in the main beam of these antennas do not adversely affect the co-channel interference levels.

The reduction in peak gain results in a reduction in the aperture size and correspondingly a reduction in the number of elements in the array. This is illustrated in Figure 22 where a 28 x 28 or 784-element array for a conventional spot beam (corresponds to Ns =1) is replaced with arrays of fewer elements as the number of sub-beams increase in the cluster. It is observed that the limit case is approached for sub-beam cluster sizes of 7, where less than 300 elements are required.

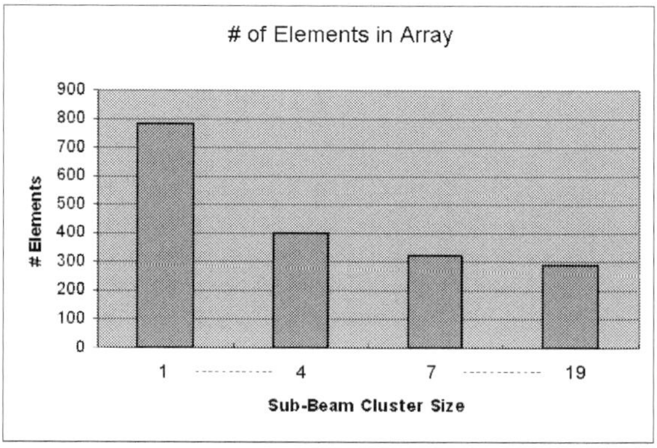


Fig. 22. Array size reduction with sub-beam partitioning

The reduction in the receive and transmit antenna aperture sizes result in fewer components. However, the generation of sub-beam clusters requires further de-multiplexing and increase in the size of the digital traffic router. Corresponding multiplexing on the transmit side is also required. However, the added complexity occurs in the software-driven processor on board the satellite, with minimal addition to the hardware.

### 3.5 Interference in Sub-beam Partitioned Coverage

The same LEO system as in section 2.5 is considered for the interference analysis of the sub-beam partitioned coverage. A comparison of beam coverage on earth for spot beam and sub-beam configuration with partition size of 3 is shown in Figure 23. The beams are overlaid on the contour plot of the antenna pattern that generates the center beam, the beam of interest for interference calculations.

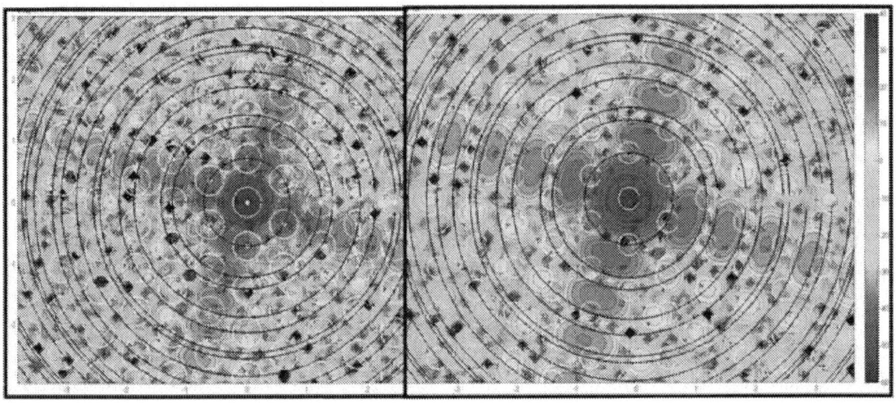

Fig. 23. Coverage on Earth - Spot beam pattern (left) versus Sub-beam pattern (right)

Sub-beam partition sizes of four and seven will be considered for the interference analysis. Figure 24 shows a zoomed in depiction of which one of the sub-beams that replace a spot beam are used for interference calculations. The spot beam to be replaced is the blue circle, while the red circles are the sub-beams for a partition size of four (left) and seven (right). The green circle is the sub-beam that will be selected in the interference analysis.

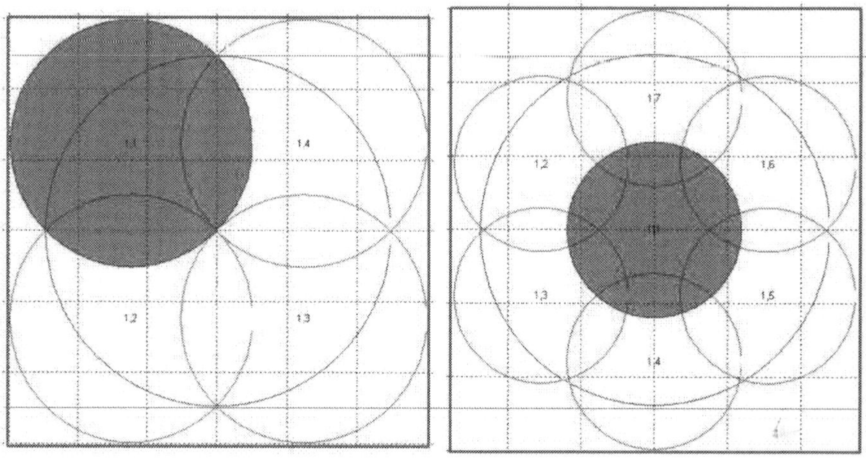

Fig. 24. Sub-beam selection for interference calcualtions, Ns = 4

The carrier to interference ratio and total interference are shown for frequency reuse of 3, 4, 7, 9 and 13 in Figure 25 for sub-beam partition sizes of four and seven. It is observed that the C/I at the center of the beam (where carrier power is maximum) are better for the spot beam configuration for large cluster sizes (i.e. N > 9). However in certain cases such as No = 3 and No = 9, the performance of the sub-beam approach is better or comparable to the spot beam configuration. It should be noted that this comparison presents a best case scenario for the spot beams as the carrier power is at its peak value. The total interference shown in Figure 25(b) indicates that sub-beams may have lower interference values relative to the spot beam cases of the same frequency re-use number. A large cluster size separates co-channel cells, and the highly directive spot beam antenna tends to have lower sidelobe levels at these separations.

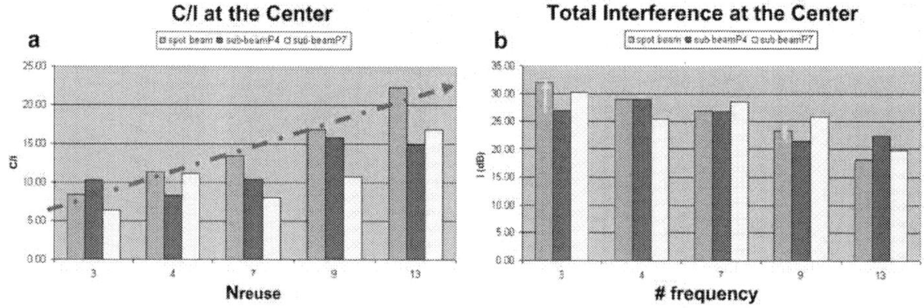


Fig. 25. Interference at the center by tier for sub-bem partition=4, 7 and frequency reuse 3, 4, 7, 9, 13. (a) C/I (b) Total interference

To better understand the sub-beam performance, a better analysis is calculating C/I as one moves within the spot beam to be replaced by the sub-beams as depicted in Figure 26. The spot beam shown at the center with the blue circle is replaced by 4 sub-beams. The C/I is calculated at discrete distances from the center (d) and at angular positions (α). The chart on the right shows the C/I for the spot beams (blue lines) and sub-beams (red lines) for different d and α values. The x-axis shows the ratio of the distance from the center to the spot beam radius, R.

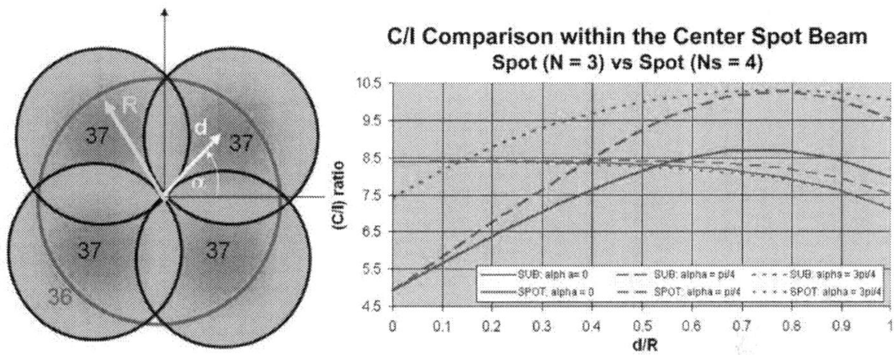

Fig. 26. C/I performance comparison within the spot beam

It is observed that partitioning the spot beam into sub-beams performs superior towards the spot beam edge because the carrier power in the sub-beam gets stronger. This is typically where the spot beam performs worst and a significant advantage for the sub-beam approach results. The main region where the spot beam seems to perform better is at the center of the beam as expected due to the high carrier power levels in the spot beam. In the regions near the beam center, it is observed that the sub-beam approach performs worst where the spot beams have the best performance. As expected, this is due to the high carrier power the spot beam has at the center. Typically the C/I value at the beam center has a higher margin as the systems are designed for the edge gain and the degraded performance of the sub-beam at the center may not pose a problem. To mitigate the relatively poor performance at the center, it would be possible to optimize the sub-beam antennas to perform better at the beam centers. If necessary a further step could be to modify the frequency allocation. However, this could result in lower system capacity, and should be attempted only if the antenna size and cost is the primary concern. Overall, the findings suggest that the sub-beam approach holds promise as its C/I is mostly not degraded compared to the spot beam approach, while its peak gain is lower and thus requires smaller antenna aperture.

## 4. Conclusion

Co-channel beam interference in multi-beam satellite communications systems was investigated particularly for the downlink. Concept of frequency reuse was explained and the role of satellite antenna size and pattern was examined. Conventional spot beam coverage and its impact on determining the antenna size on board was discussed.

A method was presented to reduce the array antenna aperture on board the satellite for multiple-spot-beam cellular coverage. The overlapped spot beams have edge gain that is determined based on link budget, coverage requirement, frequency band allocations, and corresponding channel capacity allocations. The edge gain relative to the peak gain of the spot beam determines the size of the required antenna aperture. Dividing the spot beam into a number of sub-beams that overlap at the same gain level as the spot beam but with lower gain taper over the sub-beam results in lower peak gain and consequently smaller aperture. The frequency band allocated to the spot beam is divided among the constituent sub-beams and can be assigned with enough flexibility to meet projected traffic demands.

The reduction in the antenna aperture with this approach translates into significant reductions in number of array elements, RF components, and A/D and D/A converters. These savings are obtained at the expense of more complex digital beam forming and traffic routing algorithms. Analysis has shown that in spite of the smaller aperture and the broader beams of the sub-beams, the co-channel interference between sub-beams using the same frequency segment is not adversely affected.

## 5. References

Balanis C.A., (2005) Antenna Theory: Analysis and Design, John Wiley and Sons, New York, 3rd Edition, ISBN 0471592684.

Kilic, O., and A. I. Zaghloul (2009), Antenna aperture size reduction using subbeam concept in multiple spot beam cellular satellite systems, *Radio Sci.,* 44, RS3001, doi:10.1029/2008RS004052

Mehrotra, (1994) A. *Cellular Radio Performance Engineering,* Artech House, Boston, ISBN 0890067481.

Stutzman, W.L. and Thiele, G.A., (1998) *Antenna Theory and Design,* John Wiley and Sons, New York, 2nd Edition, ISBN 0471025909.

Zaghloul, A.I., O. Kilic and A.E. Williams(2000), "Aperture Size Reduction Using Sub-Beam Design Concepts for Multiple Spot Beam Satellites," *IEEE International Symposium on Antennas and Propagation,* Salt Lake City, Utah, July 2000, pp. 534-537.

16

# Beyond life-cycle utilization of geostationary communication satellites in end-of-life

Shi Hu-Li, Han Yan-Ben, Ma Li-Hua, Pei Jun, Yin Zhi-Qiang and Ji Hai-Fu
*National Astronomical Observatories, Chinese Academy of Sciences, 20A Datun Road, Chaoyang District, Beijing 100012*
*China*

## 1. Introduction

In 2002, we proposed an idea that carrier signals with ranging codes and navigation data from ground navigation master station are transmitted in transponders of geostationary (GEO) communication satellites and then ground users receive those signals and realize passive navigation. In 2003 we began to lease transponders of GEO communication satellites to carry out experimental verification, and a success was achieved in 2005[11, 12].

The most challenging problem in the verification is that all rental GEO communication satellites are located in GEO orbit and therefore three dimensional positioning is impossible. To improve geometric configuration of satellites constellation and positioning precision, we hope that several inclined geosynchronous orbit (IGSO) satellites can be adopted except for GEO satellites to constitute the constellation. Considering the long designing period of IGSO satellite, a concept that we try to utilize IGSO satellites was proposed. Unfortunately, those efforts are in vain, and therefore we proposed the concept turning GEO communication satellites in end-of-life to IGSO by regulation.

Communication satellites running in GEO orbit with altitude of 35786 km above equator are maintained in fixed track with up to about ±0.1 degree precision and some even with better than about ±0.05 degree precision. However, because of the perturbation from other celestial bodies, the position of GEO satellites are changing relative to the Earth, and the drift of satellite's fixed position relative to the equator are constantly increasing and even exceeds more than the ITU fixed-point drift value. In order to control the satellite's drift, we must adjust the satellite's orbital position and attitude, which will inevitably consume fuel, when fuel in the satellite is consuming step by step, the satellite's service life will come to end. We call such a geostationary communications satellite "late-life" or "end-life" GEO satellite. In fact, the life of satellite's solar battery and signal transponders has a certain redundancy and it can still work for many years, just because the remaining fuel in satellite has almost run out, this satellite with heavily investment in the process of development and launch has to been abandoned. To ensure custom's business is not interrupted, satellites company must launch a new satellite to replace the GEO communication satellite in end-of-life. According to the IADC's suggestion on retiring GEO communication satellites, the satellite solar power and related equipment have to been closed and satellite's must been pulled up about 200 ~ 300 km to the grave orbit and then those retired satellites become space junk[14, 15].

According to IADC's monitoring, there have been approximately more than a thousand satellites in the vicinity of GEO orbit within about 60 km, but only about 300 satellites work normally[16]. Increasing space junks seriously affect the space environment, especially the valuable space environment near the equatorial orbit. How to extend the life of the satellite in orbit and thereby turning waste into wealth has became one of important topics for space technology researchers.

To this end, according to concept of modern design and advanced manufacturing technology, we explored the issue of reusing satellite in end-of-life[1, 9, 13]. Specifically, with use of remaining fuel resources in retired satellites in end-of-life, gradually turning GEO satellite in end-of-life to slightly inclined geostationary orbit (SIGSO) satellite by control of satellite attitude. Utilizing those SIGSO satellites to constitute a new navigation and positioning constellation is for the purpose of improving the navigation and positioning accuracy of the transmitting satellite navigation system. In addition, the abundant transponders resources in those satellite can also been used to carry out satellite communication business or open up new satellite communication services, that is to say, it is possible to develop a second life-cycle of the satellite in end-of-life.

## 2. Orbital dynamic analysis of GEO communication satellite orbit

In order to explore the issue of reusing the GEO satellite in end-of life, we must thoroughly understand and analyze the dynamic theory of GEO satellite[34].

### 2.1 Description of GEO orbit

GEO is a special orbit, on which the satellites rotating around the Earth has the same rotation cycle with Earth's rotation period and its nadir position in the Earth's surface should be fixed. Generally speaking, GEO has the following three characteristics:

(1) the shape of orbit is circular, eccentricity e = 0;
(2) orbit is located in the Earth equatorial plane, the inclination i = 0;
(3) orbit's spin cycle is same with the earth rotation cycle, $T = 2\pi / \omega_e$ = 23 hours 56 minutes 4.1 seconds; the orbit altitude is 35786 km, or semi-major axis a = 42164km, where $\omega_e$ is for the Earth's rotation angular velocity.

However, due to the Earth's non-spherical gravity, moon and sun gravitational perturbation and solar light pressure, satellite orbit elements (including a, e, i) are constantly changing and deviating from the fixed location of the satellite.

### 2.2 The relationship between deviation of satellite orbital elements and drift

Semi-major axis deviation causes long-term east-west direction drift, At t time, sub-satellite point longitude λ is:

$$\lambda = \arctan(\cos i \tan u) + \Omega - \omega_e(t - t_0) - s_{g0} \tag{2.1}$$

*Where,* $s_{g0}$ *is GMT at* $t_0$ *time; u is satellite longitude argument,* $u = \omega + f$ , namely the angle between satellite and ascending node; is perigee argument, namely the angle between orbit ascending node and perigee; $\omega$ is true anomaly, that is, angle between satellite and perigee; $\Omega$ is RAAN, namely angle between ascending node and the equinox.
When $i = 0$, $e = 0$, we have $f = M$, at t time, sub-satellite point longitude is:

$$\lambda = \omega + M + \Omega - \omega_e(t - t_0) - s_{g0} = \omega + \mathrm{M}_0 + n_e(t - t_0) + \Omega - \omega_e(t - t_0) - s_{g0} \quad (2.2)$$

In initial time $t_0$, satellite longitude $\lambda_0$ is

$$\lambda_0 = \omega + \mathrm{M}_0 + \Omega - s_{g0} \quad (2.3)$$

And then can been simplified as

$$\lambda = n_e(t - t_0) - \omega_e(t - t_0) + \lambda_0 \quad (2.4)$$

When semi-major axis $a$ has no error, we have

$$n = \omega_e \quad (2.5)$$

$$\lambda \equiv \lambda_0 \quad (2.6)$$

And when semi-major axis has error, there is

$$\Delta n = -\frac{3}{2} n_e \frac{\Delta a}{a} \quad (2.7)$$

Therefore, when a has deviation $\Delta a$, satellite drift in east-west direction is

$$\dot{\lambda} = -0.012806 \cdot \Delta a \ (^\circ/day) \quad (2.8)$$

When $\Delta a > 0$, $\dot{\lambda} < 0$, satellite drifts to west; when $\Delta a < 0$, $\dot{\lambda} > 0$ satellite drifts to east.

### 2.3 Oscillation in east-west direction caused by eccentricity deviation

The instantaneous sub-satellite point longitude can be denoted as:

$$\lambda = \lambda_0 + 2e \cdot \sin M \quad (2.9)$$

where $\lambda_0$ is central point longitude, $M$ is mean anomaly; when $e$≠0, satellite swings periodically around the center of $\lambda_0$, the largest Oscillation argument is 2$e$(rad), Oscillation cycle is equal to satellite orbit period.

### 2.4 Oscillation in north-south and east-west direction caused by Orbit inclination deviation

When $\Delta a = 0, e = 0$, if satellite orbital plane deviates from the earth equator, namely there is deviation $\Delta i$.

Satellite geocentric longitude is

$$\sin\phi = \sin\Delta i \sin u \tag{2.10}$$

While is $\Delta i$ *is* small, then

$$\phi = \Delta i \sin u \tag{2.11}$$

The former equation indicates that when $\Delta i$≠0, satellite swings in latitude direction with the period of one day, and the argument in south-north direction is $i$. $u$ is latitudinal argument, with the range from 0° to 360° in one day. When inclination is not zero, satellite will has oscillation both in south-north direction and east-west direction, and now sub-satellite point longitude is

$$\lambda = \arctan(\tan u) - \frac{i^2}{4}\sin 2u + \Omega - \omega_e(t - t_0) - s_{g0} = \lambda_0 - \frac{i^2}{4}\sin 2u \tag{2.12}$$

The former formulation denotes: when the inclination is not zero, satellite has oscillation with a half-day period in longitude direction, and argument of oscillation is $i^2/4$, the track of sub-satellite point has "8" shape which is shown in figure 2.1.

Fig. 2.1 Diagram of sub-satellite point "8" shape due to the change of GEO satellite orbital inclination

Due to various perturbations, GEO satellite position relative to Earth is not absolutely static, in the actual project implementation; it needs to continually and periodically modify its orbit[34, 35], restricting its motion in a small area around its specified location. Figure 2.2 shows that of Canada Anik-C3 satellite still has typical latitude and longitude changes after position maintenance[7].

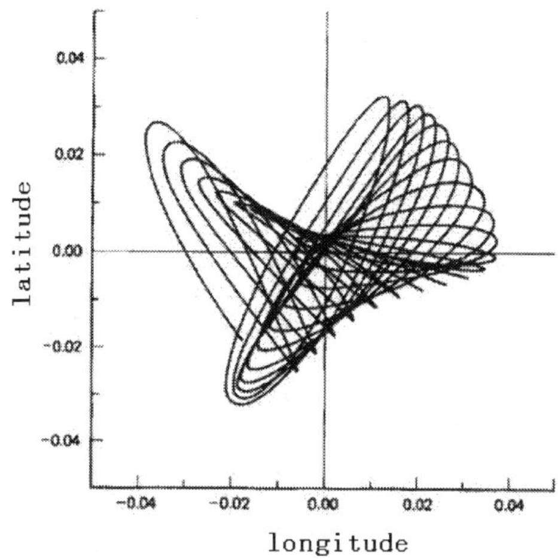


Fig. 2.2 Typical track of GEO satellite's sub-satellite point

### 2.5 GEO satellite orbit perturbations

GEO satellites are subject to various perturbations, including: additional gravitational force caused by non-spherical shape and the uneven quality of the earth, the sun and the moon's gravity and solar radiation pressure. Under the perturbations, satellite orbit no longer follows the two-body orbital motion law, the period, eccentricity, RAAN and inclination are constantly changing, and thereby causing satellite orbit drift. The details of these perturbations are as follows:

1) Earth's non-spherical gravitational perturbation

   Earth's non-spherical gravity will mainly cause changes in sub-satellite point longitude; especially Earth's equatorial ellipticity has a long-term accelerating effect on sub-satellite point longitude. The difference between semi-major axis and semi-minor axis is about 68 meters, the short axis pointed at west longitude 75 degrees and west longitude 105 degrees, the long axis point and short point axis has a 90 degrees in longitude direction. As in the geostationary orbit, semi-minor axis point is stable and the semi-major axis point is not, so the equatorial ellipticity will bring about perturbation which will cause the satellite drift back and forth near the center of semi-minor axis point in east-west direction.

Considering the first 4-order harmonic field perturbation primarily, the average longitude perturbing acceleration can be described as follows[39]:

$$\ddot{\bar{\lambda}} = 0.00168\sin 2(\lambda-\lambda_{22}) - 7.79\times10^{-5}\sin(\lambda-\lambda_{31}) + 2.34\times10^{-4}\sin 3(\lambda-\lambda_{33}) - 8.95\times10^{-6}\sin 2(\lambda-\lambda_{42}) + 1.14\times10^{-5}\sin 4(\lambda-\lambda_{44})\,°/d^2 \tag{2.13}$$

Where $\lambda_{lm}$ is defined as phase longitude[40], data of $\lambda_{lm}$ is presented in Table 2.1

Table 2.1 Data table[51] for $\lambda_{lm}$

| $\lambda_{22}$ | $\lambda_{31}$ | $\lambda_{33}$ | $\lambda_{42}$ | $\lambda_{44}$ |
|---|---|---|---|---|
| -14.91° | 7.0° | 21.1° | 31.2° | 30.65° |

In the east longitude 0° ~ 180°, the average longitude perturbing acceleration caused by the Earth's gravitational perturbation is shown in Figure 2.3. As can be seen from the diagram, average longitude drift acceleration reaches the maximum value of about -0.00197° / day$^2$ near east longitude 118° and equals to zero in the vicinity of east longitude 75° and 165°.

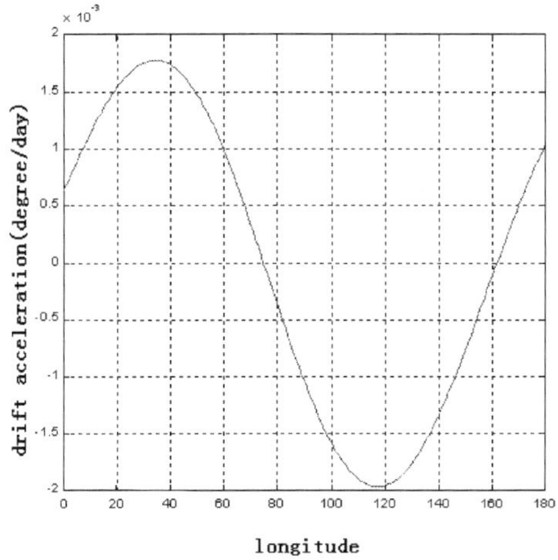


Fig. 2.3 Mean longitude perturbing acceleration caused by the Earth's gravitational perturbation

2) The sun light pressure perturbation
   Sun light pressure perturbation mainly affects eccentricity of geostationary satellites orbit. Eccentricity vector components is defined as:

$$e_x = e\cos(\Omega+\omega) \tag{2.14}$$

$$e_y = e\sin(\Omega+\omega) \tag{2.15}$$

where Ω is RAAN, ω is perigee argument.

The trajectory of eccentricity vector Endpoint forms an ellipse with one-year cycle, its semi-major axis and semi-minor axis are as follows:

$$a_e = 0.0114\frac{A}{m} \tag{2.16}$$

$$b_e = 0.0105\frac{A}{m} \tag{2.17}$$

Where A is surface area of satellite, m is satellite quality. The eccentricity vector caused by visible sun light pressure is mainly depends on the ratio of surface area to mass. Figure 2.4 shows the changes of eccentricity vector curve caused by various factors in one year. The value of eccentricity equals the distance between the point in the figure and the origin, the size of large circular curve mainly caused by the pressure of sunlight, large wavy lines caused by the lunar gravity, and small zigzag line is derived from several factors such as the Earth's gravity, solar gravity, and lunar gravity.

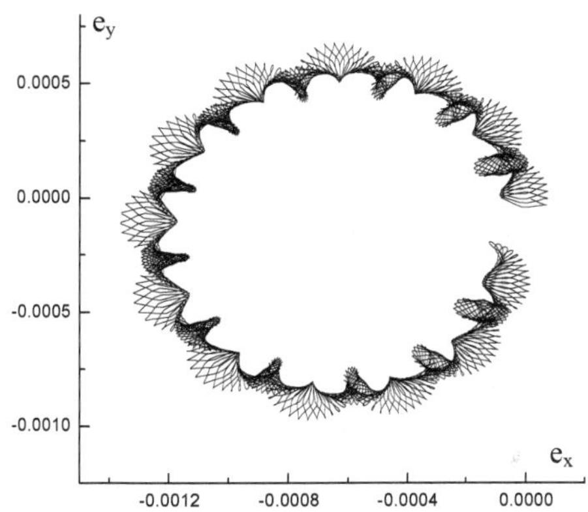


Fig. 2.4 Various perturbations' affects on eccentricity in one year

3) Sun and Moon gravitational perturbation
Sun and Moon gravitational perturbation is most important factor causing the changes of GEO orbit inclination. Orbital inclination vector (or orbital polar) is defined as the unit vector for the normal direction of orbital plane. Under the combined effect of gravity field $J_2$ items of the sun, the lunar and the earth (the earth with harmonic coefficients), the normal direction has gradually deviated from the Arctic. Orbital normal direction generally rotates around a certain direction which is in the Arctic pole plane and ecliptic pole plane, between the ecliptic pole and the Arctic pole, and has an angle of about 7.5° from Arctic pole. With a 52 years cycle, orbital inclination reached its maximum value of about 15° in about 26 years and back to zero 52 years later which means normal direction meets Arctic Pole again.

As the lunar orbit has an angle between 23.5° ± 5.14° relative to equatorial plane, the drift of satellite's inclination is changing every year, the specific value of which depends on the 18.6-year -cycle average value of lunar orbit inclination. Drift rate ranges from 0.7°/ year to 0.95° / years and its direction varies in different years as the lunar orbit pole varies. The most maximum value of drift occurred in 2006, the lunar orbit inclination reached maximum value at the same time.

Figure 2.5 is two-dimensional inclination vector diagram for the satellite, the moon and the sun. As can be seen from the figure, the angle between terrestrial pole and the ecliptic pole is 23.5°, the angle between the lunar orbital pole and ecliptic pole was 5.14°. Lunar orbital plane is not fixed, its orbital pole regress around ecliptic pole with a period of 18.6 years. If we don't maintain satellite position in north-south direction, the satellite's orbit rotates around the pole which has a 7.5° angle with terrestrial pole and a period of about 52 years. If the initial orbit inclination satellite is 0°, the orbital inclination will reach the maximum value of about 15° and then gradually decreasing with year.

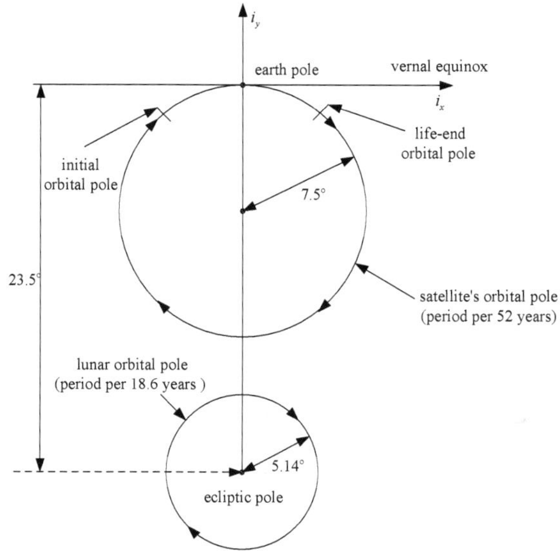


Fig. 2.5 Two dimensional diagram of GEO satellite orbit, moon's path and ecliptic

The effects on inclination vector caused by solar gravity, lunar gravity and the Earth's non-spherical gravitational perturbations vector can been defined as

$$i_x = \sin i \cos\Omega \tag{2.18}$$

$$i_y = \sin i \sin\Omega \tag{2.19}$$

Lagrangian perturbation equation defined with i is as follows[40]:

$$\frac{di_x}{dt} = -n_e(5.21 i_y - 0.099 \sin \Omega_m) \times 10^{-5}\ rad/s$$
$$\frac{di_y}{dt} = n_e(-4.93\ i_x + 0.074 \cos \Omega_m + 0.648) \times 10^{-5}\ rad/s \tag{2.20}$$

Where $n_e$ is the average angular velocity of the earth's rotation, namely:

$$n_e = \sqrt{\frac{G m_e}{r_e^3}} \tag{2.21}$$

Here $\Omega_m$ is ecliptic longitude of ascending node of lunar orbit. Based on AGI's STK, the following Table 2.2 and Table 2.3 gives the changes of i and Ω for the perturbed geostationary orbit in the situation of different RAAN, with the initial inclination 0.1° and 0.5°.

Table 2.2 Changes of orbit in varied RAAN (Ω) with initial inclination 0.1°

| Ω | | 3 monthes | 6 monthes | 1 years | 2 years | 3 years | 4 years |
|---|---|---|---|---|---|---|---|
| 0° | *i* | 0.266 | 0.58 | 1.04 | 1.89 | 2.82 | 3.79 |
| | *Ω* | 80.45 | 82.35 | 84.6 | 82.86 | 79.4 | 75.1 |
| 45 ° | *i* | 0.334 | 0.65 | 1.02 | 1.97 | 2.91 | 3.86 |
| | *Ω* | 87.07 | 85.36 | 86.3 | 83.54 | 79.6 | 75.26 |
| 90 ° | *i* | 0.368 | 0.678 | 1.05 | 2.01 | 2.96 | 3.90 |
| | *Ω* | 98.242 | 91.389 | 90.05 | 85.5 | 80.86 | 76.2 |
| 135 ° | *i* | 0.359 | 0.66 | 1.03 | 1.99 | 2.93 | 3.89 |
| | *Ω* | 110.15 | 97.73 | 94.12 | 87.60 | 82.35 | 77.36 |
| 180 ° | *i* | 0.31 | 0.60 | 0.97 | 1.93 | 2.89 | 3.83 |
| | *Ω* | 120.14 | 101.82 | 96.58 | 88.88 | 83.12 | 78.03 |
| 225 ° | *i* | 0.24 | 0.52 | 0.90 | 1.86 | 2.82 | 3.75 |
| | *Ω* | 123.24 | 100.88 | 95.29 | 88.46 | 82.92 | 77.86 |
| 270 ° | *i* | 0.18 | 0.48 | 0.85 | 1.81 | 2.39 | 3.71 |
| | *Ω* | 109.77 | 93.61 | 91.72 | 86.44 | 83.56 | 76.93 |
| 315 ° | *i* | 0.19 | 0.50 | 0.88 | 1.83 | 2.78 | 3.72 |
| | *Ω* | 86.25 | 85.15 | 86.76 | 84.13 | 80.08 | 75.77 |

Table 2.3 Changes of orbit in varied RAAN ($\Omega$) with initial inclination 0.5°

| Ω | | 3 months | 6 months | 1 years | 2 years | 3 years | 4 years |
|---|---|---|---|---|---|---|---|
| 0° | $i$ | 0.510 | 0.656 | 1.020 | 1.900 | 2.816 | 3.736 |
| | $\Omega$ | 29.48 | 43.09 | 61.70 | 70.81 | 71.149 | 69.070 |
| 45° | $i$ | 0.682 | 0.882 | 1.321 | 2.249 | 3.176 | 4.101 |
| | $\Omega$ | 63.19 | 66.41 | 73.314 | 75.52 | 73.618 | 70.536 |
| 90° | $i$ | 0.766 | 0.975 | 1.449 | 2.408 | 3.363 | 4.294 |
| | $\Omega$ | 93.168 | 89.463 | 88.211 | 83.986 | 79.487 | 74.990 |
| 135° | $i$ | 0.745 | 0.918 | 1.380 | 2.333 | 3.286 | 4.231 |
| | $\Omega$ | 122.480 | 112.562 | 103.53 | 93.111 | 86.022 | 80.131 |
| 180° | $i$ | 0.623 | 0.723 | 1.129 | 2.048 | 2.987 | 3.938 |
| | $\Omega$ | 153.281 | 135.878 | 116.831 | 99.877 | 90.641 | 83.681 |
| 225° | $i$ | 0.427 | 0.419 | 0.758 | 1.665 | 2.610 | 3.560 |
| | $\Omega$ | 190.385 | 160.25 | 122.407 | 100.05 | 90.39 | 83.450 |
| 270° | $i$ | 0.245 | 0.055 | 0.460 | 1.420 | 2.369 | 3.317 |
| | $\Omega$ | 253.177 | 199.55 | 99.09 | 89.273 | 83.618 | 78.605 |
| 315° | $i$ | 0.304 | 0.328 | 0.635 | 1.534 | 2.465 | 3.397 |
| | $\Omega$ | 341.197 | 18.153 | 62.185 | 74.94 | 74.757 | 72.170 |

As can be seen from the above results, for the geostationary orbit whose initial inclination is in the vicinity of 0°, the general trend is that its orbital inclination is to increase. However, in the different case of RAAN, there are still some slight differences in the changes of orbital inclination. Specifically, for $\Omega = 90°$, orbital inclination will directly increase, and for $\Omega = 270°$, the inclination will decrease first and then increase. $\Omega$ is to change in the direction of 90 ° with a 52-year period. to the specific, $\Omega$ will be 90° from the first beginning 0 ° after 26 years and 270° after 52 years (in 52 years, the changing law of $\Omega$ is 90° -> 0°, 360° -> 270°). This variation does not change with sub-satellite point longitude.

### 2.6 Orbit control scheme for GEO satellite

To maintain GEO satellites in the fixed position with the required accuracy, it is necessary to regulate the position of satellite's orbit periodically based on the characteristics of the orbit's change. In order to keep GEO satellites operating in the designated area, it is required to increase the velocity by jet to overcome various perturbation imposed on satellite. For GEO satellites, the velocity increment in north-south direction resulting from north-south position maintenance is to overcome satellite inclination's changes due to the sun and moon gravity, and while the velocity increment in north-south direction resulting from east-west position maintenance is to overcome the changes of eccentricity caused by solar light pressure and satellite's drift in east-west direction caused by harmonic perturbation of the Earth's gravitational field. East-west position maintenance for the GEO satellite is to change inclination by control of orbital normal direction in the vicinity of RAAN. Taking a longer time interval for East-west position maintenance into account, RAAN should be adjusted to

270 ° at the same time of controlling orbital inclination. Velocity increment for north-south position maintenance has a relative large value of about 42 m/s ~51m/s every year.
East-west position maintenance for the GEO satellite is mainly to overcome the long-term accelerated drift of average longitude caused by the tangential perturbation resulted from the Earth's non-spherical gravity. Through the control method with limit cycle[40], satellite can be restricted to do the long term limit cycle oscillatory movement near the fixed-point. Control mode of drift limit cycle will be introduced in the following via typical cycle of east-west amendments (as shown in Figure 2.6).

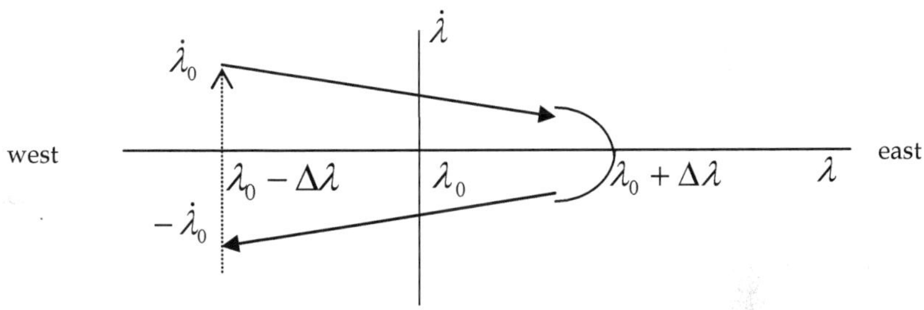


Fig. 2.6 Limit cycle of east-west position maintenance

Supposed $\lambda_0$ is sub-satellite point, $\Delta\lambda$ is the allowed longitude deviation, $\ddot{\lambda}$ is for the longitude drift acceleration which is a function of longitude (in China's space, $\ddot{\lambda} < 0$). If Satellite's longitude is $\lambda_0$ at time $t_0$, Satellite's longitude will be as followings at time t:

$$\lambda = \lambda_0 + \dot{\lambda}_0 t + \frac{1}{2}\ddot{\lambda}t^2 \quad (2.22)$$

where $\dot{\lambda}_0$ is initial drift velocity. Due to $\ddot{\lambda}$, satellite will has drift velocity. Specifically, satellite will have westward drift when $\dot{\lambda}_0 < 0$ and eastward drift when $\dot{\lambda}_0 > 0$. as in the vicinity of china area $\ddot{\lambda} < 0$, control is usually implemented in the western boundary of the limit cycle to give the satellites a long running time within the fixed-point accuracy (such as ± 0.1° the east-west direction), which will equip the satellite an eastward drift velocity . In the role of $\ddot{\lambda}$, the velocity of satellite will gradually slow down during the eastward drift until satellite drifts to the east boundary of limit cycle. At that time, the satellite will have westward drift velocity and then drift to west continuously by the effect of $\ddot{\lambda}$. When the satellite reaches the western boundary of limit cycle $\lambda_0 - \Delta\lambda$, it can be equipped with drift velocity $\dot{\lambda}_0$ by the control of executive organization and thereby drifts to the east again. A drift cycle has been formed after several repetition of former process.

Velocity increment for east-west position maintenance each year is generally less than 2 m/s, especially in the equilibrium point $\ddot{\lambda}$ will be close to zero and here the required velocity increment is also approaching zero.

### 2.7 issue of GEO orbital phase

For geostationary orbit with a certain inclination, the satellite phase we actually call refers to Greenwich sidereal time when satellite passes the RAAN. As can been seen from $\lambda = \omega + M + \Omega - s_g$, if $\omega + M = 0$, while satellite passing the RAAN, there will be, $\varphi = s_{g(M+\omega=0)} = \Omega - \lambda$. Therefore satellite phase can been denoted as $\Omega - \lambda$, where $\lambda$ is sub-satellite point longitude and $\Omega$ is RAAN. When there is no north-south position maintenance for GEO satellites, they generally have the same law for the changes of orbital inclination and RAAN. Therefore their phases more depend on the differences of fixed position. For the two satellites with 10° difference in fixed position, their phase difference will be about 10°. The change of two satellites' phase-difference can be seen from $\phi = \Omega - \lambda$ and realized by altering their $\Omega$ or $\lambda$.

## 3. Technical scheme on altering control mode of the satellite's attitude

In precious section, we have explored the characteristics of IGSO satellite orbit and its running law. In this section, technical scheme on altering control mode of satellite's attitude in end-of-life will be introduced.

### 3.1 Technical schemes on regulating GEO satellite in end-of-life into IGSO satellite

Known from the previous section, due to various perturbations such as gravity of sun and moon, GEO satellite's inclination will change. Satellite will gradually deviate from the original position and drift in both east-west and north-south direction which is more apparent. There is 0.269° increase in IGSO satellite's inclination each year caused by solar gravity and 0.478°~0.674° increase caused by lunar gravity. The inclination drifts 0.75°~0.95° in north-south direction each year under joint effects of sun and moon. The inclination changes periodically with period of 18.6 years and its largest range ability is about 14.67°. After implementing the new control model, GEO satellite will become a new kind of satellite, namely SIGSO whose inclination is between 1° and 10°.

So if there is just east-west position maintenance and no north-south position maintenance which means satellite can drift in latitudinal direction freely, the retired GEO satellite can become half-controlled SIGSO satellite. As the fuel that north-south position maintenance consume is 9 times as much as the fuel that east-west position maintenance consume, the using time of remained fuel in retired satellite can be prolonged about 10 times with the new orbit regulation model. As mentioned earlier, the satellite's electronic equipment such as solar cells, signal transponders has a relative longer service life, SIGSO satellites can service for many years in transmitting navigation and positioning satellite system. Meanwhile, due to retire GEO communications satellites are no longer bear the original communication and broadcasting tasks, they can be regulated to other required and available track in a larger range in accordance with the requirements about formation of navigation constellation. The satellite management departments regulating the obits in this way has some technical difficulties, but much easier than developing and launching new satellites, and the satellite transponder lease costs is much less than the costs of developing and launching dedicated navigation satellites. Therefore the formations of navigation constellation composed of the SIGSO satellites which are regulated from GEO satellites make retired GEO communications satellites playing a role in improving the accuracy of navigation. The usage of the retired GEO satellites can not only shorten the time and saves a lot of money to organizing a navigation system, but also delays the time of satellites into space junk which will pollute the space environment. CAPS verification system using SIGSO satellites[1, 9, 10, 13] which is deviated from regulating retired satellites has proven effective.

Of course, utilizing this kind of satellites need regulate the SIGSO satellites to required orbits and meet the provisions of the ITU about anti-interference. It also has extremely high demand of orbits determination since the orbits of satellites are the resources for navigation. CAPS project team using two-way time transfer technology and code-related dispreading demodulation makes satellite ranging accuracy better than 1 cm and orbit observation residuals less than 9 cm[3, 17] which can meet the requirement of transmitting satellite navigation and positioning system.

### 3.2 Antenna's direction and its Isotropic Radiated Power

After retired satellites regulated to the new orbit, the ground beacons should be re-established or technology of geocentric point mode should be adopted. To meet the navigation's requirement on the effective coverage area and power distribution, effective coverage area of the satellite's antenna and EIRP allocation should be altered by adjusting antenna's direction.

### 3.3 The method of changing sub-satellite point longitude $\lambda$

For more than two satellites, their intervals in longitude direction must be drawn away. Method of the operation is not complicated, as long as one of the satellites has been drifted away, the purpose of changing the fixed-point longitude could be achieved, since changes in phase is consistent with changes in longitude. About 5~6m/s velocity increment is enough for drifting satellite and drift speed is 1°/ day. giving the scope of China's territory, drift of the longitude range can be limited to 50° E ~ 170° E. when satellite is in the new position, it is also necessary to consider electromagnetic compatibility of other satellites nearby.

### 3.4 The method of changing RAAN

As described above, each satellite's orbital inclination $i$ has the same change with its RAAN when there is no north-south position maintenance. If there is some time difference in suspending the two satellite's north-south position maintenance, a certain Right Ascension difference will be retained during process of change. As is shown in Table 2.2, when the first satellite has been no north-south position maintenance for four years and the second satellite 2 years, there will be 10° difference in Right Ascension of the two satellites.

Changing the satellite's RAAN usually consumes a large amount of fuel. Taking two satellite with inclination 3° and RAAN 75° for example, velocity increment of 55m/s is required to bring about 20° difference in RAAN and velocity increment of 55m/s for 60° difference in RAAN, which is unrealistic for the huge cost. The initial inclination is about 0°

in the beginning of suspending north-south position maintenance, and therefore RAAN is meaningless and thereby setting Right Ascension difference is relatively easy. To set the Right Ascension of two satellites arbitrarily, the only thing need to do is that one of satellite has been established a non-zero angle at a certain time. For example, when satellite passes Right Ascension 90°, RAAN of satellite can be changed to 90° with north direction position maintenance. But the Right Ascension we have set will decrease generally during the following perturbed evolution, the extent of decrease depend on the inclination when setting the RAAN.

For example, RAAN of the two satellites were set to 0° and 180° respectively and inclination is controlled to 0.1°, velocity increment is about 5.5m/s and thereby consuming less fuel. Table 2.2 shows the subsequent changes, as can be seen from Table 2.2, the RAAN of the satellite with the initial RAAN 0° will change to 84.6° in a year and 79.4° in three years, and the RAAN of the satellite with the initial RAAN 180° will change to 96.6° in a year and 78° in three years. The right ascension difference will decrease to 12° in one year from the beginning 180° and to 1.4° three years later. If the inclination of the two satellites is set to 0.5°, while the RAAN is still set to 0° and 180° respectively, the reducing trend of right ascension difference will slow down.

The subsequent evolution of the situation is shown in Table 2.3, the RAAN of the satellite with the initial RAAN 0° will change to 61.7° in a year and 71.7° in three years, and the RAAN of the satellite with the initial RAAN 180° will change to 116.8° in a year and 90.6° in three years. The right ascension difference will decrease to 55° in one year from the beginning 180° and to 15.5° three years later. However, if the inclination has been set too larger, the consumption of propellant will have a substantial increase which is equal to the amount of half-year's consuming fuel and velocity increment will have to be about 27m/s. If we further increase the initial inclination, the reduction of the two satellite's RAAN difference will further slow down but the consumption of propellant will have a substantial increase, which can be seen from the table 2.3, 2.5, 2.5.

### 3.5 Conclusions

Based on the above analysis, we can obtain the following conclusion:

1. GEO satellite's orbital inclination and RAAN have a 52-year cycle of change due to solar and lunar perturbations, right ascensions reducing from about 90° to 0°, then to 270°, and at last to 270° directly. Therefore, GEO satellites without north-south position maintenance have same changes in their orbital inclination and RAAN.
2. There are two ways to change the time when the two retired satellites pass the equator: one is changing sub-satellite point longitude difference $\Delta\lambda$ between the two satellites, the other is to change the satellite's RAAN difference $\Delta\Omega$ .
3. Altering the time when the two retired satellites pass the equator by changing the sub-satellite point longitude can improve navigation performance via the improvements of satellites constellation;
4. It is easier to change the sub-satellite point longitude with control volume 5 ~ 6m/s; consider the scope of China's territory, drift in the longitude range can be limited to 50° E ~ 170° E. However, we should note that the re-coordination of the orbit is required;
5. Altering the orbital phase by the change of the RAAN can improve navigation performance actually via increasing geocentric angles of two satellites, but in case of small

inclination, geometrical distribution can't be significantly improved because the increasing amount of the geocentric angle is not more than the sum of two satellites' inclination;
6. The method that we alter the orbital phase by the change of the RAAN can be optimized in different initial conditions. But even so, the control volume has the same amount of the normal north-south position maintenance and the initial set of orbital phase differences will gradually reduce and have the same value eventually.

## 4. Contribution of SIGSO satellite for transponder satellite navigation constellation

In this section, through simulation analysis, we will discuss the SIGSO communications satellites in the transponder satellite navigation system how to affect the navigation and positioning constellation structure and improve the navigation and positioning solutions.

### 4.1 Requirements of positioning on the constellation structure

Generally, position accuracy in satellites navigation system can be represented as:

$$\sigma_x = \text{GDOP}\ \sigma_\rho \tag{4.1}$$

Where $\sigma_x = \sigma_{(x,y,z,t)}$ is positioning and time service error, $\sigma_\rho$ is pseudo-range measurement error. GDOP (Geometric Dilution of Precision) is an important index to describe the quality of the space section of a positioning system, and its value determines position accuracy[42, 49, 53, 56].

An excellent positioning system requests PDOP value to be small in effective covering region in any time. The GPS system is formed by more than thirty satellites and corresponding PDOP value in covering region is smaller than 4 or less [26].

When GEO communications satellites are used in constellation of regional navigation system, the system needs three communications satellites with the large orbital inclination (IGSO) too. Thus the coverage of the system will be of about one-third of the earth. Three such regional systems can cover the whole world.

In 2003, the problem existing in CAPS test system is that we have not appropriate IGSO satellites to apply in the system. Therefore, we propose to use GEO satellites which are going to be retired for solving the problem. That is, using new mode to control thus GEO satellites to become SIGSO satellites for navigation and positioning. However, the improvement of constellation structure when SIGSO satellites are used should be studied in depth. In this section, by calculating the distribution and variation of constellation PDOP within a certain coverage we mainly analyze quality of the constellation.

### 4.2 Improvement to PDOP value in satellites navigation system by SIGSO satellites

Based on the theoretical analysis of the constellation and PDOP changing with time, we developed a set of application software of calculation and analysis on PDOP change. For the navigation and positioning constellation includes some satellites with different orbital parameters, the software can calculate the PDOP change in any place on the ground at any time, calculate the PDOP change in the designated area within a certain period, and can produce various charts of PDOP by the time and distribution of regional areas for simulation and analysis constellation quality.

**(1) Analysis of the impact of SIGSO satellites for navigation constellation PDOP**
We will analyze the function of SIGSO satellite through studies of distribution and variation of PDOP values for various constellation structures. In simulation, the satellite number of constellation is designed to be six, including the GEO satellites and SIGSO satellites. The number of SIGSO satellite changes from one to six, and number of GEO satellite changes from five to zero correspondingly. The longitude of the satellites is 57°, 87.5°, 103°, 122°, 142° and 163° respectively, and the least of the satellite's elevation for users is 5°. The orbit inclination of SIGSO satellite is 3° and 5°. For each 2° × 2° grid point to calculate the 10-minute intervals PDOP values, and plotted PDOP distribution. If the PDOP values of a grid point is always less than 1000 in the required effective period (such as 95%, 98% or 100% of the day), the points will be called effective grid points and they will compose the effective coverage area. Effective area of the PDOP calculated respectively the minimum, maximum and average. The figures and data will be used in analysis of the function of SIGSO satellites.

**(a) Constellation including SIGSO and GEO satellites can achieve three-dimensional navigation.**
If the positioning system only uses GEO communication satellites without other assistant methods, it cannot do the 3-D positioning since these satellites are all in the earth's equator plane. The calculations show that the DOP value of this constellation is too large for navigation and positioning. For example, when using only the six GEO satellite, the PDOP value of Beijing area approximately is one million or more. Supposing an retired GEO satellite (such as at 110.5° E) has been controlled to be a SIGSO satellite with $i$ = 3°, the effective time is 95% of one day, the calculated PDOP distribution is shown in Figure 4.1. In the scope of about 60° E—160° E and about 65° S—65° N, PDOP minimum is 57.1, and average is 76.2, PDOP <60 in the center of the covered area and PDOP <70 in a considerable area. Although the effective time has not yet reached whole day, and the value of PDOP is also larger, the significant change is the 3-D positioning equation could be solved and fundamentally change the situation that the equation cannot be solved when GEO satellites are used only.

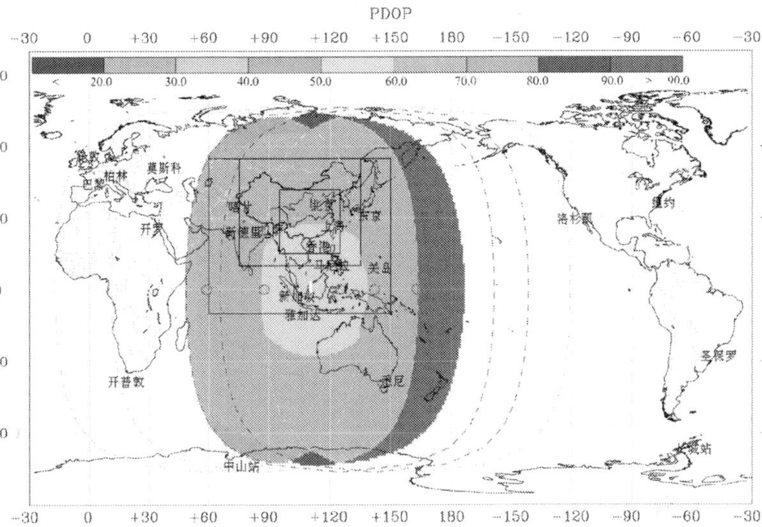


Fig. 4.1 Distribution of PDOP equal-value-area of the constellation of one IGSO ($i$=35°,$e$=0) and five GEO satellites.

Notes: Different color stands for the distribution grade levels of the PDOP daily mean and the same color area is called the equal-value-area. The circumscription value of the equal-value-area is marked on the top of the figure, such as PDOP<20 in the blue area and on the analogy of this. The red circles marked with 'G' along the equator are the ground projecting of GEO satellite, the corresponding coverage boundary are shown by dotted lines. The '8' shaped ground track of IGSO satellite is plotted with white dash-dot line, on which the red star represents the position of IGSO satellite, while those white points are the positions at every hour. The three key areas marked by the black line, they are Chinese hinterland, most of the territory and surrounding areas, and it is the same below.

**(b) Increase in the number of SIGSO satellite PDOP of constellation can be improved.**
Improvement of the PDOP value of constellation is relatively obvious along with the increasing in the number of SIGSO satellites. Based on above case, if retired GEO satellite at 142° E is also controlled to be a SIGSO satellite with $i$ = 3°, that is the combination of 4 GEO satellites and 2 SIGSO satellites is formed, the PDOP is improved obviously. In the cases of effective time is 98% of one day, the minimum reduces from 57.1 to 36.0 and the average decrease to 47.3 from 76.2 in the entire covered region, and the largish region will reach PDOP <50. Assuming an retired GEO satellite at 87.5° E is controlled to be a SIGSO satellite with $i$=3° and the combination of 3 GEO and 3 SIGSO satellites will be formed and ask the ratio of effective time to reach to 100% of one day, the PDOP distribution is shown in Figure 4.2. PDOP minimum further reduces to 15, average of the entire coverage region reduces to 28.6, the largish region reaches to PDOP <17.
Further assuming the retired GEO satellite at 125° E is controlled to be a SIGSO satellite with $i$=3° and the combination of 2 GEO and 4 SIGSO satellites will formed, the PDOP is further improved (see Figure 4.3). The minimum will reduce to 12.8, average of the entire coverage

region will reduce to 16.6, the largish region will reach to PDOP <14. If the number of controlled SIGSO satellite increases again, for example, all 6 GEO satellites are controlled to SIGSO ones with $i$ = 3°, PDOP minimum will further reduce to 11.9. For comparison, simulated PDOP values of some constellations are listed in Table 4.1 ($i$ stands the orbital inclination, T is the percentage of effective coverage time in a whole day, min, max and mean show minimum, maximum and regional averages of daily average PDOP respectively, H means using atmospheric pressure altimeters).

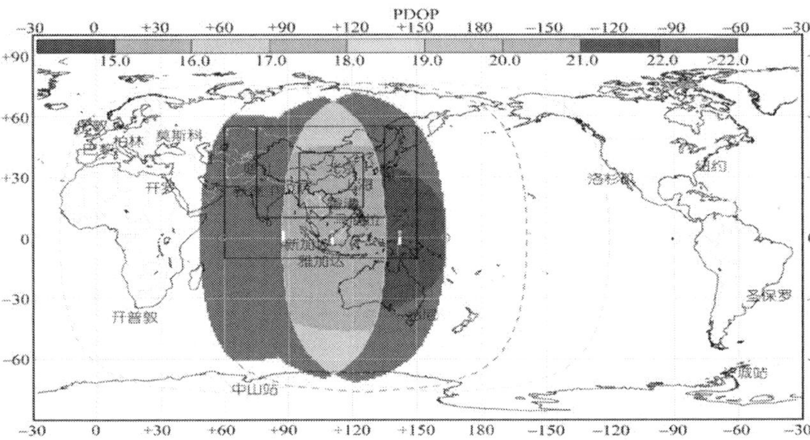


Fig. 4.2 Distribution of PDOP equal-value-area of the constellation of three GEO and three SIGSO satellites with $i$=3°.

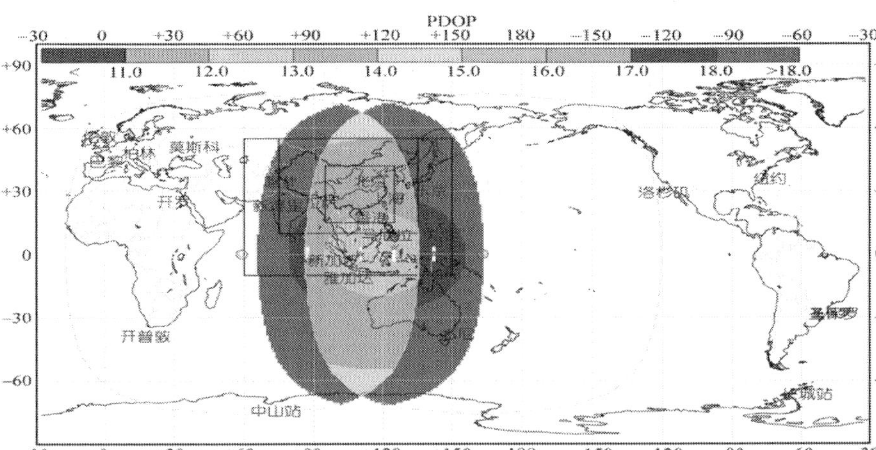


Fig. 4.3 Distribution of PDOP equal-value-area of the constellation of two GEO and four SIGSO satellites with $i$=3°.

It should be noted that PDOP maximum became larger for the two groups of using atmospheric pressure altimeters in the table 4.1. It is the reason that the effective coverage area increases and the PDOP is relatively large in the marginal parts of the effective coverage area. As the parts are smaller, the impact on the average is also small, and it will not adversely affect the largish region we care about.

Table 4.1 PDOP values of various constellations
(The minimum elevation of satellite is 5°)

**Table 4.1 PDOP values of various constellations (min elevation 5 degree)**

| constellition | $i$ | $T$/% | Min | Max | mean | 备注 |
|---|---|---|---|---|---|---|
| 5GEO+1SIGSO | 1° | 95 | 139.0 | 170.4 | 151.8 | |
| | 3° | 95 | 57.1 | 149.1 | 76.2 | |
| 4GEO+2SIGSO | 3° | 98 | 36.0 | 75.1 | 47.3 | |
| 3GEO+3SIGSO | 3° | 100 | 15.0 | 64.0 | 28.6 | |
| | 3°+H | 100 | 2.9 | 108.3 | 12.2 | |
| 2GEO+4SIGSO | 3° | 100 | 12.8 | 33.4 | 16.6 | |
| | 5° | 100 | 7.9 | 23.5 | 10.5 | |
| | 7° | 100 | 5.9 | 18.9 | 8.0 | |
| 1GEO+5SIGSO | 5° | 100 | 12.1 | 61.5 | 18.8 | satellite phase 60° |
| | 5° | 100 | 9.1 | 35.8 | 13.3 | satellite phase 75° |
| 0GEO+6SIGSO | 3° | 100 | 24.0 | 77.2 | 36.6 | |
| | 5° | 100 | 14.7 | 55.9 | 22.3 | |
| | 5°+H | 100 | 3.1 | 106.4 | 12.8 | satellite phase 60° |
| | 7° | 100 | 10.8 | 40.4 | 16.4 | satellite phase 120° |
| | 5° | 100 | 6.6 | 51.7 | 11.8 | |

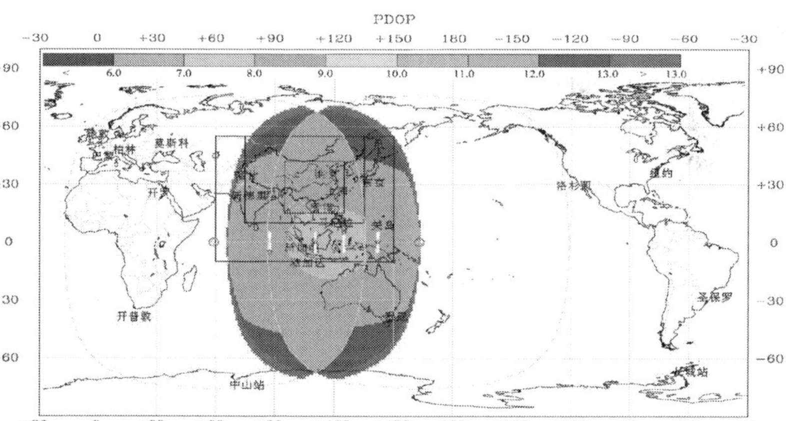


Fig. 4.4 Distribution of PDOP equal-value-area of the constellation of two GEO and four SIGSO satellites with $i$=5°.

**(c) The relatively greater inclination of the SIGSO satellite is, the greater the degree of improvement of PDOP value is.**

Simulation shows that even if only one SIGSO satellite exists and the orbital inclination *i* increases to 3°, the PDOP minimum will reduce to 57.1. Here we use the combination of two GEO and four SIGSO satellites to further display the impact of the orbital inclination on PDOP. Four SIGSO satellites locate at 87.5° E, 110.5° E, 125° E and 142° E and *i* changes from 3° to 5° and 7° for the simulation. The result of *i*=3° has been showed in Figure 4.3. The results of *i*=5° and 7°, respectively, are showed in Figure 4.4 and 4.5. The comparison of results in Table 4.1 and Figures 4.3, 4.4 and 4.5 show that PDOP value is obviously improved along with the increasing of inclination of SIGSO. The minimum PDOP reduces from 12.8 to 7.9 then to 5.9 and the average decreases from 16.6 to 10.5 and then to 8.0. At the same time, when *i* = 5°, 4/5 of Chinese domain achieves PDOP <9. The change curves of PDOP per minute in 24 hours of some representative locations are in Figure 4.6. The results show that the variation of PDOP per minute is relatively stable and the value is relatively small except for the PDOP is relatively large in north-west marginal region of China which cannot observe 163° E satellite. Using SIGSO satellites are from retired GEO satellites to form constellation and then get such improvements of PDOP that may be not imagine before. In Figure 4.5, we can see that 4/5 of Chinese domain achieves PDOP<7 when *i* = 7°. However it needs a longer time that SIGSO is controlled to be the orbital inclination of 7°, and there should be more of remaining fuel, as well as solar battery and the transponder has a long working life.

In the simulation, if three SIGSO satellites are used and their orbit inclination is about 5°, the PDOP average will be about 15.7, PDOP <12 in a majority of the satellite coverage area and a larger area of PDOP <10 appears in the middle (see Figure 4.6). When using four IGSO satellites, PDOP reduced to an average of 11.8, and the area of PDOP <8 appears in the middle (see Figure 4.7). Compared with the combination of the number of IGSO satellites of *i* = 3°, the improvement of PDOP is obvious when the SIGSO satellites are used, of course, the navigation and positioning accuracy will be improved significantly (as shown in Figure 4.4).

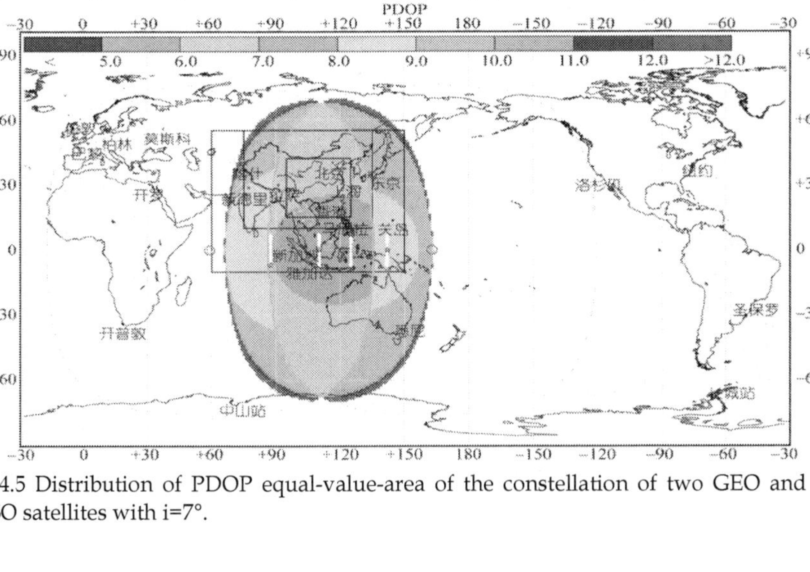


Fig. 4.5 Distribution of PDOP equal-value-area of the constellation of two GEO and four SIGSO satellites with i=7°.

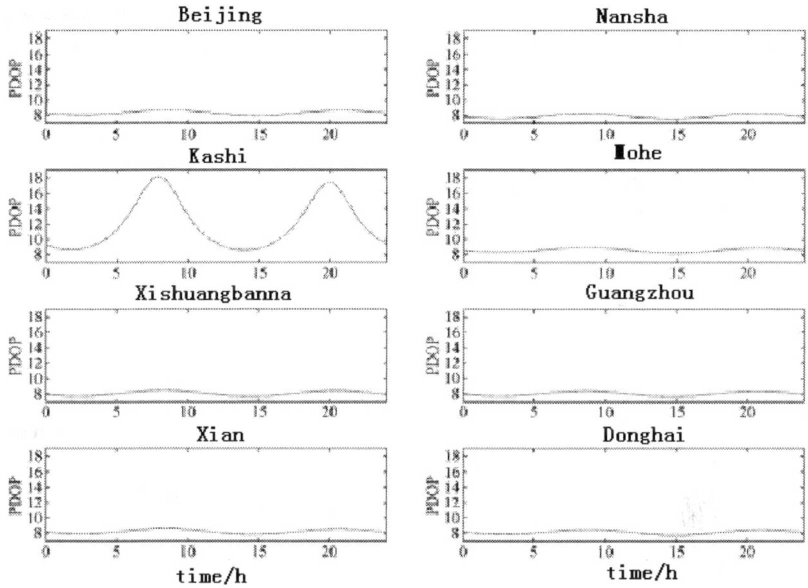


Fig. 4.6 the PDOP value per minute of a constellation with two GEO and four SIGSO (i=5°) satellites in different sites in Chinese.

In order to make better use of contribution of SIGSO satellites for improving the distribution of satellite constellation, it is needful to take into account the phase difference of the SIGSO satellites. These GEO satellites have the same ascending node when they are launched, there are two ways to change their phase. We can adjust the phase directly, of course this way will consume more fuel and it is not advisable since SIGSO satellites have a spot of fuel. Another way is to move SIGSO satellites in the longitude direction in order to increase the interval between satellites and then their phase difference will be increased and the satellite coverage will be increased.

In summary, joining of SIGSO satellites changed the situation that the three-dimensional positioning cannot be made if only use GEO satellites, the three-dimensional positioning can be actualized. More SIGSO satellites and greater orbital inclination, the improvement of PDOP is more obvious. To increase the phase difference between SIGSO satellites, can be used pull SIGSO satellite in longitude direction interval.

**(2) SIGSO satellites for improvement of PDOP value of CAPS validating system**

The coverage of CAPS focuses on Chinese land and coastal areas, considering there should be a certain redundancy for the coverage, the simulation will be extended to Chinese surrounding areas (60° E−150° E, 0° −65° N). The GEO satellites involved in CAPS are at 87.5° E, 110.5° E, 125° E and 142° E, their span in west-east direction is relatively small. We suppose to launch new GEO communication satellites at 59° E and 163° E, or retired GEO satellites are controlled to become SIGSO ones and moved there (here we mainly consider Chinese existing GEO communication satellites and the possible new ones, as well as SIGSO satellites from the retired GEO satellites). For the above-mentioned 6 satellites, they gradually enter the end of the life and become SIGSO satellites in sequence, their orbital inclination will reach to 1°, 3° and 5° etc., and minimum altitude angle of satellite is 5°. Under these assumptions, the simulation will provide PDOP value every 10 minutes, at each 1° × 1° grid (latitude × longitude) in the effective coverage area, as well as the daily average and minimum value of PDOP. For regional navigation and positioning system, we are concerned about the value of PDOP in the region and its distribution, so we draw the value of PDOP distribution in different levels. At the same time, in order to fully understand the details of the PDOP changes and further comparative analysis, we calculate PDOP changes per minute in one day and draw PDOP curve for some selected representative sites within the coverage area. Through the analysis of the change and distribution of PDOP, we can get the following results:

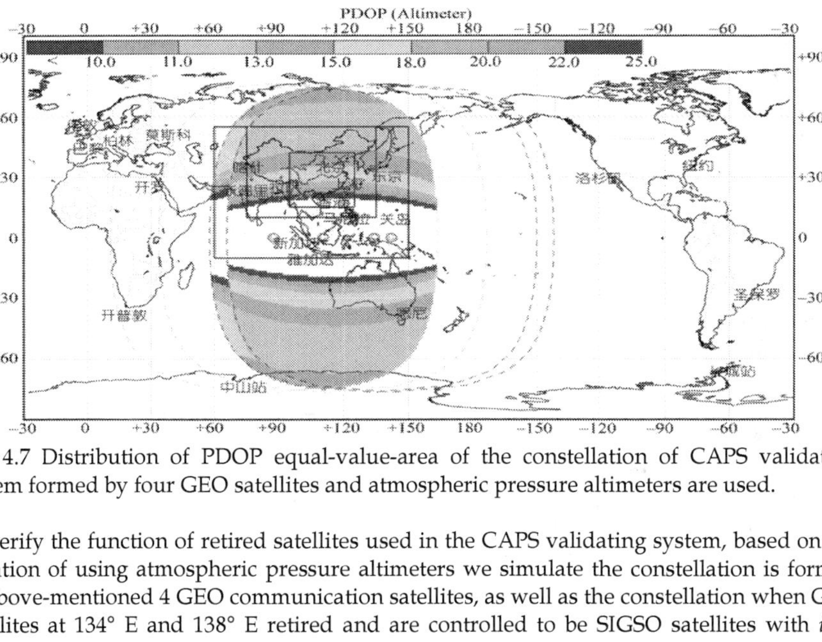


Fig. 4.7 Distribution of PDOP equal-value-area of the constellation of CAPS validating system formed by four GEO satellites and atmospheric pressure altimeters are used.

To verify the function of retired satellites used in the CAPS validating system, based on the situation of using atmospheric pressure altimeters we simulate the constellation is formed by above-mentioned 4 GEO communication satellites, as well as the constellation when GEO satellites at 134° E and 138° E retired and are controlled to be SIGSO satellites with $i$=3° respectively. The results are shown in Figures 4.7 and Figures 4.8. In absence of retired satellites, the average, minimum and maximum of PDOP value is 43.0, 10.9 and 970.8 (in Figure 4.7, the part for PDOP>25 is set to be white, and the same below). When two satellites retired and became SIGSO ones with $i$=3°, these values of PDOP are improved as 19.4, 10.9 and 131.8, respectively (see Figure 4.8). It is obvious that retired satellite is certain valuable for improving PDOP value of CAPS validating system. With their orbital inclination increases, the improvement on PDOP will be relatively more pronounced. When the orbit inclination of the two satellites is about 5°, these PDOP values are 14.7, 10.8 and 124.7 (see Figure 4.9). We could see that, although the minimum of PDOP unchanged with the increase of the orbital inclination, the change of the average is obvious, the region with smaller PDOP expanded. For example, the south boundary of the region of 11 <PDOP ≤13 is at the

north of Beijing at first, and then it extends to the southern of Beijing when $i$ = 3°, then the boundary extends to Hainan island, and the region of PDOP ≤ 11 exists in the eastern part of China when $i$ = 5°. Therefore, using relative more SIGSO satellites are from retired GEO satellites, and the orbital inclination is greater relatively, the improvement for the PDOP of CAPS validating system is relatively more pronounced.

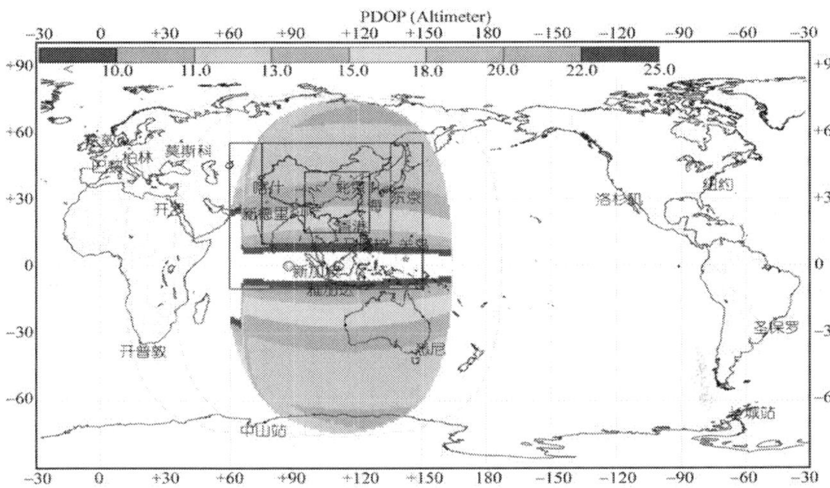


Fig. 4.8 Distribution of PDOP equal-value-area of the constellation of CAPS validating system formed by two GEO satellites and two SIGSO satellites with i=3° and atmospheric pressure altimeters are used.

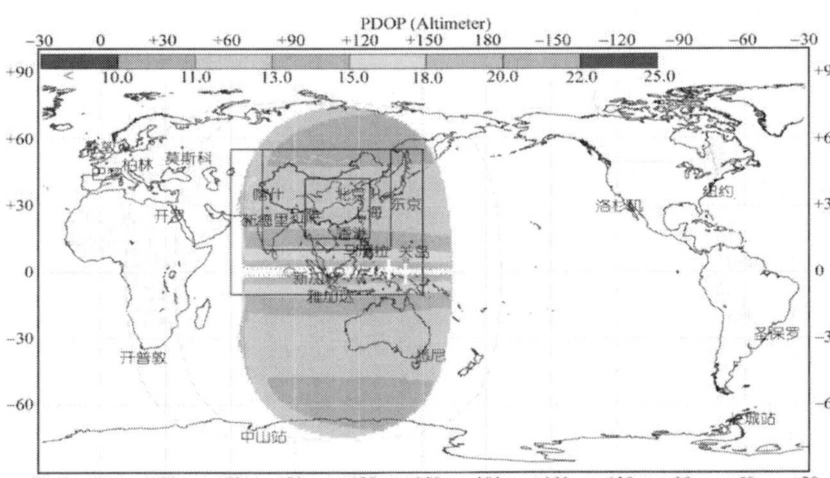


Figure 4.9 Distribution of PDOP equal-value-area of the constellation of CAPS validating system formed by two GEO satellites and two SIGSO satellites with i=5° and atmospheric pressure altimeters are used.

The results show that the retired GEO communication satellite which is controlled to be SIGSO satellite can be used to the transmitting navigation and positioning system, and 3-D navigation and positioning can be actualized. Of course, a certain number of IGSO satellites are necessary to form an ideal constellation for navigation and positioning even if the GEO and SIGSO communication satellites are together used in the constellation or all members of the constellation are SIGSO satellites. The analysis shows that controlled SIGSO satellite from retired GEO satellite can improve PDOP value of the constellation, and its function for improvement of PDOP is better than one of GEO satellite. More SIGSO satellites and greater orbital inclination, the improvement of PDOP is more obvious.

Studies showed that the CAPS needs SIGSO satellites not only lie within the range of 80° E—140° E, but also lie within the ranges of 50° E—80° E and 150° E—180° E, thus navigation accuracy and effective coverage range of the CAPS will be improved significantly. However, Chinese GEO communications satellites launched distributed in 80° E—140° E generally. Through studies, the research team adopted the way to change the position of retired GEO communication satellites and move the satellites into the ranges of 50° E—80° E and 150° E—180° E respectively, in order to form a relatively better navigation constellation. If retired GEO communication satellites do not be used, thus we need launch two GEO communications satellites to the two areas, this must take longer time and spend more fund.

### 4.3 The capability of anti-jamming

As a lot of C-band transponders in SIGSO satellites with 500/700MHz bandwidth to generate three-frequency, multi-frequency and frequency navigation signals for enhancement of corresponding ground receiving terminals can be used for user's navigation and positioning, it can overcome the limitations of using fixed navigation signal in current navigation system. The implementation procedures are to choice different and contiguous navigation frequencies which can be covered by a ground receiving antenna to convert them down to expected navigation base band signal, and then measure and compensate their converting time delay by software to achieve finally the accuracy navigation and positioning[52, 2]. As the fixed navigation frequency signals with a low power and easily interfered are used in common satellite navigation systems, frequency-hopping technology[1, 6, 7] may be applicable to anti-jamming in satellite navigation, therefore, an innovative satellite navigation system can be built up with the functions of multi-frequency and hopping for navigation, they will significantly improve anti-jamming capability and increase more functions in navigation system.

## 5. A low data rate satellite communications technique

A low data rate satellite communication system is being developed in the navigation and positioning system based GEO communications satellites as SIGSO satellites around end of life span. Such a communication system has its especially technical issues and demands, of course, and its technical difficulties.

### 5.1 Why developing this type of satellite communication techniques

By means of changing the attitude control mode, GEO communication satellite around the end of life span will drift into a small angle inclined orbit (SIGSO) satellite. Except 2-3 band transponders for navigation purpose, if remaining 21 to 22 transponders resource can be used well? They are very cheap space resources. In general, human beings demand more increasing communications, so satellite communications techniques tend to improve bandwidth efficiency and data transmission rate with higher cost-effective .Would SIGSO satellite meet these needs?

As SIGSO satellites have "8" shape shift daily movement and continue increasing drift, the communication terminal antenna must point to the satellite. It will be required to be able to control the antenna beam or antenna beam directive should be able to cover a very wide range, in particular, the composition and structure of the terminal installed on the mobile carrier will become more complex.

There are various demands of application, for example, vehicle travel monitoring carrying dangerous goods; the transmission of measured data in meteorological and hydrological monitoring stations in remote mountains, ocean and desert etc.; emergency information reporting in real-time in case of disaster; corrosion monitoring for petroleum pipeline and so on. Particularly, after GPS is being used widely, people needs increasingly location and time information transmission service which needs an innovative medium with capacities of long-distance transmission in real-time, low information transmission under a strict error code rate and stronger anti-interference. There are potentially numerous users who hopefully need cheaper device and lower using cost. Thus, a new classified satellite communications with an extra low-information rate are being developed widely.

### 5.2 Configuration of navigation and communications system

Navigation and communication system is composed of SIGSO communications satellites, master communication ground station, user terminal, data distribution and network management system. User segment is user receivers that are classified into navigation/ communication or combined receivers. User receivers in functions may include Positioning, Timing, Velocity and Communication; carrier feature for user terminals could be classified as static, dynamic and highly dynamic motion. The sorts of device include portable devices, vehicle terminal, mobile terminals and static terminals.

CAPS project now has two SIGSO satellite communications with capability of communications for users, and provide positioning and navigation services with higher positioning accuracy—similar to GPS positioning accuracy, and higher timing precision in all weather and day in service area (longitude 60° ~ 160°, latitude 20° ~ latitude 60°) user terminals are featured by short message and low data rate voice communication and compressed image transmission. CAPS communications hub station locates in Huairou region Beijing with 16 meters aperture antenna operated by automatic tracking. Test system shown in Figure 5.1.

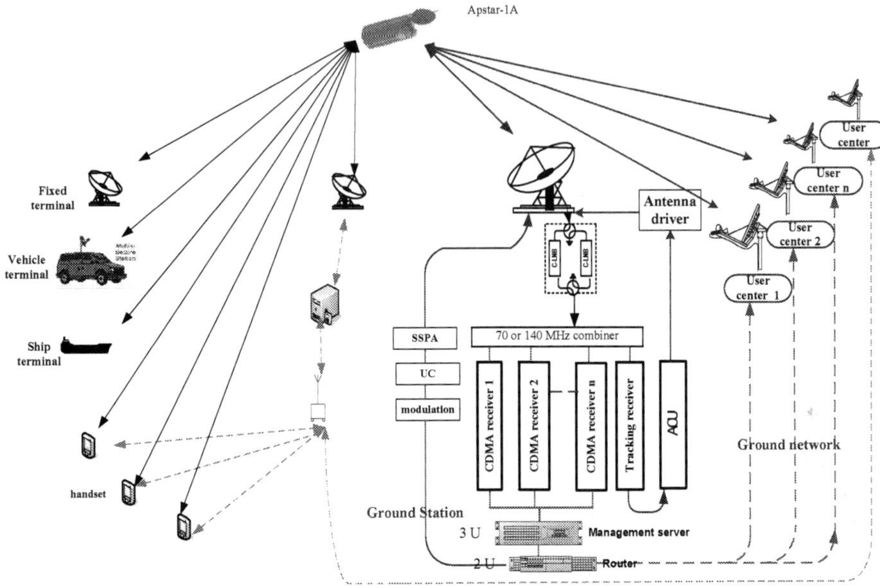


Fig. 5.1 Test principle of CAPS-based integrated communications and navigation in CAPS

### 5.3 Optimal Design of Link

A portable mobile devices with lighter weight, lower power and lower cost seems to be incompatible in a low data rate satellite communication system, development difficulties focus on user terminal with low power, miniaturized and cost-effective, especially, the design of miniaturized antenna in portable user terminal without tracking systems have to be solved, so, the design of the transmission power link must be optimized.

Power link margin is usual expressed by decibel difference between system Eb/N0 and demodulation threshold (Eb/N0). Eb/N0 obtained by the following formula

$$[E_b/N_0] = [C/T]_t - [R_b] - [k], \tag{5.1}$$

where $[C/T]_t$ (C / T) t is total link carrier power and noise power ratio, shown in (5.2) as follows:

$$\left(\frac{C}{T}\right)_t = \left[\left(\frac{C}{T}\right)_u^{-1} + \left(\frac{C}{T}\right)_d^{-1} + \left(\frac{C}{T}\right)_{co}^{-1} + \left(\frac{C}{T}\right)_{im}^{-1} + \left(\frac{C}{T}\right)_{adj}^{-1}\right]^{-1}, \tag{5.2}$$

where, $C/T_u$, $C/T_{co}$, $C/T_{im}$, $C/T_{adj}$ uplink CNR, co-channel interference CNR, inter-modulation interference CNR and adjacent satellite interference CNR, respectively, their expression are shown as follows:

$$[C/T]_u = [\text{EIRP}]_e - [L_u] - [L'] + [G/T]_s, \tag{5.3}$$

$$[C/T]_{co} = 23 + [k] + [B], \tag{5.4}$$

$$[C/T]_{im} = 20 + [k] + [B], \tag{5.5}$$

$$[C/T]_{adj} = [G_r] - 29 + 25 \times \log(2) + [k] + [B], \tag{5.6}$$

where, $L_u$ is free space path loss in uplink transmission, $L'$ are other losses, including the uplink rain attenuation, antenna pointing error in master control navigation station; $[G/T]_s$ are the gain and noise temperature ratio of satellite receiving antenna.

The three parameters $C/T_{co}$, $C/T_{im}$, $C/T_{adj}$ of total Carrier to Noise ratio $[C/T]_t$ in satellite communications system relate with carrier noise bandwidth and temperature parameters, and $C/T_{co}$ relate with the number of CDMA users working in same frequency and cross-correlation of their PN codes. If 20 groups GOLD code are selected as PN codes with 2047 code length, the above three values are generally 130 ~ 150 dBW / K in inbound communication link, the $C/T_u$ at the margin of APstar-1 coverage is about 190 dBW / K, therefore, in equation (5.2), $C/T_u$ and the $C/T_d$ are dominant. $(G/T)_s$ is about 7 dB / K at the margin of APstar coverage, if $R_b$ takes 50 bps, demodulation threshold $(E_b/N_0)_{th}$ is 10.6dB, it will be known by Eq. (5.3 ), $(C/T)_u$ value is 195.3 dBW / K, the uplink C/N is very low, and the demodulation threshold of signal to satellite is close to system threshold. Similarly, for the outbound link, as receiving antenna gain is lower, downlink or C/N is also limited power system.

In order to achieve miniaturization of user terminal, the terminal's power amplifier configuration should not be too high. Data rate is less to 50-300bps for short message transmission with antenna gain of 5-6dBi. CNR uplink signal to satellite is close to demodulation threshold with very weak signal, so a large size antenna need to be configured to center ground station in this system so that downlink CNR is not dominant to have weak signal receiving and demodulation. The link power budget in the existing APstar communications satellite are shown as follows.

The link power budget results in inbound (terminal transmitting and receiving in central ground station) and outbound (transmitting in central ground station and terminal receiving), respectively are shown in Table 5.1 and Table 5.2.

Table 5.1 Inbound link power budget in C-band mobile communication

| APstar-I 2A-50bps--single-carrier transmitting in margin region and received in Beijing | |
|---|---|
| | **Edge coverage - Beijing** |
| **A. Satellite Parameters** | **Parameters** |
| 1. Satellite Name | APstar-I |
| 2. Orbital position (deg.E) | 142 |
| 3. Transponder No. | 1A |

| | |
|---|---|
| 4. Transponder gain attenuation(dB) | 2 |
| 5. Satellite saturated effective isotropic radiated power (dBw) | 36.00 |
| 6. Satellite Saturation Flux Density (dBw/m^2) | -90.82 |
| 7. Satellite receiving antenna gain and noise temperature ratio (dB/k) | -2.00 |
| 8. Transponder Output backoff (dB) | 4.50 |
| 9. Transponder input backoff (dB) | 10.50 |
| 10. Uplink Frequency (MHz) | 5965.00 |
| 11. Downlink Frequency (MHz) | 3740.00 |
| 12. Transponder Bandwidth (MHz) | 72.00 |
| | |
| **B. Ground station parameters** | |
| 1. Transmission parameters□ | |
| a. Antenna diameter (m) | Monopole Antenna |
| b. Ground station location | Coverage |
| c. Feed losses (dB) | 1.00 |
| d. Power amplifier to feed insertion loss (dB) | 0.50 |
| e. Type of power amplifier (klystron, traveling wave tube, SSPA) | SSPA |
| f. Power amplifier output (dBw) （10W） | 10.00 |
| g. Power amplifier maximum output power (dBw) | 10.79 |
| h. Amplifier output margin (dBW) | 0.79 |
| i. Antenna efficiency (%) | 45.00 |
| l. Antenna gain (dB) | 5.70 |
| k. Antenna tracking capability | no |
| l. Type of antenna | mobile |
| | |
| 2. Transmission parameters | |
| a. Antenna diameter (m) | 16.00 |
| b. Ground station position | Beijing |
| c. Longitude (deg.E) | 116.46 |
| d. Latitude (deg.N) | 39.92 |
| e. To the satellite distance (km) | 38031.35 |
| f. Antenna elevation angle (deg) | 36.85 |
| g. Antenna Azimuth (deg) | 143.33 |
| h. Antenna efficiency (%) | 60 |
| i. Antenna gain (dBi) | 53.70 |

| | |
|---|---|
| j. Receiving system noise temperature (dBk) | 19.54 |
| k. Antenna gain and noise temperature ratio (dB/k) | 34.16 |
| l. Antenna tracking capability | Automatic |
| m. Antenna type | Fixed |
| | |
| **C. Space loss** | |
| 1. Uplink Rain Attenuation (dB) | 0.5 |
| 2. Downlink Rain Attenuation (dB) | 0.3 |
| 3. Uplink free space attenuation (dB) | 200.50 |
| 4. Downlink free-space attenuation (dB) | 195.51 |
| 5. Uplink polarization loss | 3.00 |
| | |
| **D. Carrier Parameters** | |
| 1. Carrier Type | DSSS |
| 2. Multiplexed mode | CDMA |
| 3. Modulation | BPSK |
| 4. Data Rate (bps) | 50 |
| 5. Chip rate (kbps) | 2047.00 |
| 6. Spreading frequency gain | 46.12 |
| 7. Carrier Noise Bandwidth (khz) | 2800.00 |
| 8. Carrier allocated bandwidth (khz) | 3000.00 |
| 9. Threshold $E_b/N_0$ (dB) | 10.00 |
| | |
| **E. Link Budget** | |
| 1. Uplink C/T | |
| a. Ground station effective isotropic radiated power (dBw) | 11.20 |
| b. Uplink free space loss (dB) | 200.50 |
| c. Loss of antenna pointing (dB) | 2.00 |
| d. Antenna gain per square meter (dB/m^2) | 36.96 |
| e. Satellite carrier flux density (dBw/m^2) | -141.55 |
| f. Transponder saturation flux density (dBw/m^2) | -90.82 |
| g. Carrier input backoff (dB) | 50.73 |
| h. Satellite antenna gain and noise temperature ratio (dB/k) | -2.00 |
| i. Uplink C/T (dBw/k) | -193.80 |
| | |
| 2. Downlink C/T | |

| | |
|---|---|
| a. Satellite effective isotropic radiated power saturation (dBw) | 36.00 |
| b. Carrier Output backoff (dB) | 44.73 |
| c. Carrier Downlink effective isotropic radiated power (dBw) | -8.73 |
| c. Downlink carrier per channel effective isotropic radiated power (dBw) | -19.52 |
| d. Downlink free-space loss (dB) | 195.51 |
| e. Receiving antenna pointing error (dB) | 0.30 |
| f. Antenna gain and noise temperature ratio (dB/k) | 34.16 |
| g. Downlink C/T (dBw) | -181.47 |
| | |
| 3. Co-channel interference C/T (dBw/k) | -162.31 |
| | |
| 4. Inter-modulation interference C/T (dBw/k) | -144.13 |
| | |
| 5. Satellite interference C/T ( dBw/k) | -131.90 |
| | |
| 6. Total C/T and C/N | |
| a. Total C/T (dBw/k) | -194.05 |
| b. Boltzmann constant (dBw/k-Hz) | -228.6 |
| c. Receiver Noise Bandwidth (dB-Hz) | 64.47 |
| d. Total C/N (dB) | -29.92 |
| e. Threshold $E_{bi}/N_0$ (dB) | 10 |
| f. Threshold C/N (dB) | -37.48 |
| g. Link margin (dB) | 7.56 |

Table 5.2 Outbound link power budget in C-band mobile communication

| APstar-I 2A-50bps—Beijing station transmitting and terminal received in margin region | |
|---|---|
| | |
| | **Beijing-Edge coverage** |
| **A. Satellite Parameters** | **Parameters** |
| 1. Satellite Name | Apstar-1 |
| 2. Orbital position (deg.E) | 142 |
| 3. Transponder No. | 5A |
| 4. Transponder gain attenuation(dB) | 8 |
| 5. Satellite saturated effective isotropic radiated power (dBw) | 36.00 |
| 6. Satellite Saturation Flux Density (dBw/m^2) | -86.73 |

| | |
|---|---|
| 7. Satellite receiving antenna gain and noise temperature ratio (dB/k) | -1.12 |
| 8. Transponder Output backoff (dB) | 4.50 |
| 9. Transponder input backoff (dB) | 10.50 |
| 10. Uplink Frequency (MHz) | 6105.00 |
| 11. Downlink Frequency (MHz) | 3880.00 |
| 12. Transponder Bandwidth (MHz) | 36.00 |
| | |
| **B. Ground station parameters** | |
| 1. Transmission parameters | |
| a. Antenna diameter (m) | 16.00 |
| b. Ground station location | Beijing |
| c. Longitude (deg.E) | 116.46 |
| d. Latitude (deg.N) | 39.92 |
| e. Distance to satellite (km) | 38031.4 |
| f. Elevation angle (deg) | 36.85 |
| g. Azimuth angle (deg) | 143.33 |
| h. Feed losses (dB) | 0.30 |
| i. Power amplifier to feed insertion loss (dB) | 3.00 |
| j. Type of power amplifier (klystron, traveling wave tube,SSPA) | SSPA |
| k. Power amplifier output (dBw) | 2.21 |
| l. Power amplifier maximum output power (dBw) | 20.00 |
| m. Amplifier output margin (dBW) | 17.79 |
| n. Antenna efficiency (%) | 60.00 |
| o. Antenna gain (dB) | 57.98 |
| p. Antenna tracking capability | Automatic |
| q. Type of antenna | Fixed |
| | |
| 2. Transmission parameters | |
| a. Type of antenna | Double spiral antenna |
| b. Station position | Edge coverage |
| c. Antenna gain (dBi) | 5.00 |
| d. Receiving system noise temperature (dBk) | 24.13 |
| e. Antenna gain and noise temperature ratio (dB/k) | -19.13 |
| f. Antenna tracking capability | None |
| g. Antenna type | Mobile |

| | |
|---|---|
| **C. Space loss** | |
| 1. Uplink Rain Attenuation (dB) | 0.5 |
| 2. Downlink Rain Attenuation (dB) | 0.3 |
| 3. Uplink free space attenuation (dB) | 199.77 |
| 4. Downlink free-space attenuation (dB) | 196.50 |
| 5. Uplink polarization loss | 0.00 |
| 6. Downlink polarization loss | 3.00 |
| | |
| **D. Carrier Parameters** | |
| 1. Carrier Type | DSSS |
| 2. Multiplexed mode | Single carrier |
| 3. Modulation | BPSK |
| 4. Data Rate (bps) | 50 |
| 5. Chip rate (kbps) | 2047.00 |
| 6. Spreading frequency gain | 49.13 |
| 7. Number of path | 1 |
| 8. Carrier Noise Bandwidth (khz) | 2800.00 |
| 9. Carrier allocated bandwidth (khz) | 3000.00 |
| 10. Threshold $E_b/N_0$ (dB) | 10.00 |
| | |
| **E. Link Budget** | |
| 1. Uplink C/T | |
| a. Ground station effective isotropic radiated power (dBw) | 56.89 |
| b. Uplink free space loss (dB) | 199.77 |
| c. Loss of antenna pointing (dB) | 0.30 |
| d. Antenna gain per square meter (dB/m^2) | 37.16 |
| e. Satellite carrier flux density (dBw/m^2) | -105.71 |
| f. Transponder saturation flux density (dBw/m^2) | -86.73 |
| g. Carrier input backoff (dB) | 18.98 |
| h. Satellite antenna gain and noise temperature ratio (dB/k) | -1.12 |
| i. Uplink C/T (dBw/k) | -144.80 |
| | |
| 2. Downlink C/T | |
| a. Satellite effective isotropic radiated power saturation (dBw) | 36.00 |
| b. Carrier Output backoff (dB) | 12.98 |
| c. Carrier Downlink effective isotropic radiated power (dBw) | 23.02 |

| | |
|---|---|
| c. Downlink carrier per channel effective isotropic radiated power (dBw) | 23.02 |
| d. Downlink free-space loss (dB) | 196.50 |
| e. Receiving antenna pointing error (dB) | 2.00 |
| f. Antenna gain and noise temperature ratio (dB/k) | -19.13 |
| g. Downlink C/T (dBw) | -197.92 |
| | |
| 3. Co-channel interference C/T (dBw/k) | -141.13 |
| | |
| 4. Inter-modulation interference C/T (dBw/k) | -144.13 |
| | |
| 5. Satellite interference C/T ( dBw/k) | -180.60 |
| | |
| 6. Total C/T and C/N | |
| a. Total C/T (dBw/k) | -198.00 |
| b. Boltzmann constant (dBw/k-Hz) | -228.6 |
| c. Receiver Noise Bandwidth (dB-Hz) | 64.47 |
| d. Total C/N (dB) | -33.87 |
| e. Threshold $E_{bi}/N_0$ (dB) | 10 |
| f. Threshold C/N (dB) | -40.49 |
| g. Link margin (dB) | 6.62 |

From Table 5.1 and Table 5.2, inbound and outbound margins are about 7dB, and relatively abundant, and can ensure reliable information transmission through satellite.

### 5.4 The key technologies

The following technologies are adopted to solve limited power and meet special design requirements for small terminal device, as well as SIGSO satellites operational requirements.

1) Broadband spread spectrum technology

As bandwidth resources in the SIGSO satellite are abundant, so it can be used to wideband spread-spectrum communication. The trade off using frequency bands for power solves the limited power and enhancement of C/N. the bandwidths of spread spectrum are optimized in terms of meeting the relevant ITU provisions so that we can solve adjacent satellite interference in using wide beam antenna, reducing the power flux density, and improving the system of anti-jamming capability.

2) Large size receiving antenna in ground station

The small size and wide beam antenna with lower gain and power result in power and CNR limitation in satellite transmission links, it can be solved by adjusting attenuation in the transponder, or adopting FEC technique, besides that, a practical solution is to use large size receiving antenna on ground station. Thus, CNR in the downlink cannot effect on the SNR of the whole link which can be met well.

3) Shaped wide beam antenna

As SICSO satellite drift all the time, and its daily motion is "8" shape (Figure 5.2). Such as the APstar-1 satellite drifts over ± 4°/day in north-southern direction, the azimuth and elevation of satellite observing from antenna change larger per day (Figure 5.3).

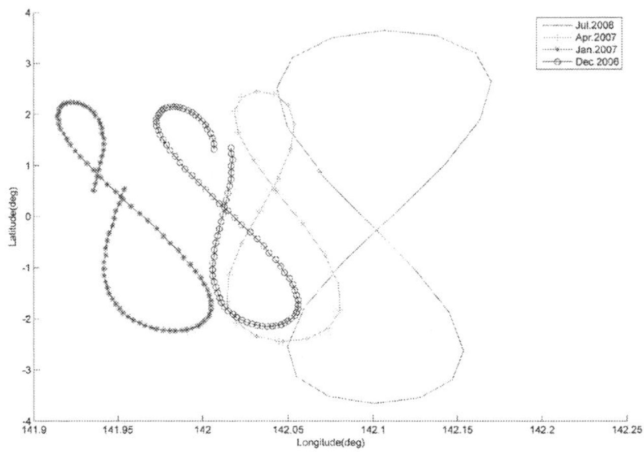


Fig. 5.2 APstar-1 track of subsatellite point

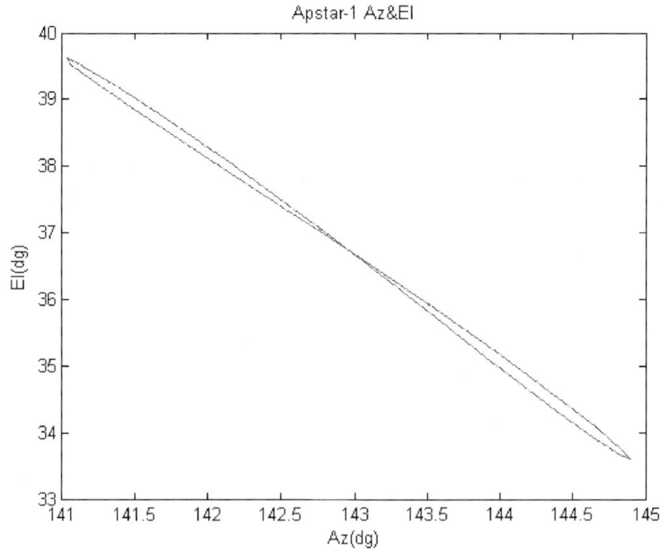


Fig. 5.3 The azimuth and elevation in Beijing earth station

In the mobile satellite communication systems, antenna is a crucial component. Antenna directive, tracking accuracy and its cost restrict the overall system performance and price in certain extent, restricted definitely the availability and promotion of the system. Therefore, we developed a monopole shaped beam antenna. Antenna pattern is shown in Figure 5.4, where it is omnidirectional and wide-beam with about 60° elevation angle, the antenna maximum direction can be designed to required direction. It totally does not need to change the antenna attitude to meet the demand for satellites drift. And even devices would be installed in the mobile vehicle, they can meet the moving dynamic requirements, so it is very simple and cheap.

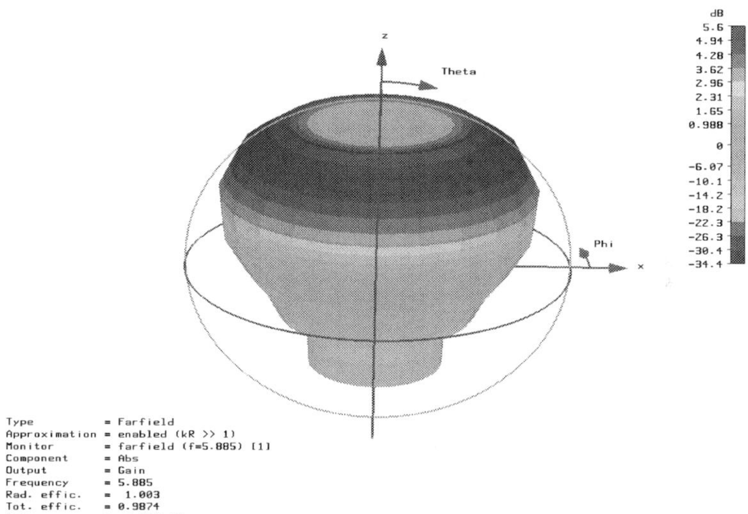


Fig. 5.4 Monopole Antenna 3-D pattern

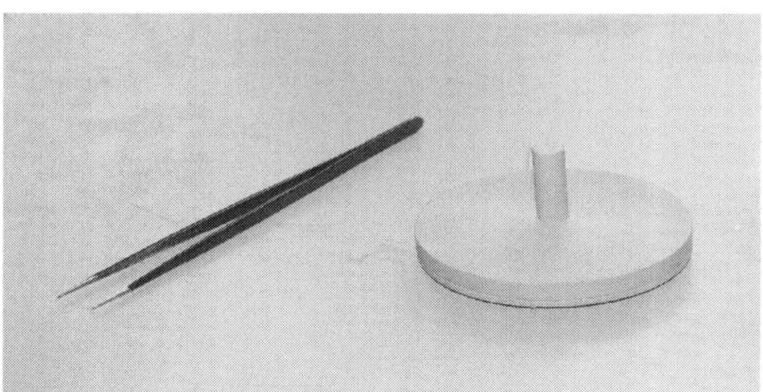

Fig. 5.5 Monopole Antenna

Terminal antenna is vertical polarization; polarization loss is relatively small between antenna and satellite. Antenna shape is shown in Figure 5.5. The disc of its bottom is reflection plate and its diameter is 6cm.

Other kind receiving antenna is double spiral antenna; its 3-D antenna pattern is shown in Figure 5.6.

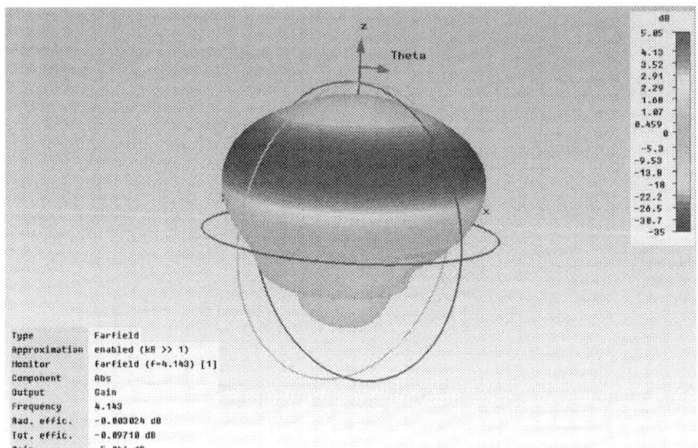


Fig. 5.6 Three-dimensional pattern of double spiral antenna

Double spiral antenna gain is relatively low around 5dBi with whip-like in (as shown in Figure 5.7).

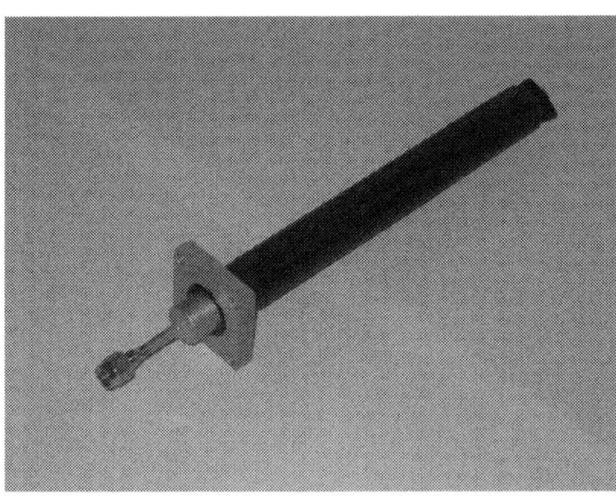

Fig. 5.7 Double spiral antenna

In order to prevent carrier signal interfere from earth stations with neighboring satellites, the International Radio Consultative Committee and the International Telecommunication Union (ITU) on the uplink station carrier axis radiation power density is strictly required; in order to prevent interference to ground microwave communications, ITU regulated international standards to allow power flux density on the ground transmitted from space communications systems. In addition, the coordination is required about above two issues by satellite operators. From Figure 5.4 and Figure 5.6 it can be seen that such antenna beam is wide, and even adjacent satellite is radiated by antenna main beam, so, the measures must be taken to avoid interference with adjacent satellite.

It is very important to three aspects mentioned above for low data rate satellite communication system design which is typically different from normal satellite communication system, of course, in addition, technologies generally used in satellite communications need to be optimized, such as the use of data compression, optimized multiple access and modulation and so on.

### 5.5. The system characteristics and advantages

1) Information exchange

As CAPS has successfully developed satellite communications technology with a low data rate, it can exchange and transmit spatial and time information. This system can provide the spatial location and time information.

2) Stronger anti-interference

As the system uses wideband PN spread spectrum system, signal will be buried in the noise below, so the signal will not be easy to be intercepted and deciphered, but also it is difficult to interfere with its transmission and has anti-interference ability.

3) Cost-effectiveness

Constellation adopts GEO communications satellite around late life span; it is prolonged 10 times life expectancy and satellite by changing attitude control with very cheap transponder. System is designed to use small devices and small antenna; equipment development costs are relatively small. Ground station can continue to use the original communication equipments and operate with a little modified equipment. So, the whole system invests less, system cost-effective and easy to build up.

### 5.6 Typical application

1) Based on the position and time service

The development of a low data rate of satellite communications technology in satellite navigation system based on communication satellite enables to communicate each other to position message and create more applications in combining of satellite navigation and position and timing service, it solve the limitations of individual navigation service in GPS, and to realize the group and center system navigation, and center. Figure 5.8 is shown terminal vehicle positioning information in the electronic map picture.

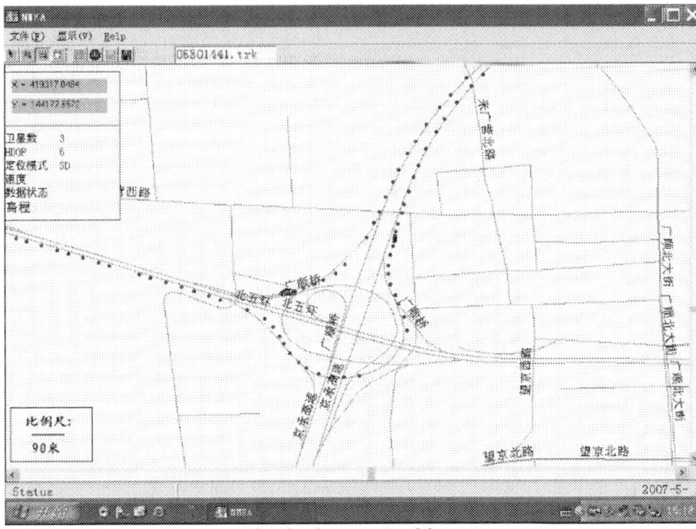


Fig. 5.8 Display on electronic map of vehicle terminal location

2) Applicative for scattered users, rare routing and small traffic communication network, data transmission in automated and unattended station.

3) Applicative for anti-terrorism, disaster relief, emergency communications, When the earthquake disaster and other emergencies happen, it is very important the timely to report the disaster. Disaster relief center also need to control the situation in real-time, directly command and deploy disaster relief teams. People need emergency communication means as facing terrorist activities, also need reliable means of communication, which requires. The project develops a low information rate of satellite communications to meet the emergency communication needs.

4) A new satellite monitoring function

Low information rate satellite communication system can send timely data to control center, for example, operating parameters of large equipments like scraper, oil pipeline corrosion data, measurements data in marine, hydrological and meteorological investigation, logistics management information of ship and vehicle, displacement transfer parameters of reservoirs, rivers, lakes, and dams and so on, contrarily, control center can also transmit instructions to the monitoring terminal in time, to change work parameters and set working status for monitoring.

Totally, the low information rate satellite communications experiments for integration of navigation and communication and multi-satellite communication applications, and to be a practical technology orient to market applications.

## 6. The application value of GEO communications satellites with extra-live span

We proposed the technological methods to reuse retired communication satellite[1], it has obtained significant results through the conceptive research, simulation and project validation, as well as practical application.

CAPS system now use Apstar-1 and Apstar-1A satellites, where orbital position of Apstar-1 satellite was adjusted from the 138° E to the current 142° E orbital position, Apstar-1A satellite from the 134° E to current 130° E. Apstar-1 and Apstar-1A satellite began to be adjusted to SIGSO satellite in August 2004 and in May 2005 respectively. They still keep orbital precision to the original ± 0.05° [19, 20] along latitudinal direction without control in longitudinal direction. Apstar-1 satellite's inclination had reached 4.1°, Apstar-1A satellite's inclination was 3.5° till March 2009. The following detailed description of the project application value:

**1) The protection of spatial environmental resources**

Research and realization of retired GEO communications satellite remaining payload fuel without abandonment can be used as constellation of satellite navigation and positioning system. they are only be maintained and adjusted along longitude in orbit spaces, the working mode can save 90% fuel keeping the new control method for CAPS and significantly extend the life of satellites, a retired communications satellite achieves to reuse, creating a development and utilization of extra life-cycle of satellite.

A new model of satellite orbit control has been study and proposes new methods and technologies of reuse retired GEO communications satellites. For example, AP-star 1 satellite reuse, the remained using period of fuel in normal control mode was 3 to 6 months while it was replaced by other new satellite in 2004. At this time, the satellite's solar cells and electronic equipment is still intact with 24 transponders still working, even six spare transponders not yet being used. In the new orbit control mode, the retired satellite became the SIGSO satellite, and has worked more than five years, the GEO communications satellite resources are ultimately used and delay the period to become space junk and pollute the space environment, it plays a very important role to protect the space environment resulting in significant social and economic benefits. In contrast, in previous years, without this new technique, the AP-star 1 satellite in end of life span would be replaced by new satellite and retire to be abandoned as space junk.

**2) The constellation of satellite navigation and positioning system, improvement of navigation and positioning accuracy**

In the new satellite control mode, the retired GEO communications satellite were regulated as SIGSO satellite to improve the geometry of the navigation constellation, reducing constellation DOP, and can further improve the navigation and positioning accuracy, therefore, it is obvious augmentation due to SIGSO satellite.

Chinese satellite navigation and positioning system (CAPS) purchased retired APstar-1 GEO communications satellite at 138° E orbit, and maneuvers to the 142° E orbital position to SIGSO satellite, it realized GEO communications satellites to reuse. A preliminary application system was configured by adding other two transponders in two GEO satellites, and barometer supplement to realize 3-D positioning. In 2005, the system has reached the GPS navigation and positioning accuracy in coarse code level, positioning accuracy of 10 m, velocity accuracy is 0.2 m / s, timing accuracy of 12 ns [2, 12]. Communication data rate 50bps ~ 1.2kbps; spread spectrum bandwidth of ± 1.5M; transmitting power 3W ~ 20W; BER 1 × 10-5 ~ 1 × 10-6; modulation is BPSK. Reusing retired satellites can extend the life span of the satellite more than 8 to 10 times, and the band-costing decreasing by about 80%. It will save investment and costing less than one-tenth of the new satellite launching to buy this satellite, or lease their transponders, this can generate significant economic benefits in terms of deploying navigation system and investment. The innovative satellite navigation and positioning system is featured lower cost, shorter deployment period and better performance.

Retired satellites often have full-band (C band) transponders, it can be used to navigation frequency change to become a multi-band satellite navigation system and find new way to improve anti-interference capability, lay the foundation for reuse retired GEO satellites. For example, currently, it is a rather large PDOP <10 region. SIGSO satellite orbital inclination increases along with regulation time (for example, when at about 7°), PDOP value is further reduced,

**3) Developing a satellite communication system with a low data rate transmission and small satellite user terminals**

GPS class satellite navigation system can provide user position, but its positioning message cannot transfer and communicate each other within the system. The satellite navigation system based communications transponders have a strong communication transmission capacity, easily achieved communicating positioning message, and enhance information value. Conventionally, we need to lease other GEO communication satellite transponders, but do need to increase the cost of transponder rental. Fortunately, we bought a retired GEO communications satellite with 24 transponders where two transponders use for navigation, in addition, using three transponders for tri-frequency measurement, so there are 21 transponders can be used for communication. Reuse of retired communications satellite is to turn waste into treasure, and make lower channel cost in this satellite communication system. Mobile Micro-Satellite Terminal( MMSAT) with wide-beam antenna are applicable to mobile communications, especially time, position and state message service in navigation system, the development of the spread spectrum communication system with low data rate and small antenna and user terminal can be used for short messaging, voice communication and transmission of static images in various fields such as traffic safety and vehicle management, calling, emergency rescue, search and rescue, emergency rescue and disaster relief, mobile TV and video broadcast. Satellite MMSAT terminal series can be formed to provide a short message and data communication services, which promote the application of satellite communications. Therefore, it is necessary to use more retied GEO communications satellite which provides a plentiful space transponder resource to give the full enhancement of combined navigation and communication.

NAOC in china has developed a class low data rate satellite communications technology with short text and data messages more than four years in vehicle monitoring and oceanic observation, recently, a class of low data rate voice communication system is developing.

## 7. Reference

1. Inventor: Shi H L, Ai G X, Chen J B, Han Y B, Shang J N, Geng J P, Ma L H, applicant: National Astronomical observatories, CAS, patent name: methods of changing retired satellites into SIGSO, patent number: 200610055909.2, date of application: October 5, 2006, date of approval: June 3, 2009
2. Inventor: Ai G X, Shi H L, Wu H T, Yan Y H, Bian Y J, Hu Y H, Li Z G, Guo J, Cai X D, applicant: National Astronomical observatories, CAS, National Time Service Center, CAS, patent name: the transmitting satellite navigation and positioning system, patent number: 20041004760264.1, date of application: June 3, 2004, date of approval: June 5, 2009
3. Inventor: Li Z G,Ai G X,Shi H L, Yang X H, applicant: National Astronomical observatories, CAS, National Time Service Center, CAS, patent name: a new method for determination of satellite orbits by transfer, patent number: 200310102197.1, date of application: December 30, 2003, date of approval: July 5,2006
4. Inventor: Shang J N, Shi H L, Lu X C, Zhou X applicant: National Astronomical observatories, CAS, National Time Service Center, CAS, patent name: integrated system of navigation and communication based on PT and status messages, patent number: 200610012056.4, date of application: May 31, 2006
5. Inventor: Shi H L, Yang P C, Liang K, Pei J, Cui J X, Ji Y F, Li S M, Ning C L applicant: National Astronomical observatories, CAS, patent name: parasitic anti-interference method and system for transmitting satellite navigation signal, patent number: 200710081260.6, date of application: May 10, 2007
6. Inventor: Sun X Y, Ai G X, Shi H L, Ji Y F, Wang X N, Pang F, Ning C L, applicant: National Astronomical observatories, CAS, patent name: method of realizing satellite navigation and positioning with different navigation frequency, patent number: 200810056250.1, date of application: January 16,2008
7. Inventor: Shi H L, Cui J X, Han Y B, Pei J, Du X H, Hu Z Q, Ji Y F, Wang M, Zhu H L, applicant: National Astronomical observatories, CAS, patent name: signal structure for integrated system of navigation and communication, patent number: 200910084033.8, date of application: May 13, 2009
8. SHI HuLi, AI GuoXiang, HAN YanBen, MA LiHua, CHEN JiBin, GENG JianPing. Multi-life cycles utilization of retired satellites. Science in China Series G: Physics, Mechanics & Astronomy, 2009, 52(3):323-327.
9. HAN YanBen, MA LiHua, QIAO QiYuan, YIN ZhiQiang,SHI HuLi, AI GuoXiang, Function of retired GEO communication satellites on improving PDOP value of CAPS, Science in China Series G: Physics, Mechanics &Astronomy, 2009, 52(3):423-433.
10. Ai G X, Shi H L, Wu H T, et al. Positioning system based satellite communication and Chinese Area Positioning System (CAPS). Chin J Astron Astrophys, 2008, 8(6): 611−635.
11. AI GuoXiang, SHI HuLi1, WU HaiTao2, LI ZhiGang2 & GUO Ji2. The principle of the positioning system based on communication satellites, Science in China Series G: Physics, Mechanics & Astronomy, 2009, 52(3):472-488.
12. Shi H L, Han Y B, Ma L H, exploration and practice on multi-life cycles utilization of GEO satellites, Satellite and Network, 2007, (8): 34-37

13. HAN YanBen, MA LiHua, QIAO QiYuan, YIN ZhiQiang, AI GuoXiang, Selection of satellite constellation framework of CAPS, Science in China Series G: Physics, Mechanics & Astronomy, 2009, 52(3):458-471.
14. CUI JunXia, SHI HuLi & CHEN JiBin. Transmission Link of CAPS Navigation and Communications system. Science in China Series G: Physics, Mechanics & Astronomy, 2009, 52(3):402-411.
15. LI Zhigang, YANG Xuhai, SHI Huli, AI Guoxiang, QIAO Rongchuan & FENG Chugang. New Method for Determination of Satellite orbit by Transfer. Science in China Series G: Physics, Mechanics & Astronomy, 2009, 52(1), in press
16. Huang Yong, Hu XiaoGong, Huang Cheng, Yang QiangWen & Jiao WenHai. Precise orbit determination of a maneu-vered GEO satellite using CAPS ranging data. Science in China Series G: Physics, Mechanics & Astronomy, 2009, 52(1), in press
17. SHI Huli & PEI Jun. The solutions of navigation observation equations for CAPS. Science in China Series G: Physics, Mechanics & Astronomy, 2009, 52(3):434-444.
18. YANG Xuhai, LI Zhigang, FENG Chugang, GUO Ji, SHI Huli, AI Guoxiang, WU Fenglie & QIAO Rongchuan. Rapid method of orbit forecasting after maneuver for geostationary satellite. Science in China Series G: Physics, Mechanics & Astronomy, 2009, 52(1), in press
19. Shi H L, Cui J X, Du X H, one kind of satellite communication technique with low information rate, China Science and Technology Achievements
20. Shi H L, Sun X Y, Li Z G, principles on the transmitting satellite navigation and positioning system, Beijing: Science Press, 2009
21. Ma L H, Han Y B, Qiao Q Y, Some questions about Satellite static positioning, Global Positioning System, 2005, 4: 9—11.
22. Ma L H, Han Y B, Qiao Q Y, calculating DOP of artificial satellites with MATLAB, Application Research of Computers (supplement), 2007, 24: 1716.
23. Deng C M, Zhuo C N, Wu T Y, Constellation of a Regional Satellite Navigation System, JOURNAL OF UNIVERSITY OF ELECTRONIC SCIENCE AND TECHNOLOGY OF CHINA, 2006, 35(5): 725—728.
24. International Telecommunication Union. Recommendation ITU-R S.484-3. Station-keeping in longitude of geostationary satellites in the fixed-satellite service Geneva: International Telecommunication Union Electronic Publishing Service.
25. International Telecommunication Union, Recommendation ITU-R S.743-1, ITU-R Recommendations on CD-ROM, Electronic Publishing Service, 2004.
26. Kaplan E D, Hegarty C J. Understanding GPS: Principle and Application. 2nd ed. Boston: Artech House, 2006.
27. Hanson J M, Evans M J, Turner R E. Designing good partial coverage satellite constellations. J Astr Sci, 1992, 20(2): 215—239.
28. IADC/WG4/WGD01, Draft IADC space debris mitigation guideline, IADC/WG4/WGD01, 2001.
29. Jing Y X, Zhang Y T, Analyzation of GEO satellite phase, CAPS Technical report. 2009, 12
30. Li Y H, principle and implementing strategy for position maintenance of GEO satellites, Journal of Spacecraft TT&C Technology, 2003, 22(4): 53-61.
31. Liu L, Hu S J, Wang X, an introduction of astrodynamics, Nanjing University Press, 2006
32. Yang J X, orbits dynamics and control of spacecraft, China Astronautic Publishing House, 1995.

33. Zhang R W, orbits dynamics and control of satellite, BEIHANG University
34. Hu S J, Chen L, Liu L, The Structure Evolution of Satellite Constellation, Acta Astronomica Sinica, 2003, 44(1): 46—54.
35. Zheng J, Chen H. Study on GEO Satellite's Re-Orbit Control Strategy in End of Life, AEROSPACE SHANGHAI, 2007, 3: 38—41
36. Wu B Y, Consideration on Inclined GEO Satellite Communication System, TELECOMMUNICATIONS SCIENCE, 2002, 11: 7—11.
37. Xu, Q F, Satellite Constellation of Local Navigation System, ENGINEERING OF SURVEYING AND MAPPING, 2001, 10(1): 1—5.
38. Liu H J, Zhang N T, CONSTELLATION DESIGN FOR REGIONAL SATELLITE POSITIONING SYSTEM, JOURNAL OF ELECTRONICS & INFORMATION TECHNOLOGY, 2001

# 17

# Planar Antennas For Satellite Communications

Jorge Sosa-Pedroza, Fabiola Martínez-Zúniga and Mauro Enciso-Aguilar
*Instituto Politécnico Nacional. Escuela Superior de Ingeniería Mecánica y Eléctrica*
*México*

## 1. Introduction

Modern Satellite Communications requires the development of very small size, low-cost, low-profile, high gain and high directivity antennas. Antennas form the link between transmitting and receiving equipment and the space propagation path. Even the general principles may apply to satellite communications antennas, imposed limitations of gain, field pattern and mainly the physical environment lead to design requirements which should be taken in account. Most satellite antennas are designed to give coverage over a specified restricted area defining restrictions over antenna gain; on the other hand payload restrictions of weight and size lead to small size and weight antennas. Especially for these characteristics, planar antennas could have wide application on communications satellites. Since the late 1970s, the international antenna community has devoted much effort to the theoretical and experimental research on microstrip and printed antennas, which offer the advantages of low profile, compatibility with integrated circuit technology and conformability to a shaped surface. The results of this research have contributed to the success of these antennas not only in military applications such as aircraft, missiles, and rockets but also in commercial areas such as mobile satellite communications, the Direct Broadcast Satellite (DBS) system, global position system (GPS) and remote sensing but their main applications are on ground transceivers. Last two decades have been especially worthwhile on planar antennas development, mainly for mobile communications, but its applications should be extended to satellite systems. Particularly for the first Mexican satellite system (Morelos I and II), primary array receiving Ku band , was a planar array to produce an elliptical footprint over Mexican territory. For the second generation of Mexican satellites, was introduced the L band circular polarization array, formed by 16 parabolic reflectors feed by crossed short dipoles producing the needed circular polarization and also an elliptical footprint. It is clear that if there were used a planar array, as those that technology has developed in the last two decades, the cost and the weight would be reduced to a fourth part. This chapter of the Satellite Communications book, is devoted to planar antennas, not only for that already in use but proposing other kind that could be applied for satellite communications. It starts with a brief description of planar antennas, their characteristics, and their applications as antenna arrays. The next part will describe actual planar antennas used in satellite communications systems and will finish with a proposal of new developments of planar antennas that could be used in the near future.

### 1.1 A Brief History

Historically microstrip antennas and consequently planar antennas, have associated with low cost and low profile, but this simplistic description is inadequate as state in (James et al.,

1989), considering that the feasibility of a low profile printed radiator has inspired the system creator with abundance of examples, as the printed paper antennas, or adaptive conformal antenna arrays and many other examples. The success on cellular telephony is based in much in the planar antenna development, as the services offered by cellular company's increase every day, planar ultra wideband antennas are an actual challenge for those working in the area. Research and engineering publications are also devoting more space to development in planar antennas; one can open any proceedings on antennas and will find at least one or two articles related with this subject. Satellite communications are not outside of this boom, especially when, as in no other industry, low weight, low profile and low cost are a daily challenge for antenna design. The invention of the microstrip concept has been attributed to many sources and the earliest include (Greig et al., 1952), and Deschamps (1953), the idea was to create a new way for connecting electronic circuits even knowing the high unwanted radiation, leading to reduce both, substrate and conducting strip. Whether the advent of the transistor influenced the rapid development of printed circuits, the main interest was the development of low cost microwave circuits (James et al., 1981). There were few proposals to use the technique for antennas in those early years, until the 1970's when the new generation of missiles create the need of low profile antennas, after that the development of microstrip and planar antennas started to fill up research publications. A key point on the design of microstrip and planar antennas, was the development of substrates, not only to define characteristics of permittivity and low loss tangent but also to make materials capable to work on extreme ambient conditions; the 80's was the decade when bigger steps were done in substrate design, however as the frequency demand increases in communication systems, the development of new materials has lead to manufacturers in tightened their specifications, and reducing substrate costs. Table 1 shows examples of materials actually used in microstrip and planar antennas.

| Type | Composition | Thickness | $\epsilon_r$ | $\tan\delta$ |
|---|---|---|---|---|
| RO3010 | PTFE Ceramic | 0.005"-0.050" | 10.2±0.30 | 0.0023 |
| RO3206 | PTFE Woven Glass | 0.005"-0.050" | 6.15±0.15 | 0.0027 |
| RO4350B | Hydrocarbon Ceramic | 0.004"-0.06" | 3.48±0.05 | 0.0037 |
| RT/Duroid 5880 | PTFE Glass Fiber | 0.005"-0.125" | 2.20±0.02 | 0.0009 |
| RT/Duroid 6006 | PTFE Ceramic | 0.005"-0.100" | 6.15±0.15 | 0.0027 |
| TMM 3 | Hydrocarbon Ceramic | 0.15"-0.125" | 3.27±0.032 | 0.0020 |
| TMM 10 | Hydrocarbon Ceramic | 0.015"-0.100" | 9.20±0.23 | 0.0022 |
| FR-4 | Woven Glass/Epoxy Resin | 0.059"-0.118" | 4.8 and 2.2 | 0.017 |

Table 1. Typical substrate materials for planar antennas

Planar antennas have influenced other fields in electromagnetics, those related with materials and the development of transmission lines; actually engineers include the feeders as a part of the design of patch antennas to get a better control over the input impedance characteristics, this is even more important for large arrays architectures where the feeders and antennas are regarded as a complete entity, emphasizing the importance of choice of array topology and the fact that feeders cannot necessarily be freely attached to printed elements. On the other hand, analytical development of antennas is no longer applied; the actual printed substrate technology uses mathematical models as the Finite Difference on Time Domain (FDTD) (Yee, 1966) or the Method of Moments (Harrington, 1992) and others, managed by computer simulators, reducing the highly mathematical analysis to a drawing in a CAD system, which determines current distribution, and after that, radiation patterns, gain, feeder impedance and other parameters of antennas and antenna arrays. Computational Electromagnetics is a field in constant development, seeking for better computer characteristics and computer architecture, as the parallel computer systems, as well for better programming efficiency.

## 2. Microstrip Antennas

Microstrip antennas are very attractive to be used as radiators in satellite communications systems because of its several interesting features like low profile, light weight, easy fabrication, robust nature, conformability to mounting hosts, compatibility with microwave monolithic integrated circuits (MMICs) and optoelectronic integrated circuits (OEICs) technologies, they are very versatile in terms of resonant frequency, polarization, radiation pattern and impedance Microstrip antennas were first introduced in the mid of 70s (Howell, 1975; Munson, 1974) since then a really intense research activity on microstrip antenna has been taken place around the world for both academic and industrial entities. The basic configuration of a microstrip patch antenna is shown in Fig. 1.

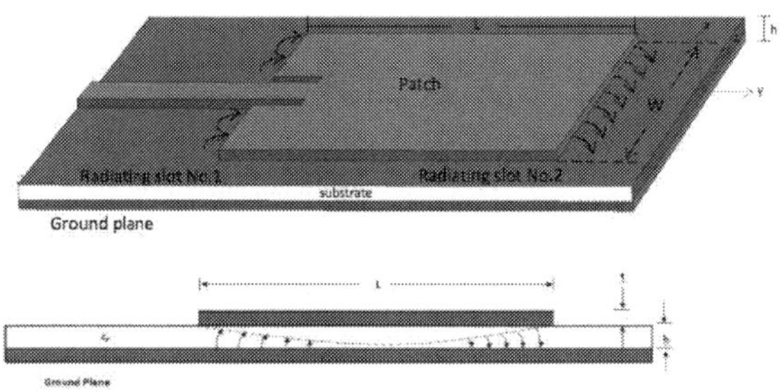


Fig. 1. Microstrip patch antenna configuration

It consist of a dielectric substrate sandwiched structure. The upper surface supports a printed conducting radiator strip or *patch* (even if a microstrip antenna is essentially different to a *patch* antenna, we will use both terms indifferently in this subsection) which is well contoured while the entire lower surface of the substrate is backed by the ground plane. The microstrip patch is designed in a such way that its maximum radiation pattern is normal to the surface of the patch (broadside radiator). This can be achieved by choosing the mode of excitation beneath the patch (Balanis, 2005). The patch radiates from fringing fields around its edges. Impedance match occurs when a patch resonates as a resonant cavity. When matched, the antenna achieves peak efficiency. For microwave applications, thinner substrates with higher dielectric constants are desirable because they require tightly bound field to decrease parasitics effects like undesired radiation and coupling. This also leads to smaller element sizes.
The shape of the patch can take several geometrical forms like square, rectangular, circular, triangular, elliptical, linear (thin strip dipole) or any other configuration. The most common patch shapes are square, rectangular, and circular due to its simplicity of design, manufacture and analysis. They are also attractive because of their radiation characteristics, especially for low cross-polarization radiation. Linear and circular polarization can be obtained by either one element or an array of microstrip antennas. Rectangular patches tend to have the largest impedance bandwidth, simply because they are larger than the other shapes. Circular and elliptical patches are slightly smaller than their rectangular counterpart and as a result have

slightly lower gain and bandwidth. One of the primary reasons the circular geometry was intensively investigated in the past was because of its symmetry allowed full-wave analysis tools utilizing a spectral domain technique. Circular patch antennas provide circularly polarized conical patterns for effective data transmission from a satellite to the earth at UHF and S-band frequencies. In the following sections we shall briefly introduce the theoretical framework of only two types of patch: rectangular and circular. Full wave methods like Method of Moments and FDTD along computational tools are discussed later in this chapter.

## 3. Main Figures of Merit of Microstrip Antennas

### 3.1 Resonant Frequency

In general, microstrip antennas are half-wavelength structures and are operated at the fundamental resonant mode $TM_{01}$ or $TM_{10}$, with a resonant frequency given by (James et al., 1989):

$$f_r = \frac{c}{2L\sqrt{\epsilon_r}} \tag{1}$$

The above expression is only valid for a rectangular microstrip antenna with a thin microwave substrate, where $c$ is the speed of light, $L$ is the patch length of the rectangular microstrip antenna, and $\epsilon_r$ is the relative permittivity of the grounded microwave substrate. From 1 it is clear that the microstrip antenna has a resonant length inversely proportional to $\sqrt{\epsilon_r}$, hence the use of a microwave substrate with a larger permittivity thus can result in a smaller physical antenna length at a fixed operating frequency.

If the open-end extension due to fringing effect is considered then the resonant frequency for a rectangular microstrip antenna becomes (Hirazawa et al., 1992; James et al., 1989):

$$f_r \cong \frac{c}{2L_{eff}\sqrt{\epsilon_{eff}}} \tag{2}$$

where:

$$L_{eff} = L\left\{1 + 0.824\frac{h}{L}\frac{(\epsilon_{eff}+0.3)(L/h+0.262)}{(\epsilon_{eff}-0.258)(L/h+0.813)}\right\} \tag{3}$$

$$\epsilon_{eff} = \frac{\epsilon_r+1}{2} + \frac{\epsilon_r-1}{2}\left(1+10\frac{L}{h}\right)^{-\frac{1}{2}} \tag{4}$$

with $h$ being the substrate thickness. If the microstrip antenna is circular then its resonant frequency is approximated by:

$$f_r \cong \frac{\chi_{11}c}{2a_{eff}\sqrt{\epsilon_r}} \tag{5}$$

where

$$a_{eff} = L\left\{1 + \frac{2h}{\pi a\epsilon_r}\ln\frac{\pi a}{2h} + 1.7726\right\}^{\frac{1}{2}} \tag{6}$$

and

$$\chi_{11} = 1.841 \tag{7}$$

$a$ being the radius of the patch.

### 3.2 Radiation Pattern

It is quite hard to talk on radiation pattern characteristics without associate it with the internal construction consideration. Here we only briefly discuss the associated internal structures that affect the pattern. Rectangular and circular are the most common shapes for patch and they radiate similar patterns. When the cavity is loaded in order to shrink its size, it radiates wider beamwidth patterns that decreases directivity. Structures that couple to coplanar patches to increase the impedance bandwidth will radiate narrower beams, but the basic patch has a wide beamwidth. If multiple coplanar patches are coupled, one can expect the pattern to narrow or vary its shape as the mixture of modes on the various patches changes over the frequency range of operation(Milligan, 2005). Radiation pattern of microstrip antennas have been investigated by several techniques including the equivalent magnetic current method (Derneryd, 1976; James et al., 1989; Long, 1978), full-wave analysis (Chang, 1989; Itoh et al., 1981), and method of moments (Agrawal et al., 1977; Chang, 1989). The former has been often used because its procedural simplicity and extensive applicability. This method considers magnetic current sources $\mathbf{K}_{m1}$ and $\mathbf{K}_{m2}$ laying at the edge or within the vicinity of the edge of the patch that can be approximated by a magnetic current loop antenna. The radiation pattern can be determined by using a Hertz or potential vector. Fig. 2 depicts the fringing electric fields around the edges of square and circular patch antennas excited in the lowest-order cavity modes. The arrow sizes indicate the magnitude of the fields. As seen the square patch exhibit nearly uniform fields along two edges called the width, and a sinusoidal evolution along the other two edges, called the resonant length. The fields vanish along a virtual electrically short-circuited plane halfway across the patches. On either side of the short-circuit plane, the fields are directed in opposite directions.

The circular patch fringing fields distribution varies as $\cos\phi$, where the angle $\phi$ along the edge is measured from the peak electric field. Magnetic currents found from the fringing electric fields can replace the electric currents located on the patch and the surrounding ground plane for pattern analysis.

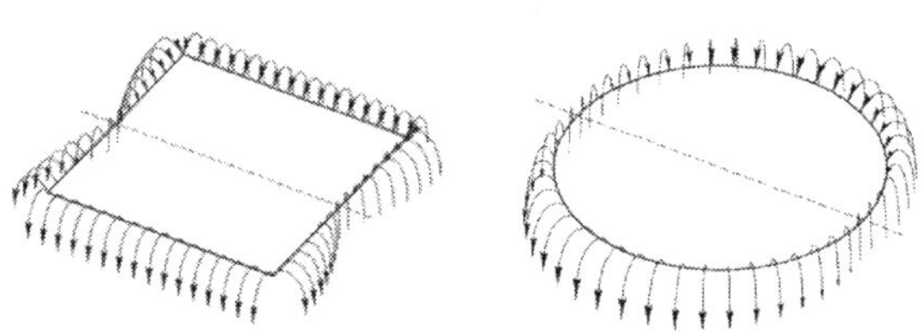

(a) Square (b) Circular

Fig. 2. Fringing electric fields around microstrip patches: (a) square; (b) circular.(After (Milligan, 2005))

The source of equivalent magnetic current corresponding to the $E_z$ field component (along the vertical direction) of a dominant mode is expressed taking the image component as $\mathbf{K}_m = 2\mathbf{E}_z \times \mathbf{n}$, $\mathbf{n}$ being a unity normal vector. The pattern is determined by the line integration of $\mathbf{K}_m$ at the rim of the patch over the adjacent source region (James et al., 1989). For a circular patch with a $TM_{110}$ dominant mode, the expressions to approximate the pattern are: (Hirazawa et al., 1992; James et al., 1989; Long, 1978):

$$E_\theta = -jK_d a\cos(\phi)\sin(m)\left\{\frac{J_1'(n)}{\cos\theta}\right\} \tag{8}$$

$$E_\phi = \frac{jK_d}{k_0}\sin(\phi)\cos(m)\left\{\frac{J_1(n)}{\cos\theta}\right\} \tag{9}$$

with

$$m = k_0 h\cos\theta;\quad n = k_0 a\sin\theta;\quad K_d = E_0\frac{e^{-jk_0R}}{R}J_1(ka) \tag{10}$$

For a rectangular patch with a dominant $TM_{100}$ mode the expression for the radiation pattern are given by:

$$E_\theta = -jK_r f(\theta,\phi)\cos\phi\left\{\frac{\epsilon_r - \sin^2\theta}{\epsilon_r - (\sin\theta\cos\phi)^2}\right\} \tag{11}$$

$$E_\phi = jK_r f(\theta,\phi)\cos\theta\sin\phi\left\{\frac{\epsilon_r}{\epsilon_r - (\sin\theta\cos\phi)^2}\right\} \tag{12}$$

with:

$$f(\theta,\phi) = \left\{\frac{\sin(m)}{m}\right\}\cos n;\ m = \frac{k_0L}{2}\sin\theta\sin\phi;\ n = \frac{k_0L}{2}\sin\theta\cos\phi;$$

$$K_r = \frac{V_0k_0L}{\pi}\frac{e^{-jk_0R}}{R};\ V_0 = E_0h \tag{13}$$

where $k_0$ is the wave number in free space, $k$ is the wave number in the substrate, $J(n), J'(n)$ first kind of Bessel function and its derivative, respectively. $E_0$ is an arbitrary constant, $h$ is the substrate thickness, $a$ is the radio of the circular patch , $L$ is the patch length of the rectangular shape.

### 3.3 Quality Factor, Radiation Efficiency, Bandwidth and Gain

The quality factor, bandwidth, and efficiency are antenna figures-of-merit, which are interrelated, and there is no complete freedom to independently optimize each one.The quality factor $Q$ of a microstrip antenna becomes one of the most important design parameters. Quality Factor is a measure of the radiation loss in the microstrip antenna. For a given substrate characteristics like $\epsilon_r$, tan$\delta$, $\sigma$ and a radiation pattern, the unloaded quality factor $Q_0$ and the radiation efficiency is (James et al., 1989; Long, 1978):

$$\frac{1}{Q_0} = \frac{1}{Q_r} + \frac{1}{Q_c} + \frac{1}{Q_d} \tag{14}$$

where

$$\eta = \frac{Q_0}{Q_r};\ Q_d = \frac{1}{tan\delta};\ Q_c = \frac{h}{\delta_s};\ \delta_s = \frac{1}{\sqrt{\pi f\mu_0\sigma}};\ Q_r = \omega\frac{W_T}{P_{rad}};\ W_T = \frac{1}{2}\epsilon\int_v |E|^2dv \tag{15}$$

and

$$P_{rad} = \frac{1}{2}\Re\left\{\int_{sh}((E\times H)^*\cdot nds)\right\} = \frac{1}{2Z_0}\int_0^{2\pi}\int_0^{\frac{\pi}{2}}(|E_\theta|^2 + |E_\phi|^2)R^2\sin\theta d\theta d\phi \tag{16}$$

$Q_r$ being the quality factor $Q$ of radiation loss, $\delta_s$ skin depth, $Q_c$ is $Q$ of conductor loss, $Q_d$ $Q$ of dielectric loss, $P_{rad}$ radiated power, $P_{in}$ input power, $Z_0$ free space intrinsic impedance, $\Re$ real part, $Sh$ sphere. Those expression agree well with measured values of $Q_0$ and $\eta$ for common design ranges as long as the substrate is not thick enough to excite the surface wave and higher order modes (Hirazawa et al., 1992). In the general case the bandwidth of the antenna is expressed as:

$$BW = \frac{(VSWR - 1)}{Q_0\sqrt{VSWR}} \tag{17}$$

Then the BW of a practical antenna is determined for a given value of $Q_0$ and $VSWR$.

### 3.4 Directivity Gain

Directivity gain can be obtained by integrating the radiation pattern over an arbitrary surface containing the antenna. A useful expression to compute the directivity gain is:

$$G_d(\phi,\theta) = \frac{|E(\theta_0,\phi_0)|^2}{\frac{1}{4\pi}\int_0^{2\pi}\int_0^{\pi}|E(\theta,\phi)|^2\sin\theta d\theta d\phi} \tag{18}$$

where

$$|E(\theta,\phi)|^2 = |E_\theta(\theta,\phi)|^2 + |E_\phi(\theta,\phi)|^2 \tag{19}$$

and $E_\theta$ and $E_\phi$ are given by (8, ), (9), (11, ) and (12, )

### 3.5 Impedance Characteristics

Impedance characteristics of a microstrip antennas has been treated mainly from two points of view: self impedance and auto-impedance. Here the basic procedures for computing the impedance characteristics are briefly described.

#### 3.5.1 Input Impedance

The Input impedance or self-impedance of a microstrip antennas has been studied by several models like Transmission Line Model (Derneryd, 1976; James et al., 1989) cavity model (Richards et al., 1979; 1981) and full wave models that includes several numerical techniques like method of moments (Gupta et al., 1988; Pozar, 1982) finite element (Volakis et al., 1998) and finite difference time domain method (FDTD) (Reineix et al., 1989; Sheen et al., 1990) among others. Transmission Line Model has been widely used because of its simplicity to describe the impedance of a general microstrip and provides a good physical insight, however it may lack in accuracy. In contrast, the cavity model is more accurate but at the same time, more complex although it also provides a good physical insight. Full wave models are in general very accurate and versatile and can manage a wide number of elements, complex patch shapes and coupling problems, but they are more complex models with less physical insight. Here we shall discuss the transmission line model only. In general, input impedance is complex and includes both a resonant an nonresonant components which is commonly reactive. Both real and imaginary parts of impedance vary with frequency. Ideally the reactance and the resistance exhibit symmetry about resonant frequency. At resonance, the reactance is equal to the average of the addition of its maximum and minimum values.

### 3.5.2 Transmission Line Model

The simplest analytical description of a rectangular microstrip patch utilizes transmission-line theory and models the patch as two parallel radiating slots (Munson, 1974). Each radiating edge of length $L$ (Fig. 3) is modeled as a narrow slot radiating into a half-space,with a slot admittance given by (Harrington, 1961)

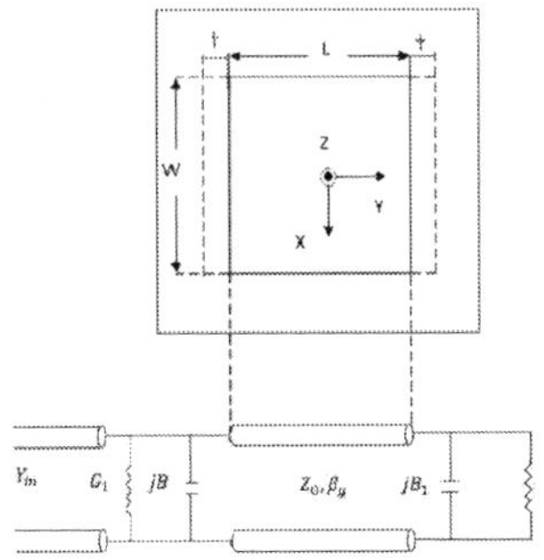


Fig. 3. Top view of a rectangular patch antenna and its associated transmission line equivalent circuit

$$G_1 + jB_1 \cong \frac{\pi q}{\lambda_0 \eta_0}[1 + j(1 - 0.636 \ln k_0 w)] \tag{20}$$

where $\lambda_0$ is the free-space wavelength, $\eta_0$ the intrinsic impedance of the free-space, $k_0 = \frac{2\pi}{\lambda_0}$ and $w$ is the slot width, approximately equal to the substrate thickness $h$. Assuming that the field remains constant along the direction parallel to the radiating edge, the characteristic impedance is given by:

$$Z_0 = \frac{1}{Y_0} = \frac{t\eta_0}{L\sqrt{\epsilon_r}} \tag{21}$$

and the propagation constant is approximated by :

$$\beta_p \cong k_f \sqrt{\epsilon_{eff}} \tag{22}$$

where $k_f$ is the propagation constant in free space. $\epsilon_{eff}$ is the effective dielectric constant given by:

$$\epsilon_{eff} = \frac{\epsilon_r + 1}{2} + \frac{\epsilon_r - 1}{2}\left(1 + 12\frac{h}{L}\right)^{-\frac{1}{2}} \tag{23}$$

The conductive component $G_1$ and the reactive component $B_1$ in (20) are related to the fringing effect and the radiation loss, and are approximated respectively by:

$$B_1 = \frac{k_f \Delta l}{Z_0} \sqrt{\epsilon_{eff}} \tag{24}$$

$$G_1 \approx \begin{cases} \frac{L^2}{90\lambda_0^2}, & L < 0.35\lambda_0 \\ \frac{L}{120\lambda_0} - \frac{1}{60\pi^2}, & 0.35\lambda_0 \leq L \leq 2\lambda_0 \\ \frac{L}{120\lambda_0}, & 2\lambda_0 < L \end{cases} \tag{25}$$

where $\Delta l$ denotes the line extension due to fringing effect. This value can be expressed by:

$$\Delta l \approx 0.412t\frac{(\epsilon_{eff}+0.3)}{(L/h+0.264)} \tag{26}$$

The input admittance of the patch antenna on Fig. 3 can be treated as two slot antennas interconnected by a transmission line having characteristic admittance and propagation constant approximated by (20) and (22):

$$Y_{in} = G_1 + jB_1 + Y_0\frac{(G_1+jB_1)+jY_0\tan(\beta_p L)}{Y_0+j(G_1+jB_1)\tan(\beta_p L)} \tag{27}$$

When the average stored magnetic and electric energies are equal then resonance occurs (Pozar, 2005), under these conditions the imaginary part of the admittance must vanish.

$$\Im\{Y_{in}\} = 0 \tag{28}$$

leading to:

$$\tan(\beta_p L) = \frac{2Y_0B_1}{G_1^2+B^2-Y_0} \tag{29}$$

This condition is used to determine the resonant frequency for a given patch length $L$ or inversely to establish the resonant leght $L$ for a given resonant frequency. The input admittance at resonance is:

$$Y_{in} = 2G_1 \tag{30}$$

### 3.6 Arrays and feed networks

Even that most of the planar antennas are low and medium gain, wide pattern antennas, its versatility and mainly its low profile make them very popular in arrays. Its use to synthesize a required pattern is a nowadays application in many communication systems; most of the microstrip arrays are designed for fixed-beam broadside applications, increasing directivity and performing other functions, as for scanning purposes; those functions are difficult to obtain with a single radiator or even other classes of antennas; often the feed network is located coplanar with the array elements. The main versatility of planar antenna arrays is that the feed network usually is part of the printed environment; the elements can be feed by a single line in a series-feed network, or by multiple lines as a corporate-feed network (Balanis, 2005), including both parallel feed and hybrid series/parallel feed. Series-feed arrays are usually built using photolithography for both the radiating elements and the feed network, reducing its application to fixed beams, linear or planar arrays with single or dual polarization. In this configuration multiple elements are arranged linearly and feed serially by a single transmission line, multiple linear arrays can then be fed either serially or parallel to form a two dimensional planar array (Lee, 1997). Series-feed line can be in-line feed when all radiators are arranged in the same line or out-of-line feed when elements are parallel each other. The series-feed array can be classified as resonant and traveling-wave. For the first kind, impedances at the transmission line junctions and the patch elements are not matched and elements are spaced multiple integers of one wavelength apart creating a broadside beam, with a very narrow resonant bandwidth, around 1%. For the traveling wave array type, transmission line and elements are matched, with spacing of one wavelength for broadside or less for off-broadside radiation. The traveling current wave over the feeding line moves almost without reflections reducing almost to zero the energy at the end of array which can be either absorbed by a matching network or reflected back to be reradiated in phase. Corporate feed arrays, also called parallel feed arrays, are more versatile and the designer has more control of the phase and amplitude of each feeding element, making them ideal for scanning phased arrays, multibeam arrays, or shaped-beam arrays.

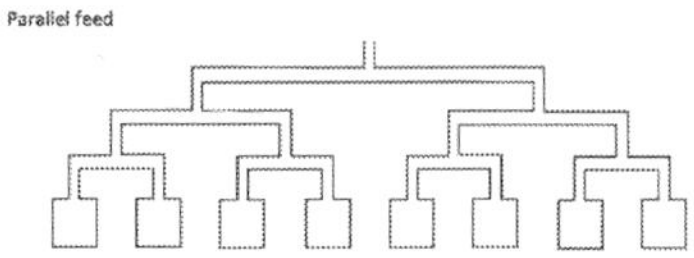


Fig. 4. Parallel feed configuration for microstrip arrays. After (Lee, 1997)

Fig. 4 shows how the patches are feeding by power division lines. Transmission lines divides into two branches and each branch divides again until it reaches the patch elements. For a broadside array all the divided lines are of the same length. A disadvantage of parallel feeding is that the insertion loss is higher than that of a series feeding, however is less affected by phase changes due frequency changes, because relative phases between all elements will remain the same. The corporate feed array can achieve a bandwidth of 15% or more, depending on the design.

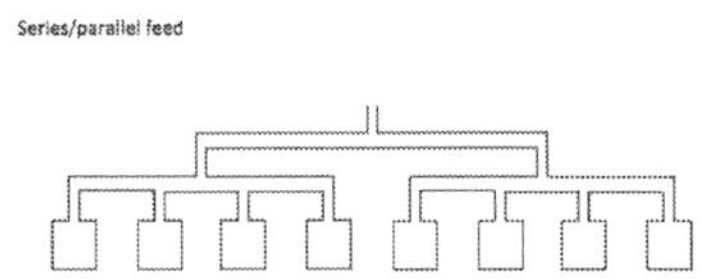


Fig. 5. Series-Parallel feed configuration for microstrip arrays. After (Lee, 1997)

The hybrid series/parallel array is shown in Fig. 5 where a combination of series and parallel feed lines are used, feeding elements in this way is possible to have a wider bandwidth than the series feed array but with a higher insertion loss due the parallel feeding, but the technique allows to make design trade-offs between bandwidth and insertion loss.
A microstrip can be designed in a single layer or multilayer configuration, decision is related with complexity and cost, sidelobes, cross polarization, bandwidth and other factors. Using a single layer reduces manufacturing costs but other characteristics are degraded. When low sidelobe or cross polarization is needed the double layer design seems to be the better choice. Fig. 6 shows a multilayer of a dual polarization antenna, where the antenna feeding is obtained with crossed slots on the ground plane and the feeding of the two polarizations is obtained using two orthogonal microstrip lines. A disadvantage of planar arrays is the influence between elements and feed lines, which affects the performance of the others, then in the design is very important to take into account effects as mutual coupling and internal reflections; coupling between elements generates surface waves within the substrate which can be eliminated using cavities in conjunction with microstrip feeding elements, but these effects are difficult to analyze for common analytical methods, therefore for accurate results should be used full wave solutions as those presented in this chapter, applied in most of the actual computational tools of present days.

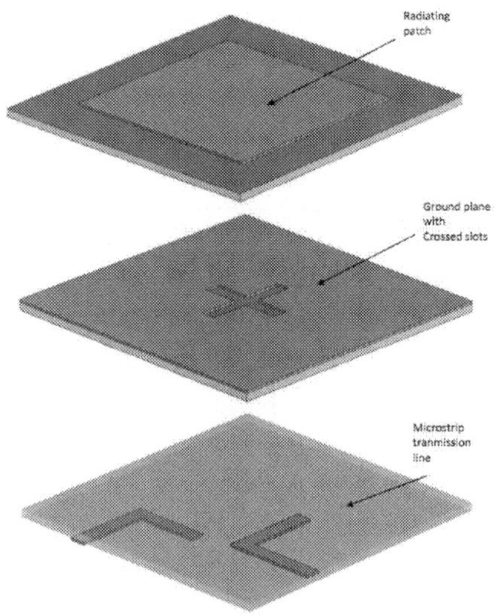


Fig. 6. Multilayer dual-polarized microstrip patch element. After (Lee, 1997)

## 4. Computational Tools

At the beginning of 80's, when the boom of planar antennas started the only available computational tools where the CAD packages, but their applications were constrained to low frequency electronic systems (Harrington, 1992) and the microwave programs were limited and expensive. The situation has changed radically since the arrival of personal computers and mainly the permanent improvement of their characteristics, making them a powerful tool for analysis of electromagnetic problems; on the other hand, the use of mathematical models had become in the development of specialized software tools widely used nowadays in the field of planar antenna analysis and design. Although the appearance of many commercial tools is a common situation every day, many researchers use their own programs; authors have been working since some years ago in development of their own software, applying it to antenna and microstrip devices (Barrera-Figueroa et al., 2007; 2009; Sosa-Pedroza et al., 2008; 2009) considering both, saving economical resources and "learning doing" mainly for academic reasons. Most of the actual work on antenna computational methods is based on solution of Maxwell Equations in integral or differential form, Method of Moments is an example and maybe the most applied for integral Methods and Finite Elements (FE) and Finite Difference on Time Domain (FDTD) for differential methods.

### 4.1 Method of Moments (MoM)

Integral Methods solve Maxwell's Equations in its integral form, describing the electromagnetic problem. As the current density on the conductor is related with the Electric Field, some equations have been derived from Maxwell equations, two examples of those are Pocklington equation and Hallén Equation, both can be studied in the literature (Balanis, 2005; Kraus et al., 2002). For an arbitrary shaped wire (Sosa-Pedroza et al., 2007) as the one shown in Fig. 7, is possible to deduce Pocklington Equation given by:

$$E_s^I = -\frac{j}{\omega\epsilon}\int_s^{\prime} I_s(s')\left[k^2\mathbf{s}\cdot\mathbf{s}+\frac{\partial^2}{\partial s\partial s'}\right]\frac{e^{-jkr-r'}}{4\pi r-r'} \tag{31}$$

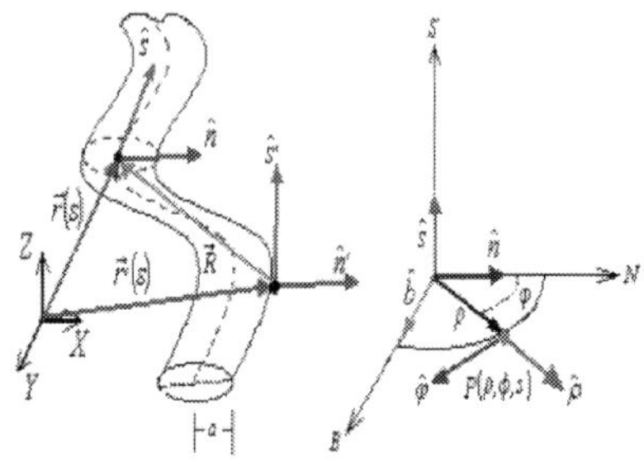

Fig. 7. Arbitrary shaped wire

$E_s^I$ is the tangential incident electric field. As it seen the unknown distribution current is inside the integral and solution is not possible. The MoM formulation, introduced by Harrington in the 60's (Harrington, 1961), is used to get a numerical solution of (31) expressing the unknown function as a linear combination of $n$ linearly independent functions, called basis functions:

$$I_s(s') = \sum_{n=1}^{N} c_n i_n(s') \tag{32}$$

where $c_n$ are the unknown coefficients to be determined. Substituting (32) into (31) results the equation with $N$ unknowns:

$$E_s^I = -\frac{j}{\omega\epsilon}\sum_{n=1}^{N} c_n \int_s w_m \int_{s'} i_n(s')\left[k^2\mathbf{s}\cdot\mathbf{s}+\frac{\partial^2}{\partial s\partial s'}\right]\frac{e^{-jkr-r'}}{4\pi r-r'} \tag{33}$$

with $m = 1, 2, ..., N$.
for a consistent equation system, is necessary to find $N$ linearly independent equations, then is used the inner product of (33) with other set of $N$ chosen linearly independent functions $w_m(s)$ named weight function, then

$$\int_s E_s^I ds = -\frac{j}{\omega\epsilon}\sum_{n=1}^{N} c_n \int_s w_m \int_{s'} i_n(s')\left[k^2\mathbf{s}\cdot\mathbf{s}+\frac{\partial^2}{\partial s\partial s'}\right]\frac{e^{-jkr-r'}}{4\pi r-r'} \tag{34}$$

Which can be reduced in a matrix form as:

$$\begin{pmatrix} Z_{11} & Z_{12} & \dots & Z_{1N} \\ Z_{21} & Z_{22} & \dots & Z_{2N} \\ \vdots & \vdots & \ddots & \vdots \\ Z_{N1} & Z_{N2} & \dots & Z_{NN} \end{pmatrix}\begin{pmatrix} c_1 \\ c_2 \\ \vdots \\ c_N \end{pmatrix} = \begin{pmatrix} v_1 \\ v_2 \\ \vdots \\ v_N \end{pmatrix} \tag{35}$$

where the elements $Z_{mn}$ are obtained from:

$$Z_{mn} = -\frac{j}{\omega\epsilon}\int_S w_m \int_{S'} i_n(s')\left[k^2\mathbf{s}\cdot\mathbf{s} + \frac{\partial^2}{\partial s \partial s'}\right]\frac{e^{-jkr-r'}}{4\pi r - r'} \tag{36}$$

and the elements $v_m$ are:

$$v_m = \int_S w_m E_s' ds \tag{37}$$

$c_n$ represent the system's unknowns. Matrices of (35) are known as impedance matrix $(Z_{mn})$, current matrix $(c_n)$, and voltage matrix $(v_m)$. The solution for (35) is:

$$(c_n) = [Z_{mn}]^{-1}(v_m) \tag{38}$$

where the inverse matrix$(Z_{mn})^{-1}$ is obtained by a numerical technique. It is important to mention that both, basis and weight functions are arbitrary functions, selected considering computational resources and time and accuracy of solution. As is seen the Kernel of the integral includes Green's function, representing the electromagnetic influence on the entire surrounding space. Solution is valid at every point of an infinite space, including the far field radiation phenomena that are vital for antenna analysis.

Integral equations are established as a multilayered Green's function, such that the background can consist of an arbitrary number of horizontal, infinitely stretched layers, containing dielectric substrates and conducting ground planes, always present on planar antennas. The main components on the antenna are replaced with equivalent surface/volume currents, appearing as the primary unknowns in the resulting integral equations, solved by MoM. From (35) we can see that the method creates a dense matrix equation which solution gives the current distribution on the environment, after that, all other antenna parameters are easily obtained.

### 4.2 Finite Difference on Time Domain (FDTD)

Differential methods solve Maxwell Equations in their differential form; most of the computational solvers use the Finite Element Method (FEM) and the Finite Difference on Time Domain (FDTD).

The FDTD method, introduced by (Yee, 1966), transforms the Differential Maxwell Equations in Finite Difference Equations, used to generate an algorithm and a program code to solve them, over a specific propagation region; it uses the Yee's cell algorithm in a central difference scheme, considering field variations in time and space as the original Maxwell Equations, the characteristics of media are also defined in the method by means of $\epsilon$, $\mu$ and $\sigma$ characteristics, the algorithm is processed by a computer to analyze behavior of EM field moving over an environment of any kind. The interacting electric and magnetic fields are given by:

$$\frac{\partial H_x}{\partial t} = \frac{1}{\mu}\left[\frac{\partial E_y}{\partial z} - \frac{\partial E_z}{\partial y} - (M_{source_x} + \sigma^* H_x)\right] \tag{39}$$

$$\frac{\partial H_y}{\partial t} = \frac{1}{\mu}\left[\frac{\partial E_z}{\partial x} - \frac{\partial E_x}{\partial z} - (M_{source_y} + \sigma^* H_y)\right] \tag{40}$$

$$\frac{\partial H_z}{\partial t} = \frac{1}{\mu}\left[\frac{\partial E_x}{\partial y} - \frac{\partial E_y}{\partial x} - (M_{source_z} + \sigma^* H_z)\right] \tag{41}$$

$$\frac{\partial E_x}{\partial t} = \frac{1}{\epsilon}\left[\frac{\partial H_z}{\partial y} - \frac{\partial H_y}{\partial z} - (J_{source_x} + \sigma E_x)\right] \tag{42}$$

$$\frac{\partial E_y}{\partial t} = \frac{1}{\epsilon}\left[\frac{\partial H_x}{\partial z} - \frac{\partial H_z}{\partial x} - (J_{source_y} + \sigma E_y)\right] \tag{43}$$

$$\frac{\partial E_z}{\partial t} = \frac{1}{\epsilon}\left[\frac{\partial H_y}{\partial x} - \frac{\partial H_x}{\partial y} - (J_{source_z} + \sigma E_z)\right] \tag{44}$$

These equations are transformed in a discrete form using the Yee algorithm, which can be solved by computational methods, as an example is presented only one of them:

$$\frac{E_x|^{n+1/2}_{i,j+1/2,k+1/2} - E_x|^{n-1/2}_{i,j+1/2,k+1/2}}{\Delta t} = \frac{1}{\epsilon_{i,j+1/2,k-1/2}} \cdot$$
$$\left\{ \frac{H_z|^{n}_{i,j+1,k+1/2} - H_z|^{n}_{i,j,k+1/2}}{\Delta y} - \frac{H_y|^{n}_{i,j+1/2,k+1} - H_y|^{n}_{i,j+1/2,k}}{\Delta z} \right.$$
$$\left. - J_{source_x}|_{i,j+1/2,k+1/2} - \sigma_{i,j+1/2,k+1/2} E_x|_{i,j-1/2,k+1/2} \right\} \tag{45}$$

Yee Algorithm discretizes both time and space, represented by parameters $n, i, j, k$ with intervals of $\Delta t$ and $\Delta$ respectively. As seen in equation (45) the media characteristics are specially considered as $\epsilon$, $\mu$ and $\sigma$ which position is defined using the $(i, j, k)$ subindex, then is possible to analyze the effects of any material at any position on the computational space. A computational code of any program language permits to know the EM behavior over the entire computational space. The FDTD and also the FE differential-equation methods are particularly suitable for modeling full three-dimensional volumes that have complex geometrical details. They are extremely efficient for smaller close-region problems involving inhomogeneous media (James et al., 1989).

### 4.3 Computational tools comparison

An excellent summary and comparison of actual available commercial software used on planar antenna analysis and design is presented in (Vasylchenko, 2009), they analyze 5 commercial tools and one "in house", comparing all of them in the analysis of planar antennas looking to guarantee the optimal use of each of the software packages, to study in detail any discrepancies between the solvers, and to assess the remaining simulation challenges. Even their work is not the first one on the theme, mentioning references strengthening their vision that, an extensive benchmark study over a large variety of solvers and for several structures has not yet been documented.

As the operation of EM solvers is based on the numerical solution of Maxwell's equations in differential or integral form, one or other influences the efficiency and accuracy and users may get the wrong impression that a given solver is automatically suited to solve any kind of problem with arbitrary precision. Comparison in the Vasylchenko work verifies the plausibility of such expectations by presenting an extensive benchmark study that focuses on the capabilities and limitations of the applied EM modeling theories that usually remain hidden from the antenna designer. The integral solvers they analyze are the one they designed in K. U. Leuven's: MAGMAS 3D, the others are IE3D from Zeland Software, FEKO from EM Software & Systems, and ADS Momentum from Agilent. On the other hand they analyze the two leading differential EM tools, HFSS from Ansoft for the finite-element method, and CST Microwave

Studio for the FDTD method. After a careful analysis, comparing results with measurement of 4 common planar antennas, their conclusion is as follows:
Classical patch antennas could be predicted by every simulation program with a deviation not beyond 1.5 %. The simulation based on MoM was inherently faster and are more attractive in price. On the other hand the FEM and FDTD are inherently able to analyze much more general structures, but require the inversion of much larger, but sparse, matrices, requiring higher memory resources. Although the calculation times were not that different at the time of experiment, they presented a reference in which it seems that dedicated inversion techniques for MoM solvers are nowadays fully in development, opening the possibility that better times can be obtained for differential equations solvers.
Proper mesh generation and a correct feeding model are two crucial issues predetermining the successful simulation in the software packages reviewed. In general, a very neat adaptive mesh refinement, implemented in Ansoft's HFSS and as an option in CST's MWS, allows better handling of a design with difficult electromagnetic coupling between its different parts. Such characteristics pertain to applications in mobile gadgets, such as the GSM antennas. Having no mesh refinement option, MoM-based programs require more careful consideration of the initial meshing. MoM solvers can provide an improvement in simulation results and time using so called edge-meshing features, while avoiding excessive meshing on the bulk of the metal structure. However the study concludes that the meshing schemes in all solvers are adequate.
Some designs, such as the GSM and UWB antennas, require finite substrate effects to be taken into account, such as diffraction from substrate edges. MoM based solvers show better convergence when a dielectric substrate is infinite, but the trend toward miniaturizing antennas diminishes the advantage of using these solvers, then they conclude that at present, differential equations programs are better suited for modeling small antennas. On the other hand(Vasylchenko, 2009) suggest that the feeding models, as implemented today in the widespread commercial 35 solvers, are probably unsatisfactory in the case of small structures with complicated electromagnetic-coupling behavior, but HFSS and CST MWS solvers are better suited to handle the problem.
As a final guideline, authors recommend the use of two different solvers, based on different theoretical methods (integral and differential), to characterize a specific device if both results are in good agreement, it is reasonable to expect that the results can be trusted, if the two results are in disagreement, a deeper investigation of the structure and its modeling is absolutely necessary.

## 5. Planar antennas on space applications

When a designer decide to use planar or microstrip antennas on a space applications should take in account three factors among those related with the inherent design of the radiator (Lee, 1997); those factors are critical and need to be considered. One is that the antenna must be able to support the high vibration produced during the launch from the Earth; acceleration can be as high as 10 Gs or more, under this conditions soldering junctions and laminating of multilayer antennas tend to breakdown, then they should be made strong enough to survive the vibration, a solution could be the use of noncontacting feeds as proximity, capacitive or aperture coupling. The second factor is related with the extreme temperature difference which can be as high as 100°C to -70°C, whether the antenna "sees" the sun or not, behind a shaded area. Under this condition, the laminating adhesive material must survive physically and electrically into this environment. Third factor is the space vacuum, as is known at low pressures, electrons are almost free to leave an electrode and move across to the opposite electrode, a phenomenon known as multipacting. For a microstrip antenna, the two electrodes are the patch and the ground plane, when the phenomenon is present reduces the capacity of power handling of the antenna then it should be designed with the proper thickness. These three factors limit the use of planar and especially microstrip antennas, nevertheless there are many examples of spacecrafts which can be mentioned: Earth Limb Measurements Satellite, Shuttle Imaging Radar, Geostar system and especially the Mars Pathfinder using a small X band microstrip antenna providing circular polarization with a peak gain of 25 dB. Antenna was constructed with a parallel feed power divider and electromagnetically coupled dipoles. The divider and the dipoles were printed on multilayer honeycomb substrates which have open vented cells for space applications.

### 5.1 Morelos: First Mexican Satellite System

Historically the first satellites using planar antennas could be the Mexican Morelos System, constructed by Hughes Aircraft Company (Satmex, 2010);. They were launched on the space Shuttle in June 17 and November 27, 1985 and they were the first in use the HS-376 platform as a hybrid satellite operating in two frequency bands (C and Ku) simultaneously. The four Ku-band channels used the planar arrays for reception only having a bandwidth of 108 MHz with a minimum effective isotropic radiated power (EIRP) of 44 dBW throughout Mexico. Transmit and receive beams in the C-band and the transmit beams in the Ku-band were created by a 1.8 m wide shared aperture grid antenna with two polarization-selective surfaces. The front surface was sensitive to horizontally polarized beams and the rear was sensitive to vertically polarized beams. Separate microwave feed networks are used for the two polarizations. Fig. 8(a) shows the spacecraft with the planar array and Fig. 8(b)the antenna and the reflector in the construction bay. Morelos Satellites were a very successful communications system; Morelos 1 exceeded his life from 9 years to 10, when it was substituted in 1996 for the first satellite of 2$^{nd}$ generation of Mexican satellites, but Morelos 2 was in operation until to 2002, almost doubling its life designed time.

### 5.2 The IRIDIUM Main Mission Antenna Concept

A commercial satellite system using planar antennas is the MOTOROLA's IRIDIUM (Schuss et al., 1990) shown in Fig. 8(c) used for personal satellite communications with a constellation of 66 satellites placed in low earth orbit, positioned in six polar orbital planes with 11 satellites plus one spare per plane. The main mission antenna (MMA), consists of three fully active phased-array panels providing the band link from the satellite to the ground user. Each phased-array panel produces 16 fixed simultaneous beams for a total of 48 beams per satellite linked to hand-held phones having low-gain antennas. The MMA radiates multiple carriers into multiple beams with high efficiency and linearity as well as being lightweight and able to function in the thermal and radiation environment of space. MMA was optimized for the highest link margin accordingly with its size and the budgeted RF power per carrier. The architecture of the MMA phased-array panel is shown in Fig. 8(d); each array consists of over 100 lightweight patch radiators, each of which is driven by a Transmitter/Receiver (T/R) module, which are in turn collectively excited by an optimized beamformer network. The beamformer network forms the 16 optimized shaped beams for both transmit and receive operation with the T/R modules maintaining a high G/T in receive operation and efficient EIRP generation for transmit operation. The satellite can receive or transmit through each beamport, providing the RF access to a particular fixed beam. In general, several or all beams

(a) The *Morelos* satellite

(b) The *Morelos* at the construction bay

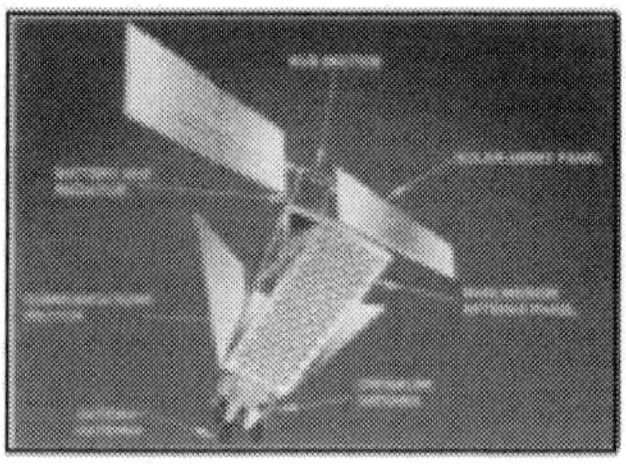

(c) IRIDIUM space vehicle (©(1999) IEEE)

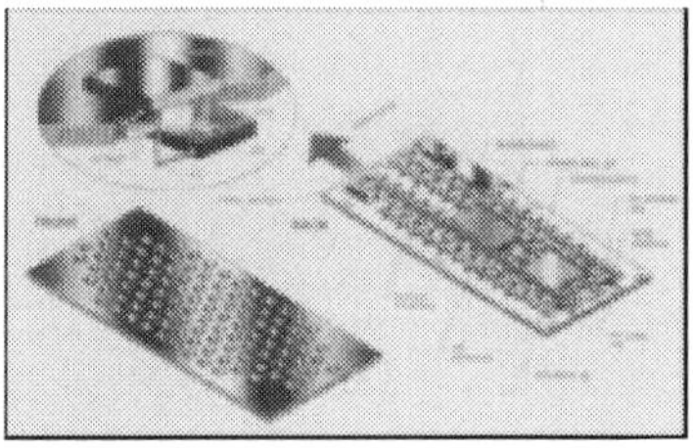

(d) MMA panel construction (©(1999) IEEE)

Fig. 8. The use of planar antennas in commercial satellites and space vehicles

can be utilized at once in either transmit or receive operation with the only limitation being the MMA capacity constraints on transmit.

### 5.2.1 Patch Radiator

(a) Bottom view of patch radiator (©(1999) IEEE)

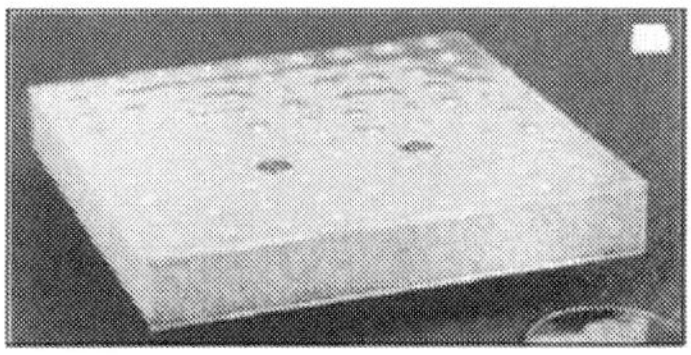

(b) Top view of patch radiator (©(1999) IEEE)

Fig. 9. Patch radiator developed for the MMA

Fig. 9(a) and Fig. 9(b), show the patch radiator developed for the MMA, which was manufactured as a separate component and bonded onto the MMA panel during array assembly; its radiator is built as one assembly and contains the matching and polarizing networks; a single 50 Ω input connector is provided on the underside of the patch for connection to the $T/R$ module. The radiator cavity is loaded with an artificial dielectric substrate whose weight is approximately one tenth that of teflon, but which has a dielectric constant of approximately two. This dielectric constraint is needed to obtain the desired scan and polarization performance of the array. The artificial dielectric also permit efficient heat radiation out the front face of the array during peak traffic loads.

### 5.3 Antennas for Modern Small Satellites

Many examples of planar antennas application are discussed in literature, but its major application could be the modern small satellites (MSS) which are revolutionizing the space industry (Gao et al., 2009). They can drastically reduce the mission cost, and can make access to space more affordable.

These modern small satellites are useful for various applications, including telecommunications, space science, Earth observation, mitigation and management of disasters (floods, fire, earthquake, etc.), in-orbit technology verification, military applications, education, and training. Typical antenna coverages ranges from low-gain hemispherical, to medium-gain antennas. The basic radiator designs used are normally helices, monopoles, patches, and patch-excited cups (PEC), depending on frequency and range, coverage requirements, and application. As antenna examples of small satellites are mentioned various monopole antennas, printed inverted-F-shaped antennas (PIFAs), microstrip-patch antennas, helices, and patch-excited cup antennas, developed for telemetry, tracking, and command in the UHF, VHF, S, C, and X bands. These antennas are simple, cheap, easy to fabricate, and have wide radiation-pattern coverage; the satellite thus does not need accurate control of attitude.

Universities have played an important role in satellites development, since the beginning of space era; professors were interested in the new research area, either as academic developers or as a part of contracts with satellite industry, but small satellites seems to be a very appropriate area to be working in by universities, due the few economical resources needed. As an example we can mention universities in Mexico, creating clusters to design small satellites; institutions as CICESE (Centro de Investigación Científica y de Educación Superior de Ensenada) in north of Mexico developing transponders and the Instituto Politécnico Nacional working with satellite structures and integration into a clean room, design of monopoles and planar antennas for satellite applications and also exploring the capabilities of new active devices as candidates for LNA amplifiers (Enciso et al., 2005). An especial mention should be make to the Universidad Nacional Autónoma de México (UNAM) which has been working towards the design of a femto satellite.

Other illustrative example is the University of Surrey, which has been developing modern small satellite technology since starting its UoSAT program in 1978. UoSAT-l, developed by Surrey, was launched in 1981. This was followed by UoSAT-2 in 1984. UoSAT-l continued to operate for eight years, while UoSAT-2 was still operational after 18 years in orbit. During the past 30 years, the University of Surrey's spinoff company, Surrey Satellite Technology Ltd. (SSTL), together with Surrey Space Centre (SSC), have successfully designed, developed and launched 32 modern small satellites for various countries around the world. (Gao et al., 2009) have a complete description of various small satellites, which are described in the next lines and figures. Fig. 10 shows a photograph of the S-band microstrip-patch antenna used at SSTL;

it employs a circular microstrip patch, fed by a 50Ω probe feed at the bottom. It can operate within a tunable frequency range of 2.0-2.5 GHz. Left-hand or right-hand circular polarization can be achieved by using a single feed combined with patch perturbation, or a 90°microstrip hybrid combined with a circular patch. It achieves a maximum gain of about 6.5 dBi, has a size of 82 x 82 x 20 mm, and a mass of less than 80 g. It can operate within -20°C to +50°C, is radiation tolerant to 50 kRad, and qualified to 50 Gs rms random vibration on three axes.

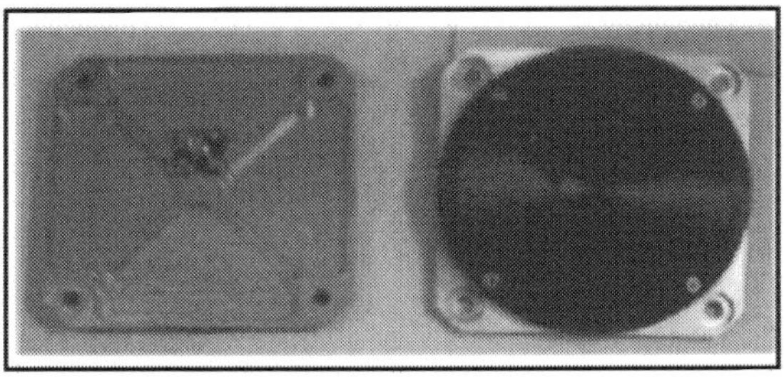

Fig. 10. An S-band patch antenna SSTL. (©(2009) IEEE)

To respond the need for single-frequency low-profile and low-weight hemispherical or near-hemispherical antennas, working at S, C, or X band, patch-excited cup antennas were developed at RUAG Aerospace Sweden. They consist of a short cylindrical cup, with a circular cross section and an exciter. The cup is excited using a stacked circular dual-patch element, or a single patch. The lower patch or the single patch is fed at one point, and the patch has two opposite perturbations for generating circular polarization. The antennas have special features to minimize their coupling to the surrounding spacecraft environment, as this is a common problem for low-gain antennas of this type, and it has an effect on the installed performance. The antenna's diameter is 60 mm for the C band antenna, and 40 mm for the X-band antenna. The mass is less than 90 g for the C-band antenna, and less than 20 g for the X-band antenna. They are both almost all metal antennas (which is a preferred property), with dielectric material only in the interface connector.

Fig. 11 shows the X-band patch-excited cup antennas that can be used for the telemetry, tracking, and command function. Fig. 12(a) shows the S-band patch-excited cup antenna, developed at Saab Space. It consists of three patches, mounted within a thin aluminum cup with a rim height of about a quarter wavelength. Two lower patches form a resonant cavity, allowing broadband or double tuning. The top patch acts as a reflector that affects the illumination of the aperture, and is used to improve the aperture efficiency. To achieve circular polarization, the lower patch is fed in phase quadrature at four points by a stripline network. It achieves a maximum gain of about 12 dBi. A patch-excited cup antenna development performed at Saab Space is the update of the antenna in Figure 6, to be used for other missions; it has a radiator tower that is modified compared to the original design. It is now an all-metal design, and has a new feed network configuration: an isolated four-point feed design, antenna is shown in Fig. 12(b).

Surrey also pioneered the use of GPS and global navigation satellite systems (GNSS) in space. A GPS receiver can provide accurate position, velocity, and time for LEO satellites. For this application, the antenna needs to be compact, low profile, able to operate at GPS frequencies in the L1 (1.575 GHz) and L2 (1.227 GHz) bands with stable performance, and produce low backward radiation towards the small satellite body.

Fig. 11. An X-band patch-exited cup antenna (©(2009) IEEE).

A medium-gain antenna, shown in Fig. 13(a), was launched on the UK-DMC satellite of SSTL for the purpose of collecting reflected GPS signals in orbit. This satellite has begun to collect reflected signals under a variety of sea conditions, and over land and ice. The antenna is a three-element, circularly polarized microstrip-patch array with a gain of 12 dBi. Antenna-design challenges remain in terms of further reducing antenna size, improving the antenna's efficiency, multi-band (L1/L2/L5 band) operation, constant phase center, multipath mitigation, etc.

Fig. 13(b) shows the patch-excited cup antenna developed at RUAG Aerospace Sweden. It consists of two patches placed in a circular cup. To obtain a stable antenna covering two GPS frequency bands (Ll, L2), the bottom patch was capacitively fed by four probes and an isolated feed network. The antenna achieved a coverage out to $80^0$ in zenith angle, and low backward radiation. The antenna's diameter is 160 mm, and the mass is 345 g. This antenna shows how shorted-annular-patch can achieve high-accuracy GPS/GNSS performance without compromising the physical constrains.

## 6. Some proposals for future applications

Spacecraft development and research never ends and antenna improvements are not the exception, even thinking that some of them were designed for other applications, always is possible to extrapolate to space applications, but antenna research and design for satellites and spacecrafts is an area of permanent expansion. Starting with airborne applications, where planar antennas have a permanent development, to meet the low profile and conformal challenges, is possible to extrapolate them to satellite systems. For airplanes as for satellite and spacecrafts, an array antenna should have good isolation, high efficiency, and ease of integration, also a simple feeding-line network with lower loss and high isolation is generally desired. Microstrip series-fed arrays have been shown to have a structure that enhances the antenna's efficiency. This is because the array feeding-line length is significantly reduced, compared to

(a) Cup antenna at RUAG (©(2009) IEEE)

(b) Medium-dowlink antennas (©(2009) IEEE)

Fig. 12. S-band patch-excited cup antenna.

(a) For the UK DMC satellite at SSTL (©(2009) IEEE)

(b) Antennas at RUAG (©(2009) IEEE).

Fig. 13. GPS antennas.

the conventional corporate feeding-line network. A planar structure with a thin and flexible substrate is a good choice, because it will not disturb the appearance of the aircraft, and can be easily integrated with electronic devices for signal processing.

### 6.1 The Shih planar antenna

An example of a planar antenna first designed for aircrafts is the dual-frequency dual-polarized array antenna presented by (Shih et al., 2009). It consists of a multilayer structure of two antennas separated on different layers, adopted for dual-band operation, working in the S band and X band frequencies. To reduce the array's volume and weight, a series-fed network is used. An ultra-thin substrate is chosen in order to make the array conformal, and the array can be easily placed on an aircraft's fuselage, or inside the aircraft.

### 6.1.1 S-band Array Design

The multilayer array structure for dual-band operation is shown in Fig. 14. The S-band antenna elements sit on the top layer, and the X-band antennas are on the bottom layer. A foam layer ($h_2$) serves as the spacer, and is sandwiched between the two substrate layers. One of the important design considerations for this multilayer dual-band array is that the S-band antenna element should be nearly transparent to the X-band antenna elements. Otherwise, the S-band element may degrade the performance of the X-band antenna. Two RTlDuroid 5880 substrates ($\epsilon_1 = \epsilon_3 = 2.2$) and a foam layer ($\epsilon_2 = 1.06$) form the multilayer structure. The thicknesses of the substrates ($h_1$ and $h_2$) are both only 0.13 mm. These ultra-thin and flexible substrates make it possible for the array to be easily attached onto the aircraft's fuselage, or installed inside the aircraft. The foam layer has a thickness of $h_2$ =1.6 mm.

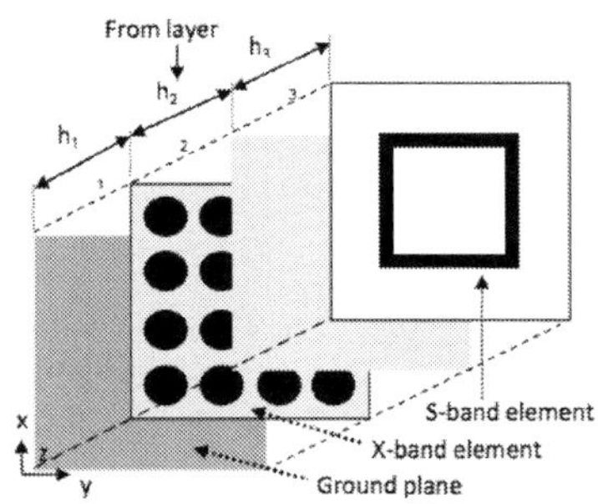


Fig. 14. The multilayer structure of dual-band dual polarized array antenna (©(2009) IEEE).

### 6.1.2 X-Band Antenna and Subarray

The X-band array uses the circular patch as its unit antenna element. The circular patches are fed with microstrip lines at the circumferential edge, as shown in Fig. 15(a) for a single circular patch, two microstrip feeding lines are used to feed the circular patch to generate two orthogonally radiating $TM_{11}$ modes for dual polarized operation. Two feed points are located at the edge of the patch, 90ą away from each other, so that the coupling between these two ports can be minimized. The port isolation also depends on the quality factor of the patch. Increasing the substrate's thickness decreases the isolation, therefore using thin substrates could improve the quality of isolation.

Fig. 15(a) shows a 4 x 8 dual-polarized X-band array. The V port and the H port are the input ports for the two orthogonal polarizations (vertical and horizontal). The array is composed of two 4 x 4 subarrays. The corporate-fed power-divider lines split the input power at each port to the subarrays. Within each subarray, the circular patches are configured into four 4 x 1 series-fed resonant type arrays, which make the total array compact and have less microstrip line losses than would a purely corporate-fed type of array. An open circuit is placed after the last patch of each 4 x 1 array. The spacing between adjacent circular-patch centers is about one guided wavelength ($\lambda_g$ =21.5 mm at 10 GHz). This is equivalent to a $360^o$ phase shift between patches, such that the main beam points to the broadside. The power coupled to each patch can also be controlled by adjusting the size of the individual patch to achieve a tapered amplitude distribution for a lower-sidelobe design.

As shown in Figure Fig. 15(b), the S-band antenna elements are printed on the top substrate, and are separated from the X-band elements by the foam layer. To reduce the blocking of

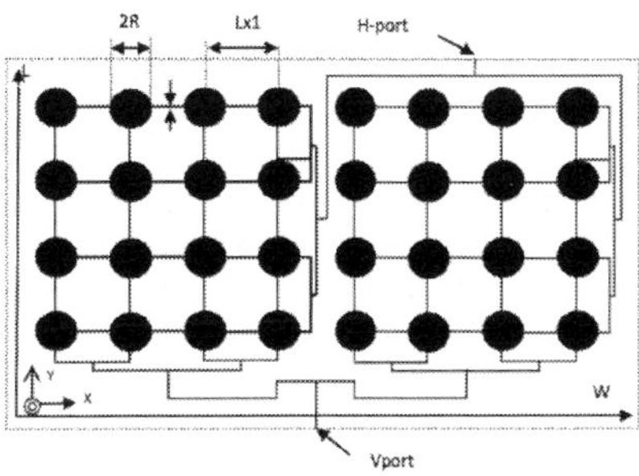


(a) X-band antenna (©(2009) IEEE)

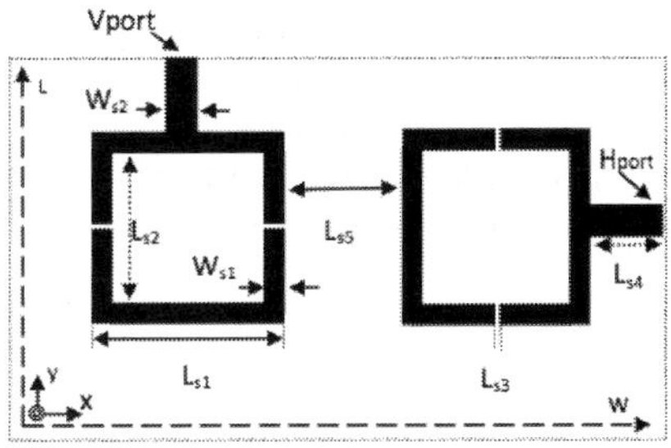


(b) S-band antenna (©(2009) IEEE)

Fig. 15. Microstrip Antenna Arrays

the radiation from the X-band elements at the bottom layer, the shape of the S band elements has to be carefully selected. A ring configuration was a good candidate, since it uses less metallization than an equivalent patch element. Here, a square-ring microstrip antenna is used as the unit element of the S-band array. Because antenna elements at both frequency bands share the same aperture, it is also preferred that the number of elements on the top layer be as small as possible, to minimize the blocking effects.

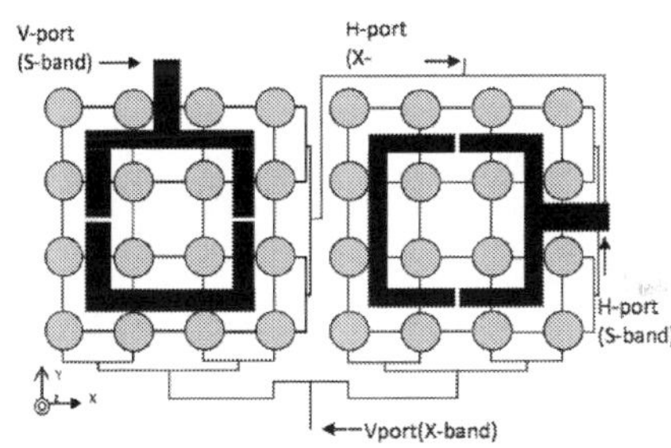


Fig. 16. Geometry of dual antenna (©(2009) IEEE)

The stacked X-band and S-band array antennas are shown in Fig. 16. As can be seen in the figure, the four sides of the square-ring element are laid out in such a way that they only cover part of the feeding lines on the bottom layer, but none of the radiating elements. Unlike an ordinary microstrip-ring antenna that has a mean circumference equal to a guided wavelength, the antenna proposed here has a mean circumference of about $2\lambda_g$ ($\lambda_g$ = 82.44mm at 3 GHz). Although the size of the proposed unit element is larger than an ordinary ring antenna, its gain is about twice as high, because of its larger radiation-aperture area. The ring is loaded by two gaps at two of its parallel sides, these make possible to achieve a 50 $\Omega$ input match at the edge of the third side without using a small value of $L_{s2}/L_{s1}$. For an edge fed microstrip ring, if a second feed line is added to the orthogonal edge, the coupling between the two feeding ports will be high. The V-port and H-port feeds are therefore placed at two individual elements, so that the coupling between the two ports can be significantly reduced. Using separate elements seems to increase the number of antenna elements within a given aperture.

However, this harmful effect could be minimized by reducing the number of elements with the use of larger-sized microstrip rings.

### 6.2 The Cross Antenna

The cross antenna is another possibility of use in spatial applications, it is a traveling wave antenna with circular polarization formed by conductors over a ground plane, proposed by (Roederer, 1990). Antenna can be constructed as a wire or printed antenna. Roederer's paper do not describe completely the antenna but it was reanalyzed by authors (Sosa-Pedroza et al., 2006).
The cross antenna is a printed structure of medium gain and circular polarization, consisting of a conductor or microstrip over a ground plane following the contour of a cross with four or more arms and a diameter of about 1.5 wavelengths. The antenna is feeding on one end by a coaxial line and finished on the other end by a load impedance, considering behavior of travelling wave. Even the antenna was primarily designed for applications in L Band (1500 MHz) mobile communications, the design and experimental characterization was made at 10 GHz and for an eight arms antenna besides original four arms antenna, showing the possibility of extrapolation for other applications as satellite communications. For the cross antenna, feed connector and load position define the right or left circular polarization; it can be used as a unique radiator or as a part of an array, a proposal is that could be used as primary antenna for parabolic reflector with wide focal length and diameter relationship. The main advantage of the cross antenna is its gain (12-15 dBi) compared with its size and weight, ideal for space communications.

Fig. 17. The cross wire antenna

| | |
|---|---|
| Arm length | $\lambda_{eff}$ |
| Arm width | $0.25\lambda_{eff}$ |
| Cross diameter | $2.5\lambda_{eff}$ |
| Wire diameter | $0.01\lambda_{eff}$ |

Table 2. Geometric characteristics of cross antenna

The power at the end of the antenna is controlled by the load impedance and is limited to a small percentage, changing the height of the conductor over the ground plane (typically $\lambda_{eff}/20$ to $\lambda_{eff}/4$) which also affects the axial rate. The bandwidth of the cross antenna is around 5% depending on the number of arms. Fig. 17 shows photograph of a 8 arm radiator,

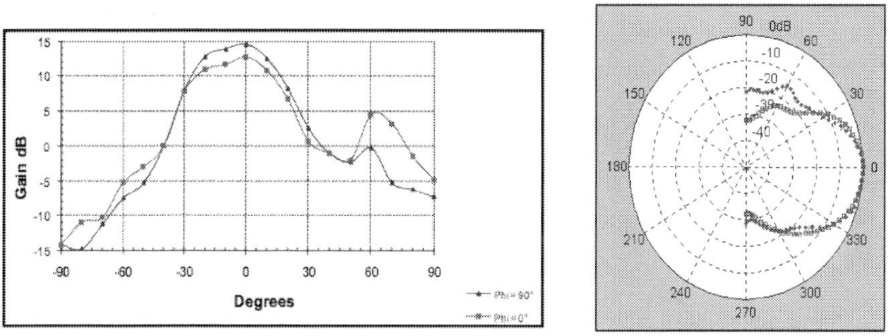


(a) Gain (b) Radiation Pattern

Fig. 18. Electrical characteristics of the Cross antenna

which was constructed both, as a microstrip antenna using a 3.6 mm thick RTDuroid with 2.3 of $\epsilon_{eff}$=2.3 and as a wire antenna using copper wire, supported over the ground plane by small Teflon fragments giving flexibility to move up the structure to analyze the effect of height over the ground plane. Table 2 shows dimensions of the antenna. On the other hand Fig. 18(a) and Fig. 18(b) show the gain and the radiation pattern respectively, for one of the antennas.

### 6.3 Rhombic cross antenna

A variation over cross antenna is a four arm rhombic cross antenna (Lucas et al., 2008), it is also a medium gain and circular polarization structure made of a conductor or strip line over a ground plane, following a rhombic contour of four branches. One end is connected to the feed line and the other is grounded by a load impedance. Antenna was analyzed using Method of Moments and constructed for experimental analysis using both, a 12 AWG wire over a ground plane and printed as a microstrip structure working in 4.2 GHz. The rhombic antenna shows a better performance compared with the four arms Roederer's antenna, with almost 15 dB gain and 1.4 dB for axial ratio. The antenna can be used in mobile communication or as primary radiator of parabolic reflectors, when circular polarization is needed. The construction repeatability is very easy as well the facility to obtain 15 dB gain in a very small antenna.

| | |
|---|---|
| A | $0.430\lambda_{eff}$ |
| B | $0.276\lambda_{eff}$ |
| C | $0.3911\lambda_{eff}$ |
| D | $1.4112\lambda_{eff}$ |

Table 3. Dimensions of rhombic antenna

The rhombic cross antenna geometry is shown in Fig. 19(a), and antenna dimensions as function of effective wavelength, are given in Table 3.

There were constructed several antennas, both wire (air dielectric) and strip line (fiber glass dielectric), the last one is shown in Fig. 19(b); wire antenna uses Teflon supports over the

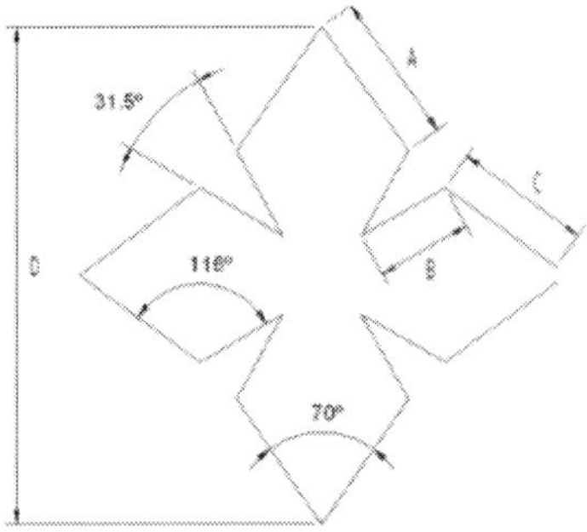


(a) Scheme of rhombic antenna

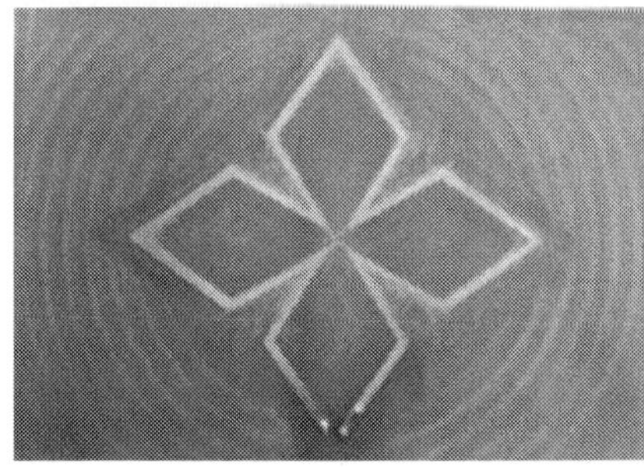

(b) Microstrip antenna

Fig. 19. Physical characteristics of Rhombic Antenna

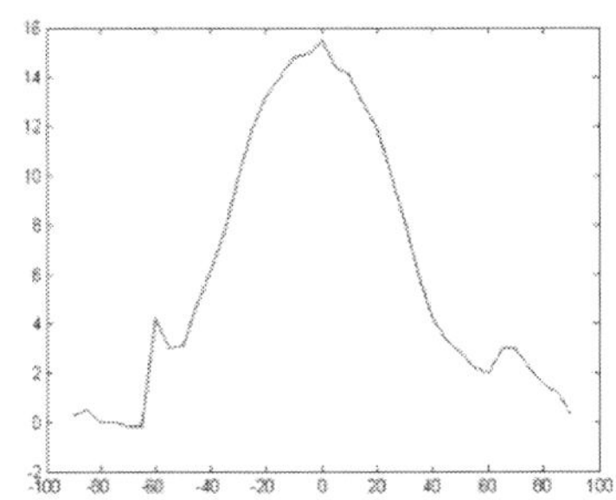

(a) Rhombic antenna gain

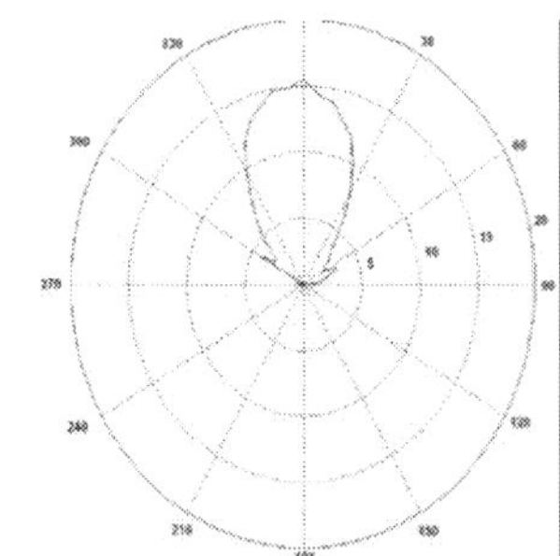

(b) Rhombic antenna field pattern

Fig. 20. Physical characteristics of Rhombic Antenna

ground plane, giving flexibility to change the height over it. Results for gain and field pattern are shown in Fig. 20(a) and Fig. 20(b) respectively, for a 50 $\Omega$ load impedance; the feed impedance is $Z = 38.6 - j56.8\Omega$ for 2.4 GHz:
Even the proposed antennas have not been used yet for spatial applications, their profiles can match for it, in frequencies ranging from L band, S band, commercial C band or X band, either as single structures or as arrays.

## 7. References

Agrawal, P. K., Bailey M. C. (1977) *An Analysis Technique for Microstrip Antennas.* IEEE Trans. on Antennas and Propagation AP-25, pp.756-759

Balanis, C.(2005).*Antenna Theory Analysis and Design*, Wiley-Interscience ISBN 0-471-66782-X

Barrera-Figueroa, V.; Sosa-Pedroza, J.; Lopez-Bonilla, J. (2007). *Numerical approach to King's analytical study for circular loop antenna*, Journal of Discrete Mathematical Sciences & Cryptography, Vol. 10, No. 1 February 2007, pp 82-92.

Barrera-Figueroa, V.; Sosa-Pedroza, J.; Lopez-Bomilla, J. (2009) *Pocklington Equation via circuit theory* Apeiron, on line Journal, Vol 16, No. 1.

Chang, K.(1989). *Handbook of Microwave and Optical Components*. Vol. 1, John Wiley & Sons, New York, USA

Derneryd, A.,G.(1975). *Linear Polarized Microstrip Antennas*. IEEE Trans. on Antennas and Propagation, AP-24, pp. 845-850.

Deschamps, G. A. (1953) Microstrip microwave antennas. *3rd USAF Symposium on Antennas.*

Enciso-Agilar, M.; Crozat, P.; Hackbarth, T; Herzog, H,; Aniel F.(2005). *Microwave Noise Performance and Modeling of SiGe-Based HFETs*. IEEE Trans. on Electron. Dev., Vol. 52, No. 11, November 2005

Gao S.; Clark K.; Unwin M.; Zackrisson J.; Shiroma W.A.; Akagi J.M.; Maynard K.; Garner P.; Boccia L.; Amendola G.; Massa G.; C. Underwood; Brenchley M.; Pointer M.; Sweeting M.N. (2009). *Antennas for Modern Small Satellites.* IEEE Antennas and Propagation Magazine (August 2009), Vol. 51 No.4, pp. 40-56. IEEE. ISSN 1045 9243/2009.

Greig D., D.; Engleman H., F. (1952) Microstrip a new transmission technique for the kilomegacycle range. *Proceedings of IRE*, No. 40.

Gupta,K.C.; Benalla, A. eds. *Microstrip Antenna Design*. Artech House, Norwood MA, USA

Harrington, R.F.,(1961). *Time-Harmonic Electromagnetic Waves*, McGraw-Hil, New York, USA

Harrington,F.,R., *Matrix Methods for Field Problems*, Proc. IEEE, Vol. 55, No. 2, Feb. 1967.

Roger F. Harrington, (1992). *Field Computation by Method of Moments*, IEEE Press Series on Electromagnetic Waves, ISBN 0 7803 1014 4, Piscataway, NJ, USA.

Hirasawa, K.; Haneishi, M. (1994). *Analysis, Design and Measurement of Small and Low Profile Antennas*, Artech House, ISBN 0-89006-486-5,USA

Howell, J., Q.(1975) *Microstrip Antennas*. IEEE IEEE Trans. on Antennas and Propagation, vol. AP22 , pp. 74-78.

Itoh, T.; Menzel,W. (1989) *A Full Wave Analysis Method for Open Microstrip Structures*. IEEE Trans. on Antennas and Propagation AP-29 , 63-68.

James, J.R., Hall, P. S. (1989). *Handbook of Microstrip Antennas*, Vol 1. Peregrinus Ltd, IEE Electromagnetic Waves Series, ISBN 0 86341 150 9 Peter, London United Kingdom.

James, J. R.; Hall, P., S.; and Wood, C. (1981). *Microstrip antenna theory and design*. IEE Peter Peregrinus.

Kraus, J.,D.; Marhefka, R.,J. (2002) *Antennas 3rd. Edition*, ISBN 0 07 232103 2, New York NY USA.

Lee H.,F.; Chen, W. (1997). *Advances in Microstrip and Printed Antennas*. John Wiley & Sons, New York, USA

Long, S., A.; Shen, L.,C.; & Morel,P.,B. (1978). *A Theory of the Circular Disk Printed Circuit Antenna*. Proc IEE, Pt. H, Vol. 125, October 1978, pp. 925-928

Lucas-Bravo A., Sosa-Pedroza J., Barrera-Figueroa V. (2008). Experimental and numerical results of a rhombic cross antenna. *5o Congreso Internacional de Ingeniería Electromecánica y de Sistemas*, pp. 1178-1182. Mexico D.F., November 2008.

Thomas, A., Milligan.(2005). *Modern Antenna Design*. John Wiley & Sons. New Yersey, USA

Munson, R., E. (1974). "Conformal Microstrip Antennas and Microstrip Phased Arrays" *IEEE Trans. Antennas and Propagation*, Vol. AP-22, January 1974, pp 74-78, ISSN 0018-926X

Pozar,D.,M. (1982) *Input impedance and mutual coupling of rectangular microstrip antennas*. IEEE Trans Antennas Propagatation. AP-30, 1191-1196.

Pozar,D.M. (2005). *Microwave Engineering*, John Wiley & Sons, USA

Reineix, A.; Jecko, B. (1989) *Analysis of Microstrip Patch Antennas Using Finite-Diference Time Domain Method*. IEEE Trans. on Antennas and Propagation AP-37 pp. 1361-1369.

Richards,W.,F.; Lo,Y., T.; Harrison, D.,D. (1979) *An improved theory for microstrip antennas* .Electron. Letters, Vol 15 , 42-44.

Richards,W.,F.; Lo,Y., T.; Harrison, D.,D. (1981) *An improved theory for microstrip antennas and applications*. IEEE Trans. on Antennas and Propagation AP-29 , 38-46.

Roederer A. (1990). *The Cross Antenna: a New Low Profile Circularly Polarized Radiator*. IEEE Transactions on Antennas and Propagation, Vol. 38, no. 5, May 1990.

Schuss J. J., Upton J., Myers B., Sikina T., Rohwer A., Makridakas P., Francois R., Wardle L., and Smith R. (1999). *The IRIDIUM Main Mission Antenna Concept*. IEEE Transactions on Antennas and Propagation, Vol. 47, No. 3, March 1999, pp. 416-424, IEEE, ISSN 0018 926X/99

Sheen, D.,M.; Ali, S.,M.;Abouzahra M.,D.; Kong J.,A. (1990). *Application of Three-Dimensional Finite-Difference Time-Domain Method to the Analysis of Planar Microstrip Circuits*. IEEE Trans. on Microwave Theo. and Tech. vol. 38, pp. 849-857

Shih-Hsun H., Yu-Jiun R., and Kai C. (2009). *A dual-Polarized Planar-Array Antenna for S-Band and X-Band Airbone Applications*. IEEE Antennas and Propagation Magazine, Vol. 51, No.4, August 2009, pp.70-78

Sosa-Pedroza J., Lucas-Bravo A., Lopez-Bonilla J. (2006). Numerical and Experimental Analysis for a Cross Antenna. *International Caribbean Congress on Devices, Circuits and Systems* pp 207-211, ISBN: 1 4244 0042 2, Playa del Carmen, Quintana Roo, Mexico, April 2006.

Sosa-Pedroza, J.; Enciso-Aguilar, M.; Benavides-Cruz, M. (2008). A 9 slots antenna designed by Chebyshev technique and modeled by Finite Difference Time Domain, *Proc. of 7th International Caribean Conference on Devices Circuits and Systems*, Cancun Quintana Roo, Mexico, April 2008, IEEE.

Sosa-Pedroza, J., Barrera-Figueroa,V., Lopez-Bonilla, J. Equidistant and non-equidistant sampling for method of moments applied to Pocklington Equation. *The 18th Annual IEEE International Symposium on Personal, Indoor and Mobile Radio Communications*, Cancun Mexico, 2007.

Sosa-Pedroza J.; Enciso-Aguilar, M.; Benavides-Cruz, M.; Nieto-Rodríguez, M.; Galaz-Larios, M. A parametric analysis of perfect matched layer model of Finite Difference Time Domain Method, *Proc. of PIERS* 2009, Moscow, Russia.

Vasylchenko, A., Schols, Y., De Raedt , W., and Vandenbosch, G. A. E. *Quality Assessment of Computational Techniques and Software Tools for Planar-Antenna Analysis*, IEEE Antennas and Propagation Magazine, Vol. 51, No.1, February 2009

Volakis, J.,L.; Chatterjee, A.; and Kempel, L.,C. (1998). *Finite element method for electromagnetics* IEEE Press, Chap. 7, 1998.

www.satmex.com. (2010)

Yee, K. S (1966). *Numerical solution of initial boundary value problem involving Maxwell's equations in isotropic media*. IEEE Trans. on Antennas and Propagation, vol. 14, pp. 302-307.

# 18

# Power and Spectral Efficient Multiuser Broadband Wireless Communication System

Santi P. Maity
*Bengal Engineering & Science University, Shibpur*
*India*

## 1. Introduction

Communication Satellite plays significant role in long distance broadband signal transmission in recent times. Development of an efficient high data rate communication system for multiusers becomes important and challenging. To meet ever-increasing demand for broadband wireless communications, a key issue that should be coped with is the scarcity of power and spectral bandwidth. In the conventional communication scenario, supporting ubiquitous users with high quality data communication services requires huge bandwidth. However, since the spectrum of nowadays is a very costly resource, further bandwidth expansion is impracticable. Thus, the use of efficient multiple access is critical. Moreover, for a given limited bandwidth and limited number of antennas, the only option that can increase data rate is to increase transmit power as suggested by Shannon channel capacity theorem. Therefore, development of power and bandwidth efficient coding, modulation, and multiple-access techniques is essential for the future wireless communication systems implemented through Satellite link.

Considering the above issues, this chapter discusses a new communication system for Satellite system that can achieve high power and spectral efficiency in broadband wireless communication. To achieve the goal, development of a new and complete communication system with high user capacity and variable data rate become essential. The system includes multicarrier code division multiple access (MC-CDMA) with peak-to-average power ratio (PAPR) reduction using channel coding, optimization in MC-CDMA, estimation of wireless channel condition and MC-DMA with multi-user detection (MUD). The chapter proposal focuses on different aspects for different part of a communication system, namely PAPR reduction in transmitter, channel estimation for design of adaptive and optimized system, multiuser detection at the receiver for increase in user capacity. The different issues are described under four broad subheadings. Prior to that a brief literature review on the above issues are also presented.

The organization of the chapter is as follows: Section 2 presents literature review, while proposed system model is described in Section 3. Design of power and spectral efficient system is described in Section 4. Performance evaluation of the system is presented in Section 5, while conclusions and scope of future works are highlighted in Section 6.

## 2. Literature Review

In this section, we present a brief literature review related to PAPR reduction, multiuser detection in CDMA, channel estimation and optimization in CDMA/MC-CDMA system. The objective of this review is to discuss the merits and the limitations of the existing works, scope of the present work and finally to compare the performance of this work with respect to the related existing works.

A growing number of techniques have been proposed in the recent literature to alleviate power efficiency problem in multicarrier system through the development of PAPR reduction methods. The approaches include simplest filtering and clipping (Ochiai & Ima; 2000), partial transmit sequence (PTS) (Lim et al; 2006), selective mapping (Yoo; 2006) to the sophisticated trellis shaping approach (Ochia; 2004). Ochiai et al (Ochia; 2004) propose a new trellis shaping design based on recursive minimization of the autocorrelation sidelobes for reducing PAPR of the bandlimited orthogonal frequency division multiplexing (OFDM) signals. The exponentially increasing complexity of the PTS scheme prevents its use for MC-CDMA. Kang et al. (Kang et al; 2005) propose an efficient PTS scheme through phase factor optimization to reduce the complexity. Natarajan et al (Natarajan; 2004) address signal compactness issues in MC-CDMA employing CI codes using the measure of crest factors.

Literature on multiuser detection in CDMA is quite rich. The optimum multiuser detector proposed in (Vedu; 1986) achieves significant performance improvement relative to single user receiver but the computational complexity increases exponentially with the number of users. This has motivated the use of low complexity linear (Lupas & Vedu; 1989) and decision driven suboptimal multiuser detection techniques. To overcome the disadvantage intrinsic to the total parallel interference cancelation (PIC), some modified versions like partial parallel interference canceling (Divsalar et al; 1989) and linear parallel interference cancelation technique (Kim & Lee; 2001 ) have been proposed. Parallel interference weighted canceler has been proposed in (Xiao & Liang; 1999) to mitigate the degrading effects of unreliable interference estimation. Two variants of PIC known as threshold PIC and block PIC have been proposed in (Thippavajjula & Natarajan; 2004) for synchronous CI/MCCDMA uplink. Block PIC is shown to provide performance gain relative to conventional PIC. Maity et al (Maity at al; 2008a) further improves bit error rate (BER) performance of diversity assisted block PIC using genetic algorithms.

A growing number of techniques have been proposed in the recent literature to estimate wireless channel parameters. Sgraja et al (Sgraja & Linder; 2003) propose an algorithm to estimate the whole channel matrix using Wiener filtering. Such an estimator requires a Wiener filter, a multiplication operation for each element of the channel matrix and matrix inversion, which increases the complexity of the receiver for large number of sub-carriers. In some systems, a known training sequence is sent by the transmitter and a training algorithm is performed by the receiver on the observed channel output and the known input to estimate the channel (Chow et al; 1991), (Ziegler & Cioffi; 1992) The deterministic least squares (DLS) channel identification algorithm in (Ziegler & Cioffi; 1992) is such a simple but widely used training approach. However, it is not suited for time varying system. In (Wang & Ray; 1999), authors propose an adaptive channel estimation algorithm using the cyclic prefix. This algorithm can adaptively track the channel variation without additional training sequences.

In (Choi et al; 2001), time-domain channel estimation followed by symbol detection based on zero forcing (ZF) and minimum mean square error (MMSE) criterion have been proposed. Pilot based channel estimation is also widely used in many algorithms. The most common pilot based channel estimation scheme is the least square (LS) method of channel estimation

(Coleri et al; 2002), (Schramm et al; 1998). Blind adaptive channel estimation, based on least mean square (LMS) for OFDM communication, is proposed in (Doukopoulos & Moustakides; 2004). Kalman filter is also used for channel estimation in (Gupta & Mehra; 2007). Finite affine projection (FAP) algorithm is used to estimate the channel response for OFDM system in (Gao et al; 2007). Gao et al (Gao et al; 2007) propose an efficient channel estimation for multiple input multiple output (MIMO) single-carrier block transmission with dual cyclic timeslot structure. An optimal channel estimation is then investigated in the minimal mean square error (MMSE) sense on the time slot basis.

This channel state information (CSI) is exploited to design optimized system and several such algorithms for CDMA and MC-CDMA systems are reported in literature. The optimized systems include adaptive algorithm to minimize the total transmitted power (Lok & Wong; 2000), signal-to-interference ratio (SIR) balanced optimum power control in CDMA (Wu; 1999), joint optimization of spreading codes and a utility based power control (Reynolds & Wang; 2003), joint rate and power control (Kim; 1999), joint transmitter-receiver optimization using multiuser detection and resource allocation for energy efficiency in wireless CDMA networks (Buzzi & Poor; 2008), using transmitter power control, receiver array processing and multiuser detection (Seo & Yang,2006) etc. The objective of the optimized methods, in general, is to minimize transmit power and maximize per user's data transmission rate.

## 3. Proposed System

MC-CDMA system was first proposed in (Yee; 1993) and is a combination of CDMA and OFDM with the spreading codes applied in frequency domain. We use carrier interferometry (CI) code as spreading code. Interferometry, a classical method in experimental physics, refers to the study of interference patterns resulting from the superpositioning of waves. CI codes of length N supports N users orthogonally and then, as system demand increases, codes can be selected to accommodate upto addition N-1 users pseudo-orthogonaly. Additionally, there is no restriction on the length N of the CI code (i.e. $N \in I$), making it more robust to the diverse requirements of wireless environments. Since the present MC-CDMA system employs CI code as spreading code, this multiple access scheme is called as CI/MC-CDMA system. In other words, the CI/MC-CDMA is an MC-CDMA scheme employing complex carrier interferometry spreading codes. Assuming that there are K users and N subcarriers in the MC-CDMA system, the CI code for the k-th user $(1 \leq k \leq K)$ is given by (Natarajan et al; 2001):

$[1, e^{j\Delta}\theta_k, ... e^{j(N-1)\Delta}\theta_k]$ where $\Delta\theta_k = 2\pi.k/N$.

In this chapter, we have discussed the operation and performance of a newly developed CI/MC-CDMA model which is a variation that is discussed in (Natarajan et al; 2001). The system proposed here is capable of supporting high capacity with reduction in PAPR as well as low BER values at the receiver. Therefore, we present a brief review of the model in (Natarajan et al; 2001) followed by our new model.

### 3.1 Transmitter Model

In CI/MC-CDMA transmitter (Natarajan et al; 2001), the incoming data $a_k[n]$ for the k-th user, is transmitted over N narrow-band sub-carriers each multiplied with an element of the k-th user spreading code. Binary phase shift keying (BPSK) modulation is assumed, i.e. $a_k[n] = \pm 1$, where $a_k[n]$ represents n-th bit of k-th user. The transmitted signal corresponding to n-th bit of the incoming data is given by

$$S(t) = \sum_{k=0}^{N-1} \sum_{i=0}^{N-1} a_k[n] exp^{j(2\pi.f_i t + i\Delta\theta_k)} p(t) \quad (1)$$

where $f_i = f_c + i.\Delta f$ is the i-th subcarrier, and p(t) is a rectangular pulse of duration $T_b$. As with the traditional MC-CDMA, the $\Delta f$'s are selected such that the carrier frequencies are orthogonal to each other, typically $\Delta f = 1/T_b$, where $T_b$ is the bit duration.

The transmitter model developed in (Natarajan et al; 2001) allows high data rate transmission for (2N-1) users employing 'N' number of sub-carriers. However, in many practical situations, users may have transmission of variable data rate. It is logical to consider variable data rate transmission in multimedia signals environment as diverse classes of traffic signals have different requirement of quality of services (QoS)(Maity et al; 2008b). We like to modify the above transmitter model by allocating all sub-carriers to the users of high data rate transmission and odd and even sub-carriers are shared alternately among the users of low data rate. In other words, we can split the sub-carriers in even and odd parts as well as the 'N' length PO-CI (pseudo-orthogonal) codes in N/2 odd and N/2 even parts. The mathematical form for the transmitted signal $S_1$(t) becomes

$$\begin{aligned} S_1(t) = [\sum_{k=0}^{N-1} \sum_{i=0}^{N-1} a_k[n] exp^{j(2\pi.f_i t + 2\pi.i.k/N)} + \sum_{k=N}^{3N/2-1} \sum_{\forall i=odd}^{N-1} a_k[n] exp^{j(2\pi.f_i t + 2\pi.i.k/N + i\Delta\phi_1)} \\ + \sum_{k=3N/2}^{2N-1} \sum_{\forall i=odd}^{N-1} a_k[n] exp^{j(2\pi.f_i t + 2\pi.(i+1).k/N + i\Delta\phi_2)} + \sum_{k=2N}^{5N/2-1} \sum_{\forall i=even}^{N-1} a_k[n] exp^{j(2\pi.f_i t + 2\pi.i.k/N + i\Delta\phi_3)} \\ + \sum_{k=5N/2}^{3N-1} \sum_{\forall i=even}^{N-1} a_k[n] exp^{j(2\pi.f_i t + 2\pi.(i+1).k/N + i\Delta\phi_4)}] p(t - n.T_b) \end{aligned} \quad (2)$$

where $\Delta\phi_1 = +\pi/2, \Delta\phi_2 = -\pi/2, \Delta\phi_3 = -\pi/N - \pi$ and $\Delta\phi_4 = -\pi/N$ indicate respective phase shift angles with respect to orthogonal CI codes. The phase shift are set in order to make the same even or odd subcarriers used by the different users through the respective spreading codes in out-of-phase. This out-of-phase condition reduces cross-correlation values among the spreading code patterns that not only leads to the reduction in PAPR values but also lower BER values at the receiver. This fact is analyzed mathematically as well as is supported by simulation results.

### 3.2 Channel Model

We also introduce here the channel model. The particular channel model needs to be considered since we will analyze later BER performance of the proposed system and its relative improvement/ degradation due to reduction in PAPR. An uplink model has been considered where all the users' transmissions are assumed to be synchronized for simplification of analysis, although, this condition may seem difficult to be valid practically. It is also assumed that every user experiences an independent propagation environment that is modeled as a slowly varying multipath channel. Multipath propagation in time translates into frequency selectivity in the frequency domain (Proakis, 1995). Frequency selectivity refers to the selectivity over the entire bandwidth of transmission and not over each subcarrier transmission. This is because $\frac{1}{T_b} \ll (\Delta f_c) \ll BW$, where $\Delta f_c$ is the coherence bandwidth and BW is the total bandwidth.

### 3.3 Receiver Design for the proposed system

In this subsection, we describe the multiuser receiver operation when all the users transmit data through all sub-carriers as developed $S(t)$ in Eq. (1). The received signal $r(t)$ for this $S(t)$ corresponds to

$$r(t) = \sum_{k=1}^{K} \sum_{i=0}^{N-1} \alpha_{ik} a_k[n] cos(2\pi . f_i t + i.\Delta\theta_k + \beta_{ik}).p(t) + \eta(t) \tag{3}$$

where $\alpha_{ik}$ is the Rayleigh fading gain and $\beta_{ik}$ is uniformly distributed phase offset of the k-th user in the i-th carrier and the symbol $\eta(t)$ represents additive white Gaussian noise (AWGN). The received signal is projected onto N-orthogonal carriers and is despread using j-th users CI code resulting in $r^j = (r_0^j, r_1^j, ....r_{N-1}^j)$, where $r_i^j$ corresponds to

$$r_i^j = \alpha_{ij}.a_j[n] + \sum_{k=1,k\neq j}^{K} \alpha_{ik} a_k[n] cos(i(\Delta\theta_k - \Delta\theta_j) + \beta_{ik} - \hat{\beta}_{ij}) + \eta_i \tag{4}$$

where $\eta_i$ is a Gaussian random variable with mean 0 and variance $N_0/2$. Exact phase and frequency synchronization for the desired user is assumed i.e.,$\hat{\beta}_{ij} = \beta_{ij}$. Now, a suitable combining strategy is used to create a decision variable $D_j$, which then enters a decision device that outputs $\hat{a}_j$. Minimum mean square error combining (MMSEC) is employed as it is shown to provide the best performance in a frequency selective fading channel (Cal & Akansu; 2000). The decision variable $D_j$ for the n-th bit is (Natarajan et al; 2001)

$$D_n^j = \sum_{i=1}^{N} r_i^j w_{ij} \tag{5}$$

where

$$w_{ij} = \frac{\alpha}{(var(a_k)A_{ij} + N_0/2)} \tag{6}$$

where $A_{ij} = \sum_{k=1}^{K} \alpha_{ik}^2 cos(i\delta\theta_k - i\delta\theta_j + \beta_{ik} - \beta_{ij})^2$ and $var(a_k) = 1$. Thus, the outputs as the single user detector for all the users generate a decision vector $D = [D^1, D^2, ...D^K]$ which is used to obtain the initial estimates of the data $\hat{a} = (\hat{a}_1, \hat{a}_2, ...\hat{a}_k)$. These initial estimates are then used to evaluate multiple access interference (MAI) experienced by each user in the interference cancelation technique.

## 4. Design of power and spectral efficient system

This section describes multi-carrier code division multiple access (MC-CDMA) with PAPR (peak-to-average power ratio) reduction using channel coding, optimized system design for number of subcarriers and users, estimation of wireless channel condition, and finally MC-CDMA with multi-user detection (MUD). The overall discussion focuses on different issues for different section of a communication system, namely PAPR reduction and optimized system design in transmitter, estimation of parameters for the channel, multiuser detection at the receiver for increase in user capacity. The different issues are described under four broad subheadings as follows.

### 4.1 MC-CDMA with PAPR reduction using channel coding

In this section, we will first define PAPR mathematically and then see whether this PAPR is related to the properties of the spreading codes in multiuser systems. The rationale behind such relation lies as each subcarrier is multiplied by chip of individual user's spreading code which is generated from phase shift for CI codes. It would not be irrelevant to mention here that high PAPR occurs due to the superpositioning of several in-phase or near in-phase subcarriers. So cross-correlation and auto-correlation values of CI codes expect to have an influence on PAPR values of the resultant multiuser signals. In MC-CDMA system, the corresponding PAPR definition per discrete-time symbol is given by

$$PAPR = \frac{\max_{0\le i\le N-1} |S(i)^2|}{E[S(i)^2]} \tag{7}$$

where $E[|S(i)|^2]$ and $\max_{0\le i\le N-1} |S(t)|^2$ denote the the average power and the peak power, respectively in one symbol interval. The power of i-th time-domain sample denoted by $P(i)$, $1 \le n \le N$, from equation (2) for complex signal is

$$\begin{aligned} P(i) &= \Big(\sum_{k=1}^{K}\sum_{i=0}^{N-1} a_k[n]c_k^i exp(j2\pi f_i t)\Big)\Big(\sum_{k=1}^{K}\sum_{i1=0}^{N-1} a_k[n]c_k^i exp(j2\pi f_i t)\Big)^* \\ &= \sum_{k=1}^{K} |a_k[n]|^2 + \sum_{k=1}^{K}\sum_{k=1,k\neq k1} a_k[n]a_{k1}[n] \sum_{i2=-(N-1)}^{N-1} Z^{(k,k1)}(i2)exp(i2\pi f_{i2}t) \\ &\quad + \sum_{k=1}^{K} a_k[n]^2 \sum_{i2=-(N-1),i2\neq 0}^{N-1} Z^{(k,k)}(i2)exp(j2\pi f_{i2}t) \end{aligned} \tag{8}$$

The symbol $a_k[n]$ indicates n-th symbol of k-th user. Here $Z^{k,k1}(i2)$ is the aperiodic crosscorrelation function of spreading codes between user k and k1,

$$Z^{(k,k1)}(i2) = \sum_{m=1}^{N-i2} c_k^m (c_{k1}^{m+i2})*$$

and $Z^{k,k}(i2)$ is the aperiodic autocorrelation function of the k-th user. It can be found that the signal power of MC-CDMA system is partially determined by the correlation property of the selected set of spreading codes. We now briefly present proposed PAPR reduction method using code optimization, phase optimization and trellis coding.

*A. Code Optimization*

As defined earlier, the CI code for the $k^{th}$ user is given by the spreading sequence,

$$\begin{aligned} \{\beta_k^0, \beta_k^1, \beta_k^2, ....\beta_k^{N-1}\} &= \{e^{j\Delta\theta_k^0}, e^{j\Delta\theta_k^1}...e^{j\Delta\theta_k^{N-1}}\} \\ &= \{e^{j(2\pi/N)0.k}, e^{j(2\pi/N)1.k}...e^{j(2\pi/N)(N-1).k}\} \end{aligned} \tag{9}$$

Here, $k\epsilon[0, 1, 2, ....K-1]$. However, low value of PAPR is achieved if we give cycle shift of the code that leads to new code. We can write the new code as

$$\begin{aligned} \text{New code} &= \{\beta_{k+shift}^0, \beta_{k+shift}^1, ....\beta_{k+shift}^{N-1}\} \\ &= \{e^{j\Delta\theta_{k+shift}^0}, e^{j\Delta\theta_{k+shift}^1}...e^{j\Delta\theta_{k+shift}^{N-1}}\} \\ &= \{e^{j(2\pi/N)0.(k+shift)}, e^{j(2\pi/N)1.(k+shift)}... \\ &\qquad e^{j(2\pi/N)(N-1).(k+shift)}\} \end{aligned} \tag{10}$$

Here 'shift' indicates the number of cycle shift which is fixed for all users for a given period of time, $[0 \leq' shift' \leq K]$. It is seen from the simulation results that for a minimum value of PAPR, the 'shift' is either [1,2,3] or [(K-1) to (K-3)]. It is also found that performance achieved by doing (K-1) to (K-3) shifting operations is same as is obtained by performing 1 to 3 times reverse shifting. This means that as far as the code optimization is concerned, code sequence for the users is shifted in forward or reverse cyclic shifting by an amount of 1 to 3 shifts.

B. Phase Optimization

To achieve lower PAPR value, we can use the iterative method where we shift the phase of each set of users with additional rotation in between $\Delta\phi_{(max)}$ and $\Delta\phi_{(min)}$ so that the condition $\Delta\phi_{(max)} > \Delta\phi > \Delta\phi_{(min)}$ is satisfied. It is also to be mentioned here that $\Delta\phi_{(max)}$ is allowed to take value of $\Delta\phi + \pi/32N$, while $\Delta\phi_{(min)}$=$\Delta\phi - \pi/32N$. Fig. 1 shows the proposed code and phase optimized variable data rate CI/MC-CDMA system with reduction in PAPR value.

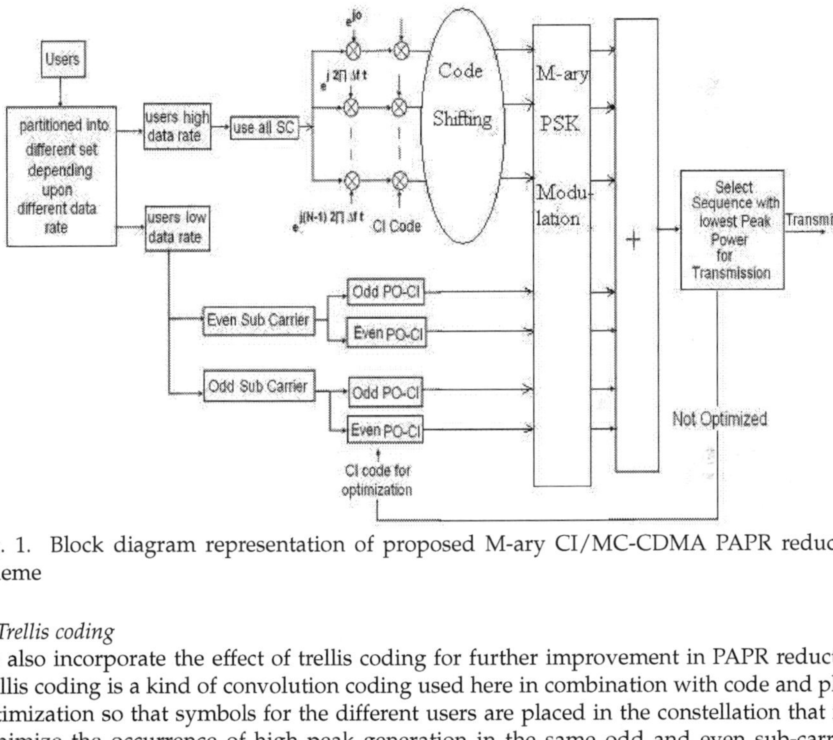


Fig. 1. Block diagram representation of proposed M-ary CI/MC-CDMA PAPR reduction scheme

*C. Trellis coding*

We also incorporate the effect of trellis coding for further improvement in PAPR reduction. Trellis coding is a kind of convolution coding used here in combination with code and phase optimization so that symbols for the different users are placed in the constellation that may minimize the occurrence of high peak generation in the same odd and even sub-carriers. Moreover, the usage of trellis coding offers the benefit of reduction in bit error rate (BER) performance at the receiver. Fig. 2 shows the simplified block diagram of Fig. 1 with trellis coding.

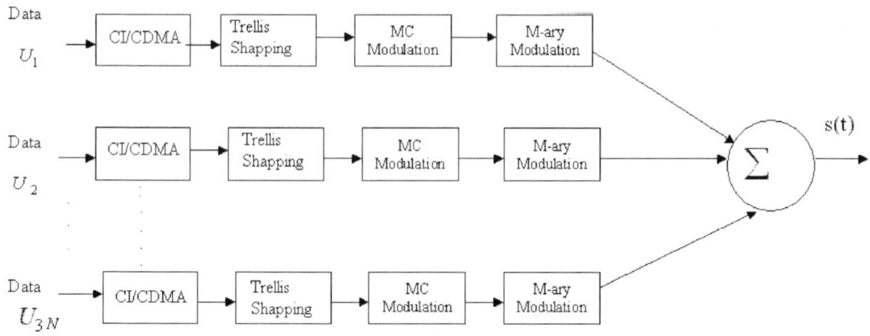


Fig. 2. Simplified block diagram representation for the proposed M-ary CI/MC-CDMA scheme with trellis coding

### 4.2 Proposed GA based channel estimation

This subsection first defines fitness function 'F' and GA based minimization is presented later.

*A. Formation of Fitness function*

We assume that actual fading gain at n-th subcarrier is $\alpha_n$ and its estimated value is denoted by $\hat{\alpha}_n$. The error in estimated value is represented by $e_n$. We can think of probability density function (pdf) of the estimation error as central Chi-square distribution (Ahamad & Assaad; 2009) and can be written as follows:

$$f(|e_n^2|) = \frac{1}{\sigma_n^2} e^{\frac{-|e_n|^2}{\sigma_n^2}} \tag{11}$$

where $\sigma_n^2$ is the variance of estimation error. So, in terms of estimated channel gains and its corresponding error terms, the expression of (4) can be written as follows:

$$\begin{aligned} r_i^j &= \hat{\alpha}_i + e_i + \sum_{k=1,k\neq j}^{K} \hat{\alpha}_i \rho_{kj} + \sum_{k=1,k\neq j}^{K} e_i \rho_{kj} + \eta_j \\ &= \hat{\alpha}_i + e_n + \sigma_I^2 + \sigma_{Ie}^2 + \sigma_N^2 \end{aligned} \tag{12}$$

The first, second, third, forth and fifth term of (12) can be designated for the $j$-user as signal term in $i-$ th subcarrier, estimated error in signal term, variance of interfere i.e. interference power due to all other users at i-th subcarrier, interference power for estimation error, and noise power at i-th subcarrier, respectively. For large number of users, the third and the forth term correspond to a random variable with normal distribution (according to central limit theorem).

We define signal-to-interference noise power ratio (SINR) corresponding to j-th user's bit at i-th subcarrier as follows:

$$(SINR)_i^j = \frac{(|\hat{\alpha}_i| + |e_i|)^2}{\sigma_I^2 + \sigma_{Ie}^2 + \sigma_{N_i}^2} \tag{13}$$

We assume SINR for all subcarriers are independent, so total SINR for j-th user's bit is

$$(SINR)^j = \sum_{i=1}^{N} (SINR)_i^j \tag{14}$$

The channel capacity corresponding to j-th user's is

$$C^j = log(1 + SINR^j)\ bits/sample \tag{15}$$

Now total channel capacity can be calculated as $C = \sum_{\forall j} C^j$.
We now define 'F' as weighted average of channel capacity and detection probability $p_e$ i.e. $F = f(C, p_e)$. It is preferable to minimize 'F', while target is to maximize C and minimize $p_e$. It is logical to express C as $C_{norm} = \frac{C_{\hat{\alpha},e}}{C_{\alpha=1}}$, that indicates normalization of the channel capacity. The symbol $C_{\alpha=1}$ corresponds to non-fading situation and obviously channel capacity with reliable decoding will be high. It is quite true that the value of $\frac{C_{\hat{\alpha},e}}{C_{\alpha=1}}$ is less than 1 but our target is to achieve this value close to 1. The $p_e$ is calculated as follows:

$$p_e = \frac{1}{K}\sum_{k=1}^{K}(b_k - \hat{b}_k) \tag{16}$$

where, K is total number of users, $b_k$ and $\hat{b}_k$ are the transmitted and the received $k$-th user bit, respectively.
The objective function 'F' can be defined as

$$F = w_1(1 - C_{norm}) + w_2 p_e = w_1(1 - \frac{C_{\hat{h},e}}{C_{h=1}}) + w_2 p_e \tag{17}$$

where $w_1$ and $w_2$ are the weight factors of channel capacity and detection reliability, respectively. Each weight factor represents how important each index is during the searching process of GA. For example, if both indices are equally important, each one should be 0.5 so that the relationship $w_1 + w_2 = 1.0$ must hold.
*B. Optimization of Fitness function using GA*
The experimental conditions of GA for the present problem are depicted as follows:
(i) size of population is 10,(ii) number of generation/iterations 100, (iii) probability of crossover per generation is 0.80, and (iv) probability of mutation per bit is 0.07.
Initialization of twenty sets of random values for $\hat{\alpha}_1, \hat{\alpha}_2, ......\hat{\alpha}_n, e_1, e_2, ......e_n$ are done. The values of $\hat{\alpha}_i$ are taken from Rayleigh distribution, while the values of $e$'s are taken from (11). Then the value of C and $p_e$ are calculated for each set. Using the procedure outlined in previous subsection, the value of fitness function 'F' is calculated for each of $\hat{\alpha}_1, \hat{\alpha}_2, ......\hat{\alpha}_n, e_1, e_2, ......e_n$ using (17). An upper bound of F value $(F_U)$ is determined based on the calculated 'F' values. The value of $(F_U)$ acts as a threshold and is adjustable. This is required so that the needful number of sets of $\hat{\alpha}_1, \hat{\alpha}_2, ......\hat{\alpha}_n, e_1, e_2, ......e_n$ values for which 'F' values lie below the $(F_U)$ are duplicated. The remaining sets having 'F' values higher than $(F_U)$ are ignored from the population. This process is done from the concept of selection of GA based algorithm. A binary string is generated through decimal-to-binary conversion for each selected set of $\hat{\alpha}_1, \hat{\alpha}_2, ......\hat{\alpha}_n, e_1, e_2, ......e_n$ value and thus a set of strings are calculated for all selected combinations. Now, crossover and mutation operations are done with above probabilities.Operations as described, when applied to the selected sets, generate a new set of $\hat{\alpha}_1, \hat{\alpha}_2, ......\hat{\alpha}_n, e_1, e_2, ......e_n$ value. This set is considered as population for next iteration/generation of the proposed GA based optimization problem. The operations are repeated for desired number of iterations or till a predefined acceptable values for C and $p_e$ are achieved.

### 4.3 Subcarrier PIC scheme

Multicarrier (MC) signal can be modeled as a vector in $N$-dimensions where $N$-mutually orthogonal subcarriers form the basis signals. Accordingly, MC-CDMA signals with high and low data rates can be modeled as vectors of different dimensions in subcarrier signal space. In MUD, CDMA signal is treated as a vector and is projected onto individual orthogonal spreading function to estimate the interference of other users for the respective user. In PIC, for a given user, CDMA composite signal is projected simultaneously to all spreading codes except the desired one and the process demands high computation cost. The multiple access interference (MAI) estimation and its subsequent cancelation in MC-CDMA with variable data rates can be implemented in the respective subcarrier component rather than projecting it onto spreading functions followed by calculation of vector magnitude (Eq. 5 and Eq. 6), as is done in case of conventional PIC method. This very simple concept is used to develop a low computation cost PIC scheme when MAI is subtracted at the sub-carriers level. Fig. 3 shows the block diagram representation of the proposed PIC method. The users are divided in two different blocks, block 1(B-1) with high data rate transmission and block-2 (B-2) with low data rate transmission.

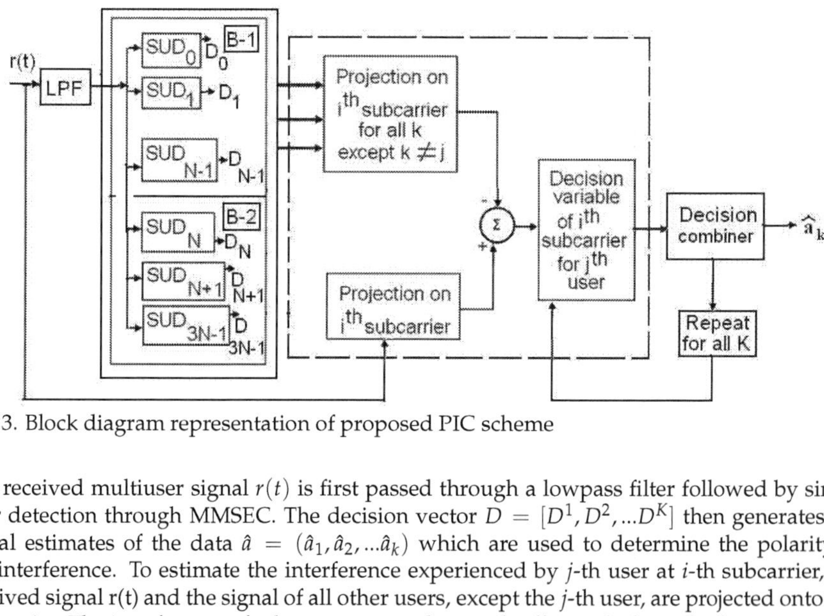


Fig. 3. Block diagram representation of proposed PIC scheme

The received multiuser signal $r(t)$ is first passed through a lowpass filter followed by single user detection through MMSEC. The decision vector $D = [D^1, D^2, ...D^K]$ then generates the initial estimates of the data $\hat{a} = (\hat{a}_1, \hat{a}_2, ...\hat{a}_k)$ which are used to determine the polarity of the interference. To estimate the interference experienced by $j$-th user at $i$-th subcarrier, the received signal r(t) and the signal of all other users, except the $j$-th user, are projected onto $i$-th subcarrier. The interference of other users are subtracted collectively from the former.
We now analyze mathematically the interference free estimation for the n-th bit of j-th user for conventional PIC, two block PIC and the proposed subcarrier PIC. The expression of the same for the conventional PIC system is

$$\hat{a}_n^j = r_n - \sum_{k=1, k \neq j}^{K} \hat{a}_n^k \tag{18}$$

The similar expression for an arbitrary j-th user belonging to strong group and when belong to weak group, respectively (of a two block PIC system) can be written as follows:

$$\hat{a}_n^j = r_n - \sum_{\forall k_1 \in S_1, k_1 \neq j} \hat{a}_n^{k_1} \tag{19}$$

$$\hat{a}_n^j = r_n - \sum_{\forall k_1 \in S_1} \hat{a}_n^{k_1} - \sum_{\forall k_2 \in S_2, k_2 \neq j} \hat{a}_n^{k_2} \tag{20}$$

where the symbols of $\hat{a}_n^j, r_n, S_1$ and $S_2$ represent n-th estimated bit of j-th user, resultant received signal for n-th bit, set of strong and weak user groups, respectively.
The interference free estimation of the $i$-th subcarrier for the $j$-th user of block $B_1$,

$$r_i^j = r_i - \sum_{\forall k_1 \in B_1, k_1 \neq j} r_i^{k_1} \tag{21}$$

while the expression for the same belonging to block $B_2$,

$$r_i^j = r_i - \sum_{\forall k_1 \in B_1} r_i^{k_1} - \sum_{\forall k_2 \in S_2, k_2 \neq j} r_i^{k_2} \tag{22}$$

where $r_i^j$ and $r_i$ indicate the i-th subcarrier component of the j-th user and the i-th subcarrier component of resultant received signal, respectively.
Now, the decision variable $D_n^j$ for the n-th bit of an arbitrary $j$-th user of block $B_1$ can be written accordingly to Eq. (5). The equation is rewritten in Eq. (22) for convenience of further analysis. The similar expression for the $j$-th user of the block $B_2$ is written in Eq.(23)

$$D_n^j = \sum_{i=1}^{N} r_i^j w_{ij} \tag{23}$$

$$D_n^j = \sum_{\forall i = odd \ or \ even}^{N} r_i^j w_{ij} \tag{24}$$

The values of $r_i^j$ in Eq. (22) and Eq. (23) can be put from Eq. (20) and Eq. (21), respectively. Mathematical analysis of Eq. (20)-Eq.(23) show that stable decision variable $D_n^j$ can be achieved for subcarrier PIC method compared to conventional PIC and block PIC. This is because interference due to other users are subtracted before forming $D_n^j$ in the case of former unlike the latter methods where $D_n^j$ is formed first and then interference is subtracted. This stable decision variable $D_n^j$ gives rise to better estimation of $\hat{a}_j$ leading to an improvement in BER through interference cancelation. Moreover, the computation cost and time requirement is less compared to the conventional PIC and other block PIC (Thippavajjula; 2004). This can be well understood considering the fact that the users of block-1, transmit data at 'q' times higher rate (say) compared to the users of block-2. So, for 'q' consecutive bits detection of high data rate users, estimated interferences for the single bit due to low data rate users can be made use. In other words, interferences estimated for low users' single bit can be applied to improve subsequent BER of high user data. It is to be mentioned here that the projection of the received signal on the mutually orthogonal subcarriers is essentially the same for both the conventional PIC and the proposed PIC. However, the interference in the former is subtracted after combining them as decision vector while in the later, interference is subtracted in subcarrier level itself. The computation complexity, for CI codes of length $N$, is $O(2N-1)$ for conventional PIC, 2 $O(N)$ for block PIC (Thippavajjula; 2004) and $O(N)$ for the proposed subcarrier PIC.

### 4.4 Optimized system design

In the present study, optimum values for N, K, and SNR are calculated based on the minimization of a fitness function '$F'$' that depends on PAPR, BER and average data rate (ADR). We first define 'F' and then GA based minimization technique is presented. Let us define the following symbols:
$P_{diff}$: Difference in threshold PAPR and actual PAPR
$B_{diff}$:Difference in threshold BER and actual BER
$ADR$:Average data rate which is normalized with respect to the collective data of all users transmitting at high data rates.
The predefined values of PAPR and BER indicate threshold of the permissible values for the respective parameters. It is to be noted that we have considered $P_{diff}$ and $B_{diff}$, instead of considering their actual values, while forming the objective function and its subsequent minimization process is performed. This is done so as these two measures are related in a conflicting manner with their associated parameters and the simultaneous minimization of both of these is not optimally achievable with respect to the overall optimal system design is concerned. So the objective is to reduce $P_{diff}$ and $B_{diff}$ whether their actual values are below or above the threshold value. The fitness function '$F'$' is defined as follows:

$$F = w_1 P_{diff} + w_2 B_{diff} + w_3(1 - ADR) \tag{25}$$

where $w_1$,$w_2$ and $w_3$ are defined as follows:

$$w_1 = \frac{B_{diff} \times (1 - ADR)}{(P_{diff} + B_{diff})(1 - ADR) + B_{diff}P_{diff}} \tag{26}$$

$$w_2 = \frac{P_{diff} \times (1 - ADR)}{(P_{diff} + B_{diff})(1 - ADR) + B_{diff}P_{diff}} \tag{27}$$

$$w_3 = \frac{P_{diff} \times B_{diff}}{(P_{diff} + B_{diff})(1 - ADR) + B_{diff}P_{diff}} \tag{28}$$

so that $0 < w_1,w_2,w_3 < 1$ and $w_1+w_2+w_3 = 1$ relation must hold.
The values of $w_1$,$w_2$ and $w_3$ when put in Eq. (24), the expression of 'F' becomes

$$F = \frac{3P_{diff} \times B_{diff} \times (1 - ADR)}{(P_{diff} + B_{diff})(1 - ADR) + B_{diff}P_{diff}} \tag{29}$$

We now describe very briefly GA based optimization process. Initialization of twenty sets of random $\{N, K, SNR\}$ values within the predefined range are done. The values for $P_{diff}$, $B_{diff}$ and $ADR$ as well as the value of fitness function 'F' for each set of {N,K,SNR} are then calculated. A predefined threshold ($F_U$) value of 'F' is assigned to identify the fit parameter sets. The particular sets of {N,K,SNR} for which 'F' values lie below the $F_U$ are duplicated and the other sets are ignored from the population following the spirit of GA. Now, crossover and mutation operations are done on the strings. The experimental conditions of GA for the present problem are depicted as, (i) size of population is 20, (ii) probability of crossover per generation is 0.80, and probability of mutation per bit is 0.09, user capacity i.e. number of users vary from 15-75, number of subcarriers range from 10-25, value of SNR range from 7-14 dB. The operations are done for the desired number of iterations or till a predefined acceptable values for PAPR and BER are achieved for optimal values of $N$, $K$ and SNR.

## 5. Performance Evaluation

In this section, we present performance analysis of PAPR reduction, channel estimation, subcarrier PIC for improved receiver operation and finally optimization system for power and spectral efficient system design.

### 5.1 Performance for PAPR reduction

This subsection presents the performance of the proposed PAPR reduction technique in synchronous M-ary CI/MC-CDMA system. As figures of merit, we evaluate the complementary cumulative distribution function (CCDF) of PAPR under various phase modulation. We also show through simulation results that the proposed CI/MC-CDMA method offers lower PAPR and BER values compared to conventional CI/MC-CDMA system (Natarajan et al; 2001) and other recent work [12]. We first show the cross-correlation values for the different user's spreading codes under the phases as described in equation (7). The cross correlation (CC) between user k's signature waveform $c_k(t)$[created using the spreading sequence (1, $e^{j\Delta}\theta_k$ ,... $e^{j(N-1)\Delta}\theta_k$) and user j's signature waveform [created via the spreading sequence (1, $e^{j\Delta}\theta_j$ ,... $e^{j(N-1)\Delta}\theta_j$) can be shown as follows

$$R_{k,j}(\tau) = \frac{1}{2\Delta f} \sum_{i=0}^{i=N-1} cos(i(2\pi.\Delta f \tau)) \tag{30}$$

$$R_{k,j}(\tau) = \frac{1}{2\Delta f} \frac{sin(1/2N.2\pi.\Delta f.\tau)}{sin(1/2.2\pi.\Delta f.\tau)} cos(\frac{(N-1)}{2} 2\pi\Delta f \tau) \tag{31}$$

where $\tau = \Delta t_k - \Delta t_j = (\Delta\theta_k - \Delta\theta_j)/2\pi.\Delta f$. Table 1 shows the cross-correlation values for some arbitrary pair of spreading codes under various phase shifts. Numerical values show that $R_{j,k}$ values are reduced significantly for the proposed phase shifts of the code patterns.

We now show cumulative complementary distribution function (CCDF) of PAPR values for different $M$-ary modulations with separate code and phase optimization of the spreading codes in Fig. 4 and as well as with the combined effect of code and phase optimization in Fig. 5. It is also to be mentioned here that the proposed PAPR reduction algorithm is very fast. It takes approximately 2.25 second for 10 time iteration in MATLAB 7 platform running on a Pentium IV 2.79 GHz, 512 MB RAM PC system. The graphical representation shows that best PAPR reduction performance is achieved for the proposed algorithm with QPSK system, while the worst PAPR reduction performance is shown in BPSK performance. Similarly, PAPR reduction performance of 16-ary PSK is comparatively better than 8-ary PSK system. The results obtained are quite consistent with the cross-correlation values shown in the Table 1 for any arbitrary pair of code patterns and different M-ary modulation. While phase quadrature in QPSK system is not much influenced by the polarity of the symbols, but in case of BPSK modulation the polarity of symbols makes superimposition of subcarriers for in-phase that leads to an increase in PAPR value. It is interesting to see that for BPSK, QPSK and 8-ary PSK

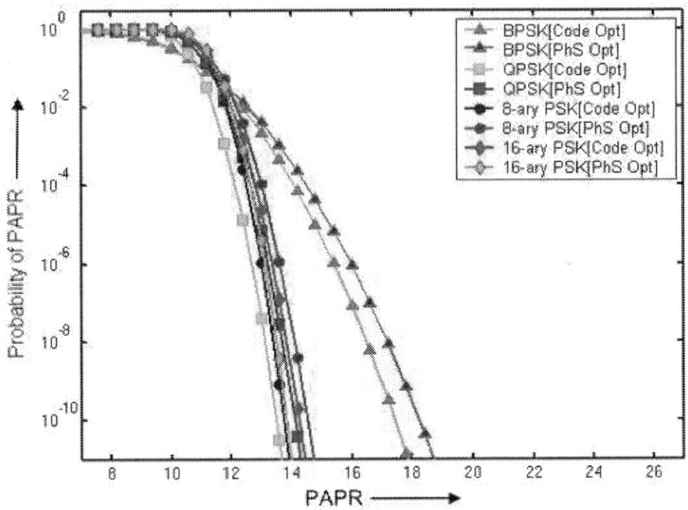


Fig. 4. CCDF of PAPR for N=8 and K=24 with code and phase optimization separately

code optimizations are more effective for reduction in PAPR values. On the other hand, for 16-ary PSK, phase optimization is more effective. Significant improvement in PAPR reduction is possible to achieve by exploiting the combined effect of code and phase optimization. Fig. 5 shows that PAPR values are not only reduced but also the probability for the PAPR values over a particular threshold value also become low. Fig. 6 shows improvement in PAPR using trellis coding.

We also test PAPR reduction performance of the proposed system in real time system of Agilent made N5182A MXG RF vector signal generator (VSG). The signal characteristics is 100 kHz to 3 GHz, +23 dBm output power at 1GHz = -121 dBc/Hz (typ), phase noise at 1 GHz and 20 kHz offset =1.2 ms switching speed in SCPI mode; $\leq$ 900 micro-second simultaneous frequency, amplitude, and waveform switching in list mode. Fig. 7 shows CCDF for PAPR reduction when the proposed algorithm is generated in VSG before code and phase optimization while the same is shown in Fig. 8 after code and phase optimization. The results show that due to optimization peak power is not only reduced from 6.37 dB to 4.06 dB, but also probability for attaining 0dB value also increases from 37% to 46 %. Similarly, 10% probability is found for PAPR value with 3.52dB, while after optimization this value reduces to 3.06dB. So theoretical performance improvement of PAPR reduction is also verified in hardware system indicating the importance of implementation of the proposed algorithm in practical system.

PAPR reduction for the proposed algorithm is also verified in VSG for 8-ary PSK modulation. Results are shown in Fig. 9 and Fig. 10. The results show that due to optimization not only peak power is reduced from 7.27 dB to 6.13 dB, but also probability for attaining 0dB value increases from 29% to 35 %. Similarly, 10% probability is found for PAPR value with 3.35dB, while after optimization this value reduces to 3.00dB. Verification results once again support the theoretical results that PAPR reduction performance for the proposed algorithm with 8-ary and 16-ary modulation system lie between performance for BPSK and QPSK modulation.

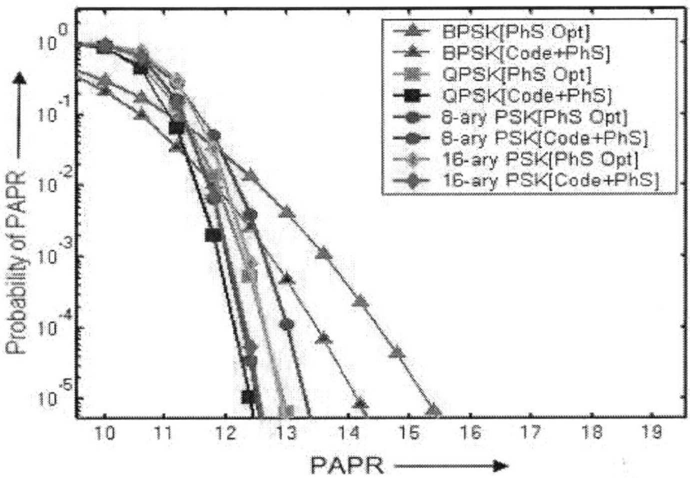


Fig. 5. CCDF of PAPR for N=8 and K=24 with combined code and phase optimization

### 5.2 Performance evaluation for channel estimation

This subsection represents performance analysis of the proposed channel estimation method. We have evaluated BER for different number of users at SNR=7dB and the number of subcarriers N=10. Fig. 11 shows that BER values corresponding to the estimated channel parameters become more closer to the values corresponding to the actual values of the channel parameters. Simulation results show that for N=10, difference in BER value corresponding to the actual and the estimated channel parameters become of the order of 0.085 for the number of users 20, and is of the order of 0.055 for the number of users 30. We also test the effect of the number of iterations/generations on the estimated channel parameters in terms of detection reliability and channel capacity. Fig. 12 shows that BER values increase with the increase of number of users but decreases with the number of generations. At the same time normalized (maximum values is 1) data transmission capacity is increased significantly with the number of generations which is shown in Fig. 13.

We also compare performance of the proposed method with the method in (Doukopoulos & Moustakides; 2004). For compatibility, orthogonal frequency division multiplexing (OFDM) system of (Doukopoulos & Moustakides; 2004) is designed as MC-CDMA system for the same number of subcarriers N=10 and simulation is performed at SNR=7 dB. Simulation results shown in Fig. 14 reflects the fact that the proposed channel estimation offers better BER performance due to the optimal selection of GA. The computation complexity for the proposed method is $O(n)$, while the same for LMS method in (Doukopoulos & Moustakides; 2004) is $O(n^2)$.

### 5.3 Performance evaluation of subcarrier PIC

Fig. 15 shows BER performance of the proposed scheme at 14 dB SNR and with N=16 subcarriers for 3N users system where N number of users transmit high data rate and 2N number of users transmit low data rate. The graphical result shows that when capacity is low (less

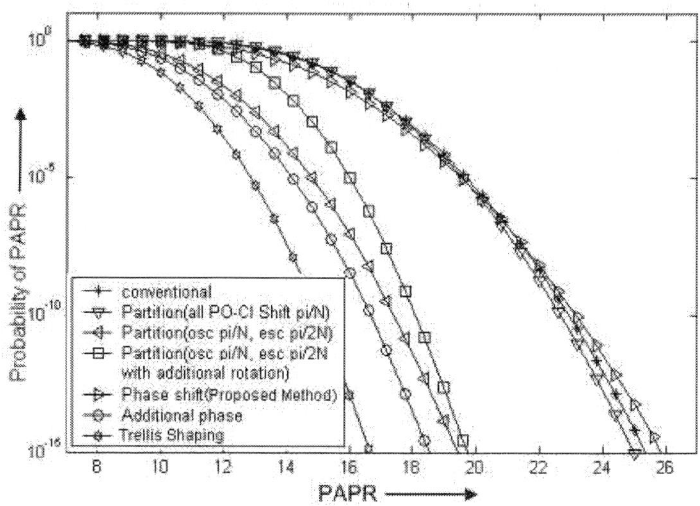


Fig. 6. CCDF of PAPR for N=8 and K=24 using trellis coding

than 20 users), MMSEC receiver of conventional CI/MC-CDMA system (shown as Con. MM-SEC in the figure) performs significantly better compared to MMSEC receiver of the proposed system (shown as MMSEC in the figure). This is due to the fact that all users in the former transmit data using all sub-carriers while some of the users in the latter transmits data using either odd or even sub-carriers. However, significant improvement in BER performance is achieved in the latter case compared to the former when the number of users are gradually increasing over 20. Further significant improvement in BER performance can be achieved after different stages of the proposed subcarrier PIC scheme. Simulation results show that the proposed subcarrier PIC scheme after third stage iteration, can support the number of users three times the number of sub-carriers with BER of the order of 0.0428 (0.11071 for (Natarajan et al; 2001)), and it can support users upto four times the number of sub-carriers with BER of the order of 0.1350 (0.2193 for (Natarajan et al; 2001)).

We also test BER performance of the proposed scheme with the increase of number of users transmitting at high data rate. Fig. 16 shows BER performance of the proposed scheme at SNR 14 dB with N=16 sub-carriers for 2.5N users system. Here 1.5 N number of users transmit at high data rate and N number of users transmit low data rate. Simulation results show that the system supports the number of users two-and-half and three times the number of sub-carriers with BER values of 0.0591 and 0.104, respectively after three stage iterations. The relative degradation in BER performance for 2.5 N system, over 3N system, based on the proposed subcarrier PIC is due to the increase in overall data transmission rate for the former compared to the latter. The over all data transmission rate is 0.7742 times for 3N user and 0.9032 times for 2.5 N system with respect to the conventional CI/MC-CDMA system (Natarajan et al; 2001). The numerical values specified here for data transmission rate is obtained when transmission rate between high and low data rate user differs by a factor of 4.

We also compare BER performance of the proposed subcarrier PIC scheme (SCPIC) and block PIC (BPIC) scheme (Thippavajjula; 2004). The results are reported with the performance of

| Arbitrary code pairs | Diff. PSK Sys. | Two Ort. cod. (j,k) | Orth.Code (j) $\pm\pi/M$ Shift. codes | + $\pi/M$ & -$\pi/M$.shift. codes |
|---|---|---|---|---|
| 1st pair | BPSK | 0.3953 | 0.3750 | 0.3536 |
| | QPSK | 0.2812 | 0.2789 | 0.2542 |
| | 8- PSK | 0.3162 | 0.3149 | 0.3020 |
| | 16- PSK | 0.3617 | 0.3411 | 0.3211 |
| 2nd pair | BPSK | 0.3953 | 0.3750 | 0.3536 |
| | QPSK | 0.2834 | 0.2791 | 0.2590 |
| | 8- PSK | 0.3179 | 0.3151 | 0.3001 |
| | 16- PSK | 0.3619 | 0.3459 | 0.3291 |
| 3rd pair | BPSK | 0.3953 | 0.3750 | 0.3536 |
| | QPSK | 0.2834 | 0.2791 | 0.2590 |
| | 8- PSK | 0.3179 | 0.3151 | 0.3001 |
| | 16- PSK | 0.3619 | 0.3459 | 0.3291 |
| 4rth pair | BPSK | 0.3962 | 0.3690 | 0.3542 |
| | QPSK | 0.2842 | 0.2709 | 0.2581 |
| | 8- PSK | 0.3287 | 0.3124 | 0.3054 |
| | 16- PSK | 0.3754 | 0.3589 | 0.3205 |

Table 1. Cross correlation values for arbitrary code pair

| Sl. no | N | K | SNR | PAPR | BER | ADR |
|---|---|---|---|---|---|---|
| 1 | 22 | 64 | 10 | 10.2649 | 0.0482 | 0.7444 |
| 2 | 24 | 70 | 10 | 11.4017 | 0.0450 | 0.6700 |
| 3 | 21 | 60 | 14 | 10.0973 | 0.0125 | 0.8100 |
| 4 | 23 | 68 | 14 | 11.4012 | 0.0650 | 0.5000 |
| 5 | 18 | 52 | 11 | 10.4698 | 0.0464 | 0.5577 |
| 6 | 21 | 65 | 12 | 10.8132 | 0.0414 | 0.5407 |
| 7 | 23 | 67 | 12 | 11.7433 | 0.0701 | 0.5135 |
| 8 | 19 | 57 | 14 | 10.066 | 0.0422 | 0.5716 |

Table 2. GA based optimization for the proposed method

(Natarajan et al; 2001) through interference cancelation (IC). Fig. 17 shows that BER performance of the proposed subcarrier PIC scheme is significantly better compared to that of block PIC scheme and needless to mention its superior BER performance compared to MMSE scheme of (Natarajan et al; 2001). The performance improvement for the proposed subcarrier PIC is due to the twofold advantages in interference cancelation. Since the high data rate transmission uses all sub-carriers, the data can be decoded with greater reliability and interference due to these users can be estimated with greater accuracy. This interference when subtracted from the resultant received signal improves detection performance of the low data rate users. On the other hand, low data rate transmission uses alternate sub-carriers, so subcarriers of high data rate users experience less interference that leads to an improvement in BER performance of the latter. This cumulative effect on BER performance in multistage interference cancelation significantly improves overall BER performance of the proposed system unlike to that of the block PIC scheme in (Thippavajjula; 2004). In block PIC, in any stage

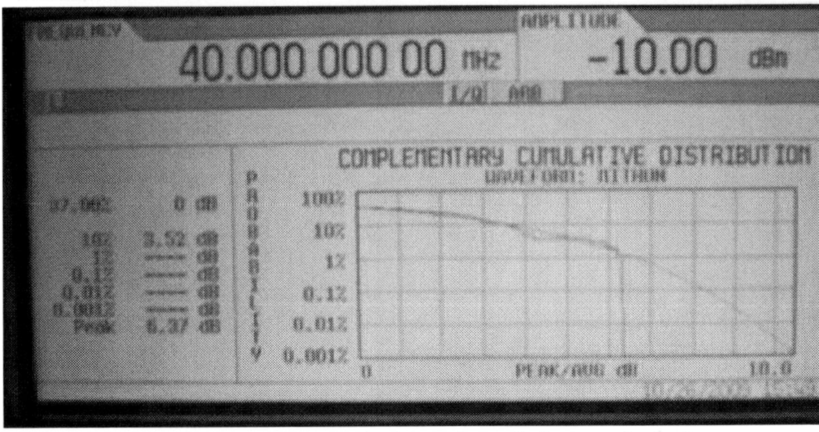


Fig. 7. CCDF of PAPR in QPSK for N=24 before code and phase optimization

| Sl. no | N | K | SNR | PAPR | BER | ADR |
|---|---|---|---|---|---|---|
| 1 | 25 | 54 | 14 | 18.6040 | 0.0770 | 1.000 |
| 2 | 23 | 48 | 14 | 16.0504 | 0.0691 | 1.0000 |
| 3 | 21 | 44 | 13 | 14.1088 | 0.0805 | 1.0000 |
| 4 | 20 | 35 | 10 | 13.1250 | 0.0691 | 1.0000 |
| 5 | 22 | 43 | 12 | 15.3944 | 0.0714 | 1.0000 |
| 6 | 18 | 33 | 11 | 13.1597 | 0.0670 | 1.0000 |
| 7 | 24 | 46 | 10 | 16.9413 | 0.0683 | 1.0000 |

Table 3. GA based optimization in

of interference cancelation, weak users data noway benefits BER performance of strong users data unlike the proposed subcarrier PIC scheme.

Fig. 18 shows graphical representation for BER performance with the number of users for subcarrier parallel interference cancelation (PIC) (Maity & Mukherjee; 2009), code and subcarrier PIC and trellis coded system for 'N'=16 and SNR=14 dB. It is found that trellis coded system provides significant improvement in BER performance even at less number of interference cancelation stage compared to the same for higher stage interference cancelation of subcarrier PIC and combined code & phase PIC.

### 5.4 Performance evaluation of optimized system

Table 2 and Table 3 show the performance of the optimization for the proposed system and CI/MC-CDMA system in (Natarajan et al; 2001), respectively. Simulation results clearly specify the importance of the optimization problem. The values of PAPR, BER and ADR for both the optimized systems are quite consistent for the particular combinations of $K$, $N$ and SNR values i.e. large $N$ values offer lower BER and increased data transmission rate, while large K values yield increased BER. The values of SNR have both way effect on BER performance in multiuser communication system. As a matter of fact, a set of $N$, $K$, SNR values are (at least) near optimal for the set of PAPR, BER and ADR values with respect to the status of the wireless channel condition. For example, if we see the results depicted in 4th row (Sl. no 3)

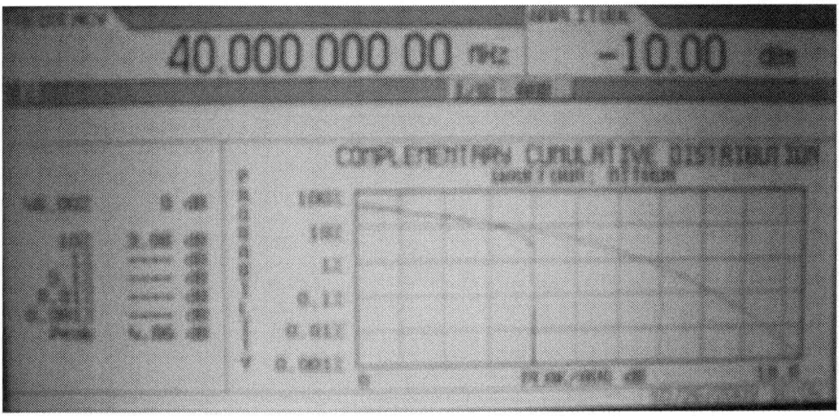


Fig. 8. CCDF of PAPR in QPSK for N=24 after code and phase optimization

of Table 2 and Table 3, for $N$=21, and for similar SNR values (14 dB for the proposed system and 13 dB for (Natarajan et al; 2001)), PAPR values of proposed system is lower compared to (Natarajan et al; 2001) due to improved PAPR reduction performance for the proposed system. At the same time a significant improvement in BER is achieved for our method due to novelty of the proposed subcarrier scheme, even at nearly 1.5 times increase in user capacity. Similar explanation is applicable for other set of results in Table 2 and Table 3.

## 6. Conclusion

This chapter discusses a new model of high capacity CI/MC-CDMA system with variable data rates along with simple, fast and efficient PAPR reduction at transmitter and subcarrier PIC scheme at receiver. PAPR reduction is achieved through phase shift of pseudo-orthogonal codes with respect to the orthogonal spreading codes assigned for low and high data rate transmission, respectively. The algorithm has been extended for $M$-ary PSK system. Significant reduction in PAPR is achieved with combined code and phase optimization in conjunction with trellis coding. Simulation results show that code optimization is more effective for PAPR reduction in BPSK, Q-PSK and 8-ary PSK while phase optimization is effective for the same in case of 16-ary PSK. In the receiver, a simple, fast and efficient subcarrier PIC scheme is proposed. BER performance of the proposed method not only shows improved result compared to the conventional PIC and block PIC system but also requires low computation complexity. The scope of usage of genetic algorithms for the estimation of channel parameters for the proposed MC-CDMA system is then explored. The results reported here show that with the increase of number of users, BER values corresponding to the estimated parameters closely follow to that of BER values obtained for actual parameters values. Simulation results also show that with the increase of number of generations both BER values decrease and channel capacity increases. Finally, GA based optimized system is designed to achieve acceptable values of PAPR, BER and ADR for optimal set of the number of users, the number of subcarriers and SNR values based on the status of the wireless channel.The prposed system may be used as a potential multiple access with broadband data transmission for both uplink and downlink satellite system in conjunction with mobile communication.

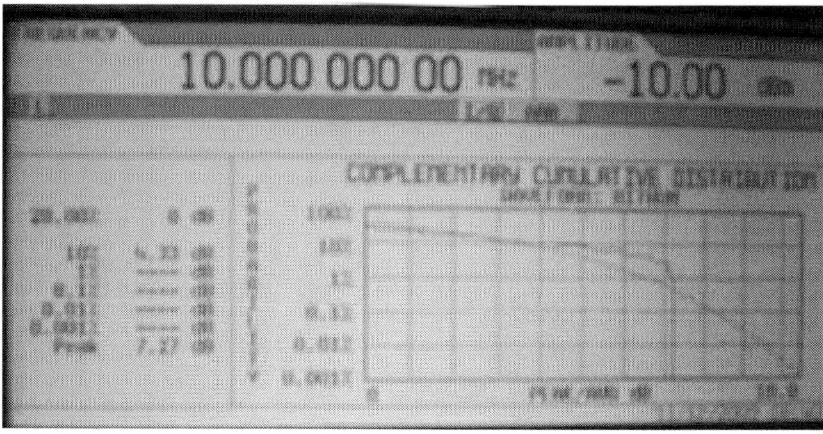


Fig. 9. CCDF of PAPR in 8-ary PSK for N=24 before code and phase optimization

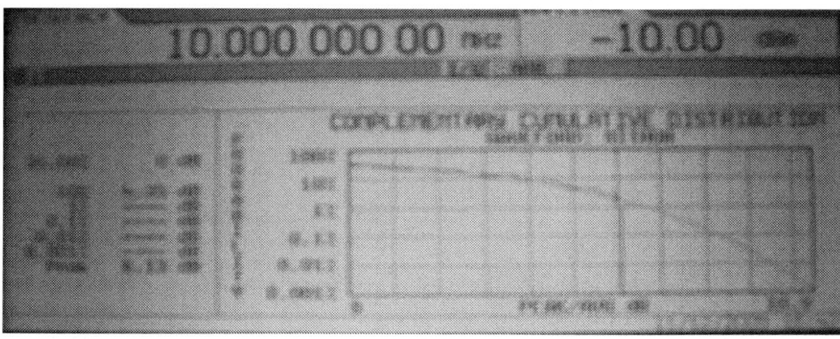


Fig. 10. CCDF of PAPR in 8-ary PSK for N=24 after code and phase optimization

**Acknowledgment**

The author acknowledge financial support for the project on "Development of high power and spectral efficiency multiuser system for broadband wireless communication" funded by Ministry of Communication and Information Technology, Govt. of India vide administrative approval no. 13(2)/2008-CC & BT dated 31.03.2008.

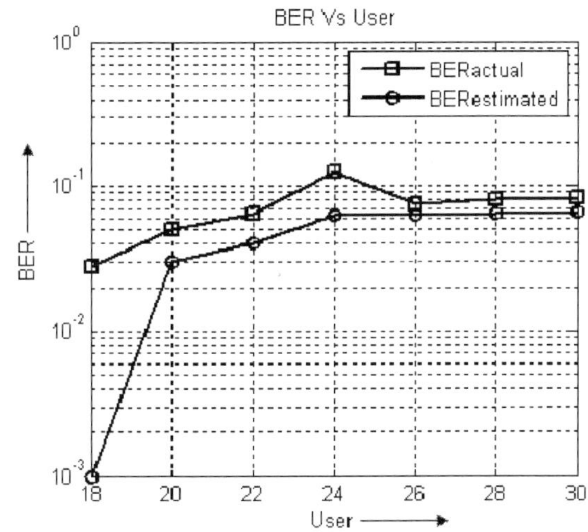


Fig. 11. BER comparison for estimated and actual channel parameters

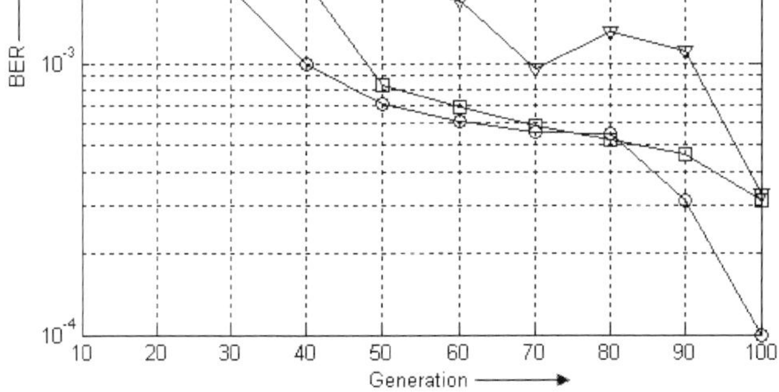


Fig. 12. BER performance with the number of generations

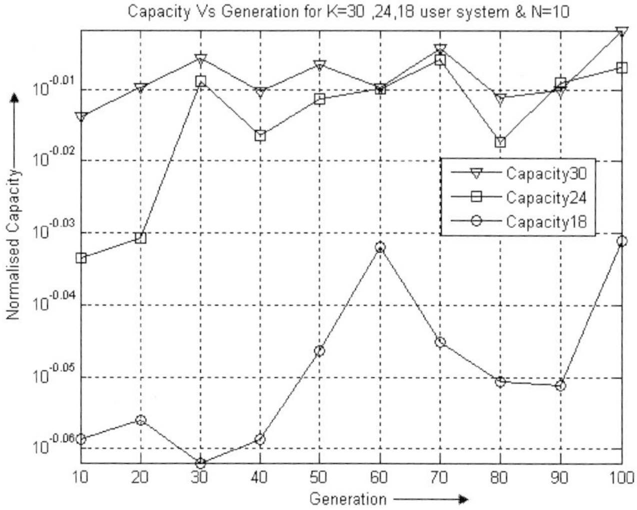


Fig. 13. Channel capacity with number of generations

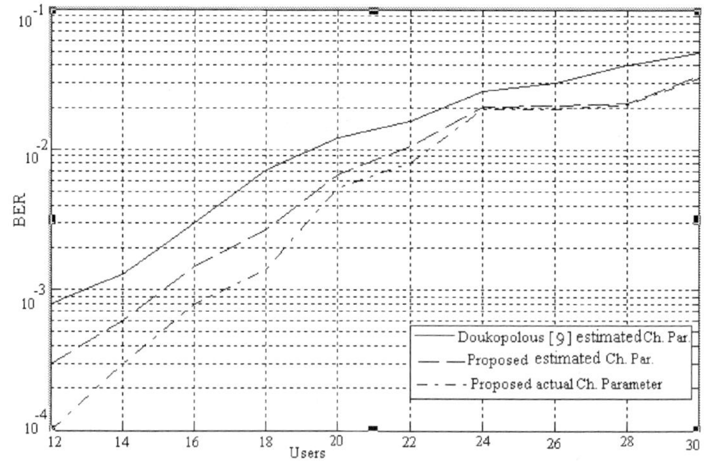


Fig. 14. Comparison of BER performance through channel estimation using N=10 and SNR=7dB

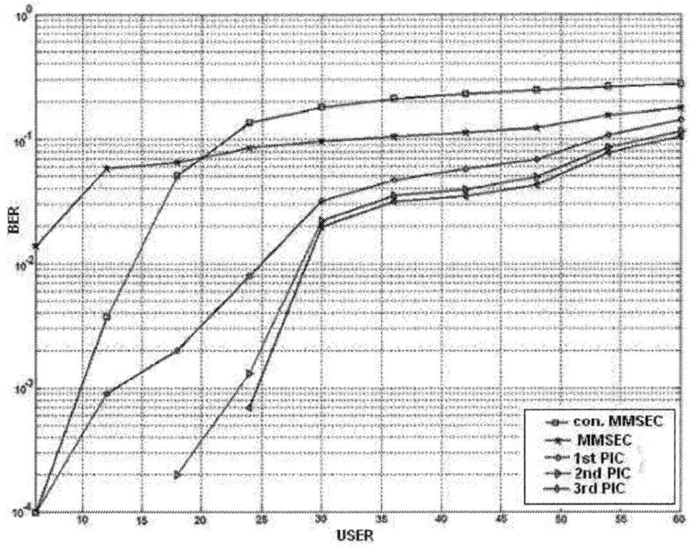


Fig. 15. Performance of subcarrier PIC scheme for 3N user system

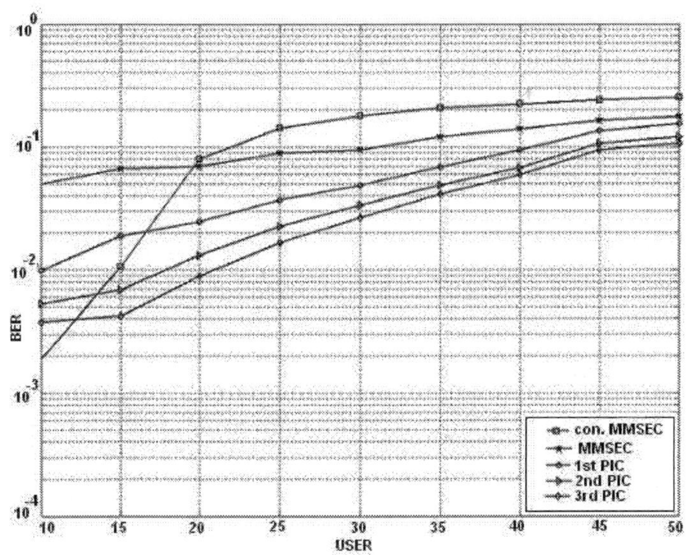


Fig. 16. Performance of subcarrier PIC scheme for 2.5N users system

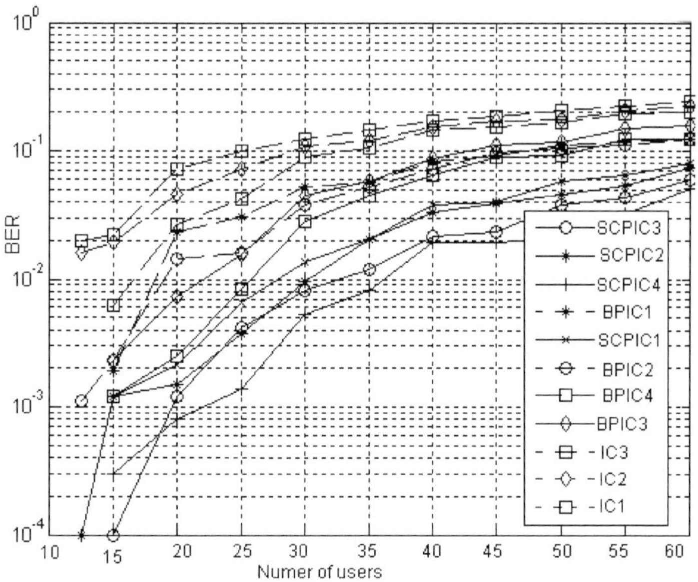


Fig. 17. Performance comparison of subcarrier PIC & block PIC schemes for 3N users system

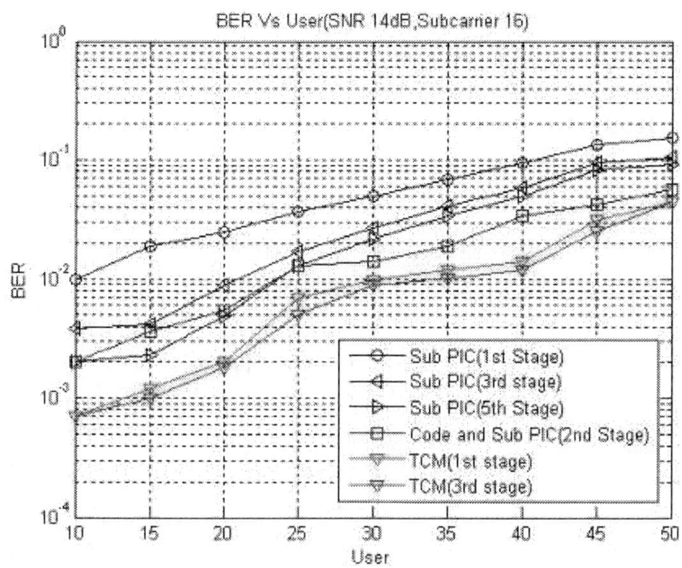


Fig. 18. BER performance for subcarrier PIC, Code and subcarrier PIC and trellis coded system for N=8 and K=24

## 7. References


Ochiai H. and Ima H.(2000). Performance of the deliberate clipping with adaptive symbol selection for strictly bandlimited OFDM systems. *IEEE Journal on Selected Areas in Communications*, Vol. 18, No. 11, (2000) (2270-2277)

Lim D. W., Heo S. J., No J. S., and Chung, H. A New PTS OFDM Scheme with Low Complexity for PAPR Reduction. *IEEE Tran. Broadcasting* Vol. 52, No. 1, (2006)(77-82).

Yoo, S., Yoon, S., Kim S. Y. and Song, I. A novel PAPR reduction scheme for OFDM systems: Selective Mapping of Partial Tones (SMOPT). *IEEE Trans. on Consumer Electronics*, Vol. 52, No. 1, (2006) (40-43)

Ochia, H. A novel trellis-shaping design with both peak and average power reduction for OFDM systems. *IEEE Trns. on Communication*, Vol. 52, No. 11, (2004)(1916-1926).

Kang, K., Kim, S., Ahn, D. and Lee, H.J. Efficient PAPR reduction scheme for satellite MC-CDMA systems, *IEE Proc. on Communication*, Vol. 152, No. 5, (2005)(697-703).

Vedu, S. Minimum probability of error for asynchronous gaussian multiple access channels. *IEEE Transactions on Inform. Theory*, Vol. 32, (1986)(85-96)

Lupas, R. and Verdu, S. Linear multiuser detectors for synchronous code division multiple access channels. *IEEE Transactions on Inform. Theory*, Vol. 35,(1989)(123-136)

Divsalar, D., Simon, M. K. and Raphaeli, D. Improved Parallel Interference Cancelation for CDMA. *IEEE Trans. Communication*, Vol. 46, No. 2 (Feb 1998)(258-268)

Kim, S. and Lee, J. H. Performance of iterative multiuser detection with a partial PIC detector and serial concatenated codes, *IEEE VTS 54th Vehicular Technology Conference*, Vol. 1, pp.487-491, 2001.

Xiao, L. and Liang, Q. The study of parallel interference weighted canceler multiuser detection, *IEEE VTS 50th Vehicular Technology Conference*, Vol. 5,pp.3009-3013, 1999.

Thippavajjula, V. and Natarajan, B. Parallel interference cancelation techniques for synchronous carrier interferometry/MC-CDMA uplink, *IEEE Vehicular Technology conference*,pp.399-403, 2004.

Maity, S. P., Hati, S. and Maity, S. Diversity Assisted Block PIC for Synchronous CI/MC-CDMA Uplink Systems Using Genetic Algorithm, *Proceedings of the third IEEE International Conf. on Industrial and Information System*, Indian Institute of Technology, Kharagpur, India, (December 2008).

Sgraja, C. and Linder, J. Estimation of Rapid Time- Variant Channels for OFDM using Wiener Filtering, Proc. IEEE Int. Conf.on Comm., Vol. 4, pp. 2390-95, 2003.

Chow, J. S., Tu, J. C. and Cioffi, J. M. A discrete multitone transceiver system for HDSL application. *IEEE J. Select.Areas Communication*, Vol. 9,(Aug. 1991)(895Ű-908).

Ziegler, R. A. and Cioffi, J. M. Estimation of time-varying digital radio channel. *IEEE Trans. Veh. Tech.*, Vol. 41, (1992)(134Ű-151).

Wang, X. and Ray Liu, K. J. Adaptive channel estimation using cyclic prefix in multicarrier modulation system. *IEEE Commun. Lett.*, Vol. 3, No. 10, (1999)(291-293).

Choi, Y. S., Voltz, P. J. and Cassara, F. A. On channel estimation and detection for multicarrier signals in fast and selective Rayleigh fading channels. *IEEE Trans. on Communication*, Vol. 49, No. 8,(2001)(1375-1387).

S. Coleri, M. Ergen and A. Puri, A study of channel estimation in OFDM systems, IEEE Globecom, 2002.

P. Schramm and R. Mullar, Pilot symbol assisted on Rayleigh fading channels with diversity: Performance analysis and parameter optimization. *IEEE Trans. on Communication*, Vol. 46, No. 12, (1998)(1560-1563).

Doukopoulos X. G. and Moustakides, G. V. Blind adaptive channel estimation in OFDM systems, *Proc. Of IEEE ICC*, Vol. 4, (2004)(20-24).

Gupta, P. and Mehra, D. K. Kalman filter based equalization for ICI suppression in High mobility OFDM systems, *Proc. of 13th National Conf. on Commun.*, (NCC-07), IIT Kanpur, pp.21-25, 2007.

Ramesh, C., Jawakar P. K., and Vaidehi, V. Pilot based adaptive channel estimation for OFDM system using GS FAP algorithm, *Proc. of 12th National Conf. on Commun. (NCC-2006)*, IIT Delhi, pp. 94-98, 2006.

Gao, X., Jiang, B.,You, X., Pan, Z., Xue, Y. and Schulz, E. Efficient Channel Estimation for MIMO Single-Carrier Block Transmission With Dual Cyclic Timeslot Structure. *IEEE Trans. on Communications*, Vol. 55, no. 11, (November 2007), (2210-2223).

Lok, T. M. and Wong, T. F. Transmitter and Receiver Optimization in Multicarrier CDMA Systems. *IEEE Transaction on Communication*, (2000)(1197-1207).

Wu, Q. Performance of optimum transmitter power control in CDMA cellular mobile systems. *IEEE Transaction on Vehicular Tech.*, Vol. 48, (1999).

Reynolds, D. and Wang, X. Adaptive transmitter optimization for blind and group-blind multiuser detection. *IEEE Trans. on Signal Proc.*, Vol. 51, (2003)(825-38).

Kim, D. Rate-regulated power control for supporting flexible transmission in future CDMA mobile networks. *IEEE Journal on Selected Areas Commmunications*, Vol. 17, (1999)(968-977).

Buzzi, S. and Poor, H. V. Joint Transmitter and Receiver Optimization for Energy-Efficient CDMA Communications. *IEEE Journal Selected Areas Communication -Special issue on multiuser detection for adv. commun. and networks*, Vol. 26,(Apr. 2008)(pp. 459-472).

Seo, K. and Yang, L. Joint transceiver optimization in MC-CDMA systems exploiting multipath and spectral density, *IEEE GLOBECOM Proceedings*, pp. 1-5, 2006.

Yee, N., Linnartz, J. P. and Fettweis, G. BER of multicarrier CDMA in indoor wireless networks. *IEEE Trans. on Commun.*, Vol. 1, (1993)(2817-2821).

Natarajan, B., Nassar, C.R., Shattil, S., Michelini, M. and Wu, Z. High performance mc-cdma via carrier interferometry codes. *IEEE Transactions on Vehicular Technology*, Vol. 50, (2001)(1344-1353).

Maity, S. P., Hati, S., Maity, S. and Mandal, M. K. Transmitter Optimization in Diversity Assisted Synchronous CI/MC-CDMA Uplink Systems Using Genetic Algorithm, *24th IEEE Queen's Biennial Symposium on Communications (QBSC-2008)*, Canada, pp. 62-67, 2008.

Proakis, J.G. *Digital communications*, 3rd Ed,McGraw-Hill, 1995.

Cal, X. and Akansu, A. N. Multicarrier CDMA systems with transmit diversity. *IEEE Transactions on Vehicular Technology*, Vol. 2, (2000)(658-661)

Ahamad, A. and Ali Assaad, M. Margin adaptive resources allocation in downlink OFDMA system with outdated channel state information, *PIMRC 2009*, Japan.

Doukopoulos, X. G. and Moustakides, G. V. Blind adaptive channel estimation in OFDM systems, *Proc. Of IEEE ICC*, Vol. 4, pp. 20-24, 2004.

Maity S. P. and Mukherjee M., Subcarrier PIC scheme for high capacity CI/MC-CDMA System with Variable Data Rates, IEEE Mobeile WiMAX'09, July 9-10, Napa Valley, California, pp. 135-140, 2009.

# Quantum Based Information Transfer in Satellite Communication

Laszlo Bacsardi and Sandor Imre
*Department of Telecommunications, Budapest University of Technology and Economics*
*Hungary*

## 1. Introduction

The first electronic computer, the ENIAC (Electronic Numerical Integrator And Computer) was developed in 1943 at the University of Pennsylvania to calculate artillery firing tables. It contained around 17500 vacuum tubes and it weighed about 27 tonnes. Since that we construct smaller and smaller computers from year to year, whose performance is becoming higher and higher. Gordon Moore, co-founder of the Intel Corporation examined the number of transistors that can be placed inexpensively on an integrated circuit in 1965. He found that this number had doubled every second year. In his original paper he examined the time interval between 1958 and 1965. However, the trend - called Moore-law - has continued more than half a century and is not expected to stop in the next five years (Moore, 1965). The law is represented on Fig. 1.

Capabilities of many electronic devices are linked to the Moore-law, for example processor speed, memory capacity etc. We can observe a continuing size decreasing in the field of integrated circuits as well. The growth in the performance of the processor is due to putting more and more transistors on the microchip of same size. This requires smaller and smaller transistors, which can be achieved if we are able to draw thinner and thinner lines onto the surface of a semiconductor disk. The big question is how long this trend can continue? We will reach the limit of our technology and won't be able to place more transistors on an integrated circuit. Researches offer different solutions for this problem like using parallel computers, DNS-technology or informatics based on quantum mechanics. Why quantum mechanics? If we want to place more transistors on an integrated circuit of a given size, the size of transistors have to be decreased. At a point we will cross the line to the world of the atoms. In that world the classical Ebers-Moll equals are not valid anymore, and quantum mechanical equals have to be used instead. Informatics based on quantum mechanical models is called quantum informatics.

In the last years, quantum theory has appeared in satellite communications offering answers for some of nowadays' technical questions. Although quantum computers are going to be the tools of the far future, there exist already algorithms to solve problems which are very difficult to be solved by traditional computers (Imre & Ferenc, 2005).

The quantum informatics can play a key role in the field of cryptography. In present classical cryptographic methods, the key exchange is generally based on public key methods. The security of modern cryptographic methods like asymmetric cryptography, relies heavily on the problem of factoring integers. In the future, if quantum computers become reality, any information exchange using current classical cryptographic schemes will be immediately insecure. Current classical cryptographic methods are not able to guarantee long-term security. Other cryptographic methods, with absolute security must be applied in the future (Gyongyosi & Imre, 2009). The quantum cryptography gives better solutions for communication problems than the classical cryptographic methods.

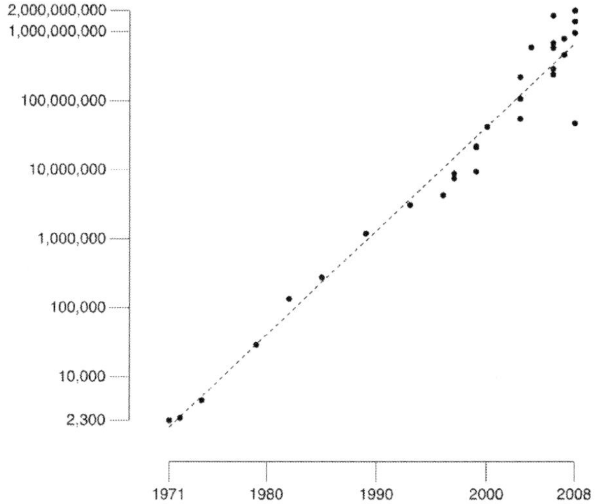


Fig. 1. One representation of the Moore-law. Horizontally the years, vertically the number of transistors in a CPU are represented. The points are for different CPU's between 1971 and 2008. The dashed line represents the Moore-law.

One of the interesting communication problems is how we can distribute a secret key for a secure communication between different parties. This is the so-called key distribution. The free-space Quantum Key Distribution (QKD) has a 16-year-old history. The first quantum cryptography protocol, the BB84 was introduced in 1984 and offered a solution for secure key distribution based on quantum theory principles like No Cloning Theorem.

The free-space quantum communications can be extended to ground-to-satellite or satellite-satellite quantum communications, which could be an ideal application for global quantum cryptography (Bacsardi, 2005).

One of the primary requirements of long-distance and free-space quantum communications is the capability of the effective transmission of quantum states in non-ideal, noisy environments. The free-space and satellite quantum channels are possible ways to increase significantly the distance limit of current quantum communication systems. To exploit the advantages of free-space quantum channels, it will be necessary to use space and satellite technology. The free space optical technology has been combined successfully with entangled pairs and satellite communications.

One of the main advantages of the usage of space for future quantum communication is the loss-free and distortion-free optical communication. In space, communication between

satellites can exploit the advantages of vacuum, where the noise of the channel can be negligible. Entanglement can be used in satellite communication to enhance the security level of key agreement process, and to realize a more secure communication compared to faint pulse quantum-key distribution technology.
This chapter is organized as follows. At first we introduce basics of quantum computing (Section 2) and quantum communication (Section 3). In Section 4, we discuss the possible connections between quantum and satellite communication including different approaches for quantum based information transfer in satellite communication, which can help to establish a secure communication link. Section 5 introduces our solutions with zero redundancy error correction which can help to establish an efficient communication link.

## 2. What is Quantum Computing?

### 2.1 Short Introduction to Quantum Informatics

From the viewpoint of quantum informatics the traditionally used communication methods are called classical methods. Communication algorithms based on classical methods are called classical algorithms. Quantum research started more than 25 years ago, and a lot of interesting results has been published since that. Although Deutsch has published the theoretical plan of a quantum computer, until now it hasn't been possible to build a real working quantum computer. Researches have had a lot of success in this area, and a lot of interesting physical implementation has been demonstrated. However, quantum informatics could not play a key role because quantum based algorithms are impossible to use without a working computer. These algorithms are very different from classical ones. Their properties have advantages in factoring, encrypting messages or creating unbreakable cryptography methods. Such solutions can be bought for commercial use from different quantum companies like id Quantique, MagiQ Technologies, Quintessence Labs.

The mathematical background of Quantum Informatics can be described by four postulates. In the first postulate the state space is defined. The second axiom describes the evolution of a closed system. The third postulate deals with measurements to create connection between quantum and classical world. In the fourth one composite systems are specified (Nielsen & Chuang, 2000).

*1st postulate.* The actual state of any closed physical system can be described by means of a so-called state vector **v** having complex coefficients and unit length in a Hilbert space *V*, i.e. a complex linear vector space equipped with an inner product.

*2nd postulate.* The evolution of any closed physical system in time can be characterized by means of unitary transforms depending only on the starting and finishing time of evolution.

*3rd postulate.* Let *X* be the set of possible results of the measurement. A quantum measurement can be described by means of a set of corresponding measurement operators .

$$M = \{\mathbf{M}_x\}, x \in X, \mathbf{M}_x \in \mathrm{H} \tag{1}$$

The operators should satisfy the completeness relation:

$$\sum_x \mathbf{M}_x^T \mathbf{M}_x = \mathbf{I} \tag{2}$$

The probability of measuring $x$ if the system is in state $|\varphi\rangle$ can be calculated as

$$p_x = \langle \varphi | \mathbf{M}_x^T \mathbf{M}_x | \varphi \rangle \tag{3}$$

The state of system after measurement is the following:

$$|\varphi\rangle = \frac{\mathbf{M}_x |\varphi\rangle}{\sqrt{p_x}} \tag{4}$$

*4th postulate.* The state space of a composite physical system $W$ can be determined using the tensor product of the individual system $V$ and $Y$:

$$W = V \otimes Y \tag{5}$$

**2.2 Quantum bits**

In classical information theory, the smallest unit is the bit. In digital computers, the voltage between the plates of a capacitor represents a bit of information: a charged capacitor denotes bit value 1 and an uncharged capacitor bit value 0. The smallest unit of the quantum informatics is the quantum bit (or qbit). One bit of information can be encoded using two different polarisations of light or two different electronic states of an atom. However, if we choose an atom as a physical bit, then apart from the two distinct electronic states the atom can be also prepared in a coherent superposition of the two states according to the rules of quantum mechanics. Therefore the atom is both in state 0 and state 1. Quantum computers use quantum states which can be in a superposition of many different numbers at the same time. In long distance communication photons are used as carriers of quantum bits. The channel can be a wired optical cable or the free-space. The problem is caused by No Cloning Theorem (NTC). According to NCT, copies can not be made of a non classical state, which means it is impossible to copy an electron spin based quantum bit to a photon based quantum bit without destroying the original quantum bit (Wootters & Zurek, 1982).

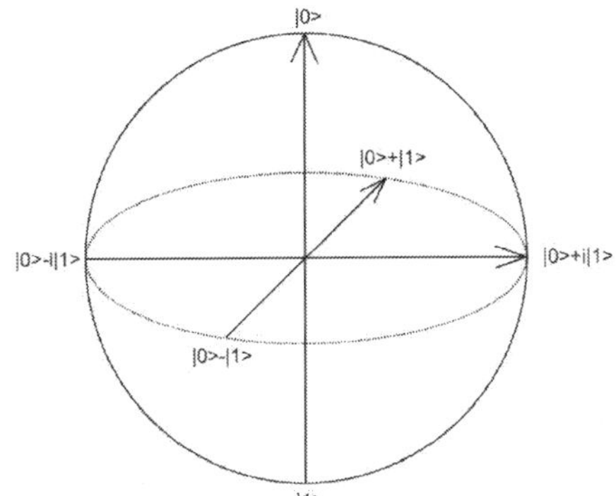


Fig. 2. Bloch sphere - a special visual representation of a quantum bit

A simple quantum system is a half-state of the two-level spin. Its basic states, spin-down $|\downarrow>$ and spin-up $|\uparrow>$, may be relabelled to represent binary zero and one, i.e. $|0>$ and $|1>$, respectively. The state of a single such particle is described by the wave function $|\psi> = \lambda |0> + \beta |1>$. The squares of the complex coefficients - $|\lambda|^2$ and $|\beta|^2$ - represent the probabilities for finding the particle in the corresponding states. The representation of a two dimensional quantum bit can be seen in Fig.2.
For example, $|\psi> = 0.6 |0> + 0.8 |1>$ means that we get 0 as result after the measurement with probability of 0.6, and we get 1 as result after the measurement with probability of 0.8. Generalizing this to a set of $k$ spin-1/2 particles we find that there are now $2^k$ basis states which equals to $2k$ possible bit-strings of length $k$ (Nielsen & Chuang, 2000).

### 2.3 Quantum algorithms

ased on the postulates of quantum informatics quantum gates can be created, which perform a typical operation and/or transformation like identity, rotation, controlled NOT etc. A quantum gate can be described with its result or with its transformation matrix.
Some important gates are the following

$$I \equiv |0\rangle\langle 0| + |1\rangle\langle 1| \quad (6)$$

$$X \equiv |0\rangle\langle 1| + |1\rangle\langle 0| \quad (7)$$

$$\mathbf{Y} = \begin{bmatrix} 0 & -i \\ i & 0 \end{bmatrix} \quad (8)$$

$$\mathbf{Z} = \begin{bmatrix} 1 & 0 \\ 0 & -1 \end{bmatrix} \quad (9)$$

$$\mathbf{H} = \frac{1}{\sqrt{2}} \begin{bmatrix} 1 & 1 \\ 1 & -1 \end{bmatrix} \quad (10)$$

where I is the identity transformation, X is the bit flip, Z is the phase flip, Y exchanges the probability amplitudes multiplied by $j$, and H is the Hadamard transformation.

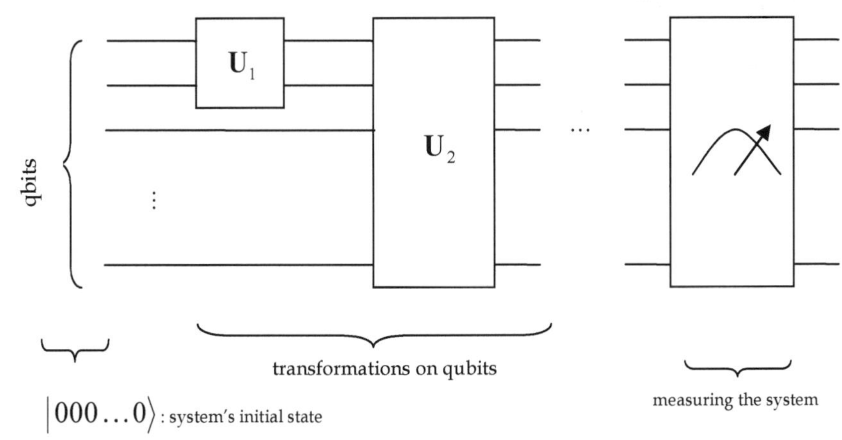


Fig. 3. General model of a quantum circuit

From the quantum gates, quantum circuits can be built. The quantum circuit is equivalent to the quantum algorithm, therefore quantum algorithms can be described by quantum circuits. The general model of a quantum circuit is illustrated in Fig.3. A quantum computer manipulates qbits by executing a series of quantum gates, each being unitary transformation acting on a single qbit or pair of qbits. At the end of the process, a classical result like 0 or 1 is determined by a measurement.
Let us create a quantum circuit like in Fig. 4. H is for Hadamard transformation, and the other gate is the Controlled-NOT gate (Imre & Ferenc, 2005). The final state of the system is

$$|\psi\rangle = \mathbf{CNOT} \cdot (\mathbf{H} \otimes \mathbf{I})(|0\rangle \otimes |0\rangle) = \frac{1}{\sqrt{2}}(|00\rangle + |11\rangle) \tag{11}$$

If we try to decompose the $|\psi\rangle = |00\rangle + |11\rangle$, our effort provides an interesting result because individual one-qbit states do not exist . States whose decomposition comprise one-qbit states are called product states while qbit and qregisters bounded together by a special phenomenon are referred to as entangled states. There are four distinguished entangled pairs called EPR pairs (named after Einstein, Podolsky and Rosen) or the Bell states.

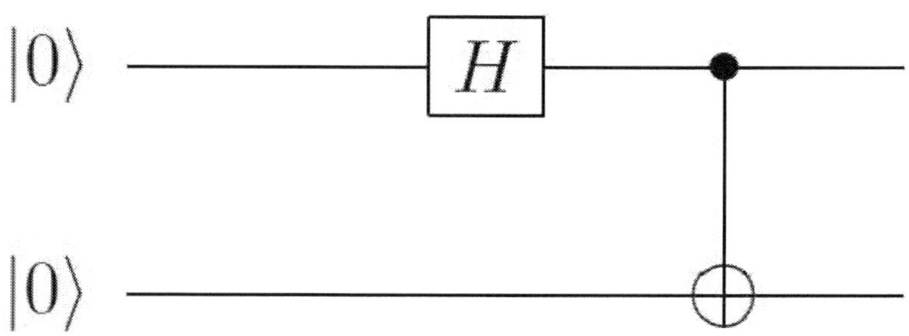


Fig. 4. EPR pairs generator quantum circuit

There are many interesting quantum algorithms, we would like to highlight some of them (Imre & Ferenc, 2005). In the quantum teleportation we use entangled pairs to transport information from point A to point B. The search in an unsorted database is faster with the Grover-algorithm than with the classical methods. We can break and decrypt the keys based on RSA algorithm with the help of Shor-algorithm. With the so-called superdense coding algorithm, two classical bits can be sent over a channel with only one quantum bit, using EPR pairs. This algorithm is illustrated in Fig.5.
From the viewpoint of the communication system, a secure key distribution is essential. In present classical cryptographic methods, the key exchange is generally based on public key methods. The security of modern cryptographic methods like asymmetric cryptography, relies heavily on the problem of factoring integers. In the future, if quantum computers become reality, any information exchange using current classical cryptographic schemes will be immediately insecure. Current classical cryptographic methods are not able to guarantee long-term security.

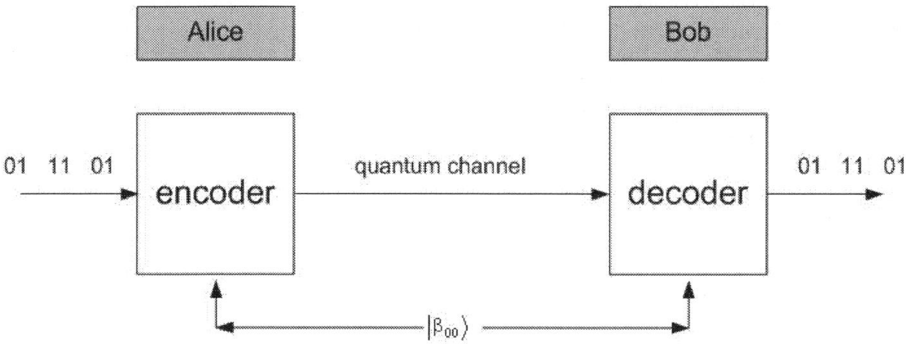


Fig. 5. In the superdense coding scenario we can send two classical bits over a channel with only one quantum bit.

## 3. Quantum Computing and Key Distribution

Quantum cryptography provides new ways to transmit information securely, using the fundamental principles of quantum-mechanics. As classical cryptography uses and manipulates classical bits, quantum cryptography does the same with qbits to realize provable absolute secure communication (Gyongyosi & Imre, 2009). In quantum cryptographic schemes, the secret information is not encoded directly into the quantum states, the qbits are used only to generate a secret cryptographic key, shared between two legal parties, called Alice and Bob. The main idea behind the quantum cryptographic protocols was the absolute secure key distribution. This is why we rather call these cryptographic methods as Quantum Key Distribution (QKD) systems.

These systems can provide unbreakable quantum communication. Therefore the QKD schemes may have a big relevance in future's space communication. There exist several fields, where private information must be sent through a free-space quantum communication channel. The QKD schemes solve the problem of key exchange, however some security steps are integrated after the process of quantum-based key exchange to provide provable safeness. Using quantum cryptography, the safety of future's space communication against various forms of possible attacks can be guaranteed.

The first QKD protocol was the BB84 protocol. The inventors of the BB84 protocol were C. H. Bennett and G. Brassard, and they published it in 1984 (Bennett & Brassard, 1984). C. H. Bennett in 1992 published the B92 protocol, the simplified version of the four state BB84 protocol (Bennet et al., 1992). The B92 protocol based on the similar principles of quantum mechanics like the BB84, however it uses only two polarization states.

The QKD schemes can solve the problem of key exchange and help to support sending private information through a free-space quantum channel. Using quantum cryptography, the safety of future's space communication against various forms of possible attacks can be guaranteed.

The QKD protocols are symmetric cryptographic systems. In these protocols, the action of encryption transforms the original message into an encoded message. The reverse operation of encryption is called decryption, which decodes the encoded message into the original

message. The decryption operation is based on a secret key, which is a symmetric key, hence the sender and the receiver have the same one to encode and decode the given message.
In the BB84 protocol Alice draws two random series of bits. One of these are the bits for the key, the second one is for the basis of measurement. If she has 0, she sends the $|0\rangle$ state, in other case she sends $|1\rangle$ state. If the base of the measurement is 1, than she performs a Hadamard transformation on the qubit. She sends the qubit to Bob. Bob draws a random series of bits. If he has 1 bit, he performs a Hadamard transformation on the received qbit. If he has a 0 bit, he performs an Identity transformation. Bob measures the quantum bit. Alice and Bob reconcile the base of their measurement on a public channel. They delete the quantum bits on which they used different measurement base. If the channel is noiseless and there is no eavesdropper in the channel, then they have the same series of qbits. To get ensured, they conciliate some selected bits.
If Eve wants to eavesdrop the channel, then she must measure Alice's qbits with a random selected measurement base, and send new qbits. However, if Alice and Bob conciliate their bits, they will notice the errors occurred by Eve's random selection. If they conciliate *N* bits, then they can find Eve's present Eve (?) with $1-\left(\frac{3}{4}\right)^N$ probability.

## 4. Quantum Satellite Communications

The quantum cryptography means distributing secret keys over a public communication channel, and not encrypting and decrypting of qbits like in the classical cryptography. The main task of the QKD protocols is to achieve an absolute secure key exchange process on the quantum channel.
Currently we use quantum algorithms in space communication mostly to secure key exchange. The values of quantum bits are encoded by photon polarizations. For the communication multilaser sources and optical receiver units are used. The current satellite and free-space schemes use weak laser pulses, instead of single photon communication. The polarization states of photons are related to the logical values of quantum states, and the particles of special photon-pairs can form entanglement (Lo & Chau, 1999).
The free-space QKD was first introduced over an optical path of about 30 cm in 1991. Since that several demonstrations (like indoor optical paths of 205 m and outdoor optical paths of 75 m) increased the utility of QKD by extending it to line-of-site laser communications systems during the 90's. In 1998, a research group at Los Alamos National Laboratory, New Mexico, USA developed a free-space QKD over outdoor optical paths of up to 950 m under night time conditions (Buttler et al., 1998). Four years later, in 2002 the same laboratory demonstrated that free-space QKD is possible by daylight as well as at night. The optical path had a length of appr. 10 km (Hughes et al., 2002). In 2006, an European research group reached the distance of 144 km (Schmitt-Manderbach et al., 2007). Space quantum communication systems have been implemented over a free-space distance of 23.4 km in Germany. The proposed satellite based quantum key distribution scheme used optical devices, improved spatial filtering and narrowband passfilter to control the transmitter lasers. In this implementation, the error rate was below 6%, and the polarization preparation devices and analysis modules were stable.

Establishing a successful free-space communication is not simple. The errors would arise from background photons collected at the satellite. The background rate depends on full or new moon: the error rate will be dominated by background photons during full moon periods, and by detector noise during new moon. During the daytime orbits the background radiance would be much larger

According to our models the quantum computing algorithms can be used to affirm our free-space communication in four different ways, which are the open-air communications, earth-satellite communications, satellite broadcast and inter-satellite communications (Bacsardi, 2007).

1. *Open-air communications:* a 'horizontal' communication channel is used. The communication happens below 100km height.

2. *Earth-satellite communications:* the communications take place through greater heights than in the Open-air communication, usually between 300 and 800 km altitude. Signal encoding and decoding can be used to produce quantum error correction that allows operation in noisy environment.

3. *Satellite broadcast:* the broadcast satellites are in orbit at 36,000 km using 27 MHz frequency for signalling. The quantum algorithms can play an important role to improve the effective bandwidth, thus the brand is better utilized as in traditional cases.

4. *Inter-satellite communication:* Equals the satellite-to-satellite communications. In this case the channel is the free-space. Any kind of coding and encoding can be used, to increase stability.

The future's free-space quantum communication should be able to realize secret key distribution over long distances using low, median and geostationary satellite system. The QKD can be extended to achieve global quantum data protection. The future free-space satellite QKD schemes can be applied in two schemes [24, 25]:

1. Global quantum key distribution based on satellite-to-ground faint pulse quantum communication,

2. Simultaneous key generation between ground stations using EPR-based QKD scheme.

In Fig. 6 we illustrated the schematic of a global key exchange QKD system in free-space. The global key free-space QKD scheme uses symmetric key encoding at the space satellite modules, and the decoding process is implemented at ground level. According to the security requirements of free-space communication, the global key free-space method requires the generation of new keys regularly. In this scheme, QKD is applied during the key upload to satellites, which (?) key will be used by the satellite module to data scrambling. The satellites scramble the transmitted data to the licensed users only, and in this process, a QKD based key can be applied efficiently to ensure the advantages of quantum mechanics.

The global key exchange based QKD scheme uses faint pulse quantum cryptography. The bitrate of the key agreement process is about 1000 bit/sec. The tested implementations have worked between 600 km and 2000 km distances. The global key exchanging satellite QKD systems can be implemented with more than one ground receiving stations for a single satellite module. The exchanged keys can be used by ground stations on long distances (Koashi et al., 2008).

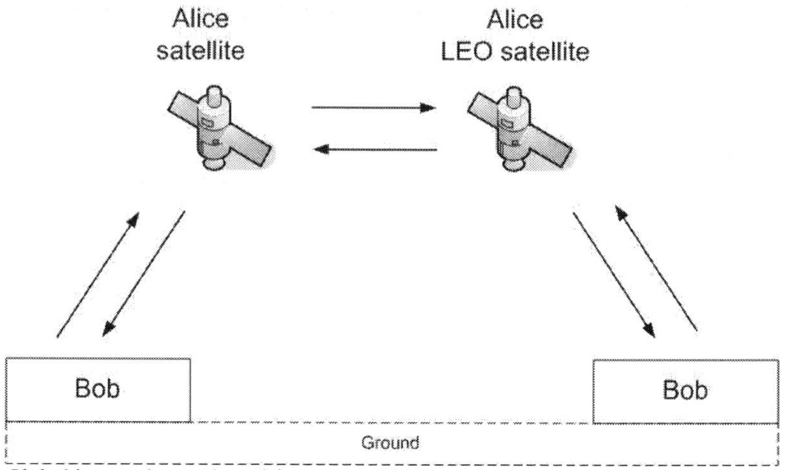

Fig. 6. Global key exchange based free-space QKD

We illustrated a satellite-to-ground free-space QKD scheme in. Fig 7. This method uses entangled photons. The bitrate of the key agreement process in the current implementations is greater than 100 bit/s.

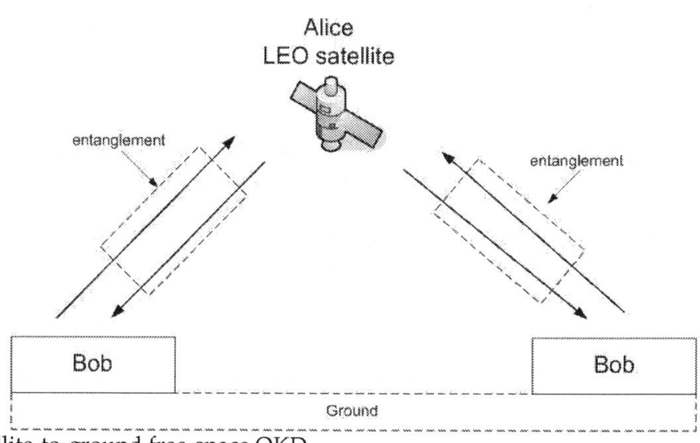

Fig. 7. Satellite-to-ground free-space QKD

The secret key agreement process requires two optical paths from the space module to the ground communication module, which increments the error rate in the current solutions. Another drawback of polarization based satellite-to-ground QKDs, that Alice and Bob rotate to each other according to the movement of the Earth, thus it is necessary to continuously compensate this movement in polarization detection (Schmitt-Manderbach et al., 2007). This kind of cryptography uses EPR-based QKD scheme. There are still many challenges in practical implementation to use entanglement. However, the free-space EPR technology can be implemented easily in current free-space QKD realizations (Ho et al., 2008; Ling et al., 2008; Tsurumaru & Tamaki, 2008).

## 5. Redundancy-free Quantum Channels

### 5.1. Importance of Redundancy-free Quantum Channels

In the classical communication we need a channel coding to handle the errors appearing in a communication channel. The error correction can be performed only with the help of redundancy in the classical system. The error correction capabilities are required for any large scale computation and communication. In these classical systems the simplest form to give redundancy to the communication is to encode the bits more than once. However, the quantum communication can be performed by more complex strategies.

In quantum computing the classical error coding methods could not be used because of the following three reasons (Nielsen & Chuang, 2000):

1. The errors are continuous. The errors can result either amplitude or phase decoherence. Moreover both errors have complex coefficients. This means that their co-domains are continuous.
2. Through the No Cloning Theorem a simple copy-based redundancy is not possible.
3. Problems occur at the measurement of the transmitted states. For the error correction the type of error has to be known. But if the quantum bits are measured for determination of the failure, then the original bits are lost.

Many quantum error correction methods have been introduced to overcome these problems and to try to handle the limitations of quantum theory principles. Since the quantum states cannot be cloned perfectly, or cannot be measured nondestructively, a simple copy-based solution is not possible. In these proposals some redundancy is required for successful error correction. However, best would to implement a redundancy-free proposal.

The redundancy-free solutions could be very useful in the long-distance free-space communication, because there would be no need to use redundant error correction codes as nowadays. With the redundancy-free techniques the effective capacity of the satellite link could also be increased

In our solution we would like to provide error correction by sending certain amount of qbits over a noisy quantum channel. The qbits are independent. Each contains information that needs to be processed. We show two different redundancy-free solutions. In the first one the noise of the quantum channel is modeled by a rotation angle. In the second one we consider the redundancy-free implementation of a unitary error correcting operator $\mathcal{R}_\theta$.

### 5.2. Redundancy-free Quantum Channel – Solution 1.

Our initial assumption is that the channel rotates the qbit with a $\omega$ degree which is considered to be constant. We wish to create a system where error correction is possible. To achieve this we mix the qbits and send them over the channel, as shown in Fig. 8.

We use $n$ long qbits for the communication so $2^n = N$, where $n$ is the length of the qbits and $N$ is the size of the space. If we code the classical states into the eigenvectors of the $U$ matrix, we can construct a description which leads to a redundancy-free solution With the appropriate selection of the matrix $A$, we can restore one quantum bit sent over the channel without any other (redundant) information.

Because $U$ is unitary, it can be written in the following form

$$U = \sum_i \lambda_i |u_i\rangle, \qquad (12)$$

where $\lambda_i$, $|u_i\rangle$ are the eigenvalues and the eigenvectors of matrix $U$ and

$$\lambda_n = e^{j\alpha_n}. \qquad (13)$$

Because $U$ is unitary, it acts on each qbits and changes it as

$$|\psi\rangle = U|\psi_k\rangle. \qquad (14)$$

Using the eigenvalues we get the following matrix for $U$

$$U_k = \begin{bmatrix} e^{j\alpha_{k1}} & 0 \\ 0 & e^{j\alpha_{k2}} \end{bmatrix}. \qquad (15)$$

For the eigenvalues, we have two cases. In the first case we can describe $U$ as

$$U = \begin{bmatrix} e^{j+\alpha} & 0 \\ 0 & e^{j+\alpha} \end{bmatrix} \otimes \begin{bmatrix} e^{j+\alpha} & 0 \\ 0 & e^{j+\alpha} \end{bmatrix} = \begin{bmatrix} e^{j2\alpha} & 0 \\ 0 & e^{j2\alpha} \end{bmatrix}, \qquad (16)$$

In the second case we can describe $U$ as

$$U = \begin{bmatrix} e^{j+\alpha} & 0 \\ 0 & e^{j+\alpha} \end{bmatrix} \otimes \begin{bmatrix} e^{j-\alpha} & 0 \\ 0 & e^{j-\alpha} \end{bmatrix} = \begin{bmatrix} 1 & 0 \\ 0 & 1 \end{bmatrix} = I. \qquad (17)$$

This description leads to a redundancy-free solution because the classical states are coded into the eigenvectors of the $U$ matrix. The eigenvalues can be written in the form shown in (13) in case of a unitary transformation. The whole mathematical background of the algorithm is described in (Bacsardi et al., 2009).

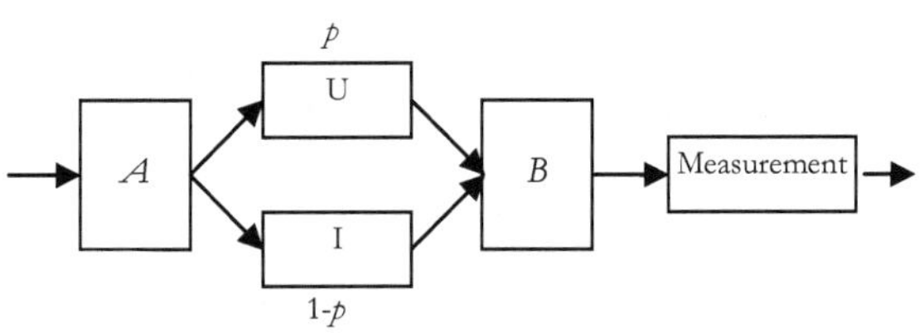


Fig. 8. Our channel model for redundancy free-error correction.

### 5.3. Redundancy-free Quantum Channel – Solution 2.

In this solution we consider the redundancy-free implementation of a unitary error correcting operator $\mathcal{R}_\theta$. The solution achieves the redundancy-free quantum communication using local unitary operations and unitary matrices. The error correcting operator is illustrated on Fig. 9.

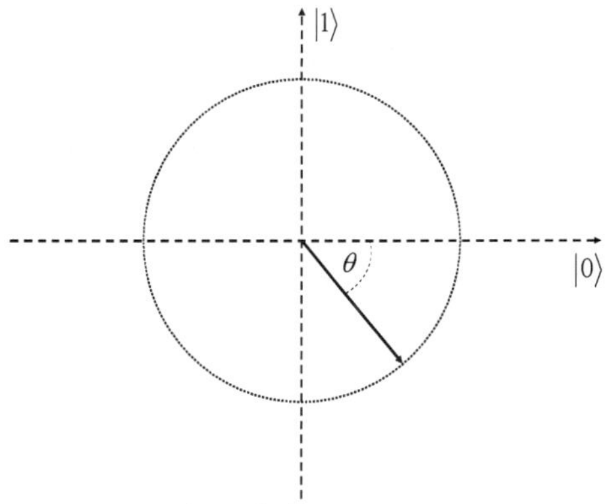


Fig. 9. Redundancy-free error correction, circle

The error of the satellite quantum channel can be modeled by a unitary rotation $\mathcal{R}_\theta^\dagger$, thus the error of the satellite quantum channel can be expressed as an angle: $\theta_i \in [0, 2\pi)$. At the beginning of the communication, Alice sends her quantum state $|\psi_A>$ on the quantum channel, which transforms it to $|d\rangle = \mathcal{R}_\theta^\dagger(\psi_A)$ with given probability $p$. The error of the quantum channel is denoted by $\mathcal{R}_\theta^\dagger$. In the error-correcting process while reading the sent quantum state, Bob doesn't know the properties of the noise on the quantum channel. In the redundancy-free coding mechanism, Alice's initial state is $|\psi_A>$, the correction transformation denoted by $\mathcal{R}_\theta$. Bob uses a CNOT to correct the error of the quantum channel. In order to read the sent quantum bits correctly, Bob must rotate the $i$-th data quantum bit by the angle $\theta_i$ in the opposite direction of that rotated by the error of the quantum channel. The method is illustrated on Fig. 10.

Bob has a chance not greater than $\varepsilon = \sin^2(\theta_i)$ to correct the sent states, because he doesn't know the original rotation angle $\theta_i$ of the quantum channel's error on the $i$-th sent qbit. The rotation operation $\mathcal{R}_\theta$ of the error correcting mechanism can be given by the angle $|\theta\rangle$, where

$$|\theta\rangle = \frac{1}{\sqrt{2}}\left(e^{i\frac{\theta}{2}}|0\rangle + e^{-i\frac{\theta}{2}}|1\rangle\right) \tag{18}$$

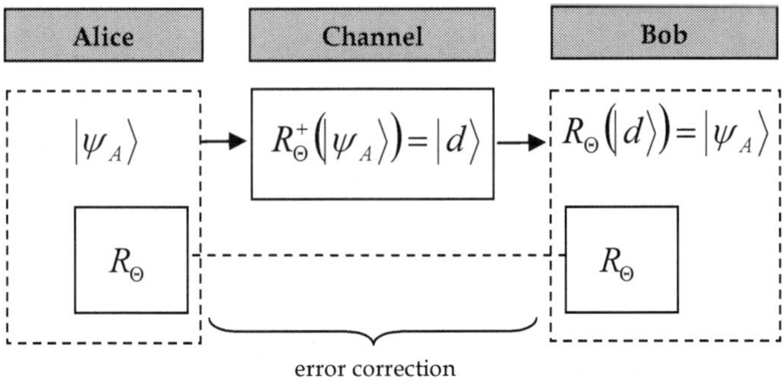


Fig. 10. Redundancy-free error correction, solution 2.

The error-correcting method consists of a control qbit, which corresponds to the modified qbit $|d\rangle$, and a target qbit, which is equal to the error-correction angle state $|\theta\rangle$. To correct state $|d\rangle$ to ket( $\psi_A$ ), Bob uses a simple CNOT transformation, thus our state is transformed to

$$|d\rangle \otimes |\theta\rangle \rightarrow \frac{1}{\sqrt{2}} \left( \mathcal{R}_\theta |d\rangle \otimes |0\rangle + \mathcal{R}_\theta^\dagger |d\rangle \otimes |1\rangle \right), \tag{19}$$

and therefore a projective measurement in the $\{|0\rangle, |1\rangle\}$ base of the correction-state $|\theta\rangle$ will make the modified qbit $|d\rangle$ collapse either into the desired state $\mathcal{R}_\theta |d\rangle$ or into the wrong state $\mathcal{R}_\theta^\dagger |d\rangle$. Bob cannot determine the received state exactly, since he does not know the angle of the error $\theta_i$. In this phase, Bob can not be sure whether the *i*-th quantum state $|d_i\rangle$ is identical to the original sent state $|\psi_i\rangle$ or not. The whole algorithm is described in (Bacsardi et al., 2009).

## 6. In the near future

Although the free-space QKD seems to be a tool for the future, the QKD implementations include some challenges, like the extremely high precision of optical devices, the nano-second time interval synchronization of key exchange process, and the generation of entangled photons for long distances (Ho et al., 2009; Ling et al., 2008). The future free-space QKD applications can include direct communication between satellites, where the secret quantum key agreement process will be achieved between satellites in the space. The current low, medium and geostationary earth orbit satellites between the range of 1000 km and 35,000 km can be used to achieve secure future space quantum communication (Schmitt-Manderbach et al., 2007). The fiber based results are promising as well. The Japanese Researches at National Institute of Information and Communications Technology

(NICT) reported a QKD field test through a 97 km installed fiber. Their experiments established the clock transmission technique for long distance QKD without degrading the quantum signal, using an optical amplifier for the clock signal (Tanaka et al., 2008). The first experimental implementation of a scheme for single-photon exchange between an Earth-based satellite was reported by Italian researches. They have built such an experiment which has a single photon source on a satellite, exploiting the telescope at the Matera Laser Ranging Observatory to detect the transmitted photons. Returning signals were detected from a low-earth orbit geodetic satellite (Ajisai), whose orbit has a perigee height of 1485 km.

Space-QUEST (Quantum Entanglement in Space Experiments) is a promising project. This project is led by a European research consortium and supported by the European Life and Physical Sciences in Space Program of ESA. The aim is to make a quantum communication space-to-Earth experiment from the International Space Station According to their mission scope, bringing quantum entanglement for the first time in the space environment will open a new range of fundamental physical experiments (Ursin et al., 2009).

## 7. Conclusions

In the next years we have to handle the problems originated from the decreasing size of microchips. As everyday life is getting faster it requires researchers and engineers to respond to the growing needs. In telecommunication applications one of the most crucial issue to solve is how to transmit as many bytes at a given frequency in a second as possible. One way to determine the system's efficiency is to measure the number of bytes sent per second.

In this chapter we presented the base of quantum mechanic based communication. We showed that there exist already algorithms to solve problems which are very difficult to be solved by traditional computers. The quantum theory has appeared in satellite communications offering answers for some of nowadays' technical questions. We introduced two solutions which can be useful in redundancy-free communications. One of the primary requirements of long-distance and free-space quantum communication is the capability of the effective transmission of quantum states in non-ideal, noisy environments. The free-space and satellite quantum channel could be the way to increase significantly the distance limit of current quantum communication systems. The current earthbound free-space quantum channels have the advantage in that they can be combined with satellite quantum communication. In the future, we will be able to overcome the current distance limits in quantum communication by transmitting EPR-states from space to Earth.

## 8. References

Bacsardi, L. (2005). Using Quantum Computing Algorithms in Future Satellite Communication. *Acta Astronautica*, Vol 57., No. (2005) 224-229, ISSN 0094-5765

Bacsardi, L. (2007). Satellite communication over quantum channel. *Acta Astronautica*. Vol 61. No. 1-6. (June-August 2007), 151-159, ISSN 0094-5765

Bacsardi, L.; Gyongyosi, L. & Imre, S. (2009). Solutions for Redundancy-free Error Correction in Quantum Channel, *Proceeding of the QuantumComm 2009*, Vico Equenso, Italy, October 2009, ICST, Gent

Bennett, C. H. & Brassard, G. (1984). Quantum cryptography: Public key distribution and coin tossing. *Proceedings of the IEEE International Conference on Computers, Systems and Signal Processing*, Bangalore, India, December 1984, IEEE, New York

Bennett, C. H.; Brassard, G. & Ekert, A.K. (1992). Quantum cryptography. *Scientific American.* Vol. 267(4) Issue. 50 (October 1992), ISSN 0036-8733

Buttler, W. T.; Hughes, R. J.; Kwiat, P. G. ; Lamoreaux, S. K.; Luther, G. G.; Morgan, G. L.; Nordholt, J. E.;Peterson, C. G. & Simmons, C. M. (1998) Practical free-space quantum key distribution over 1 km. *Phys. Rev. Lett.* Vol 81., October 1998, ISSN 1079-711

Gyongyosi, L & Imre, S. Fidelity Analysis of Quantum Cloning Based Attacks in Quantum Cryptography, *Proceedings of the 10th International Conference on Telecommunications,* Zagreb, Croatia, 2009

Ho, C; Lamas-Linares, A & Kurtsiefer, C (2009). Clock synchronization by remote detection of correlated photon pairs. *New Journal of Physics* Vol. 11, April 2009, ISSN 1367-2630 (2009)

Hughes, R. J.; Nordholt, J. E; Derkacs, D. & Peterson, C. G (2002). Practical free-space quantum key distribution over 10 km in daylight and at night. *New Journal of Physics*. Vol. 4., Issue 1, Jul 2002, ISSN 1367-2630

Imre, S. & Ferenc, B. (2005) *Quantum Computing and Communications: An Engineering Approach.* Wiley, ISBN 9780470869024

Ling, A. ; Peloso, M.P.; Marcikic, I.; Scarani, V. ; Lamas-Linares, A. & Kurtsiefer:, C. (2008). Experimental quantum key distribution based on a Bell test. *Physical Review A.* Vol.78 , Issue 2, 2008, ISSN 1050-2947 (2008)

Lo, H.-K. & Chau, H.F. (1999). Unconditional security of quantum key distribution over arbitrarily long distances. *Science* Vol. 283 No. 2050 (March 1999), ISSN 0036-8075

Koashi, M. ; Adachi, Y. ; Yamamoto, T. & Imoto, N. (2008). Security of Entanglement-Based Quantum Key Distribution with Practical Detectors. http://arxiv.org/abs/0804.0891, 2008.

Nielsen, M.A. & Chuang, I.L. (2000). Quantum Computation and Quantum Information. Cambridge University Press, ISBN 9780521635035, Cambridge

Schmitt-Manderbach, T., Weier, H; Fürst, M; Ursin,R; Tiefenbacher, F; Scheidl, T; Perdigues, J.; Sodnik, Z.; Kurtsiefer, C.; Rarity, J. G., Zeilinger, A.; & Weinfurter, H (2007). Experimental Demonstration of Free-Space Decoy-State Quantum Key Distribution over 144 km, *Physical Review Letters,* Vol. 98, Issue 1, 2007, ISSN 1079-7114

Tanaka, A; Fujiwara, M.; Nam, S. W.; Nambu, Y.; Takahashi, S. ; Maeda, W. ; Yoshino, K.; Miki, S.; Baek, B.; Wang, Z.; Tajima, A.; Sasaki, M. & Tomita, A. (2008) Ultra fast quantum key distribution over a 97 km installed telecom fiber with wavelength division multiplexing clock synchronization. *Optics Express* Vol. 16, Issue 15, July 2008, ISSN 1094-4087

Tsurumaru, T. & Tamaki, K. (2008). Security Proof for QKD Systems with Threshold Detectors. http://arxiv.org/abs/0803.4226, 2008.

Ursin, R. et al. (2009), Space-QUEST. Experiments with quantum entanglement in space. *Europhysics News* Vol. 40, Issue 3, 2009, ISSN 0531-7479

Wootters, W. K. & Zurek, W. H. (1982) A single quantum cannot be cloned, *Nature* Vol. 299 No. 802–803, 1982, ISSN 0028-0836

# 20

# Satellite coverage optimization problems with shaped reflector antennas

Adriano C. Lisboa and Douglas A. G. Vieira
*ENACOM - Handcrafted Technologies*
*Brazil*

Rodney R. Saldanha
*Universidade Federal de Minas Gerais*
*Brazil*

This chapter is devoted to optimization formulations of satellite coverage problems and their solution. They are specially useful in satellite broadcasting applications, where the information goes from one to many. Due to distance from target and power supply constraints, highly directive efficient reflector antennas are the best choice. There are many possible optimization formulations for this problem, depending fundamentally on specifications and degrees of freedom in the system. Some specific coverage problems considering broadband, radiation pattern constraints and interference are presented, where the degrees of freedom are purely geometric.

## 1. Satellite coverage problems

A satellite coverage problem queries for antennas in geostationary orbit of a planet, as shown in the example in Fig. 1 for the Brazilian territory, whose radiation pattern over a target is as close as possible to a specification.

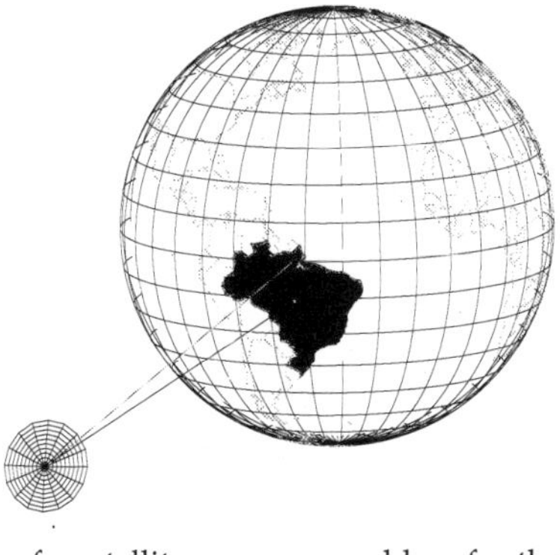

Fig. 1. Graphical illustration of a satellite coverage problem for the Brazilian territory.

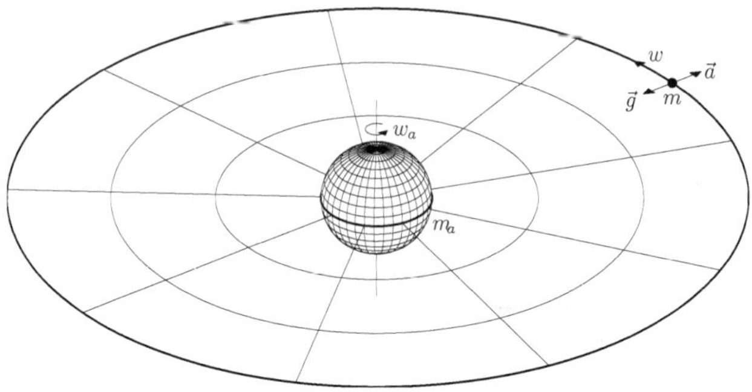


Fig. 2. Geostationary orbit (true scale) around a planet of mass $m_a$ and angular speed $w_a$.

### 1.1 Satellite's orbit

Geosynchronous orbit is a requirement because it is considered a fixed target on the planet surface. Geostationary orbit (a special case of geosynchronous orbit) is a further requirement to reduce orbital station-keeping of the satellite. Geostationary means that the satellite centripetal force $m\vec{a}$ and the planet gravitational force $m\vec{g}$ are equal in modulus and opposite in direction (see Fig. 2), i.e. $m\vec{a} = -m\vec{g}$, where $m$ is the satellite mass, $\vec{a}$ is the satellite centripetal acceleration and $\vec{g}$ is the planet gravitational acceleration. The direction constraint implies that the satellite is above the planet's equator. Using the notation $\vec{a} = a\hat{a}$ to denote the modulus $a = |\vec{a}|$ and unit direction $\hat{a} = \vec{a}/|\vec{a}|$ of a vector $\vec{a}$, the modulus constraint implies $ma = mg$, where

$$a = w^2 r \tag{1}$$

$$g = \frac{Gm_a}{r^2} \tag{2}$$

where $w$ is the satellite angular speed, $G$ is the gravitational constant ($G \approx 6.674 \times 10^{-11}\text{m}^3/\text{kg}/\text{s}^2$), $m_a$ is the planet mass ($m_a \approx 5.974 \times 10^{24}\text{kg}$ for Earth), and $r$ is the distance from the planet center to the satellite, so that the geostationary orbit radius is given by

$$r_s = \sqrt[3]{\frac{Gm_a}{w_a^2}} \tag{3}$$

where $w_a$ is the angular speed of the planet's rotation around its own axis ($w_a \approx 7.292 \times 10^{-5}\text{rad}/\text{s}$ for Earth). The geostationary orbit radius depends only on planet properties. For Earth it is about $42.2 \times 10^6\text{m}$.

### 1.2 Coordinate systems

A very important, yet simple, point in the mathematical model of the satellite is the definition of coordinate systems. In the context of satellite coverage problems, they can be defined by taking the planet, the antenna reflector or the antenna feed as reference. For now, consider planet coordinate system as any coordinate system that takes the planet center as its origin and the planet rotation axis as its $z$-axis.

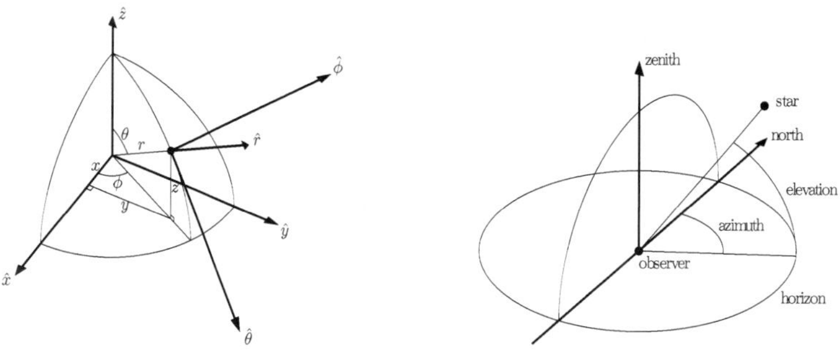


(a) (b)

Fig. 3. Spherical coordinate system (a) and hoizontal coordinate system (b).

Denote $\vec{p} = x_p\hat{x} + y_p\hat{y} + z_p\hat{z}$ a point given in Cartesian coordinate system, and $\vec{p} = r_p\hat{r} + \theta_p\hat{\theta} + \phi_p\hat{\phi}$ a point given in spherical coordinate system (see Fig. 3(a)), so that

$$\begin{aligned} x_p &= r_p \sin\theta_p \cos\phi_p \\ y_p &= r_p \sin\theta_p \sin\phi_p \\ z_p &= r_p \cos\theta_p \end{aligned} \tag{4}$$

Denote $\vec{a} = a_x\hat{x} + a_y\hat{y} + a_z\hat{z}$ a vector given in Cartesian coordinate system, and $\vec{a} = a_r\hat{r} + a_\theta\hat{\theta} + a_\phi\hat{\phi}$ a vector given in spherical coordinate system, so that

$$\begin{bmatrix} a_x \\ a_y \\ a_z \end{bmatrix} = \begin{bmatrix} \sin\theta\cos\phi & \cos\theta\cos\phi & -\sin\phi \\ \sin\theta\sin\phi & \cos\theta\sin\phi & \cos\phi \\ \cos\theta & -\sin\theta & 0 \end{bmatrix} \begin{bmatrix} a_r \\ a_\theta \\ a_\phi \end{bmatrix} \tag{5}$$

and, since the transformation matrix $U$ is unitary (i.e. $U^{-1} = U^T$),

$$\begin{bmatrix} a_r \\ a_\theta \\ a_\phi \end{bmatrix} = \begin{bmatrix} \sin\theta\cos\phi & \sin\theta\sin\phi & \cos\theta \\ \cos\theta\cos\phi & \cos\theta\sin\phi & -\sin\theta \\ -\sin\phi & \cos\phi & 0 \end{bmatrix} \begin{bmatrix} a_x \\ a_y \\ a_z \end{bmatrix} \tag{6}$$

where it is important to note that the vector components depends on the angular components $\theta$ and $\phi$ of the point where the vector itself is defined. Coherently to this, a vector field (e.g. electric field $\vec{E}$ and current density $\vec{J}$) is a function of the position where each vector is defined (e.g. $\vec{E}(\vec{r})$ and $\vec{J}(\vec{r}')$, where $\vec{r}$ is an observer point and $\vec{r}'$ is a source point). Also because the transformation matrix is unitary, the modulus of a vector does not depend on which coordinate system it is evaluated: just take the square root of the sum of its squared components.

The horizontal coordinate system, as shown in Fig. 3(b), is slightly different. It is a kind of celestial coordinate system, where positions are mapped on a sphere centered in the observer, originally taking the north pole of Earth as origin and the zenith as reference. It is defined by two components: azimuth (horizontal angular distance to the north) and elevation (vertical angular distance to the ground). It is very convenient to map deviations around a direction and it will be used to map radiation patterns where the observer is the satellite.

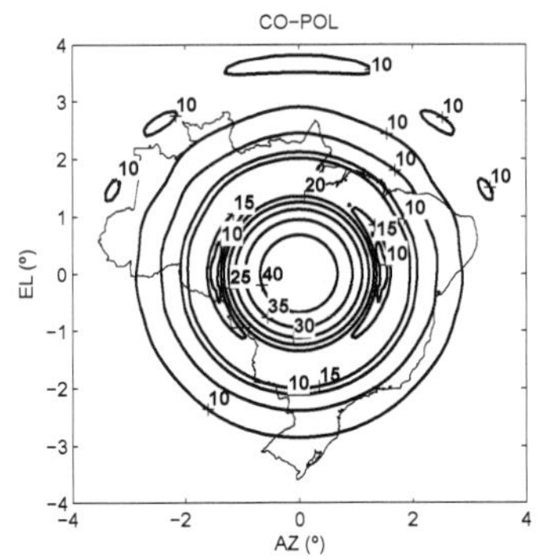


Fig. 4. Radiation pattern of a parabolic antenna illuminating the Brazilian territory. It is shown the level sets of the co-polarization gain (dBi) relative to an isotropic radiation in each direction given in reflector horizontal coordinate system.

### 1.3 Degrees of freedom in satellite's position and direction

Considering a geostationary orbit of radius $r_s$ relative to the planet center, the remaining degree of freedom in the satellite position given in planet spherical coordinate system

$$\vec{p} = r_s\hat{r} + \frac{\pi}{2}\hat{\theta} + \phi_p\hat{\phi} \tag{7}$$

is its angular position $\phi_p$ (it defines a location on the geostationary circle). The satellite is directed to a point on the planet surface

$$\vec{t} = r_a\hat{r} + \theta_t\hat{\theta} + \phi_t\hat{\phi} \tag{8}$$

where $r_a$ is the planet radius ($r_a \approx 6.4 \times 10^6$m for Earth). The center of mass $\vec{c}$ of the target is a good reference, as shown in Fig. 1, so that $\theta_t = \theta_c$ and $\phi_t = \phi_p = \phi_c$. However, the degrees of freedom $\phi_p$, $\theta_t$ and $\phi_t$ may be let as optimization variables for the satellite coverage problem.

### 1.4 Radiation pattern specification

A typical radiation pattern of a classical reflector antenna is shown in Fig. 4, where the target is the Brazilian territory. This highly directive pattern is desired for one-to-one communications (i.e. when target is a point), but can be improved when one-to-many communications (i.e. when target is an area) are considered. The most fair radiation pattern for broadcasting is the one that uniformly illuminates the target area, and only that. However different types of coverage can also be specified by gain lower or upper bounds, energy confinement inside target, radiation polarization or broadband operation.

## 2. Electromagnetic model

### 2.1 Antenna type selection

Since energy must be collimated within a few degrees around its main radiation direction, reflector antennas are very suitable and the natural choice.

A relector antenna is composed by feed and reflector, as shown in Fig. 5. The main function of the feed is to supply energy. The main function of the reflector is to collimate energy.

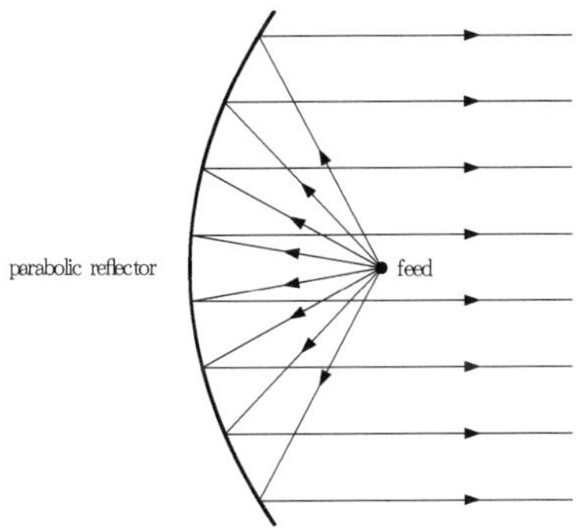


(a)

Fig. 5. Parabolic reflector antenna (a).

### 2.1.1 Parabolic reflector antenna

A point $\vec{b} = (z, \rho)$ belongs to a parabola, as shown in Fig. 6, when it is at the same distance from the focus $\vec{a} = (F, 0)$ and its projection $\vec{c} = (-F, \rho)$ onto the directrix $z = -F$, which leads to the implicit equation

$$z = \frac{\rho^2}{4F} = \frac{x^2 + y^2}{4F} \tag{9}$$

where $F$ is called the focal distance.
The unit normal vector to the parabola can be written as

$$\hat{n} = \frac{2F\hat{z} - \rho\hat{\rho}}{\sqrt{4F^2 + \rho^2}} = \frac{2F\hat{z} - x\hat{x} - y\hat{y}}{\sqrt{4F^2 + x^2 + y^2}} \tag{10}$$

The angle of the normal vector $\hat{n}$ to $\hat{z}$ is

$$\theta = \arctan\left(\frac{|\rho|}{2F}\right) = \arctan\left(\frac{\sqrt{x^2 + y^2}}{2F}\right) \tag{11}$$

The direction of a ray from the focus to a point on the parabola can be written as

$$\hat{d} = \frac{(z - F)\hat{z} + \rho\hat{\rho}}{z + F} \tag{12}$$

so that

$$\begin{aligned} -\hat{d} \cdot \hat{n} &= \frac{-2zF + 2F^2 + \rho^2}{(z + F)\sqrt{4F^2 + \rho^2}} = \frac{-2zF + 2F^2 + 4zF}{(z + F)\sqrt{4F^2 + \rho^2}} \\ &= \frac{2F}{\sqrt{4F^2 + \rho^2}} \\ &= \hat{n} \cdot \hat{z} \\ &= \cos\theta \end{aligned} \tag{13}$$

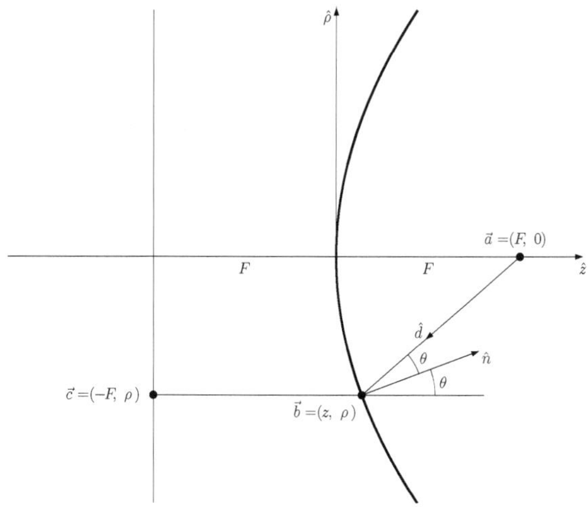


Fig. 6. Parabola with focus $\vec{a} = (F, 0)$ and directrix $z = -F$, defined by points $\vec{b} = (z, \rho)$.

Hence, the angle between $-\hat{d}$ and $\hat{n}$ is the same as the angle between $\hat{n}$ and $\hat{z}$. If the reflection is perfect (e.g. for electromagnetic waves reflector must be a perfect electric conductor), then $\theta$ is the angle of reflection of the ray, which equals the angle of incidence, so that all reflected rays go towards $\hat{z}$.

A last notable feature of parabolic reflectors is that the distance traveled by two reflected rays originated from the focus up to a given $z = z_a$ plane are exactly the same and valued to

$$\begin{aligned} \left|\vec{b} - \vec{a}\right| + z_a - z &= \left|\vec{b} - \vec{c}\right| + z_a - z \\ &= F + z + z_a - z \\ &= F + z_a \end{aligned} \tag{14}$$

Therefore, if all rays from the feed are in phase, then they will still be in phase after reflecting and reaching $z = z_a$. This is highly desired, considering that an aperture (i.e. a bounded region at a given $z = z_a$) is most directive when the field distribution on it is uniform in phase and amplitude. Unfortunately, the same does not hold true for the amplitude, since the dispersive behavior from the feed up to the reflector attenuates non-uniformly each ray according to the respective distance $F + z$.

Fig. 7 shows a parabolic reflector antenna with circular aperture of radius $R$. The reflector center was displaced $H$ towards $\hat{y}$ together with a feed tilt of $\tau$ in the $x = 0$ plane, so that the feed do not block reflected waves.

### 2.2 Geometric parameterization of the antenna

In optimization, a parameterization that considers the expected configurations to meet specified requirements reduce the search space and, hence, the computational effort to solve the optimization problem.

For a single reflector antenna, the expected shape is nearly parabolic, convex, continuous and smooth. A suitable parameterization would be a set of basis functions capable to represent

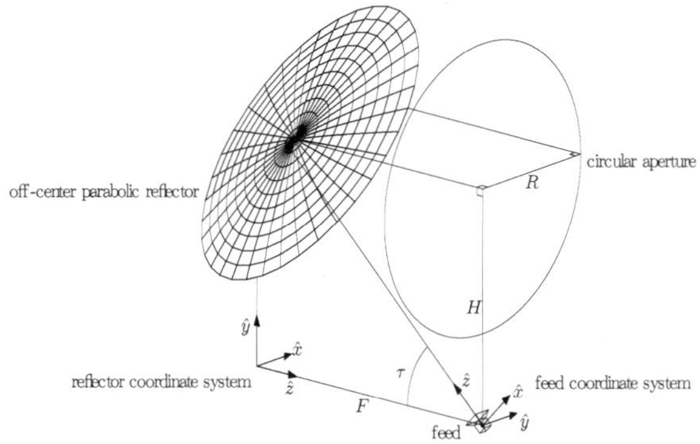


Fig. 7. Off-center parabolic reflector antenna.

any convex function, just like Fourier series are for periodic functions. Unfortunately, this is still an open problem. A parameterization that takes into consideration the expected shapes for a single reflector antenna is based on four triangular Bézier patches (Lisboa et al., 2006) to represent the reflector shape. Bézier patches are very common in computer aided geometric design. The shape of the object they represent is controlled by the position of vertices of a grid, called control points.

The feed position and direction can represent up to 5 geometric degrees of freedom for the antenna.

### 2.3 Electromagnetic analysis

In optimization, the antenna must be simulated (analysed) so that specified measures can be evaluated. These measures are evaluated at the so called observer points, denoted by $\vec{r}$. In spherical coordinates, observer points are written as $\vec{r} = r\hat{r} + \theta\hat{\theta} + \phi\hat{\phi}$. Analogously, source points are denoted by $\vec{r}'$.

Some key points concerning the electromagnetic model of the reflector antenna are depicted next.

#### 2.3.1 Harmonic fields

A very useful simplification in field propagation formulation is to consider harmonic components of it. This assumption allows to express the instantaneous field as the product of two terms: one depending only on space (i.e. amplitude, phase and direction at some spacial position) and another one depending only on time. The instantaneous field relative to a single harmonic component can be written as

$$\begin{aligned}\vec{E}(\vec{r},t) &= \mathrm{Re}\left(\vec{E}(\vec{r})\,\mathrm{e}^{jwt}\right)\\ &= \mathrm{Re}(\vec{E}(\vec{r}))\cos(wt) - \mathrm{Im}(\vec{E}(\vec{r}))\sin(wt)\\ &= \frac{1}{2}\left(\vec{E}(\vec{r})\,\mathrm{e}^{jwt} + \vec{E}^{*}(\vec{r})\,\mathrm{e}^{-jwt}\right)\end{aligned} \tag{15}$$

where $t$ is the time, $\vec{E}(\vec{r})$ is an harmonic field at angular frequency $w = 2\pi v$, $v$ is the frequency, and $j$ is the imaginary unit. Since any time independent phase angle or attenuation could be put into $\vec{E}(\vec{r})$ (i.e. $A\,e^{jwt+\alpha+j\beta} = A\,e^{\alpha}\,e^{j\beta}\,e^{jwt} = A'\,e^{jwt}$), they were set to zero. Analyzing field distributions at a single frequency (harmonic) is enough in many applications and, even if it is not, it may be possible to formulate a field distribution as a sum of harmonic components.

### 2.3.2 Feed model

The feed of a reflector antenna must radiate the energy that will be collimated by the reflector. It must be directive in order to waste less energy by missing the reflector.
One of the most simple mathematical feed model is the raised cosine, which can be written in feed coordinate system as

$$\vec{E}(\vec{r}) = \begin{cases} \frac{e^{-jkr}}{r} \cos^q \theta \left(\cos\phi\hat{\theta} - \sin\phi\hat{\phi}\right), & \theta \leq 90^\circ \\ 0, & \theta > 90^\circ \end{cases} \tag{16}$$

where $q$ is an exponent that controls the feed directivity, $k = w\sqrt{\mu\epsilon}$ is the wavenumber, $\mu$ and $\epsilon$ are the magnetic permeability and electric permittivity of free space, and $j$ is the imaginary unit. Its simplicity allows some analytical analysis and its radiation pattern approximate real world feeds.
The raised cosine formulation is a typical far field equation: no field component towards $\hat{r}$, phase factor $e^{-jkr}$, and attenuation factor $1/r$. The remaining terms are related to field strength and polarization in each direction defined by $\phi$ and $\theta$. Given the electric field $\vec{E}(\vec{r})$ in free space at $r >> 1/k$, the magnetic field $\vec{H}(\vec{r})$ is given by

$$\vec{H}(\vec{r}) = \frac{1}{\eta}\hat{r} \times \vec{E}(\vec{r}) \tag{17}$$

where $\eta$ is the intrinsic impedance of free space.

### 2.3.3 Radiated power

The electromagnetic power density is given by the Poynting vector

$$\begin{aligned} \vec{W}(\vec{r},t) &= \vec{E}(\vec{r},t) \times \vec{H}(\vec{r},t) = \frac{1}{2}\left(\vec{E}(\vec{r})\,e^{jwt} + \vec{E}^*(\vec{r})\,e^{-jwt}\right) \times \frac{1}{2}\left(\vec{H}(\vec{r})\,e^{jwt} + \vec{H}^*(\vec{r})\,e^{-jwt}\right) \\ &= \frac{1}{2}\,\mathrm{Re}\left(\vec{E}(\vec{r}) \times \vec{H}^*(\vec{r})\right) + \frac{1}{2}\,\mathrm{Re}\left(\vec{E}(\vec{r}) \times \vec{H}(\vec{r})e^{j2wt}\right) \end{aligned} \tag{18}$$

where the first term independs on time $t$ (mean power density) and the second term oscillates at $2w$.
The radiated power in free space of a source can be easier found by integrating the power density over a sphere in the far field region, where no field component towards $\hat{r}$ is a good approximation. It can be written as

$$P_{rad} = \frac{1}{2}\oint_s \mathrm{Re}\left(\vec{E}(\vec{r}) \times \vec{H}^*(\vec{r})\right) \cdot d\vec{s} = \frac{1}{2\eta}\int_0^{\pi}\int_0^{2\pi} \left|\vec{E}(\vec{r})\right|^2 r^2 \sin\theta d\phi d\theta \tag{19}$$

where $s$ is an sphere of large radius and $\eta$ is the intrinsic impedance of free space.
The radiated power of a raised cosine feed (16) is

$$\begin{aligned} P_{rad} &= \frac{1}{2\eta}\int_0^{\pi/2}\int_0^{2\pi} \frac{\cos^{2q}\theta}{r^2} r^2 \sin\theta d\phi d\theta = \frac{\pi}{\eta}\int_0^{\pi/2} \cos^{2q}\theta \sin\theta d\theta \\ &= \frac{\pi}{\eta(2q+1)} \end{aligned} \tag{20}$$

where $q$ is the raised cosine exponent.

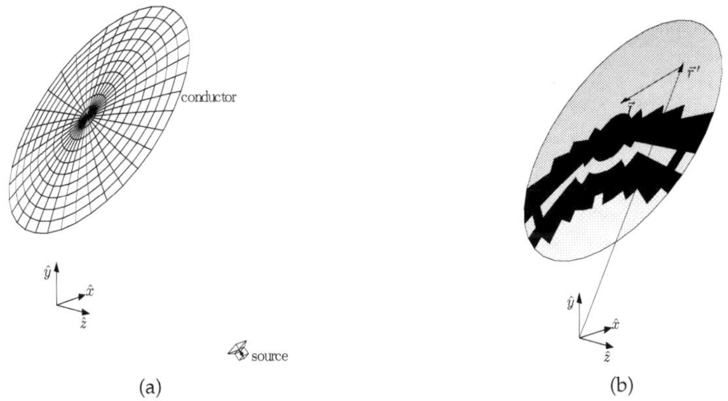


Fig. 8. Equivalence theorem: any field sources and scatters (a) can be replaced by equivalent current distributions (b).

#### 2.3.4 Equivalent current sources

A common way to analyze reflector antennas is to replace the feed and reflectors (scatterers) by equivalent current sources, which is supported by the equivalence theorem (see Fig. 8). For a single, smooth and perfect electric conducting reflector, the physical optics approximation is accurate and can be written as

$$\vec{J}(\vec{r'}) \approx 2\hat{n}(\vec{r'}) \times \vec{H}_i(\vec{r'}) \tag{21}$$

where $\hat{n}$ is the unit normal vector at the source point $\vec{r}'$, and $\vec{H}_i$ is the incident magnetic field at the source point $\vec{r}'$. Multiple reflectors may interact with each other, so that more complex procedures (e.g. method of moments) may have to be considered in order to determine accurate equivalent current sources.

#### 2.3.5 Far field

Given a current density source $\vec{J}(\vec{r}')$ over a surface $s'$ in free space as shown in Fig. 9, the electric field can be accurately approximated for $r >> 1/k$ by the integral

$$\vec{E}(\vec{r}) = -j\frac{k\eta}{4\pi}\frac{\mathrm{e}^{-jkr}}{r}\int_{s'}\left[\vec{J}(\vec{r}') - \left(\vec{J}(\vec{r}')\cdot\hat{r}\right)\hat{r}\right]\mathrm{e}^{jk\vec{r'}\cdot\hat{r}}\,ds' \tag{22}$$

where $k = w\sqrt{\mu\epsilon}$ is the wavenumber, $\mu$ and $\epsilon$ are the magnetic permeability and electric permittivity of free space, $\eta$ is the intrinsic impedance of free space and $j$ is the imaginary unit.

The far field integral can be evaluated by standard numerical methods (e.g. trapezoidal rule or Gaussian quadrature).

#### 2.3.6 Definition of polarization

Polarization of a radiation is related to the orientation of the electric field. The third definition of Ludwig (1973) is very commonly used. It can be written as

$$E_{co}(\vec{r}) = \cos\phi E_\theta - \sin\phi E_\phi \tag{23}$$

$$E_{cx}(\vec{r}) = \sin\phi E_\theta + \cos\phi E_\phi \tag{24}$$

Fig. 9. Farfield $\vec{E}$ at observer point $\vec{r}$ relative to a current distribution $\vec{J}$ at source points $\vec{r}'$ over $s'$.

Note that the definition of co and cross polarization must be coherent to the feed radiation. For the raised cosine model (16), the field components in each polarization are

$$E_{co}(\vec{r}) = \frac{\cos^q \theta \, e^{-jkr}}{r} \tag{25}$$

$$E_{cx}(\vec{r}) = 0 \tag{26}$$

for $\theta \leq 90°$. Unfortunately, reflectors may insert cross polarized components into the collimated field, even when the source does not radiate them.

### 2.3.7 Gain relative to an isotropic radiation

An isotropic radiation is the one that has the same intensity regardless of the direction. Hence, if an isotropic antenna radiates a power $P_{rad}$, the power density at a distance $r$ away from it is

$$\bar{W}(\vec{r}) = \frac{P_{rad}}{4\pi r^2} \hat{r} \tag{27}$$

considering that $P_{rad}$ is uniformly distributed over the area of an sphere of radius $r$.
The power density relative to an electric field $\vec{E}$ in free space at $r >> 1/k$ away from the source is given by

$$\begin{aligned} \vec{W}(\vec{r}) &= \frac{1}{2} \operatorname{Re}\left(\vec{E}(\vec{r}) \times \vec{H}^*(\vec{r})\right) = \frac{1}{2\eta} \operatorname{Re}\left(\vec{E}(\vec{r}) \times \hat{r} \times \vec{E}^*(\vec{r})\right) \\ &= \frac{\left|\vec{E}(\vec{r})\right|^2}{2\eta} \hat{r} \end{aligned} \tag{28}$$

The gain relative to an isotropic radiation can be written as

$$G(\vec{r}) = \frac{W(\vec{r})}{\bar{W}(\vec{r})} = \frac{2\pi r^2 \left|\vec{E}(\vec{r})\right|^2}{\eta P_{rad}} \tag{29}$$

where $P_{rad}$ is the feed radiated power and $\eta$ is the intrinsic impedance of free space. When the gain is given in decibels (i.e. $10 \log_{10} G$), the unit is called dBi to indicate that it is relative to an isotropic radiation.

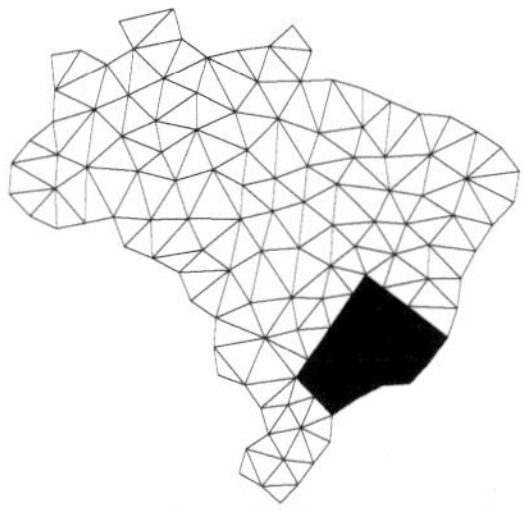

Fig. 10. Mesh over Brazilian territory used to define sample observer points.

## 3. Optimization problem formulations and solutions

### 3.1 Sampling of observer points

To formulate optimization problems, sample observer points $\vec{r}_i \in \mathbb{P}, i = 1, ..., n_r$, are uniformly spread all over the target. Fig. 10 shows a target instance, the Brazilian territory, where a triangular mesh is used to support the sampling. The antenna gain is evaluated in the vertices $\mathbb{V} \subset \mathbb{P}$ for integrative measures (e.g. energy confinement) and in the centroids of triangles $\mathbb{B} \subset \mathbb{P}$ for average measures (e.g. coverage). Subsets of samples can be differentiated to allow distinct specifications.

### 3.2 Useful measures

#### 3.2.1 For coverage

A good measure of coverage is the average gain observed at the sample points. Considering the gain given in dB and that a maximum coverage is desired, an implicit uniformity measure is imbued in the average value.

#### 3.2.2 For energy confinement

Ideally, coverage also means energy confinement since illuminating a target implicitly means only that target. However, due to physical constraints, transition to outside regions is not discontinuous, so it is meaningful to formulate apart an energy confinement measure. The energy ratio inside the target can be numerically computed using the supporting mesh for sample observer points.

### 3.3 Instance formulations of satellite coverage problems

The following instances of satellite coverage problems consider the Brazilian territory as target of a single feed single reflector antenna. The reference reflector is an off-center parabolic with circular aperture of radius $R = 0.762$m, focal distance $F = 1.506$m, and offset $H = 1.245$m along $\hat{y}$ (Duan & Rahmat-Samii, 1995). The feed is modeled by a raised cosine radiation with exponent $q = 14.28$ at frequency $\nu = 11.95$GHz, whose reference position is the reflector focus and reference direction is $\tau = 42.77°$ above the $-z$-axis on the $x = 0$ plane (Duan & Rahmat-Samii, 1995). The satellite (reflector) is directed to the center of mass of the target. The reflector shape is parameterized by four triangular Bézier patches and the feed is parameterized by its position and direction on the plane $x = 0$ of the reflector coordinate system (Lisboa et al., 2006). The number of design variables is $n = 38$ and they compose the respective vector of design variables $x \in \mathbb{R}^n$. The sampling is supported by a mesh with 125 vertices $\mathbb{V}$ and 190 triangle centroids $\mathbb{B}$, where 18 triangle centroids are labeled as southeast region $\mathbb{B}'$.

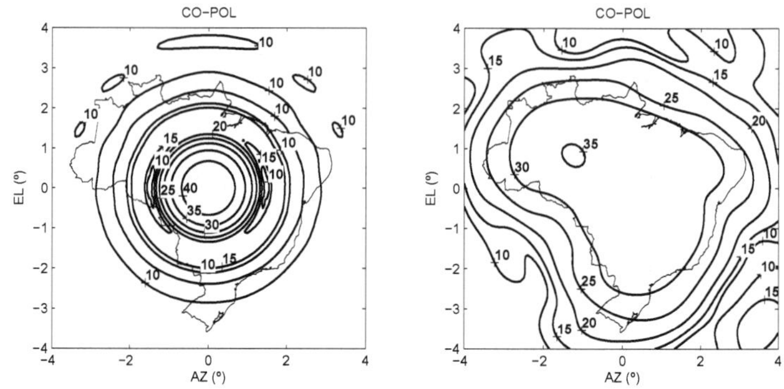


(a) (b)

Fig. 11. Radiation patterns of the reference parabolic antenna (a) and of the optimal antenna for maximum uniform coverage (b).

Some instances of satellite coverage problems are shown next. Each one is intended to tackle a specific point on design of reflector antennas. Hence, the formulation techniques can be combined to produce more complete optimization problem formulations.

### 3.3.1 Maximum uniform coverage

The most simple satellite coverage problem aims to uniformly cover the target with the maximum energy possible. It can be formulated as the monoobjective optimization problem (Duan & Rahmat-Samii, 1995)

$$\text{minimize} \quad f(x) = -\frac{1}{n_b}\sum_{i=1}^{n_b} 10\log_{10} G(\vec{r}_i, x) \tag{30}$$

$$\text{subject to} \quad x_{\min} \leq x \leq x_{\max} \tag{31}$$

where $x \in \mathbb{R}^n$ is the vector of design variables, $G$ is the antenna gain at an observer point $\vec{r}$, and $\vec{r}_i \in \mathbb{B}$ denotes an observer point derived from triangle centroids.
The optimization algorithm iteratively queries the oracle of the optimization problem (antenna simulator in this case) with a vector of design variables $x_k$ at iteration $k$, which in turn must answer with the respective values of the optimization functions at $x_k$ (only $f(x_k)$ in this case). Since the oracle query is expensive, the fewer iterations required for convergence, the better. Considering that the optimal antenna must still collimate energy since the target is far away, a highly directive classical reflector antenna is a good starting point.
Fig. 12 shows the optimal radiation pattern for maximum uniform coverage of the Brazilian territory (Lisboa et al., 2006). It was achieved by a search direction optimization algorithm with theoretical guarantees on improving the objective functions every iteration (Vieira, Takahashi & Saldanha, 2010), starting from a classical off-center parabolic reflector antenna.

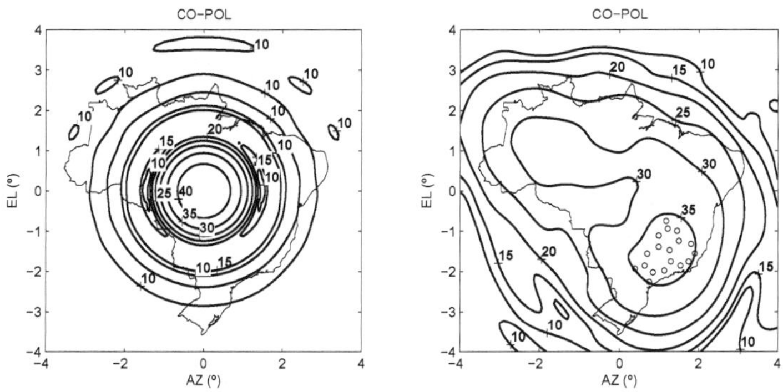


(a) (b)

Fig. 12. Radiation patterns of the reference parabolic antenna (a) and of the optimal antenna for maximum uniform coverage constrained to lower bound of 35dBi in the southeast region (b).

#### 3.3.1.1 Minimum gain constraints

To specify lower bounds of gain at subregions of the target, a subset of sample points can be used. It can be formulated as the monoobjective optimization problem (Vieira, Lisboa & Saldanha, 2010)

$$\text{minimize} \quad f(x) = -\frac{1}{n_b}\sum_{i=1}^{n_b} 10\log_{10} G(\vec{r}_i, x) \tag{32}$$

$$\text{subject to} \quad g_j(x) = G_{\min,i} - G(\vec{r}_i, x) \leq 0, \ \ \forall \vec{r}_i \in \mathbb{B}' \subset \mathbb{B} \tag{33}$$

$$x_{\min} \leq x \leq x_{\max} \tag{34}$$

where $x \in \mathbb{R}^n$ is the vector of design variables, $G$ is the antenna gain at an observer point $\vec{r}$, $\vec{r}_i \in \mathbb{B}$ denotes an observer point derived from triangle centroids, and $G_{\min,i}$ is the desired minimum gain at $\vec{r}_i \in \mathbb{B}'$. Each sample observer point inside the subregion $\mathbb{B}'$ defines one constraint.

Fig. 12 shows the optimal radiation pattern for maximum coverage with a minimum gain 35dBi in the southeast region (Vieira, Lisboa & Saldanha, 2010), leading to a problem with 18 constraint functions. It was achieved by an ellipsoid method with an improved constraint functions treatment (Shor, 1977; Vieira, Lisboa & Saldanha, 2010; Yudin & Nemirovsky, 1977), starting from a classical off-center parabolic reflector antenna.

### 3.3.2 Maximum energy confinement

The search for maximum energy confinement can be formulated as the monoobjective optimization problem (Lisboa et al., 2008)

$$\text{minimize} \quad f(x) = -\sum_{i=1}^{n_b} \frac{A_i}{4\pi r_a^2} \frac{1}{3} \sum_{j=1}^{3} G(\vec{r}_{ij}, x) \tag{35}$$

$$\text{subject to} \quad x_{\min} \leq x \leq x_{\max} \tag{36}$$

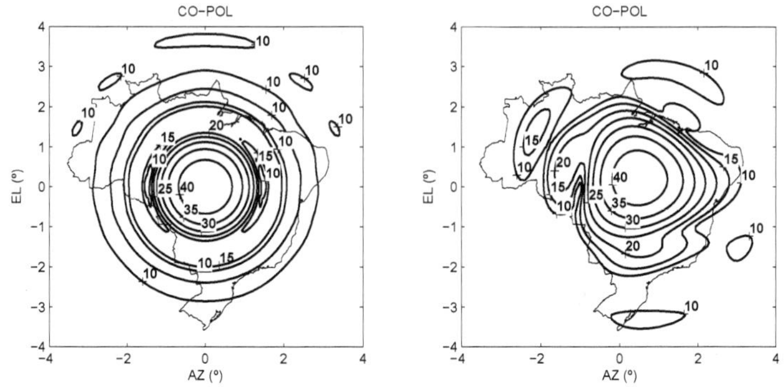


(a) (b)

Fig. 13. Radiation patterns of the reference parabolic antenna (a) and of the optimal antenna for maximum energy confinement inside target area (b).

where $x \in \mathbb{R}^n$ is the vector of design variables, $G$ is the antenna gain at an observer point $\vec{r}$, $A$ is the area of triangles, and $\vec{r}_{ij} \in \mathbb{V}$ is used to denote the $j$-th vertex of the $i$-th triangle.
Fig. 13 shows the optimal radiation pattern for maximum energy confinement inside target (Lisboa et al., 2008). It was achieved by a search direction optimization algorithm with theoretical guarantees on improving the objective functions every iteration (Vieira, Takahashi & Saldanha, 2010), starting from a classical off-center parabolic reflector antenna.

### 3.3.3 Maximum coverage and energy confinement

Optimal tradeoffs between maximum coverage and energy confinement can be formulated as the solution of the multiobjective optimization problem (Lisboa et al., 2008)

$$\text{minimize} \quad f(x) = \begin{bmatrix} -\dfrac{1}{n_b}\displaystyle\sum_{i=1}^{n_b} 10\log_{10} G(\vec{r}_i, x) \\ -\displaystyle\sum_{i=1}^{n_b} \dfrac{A_i}{4\pi r_a^2}\dfrac{1}{3}\sum_{j=1}^{3} G(\vec{r}_{ij}, x) \end{bmatrix} \tag{37}$$

$$\text{subject to} \quad x_{\min} \leq x \leq x_{\max} \tag{38}$$

where $x \in \mathbb{R}^n$ is the vector of design variables, $G$ is the antenna gain at an observer point $\vec{r}$, $A$ is the area of triangles, $\vec{r}_i \in \mathbb{B}$ denotes an observer point derived from triangle centroids, and $\vec{r}_{ij} \in \mathbb{V}$ is used to denote the $j$-th vertex of the $i$-th triangle.
Fig. 15 shows the optimal Pareto front for coverage and energy confinement for the Brazilian territory, whose respective radiation patterns of marked points are shown, from maximum coverage to maximum confinement, in Fig. 14(a), 14(b), 14(c) and 14(d). It was achieved by launching several times a search direction optimization algorithm with theoretical guarantees on improving all the objective functions every iteration (Vieira, Takahashi & Saldanha, 2010), starting from a classical off-center parabolic reflector antenna.

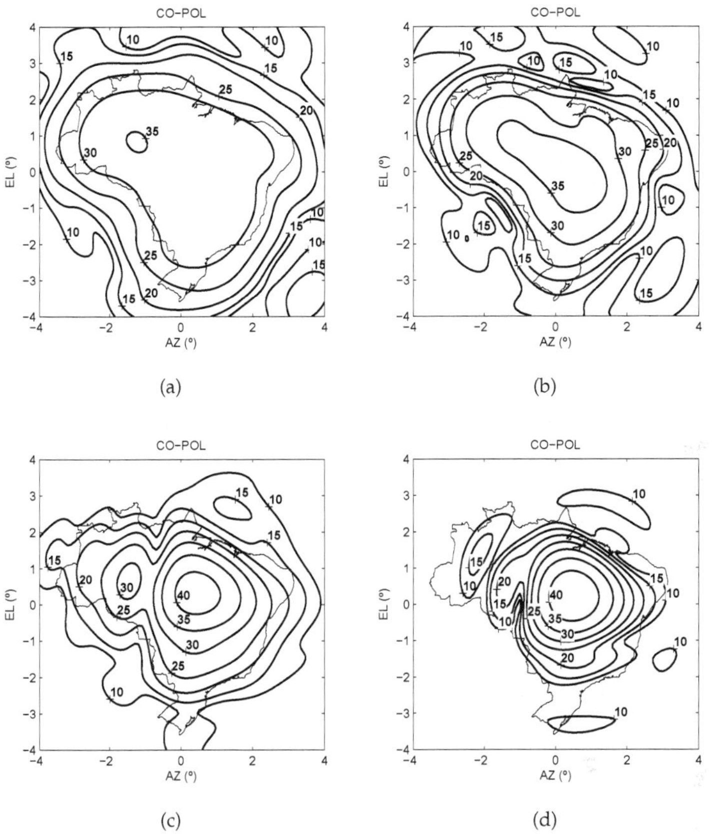


Fig. 14. Radiation patterns for maximum coverage (a), intermediate tradeoff for coverage and energy confinement (b-c), and for maximum energy confinement (d).

### 3.3.4 Broadband operation

To cope with broadband specifications, sample frequencies can be chosen inside the range so that the set of objective functions is considered for each sample frequency (Lisboa et al., 2006; 2008). For example, 5 sample frequencies for coverage and energy confinement lead to an optimization problem with 10 objective functions.

Another possible formulation is to take the worst case of each objective function inside the frequency range, which can be even more meaningful in practice. However, the convergence rate will be slower, considering that optimization algorithms that can directly cope with multiple objectives without any scalarization procedure (Vieira, Lisboa & Saldanha, 2010; Vieira, Takahashi & Saldanha, 2010) are faster when the number of objective functions increases. A final remark in this sense, is that the number of sample frequencies can be increased until it surpasses the number of design variables. From this point on, the Pareto set is likely to have

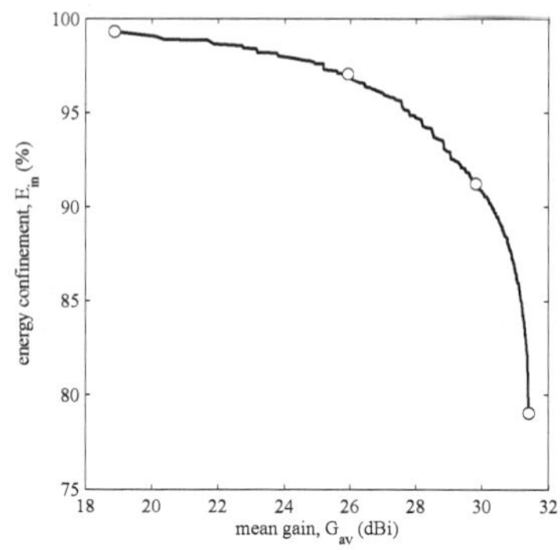


Fig. 15. Pareto front for coverage and energy confinement.

the same dimension of the search space, making numerically much easier to assert if a point is a local optimum.

**Acknowledgements**

This work was supported by FAPEMIG, CAPES and CNPq, Brazil.

## 4. References

Duan, D.-W. & Rahmat-Samii, Y. (1995). A generalized diffraction synthesis technique for high performance reflector antennas, *IEEE Transactions on Antennas and Propagation* **43**(1): 27–39.

Lisboa, A. C., Vieira, D. A. G., Vasconcelos, J. A., Saldanha, R. R. & Takahashi, R. H. C. (2006). Multi-objective shape optimization of broad-band reflector antennas using the cone of efficient directions algorithm, *IEEE Transactions on Magnetics* **42**: 1223–1226.

Lisboa, A. C., Vieira, D. A. G., Vasconcelos, J. A., Saldanha, R. R. & Takahashi, R. H. C. (2008). Decreasing interference in satellite broadband communication systems using modeled reflector antennas, *IEEE Transactions on Magnetics* **44**: 958–961.

Ludwig, A. C. (1973). The definition of cross polarization, *IEEE Transactions on Antennas and Propagation* **21**(1): 116–119.

Shor, N. Z. (1977). Cut-off method with space extension in convex programming problems, *Cybernetics* **12**: 94–96.

Vieira, D. A. G., Lisboa, A. C. & Saldanha, R. R. (2010). An enhanced ellipsoid method for electromagnetic devices optimisation and design, *IEEE Transactions on Magnetics* **(accepted for publication)**.

Vieira, D. A. G., Takahashi, R. H. C. & Saldanha, R. R. (2010). Multicriteria optimization with a multiobjective golden section line search, *Mathematical Programming* **1-31**.

Yudin, D. B. & Nemirovsky, A. S. (1977). Informational complexity and effective methods for solving convex extremum problems, *Matekon* **13**: 24–45.

# Satellite Laser Communication With Widely Dispersed Ground Stations

Paul Christopher
*Leesburg, VA 20175*
*pfchristop@aol.com*

## Abstract

We use a cloud attenuation model derived from key Italsat results to find laser cloud attenuation for satellite-ground links. Gaseous attenuation is included for 2- 10 micron attenuation with the aid of a MODTRAN model. New silicon laser developments at 3 microns allow special 10 and 3 micron comparisons. Ten microns are emphasized. High worldwide laser attenuation and relatively low link availability are indicated for latitudes south of Washington, DC.

Link availability is raised with suitable Northern Latitude satellites, such as Earth Observation Satellites. Soviet cloud correlation results indicate that link availability would be raised to acceptable levels with ground sites separated by 100-390 km.

## 1. Introduction

The Jet Propulsion Laboratory and Kim (1,2) have recognized that optical communication near one micron wavelength can offer significant broadband advantages for satellite to ground communication links. They recognized that optical advantages over conventional satellite frequencies at 12-14 GHz included small size, low cost convenient signal processing, and freedom from government frequency regulations. Laser communication links also hold the promise of high data rate. Satellite-ground links would, however, be complicated by propagation loss and sporadic availability.

Chu and Hogg (3) allowed us to note (4) that 10 micron lasers would have less atmospheric attenuation than 1 micron lasers. The 10 micron attenuation (function of longitude, latitude, and link availability) is available as a Mathematica function (4). The results can be shown as zenith attenuation maps, and we emphasize modest 80% availability here for reasonable link attenuation. We include attenuation, size, and cost features here for a fuller comparison of lasers for satellite-ground links. We combine the Chu and Hogg cloud attenuation results with the discrete atmospheric windows shown by MODTRAN (5). These windows include favorable observations near 2, 3, and 10 microns.

Laser links might be especially useful for sun synchronous Earth Observation Satellites. Landsat 4 and 5 missions were precursors to the valuable missions which may occur for EOS, but we reconsider the 701 km Landsat altitude. Large clouds of space debris were released by the Chinese antisatellite test of January 2007. The average altitude of the debris is near 800km, and we suggest new sun synchronous orbits near 1000 km. We show a key equation for required sun synchronous inclination (Appendix A) as a function of semi major axis, and we show the altered inclinations for the new altitudes. We show figures of the new earth observation orbits.

The Chu and Hogg cloud attenuation implied notable advantages for 10 micron lasers relative to the shorter wavelengths. In addition, we use a MODTRAN model to show atmospheric transmittance regions between 1 to 10.6 microns. The combination of Chu and Hogg cloud attenuation with MODTRAN gaseous attenuation is used as a basis for comparisons of laser links. The combination often favors 3 and 10 micron links. The attenuation results are also found to be a strong function of Latitude and Longitude.

Worldwide attenuation plots are emphasized for the 10 and 3 micron regions. The shorter wavelengths are indicated to become competitive with 10 micron attenuation at Latitudes greater than 50 degrees north. Communication links at Northern latitudes are especially interesting for EOS because downlinks are concentrated near the poles. The polar EOS tends to have successive repeat passes over Northern ground stations.

The shorter wavelengths also have some intrinsic advantages in size and system cost. Satellite system cost studies as Teichmann (6,7,8) tend to place high emphasis on system antenna gain. More than half of the system cost for optimum system studies was implied to reside in the antenna subsystem to achieve high gain. The shorter wavelength lasers would overcome part of their attenuation disadvantages by achieving high gain at small apertures. We discuss the system cost relations of lasers in the 2 to 10 micron region, and find that 3 micron lasers may be competitive with 10u EOS downlinks at high latitudes.

## 2. Recent Developments

### Recent Silicon Laser Developments

We applied the Chu and Hogg insights to the cloud water content which can be derived from key millimeter wave results of Barbaliscia et al (9,10) by analytical methods (11,12). Ten micron communication links appeared attractive at mid and high latitudes. Also, Frank Hanson (13) had fascinating comparisons of 10 microns and 3 micron links.

Since then (14), important new silicon laser developments have been proceeding rapidly. Bahram Jalali observed that silicon allowed photon-phonon exchange (the Raman Effect) to proceed better than in other materials. He has noted that silicon lasers are promising for 2.3 to 7 micron radiation. Fig. 2-1 is a schematic of the Jalali amplitude as the radiation traverses a silicon light pipe. The amplitude builds rapidly along the length of the silicon.

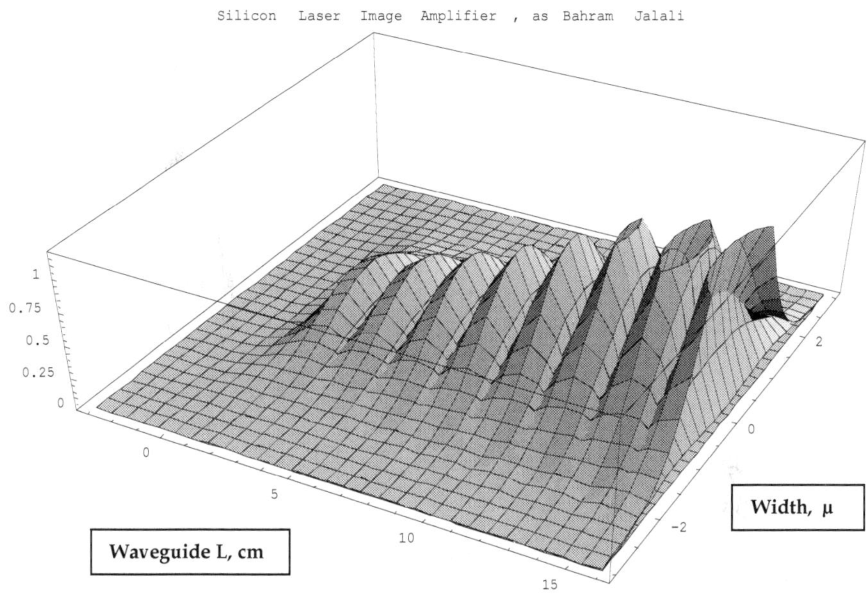


Fig. 2-1. Amplitude v. Length for Silicon Laser, after Jalali May10—Jalali2.nb.

The effectiveness and reliability of the silicon lasers in the estimated 2.3- 7 micron range is a strong inducement for us to include this range of wavelengths for Earth Observation Satellite laser link studies.

## Intense Satellite Debris Field Developments

A Chinese anti-satellite test in January 2007 left an enlarged debris field between centered near 800 km altitude. It stretches from 700 km to over 900 km altitude. In addition, a Russian satellite hit a valuable Iridium satellite in 2009 at lower altitude. Therefore, we alter a useful 701 km Landsat orbit to nearly 1000 km altitude. The inclination must also be altered to keep the useful sun-synchronous features. Fig. 2-2 indicates the mean debris field, and Fig. 2-3 the higher orbit required to clear the debris. Appendix A discusses the change in inclination needed for the higher orbit.

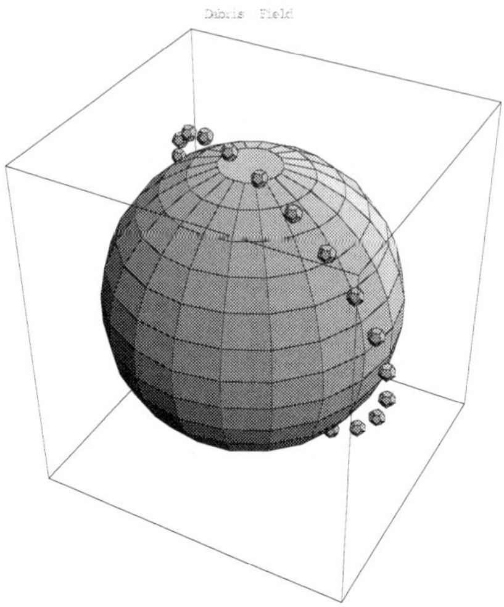

Fig. 2-2 ASAT Debris Field, Mean Altitude 800 km

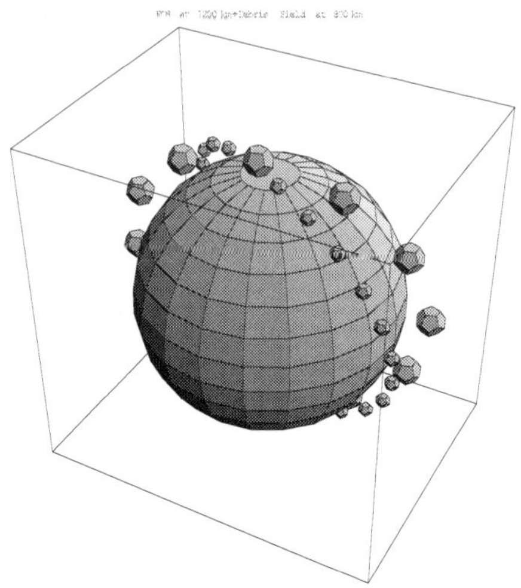

Fig. 2-3 EOS Higher (1000km) than Debris Field

Mathematica Apr27—EOScx.nb.

Valuable EOS views of Earth include the endangered Arctic region (Fig. 2-4). The coastlines are shown in relief so that visibility regions can be found quickly.

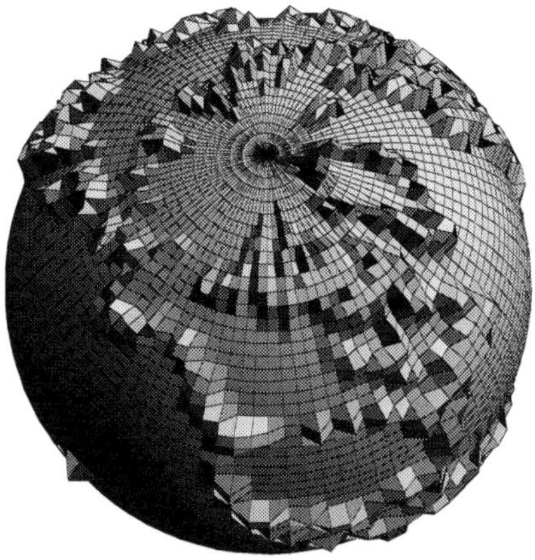

Fig. 2-4 EOS View of Canada, Alaska, Arctic Ocean Jan8ConstEOS--.nb

## 3. Laser Propagation Comparisons

Four decades ago, Chu and Hogg attempted to compare optical communication with infrared communication at 10.6 microns for terrestrial communication. They found that 10 micron signals propagated much better through fog than did the optical communication.

Fig. 3-1 indicates the kind of signal loss they anticipated over a wide range of frequencies. The upper curve indicates loss through 1 km of fog with 0.1 gm/m3 liquid water density, vs Frequency on the abscissa.
Note that 10 microns corresponds to 30 THz and the 1 micron region corresponds to 300THz. The fog comparison for this figure indicates about 60 dB and 200 dB loss, respectively, for the two frequencies.

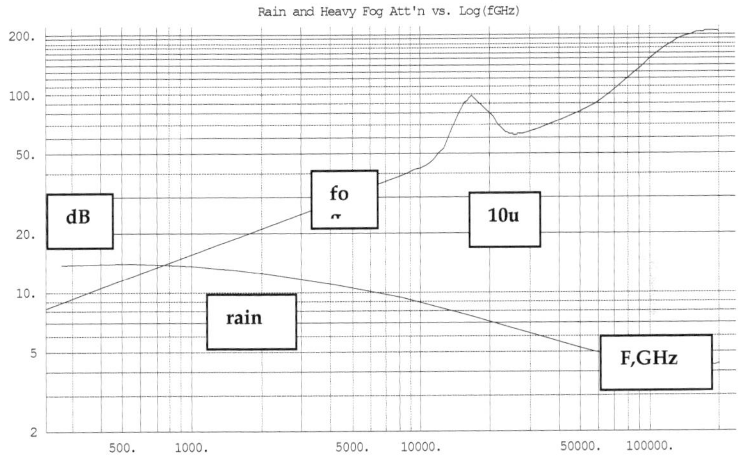


Fig. 3-1 Chu and Hogg's Fog (top) and Rain Attenuation (dB) v. Frequency
Note, 10u at 30,000 GHz and 1u at 300,000 GHz

Cloud attenuation can be estimated as a function of worldwide location and desired availability. An 80% availability was used for 10 micron propagation in order to keep propagation loss to reasonable levels. A form of Quad-Diversity was chosen for the ground site to get desired overall link availability. Four sites on the order of 15 km apart were used to approach 98-99% availability. The diversity section below will indicate even higher link availability at hundreds of kilometer separation.
We also add gaseous attenuation here for completeness. A smoothed MODTRAN result is seen as Fig. 3-2. It includes gaseous attenuation for a moderate atmosphere.

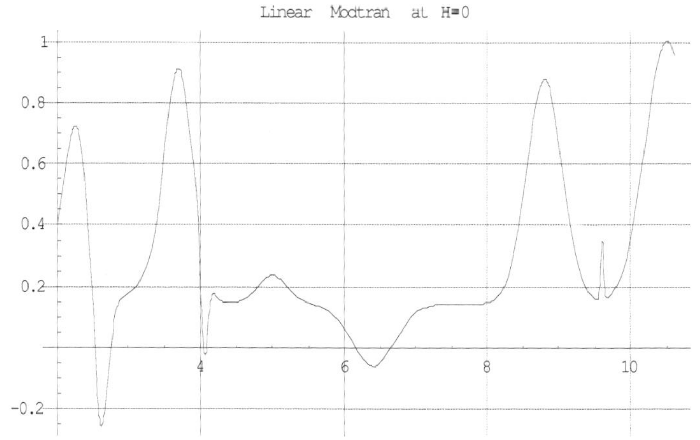


Fig. 3-2 Smoothed MODTRAN Transmission v. Wavelength, Microns

The MODTRAN model can be combined with the cloud zenith attenuation model indicate low attenuation at 10u and 3 u.

Variations in attenuation for the 10 micron and 3 micron maps may be seen as Figs. 3-3, -4.

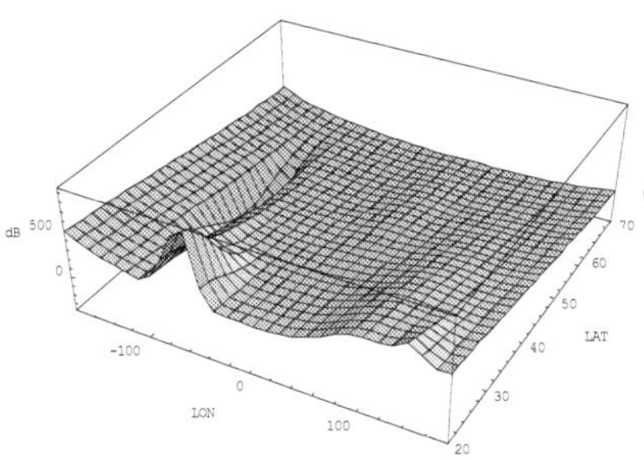


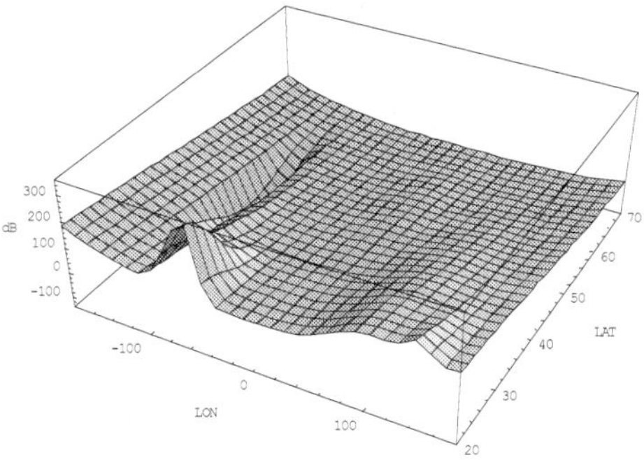


Fig. 3-3 10 Micron Attenuation Map Fig. 3-4 3 Micron Attenuation Map

80% Availability

The ten micron cloud attenuation varies, as clouds form and dissipate. The attenuation may be described by an exponential probability density function, with high attenuation at the $90^{th}$ percentile level. For convenient probability calculations, this may be referred to as the 10 percent exceedance level. Fig. 3-5 shows the intense attenuation that a ground observer would experience by looking straight up (zenith), looking through the clouds. An observer near Miami may experience hundreds of dB attenuation.

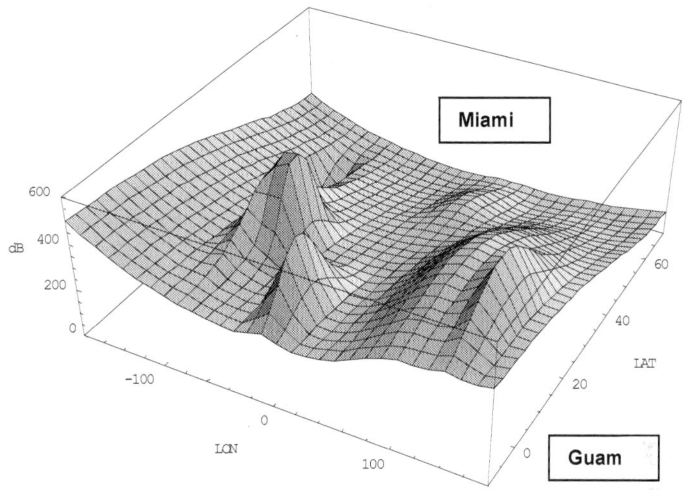


Fig. 3-5 10u Zenith Attenuation in Northern Hemisphere 90% Availability [PR=.1] Mathematica Jan22010--.nb.

Many successful systems would require attenuation less than 10 or 15 dB. Lower link availability, as Fig. 3-6, would be required. Fig 3-6 indicates zenith attenuation at 80% availability, and many important ground sites in northern New England (note Boston 42 deg North) may enjoy reasonable attenuation as 10 dB.

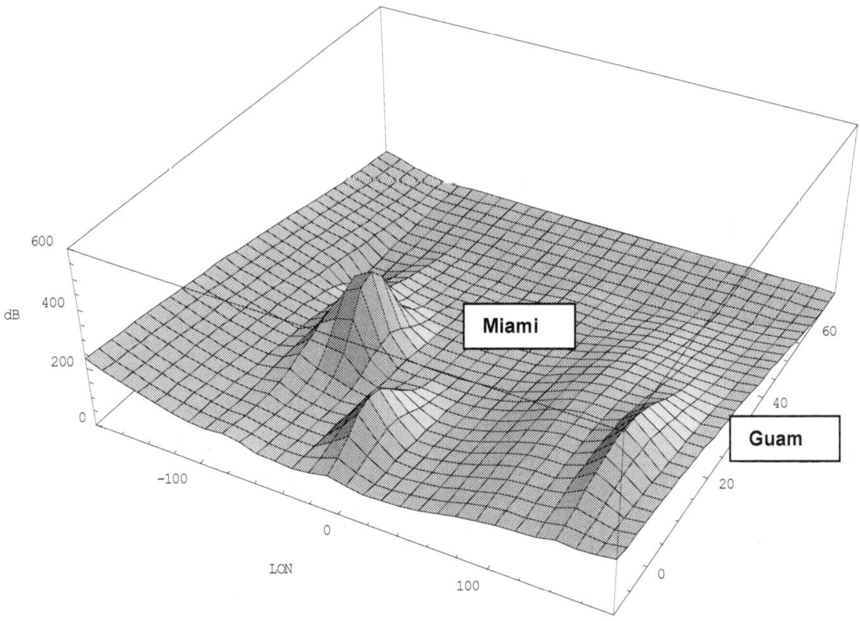


Fig. 3-6 10u Zenith Attenuation in Northern Hemisphere
80% Availability[PR=.2] Mathematica Jan22010--.nb .

Zenith attenuation at 10 microns is notably less than attenuation at 3u (Fig 3-7).

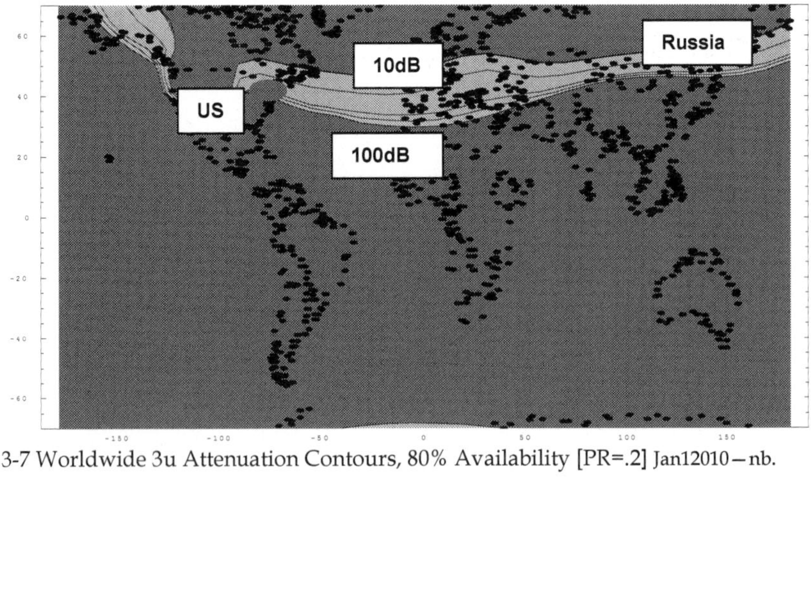


Fig. 3-7 Worldwide 3u Attenuation Contours, 80% Availability [PR=.2] Jan12010 – nb.

The modest 10 micron attenuation at 80% availability allows attractive communication sites to appear, as Bangor Maine near 10 dB loss and 50dB near Washington DC, as in Fig 3-8. Difficult or untenable attenuation as 80-100 dB is indicated near Charleston, SC.

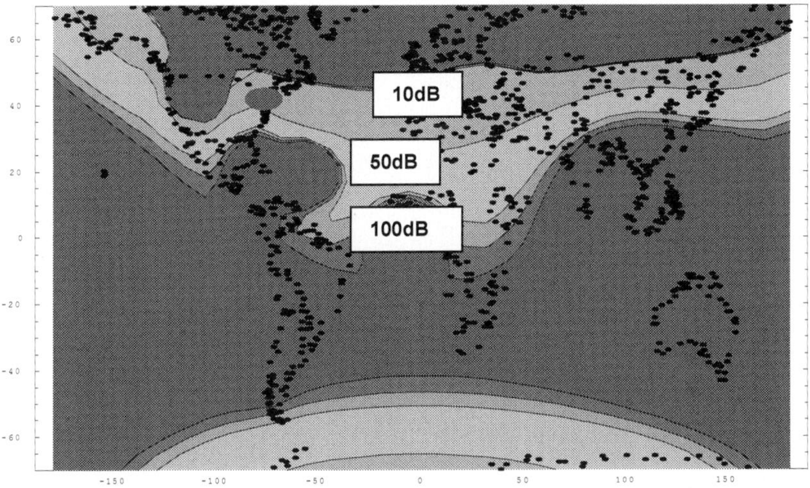


Fig. 3-8 Worldwide 10u Attenuation Contours, 80% Availability [PR=.2] Jan12010 – nb.

The asymmetry in the attenuation contours between the northern and southern hemispheres is apparent in Fig. 3-8. In addition, the attenuation of the southern hemisphere has not been as thoroughly studied as the northern hemisphere.

The 80% attenuation can offer interesting ground sites, but if we let availability standards slip even further to 70%, the area of ground sites expands sharply, as in Figs. 3-9 and -10. The areas east of Cuba and off West Africa are the only intense propagation loss regions remaining at both 3u and 10u in Figs. 3-9 and -10, respectively.

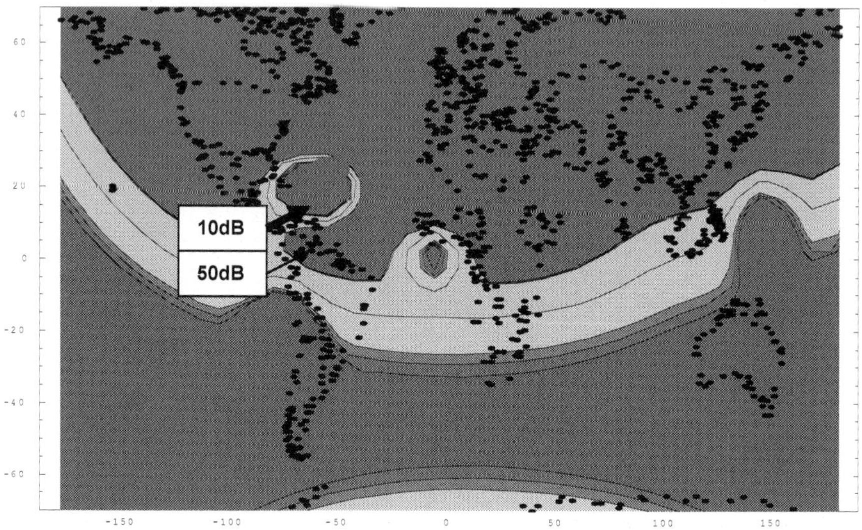


Fig. 3-9 Worldwide 3u Attenuation Contours, 70% Availability [PR=.3]

The attractive coverage for 3u at 70% availability extends even further into the Southern Hemisphere for 10u, as in Fig 3-10. Appendix B lists the 10u attenuation equation.

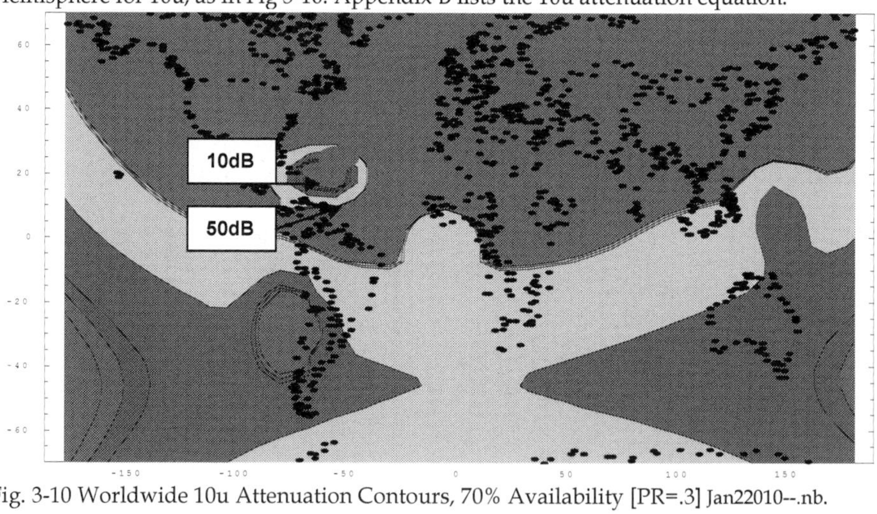


Fig. 3-10 Worldwide 10u Attenuation Contours, 70% Availability [PR=.3] Jan22010--.nb.

Of course, the limited single link availability we have discussed here is not acceptable for most important communication applications. Key Japanese tests (15) of 1u laser links addressed the question of extending the observed 20% single link availability. They placed laser receiver sites throughout the country, and were delightfully surprised that availability increased to greater than 90%. The Soviets may have anticipated this sharp rise in availability with cloud correlation studies in the early '70s. We will address the Soviet cloud correlation function in Section 4, to imply much better availability when 100-390 km site separation is available.

The ratio of 3micron/ 10micron attenuation can also be found. The ratio appears useful for estimating where 3 micron lasers could be seriously considered as an alternative to 10 microns. Fig. 3-11 uses the ratio, and draws contours at locations where 3 micron lasers are estimated to be competitive. Regions north of the contours of Fig. 3-6 may be considered as reasonably competitive 3 micron regions. Note that regions north of Labrador and Denmark appear to be a demarcation for 3 micron laser use.

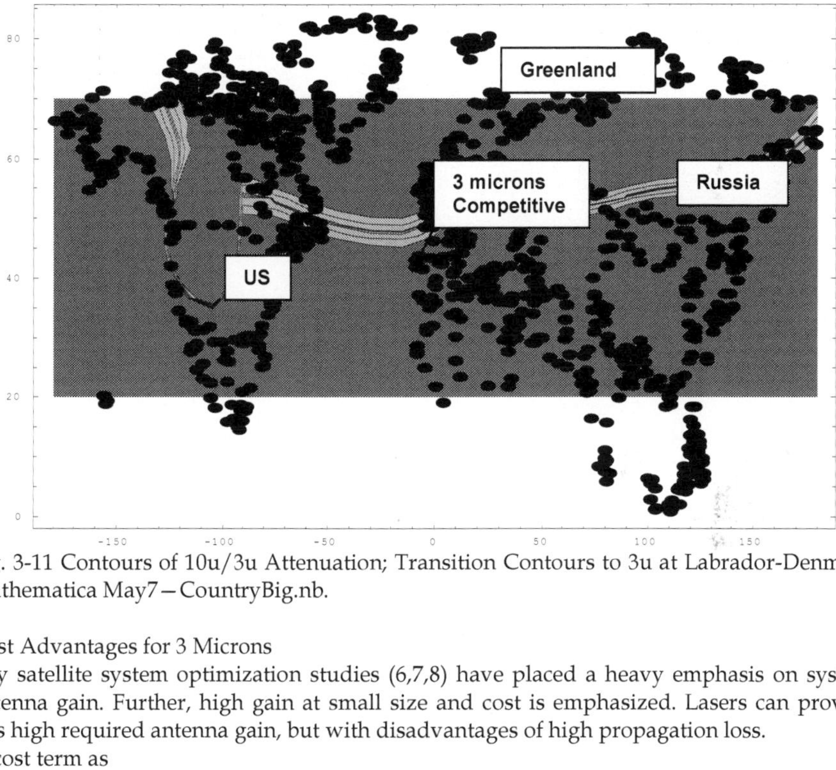


Fig. 3-11 Contours of 10u/3u Attenuation; Transition Contours to 3u at Labrador-Denmark Mathematica May7—CountryBig.nb.

Cost Advantages for 3 Microns

Key satellite system optimization studies (6,7,8) have placed a heavy emphasis on system antenna gain. Further, high gain at small size and cost is emphasized. Lasers can provide this high required antenna gain, but with disadvantages of high propagation loss.

A cost term as

$$\text{Cost/Bit}=\left(\frac{\text{loss}}{f^2}\right)^{1/3}$$

was found (5) to include both the loss disadvantages and the gain advantages of a satellite system. This is also closely related to a method of minimizing (Loss,dB - 20 LogF) for millimeter wave system. This cost per information bit method can be applied to Fig. 3-6 to find the transition line moves slightly south for the system cost comparisons, as Fig. 3-12.

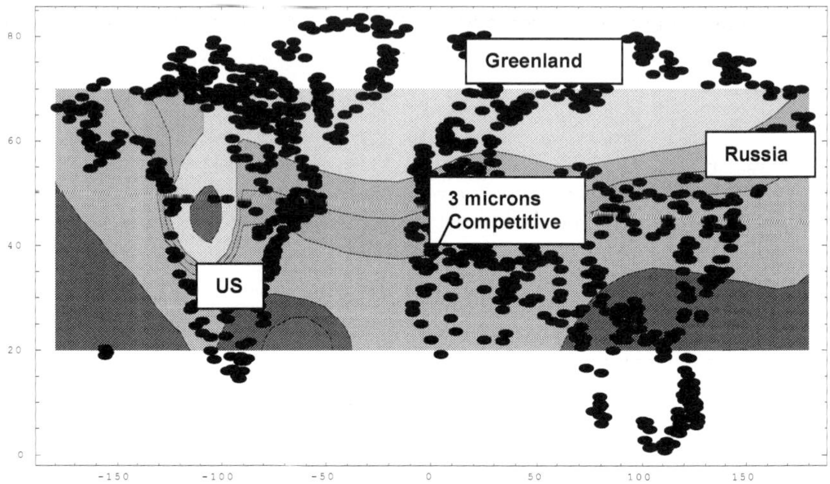


Fig. 3-12 Estimated Cost Contours for 3 Micron/ 10 Micron Comparisons

A transition region to 3 microns is now indicated to move south from Labrador to Northern Maine.

#### Intense 10u Attenuation at 90% Availability

The modest 80% availability attenuation of Figs 3-8 would require ground site diversity for useful communication. Alternatively, one might consider 90% availability for single sites (Fig. 3-13).

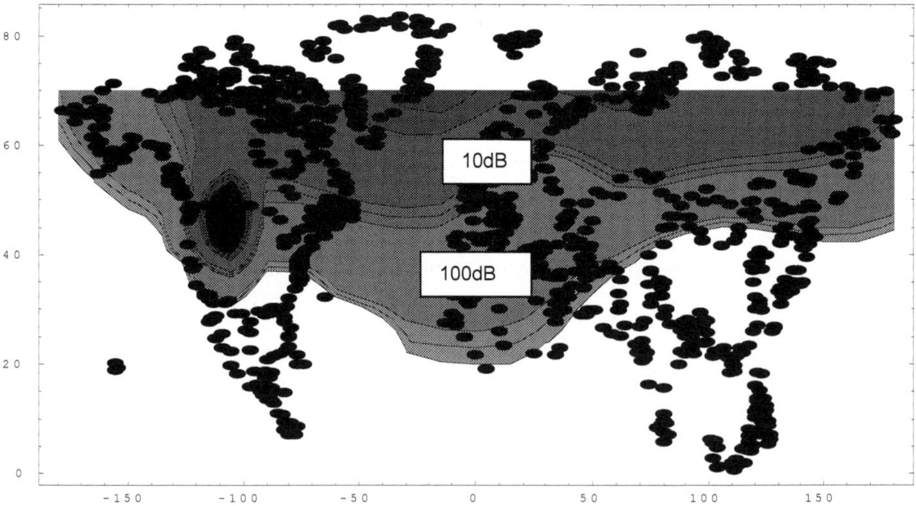


Fig. 3-13 Intense 10 Micron Attenuation Contours for 90% Availability May17-PR1.nb

Severe 100dB zenith attenuation contours are indicated at Labrador and Denmark if 90% availability is required for single links.

## 4. Diversity Analysis

Boldyrev and Tulupov (16) developed cloud correlation functions to describe the cloudy/ clear relations between two separated ground stations. The correlation functions were compiled from many satellite observations, with large scale characteristics for hurricanes and much smaller sizes for more typical observations. Fig. 4-1 shows correlation function results for a number of different seasons.

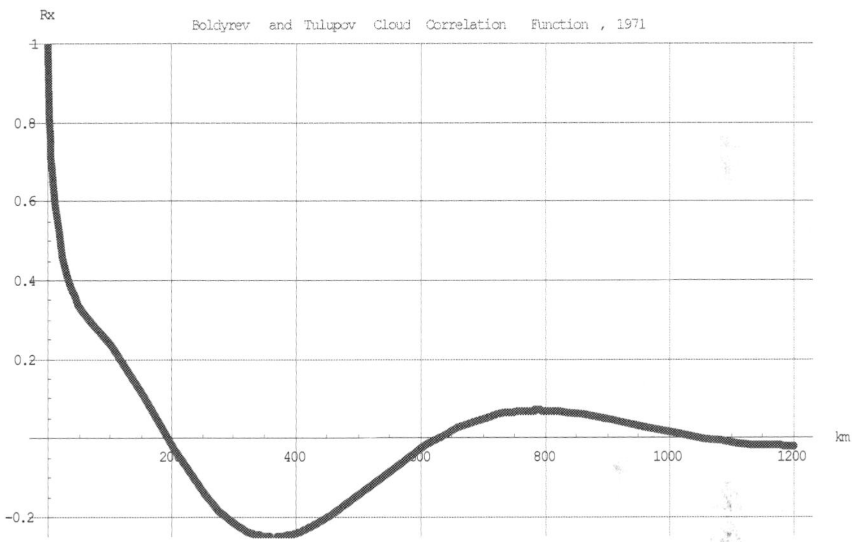


Fig. 4-1 Boldyrev and Tulupov Cloud Correlation Function

Boldyrev and Tulupov recognized that a correlation for all clouds could be represented conveniently by

$$\text{Cloud correlation}=0.2e^{-x}+e^{-0.036x}-e^{-0.15x}+0.8e^{-0.003x}\text{Cos}[0.0075x] \qquad (4\text{-}1)$$

where x= separation, km

Their insight into the importance of the correlation function will next be seen with a description of Pierce's bivariate exponential probability density function.

Boldyrev's function can be used as a direct input into a correlated bivariate exponential pdf, and then to develop quantitative results for diversity advantages. We use the form of the correlated exponential pdf as:

$$p(A1, A2) = \frac{e^{-\frac{A1+A2}{B(1-r)}}\, p\, I_0\left(2\sqrt{\frac{A1\,A2\,r}{B^2(1-r)^2}}\right)}{B^2(1-r)} \qquad (4\text{-}2)$$

where A1 and A2 are the rain attenuation (dB) at the 2 sites
r= correlation between sites, found from Boldyrev
B=standard deviation of the univariate pdf, and
p= conditional probability that rain attenuation is observed; as 0.80

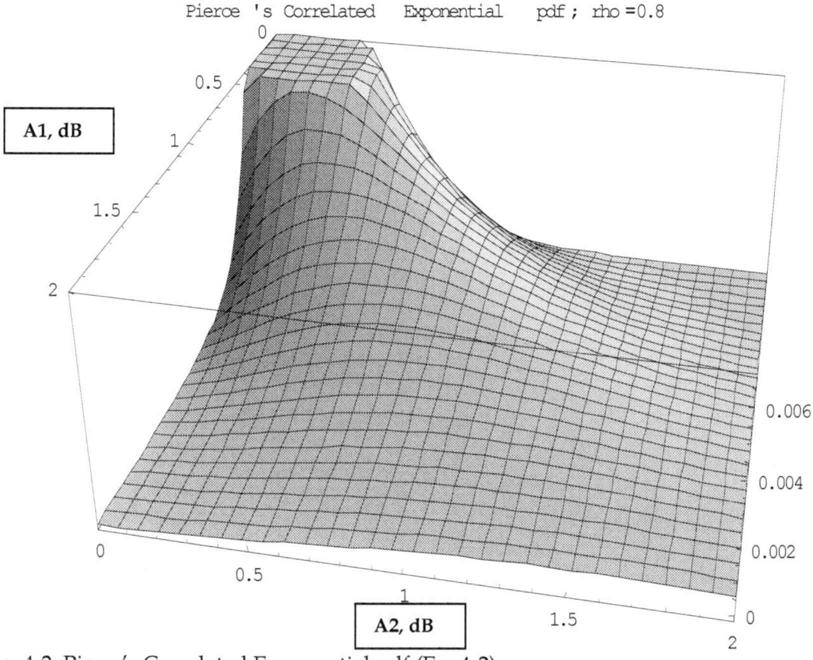


Fig. 4-2. Pierce's Correlated Exponential pdf (Eq. 4-2)

The double integral on the density function should be evaluated to find the joint probability of exceeding arbitrary attenuation levels. The probability of both sites having attenuation greater than AR (dB) may be functionally shown as Eq. 4-3.

P[A1>AR, A2>AR]=

$$\frac{p \int_{AR}^{\infty} \int_{AR}^{\infty} e^{-\frac{A1+A2}{B(1-r)}} I_0\left(2\sqrt{\frac{A1\,A2\,r}{B^2(1-r)^2}}\right) dA2\, dA1}{B^2(1-r)} \qquad (4\text{-}3)$$

The low values of the Boldyrev correlation function (r) will be key to finding low probability of attenuation AR. System operators will recognize AR (dB) as the rain attenuation available with switched diversity, when they can choose the site with the least rain attenuation.
The integral has not yielded to attempts to integrate it exactly, and it turns out to be a very lengthy numerical integration. An upper limit is chosen as (10 B) rather than infinity. The exceedance probability (Eq. 2-3) is abbreviated as (pr) below.
The result of the double integration can be found as Figures 4-3 and 4-4. Fig. 4-3 shows exceedance probability (PR=Log10(pr)) v. normalized (AR/B) and correlation coefficient r. Exceedance probability contours may also be shown as Fig. 4-4 for 0<r<1.

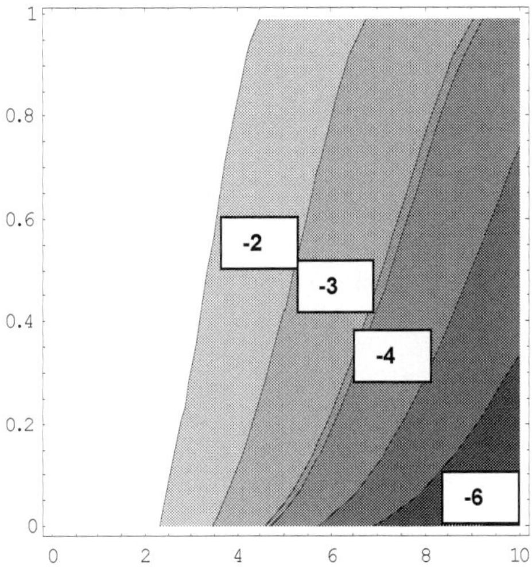


Fig. 4-3. PR v. (AR/B), r

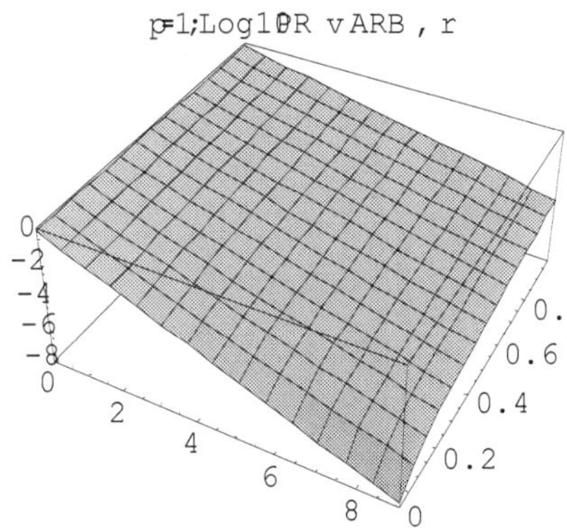


Fig. 4-4 PR Contours v. (AR/B), r

The bivariate function of Fig. 4-3 is defined for all possible weather events, even clear weather at both ground sites. Instead, the exponential density function should be defined for rain events in the 1% to 0.1% range: This is where the exponential pdf has the most relevance. We note again that PR= -0.1 corresponds to 90% availability.

The Boldyrev function offers even more promise for ground diversity, however.

The integral can be evaluated numerically even for negative correlation, as occurs at 390 km for the Boldyrev function. Sharp new benefits can be found in the results of Fig 4-5, with the negative correlation in the 200km-390 km interval as specified by Boldyrev.

The joint exceedance probability is seen to take a quantum step down in the 200-390 km region. This means that if one ground site has bad weather, another ground site 200-390 km away probably has good weather. Indeed, one can test this by experience: if you get a blizzard, your cousin 300 km away is likely to have brilliant blue skies. Two widely separated ground sites are not just better than independent sites, but are much better. Fig. 4-5 extends 4-4 to negative correlation, and the sharp change in PR can be seen at correlation r= -0.2.

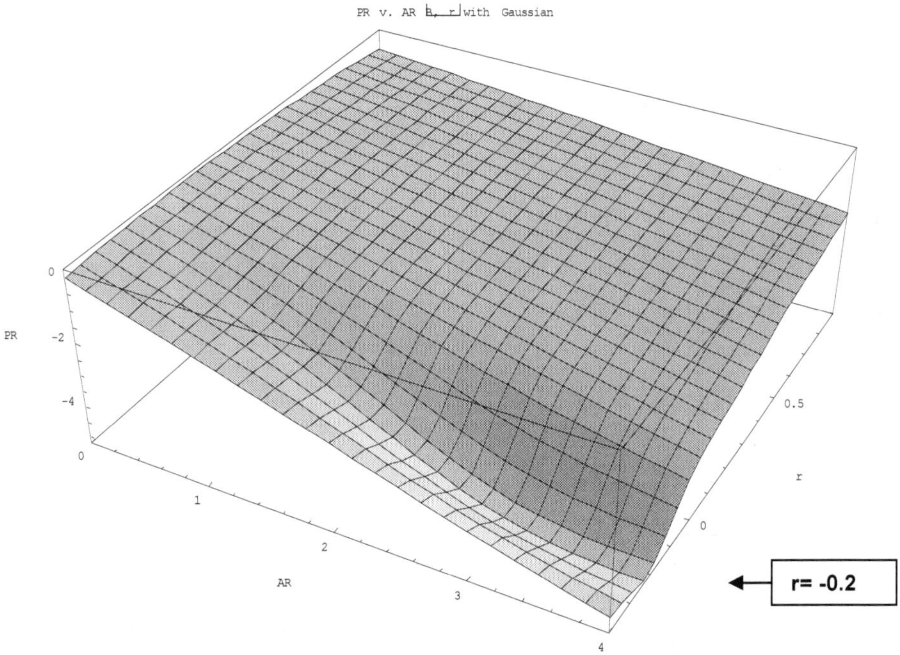


Fig. 4-5 PR v. Single Link Attenuation AR, Cloud Correlation r Mathematica Jan18PROBlistb--.nb.

A short approximation for the exceedance probability PR can be found as:
Log10[Exceedance]=

$$0{,}131989-0{,}861214\,AR-0{,}0113578\,AR^2-1{,}69174\,e^{-\frac{1}{2}(3-AR)^2-50.(-0.2-r)^2}+0{,}0383318r+0{,}603411\,ARr-0{,}651209r^2 \quad (4\text{-}4)$$

Where AR is single link attenuation(dB), with standard deviation B=1
r= cloud correlation coefficient

The probability contours may be seen more clearly as Fig. 4-6.

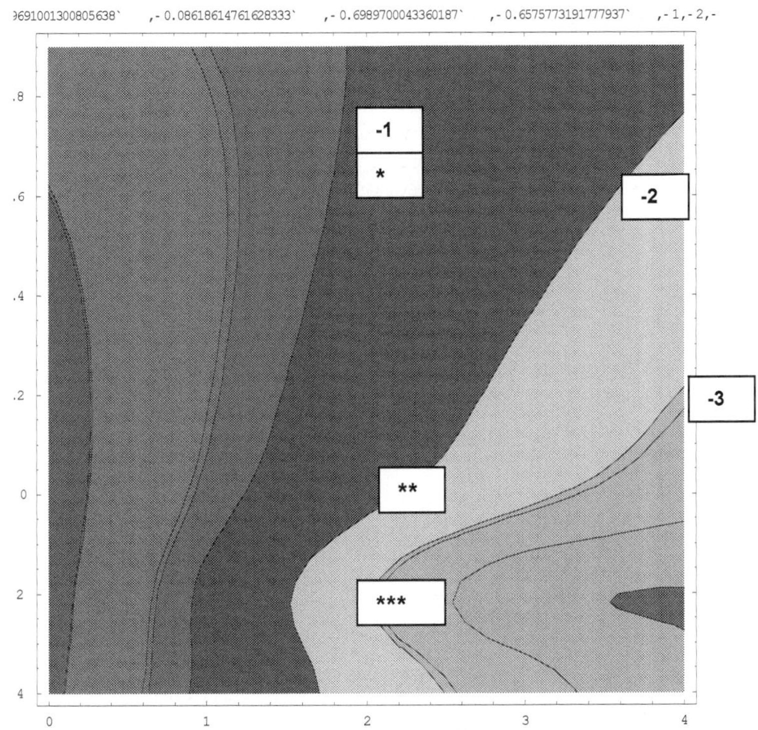


Fig. 4-6 Probability Contours v. AR/B, Correlation Coefficient r
Mathematica Jan18PROBlistb--.nb.

The PR contours can be checked immediately by comparing the PR=-1 and PR=-2 curves.
At constant attenuation AR, point (*) at correlation=1 corresponds to point (**) at correlation=0.
This is necessary because 0.1 exceedance must correspond to 0.01 at zero correlation. The benefits of negative correlation, as seen by Boldyrev, are seen by further dropping to point 3 (***) at correlation= -0.2. Point 3 shows much better probability, approaching PR=-3 or 0.001 exceedance probability, or 99.9% availability.

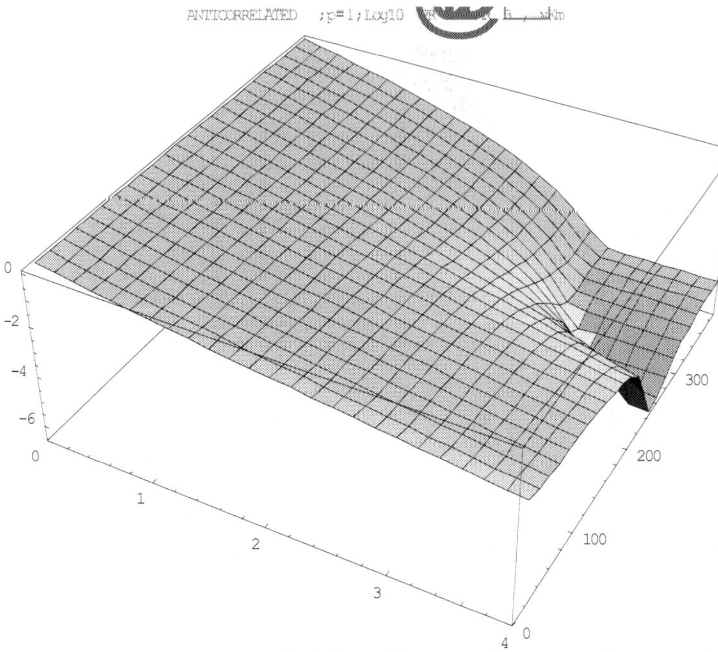


Fig. 4-7 PR v. Attenuation, Distance (km) by Boldyrev and Pierce Mathematica Jan122010–Boldeyrev.nb

Fig. 4-7 imposes the Boldyrev correlation function as a function of distance. The cloud correlation function is 1 at x=0 and drops to 0 at x=200 km. It then drops to negative values for 200<x<390 km, and exceedance probability PR drops abruptly to give clear advantages over independent sites. With independent sites (correlation=0) a single site outage probability of 20% would drop to 4%, or 96% availability. Site separation as 300- 390 km would imply even better availability, from further analysis of Fig. 4-6.

The 80% exceedance (20% availability) of the Japanese experiments would be implied to improve to nearly 85% availability, with an approximate solution of Fig 4-7. The approximate solution is seen as the contours of Fig 4-8. The Japanese experimenters were delighted with results near 90% availability for widely separated ground stations in Japan.

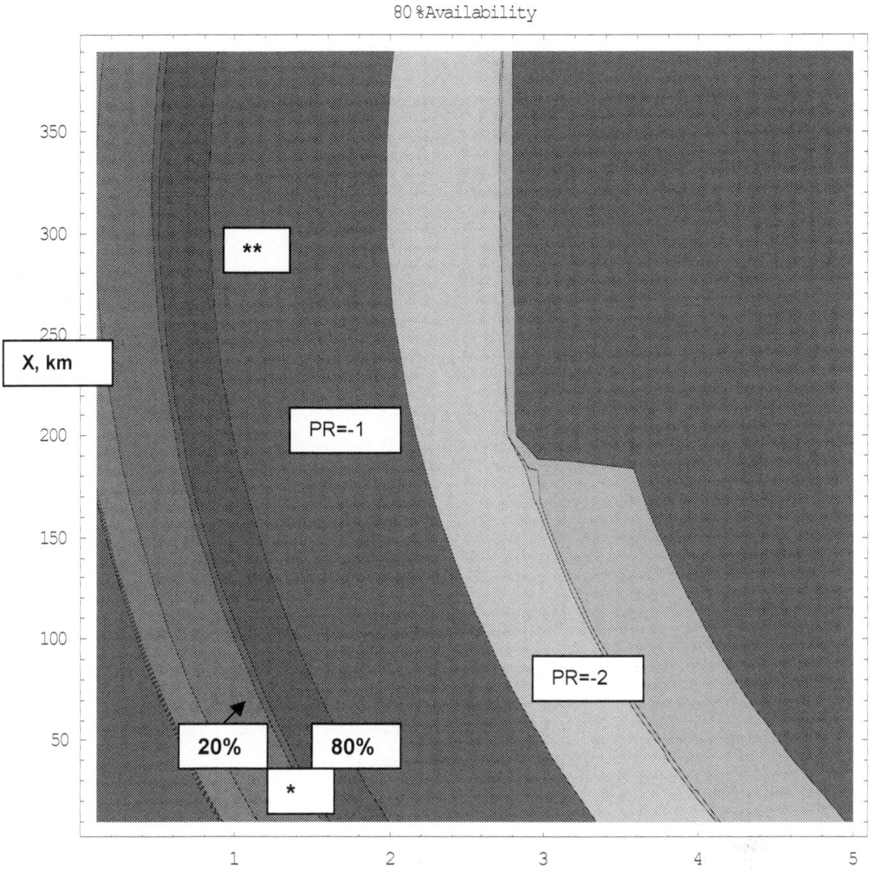


Fig. 4-8 Approximate Availability Contours v. Attenuation, Distance
Compare *20% Availability at 0km with **85% at 300 km

## 6. Summary

We applied Chu and Hogg's key cloud attenuation results to Italsat observations (9) and cloud water analysis (11) to estimate worldwide laser attenuation for satellite- ground communication links. Section 3 gave worldwide zenith attenuation estimates for 80% availability, 90% availability, and 70% availability. The equation for cloud attenuation at 70% availability is given in Appendix B. We` emphasized 10 micron wavelengths and 80% availability for modest attenuation. Reasonable availability was suggested by using Boldyrev and Tulupov's negative cloud correlation at large distances, as 200 - 390 km. Section 4 outlined the analysis and anticipated results for dual switched ground sites. Fig. 4-8 anticipated the encouraging current Japanese (15) results: 20% single site availability growing to over 85% at 300 km site separation. Eighty percent availability and zero correlation at 200 km were indicated to give 96% availability for the dual sites, and 390 km was indicated as notably better. Even better performance should be expected for quad diversity.

We addressed possible 10 micron laser reliability issues by considering new silicon lasers in the 3 -10 micron range. The cloud attenuation model (1,2) and a smoothed MODTRAN atmospheric transmission model indicated the 3 micron space-ground attenuation as too high except in regions north of Labrador-Denmark. The 3 micron lasers might be well suited to small Earth Observation Satellites with widely dispersed ground sites at far northerly latitudes.

Laser space-ground links would be similarly attractive for Molniya satellites and temperate and Arctic ground locations. Indeed, Molniya satellites may avoid a difficult acquisition concern intrinsic to the fast moving EOS. The Molniya satellites would have an additional concern: attenuation would be slightly higher than the zenith attenuation shown here. This has been addressed at the Ka Conference in Naples (4).

## Acknowledgements

Daniel Brandel of ITT noted concerns for 10 micron CO2 laser reliability. David Brown of Wolfram Research (17) gave key advice on Mathematica worldwide mapping functions. Analytical Graphics (AGI) gave excellent descriptions of the Chinese ASAT debris.

## 7. Reference

K.E. Wilson et al, "Overview of the Ground-to-Orbit Lasercom Demonstration," Free-Space Laser Comm. Tech IX, SPIE, Jan. 1997. The GOLD experiment was intended for ground to GEO communication with the Japanese ETS-VI satellite. Instead, the ETS-VI was stuck in the transfer orbit and only partial data was collected. The 1.024 MB/s links showed 10E-4 BER on the uplink and 10E-5 BER on the downlink.

I. Kim et al., "(Very) Preliminary Results of the STRV-2 Satellite-to-ground Lasercom Experiment," Free-Space Laser Comm. Tech. XII, SPIE, vol. 3932, Jan. 2000.

T.S.Chu and D.C.Hogg, "Effects of Precipitation at 0.63, 3.5, and 10.6 Microns", Bell System Technical Journal, May-June 1968

Paul Christopher, "Considerations for Lunar Relays with Millimeter Wave and IR Communication," Ka and Broadband Conference, Naples, Sept. 2006.

Report of the APS Study Group of boost- phase intercept systems for national missile defense. American Physical Society, NY, July 15, 2003.

T. Teichmann, "Some Elementary Considerations of Satellite Earth Communication Systems," IRE Trans. Space Electronics and Telemetry, VOL SET-7, June 1960.

P. Christopher, "Optimum Space-Ground Tradeoffs for Millimeter Wave Satellite Communication," Proc. Int'l Mobile Satellite Conference, Pasadena CA, June 1997.

Richard Orr, Leonard Schuchman, Proc. Ka and Broadband Conference, Naples, Italy, Sept. 2006.

F. Barbaliscia, M. Boumis, A. Martellucci, "World Wide Maps of Non Rainy Attenuation for Low-Margin Satcom Systems Operating in SHF/EHF Bands," Ka Band Conference, Sept. 1998.

F. Barbaliscia, M. Boumis, A. Martellucci, "Characterization of Atmospheric Attenuation in the Absence of Rain in Europe in SHF/EHF Bands for VSAT Satcom Systems Applications," Ka Band Conference, Sorrento, Italy, Sept. 1997.
Paul Christopher, "World Wide Millimeter Wave Attenuation Functions from Barbaliscia's 49/22 GHz Observations," Ka Band Conference, Taormina Sicily, Oct. 1999.
Paul Christopher, "Millimeter Waves for Broadband Satellite Communication, 75- 98 GHz;" Proc. Ka and Broadband Conference, Isle of Ischia, Oct. 2003
Frank Hanson, "Coherent laser radar performance in littoral environments- a statistical analysis based on weather observations," Optical Engineering, Vol. 39 No. 11, Nov. 2000, 3044-3052.
Bahram Jalali, "Making Silicon Lase," Scientific American, pages 58-65, February 2007.
Yoshihisa Takayama, personal communication on Japanese Satellite Laser Experiments, at Ka and Broadband Conference, Cagliari, Sicily, Sept. 2009.
Boldyrev and Tulupov, Cloud Correlation Function, COSPAR Space Research XI, Leningrad USSR 20-29 May 1970, Vol. 1 Akademie-Verlag Berlin, 1971.
Michael Trott, The Mathematica Guidebook for Symbolics, Springer, New York, NY, 2006.

## Appendix A Required Inclination for Higher EOS Orbit

The typical Landsat 4 orbit, at 701 km altitude, and 98.4 degrees inclination, must be altered for higher altitudes. The method of Osculating Elements shows the rate of change of Right Ascension as

$$\frac{3\,J2\,\sqrt{\frac{mu}{a}}\,r^2\,(5\cos^2(i)-1)}{4\,a^3}$$

with i as the inclination, a as the semimajor axis, and mu the Earth gravitational constant.

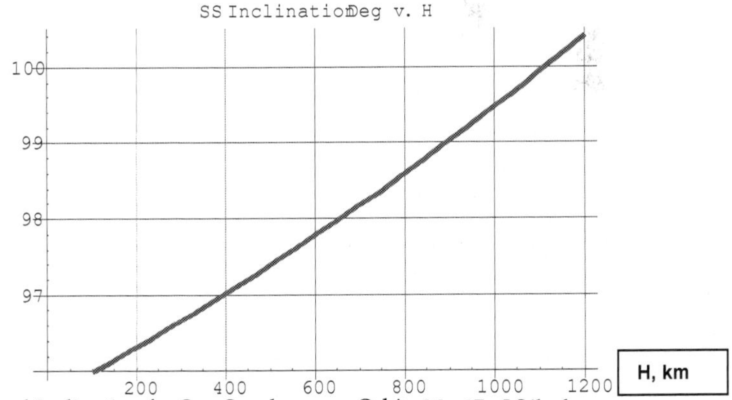


Fig. A-1 Required Inclination for Sun Synchronous Orbits May17–LS4b.nb .
The shift from the Landsat4 altitude to 1000 km is seen to require almost 99.5 deg inclination.

## Appendix B 10 Micron Zenith Attenuation, dB PR=.3

This Mathematica program is available from the author. It gives zenith attenuation maps at 70% availability.

Zenith Attenuation [70%]=

Out[4]=

$$
\begin{aligned}
& 4.60285 - \\
& 694.517 \Big(-0.764706 \Big(-0.0878644 - 0.352549 \Big(2.53967 + 0.66746\, e^{-\frac{1}{200}(-10+\mathrm{LAT})^2-\frac{1}{200}(-145+\mathrm{LON})^2} + 1.12036\, e^{-\frac{1}{800}(-18+\mathrm{LAT})^2-\frac{(-102+\mathrm{LON})^2}{1800}} + \\
& 0.70296\, e^{-\frac{1}{200}(-52+\mathrm{LAT})^2-\frac{1}{800}(-28+\mathrm{LON})^2} - 1.28258\, e^{-\frac{\mathrm{LAT}^2}{1800}-\frac{\mathrm{LON}^2}{180000}} - 1.22726\, e^{-\frac{1}{72}(-2+\mathrm{LAT})^2-\frac{1}{200}(7+\mathrm{LON})^2} - \\
& 1.92779\, e^{-\frac{1}{200}(28+\mathrm{LAT})^2-\frac{1}{200}(77+\mathrm{LON})^2} + 0.865964\, e^{-\frac{1}{200}(-20+\mathrm{LAT})^2-\frac{1}{128}(82+\mathrm{LON})^2} - 0.0909198\, e^{-\frac{1}{98}(-45+\mathrm{LAT})^2-\frac{1}{128}(110+\mathrm{LON})^2} + \\
& 0.00128521\,\mathrm{LAT} - 0.000975417\,\mathrm{LAT}^2 + 2.23192\times10^{-7}\,\mathrm{LAT}^3 + 1.03045\times10^{-7}\,\mathrm{LAT}^4 - 0.000416968\,\mathrm{LON} - 2.18586\times10^{-6}\,\mathrm{LAT}\,\mathrm{LON} + \\
& 2.44557\times10^{-7}\,\mathrm{LAT}^2\,\mathrm{LON} + 0.000011497\,\mathrm{LON}^2 + 1.0003\times10^{-8}\,\mathrm{LAT}\,\mathrm{LON}^2 - 1.38211\times10^{-8}\,\mathrm{LON}^3 - 1.36816\times10^{-10}\,\mathrm{LON}^4\Big) - \\
& 1.20397 \Big(0.132924 + 0.140592 \Big(2.53967 + 0.66746\, e^{-\frac{1}{200}(-10+\mathrm{LAT})^2-\frac{1}{200}(-145+\mathrm{LON})^2} + 1.12036\, e^{-\frac{1}{800}(-18+\mathrm{LAT})^2-\frac{(-102+\mathrm{LON})^2}{1800}} - \\
& 0.70296\, e^{-\frac{1}{200}(-52+\mathrm{LAT})^2-\frac{1}{800}(-28+\mathrm{LON})^2} - 1.28258\, e^{-\frac{\mathrm{LAT}^2}{1800}-\frac{\mathrm{LON}^2}{180000}} + 1.22726\, e^{-\frac{1}{72}(-2+\mathrm{LAT})^2-\frac{1}{200}(7+\mathrm{LON})^2} - 1.92779 \\
& e^{-\frac{1}{200}(28+\mathrm{LAT})^2-\frac{1}{200}(77+\mathrm{LON})^2} + 0.865964\, e^{-\frac{1}{200}(-20+\mathrm{LAT})^2-\frac{1}{128}(82+\mathrm{LON})^2} - 0.0909198\, e^{-\frac{1}{98}(-45+\mathrm{LAT})^2-\frac{1}{128}(110+\mathrm{LON})^2} + 0.00128521 \\
& \mathrm{LAT} - 0.000975417\,\mathrm{LAT}^2 + 2.23192\times10^{-7}\,\mathrm{LAT}^3 + 1.03045\times10^{-7}\,\mathrm{LAT}^4 - 0.000416968\,\mathrm{LON} - 2.18586\times10^{-6}\,\mathrm{LAT}\,\mathrm{LON} + \\
& 2.44557\times10^{-7}\,\mathrm{LAT}^2\,\mathrm{LON} + 0.000011497\,\mathrm{LON}^2 + 1.0003\times10^{-8}\,\mathrm{LAT}\,\mathrm{LON}^2 - 1.38211\times10^{-8}\,\mathrm{LON}^3 - 1.36816\times10^{-10}\,\mathrm{LON}^4\Big)\Big)\Big) + \\
& 1. \Big(-0.919661 - 0.139753 \Big(3.84195 + 1.49573\, e^{-\frac{1}{200}(-10+\mathrm{LAT})^2-\frac{1}{200}(-145+\mathrm{LON})^2} + 2.04653\, e^{-\frac{1}{800}(-18+\mathrm{LAT})^2-\frac{(-102+\mathrm{LON})^2}{1800}} + \\
& 1.16176\, e^{-\frac{1}{200}(-52+\mathrm{LAT})^2-\frac{1}{800}(-28+\mathrm{LON})^2} - 2.51927\, e^{-\frac{\mathrm{LAT}^2}{1800}-\frac{\mathrm{LON}^2}{180000}} + 2.46395\, e^{-\frac{1}{72}(-2+\mathrm{LAT})^2-\frac{1}{200}(7+\mathrm{LON})^2} + \\
& 1.41489\, e^{-\frac{1}{50}(-20+\mathrm{LAT})^2-\frac{1}{200}(60+\mathrm{LON})^2} - 2.25769\, e^{-\frac{1}{200}(28+\mathrm{LAT})^2-\frac{1}{200}(77+\mathrm{LON})^2} + 1.69814\, e^{-\frac{1}{200}(-20+\mathrm{LAT})^2-\frac{1}{128}(82+\mathrm{LON})^2} - \\
& 1.07578\, e^{-\frac{1}{98}(-45+\mathrm{LAT})^2-\frac{1}{128}(110+\mathrm{LON})^2} + 0.000919687\,\mathrm{LAT} - 0.00176137\,\mathrm{LAT}^2 - 8.75845\times10^{-7}\,\mathrm{LAT}^3 - \\
& 1.79394\times10^{-7}\,\mathrm{LAT}^4 - 0.000725197\,\mathrm{LON} - 6.69472\times10^{-6}\,\mathrm{LAT}\,\mathrm{LON} - 3.25314\times10^{-7}\,\mathrm{LAT}^2\,\mathrm{LON} + 0.0000241438\,\mathrm{LON}^2 - \\
& 3.07508\times10^{-9}\,\mathrm{LAT}\,\mathrm{LON}^2 - 3.34843\times10^{-8}\,\mathrm{LON}^3 - 4.24764\times10^{-11}\,\mathrm{LON}^4\Big) + \frac{1}{(293. - 0.25\,(-40. - \mathrm{LAT}))^2} \\
& \Big(85849. \Big(1.01966 + 0.139753 \Big(3.84195 + 1.49573\, e^{-\frac{1}{200}(-10+\mathrm{LAT})^2-\frac{1}{200}(-145+\mathrm{LON})^2} + 2.04653\, e^{-\frac{1}{800}(-18+\mathrm{LAT})^2-\frac{(-102+\mathrm{LON})^2}{1800}} - \\
& 1.16178\, e^{-\frac{1}{200}(-52+\mathrm{LAT})^2-\frac{1}{800}(-28+\mathrm{LON})^2} - 2.51927\, e^{-\frac{\mathrm{LAT}^2}{1800}-\frac{\mathrm{LON}^2}{180000}} + 2.46395\, e^{-\frac{1}{72}(-2+\mathrm{LAT})^2-\frac{1}{200}(7+\mathrm{LON})^2} - \\
& 1.41489\, e^{-\frac{1}{50}(-20+\mathrm{LAT})^2-\frac{1}{200}(60+\mathrm{LON})^2} - 2.25769\, e^{-\frac{1}{200}(28+\mathrm{LAT})^2-\frac{1}{200}(77+\mathrm{LON})^2} - 1.69814\, e^{-\frac{1}{200}(-20+\mathrm{LAT})^2-\frac{1}{128}(82+\mathrm{LON})^2} - \\
& 1.07578\, e^{-\frac{1}{98}(-45+\mathrm{LAT})^2-\frac{1}{128}(110+\mathrm{LON})^2} + 0.000919687\,\mathrm{LAT} - 0.00176137\,\mathrm{LAT}^2 + 8.75845\times10^{-7}\,\mathrm{LAT}^3 - \\
& 1.79394\times10^{-7}\,\mathrm{LAT}^4 - 0.000725197\,\mathrm{LON} - 6.69472\times10^{-6}\,\mathrm{LAT}\,\mathrm{LON} + 3.25314\times10^{-7}\,\mathrm{LAT}^2\,\mathrm{LON} + \\
& 0.0000241438\,\mathrm{LON}^2 - 3.07508\times10^{-9}\,\mathrm{LAT}\,\mathrm{LON}^2 - 3.34843\times10^{-8}\,\mathrm{LON}^3 - 4.24764\times10^{-11}\,\mathrm{LON}^4\Big)\Big)\Big) - \\
& 1.20397 \Big(0.332309 - 0.186801 \Big(3.84195 - 1.49573\, e^{-\frac{1}{200}(-10+\mathrm{LAT})^2-\frac{1}{200}(-145+\mathrm{LON})^2} + 2.04653\, e^{-\frac{1}{800}(-18+\mathrm{LAT})^2-\frac{(-102+\mathrm{LON})^2}{1800}} + \\
& 1.16178\, e^{-\frac{1}{200}(-52+\mathrm{LAT})^2-\frac{1}{800}(-28+\mathrm{LON})^2} - 2.51927\, e^{-\frac{\mathrm{LAT}^2}{1800}-\frac{\mathrm{LON}^2}{180000}} + 2.46395\, e^{-\frac{1}{72}(-2+\mathrm{LAT})^2-\frac{1}{200}(7+\mathrm{LON})^2} + \\
& 1.41489\, e^{-\frac{1}{50}(-20+\mathrm{LAT})^2-\frac{1}{200}(60+\mathrm{LON})^2} - 2.25769\, e^{-\frac{1}{200}(28+\mathrm{LAT})^2-\frac{1}{200}(77+\mathrm{LON})^2} + 1.69814\, e^{-\frac{1}{200}(-20+\mathrm{LAT})^2-\frac{1}{128}(82+\mathrm{LON})^2} - \\
& 1.07578\, e^{-\frac{1}{98}(-45+\mathrm{LAT})^2-\frac{1}{128}(110+\mathrm{LON})^2} + 0.000919687\,\mathrm{LAT} - 0.00176137\,\mathrm{LAT}^2 + 8.75845\times10^{-7}\,\mathrm{LAT}^3 - \\
& 1.79394\times10^{-7}\,\mathrm{LAT}^4 - 0.000725197\,\mathrm{LON} - 6.69472\times10^{-6}\,\mathrm{LAT}\,\mathrm{LON} + 3.25314\times10^{-7}\,\mathrm{LAT}^2\,\mathrm{LON} + \\
& 0.0000241438\,\mathrm{LON}^2 + 3.07508\times10^{-9}\,\mathrm{LAT}\,\mathrm{LON}^2 - 3.34843\times10^{-8}\,\mathrm{LON}^3 - 4.24764\times10^{-11}\,\mathrm{LON}^4\Big)\Big)\Big)\Big)
\end{aligned}
$$

dB

(B-1)

Where LON= Longitude, deg East

LAT= N. Latitude

Mathematica [Jan2—Point3.nb]

# Satellite Motion

Miljenko Solarić
*University of Zagreb - Faculty of Geodesy*
*Croatia*

## 1. Introduction

*What is satellite?*

The word satellite is coming from the Latin language (Latin *satelles* - escort, companion). Satellites are objects that rotate around the planets under the influence of the gravitational force. For example, the Moon is natural satellite of the Earth.

*What is the artificial Earth satellite?*

The artificial Earth satellites are artificial objects which are launched into orbit around the Earth by a rocket vehicle. This kind of satellites was named, the Human-Made Earth Satellites.

*How do rockets function?*

Aeroplanes work on the principle of buoyancy difference on their wings. This is the reason why aeroplanes can fly only in the air but not in the vacuum. Thus, an airplane cannot be used for launching satellites in their orbit around the Earth (Fig. 1).

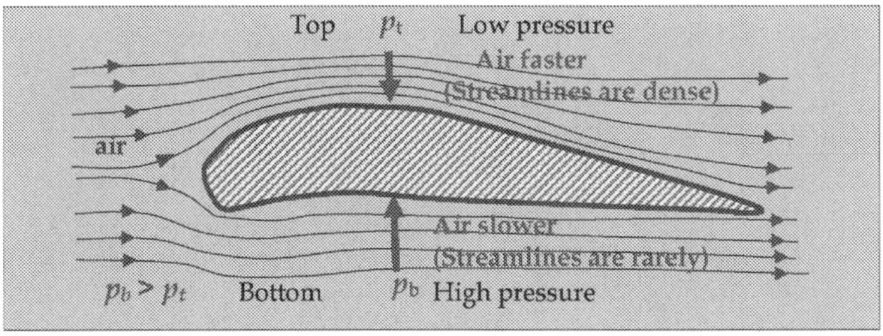


Fig. 1. When the aeroplanes have velocity in the air on the bottom of his wings they have higher pressure then on the top of wings. This difference of pressure is giving the force of buoyancy and the aeroplanes can fly.

It is also not possible to launch a human-made satellite into the orbit around the Earth with a cannon or a gun because a cannon-ball has the velocity of about 0.5 km/s. It means that this velocity of cannon-ball is about 15 times smaller than the first cosmic velocity (7.9 km/s). So for it has been possible to launch a satellite into an orbit around the Earth where there is vacuum is possible only with using rockets.

The word *rocket* comes from the Italian *Rocchetta* (i.e. *little fuse*), a name of a small firecracker. It is commonly accepted that the first recorded use of a rocket in battle was by the Chinese

in 1232 against the Mongol hordes at Kai Feng Fu. The Mongols were the first to have applied rocket technology in Europe as they conquered some parts of China and of Russia, Eastern and Central Europe.
Konstantin Tsiolkovsky (1857-1935) (Fig. 2) from the Imperial Russia and after from the Soviet Union published the first serious scientific work on space travels titled *"The exploration of Cosmic Space by Means of Reaction Devices"* in 1903. He is considered by many to be the *father of theoretical astronautics.* He also advocated the use of liquid hydrogen and oxygen for propellant, calculating their maximum exhaust velocity. His work inspired further research, experimentation and the formation of Society for Studies of Interplanetary Travel in 1924. Also in 1924, Tsiolkovsky wrote about multi-stage rockets, in *"Cosmic Rocket Trains"*.

Fig. 2. Konstantin Tsiolkovsky

Fig. 3. Robert Goddard

Fig. 4. Sergei Korolev

Fig. 5. Wernher von Braun

In the USA, Robert Goddard (Fig. 3) began a serious analysis of rockets in 1912. It can thus be concluded that conventional solid-fuel rockets needed to be improved in three ways. One of them is that rockets could be arranged in stages. He also independently developed the mathematics of rocket flight. For his ideas, careful research, and great vision, Goddard was called the *father of modern astronautics.*
After the World War II in the USA Wernher von Braun (Fig. 5) and Sergei Korolev (Fig. 4) in the Soviet Union were the leaders in the advancing rockets technology.
The operational principle a rocket can be explained by means of a balloon (Fig. 6).
In a balloon the pressure of gas is practically equal on all sides (Fig. 6 a). When this balloon has an aperture then a particle of gas on this aperture will under pressure of gas be thrown out with velocity $v_g$ (Fig. 6 b). The pressure in the balloon in opposite direction of the aperture will produce the pressure on the balloon and will give it the velocity $v_B$. The pressures in the other directions will be mutually cancelled in opposite direction. The rocket operates on this principle.

A rocket travelling in vacuum is accelerated by the high-velocity expulsion of a small part of its mass (gas). The Fig. 7. represents a rocket with the situations before and after the explosion. This is closed material system and for this system linear momentum needs to be conserved. So we can say that the momentum in the beginning position ($M \cdot \mathbf{v}$) is equal to momentum of this system after the explosion when a particle with the mass ($dm$) is be thrown out with velocity ($\mathbf{u}_{rel}$) in the opposite direction of this rocket velocity.

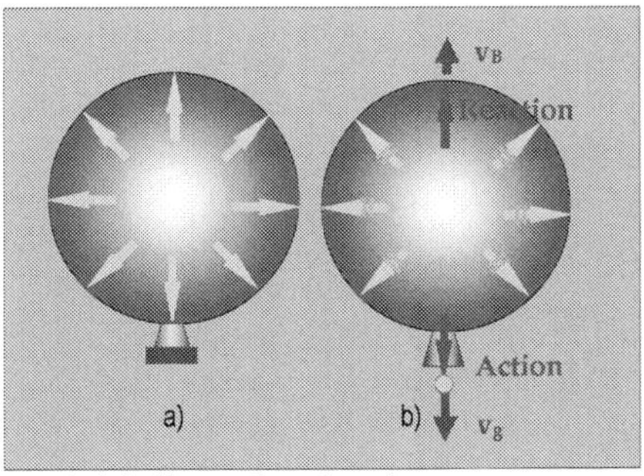


Fig. 6. A balloon closed under gas pressure and after opening an aperture.

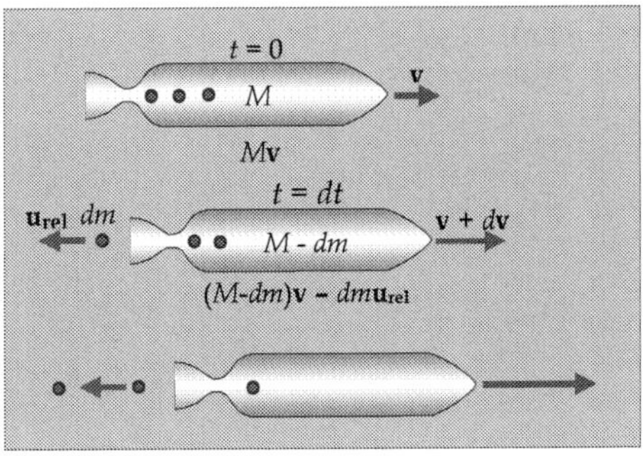


Fig. 7. The principle of a rocket operation.

It means that the momentum for a particle is negative ($-dm \cdot \mathbf{u}_{rel}$) and we can write the equation (Carton, 1965) and (Danby, 1989):

$$M \cdot \mathbf{v} = (M - dm) \cdot (\mathbf{v} + d\mathbf{v}) - dm \cdot \mathbf{u}_{rel} \,. \tag{1}$$

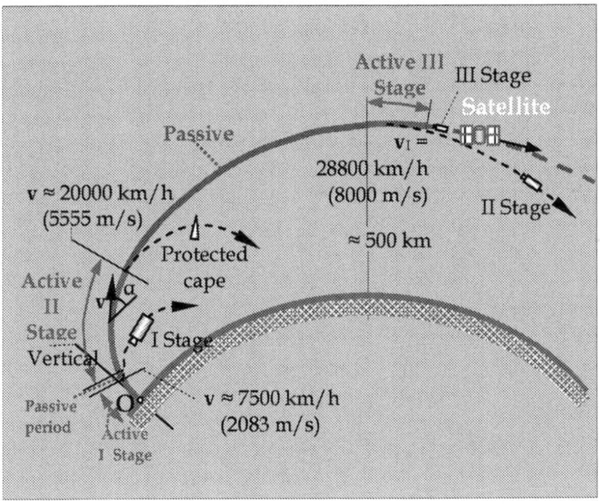


Fig. 8. Imagery of launching a satellite by rocket with the three stages.

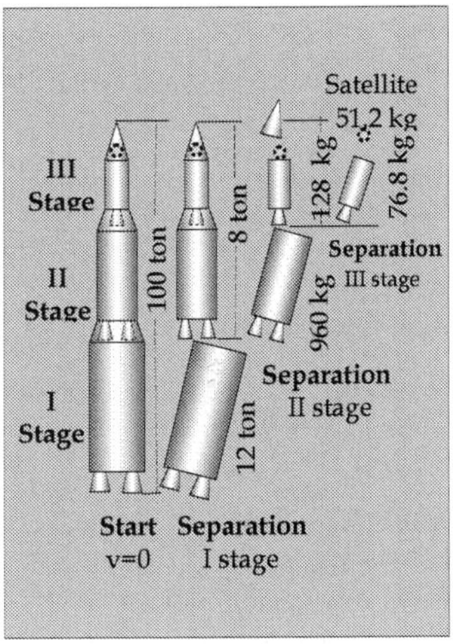


Fig. 9. For example during the start a three stages rocket may have the mass 100 t but mass of a space vehicle will be only 51.2 kg.

It follows from this equation that the velocity of rocket increased for the elementary magnitude

$$d\mathbf{v} = \frac{dm(\mathbf{v} + \mathbf{u}_{rel})}{M - dm}. \qquad (2)$$

From this equation it is possible to see that the increase of the rocket velocity is larger if the velocity $\mathbf{u}_{rel}$ of the particles of gas is maximally the greater when the mass of particle $dm$ has some magnitude. This is the reason why the constructors of rockets like to make rockets with very high (maximal) velocity of the particles (gas) of the rocket.
Usual rockets have vertical start. Longer delaying of rockets in the Earth gravitation field causes the loss of velocity but also to big thrust during the start is not suitable. So rockets are usually made in same stages (Fig. 8 and 9). During the starts of rockets the consummation of fuel is very large so that a satellite or a space vehicle enters into the orbit with a mass practically next to nothing (see for example Table 1).

Table 1. Data on initial mass of a rocket in a start, spend fuel, thrown parts for a launching space vehicle of the mass 51.2 kg

| Stage | 1 | 2 | 3 |
|---|---|---|---|
| Initial mass | 100 t | 8 t | 640 kg |
| Spend fuel | 80 t | 6.4 t | 512 kg |
| Thrown part | - | 12 t | 960 kg |
| Definitely mass | 20 t | 1.6 t | 128 kg |
| Velocity | 3860 (m/s) | 7720 (m/s) | 11580 (m/s) |

## 2. Planet Motion

Until the 17th century people were thinking that the Sun and planets are rotated around the Earth by circles. Such opinions were practically usual until Johann Kepler.

### 2.1 Kepler's Laws of Planetary Motion

Johann Kepler (1546 -1601) discovered the laws of planetary motion empirically from Tycho Brahe's (1546 -1601) astronomical observations of the planet Mars. The first and the second laws he published in *Astronomia Nova (New Astronomy)* in 1609, and the third law in *Harmonices mundi libri V (Harmony of the World)* in 1619.

***a) Kepler's First Law of Planetary Motion***

This law can be expressed as follows:

***The path of each planet describes an ellipse with the Sun located at one of its foci.***
***(The Law of Ellipse)***

This first Kepler's Law (Fig. 10 and 11) is sometimes referred to as *the law of ellipse* because planets are orbiting around the Sun in a path described as an ellipse. An ellipse is a special curve in which the sum of the distances from every point on the curve to two other points (foci $F_1$ and $F_2$) is a constant.

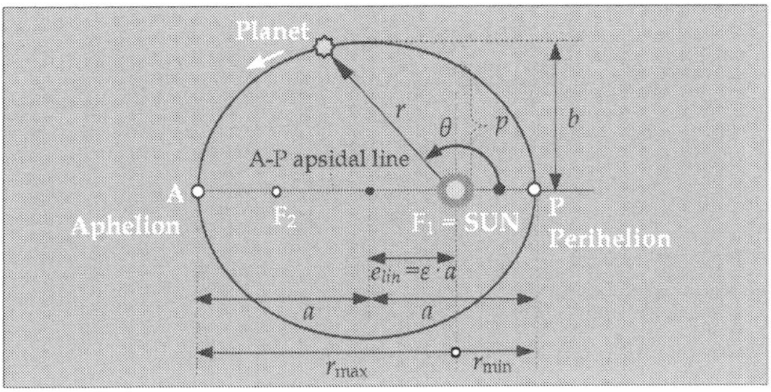


Fig. 10. The Orbit of a planet is an ellipse and its elements.

Before Kepler the Greek astronomer Ptolemy and many others after him were thinking that the Sun and planets travel in circles around the Earth. The ellipse can be mathematically expressed in the polar coordinate system by this equation:

$$r = \frac{p}{1+\varepsilon\cos\theta}, \tag{3}$$

where $(r, \theta)$ are heliocentric polar coordinates for the orbit of planet ($r$ - the distance between the Sun and a planet, $\theta$ - angle from the perihelion to the planet as seen from the Sun, respectively known as *the true anomaly*), $p$ is the semi-latus rectum, and $\varepsilon$ is the numerical eccentricity.

At $\theta = 0°$ the minimum distance is equal to

$$r_{min} = \frac{p}{1+\varepsilon}. \tag{4}$$

At $\theta = 90°$ the distance is equal $p$.

At $\theta = 180°$ the maximum distance is

$$r_{max} = \frac{p}{1-\varepsilon}. \tag{5}$$

The semi-major axis is the arithmetic mean between $r_{min}$ and $r_{max}$:

$$a = \frac{r_{max}+r_{min}}{2} = \frac{p}{1-\varepsilon^2}. \tag{6}$$

The semi-minor axis is the geometric mean between $r_{min}$ and $r_{max}$:

$$b = \sqrt{r_{min} \cdot r_{max}} = \frac{p}{\sqrt{1-\varepsilon^2}} = a\sqrt{1-\varepsilon^2}\,. \quad (7)$$

The semi-latus rectum $p$ is equal to

$$p = \frac{b^2}{a}\,. \quad (8)$$

The area $A$ of an ellipse is

$$A = \pi ab\,. \quad (9)$$

In the special case when $\varepsilon = 0$ then an ellipse turns into a circle where $r = p = r_{min} = r_{max} = a = b$ and $A = \pi r^2$.
Using ellipse-related equations Kepler's procedure for calculating heliocentric polar coordinates $r$, $\theta$, for planetary position as a function of the time $t$ from Perihelion, and the orbital period $P$, follows four steps:

1. Compute the *mean anomaly* $M_a$ from the equation $M_a = \frac{2\pi t}{P}$. (10)
2. Compute the *eccentric anomaly* $E$ by numerically solving Kepler's equation:

$$M_a = E - \varepsilon \sin E\,. \quad (11)$$

3. Compute the *true anomaly* $\theta$ by the equation $\tan\frac{\theta}{2} = \sqrt{\frac{1+\varepsilon}{1-\varepsilon}}\tan\frac{E}{2}$. (12)
4. Compute the *heliocentric distance* $r$ from the equation $r = \frac{p}{1+\varepsilon\cos\theta}$. (13)

For the circle $\varepsilon = 0$ we have simple dependence $\theta = E = M_a$.

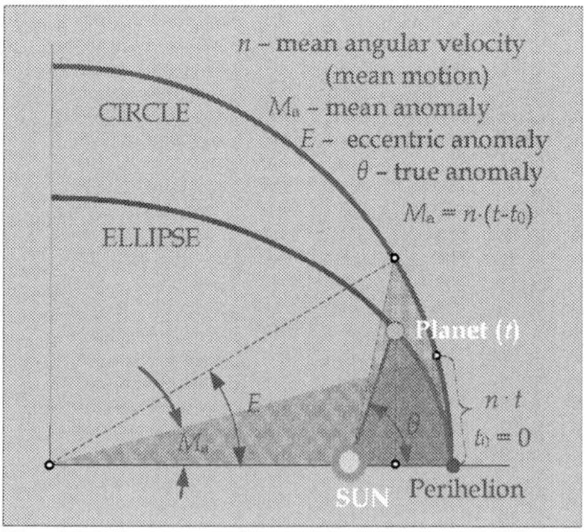


Fig. 11. Elements of parameters of a satellite orbit.

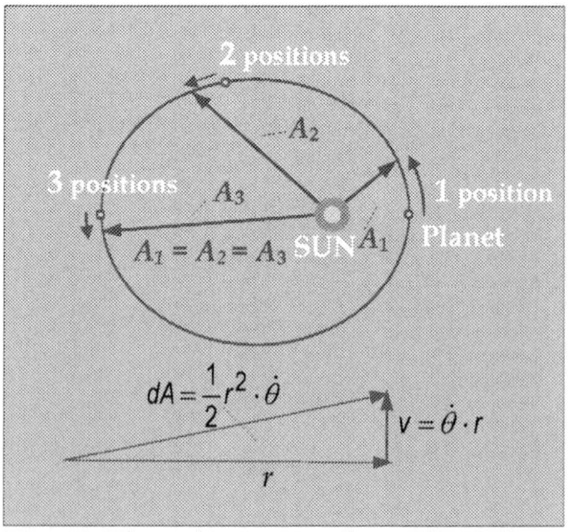


Fig. 12. The radius vector drawn from the Sun to a planet covers equal areas in equal times.

*b) The Second Kepler's Law of Planetary Motion*
This law can be expressed as follows:

***The radius vector drawn from the Sun to a planet covers equal areas in equal times. (The Law of equal areas)***

Mathematically this law can be expressed with the equation:

$$\frac{d}{dt}\left(\frac{1}{2}r^2\dot{\theta}\right)=0, \qquad (14)$$

where $\dot{\theta}$ is angular velocity of true anomaly, $\frac{1}{2}r^2\dot{\theta}$ is the *"areal velocity"* that the radius vector $r$ drawn from the Sun to the planet sweeps in one second (Fig. 12).

From this law it follows that the speed at which any planet moves through space is continuously by changing. A planet moves most quickly when it's closer to the Sun and more slowly when it is further from the Sun.

*c) The Third Kepler's Law of Planetary Motion*
This law can be expressed as follows:

***The squares of the periodic times of the planets are proportional to the cubes of the semi-major axes of their orbits. (The harmonic law)***

This law is giving the relationship between the distance of planets from the Sun and their orbital periods. Mathematically and symbolically it's possible to express as follows:

$$P^2 \,\alpha\, a^3 ,$$

where $P$ is the orbital period of circulate planet around the Sun and $a$ is the semi-major axis of this orbit. Because this proportionality is the same for any planet which rotates around the Sun it's possible to write the next equation:

$$\frac{P^2_{planet1}}{a^3_{planet1}}=\frac{P^2_{planet2}}{a^3_{planet2}}, \quad \text{namely} \quad \frac{P^2_{planet1}}{P^2_{planet2}}=\frac{a^3_{planet1}}{a^3_{planet2}}. \qquad (15)$$

## 3. The Physical Laws of Motions

Sir Isaac Newton's formulated three fundamental laws of the classical mechanics and the law of gravitation in his great work *Philosophieæ Naturalis (Principia Mathematica)* published on July 5, 1687. Before Isaac Newton the great contribution to the advance of mechanic was given by Galileo, Kepler and Huygens.

### 3.1 The First Law of Motion – Law of Inertia

This law can be expressed as follows:

***Everybody persists in its state of being at rest or moving uniformly straight forward, except insofar as it is compelled to change its state by force impressed.***

Newton's first law of motion is also called *the law of inertia*. It states that if the vector sum of all forces acting on an object is zero, then the acceleration of the object is zero and its velocity is constant. Consequently:

- An object that is at rest will stay at rest until a balanced force acts upon it.
- An object that is in motion will not change its velocity until a balanced force acts upon it.

***If the resultant force acting on a particle is zero, the particle will remain at rest (if originally at rest) or will move with constant speed in a straight line (if originally in motion).***

### 3.2 The Second Law of Motion – Law of Force

This law can be expressed as follows:

***Force equals mass times acceleration.***
***If the resultant force acting on a particle is not zero, the particle will have acceleration proportional to the magnitude of the resultant and in the direction of this resultant force.***

This law may be expressed by the equation:

$$\mathbf{F} = m\mathbf{a}\,, \tag{16}$$

where **F** is the vector of force, $m$ the mass of particle and **a** is the vector of acceleration (Fig. 13. a).

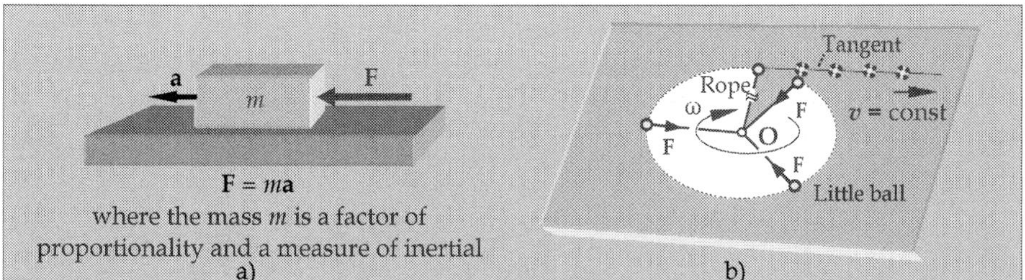


Fig. 13. a) An acceleration **a** of a free body on a horizontal plane under influence of a force **F**, b) In the rotation when the rope is broken, a little ball shall start moving with constant velocity along the line of tangent in the horizontal plane.

Really, this is differential equation which represents a basic equation of motion or basic equation of dynamic.
Alternatively this law can be expressed by the equation:

$$\mathbf{F} = \frac{d}{dt}(m\mathbf{v})\,, \tag{17}$$

where the product $m\mathbf{v}$ is the momentum: $m$ - the mass of particle and $\mathbf{v}$ - the velocity. So, we can say:

*The force is equal to the time derivative of the body's momentum.*

### 3.3 The Third Law – Law of action and reaction

This law can be expressed in the following way:

*To every action there is an equal by magnitude and opposite reaction* (Fig. 14. a).

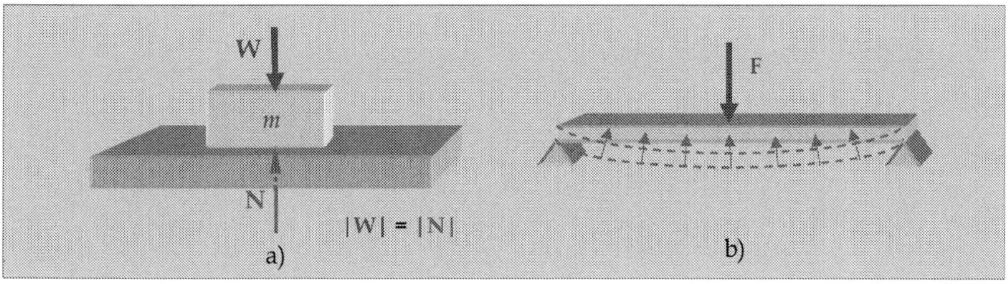


Fig. 14. a) Under the influence of the weight **W** of a body, a normal reaction of its support occurs, b) A beam under loading by a force **F** will be deformed as reaction to an active force **F**.

This law can be also expressed:

*The forces of action and reaction between bodies in contact have the same magnitude, same line of action and opposite sense.*

### 3.4 The Law of Gravitation – The Law of Universal Gravitation

Isaac Newton stated that two particles at the distance $r$ from each other and, respectively, of mass $M$ and $m$, attract each other with equal and opposite forces $\mathbf{F}_1$ and $\mathbf{F}_2$ directed along the line joining the particles (Fig. 15). The common magnitude $F$ of these two forces is:

$$F = G\frac{Mm}{r^2}, \tag{18}$$

where G is the *universal constant of gravitation* $G \approx 6.67428 \times 10^{-11}\ m^3kg^{-1}s^{-2}$),or approximately $3.44{\times}10^{-8} ft^4/lb\text{-}sec^4$ in British gravitational system of units (Beer & Johnston 1962).

The force of attraction exerted by the Earth on body of the mass $m$ located on or near its surface is defined as the weight (W) of the body (Fig. 16)

$$W = mg, \tag{19}$$

where $g$ is the acceleration of gravity, being also the acceleration of force of weight.

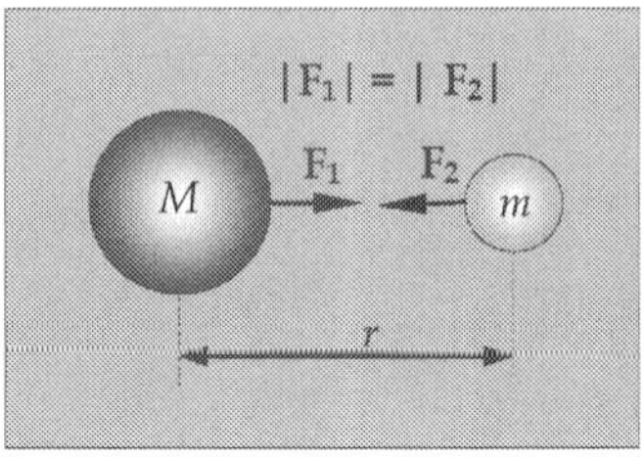


Fig. 15. Newton's Law of the universal gravitation.

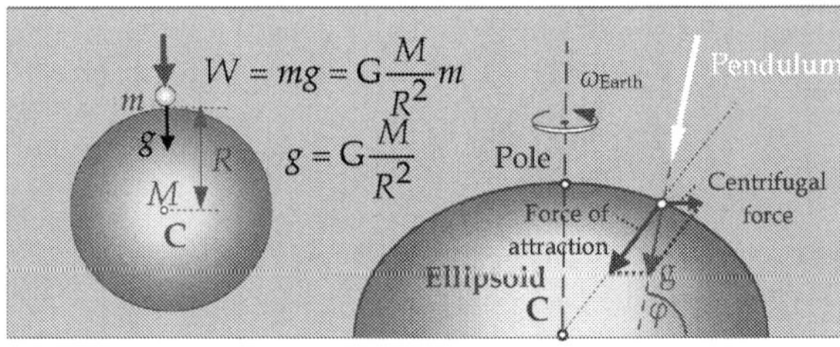


Fig. 16. The weight of a body on the surface of the Earth and the influence of centrifugal forces.

Because this force is really the force of universal gravitation it's possible to say

$$W = mg = G\frac{M}{R^2}m\,. \tag{20}$$

From this equation it follows that the acceleration of gravity is

$$g = \frac{GM}{R^2}\,. \tag{21}$$

The Earth is not truly spherical so the distance $R$ from the centre of the Earth depends on the point selected on its surface. This will be the reason why the weight of the same body will also not be the same weight on different geographical latitude and altitude of the considered point. For more accurate definition of the weight of a body it's necessary to include a component representing the centrifugal force due to the rotation of the Earth. So, the values of $g$ for a body in rest at the sea level vary from 9.780 $m/s^2$ (32.09 $ft/s^2$) at the equator to 9.832 $m/s^2$ (32.26 $ft/s^2$) at the poles.

### 3.5 D'Alembert's Principle

Jean le Rond d' Alembert (1717-1783) postulated the principle called by his name from the basic equation of dynamic:

$$\mathbf{F} = m\cdot\mathbf{a}\,. \tag{22}$$

This equation can be written in this form:

$$\mathbf{F} + (-m\cdot\mathbf{a}) = 0\,. \tag{23}$$

From this equation the magnitude $(-m\cdot\mathbf{a})$ is called *inertial force*. So, this equation represents an equation of fictive equilibrium where **F** represents resultant of all active and reactive forces, and inertial force which has the magnitude $m{\cdot}a$, but in opposite direction of the acceleration **a**. This equation is named *equation of dynamic equilibrium.*

For example, when a body is rotating in a circle with constant velocity $v$ restrained by a rope length $R$ then centrifugal force appears (Fig. 17).

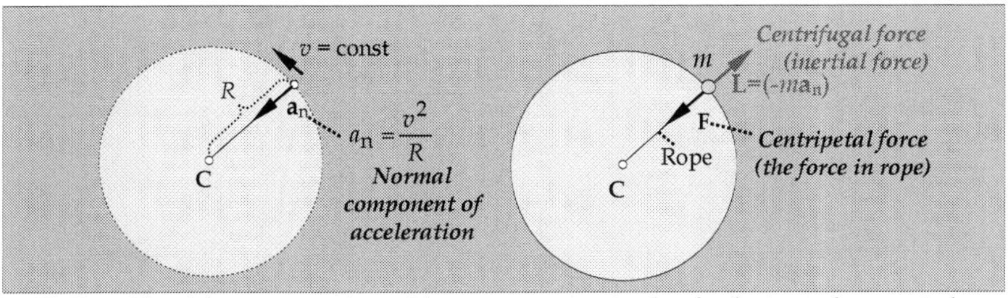


Fig. 17. Centrifugal force at rotation with constant magnitude of velocity $v$ along a circle, and then centrifugal force **L** appears.

### 3.6 Potential due to a Spherical Shell

A basic result proved by Isaac Newton is that spherical shell which is homogeneous (with constant density) attracts an exterior point with mass $m$=1 as if all of the mass $M$ of the spherical shell concentrated at its centre C (Fig. 18). This is the same, as if we have homogenous concentric layers but with different densities and whole masses $M$ then an exterior mass point $m$ attracts as if all of the mass $M$ of the spheres was concentrated at its centre (Fig. 19).

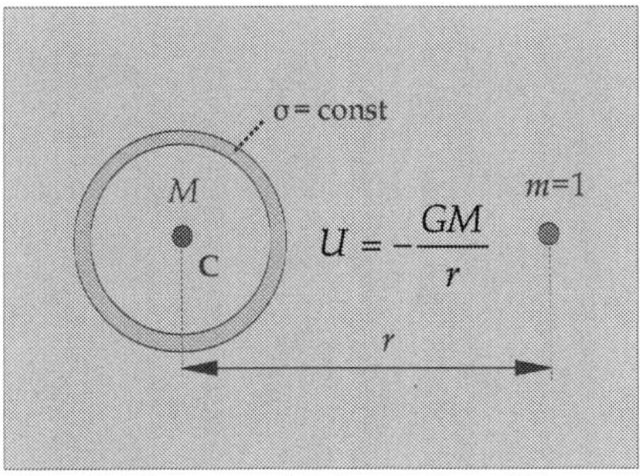


Fig. 18. The potential due to the solid spherical shell.

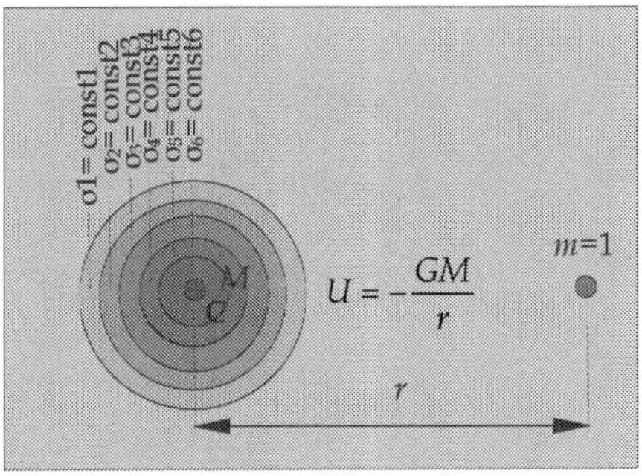


Fig. 19. The potential due to the concentric solid homogeneous spherical shells.

This fundamental result allows us to consider that the attraction between the Earth and the Sun, for example, to be equivalent to that between two mass points.
So we can say:

***The solid sphere of constant density attracts an exterior unit mass though all of its masses were concentrated at the centre.***

The potential, therefore, due to a spherical body homogeneous in concentric layers, for a point outside the sphere is

$$U = -\frac{GM}{r}, \tag{24}$$

where $r$ is the distance from the point with mass $m$ to the centre C of the mass of the homogeneous sphere or to the centre C of the concentric homogeneous spheres.

## 4. Determination of Orbits

Jacques Philippe Marie Binet (1786-1856) derived the differential equation in the polar coordinate system of the motion free material particle under action of the central force when areal velocity by the second Kepler's law is constant. This Binet's differential equation can be put down in writing

$$-mC^2u^2\left[\frac{d^2u}{d\theta^2} + u\right] = F_{rad}, \tag{25}$$

where is $u = \frac{1}{r}$, $C$ - double areal velocity ($C = \dot{\theta}r^2$), $F_{rad}$ gravitation force of the central body with mass $M$ on the free particle with mass $m$ ($F_{rad} = -GMmu^2$), where minus sign indicates that this is an attracting force, and plus sign stands for the repulsive force.
This differential equation is equation of free particle motion in a plane displayed in the polar coordinate system. Thus, inhomogeneous differential equation is obtained

$$\underbrace{\left[\frac{d^2u}{d\theta^2} + u\right]}_{\text{Homogeneous Part}} = \underbrace{\frac{GM}{C^2}}_{\text{Inhomogeneous Part}}. \tag{26}$$

The solution for the homogeneous part of this equation is

$$u_1 = B\cos(\theta - \theta_0), \tag{27}$$

where $B$ and $\theta_0$ are the constants of integration. Choosing the polar axis so that $\theta_0 = 0$ we can write

$$u_1 = B\cos(\theta) \tag{28}$$

and for the inhomogeneous part of the equation

$$u_2 = \frac{GM}{C^2} = \frac{1}{p}. \tag{29}$$

The solution of this inhomogeneous differential equation (26) is

$$u = u_2 + u_1 = \frac{1}{p}\left[1 + \frac{C^2}{GM} B\cos(\theta)\right]. \tag{30}$$

The equation for the ellipse and for the other conic section in the polar coordinate system can be written in the form

$$\frac{1}{r} = \frac{1}{p}[1 + \varepsilon\cos\theta]. \tag{31}$$

After comparing the equations (30) and (31) it is possible to see that the equation (30) truly represents the equation of a conic section. The product of the constants *B* and $C^2/(GM)$ defines the *eccentricity* ε of the conic section. So it can be expressed by the equation (Beer & Johnston, 1962)

$$\varepsilon = \frac{B}{\frac{GM}{C^2}} = \frac{BC^2}{GM}. \tag{32}$$

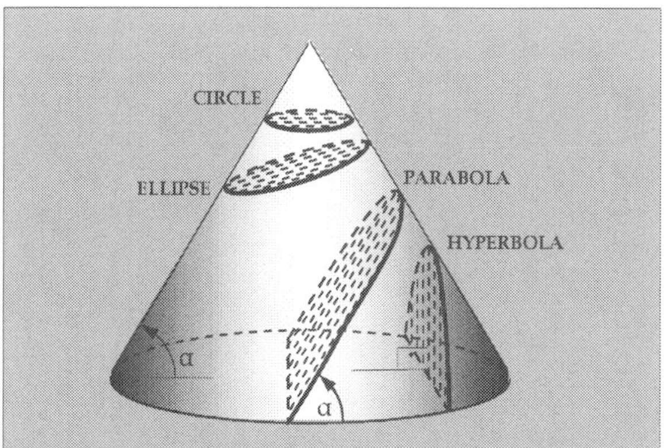


Fig. 20. The conic sections: circle, ellipse, parabola and hyperbola.

Four cases may be distinguished for different eccentricities (Fig. 20).
1) The conic section is a *circle* when is ε = 0.
2) The conic section is an *ellipse* when 0 < ε < 1.
3) The conic section is a *parabola* when ε = 1.
4) The conic section is a *hyperbola* when ε > 1.
Of cause for the planets and for the satellites orbits can be only circulars or ellipses.

## 5. The Two-Body Problem

It is possible to investigate the motion of two bodies that are only under their mutual attraction. It can also be assumed that the bodies are symmetrical and homogeneous and that they can be considered to be point masses. So we can do analysis of the motion of planets and the Sun.

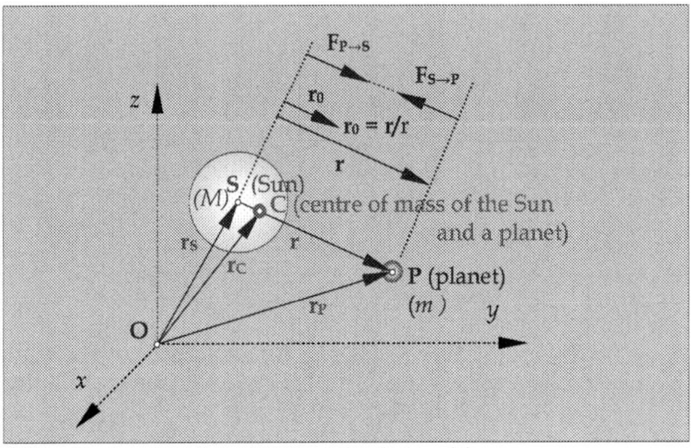


Fig. 21. Motion of the Sun and a planet in two-body problem.

The differential equation of the Sun motion (Fig. 21) is

$$M\frac{d^2\mathbf{r}_S}{dt^2} = +G\frac{Mm}{r^2}\frac{\mathbf{r}}{r}. \tag{33}$$

The sign + is because the force $\mathbf{F}_{P\to S}$ has the same orientation as the vector $\mathbf{r}_0$.
The differential equation of the planet motion is

$$m\frac{d^2\mathbf{r}_P}{dt^2} = -G\frac{Mm}{r^2}\frac{\mathbf{r}}{r}. \tag{34}$$

The sign - is because the force $\mathbf{F}_{S\to P}$ has the same orientation as the vector $\mathbf{r}_0$.
After summing up the equations (33) and (34) we can write

$$M\frac{d^2\mathbf{r}_S}{dt^2} + m\frac{d^2\mathbf{r}_P}{dt^2} = 0 \quad \text{or} \quad \frac{d^2}{dt^2}\left(M\mathbf{r}_S + m\mathbf{r}_P\right) = 0. \tag{35}$$

From the static it is known that the sum of the moment forces is equal to the moment of resultant. So we can say that the sum of the moment masses is equal to the moment of resultant mass. Now it is possible to write

$$M\mathbf{r}_S + m\mathbf{r}_P = (M+m)\mathbf{r}_C. \tag{36}$$

After the first and the second derivation we have

$$\frac{d}{dt}\left(M\mathbf{r}_S + m\mathbf{r}_P\right) = \frac{d}{dt}\left[(M+m)\mathbf{r}_C\right] \qquad \text{and}$$

$$\frac{d^2}{dt^2}\left(M\mathbf{r}_S + m\mathbf{r}_P\right) = \frac{d^2}{dt^2}\left[(M+m)\mathbf{r}_C\right] = \frac{d^2\mathbf{r}_C}{dt^2}(M+m). \tag{37}$$

From the equations (35) and (37) it follows

$$\frac{d^2\mathbf{r}_C}{dt^2}(M+m) = 0 \quad \text{or} \quad \frac{d^2\mathbf{r}_C}{dt^2} = 0. \tag{38}$$

Therefore the centre of this material system has no acceleration, namely this material system is in the inertial motion with the possibility to move with constant velocity $\mathbf{v}_C$, or remain at rest.

The equation (33) may be multiplied by $m$, and equation (34) by $M$ and after subtracting the equation (33) from (34) we can write

$$Mm\left[\frac{d^2\mathbf{r}_P}{dt^2} - \frac{d^2\mathbf{r}_S}{dt^2}\right] = -G\frac{mM}{r^2}\frac{\mathbf{r}}{r}(M+m), \tag{39}$$

$$Mm\frac{d^2}{dt^2}\left[\mathbf{r}_P - \mathbf{r}_S\right] = -G\frac{mM}{r^2}\frac{\mathbf{r}}{r}(M+m) \tag{40}$$

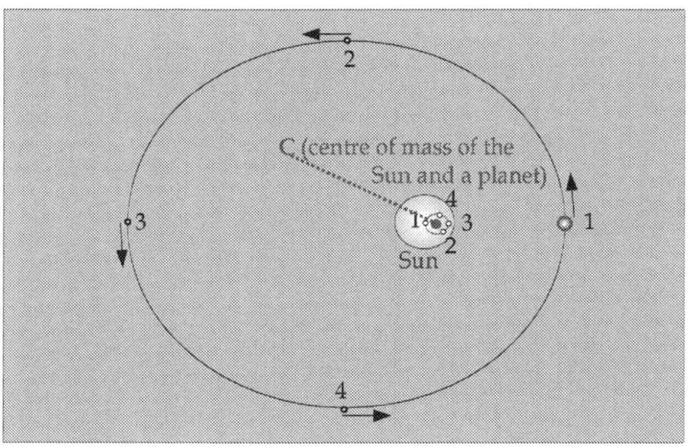


Fig. 22. The Sun is moving also around the centre C of mass of the Sun and a planet by a small ellipse and a planet is moving about the same centre C by the bigger ellipse, not around the geometrical centre of Sun.

After dividing the equation (40) by $M$ and taking from Fig. 21 that $\mathbf{r}=\mathbf{r}_P-\mathbf{r}_S$ we can write

$$m\frac{d^2\mathbf{r}}{dt^2} = -G\frac{m(M+m)}{r^2}\frac{\mathbf{r}}{r}. \tag{41}$$

This is differential equation (41) of the planet motion when taken into account and the planet acting on the Sun. It is easy to prove that this planet is really rotating around the centre of mass (C) of the Sun and the planet. Also the Sun's geometrical center is rotating by the small ellipse around the centre of mass (C) (Fig. 22).
Hence:

***The planet is rotating around the centre of masses (C) of the Sun and the planet by the bigger ellipse. The geometrical centre of the Sun also rotates around the centre of masses (C) by a small ellipse.***

## 6. Satellite Motion

The problem, of two bodies is solved exactly in the celestial mechanics, but only in the special case if both bodies are having small dimensions, i.e. if the Sun and a planet can be thought of as particles. In this special case the motions of particles around the body with finite dimensions is also included, if this body with finite dimensions has the central spherical field of forces. (For example, as a homogeneous ball (Fig. 18) or concentric solid homogeneous spherical shells with different densities (Fig. 19)). Just because our Earth is not a ball and with homogeneous masses some discrepancies appear at satellite motions around the Earth from the exact solutions of two bodies when we imagine whole mass of the Earth as concentrated in it the centre of mass. For the solution of the problem of the motion of bodies (two particles) exactly valuable are three Kepler's laws from which fallow that the satellite would be moving constantly in the same plane by the ellipse with constant areal velocity.

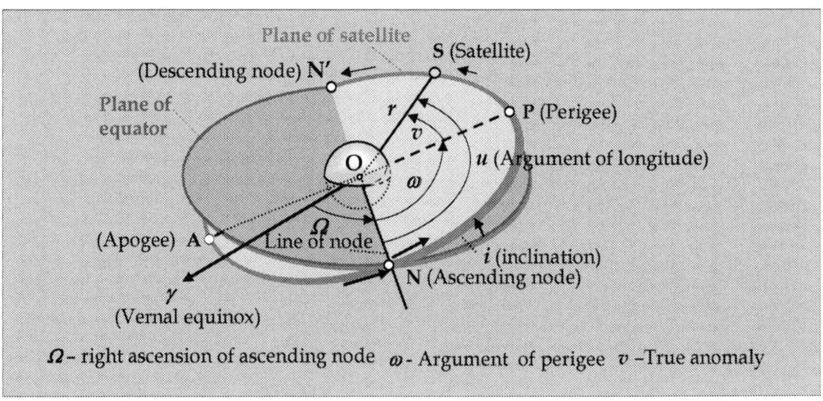


Fig. 23. Keplerian orbital parameters.

The positions of satellites are determined with six Keplerian orbital parameters: $\Omega$, $i$, $\omega$, $a$, $e$ and $v$ or $t$ (Fig. 23):

- *The orientation of orbits* in space is determined by:

$\Omega$ - *the right ascension of ascending node* (the angle measured in the equator plane between the directions to the vernal equinox and ascending node N where the satellite crosses equatorial plane from the south to the north celestial sphere), $i$ – *the inclination of orbit,* (the angle between the equatorial plane and orbital plane) and $\omega$ – *the argument of perigee* (the angle between the ascending node and the direction to perigee (as the nearest point of satellite)).

- *The dimensions of orbit* are determined by: $a$ - *the semi-major axis* and $\varepsilon$ – *the numerical eccentricity* of an ellipse.
- *The position of satellite on its orbits* is determined by: $v$ - *the true anomaly* (as the angle between the directions to perigee and instantaneous position of satellite) or by $t$ - the difference of time in instantaneous position and the time in perigee.

***All Kepler's laws and Newton's laws for a planet motion are valued also for the Earth's satellites motion but at satellites there are some more perturbations.***

## 6.1 Required Velocity for a Satellite

A body will be a satellite in a circular orbit around the Earth if it has velocity in the horizontal line so that centrifugal force is equal to centripetal force which is produced by the Earth's gravitation attraction (Fig. 24). So it can be expressed with the equation:

$$m\frac{v_I^2}{R+H_S} = G\frac{mM}{(R+H_S)^2}, \tag{42}$$

where: $m$ - mass of a satellite, $M$ – the Earth's mass, $R$ - radius of the Earth, $H_S$ - altitude of a satellite above the surface of the Earth, G constant of universal gravitation and $v_I$ velocity of a body which will become the satellite.

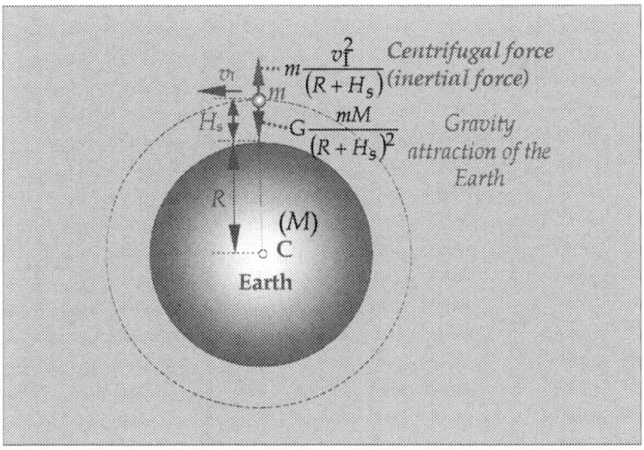


Fig. 24. On the satellite in orbit act the Earth's gravity attraction and the centrifugal force.

From this equation (42) next the equation follows

$$v_I = \sqrt{\frac{GM}{R+H_s}}. \tag{43}$$

The equation (21) can be expressed as

$$GM = gR^2 \tag{44}$$

and after including this equation (44) into (43) we have

$$v_{\mathrm{I}} = \sqrt{\frac{gR^2}{R + H_{\mathrm{S}}}}\,. \tag{45}$$

This velocity $v_{\mathrm{I}}$ is sometimes called the *first cosmic velocity*, see Table 2.

Table 2. The first cosmic velocity depending on the heights over the Earth.

| The heights over the Earth (km) | 200 | 300 | 400 | 500 | 1000 | 3000 | 5000 |
|---|---|---|---|---|---|---|---|
| Velocity $v_{\mathrm{I}}$ (m/s) | 7791 | 7732 | 7675 | 7619 | 7356 | 6525 | 5924 |

### 6.2 Period of Circulation a Satellite around the Earth by a Circle

Since a satellite makes a circle $2\pi(R + H_{\mathrm{S}})$ long around the Earth in a time $P$ (period) and because its velocity is constant then the equation (45) can be expressed as

$$2\pi(R + H_{\mathrm{s}}) = \sqrt{\frac{gR^2}{R + H_{\mathrm{s}}}}\,P. \tag{46}$$

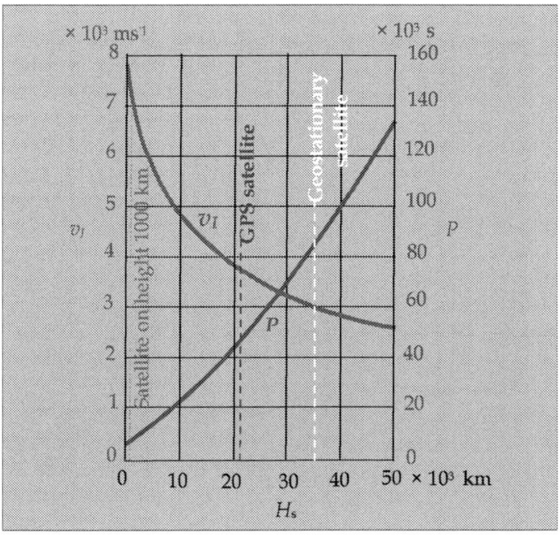


Fig. 25. Dependence the velocity $v_{\mathrm{I}}$ on the circular orbit and the period $P$ of satellite height over the Earth's surface.

From this equation it follows

$$P = 2\pi\sqrt{\frac{(R + H_S)^3}{gR^2}} \, . \tag{47}$$

With the help of the equations (45) and (47) graphs on the Fig. 25 are constructed.

### 6.3 Escape Velocity

The escape velocity is minimum velocity of a body that a spacecraft must have in order to escape from the Earth's gravitation field or of other celestial bodies.
In order to launch a spacecraft into the space out of the Earth's gravitation field, it's necessary to do mechanical work for lifting a spacecraft from $r=R$ to $r=\infty$ (Fig. 26).

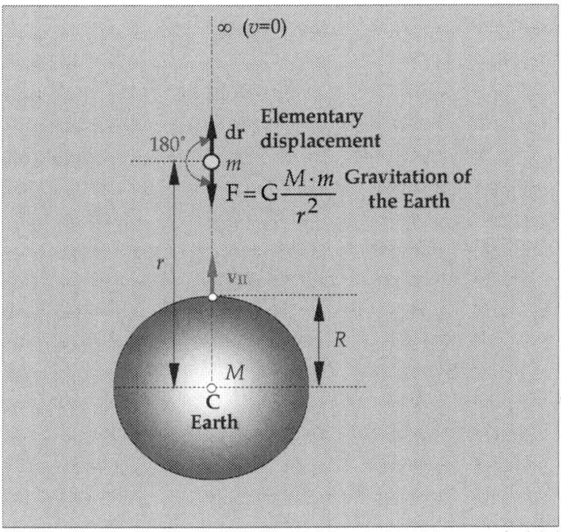


Fig. 26. Launching a spacecraft in the space out of Earth gravitation field (the second cosmic velocity).

The elementary work of the gravitation force **F** during the displacement $d\mathbf{r}$ is defined as the scalar quantity

$$dW = F \cdot dr \cos 180^\circ = -F \cdot dr \, . \tag{47}$$

For the work needed by the a spacecraft from R to the infinite $\infty$ we need to do the integration

$$W = -GMm\int_R^\infty \frac{dr}{r^2} = -GMm\frac{1}{R} \, . \tag{48}$$

The kinetic energy of a particle at the end minus the kinetic energy at the start is equal to the work, so we have

$$0-\frac{mv_{\mathrm{II}}^2}{2}=-GMm\frac{1}{R}, \quad \text{namely} \quad v_{\mathrm{II}}=\sqrt{2GM\frac{1}{R}}. \tag{49}$$

Since is $GM - gR^2$, according to the equation (44), the equation (49) can be written as

$$v_{\mathrm{II}}=\sqrt{2gR}. \tag{50}$$

When a body has the velocity $v_{\mathrm{II}}$ then it will abandon gravitation field of the bigger body. This velocity $v_{\mathrm{II}}$ is called by the some people the *second cosmic velocity.*

From the equation (50) for the Earth it follows that the second cosmic velocity is 11.2 km/s (6.96 mi/second) and in this case spacecrafts will abandon gravitation field of the Earth.

### 6.4 The Potential of a Body with Irregular Disposition of Mass

The elementary potential of an attraction force of a parcel $P_1$ with mass d$M$ in the point P (Fig. 27) is equal

$$dU=G\frac{dM}{\Delta}. \tag{51}$$

where: $G$ - the universal gravitational constant and $\Delta$ the distance point P of $P_1$.

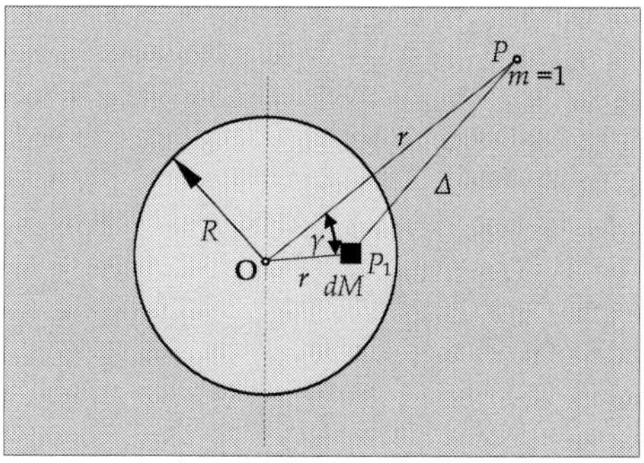


Fig. 27. The elementary potential of an attraction force of a body with arbitrary disposition of mass.

For a body with finite dimensions and with arbitrary disposition of mass it can be written in this form

$$U=G\int_M\frac{dM}{\Delta}. \tag{52}$$

This equation can be expressed by spherical functions in the form

$$U=G\frac{M}{r}\left[1+\sum_{n=2}^{\infty}\sum_{m=0}^{n}\left(\frac{R}{r}\right)^n P_{nm}(\sin\phi)\left(c_{nm}\cos m\lambda+s_{nm}\sin m\lambda\right)\right], \tag{53}$$

where the potential is evaluated at a point whose geocentric spherical coordinates are: $r$ - the geocentric distance, $\varphi$ - the geocentric latitude, $\lambda$ - the longitude measured eastwards and $R$ - the Earth's equatorial radius, $c_{nm}$ and $s_{nm}$ - the coefficients of the Earth's harmonics. The origin of the coordinate system is in the centre of Earth's mass, the geocentric longitude $\lambda$ is calculated starting from Greenwich meridian and can have the values $0 \le \lambda \le 2\pi$, the geocentric latitude $\varphi$ can have the values $(+\pi/2) \ge \varphi \ge (-\pi/2)$. In the equation (53) the expression $P_{nm}$ (sin $\varphi$), is spherical function of sin $\varphi$ which is determined by this expression

$$P_{nm}(\sin\phi) = \left(1-\sin^2\phi\right)^{m/2} \frac{d^m P_{n0}(\sin\phi)}{d(\sin\phi)^m}, \tag{54}$$

where is

$$P_{n0}(\sin\phi) = \frac{1}{2^n n!}\frac{d^n\left(\sin^2\phi - 1\right)^n}{d(\sin\phi)^n}. \tag{55}$$

The functions:

- $P_{n0}(\sin\varphi)$ are the *zonal spherical function* or shorter *zonal harmonics* (Fig. 28. a),
- $P_{nn}(\sin\varphi)\cos n\lambda$ and $P_{nn}(\sin\varphi)\sin n\lambda$ are *sectorial spherical harmonics* or sectorial harmonics (Fig. 28. b) and
- $P_{nm}(\sin\varphi)\cos m\lambda$ and $P_{nm}(\sin\varphi)\sin m\lambda$ are *tesseral spherical harmonics* or tesseral harmonics (Fig. 28. c).

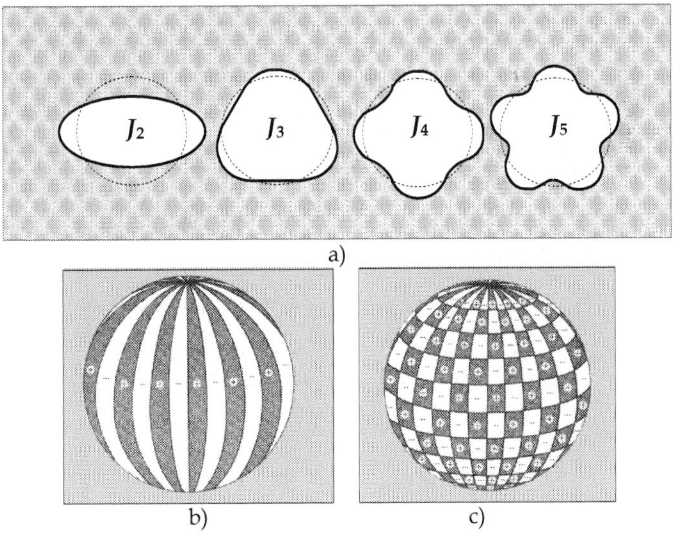


Fig. 28. Images of influences a) zonal hamonics, b) sectorial harmonics and c) tesseral harmonics.

If we suppose that the disposition of the mass in interior of the Earth is symmetrical with respect to the rotation axis than we obtain a somewhat simpler equation:

$$U = G\frac{M}{r}\left[1+\sum_{n=2}^{\infty} J_n\left(\frac{R}{r}\right)^n P_{n0}\left(\sin\phi\right)\right], \tag{56}$$

where $J_n$=-$c_{n0}$. Hence, than only zonal harmonics remain. The asymmetry of the north and of the south hemisphere is characterised by the coefficients of the odd zonal harmonics. If we suppose that the Earth is symmetrical in respect of the equator than will coefficients of the odd zonal harmonics disappear. So, it can be expressed with the equation

$$U = G\frac{M}{r}\left[1-J_2\left(\frac{R}{r}\right)^2 P_{20}\left(\sin\phi\right)-J_4\left(\frac{R}{r}\right)^4 P_{40}\left(\sin\phi\right)-....\right]. \tag{57}$$

## 6.5 Influence of the Earth's Gravitation on Movement of Artificial Satellites

Because the Earth is not a ball, i. e. not homogeneous ball, and also because it does not consist of concentric homogeneous spherical shells, the deviations of satellites orbits from an ellipse which corresponds to Kepler's laws will appear. Consequently to the second Newton's law can be written following the differential equations of satellites motion in rectangular coordinate system:

$$\ddot{x} = \frac{\partial U}{\partial x}, \quad \ddot{y} = \frac{\partial U}{\partial y}, \quad \ddot{z} = \frac{\partial U}{\partial z}. \tag{58}$$

Only the gravitation force of the Earth is taken into account but the other influences are neglected (the resistances of the air, the attraction of the Sun and the Moon, the radiation pressure of the Sun and so on). In this case the force of disturbance is

$$R^* = U - \frac{GM}{r} \tag{59}$$

and the equations (58) can be written in the form

$$\frac{d^2x}{dt^2}+\frac{\mu x}{r^3}=\frac{\partial R^*}{\partial x}, \qquad \frac{d^2y}{dt^2}+\frac{\mu y}{r^3}=\frac{\partial R^*}{\partial y}, \qquad \frac{d^2z}{dt^2}+\frac{\mu z}{r^3}=\frac{\partial R^*}{\partial z}, \tag{60}$$

where $\mu$=$G(1+M)$. The solution of these systems of differential equations and replacing the coordinates by the elliptical elements yields the parameters of orbit depending on time and the position of satellite.

Very complicated equations are obtained that can be found in the professional experts literature (Kozai 1959 and 1966), (McCuskey 1963). They are made of the starting or the mean magnitude of orbital parameters, their secular perturbations (which are linearly progressive in the time), the long-periodic and the short-periodic perturbations (Fig. 29) (Escobal, (1965).

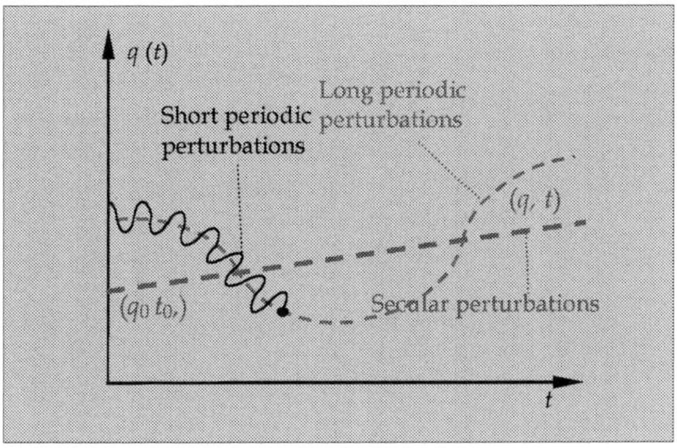


Fig. 29. The typical changes of orbital elements $\omega$, $\Omega$ and $M_0$ by time.

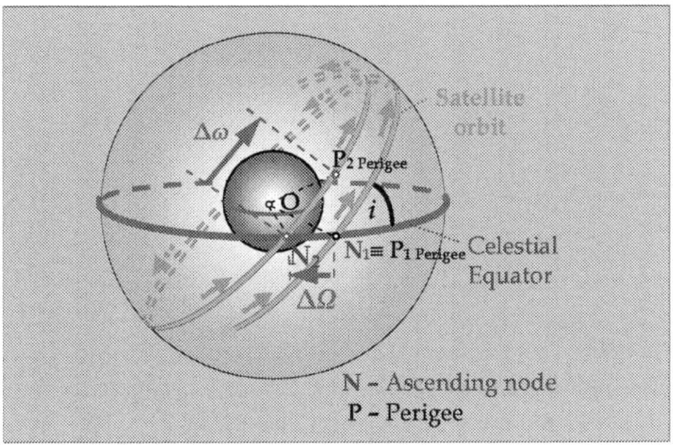


Fig. 30. Perturbations of the ascending node and perigee.

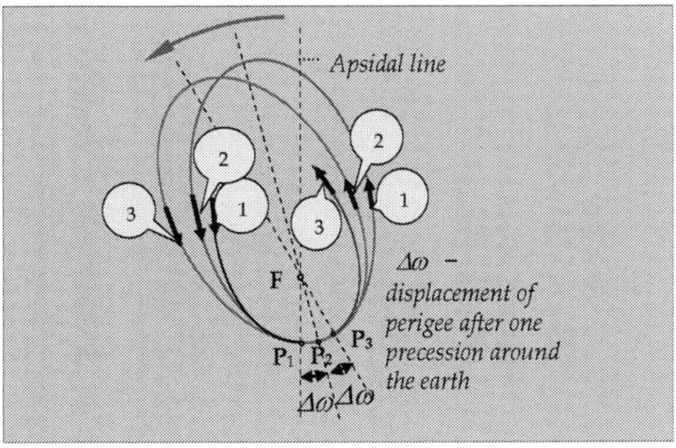


Fig. 31. Perturbations of the perigee P.

The parameter $a$ has only short periodic perturbations, the parameters $\varepsilon$ and $i$ have long and short periodic perturbations and $\omega$, $\Omega$ and $M_a$ have secular, long and short periodic perturbations (Hofmann-Wellenhof et al. 1994) and (Kaula 1966).
From this solutions it follow that the ascending node will change position from the point $N_1$ in the point $N_2$ after one satellite precession around the Earth and so the longitude of node will dislocate for the angle $\Delta\Omega$ (Fig. 30). This phenomenon is named the precession. Also the point of perigee $N_1= P_1$ will change position in the point $P_2$ (Fig. 30 and Fig. 31) and thus, it will arise the argument of perigee for the angle $\Delta\omega$. Also the apsidal line will rotate.

The secular perturbation of node longitude and the argument of perigee can be determined from approximate equations

$$\dot{\Omega} = -\frac{3}{2} n a_E^2 \frac{\cos i}{a^2\left(1-\varepsilon^2\right)^2} J_2 + \ldots\ldots , \tag{61}$$

$$\dot{\omega} = \frac{3}{4} n a_E^2 \frac{5\cos^2 i - 1}{a^2\left(1-\varepsilon^2\right)^2} J_2 + \ldots\ldots , \tag{62}$$

where are: $\dot{\Omega}$ and $\dot{\omega}$ are the temporal changes of angles in the degrees per day, $n$ - the mean angular satellite velocity ($n = 2\pi / P = \sqrt{GM/a^3}$, $G$ - the universal constant of Gravitation, $M$ - the Earth's mass), $a_E$ - the semi-major axis of the Earth, $a$ - the semi-major axis of satellite orbit, $\varepsilon$ - the numerical eccentricity of elliptic orbit, $i$ - the inclination of orbital plane and $J_2$ - the zonal coefficient of Earth's harmonics. This coefficient is about 500 and more time bigger than other coefficients and characterizes the flatness on poles and the bulginess on equator, which is caused by the Earth's rotation.
From the equation (61) it is possible to make the decision that $\dot{\Omega}$ =0°/day for $i$ = 90° when satellites are in polar orbits. Then the ascending node will not move. From the eqution (61) it also thus follows that $\dot{\Omega}$ has maximal value for $i$ = 0° when the satellite rotates in the equatorial plane.
From the equation (62) it is possible to see that $\dot{\omega}$ =0°/day for $\cos^2 i$ = 0.2 and thus, it is obtained that the inclination of orbit is $i$ = 63.4° or $i$ = 166.6°. The maximal value of $\dot{\omega}$ is when is $i$ = 90° that is when the satellite rotates in the polar orbit.
In the equations (61) and (62) there is ratio $a_E^2/a^2$ which is in every case less than 1. The satellites with bigger distance from the Earth's centre have the lower value of ratio $a_E^2/a^2$ and then the dislocations of nodes and apsidal lines will be much smaller when the satellites are farther from the Earth.
The zonal coefficient $J_2$ has the largest influence on satellites motion. The other coefficients have considerably smaller influence but in spite of this fact the artificial satellites made possible to determine the true shape of the Earth that is the geoid (Fig. 32). For this purpose usually the low satellites were used (altitudes about 1000 km), but before calculation we are obliged to remove the other influences which is a very complicated mode of proceedings. In the recent time the satellite GOCE was launched for the determination Earth's gravitation field with the very high accuracy.

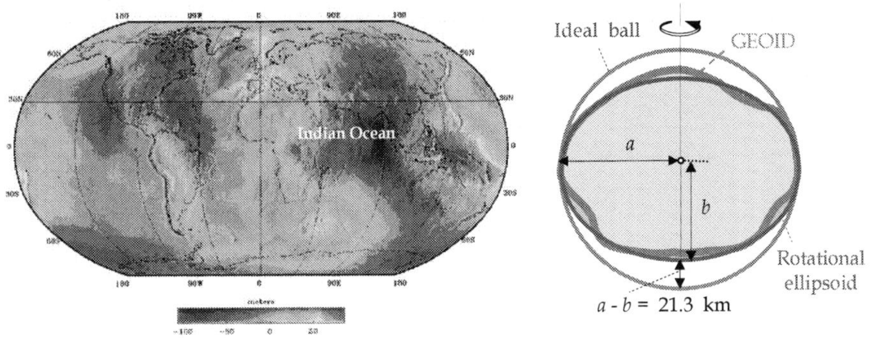


Fig. 32 The geoid is the true shape of the Earth - the maximal elevation over the reference ellipsoid is 85.4 m and the largest hollow 107.0 m in Indian Ocean (Geoid 050 NASA).

### 6.6 Influence of the Atmosphere on the Satellite Motion

The influence of atmosphere on the satellite motions is bigger if the distance of satellite from the Earth's surface is smaller. This effect is the biggest immediately after the effects of Earth's flatness for the low Earth satellites. The force of satellite retardation ($F_0$) can be determined from the equation

$$F_0 = -0.5 C_D \rho v_{rel}^2 \frac{A}{m}, \tag{63}$$

where: $C_D$ - the aerodynamic coefficient, $\rho$ - the density of atmosphere, $v_{rel}$ - the relative velocity of a satellite in respect to the atmosphere, $A$ - the area of satellite cross section and $m$ - the mass of a satellite.

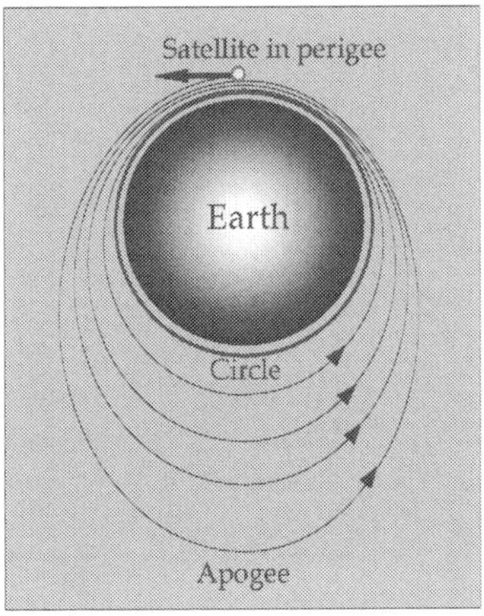


Fig. 33. Image of influence the resistance

The density of atmosphere is changeable and depends of altitudes but bigger changes caused by the activity of the Sun (phenomenon of the Sun spot). Because the resistance of the air depends on its density which is changeable and depends on more factors, this influence can be taken in account with more limited accuracy. The eccentricity of elliptic orbits in low orbits is progressively smaller (Fig. 33).
From the parameters of satellites motions in the low Earth orbit the effect of air resistance can be removed only with limited accuracy.

### 6.7 The Influence of Gravitation of Moon and Sun

The gravitation influence of the Moon and the Sun can be eliminated by equations which are obtained solving the differential equations of satellite motion in which the functions of the Moon and the Sun attractions forces are inserted.
Secular motions due to the lunar and solar attractions are denoted by $\Delta\omega_{LS}$ and $\Delta\Omega_{LS}$ are expressed as,

$$\Delta\omega_{LS} = \frac{1}{2}\psi\left(-1+5\theta_i^2+\varepsilon^2\right) \quad \text{and} \quad \Delta\Omega_{LS} = \Psi\left((1+\frac{1}{2}\varepsilon^2\right), \tag{64}$$

where: $\varepsilon$ is the numerical eccentricity of the satellite orbit at the beginning for $t = 0$,
$\theta_i = \cos(i)$ for inclination of the satellite orbit $i$ at beginning for $t = 0$,

$$\Psi = \frac{3}{4}\frac{1}{n^2\sqrt{1-\varepsilon^2}}\left[n_{\odot}^2\left(1-\frac{3}{2}\sin^2\zeta\right)+n_{\Diamond}^2 m_{\Diamond}\left\{1-\frac{3}{4}\sin^2 J\left(1+\cos^2\zeta\right)-\frac{3}{2}\sin^2\zeta\cos^2 J\right\}\right]. \tag{65}$$

In this expression $n_{\odot}$ and $n_{\Diamond}$ are mean motions of the Sun and the Moon, respectively, $m_{\Diamond}$ is the mass of the Moon expressed in units of the mass of the Earth, $\zeta$ is then obliquity, that is the inclination between the ecliptic and the equator, and $J$ is the inclination of the lunar orbit with reference to the ecliptic and is about 5°. When $n$ is given in degrees per day, $\Psi$, the dimensionless quantity, takes the following value (Kozai 1959 and Kozai 1966):

$$\Psi = \frac{1.762}{n^2\sqrt{1-\varepsilon^2}}. \tag{66}$$

### 6.8 Tidal Effects

The Sun and the Moon deform the Earth by their tidal forces not only in the ocean but also in the body (the land). In this manner the Earth's shape is temporally changeable and the tidal effects are apparent. Merson, R. H.; Kaula, W. M.; Newton, R. R. and Kozai, Y. are the first who examined the tidal effects on the motion of polar Earth's satellites and they found out that there are small changes in inclinations of the polar satellites orbits but this results were not consistent. Today the instrument for determining the positions of satellites are more precise and this effect needs to be taken into account for better determination parameters of satellite's orbit (Hofmann-Wellenhof et al. 1994).

### 6.9 Solar Radiation Pressure (The Pressure of Light)

In the first time when the first satellites were launched in the Earth's orbit solar radiation (the pressure of light) was neglected. The motions examinations of the satellite "Vangard 1" indicated that there are some discrepancies. These perturbations were possible to explain by the pressure of light. The pressure of light on the Earth's surface is very tiny and really practically zero, but in the orbit this effect can change position of satellite especially if the mass of satellite is small and its effective cross section is very big. So this effect could be studied better at the satellite EHO 1 which really was a weightless balloon with small mass and big cross section. The module of acceleration "$a_a$" of a satellite under influence of the solar radiation pressure can be determined by the equation

$$a_a = P_r \frac{A}{m}. \tag{67}$$

where: $P_r$ - the force of light pressure on the unit of area, $A$ – the area of satellite cross section and $m$ – the mass of satellite.

It should be noted here that the satellites enter into shadow of the Earth and so the problems with the pressure of light are more complex, although small.
The solar radiation pressure which is reflected back from the Earth's surface causes an effect called albedo. The pressure of reflected light also affects on the satellites but this effect can be for the most number of cases practically neglected. From the perturbations of Earth's satellite motion it was determined in the paper (Sehnal 1978) which parts of the Earth's surface give bigger or smaller albedo.
The solar radiation pressure which is reflected to the Earth's satellite from the Moon also affects on the Earth's satellite but this effect is smaller than this of Earth's albedo.

### 6.10 The other influences

- The low satellites have very big velocities something more of 7 km/s but this velocities are relatively small comparing to the velocity of light (300,000 km/s). Meanwhile, this *relativistic effect* needs also to be taken in the accurate calculation of parameters of the satellite orbits. This effect can cause the perturbing acceleration results in an order of $3 \cdot 10^{-10}$ ms$^{-2}$.
- The influences of planets gravitations can be neglected because the perturbation of planets is really too small.
- On the satellites there are different electronic equipments which have proper magnetic and electric fields. Since the satellites are in the motion in the Earth's the magnetic field the perturbation forces appear which are provoked by the acting of this field. Such effects are very small and poorly examined.

### 6.11 Sorts of Satellite Orbits

Satellites are launched into orbits, which is to say that they are shot up into the sky on rockets to get them up above the atmosphere where there is no friction. The idea is to get them flying so fast, that when they fall back to the Earth, they fall towards the Earth at the same rate the Earth's surface falls away from them. When an object's path around the Earth "trajectory" matches the Earth curvature, the object is said to be "in orbit".

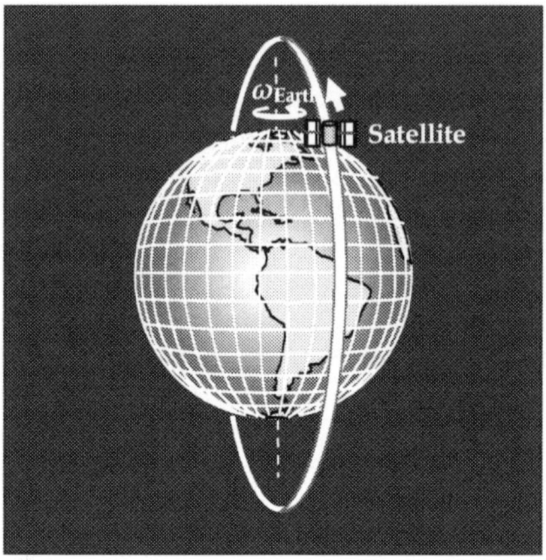


Fig. 34 Polar orbit of a satellite.

The orbits are defined by 3 factors:

▪ The first is the ***shape of the orbit,*** which can be

- *circular* (with equal distance from the Earth's centre) or
- *elliptic* (an oval shaped orbit with the *perigee* as the closet point to the Earth and the *apogee* as the farthest point from the Earth).

▪ The second is the ***altitude of the orbit.*** The altitude is constant for a circular orbit but changes constantly for an elliptical orbit.

▪ The third factor is the ***angle of the inclination,*** the angle which plane of satellite orbit makes with the plane of equator. An orbit when a satellite is passing over the Earth's poles or close to it has a large angle of inclination ($i$ = 90° or near 90°). These satellites are called the *polar satellites* (Fig. 34). An orbit that makes the satellite stay close to the equator has a small angle of inclination ($i$ = 0° or near 0°). These satellites are called the *equatorial satellites* (Fig. 35).

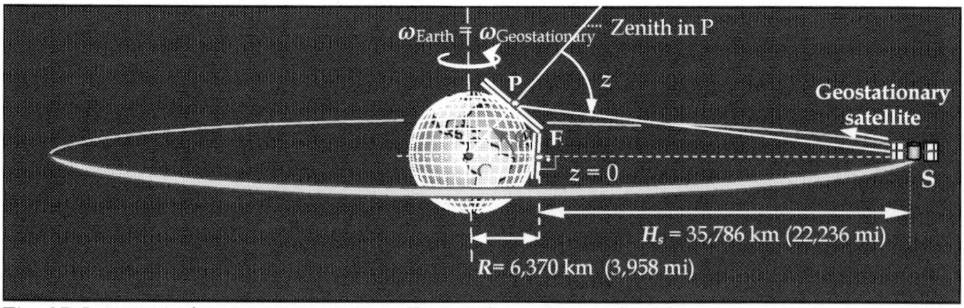


Fig. 35. Imagery of a geostationary satellite in the equatorial orbit.

□ ***Low Earth Orbits* (LEO)**

Satellites in low Earth orbit (LEO) orbit around the Earth at altitudes of less than 2000 km (1240 miles). Communication satellites in LEO can give clearer surveillance images and require much less power to transmit their data to the Earth.

For example by the *Iridium satellite network* voice and data messages can be routed anywhere in the world. This Iridium satellite constellation was originally planned to have 77 active satellites and was redesigned to on the 66 active satellites of day. These satellites are in six polar low Earth's orbital plane at a height of approximately 781 km (485 miles) and have inclinations $i$ = 86,4°. Each satellite completely orbits the Earth in 100 minutes travelling. From the horizon to the horizon it takes 10 minutes. As satellites move out of view from the satellite phone user the call is handed over the next satellite coming into view (URL 4) and (URL 5).
The ***apparent trajectory*** for LEO satellites in their circular orbits or elliptic orbits for a motionless spectator on the Earth looks like as a straight line or a mild curved line.

□ ***Medium Earth Orbits*** **(MEO)**
Satellites in medium Earth orbit (LEO) circling around the Earth are at altitudes of about 2000 km (1240 miles) to just below geosynchronous orbit at 35,786 km (22,240 miles). These kinds of orbits are known as *intermediate circular orbits.*

□ ***Geosynchronous satellites***
The geosynchronous satellites can be: the geostationary satellites and the 24-hours Earth synchronous satellites.

- ***Geostationary satellites*** **(GEO)**
A satellite in geostationary orbit circles the Earth in exactly 1 day and is placed above the equator. The geostationary satellites have angles of inclination of orbits equal 0° and they are at the elevation of approximately 35,786 km (22,236 miles) above the ground of Earth (Fig. 35). These satellites have the same directions of rotation as the Earth and so follow the Earth's rotation. For one full orbit they need 23 hours, 56 minutes, 4.091 seconds (one sidereal day). As a result the satellite seems to the man who is standing on the Earth that the satellite (GEO) is all the time in the same position without motion.

The geostationary satellites are usually launched as the first in an elliptical orbit (Fig. 36) and after when the satellite is in the apogee of this orbit it obtains velocity necessary for maintaining the height over the Earth.

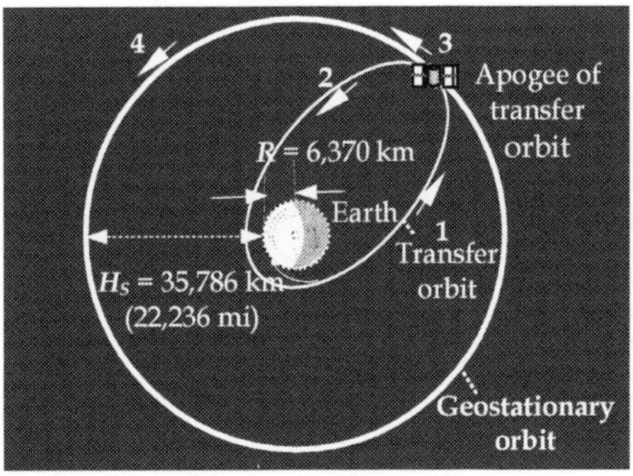


Fig. 36. Launching a satellite in the geostationary orbit with the help of an elliptic transfer orbit (Bazjanac 1977).

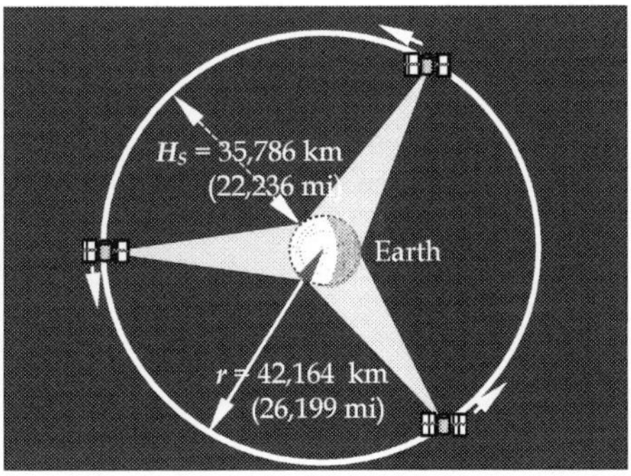


Fig. 37. All the Earth's surface is covered with only three geostationary satellites.

These satellites are usually used for communications, satellite TV, meteorology and remote sensing. With only three geostationary satellites it is practically possible to cover all the Earth's surface (Fig. 37) from the equator to the geographical latitudes +81.3° and - 81.3°.
A geostationary satellite can be immovable in the zenith only in a point on the equator, but in the point on the Earth with higher latitudes it is not possible to see immovable satellites in the zenith. This means that a satellite cannot be geostationary in the zenith at any point on the Earth with higher geographic latitudes. The planes of satellites need to pass through the centre of Earth's mass and this is the reason why the satellites cannot orbit around the Earth in the plane of parallel and to be all the time immovable in the zenith of this point on the parallel.

- ***Geosynchronous 24 Hour Earth Satellites***

These kinds of satellites have the same orbital period as the geostationary satellites but their orbital plane is not placed in the equator plane, namely they have an inclination greater than 0°. This orbit of 24-hour Earth satellite in reality may be a circular or an elliptic.

***a) Apparent Motion of 24-Hour Earth Satellites in Circular Orbits***

The 24-hour satellites in the circular orbits and under certain inclination angles $i$ in relation to equatorial plane can be described as a *number eight* on the sky for the "standing spectator" on the Earth. This apparent motion of 24-hour satellites is caused by combining rotation of the Earth around its axis and rotation of the satellite around the Earth around the axis inclined under the angle $i$ in relation to the equatorial plane.
This occurrence can be best explained by Fig. 38. In the initial moment $t = 0$ let the 24-hour satellite be in the equatorial plane. After the time $t$, the Earth will turn around its rotation axis for the angle $\beta_{Earth}$. At the same time $t$ 24-hour satellites will rotate around their orbit for the equal angle $\beta_{Sa}$. Namely, angle speed of the Earth $\omega_{Earth}$ and angle speed of 24-hour satellite $\omega_{Sa}$ are equal, so the banking angles of longitude plane and of 24-hour satellite in its plane will be equal, that is $\beta_{Earth} = \beta_{Sa}$.
However, these inclination angles $\beta_{Earth}$ and $\beta_{Sa}$ do not lie in the same plane, but their planes close between themselves an inclination angle $i$ of the 24-hour satellite orbit. This is why the projection of banking angle of the satellite in the plane of the satellite's orbit $\beta_{Sa}$ will be projected in the equatorial plane in the S' spot, e.g. it will be shorter from $\beta_{Earth}$. This means that a spectator on the Earth will at first see 24-hour satellite lapses on the west side. It will remain this way until the plane of the Earth's longitude does not turn for 90°, when the 24-hour satellite appears in that displaced longitude's plane.
In order to mathematically determine this apparent motion of 24-hour satellites, one can observe the spherical triangle $E_1SS'$ and calculate the size of the projection of the spherical side angle $\beta_{Sa}$ in the equatorial plane, which has been marked with letter S' on the Fig. 38. In this way we get:

$$s = \arctan\left[\tan(\beta_{Sa}) \cdot \cos(i)\right]. \tag{68}$$

So the lagging angle of the 24-hour satellite in the equatorial plane is:

$$\Delta\beta = \beta_{Earth} - s. \tag{69}$$

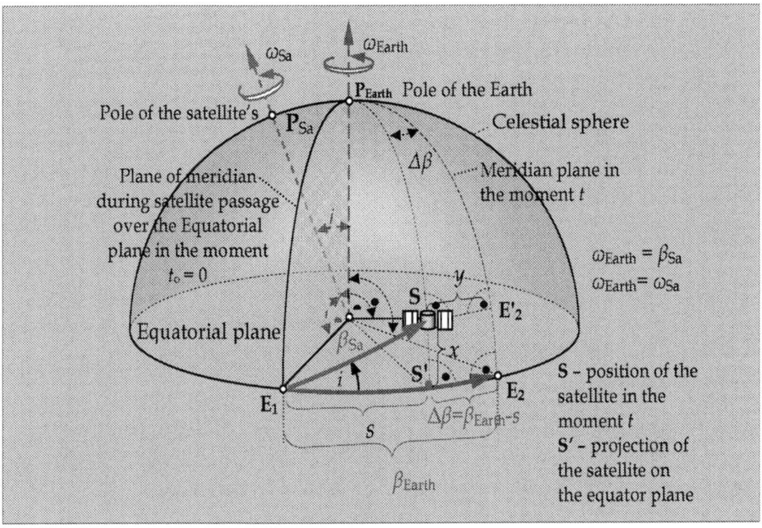


Fig. 38. Graphic illustration of why the spectator on the Earth sees relatively (apparent) motion of 24-hour satellites in circular orbit in relation to "still" Earth.

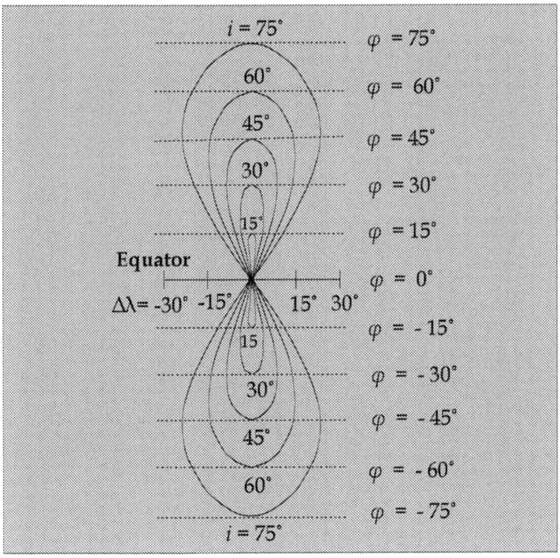


Fig. 39. Graphical illustration of apparent orbits of relative motion of 24-hour satellites on their circular orbit for different inclinations of their orbits, as a spectator on the Earth sees them.

Now it can be calculated, from the very same spherical triangle $E_1SS'$ the value of angle distance $x$ from the equatorial plane and the plane the satellite passes through. In this way, we get:

$$x = \arctan[\sin(s) \cdot \tan(i)]. \tag{70}$$

The coordinate $y$ expressed in degrees can be calculated from the spherical triangle $P_{Earth}SE'_2$. We get:

$$y = \arctan[\cos(x) \cdot \tan(\Delta\beta)]. \tag{71}$$

For different inclinations $i$ of the satellite's orbit, with the help of equations (68), (69), (70) and (71), angle values of $x$ and $y$ were calculated for drawing of graphical display of orbits of apparent (relative) motion of 24-hour satellites, as it is seen on the Earth, which is shown in Fig. 39 (Solarić, 2007).

From the chart in the Fig. 39, it is obvious that 24-hour satellite in its apparent relative motion reaches latitudes that are equal to their orbital inclinations. Besides, these *symmetrical eights* become greater when the inclination $i$ of the satellite's orbit are greater.

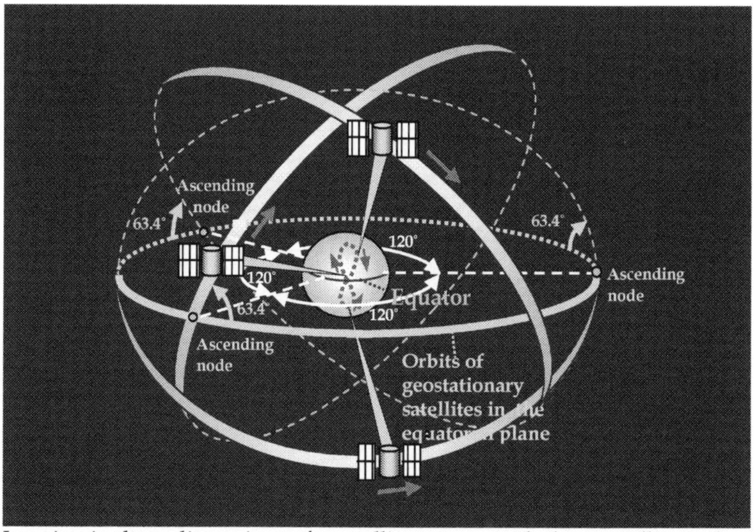


Fig. 40. Imaging in three dimensions of a satellite system with three satellite in three orbital planes.

***b) Apparent Motion of 24-Hours Earth Satellites in Elliptic Orbits***

The geosynchronous satellites on elliptical orbits with a high inclination (usually near 63.4°) and an orbital period of one sidereal day (23 hours, 56 minutes, 4.091 seconds) called *Tundra* (URL 7). A satellite placed in this orbit spends most of its time over a chosen area of the Earth.

From the second Kepler's law that satellite's velocity is lowest in the apogee, the greatest in the perigee, while the velocity of the satellite in any other point of elliptical orbit is between these values, that is: $v_{perigee} > v > v_{apogee}$. So the satellites in the apogee point of an elliptic orbits spend the most time, and because they have usually the critical inclinations $i$ = 63.4° they have not the secular perturbation of the argument of perigee caused by the Earth's equatorial bulge.

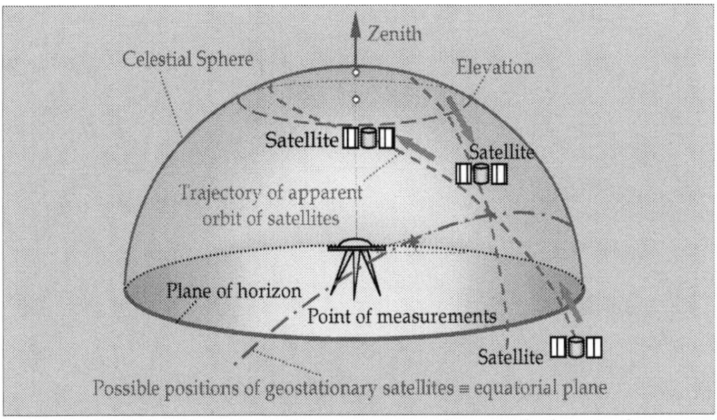


Fig. 41. View of celestial sphere on the point on the north hemisphere with apparent (relative) orbits a satellite system with the 24-hour satellites in three different orbits, as the standing spectator sees them. The possible positions of geostationary satellites have been represented as well.

The apparent ground track of a satellite in *a tundra orbit* is a closed *"asymmetrical- eight"* lemniscate's.

The Sirius Satellite Radio (the USA) uses the tundra orbits in three orbital planes with the inclination $i$ = 63.4° and the eccentricity $\varepsilon$ = 0.2684 so that their ascending nodes in the equatorial plane are moved on the longitude for 120°. Also the difference of mean anomaly of each satellite is 120°. When one satellites moves out of position, another has passed perigee and is ready to take functions over the territory of the USA and the Canada. Every satellite of these satellites is every day 16 hours over the north hemisphere and 8 hours over the south hemisphere. (Simile as at Fig. 40.)

In order to numerically determine apparent trajectory of tundra orbits with elliptical orbit, as a "standing" spectator sees it on the Earth, one should:

- project position of a satellite from the elliptical orbit to sphere,
- out of these coordinates from the sphere, according to equations (68) - (71), determine coordinates $x$ and $y$ the point of tundra orbits, expressed in degrees,
- graphically demonstrate sequential positions of satellite's apparent orbit.

Japan decided to set up their satellite communication and positioning system called *"Quasi Zenith Satellite system"* (abbreviated QZSS) or in Japanese "Jun-Ten-Cho". As expected this satellite system will up to schedule operate in 2013 (URL 8).

It is being anticipated that 3 of these 24-hour QZSS satellites would be substituting every 8 hours on the near zenith above Japan. In this way, there will always be one satellite approximately in zenith, and another QZSS satellite would be on somewhat greater zenith distance. The third QZSS satellite would then be somewhere on the south, so it can be for shorter time under the horizon in Tokyo. This could be done in the way that every QZSS satellite has its own orbital plane and that their ascending nodes in the equatorial plane are moved on the longitude for 120° (Fig. 42). The Japanese called this satellite system quasi-zenith because there will always be one QZSS satellite near the zenith.

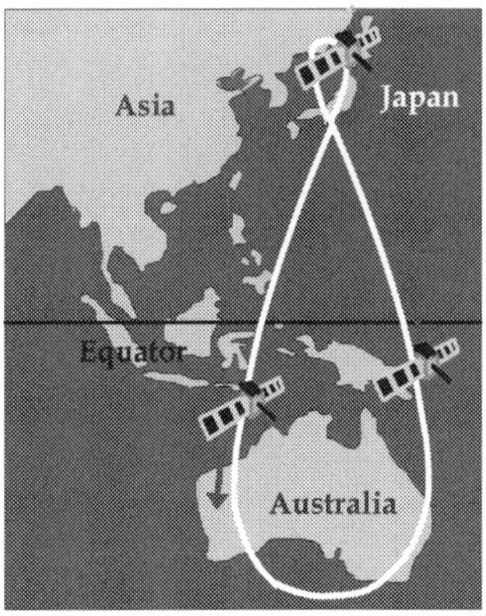


Fig. 42. Apparent orbits of three planned 24-hour QZSS-satellites with inclination $i = 45°$ and eccentricity $e = 0.099$.

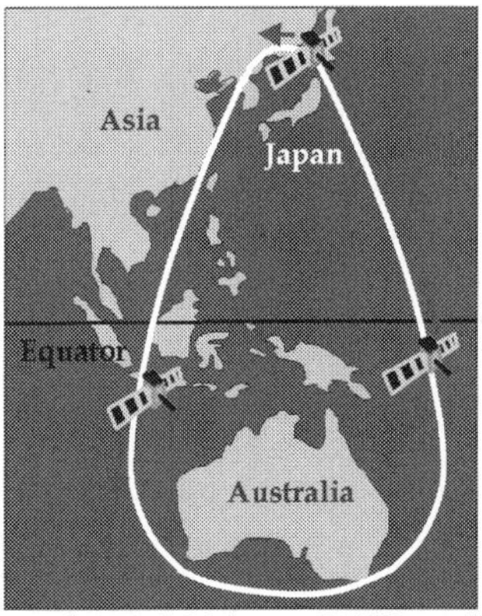


Fig. 43. Apparent orbits of three planned 24-hour QZSS-satellites with inclination $i = 42.5°$ and eccentricity $e = 0.21$.

In the article (Petrovski et al. 2003), experiments for possible usage of 24-hour satellites' orbits with different eccentricities $\varepsilon$ and for different inclinations of orbits' planes $i$, have been conducted.

Orbit of 24-hour satellite will have different form for different values of eccentricities of apparent motion:

a) For the inclination of orbital plane $i$ = 45° and eccentricity $\varepsilon$ = 0.099 apparent orbit of 24-hour satellite will have the shape of *asymmetrical eight* (Fig. 42), while they are symmetrical with the circular orbits of 24-hour satellites.
b) For the grater eccentricity of elliptical orbit, apparent orbit of 24-hour satellite motion will no longer have the shape of an asymmetrical eight, but it will assume the shape of *rain drops*. It can be seen in Fig. 43. for the eccentricity of elliptical orbits of 24-hour satellites $\varepsilon$ = 0.21 and orbital inclination $i$ = 45°. Apparent motion orbit of QZSS satellite will expand even more for somewhat greater eccentricity $\varepsilon$. This kind of orbit of a QZSS satellite would be appropriate for an area wider than Japan, so, for this case, possibilities with 3 and 4 QZSS satellites have been tested.
Clear spatial image could also be constructed for this case, when QZSS has three QZSS satellites in elliptical orbits, as it is demonstrated in Fig. 40, for circular orbits of QZSS satellite.

□ ***Semi - Synchronous Orbit***
The satellite in the semi-synchronous orbit has the period of the 0.5 sidereal day (about 11 hours, 58 minutes) of travelling about the Earth. A Molniya orbit is a type of highly elliptical orbit with an inclination of 63.4° and an orbital period of precisely one half of sidereal day. Soviet/Russian communications satellites use these kinds of orbits which have been using this type of orbit since the mid 1960s. (Molniya (Russian mean: "lightning")). A satellite in a Molniya orbit spends most of its time over the Earth as a result of "apogee dwell" (Fig. 44). Only three such satellites are needed to give 24 hour coverage.
Also the satellites from the global position system (GPS) have the semi-synchronous period of orbits around the Earth.

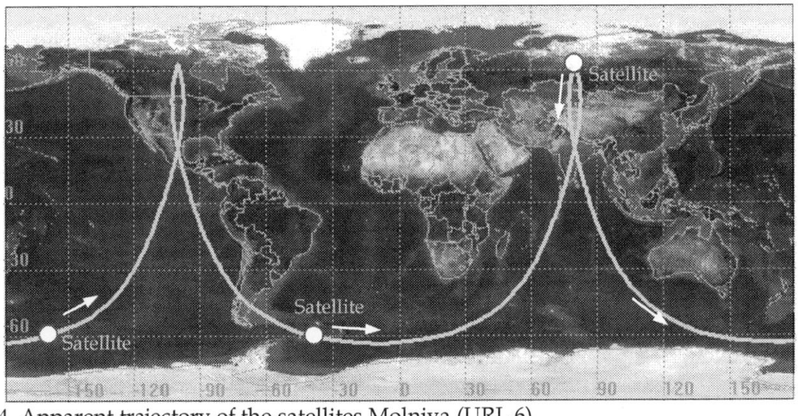

Fig. 44. Apparent trajectory of the satellites Molniya (URL 6).

□ ***Sun - Synchronous Orbit (SO)***
A Sun-synchronous orbit of the Earth's satellite is a geometric orbit which combines altitude and inclination in such a way that an object on that orbit ascends or descends over any given point of the Earth's surface at the same local mean solar time. In this way the Earth's surface illumination angle will be nearly in the same time.

Because the Earth is bulgy on the equator it follows from the equation (61) that the right ascension of ascending node has a constant negative velocity. In this way the plane of satellite orbit does not remain in the same position. So the ascending node is moving for approximately one degree per day which is equal to the angle of the Earth's shift during one day on the Earth orbit around the Sun.

Typical Sun-synchronous orbits are about 600-800 km in altitude, with periods in the 96-100 minute range, and inclinations of around 98°. Variations on this type of orbit are possible; a satellite could have a highly eccentric sun-synchronous orbit.

## 7. References

Bazjanac, D. (1977): Seminar satellites - servants from space, Sveučilišna naklada Liber, Zagreb, Croatia

Beer, F. P. & Johnston, R. E. (1962): Mechanics for Engineers (Second Edition), McGraw-Hill Book Company, Inc., New York, San Francisco, Toronto, London

Carton, D. S. (1965): The propulsion and motion of rigid rocket - propellend launch vehicles, In: Dynamic of Rockets and Satellites, Groves, G. V. (Ed.), 1-30, North-Holland Publishing Company, Amsterdam

Danby, J. M. A. (1989): Fundamentals of Celestial Mechanics, Second Edition, Revised & Enlarged, Willmann-Bell, Inc. Richmond, Virginia, USA

Escobal, P. R. (1965): Methods of Orbit Determination, John Wiley & Sons, Inc., New York, London, Sydney

Hofmann-Wellenhof, B.; Lichtenegger, H. & Collins, J. (1994): GPS Theory and Practice, Spring-Verlag, Wien, New York.

Groves, G. V. (1965): Dynamic of Rockets and Satellites, North-Holland Publishing Company, Amsterdam

Kaula, W. M. (1966): Theory of Satellite Geodesy, Blaisdell Publishing Company, Waltham, Massachusetts, Toronto, London

Kozai, Y. (1959): Smithsonian Astrophysical Observatory Special Report No 22, 10.

Kozai, Y. (1966): The Earth Gravitational Potential Derived from Satellite Motion, Space Science Reviews 5, 818-879, D. Reidel Publishing Company, Dordrecht-Holland.

McCuskey, S. W. (1963): Introduction to Celestial Mechanics, Addison-Wesley Publishing Company, Inc. Reading, Massachusetts, Palo Alto, London

Petrovski, I.; Makoto, I.; Hideyuki, T.; Hitoshi, K.; Toshio, F.; Masayuki, S.; Toshiyuki, T. & Hiroaki, M. (2003): QZSS - Japan's New Integrated Communication and Position - Service for Mobile Users, GPS Word, June 2003, pp. 24-29.

Sehnal, L. (1978): Spherical Harmonic Analysis of the Earth's Albedo, Obserwacje Sztucznych Satelitów Ziemi, 18, 315-322

Solarić, M. (2007): Japan's Quasi-Zenith Communication and Position Satellite System, Cartography and Geoinformation, Vol 6, Special issue, June 2007, pp. 234-245.

URL 1: Robert Goddard
http://en.wikipedia.org/wiki/File:Dr._Robert_H._Goddard_-_GPN-2002-0001.31.jpg (12.4.2010)

URL 2: Sergei Korolov
http://www.centennialofflight.gov/essaySPACEFLIGHT/korolev/SP5G1.htm (12.4.2010)

URL 3: Wernher von Braun:
http://en.wikipedia.org/wiki/File:Wernher_von_Braun_crop.jpg (12.4.2010)
URL 4: How the Iridium Network Works
http://www.satphoneusa.com/iridium/network.html (20.1.2010)
URL 5: Iridium satellite constellation
http://en.wikipedia.org/wiki/Iridium_satellite_constelation (20.1.2010)
URL 6: Molniya orbit http://en.wikipedia.org/wiki/File:Molniya.jpg (23.1.2010)
URL 7: Tundra orbit http://en.wikipedia.org/wiki/Tundra_orbit (25.1.2010)
URL 8: Quasi-Zenith Satellite System
http://en.wikipedia.org/wiki/Quasi-Zenith_Satellite_System (18.4.2010)

# 23

# System Aspects of Active Phased Arrays[1]

Amir I. Zaghloul[1,2], Ozlem Kilic[3] and Eric C. Kohls[4]
[1]*Virginia Polytechnic Institute and State University*
[2]*US Army Research Laboratory*
[3]*The Catholic University of America*
[4]*Lockheed Martin Corporation*
*USA*

## 1. Introduction

Demands in satellite communication systems for large number of spot beams with high beam-to-beam isolation and the need for flexibility in steering the beams impose requirements that are easier met with active phased arrays. The advances in MMIC and digital technologies played a major role in realizing these arrays for such systems. However, the added requirements and active array characteristics introduce a set of transmission impairments in the satellite link that did not necessarily exist in the conventional reflector-based satellite systems.

The design of the antennas for multiple beam communication systems depends on the beam definition, which in turn is a function of the system capacity and projected traffic patterns. Several systems require a large number of fixed narrow spot beams to cover the service area. A single reflector with a large number of feeds may provide such coverage. However, if the scanning loss is excessive due to the large number of beams scanned in one direction, multiple reflectors may be required. Alternatively, single or multiple phased arrays may be used. The phased array may have lower scan loss and a single phased array can handle a large number of beams. The phased array solution also offers higher reliability due to the use of the beam former versus the switching arrangement in a focal-region-fed reflector antenna. For a small number of beams at a time, a microwave beam former represents a simple and attractive solution, while if the number of beams is large a digital beam former is a more viable alternative. In many applications the choice between the microwave and the digital beam formers becomes a system issue.

Active phased arrays can also meet high flexibility requirements that have become the feature of several satellite communications systems. Among the flexibility requirements are the ability to form multiple beams, provide power sharing among beams through distributed amplification, and rapidly reconfigure and/or repoint the beams. The

---

requirements also include high reliability and the ability to readjust the feed or array coefficients to compensate for component failures. The conventional drawbacks of active antennas pertaining to mass, non-uniformity of performance of a large number of elements, and reliability have been largely overcome by the advances in monolithic microwave integrated circuits (MMICs), printed-circuit antennas and digital beam forming networks. The MMICs allow the beam-forming network, which may include signal-conditioning circuitry such as MMIC phase shifters and attenuators, to be incorporated in a lightweight transmission medium. The resulting antenna configurations have the potential for dramatic reductions in mass while providing increased efficiency, flexibility, and capacity. Similar features apply also to digital beam formers.

The satellite industry conducted evolutionary research and development programs to assess the feasibility of implementing MMIC-driven active phased-array antennas in high-capacity communications systems. A number of advanced concepts employing active phased arrays have been investigated. Among the early developments are Ku-band and X-band arrays that were sponsored by INTELSAT, COMSAT Corporation and the US Air Force (Zaghloul et al., 1994), (Zaghloul, 1996). Another array was also developed at COMSAT Laboratories at C-band to produce eight simultaneous and independently shaped and steerable beams. Extensive development was performed in parallel at Lockheed Martin Commercial Satellite Systems and accomplished high level of integration in active phased arrays (Jacomb & Lier, 2000). The design uses modular concepts to increase the design flexibility and afford efficient in-orbit operation using MMIC components and high-efficiency SSPAs. The characteristics of the communications systems that use phased arrays have been measured and analyzed to show the effects of the array's signal formation mechanisms on the transmission parameters. The following section reviews some of the reported development of these active phased arrays. The transmission impairment mechanisms associated with such systems are subsequently addressed. Other and similar characterizations of the transmission mechanisms, interference parameters and bit error analyses were reported (Maalouf et al., 1998), (Purdy & Lier, 2000), (Lier & Maalouf, 2003), (Maalouf & Lier, 2004). Of special interest is the analysis of intermodulation components that result from multi-carrier operation of active phased arrays (Sandrin, 1973), (Lier & Charrette, 2005). A reliability model to support the design of active phased arrays in satellite communications systems was also reported (Ruggieri, 1997).

## 2. Early Developments of Active Phased Arrays for Satellite Systems

### 2.1 Ku-Band Active Phased Array

The first example is a Ku-band array that consists of 24 active radiating elements, each equipped with an ortho-mode transducer (OMT) for dual-linear polarization (Zaghloul et al., 1994). Figure 1 shows a block diagram of the array components and Figure 2 shows the array assembly. The radiating elements are dual-linearly polarized square horns of 3-$\lambda$ aperture each; a size chosen to reduce the number of elements in the overall aperture, reduce the mutual couplings between the elements, and eliminate the grating lobes within the $\pm 9°$ field of view of the geostationary satellite.

To form a number of simultaneous and independent beams, a beam-forming matrix (BFM) is used. Directly behind each element is a 2-W Solid-State Power Amplifier (SSPA). The SSPAs are located in separate housings that are mechanically cooled by a liquid cooling loop. The outputs of the amplifiers are probe coupled to the input of the OMTs. The inputs to the amplifiers are connected to the respective outputs of the BFM, which is capable of forming four simultaneous beams. Each beam is independently steerable through the use of digital phase shifters located in the BFM. The logic circuitry necessary to drive and store the respective states of the phase shifters is also located within the BFM. The SSPA and BFM designs both make extensive use of MMICs. Phase-matched RF cables provide the signal path from the outputs of the BFM to the inputs of the amplifiers.

The BFM consists of four 1-to-24-way power dividers and 24 four-way power combiners. Each transmission path in the BFM consists of an MMIC 5-bit phase-shifter module, input and output distribution networks. A total of 96 MMIC phase-shifter modules are located in the BFM housing. A modular architecture was selected for implementation of the MMIC-populated BFM in the form of 4-input, 8-output shelves, corresponding to the four beams and eight elements, respectively. The three shelves are identical in assembly and performance. Each shelf shown in Figure 3 contains four 1-to-8-way dividers, eight 4-to-1-way combiners, thirty-two MMIC phase-shifter modules and associated control electronics. Four 1-to-3-way dividers distribute the signals for each beam to the three shelves. The input plane of the BFM shelf contains four eight-way Wilkinson dividers and MMIC phase shifter modules. The MMIC phase-shifters are assembled on separate carriers for individual testing. The output plane of the BFM shelf also contains the phase-shifter driver and level-shifter electronics that are mounted on multilayer alumina boards. The ground plane in the middle serves as an RF and DC ground and also provides isolation between input and output planes.

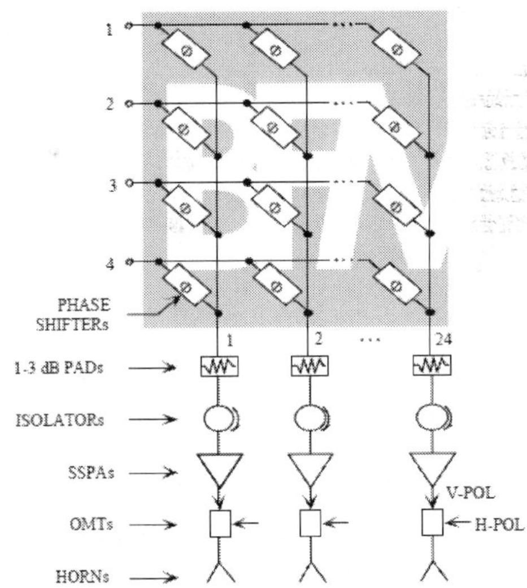


Fig. 1. Block diagram of multiple-beam active phased array

Fig. 2. Ku-band Active Phased Array

Fig. 3. 4x8 Shelf in Ku-band BFM

This design approach for the BFM provides several advantages including modularity, reproducible assembly and performance. Broadband Wilkinson dividers and combiners provide inherently good amplitude and phase balance between the paths. Furthermore, path-to-path insertion loss variations can be minimized by the appropriate selection of MMIC phase shifter modules so that low loss phase shifters are matched with high loss paths, and vice versa.

The power amplifier design combines MMIC and quasi-monolithic technology to achieve optimum DC-to-RF conversion and linearity, while simultaneously meeting output power and size requirements. The amplifiers are designed to be integrated directly in back of the array to minimize total output losses.

## 2.2 X-Band High-Power Active Phased Array

A study performed at COMSAT Laboratories addressed a set of requirements for a future DSCS system (Zaghloul, 1996). A key antenna specification is to transmit four beams simultaneously. These beams are to be shaped and steered independently, and can range in size from the minimum of 2° to the full earth coverage of 17°. An EIRP of 50 dBW is required for the 2° beams, and the overall antenna mass and DC power consumption are to be minimized with goals of 200 lb. and 400 Watts, respectively.

Fig. 4. Demonstration X-band array for DSCS system

A proof-of-concept array was developed as a subarray of such an antenna, and aims at demonstrating the most critical components required of a full-up active array antenna. This demonstration array, shown in Figure 4, utilized a total of eight radiating elements to form two independently steerable and reconfigurable beams. A key component developed for the program is the Beam Forming Matrix (BFM) whose function is to shape and steer the beams. The BFM implements modular MMIC technology in its phase shifters and attenuators which provide highly repeatable and uniform performance that is critical for this application. The lightweight radiating elements are composed of scalar ring horns and pin-polarizers that together provide high polarization purity required in a dual-polarization system should that option be exercised. Another key development is the highly efficient 2 Watt power amplifiers which are positioned within the transmit modules behind each radiating element.

### 2.3 C-Band Multiple-Beam Active Phased Array

The linearly polarized Ku-band array described above demonstrated the concept of active phased arrays for communications systems. It used large pyramidal horns as the radiating elements. One of the most desirable features, which have to be achieved, is lightweight. Because the size of the array can be large, especially at lower frequencies, the need for compact, lightweight radiating elements became paramount. The next array in this development series used printed-circuit patch radiators in order to achieve this goal. The array, shown in Figure 5, was designed at C-band (3.6-4.2 GHz) and featured printed-circuit radiators, a highly modular approach for the beam-forming matrix as well as for the active aperture, and an integrated thermal control system (Zaghloul et al., 1994). The array consists of 69 electromagnetically coupled patch (EMCP) elements, each producing two orthogonal circular polarizations. The elements are fed with 2-Watt SSPAs that are integrated in the array structure. The BFM provides independent control of eight simultaneous beams.

Fig. 5. C-Band active phased array

The array active elements demonstrate a high level of integration. The integrated element includes the EMCP radiating element, redundant SSPAs, MMIC gain blocks, redundancy switches, and monitoring circuit at the SSPA output. The SSPA consists of a high-efficiency power amplifier preceded by multistage MMIC preamplifiers and an MMIC linearizer to predistort the signal such that high linearity is achieved simultaneously with the required efficiency.

The active array is controlled by an 8X69 BFM, which consists of nine 8x8, shelves and accommodates up to eight beams connected to a maximum of 72 elements. For the 69-element array, three outputs are match terminated. The modular architecture of the Ku-band BFM described above has been further enhanced by implementing the C-band BFM shelf with eight 1x8 BFM modules. Each of these modules is realized by modular MMIC packages, each containing phase and amplitude control elements and associated digital control for each BFM cross point. This additional modularity further offers improved design flexibility for larger size BFMs, ease of repair and reduced manufacturing costs. The 8 x 8 shelf, shown in Figure 6, is composed of eight 1x8 BFM modules on the topside and eight 8-way power combiners on the bottom surface. Each 1x8 BFM module is an independently testable unit and includes the control circuits for eight MMIC packages and also provides the RF signal paths from the top to the bottom of the BFM shelf. Each 1x8 module consists of an eight-way power divider, eight MMIC packages and the digital control circuits. The most basic building block of the BFM is the MMIC package, shown in Figure 7, which contains an MMIC 5-bit phase-shifter, a 5-bit MMIC attenuator and three digital control ICs. This MMIC package can be individually tested and hermetically sealed to provide a rugged design.

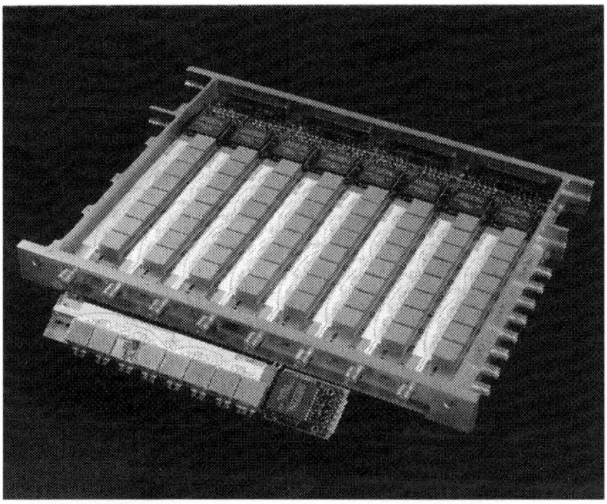

Fig. 6. 8 X 8 Shelf in C-band BFM

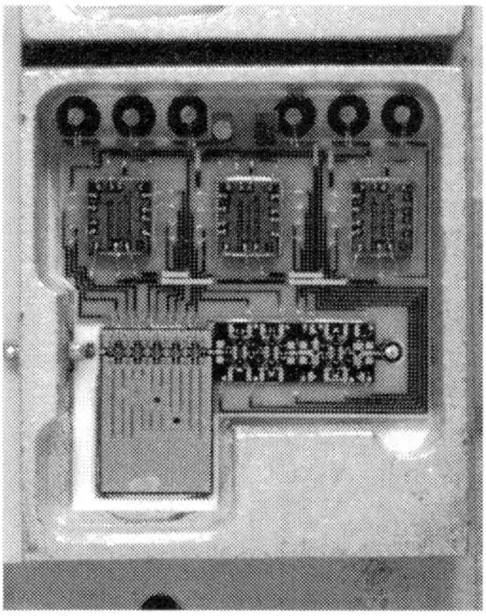

Fig. 7. Basic C-band MMIC module

## 3. Evaluation of Transmission Parameters

This section presents evaluations of three transmission parameters through analysis and/or measurements: radiation patterns of intermodulation products, bit error ratios and co-channel interference between beams.

### 3.1 Far-field radiation pattern measurements of intermodulation products

Intermodulation products (IMP) result from the nonlinearities in the power amplifiers that are used in the active phased arrays. The distributed amplifiers are shared between all channels that are transmitted through all elements in the array. Due to the periodicity of the array structure and the nature of the intermodulation product generation, the location of such undesired and interference-like radiation can be predicted. The intermodulation products generated by a pair of carriers whose frequencies are $f_1$ and $f_2$ form distinctive beams with peak positions that are the vector sums of the carriers' positions. The frequency of the intermodulation product is defined by:

$$f_{third} = 2*f_2 - f_1$$

The direction of the intermodulation product beam can be predicted by (Sandrin, 1973):

$$\bar{k}_{third} = 2*\bar{k}_2 - \bar{k}_1$$

where:

$$\bar{k} = 2\pi\frac{d}{\lambda}\left(\cos\varphi\ \sin\theta\ \hat{x} + \sin\varphi\ \sin\theta\ \hat{y}\right)$$

is the direction vector of the beam. The levels of the intermodulation products can be estimated knowing the nonlinear input/output characteristics of the amplifiers. The nonlinearities are represented in Bessel function expansion series that leads to power level evaluations (Fuenzalida et al., 1994). Typical amplifier characteristics are shown in Figure 8. The output power levels are plotted against the input back-off (IBO). The IBO is the back-off of the amplifier total input power level that corresponds to the output back-off (OBO), which is defined as the ratio of the peak single-carrier output power to the total multi-carrier output power at the operating point.

Multiple-beam intermodulation measurements for the 24-element Ku-band array described above, and shown in Figures 1 and 2, were performed in an anechoic chamber, where the antenna under test operated in transmit mode at relatively high power levels. Array control, measurement control, and data collection proceeded in essentially the same manner as during regular single carrier pattern measurements (Ekelman et al., 1994). The signal sources were composed of a synthesizer driving a 20-Watt Traveling Wave Tube Amplifier (TWTA) through a band-pass filter and 10-dB directional coupler. Three input signals were coupled off at the BFM inputs, and drive levels were continually monitored for the duration of the measurement. The drive level for all carriers was 33 dBm so as to provide the maximum drive into the SSPA's without incurring gain compression at the phase shifters in the BFM. Carrier frequencies and beam locations were chosen as shown in Figure 9 and Table 1, and designated as frequencies a, b and c.

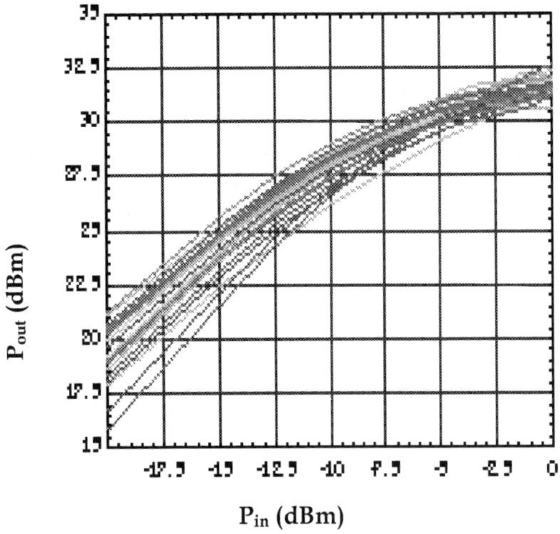


Fig. 8. Transfer characteristics of SSPA models

| Signal Type | Frequency (GHz) | Generator Polynomial | Location (Θ°, Φ°) |
|---|---|---|---|
| IMP | 11.67 | 2 • a - c | (0°, 4°) |
| IMP | 11.72 | a + b - c | (4°, 4°) |
| IMP | 11.77 | 2 • b - c | (8°, 4°) |
| IMP | 11.80 | 2 • a - b | (-4°, 0°) |
| Carrier | 11.85 | a | (0°, 0°) |
| Carrier | 11.90 | b | (4°, 0°) |
| IMP | 11.95 | 2 • b - a | (8°, 0°) |
| IMP | 11.98 | c + a - b | (-4°, 4°) |
| Carrier | 12.03 | c | (0°, 4°) |
| IMP | 12.08 | b + c - a | (4°, 4°) |
| IMP | 12.16 | 2 • c - b | (-4°, 8°) |
| IMP | 12.21 | 2 • c - a | (0°, 8°) |

Table 1 Carrier and Intermodulation Parameters

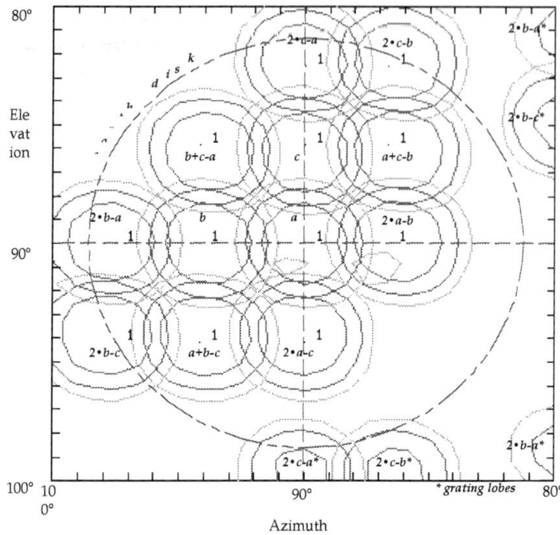


**Contour s**

**(Rel. dB)**

1 = 0.00
2 = -3.00
3 = -6.00
4 = -9.00

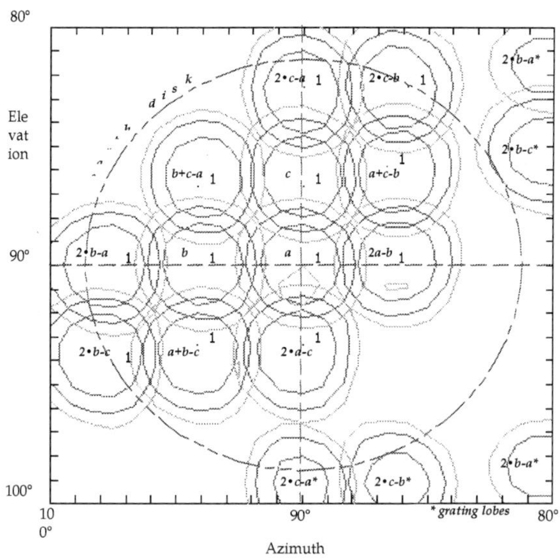


Fig. 9. Radiation far field intermodulation contours, Top: predicted, Bottom: measured

Predicted third order IMP locations are shown in Figure 9 (top). Pattern contours were measured at the 3 carrier frequencies and the 9 in-band 3rd-order IMP frequencies, and are shown in Figure 9 (bottom). The carrier and intermodulation component locations are listed in Table 1. Measured pattern contours demonstrate that the predicted patterns are accurate in regard to the beam shapes and locations.

### 3.2 Bit-Error Ratio Measurements

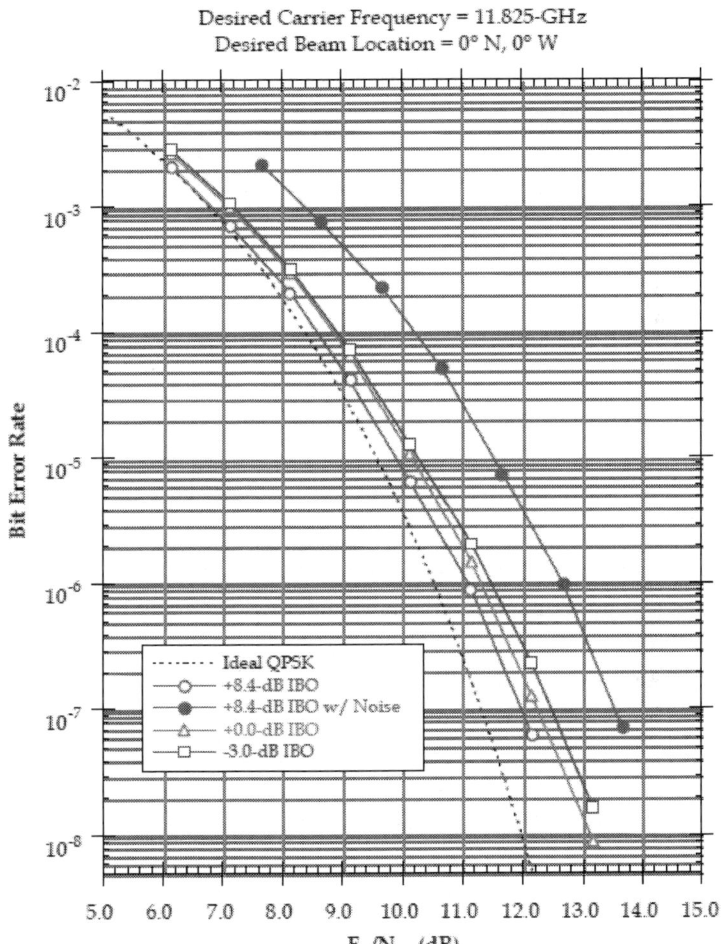


Fig. 10. BER for single carrier

The bit-error ratio (BER) measurements of the same Ku-band active phased array described above were performed under different operating conditions: single channel and multi-channel operations at different power levels in one beam with and without traffic loading in a different beam (Kohls et al., 1995). In order to simulate the traffic loading represented in multitude of populated signal channels, a single white noise source was connected to one of the BFM input ports. The noise-loading beam was scanned to a position isolated from the beam under test. When in operation, the drive level of the noise-loading channel was maintained at a level equivalent to a fully populated transponder bank. At such a drive level, it is possible to observe the effects of driving the SSPA's into saturation.

The purpose of the single channel measurements is to assess independently the impact of noise loading and carrier input drive level on BER performance. Figure 10 shows the results of the single channel measurements where four operating conditions are investigated: the desired carrier at -3.0, 0.0, and +8.4 input back-off (IBO) levels with no noise loading, and the desired carrier at +8.4-dB IBO in the existence of a noise loading at a level of +0.7-dB IBO. The dotted line indicates the theoretical BER limit for a QPSK signal. The changes in the BER characteristics as the channel drive level is increased indicate the degradation that is attributable to the non-linear distortion effects as the SSPA's are driven towards saturation. The difference between the curves with and without the noise loading indicates that the degradation is primarily attributable to carrier suppression as most of the available SSPA power is expended in the loading channel. As the traffic or the number of carriers represented by the noise loading increase, the relative level of the carrier under test decreases within the available SSPA power. This causes the carrier suppression and the corresponding degradation in the BER performance. The same degradation due to non-linear distortion is still present in the unloaded case, as in the +0.0-dB back-off case, but the impact is secondary by comparison.

The purpose of the multi-channel measurements is to assess the impact of adjacent channel interference on BER performance, while the adjacent-channel beam is either in the same location as, or scanned away from, the desired-channel beam. It is also the purpose of the multi-channel measurements to assess the impact of co-channel interference on BER performance, while the co-channel interferer beam is scanned away from the desired-channel beam. All multi-channel measurements were performed with and without noise loading. Figure 11 shows the results of one set of multiple channel measurements. In this case, the BER of the desired channel was measured with the desired channel set at a +8.4-dB IBO drive level. The BER was then re-measured after the addition of the interfering channel, which was also set at a +8.4-dB back-off drive level. Finally, the desired channel BER was measured with the addition of the noise-loading channel, which was set to +1.5-dB back off. Similar to the single channel case, the most degradation occurs due to the carrier suppression as a result of the noise loading.

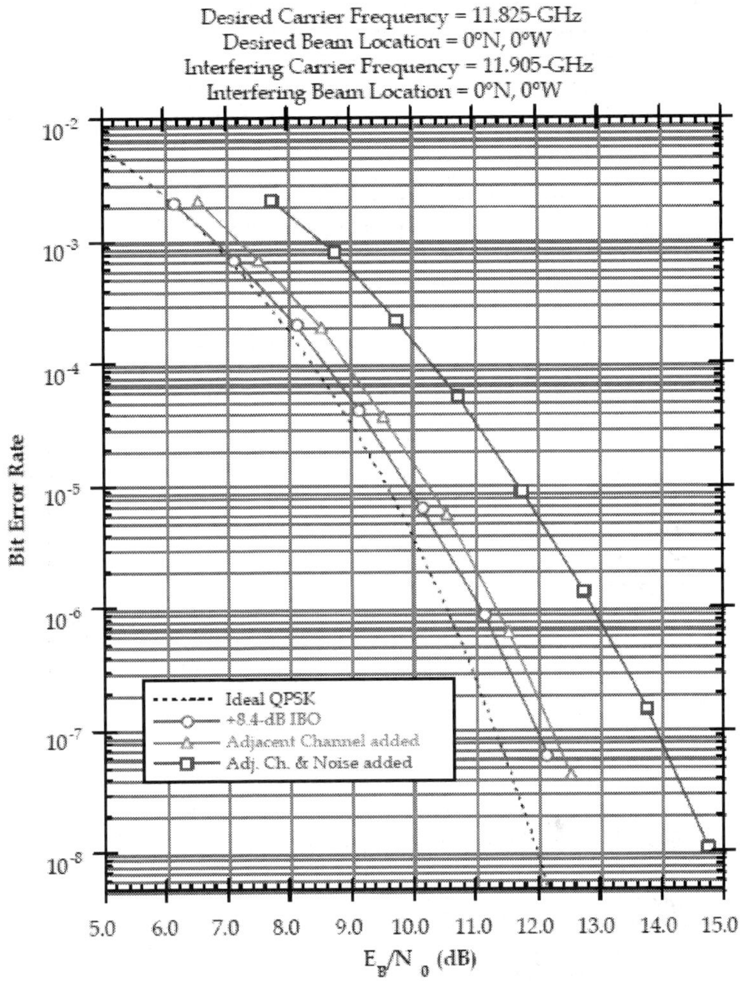


Fig. 11. BER for a carrier with on-axis adjacent channel interferer

### 3.3 Co-Channel Beam Interference

In cellular satellite communication systems employing multiple beam antennas, mutual interference among beams operating at the same frequency can be severe especially for systems that use a large number of beams. The co-channel interference depends on the coverage requirements, antenna performance for both user terminals and satellite payload, and user traffic within each cell. The carrier-to-interference ratio (CIR), which is the ratio of the power levels of the desired signal and the aggregate interference, is used to quantify this effect. Multiple parameters affect the CIR, some of which are readily predictable from the system parameters (such as the total number of co-channel beams, and their separation), while the others need to be determined by statistical means (such as the antenna pointing error and user traffic).

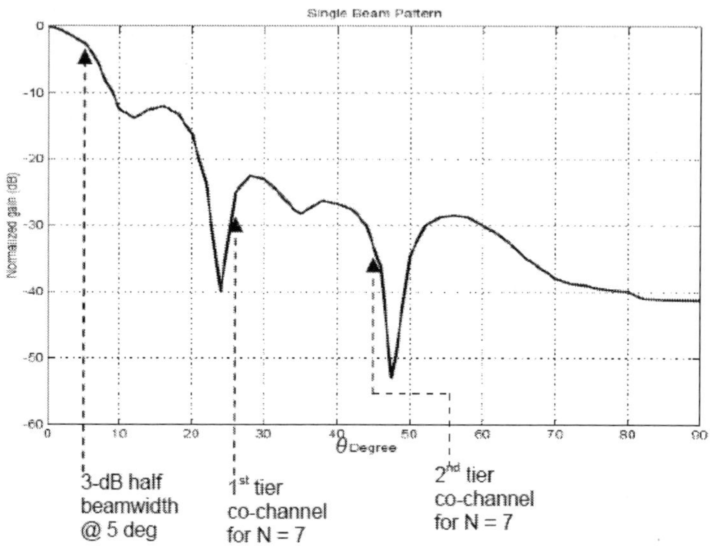


Fig. 12. Satellite antenna pattern used in interference calculations

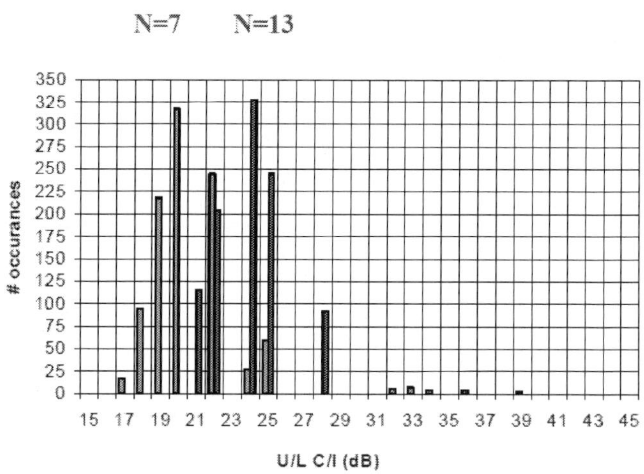


Fig. 13. Example of simulation for uplink interference

The downlink CIR is measured at the user terminal and is a function of the payload antenna characteristics and the location of the user terminal within the beam. The uplink CIR is measured at the payload and is a function of the user traffic within all co-channel beams at a given time, as well as their locations within the beam (Kilic et al., 1999; Chhabra, et al., 2005; Chhabra et al., 2009). To demonstrate these interferences, array patterns for the satellite antenna are assumed based on large apertures consistent with the cell size. Designs similar to the ones described above can be used with the aperture distribution optimized to minimize the interference. In the results to follow, the pattern shown in Figure 12 was used to represent the relative gain function of all cellular satellite beams in a cellular low-altitude system. The pattern is based on an actual satellite antenna design provided by a satellite manufacturer. The figure shows the location of the co-channel interference beams for a hexagonal cellular system with a frequency re-use pattern of N = 7. The antenna pattern can be optimized further to position the pattern nulls at the center of the interfering cells.

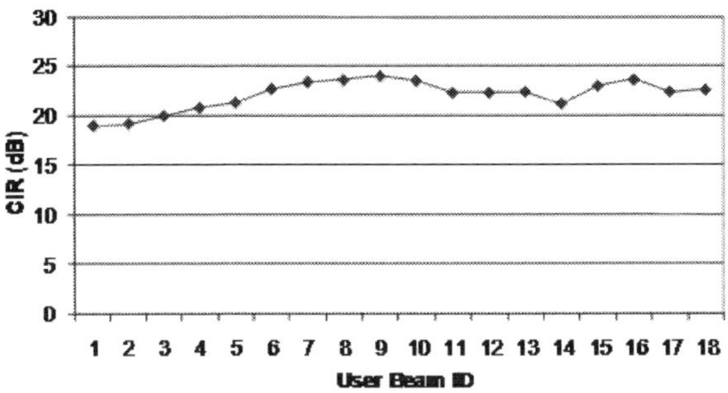


Fig. 14. Example of simulation for downlink interference

Figure 13 demonstrates a statistical histogram for the uplink CIR for two cases where the number of frequency segments in the frequency re-use pattern N is 7 and 13, respectively. The uplink CIR is calculated hundreds of times with uniform statistical distribution of the interfering users within the co-channel beams. The number of times the CIR falls within certain range (# of occurrences) is recorded for different ranges and the results constitute the statistical histogram. As the number of frequency segments increases, the number of interfering beams that use the same frequency segment decreases, leading to higher levels of CIR. Figure 14 shows the downlink CIR levels at different beams that are identified in an ascending order starting from the coverage center. All users in these examples are assumed to be located at the beam center. As Figure 13 suggests, the uplink CIR can vary as much as 12 dB for N = 7 and 6 dB for N = 13, depending on the user traffic within each beam. The downlink CIR, on the other hand, is more stable due the fact that all co-channel interferers are always on. However, depending on the location of the beam in the coverage area, the variation in downlink CIR can be as much as 5 dB as the simulation for the case shown in Figure 14 indicates.

## 4. Conclusion

Active phased arrays provide significant advantages in satellite communications systems that require the generation of a large number of beams. The advantages include the flexibility in generating the beams, sharing of power between beams and achieving a large number of frequencies re-use. These advantages come at a cost at the system level performance that has to be studied and can be minimized. The system-level performance of active phased arrays, which employ distributed amplifiers across the array aperture can be accessed through measurements and computer simulation. The primary parameter in identifying the goodness of the system is the BER, which is a function of the net carrier to noise and interference ratio (CNIR).

In a multiple beam phased array system, intermodulation products and co-channel interference are two key contributors to the degradation of the CNIR. The far-field intermodulation radiation follows certain patterns, which can be predicted using software tools that are verifiable by measurements. The intermodulation radiation degrades the CNIR if it falls within certain frequency bands in the directions of the beams that use those bands. Judicial choices of the operating frequencies and associated beams can reduce or mitigate such degradation.

The two components of the co-channel interferences, downlink and uplink, are functions of both deterministic parameters of the system, such as the total number of beams, antenna performance and frequency re-use scheme, as well as stochastic parameters, such as the traffic pattern, location and equipment pointing errors of the users. While these effects can cause significant variations in the co-channel CIR, the system can be designed based on the worst-case predictions in order to offer higher levels of margin.

An associated phenomenon is the BER degradation due to carrier suppression as a result of traffic loading. This exceeds the degradation that results from driving the desired channel in the non-linear region of the amplifier or at saturation. It also exceeds the degradation that results from adjacent channel or co-channel interference. This type of BER degradation can be controlled by proper loading of the satellite channels.

## 5. References

Chhabra, S.; Zaghloul, A.I. & Kilic, O. (2005). Co-Channel Interference in Satellite-Based Cellular Communication Systems, *Proceedings of the URSI General Assembly*, New Delhi, India, October 2005.

Chhabra, S.; Kilic, O. & Zaghloul, A.I. (2009). *Co-Channel Interference in Satellite Links Utilizing Frequency Reuse,* VDM Publisher, ISBN 978-3-639-17714-5, Saarbrucken, Germany.

Ekelman, E.P.; Kohls, E.C.; Zaghloul, A.I. & Assal, F.T. (1994). Measured Performance of a Ku-Band Multibeam High-Power Phased-Array, *Proceedings of the IEEE International Symposium on Antennas and Propagation,* pp. 844-847, Seattle, Washington, June 1994.

Fuenzalida, J.C., Shimbo, O. & Cook, W.L. (1973). Time-Domain Analysis of Intermodulation Effects Caused by Nonlinear Amplifiers, *COMSAT Technical Review*, 1973.

Jacomb-Hood, A. & Lier, E. (1999). Multibeam Active Phased Arrays for Communications Satellites, *IEEE Microwave Magazine*, pp. 40-47, December 2000.

Kilic, O.; Zaghloul, A.I.; Thai, M. & Nguyen, L. (1999). Beam Interference Modeling in Cellular Satellite Communication Systems, *Proceedings of the URSI General Assembly*, Toronto, Canada, 1999.

Kohls, E.; Ekelman, E.; Zaghloul, A.I. & Assal, F. (1995). Intermodulation and BER Performance of a Ku-Band High Power Active Phased Array, *Proceedings of the IEEE International Symposium on Antennas and Propagation*, pp. 1404-1408, Newport Beach, California, June 1995.

Lier, E.; Purdy, D. & Maalouf, K. (2003). Study of Deployed and Modular Active Phased Array Multibeam Satellite Antennas, *IEEE Antennas and Propagation Magazine*, vol. AP-45, no. 5, pp. 34-45, October 2003.

Lier, E. & Cherrette, A. (2005). An Intermodulation Suppression Technique for Transmit Actrive Phased Array Satellite Antennas with Multiple Shaped Beams, *IEEE Transactions on Antennas and Propagation*, vol. AP-53, no. 5, pp. 1853-1858, May 2005.

Maalouf, K.; Gaus, Jr., R.; Sowelam, S. & Jacomb-Hood, A. (1998). Error Rate Estimation in a Multi-Channel Active Phased Array," *Proceedings of the IEEE International Conference on Communications*, pp. 402-406, June 1998.

Maalouf, K. & Lier, E. (2004). Theoretical and Experimental Study of Interference in Multibeam Active Phased Array Transmit Antenna for Satellite Communications, *IEEE Transactions on Antennas and Propagation*, vol. AP-52, no. 2, pp. 578-592, February 2004.

Purdy, D. & Lier, E. (2000). "EIRP and C/I Analysis of Multibeam Active Phased Array Transmit Antennas for Communication Satellites," *Proceedings of the IEEE International Conference on Phased Array Systems and Technology*, Dana Point, CA, pp. 203-206, May 2000.

Ruggieri, M. (1997). A (Design-Oriented) Reliability Model for Active Phased Arrays in Satellite Communication Systems, *IEEE Aerospace and Electronic Systems Magazine*, vol. 12, no. 10, pp. 23-29, October 1997.

Sandrin, W. (1973). Spatial Distribution of Intermodulation Products in Active Phased Array Antennas, *IEEE Transactions on Antennas and Propagation*, vol. AP-26, no. 6, pp. 864-868, November 1973.

Zaghloul, A.I.; Gupta, R.K.; Assal, F.T.; Martin, J.E. & Wernek, S.B. (1994). Active Phased-Array Development at C- and Ku-Band, *Proceedings of the 15th AIAA International Communications Satellite Systems Conference*, pp. 1683-1691, San Diego, California, February - March, 1994.

Zaghloul, A.I. (1996). X-Band Active Transmit Phased Array, *US Air Force Report*, RL-TR-96-164, September 1996.